HORMONES

SECOND EDITION

HORMONES

SECOND EDITION

Anthony W. Norman
University of California, Riverside

Gerald Litwack
Thomas Jefferson University
Philadelphia, Pennsylvania

Academic Press

San Diego London Boston New York Sydney Tokyo Toronto

Front cover: The crystal structure of the rat I1 thyroid hormone receptor hormone-binding domain bound with hormone has been determined at a 2.0 Å resolution via X-ray crystallographic procedures (Wagner, R. L., Apriletti, J. W., McGrath, M. E., West, B. L., Baxter, J. D., and Fletterick, R. J. (1995). A structural role for hormone in the thyroid hormone receptor. *Nature* **378**, 690–697). This is a schematic representation of the secondary structure elements as well as that of the hormone. The hormone-binding domain of the thyroid receptor is composed of 255 amino acid residues which represent residues 156–410 in the intact thyroid receptor (see Figures 6-18 and 6-19).

The ribbon drawing traces the paths of the polypeptides of the 12 I-helices beginning with the color red (for I-helix$_1$) at the *N*-terminus and progressing down the the colors of the visible spectrum to the color royal blue (for I-helix$_{12}$) at the *C*-terminus. The four strands of the J-sheet are indicated by the color yellow and are designated as S1 ‡ S4. The hormone (purple) is depicted as a space-filling model. Background information and additional details are presented in Chapter 6.

This book is printed on acid-free paper. ∞

Academic Press
a division of Harcourt Brace & Company
525 B Street, Suite 1900, San Diego, California 92101-4495, USA
http://www.apnet.com

Academic Press Limited
24-28 Oval Road, London NW1 7DX, UK
http://www.hbuk.co.uk/ap/

Library of Congress Card Catalog Number: 97-80235

International Standard Book Number: 0-12-521441-3

PRINTED IN THE UNITED STATES OF AMERICA
99 00 01 02 MM 9 8 7 6 5 4 3 2

Contents

CHAPTER

3

Hypothalamic Releasing Hormones

CHAPTER

4

Posterior Pituitary Hormones

CHAPTER

5

Anterior Pituitary Hormones

CHAPTER

6

Thyroid Hormones

CHAPTER

7

Pancreatic Hormones: Insulin and Glucagon

CHAPTER

8

Gastrointestinal Hormones

CHAPTER

9

Calcium-Regulating Hormones: Vitamin D, Parathyroid Hormone, and Calcitonin

CHAPTER

10

Adrenal Corticoids

CHAPTER

11

Hormones of the Adrenal Medulla

CHAPTER

12

Androgens

CHAPTER

13

Estrogens and Progestins

CHAPTER

14

Hormones of Pregnancy and Lactation

CHAPTER

15

Hormones Related to the Kidney and Cardiovascular System

CHAPTER

16

Prostaglandins

CHAPTER

17

Thymus Hormones

CHAPTER

18

Pineal Hormones

CHAPTER

19

Cell Growth Factors

CHAPTER

20

Hormones and Cancer

Preface to the First Edition

The past decades have brought startling advances in our understanding of endocrinology. Of paramount importance is the large increase in the number of legitimate hormones, which now number more than 100, as well as the application of the modern concepts and methodologies of biochemistry and molecular biology to endocrinological research. It is now feasible to approach virtually all classical topics in endocrinology at the cellular and molecular levels.

This book provides a comprehensive treatment of human hormones viewed in the light of modern theories of hormone action and in the context of our current understanding of subcellular and cellular architecture and classical organ physiology. The book is intended for use by first-year medical students, graduate students, and advanced undergraduates in the biological sciences. Also, physicians-in-training should be cognizant of new insights into the etiology of endocrine-related diseases and appreciative of the contribution of basic science to the development of new treatments which are possible through the application of molecular biology and biochemistry to the classic domain of endocrinology. For example, who could have predicted a decade ago that hormonal receptors, or components of hormonal receptors, as well as key cell growth factors could be associated with oncogenes and the cellular expression of certain forms of cancer? Increasingly, medical school curricula are being revised to include a significant coverage of molecular endocrinology. The curriculum for advanced undergraduate biology majors is also being expanded to include molecular endocrinology. Graduate students in medical or biological sciences, including immunology, entomology, genetics, anatomy, physiology, and biochemistry, will inevitably encounter, either in the classroom or in excursions into the modern scientific literature, the contributions and impact of modern endocrinology. It is hoped that this book will fill the void that currently exists for resource materials for teaching cellular and molecular endocrinology and that it will be employed as an equal partner with most standard biochemistry textbooks to provide a comprehensive and balanced coverage of this realm of biology.

Our book presumes that the reader will have been exposed in detail to the fundamental areas of biochemistry, including enzymology and the structure and function of macromolecules and the other bioorganic substances of intermediary metabolism, as well as to selected topics in molecular biology. In addition, an understanding of cell biology, cellular and subcellular organization, and mammalian physiology will be useful. It is the tetrad of biochemistry, molecular biology, and cell and organ physiology that provides the principles and biological "facts of life" that are critical to our modern understanding of hormones.

The book provides two introductory chapters followed by seventeen chapters on selected endocrinological topics pertinent to humans. The first chapter presents the first principles of hormone action. These include a discussion of the structural and functional classification of hormones and a detailed presentation of current general theories of mechanisms of hormone action at both the cellular and subcellular levels. Chapter 2 provides a detailed presentation of the seven classes of steroid hormones and their chemistry, biosynthesis, and metabolism. These introductory chapters are followed by sixteen chapters that address either a classical endocrine system—e.g., hypothalamic hormones (Chapter 3), posterior pituitary hormones (Chapter 4), anterior pituitary hormones (Chapter 5), thyroid hormones (Chapter 6), pancreatic hormones (Chapter 7), gastrointestinal hormones (Chapter 8), calcium-regulating hormones (Chapter 9), adrenal corticoids (Chapter 10), hormones of the adrenal medulla (Chapter 11), androgens (Chapter 12), estrogens and progestins (Chapter 13), and hormones of pregnancy and lactation (Chapter 14)—or newer domains of hormone action which are now essential to a comprehensive understanding of hormone action, including prostaglandins (Chapter 16), thymus hormones (Chapter 17), and pineal hormones (Chapter 18). The book concludes (Chapter 19) with a presentation of hormones of the future, i.e., cell growth factors. Chapter 15, Hormones of the Kidney, of necessity is not devoted exclusively to a single hormone system; it focuses on the hormones, excluding 1,25-dihydroxy-vitamin D (Chapter 9), which are made in (erythropoietin, kallikreins) or which act (aldosterone, vasopressin) on the kidney.

Each of the last seventeen chapters is organized in parallel fashion. Thus, each chapter has the following sections: (a) introduction; (b) human anatomical–morphological relationships; (c) chemistry; (d) biochemistry; (e) biological and molecular action of the hormone(s); (f) clinical aspects; and (g) reference citations. The clinical aspects section is not intended to be comprehensive, but rather to provide for the medical student an introduction/resume of key disease states and contemporary medical problems related to the hormone(s) deficiency or excess. In addition, Appendixes C and D provide insight into the definition and incidence rate of prominent endocrine-related disease states. Each chapter is highly illustrated with respect both to human physiology and anatomy and to the details and models of hormone action. Each chapter culminates with a listing of key reference citations, including books and review articles as well as recent research papers.

The book also contains eight appendixes: (A) a table listing over 100 hormones; (B) a table of the blood concentrations of major hormones; (C) a list of prominent endocrine disorders; (D) a table of the rate of incidence of principal endocrine disease diagnoses; (E) a tabulation of Nobel Prizes awarded in endocrinology and related areas; (F) a table of the genetic code; (G) a table of the three-letter and single-letter abbreviations for amino acids; and (H) a tabulation of the units of scientific measurement.

A major challenge to modern publication techniques as well as to authors is the ability of scientists to obtain the primary amino acid sequence of large proteins/hormones as well as the complementary DNA sequence (cDNA) of nucleotides and the genomic sequence of nucleotides for proteins of interest. Thus, modern triumphs of molecular endocrinology include the primary amino acid sequence of prolactin (199 amino acid residues), the cDNA sequence of the steroid receptors for glucocorticoids (4800 nucleotides) and estradiol (3600 nucleotides), as well as the genomic organization of several hormones, e.g., insulin (1720 nucleotides). The dilemma to the authors was whether to include such extensive and detailed information in a volume that is intended for use as a textbook. Clearly, most students will not study protein amino acid sequences or cDNA sequence at the individual amino acid or nucleotide level. Yet we feel it instructive for the student to realize and appreciate the intrinsic complexity and detail of information pertaining to hormones which molecular endocrinologists can now almost routinely achieve. Accordingly, the authors have included many sequences of large peptide hormones as well as cDNA sequences. We have chosen in some instances to limit their format to a single-page "miniprint" rather than to extend their presentation over two or three pages. Interested readers should utilize a magnifying or "reading" glass to facilitate their study (e.g., see Figs. 5-4, 5-7, 7-13, 10-16A, 10-20B, and 13-15).

The text is related to biochemical endocrinology courses we have taught to first-year medical students and graduate students at the University of California at Riverside and Temple University. The authors hope that the uniform organization of the chapters and the subdivision of topics within each chapter will allow instructors to select the level of coverage they require from a comprehensive presentation to one focusing on only the subcellular mode of action of the hormones.

We would like to acknowledge the students at UC Riverside in Biochemistry 120 who used a draft of this textbook. From their comments and from our co-instructors, Professors H. L. Henry and R. A. Luben, we received much useful feedback.

In addition we thank our professional colleagues who individually read and critiqued the various chapters. These include Julius Axelrod (18), Om P. Bahl (5), John D. Baxter (6, 10), Esther Breslow (4), Josiah Brown (6), Ralph A. Bradshaw (19), P. Michael Conn (3), Michael P. Czech (19), Leonard J. Deftos (9), Isidore Edelman (15), John H. Exton (11), H. Hugh Fudenberg (17), W. F. Ganong (3), Jack Geller (12), Allan L. Goldstein (17), Jack Gorski (1), Oscar Hechter (4), Bernard L. Horecker (17), Benita S. Katzenellenbogen (1), Leonard D. Kohn (5), William E. M. Lands (16), Joseph Larner (7), J. B. Lee (16), Robert J. Lefkowitz (11), Choh Hao Li (5), Marc E. Lippman (13), Walter Lovenberg (18), Joseph Meites (3), R. Curtis Morris (9), Allan Munck (10), William D. Odell (14), Jack H. Oppenheimer (6), Peter W. Ramwell (16), Russel J. Reiter (18), Herbert H. Samuels (1, 6), David A. Sirbasku (19), Melvin S. Soloff (4), Donald F. Steiner (7), E. Brad Thompson (10), Sidney Udenfriend (11), Larry Vickery (2), John H. Walsh (8), Owen N. Witte (15), and Richard J. Wurtman (18). We thank Dr. A. Geoffrey Norman, Dr. Valerie Leathers, and Dr. Helen L. Henry, who all read major portions of the manuscript and provided essential feedback. If our book has merit it is due largely to the contributions of our scientific mentors and colleagues as well as those individuals who read various sections of the manuscript. Shortcomings still remaining are, of course, completely the responsibility of the authors.

We are also indebted to numerous authors and publishers for permission to reproduce figures and tables that were originally published elsewhere. In particular we recognize the masterful skill of Dr. Thomas Lentz and thank the publisher W. B. Saunders for allowing us to utilize numerous of his outstanding line drawings (based on electron micrographs) taken from his book *Cell Fine Structure.* Thanks are also due to Dr. Laurie Paavola of the Department of Anatomy, Temple University School of Medicine, who allowed us to reproduce several of her own drawings of various cells and their morphological components in relation to hormone function.

The authors have been privileged to utilize the skills of Ms. Patti Mote, a scientific illustrator, who prepared original line drawings or anatomical ink washes for approximately 120 of the 442 figures published in the book. Throughout the entire publication process, Ms. Mote was always a positive force striving for simplicity, accuracy, and clarity in our illustrations.

Also, the authors have been fortunate to have the services of a series of dedicated, loyal, and patient secretaries over the 6 years in which the typescript was being written or revised. In the early phases this included Ms. Sherry Bataglia, Ms. Cassie Wooten, Ms. Marcia Johnson, Ms. Pam Moore, and Ms. Jerri Flagg, and in the latter phases Mrs. Sharon Herbert, Mrs. Ann Hall, Mrs. Grace Jones, and Mrs. Lean Gill (who retyped several cDNA sequences). We thank you all! The authors are also indebted to the production staff of Academic Press for their professional assistance during this project.

Finally, we express appreciation to our families in Philadelphia (Ellie, Geoffrey, Kate, Claudia, David, and Debbie) and in Riverside (Helen, Thea, Jacqueline, and Derek) who were most understanding of the long hours and absence from home that were required to produce the final manuscript.

Anthony W. Norman
Gerald Litwack

Preface to the Second Edition

This second edition of *Hormones* is intended to provide a comprehensive treatment of the hormones of humans viewed from the context of modern theories of their action in the framework of our current understanding of the molecular, subcellular, and cellular architecture as well as classical organ physiology. The book is intended for use by graduate students and advanced undergraduates in the biological sciences and, as well, by first-year medical students. Increasingly, medical school curricula are being revised to include a significant component of molecular endocrinology. Also physicians-in-training should be cognizant of the newer insights into the etiology of endocrine-related diseases that are possible as a consequence of the application of molecular biology and biochemistry to the classic domain of endocrinology. *Hormones* also will be used to varying extents by researchers at all levels.

The first edition of *Hormones* was published in 1987, and at that time the authors wrote: "The last decades have seen startling advances in our understanding of the science of endocrinology. Of equal importance is the large increase in the number of legitimate hormones (which now number more than 150) as well as the application of the modern concepts and methodologies of biochemistry and molecular biology to endocrinology research frontiers. Thus it is now feasible to approach all classical endocrinology topics at the cellular and molecular level."

Although only 11 years have passed since the authors finalized the manuscript for the first edition of *Hormones,* the pace of discovery of new hormones and the development of an in-depth molecular understanding of older hormones and endocrine systems have continued unabated. The current understanding of endocrinology in 1997 is so different from that in 1986 that this second edition of *Hormones* now has one new chapter (Chapter 20, concerning hormones and cancer) describes many hormones discovered or biochemically characterized over the past decade (including leptin, interleukins 3–13, inhibins, activins, endothelins, lipoxins, leukemia inhibitory factor, and stem cell factor). It also has an increased page size (from 6×9 to 8.5×11 inches). Also, this second edition has a total of 449 figures, 270 of which are either new or significantly changed.

Similar to the teaching of medical students, the curriculum for advanced undergraduate biology majors increasingly includes exposure to both the principles of molecular endocrinology and its research frontiers. Finally, graduate students in all aspects of medical or biological science (immunology, genetics, anatomy, physiology, biochemistry, entomology, etc.) will inevitably encounter, either in the classroom or in excursions into the modern scientific literature, the contributions to and impact on fundamental cellular processes of modern endocrinology.

It is hoped that this revised textbook will fill the void of resource materials that currently exist for teaching cellular and molecular endocrinology. We visualize that this book could be employed as an equal partner with most of the available standard biochemistry textbooks to provide a comprehensive and balanced coverage of this realm of biology.

Our book presumes that the reader will have been exposed in detail to the areas of knowledge fundamental to biochemistry, including enzymology and the structure and function of macromolecules and the other bioorganic substances of intermediary metabolism, as well as to selected topics of molecular biology. In addition, an understanding of cell biology, cellular and subcellular organization, and mammalian physiology will be useful. It is the pentad of biochemistry, structural biology, molecular biology, and cell and organ physiology that forms the foundation for the principles and biological "facts of life" that are critical to the development of our modern understanding of hormones.

The book is organized to provide two introductory chapters followed by 18 chapters on selected endocrinology topics pertinent to humans. The opening chapter is concerned with a delineation of the first principles of hormone action. These include a discussion of the structural and functional classification of hormones and a detailed presentation of current general theories of mechanisms of hormone action at both the cellular and the subcellular level. In this second edition this includes the addition of information on the families of hormone receptors (steroid and growth factors) and a detailed introduction to the topic of signal transduction, which describes how the chemical message of the hormone is transduced to generate specific biological responses. Chapter 2 provides a detailed presentation of the seven classes of steroid hormones and their chemistry, biosynthesis, and metabolism. These introductory chapters are then followed by 16 chapters that individually address either a classical endocrine system—for example, hypothalamic hormones (Chapter 3), posterior pituitary hormones (Chapter 4), anterior pituitary hormones (Chapter 5), thyroid hormones (Chapter 6), pancreatic hormones (Chapter 7), gastrointestinal hormones (Chapter 8), calcium-regulating hormones (Chapter 9), adrenal corticoids (Chapter 10), hormones of the adrenal medulla (Chapter 11), androgens (Chapter 12), estrogens and progestins (Chapter 13), hormones of pregnancy and lactation (Chapter 14)—or newer domains of hormone action which are now essential to a comprehensive understanding of hormone action—such as prostaglandins (Chapter 16), thymus hormones (Chapter 17), pineal hormones (Chapter 18), and cell growth factors with reference to oncogenes (Chapter 19). The book concludes with a new chapter (20), "Hormones and Cancer." Chapter 15, "Hormones of the Kidney," of necessity is not devoted exclusively to a single hormonal system; it focuses on the several hormones, excluding 1,25-dihydroxyvitamin D (Chapter 9), that are made in (erythropoietin, kallikreins) or that act (aldosterone, vasopressin) on the kidney, as well as focusing on endothelins, nitric oxide and atrial natriuretic hormone (ANF), which collaborate to regulate plasma fluid volume, plasma Na^+ concentration, and blood pressure: Revised Chapter 15 also provides an introduction to hematopoiesis.

Each of the last 17 chapters individually focuses on separate vertebrate hormonal systems. Each is organized in parallel fashion with the following sections: (a) introduction; (b) human anatomical–morphological relationships pertinent to the topic under discussion; (c) chemistry of; (d) biochemistry of and (e) biological and molecular action of the hormone(s); (f) clinical aspects; and (g) reference citations. The clinical aspects section is intended to be brief in order to provide an introduction / resume for the medical student of key disease states and contemporary medical problems related to the hormone(s) deficiency or excess. Further, each chapter is highly illustrated with respect to human physiology and anatomy and also with schematic figures describing the details and models of hormone action. Each chapter culminates with a listing of key reference citations, including books and review articles as well as pertinent recent research citations. These are intended to provide the interested reader with an entry into the scientific literature (books, review articles, or research papers) of interest.

Also included in the book are six appendices: (A) an extensive table listing over 133 known hormones in higher mammals and humans; (B) a table of the blood concentrations of major human hormones; (C) a list of prominent endocrine disorders; (D) a table of the genetic code; (E) a table of the three-letter and single-letter abbreviation(s) for amino acids; and (F) a tabulation of the units of scientific measurement in biological systems. It is hoped that these appendices will provide the student and other readers a ready reference source to facilitate assimilation of the endocrinology topics in this book.

On the whole the text is related to biochemical endocrinology courses we have taught to first-year medical students and graduate students at the University of California at Riverside, Temple University, and Jefferson Medical College. The authors hope that the uniform organizational framework of the chapters as well as the division of chapter topics into separate hormone systems will allow instructors to selectively identify varying levels of coverage. Thus it should be possible to prepare a teaching syllabus that is comprehensive or one that focuses on only the subcellular mode of action of selected hormones divorced from a detailed understanding of their human anatomy and physiology.

We are indebted to numerous authors and publishers for permission to reproduce figures and tables that were originally published elsewhere. In particular we thank Professors John Baxter and Robert Fletterick of the University of California at San Francisco for permission to use the ribbon drawing of the 12 α-helices of the thyroid hormone receptor that is on the cover of *Hormones*. We also acknowledge the students in Biochemistry 120 at UC Riverside who utilized a draft of this revised textbook. From their comments and those of our co-instructors, Drs. Helen L. Henry and Richard A. Luben, we received many useful ideas.

The authors have been privileged to utilize the scientific illustration skills of Ms. Patti Mote (first edition) and Ms. Karin Christensen (second edition), who prepared original line drawings or anatomical ink washes of approximately 80% of the figures published in the book. Throughout the entire publication process, both Ms. Mote and Ms. Christensen were always a positive force striving for simplicity, accuracy, and clarity in our illustrations.

Also, the authors have been fortunate to have the services of loyal and patient secretaries over the three-year interval that the typescript for the second edition was being written or revised: Ms. Marian Herbert in Riverside and Mrs. Kathleen Whittock in Philadelphia.

If our book has merit, it is due largely to the contributions of our scientific mentors, colleagues, and students, as well as to those individuals who read various sections of the manuscript. Shortcomings still remaining are, of course, completely the responsibility of the authors.

Finally, we express appreciation to our families in Riverside (Helen, Derek, Jacqueline, and Thea) and Philadelphia (Ellie, Geoffrey, Kate, Claudia, David, and Deborah), who were most understanding of the long hours and absences from home that were required to produce the new manuscript for this second edition of *Hormones*.

Anthony W. Norman
Gerald Litwack

CHAPTER

1

General Considerations of Hormones

I. OVERVIEW OF HORMONES, RECEPTORS, AND ENDOCRINOLOGY

A. Introduction

The domain of endocrinology represents a scientific description of how, in a higher organism, cell A communicates with cell B via the sending of chemical messengers termed "hormones." A detailed understanding of a particular endocrine system likely will include an understanding of the following: (a) the anatomical description of cells A and B and their immediate environment (are they part of a gland?), as well as the distance of separation of A from B; (b) the chemical structure of the messenger, H (the hormone); (c) the details of the biosynthesis of the hormone by cell A; (d) the mode of transfer of H from cell A to cell B; (e) the detailed mechanism by which cell B uses receptors to detect the presence of H; (f) how cell B transduces the presence of H to initiate and sustain a biological response; and (g) how cell B communicates via a feedback loop with cell A to indicate the adequate presence of the hormone.

The study of endocrinology over the past century quite naturally has been dependent upon the scientific methodologies available to probe the various endocrine systems. Thus, in the interval 1900–1960, endocrinology was largely pursued at the physiological level. This resulted in the discovery of approximately 25 hormones; the detail of specification of the hormone structure was usually inversely proportional to the size of the hormone. Accordingly, the complete structure of thyroxin (molecular weight 770) was defined in 1926, while the sequence and structure of the small protein hormone insulin would not be obtained until 1953 (amino acid sequence) and 1969 (three-dimensional structure).

The biochemical era of endocrinology began in approximately 1955–1960 and extends to the present time. The availability of radioactive isotopes of carbon (^{14}C), hydrogen (^{3}H), phosphorus (^{32}P), etc., coupled with advances in chemical methodology (chromatography, mass spectrometry, nuclear magnetic resonance spectroscopy (NMR), and X-ray crystallography), has and continues to permit the detection and chemical characterization of minute quantities (nanograms or picograms) of new hormones and the characterization of the many receptors. Now in the 1990s we are experiencing the cellular and molecular biological era of endocrinology. With the advent of scanning electron microscopes and confocal microscopy with the capability for real time imaging of living cells, coupled with the application of the plethora of molecular biological procedures and manipulations, the science of endocrinology is flourishing. It is not clear how soon the apex will be realized. Thus, today we know of the existence of over 130 hormones (see Appendix A for a tabulation), and scientists are actively engaged in categorizing and defining the sphere of influence and molecular mode of action of each hormone. Undoubtedly there are still additional hormones waiting to be discovered.

The terms "endocrine" and "hormone" were both derived from the Greek, "endokrinein" and "hormaein," respectively. The original use of hormone implied an agent that can excite or arouse; in the modern era a hormone is perceived to be a defined chemical entity that is secreted or produced by specific glands or cells and that acts as a chemical messenger or signal molecule. A hallmark of hormones is their potency. A typical concentration of hormone in the blood may be $(1–100) \times 10^{-10}$ *M*. The term "endocrine" originally implied a glandular secretion of hormonal products; in the modern era it is appreciated that single cells or clusters of cells that are not anatomically definable as a gland can also secrete or produce hormones.

Finally, the cellular constitutent that is the immediate recipient of the "information unit" or message of a hormone is the receptor. The biochemical organization of receptors is diverse but each receptor is, in general, structurally organized so that it can specifically recognize and interact with its own cognate hormone (not unlike the lock and key model of the substrate–enzyme intermediate). Because of the low circulating concentrations of the hormones, the receptor must have a very efficient "capture" mechanism for its hormone. As a consequence of the receptor–hormone interaction (however transient it may be), signal transduction occurs and a specific biological response(s) is generated within and, in some instances, around the target cell, i.e., the cell responds to the presence of the hormone. An overview of a traditional glandular secretion of a hormone and its transport through the bloodstream to a distal target organ is presented in Figure 1-1.

The objective of this book is to provide a status report on the field of human hormones, viewed in the light of our current understanding of cellular and subcellular architecture, as well as the molecular details of their mode of action.

We now realize that the domain of endocrinology has become a science of signaling or communication mechanisms involving substances in a much broader category than previously connoted by the term "endocrine hormones." Logically, any substance that operates at the cellular level, generated either externally or internally, which conveys to that cell a message to stop, start, or modulate a cellular process, will come under the perview of modern endocrinology. In recognition of this fact, we have preferred to entitle this book *Hormones* rather than *Endocrinology,* since the latter term refers to the limited category of substances operating over long distances within the organism.

This chapter presents the first principles of hormone action. These include a discussion of the structural and functional classification of hormones and a detailed presentation of current theories of mechanisms of hormone action at both the cellular and subcellular levels. The remaining chapters then focus on individual classical endocrine systems (posterior pituitary hormones, pancreatic hormones, Ca^{2+} regulatory hormones, etc.) or newer domains of hormone action that are essential to a comprehensive understanding of the domain of hormone action (e.g., prostaglandins, cancer, pineal hormones).

B. Classes of Hormone Molecules

Originally there were perceived to be three major classes of hormones based upon their chemical structure. These are the peptide and protein hormones, the steroid hormones, and the amino acid-related hormones. Examples of human hormones for these categories are given in Table 1-1 (and include examples from the endocrine, paracrine, autocrine, and neurotransmitter systems). As the discovery of new hormones continues, it is apparent that it is not feasible to categorize all hormones into precisely these three families; e.g., the prostaglandins are derived from fatty acids. Appendix A provides a tabulation of the ≈130 hormones known at the present time, and the trivial name, source, and principal biological action are listed for each hormone.

C. Classification of Hormonal Communication Systems

As stated earlier, hormones are chemical messengers that send a signal within a physiological system from point A to point B. There are four identifiable hormonal communication systems; each has a differ-

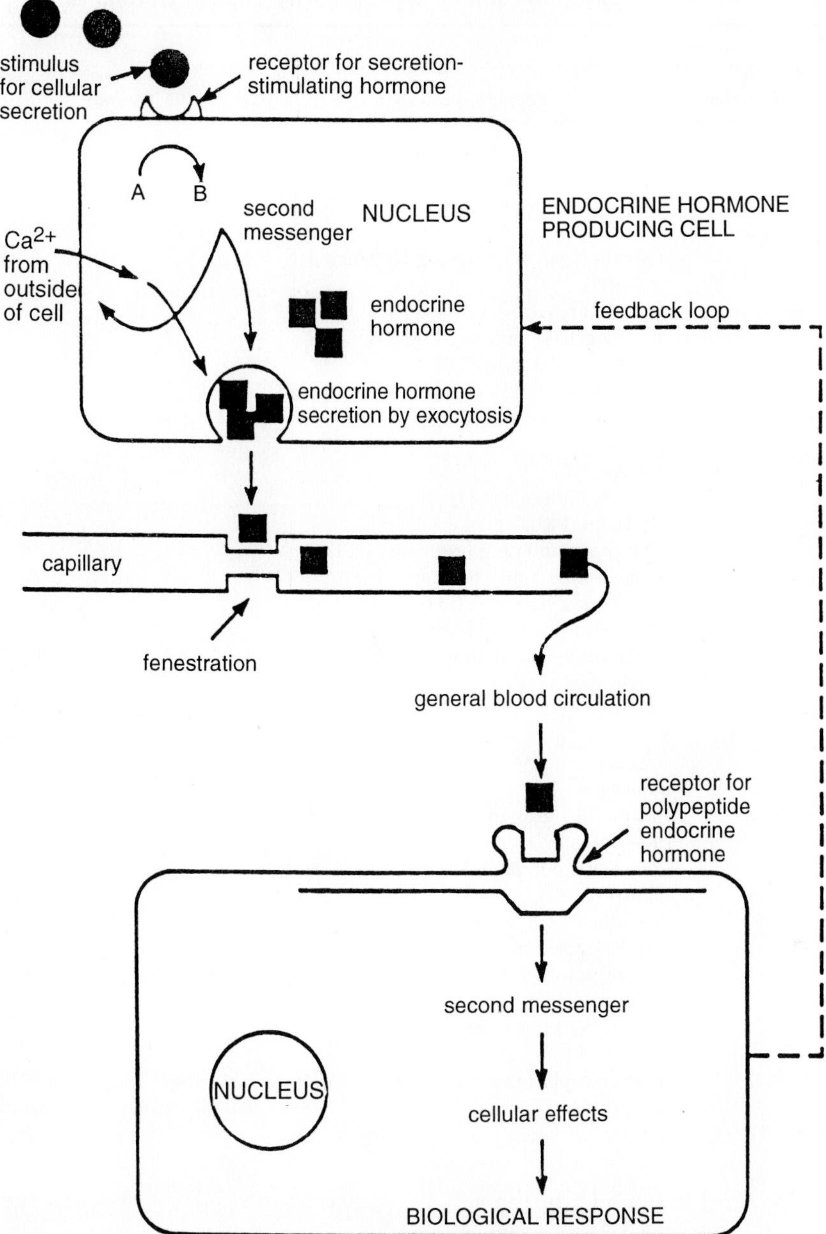

FIGURE 1-1 Overview of traditional endocrine system. The endocrine hormone producing cell is stimulated to biosynthesize the hormone (■) as a consequence of the balance of incoming physiological signals represented by (i) the stimulus for cellular secretion (●) and (ii) the negative feedback loop. The secreted hormone (■) travels through the circulatory system to distal target cells. Thus, the hormone (■) interacts with a specific receptor to initiate the generation of a specific biological response(s). As a consequence of the pressure of the biological response(s), a chemical signal is sent back to the endocrine hormone producing cell via the "negative feedback loop" to indicate the presence of the hormone.

TABLE 1-1 Categorization of Representative Human Hormones[a]

Hormone system	Chapter	Peptide–protein	Steroid	Amino acid or fatty acid derived
Hypothalamic regulating	3	Thyrotropin releasing hormone (TRH) Gonadotropin releasing hormone (GnRH) Corticotropin releasing hormone (CRH or CRF) Growth hormone releasing hormone (GRH) Prolactin releasing factor (PRF) Melanocyte-stimulating hormone releasing factor (MRF) Melanocyte-stimulating hormone-release inhibitory factor (MIF)		
Posterior pituitary	4	Oxytocin Vasopressin		
Anterior pituitary	5	Growth hormone (GH) Prolactin (PRL) Thyroid-stimulating hormone (TSH) Follicle-stimulating hormone (FSH) Luteinizing hormone (LH) Adrenocorticotropic hormone (ACTH) Melanocyte-stimulating hormone (MSH) Lipotropin		
Thyroid hormone	6			Thyroxine (T_4) Triiodothyronine (T_3)
Pancreatic hormone	7	Insulin Glucagon Insulin-like growth factors (IGFs) Somatostatin Pancreatic polypeptide		
Gastrointestinal	8	Gastrin Secretin Cholecystokinin		
Calcium regulating	9	Parathyroid (PTH) Calcitonin (CT) Parathyroid hormone-related peptide (PTHrP)	$1,25(OH)_2D_3$	
Adrenal corticoids	10	Corticotropin-releasing factor (ACTH) Angiotensin	Cortisol Aldosterone	
Adrenal medulla	11			Epinephrine Norepinephrine
Androgens (the male)	12	Luteinizing hormone (LH) Follicle-stimulating hormone (FSH) Inhibin	Testosterone Dehydroepiandrosterone (DHEA)	
Estrogens and progestin (the female)	13/14	Follicle-stimulating hormone (FSH) Luteinizing hormone (LH) Prolactin Human chorionic gonadotropin (HCG) Relaxin Oxytocin Inhibin	Estradiol Progesterone	
	14	Human placental lactogen (hPL)		
Plasma volume and sodium regulating	15	Atrial natriuretic factor (ANF) Erythropoietin Kallikreins Kinins Renin	Aldosterone	
Prostaglandins	16			Prostaglandins
Thymus	17	Thymosin α_1 Thymosin α_4		

(*continues*)

TABLE 1-1—*Continued*

Hormone system	Chapter	Peptide–protein	Steroid	Amino acid or fatty acid derived
Pineal	18			Melatonin Serotonin
Cell growth factors	19	Epidermal growth factor (EGF) Nerve growth factor (NGF) Platelet-derived growth factor (PDGF) Insulin-like growth factors (IGFs) Fibroblast growth factor (FGF)		

[a] Appendix A provides a comprehensive summary of most human hormones.

ent anatomical relationship between point A and point B.

1. *Systemic*

The classic systemic endocrine system is described in Figure 1-1. The hormone is biosynthesized and stored within specific cells associated with an anatomically defined endocrine gland. The hormone is not released until receipt of an appropriate physiological signal, which may take the form of either a change in the concentration of some component in the blood (e.g., [Ca^{2+}], [glucose], [K^+]) or the delivery of a neural signal. Once released into the bloodstream, the hormone can travel to a distant cellular target, frequently the hormone (particularly if it is a hydrophobic molecule like

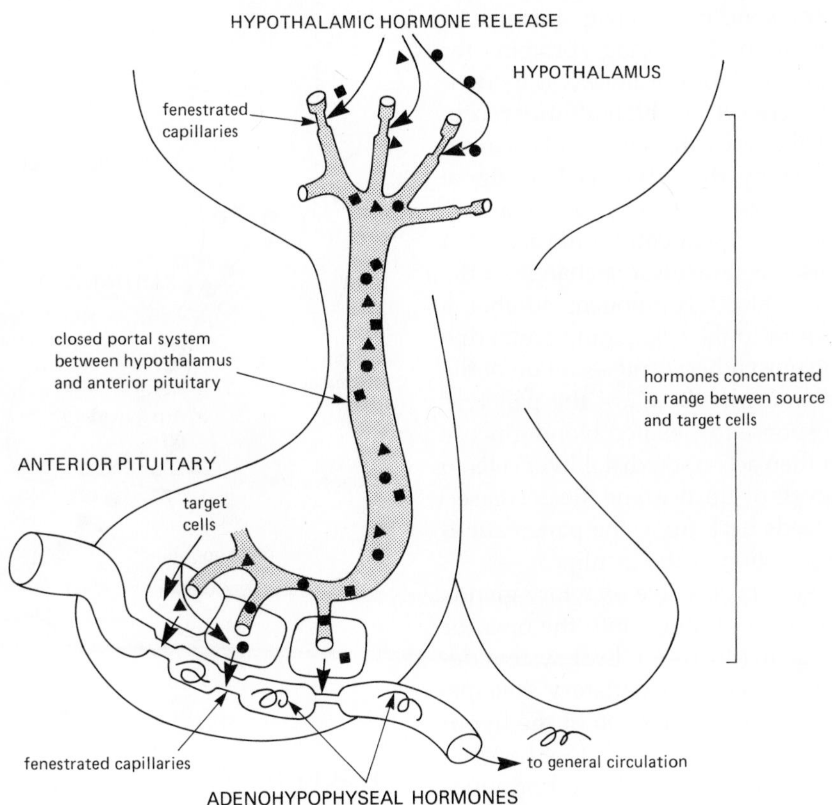

FIGURE 1-2 Example of an endocrine system that involves a closed portal system that effects conservation of hormone concentrations between the source secreting system and the target receptor-containing cells.

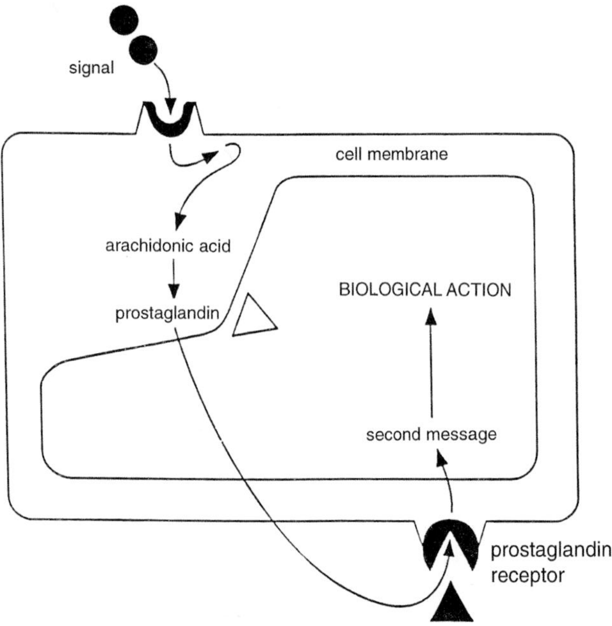

FIGURE 1-3 Example of an autocrine system operating on the same cell that released it. The active hormone also could reach its receptor from the cell interior without having been secreted first.

the steroid hormones) may be carried in the bloodstream by a specific transport protein [e.g., transcortin for cortisol, the vitamin D-binding protein for $1,25(OH)_2D_3$]. The target cell for hormone X is by definition a cell that possesses specific high-affinity receptors for hormone X. Then, as a consequence of accumulation of the hormone by the target cell, a signal transduction process ensues and a specific set of biological responses is generated. Frequently some aspect of the biological responses will result in a change in the concentration of some blood component so that a "feedback" signal is sent to the originating endocrine gland to diminish the biosynthesis and secretion of the hormone. Thus, the endocrine B cell of the pancreas secretes insulin in response to elevated blood glucose levels. The glucagon then acts upon distal liver cells to decrease the blood levels of glucose, and the decreased blood glucose then feeds back upon the pancreatic B cell to diminish the secretion of the insulin.

Classic endocrine systems involve secretory glands that release their hormone products into the open or general circulatory system (the B-cell/liver system described earlier), as well as closed circulatory systems. An example of the latter is the secretion of the hypothalamic releasing hormones into a closed portal system, which ensures that most of these hormones will be delivered to the anterior pituitary, which contains the target cells for these releasing hormones; see Figure 1-2.

2. *Paracrine*

In the paracrine system, the distance from secretion point A to target cell point B is sharply reduced. Here cell A biosynthesizes and secretes hormone X. Hormone X then diffuses from the secreting cell A to immediately adjacent target cells B. An example of a paracrine system is the production of testosterone by the interstitial Leydig cells of the testes, which then diffuses into the immediately adjacent seminiferous tubules.

3. *Autocrine*

A variation of the paracrine system is when the hormone-secreting cell and the target receptor-containing cell are one and the same; this is an autocrine system. Figure 1-3 represents a model of a paracrine and an autocrine endocrine system. Examples of autocrine hormones would be the prostaglandins and some of their relatives, such as the thromboxanes, leukotrienes, and lipoxins.

4. *Neurotransmitters*

Over the past decade, it has been increasingly accepted as a consequence of mechanistic studies that neurotransmitters are really hormones, i.e., they are

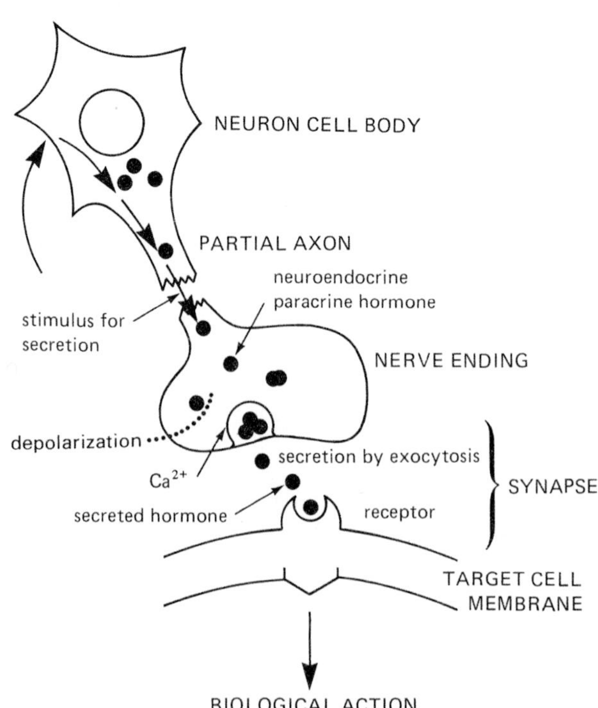

FIGURE 1-4 Example of a neurohormone operating in a synapse. In the case of some amine hormones or neurotransmitters, their synthesis may be confined to the nerve ending, whereas other substances are synthesized in the cell body and transported to the nerve ending, as shown here.

chemical messengers. Thus, a neuron normally innervates a single cell. While the electrical signal of the neuron may travel a long anatomical distance over the axon, the actual electrical "message" is transformed into a chemical mediator, the neurotransmitter, which is secreted by the axon. The neurotransmitter then diffuses locally across the synapse to the adjacent receptor cell. Thus, neurotransmitters, such as acetylcholine and norepinephrine, may be thought of as paracrine hormones. An example of a neurohormone operating in a synapse is given in Figure 1-4.

D. General Concept of a Hormone Receptor

Hormones effect delivery of their "message" to target cells by interaction with a cellular constituent, namely, the receptor. All receptors have two key components: (a) a ligand-binding domain that noncovalently but stereospecifically binds the correct hormone for that receptor and (b) an effector domain that "recognizes" the presence of the hormone bound to the ligand domain and that then initiates the generation of the biological response(s). The "recognition" process that occurs between the ligand-binding domain and the effector domain is achieved by a coupling or transduction process believed for many receptors to be an allosteric conformational change. Thus, binding of the hormone to the ligand domain results in subtle but critically important changes in the environment of the effector site, so that the signal transduction process is initiated (see Figure 1–5). The competent hormone effector site may then interact with other cellular constituents to complete the signal transduction process.

Those steroid hormone receptors that are present in the cytosolic and nuclear compartments of the cell interact with DNA. Peptide and protein hormones usually (but not always) interact with receptors present in the outer membrane of the cell. Then, depending upon the system, the hormone–receptor complex will physically interact with other cellular constituents such as G proteins (discussed later in this chapter) and/or ion channels. As a consequence of occupancy of the membrane receptor, a variety of intracellular second messengers may be generated inside the cell; these second messengers, e.g., cyclic AMP, cyclic GMP, Ca^{2+}, diacylglycerol (DAG), inositol triphosphate (IP_3), then move elsewhere inside the cell and complete the transfer of the "message" to still other cellular constituents, which then produce the biological response "signaled" by the initiating hormone.

Receptors are largely composed of proteins, but they can contain secondary modifications of carbohydrate and, in addition, may be selectively imbedded within a lipid membrane bilayer. Receptors can also become phosphorylated or myristoylated (addition

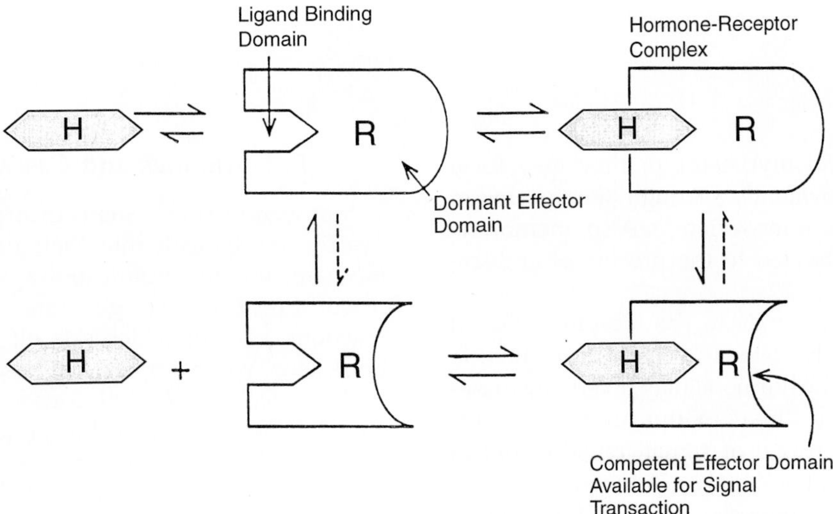

FIGURE 1-5 Schematic model of a generic receptor for a hormone. The ligand-binding domains and the effector domains are indicated. Upon binding of the hormone (H), a conformational change occurs in the receptor, which converts a dormant effector domain into a competent effector domain (i.e., able to initiate signal transduction). The whole process of hormone binding and generation of a competent receptor followed by dissociation of the hormone is believed to operate in a cyclic fashion.

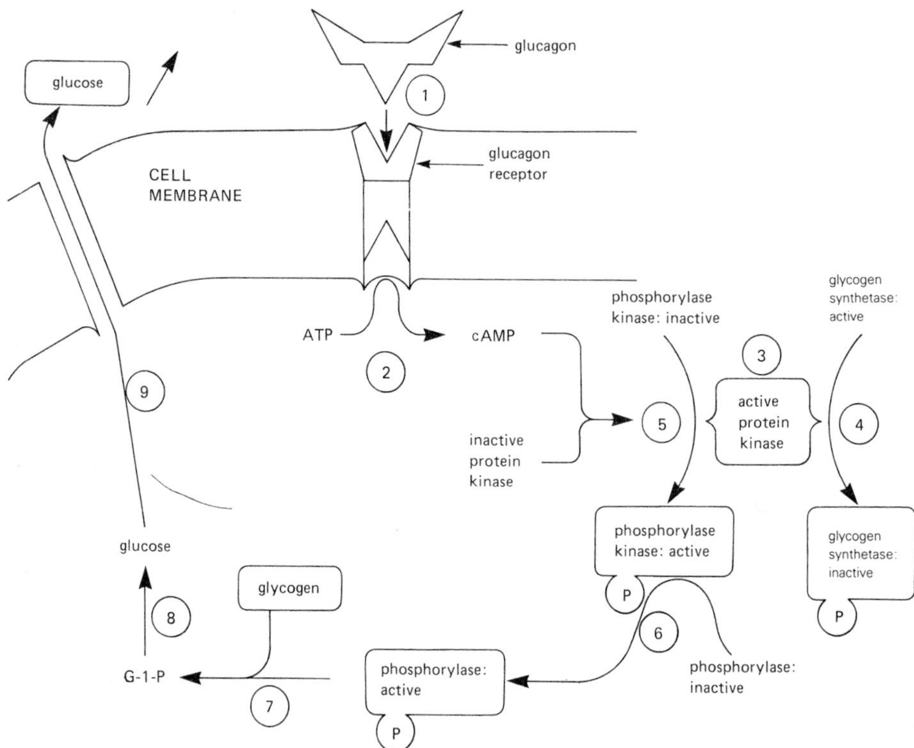

FIGURE 1-6 Cascade mechanism leading from the cell surface hormonal signal to the cellular metabolic response: glucagon and glycogenolysis. Glucagon combines with its cell membrane receptor (1), which stimulates the activity of adenylate cyclase, possibly mediated by a transducing element, on the cytoplasmic side of the membrane (2). The resulting increased level of cyclic AMP activates a protein kinase (3) by the mechanism shown in Fig. 1-37. The protein kinase subunits catalyze the phosphorylation of inactive phosphorylase kinase (5) and active glycogen synthetase (4) to produce the phosphorylated inactive form (6), which stimulates glycogenolysis (7) to form glucose 1-phosphate (G-1-P), which is further metabolized to glucose (8). Glucose is transported to the extracellular space and into general circulation (9). This represents a cascade system because each stimulated step after hormone binding is accomplished by an enzyme that can turn over multiple substrate molecules.

of the C-14 fatty acid myristate), or they may form oligomers, either covalently through the formation of disulfide bonds or noncovalently. Also, membrane receptors can be subjected to the process of endocytosis.

Since the hormone binds to the receptor's ligand domain noncovalently, after the signal transduction process is complete the hormone may reversibly dissociate, thus generating "free" or unliganded receptor available for a new round of signal transduction. In reality, after one round of signal transduction, inactivation or degradation of both the hormone (via catabolism or metabolism in the case of steroid hormones) and the receptor (endocytosis or phosphorylation) may ensue. Thus, there is a continuing need for the renewal (biosynthesis and secretion) of the hormone by the endocrine gland and of the receptor by the population of target cells.

E. Hormones and Cascade Systems

Many endocrine systems incorporate some form of cascade mechanism into their operation. A cascade mechanism is an amplification system where an initial reaction results in the generation of multiple second reactions, each of which sets off multiple third reactions, etc. Two examples will be presented.

A classical biochemical cascade mechanism is generated by the action of a hormone, such as glucagon acting at the cell membrane to produce an increase in cyclic AMP (see Figure 1-6). The cascade may be visualized in terms of alterations of cellular metabolism, which generate the ultimate cellular response, stimulation of glycogenolysis to generate glucose for export to the extracellular space, and the general circulating system, as in the example given in Figure 1-6.

Another important endocrine cascade system involves the central nervous system (CNS), the hypothalamus, pituitary, and distal endocrine secreting glands, and final target issues (see Figure 1-7).

The cascade effect may be produced by a single event or signal in the external or internal environment. By either electrical or chemical transmission, a signal is sent to the limbic system and then to the hypothalamus, resulting in the secretion of a releasing hormone into the closed portal system connecting the hypothalamus and anterior pituitary. Releasing hormones may be secreted in nanogram amounts and have half-lives of about 3–7 min. In turn they signal the release of the appropriate anterior pituitary hormones, which may be secreted in microgram amounts with half-lives on the order of 20 min or longer. The anterior pituitary hormone elicits the secretion of the ultimate hormone, which may be secreted in many micrograms or milligram amounts and may be fairly stable. Thus, in terms of increasing stability of hormones as one proceeds down the cascade, together with increasing amounts of hormones elaborated down the cascade, amplification of a single event at the outset could involve a factor of thousands to a millionfold. The ultimate hormone will operate on its receptors in many cell types, and these hormone–receptor complexes may stimulate the appearance of many phenotypes, augmenting the amplification built into the cascade system even further. Although not all signaling mechanisms involve this entire system, a good many do, as will be illustrated in later chapters.

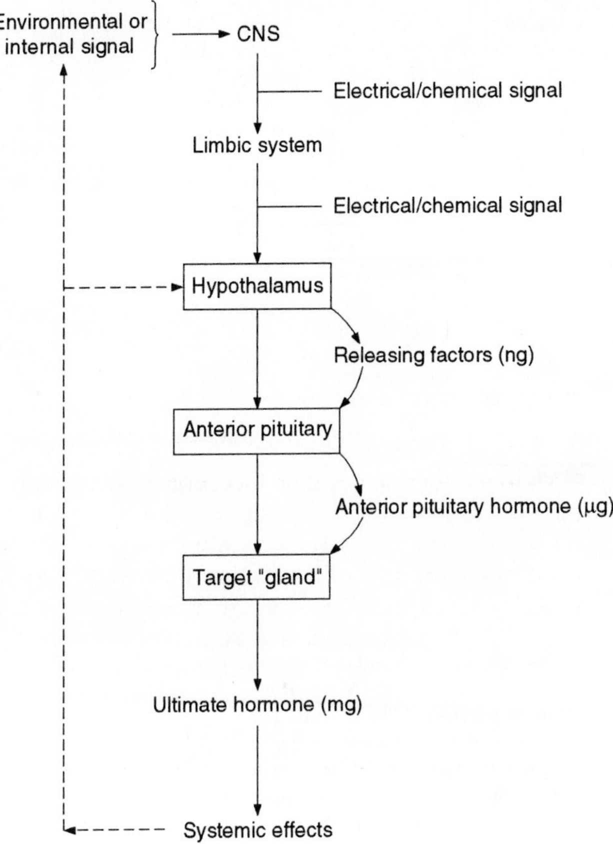

FIGURE 1-7 Hormonal cascade of signals from CNS to ultimate hormone. The target "gland" refers to the last hormone-producing tissue in the cascade that is stimulated by an appropriate anterior pituitary hormone. The dashed line indicates a negative feedback loop.

II. BIOSYNTHESIS OF HORMONES

A. Introduction

The biosynthesis of all hormones largely occurs in specialized cells usually present in endocrine glands, which have the genetic phenotype to permit the correct orderly production of the hormone in question. Chapter 2 describes the production of the steroid hormones, while Chapters 6, 11, and 16 describe the biosynthesis of the thyroid hormones, epinephrine, and prostaglandins, respectively. The next section of this chapter will provide a general description of the biosynthesis of peptide and protein hormones.

B. Biosynthesis of Peptide and Protein Hormones

Each of the over 125 protein and peptide hormones is biosynthesized in specific cells, whose phenotype results in the orderly operation of the processes of gene transcription and mRNA translation. Figure 1-8 summarizes the many steps involved with transcription and translation. In the field of molecular biology, complementary DNA (cDNA) cloning technology, rapid sequencing of DNA, the use of PCR (polymerase chain reaction) methodology to amplify rare DNA sequences, and the availability of immunological (highly specific monoclonal or polyclonal antibodies) and nucleic acid probes have all contributed enormously to the generation of our current understanding of these fundamental biological responses.

Genes for polypeptide hormones contain the information to generate the specific amino acid sequence of the hormone and the control elements upstream from the start of the transcriptionally active sequence. However, the old adage, "one gene for one protein," is no longer absolutely true. There are circumstances where one gene can encode information for multiple hormones or multiple copies of the same hormone (see the following).

Figure 1-9 compares the details of the gene for procorticotropin releasing hormone (CRH), where one

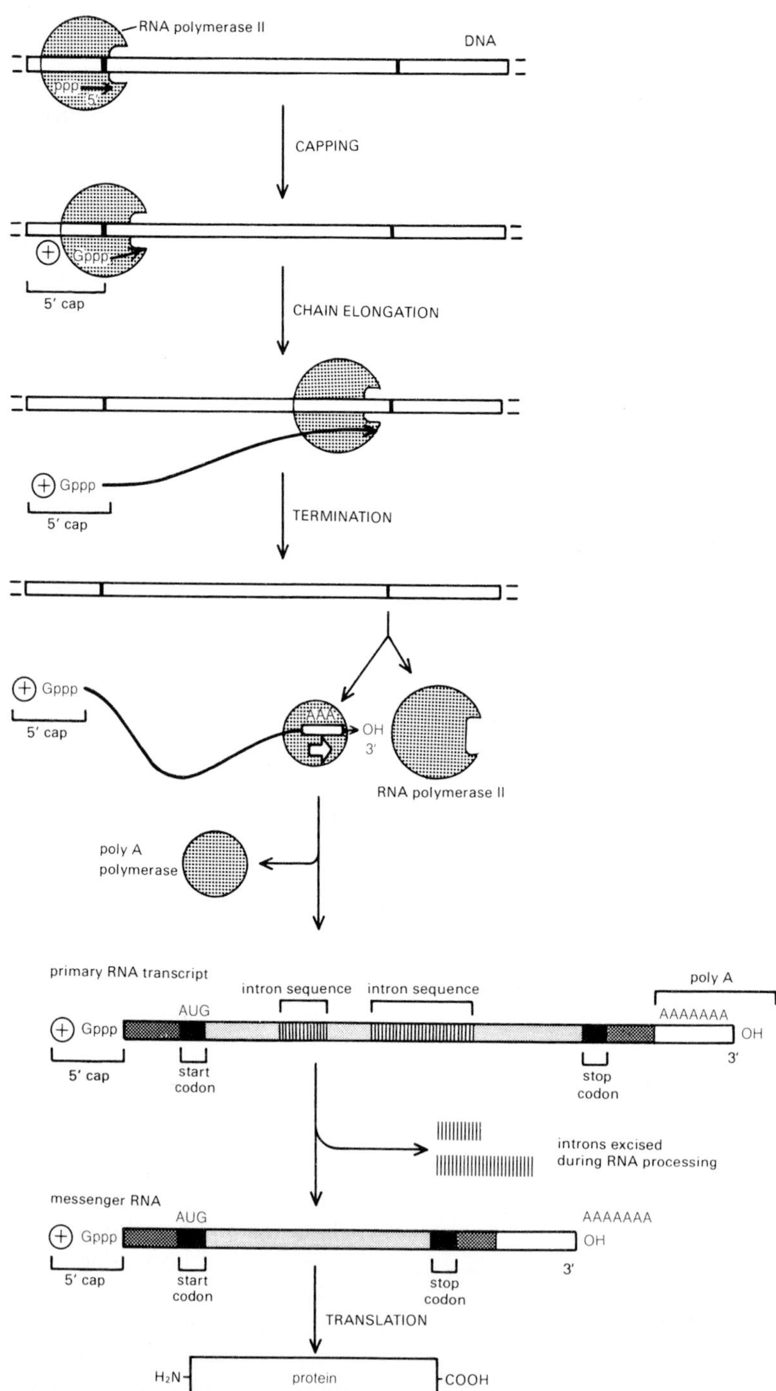

FIGURE 1-8 Overview of the processes of transcription (top) and translation (bottom). The transcriptional process consists of synthesis of RNA by RNA polymerase II, capping at the 5′ end, and addition of poly(A) at the 3′ end. The positive charge at the cap is the result of methylation of the nitrogen at the 7-position of guanine. The mRNA is further processed by excision of the intron, leading to the mature mRNA that is translated in the cytoplasm. Reproduced from Alberts B, Bray D, Lewis J, Raft, M, Roberts K, and Watson, J. D. (1983). "Molecular Biology of the Cell." Garland Publishing, New York.

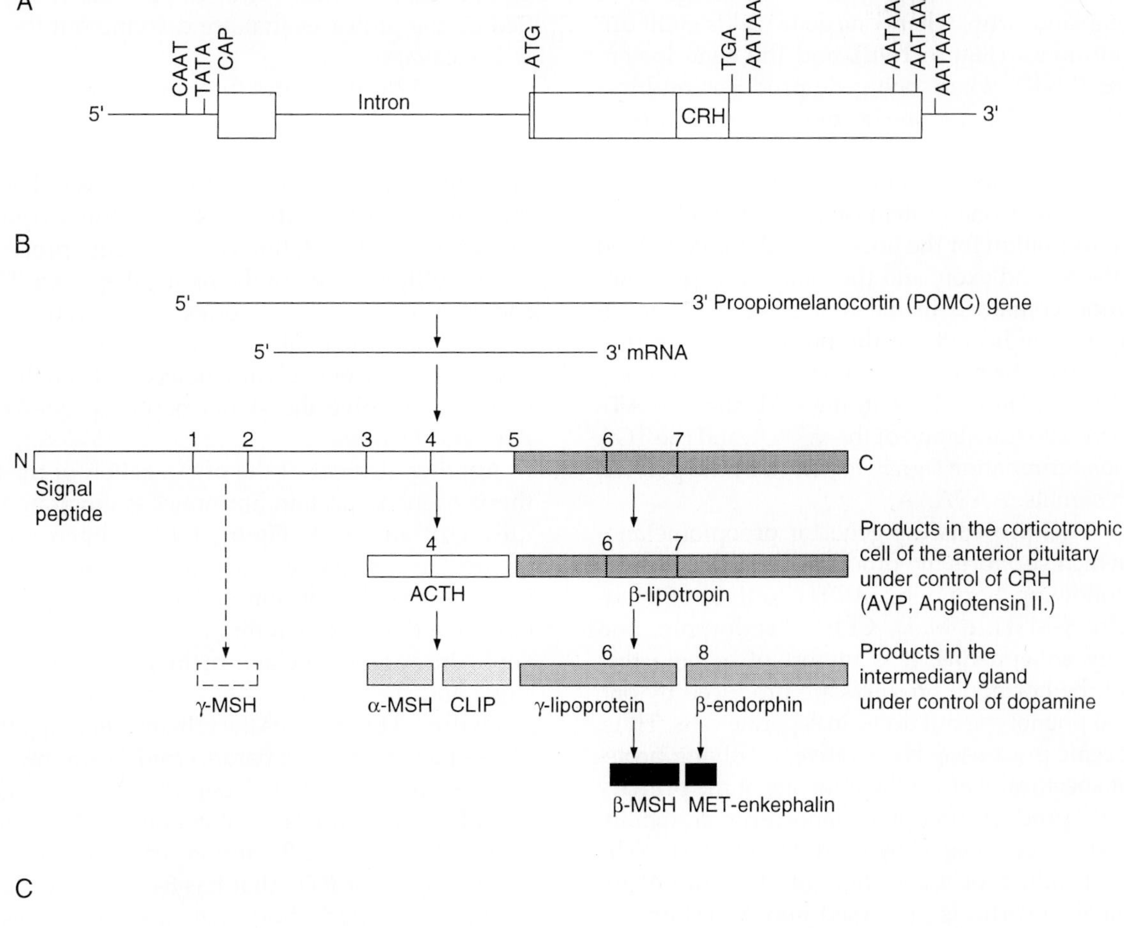

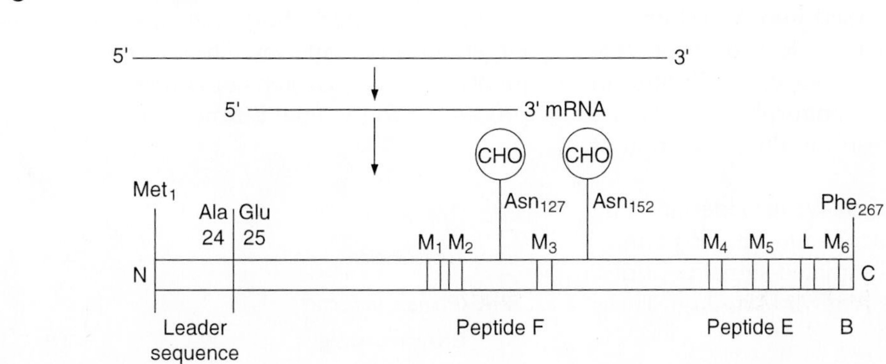

FIGURE 1-9 (A) Nucleic acid sequence for rat *proCRH* genes. A schematic representation of the rat *proCRH* gene. The exons are shown as blocks and the intron is shown by a line. The TATA and CAAT sequences, putative cap site, translation initiator ATG, translation terminator UGA, and poly(A) addition signals (AATAAA) are indicated. The location of the CRH peptide is indicated by CRH. Redrawn from Thompson, R. D., Seasholz, A. F., and Herbert, E. (1988). *Mol. Endocrinol.* **1,** 363. (B) Proopiomelanocortin is a polypeptide product encoded by a single gene. The dark vertical bars represent proteolytic cleavage sites for specific enzymes. The cleavage sites are Arg-Arg, Lys-Arg, or Lys-Lys. Some specificity also may be conferred by neighboring amino acid residues. In the corticotrophic cell of the anterior pituitary, enzymes are present that cleave at sites 3 and 5, releasing the major products ACTH and β-lipotropin into the general circulation. In the pars intermedia, especially in vertebrates below humans, these products are further cleaved at major sites 4, 6, and 7 to release α-MSH, CLIP, γ-lipotropin, and β-endorphin into the general circulation. Some β-lipotropin arising in the corticotroph may be further degraded to form β-endorphin. These two cell types appear to be under separate controls. The corticotrophic cell of the anterior pituitary is under the positive control of the CRH and its auxiliary helpers, arginine vasopressin (AVP) and angiotensin II. AVP by itself does not release ACTH, but enhances the action of CRH in this process. The products of the intermediary pituitary, α-MSH, CLIP (corticotropin-like intermediary peptide), γ-lipotropin, and β-endorphin, are under the positive control of dopamine, rather than CRH, for release. Obviously different proteases must exist in these different cell types in order to generate a specific array of hormonal products. β-Endorphin also contains a pentapeptide, enkephalin. (C) Model of the enkephalin precursor. The distribution of Met-enkephalin sequences (M_1–M_6 and Leu-enkephalin (L) is presented. Redrawn from Comb, M, Seeburg, P. H., Adelman, J., Eiden, L., and Herbert, E. (1982). *Nature* (*London*) **295,** 663.

gene yields one hormone (Figure 1-9A), the gene for proopiomelanocortin, where one gene yields eight different hormones (Figure 1-9B), and the gene for enkephalins (ENK), where one gene produces multiple copies of the mRNA for one hormone, met-ENK (Figure 1-9C).

Certainly the most general circumstance is where one gene encodes one protein (hormone). In Figure 1-9A, the information for the hormone CRH is contained only in the second exon, and the amino acid sequence information contained in the first exon is never expressed. Also indicated are the positions of the promoter region of the gene containing the TATA box and the CAAT regulatory element, the CAP site, the AT6 signal to initiate translation of the mRNA, and the TGA translation termination signal, followed by the poly(A) addition signals, AATAAA.

Figure 1-9B presents the gene for proopiomelanocortin, which supports the production in the pituitary of the following hormones: ACTH, β-lipotropin, α-lipotropin, γ-MSH, α-MSH, CLIP, β-endorphin, and potentially enkephalins. It is important to note that not all of the hormone products are produced by one single-cell phenotype but occur in separate cells. Thus, a cell-specific processing is operative, involving proteases that specifically affect the cleavage of the primary translation product proopiomelanocortin at the indicated cleavage sites shown in Figure 1-9B. The principal product of translation of the mRNA for proopiomelanocortin is processed into ACTH and β-lipotropin. The ACTH can then be cleaved to generate α-MSH and CLIP, while the cleavage of β-lipotropin generates α-lipotropin and β-endorphin. Table 1-2 summarizes which stimuli result in the generation of which hormone products.

Each of the specific sites of cleavage is identified by the presence in the amino acid sequence of pairs of basic amino acids that are designated signal peptides, usually either Lys-Lys, Lys-Arg, or Arg-Arg. These pairs of dibasic amino acids can be specifically identified by the proteases that are carrying out the proteolytic cleavage.

TABLE 1-2 Summary of Stimuli and Products of Proopiomelanocortin[a]

Cell type	Corticotroph	Pars intermedia
Stimulus	CRH	Dopamine
Auxiliary stimulus	AVP, AII	
Major products	ACTH β-lipotropin (β-endorphin)	α-MSH, CLIP, γ-lipotropin, β-endorphin

[a] CRH, corticotropin releasing hormone; AVP, arginine vasopressin; AII, angiotensin II; ACTH, adrenocorticotropin; α-MSH, α-melanocyte-stimulating hormone; CLIP, corticotropin-like intermediary peptide.

Figure 1-9C illustrates the gene organization for enkephalins that are produced by the chromaffin cells of the adrenal medulla. A typical enkephalin comprises only eight amino acids, yet it is still biosynthesized by the traditional procedures of gene transcription and mRNA translation followed by specific processing to yield multiple copies of the final enkephalin. The gene shown in Figure 1-8C encodes six separate coding regions for the enkephalin. Again, the processing sites to release the final enkephalin molecules from the protein precursor involve the signal peptides Lys-Arg, Arg-Arg, and Lys-Lys.

Another element of the organization of the biosynthesis of many protein hormones is illustrated in Figures 1-10 and 1-11. Figure 1-10 enumerates the sequence of steps employed for many secreted protein–peptide hormones, e.g., insulin, parathyroid hormone (PTH), calcitonin, somatostatin, gastrin, enkephalins, proopiomelanocortin, glucagon, neurophysins I and II.

Figure 1-11 illustrates the scheme depicting the steps of the biosynthesis of parathyroid hormone taken as a representative secreted peptide hormone. One can identify a pre-pro-PTH that contains 115 amino acids, a pro-PTH that has 93 amino acids, and the mature secreted form of PTH that has 84 amino acids.

Secreted peptide hormones are biosynthesized as larger precursor species. These precursor species can involve one or two steps of conversion by proteolytic processing to the final hormone; if it is one step, then

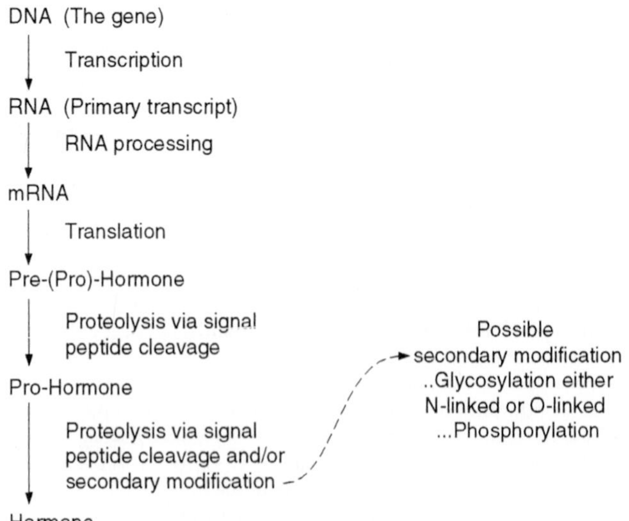

FIGURE 1-10 Schematic illustration of the steps in the biosynthesis of a peptide hormone.

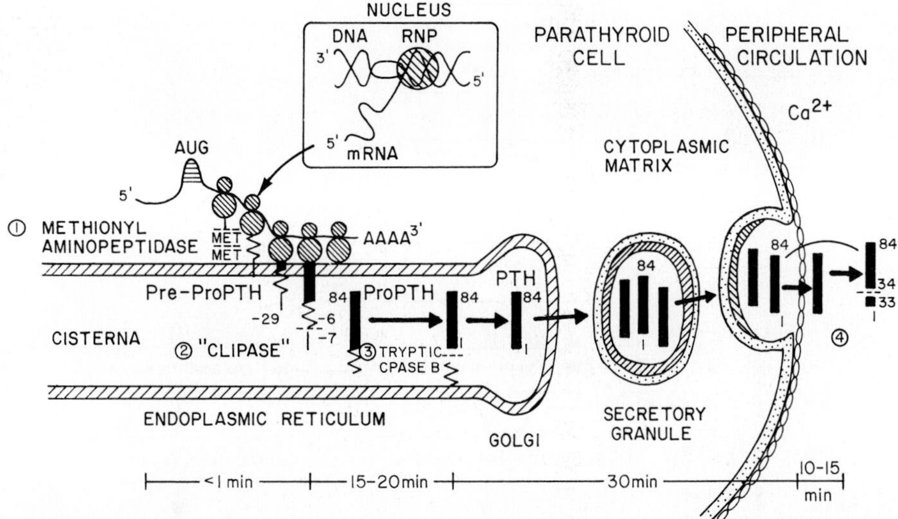

FIGURE 1-11 Scheme depicting the proposed intracellular pathway of the biosynthesis of parathyroid hormone. Preproparathyroid hormone (pre-pro-PTH), an initial product of synthesis on the ribosomes, is converted to proparathyroid hormone (pro-PTH) by the removal of (1) the NH_2-terminal methionyl residues and (2) the NH_2-terminal sequence (−29 through −7) of 23 amino acids during or within seconds after synthesis, respectively. The conversion of pre-pro-PTH probably occurs during transport of the polypeptide into the cisterna of the rough endoplasmic reticulum. By 20 min after synthesis, pro-PTH reaches the Golgi region and is converted into PTH by (3) removal of the NH_2-terminal hexapeptide. PTH is stored in the secondary granule until it is released into circulation in response to a fall in the blood concentration of calcium. The times needed for these events are given below the scheme. Modified from Habener, J. F., Kemper, B. W., Rich, A., and Potts, J. T. (1977). Biosynthesis of parathyroid hormone. *Recent Prog. Horm. Res.* **33,** 287.

a prohormone is the immediate product of the translation process, while if two steps are involved, then a pre-pro version of the hormone is the immediate translation product. The function of the pre sequence or pre-pro sequence is to facilitate the insertion of the nascent peptide into the membrane of the rough endoplasmic reticulum (RER) immediately followed by movement of the entire peptide into the cisterna of

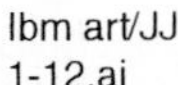

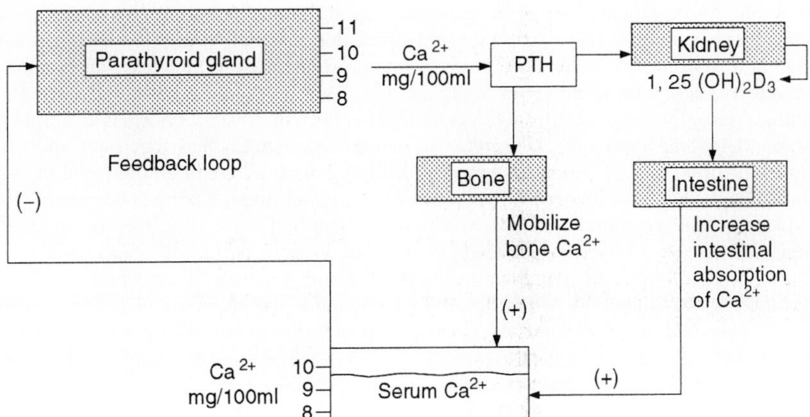

FIGURE 1-12 Schematic diagram of a peripheral feedback system analogous to that utilized by PTH and $1,25(OH)_2D_3$ to maintain a normal serum Ca^{2+} level in the range 9.5–10.5 mg/100 ml of serum.

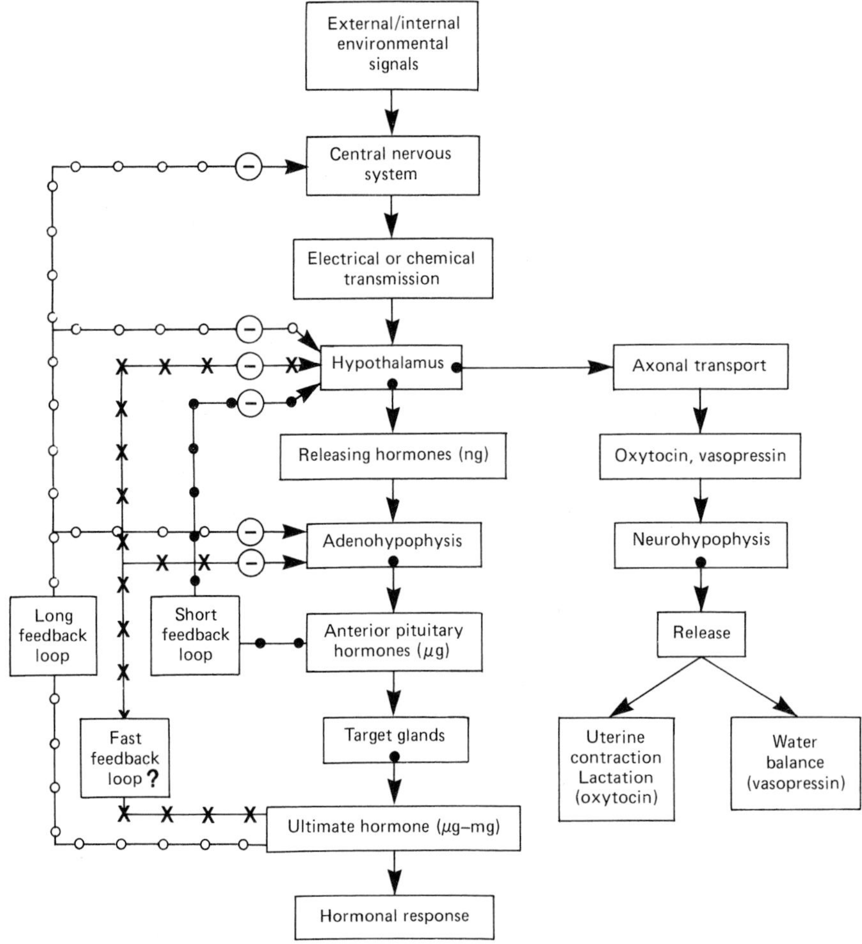

FIGURE 1-13 Schematic diagram of a hormone system involving a peripheral feedback mechanism. Many hormonal systems involve the hypothalamus. This figure shows a cascade of hormonal signals starting with an external or internal environmental signal. This is transmitted first to the central nervous system and may involve components of the limbic system, such as the hippocampus and amygdala. These structures innervate the hypothalamus in a specific region, which responds with the secretion of a specific hormone, usually in nanogram amounts. Releasing hormones are transported down a closed portal system connecting the hypothalamus and the anterior pituitary. Peptide releasing hormones pass the blood–brain barrier at either end through fenestrations. A specific releasing hormone binds to a specific anterior pituitary cell membrane receptor and causes the secretion of specific anterior pituitary hormones, usually in microgram amounts. These access general circulation through fenestrated local capillaries and bind to specific target gland receptors. The interactions trigger the release of the ultimate hormone in microgram to milligram daily amounts, which cause the hormonal response by binding to receptors in several distal target tissues. In effect, this overall system is an amplifying cascade. Releasing hormones are secreted in nanogram amounts, and they have short half-lives on the order of a few minutes. Anterior pituitary hormones are often produced in microgram amounts and have longer half-lives than releasing hormones. Ultimate hormones can be produced in milligram amounts daily, with much longer half-lives. Thus, the system taken as a whole constitutes an enormous cascade mechanism. With respect to differences in the mass of hormone produced, from hypothalamus to target gland, the range is nanogram to milligram or as much as a millionfold. When the ultimate hormone has receptors in nearly every cell type, it is possible to affect the body chemistry of virtually every cell by a single initiating environmental signal. Consequently, the organism is in intimate association with the external environment, a fact that we tend to underemphasize. Arrows with a black dot at their origin indicate a secretory process.

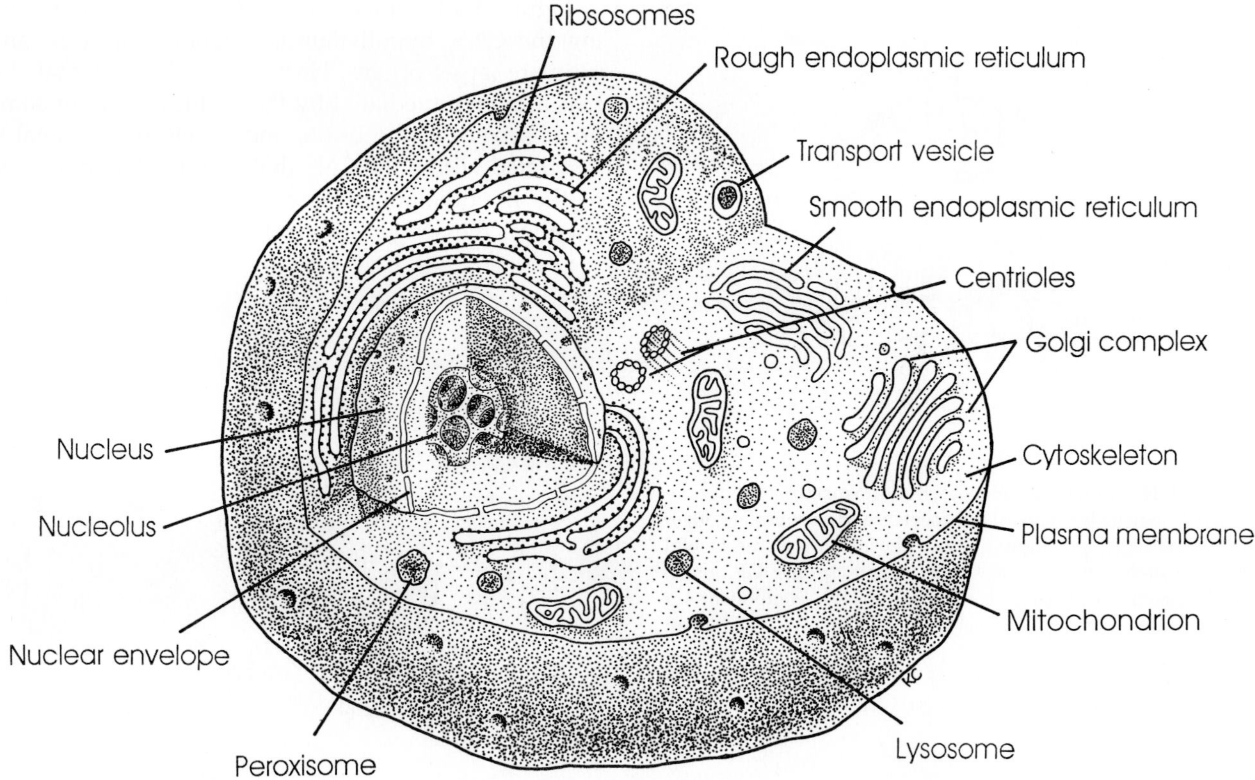

FIGURE 1-14 Schematic illustration of a representative animal cell. The important structural aspects of the cell are identified.

the RER. A signal recognition particle (SRP) protein attaches to the signal peptide (the N-terminus region of the nascent peptide) and simultaneously attaches to a SRP receptor in the membrane of the rough endoplasmic reticulum. This facilitates the insertion process of the growing peptide and the efficient accumulation of the pre-(pro)hormone inside the RER. Next, specific peptidases cleave off the pre form (23 amino acids) of pre-pro-PTH. The resulting pro-PTH (92 amino acids) is then cleaved by another specific peptidase so that the mature 1-84 amino acid PTH is packaged into a secretory granule, where it is stored in the chief cell of the parathyroid gland. The PTH is then released from the cell by the process of exocytosis, which results in the contents of the secretory gland being released into the general circulatory system.

C. Regulation of Hormone Secretion

The production of all hormones is regulated by the homeostatic mechanisms operative in that particular endocrine system. The secretion or release of the hor-

TABLE 1-3 Lipid Compositions of Different Cell Membranes (% of Lipid Weight)[a]

Lipid	Liver plasma membranes	Erythrocyte membranes	Mitochondria	Endoplasmic reticulum (ER)
Cholesterol	17	23	3	6
Phosphatidylethanolamine	7	18	35	17
Phosphatidylserine	4	7	2	5
Phosphatidylcholine	24	17	39	40
Sphingomyelin	19	18	0	5
Glycolipids	7	3	Trace	Trace
Others	22	13	21	27

[a] This table was adapted with permission from B. Alberts, D. Bray, J. Lewis, M. Raff, K. Roberts, and J. D. Watson (1983). "Molecular Biology of the Cell." Garland Publishing Co., New York.

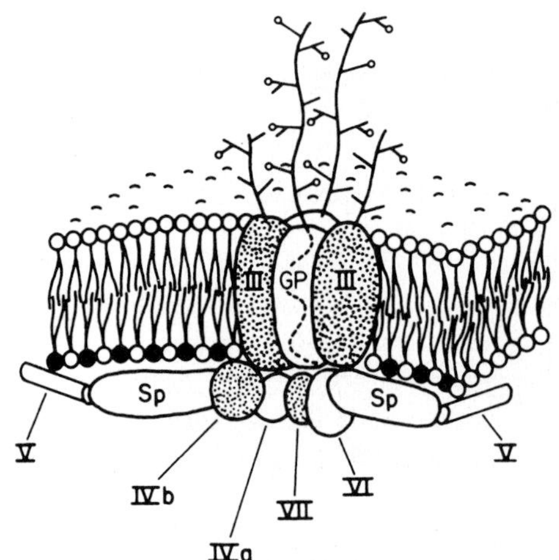

FIGURE 1-15 Model of human erythrocyte membrane. Speculative model based on the human erythrocyte major glycoprotein oligomeric complex. Phosphatidylserine and phosphatidylglycerol (solid circles) are distributed asymmetrically. Abbreviations: GP, glycophorin; III, component 3; IVa, component 4.1; IVb, component 4.2; V, component 5 or actin; VI, component 6 or GP-3-D; VII, component 7; Sp, spectrin. Reproduced from Nicholson, G. L. (1982). *In* "Biological Regulation and Development" (R. F. Goldberger and K. R. Yamamoto, eds.), Vol. 3A, p. 222. Plenum, New York.

mone is normally (in the absence of an endocrine disease related to hormone secretion) strictly related to the organism's perceived "need" for the biological response(s) generated by the hormone in question. The "need" for any hormone will be determined by the physiological set point for that endocrine system. Thus, a characteristic feature of most endocrine systems is the existence of a feedback loop that limits or regulates the secretion of the hormonal messenger. There are two general categories of endocrine feedback systems: those involving the CNS and the hypothalamus and those where the function achieved by the hormone (elevated blood glucose, elevated serum Ca^{2+}, etc.) directly feeds back upon the endocrine gland that secretes the hormones.

Figure 1-12 depicts a schematic diagram of a peripheral feedback system analogous to that utilized by PTH and $1,25(OH)_2D_3$ to maintain a normal serum Ca^{2+} level in the range 9.5–10.5 mg/100 ml of serum. When serum Ca^{2+} falls, PTH secretion is stimulated and the secreted PTH initiates, at distal target organs, a biological response(s) that elevates serum Ca^{2+}. In the event that the serum Ca^{2+} rises above 10.5 mg/100 ml of serum, there is a feedback signal to the parathyroid gland that represses the secretion of PTH (see Chapter 9 for additional details).

Figure 1-13 depicts an endocrine system involving the CNS, hypothalamus, adenohypophysis, and ultimate target organs. Here, external or internal signals are often mediated by the limbic system or some other system of the brain, and an electrical signal is generated within the CNS that is ultimately delivered to the hypothalamus, where it is transduced into a chemical messenger—the releasing hormones of the

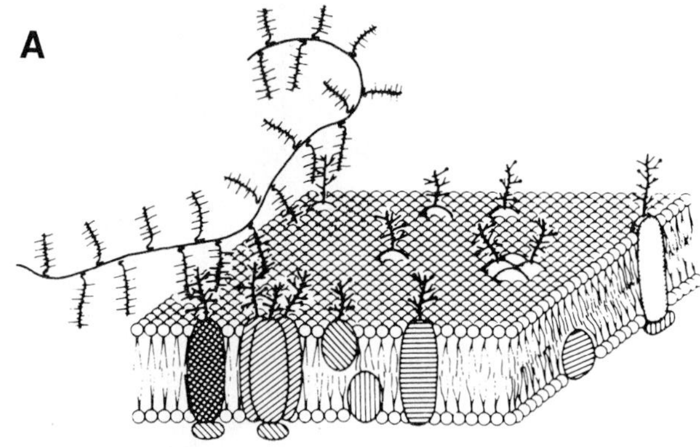

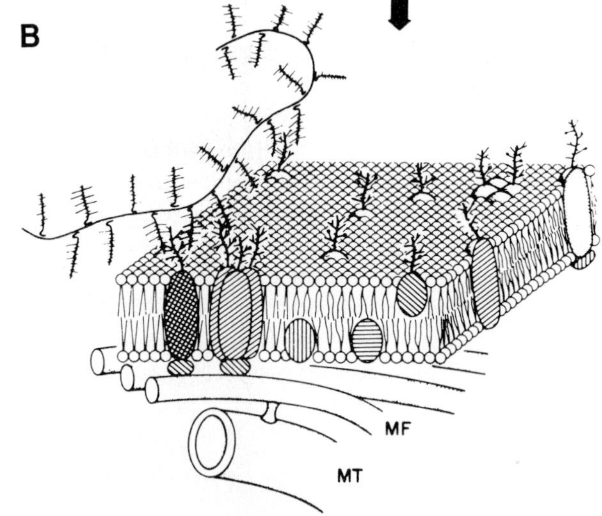

FIGURE 1-16 Model of a fluid mosaic membrane showing distribution and mobility of cell membrane receptors. (A) Integral transmembrane glycoproteins occur in a domain of the membrane of different lipid composition; a glycosaminoglycan molecule is shown associated with the carbohydrate side chains of the glycoproteins. (B) Aggregation of some of the glycoproteins stimulates the attachment of cytoskeletal components: microfilaments (MF) and microtubules (MT) to peripheral membrane components on the cytoplasmic surface are shown. Other unattached integral membrane proteins are capable of lateral motion. Reproduced from Nicholson, G. L. (1982). *In* "Biological Regulation and Development" (R. F. Goldberger and K. R. Yamamoto, eds.), Vol. 3A, p. 225. Plenum, New York.

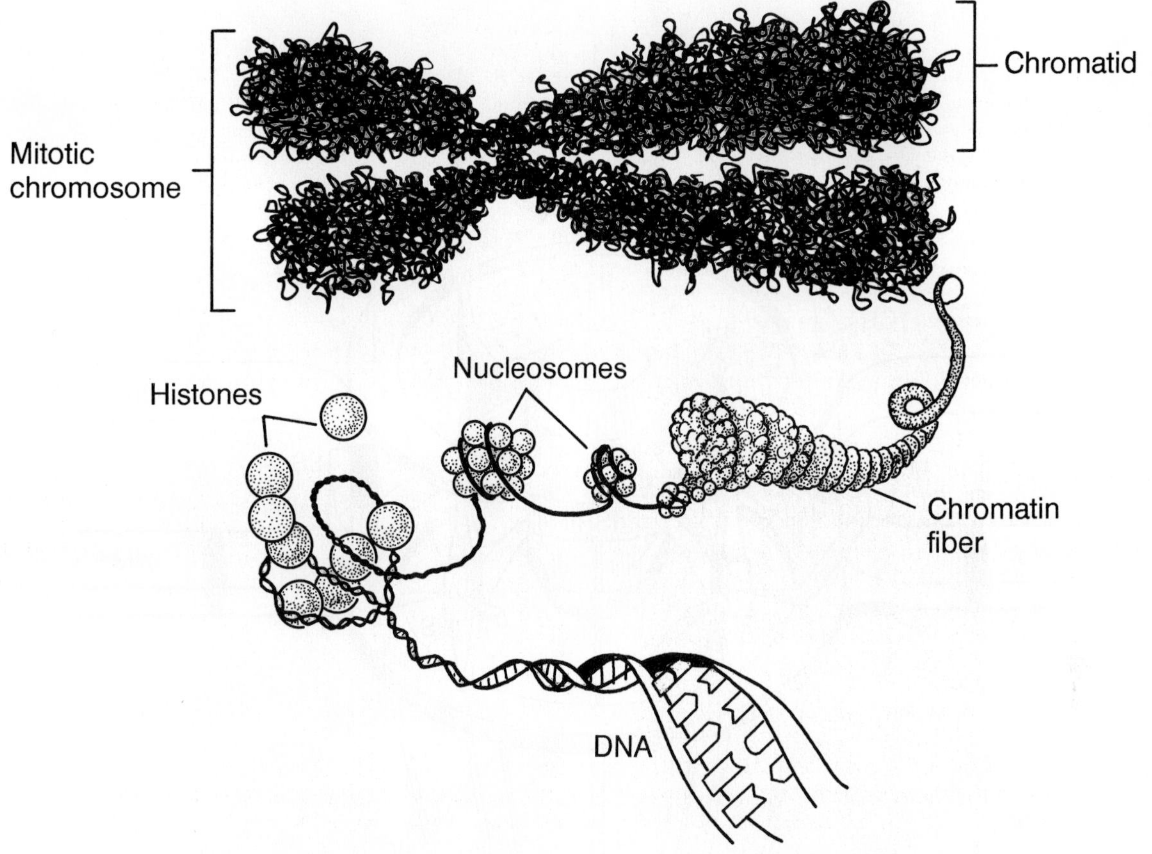

FIGURE 1-17. Schematic model of a chromosome. The DNA double helix is wrapped around a core of histone proteins to form a nucleosome; altogether the DNA of a single human chromosome is estimated to have approximately 1 million nucleosomes. The nucleosomes, in turn, aggregate to create a very ordered structure termed the "chromatin fiber." The resulting chromatin fibers, which are approximately 30 nm in diameter, condense to create clustered, looped regions that create the chromosome. Each chromosome is composed of two chromatids. This figure was adapted from Becker, W. M., and Deamer, D. W. (1991). "The World of the Cell," 2nd edition. The Benjamin/Cummings Publishing Co., Menlo Park, CA.

adenohypophysis (see the legend to Figure 1-13 for additional detail).

D. Oncogenes and Hormones

It has become clear that there is a close link between endocrinology and certain forms of cancer. A cancer cell may be described as a cell that has escaped the normal constraints of physiological regulation that maintain the function of that cell within the normal boundaries of the homeostatic state. The "escape" can involve the oncogenic production by the cancer cell of excess or altered forms of a hormone or its receptor. An oncogene is a gene related to cancer. The presence of this oncogenic hormone and/or oncogenic receptor can then totally distort the regulated state of the cell; this may contribute to the aggressive properties of a tumor. This is particularly so if the oncogene product is an analog of a cellular growth factor that stimulates growth of the tumor. These complex details are considered further in Chapter 20.

III. HORMONE RECEPTORS

A. Cellular Organization

In order to describe the complex details of hormonal receptors and how they generate biological responses, it is appropriate to present a brief review of cellular organization. A typical animal cell is shown in Figure 1-14.

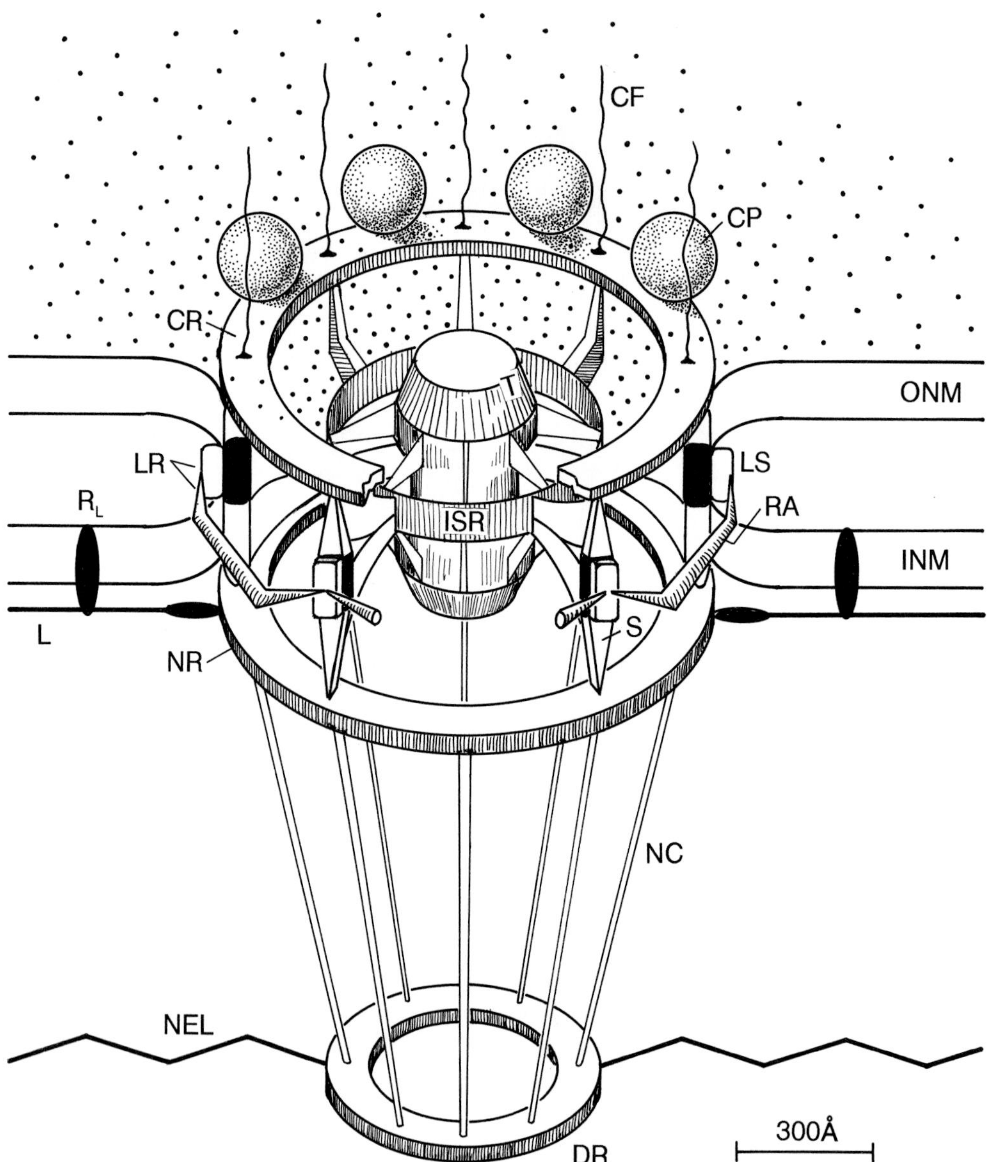

FIGURE 1-18 Three-dimensional ribbon representation of the nuclear pore complex. For clarity, the dimensions of certain features have been altered; for example, the inner spoke ring (ISR) is disproportionately short and thin. Abbreviations: CR, cytoplasmic ring; NR, nucleoplasmic ring; LR, lumenal ring; S, spokes; LS, lumenal spoke domain; RA, radial arm dimers; T, transporter, CP, cytoplasmic particles; CF, cytoplasmic filaments; NC, nucleoplasmic cage; DR, distal ring; NEL, nuclear envelope lattice; ONM, outer nuclear membrane; INM, inner nuclear membrane; L, nuclear lamina; R_L, lamin receptor. Modified with permission from Akey, C. W., and Radermarcher, M. (1993). Architecture of the *Xenopus* nuclear pore complex revealed by three-dimension cryo-electron microscopy. *J. Cell. Biol.* **122,** 1–19.

1. Cell Membranes

Although the precise content of substances that comprise the cell membrane differs in different cell types, many components are common to all membranes. These are lipids (including phospholipids and glycolipids), proteins, and glycoproteins. The cell membrane that encloses the cell resembles the internal membranes, such as those associated with the nucleus, mitochondria, and microsomes. The approximate lipid compositions of different cell membranes are presented in Table 1-3.

The lipids consist of two portions: a head group and a tail. The head groups are charged (hydrophilic), while the tails are uncharged (hydrophobic). They form bilayers spontaneously oriented, so that the heads are on the water side and the tails are internal. This resembles the basic structure of the cell membrane and provides for a negative charge on the head side of the membrane. This arrangement is pictured in Figure 1-15. Included in the membrane are various proteins, some of which are receptors, ions and other small molecule transporters, and enzymes. The specific proteins vary from cell type to cell type depending on the special membrane functions required by a given cell. Some of the proteinaceous constituents span the membrane so as to make contact with both the extracellular space and the cytoplasm, while others do not but can translocate to either side. Proteins that span the membrane are called intrinsic or integral proteins. Extrinsic proteins are those associated with the periphery of the membrane but that may not be associated directly with the phospholipid layer. The cell membrane interacts with the fibrillar and tubular structures in the cytoplasm, facilitating dynamic movements of portions of the membrane into the cell (e.g., during internalization of ligand–receptor complexes). Thus, membrane components may connect with proteins called spectrin, actin, and others, partly visualized in Figure 1-15.

2. *Fluid Cell Membrane Model*

In the context of the model just described, it is believed that various conditions, such as changes in the metabolism of phospholipids and responses to various signals via receptor interactions, can lead to changes in fluidity of the membrane that allow proteins to move within the membrane. This movement can generate productive interactions that can result in the stimulation of enzymatic activities on the cytoplasmic side of the cell membrane or other events. A model showing this type of motion is presented in Figure 1-16.

3. *Nuclear Organization*

In eukaryotes, the nucleus and cytoplasm are separated compartmentally, except during interphase. The nucleus is surrounded by a nuclear envelope, which comprises two membranes separated by a small space. Also present in this nuclear membrane are pores (see the next section).

Inside the nucleus are the nucleolus and the chromatin. The nucleolus is where the DNA that encodes ribosomal RNA is continuously transcribed. The chromatin comprises DNA and tightly associated basic proteins (histones) that represent the chromosomes; the individual chromosomes are only visible during the time of cell division. The DNA and histones interact to create nucleosomes. Figure 1-17 presents a schematic model of the organization of a chromosome which includes in, ascending hierarchical order, DNA + histones, nucleosomes, chromatin fiber, and chromatid.

4. *Nucleus–Cytoplasm Transport*

Communication between the nuclear and cytoplasmic compartments is important for a wide variety of cellular activities. These can include the transfer of enzymes synthesized in the ribosomes that are essential in the nucleus for DNA replication, repair, and transcription. In addition, mRNA precursors and associated proteins move out of the nucleus. Also, some steroid receptors move from the cytoplasm to the nucleus via the nuclear pores.

Evidence has been presented for the existence of specialized communication pathways, termed the nuclear pore complex (NPC). The NPC is responsible for mediating molecular trafficking between the nuclear and cytoplasmic compartments. The NPC is an octagonally symmetrical cylinder of $\approx$80 nm length $\times$ $\approx$100 nm diameter; its molecular weight has been estimated by scanning electron microscopy to be $\approx 1.25 \times 10^8$. The NPC comprises a family of proteins known as nucleoporins. Figure 1-18 presents a schematic model of a nuclear pore complex.

B. Membrane Receptors

Peptide hormones and growth factors typically bind to their cognate receptors present in the plasma membrane of their target cell(s). Once the membrane recep-

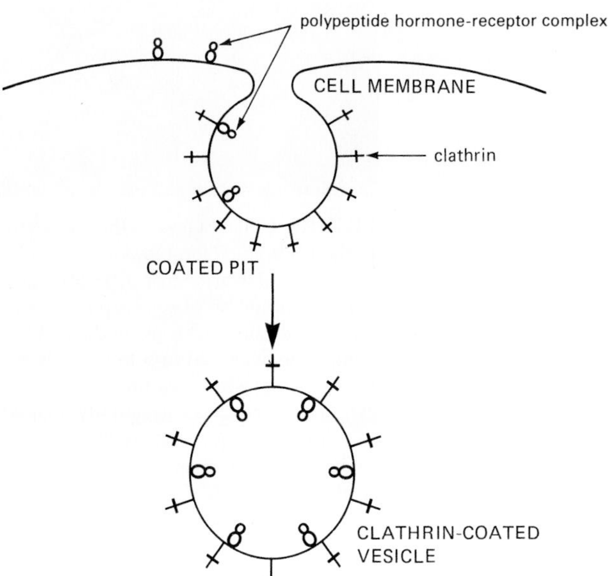

FIGURE 1-19 Diagram of coated pit and vesicle.

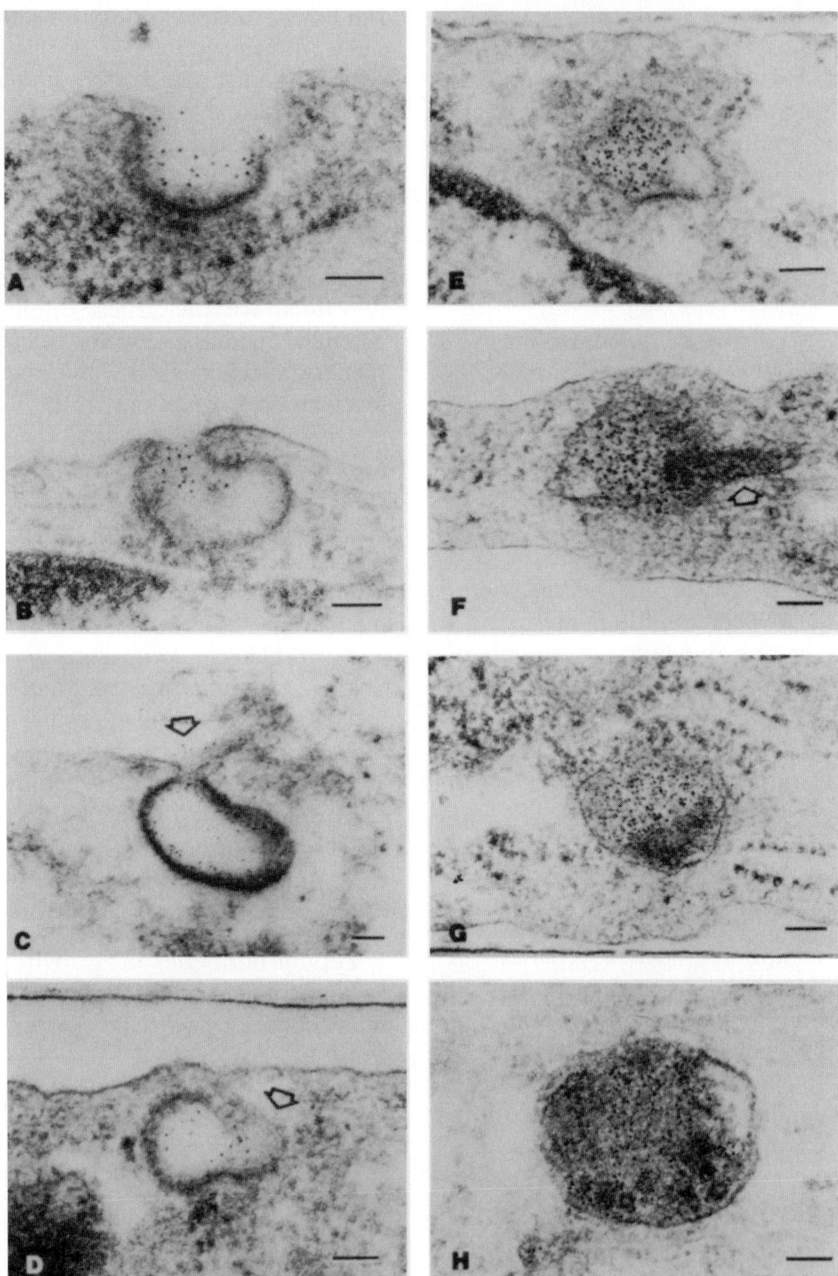

FIGURE 1-20 Stages in the endocytosis of LDL ferritin and its subsequent appearance in the lysosome. Fibroblasts were incubated with 47.5 μg/ml LDL ferritin for 2 hr at 4°C, washed, and warmed at 37°C for various times (bar = 1000 Å): (A) typical coated pit; (B) transformation of coated pit into endocytotic vesicle; (C) formation of a coated vesicle (arrow indicates that some of the LDL ferritin is left on the surface of the cell as the plasma membrane begins to fuse to form the vesicle); (D) fully formed vesicle (arrow shows the loss of cytoplasmic coat on one side); (E) endocytotic vesicle that has completely lost the cytoplasmic coat; (F) irregularly shaped endocytotic vesicle (arrow shows the region of increased electron density with lumen); (G) similar to (F) with more electron-dense material in the lumen; (H) secondary lysosome containing LDL ferritin. Reproduced with permission from Anderson, R. G. W., Brown, M. S., and Goldstein, J. L. (1977). Role of the coated endocytotic vesicle in the uptake of receptor-bound low-density lipoprotein in human fibroblasts. *Cell* **10,** 351–364. Copyright © 1977 by MIT, Cambridge, MA.

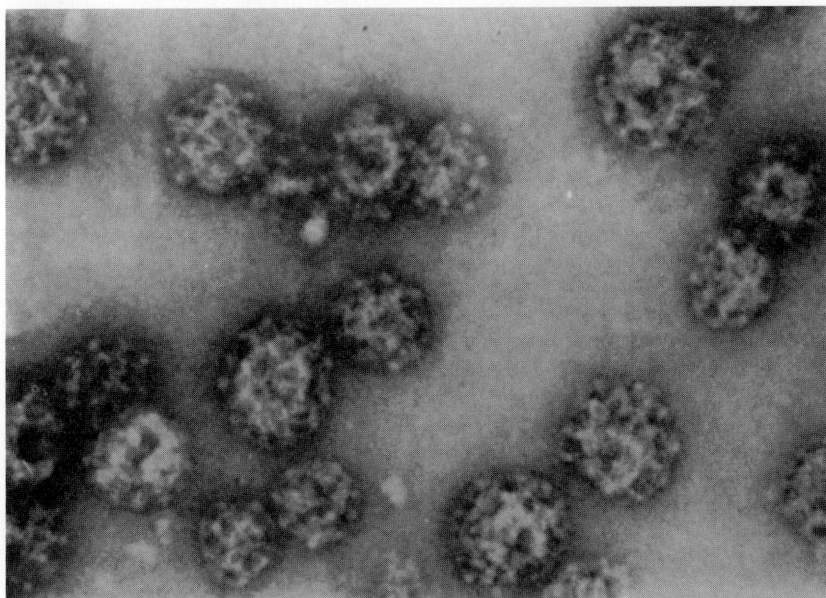

FIGURE 1-21 Electron micrograph of isolated coated vesicles from adrenal medulla magnified 125,000 times. Reproduced from Pearse, B. M. F., and Bretscher, M. S. (1981). Membrane recycling by coated vesicles. *Annu. Rev. Biochem.* **50,** 85–101. Copyright © 1981 Annual Reviews Inc., Palo Alto, CA.

tor–peptide hormone complex is formed, there are two possible resulting scenarios: (a) the ligand–receptor complex may be internalized into the cell or (b) the ligand–receptor complex may generate its signal transduction event(s) on the inside of the cell while maintaining its position in the cell membrane. These alternatives are discussed in the following sections.

1. Membrane Receptors That Are Internalized

The internalization of the ligand–receptor complex may function to allow the release of the intact hormone or derived peptide fragments on the inside of the cell. Among the possibilities are the following: a means to degrade the ligand after it has promoted some action (or not) at the cell membrane; a means to degrade the receptor; a means to degrade the ligand to a product that is active in the cell's interior or to liberate an unaltered ligand inside the cell to bind to interior receptors (e.g., on the nuclear membrane) to produce a mitogenic effect; a means to degrade the receptor to an active fragment that would operate within the cell; and finally, and perhaps most importantly, it may be the off-signal for activity at the cell membrane level. Hormones have immediate effects at the cell membrane and in some cases have long-term effects through processes generated at the membrane or after internalization.

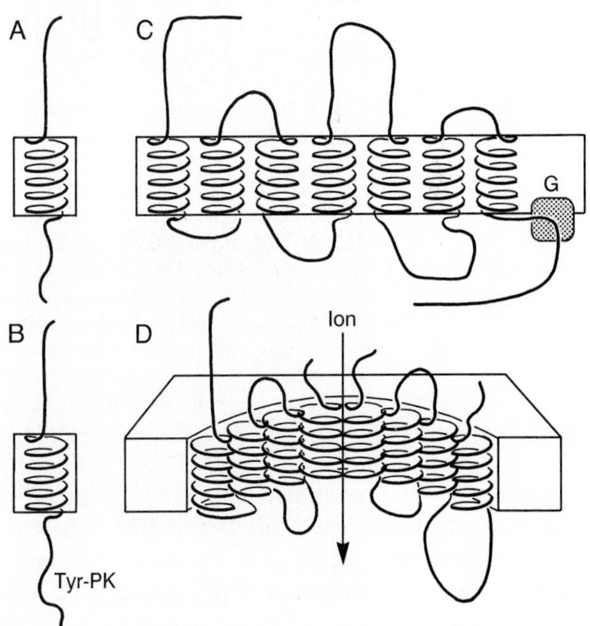

FIGURE 1-22 Four classes of membrane-spanning hormone receptors: (A) receptor with a single-membrane-spanning component and without a tyrosine kinase component; (B) receptor with a single-membrane-spanning component involving a tyrosine kinase component inside the cell; (C) receptor with seven-membrane-spanning α-helical regions that are coupled to a separate G protein facing the interior of the cell; (D) receptor with part of a membrane-spanning ion channel.

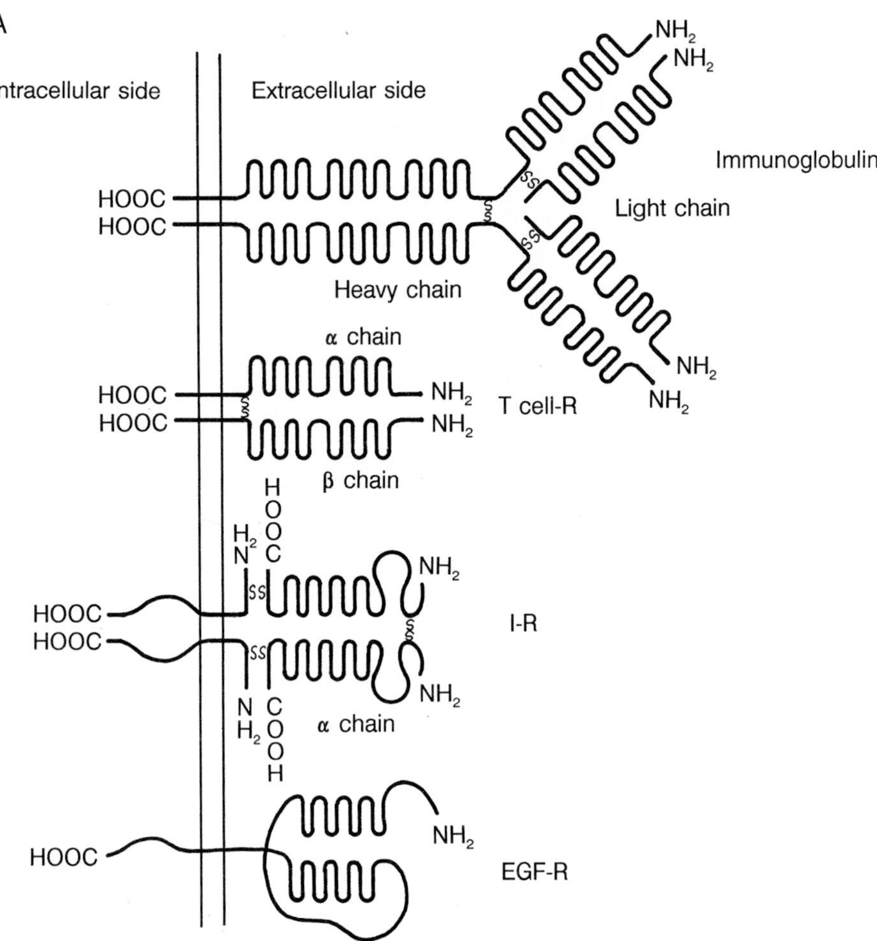

FIGURE 1-23 Example of hormone single-membrane-spanning receptors (B) and illustration of the structural homology of some of them to immunoglobulins (IgG) (A). The disulfide, S–S, bonds are indicated. (A) Structural comparison between membrane receptors and immunoglobulin. In this very schematic drawing, the overall structures of two hormone receptors (EGF-R and I-R), the T-cell receptor, and an immunoglobulin are compared. Only S–S bonds between different chains have been highlighted. Repeated loops represent cysteine-rich regions. (Hood, L., Kronenberg, M., and Hunkapiller, T. (1985). T cell antigen receptors and the immunoglobulin supergene family. *Cell* **40,** 225–220.) (B) Single-membrane-spanning receptors. The extracellular portion of these receptors contains the hormone-binding site. Tyrosine protein kinase domains are shown as open boxes, with insertions in the case of PDGF-R and CSF-I-R. Dots represent cysteines, and cross-hatched boxes represent the cysteine-rich domains. All N-terminal extremities are extracellular. All proteins are drawn to the same scale.

In terms of a polypeptide hormone that produces a mitogenic effect on a cell, one possibility is to view the internalization process as a means to generate a mitogenically active fragment or phosphorylated form of the hormone or its undegraded structure inside the cell, so that it can interact in some way (receptor) with the nucleus to increase the rate of DNA replication. Perhaps some of these intracellular products can lead to activators of DNA polymerase α, the replicative enzyme.

Internalization has been shown to apply to insulin–receptor complexes, EGF–receptor complexes, nerve growth factor (NGF) and thyroid stimulating hormone, and others. This process is referred to as "adsorptive" endocytosis. Polypeptides bind to receptors in coated pits that are indented sites on the plasma membrane. The coated pit invaginates into the cytoplasm, forming a coated vesicle containing hormone–receptor complexes. The vesicles then shed their coats and fuse with each other, and either then the receptors are returned to the cell surface after subsequent fusion with the Golgi apparatus or they fuse with lysosomes and the contents may be degraded. Formation of a coated vesicle is shown in Figure 1-19.

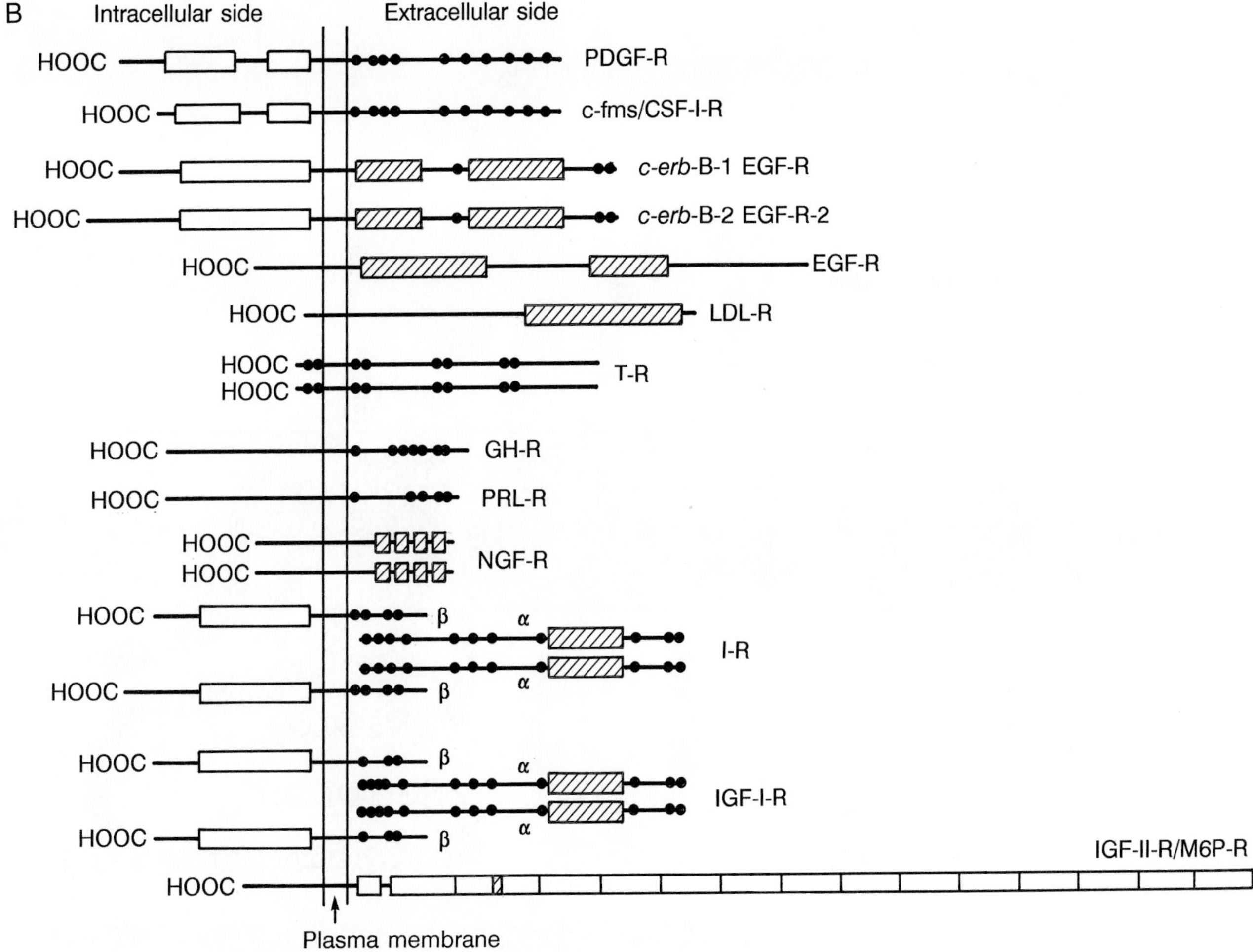

FIGURE 1-23—*Continued*

Photographs of a coated pit and an endocytotic vesicle are shown in Figure 1-20. Recent evidence suggests that the coated vesicle forming is a molecular filter, accounting for the selectivity of the contents. Diameters of coated vesicles are in the range 50–150 nm. Coated vesicles from adrenal medulla are shown in Figure 1-21. The major protein component of the coated vesicle is clathrin, a nonglycosylated protein of molecular weight 180,000. Its primary sequence appears to be highly conserved. Clathrin makes up 70% of the total protein of coated vesicles; 5% is made up of polypeptides of molecular weight about 35,000, while various other polypeptides of molecular weight 50,000–100,000 are present. Coated vesicles have a remarkable surface structure resembling a lattice comprising hexagons or pentagons. Three clathrins probably contribute to each polyhedral vertex and two clathrins contribute to each edge. Thus, the smallest structures would contain 12 pentagons plus 4–8 hexagons with 84 or 108 clathrin molecules. A 200-nm diameter coated vesicle would have about 1000 clathrin molecules. Clathrin can form flexible lattices to act as scaffolds to form vesicles during budding. Actual electron micrographs of the budding vesicle are shown in Figure 1-21. By this mechanism, the ligands in the vesicle are delivered to the lysosome and degraded. Some specially active degradation products or the intact hormone itself may emerge to promote the long-term effects of the original polypeptide hormone. Alternatively, long-term effects may occur through the activities of second messenger molecules generated by action at the cell membrane.

2. *Membrane Receptors That Are Not Internalized*

The membrane receptors that are not internalized as a consequence of binding their cognate hormone all span the outer membrane of the cell in which they reside. These receptors have three clearly identifiable domains: the extracellular component, the membrane-

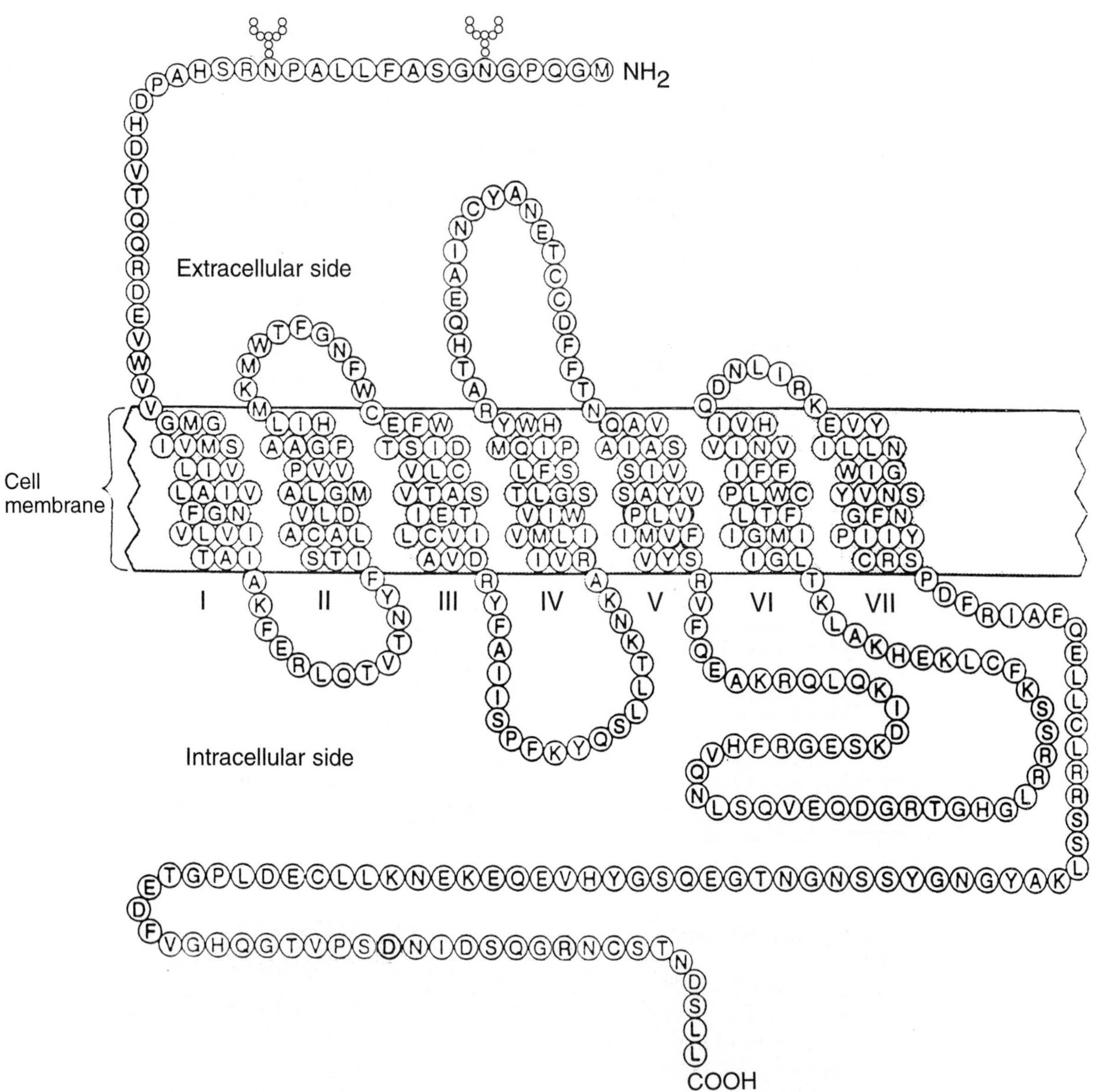

FIGURE 1-24 Proposed model for the insertion of the β_2-adrenergic receptor (AR) in the cell membrane. The model is based on hydropathicity analysis of the human β_2AR. The standard one-letter codes for amino acid residues are used (See Appendix E). Hydrophobic domains are represented as transmembrane helices. Noted are the potential sites of N-linked glycosylation. Redrawn from Kobilka, B. K. *et al.* (1987). *Proc. Natl. Acad. Sci. USA* **84,** 46.

spanning component, and the intracellular component. Each domain has its own unique biochemical properties reflecting its location and function. Frequently the membrane receptor comprises a single polypeptide chain where the N-terminus lies outside the cell and the C-terminus lies inside the cell, but some receptors may comprise subunits, i.e., ion channels.

The N-terminal portion of the membrane receptor usually possesses the ligand-binding domain, which will be defined by the primary amino acid composition of the peptide as well as by the resulting protein secondary structure. Sometimes the ligand-binding domain is determined by several membrane-spanning domains. Frequently, on the extracellular domain, there are glycosylation sites involving Asn-X-Ser/Thr for the attachment of N-linked oligosaccharides. The function of the oligosaccharide moieties is not known, but they could play a role in defining the ligand-

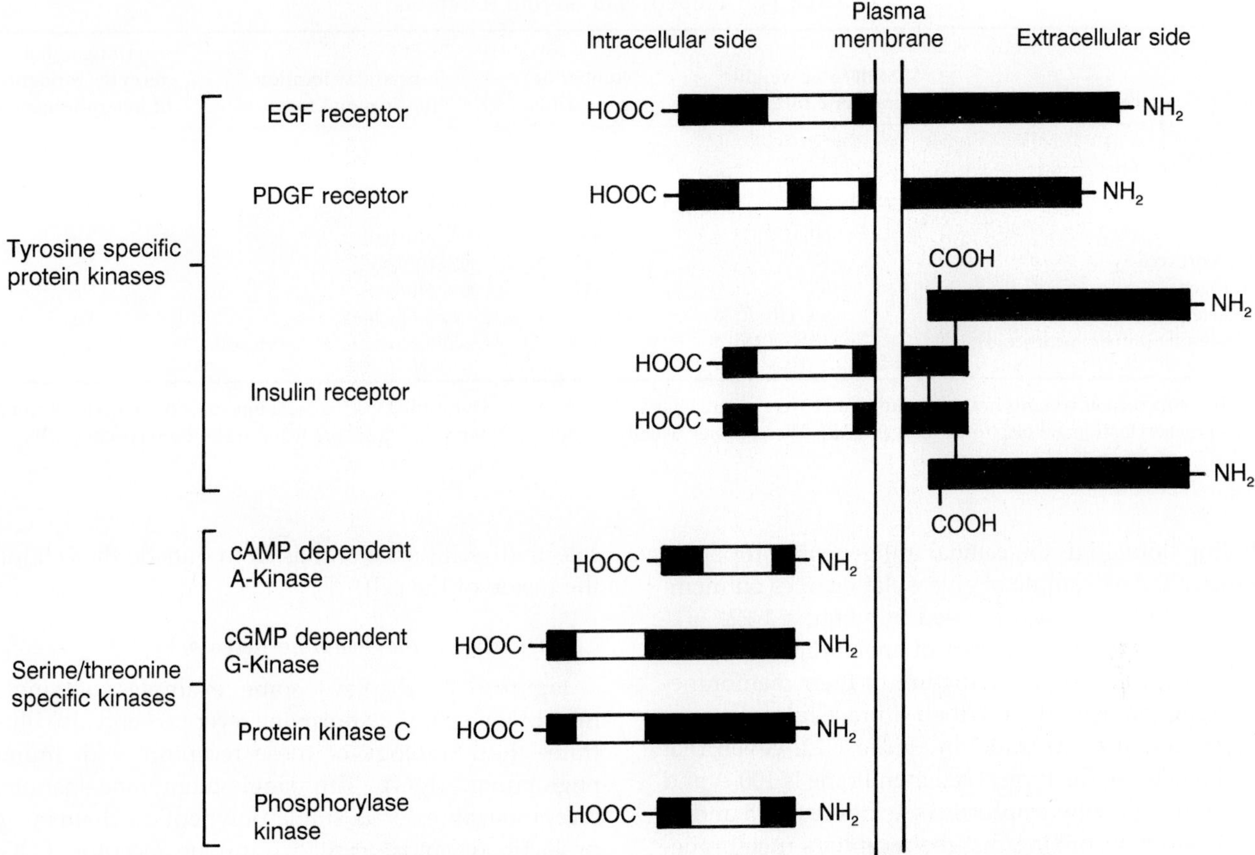

FIGURE 1-25 Protein kinases illustrating the size and location of their catalytic domain. In each case, the catalytic domain (open frame) is about 250 amino acid residues long. The regulatory subunits normally associated with protein kinase A (PKA) and phosphorylase kinase are not shown. Redrawn from Alberts, B., Bray, D., Lewis, J., Raff, M., Roberts, K. and Watson, J. D. (1983). "Molecular Biology of the Cell." 2nd ed., p. 707. Garland Publishing, New York.

TABLE 1-4 Hormone-Related Membrane-Spanning Receptors[a]

Single-membrane-spanning receptor			
Without Tyr-PK activity	**With Tyr-PK activity**	**Multiple-membrane-spanning receptors coupled to G protein**	**Hormone-gated ion channels**
Prolactin	Insulin receptor	Acetylcholine receptor	γ-Aminobutyric acid (GABA)
Insulin-like growth factors (IGF-1) (IGF-2)	Platelet-derived growth factor (PDGF-R)	Parathyroid hormone	
		Calcitonin	
Nerve growth factor (NGF)		Luteinizing hormone (LH)	IP_3 receptor (endoplasmic reticulum)
Growth hormone		Human chorionic gonadotropin (HCG)	
		Rhodopsin (the light receptor)	
		Thyrotropin releasing hormone (TRH) receptor	

[a] Note: This is only a partial listing.

TABLE 1-5 Properties of Steroid Receptors

Receptor	Molecular weight[a] (monomeric form) (×10³)	Number of amino acids	Subcellular location (unliganded form)	Unliganded receptor formation of hetero-oligomers
Thyroxine	~46	395	Exclusively nucleus	No
Retinoic acid	~47	432	~95% Nucleus	No
$1,25(OH)_2D_3$	50	427	75% Nucleus, 25% cytosol	?
Estrogen	60	595	95% Nucleus	Yes
Glucocorticoid	~94	777	90% Cytosol	Yes
Androgen	95	917	90% Nucleus	Yes
Progesterone	99, 110	933	~95% Nucleus	Yes
Aldosterone (mineralocorticoid)	~115	984	~40% Nucleus, 60% cytosol	Yes

[a] These molecular weights are approximate and were usually derived from values determined via the technique of sodium dodecyl sulfate (SDS) polyacrylamide gel electrophoresis (PAGE). These values usually do not agree with the value predicted from the respective cDNAs.

binding domain or the cellular antigenicity properties of the cell. An example of glycosylation sites on membrane receptors is incorporated into Figure 1-22.

There are four categories of membrane receptors that are defined by the structure of their membrane-spanning domain(s) and their intracellular components; these are illustrated in Figure 1-22. Given that the diameter of the typical cell membrane is 100 Å and that it is largely hydrophobic (see Figures 1-15 and 1-16), it is not surprising that the receptor's membrane-spanning components largely consist of hydrophobic and noncharged amino acids. It requires 20–25 amino acid residues organized into an α-helix to cross the membrane once. Membrane receptors have been found to consist of those that span the membrane once (Figure 1-22A and 1-22B) or seven times (Figure 1-22C) or of multiple subunits, each of which spans the membrane seven times so as to create a ligand-gated ion channel (Figure 1-22D). The ion channel operates to effect the selective admittance of ions from outside the cell into the inside of the cell.

a. Single Membrane-Spanning Receptors

Figure 1-23 displays some examples of single-membrane-spanning hormone receptors and also illustrates the homology of these receptors with immunoglobulins (IgG). The single-membrane-spanning receptors can exist as single polypeptide chains (e.g., prolactin receptor, growth hormone receptor, PDGF receptor), dimers of noncovalently bonded single polypeptide receptors [e.g., nerve growth factor (NGF)], or tetramers of four single peptide chains joined together by disulfide bonds [e.g., the insulin receptor, insulin-like growth factor-1 receptor (GF-I)]. Some membrane receptors dimerize to become active.

Figure 1-23B demonstrates the structural homology between the insulin and IGF-I receptors with the immunoglobulins. Each receptor comprises four separate

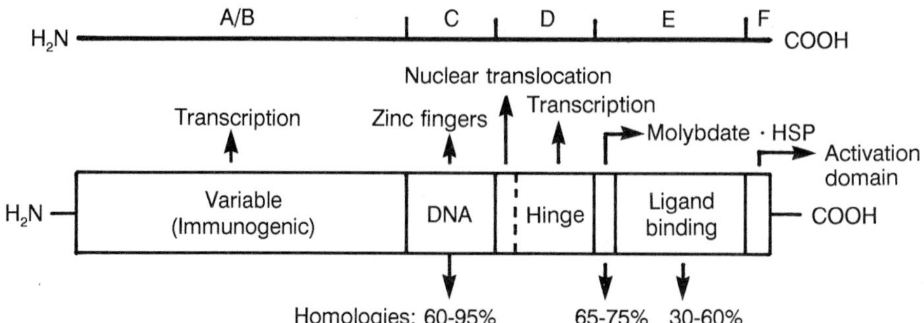

FIGURE 1-26 Model of a typical steroid hormone receptor illustrating the six functional domains A → F. The results are derived from studies on cDNA of the various steroid receptors (see Figure 1-27) in various laboratories, especially those of R. Evans, K. Yamamoto, and B. O'Malley. [This figure was modified from Tsai, M.-J., and O'Malley, B. W. (1994). Molecular mechanisms of action of steroid/thyroid receptor superfamily members. *Ann. Rev. Biochem.* **63,** 451–486.]

polypeptide chains joined by three disulfide bonds; two of the polypeptide chains span the cell membrane (to effect the signal transduction for the hormone receptors) and two polypeptide chains lie entirely outside the cell. All of the proteins–receptors described in Figure 1-23 have regions that are rich in cysteine residues (indicated by the shaded regions).

b. Multiple Membrane-Spanning Receptors

A more detailed depiction of the insertion of a seven-membrane-spanning receptor is given in Figure 1-24. This illustrates both the beauty of the organizational detail as well as the complexity of this family of receptors.

The intracellular components of membrane receptors, as suggested in Figure 1-22, are concerned with the signal transduction component of the receptor. The single-membrane-spanning class of receptors frequently has an enzymatic activity associated with the hydrophilic peptide moiety extending into the interior of the cell. This can take the form of a tyrosine protein kinase. Thus, when the receptor is occupied, the kinase activity becomes activated, and substrates for the kinase inside the cell become phosphorylated on tyrosine residues (see Figure 1-22B). The Tyr-PK activity is created by the secondary structure generated by a span of approximately 250 amino acids. Frequently, at the C-terminal domain of this peptide region is a tyrosine residue that can become autophosphorylated by the Tyr-PK activity of the receptor. Figure 1-25 illustrates the size and location of catalytic activities of membrane-spanning receptors and some related intracellular enzymes, like protein kinase C (PKC) or phosphorylase kinase.

The multiple-membrane-spanning receptors also belong to a family of membrane proteins that function as ion pumps, e.g., Ca^{2+} ATPase, $Na^{+}K^{+}$ ATPase, or H^{+},K^{+} ATPase. These enzymes are also composed of seven-membrane-spanning regions, which display sequence similarity to the hormonal receptors but lack the ligand-binding domain.

As illustrated in Figure 1-22C, the receptors with seven-membrane-spanning components interact with separate membrane components known as G proteins (these will be discussed in detail later in this chapter) to effect the signal transduction process. Finally, the intracellular components of all four classes of membrane receptors may contain serine or threonine residues that can become phosphorylated; the presence of the protein-bound phosphate then, in some fashion, can alter the properties of the signal transduction process. Table 1-4 tabulates some examples of hormone membrane receptors.

c. Membrane Receptors for Steroid Hormones

The great bulk of receptors for steroid hormones has been shown unequivocally to exist in the nuclear and cytoplasmic compartments of the cell. It is an area of continuing research and debate as to whether receptors for steroid hormones exist in the outer cell membrane. A membrane receptor for progesterone has been described in the cell membrane of the *Xenopus laevis* oocytes, and it has been functionally implicated in the process of meiosis. Also, there are indications that there may be a membrane receptor for 1,25-dihydroxy vitamin D_3 in the basal lateral membrane of the chicken duodenum that is involved in the process of intestinal calcium absorption.

C. Intracellular Receptors

1. Steroid Receptors

The receptors for all of the steroid hormones, $1\alpha,25(OH)_2$-vitamin D_3, thyroid hormone, and retinoic acid are all proteins that exist exclusively in the cell nucleus or are partitioned between the cytoplasm and nucleus. Each of these receptor proteins function as

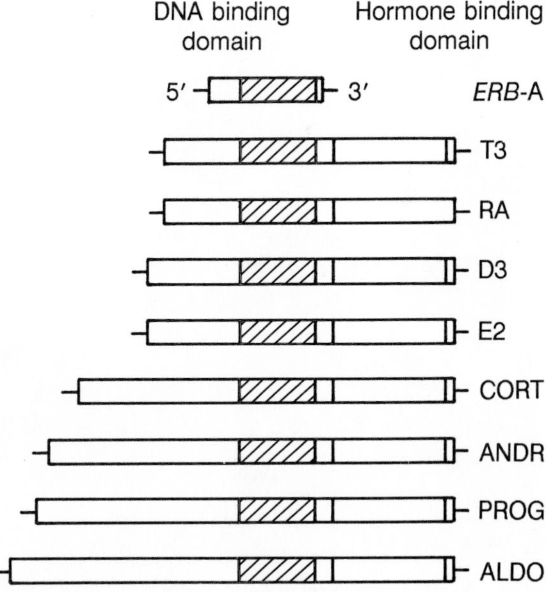

FIGURE 1-27 Steroid receptor gene superfamily. Abbreviations: T3, triiodothyronine; RA, retinoic acid; D3, 1,25-dihydroxyvitamin D_3; E2, estradiol; CORT, cortisol; ANDR, androgen; PROG, progesterone; ALDO, aldosterone. The N-terminal domain (open rectangle to the left of the DNA-binding domain) represents the sequences of least homology among the receptors. This figure roughly shows the relative sizes of these receptors. The information is derived from the laboratories of R. Evans, K. Yamamoto, P. Chambon, and others. In some cases there is high homology in DNA-binding domains and lower homologies in ligand-binding domains.

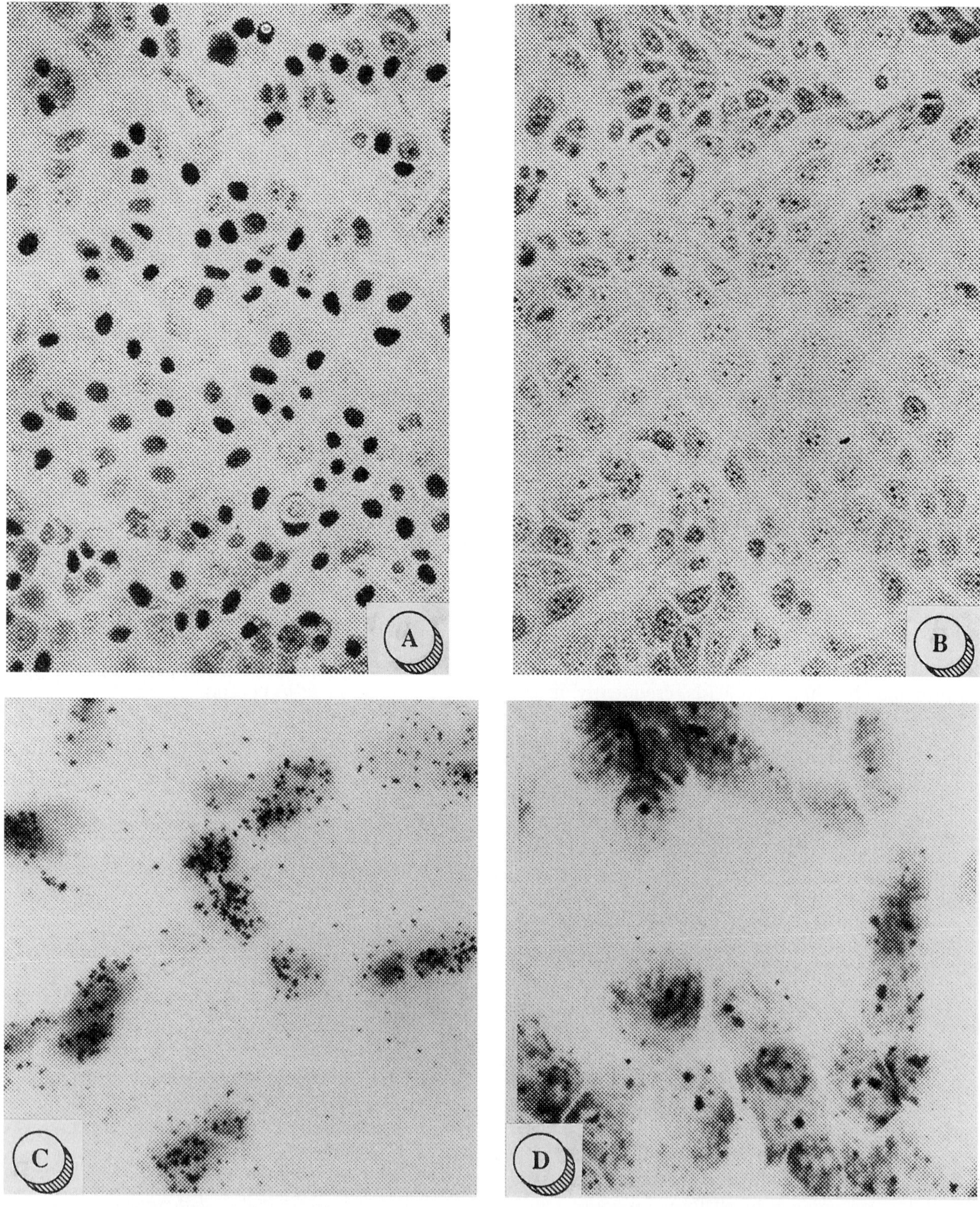

FIGURE 1-28 (A and B) Demonstration that unoccupied steroid hormone receptors may be associated with the cell nucleus. In this study the unoccupied estrogen receptor in MCF-7 human breast cancer cells was visualized through application of monoclonal antibodies which are specific for the estrogen receptor; the cells shown in B were treated identically except that a nonimmune monoclonal antibody was substituted. [Reproduced with permission from Figures 3A and B of King, W. J., and Greene, G. L. (1984). Localization of estrogen receptor in MCF-7 human breast cancer cells: Monoclonal antibodies localize estrogen receptor in the nuclei of target cells. *Nature* **307,** 745–747. (C and D) Demonstration that occupied steroid hormone receptors are localized in the cell nucleus. This panel is an autoradiogram of MCF-7 human breast cancer

DNA-binding proteins, where they regulate the expression of genes related to the biological response of the hormone in question (described in detail later in this chapter).

Determination of the exact site of the subcellular location of the unliganded steroid receptors has been a subject of extensive research. While all of the steroid receptors have a high degree of structural homology there are still some key differences in biochemical properties, which result in differing proportions of the unoccupied receptor being present in the cytosol or nuclear compartments; some of the properties of the members of the family of steroid receptors are summarized in Table 1-5.

The unliganded forms of many of the receptors form hetero-oligomers that include one or more receptor molecules and a dimer of the 90-kDa heat shock protein (HSP), as well as other proteins (e.g., see Chapter 10). The function of the interaction of the unoccupied receptors with the HSP may be to occlude the DNA-binding domain of the unoccupied receptor and prevent inappropriate interaction with the DNA. Interaction with other proteins through HSP-90 may locate the receptor to the cytoskeleton. After the steroid receptor binds ligand, the receptor dissociates from the HSP and undergoes some form of conformational change that will optimize specific interaction with regions of DNA (see Figure 1-26).

Figure 1-26 presents a generic model of a typical steroid receptor. The protein consists of a single polypeptide chain divided into six domains. Starting at the N-terminus, the domains are, respectively, (i, ii) the A/B or variable (sequence and length) domain; (iii) the C or DNA binding domain which contains two zinc fingers; (iv) the D or variable hinge region; (v) the E or ligand binding domain; and (vi) the F or variable domain. The functioning of these domains will be presented later under the topic of mechanism of steroid action.

As shown in Fig. 1-27, there is a striking structural homology between all of the domains of the cell of the members of the steroid receptor gene superfamily; clearly, all of these proteins are related evolutionarily.

FIGURE 1-28—*Continued* cells incubated for 2 hr with high specific activity tritiated estradiol at 37°C for 2 hr. (C) 1×10^{-9} *M*) at 0°C and (D) at 37°C. The predominant localization is nuclear at both 0 and 37°C. Large black dots seen in the nuclei of unlabeled cells are nucleoli. Stained with methylgreen-pyronin, 2-week exposure, 4-μm sections, ×900. [Reproduced by permission from Figure 4A of Sheridan, P. J., Buchanan, J. M., Anselmo, F. C., and Martin, P. M. (1979). Equilibrium: The intracellular distribution of steroid receptors. *Nature* **282,** 579–582.

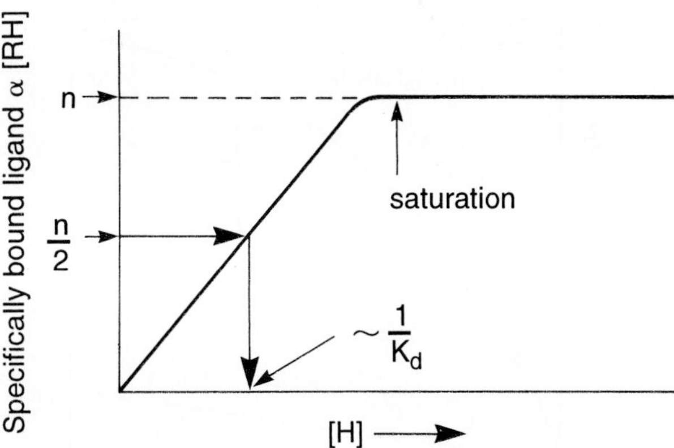

FIGURE 1-29 Progress curve for determining specific hormone binding to a receptor.

The ancestor to these proteins may be the oncogene product v-erbA or c-erbA. v-erbA is an oncogene product that binds to DNA but has no ligand-binding domain. Thus, v-erbA can constitutively activate a subset of genes that would normally be dependent upon the presence of hormone to occupy the ligand-binding domain. It is interesting that the thyroid hormone and retinoic acid receptors, as well as the hormonally active form of vitamin D_3, $1,25(OH)_2D_3$, are members of this family of hormone receptors, since the structures of their ligands are not classically steroidal in nature.

2. *Peptide–Protein Hormone Receptor*

A number of protein hormones and, in some instances, their receptors, such as nerve growth factor (NGF), epidermal growth factor (EGF), and insulin, have been found by autoradiography experiments (employment of radioactive preparations of the hormone) to be localized in the nucleus or nuclear membrane of their target cells. It is not yet clear whether these hormone–receptor complexes are functional; they could have arrived at their intracellular location by the process of endocytosis, as described in Figure 1-20.

D. Measurement of Hormone–Receptor Interactions

Historically there have been two different techniques employed for studying the interaction of hormones with their receptors: autoradiography, which determines localization of the hormone under both *in vivo* and *in vitro* conditions, and saturation analysis, which is carried out under *in vitro* conditions. Both techniques are dependent upon access to radioactive preparations of the hormone under study with a high specific activity.

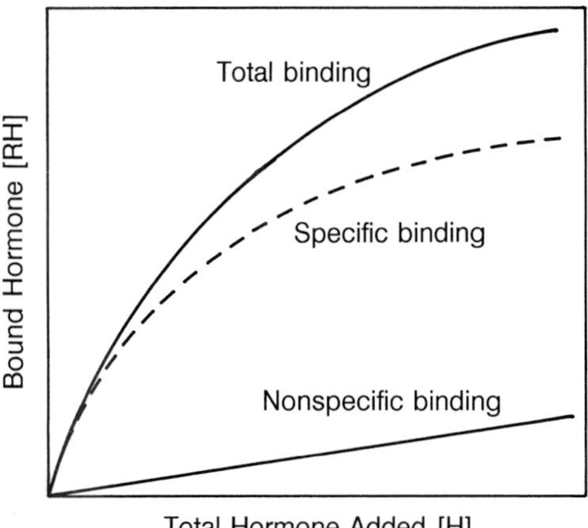

FIGURE 1-30 Typical plot showing specific hormone binding to a receptor.

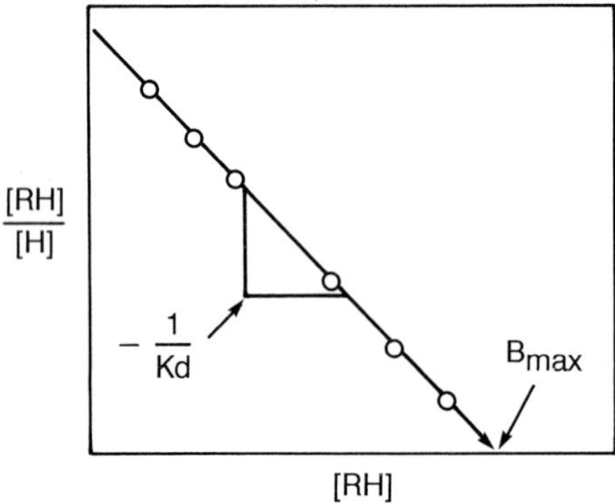

FIGURE 1-31 Typical plot of a Scatchard analysis of specific binding of hormone to its receptor.

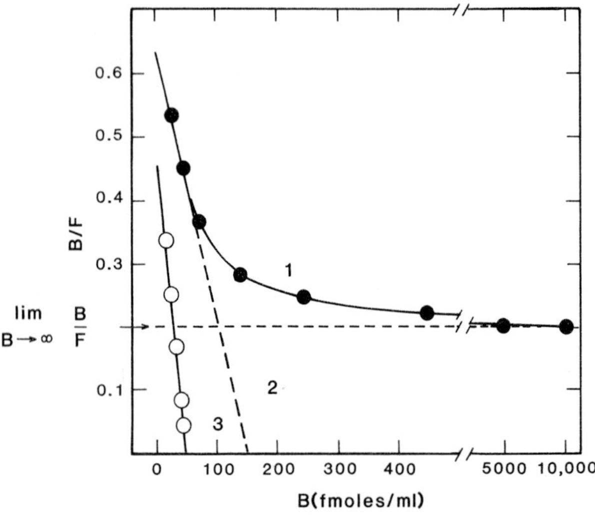

FIGURE 1-32 Scatchard analysis of curves representing two binding components. Curve 1: Scatchard plots of total binding. Curve 2: Linear extrapolation of the high-affinity component that includes contribution from the low-affinity component. Curve 3: Specific binding of the high-affinity component after removal of the nonspecific contribution, representing the true high-affinity curve. [Redrawn from Chamness, G. C. and McGuire, W. L. (1975). Scatchard plots: Common errors in correction and interpretation. *Steroids* **26,** 538.

1. *Autoradiogoraphic Localization of Hormones*

In a typical experiment, a physiological dose of radioactive hormone, usually labeled with either tritium or carbon-14, is administered to an animal. At the appropriate time, the tissues of interest are collected, fixed, sectioned, and exposed to film for extensive intervals of time. Figures 1-28C and 1-28D displays the results of administration of tritiated estrogen to MCF-7 human breast cancer cells; the results clearly indicate that the steroid hormone–receptor complex is associated with the nucleus of the cell. In Figures 1-28A and 1-28B, the localization of the unoccupied and occupied estrogen receptors was studied through the application of fluorescent monoclonal antibodies to the estrogen receptor. It is apparent that both the unoccupied and occupied forms of the estrogen receptor are found in the nucleus of the target cell.

2. *Saturation Analysis*

A very informative technique for studying the interaction of radioactive preparations of hormone with homogenates of target cells (which contain highly impure receptor) or with partially purified or purified receptor preparations is the procedure of saturation analysis. Since the cognate receptor has a highly specific ligand-binding domain, under usual incubation conditions the noncovalent hormone–receptor complex is formed rapidly. The interaction of the hormone (H) and the unoccupied receptor (R) to form receptor–hormone complex (RH) can be formally described in the following mathematical fashion:

$$[H] + [R] \underset{k_{-1}}{\overset{k_{+1}}{\rightleftarrows}} [HR]. \tag{1}$$

The association of H and R to yield HR and the dissociation of HR into H and R are readily reversible processes, i.e., it is a dynamic equilibrium since the hormone does not become covalently bound to the receptor. Thus, the equilibrium can be expressed in terms of the association constant, K_a, which is mathematically equivalent to 1/dissociation constant (K_d):

$$K_a = \frac{[RH]}{[H][R]} = \frac{k_{+1}}{k_{-1}} = 1/K_d. \tag{2}$$

The individual rate constants k_{+1} and k_{-1} numerically describe the rates of the forward (on-rate) and backward (off-rate) reactions, respectively, as written in Eq. (1). Experimentally, the approach to equilibrium can be followed by a progress curve of binding that reaches saturation (see Figure 1-29). A saturating amount of hormone is determined by using variable amounts of

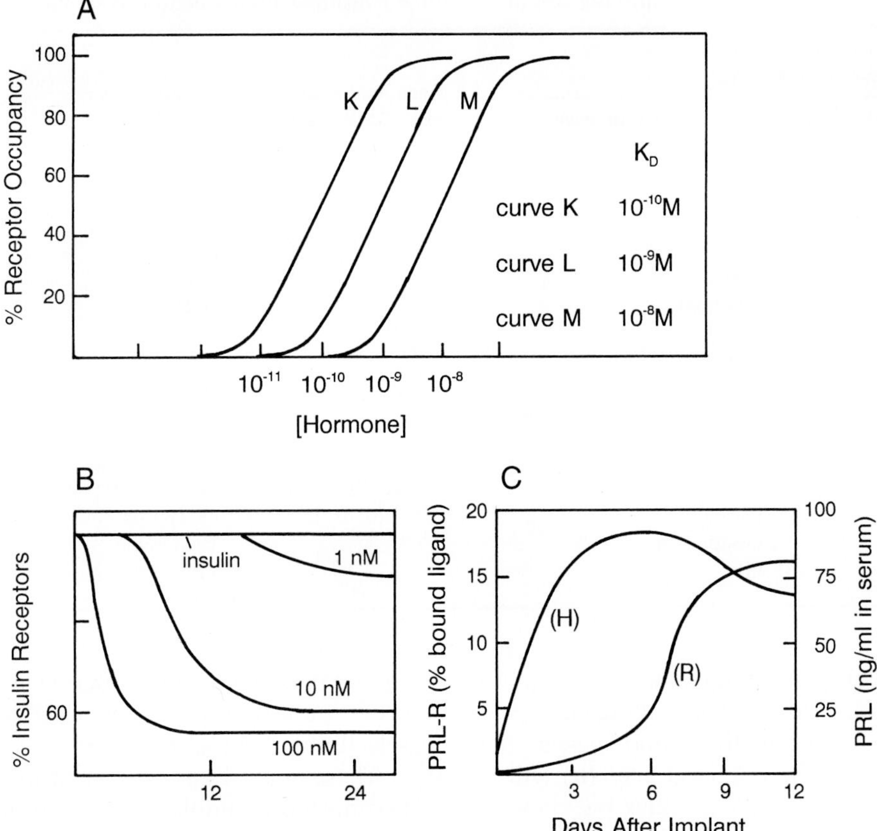

FIGURE 1-33 Regulation of receptor performance, the K_d (A), and receptor concentration via either up-regulation (B) or down-regulation (C). In A, three saturation curves for a receptor are shown; the K_d's for binding curves K, L, and M are 10^{-10}, 10^{-9}, and 10^{-8} *M*, respectively. If the prevailing hormone concentration bathing the target cell is 10^{-9} *M*, then depending upon the affinity of the receptor for its ligand, the level of receptor occupancy can vary from 80% (curve K) to 50% (curve L) to 20% (curve M). In B, homologous down-regulation of the insulin receptor is displayed; cultured lymphocytes were exposed to various concentrations of hormone, and total insulin receptor levels were determined. Gavin, J. R., Roth, J., Neville, D. M., De Meyts, P., and Buelle, D. N. (1974). Insulin-dependent regulation of insulin receptor concentration: A direct demonstration in cell culture. *Proc. Natl. Acad. Sci. USA* **71** (84–88). In C homologous up-regulation of the prolactin receptor (PRL) is displayed. Hypophysectomized rats were implanted with pituitaries under the kidney capsule. The increase of the hormone in serum preceded the increase of the receptor in liver. (Posner, B. I., Kelly, P. A., and Friesen, H. G. (1975). Prolactin receptors in rat liver: Possible induction by prolactin. *Science* **188**, 57–59.

free hormone and measuring the amount bound with some convenient assay. The half-maximal value of a plot of receptor-bound hormone (ordinate) versus total free hormone concentration (abscissa) approximates the dissociation constant, which will have a specific hormone concentration in molarity as its value. Hormone bound to receptor is corrected for nonspecific binding of the hormone to membranes. This can be measured conveniently if the hormone is radiolabeled by measuring receptor plus labeled hormone (radioactive) and receptor plus radioactive hormone after the addition of an excess (100–1000 times) of unlabeled hormone. The excess unlabeled hormone will displace the high-affinity hormone-binding sites but not the low-affinity nonspecific binding sites. Thus, when the "radioactive plus nonradioactive" curve is subtracted from the "radioactive" curve (Figure 1-30), an intermediate curve will represent specific binding to receptor. This is of critical importance when the receptor is measured in a system containing other proteins. As an approximation, 20 times the K_d value of hormone is usually enough to saturate the receptor.

Most measurements of K_d are made by using Scatchard analysis, which is a manipulation of the equilibrium equation. The equation can be developed by a number of routes, but can be envisioned from mass

TABLE 1-6 Membrane-Bound Receptor Hormone Transduction Systems

Hormones acting via utilization of a cAMP system		Hormones acting via a cGMP system	Hormones acting via IP_3 pathway	Hormones acting via ion channels
cAMP decrease	cAMP increase			
Bradykinin	ACTH	Atrial natriuretic hormone	ACTH	Acetylcholine
Norepinephrine (α_2-adrenergic)	α-MSH		Angiotensin II/III	Nicotinic receptor (Na^+ channel)
Somatostatin	Calcitonin		Cholecystokinin	Nitric oxide (NO) (Ca^{2+} channel)
	FSH		EGF	
	Gastrin		Epinephrine	
	Glucagon		GnRH	
	GnRH		Histamine (HI)	
	Histamine (H_2)		NGF	
	LH		Norepinephrine (α_1 adrenergic)	
	Norepinephrine (β_1 adrenergic)		PGF	
	PGE_1		Thromboxanes	
	PTH		TRH	
	Secretin		TSH	
	TSH		Vasopressin	
	Vasopressin		VIP	
	VIP			

action analysis of the preceding equation. At equilibrium, the total possible number of binding sites (B_{max}) is equal to the unbound plus the bound sites, so that $B_{max} = [R] + [RH]$, and the unbound sites (R) will be equal to $[R] = B_{max} - [RH]$. To consider the sites left unbound in the reaction, the equilibrium equation becomes

$$K_a = \frac{[RH]}{[H](B_{max} - [RH])}. \qquad (3)$$

Thus,

$$\frac{\text{bound}}{\text{free}} = \frac{[RH]}{[H]} = K_a(B_{max} - [RH]) = \frac{1}{K_d}(B_{max} - [RH]). \qquad (4)$$

The Scatchard plot of bound/free = [RH]/[H] on the ordinate versus bound [RH] on the abscissa yields a straight line, as shown in Figure 1-31. When the line is extrapolated to the abscissa, the intercept gives the value of B_{max} (the total number of specific receptor-binding sites). The slope of the negative straight line is $-K_a$ or $-1/K_d$.

The K_d values for steroid receptors typically fall in the range of 10^{-10}–10^{-8} *M*. This very low number is a reflection of how far to the right Eq. (1) lies. This is to say, in a mixture of H and R, there is virtually no free hormone and most exists as HR. Also, the low K_d value is a testament to the three-dimensional organization of the ligand-binding domain, which very effectively interacts to "capture" the ligand. These interactions are generally marked by a high degree of specificity so that both parameters describe interactions of a high order, indicating the uniqueness of receptors and the selectivity of signal reception.

These analyses are sufficient for most systems, but become more complex when there are two components in the Scatchard plot. In this case, the straight line usually bends as it approaches the abscissa, and a second phase is observed somewhat asymptotic to the abscissa while still retaining a negative slope (Figure 1-32). In order to obtain the true value of K_d for the steeper, higher affinity sites, the low-affinity curve must be subtracted from the first set, which also corrects the extrapolated value of B_{max}. From these analyses, information is concluded on the K_d: the number of classes of binding sites (usually one or two) and the maximal number of high-affinity receptor sites (receptor number) in the system.

E. Regulation of Hormone Receptors

The homeostatic set point of any given endocrine system is determined not only by the balance of hormone secretion in relation to the generation of the desired biological response(s) but also through regulation of the receptor performance (K_d) and receptor levels in the various target tissues.

In some hormonal receptor systems, e.g., the insulin receptor or the estrogen receptor, the affinity of the receptor for the cognate hormone may be changed in accordance with physiological circumstances, i.e., the K_d is either increased or decreased. The most usual explanation for this change is via a mechanism of nega-

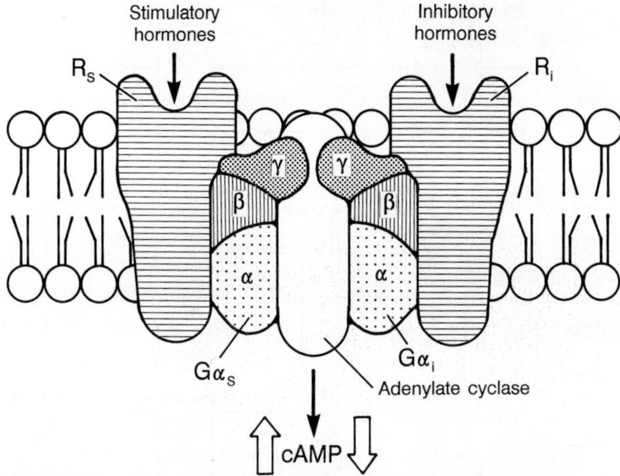

FIGURE 1-34 Schematic diagram of occupied membrane receptors interactive with G proteins and adenyl cyclase. Adenylate cyclase is responsible for the conversion of ATP to cAMP. The occupancy of R_s by stimulatory hormones stimulates adenylate cyclase via formation of an active dissociated G_{α_s} subunit. The occupancy of R_i by inhibitory hormones results in the formation of an "active" G_{α_i} complex and a concomitant reduction in cyclizing activity. The fate of β- and γ-subunits in these dissociation reactions is not yet known. Abbreviations: G_{α_s}, stimulatory hormone receptor, G_{α_i}, inhibitory hormone receptor.

tive cooperativity. In the absence of cooperativity, the population of the receptor's hormone-binding sites behaves identically and independently, but in the instance of cooperativity (either positive or negative), the properties of the ligand domain change as the level of receptor occupancy or receptor concentration changes (Figure 1-33A). Thus, the K_d of the receptor in question may be changed via the presence of a second allosteric binding site, which, when it is occupied, causes a conformational change in the receptor that, in turn, results in a small change in the environment of the hormone-binding domain. Potential ligands for the second allosteric binding site include Ca^{2+} and GTP (for those receptors that interact with G proteins) and changes in local pH; also, phosphorylation of membrane receptors is known to change the K_d for hormone binding. Alternatively, the conformational change in the hormone-binding domain may be affected through the formation of receptor dimers. Here the physical contact between the two receptor molecules generates a conformational change that can alter the K_d for cognate hormone binding.

The concentration of a specific receptor in a given target cell can also be regulated by the host organism. In a normal target cell, the number of receptor molecules has been found to vary from approximately 1000 to 50,000 per cell. But for any given receptor, the concentration of the functional receptor in any given cell may, under some circumstances, be increased (upregulation) or decreased (down-regulation) (see Figures 1-33B and 1-33C). Obviously, large changes in receptor concentration could dramatically increase or decrease the ability of that target cell to generate biological responses. A variety of mechanisms exist for changing functional receptor concentration; these include increased rate of biosynthesis (by a hormonally induced increase in gene transcription for that receptor or changes in the rate of receptor catabolism (by endocytosis or proteolysis). When the change in receptor concentration is effected by the cognate hormone, the process is referred to as homologous regulation, and when the change is effected by other regulatory factors the process is referred to as heterologous regulation.

IV. MECHANISM OF HORMONAL ACTION

A. Hormone Entry into Target Cells

1. Peptide-Internalized Hormones

Most peptide hormones do not require entry into the cytoplasm of their target cells to initiate a biological response since the ligand-binding domain of the receptor exists on the outer surface of the target cell (see Figure 1-22). As discussed in relation to Figure 1-20, some peptide hormones (e.g., insulin and nerve growth factor), when bound to their receptor, are subjected to endocytosis and are moved into the interior of the cell.

2. Steroid Hormones

As discussed in Chapter 2, most steroid hormones have limited solubility in plasma due to their intrinsic hydrophobic character; accordingly, each steroid hormone is carried through the plasma compartment by specific plasma transport proteins (PTP). Each transport protein has a specific ligand-binding domain for

TABLE 1-7 Signal Transduction by G Proteins

	Product transducer	
Activity	**Stimulatory**	**Inhibitory**
Adenylate cyclase	cAMP, G_s ($\alpha_s\beta\gamma$)	G_I ($\alpha_i\beta\gamma$)
Phospholipase C	IP_3, DAG G_p ($\alpha_p\beta\gamma$)[a]	G_I ($\alpha_i\beta\gamma$)
cGMP phosphodiesterase	5′-GMP	
Ca^{2+} channel	Open channel G_c ($\alpha_c\beta\gamma$)[a]	
K^+ channel	Open channel G_k ($\alpha_k\beta\gamma$)[a]	

[a] The stimulatory G proteins that activate the phospholipase, Ca^{2+} channel, and K^+ channel are designated G_p, G_c, and G_k, respectively.

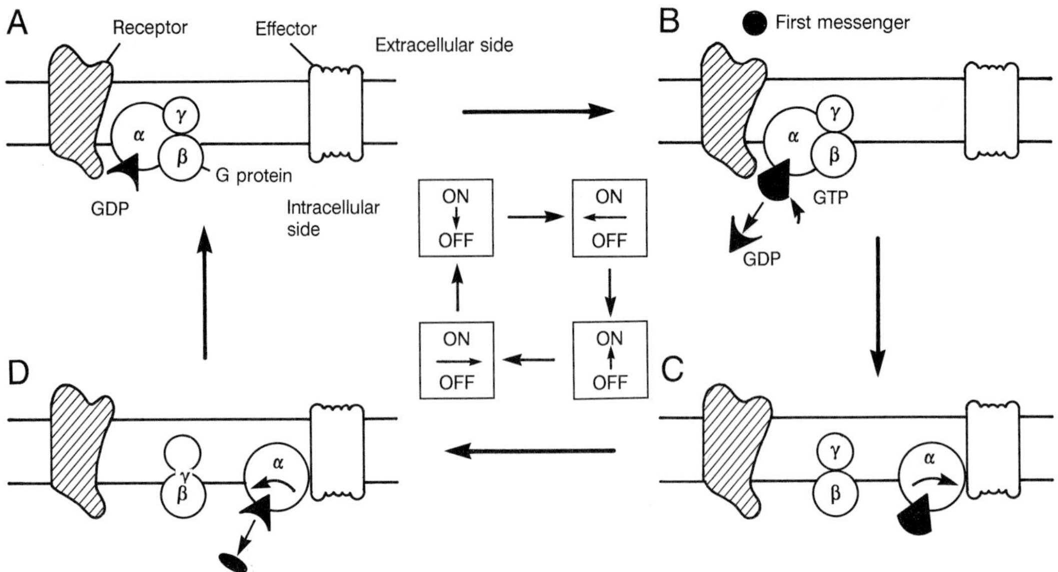

FIGURE 1-35 Model of how G proteins switch effectors on and off. (A) In their resting state, G proteins that consist of α, β, and γ-subunits are bound by the nucleotide guanosine diphosphate (GDP) and have no contact with receptors. (B) When a hormone or other first messenger binds to a receptor, the receptor causes the G protein to exchange GDP for the nucleotide guanosine triphosphate (GTP), which activates the G protein. (C) The G protein then dissociates, after which the GTP-bound α-subunit diffuses along the membrane and binds to an effector, activating it. The switch is "on." (D) After a few seconds, the α-subunit converts GTP to GDP, thereby inactivating itself. The α-subunit will then reassociate with the β–γ complex. [Modified with permission from the article by Linder, M. E., Gilman, A. G. (1992). G proteins. *Sci. Am.* July, p. 59.]

its cognate hormone. These ligand domains display little amino acid sequence homology with the ligand binding of the cognate receptors. Nevertheless, the PTP ligand-binding domain also displays a high affinity for its ligand: usually the K_d for the PTP ligand is 10–100× lower than the K_d of the hormone's cognate receptor.

Since the hormone does not form a covalent linkage with the PTP, the interaction of the hormone (H) with PTP is governed by mass action considerations:

$$[\mathrm{H}] + [\mathrm{PTP}] \underset{k_{-1}}{\overset{k_{+1}}{\rightleftharpoons}} [\mathrm{H{\bullet}PTP}], \quad (5)$$

and the formation, k_{+1}, and dissociation, k_{-1}, steps for $\mathrm{H} \cdot \mathrm{PTP}$ are readily reversible. The equation for K_d is

$$K_d = \frac{k_{-1}}{k_{+1}} = \frac{[\mathrm{H}][\mathrm{PTP}]}{[\mathrm{H{\bullet}PTP}]}. \quad (6)$$

For some endocrine systems, the concentration of the plasma transport protein can be subject to physiological regulation, i.e., the concentration of PTP can be either increased or decreased. Thus, changes in the amount of PTP can, in principle, alter the amount of "free" H, as described by the relationships of Eq. (6).

The current view is that it is the "free" form of steroid hormones and not the conjugate of the hormone with its PTP that enters target cells to begin the sequence of steps that results in the generation of a biological response. Entry of steroid hormones through the target cell membrane may occur by a process of free diffusion, but the exact mechanism is still not known.

B. Receptor-Mediated Transmembrane Signaling

This section of Chapter 1 presents current understanding of the several pathways of signal transduction involving membrane-based receptors. These include receptors that couple to (a) adenylate cyclase, (b) guanylate cyclase, (c) the phosphoinositide pathway, (d) the utilization of intracellular Ca^{2+} as a second messenger, and (e) ion channels. These topics will be presented sequentially. Table 1-6 tabulates the hormones that activate the various specific signal transduction pathways. A given hormone–receptor system may utilize more than one signaling pathway.

1. G-Proteins

A key participant in all of these signal transduction pathways is a family of G proteins. When the membrane receptor becomes occupied, the resulting confor-

TABLE 1-8 Pharmacological Agents That Interact with G Proteins

Agent	Consequence of action
Cholera toxin	Increased production of cAMP by adenylate cyclase
Pertussis toxin	Blockage of action of hormone working through G_i
NaF	Inactivation of both G_s and G_i

mational change results in the receptor coming into contact with a G protein; an example is presented in Figures 1-34 and 1-36. The G proteins consist of an oligomer with three subunits, designated α, β, and γ, with a total molecular weight of approximately 80,000–95,000 ($\alpha \approx 45{,}000$, $\beta \approx 35{,}000$, $\gamma \approx 5000$). Functionally the G proteins are the true signal transducers since they respond to the occupancy of the receptor and modulate the activity of the catalytic subunit of the membrane-bound adenylate cyclase (AC) enzyme. There are two functions of the α-subunit: $G_{\alpha s}$, which activates the AC, and $G_{\alpha i}$, which inhibits the activity of AC. Both $G_{\alpha s}$ and $G_{\alpha i}$ are activated by guanosine triphosphate (GTP) and both also function as a GTPase. The GTPase activity endows the G proteins with a turn-off mechanism. The functions of the β- and γ-

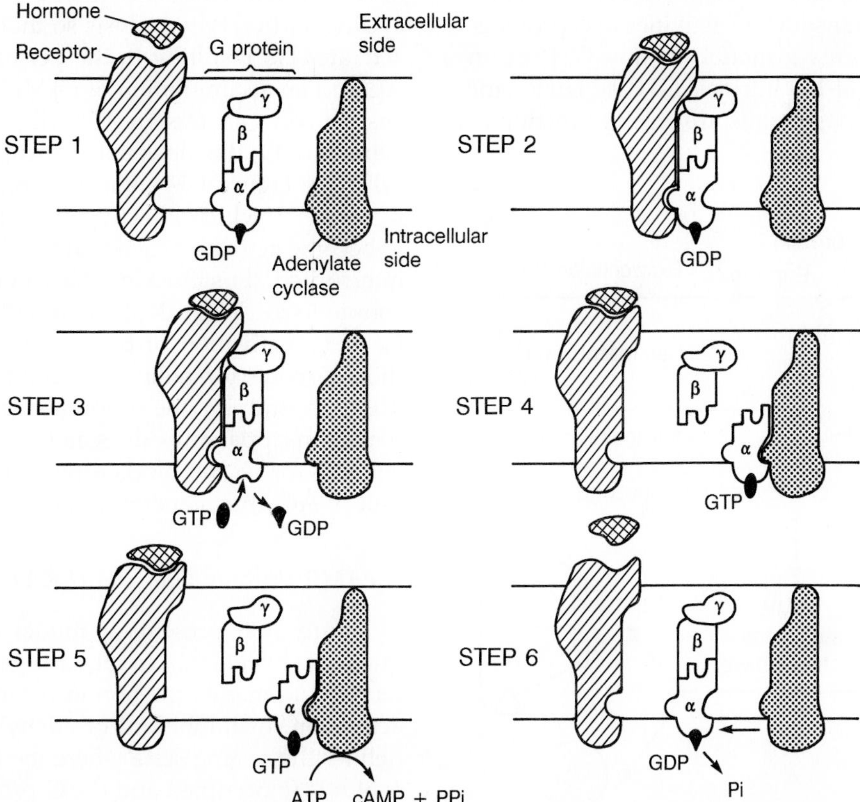

FIGURE 1-36 Membrane receptor activation of adenylate cyclase by binding of a hormone to its receptor. The cell membrane is depicted, which contains on its outer surface a receptor protein for a hormone. On the inside surface of the membrane is the adenylate cyclase protein and the transducer protein G. In the resting state, GDP is bound to the α-subunit of the G protein. When a hormone binds to the receptor, a conformational change occurs (step 1). The activated receptor binds to the G protein (step 2), which activates the latter so that it releases GDP and binds GTP, causing the α-subunit and the complex of β- and γ-subunits to dissociate (step 3). Free G_α subunit binds to the adenylate cyclase and activates it so that it catalyzes the synthesis of cAMP from ATP (step 4); this step may involve a conformational change in G_α. When GTP is hydrolyzed to GDP, a reaction most likely catalyzed by G_α itself, G_α is no longer able to activate adenylate cyclase (step 5), and G_α and $G_{\beta,\gamma}$ reassociate. The hormone dissociates from the receptor and the system returns to its resting state. [Redrawn from Darnell, J., Lodish, H., and Baltimore, D., (1986). "Molecular and Cellular Biology", p. 692. *Mol. Cell. Biol.*, New York: Scientific American Books, Inc., New York, with the modification indicating that the α-subunit may actually contact the receptor.]

TABLE 1-9 Intracellular Second Messengers

Second messenger	Abbreviation
Cyclic AMP	cAMP
Cyclic GMP	cGMP
Calcium	Ca^{2+}
Diacylglycerol	DAG
Inositol 1,4,5-triphosphate	IP_3

subunits are not so clearly defined; they likely participate in transferring the signal transduction from the occupied receptor to the $G_{\alpha s}$ and $G_{\alpha i}$ subunits. There are also comparable forms of G_s that activate the phosphodiesterase and membrane channels. Table 1-7 summarizes the signal transduction activities of G proteins.

Figure 1-35 presents a model of how GTP as an activating ligand and the intrinsic GTPase enzymatic activity of the α_s or α_i-subunits are able to function as a switch to sequentially turn on and turn off the signal transduction actions of α. This represents the essence of the signal transduction process. Thus, the alternate binding of GTP induced by the hormone occupancy of its receptor, followed by the movement of α within the membrane bilayer to AC, followed by a slow appearance of GTPase activity resulting in the conversion of α-GTP (active) $\rightarrow$ α-GDP (inactive) describes the molecular details of the transduction process.

It has been possible to biochemically characterize G_S and G_i through the use of cholera toxin, pertussis toxin, and the reagent NaF; the effects of the consequences of their presence on adenylate cyclase activity are summarized in Table 1-8. Cholera toxin transfers an ADP-ribosyl moiety from NAD^+ to the α-subunit of G_S, which has the consequence of interrupting the G_S cycling and GTP hydrolysis so that GTP permanently activates G_S, resulting in the permanent activation of AC and an accumulation of cAMP. In contrast, pertussis toxin transfers an ADP-ribosyl moiety to the α-subunit of G_i; this does not interrupt the hydrolysis of GTP (see Figure 1-34), but does reduce the affinity of G_i for GTP, which effectively blocks the ability of G_i to become activated such that the G_i-protein cycling is interrupted, thus blocking the action of hormones that couple to G_i. When NaF is added to preparations of G_S or G_i, activation of both proteins ensues. Finally, the diterpene, forskolin, has been found to have great utility in studying the properties of adenylate cyclase. Forskolin apparently does not interact directly with either G_S or G_i, but binds directly to the catalytic subunit of AC, thus generating permanent activation.

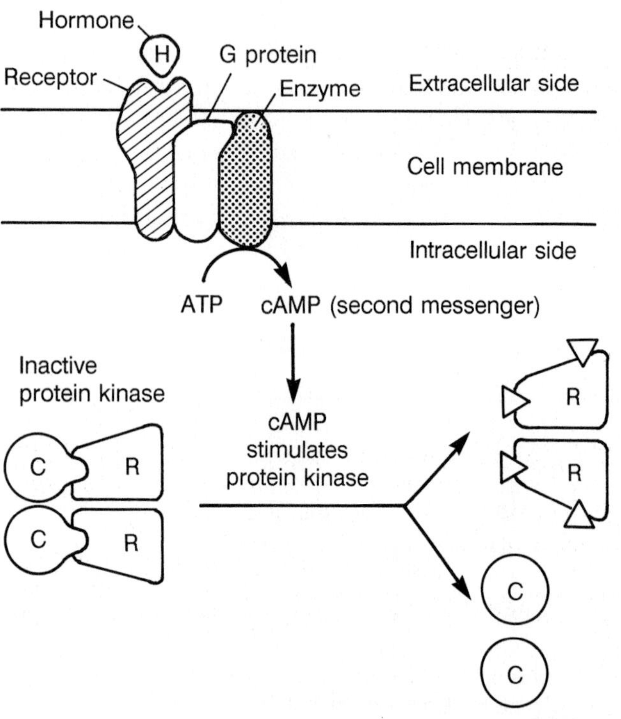

FIGURE 1-37 Activation of protein kinase A. Binding of hormone (H) to the membrane-bound receptor on the outside of the cell results in a signal transduction using the G protein to the inside of the cell so as to catalyze the enzymatic conversion of ATP into cAMP (second messenger). The cAMP then diffuses to the soluble tetrameric inactive protein kinase A moiety. The tetramer dissociates to generate two active catalytic (C) subunits. These kinase subunits then affect the phosphorylation of other proteins involved in the generation of a biological response related to the initiating hormone. The triangles bound to the R–R dimer represent molecules of cAMP.

2. *cAMP Protein Kinase A (PKA)*

Figure 1-36 presents a model describing the sequence of events that ensues after a hormone occupies its cognate membrane-bound receptor, leading to the activation (or inhibition) of adenylate cyclase. Step 1 defines the resting state where the transmembrane receptor is unoccupied and the G-protein trimeric oligomer is not in physical contact with either the receptor or the adenylate cyclase. Also note that the α-subunit has a bound GDP molecule. As a consequence of hormone occupancy of the receptor's ligand-binding domain (step 2), a conformational change ensues, the end result of which is the probable physical interaction of the α-, β-, and γ-subunits with the unoccupied receptor. As a consequence, the affinity of the α-subunit for GDP is decreased while its affinity for GTP is increased (step 3). Then, in accordance with the model of Figure 1-36, the α-subunit with bound GTP moves within the domain of the membrane lipid bilayer (step 4) to interact physically with the membrane-bound adenyl-

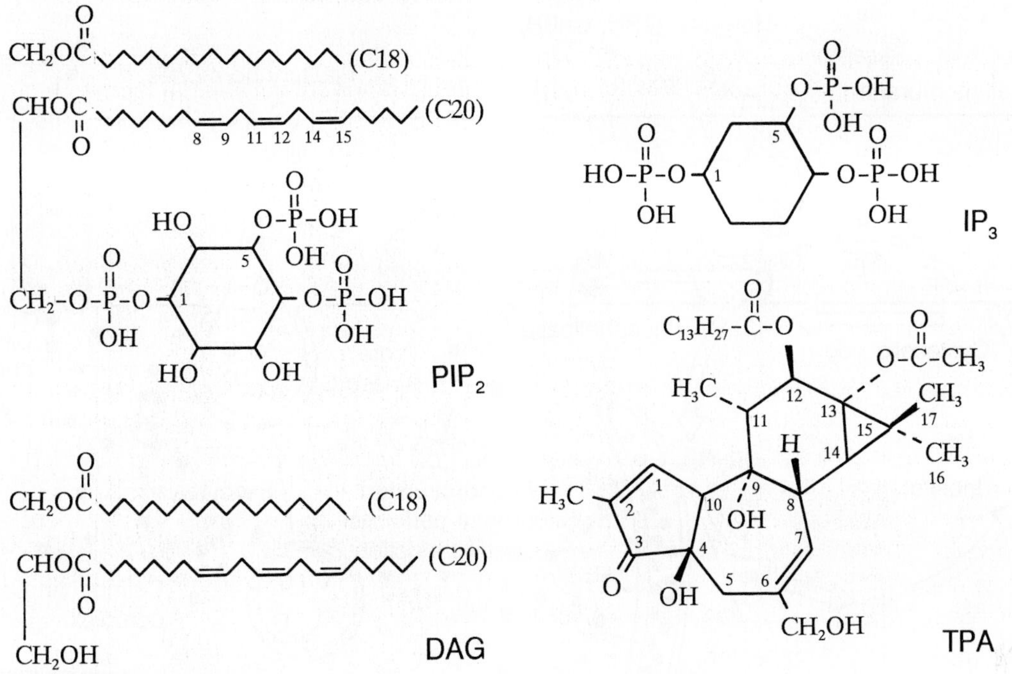

FIGURE 1-38 Structures of the three main components of the phosphoinositide pathway: phosphoinositol-4,5-diphosphate (PIP_2), diacylglycenol (DAG), and 1,4,5-inositol triphosphate (IP_3). Note that the tumor promoter TPA (12-*O*-tetradecylphorbol 13-acetate) contains a DAG-like structure. [Modified from stereochemistry and numbering of *myo*-inositol. Parthasaranthy, R., and Eisenberg, Jr., F. (1986). The inositol phospholipids: A stereochemical view of biological activity. *Biochem. J.* **235,** 313–322.]

ate cyclase. This activates the catalytic activity of the adenylate cyclase so that the substrate ATP is converted into the product cAMP + PP_i (step 5). Finally, after a short time interval, the quiescent GTPase activity of the activated α-subunit becomes active (step 6) so that the GTP is hydrolyzed to bound GDP and free inorganic phosphate; this effectively turns off the adenylate cyclase activity. Accordingly, the α-subunit returns to its original state in a complex of α, β, and γ, as depicted in step 1. The same series of events is followed where a G_i subunit is coupled to the adenylate cyclase, but the end result is reduction in the activity of the adenylate cyclases.

The cAMP produced as a consequence of the G-protein activation of adenylate cyclase functions as a second messenger. Second messengers function as intracellular messengers for the primary signaling moiety, the hormone. Table 1-9 tabulates the known second messengers.

Cyclic AMP activates protein kinase A (see Figure 1-37). Protein kinase A is a tetramer, comprising two equivalent regulatory subunits (R) and two equivalent catalytic subunits (C); the tetrameric form of protein kinase A is not able to function catalytically. Four molecules of cAMP bind to the two R subunits, which provokes the dissociation of the tetramer into a dimer of regulatory subunits and monomeric free catalytically active C subunits. The liberated C subunits function as an active kinase to phosphorylate proteins, which thereby amplifies the initial hormonal signal within the target cell. There are a large number of proteins that can be phosphorylated by PKA, including enzymes of glycolysis and glycogenolysis, chromatin proteins including histones, and ribosomal proteins. The actions of ACTH and TSH on their target cells in the *zona fasciculata* of the adrenal cortex and in the thyroid, respectively, which stimulate the release of cortisol and thyroxine, are known to use the PKA pathway and possibly other pathways. Thus, there is potentially a wide array of cellular constituents where properties can be influenced by the hormonal activation of PKA.

3. *cGMP Protein Kinase*

A cyclic GMP protein kinase G (PKG) has been identified. The receptor for atrial natriuretic factor or atriopeptin is a membrane receptor with a single-membrane-spanning domain whose extracellular ligand-binding domain binds its cognate ligand ANF and whose intracellular domain is guanylate cyclase: see Chapter 15. Occupancy of the receptor by ANF leads to activation of the guanyl cyclase and generation

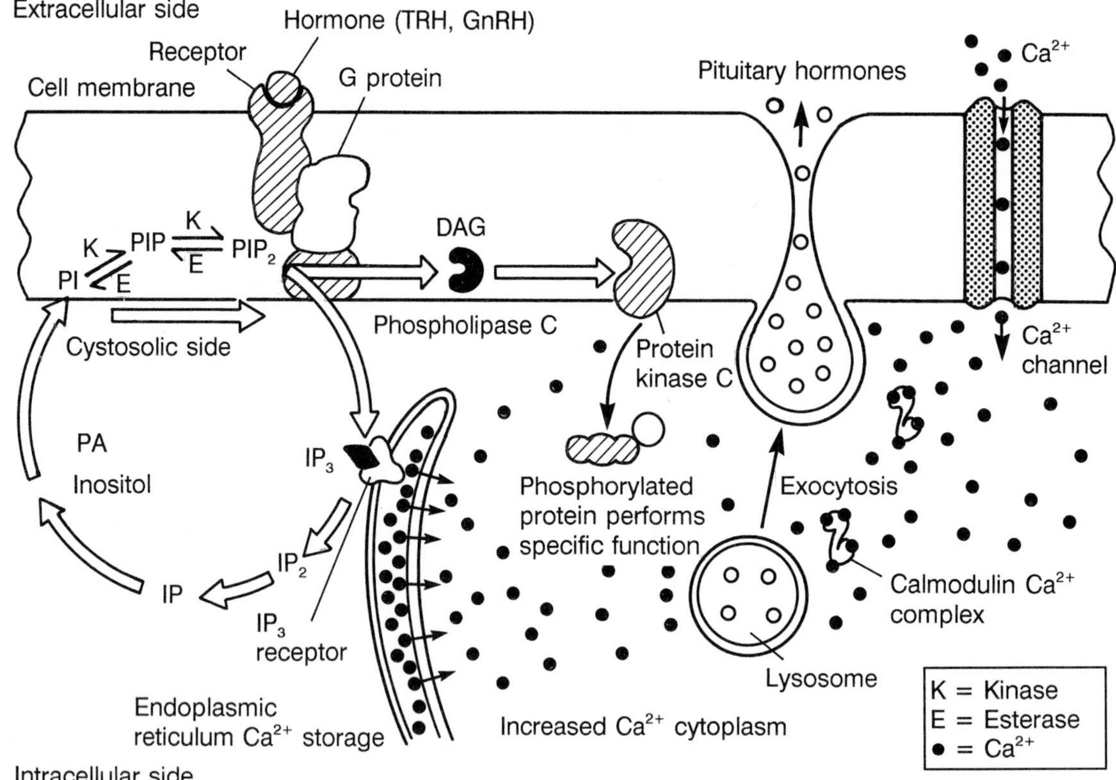

FIGURE 1-39 Overview of the hormonal signaling through the phosphatidylinositol system generating the second messengers, inositol 1,4,5-trisphosphate (IP_3), diacylglycerol (DAG), and Ca^{2+}. The action of IP_3 is to increase cytoplasmic Ca^{2+} levels by a receptor-mediated event in the cellular calcium store. Abbreviations: DAG, diacylglycerol; PA, phosphatidic acid; IP, inositol phosphate; IP_2, inositol diphosphate; IP_3, inositol 1,4,5-triphosphate; IP_4, inositol 1,3,4,5-tetraphosphate; PIP, phosphatidylinositol phosphate; PIP_2, phosphatidylinositol 4,5,-diphosphate; K, kinase; E, esterase.

of cGMP, which activates PKG and, in turn, further phosphorylates a subset of cellular proteins, producing the desired biological responses.

4. Phospholipase C Pathway (Ca^{2+}) as a Second Messenger)

A second major signal transduction pathway for membrane-bound receptors coupled to G proteins is the IP_3 phospholipase C pathway. Here, signal transduction of the α_c-subunit results in the activation of a membrane-bound phospholipase C. This enzyme catalyzes the hydrolysis of a membrane-bound substrate phosphatidylinositol 4,5-diphosphate (PIP_2) to form two second messengers, diacylglyceride (DAG) and inositol 1,4,5-triphosphate (IP_3). Figure 1-38 shows the structure of the tumor promoter TPA (12-*O*-tetradecylphorbol 13-acetate). TPA functions as a pharmacological agent for PKC in the same way that forskolin does for PKA, in that it mediates the full activity of PKC and binds directly to the enzyme.

The detailed operation of the phospholipase C activation of the phosphatidylinositol (IP_3) and diacylglycerol pathways is summarized in Figure 1-39. The second messengers IP_3 and DAG each have their own separate spheres of influence. The released IP_3 diffuses from the cell membrane to its own receptor located in the endoplasmic reticulum. The IP_3 receptor belongs to the class of multiple-membrane-spanning receptors (see Table 1-4) that are coupled to the opening of a Ca^{2+} channel. In this instance, the membrane receptor is located on an interior cell membrane and the ion channel communicates across the membrane of the endoplasmic reticulum. Ca^{+} ions are sequestered inside the endoplasmic reticulum, and when the IP_3 occupies the receptor, the Ca^{2+} channel opens, releasing the Ca^{2+} ions and thereby increasing the cytosolic concentration of Ca^{2+}.

This elevated level of Ca^{2+} also functions as a second messenger. As suggested in Figure 1-39, the increased intracellular Ca^{2+} may activate the process of exocytosis, which can explain the Ca^{2+} secretion coupling mechanism often associated with the secretion of peptide hormones by their secretory glands, e.g., pituitary hormones, insulin. Alternatively, the elevated extracellular Ca^{2+} may bind to calmodulin. Calmodulin can function as a Ca^{2+} receptor in the cell and may function to maintain intracellular Ca^{2+} in the range of 10^{-7} *M*. Calmodulin is present in the cytoplasmic compartment of virtually all cells of higher organisms. It binds four Ca^{2+} ions tightly ($K_d \sim 10^{-8}$ *M*) and then undergoes a conformational change so that it interacts with a subset of Ca^{2+}-regulated enzymes, e.g., phosphorylase kinase *a* and phosphorylase kinase *b*, which in turn amplify the initiating hormone signal.

After IP_3 is released from its intracellular membrane receptor, it is metabolized by stepwise removal of the three phosphate groups to generate free inositol. This inositol is conjugated with phosphatidic acid (PA) to yield a membrane-associated phosphatidylinositol (PI). The PI is then phosphorylated twice by a kinase to yield membrane-bound PIP_2, which is available to undergo another round of hydrolysis and reformation of the second messengers DAG and IP_3.

The other second messenger, diacylglycerol (DAG), binds to an allosteric binding site on protein kinase C, which, in the presence of phosphatidylserine and Ca^{2+}, becomes activated so that still another subset of cellular proteins becomes phosphorylated. The enzyme PKC consists of two domains, a catalytic domain and a regulatory domain, which can be separated from one another by site-specific proteolysis. The free catalytic subunit, formerly called protein kinase M, can phosphorylate substrate proteins independent of the presence of the regulatory subunit. The regulatory subunit domain contains two Zn^{+} fingers, which are usually considered to be hallmarks of DNA-binding proteins (see the next section); this activity has not yet been demonstrated for PKC.

6. Ion Channels

Ion channels present in both the outer cell membrane and the sarcoplasmic reticulum may be subject to regulation by G proteins. There are two classes of membrane channels: those that are opened by membrane depolarization (i.e., they are voltage gated) and

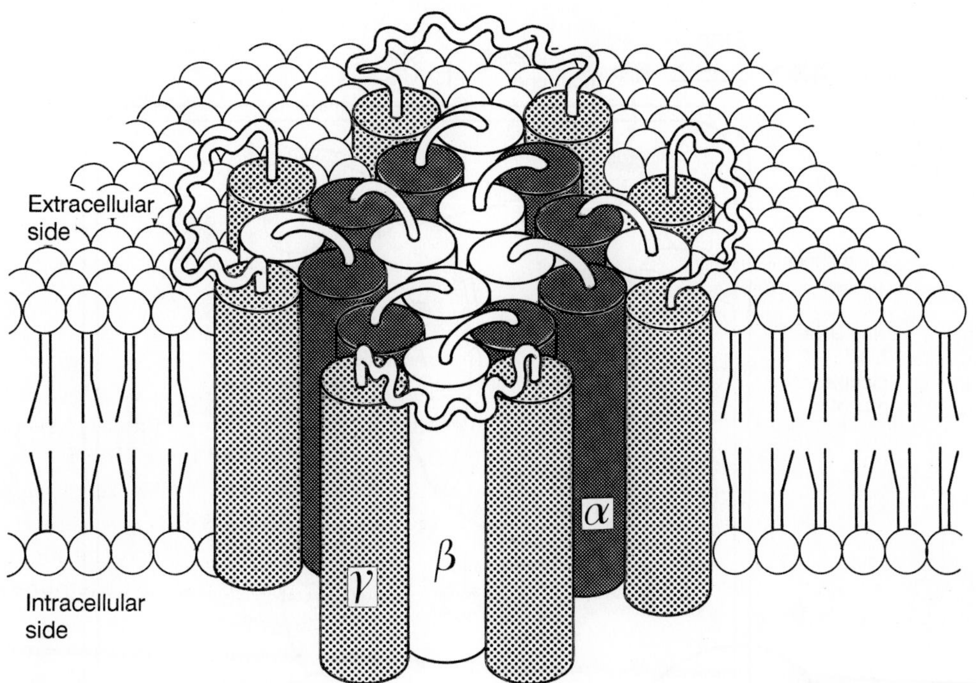

FIGURE 1-40 Schematic model for the dihydropyridine (DHP) sensitive calcium channel from skeletal muscle. The DHP channel comprises α (165 kDa), β (55 kDa), and γ (32 kDa) subunits. The subunits have been reconstituted so as to form functional Ca^{2+} channels. The α-polypeptide is the principal transmembrane subunit of the channel and forms the ion-conducting pore; it binds the ligand DHP, or phenylalkylamines and benzodiazapines, and is readily phosphorylated *in vitro* at SER687 by protein kinase A. [Information derived from Tanabe, T. *et al.* (1987); *Nature* **328,** 313–318. Catterall, W. A. (1988). *Science* **242,** 50–56; Curtis, B. M., and Catterall, W. (1984). *Biochemistry* **23,** 2113–2125.]

those that are coupled via specific G proteins to a membrane-bound receptor. Thus, the αC- and αK-isoforms, respectively, transduce the opening of a Ca^{2+} or K^+ channel in the outer cell membrane. The ion specificity of a channel is believed to be determined by the amino acid composition and sequence present in the seven-membrane-spanning domain that comprises the channel. Figure 1-40 presents a schematic model of the dihydropyridine-sensitive Ca^{2+} from skeletal muscle.

C. Steroid Hormone Regulation of Gene Transcription

Steroid hormones, including 1,25-dihydroxy vitamin D_3, thyroid hormones, and retinoic acid, generate biological responses primarily by stimulating the synthesis of mRNA at the level of the initiation of gene transcription. Basically their receptors may be considered to be transcription factors. Figure 1-41 summarizes, in a generic fashion, the sequence of steps that ensues after the arrival of the hormone at its target cells, which involves the participation of the receptor.

In step 1, the hormone (H) dissociates from the plasma transport protein. In step 2, the hormone enters the target cell by diffusing through the outer cell membrane. Then, depending upon the subcellular localization of the unoccupied receptor (see Table 1-5), the hormone will either interact with the receptor in the cytoplasmic compartment (glucocorticoid and aldosterone receptors) or continue through the cytoplasm and cross the perinuclear membrane to interact with a receptor in the nucleus (thyroid, estrogen, progesterone, retinoic acid, and $1,25(OH)_2D_3$ receptors).

As discussed earlier, the unoccupied form of the steroid receptor is believed to exist in the cell as a conjugate with a dimer of the 90-kDa heat shock protein (HSP); in Figure 1-41 the HSP dimer is indicated by a pair of ovals. It has been suggested that the HSP's function is to occlude the DNA-binding domain of the receptor (see Figure 1-23), which prevents the unoccupied receptor from binding to DNA.

In steps 3 and 4, the receptor becomes transformed or activated as a consequence of release of the 90-kDa HSP and binding of its cognate ligand, so that the DNA-binding domain is now exposed. In step 5, the

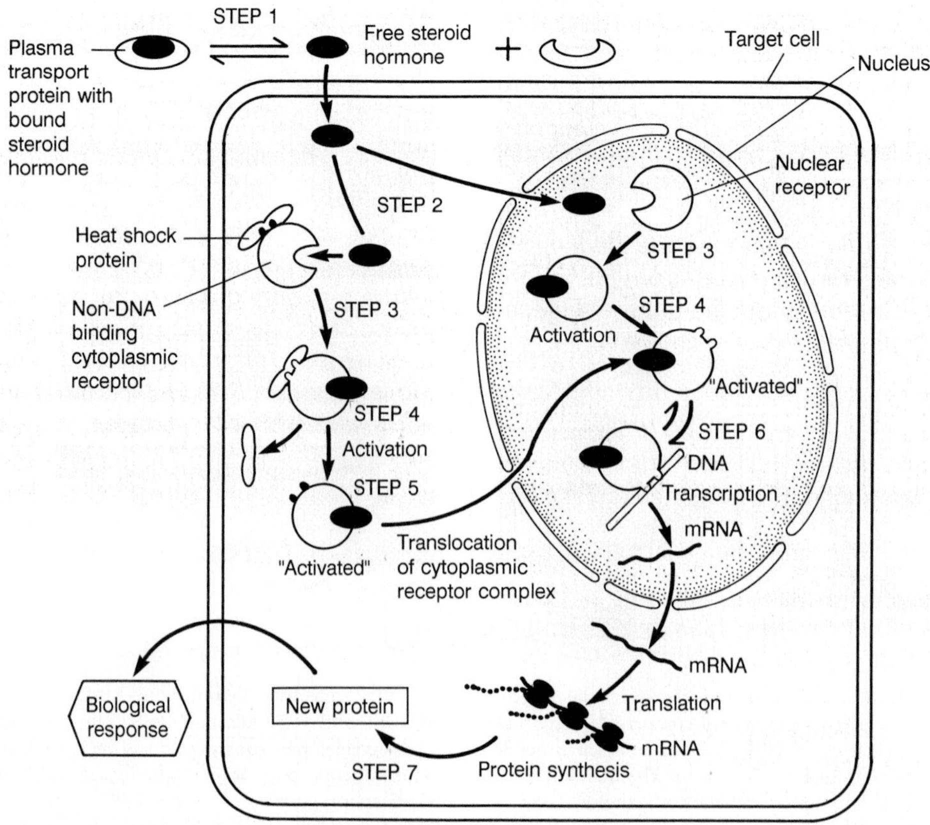

FIGURE 1-41 Model of steroid hormone mode of action. A detailed description is given in the text.

cytoplasmic activated receptor translocates to the nucleus, probably through a nucleopore. Then in step 6, the activated receptor, or more likely a homodimer of the receptor, seeks out the correct sequence of DNA that will allow it to form a high-affinity complex between the receptor and the hormone response elements (HRE) of the promoters of a selected set of genes and also any required transcription factors. As a consequence of the activated receptor binding to the promoters, either induction or repression of that gene will occur, leading to more or less of the mRNA coded for by that gene. The newly transcribed mRNAs are translocated to the cytoplasm where they become incorporated into polysomes and undergo translation (step 7). In the final step 8, the increased or decreased amount of new proteins generates more or less of the biological response(s) dictated by that hormone in that target cell. For any given hormone, an incredible array of biological responses can be modulated depending upon the phenotype of the target cell that possesses the cognate receptor. In any given target cell phenotype, only a small subset of genes will have their DNA chromatin in an active or "open" configuration. Thus, while a hormone may modulate as many as 300 genes in a given organism, in a specific target cell perhaps only a few genes will be available for regulation.

Consensus DNA sequences, known as hormone response elements (HRE), which precisely define the specific acceptor sites for the steroid hormone receptors, are now known; see Table 1-10.

On the basis of the regions of maximum sequence homology, the steroid receptors may be classified into two subfamilies: one including the glucocorticoid, progesterone, androgen, and mineralocorticoid receptors and the other group comprising the receptors for estrogen, thyroid hormone, retinoic acid, and $1,25(OH)_2D_3$. The primary difference between the two subfamilies is the presence of two discriminatory amino acids in the knuckle region of the first zinc finger, while members within a subfamily are distinguished by the presence of unique amino acids in the stem of the second zinc finger. Also within the same subfamily, the TR and VDR appear to share strong homology in most of the functional domains defined so far, which may reflect their close evolutionary relationship. It is speculated that the human VDR, which is smaller than the TR by 75 amino acids, may have evolved prior to the TR and perhaps gave rise to the latter by gene duplication. The model of a steroid hormone receptor is presented in Figure 1-26. In the center of the molecule is the DNA-binding domain. Surprisingly, among all the members of this superfamily of proteins, there is a 60–95% amino acid homology; in spite of this sequence similarity,

TABLE 1-10 Hormone Response Elements (HRE) for Subgroups of the Steroid Hormone Receptor Superfamily[a]

Subgroup	Consensus HRE	Critical amino acids for HRE specificity: Proximal box	Distal box
Subgroup A			
GR	AGAACAnnnTGTTCT	$\overset{*}{c}\overset{*}{G}\overset{*}{S}ckV$	$\overset{*}{A}\overset{*}{G}\overset{*}{R}\overset{*}{N}\overset{*}{D}$
PR	AGAACAnnnTGTTCT	$\overset{*}{c}\overset{*}{G}\overset{*}{S}ckV$	$\overset{*}{A}\overset{*}{G}\overset{*}{R}\overset{*}{N}\overset{*}{D}$
AR	AGAACAnnnTGTTCT	$\overset{*}{c}\overset{*}{G}\overset{*}{S}ckV$	$\overset{*}{A}\overset{*}{G}\overset{*}{R}\overset{*}{N}\overset{*}{D}$
MR	AGAACAnnnTGTTCT	$\overset{*}{c}\overset{*}{G}\overset{*}{S}ckV$	$\overset{*}{A}\overset{*}{G}\overset{*}{R}\overset{*}{N}\overset{*}{D}$
Subgroup B			
ER	AGGTCAnnnTGACCT	$\overset{*}{c}\overset{*}{E}\overset{*}{G}ckA$	PATNQ
VDR	AGGTCAnnnTGACCT	$\overset{*}{c}\overset{*}{E}\overset{*}{G}ckG$	PFNGD
TR-β	AGGTCAnnnnTGACCT	$\overset{*}{c}\overset{*}{E}\overset{*}{G}ckG$	KYEGK
RAR-α,β	AGGTCAnnnnnTGACCT	$\overset{*}{c}\overset{*}{E}\overset{*}{G}ckG$	HRDKN

[a] The glucocorticoid receptor (GR) is the prototype of subfamily A, while the estrogen receptor (ER) is the prototype of subfamily B. For steroid receptors, the HRE regions are known to exist in one of three orientations: (i) direct repeats; (ii) inverted repeats; or (iii) everted repeat sequences; the repeat sequences are normally separated by $1 \rightarrow 6$ (n) spacer nucleotides. Also, the HRE sequences actually present in the promoters of various hormone-regulated genes frequently display variation from the consensus HRE presented in this table. The consensus hormone responsive elements (HREs) are oriented 5′ to 3′, and the nonspecific "gap" nucleotide is denoted by n, followed by the second half-site. For the proximal and distal box regions of the receptor, the amino acids are represented by single-letter codes; amino acids that are highly conserved among species are capitalized. The discriminatory amino acids in the proximal region of the first zinc (see Figure 1-42) and in the distal region of the second zinc finger are marked with an asterisk. Reproduced from K. E. Lowe, A. Maiyar, and A. W. Norman "Vitamin-D Mediated Gene Expression," *Critical Reviews in Eukaryotic Gene Expression* **2**(1), 65–109 (1992).

each receptor still displays very high specificity with regard to the specific genes modulated.

Within the DNA-binding domain of the receptor exist two so-called zinc fingers (see Figure 1-42). These zinc fingers form the principal interface between the DNA-binding domain of the receptor and the specific nucleotides of the HRE, which constitute the acceptor site (see Table 1-10). In a zinc finger domain, the Zn^+ atom is coordinated to four amino acid side chains, usually two histidine and two cysteine residues. In the primary structure for the Zn^+ finger motif, the two fingers are separated from one another by approximately 12 amino acid residues. Figure 1-42C illustrates the subtle but important differences in the Zn^+ finger of the HRE of the glucocorticoid receptor versus that of the estrogen receptor. There are examples of single amino acid mutations that generate a crippled or totally inactive receptor. Thus, in the $1,25(OH)_2D_3$ receptor,

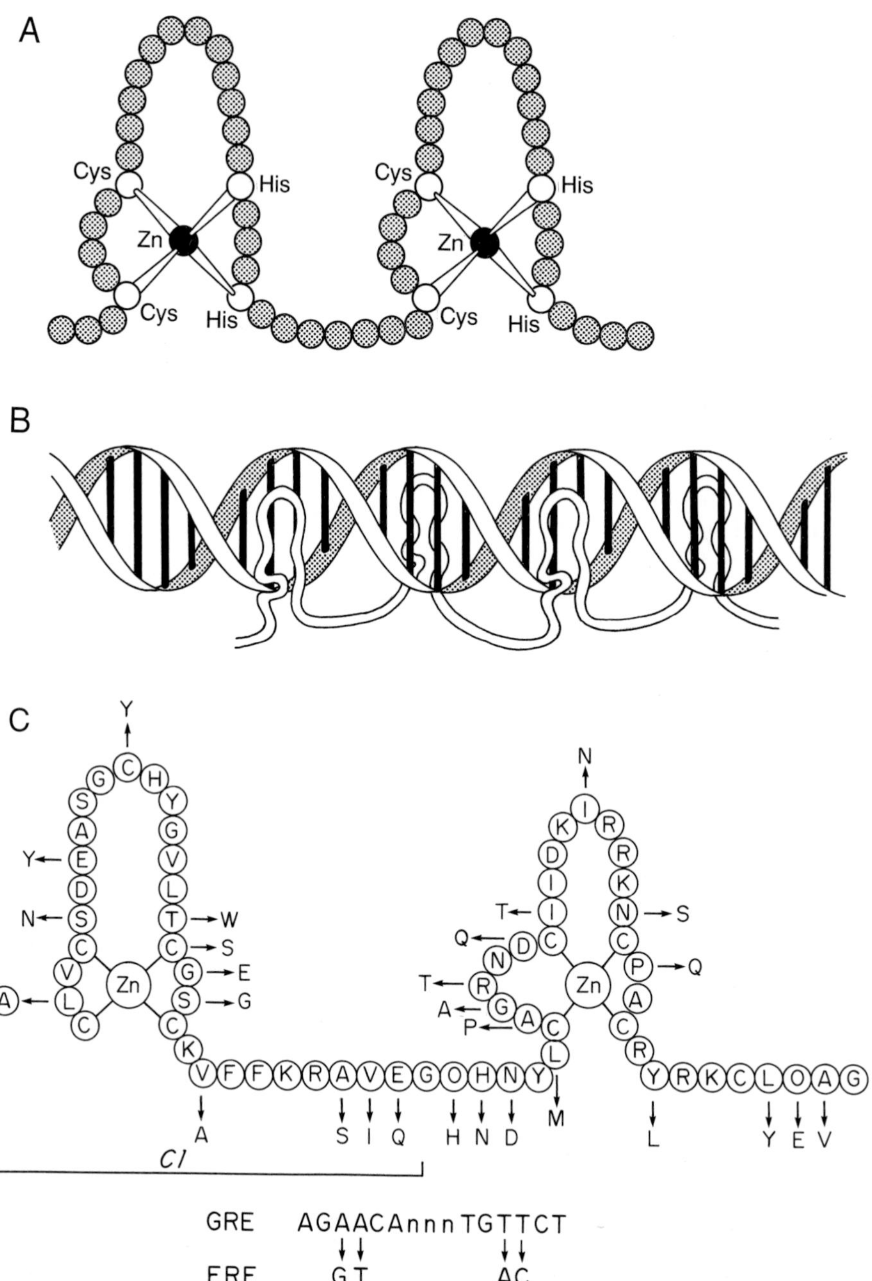

FIGURE 1-42 Zinc finger motifs present in hormone response elements of steroid receptors. (A) Generic representation of a Zn^{+} finger. A highly schematic model for the general conformation of the DNA-binding domain is shown with each amino acid represented by a sphere; the polypeptide chain in each finger is actually thought to be folded into a complex globular conformation. (B) Schematic view of how four such zinc fingers might bind to a specific DNA sequence. Each zinc finger is postulated to recognize a specific sequence of about five nucleotide pairs. [Adapted from Klug, A., and Rhodes, D. (1987). *Trends Biochem. Sci.* **12,** 464.] (C) Comparison of the amino acid sequence of the Zn^{+} finger of the glucocorticoid and estrogen receptors. The one-letter amino acid codes of the glucocorticoid zinc fingers are shown in the circles. The arrows indicate the changes in the zinc finger in the amino acid sequence of the estrogen receptor. Also shown at the bottom is a palindromic GRE with arrows indicating base pair changes converting it into an ERE. Big arrows emphasize the palindromic nature of the response element [Dahlman-Wright, K., Wright, A., Carlstedt-Duke, J., and Gustafsson, J.-A. (1992). *J. Steroid Biochem. Mol. Biol.* **42,** 131–139.]

conversion of the nucleotide sequence from GGC to GAC results in the mutation at position 30 of a glycine to an aspartic acid residue, which results in an inactive receptor. If an individual is homozygous for this mutation, then he/she will suffer from vitamin D-resistant rickets, type II (see Chapter 9).

Models of how zinc finger proteins interact specifically with DNA have been determined through X-ray crystallographic analysis and ^{1}H NMR (protein nuclear magnetic resonance spectroscopy) (see Figure 1-43). In this model, two occupied receptors, with the α-helix of the receptor containing Zn^{+} finger, interact with the major groove of the DNA, which contains the promoter

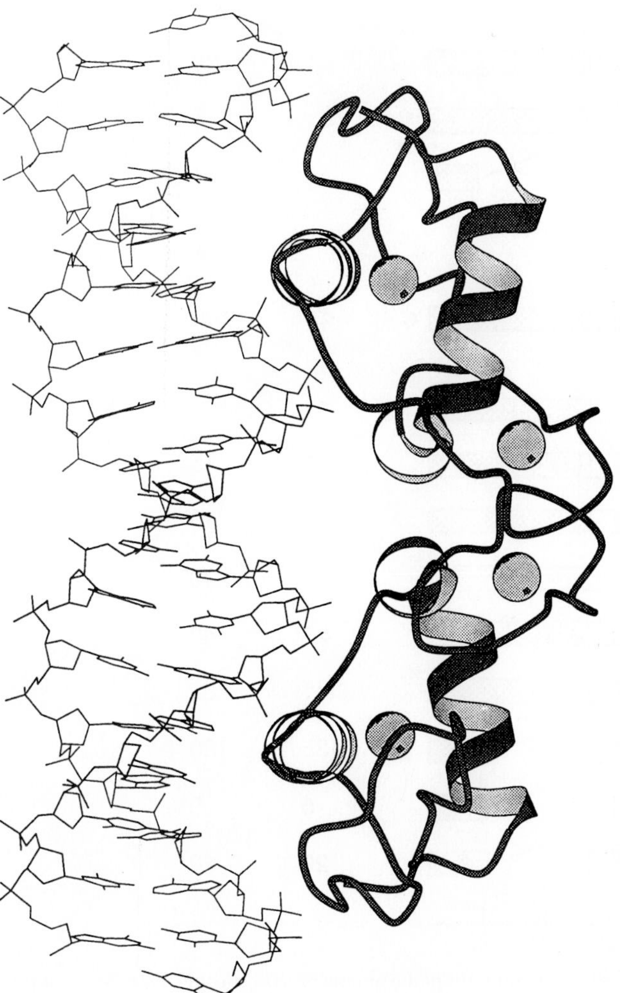

FIGURE 1-43 Three-dimensional structure of the DNA-binding domain (DBD) of two estrogen receptors interacting in the major groove of DNA. The zinc ions of the zinc fingers of the DBD are shown by spheres. The recognition helix lies in successive major grooves of the DNA. Modified with permission from Luisi, Jr., B. F., Schwabe, W. R., and Freedman, L. P. (1994). The steroid/nuclear receptors: From three-dimensional structure to complex function. *Vitam. Horm.* **49,** 1–47.

of the gene being modulated. A dimer of the activated receptor with its four zinc fingers in the α-helix interacts with the major groove, i.e., the DNA that contains the promoter of the gene being modulated.

The X-ray structure of the ligand binding domain of both the human retinoic acid RAR_{α} receptor and the human thyroid receptor have been determined; the structural features were remarkably similar, perhaps reflecting that both receptors belong to the same steroid receptor subfamily (see Table 1-10). Both structures revealed that the ligand is buried in a hydrophobic core. The carboxy-terminal activation domain F (see Figure 1-26) of the thyroid receptor was found to form an amphipathic α-helix, with its hydrophobic face constituting part of the hormone binding cavity (see front cover of this book).

One important approach to studying both the detailed structure and function of steroid receptors and the specificity of their interactions with acceptor DNA, as well as the specificity of their ligand-binding domains, has been to use the techniques of molecular cloning. Thus, it has been possible to construct chimeric receptors where, for example, a progesterone ligand-binding domain is substituted for the estrogen-binding domain of the estrogen receptor; in this instance, progesterone will activate the estrogen receptor. The specificity of the acceptor DNA sequence and the exact location of the HRE on the promoter in relation to the TATA box or start signal can also be determined. An example of this molecular biological approach to understanding steroid hormone action is presented in Figure 1-44.

The critical but complex details of how the physical presence of the occupied receptor interacting with the HRE of the appropriate gene and other transcription factors results in specific activation or repression of the gene in question remain to be elucidated. As suggested earlier, it appears that it is a homodimeric form rather than a monomeric form of the receptor that interacts with the hormone response elements. Thus, each of the receptor subunits likely interacts with one-half of the consensus nucleotide sequence of the HRE elements listed in Table 1-10.

Still another level of functional complexity concerns the formation of heterodimers by some of the steroid receptors. Thus, evidence has been presented that the $1,25(OH)_2D_3$ receptor and thyroxine and retinoic acid receptors can form functional heterodimers with themselves and, in some instances, with other transcription factors or oncogene products, e.g., c-fos and c-jun. Figure 1-45 summarizes the surprising level of complexity of known homodimeric and heterodimeric interactions

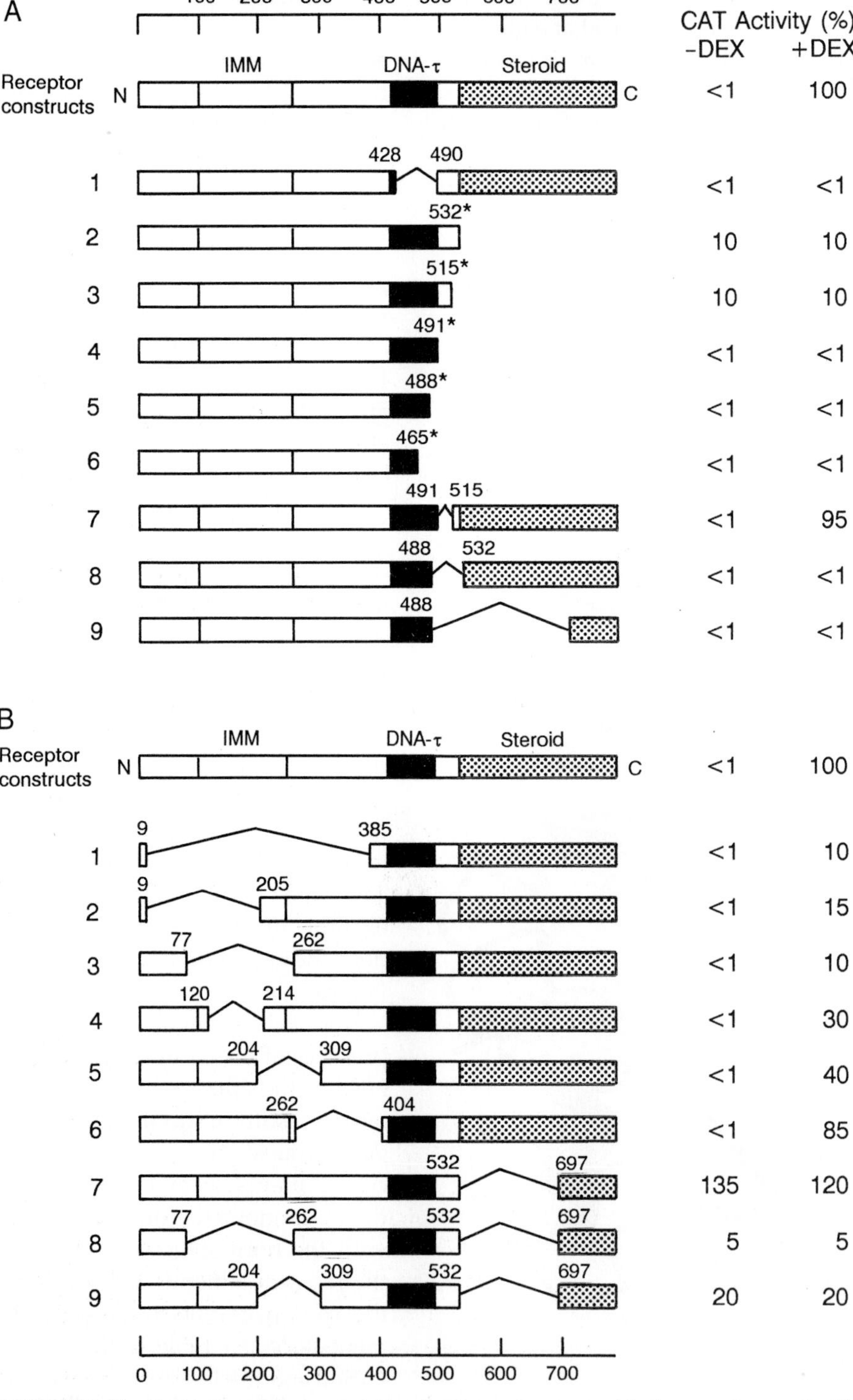

FIGURE 1-44 Evaluation of the transcriptional activity of mutants of the human glucocorticoid receptor (A), the carboxy-terminal end, and (B) the amino-terminal end of the receptor. The top of each panel shows the intact glucocorticoid receptor with 777 amino acids. Each of the domains is indicated: open box (IMM), the immunogenic region; solid black rectangles, the DNA-binding domain; stipples, the ligand-binding domain. The receptor constructs of panels A and B are prepared via site-directed mutagenesis. In (A), where truncation mutants are evaluated, the numbers with asterisks indicate the last amino acid of the receptor before the COOH-terminal nonsense peptide. In (B), where deletion mutants are evaluated, the numbers indicate amino acids bounding the deletion. These mutant receptors were constructed via standard, but complex,

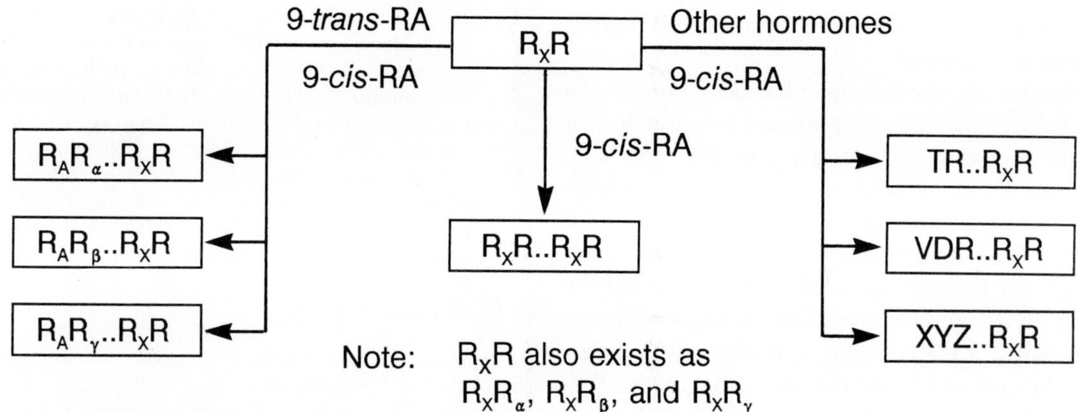

FIGURE 1-45 Illustration of homodimer and heterodimer interactions of the retinoic acid receptors. The retinoic acid receptors (RAR) exist in α_1, β_1, and γ_1 isoforms and in the more distantly related RXR gene. Thus, four functional forms of the retinoic acid receptor are generated: RAR_α, RAR_β, and RARγ, which all preferentially bind as ligands to the *trans* form of retinoic acid, 9 *trans*-RA, and the RXR_α, RXR_β, and RXRγ receptors, which bind to the *cis* form of retinoic acid, 9-*cis*-RA. Interestingly, while RAR_α, RAR_β, and RAR_γ are only found in vertebrates, the RXR isoforms has been conserved from *Drosophila* to humans. The RXR receptor isoform species can interact with itself to generate the homodimer RXR..RXR; the receptor ligand is exclusively 9-cis-RA. The RXR with its ligand 9-cis-RA can form heterodimers with each of the RAR species (which have 9-trans-RA as a ligand). Finally, RXR can also form heterodimers with the thyroxine (TR), $1,25(OH)_2D_3$ (VDR), or (postulated) receptors (indicated as XYZ). Collectively, this model suggests that RXR receptors have multiple central functions that allow the selective control of several nuclear hormone receptors as a consequence of heterodimer formation. [Modified from Zhang, X. and Pfahl, M. (1993). Hetero- and homodimeric receptors in thyroid hormone and vitamin A action. *Receptor* **3,** 183–191.]

FIGURE 1-44—*Continued* molecular biological manipulations and inserted in plasmids. CV-1 cells that have no constitutive expression of steroid receptors were transfected with the receptor construct incorporated into a plasmid. The activity of the receptor constructs was monitored by their ability to induce transcription from a mouse mammary tumor virus (MMTV)–chloramphenicol acetyl transacetylase (CAT) fusion gene also resident in the CV-I cells. The MMTV is known to have a specific HRE for the human glucocorticoid receptor; thus, a functional glucocorticoid receptor construct would activate the MMTV–CAT gene. This is expressed in each panel as CAT activity (%). In both (A) and (B), the CAT activity of the CV-I cells containing each receptor construct was determined in the presence (+DEX) and absence (−DEX) of dexamethasone. In A and B the constructs are numbered from top to bottom 1 → 9. In (A), the essential nature of the DNA-binding domain is demonstrated; compare the CAT activity of construct (1) with the intact receptor; constructs (2) and (3) suggest, in the absence of the ligand-binding domain, that the receptor can function constitutively. Construct (9) demonstrates that a full ligand-binding domain is necessary for full hormone-dependent CAT activity. In (B), constructs (1–6) demonstrate the contribution of the IMM domain to creating a fully functional receptor. Construct (7) is analogous to construct (9) of (B), except that the region deleted is smaller (leaving the domain labeled λ); thus, the panel A construct (7) is a fully active version of the receptor that can function even in the absence of ligand. [Reproduced with permission from Hollenberg, S. M., Giguere, V., Segui, P., and Evans, R. M. (1987). Co-localization of DNA-binding and transcriptional activation functions in the human glucocorticoid receptors. *Cell* **49,** 39–46.]

of the retinoic acid receptors. The structure of the ligand of the RXR receptor, namely, 9-*cis*-retinoic acid, is given in Figure 1-46. It is not yet possible to understand in detail the outcome of the operation of these receptor heterodimers, although it seems likely that

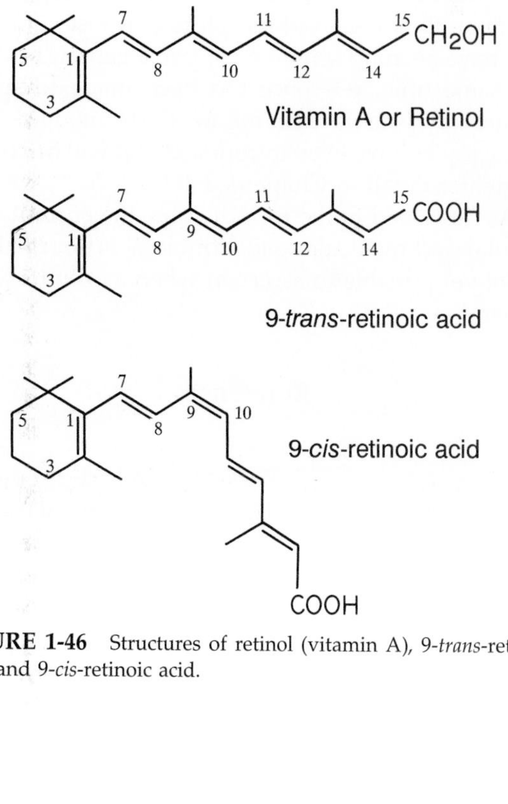

Vitamin A or Retinol

9-*trans*-retinoic acid

9-*cis*-retinoic acid

FIGURE 1-46 Structures of retinol (vitamin A), 9-*trans*-retinoic acid, and 9-*cis*-retinoic acid.

their formation allows even more sophisticated regulation of gene transcription.

Considerable intellectual effort has been invested to evaluate the evolutionary origins and relationship of the nuclear receptor gene family and their possible relationships to other nuclear transcription factors. Through the use of protein sequence comparisons of the DNA-binding domain (≈75–77 amino acids long; see Figure 1-27), the current conclusion is that all of the DNA-binding domains evolved from a single common precursor that was the ancestor of arthropods and vertebrates. With respect to the two zinc fingers (see Figure 1-42), the prevailing view is that they evolved as single, separate structural units rather than by duplication of a single ancestral zinc finger sequence. Thus, it appears that the steroid nuclear receptors do not share along with other transcription factors, or other zinc finger proteins, a common ancestral protein.

Our understanding of small molecule (steroid hormone, retinoic acid, and thyroid hormone) receptor regulation of gene transcription is still emerging. It is to be anticipated that there will be significant advances in the future involving how nuclear transcription factors, hormone receptors, etc. function to achieve regulation of gene transcription.

V. SUMMARY

This chapter has presented an overview of our current understanding of the mechanism of hormone action. The several models of membrane receptors and the operation of steroid receptors as transcription factors have been presented mostly in a generic sense. At the same time, reference has been made to specific hormonal systems that follow these models. Thus, many topics have been introduced that will be covered in greater detail in Chapters 3–20.

As described in the Introduction, the current era of cellular and molecular endocrinology is flourishing. It is not yet possible to ascertain when its zenith will be at hand.

References

A. Books

Alberts, B., Bray, D., Lewis, J., Raff, M., Roberts, K., and Watson, J. D. (1983). "Molecular Biology of the Cell." Garland Publishing Co., New York.

Becker, W. M., and Deamer, D. W. (1991). "The World of the Cell," 2nd ed., Figure 13-20. The Benjamin/Cummings Publishing Co., Menlo Park, CA. pp. 1–601.

Bolander, F. F. (1994). "Molecular Endocrinology," 2nd ed. pp. 1–601. Academic Press, New York.

Garrett, R. H., and Girsham, C. M. (1995). "Molecular Aspects of Cell Biology," pp. 1–1100. Harcourt Brace, New York.

Lehninger, A. L., Nelson, D. L., and Cox, M. M. (1993). "Principles of Biochemistry," 2nd ed., pp. 1–1103. Worth Publishers, New York.

Lodish, H., Baltimore, D., Berk, Z., Zipursky, S. L., Matsudaira, P., and Darnell, J. (1995). "Molecular Cell Biology," 3rd ed., pp. 1–1341. Scientific American Books, New York.

Nishizuka, Y., Tanaka, C., and Endo, M., eds. (1990). "The Biology and Medicine of Signal Transduction," Advances Second Messenger Phosphorylation Research, Vol. 24.

B. Review Articles

Agutter, P. S., and Prochnow, D. (1994). Nucleoplasmic transport. *Biochem. J.* **300,** 609–618.

Amero, S. A., Kretsinger, R. H., Moncrief, N. D., Yamamoto, K., R., and Pearson, W. R. (1992). *Mol. Endocrinol.* **6,** 3–7.

Berridge, M. J., and Irvine, R. F. (1989). Inositol phosphates and cell signaling. *Nature* **341,** 197–205.

Carson-Jurica, M. A., Schrader, W. T., and O'Malley, B. W. (1990). Steroid receptor family: structure and functions. *Endocrine Rev.* **11,** 201–220.

Catterall, W. A. (1988). Structural and function of voltage-sensitive ion channels. *Science* **252,** 50–61.

Karin, M. (1992). Signal transduction from cell surface to nucleus in development and disease. *FASEB J.* **6,** 2581–2590.

Kaziro, Y., Itoh, J., Kozasa, T., Nakafuku, M., and Satoh, T. (1991). Structure and function of signal-transducing GTP-binding proteins. *Annu. Rev. Biochem.* **60,** 349–400.

Luisi, B. F., Schwabe, J. W., and Freedman, L. P. (1994). The steroid/nuclear receptors: from three-dimensional structures to complex function. *Vitamins Hormones* **49,** 1–47.

Montal, M. (1990). Molecular anatomy and molecular designs of channel proteins. *FASEB J.* **4,** 2623–2635.

Nemere, I., Zhou, L.-X., and Norman, A. W. (1993). Nontranscriptional effects of steroid hormones. *Receptor* **314,** 277–291.

O'Malley, B. W., Tsai, S. Y., Bagchi, M., Weigel, N. L., Schrader, W. T., and Tsai, M.-J. (1991). Molecular mechanism of action of steroid hormones receptor. *Recent Prog. Horm. Res.* **47,** 1–26.

Pante, N., and Aebi, U. (1994). Towards understanding the three-dimensional structure of the nuclear pore complex at the molecular level. *Curr. Opin. Struct. Biol.* **4,** 187–196.

Rut, A. R., Hewison, M., Kristjansson, K., Luis, B., Hughes, M. R., and O'Riordan, J. L. H. (1994). Two mutations causing vitamin D resistant rickets: modeling on the basis of steroid hormone receptor DNA-binding domain crystal structures. *Clin. Endocrinol.* **41,** 581–590.

Taylor, C., and Marshall, I. (1992). Calcium and inositol-1,4,5-triphosphate receptors: A complex relationship. *Trends Biochem. Sci.* **17,** 403–407.

Ulrich, A., and Schlessinger, J. (1990). Signal transduction by receptors with tyrosine kinase activity. *Cell* **61,** 203–212.

C. Research Papers

Anderson, R. G. W., Brown, M. S., and Goldstein, J. L. (1977). Role of the coated endocytotic vesicle in the uptake of receptor-bound low-density lipoprotein in human fibroblasts. *Cell* **10,** 351–364.

Bourguet, W., Ruff, M., Chambon, P., Gronemeyer, H., and Moras, D. (1995). Crystal structure of the ligand-binding domain of the human nuclear receptor RXR-α. *Nature* **375,** 377–382.

Devos, A., Uhsch, M., and Kossiakoff, A. (1992). Human growth hormone and extracellular domain of its receptor. *Science* **255,** 306–312.

Hard, T., Kellenbach, E., Boelens, R., Maler, B. A., Dahlman, K., Freedman, L. P., Carlstedt-Duke, J., Yamamoto, K. R., Gustafsson, J.-A., and Kaptein, R. (1990). Solution structure of the glucocorticoid receptor DNA binding domain. *Science* **249,** 157–160.

Hughes, M. R., Malloy, P. J., Kieback, D. G., Kesterson, R. A., Pike, J. W., Feldman, D., and O'Malley, B. W. (1988). Point mutations in the human vitamin D receptor gene associated with hypocalcemic rickets. *Science* **242,** 702–1705.

Luisi, B. F., Xu, W. X., Otwinowski, Z., Freedman, L. P., Yamamoto, K. R., and Sigler, P. B. (1991). Crystallographic analysis of the interaction of the glucocorticoid receptor with DNA. *Nature* **352,** 497–505.

Mangelsdorf, D. J., Borgmeyer, U., Heyman, R. A., Zhou, J. Y., Ong, E. S., Oro, A. E., Kakuzuka, A., and Evans, R. M. (1992). Characterization of three RXR genes that mediate the action of 9-*cis*-retinoic acid. *Genes Dev.* **6,** 329–344.

Rhee, S., and Choi, K. (1992). Regulation of inositol phospholipid specific phospholipase C isoenzymes. *J. Biol. Chem.* **267,** 12393–12396.

Snyder, S. H. (1992). Nitric oxide: first in a class of neurotransmitters. *Science* **257,** 494–496.

CHAPTER

2

Steroid Hormones: Chemistry, Biosynthesis, and Metabolism

I. INTRODUCTION

This chapter deals with the structural chemistry and biosynthetic pathways of the major classes of steroid hormones. All have a complicated structure of fused rings, which can be modified by functional group substitution at many points. Furthermore, the presence of asymmetric carbon atoms introduces steric modifications and isomeric possibilities. The reader will find it prudent first to grasp the essential features of the steroid structures and relationships before attempting to delve into the discussion of specific hormonal activities in later chapters. Then, when so doing, it may be helpful to turn back to the appropriate portion of this chapter to heighten understanding of the structures of the hormones under review.

A. General Comments

The first steroid hormone, estrone, was isolated in 1929 at a time before the characteristic ring structure of the steroid nucleus had been elucidated. Today well over 225 naturally occurring steroids have been isolated and chemically characterized. In addition, an uncountable number of steroids and steroid analogs have been chemically synthesized.

All steroids belong to the chemical class of substances known as terpenoids or terpenes. Other biologically important terpenoid compounds include the plant hormones gibberellic acid and abscisic acid, insect hormone (juvenile hormone), farnesol (a plant oil), and plant-produced isoprenoids, which include carotene (a precursor of vitamin A), ubiquinone (a vitamin K analog), plastoquinones (participants in photosynthesis), and natural rubber. All terpenoids have in common the same two C_5H_8 isoprene precursors employed for their biosynthesis, namely, isopentenyl pyrophosphate and dimethylallyl pyrophosphate. These structural relationships are summarized in Figure 2-1.

B. Historical Perspective

The development of our modern understanding of hormones and the science of endocrinology has closely paralleled studies on the isolation, chemical characterization, and synthesis of steroids and the subsequent elucidation of their pathways of biosynthesis and catabolism. However, the foundation of many of these developments with steroid hormones is to be found in a lengthy series of papers that appeared in the late 1920s and early 1930s from Professor A. Windaus' laboratory in Gottingen, Germany, and led to the structural determination of cholesterol. This was an extraordinarily challenging problem given the limitation that the techniques of nuclear magnetic resonance spectroscopy (NMR), mass spectrometry, and ultraviolet (UV) and infrared (IR) spectroscopy were not available at that time. Instead, the structure was determined through elaborate classical organic chemistry manipulations, which involved the conversion of the compound under study to known reference compounds. At the present time, application of the powerful separation techniques of high-performance liquid chromatog-

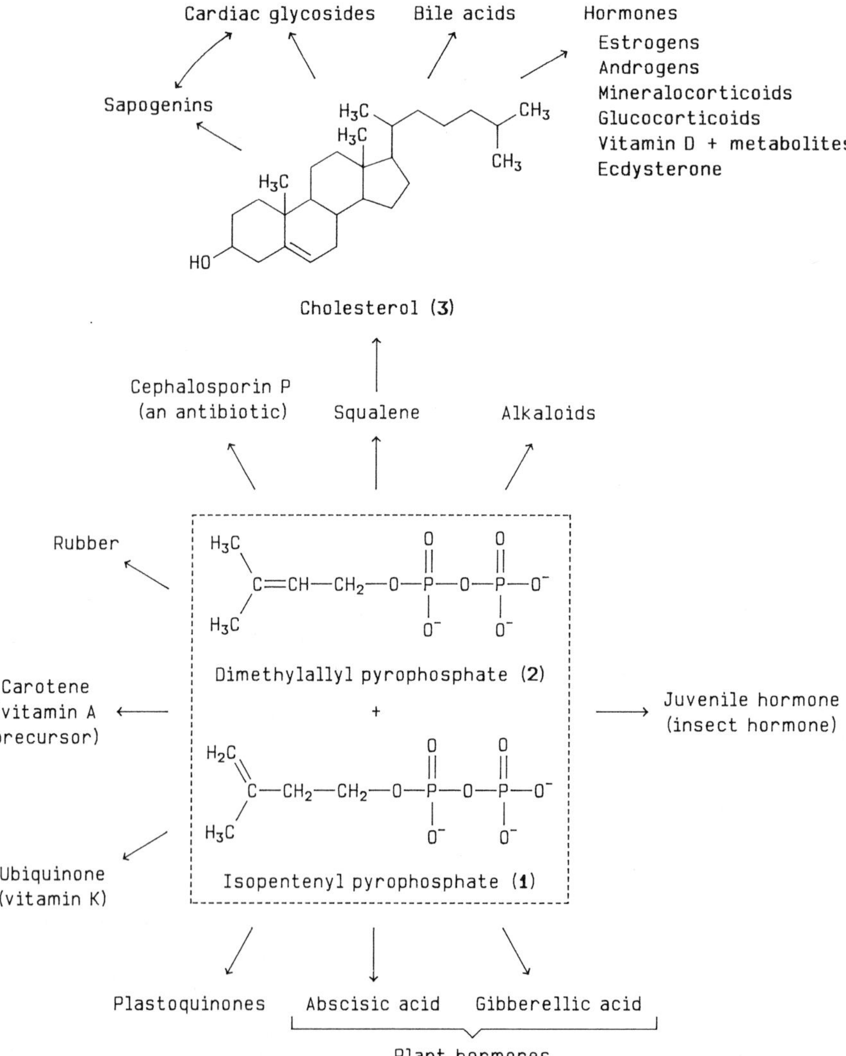

FIGURE 2-1 Representative isoprenoids that may be derived from isopentenyl pyrophosphate and dimethylallyl pyrophosphate.

raphy (HPLC) or gas chromatography, combined with the use of continuous on-line monitoring by mass spectrometry with computer-assisted data storage and analysis, frequently permit unequivocal structural determinations on impure samples that contain less than 1 ng of the steroid of interest.

An equally important contribution to our present understanding of the biochemistry of steroids was the introduction and general availability of radioactively labeled compounds. Radioactive steroids offer two major advantages: The presence of the radioactive label (a) provides a significant increase in the sensitivity of detection of the steroid under study (prior to the advent of radioactive steroids, investigators relied upon colorimetric or bioassay procedures to quantitate the steroid of interest) and (b) allows the investigator to detect, from either *in vivo* whole animal or *in vitro* experiments with perfused organs, tissue slices, cell suspensions, cell homogenates, or purified enzyme preparations, the presence of new compounds that otherwise would be unappreciated. Thus, it was through the application of radioisotope techniques, modern procedures of chromatography, and structure determination that whole categories of new steroid hormones were discovered [for example, vitamin D metabolites (Chapter 9) and catechol estrogens (Chapter 13)].

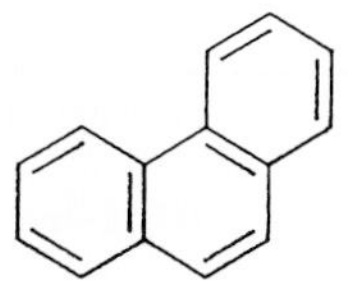

Phenanthrene (**4**)

Cyclopentanoperhydrophenanthrene (sterane) (**5**)

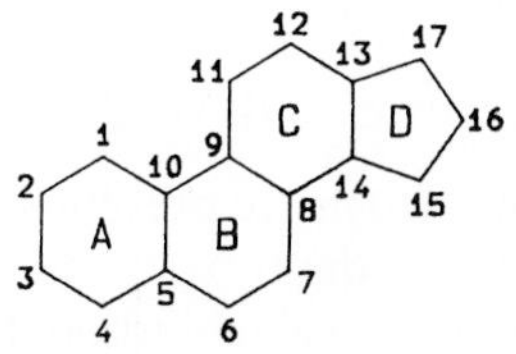

Sterane (**6**)

FIGURE 2-2 Parent ring structures of steroids.

II. CHEMISTRY OF STEROIDS

A. Basic Ring Structure

Steroids are derived from a phenanthrene ring structure (structure **4**) to which a pentano ring has been attached; this yields in the completely hydrogenated form, cyclopentanoperhydrophenanthrene, or the sterane ring structure (structure **6**, Figure 2-2).

Steroid structures are not normally written with all the carbon and hydrogen atoms as illustrated in **5** of Figure 2-2; instead, the shorthand notation as presented for sterane (**6**, Figure 2-2) is usually employed. In this representation the hydrogen atoms are not indicated, and unless specified otherwise it is assumed that the cyclohexane (A, B, and C) or cyclopentane (D) ring is fully reduced, that is, each carbon has its full complement of carbon and/or hydrogen bonds. Also, indicated in structure **6** is the standard numbering system for all of the carbon atoms in the four rings. The three six-carbon cyclohexane rings are designated A, B, and C, and the five-carbon cyclopentane ring is denoted as the D ring.

B. Classes of Steroids

In mammalian systems, there are six families of steroid hormones that can be classified on both a structural and a biological (hormonal) basis. They are the estrogens (female sex steroids), androgens (male sex steroids), progestins, mineralocorticoids, glucocorticoids, and vitamin D with its daughter metabolites. Also, the bile acids are structurally related to cholesterol and thus constitute a seventh member of the steroid family. All of these steroids are biologically derived from cholesterol. Table 2-1 summarizes some fundamental relationships of these principal mammalian classes of steroids.

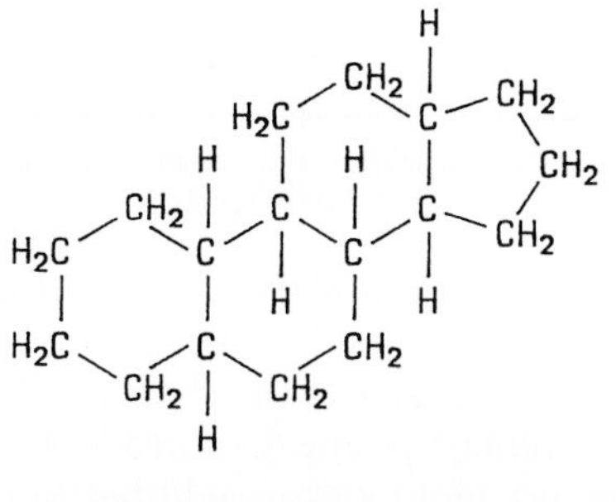

The parent ring structure for cholesterol is the fully saturated ring structure cholestane (**7**, Figure 2-3). Cholestane, which has 27 carbons, differs from sterane (**6**, Figure 2-2) by the addition of an eight-carbon side chain on carbon-17 of ring D and the presence of two angular methyl groups at the junctures of the A:B (carbon-10) and C:D rings (carbon-13). The cholestane ring structure also gives rise to the parent ring structures for the six classes of mammalian steroids and the bile acids, as summarized in Table 2-1. The parent ring compounds are the completely saturated ring structures pregnane (**8**), androstane (**9**), estrane (**10**), and cholane (**11**), of which is structurally related to cholestane. These relationships are depicted in Figure 2-3.

TABLE 2-1 Classes of Steroids

Steroid class	Principal active steroid in humans[a]	Number of carbon atoms	Parent ring structure[a]
Estrogens	Estradiol	18	Estrane
Androgens	Testosterone	19	Androstane
Progestins	Progesterone	21	Pregnane
Glucocorticoids	Cortisol	21	Pregnane
Mineralocorticoids	Aldosterone	21	Pregnane
Vitamin D steroids	1,25-Dihydroxyvitamin D_3	27	Cholestane
Bile acids	Cholic acid	24	Cholane

[a] The parent ring structures and active steroid are given in Figure 2-3.

The parent ring structures are used as the stem term in constructing the formal nomenclature of any steroid (will be discussed later in the chapter).

The biosynthetic pathway of production of each of these general steroid classes will be presented separately later in this chapter. A discussion of their hormonal and biochemical aspects will appear later in individual chapters.

C. Structural Modification

The basic ring structures presented in Figure 2-3 can be subjected to a wide array of modifications by the introduction of hydroxyl or carbonyl substituents and unsaturation (double or triple bonds). In addition, heteroatoms such as nitrogen or sulfur can replace the ring carbons, and halogens and sulfhydryl or amino groups may replace steroid hydroxyl moieties. Furthermore, the ring size can be expanded or contracted by the addition or removal of carbon atoms. The consequences of these structural modifications are designated by application of the standard organic nomenclature conventions for steroids. The pertinent aspects of this system are summarized in Table 2-2. Prefixes and suffixes are used to indicate the type of structural modification. Any number of prefixes may be employed (each with its own appropriate carbon number and specified in the order of decreasing preference of acid, lactone, ester, aldehyde, ketone, alcohol, amine, and ether); however, only one suffix is permitted.

Table 2-3 tabulates the systematic and trivial names of many common steroids. All of these formal names are devised in accordance with the official nomenclature rules for steroids laid down by the International Union of Pure and Applied Chemistry (IUPAC).[1]

D. Asymmetric Carbons

An important structural feature of any steroid is recognition of the presence of asymmetric carbon atoms and designation in the formal nomenclature of the structural isomer that is present. Thus, reduction of pregnan-3-one to the corresponding 3-alcohol will produce two epimeric steroids (see Figure 2-4). The resulting hydroxyl may be above the plane of the A ring and is so designated on the structure by a solid line; it is referred to as a 3β-ol. The epimer, or 3α-ol, has the hydroxyl below the plane of the A ring and is so designated by a dotted line for the C · · · OH bond. If the α- or β-orientation of a substituent group is not known, it is designated with a wavy line (C $\sim$ OH).

Another locus where asymmetric carbon atoms play an important role in steroid structure determination is the junction between each of the A, B, C, and D rings. Figure 2-5 illustrates these relationships for cholestanol and coprostanol. Thus, in the 5α-form, the 19-methyl and α-hydrogen on carbon-5 are on opposite sides of the plane of the A:B ring; this is referred to as a *trans* fusion. When the 19-methyl and β-hydrogen on carbon-5 are on the same side of the A:B ring fusion, this is denoted *cis* fusion. In the instance of *cis* fusion of the A:B rings, the steroid structure can no longer be drawn in one plane (as in **24**). Thus, in all 5β-steroid structures that have *cis* fusion between rings A and B, the A ring is bent into a second plane that is at approximately a right angle to the B:C:D rings (see **24** of Figure 2-5).

Thus, each of the ring junction carbons is potentially asymmetric, and the naturally occurring steroid will have only one of the two possible orientations at each ring junction. Although there are two families of naturally occurring steroids with either *cis* or *trans* fusion of the A:B rings, it is known that the ring fusions of B:C and C:D in virtually all naturally occurring steroids are *trans*.

In the estrogen steroid series in which the A ring is aromatic, there is no *cis–trans* isomerism possible at carbon-5 and -10. Also, as will become apparent upon consideration of the metabolism of many of the steroids that contain a 4-ene-3-one structure in ring A (see **12–15**, Figure 2-3), a family of structural isomers may be produced. Thus, when biological reduction of the 4-ene occurs, two dihydro products will arise, one with

[1] The IUPAC definitive rules of steroid nomenclature are presented in full in *Pure & Applied Chemistry* **31**, 285–322 (1972) or *Biochemistry* **10**, 4994–4995 (1971).

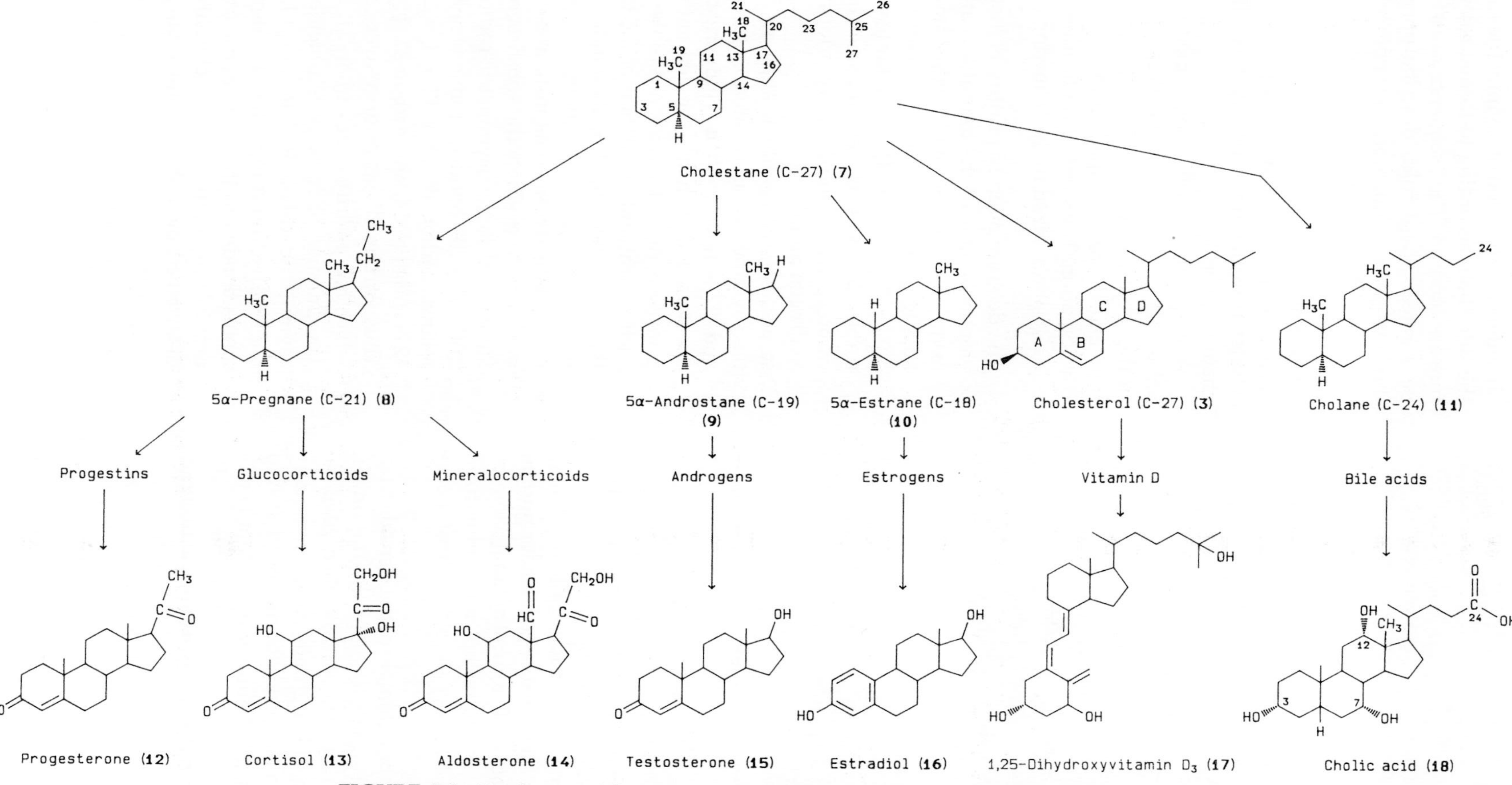

FIGURE 2-3 Family tree of the seven principal classes of steroids (bottom row) that are structurally derived from the parent cholestane (top row).

TABLE 2-2 Steroid Nomenclature Conventions

Modification	Prefix	Suffix
Hydroxyl group (OH)	Hydroxy-	-ol
Hydroxyl above plane of ring	β-Hydroxy	–
Hydroxyl below plane of ring	α-Hydroxy	–
Keto or carbonyl group (C=O)	Oxo-	-one
Aldehyde (CHO)	–	-al
Carboxylic acid (COOH)	Carboxy-	-oic acid
Double bond (C=C)	–	-ene
Triple bond (C≡C)	–	-yne
Saturated ring system	–	-ane
One less carbon atom	Nor-	–
One additional carbon atom	Homo-	–
One additional oxygenation	Oxo-	–
One less oxygen atom	Deoxy-	–
Two additional hydrogen atoms	Dihydro-	–
Two less hydrogen atoms	Dehydro-	–
Two groups on same sides of plane	*cis-*	–
Two groups on opposites sides of plane	*trans-*	–
Other ring forms (rings A and B *trans,* as in allopregnane)	Allo-	–
Opening of a ring (as in vitamin D)	Seco-	–
Conversion at a numbered carbon from conventional orientation (as in epicholesterol or 3α-cholesterol)	Epi-	–

A:B *cis* fusion and one with A:B *trans* fusion. Two additional steroids will be generated by reduction of the 3-oxo group, giving rise to a total of four possible tetrahydro products of the 3-oxo-4-ene. Under normal biological circumstances, all four structural isomers can be detected. These relationships are summarized in Figure 2-6 for the male sex steroid androst-4-ene-3,17-dione.

The side chain is a third domain of the steroid structure where asymmetry considerations are important. Historically, interest first centered on carbon-20 of the cholesterol side chain, although side chain asymmetry is also now known to be crucial for the insect steroid hormone ecdysterone, for a number of vitamin D metabolites, and in the production of many bile acids. While the α–β notation is satisfactory for the designation of substituents of the A, B, C, and D ring structures, this terminology is not applicable to the side chain. This is because there is free rotation of the side chain at the carbon-17–carbon-20 bond; thus, the side chain may assume a number of orientations in relation to the ring structure (see Figure 9-7A).

The chemical determination and designation of the absolute configuration of asymmetric carbon atoms on the side chain according to formal rules of nomenclature are complex. The "sequence rules" of Cahn[2] must be applied. These rules describe operational procedures to generate unambiguous nomenclature specifications of the absolute configuration of all chemical compounds whether they be steroids, sugars, amino acids, thiopolymers, etc. A detailed consideration of these rules is beyond the scope of this book; however, a brief consideration applicable to the steroid side chain is given.

Asymmetric carbon atoms joined to four different substituents can be designated as $C_{a,b,c,d}$. Application of the sequence rule requires determination of the a, b, c, and d groups joined to the carbon atom under consideration. The steps of application of the sequence rule are summarized in the following (see Figure 2-7):

1. The determination of priority is first based on the atomic weights of the four atoms attached to the asymmetric carbon, with higher atomic weights having priority over lower atomic weights (e.g., $^{16}O > ^{14}C > ^{12}C > H$). Thus, for asymmetric carbon-20 on the cholesterol side chain, the three carbons, 17, 21, and 22, have priority over the hydrogen atom.
2. In instances where identical atoms are bonded to the asymmetric carbon atom (e.g., carbon-17, -21, and -22 bonded to the asymmetric carbon-20 of the steroid side chain), the determination of priority is made by reference to the next atom in each radical and evaluation of the total sum of the atomic weights directly bonded to it. Again, the carbon with higher atomic weight claims higher priority. In some instances it may be necessary to evaluate the substituents on a third atom distal to the asymmetric carbon. The sums of the directly bonded atomic weights are as follows: carbon-17, $12 + 12 + 1 = 25$; carbon-22, $12 + 1 + 1 = 14$; and carbon-21, $1 + 1 + 1 = 3$. Thus, carbon-17 has priority over carbon-22, which has priority over carbon-21.
3. The chirality, or right- or left-handedness of the asymmetric carbon, is determined by visualizing the tetrahedron comprising the four substituent atoms bonded to the asymmetric carbon (see the example in Figure 2-7). The chirality is determined by looking at the atom with lowest priority (atom d) through the center of the opposite triangular face. Thus, the tetrahedron around the asymmetric carbon under

[2] The sequence rule is described in detail by Cahn, Ingold, and Prelog (1960); a simplified description is given by Cahn (1964).

TABLE 2-3 Trivial and Systematic Names of Some Common Steroids

Trivial name	Systematic name
Aldosterone	18,11-Hemiacetal of 11β,21-dihydroxy-3,20-dioxopregn-4-en-18-al
Androstenedione	Androst-4-ene-3,17-dione
Androsterone	3α-Hydroxy-5α-androstan-17-one
Cholesterol	Cholest-5-en-3β-ol
Cholic acid	3α,7α,12α-Trihydroxy-5β-cholan-24-oic acid
Corticosterone	11β,21-Dihydroxypregn-4-ene-3,20-dione
Cortisol	11β,17,21-Trihydroxypregn-4-ene-3,20-dione
Cortisone	17,21-Dihydroxypregn-4-ene-3,11,20-trione
Dehydroepiandrosterone	3β-Hydroxy-5-androsten-17-one
Deoxycorticosterone	21-Hydroxypregn-4-ene-3,20-dione
Ergosterol	5,7,22-Ergostatrien-3β-ol
Estrone	3-Hydroxyestra-1,3,5(10)-trien-17-one
Estriol	Estra-1,3,5(10)-triene,3,16α,17β-triol
Etiocholanolone	3α-Hydroxy-5β-androstan-17-one
Lanosterol	8,24-Lanostadien-3β-ol
Lithocholic acid	3α-Hydroxy-5β-cholan-24-oic acid
Progesterone	Pregn-4-ene-3,20-dione
Testosterone	17β-Hydroxyandrost-4-en-3-one
Vitamin D_3 (cholecalciferol)	9,10-Seco-5,7,10(19)-cholesta-trien-3β-ol
Vitamin D_2 (ergocalciferol)	9,10-Seco-5,7,10(19),22-ergostatetraen-3β-ol

consideration should be oriented so that the group of lowest priority (d) points away from the viewer. Then, if the other three atoms, a, b, and c, are arranged in clockwise order, the asymmetry is designated rectus (*R*) (*right*) or right-handed. Conversely, if the a, b, and c atoms are arranged in counterclockwise order, the asymmetry is designated sinister (*S*) (*left*) or left-handed. Two examples of the application of the sequence rule are provided in Figure 2-7.

Figure 2-8 identifies all of the asymmetric carbon atoms of cholesterol. In the parent sterane ring structure (**6,** Figure 2-2) there are six asymmetric carbons. Introduction of the $\Delta^{5,6}$-double bond deletes one asymmetric center, whereas addition of the eight-carbon side chain adds one asymmetric carbon at position 17. Carbon-20 of the side chain is also asymmetric. Finally, introduction of a hydroxyl group on carbon-3 creates still another asymmetric center. Thus, there are a total of eight asymmetric carbons or $2^8 = 256$ possible structural isomers. Considering that cholesterol is the most prevalent naturally occurring steroid, it is an impressive testament to evolutionary events and to the specificity of the many enzymes involved in the biosynthesis of cholesterol that only one major sterol product is present in mammalian systems.

E. Conformation of Steroids

The steroid nucleus, sterane, is composed of three cyclohexane rings and one cyclopentane ring. The six carbon atoms of a cyclohexane ring are not fixed rigidly in space, but are capable of interchanging through turning and twisting between several structural arrangements in space called conformations. The two principal conformations of a cyclohexane ring are the chair (**32**) and boat (**33**) forms (see Figure 2-9).

Each of the two substituent groups on the six carbon atoms of the cyclohexane ring may exist in either the general plane of the ring and are designated as equatorial (e) or a plane perpendicular to the ring plane and are designated as axial (a). For the equatorial bonds, it is possible to superimpose on the equatorial notation an indication of whether they are below (α) or above (β) the general plane of the ring. Cyclohexane is highly conformationally mobile, interchanging between the boat and chair forms many thousands of times per second. The most stable form of the cyclohexane ring is the chair form; in this conformer there is a greater interatomic distance between the equatorial and axial hydrogens than in the boat form. Figure 2-5 illustrates the nature of all of the equatorial (e) and axial (a) hydrogens on the cholestane and coprostane ring structures.

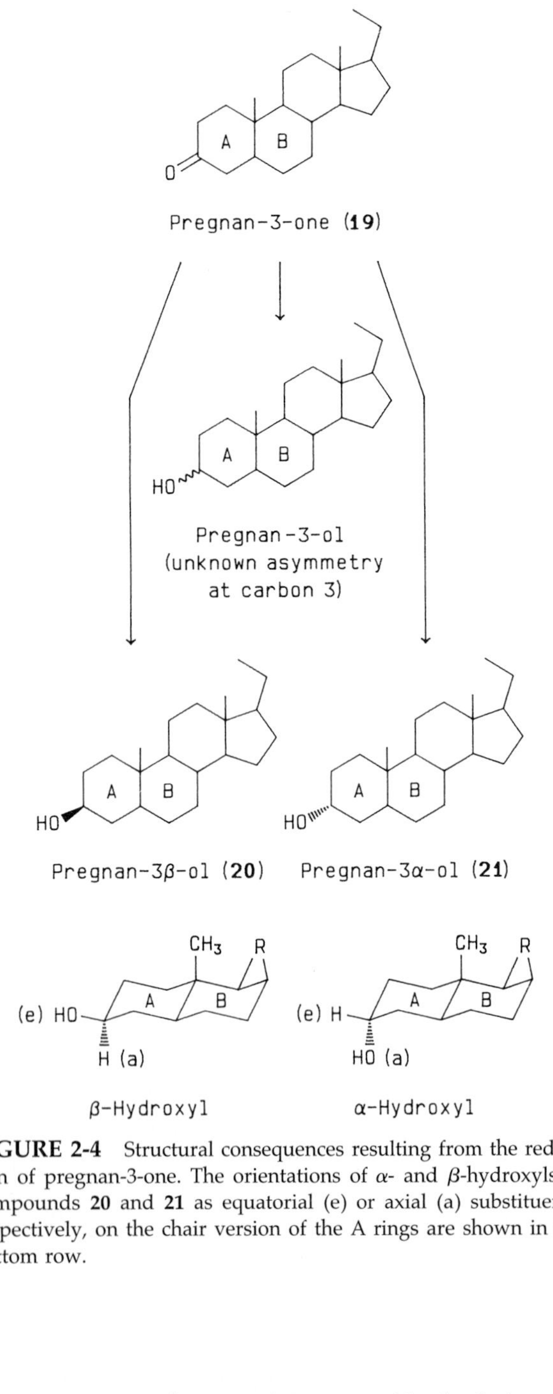

Pregnan-3-one (**19**)

Pregnan-3-ol (unknown asymmetry at carbon 3)

Pregnan-3β-ol (**20**) Pregnan-3α-ol (**21**)

β-Hydroxyl α-Hydroxyl

FIGURE 2-4 Structural consequences resulting from the reduction of pregnan-3-one. The orientations of α- and β-hydroxyls of compounds **20** and **21** as equatorial (e) or axial (a) substituents, respectively, on the chair version of the A rings are shown in the bottom row.

As indicated, the B and C rings of both cholestane and coprostane are locked into chair conformations (see Figure 2-5). Although, in principle, the A ring of both these steroids is free to interchange between the boat and chair representations, the chair form is believed to be much more favored. In the case of vitamin D steroids, which do not have an intact B ring due to the breakage of the carbon-9–carbon-10 bond (they are termed seco steroids), the A ring is much more conformationally mobile than that of the usual cholesterol-derived steroids (see Figure 9-7).

An important point for the reader to consider is that the usual structural representation given for steroids (e.g., see Figures 2-2, 2-3, and 2-4) provides no clear designation of either the three-dimensional geometry or the space-filling aspects of the electron orbitals associated with each atom involved in the formation of the requisite bonds. A comparison for cholestanol is given in Figure 2-10 of the planar representation (A), the planar conformational model (B), a Dreiding three-dimensional model emphasizing bond angles (C), and a Corey–Pauling three-dimensional space-filling model (D). Certainly the space-filling molecular representation most closely approximates the reality of the biologically relevant form of the steroid.

The approach of steroid conformational analysis has been of great value to the organic chemist as a tool to predict or understand the course of synthetic organic chemical reactions. It is also to be expected that conformational considerations will play an increasingly useful role in the understanding of steroid hormone–receptor interactions. Steroid receptors are known to have very high ligand specificities; see for example, Table 9-4, which illustrates the structural preferences of the nuclear receptor for 1α,25$(OH)_2D_3$. It would be surprising, therefore, if the intimate local structure of the receptor's ligand-binding site did not have the capability of distinguishing and discriminating between the various conformational forms of the same steroid hormone.

F. Other Steroid Structures

Figure 2-11 illustrates the structures of several other biologically significant steroids. Ecdysone (**35**) is the principal insect steroid hormone. Dexamethasone (**36**) is a potent artificial glucocorticoid. Prednisolone (**37**) is a commonly employed antiinflammatory steroid. The presently employed orally active contraceptive steroids do not contain any naturally occurring estrogens or progestins. The structure of the abortion-inducing drug RU-486, also known as mifepristone, is shown (**41**); it is structurally related to progesterone (**12**). Two major classes of synthetic progestins are employed: (i) derivatives of 19-nortestosterone, such as norethindrone (**42**) or norethynodrel (**43**), and (ii) derivatives of 17α-acetoxyprogesterone such as medroxyprogesterone acetate (**44**). These progestins are administered in combination with varying doses of two synthetic estrogens, either ethinylestradiol or ethinylestradiol 3-methyl ether (**45**). Each of these compounds has an ethinyl group on carbon-17, which enhances their oral activity. The bile acid structures in Figure 2-11 are those of the principal forms present in humans.

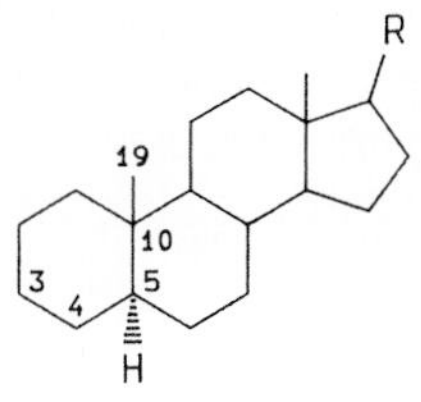

5α-Cholestane (**7**)

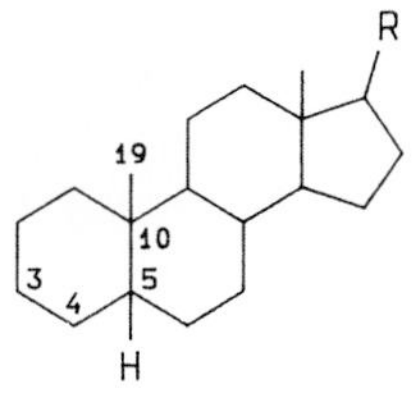

5β-Cholestane (**22**)

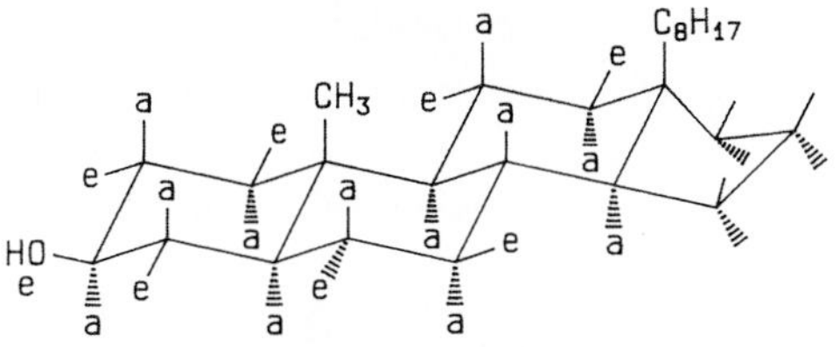

Cholestanol (**23**)
trans-union
A:B rings

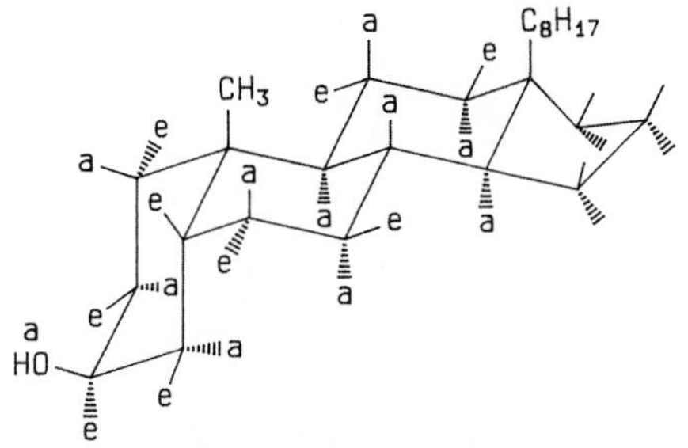

Coprostanol (**24**)
cis-union
A:B rings

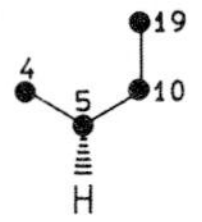

trans

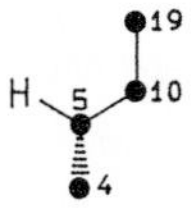

cis

FIGURE 2-5 Structural relationships resulting from *cis* or *trans* A:B ring fusion in two typical steroids. In 5α-cholestane and cholestanol the A:B ring fusion is *trans,* while in 5β-cholestane and coprostanol the A:B ring fusion is *cis.* The orientation of substituents around carbon-5 for the *cis* and *trans* circumstances are illustrated in the bottom row (●–● indicates carbon–carbon bonds).

III. BIOSYNTHESIS OF CHOLESTEROL

A. General Comments

The most commonly occurring steroid is cholesterol (**3**). Cholesterol is present in practically all living organisms, including blue-green algae and bacteria. The levels of cholesterol in plant tissues are low, except in some pollen and seed oil. Animal products are rich sources of cholesterol; cholesterol is present in high concentrations in the myelin sheath and in almost pure form in gallstones. Above-average levels can be found in skin, sperm cells, and egg yolk. Virtually all cell membranes in higher animals include cholesterol as an integral component.

A range of 180–260 mg of total cholesterol per 100 m of serum is generally accepted as the normal level for the American adult. Although there is little esterified cholesterol in the peripheral tissues, 70–75% of the cholesterol of plasma is esterified.

Most organisms have an enzymatic capability to biosynthesize cholesterol. Notable exceptions include protozoa, fungi, arthropods, annelids, mollusks, sea urchins, and sharks. However, many organisms will incorporate cholesterol into their cellular membrane if it is provided as a dietary source. In mammalian tissues the principal sites of production of cholesterol are the skin, liver, and intestinal mucosa; measurable biosynthetic activity is also found in lung, kidney, adrenal, gonad, muscle, brain, and adipose tissue.

B. Pathway of Production

Because cholesterol is the precursor of the seven classes of steroids summarized in Table 2-1, it is useful to consider the biosynthetic pathway of production of cholesterol. The pioneering work of R. Schoenheimer, D. R. Rittenburg, and K. Bloch established that all 27 carbons of cholesterol are derived from acetate. In this remarkable biosynthetic process, 18 molecules of acetate are processed by at least 27 enzymes to produce the final product, cholesterol. The sources of all the carbon atoms of cholesterol in relation to the carboxyl

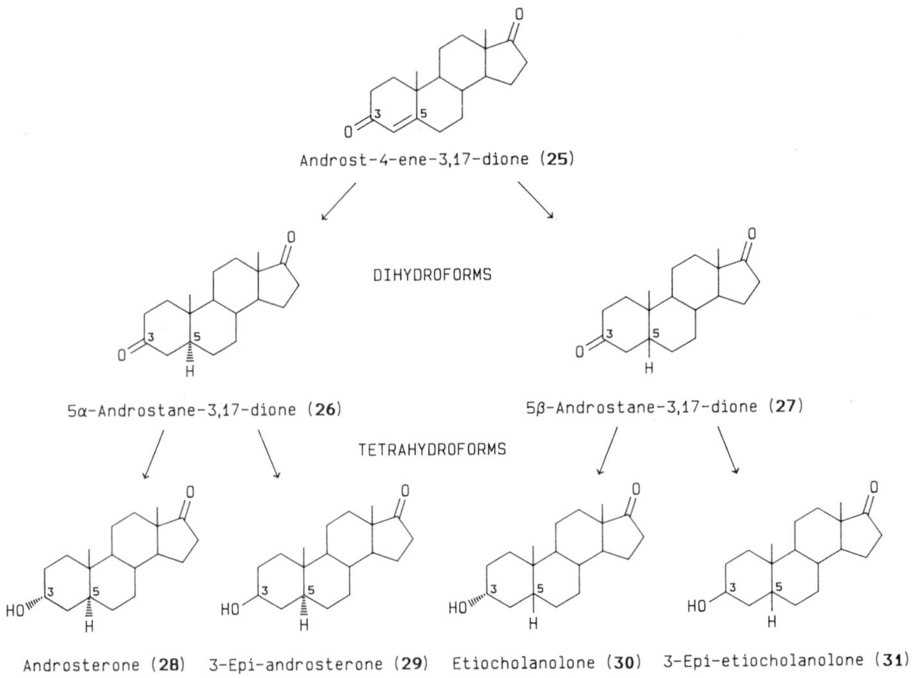

Androst-4-ene-3,17-dione (**25**)

DIHYDROFORMS

5α-Androstane-3,17-dione (**26**)

5β-Androstane-3,17-dione (**27**)

TETRAHYDROFORMS

Androsterone (**28**) 3-Epi-androsterone (**29**) Etiocholanolone (**30**) 3-Epi-etiocholanolone (**31**)

FIGURE 2-6 Structural isomers derived by reduction of the $\Delta^{4,5}$-double bond and the 3-oxo function of androst-4-ene-3,17-dione.

and methyl carbons of the starting acetate are given in Figure 2-12.

The overall biosynthetic pathway of cholesterol can be subdivided into four steps: (1) the formation of mevalonic acid (six carbons) from three molecules of acetate; (2) the conversion of six molecules of mevalonic acid through a series of phosphorylated intermediates into the hydrocarbon squalene (30 carbons); (3) the oxidation and cyclization of squalene into lanosterol, the first cyclic sterol precursor; and (4) the processing of lanosterol (to remove three methyl groups and rearrange double bonds) to yield cholesterol (27 carbons). These steps are outlined in the following sections.

1. Conversion of Acetate to Mevalonate

As summarized in Figure 2-13, mevalonic acid is derived by the condensation of three molecules of acetyl-CoA, which pass through the key intermediate 3-hydroxy-3 methylglutaryl-CoA (HMG-CoA). This latter compound is reduced by NADPH to yield mevalonic acid. As will be discussed later, the enzyme HMG-CoA reductase is an important control point in regulating the biosynthesis of cholesterol. Once the product, mevalonic acid, is produced, it is virtually irreversibly committed to conversion to cholesterol.

2. Conversion of Mevalonic Acid into Squalene

Figure 2-14 summarizes the series of eight enzymatic steps necessary to convert six molecules of mevalonic acid into the 30-carbon squalene plus six CO_2 molecules. Mevalonic acid is sequentially phosphorylated by 3 mol of ATP and then decarboxylated to yield the key isoprene building block, 3-isopentenyl pyrophosphate. The nucleophilic reagent, isopentenyl pyrophosphate, is then subjected to isomerization into the electrophilic (as a consequence of its double-bond position and its esterification with a strong acid) dimethylallyl pyrophosphate. These latter two 5-carbon units are then condensed by a "head-to-tail" mechanism, yielding the 10-carbon intermediate geranyl pyrophosphate. Geranyl pyrophosphate is then condensed in a second head-to-tail step with isopentenyl pyrophosphate, yielding the 15-carbon farnesyl pyrophosphate. Finally, two farnesyl pyrophosphate units are reductively condensed in a "nose-to-nose" mechanism to yield first presqualene pyrophosphate (which contains a cyclopropane ring) and then finally, after reduction by NADPH, the symmetric squalene (**56**).

3. Conversion of Squalene into Lanosterol

As summarized in Figure 2-15, the final steps of cholesterol biosynthesis involve the conversion of squalene into lanosterol. E. E. van Tamelen chemically

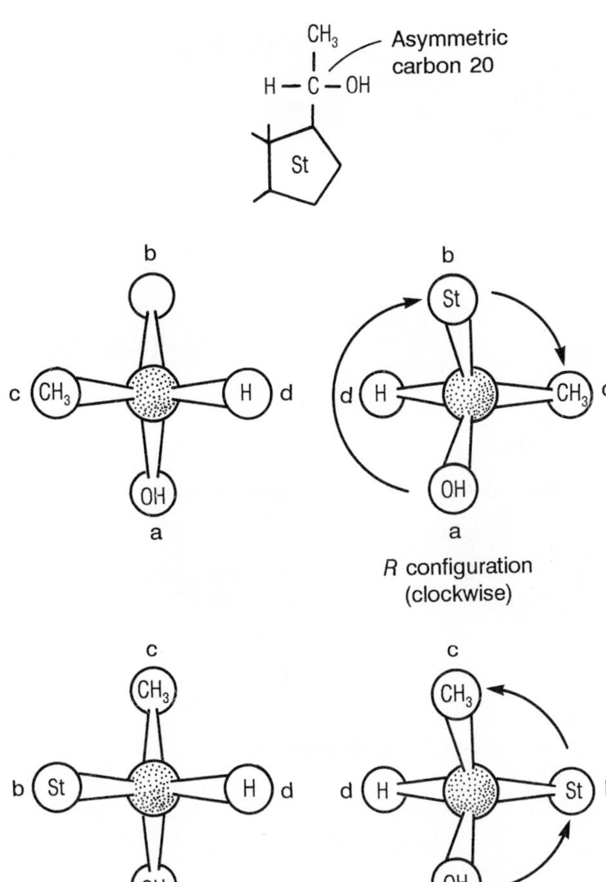

FIGURE 2-7 Application of the sequence rules of Cahn to determine the *R* (rectus) and *S* (sinister) asymmetry for carbon-20 present in the two-carbon steroid side chain. The asymmetric carbon being evaluated is indicated by the arrow; St indicates the remainder of the steroid nucleus.

synthesized 2,3-oxidosqualene and studied its spontaneous (nonenzymatic) cyclization into lanosterol. The initial step starts with the reactive species squalene epoxide, where the double bond between carbon-6

HO

Cholesterol (**3**)

FIGURE 2-8 Asymmetric carbons of cholesterol. The asymmetric carbons are indicated by ●. There are 2^8 or 256 structural isomers of cholesterol.

Chair **32** Boat **33**

FIGURE 2-9 Principal conformational representations of cyclohexane.

and -7 attacks the epoxide to form the first ring and generate a carbonium ion at carbon-6. This ion is attacked by the double bond between carbon-10 and -11 to form the second ring and generate a new carbonium ion at carbon-10. This sequence continues until all four rings are formed, followed by migration of the methyl carbons on carbon-13 and -14 to yield the first steroid product, lanosterol (**59**).

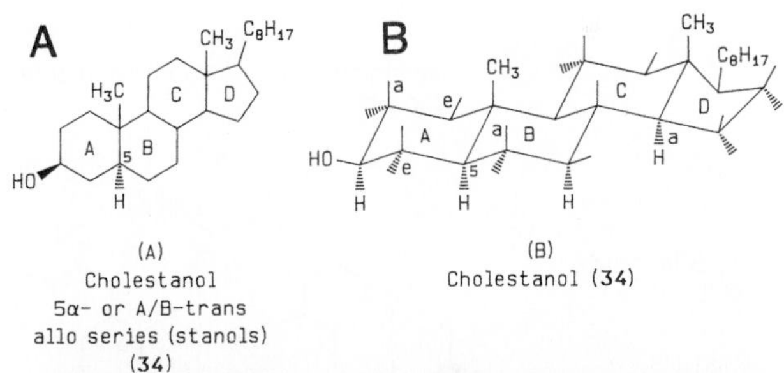

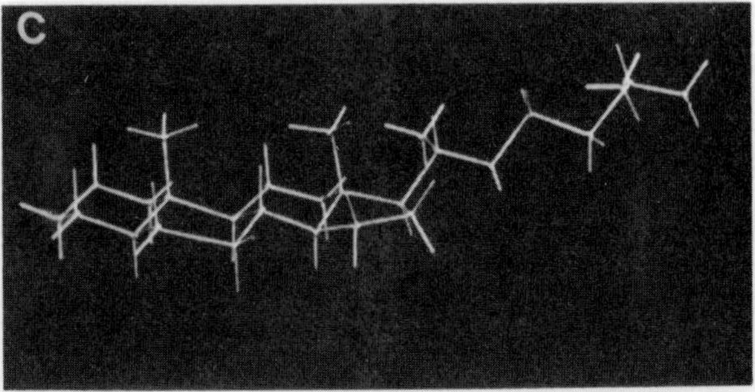

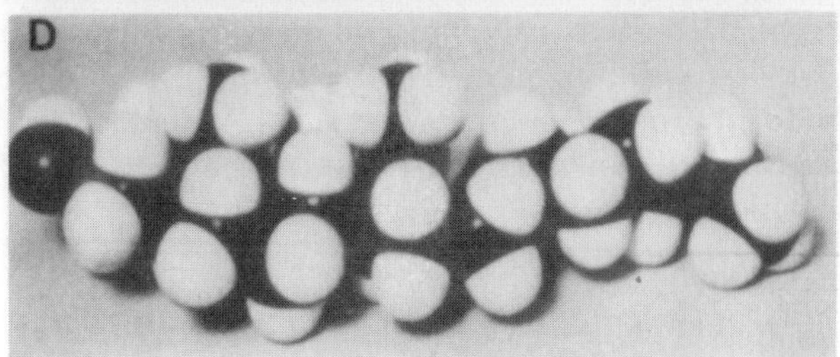

FIGURE 2-10 Structural representations of cholestanol: (A) typical two-dimensional structure; (B) planar conformational model; (C) a Dreiding model emphasizing bond angles and interatomic distances; (D) a Corey–Pauling space-filling model.

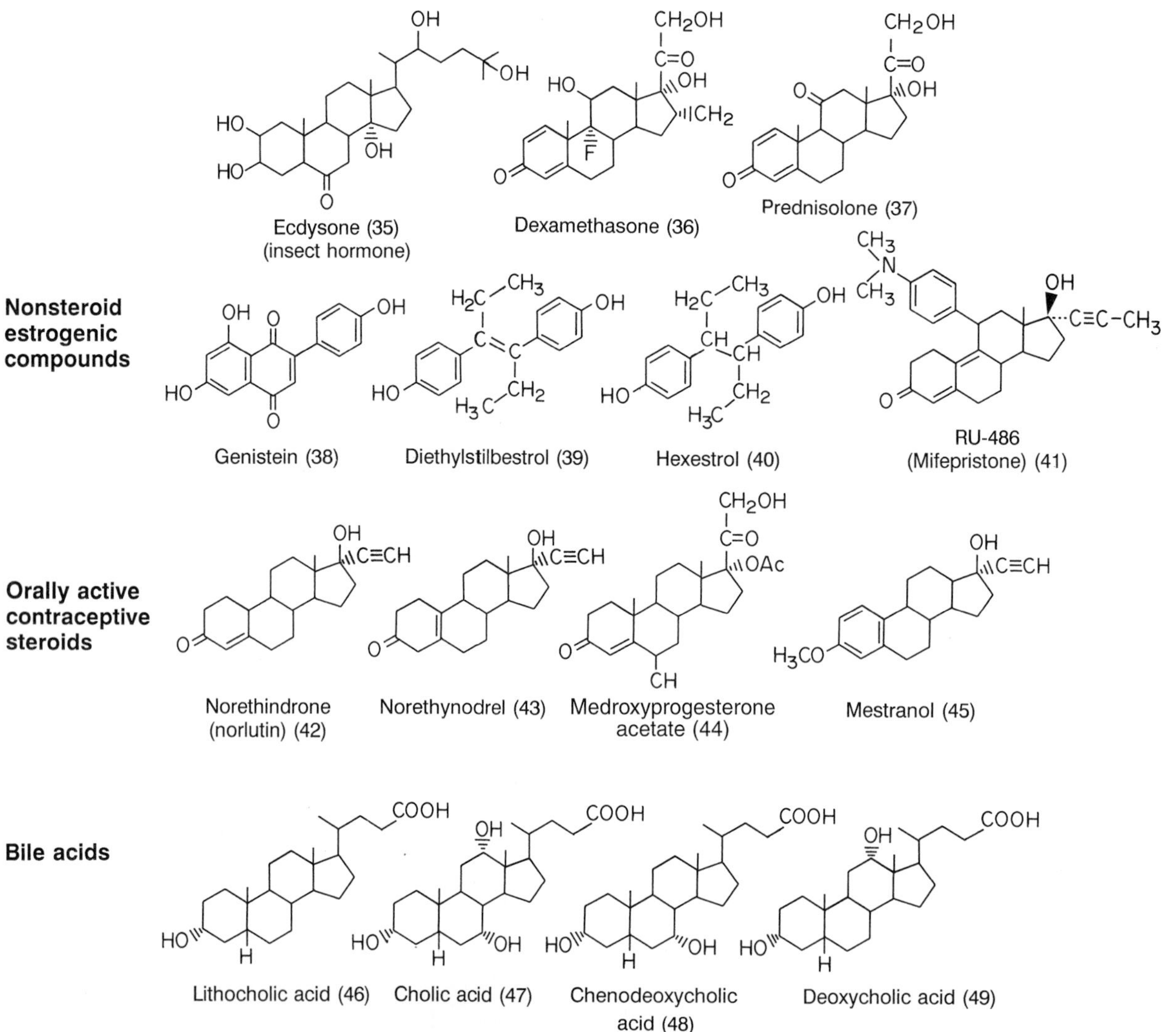

FIGURE 2-11 Structures of other biologically significant steroids. Top row: The insect hormone, ecdysone, and two synthetic glucocorticoids. Second row: Nonsteroid estrogenic compounds and the abortifacient RU-486. Third row: Orally active contraceptive steroids. Bottom row: Common bile acids.

4. Metabolism of Lanosterol to Cholesterol

Three separate enzyme-catalyzed steps are required to convert lanosterol into cholesterol. These include (a) oxidative removal of the two methyl groups at carbon-4 and the single methyl groups at carbon-14; (b) reduction of the side chain carbon-24 double bond; and (c) movement of the Δ^8-double bond to the Δ^4-position.

C. Role of Sterol Carrier Protein

The early precursors of sterol biosynthesis are all water soluble, but after the production of squalene the cholesterol precursors become markedly water insoluble. The activity of the microsomal enzymes between squalene and cholesterol has been found to be stimulated by the addition of sterol carrier protein (SCP). SCPs are heat stable, with molecular weights around 16,000.

Three sterol carrier proteins have been isolated. SCP_1 enhances the conversion of squalene, but not cholesta-4,7-dien-3β-ol (7-dehydrocholesterol), to cholesterol. SCP_2 is active in the conversion of 7-dehydrocholesterol to cholesterol, but is inactive with squalene. SCP_3 is required for the conversion of 4,4-dimethylcholest-8-3β-ol to polar precursors of cholesterol.

D. Regulation of Cholesterol Biosynthesis

The level of total body cholesterol is determined by a complex interplay of dietarily available cholesterol, *de novo* synthesis of cholesterol, and excretion of choles-

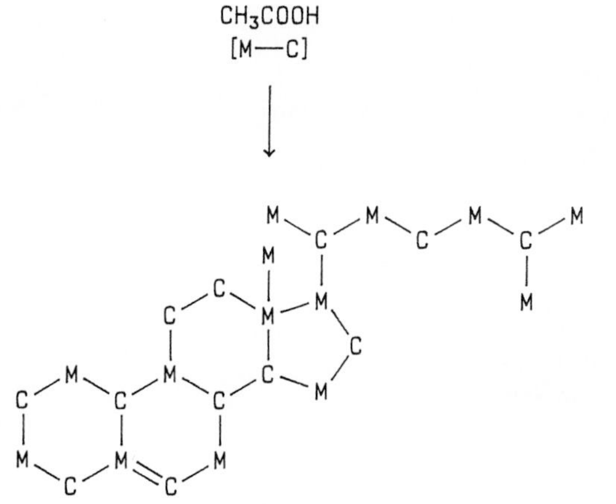

FIGURE 2-12 Source of the carbon atoms of cholesterol. The different carbon atoms are shown to be derived from either the methyl (M) carbon atom or the carboxyl (C) group of acetate.

terol and bile salts. These relationships are summarized in Figure 2-16.

In an average 70-kg man, ≈0.8–1.4 g of cholesterol is turned over per day. The daily ingestion of cholesterol in Western countries is 0.5–2.0 g; the efficiency of intestinal absorption ranges from 30 to 50%. However, it is known that dietary cholesterol is not the exclusive source of the bodily pool of cholesterol. The liver and intestine together account for more than 60% of the body's daily synthesis of this sterol. When considering the turnover of body cholesterol, it has been suggested that there are two principal components. Pool A, which comprises 30–35% of the total body cholesterol, consists of the cholesterol in liver, bile, plasma, erythrocytes, and intestine. Pool B constitutes the remaining 65–70% of the exchangeable body cholesterol and principally comprises the cholesterol in the skin, adipose tissue, and skeleton. It is now well established that hepatic cholesterol biosynthesis can be greatly reduced by the ingestion of a cholesterol-rich diet. This prevents excessive cholesterol accumulation through feedback mechanisms operative at the level of the HMG-CoA reductase enzyme. HMG-CoA reductase is the rate-limiting step in the conversion of acetate into cholesterol and thus is a logical site for feedback regulation. Dietary cholesterol does not function as an allosteric modulator directly on the HMG-CoA reductase; instead, it lowers the steady-state level of HMG-CoA reductase by decreasing the rate of biosynthesis of the enzyme without affecting its rate of enzyme degradation. Cholesterol SCP may also affect HMG-CoA reductase by a similar mechanism. The rate of hepatic cholesterol biosynthesis exhibits a marked diurnal variation caused by changes in the level of HMG-CoA reductase.

A wide variety of hormones and some dietary factors have been shown to modulate the biosynthesis of cholesterol (see Table 2-4). Most of these agents are believed to act on HMG-CoA reductase. The steady-state level of HMG-CoA reductase is affected by a complex interplay between insulin, glucagon, triiodothyronine (T_3), and growth hormone. A precise understanding of the mechanisms whereby all of these hormones modulate sterol biosynthesis is not

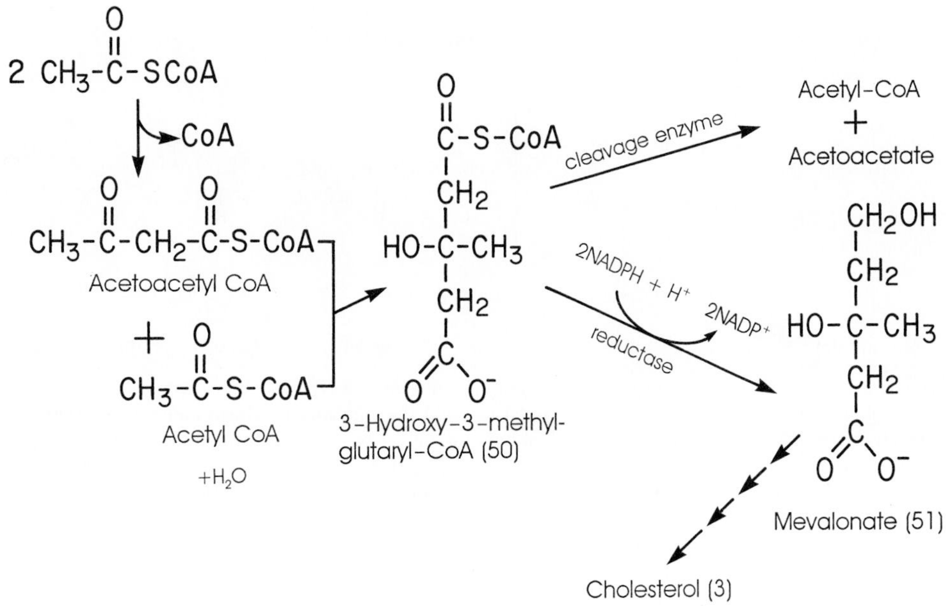

FIGURE 2-13 Biosynthesis of 3-hydroxy-3-methylglutaryl-CoA and indication of its metabolic fate. Six molecules of mevalonic acid are ultimately condensed to create one molecule of squalene.

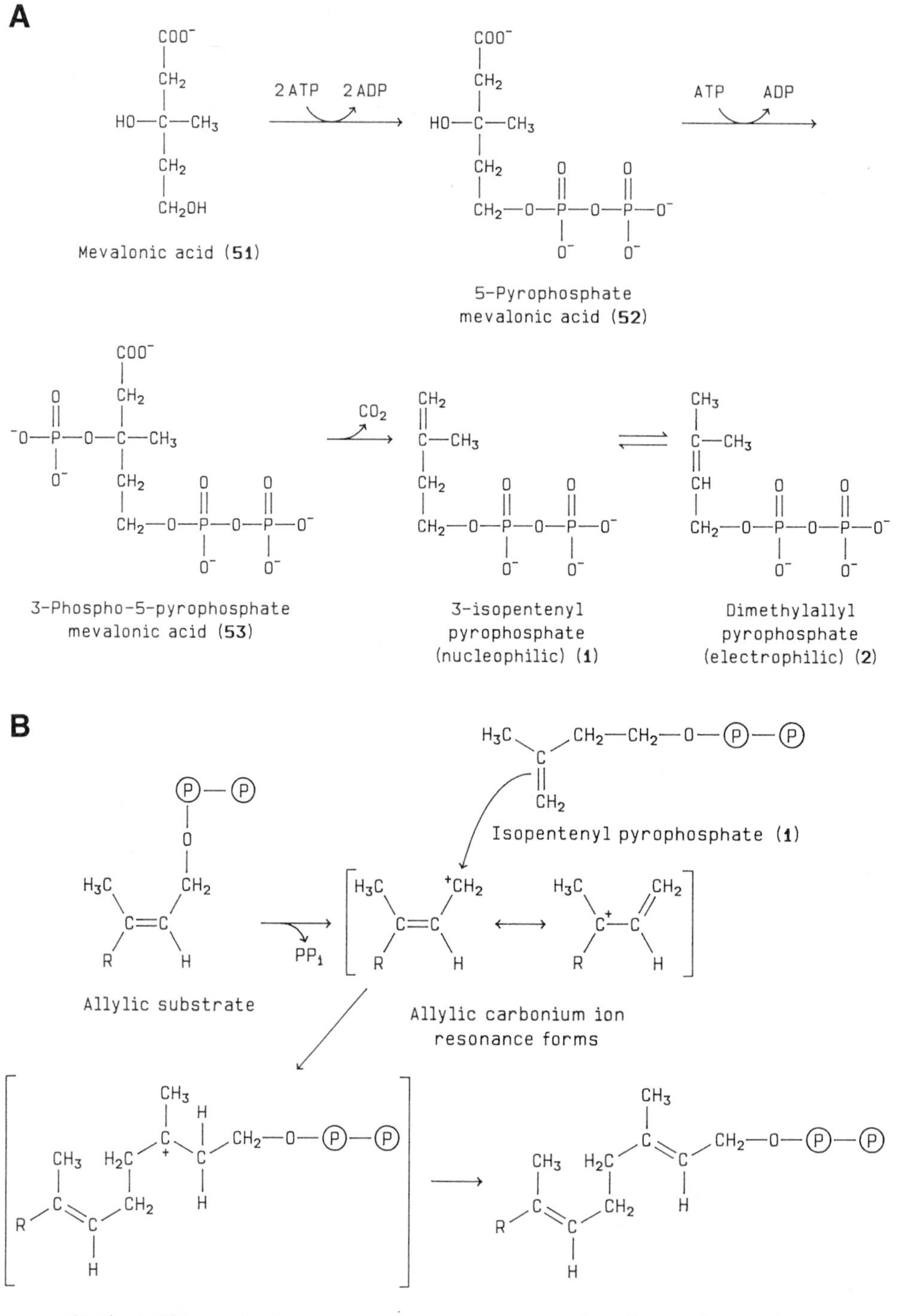

A

Mevalonic acid (51)

2 ATP 2 ADP

5-Pyrophosphate mevalonic acid (52)

ATP ADP

3-Phospho-5-pyrophosphate mevalonic acid (53)

CO_2

3-isopentenyl pyrophosphate (nucleophilic) (1)

Dimethylallyl pyrophosphate (electrophilic) (2)

B

Isopentenyl pyrophosphate (1)

Allylic substrate

PP_i

Allylic carbonium ion resonance forms

Condensation carbonium ion

Geranyl or farnesyl pyrophosphate

FIGURE 2-14 (A) Conversion of mevalonic acid **(51)** into the intermediate 3-phosphomevalonic acid 5-pyrophosphate **(53)**, which is then decarboxylated to yield the nucleophilic 3-isopentenyl pyrophosphate **(1)**. Structure **1** is then isomerized into equivalent amounts of the electrophilic dimethylallyl pyrophosphate **(2)**. (B) Proposed mechanism for the key "head-to-tail" joining of the nucleophilic isopentenyl pyrophosphate to the allylic electrophilic "acceptor," which may be either dimethylallyl pyrophosphate **(2)** or geranyl pyrophosphate **(53A)**. An encircled P indicates phosphate. (C) Pathway of reactions leading from isopentenyl pyrophosphate + dimethylallyl pyrophosphate to squalene **(56)**. A key reaction is the "nose-to-nose" reaction of *trans,trans*-farnesyl pyrophosphate **(54)** to yield presqualene pyrophosphate **(55)**, which, after reduction by NADPH, produces squalene **(56)**.

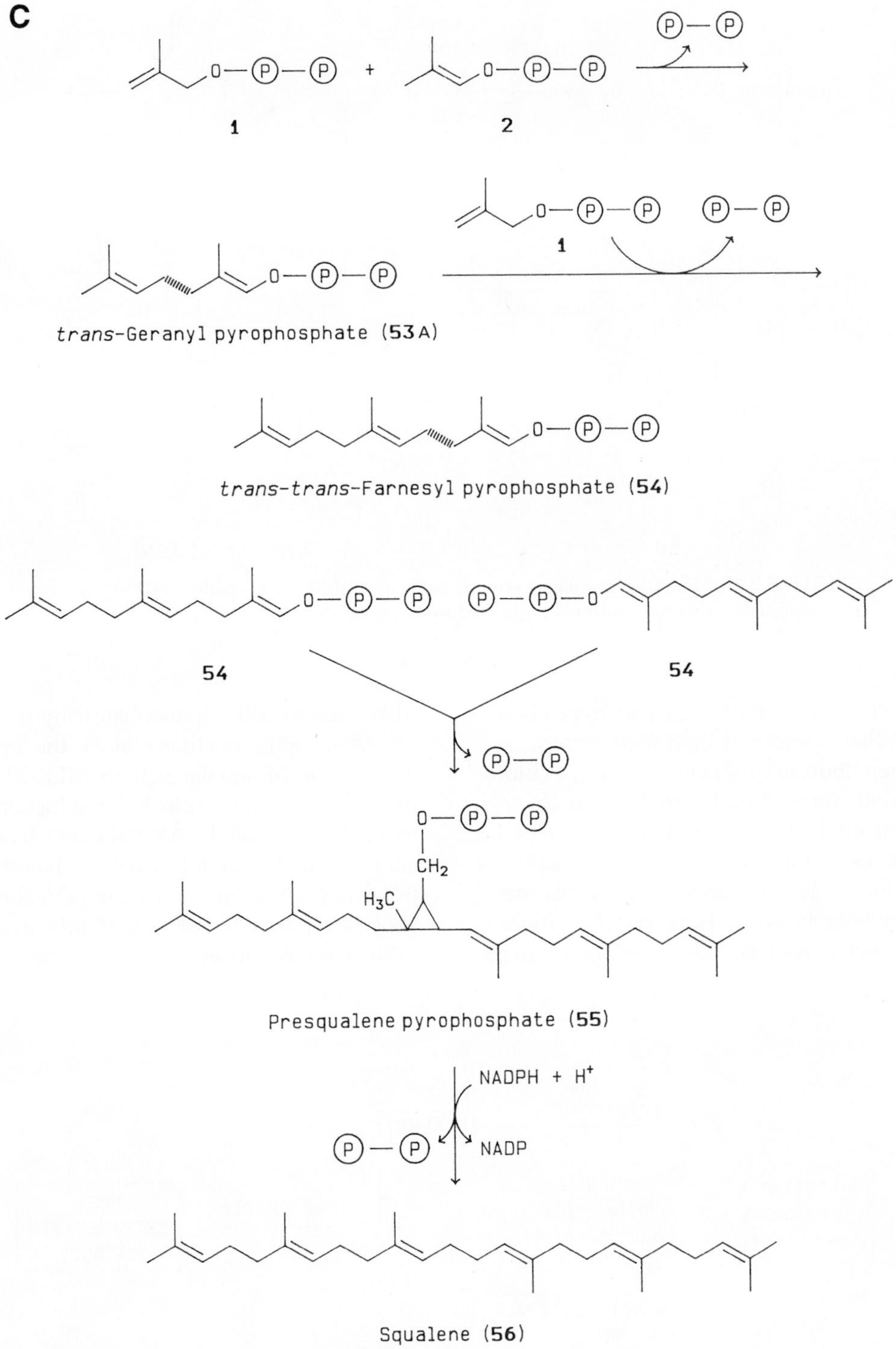

C

1 2

trans-Geranyl pyrophosphate (**53**A)

trans-trans-Farnesyl pyrophosphate (**54**)

54 54

Presqualene pyrophosphate (**55**)

NADPH + H^+

NADP

Squalene (**56**)

FIGURE 2-14 — *Continued*

yet available, but these effects certainly emphasize an intriguing interplay between many hormone systems and the production of cholesterol, which is the central precursor of the steroid hormones.

E. Inhibitors of Cholesterol Biosynthesis

Figure 2-17 presents the structures of some known inhibitors of cholesterol metabolism. These compounds have been developed in an effort to clinically control the

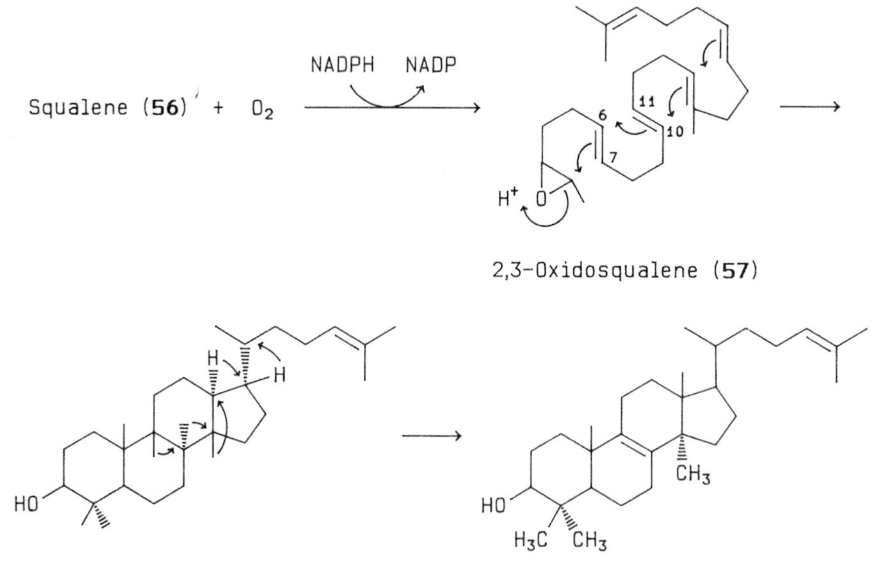

NADPH NADP

Squalene (**56**) + O_2 ⟶

2,3-Oxidosqualene (**57**)

58 ⟶ Lanosterol (**59**)

FIGURE 2-15 Proposed mechanism for the cyclization of 2,3-oxidosqualene (**57**) to produce lanosterol (**59**). This mechanism is postulated from model organic reactions.

biosynthesis of cholesterol, particularly in hypercholesterolemia, a familial disorder of lipid metabolism.

Sodium 2-phenylbutyrate (**62**) and sodium (*p*-chlorophenox)isobutyrate (or clofibrate) (**63**) are both aryl-substituted short chain fatty acids that interfere with the acylation of coenzyme A and block the initiation of cholesterol synthesis. Benzmalacene (**64**) blocks sterol synthesis somewhere between isopentenyl pyrophosphate and farnesyl pyrophosphate. Analogs of tris[2-(diethylamino)ethyl]phosphate trihydrochloride, such as SKF-7997 (**65**), effectively block the cyclization of the 2,3-epoxide of squalene. Both MER-29 (**66**) and 22,25-diazacholestanol (**67**) block the reduction of desmosterol to cholesterol, while AY-9944 (**68**) blocks the conversion of 7-dehydrocholesterol to cholesterol. Lovastatin (**69**) has been approved by the FDA for use as a serum cholesterol-lowering agent; it acts as an inhibitor of HMG-CoA reductase.

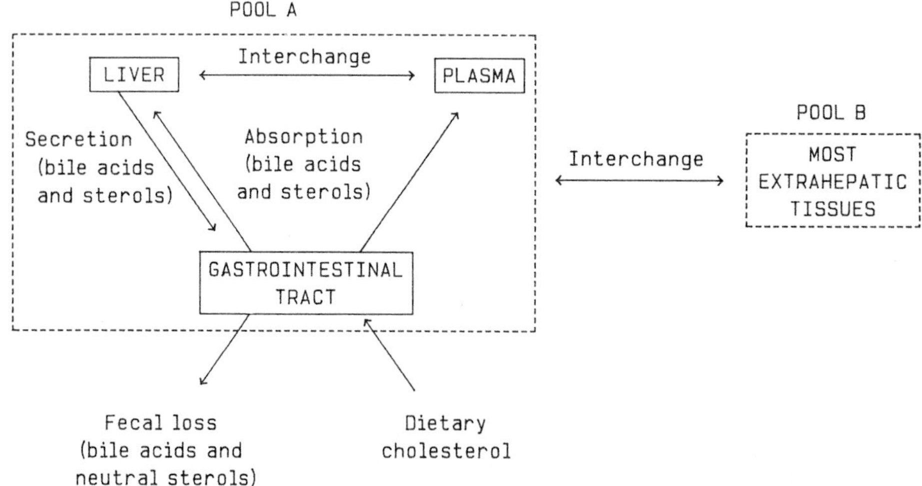

FIGURE 2-16 Diagrammatic representation of body pools of cholesterol. Modified from a similar figure by Goad, L. J. (1975). *In* "Biochemistry of Steroid Hormones" (H. L. J. Makin, ed.), p. 34. Blackwell, Oxford.

TABLE 2-4 Factors Influencing Cholesterol Synthesis in the Liver

Decrease synthesis	Increase synthesis
Hypophysectomy	Growth hormone
Diabetes	Insulin
Glucocorticoids	Thyroid hormone
Glucagon	Catecholamines
Estrogen	–
Male	Female
Fasting	Feeding
Low-fat diet	Dietary fat
Cholesterol	Cholestyramine
Bile acids	

IV. BIOSYNTHESIS OF STEROIDS

A. Introduction

The principal tissues of synthesis of the five classical classes of steroid hormones (estrogens, androgens, progestins, glucocorticoids, and mineralocorticoids) are the adrenal cortex, ovaries, and testes. Also, during pregnancy, the fetal–placental unit can serve as a source of estrogen and some other hormones. The sites of synthesis of the sixth class of steroid hormones, the prohormone vitamin D_3 and its metabolites, are the skin, liver, and kidney (see Section IV.F). The bile acids, which are the seventh important structural class of mammalian steroids, have no known hormonal activity; they are principally synthesized in the liver (see Section IV.G).

Figure 2-18 outlines the general framework of the metabolic pathways that are used to convert cholesterol into steroid hormones; the major steps are listed, as are the tissues where the transformations occur.

Table 2-5 summarizes the tissue-specific expression of the cytochrome P450 enzymes involved with the production of the classical steroid hormones. A total of six P450 enzymes is associated with the production of the classical steroid hormones produced by the adrenal cortex, ovaries, and testes.

Figure 2-19 summarizes the peptide hormones that stimulate the biosynthesis of the six steroid hormones.

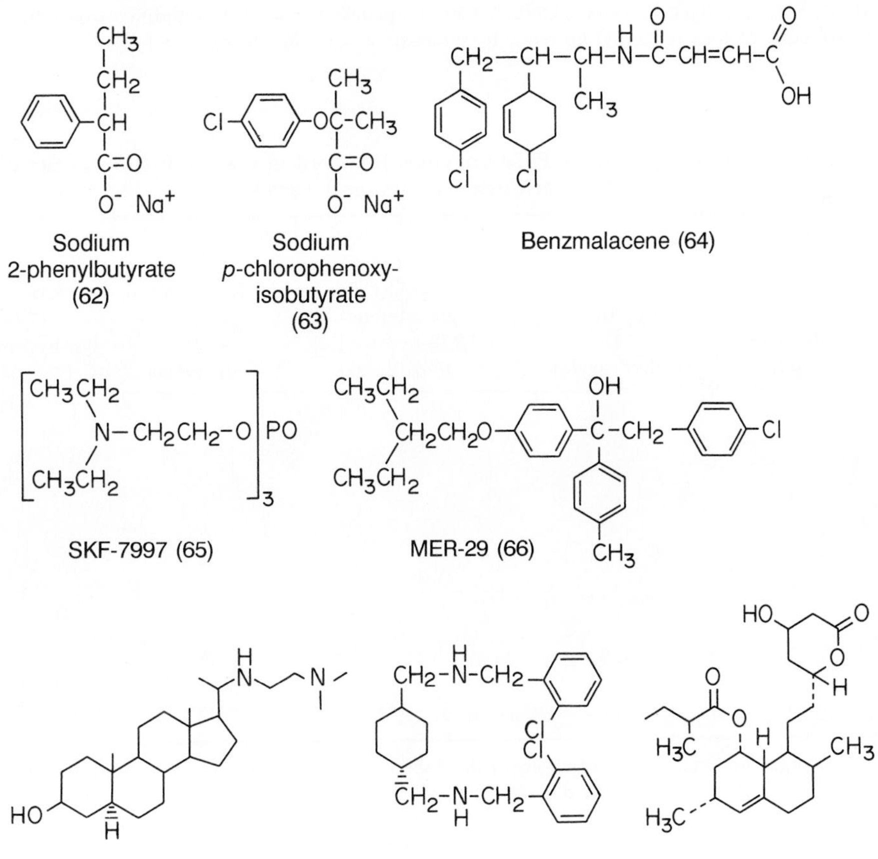

Sodium 2-phenylbutyrate (62)

Sodium *p*-chlorophenoxy-isobutyrate (63)

Benzmalacene (64)

SKF-7997 (65)

MER-29 (66)

22,25-Diazacholestanol (67)

AY-9944 (68)

Lovastatin (69)

FIGURE 2-17 Inhibitors of cholesterol biosynthesis.

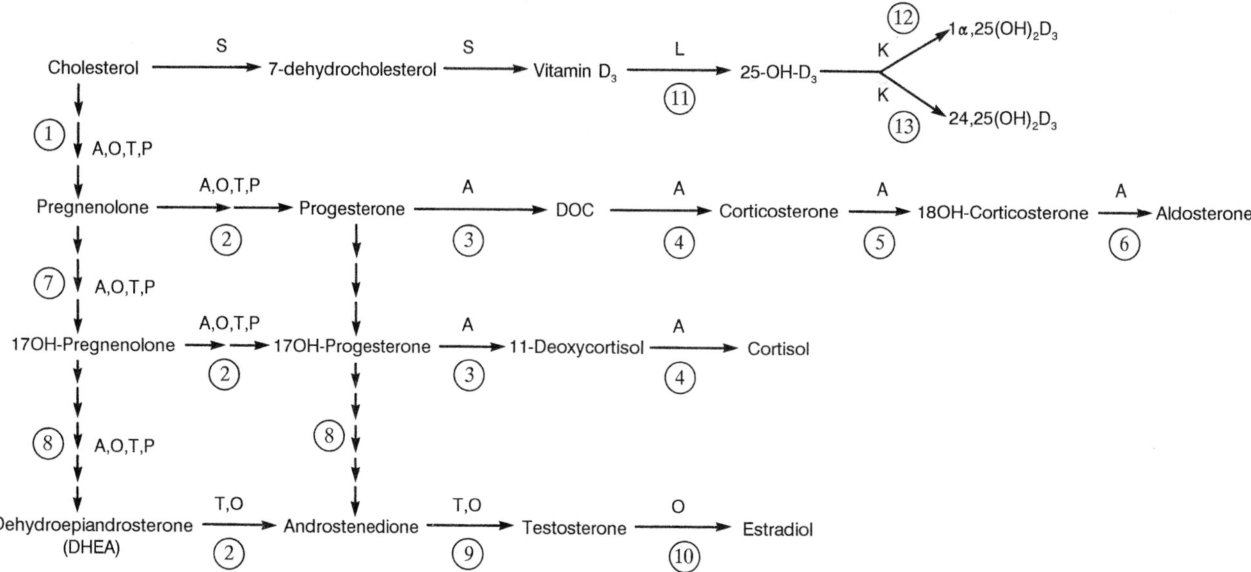

FIGURE 2-18 Principal pathways of human steroid hormone synthesis. The details of the enzyme reactions (circled numbers) are summarized in Table 2-6. Reactions ①, ②, ⑦, and ⑧ are found universally in the adrenals, ovary, testes, and placenta. Reactions ⑨ and ⑩ are found principally in the testes and ovaries. Reactions ⑪ → ⑬ describe the pathway of production from cholesterol of the prohormone vitamin D and its hormonal metabolite products 1α,25$(OH)_2D_3$ and 24,25$(OH)_2D_3$. Reaction ⑪: P450c25 present in the mitochondrial and/or endoplasmic reticulum of the liver. Reactions ⑫ and ⑬: P450c1α and P450c24 present in the mitochondria of the kidney proximal tubule. Tissue codes: A, adrenals; O, ovaries; T, testes; P, placenta; S, skin; L, liver; K, kidney; DHEA, dehydroepiandrosterone **(77)**. [Modified from Miller, W. L. (1988). Molecular biology of steroid hormone biosynthesis. *Endocr. Rev.* **9**, 295–318.]

TABLE 2-5 Tissue Expression of Cytochrome P450 Enzymes Involved in the Production of Steroid Hormones from Adrenal Cortex, Ovary, and Testes[a]

Tissue	Reaction ① cholesterol side chain cleavage (scc)	Reaction ④ 11β-hydroxylase	Reactions ④⑤⑥ aldosterone (11β,18-hydroxylase 18-oxidase)	Reaction ③ C-21 hydroxylase	Reactions ⑦⑧ 17α-hydroxylase/ C-17–C-20-lyase	Reaction ⑩ estrogen (aromatase)
Adrenals						
Zona glomerulosa	+	−	+	+	−	−
Zona fasciculata	+	+	−	+	+	−
Zona reticularis	+	+	−	+	+	−
Testes						
Sertoli cells	−	−	−	−	−	±
Leydig cells	+	−	−	−	+	−
Ovary						
Glomerulosa	+	−	−	−	−	±
Theca	+	−	−	−	+	−
Corpus luteum	+	−	−	−	−	±

[a] This table describes the organ and cellular distributions of the P450 enzymes concerned with the production of adrenal-, ovary-, and testes-derived steroid hormones. The data are largely derived from *in situ* hybridization studies employing cDNA probes from molecular biological studies on the indicated P450. The details of the enzyme reactions (circled numbers) are summarized in Table 2-6. Modified from Omara, T. (1994). Proceedings of the IX International Congress on Hormonal Steroids, Dallas, Texas. [Abstract P106]

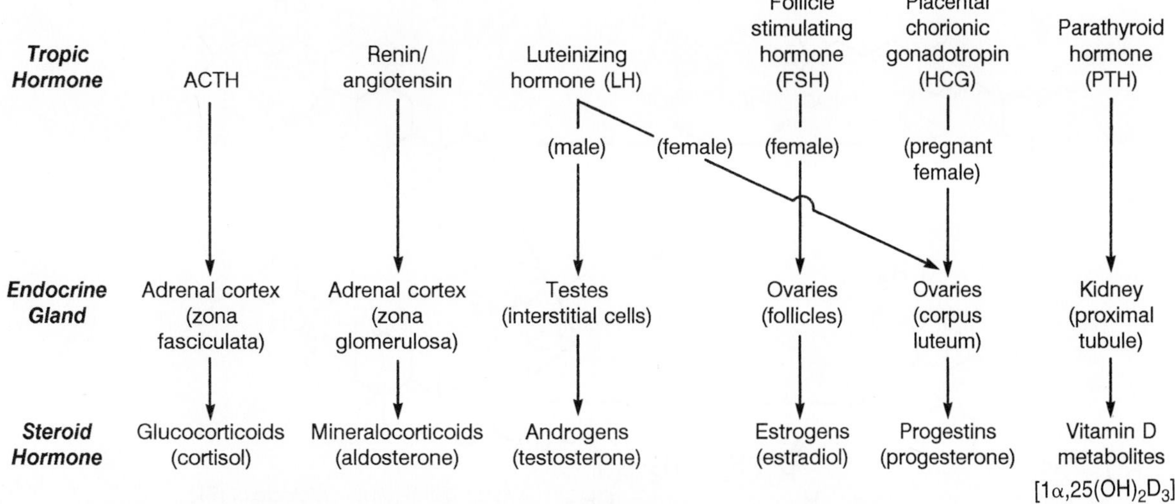

FIGURE 2-19 Relationship between stimulatory peptide hormones and the production of steroid hormones. For each of the six classes of steroid hormones, the specific stimulatory or tropic peptide hormone is indicated as well as specification of the organ or cell upon which it acts. Also, in males, FSH is known to act upon the Sertoli cells of the seminiferous tubules to increase the production not of a steroid hormone but of sperm proteins, androgen-binding protein, and inhibin, a peptide hormone.

It is important to appreciate that the production of each steroid hormone is dependent upon the stimulation of its cells of origin by a specific tropic (stimulatory) peptide hormone.

The purpose of the following section is to review the metabolic pathways by which the six classes of steroid hormones are produced. All discussions of the endocrine systems for each of the steroid hormone classes as well as of their biological properties and modes of action are deferred to subsequent chapters in this book.

B. Biosynthesis of Pregnenolone and Progestins

Figure 2-20 illustrates the various steps involved in the production of progesterone. As indicated in Figure 2-18, the conversion of cholesterol into pregnenolone and then progesterone is a common pathway leading to the production of the five classes of steroids.

The conversion of cholesterol into pregnenolone is carried out by one P450-containing enzyme known as the cholesterol side chain cleavage enzyme or P450scc. Once pregnenolone (**72**) is produced from cholesterol (**3**), it may undergo one of two conversions: it may undergo 17α-hydroxylation, ultimately leading in the adrenals to cortisol (**13**), or it may be converted to progesterone (**12**), which is a significant steroid hormone produced by the *corpus luteum* and placenta. Progesterone is also an intermediate in the synthesis of aldosterone in the adrenals. The two-step transformation of pregnenolone into progesterone is also catalyzed by one enzyme that carries out the 3β-steroid dehydrogenase oxidation of the 3β-OH and also the movement of the Δ^5,Δ^4-double bond (isomerase).

The principal progestational steroid in humans is progesterone. It is produced by both the corpus luteum of the ovaries and the placenta. The physiological actions of progesterone are described in Chapter 13.

C. Biosynthesis by the Adrenal Cortex of Glucocorticoids, Mineralocorticoids, and Some Androgens

Figure 2-21 presents the structures of the adrenal cortex steroids that lead to the biosynthesis of glucocorticoids, mineralocorticoids, and the androgen dehydroepiandrosterone (DHEA). As discussed in Chapter 10 and Table 2-6, the adrenal cortex is anatomically divided into three zones that produce different major classes of the adrenal steroids; these are the *zona glomerulosa* (mineralocorticoids), *zona fasciculata* (glucocorticoids), and *zona reticularis* (androgens). Over 45 steroids have been isolated and chemically characterized from adrenal gland extracts. The 21-carbon corticosteroids include the glucocorticoids and the mineralocorticoids; both subclasses are produced by the adrenal cortex (see Chapter 10). The corticosteroids are characterized by (i) an oxo group at carbon-3 and a double bond at carbon-4; (ii) a two-carbon side chain

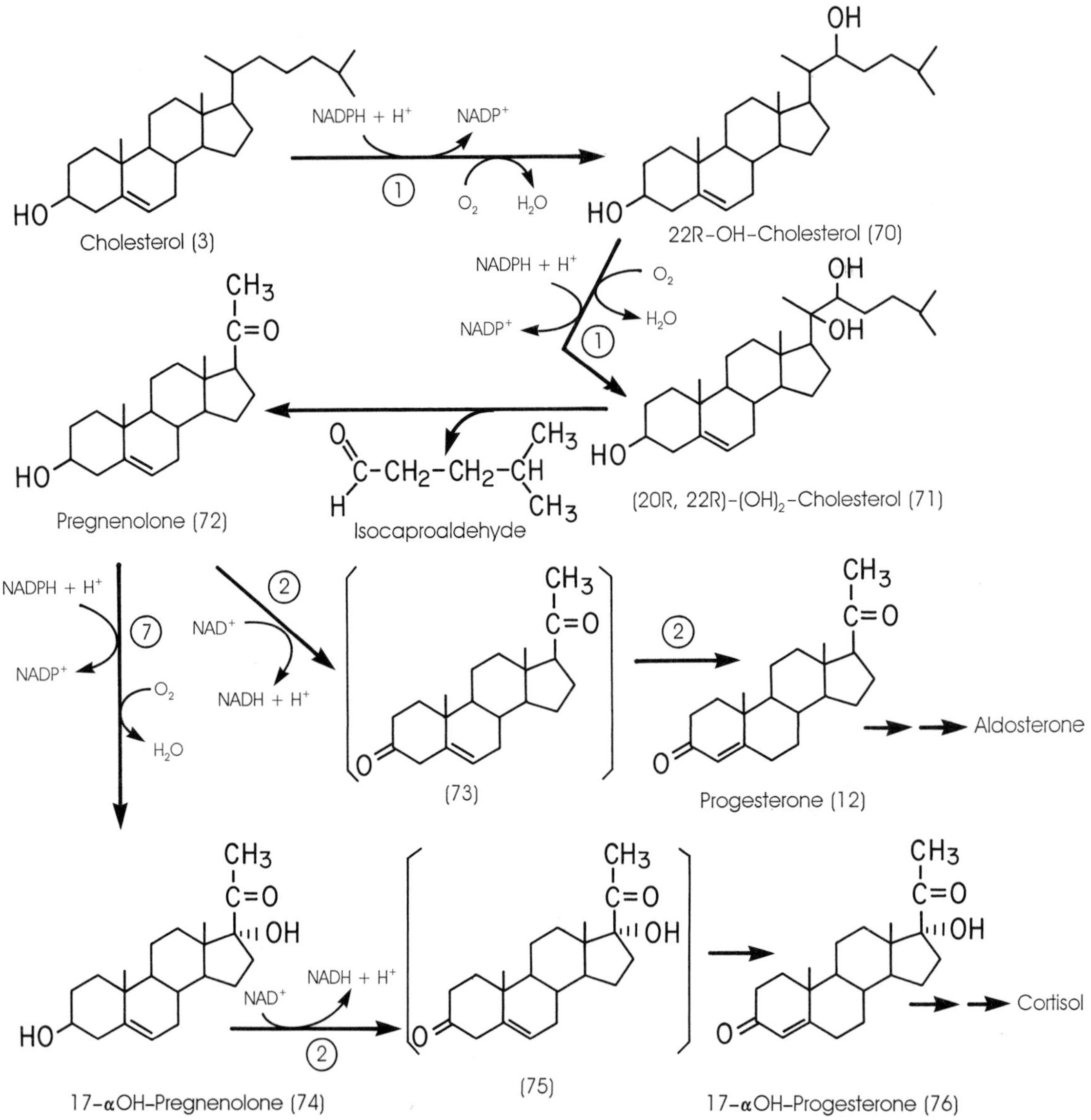

FIGURE 2-20 Biosynthesis of pregnenolone, 17α-pregnenolone, and progesterone. The steps of metabolism are numbered (in circles) according to the reactions described in Figure 2-18 and Table 2-6. The cholesterol side chain cleavage enzyme (csc) carries out three enzymatic steps, comprising two successive hydroxylations followed by scission of the C-20–C-22 carbon–carbon bond. Similarly, the steps marked (2) are mediated by one enzyme that carries out a successive dehydrogenation followed by isomerization.

on carbon-17; and (iii) an oxo group at carbon-20 and a hydroxyl on carbon-21. Glucocorticoids are characterized by (iv) the presence or absence of hydroxyls at both carbon-11 and carbon-17. The principal glucocorticoid in humans is cortisol. Mineralocorticoids are characterized by (v) a hydroxyl at carbon-11 and (vi) carbon-18 oxidized to an aldehyde.

The principal mineralocorticoid in humans is aldosterone. As a consequence of its carbon-18 aldehyde moiety, aldosterone can form a five-membered hemiacetal ring utilizing the C-18 aldehyde with the carbon-

FIGURE 2-21 Pathways of glucocorticoid, mineralocorticoid, and dehydroepiandrosterone (DHEA) biosynthesis in the adrenal cortex. The main steps of metabolism are numbered (in circles) according to the reactions described in Figure 2-18 and Table 2-6. The boldface arrows indicate the major pathways of biosynthesis of cortisol and aldosterone; however, there is no single exclusive pathway. The alternate possibilities are indicated by regular arrows. Under some circumstances, the androgen DHEA can also be produced by the adrenals (see Chapter 10).

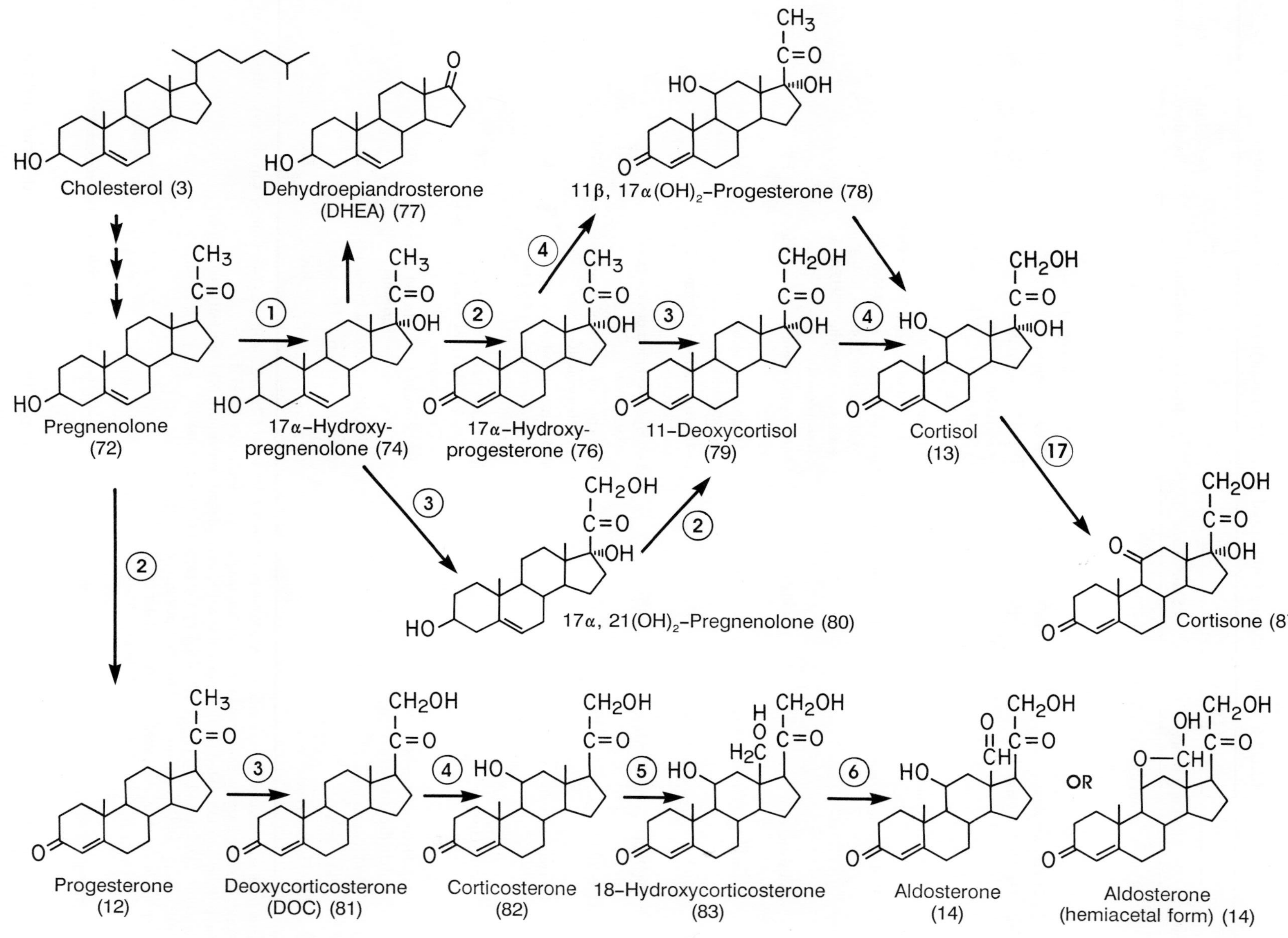
Cholesterol (3)
Dehydroepiandrosterone (DHEA) (77)
11β, 17α(OH)₂-Progesterone (78)
Pregnenolone (72)
17α-Hydroxy-pregnenolone (74)
17α-Hydroxy-progesterone (76)
11-Deoxycortisol (79)
Cortisol (13)
17α, 21(OH)₂-Pregnenolone (80)
Cortisone (87)
Progesterone (12)
Deoxycorticosterone (DOC) (81)
Corticosterone (82)
18-Hydroxycorticosterone (83)
Aldosterone (14)
OR
Aldosterone (hemiacetal form) (14)

TABLE 2-6 Key Enzymes Concerned with the Production of Steroid Hormones

Enzyme reaction[a]	Enzyme: Long name	Enzyme: Short name	Reaction(s) mediated	Subcellular location[b]	Gene location[e]
①	P450 side chain cleavage; formerly known as desmolase	P450scc	20*R*-hydroxylation; 22*R*-hydroxylation; scission of C-20–C-22 carbon bond	M	15
②	3β-Steroid dehydrogenase/Δ^5,Δ^4-isomerase	Isom	Oxidizes 3β-OH to oxo group; isomerizes Δ^5-ene to Δ^4-ene	ER	
③	21-Hydroxylase	P450c21	21-Hydroxylation	ER	6p
④⑤⑥	11β-Hydroxylase	P450c11[c]	④ 11β-Hydroxylation ⑤ 18-Hydroxylation ⑥ Oxidation of C-18 hydroxyl to an aldehyde	M	8q
⑦⑧	17α-Hydroxylase/C-17–C-20 lyase	P450c17[d]	⑦ 17α-Hydroxylation ⑧ Scission of C-17–C-20 carbon bond	ER	10
⑨	17-Ketosteroid reductase	17-Oxidoreductase	Reduces 17-oxo to a 17β-hydroxyl, as in androstenedione → testosterone; estrone → estradiol; DHEA → androstenediol	ER	
⑩	Aromatase	P450aro	Mediates the aromitization of the A ring	ER	15q 21.1
⑪	Vitamin D 25-hydroxylase	P450c25	25-Hydroxylation of vitamin D secosteroids	M,ER	
⑫	25(OH)D_3 1α-hydroxylase	P450c1	1α-Hydroxylation of vitamin D secosteroids	M	
⑬	25(OH)D_3 24-hydroxylase	P450c24	24*R*-Hydroxylation of vitamin D secosteroids	M	
⑭	5α-Reductase		Reduction of Δ^5-ene as in testosterone → dihydrotestosterone	NM	
⑮	Ferredoxin oxidoreductase		In the mitochondria transfers electrons from NADPH to ferredoxin	M	17 con → q25
⑯	P450 reductase		In the endoplasmic reticulum transfers electrons from NADPH to a P450 enzyme	ER	
⑰	11β-Hydroxysteroid dehydrogenase		Oxidation of 11β-hydroxyl to an oxo as in cortisol → cortisone	ER	
⑱	12α-Hydroxylase		12α-Hydroxylation of bile acids	ER	
⑲	6β-Hydroxylase		6β-Hydroxylation of bile acids	ER	
⑳	7α-Hydroxylase		7α-Hydroxylation of bile acids	ER	

[a] This table summarizes the key enzyme reactions in Figures 2-18, 2-20, 2-21, 2-22, and 2-23 (see numbers in circles). Note that, in some instances, the same enzyme catalyzes more than one reaction. (See footnotes c and d.)

[b] Subcellular location coded as M, mitochondria; ER, endoplasmic reticulum; NM, nuclear membrane.

[c] Reactions ④, ⑤, and ⑥ can be mediated by the same enzyme (see text).

[d] Reactions ⑦ and ⑧ are mediated by the same enzyme.

[e] Portions of the data were abstracted from Miller (1988).

11 hydroxyl or alternatively with carbon-21 to creates a six-membered hemiacetal ring.

As indicated in Figure 2-21, the adrenal cortex, particularly the zona reticularis, also has the enzymatic capability to produce steroids with mild androgenic activity. The principal adrenal androgens are androsterone, 4-androstene-3,17-dione (**25**), dehydroepiandrosterone or DHEA (**77**), and 3β,11β-dihydroxy-4-androsten-17-dione. The production of these and related steroids may increase in some instances of adrenal tumors; women with such tumors may develop secondary male characteristics, including beard growth and hirsutism.

D. Biosynthesis of Androgens

The androgens are all steroids with 19 carbons. They are produced in males in the testes and in females by the ovaries and placenta. Also, as mentioned, the adrenal cortex can under some circumstances produce steroids with weak, but physiologically significant androgen activity. The androgens in humans are characterized by (i) the absence of the two-carbon side chain on carbon-17 and (ii) the presence of an oxygen function on both carbon-3 and carbon-17. The major naturally occurring steroids with androgenic activity (in decreasing order of relative potency) are 5α-dihydrotestosterone (**86**) (150–200%), testosterone (**15**) (100%), androstanediol (**85**) (65%), androst-4-ene-3,17-dione (**25**) (25%), androsterone (10%), and dehydroepiandrosterone (**77**) (10%).

Two general metabolic pathways lead from pregnenolone to testosterone; they are, respectively, either the Δ^5- or Δ^4-pathway (see Figure 2-22). Steroid intermediates on the Δ^5-pathway can be converted to the corresponding steroid on the Δ^4-pathway by oxidation of the 3β-hydroxyl to a (oxo) ketone (3β-steroid dehydrogenase) followed by migration of the double bond from $C_{5\text{-}6}$ to $C_{4\text{-}5}$ (Δ^5, Δ^4-isomerase) (see reaction 2, Table 2-6).

The hormonally active form of testosterone in males is believed to be 5α-dihydrotestosterone (5α-DHT). There is evidence for the production of 5α-DHT by the testes, skin, and submaxillary glands, but it is formed

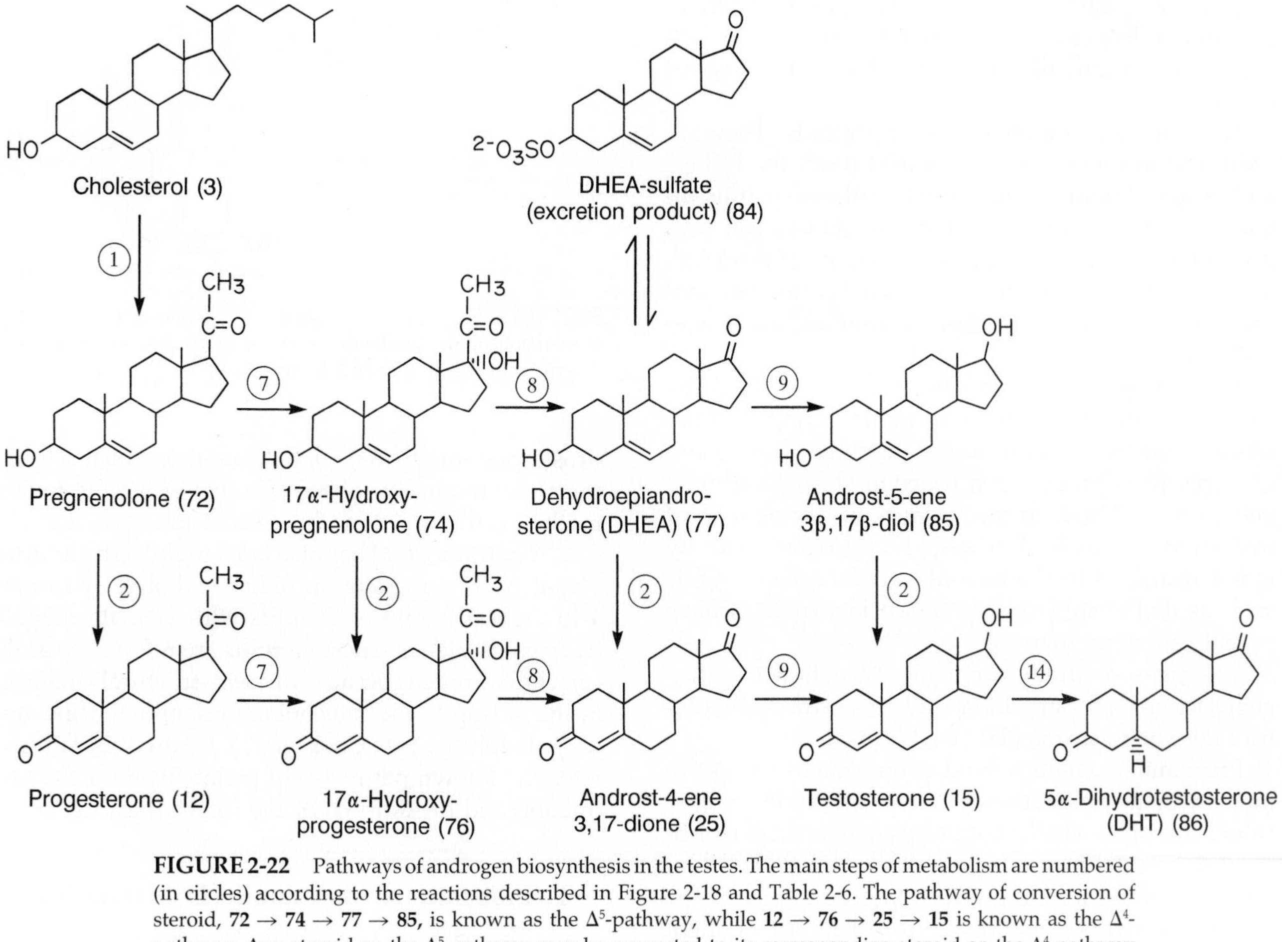

Cholesterol (3)

DHEA-sulfate (excretion product) (84)

Pregnenolone (72)

17α-Hydroxy-pregnenolone (74)

Dehydroepiandro-sterone (DHEA) (77)

Androst-5-ene 3β,17β-diol (85)

Progesterone (12)

17α-Hydroxy-progesterone (76)

Androst-4-ene 3,17-dione (25)

Testosterone (15)

5α-Dihydrotestosterone (DHT) (86)

FIGURE 2-22 Pathways of androgen biosynthesis in the testes. The main steps of metabolism are numbered (in circles) according to the reactions described in Figure 2-18 and Table 2-6. The pathway of conversion of steroid, **72** → **74** → **77** → **85**, is known as the Δ^5-pathway, while **12** → **76** → **25** → **15** is known as the Δ^4-pathway. Any steroid on the Δ^5-pathway may be converted to its corresponding steroid on the Δ^4-pathway by the sequential action of the multienzyme, the 3β-steroid dehydrogenase/Δ^5,Δ^4-isomerase (reaction 2). The conversion of testosterone to DHT (reaction 14) occurs in peripheral target tissues such as genital skin and hair follicles.

especially in androgen target glands such as the prostate. The physiological actions of the androgens are discussed in Chapter 12.

In females, in the follicular tissue there is evidence for both the Δ^5- and Δ^4-pathways of androgen production (as outlined in Figure 2-22). Pregnenolone apparently is a more efficient precursor of carbon-19 steroids than progesterone in ovaries without corpora lutea.

In the zona reticularis of the human adrenal cortex, there is also evidence for both the Δ^5- and Δ^4-pathways. However, the principal pathway appears to be a combination of both.

Pregnenolone → 17α-hydroxypregnenolone →
→ 17α-hydroxyprogesterone →
→ androst-4-ene-3,17-dione → testosterone

E. Biosynthesis of Estrogens

The several metabolic pathways for the conversion of either androst-4-ene-3,17-dione (**25**) or testosterone (**15**) into estrogens are summarized in Figure 2-23 (a unique feature of this conversion is the loss of carbon-19).

The estrogens are all 18-carbon steroids. They are produced in females in the ovaries (both the follicle and corpus luteum) and the fetal–placental unit. In males, the testes under some circumstances can produce physiologically significant amounts of estradiol. In both males and females, the adrenal cortex can generate small quantities of estrone from androst-4-ene-3,17-dione.

The estrogens in humans are characterized by (i) loss of carbon-19, (ii) an aromatic A ring, (iii) absence of the two-carbon side chain on carbon-17, and (iv) presence of an oxygen function at both carbon-3 and carbon-17 and, in the instance of estriol, a third oxygen at carbon-16. Intriguingly, estrogenic activity is not restricted to the steroid structure; compounds such as diethylstilbestrol (**39**) (see Figure 2-11) have potent estrogenic activity.

The major naturally occurring steroids with estrogenic activity are estra-3,17β-diol (**16**), estra-3,16α,17β-triol (**89**), and estrone (**88**) (see Figure 2-23).

Pregnancy is characterized in humans by a massive increase in the production of both progesterone and estrogen. In contrast to the nonpregnant woman, the principal active estrogen of pregnancy is estriol (**89**). The increase in progesterone production occurs only in the placenta, whereas the production of estriol is dependent on the combined activity of the placenta and the fetal adrenals and liver. These relationships are summarized in Figures 14-9, 14-10, and 14-11. The important role of the fetal adrenal gland as a source of dehydroepiandrosterone sulfate (**84**) (DHEA sulfate), which acts as a precursor for the estriol (**88**) of both the placenta and the mother, is discussed in Chapters 10 and 14.

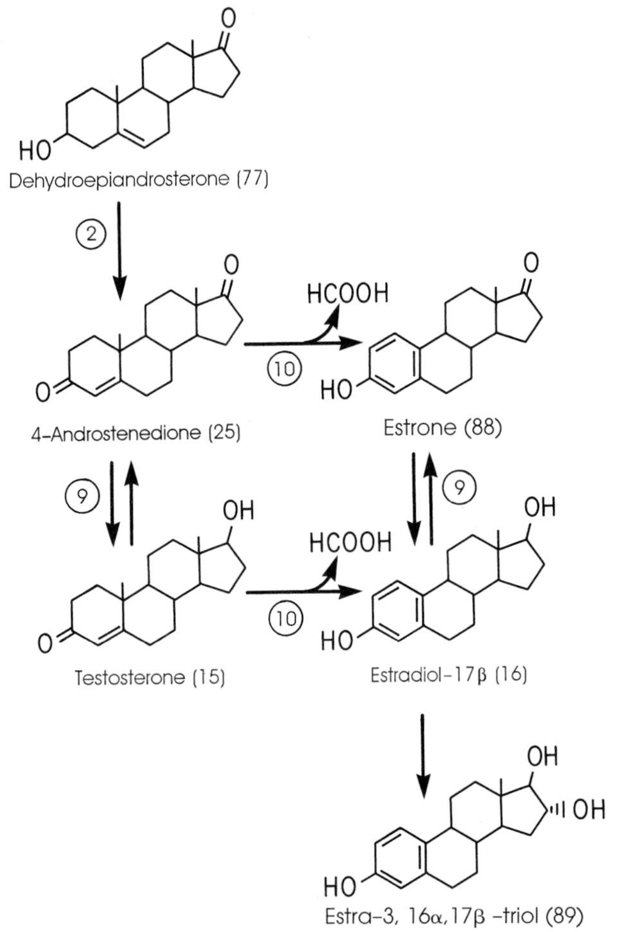

FIGURE 2-23 Pathway of estrogen biosynthesis. The main steps of metabolism are numbered (in circles) according to the reactions described in Figure 2-18 and Table 2-6.

17β-Estradiol (**16**) has also been found to be hydroxylated by brain tissue on either carbon-2 or carbon-4 to yield a family of steroids with vicinal[3] phenolic hydroxyl moieties. These steroids are referred to as the catechol estrogens because of their structural similarity in the A ring to the catecholamines, epinephrine, and norepinephrine (see Chapter 11). Figure 2-24 summarizes the known pathways of metabolism for the production and breakdown of the catechol estrogens.

F. Biosynthesis of Vitamin D Metabolites

Vitamin D_3 is chemically closely allied to the classical steroid hormones. Technically it is a secosteroid; that is, due to the breakage of the carbon-9, carbon-10

[3] Vicinal hydroxyls are hydroxyl groups on adjacent carbons.

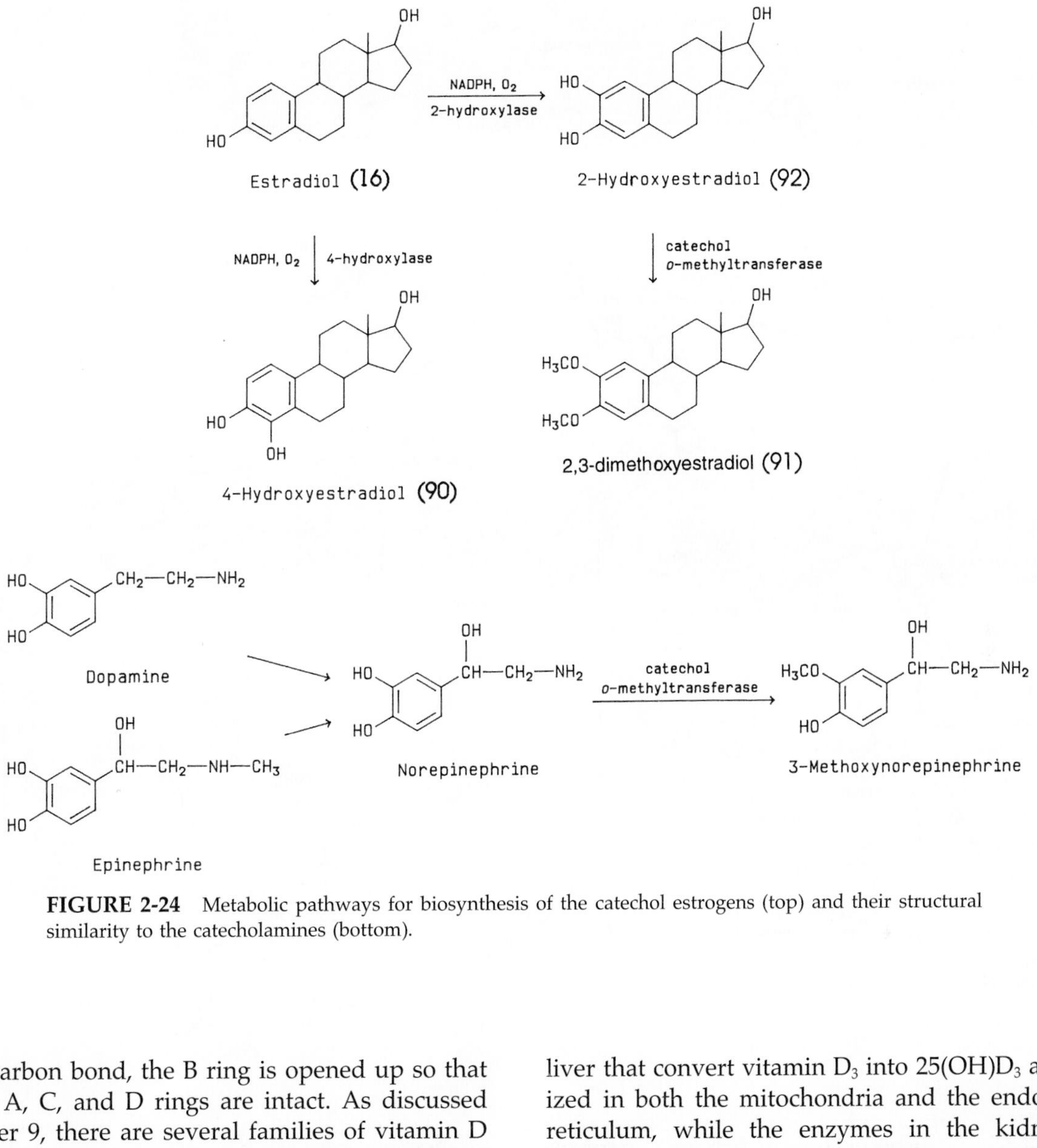

FIGURE 2-24 Metabolic pathways for biosynthesis of the catechol estrogens (top) and their structural similarity to the catecholamines (bottom).

carbon–carbon bond, the B ring is opened up so that only the A, C, and D rings are intact. As discussed in Chapter 9, there are several families of vitamin D steroids (i.e., vitamin D_2, vitamin D_3, vitamin D_4) that depend on the structure of the secosteroid side chain. When the side chain is identical to that of cholesterol, then it is the naturally occurring form and belongs to the vitamin D_3 family.

Vitamin D_3 can be obtained dietarily or produced photochemically by sunlight from a precursor, 7-dehydrocholesterol, present in the skin. As documented in Fig. 2-25, vitamin D_3 is known to be metabolized into a family of daughter metabolites.

The hormonally active forms that produce the spectrum of biological responses attributable to vitamin D_3 are 1,25-dihydroxyvitamin D_3 [1,25$(OH)_2D_3$] and 24,25-dihydroxyvitamin D_3 [24,25$(OH)_2D_3$]. The principal form of vitamin D_3 present in blood is 25-hydroxyvitamin D_3 [25$(OH)D_3$]. The enzymes in the liver that convert vitamin D_3 into 25$(OH)D_3$ are localized in both the mitochondria and the endoplasmic reticulum, while the enzymes in the kidney that catalyze the conversion of 25$(OH)D_3$ into 1,25$(OH)_2D_3$ or 24,25$(OH)_2D_3$ are located in the mitochondria of the proximal kidney tubule. All three enzymes contain cytochrome P450-type mixed-function oxidases. The other metabolites shown in Figure 2-25 are believed to be catabolites of the hormonally active forms.

G. Biosynthesis of Bile Acids

The conversion of cholesterol into bile acids takes place largely in the liver. In most mammals cholic acid (**47**) and chenodeoxycholic acid (**48**) (see Figure 2-11) are the principal products. Before excretion into the bile, the carbon-24 carboxyl of both steroids is conju-

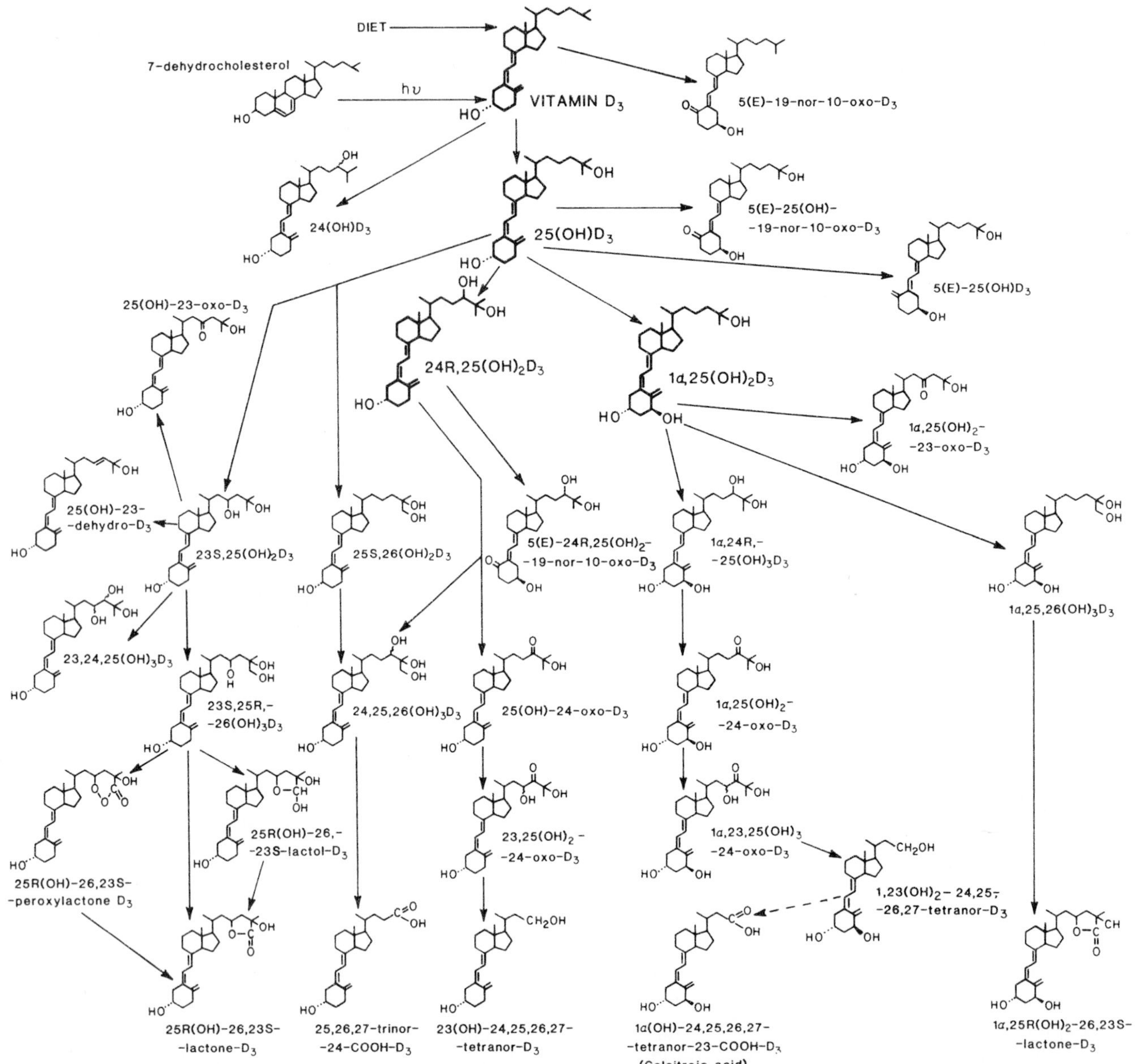

FIGURE 2-25 Metabolic pathway for the metabolism of vitamin D_3 (cholecalciferol) into its daughter metabolites. The secosteroids presented in boldface are the physiologically relevant vitamin D compounds, while the other structures are believed to be catabolites of vitamin D_3 (see Chapter 9).

gated with the amino group of the amino acids taurine or glycine.

A key step in the production of the bile acids is the cleavage of the hydrocarbon side chain between carbon-24 and carbon-25. This does not occur by the lyase-type reactions associated with cholesterol side chain cleavage or production of androst-4-ene-3,17-dione (**25**). Instead, the removal of the three terminal carbon atoms is believed to proceed by a β-oxidation mechanism similar to that occurring in fatty acid catabolism.

Additional key steps in the biosynthesis of bile acids involve (i) microsome-mediated hydroxylations at carbon-7 and carbon-12, (ii) epimerization of the 3β-hydroxyl to a 3α-orientation (see Figure 2-11) and (iii) reduction of the Δ^5-double bond so that the A:B ring junction is *cis*, as in 5β-cholestane (**22**) (see Figure 2-5).

V. PROPERTIES OF ENZYMES INVOLVED IN STEROID METABOLISM

A. Introduction

Hydroxylation reactions play an important role in the metabolic pathways of many steroids, both those producing hormones as well as those generating bile acids. The cleavage of the cholesterol side chain necessitates the hydroxylation of both carbon-22 and carbon-20, while the production of testosterone necessitates hydroxylation of progesterone at carbon-17 followed by cleavage of the carbon-17–carbon-20 bond. In the biosynthesis of cortisol, three hydroxylases, the 17α-, the 21-, and the 11β-hydroxylases, are required. The conversion of deoxycorticosterone to aldosterone would seemingly involve three separate enzymes, which sequentially hydroxylate carbon-11, then carbon-18, followed by oxidation of the carbon-18 alcohol to an aldehyde. Surprisingly, it has been learned from molecular biological cloning that one enzyme carries out all three steps.

Table 2-6 tabulates the key enzymes concerned with the production of the steroid hormones and enumerates the details of the reactions mediated. The presentation includes the short- and long-form names of the enzyme, a specification of the reaction(s) catalyzed, the subcellular location of the enzymatic activity, and the gene location. Subsequent sections of this chapter focus on a brief overview of particularly key enzymes.

B. P450 Steroid Hydroxylases

Cytochrome P450 enzymes are responsible for the oxidative metabolism of prostaglandins, fatty acids, biogenic amines, plant metabolites, and steroids, as well as chemical carcinogens, mutagens, many drugs, and other environmental pollutants. The following paragraphs present a brief description of the important steroid P450 enzymes and their electron transport chain partners.

1. *Cytochrome P450 Electron Transport Chains*

All of these seemingly diverse hydroxylase reactions are mediated by a family of homologous enzymes known as the cytochrome P450 hydroxylases. Cytochrome P450 is a generic term used to describe a large family of oxidative enzymes;[4] the hallmark of each P450 enzyme is the presence of a protoporphyrin ring group. The term "P450" is derived from early spectroscopic studies of enzymes possessing a reddish pigment, which all displayed a characteristic shift in the Soret absorbance peak from 420 to 450 nm upon reduction with carbon monoxide. In each hydroxylase enzyme, the general oxidation–reduction reaction catalyzed is as shown in Figure 2-26.

In principle, the oxygen atom of the hydroxyl group could be derived from either H_2O or molecular oxygen. But as indicated in the equation, it is known that the hydroxyl oxygen is derived exclusively from molecular oxygen in all steroid cytochrome P450-containing hydroxylases. As a class these enzymes are designated as mixed-function oxidases. The term "mixed function" indicates that one atom of the diatomic substrate oxygen ends up in the steroid (as part of the hydroxyl) and the other ends up as part of water. This result has been confirmed by utilization of $^{18}O_2$ and the subsequent appearance of ^{18}O in the steroid and in $H_2{}^{18}O$.

The key family of enzymes mediating the introduction of the oxygen atom into the steroid nucleus are proteins that have at their catalytic center a cytochrome moiety as a prosthetic group. This cytochrome moiety is structurally analogous to the hemoprotein cytochromes of the electron transport chain present in mitochondria or hemoglobin; all contain some kind of covalently bound protoporphyrin ring coordinately bound to one atom of iron, which can be reversibly oxidized and reduced (Fe^{2+}/Fe^{3+}). Many cytochrome P450 enzymes have been isolated and purified; their molecular weights are around 50,000–60,000. The terminology P450 is uniquely characteristic of cytochromes that mediate steroid and other alkyl hydroxylation reactions. As a class, most of these enzymes are subject to inhibition by the presence of carbon monoxide. The CO coordinates to the Fe^{2+}, thus blocking the essential cyclical oxidation and reduction that are associated with electron transfer. This inhibition can then be selectively reversed by light with a wavelength of 450 nm. Cytochrome P450 enzymes are known to be present in the liver, adrenal cortex, ovary, testis, kidney, placenta, lungs, and intestinal mucosa (see Tables 2-5 and 2-6).

Virtually all steroid hydroxylases are membrane bound and are present in either the mitochondria or endoplasmic reticulum of the cell. Table 2-6 tabulates the subcellular localization of the cytochrome P450 steroid hydroxylases. Comparison of the subcellular lo-

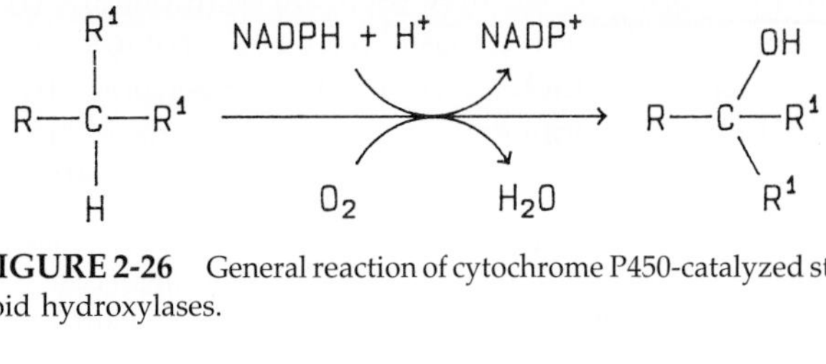

$$\mathrm{R{-}C(R^1)(H){-}R^1} \xrightarrow[\mathrm{O_2}\ \to\ \mathrm{H_2O}]{\mathrm{NADPH + H^+}\ \to\ \mathrm{NADP^+}} \mathrm{R{-}C(OH)(R^1){-}R^1}$$

FIGURE 2-26 General reaction of cytochrome P450-catalyzed steroid hydroxylases.

[4] The Nomenclature Committee of the International Union of Biochemistry (IUB) prefers the term "hemethiolate protein" instead of "cytochrome" for P450.

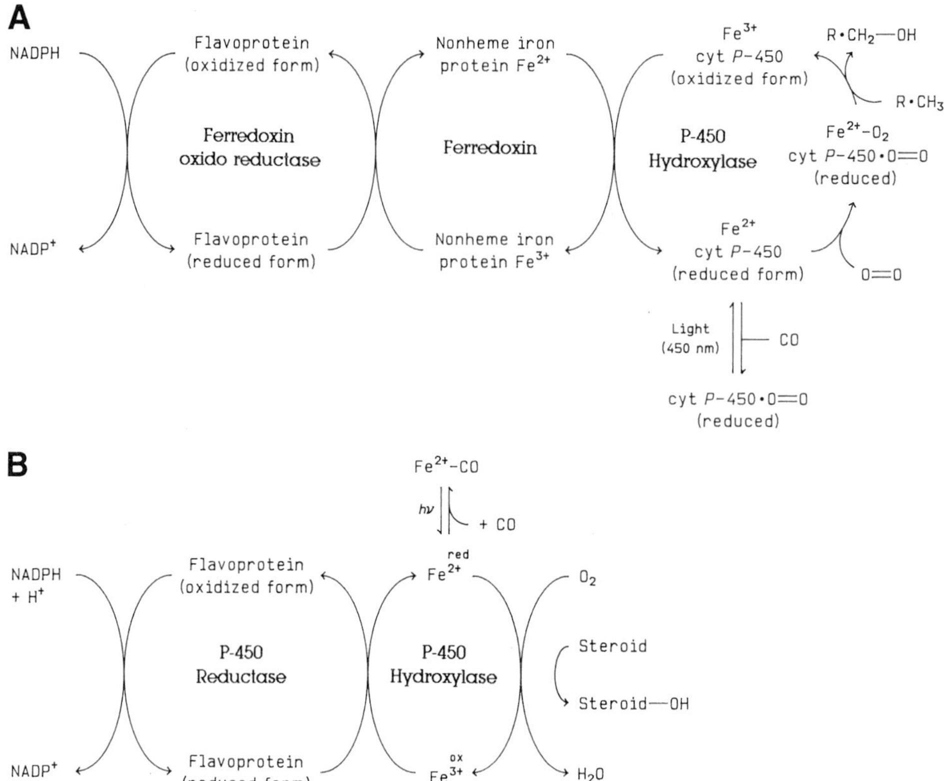

FIGURE 2-27 Electron transport chains for sterol hydroxylases. (A) Mitochondrial hydroxylases showing the participation of the three protein components, ferredoxin oxidoreductase, ferredoxin, and a P450 hydroxylase. (B) Endoplasmic reticulum hydroxylases showing the participation of the two protein components, P450 reductase and a P450 hydroxylase.

calization of the various hydroxylase enzymes with the sequence of steroid movement through a metabolic pathway indicates the important role of cellular compartmentalization. Thus, in the conversion of cholesterol into cortisol in the adrenal cortex, the steroid must move sequentially from the mitochondria (side chain cleavage) to the microsomes (17α- and 21-hydroxylation) and then back to the mitochondria (11β-hydroxylation).

Depending upon whether the cytochrome P450 hydroxylase is localized in the mitochondria or microsomes, there are two slightly different electron transport chains that function to transfer a pair of electrons from NADPH to the cytochrome P450 enzyme. These are presented in Figure 2-27. In the mitochondrial system there are at least three separate components: (i) a flavoprotein dehydrogenase known as ferredoxin oxidoreductase, which accepts the electrons from NADPH; (ii) a nonheme iron protein, termed ferredoxin[5] (molecular weight 13,000), which accepts the electrons from the flavoprotein and transfers them to (iii) the cytochrome P450 hydroxylase protein. The microsomal system lacks a ferredoxin component; thus, the flavoprotein dehydrogenase transfers electrons directly to the cytochrome P450 enzyme.

[5] In the older literature, this nonheme iron protein was discovered in the adrenals and termed adrenodoxin. With the appreciation of the generic functioning of this heme protein in mitochondrial P450 hydroxylation in many tissues it is now referred to as ferredoxin.

The most important component of the electron transport chain is the cytochrome P450 protein, in that it determines the substrate specificity and dictates the precise site of hydroxylation. Each P450 hydroxylase has a substrate-binding domain that is comparable to the ligand-binding domains of receptors and plasma transport proteins for steroid hormones (see Section VI.B). Basically, the three-dimensional structure of the substrate-binding domain of a P450 hydroxylase determines which substrates are able to acquire a new hydroxyl group.

Figure 2-28 presents a scheme describing the complex series of steps associated with electron transfer and hydroxylation of steroids.

2. *Ferredoxin*

Ferredoxin, which has ≈116 amino acids (12 kDa), is synthesized as a larger preproprotein having both carboxy-terminal (≈60 amino acids) and amino-

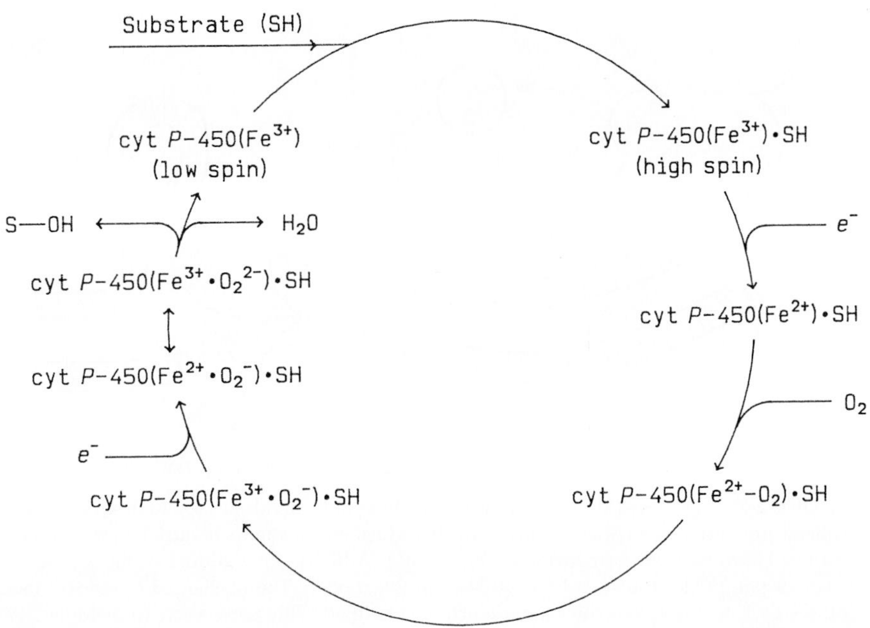

FIGURE 2-28 Proposed scheme for the cyclic reduction and oxidation of cytochrome P450 in biological hydroxylations. The process of binding of the substrates, a steroid, and molecular O_2 by the cytochrome P450 enzyme and transfer of the electrons is complex. The key participant is the hexacoordinate iron ion, which is bound to the planar tetrapyrrolic heme prosthetic group. Four of the iron valencies are taken up by binding with the four nitrogen atoms of the tetrapyrrole; the remaining two valencies are utilized to coordinate with other ligands (cysteinate and molecular oxygen) at right angles to the plane of the pyrrole ring structure. The strength or weakness of the bound ligands with respect to the nature of the uncharged or charged moiety bonding to the metal ion can greatly alter the physical properties of the hemoprotein. This alteration has been termed "low spin" or "high spin." Modified from Boyd, G. S. (1973). *In* "Biological Hydroxylation Mechanisms" (G. S. Boyd and R. M. S. Smellie, eds.), p. 1. Academic Press, New York. Copyright © 1973 The Biochemical Society, London.

terminal extensions. The N-terminal extension apparently serves as a leader sequence to facilitate the entrance of the peptide into the mitochondria. The ferredoxin protein acts as a shuttle accepting electrons from ferredoxin oxidoreductase, which then diffuses in the mitochondrial matrix to a P450 hydroxylase (see Figure 2-29) where it donates a pair of electrons.

3. *Ferredoxin Oxidoreductase*

Ferredoxin oxidoreductase (reaction 15 in Table 2-6) is an inner mitochondrial membrane-bound flavoprotein with a molecular weight of 51,100. It is responsible for the transfer of electrons from NADPH to ferredoxin and is widely expressed in many human tissues.

4. *Cholesterol Side Chain Cleavage Enzyme*

The two enzymes that catalyze the cleavage of the side chain of cholesterol (**3**) [to yield pregnenolone (**72**)] and that of 17α-hydroxypregnenolone (**74**) [to yield androst-4-ene-3,17-dione (**25**)] are known as lyases. Lyases are enzymes that cleave bonds between carbons bearing vicinal hydroxyls. The properties of these two reactions will be discussed separately.

The cholesterol side chain cleavage enzyme (P450scc) (reaction 1 in Table 2-6) carries out three successive enzymatic steps. These involve the 20*R*- and 22*R*-hydroxylations followed by cleavage of the side chain on the two carbons bearing the newly incorporated hydroxyl group. Each of these steps is an oxidation–reduction reaction requiring a pair of electrons, which is supplied by NADPH. The P450scc has been cloned and expressed. The protein subunit molecular weight is 49,000, but the functional P450scc contains 16 subunits, creating a multimer of 850,000 that is associated with the inner mitochondrial membrane. The P450scc from the human adrenal and testes have identical sequences.

5. *17α-Hydroxylase/C-17–C-20 Lyase*

Another enzyme that carries out both a hydroxylation reaction and cleavage of a carbon–carbon bond is the P450c17 (reactions 7 and 8 in Table 2-6). The P450c17 enzymes from both the adrenals and testes

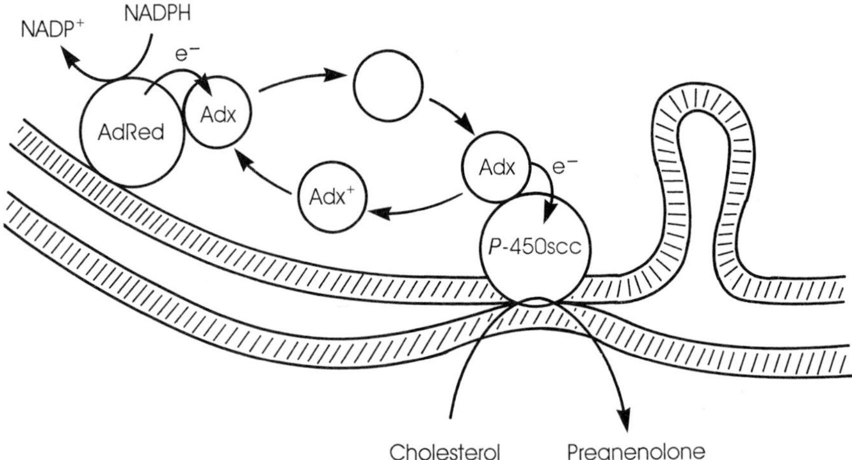

FIGURE 2-29 Proposed shuttle role of ferredoxin in mitochondrial P450 hydroxylases. The nonheme iron ferredoxin (Adx$^+$), which is freely diffusible in the mitochondrial matrix, accepts a pair of electrons from ferredoxin oxidoreductase (AdRed). After diffusion, the ferredoxin donates a pair of electrons to a P450 (P450scc in this model). The uncharged ferredoxin then diffuses back to the ferredoxin oxidoreductase. Modified with permission from Miller, W. (1988). Molecular biology of steroid hormone synthesis. *Endocr. Rev.* **9**, 295–318.

have been cloned and expressed and found to be identical proteins. The enzyme P450c17 in the adrenals is involved in the production of cortisol (**13**) (see Figure 2-21), which has a 17α-hydroxyl, while the P450c17 in the testes is involved in the production of testosterone (**15**) and androst-5-ene-3β,17-diol (**85**) (see Figure 2-22). This suggests the existence of a cellular regulatory mechanism in the adrenals to suppress the P450c17 catalytic ability for C-17–C-20 side chain cleavage.

6. 11β-Hydroxylase

The P450c11 is still another example of one enzyme that can catalyze more than one enzymatic reaction; this single P450 has 11β-hydroxylase, 18-hydroxylase, and 18-oxidase activities (reactions 4–6 in Table 2-6). Highly purified P450c11 (molecular weight 48,000) from bovine adrenal zona glomerulosa and zona fasciculata mitochondria displayed all three enzyme activities and were indistinguishable enzymatically, by amino acid composition, and immunologically. But intact mitochondria from the adrenal zona fasciculata lacks aldehyde synthetase activity, which suggests the presence of cell-specific factors that suppress the 18-hydroxylase activity. In addition, it is known that there are at least two genes for P450c11; one gene codes for the functional 11β-hydroxylase in the zona reticularis, while the second gene codes for the aldosterone synthesis capability of the zona glomerulosa. The existence of two genes is consistent with known human clinical deficiency states associated with congenital adrenal hyperplasia (see Chapter 10).

7. 21-Hydroxylase

The 21-hydroxylation essential to the production of cortisol (reaction 3 in Table 2-6) is mediated by P450c21, which is found in the endoplasmic reticulum. Evidence has been presented that humans have two genes coding for P450c21. The details of the expression of these genes are clinically important since about 1 in every 7000 individuals has an autosomally recessive genetic lesion in this enzyme. A detailed discussion of this topic is beyond the scope of this chapter.

8. Aromatase

The aromatization of the A ring of either testosterone (**15**) to yield estradiol (**16**) or androst-4-ene-3,17-dione (**25**) to yield estrone (**88**) (see Figure 2-23 and reaction 10 of Table 2-6) is mediated by P450aro. This enzyme is present in the endoplasmic reticulum of the ovary and placenta and also occurs to a limited extent in brain and adipose tissues. Figure 2-30 presents a proposed mechanism for the complex aromatization process. To date, two genes for human P450aro have been detected.

9. Ad4-Binding Protein

One of the features of the tissue expression of P450 steroid-metabolizing enzymes is their specific appearance in only steroidogenic tissues (see Table 2-5). Evi-

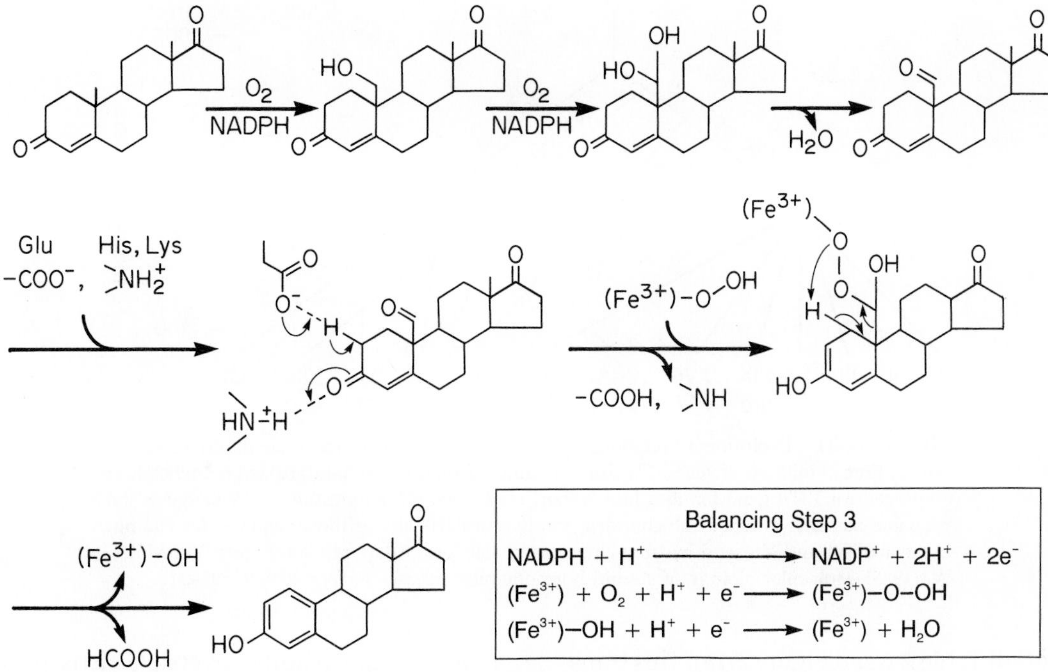

Balancing Step 3

$NADPH + H^+ \longrightarrow NADP^+ + 2H^+ + 2e^-$

$(Fe^{3+}) + O_2 + H^+ + e^- \longrightarrow (Fe^{3+})\text{–O–OH}$

$(Fe^{3+})\text{–OH} + H^+ + e^- \longrightarrow (Fe^{3+}) + H_2O$

FIGURE 2-30 Proposed mechanism for the aromatization reaction mediated by P450aro. The reactions require an appropriate A ring as substrate (either testosterone or androst-4-ene-3,17-dione), 3 mol of NADPH, and three molecules of O_2. The accumulated evidence indicates that the initial hydroxylation occurs on carbon-19, followed immediately by a second hydroxylation on carbon-19 to yield a *gem*-diol, which fragments into a C-19 aldehyde. The third oxidative reaction is proposed to involve a peroxidative attack on the C-19 carboxyl group, as indicated previously. Of the three oxygen insertions, only the third is shown in detail. The model then proposes the participation of a COO^- from glutamate and an NH_2^+ from lysine or histidine of the aromatase to facilitate 2β-hydrogen elimination, followed by 1β-hydrogen elimination and enolization of the 3-ketone, resulting in the production of an aromatic A ring. Adapted with permission from Simpson, E. R. *et al.* (1994). Aromatase cytochrome P-450, the enzyme responsible for estrogen biosynthesis. *Endocr. Rev.* **15,** 342–355.

dence has accumulated to support the existence of specific regulatory elements in the promoter regions of the genes of these steroid-metabolizing enzymes. For example, the promoter of P450c11 coding has been found to contain multiple *cis*-acting elements, termed Ad1, Ad2, Ad3, Ad4, Ad5, and Ad6. The Ad4BP motif has also been found in the promoter of other steroid hydroxylases. The Ad4-binding protein (Ad4BP) has been cloned and sequenced; it codes for a protein of 51,000 that has a zinc finger domain analogous to the zinc finger of steroid receptors and belongs to the steroid receptor superfamily (see Figure 1-42). It seems likely that Ad4BP plays an important role in the regulation of gene expression of selective P450 enzymes.

10. Cytochrome P450 Gene Families

To date there are at least 221 known P450 genes that have been described in 31 eukaryotes (including 11 mammalian and 3 plant species) and 11 prokaryotes. All of these proteins are encoded by a superfamily of P450 genes that have their origins more than 3.5 billion years ago. This superfamily comprises at least 10 separate gene families; a diagram of the evolutionary relationships of the P450 genes is provided in Figure 2-31. This figure illustrates that despite the similarity of biological roles and steroidal substrates of the four adrenal steroidogenic enzymes located in the endoplasmic reticulum, P450aro, P450c17, P450c21A, and P450c21B belong to four different gene families that are only quite distantly related to the two mitochondrial P450 enzymes, P450scc and P450c11. It appears that the highly coordinated pathways of steroid metabolism (see Figure 2-18) represent the convergence of enzymatic functions that arose at grossly different times in evolution.

C. Other Enzymes of Steroid Metabolism

1. 3β-Hydroxy Dehydrogenase/Δ^5,Δ^4-Isomerase

A key enzymatic transformation in the biosynthetic pathway of five steroid hormones is the conversion of the 5-ene 3β-hydroxysteroid to a 4-ene 3-oxosteroid

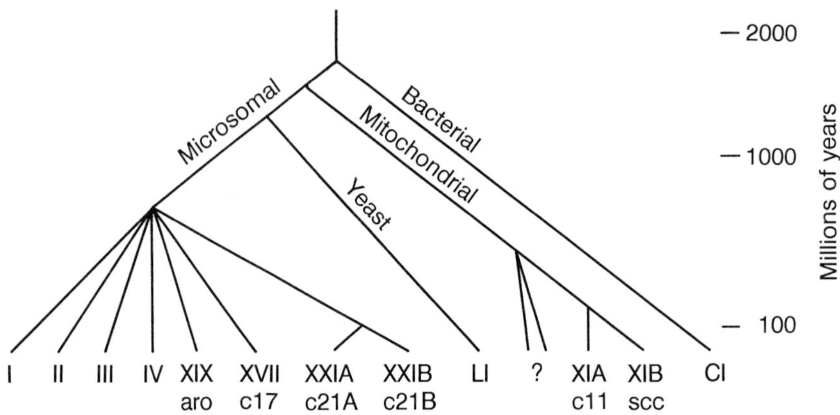

FIGURE 2-31 Evolutionary relationship of P450 genes. The vertical scale describes evolutionary time in millions of years. The Roman numerals refer to the standardized nomenclature employed for P450 gene families (see Nebert *et al.*, 1987). The question mark suggests the probable existence of other mitochondrial families of P450 such as those required for vitamin D metabolism and bile acid hydroxylation (see Table 2-6). Adapted from Figure 7 of Miller, W. (1988). Molecular biology of steroid hormone biosynthesis. *Endocr. Rev.* **9,** 295–318.

(see reaction 2 of Table 2-6 and Figure 2-18). This transformation is mediated by one enzyme located in the endoplasmic reticulum that carries out the two steps. The same enzyme is apparently involved in the adrenals, ovaries, testes, and placenta.

The 3β-hydroxysteroid dehydrogenase Δ^5,Δ^4-isomerase has an obligatory cofactor NAD^+ and catalyzes the reactions shown in Figure 2-32. The enzyme has been cloned and expressed; evidence indicates the expression of multiple forms of the 3β-hydroxy dehydrogenase / Δ^5,Δ^4-isomerase.

2. 5α-Reductase

The conversion of testosterone (**15**) to dihydrotestosterone (**86**) (see Figure 2-22 and reaction 14 of Table 2-6) is mediated by the 5α-reductase. 5α-Reductase is an NADPH-dependent non-P450 enzyme that is present in the nuclear membrane of some target cells for testosterone, e.g., genital skin and hair follicles.

3. 17-Ketosteroid Reductase

17-Ketosteroid reductase is a non-P450 enzyme present in the endoplasmic reticulum; it requires NADPH (see reaction 9 of Table 2-6 and Figures 2-22 and 2-23). The catalyzed reactions of the 17-ketosteroid reductase are readily reversible. It is present in the testes, ovaries, and placenta.

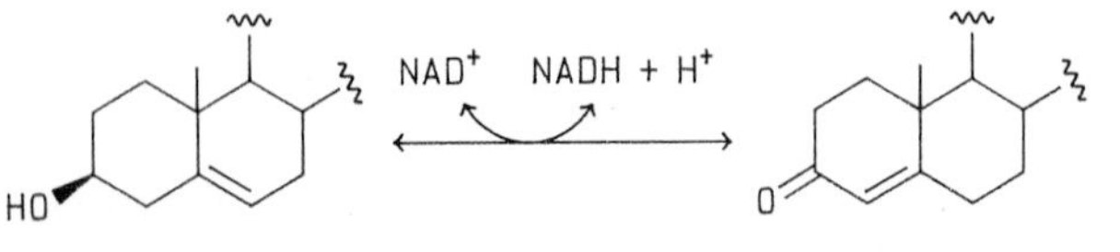

HO — NAD^+ → $NADH + H^+$ — O

FIGURE 2-32 Reaction catalyzed by the 3β-hydroxysteroid dehydrogenase, Δ^5,Δ^4-isomerase.

4. 11β-Steroid Dehydrogenase

The enzyme 11β-steroid dehydrogenase (see reaction 17 in Table 2-6 and Figure 2-21) is responsible for the NADPH-dependent reduction of the 11β-hydroxyl on cortisol (**13**) to yield cortisone (**93**). This enzyme is present in kidney, liver, brain, and lung. One significant contribution of this enzyme in the kidneys is to protect the aldosterone receptor from inappropriate exposure to cortisol, which is an effective ligand. Cortisone, in contrast, is a weak ligand for the aldosterone receptor (see Chapter 15).

5. Steroid Acute Regulatory Protein (StAR)

The first step in classical steroid hormone biosynthesis involves the conversion of cholesterol into pregnenolone by the cholesterol side chain cleavage enzyme P450scc. Although the side chain cleavage reaction is known to be the rate-limiting step for adrenal and gonadal steroid hormones, it is not the catalytic process of the enzyme that is rate limiting but, instead, the delivery of the substrate cholesterol from the cytoplasm across the outer mitochondrial membrane to the inner mitochondrial membrane site of the P450scc, which has been found to be the rate-limiting step.

A specific cholesterol transport protein, known as steroid acute regulatory protein (StAR), has been implicated as being an essential mediator of cholesterol delivery and the consequent activation of P450scc (see Figure 2-33). StAR is biosynthesized as a 285-amino

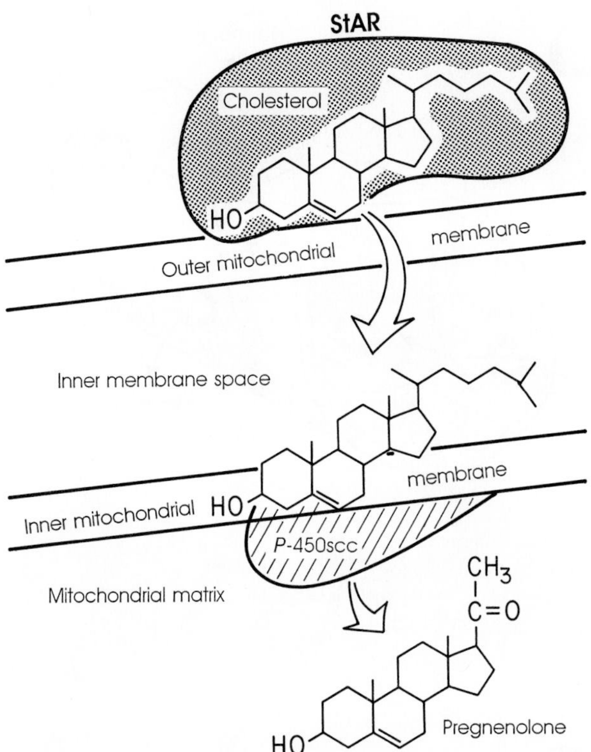

FIGURE 2-33 Proposed model for the role of the steroid acute regulatory (StAR) protein in cholesterol side chain cleavage. StAR is believed to facilitate the transfer of the hydrophobic cholesterol molecule across both the outer mitochondrial membrane and the inner membrane space to the inner mitochondrial membrane, which is the location of the P450scc where metabolism to pregnenolone occurs. [Adapted from Waterman, M. R. (1955). A Rising StAR: An essential role in cholesterol transport. *Science* **267,** 1780–1781.]

acid protein with a mitochondrial targeting sequence of 25 amino acid residues, which is cleaved from the NH_2-terminus after it is transported into the mitochondria.

The most convincing evidence for the essential nature of StAR was obtained from molecular evaluation of the defect associated with the disease known as lipoid congenital adrenal hyperplasia (CAH). (See Chapter 10 for a discussion of the autosomal recessive diseases that result from a blockade in several of the steps of steroid biosynthesis.) In lipoid congenital adrenal hyperplasia, which is characterized by a deficiency of both adrenal and gonadal steroid hormones, there is a mutation in the StAR gene, so that the StAR protein is nonfunctional due to premature stop codons. Lipoid CAH is the only steroid hormone synthesis disorder that is not a result of a mutation in the gene for a steroidogenic enzyme.

VI. PLASMA TRANSPORT, CATABOLISM, AND EXCRETION OF STEROID HORMONES

A. General Comments

The hormonally active form of most steroids is generally the molecular species released from the endocrine gland and transported systemically to various distal target tissues. A target tissue is defined as one that has stereospecific receptors permitting the accumulation of the steroid in the target tissue against a concentration gradient. This in turn permits generation of the appropriate biological response in that target tissue for the specific steroid in question. Thus, a key determinant of the ability of a target tissue to bind the steroid hormone is the hormone's actual blood concentration. The concentration of a steroid in the plasma at any particular time depends on three factors: (i) the rate at which the steroid is biosynthesized and enters the body pools; (ii) the rate at which the steroid is biologically inactivated by catabolism and removed from body pools; and (iii) the "tightness" of binding of the steroid to its plasma carrier protein.

Previous sections of this chapter have considered in detail the metabolic pathways of production of many steroid hormones; subsequent chapters will discuss in detail the regulation of steroid metabolism. The remaining sections of this chapter will briefly consider the plasma transport proteins for steroids and general

TABLE 2-7 Plasma Transport Proteins for Steroid, Thyroxine, and Retinoid Hormones

Plasma protein	Abbreviation	Principal steroids bound	Molecular weight ($\times 10^3$)
Vitamin D-binding protein	DBP	Vitamin D_3, $25(OH)D_3$, $1,25(OH)_2D_3$, $24,25(OH)_2D_3$	50 (glycoprotein)
Corticosteroid-binding globulin (transcortin)	CBG	Glucocorticoids, progesterone	52 (glycoprotein)
Sex hormone-binding globulin	SHBG	Testosterone, estradiol	A dimer, each subunit 42
Thyroxine-binding globulin	TBG	Thryoxine (T_4), triiodothyronine (T_3)	63 (glycoprotein)
Retinol-binding protein	RBP	Retinol	41

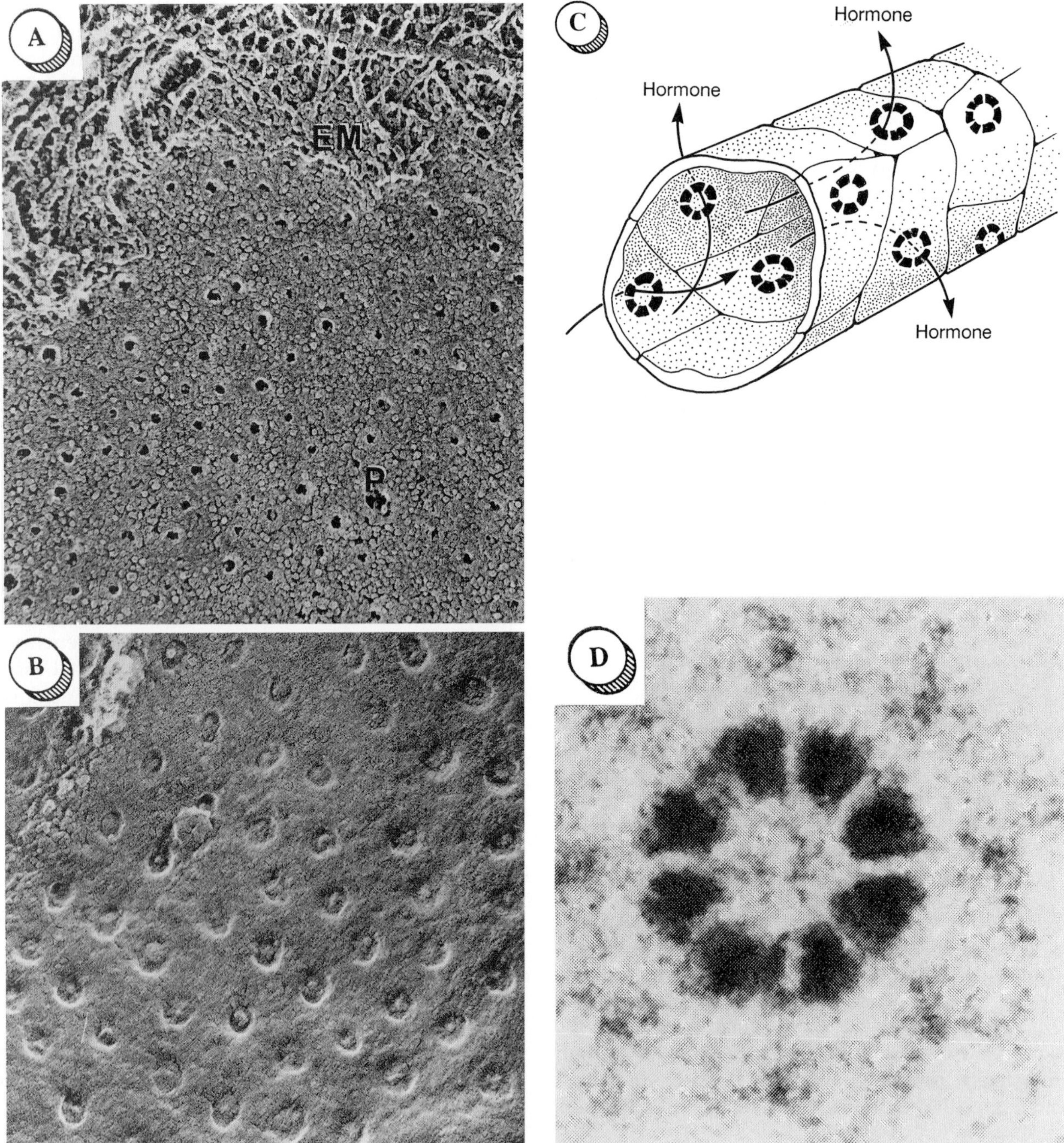

FIGURE 2-34 Example of capillary wall fenestration. The existence of fenestrations or windows in the capillary wall may allow plasma steroid transport proteins to exit the circulatory system and approach the outer cell membrane of the target cell for the steroid hormone in question. (A) Plasmalemmal vesicular openings in the P-face (P) of a diaphragm muscle capillary (×60,000). (B) Lightly etched luminal surface from a continuous capillary in the kidney medulla displays small depressions containing a central particle similar to surface views of muscle capillaries (×80,000). (C) Diagram representing the possible processes of exit or entry of hormones. (D) A fenestrated diaphragm in the endothelium of an adrenal cortex capillary. Note the eight, dark, wedge-shaped communicatory channels. Adapted with permission from Figures 2 and 6 of Bearer, E., and Orci, L. (1985). Endothelial fenestral diaphragms: A quick-freeze, deep-etch study. *J. Cell Biol.* **100,** 418–428.

pathways of steroid hormone inactivation and excretion.

B. Plasma Transport of Steroid and Thyroid Hormones

The steroid hormones are transported from their sites of biosynthesis to their target organs through the blood compartment. Due to the intrinsic low water solubility of all the steroid hormones, their transport is effected by a family of plasma transport proteins (see Table 2-7). Each plasma transport protein has a specific ligand-binding domain that allows the high-affinity binding of their cognate hormones. The exception is aldosterone, which is believed to circulate as the free steroid in the plasma compartment. Although all five plasma transport proteins are synthesized in the liver, they have no amino acid sequence homology. Also, there is no apparent sequence homology between the ligand-binding domain of the plasma transport protein and either the nuclear–cytosol receptor's ligand-binding domain or the substrate-bindiing domain of the P450 enzyme that generated the steroid.

In the plasma compartment, the steroid hormones move through the circulatory system bound to their transport protein. However, an important issue concerns the details of the mode of delivery of steroid hormones to their target cells. Since the "free" form of the steroid hormone is believed to be the form of steroid movement across the outer cell membrane of a target cell, it had been postulated that the steroid ligand bound to a plasma transport protein dissociates and diffuses first through the capillary wall and then through the outer wall membrane of target cells. However, as illustrated in Figure 2-34, it is apparent that the endothelial wall of capillaries contains fenestrations. Thus, it is conceivable that the plasma steroid transport protein could exit the capillary bed through a fenestration and move immediately adjacent ("dock?") to the outer cell membrane of the appropriate target cell for the steroid hormone in question.

Each plasma transport protein binds its ligand(s) with high affinity, the K_d values fall in the range of $(5–500) \times 10^{-8}$ *M*. In general, for any given hormone, the K_d of the hormone for its target organ receptor is 10–100× tighter than for its plasma transport protein.

C. Measurements of Rates of Secretion and Metabolic Clearance

The plasma concentration of a steroid is determined by the balance between biosynthesis and bioinactivation. With the ready availability of high specific activity

TABLE 2-8 Mean Secretion Rates, Plasma Concentrations, and Metabolic Clearance Rates (MCR) of Various Steroids[a]

Steroid	Secretion rate (mg/day)		Plasma concentration (μg/100 ml)		MCR (liters of plasma/day)	
	Men	Women	Men	Women	Men	Women
Cortisol	20	17	12		200	
Deoxycorticosterone (DOC)	0.24	0.5	0.024			
Aldosterone	0.19	0.14	0.0068			1630
Pregnenolone	9				1050	
Progesterone	0.6	2.9	0.03	0.14, 105	2920	
Testosterone	6.9	0.35	0.7	0.05	980	760
5α-Dihydrotestosterone (DHT)	0.32	0.075			500	
Androst-4-ene-3,17-dione	1.9	3.4	0.08	0.2	2300	
Dehydroepiandrosterone (DHEA)	3.0	0.7	0.50	0.48	950	
Dehydroepinandrosterone sulfate (DHEA)	15	10				
Estrone	0.11	0.11 (FP)[b] 0.15 (LP	0.036	0.004 (FP) 0.015 (Ov)	2300	1750
Estrone sulfate	0.077	0.10 (FP)	0.072	0.05 (FP) 0.31 (Ov) 0.22 (LP_2)	167	146
17β-Estradiol	0.06	0.12 (FP) 0.20 (LP)	0.0023	0.003 (FP) 0.057 (Ov) 0.04 (LP_2)	1700	1055

[a] This table was abstracted from Makin, H. L. ed. (1975). "Biochemistry of Steroid Hormones." William Clowes & Sons, Ltd., London.
[b] FP, follicular phase; Ov, ovulatory peak; LP, luteal phase; LP_2, secondary rise in LP following fall from value at Ov.

radioactive steroids, it has been possible to devise techniques of "compartmental" analysis that allow calculation of the secretion rate of steroids, as well as of their rate of disappearance from the plasma.

The metabolic clearance rate (MCR) of a steroid is the rate at which it is irreversibly inactivated. It is defined as "the volume of plasma cleared per unit of time." The plasma concentration (C) of any steroid will be determined by the balance between the rate of secretion (SR) and the MCR:

$$C = \frac{\text{SR}}{\text{MCR}} \cdot \frac{\text{nmol}}{\text{liter}} = \frac{(\text{nmol/hr})}{(\text{liters/hr})}.$$

Frequently, the MCR for steroids approximates the plasma flow through the liver (1500 liters/day in humans).

The half-life of a hormone is defined as the time required to reduce the circulating plasma concentration of the steroid by one-half (in the absence of new secretion). There is a reciprocal relationship between half-life and MCR. As the MCR increases, the half-life is reduced, and as the MCR is reduced, the half-life increases. Since catabolism of a steroid hormone occurs on the "free" form of the steroid, the relatively high K_d of the plasma transport protein for a steroid will effectively reduce its ready accessibility for clearance and thereby increase its half-life.

Table 2-8 summarizes blood concentration, secretion rate, and metabolic clearance rate for a number of important steroid hormones. These data emphasize the complex interplay between biosynthesis and biodegradation, which permits an organism to regulate the blood concentrations of these potent hormonal

Steriod class	Starting steroid	Inactivation steps	A:B ring junction	Steriod structure representations of excreted product	Principal conjugate present[a]
Progestins	Progesterone	1. Reduction of C-20 2a. Reduction of 4-ene-3-one or 2b. 3β-steriod dehydrogenase	(cis)	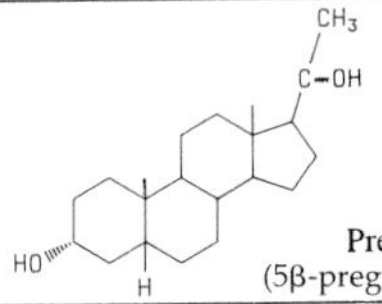 Pregnanediol (5β-pregnane-3α,20-diol)	G[a]
Estrogens	Estradiol	1. Oxidation of 17β-OH 2. Hydroxylation at C-2 with subsequent methylation 3. Further hydroxylation or ketone formation at a variety of positions, e.g., C-6, C-7, C-14, C-15, C-16, C-18		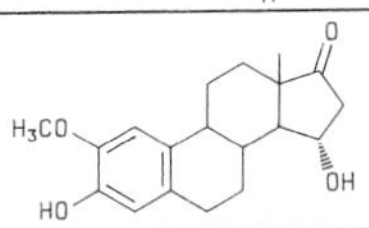 One of many possible compounds	G
Androgens	Testosterone	1. Reduction of 4-ene-3-one 2. Oxidation of C-17 oxo	(cis and trans)	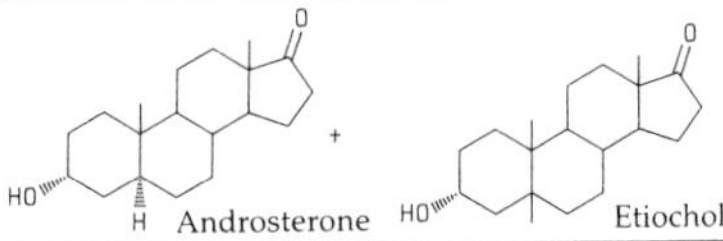 Androsterone + Etiocholanolone	G, S[a]
Glucocorticoids	Cortisol	1. Reduction of 4-ene-3-one 2. Reduction of 20-oxo group 3. Side chain cleavage	(trans)	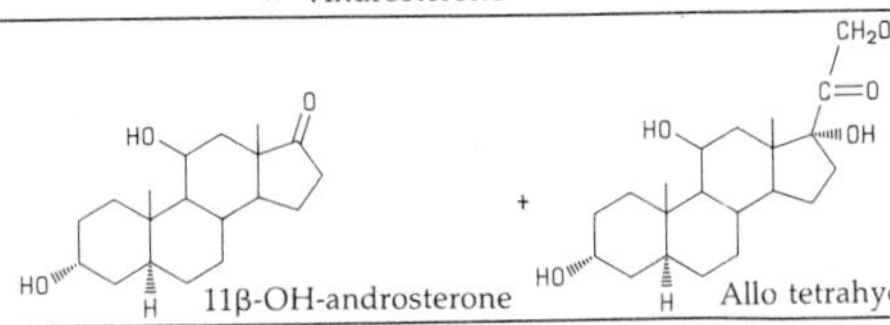 11β-OH-androsterone + Allo tetrahydrocortisone	G
Mineralocorticoids	Aldosterone	1. Reduction of 4-ene-3-one	(trans)	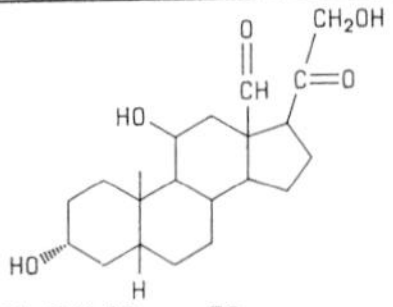 3α,11β,21-OH-20-oxo-5β-pregnane-18-al	G
Vitamin D metabolites	$1,25(OH)_2D_3$	1. Side chain cleavage between C-23 and C-24	—	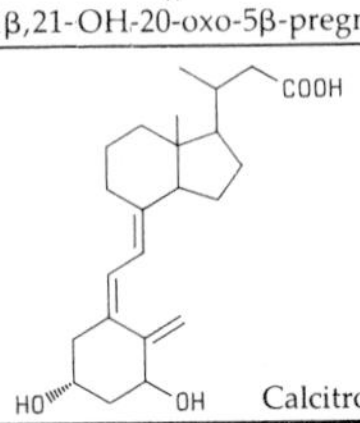 Calcitroic acid	?

[a]G, Glucuronide; S, sulfate.

FIGURE 2-35 Excretion pathways for steroid hormones.

[a] SO_4^{2-} + ATP $\xrightarrow[\text{sulfurylase}]{\text{ATP}}$ APS (adenosine 5'–phosphate) + pyrophosphate

[b] APS + ATP $\xrightarrow{\text{APS kinase}}$ PAPS + ADP

[c] PAPS + steroid—OH ⟶ steroid—O—SO_3^- + H^- + PAP [3', 5' phosphoadenosine]

FIGURE 2-36 Enzymatic steps involved with the production of "active" sulfate (PAPS).

agents. For a detailed consideration of this topic, the interested reader is referred to Tait and Horton (1966).

D. Inactivation of Steroid Hormones

Steroids are quite hydrophobic compounds, and many of the catabolic mechanisms not only inactivate the steroid hormone [i.e., markedly reduce its affinity for its receptor(s)] but also make the molecule more hydrophilic, thereby increasing its water solubility. The catabolic reactions occur mainly, although not exclusively, in the liver and are oxidative in nature. A significant increase in water solubility is effected by conjugation of the steroids with either sulfate or glucuronides; these steroid conjugates are excreted in large quantities in the urine.

Figure 2-35 summarizes some of the major excretory forms of these hormones for the six classes of steroids. The excreted species is a mixture of polyhydroxyl forms and the indicated glucuronide or sulfate. The sulfokinase enzymes have been shown to be present in the cytoplasm of the placenta, testes, adrenal cortex, and liver. These enzymes utilize "active sulfate" or phosphoadenosine phosphosulfate (PAPS) as substrate and catalyze reactions (shown in Figure 2-36).

The glucuronyl transferases that are localized in the liver endoplasmic reticulum use uridine–diphosphoglucuronic acid (UDPGA) as a substrate and catalyze the following reaction:

$$\text{UDPGA} + \text{sterol} \rightarrow \text{steroid glucuronide} + \text{UDP}.$$

References

A. Books

Dorfman, R. I., and Ungar, F. (1965). "Metabolism of Steroid Hormones." Academic Press, New York.

Fieser, L., and Fieser, M. (1959). "Steroids." Van Nostrand-Reinhold, Princeton, NJ.

Makin, H. L. (1975). "Biochemistry of Steroid Hormones." Blackwell, Oxford, UK.

B. Review Articles

Barton, D. H. R., and Cookson, R. C. (1956). Steroid conformational analysis. *Q. Rev. Chem. Soc.* **10,** 44–53.

Bloch, K. (1965). The biological synthesis of cholesterol. *Science* **150,** 19–23.

Brown, M. S., and Goldstein, J. L. (1976). Receptor-mediated control of cholesterol metabolism. *Science* **191,** 150–154.

IUPAC Rules of Steroid Nomenclature (1972). *Pure Appl. Chem.* **31,** 285–322.

Miller, W. L. (1988). Molecular biology of steroid hormone synthesis. *Endocrine Rev.* **9,** 295–318.

Nebert, D. W., Adesnik, M., Coon, M. J., Estabrook, R. W., Gonzales, J., Guengerich, F. P., Gunsalus, I. C., Johnson, E. F., Kemper, B., Levin, W., Phillips, I. R., Sato, R., and Waterman, M. R. (1987). The P450 gene superfamily: Recommended nomenclature. *DNA Cell Biol.* **6,** 1–51.

Popjak, G., and Cornforth, J. W. (1960). The biosynthesis of cholesterol. *Adv. Enzymol.* **22,** 281–335.

Rosner, W. (1990). The functions of corticosteroid-binding globulin and sex hormone-binding globulin: Recent advances. *Endocrine Rev.* **11,** 80–91.

Simpson, E. R., Mahendroo, M. S., Means, G. D., Kilgore, M. W., Hinshelwood, M. M., Graham-Lorenc, S., Amarneh, B., Ito, Y., Fisher, C. R., Michael, M. D., Mendelson, C. R., and Bulun, S. E. (1994). Aromatase cytochrome P450, the enzyme responsible for estrogen biosynthesis. *Endocrine Rev.* **15,** 342–355.

Tait, J. F. (1963). The use of isotopic steroids for the measurement of production rates *in vivo*. 1. *Clin. Endocrinol. Metab.* **23,** 1285–1297.

Tait, J. F., and Horton, R. (1966). Steroid compartmental analysis. *In* "Steroid Dynamics" (G. Pincus, J. F. Tait, and T. Nakao, eds.), p. 393. Academic Press, New York.

White, P. C., Curnow, K. M., and Pascoe, L. (1994). Disorders of steroid 11β-hydroxylase isoenzymes. *Endocrine Rev.* **15,** 421–438.

C. Research Papers

Bearer, E., and Orci, L. (1985). Endothelial fenestral diaphragms: A quick-freeze, deep etch study. *J. Cell Biol.* **100,** 418–428.

Black, S. M., Harikrishna, J. A., Szklarz, G. D., and Miller, W. L. (1994). The mitochondrial environment is required for activity of the cholesterol side-chain cleavage enzyme, cytochrome P450scc. *Proc. Natl. Acad. Sci. USA* **91,** 7247–7251.

Black, S. M., Szklarz, G. D., Harikrishna, J. A., Lin, D., Wolf, R., and Miller, W. L. (1993). Regulation of proteins in the cholesterol side-chain cleavage system in JEG-3 and Y-1 cells. *Endocrinology* **132,** 539–545.

Cahn, R. S., Ingold, C. K., and Prelog, V. (1966). Specification of molecular chirality. *Angew. Chem., Int. Ed. Engl.* **5,** 385–415. (This paper describes the "sequence rule" for determination of *R* vs *S* asymmetry.)

Clark, B. J., Wells, J., King, S. R., and Stocco, D. M. (1994). The purification, cloning, and expression of a novel luteinizing hormone-induced mitochondrial protein in MA-10 mouse Leydig tumor cells, characterization of the steroidogenic acute regulatory protein (StAR). *J. Biol. Chem.* **269,** 28314–28322.

Honda, S., Morohashi, K., Nomura, M., Takeya, H., Kitajima, M., and Omura, T. (1993). Ad4BP regulating steroidogenic P-450 gene is a member of steroid hormone receptor superfamily. *J. Biol. Chem.* **268,** 7494–7502.

Lin, D., Sugawara, T., Strauss, J. F., III, Clark, B. J., Stocco, D. M., Saenger, P., Rogol, A., and Miller, W. L. (1995). Role of steroidogenic acute regulatory protein in adrenal and gonadal steroidogenesis. *Science* **267,** 1828–1830.

Nomura, M., Morohashi, K., Kirita, S., Nonaka, Y., Okamoto, M., Nawata, H., and Omura, T. (1993). Three forms of rat CYP11B genes: 11β-Hydroxylase gene, aldosterone synthase gene, and a novel gene. *J. Biochem.* (*Tokyo*) **113,** 144–152.

Van Tamelen, E. E., Leopold, E. J., Marason, S. A., and Walspe, H. R. (1982). Action of 2,3-oxidosqualene-lanosterol cyclase on 15-nor-18,19-dihydro-2,3-oxidosqualene. *J. Am. Chem. Soc.* **104,** 6479–6480.

CHAPTER

3

Hypothalamic Releasing Hormones

I. INTRODUCTION

The pituitary gland has often been called "the master gland" because so many important functions are governed by the hormones produced there. The hypothalamic releasing hormones function to control the secretions of the pituitary hormones, mainly the anterior pituitary hormones, either positively or negatively. The releasing hormones can be thought of as the main stimuli to generate specific release of pituitary hormones, although we will learn that there are many other factors that are being found to contribute to this control. The function of the releasing hormones can be put into context by briefly considering the overall hormonal or humoral cascade system, starting with the central nervous system. This cascade is seen in overview in Figure 3-1. There are many signals originating either outside or inside the body that are mediated by the central nervous system. Thus, many changes in the environment ultimately can stimulate the secretion of releasing hormones, which produce effects in the body to adapt to the change. Likewise, signals from within the organism, such as those from internal cycling systems, also will trigger this cascade system so that end product hormones from distant target glands can be released. Thus, in Figure 3-1 it is seen that the cascade starts with signals, either from the outside environment or from within the body. These signals may be filtered and transmitted through central nervous system neurons, whose nerve endings may activate other neurons in the hypothalamus. External or internal signals are often mediated by the limbic system or some other system of the brain and may result in the generation of electrical or chemical signals to the hypothalamus. Such signals may produce depolarization of nerve endings, affecting the secretion of releasing hormones. These hormones gain access to local blood circulation through fenestrations (thin walls) in small vessels and are carried to the adenohypophysis (anterior pituitary) by the hypothalamic–hypophyseal portal system shown in Figure 3-2. The releasing hormone binds to a specific receptor on a particular cell membrane in the anterior pituitary. The binding sets off a chain of events, culminating in an elevation of intracellular calcium ions. A release of preformed anterior pituitary hormone follows. This

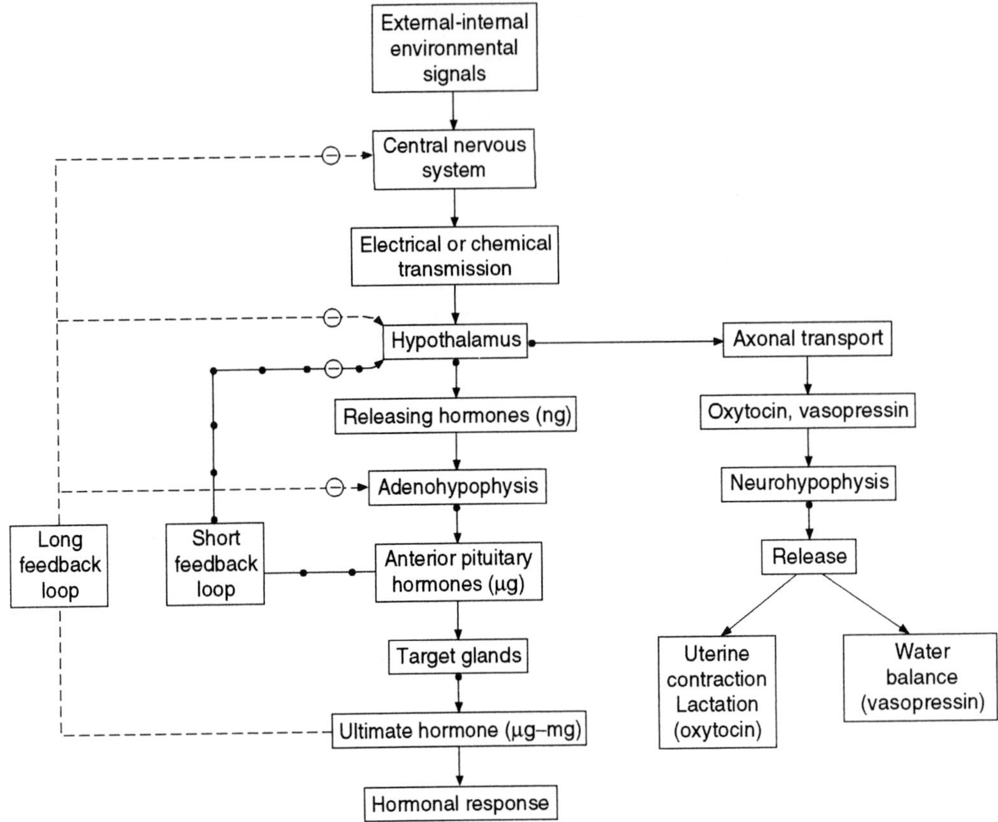

FIGURE 3-1 Many hormonal systems involve the hypothalamus. This figure shows a cascade of hormonal signals starting with an external or internal environmental signal. This is transmitted first to the CNS and may involve components of the limbic system, such as the hippocampus and amygdala. These structures innervate the hypothalamus in a specific region, which responds with secretion of a specific releasing hormone, usually in nanogram amounts. Releasing hormones are transported down a closed portal system connecting the hypothalamus and the anterior pituitary cell membrane receptor and cause the secretion of specific anterior pituitary hormones, usually in microgram amounts. These access the general circulation through fenestrated local capillaries and bind to specific target gland receptors. The interactions trigger the release of an ultimate hormone in microgram to milligram daily amounts, which generates the hormonal response by binding to receptors in several target tissues. In effect, this overall system is an amplifying cascade. Releasing hormones are secreted in nanogram amounts, and they have short half-lives on the order of a few minutes. Anterior pituitary hormones are often produced in microgram amounts and have longer half-lives than releasing hormones. Ultimate hormones can be produced in milligram daily amounts with much longer half-lives. Thus, the products of mass × half-life constitute an amplifying cascade mechanism. With respect to differences in the mass of hormones produced from hypothalamus to target gland, the range is nanograms to milligrams, or as much as 1 million-fold. When the ultimate hormone has receptors in nearly every cell type, it is possible to affect the body chemistry of virtually every cell by a single environmental signal. Consequently, the organism is in intimate association with the external environment, a fact that we tend to under emphasize. Arrows with a black dot at the origin indicate a secretory process. Long arrows studded with black or open dots indicate feedback pathways. Reproduced from "Textbook of Biochemistry" (Devlin, T. M., ed.), 3rd edition. Wiley-Liss, New York (1992).

hormone is released into the bloodstream, gaining access to the circulation through fenestrations in nearby blood vessels. Subsequently, it binds to a specific receptor on the cell membrane of a target gland cell. Within the target cell, a chain of events similar to those following the binding of releasing hormone to anterior pituitary cell occurs, culminating in the release of the ultimate hormone. The ultimate hormone has receptors in many tissues and produces the systemic hormonal response.

Hypothalamic neurons synthesize and release the releasing hormones and the release-inhibiting hormones. The signals to the hypothalamus may be either electrical or chemical or both, and often amine neurotransmitters are involved. The locations of the hypothalamic peptidergic neurons that secrete the releasing

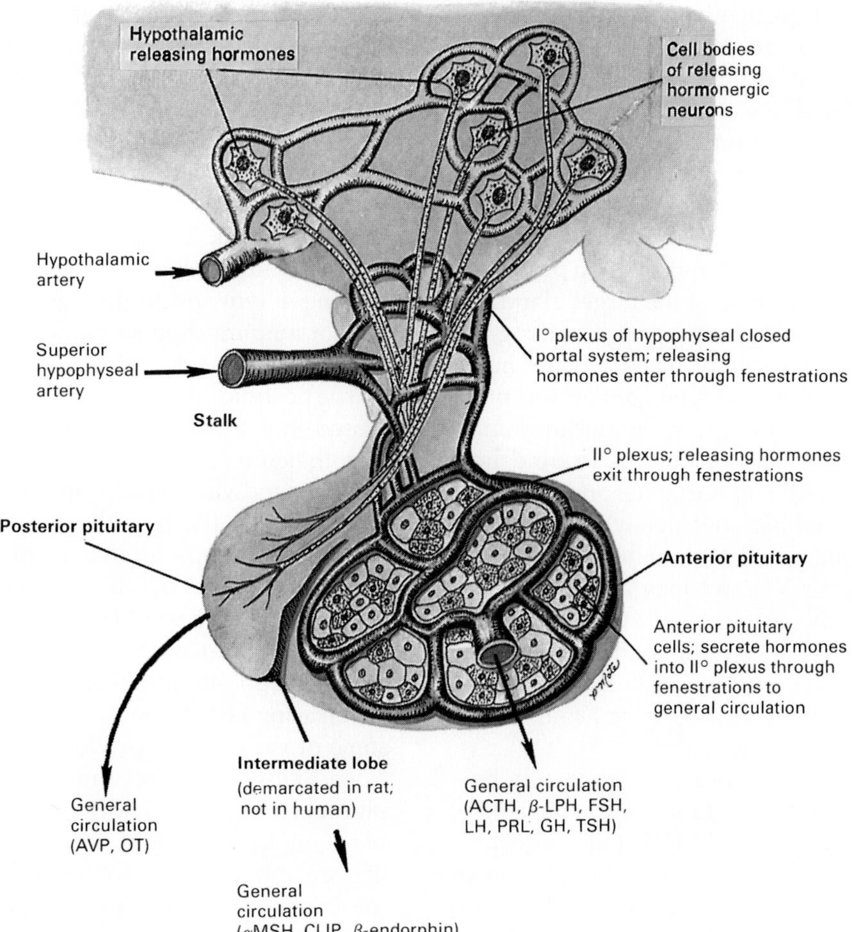

FIGURE 3-2 Diagram showing the hypothalamus with nuclei in various places in which the hypothalamic releasing hormones are synthesized. Also shown is the major vascular network consisting of a primary plexus, where releasing hormones enter circulation through fenestrated vessels, and the secondary plexus in the anterior pituitary, where the releasing hormones are transported out of circulation again through fenestrations in the vessels to the region of the anterior pituitary target cells. This figure also shows the resultant effects of the actions of the hypothalamic releasing hormones, causing the release into general circulation of the anterior pituitary hormones.

hormones will be summarized later. Once stimulation to a specific region of the hypothalamus has occurred, a specific releasing hormone will be secreted from a neuronal nerve ending, often in the area of the primary plexus of the closed portal system connecting the hypothalamus with the anterior pituitary. The primary plexus is a group of capillaries with fenestrations or functional "openings" that allow the passage of releasing hormone peptides that range from 3 to more than 40 amino acids in size into the closed portal circulation. In addition to the peptidic releasing hormones, dopamine may function in a primary or secondary role as a regulator of prolactin release, and dopamine may affect the release of other pituitary hormones. Once inside the plexus, the releasing hormones circulate down the portal vessel to the secondary plexus, which, analogously to the primary plexus, allows for outward passage of the releasing hormone peptides via fenestrations into the extracellular space in the vicinity of the anterior pituitary cells containing the anterior pituitary hormones. Specific receptors for the releasing hormones are located on the surface of specific cells of the anterior pituitary. After binding of the specific releasing hormone to its cognate receptor, signal transduction events take place that culminate in the release of the appropriate pituitary hormone(s). The closed portal circulation is needed to maintain the concentration of the releasing hormones, which are secreted in nanogram amounts from the hypothalamus following the central nervous system signal. The appropriate anterior pituitary hormone released in response to the action of the releasing hormone is secreted into the extracellular space and subsequently gains access to the general circulation; here fenestrations in local small vessels are also needed to pass the fairly large hormonal proteins. These are secreted in microgram amounts and then circulate until they reach a distant target cell containing a specific receptor in its cell membrane for the anterior pituitary hormone. The interaction of pituitary hormone and cognate receptor on the target cell is followed by signal transduction processes, which culminate in the release of the target gland cell hormone, usually in milligram or high microgram quantities. The terminal hormone then circulates in the blood until it finds target cells that contain receptors either in the cell membrane (polypeptide hormones) or in the cellular cytoplasm or nucleus (steroid hormones). The anterior pituitary hormones and the terminal hormones both feed back negatively to inhibit further release of pituitary and hypothalamic hormones in the cascade (Figure 3-1) and may also shut down the initial signaling mechanism in the central nervous system. When the cycle has been completed and the terminal hormone has generated its cellular and systemic effects, the entire system is once again poised for a subsequent round of activity.

The anterior pituitary (trophic) hormone may be a growth factor for the terminal gland cells. For example, the anterior pituitary hormone, ACTH (adrenocorticotropic hormone), in addition to causing the subsequent secretion of cortisol from the adrenal gland, is also required for the survival of the cells of the adrenal (zona fasciculata cells) that produce cortisol. Thus, if a patient is treated with pharmacological doses of cortisol, there is the risk of destroying adrenal gland cells by way of the negative feedback effects of the administered cortisol (or other glucocorticoid) on the anterior pituitary and higher functions in the cascade (Figure 3-1). As a result of the negative feedback effects of administered cortisol, the circulating levels of ACTH will fall dramatically. A fall in the circulating level of ACTH would deprive adrenal cortical cells of their growth factor, resulting in the destruction of the cortisol-producing cells. It is for this reason that glucocorticoid therapy is usually given on alternate days with the expectation that the negative feedback on ACTH release will recover during the time the treatment is discontinued and the cortisol-producing adrenal cells will survive. It is not certain whether this protection always occurs, especially when large amounts of the drug are given over a long period.

Proceeding down the cascade (Figure 3-1) from the central nervous system to the terminal hormone, there are increasing masses of the hormones released. The releasing hormones from the hypothalamus are secreted in nanogram amounts, the anterior pituitary hormones in microgram amounts, and the terminal hormone in microgram to milligram amounts. In addition, the stability of these hormones, proceeding down the cascade, seems to increase, as the releasing hormones have half-lives in the range of 3–10 min, the anterior pituitary hormones have half-lives of 20 min or longer, and the terminal hormones are usually quite stable in blood. The product of the increasing mass and increasing stability of the hormones produced proceeding downward in this system suggests a high degree of amplification stemming from a unique event, from either the external or internal environment.

The posterior pituitary hormones, which will be discussed in a subsequent chapter, also derive from the hypothalamus (Figure 3-1). The main hormones are oxytocin and vasopressin, and they are synthesized in cell bodies in the hypothalamus and transported to the nerve endings located in the posterior pituitary awaiting release, which is triggered by appropriate signals to be discussed later. Of particular interest are suggestions that oxytocin may be degraded to form active compounds and that some of these may function as releasing factors. For this to be true, the axons from some of the neurons producing oxytocin would have to end close to the secondary plexus of the anterior pituitary. Also, it is known that arginine vasopressin plays an auxiliary role in the release of ACTH from the anterior pituitary in conjunction with CRH, and the axons from vasopressinergic neurons would be expected to end in the same region, in addition to the posterior pituitary. The main roles of oxytocin and vasopressin in regulating uterine contraction and lactation and water balance, respectively, will be discussed later.

Unfortunately, modern literature has introduced a plethora of new names for the releasing hormones. Such names as gonadorelin, buserelin, goserelin, leuprolide, nafarelin, and triptorelin refer to GnRH. Protirelin refers to PRF/TRH. Somatoliberin and somatocrinin refer to GRH, and corticoliberin refers to CRH. For simplicity, the contractions are used as well as their equivalents, e.g., corticotropin-releasing hormone or CRH.

II. ANATOMICAL, MORPHOLOGICAL, AND PHYSIOLOGICAL RELATIONSHIPS

A. Hypothalamus

As seen from Figure 3-3, the hypothalamus is located below the third ventricle of the brain, just above the median eminence. The functions of the hypothalamus are intimately connected to the pituitary, which

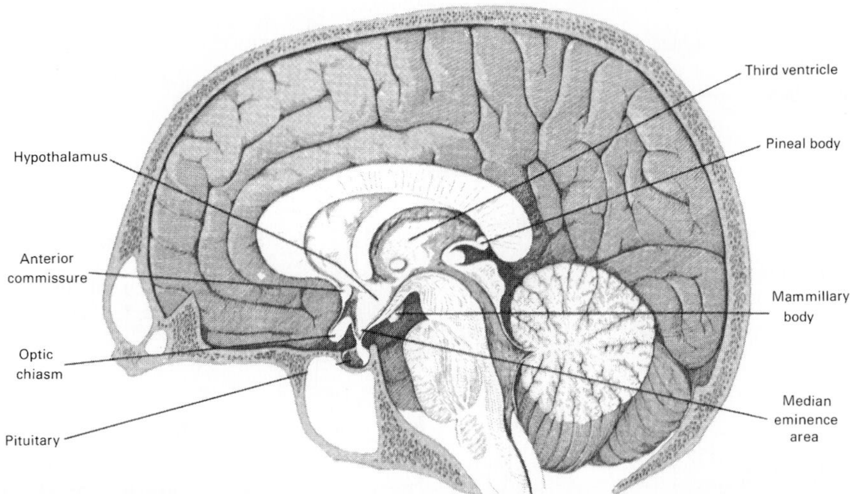

FIGURE 3-3 Lateral view of the brain showing the relationship of the pituitary gland to the hypothalamus. Reproduced from Krieger, D. T. (1980). The hypothalamus and neuroendocrinology. *In* "Neuroendocrinology" (D. T. Krieger and J. C. Hughes, eds.), pp. 3–12. Sinauer Associates, Sunderland, Massachusetts.

lies below the hypothalamus at the end of the delicate infundibular stalk.

In the process of development of the brain, a structure known as the diencephalon appears. It is referred to as the second division of the brain and is a relay center for the cerebral hemispheres. It has a central cavity known as the third ventricle (Figure 3-3). The diencephalon, the thalamus, is a relay center for tracts connecting the cerebral hemisperes. The hypothalamus contains optic chiasma where the optic nerves cross at their entrance to the brain, tuber cinerium, infundibulum, hypophysis, and mammillary region. The tuber cinerium contains gray matter behind the optic chiasma involved with olfaction. It continues ventrally as the infundibular stalk whose cavity is an extension of the third ventricle connecting the pituitary (see Chapter 5).

B. Blood Supply of Hypothalamus

Small branches of the anterior cerebral artery and the posterior communicating artery extend to form a network within the hypothalamic region. The supraoptic and paraventricular nuclei are situated in a dense capillary network supplied by branches of the supraoptic–paraventricular arteries. Small branches from the superior hypophyseal artery also serve the supraoptic nucleus. The nuclei of the tuber cinerium have a less dense vasculature, partly supplied by the posterior communicating arteries. Thus, two large pairs of arteries supply this region. There are no direct connections between vessels of the hypothalamic nuclei and the pituitary. The hypophysis is supplied by the superior and inferior hypophyseal arteries, which are branches of the internal carotid (Figure 3-2). The vasculature of the hypothalamus and pituitary is shown vividly in Figure 3-4.

C. Electrical Connections from the Brain to the Hypothalamus

The systems and circuitry between the brain and the hypothalamus (afferent) and the electrical connections from the hypothalamus to other structrures (efferent) are extremely complex and probably unnecessary to the subject matter of this book. Naturally, many of the electrical signals conducted by these fibers will either stimulate or depress the release of releasing hormones from the nerve endings of the peptidergic nerve terminals in which they are stored.

D. The Neuron

Neurons, whether they manufacture small molecular neurotransmitters such as norepinephrine or polypeptides, have the general structure diagrammed in Figures 3-5 and 3-6. The cell body is the site of synthesis and packaging of the releasing hormones. Some of the amine neurotransmitters may be synthesized in the nerve endings. The secretory granules are transported down the axon, which in some cases may be extremely long, into the nerve ending where a signal will cause the exocytotic release of the granules (defined in Chapter 4). The signal can be electrical, presumably trans-

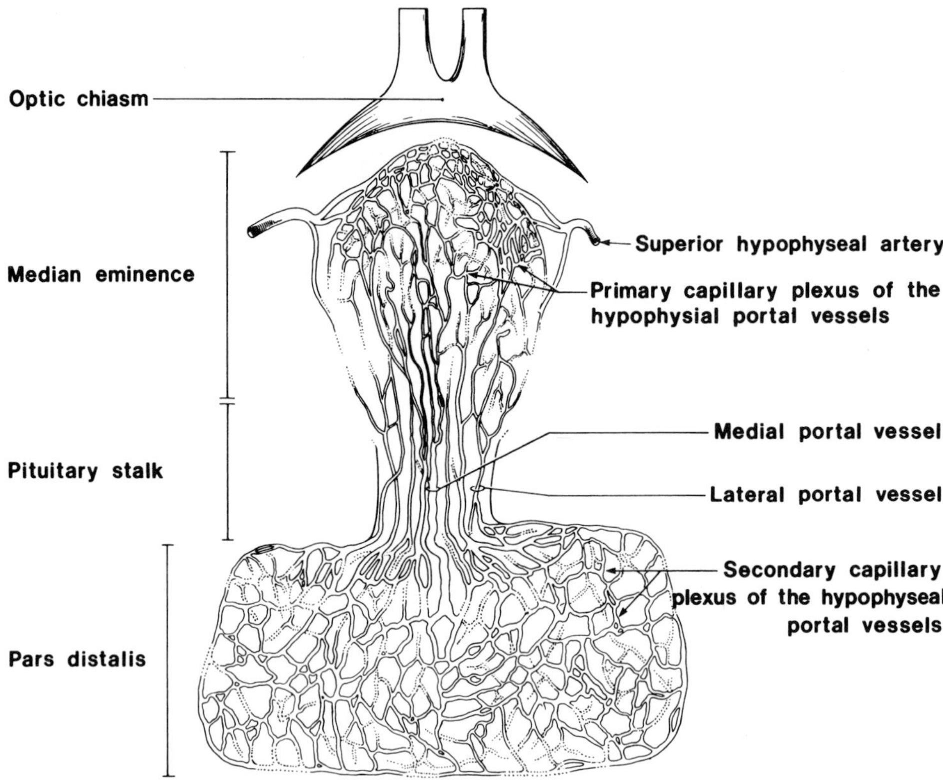

FIGURE 3-4 Schematic representation of the hypothalamic–hypophyseal portal vasculature of the rat. Reproduced from Porter, J. C. et al. (1983). *Vitam. Horm.* **40;** 145–174 with permission.

mitted through fibers traveling the length of the axon and causing depolarization at the nerve ending and Ca^{2+} uptake there, which appears to be essential in the exocytosis process.

The nerve endings may also be regulated by an interneuron forming a synapse with the nerve ending, as shown in Figure 3-7. In this case, receptors for the neurotransmitter, such as enkephalin in this example, would be located on the surface of the nerve ending in the synaptic region. Through a second messenger, an action would occur in the nerve ending that influences, positively or negatively, the release of the secretory material stored there.

The targets of the releasing hormones are the cells of the adenohypophysis in the anterior pituitary, which secrete the anterior pituitary hormones. The anatomy and cellular characteristics of the anterior pituitary are presented in Chapter 5.

III. CHEMISTRY

The releasing hormones that have been sequenced will be referred to as hormones, whereas the others, whose activity is measurable but whose primary structure is unknown, will be referred to as "factors." There are two kinds of releasing hormone systems in the hypothalamus: one in which a single releasing hormone appears to positively control the secretion of a given anterior pituitary hormone, and the other in which a pair of releasing hormones, one acting positively and one acting negatively, modulates the secretion of a specific anterior pituitary hormone (Table 3-1). As time goes on, negative controlling factors are discovered, so that in fact, all anterior pituitary hormones may be controlled ultimately in their secretion by a primary set of positive and negative releasing hormones.

In the single-controlling set are thyrotropin-releasing hormone (TRH), gonadotropin-releasing hormone (GnRH), and corticotropin-releasing hormone (CRH). All of these act positively on the release of thyroid-stimulating hormone from thyrotropic cells of the anterior pituitary (TRH), prolactin from lactotrophic cells of the anterior pituitary (PRL), luteinizing hormone (LH) from luteotrophic cells of the anterior pituitary (GnRH), follicle-stimulating hormone (FSH) from folliculotrophic cells of the anterior pituitary (GnRH), and finally adrenocorticotropin from corticotrophic cells of the anterior pituitary (CRH). Some evidence suggests that LH and FSH can be secreted from the same cell, that is, that luteotrophs and folliculotrophs may be one

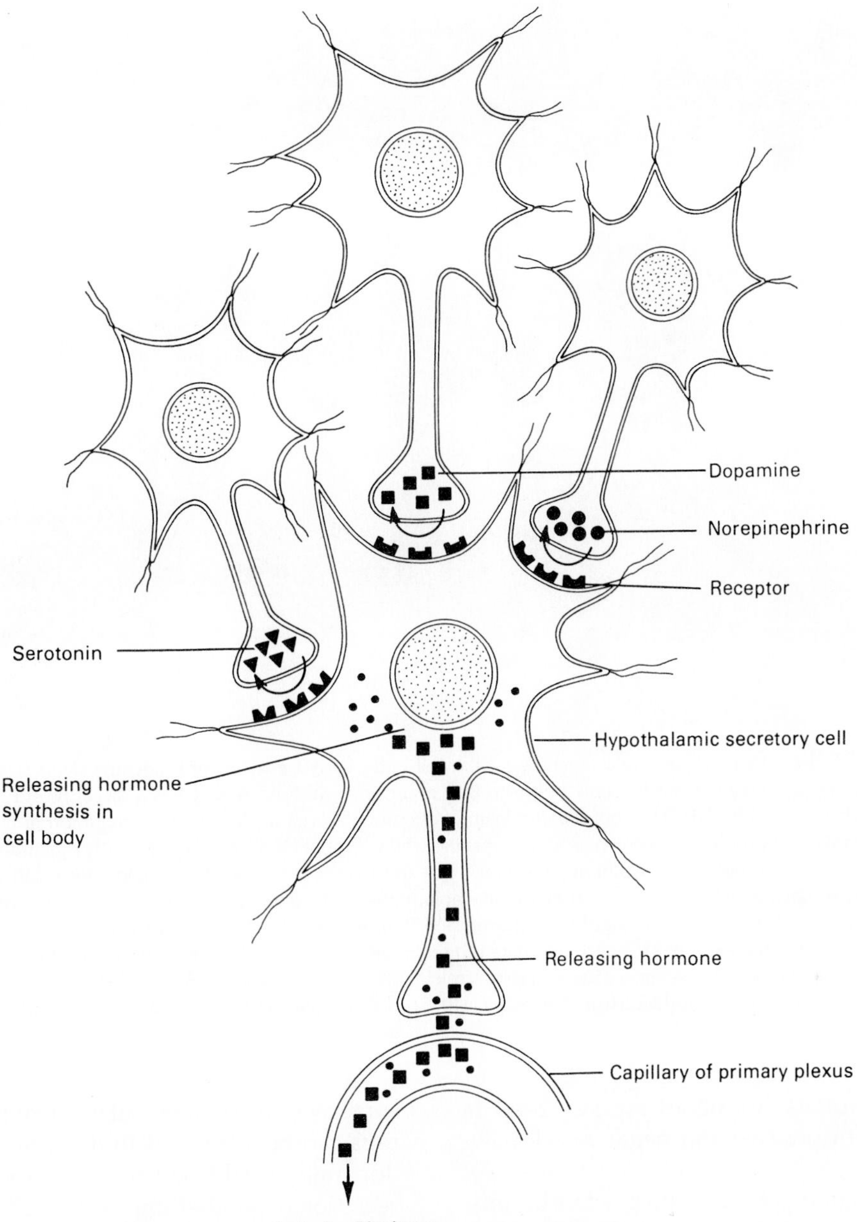

FIGURE 3-5 Drawing of a neuron (center) that is synthesizing and secreting a releasing hormone and its regulation by aminergic neurons (interneurons), which in this case secrete serotonin, dopamine, and norepinephrine. The regulatory interneurons synapse with the cell body in this illustration. The catecholamines are usually synthesized in the nerve ending. The releasing hormones and precursors are synthesized in the cell body and transported down the axon to the nerve terminal where the signal is awaited for release. Afterward, the releasing hormone enters the closed portal circulation through capillary fenestrations in the primary plexus. After transport down the portal vessel, the hormone empties from the secondary plexus into the anterior pituitary. This figure was redrawn and modified from Frohman, L. A. (1980). Neurotransmitters as regulators of endocrine function. *In* "Neuroendocrinology" (D. T. Krieger and J. C. Hughes, eds.), pp. 44–58. Sinauer Associates, Sunderland, Massachusetts.

and the same or that both types of cells may exist. In the set of dual-controlling releasing hormones, there are growth hormone-releasing hormone (GRH) and growth hormone release-inhibiting hormone (GIH or somatostatin) and prolactin-releasing factor (PRF) and prolactin release-inhibiting factor (PIF). The structures of the releasing hormones with known sequences are shown in Figure 3-8. Ultimately, it may be discovered

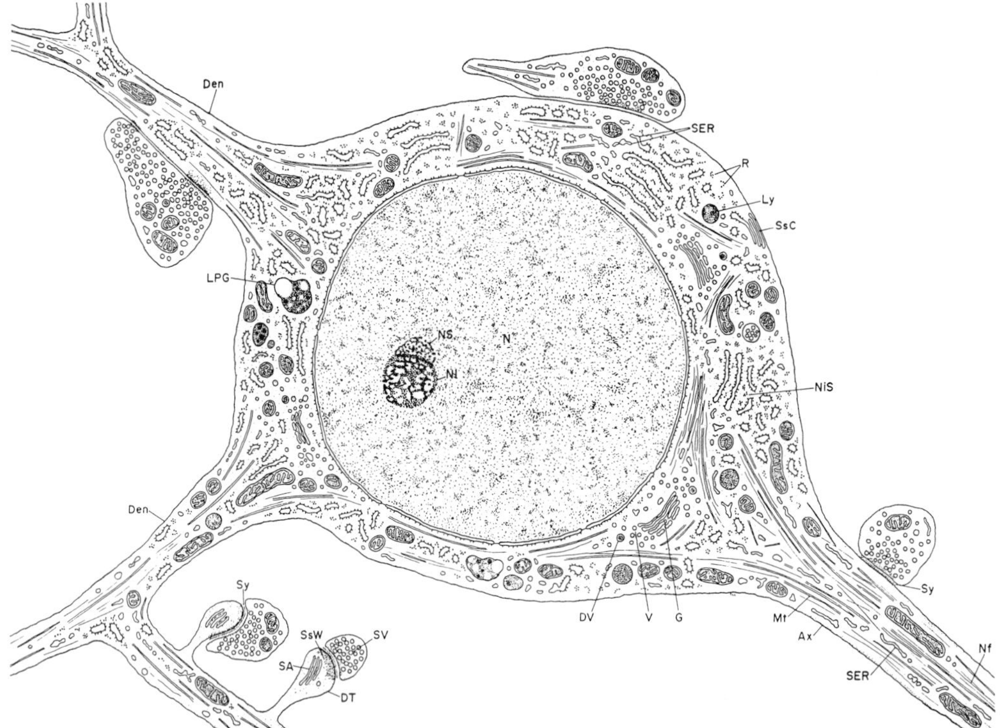

FIGURE 3-6 Drawing of a neuron showing the cell body with a number of branching dendrites (Den). The axon (Ax) extends from the pole opposite that giving rise to dendrites. It has a large, round nucleus (N) with a nucleolus (NI). The rough endoplasmic reticulum is organized into an aggregation of cisternae studded with ribosomes (R), known as Nissl substance (NiS). Smooth-surfaced endoplasmic reticulum (SER) is extensive and tubular. The Golgi apparatus (G) is a short stack of flattened cisternae with small vesicles (V); dense core vesicles (DV) may contain catecholamines. Lysosomes (Ly) are numerous. Intermediate stages between lysosomes and lipofuscin granules (LPG) are found. Microtubules (Mt) course from the perikaryon into the axon (Ax). Neurofilaments (Nf) and smooth endoplasmic reticulum run into the axon. Other abbreviations: Sy, synapse; SV, synaptic vesicle; SA, spine apparatus, SsW, subsynaptic web; SsC, subsurface cisternae. Reproduced from Lentz, T. L. (1971). "Cell Fine Structure." Saunders, Philadelphia, Pennsylvania.

that all anterior pituitary hormones are governed in their release by both positive and negative releasing factors.

GRH is a polypeptide. β-Endorphin is active in causing the release of growth hormone and PRL and in inhibiting the release of gonadotropins and TSH. β-Endorphin arises from the proopiomelanocortin precursor in corticotrophic cells of the anterior pituitary. It is a cleavage product of β-lipotropin, which is directly translated to proopiomelanocortin. It is possible that β-endorphin could stimulate GH release indirectly by increasing the release of GRH or inhibiting the release of somatostatin. A peptide and its partial degradation products have been discovered by the R. Guillemin group in human pancreatic carcinoma. This 44-amino acid peptide possesses growth hormone-releasing activity and appears to be human GRH. Its structure is shown in Figure 3-8.

PIF is closely related to dopamine, and this neurotransmitter may be identical to PIF. However, there is some evidence that not all PIF activity in the hypothalamus can be attributed to dopamine. Again, however, dopamine could be acting to modify (stimulate) the secretion of another unique PIF. The structure of dopamine is

HO HO NH_2

CRH, the corticotrophic releasing hormone, is a polypeptide, and there is an auxiliary hormone to CRH, which is vasopressin. Vasopressin is synthesized in the hypothalamus and can release ACTH when injected directly into the brain, but depends upon CRH; it stimulates the activity of CRH. The structure of vasopressin is given in Chapter 4.

α-MSH normally derives from the intermediate pituitary cells under the control of dopamine, not CRH. There may be certain disease conditions leading to abnormally produced α-MSH, perhaps from the unex-

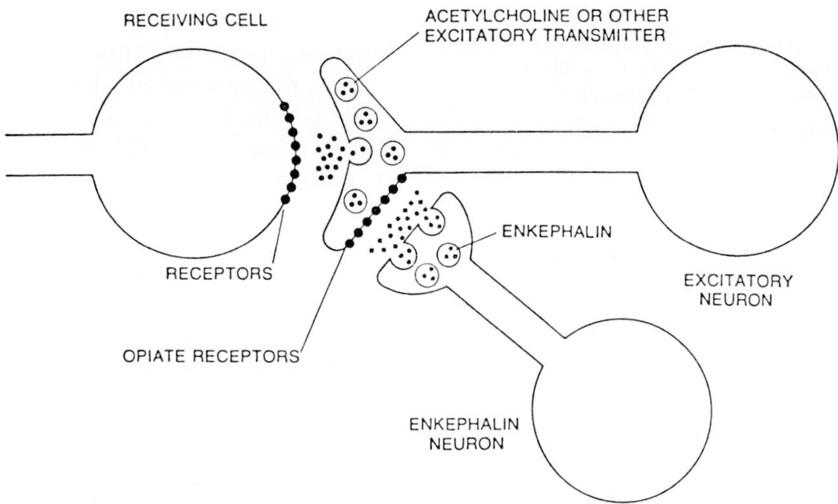

FIGURE 3-7 Diagram of a neuron (in this case, cholinergic) whose secretion is regulated by an enkephalinergic interneuron interacting at the nerve ending to modify secretion as mediated by opiate receptors. Reproduced from Agnew, D. (1970). *City of Hope Quart.* **8,** 6–10.

pected breakdown of ACTH derived from the corticotroph. Proteolytic cleavage products of oxytocin and vasopressin have been shown to result from the action of proteolytic enzymes present in brain synaptic membrane preparations. Some of these products may be active in the release of various pituitary hormones. It is presumed that some or all of these products will also have neurotransmitter or other direct functions in the CNS. In contrast to known brain aminopeptidases, synaptic (nerve ending) membrane-associated enzymes can cleave oxytocin or vasopressin without a prior requirement for reduction of the disulfide bridge. Releasing hormones (excluding GRH and CRH, which need more information) are not synthesized from mRNAs directly on polysomes, but rather evolve through the cleavage of proproteins of considerably larger size. This is also true of some anterior pituitary hormones, such as ACTH and β-lipotropin, and their potentially active breakdown products that derive from a single protein precursor, proopiomelanocortin. In humans, α-MSH derives from the intermediate pituitary cells from proopiomelanocortin under dopamine

TABLE 3-1 Well-Characterized Hypothalamic Releasing Hormones–Factors[a]

	Releases anterior pituitary hormone(s)	Inhibits release of anterior pituitary hormone(s)
Hypothalamic positive-acting single hormone		
Thyrotropin-releasing hormone (TRH, protirelin)	TSH, PRL	
Gonadotropin-releasing hormone (GnRH, gonadorelin, buserelin, goserelin, leuprolide, nafarelin, triptorelin)	LH, FSH (same cell)	
Corticotropin-releasing hormone (CRH, corticoliberin)	ACTH, β-lipotropin, β-endorphin	
Dual-acting releasing hormone		
Growth hormone-releasing hormone (somatocrinin, GRH, somatoliberin)	GH	
Growth hormone release-inhibiting hormone (somatostatin, GIH, SRIF)		GH
Prolactin-releasing factor (PRF protirelin)[b]	PRL	
Prolactin release-inhibiting factor (PIF)[b]		PRL

[a] In humans, α-MSH is probably released from pars intermedia-like cells under the positive action of dopamine, not CRH. In these cells, the ACTH moiety is further broken down to α-MSH and CLIP, and these hormones are secreted together with β-endorphin (breakdown product of β-lipotropin).

[b] These activities may derive from the neurointermediate lobe (at least in higher animals). PRF (a polypeptide of MW less than 5000) does not appear to be oxytocin, AVP, or β-END); however, PIF could be related to oxytocin.

TRH
(Thyrotropin-releasing hormone releases TSH and PRL; it is not clear whether there is a separate PRF. PIF may be dopamine or some other hormone.)

pGlu*-His-Pro-NH_2**

GnRH
(Gonadotropic releasing hormone or LHRH; releases LH and FSH, often from the same cell; residue 3, Trp, is important in association with the receptor.)

1 3 10
pGlu-H-Trp-S***-Y-G-L-R-P-Gly-NH_2

GIH or SRIF
(Somatotropin release-inhibiting factor or somatostatin; inhibits release of GH and affects release of pancreatic and other hormones.)

1 14
A-G-C-K-N-F-F-W-K-T-F-T-S-C

CRH
(Corticotropin releasing hormone; releases ACTH and β-lipotropin (β-LTH); some β-LTH forms β-endorphin.)

1 20
S-Q-E-P-***P-I-S-L-D-L-T-F-H-L-L-R-E-V-L***-E-M-T-K-A-D-Q-L-A
5-19 important for binding to receptor
30 40 41
Q-Q-A-H-S-N-R-K-L-L-D-I-Ala-NH_2

GRH
(Growth hormone releasing hormone; somatocrinin; releases growth hormone)

1 10 20
Y-A-D-A-I-F-T-N-S-Y-R-K-V-L-G-Q-L-S-A-R-K-L-L-Q-D-I-M-S-
30 40 44
R-Q-Q-G-E-S-N-Q-E-R-G-A-R-A-R-Leu-NH_2

*Denotes pyroglutamyl

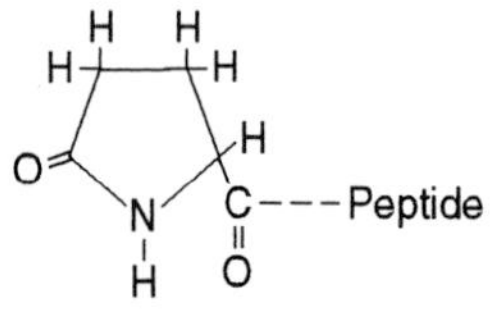

**Denotes amide of C-terminal amino acid; prolinamide in TRH, glycinamide in GnRH, alanylamide in CRH or leucinamide in GRH.

FIGURE 3-8 Sequences of known hypothalamic releasing and release-inhibiting hormones (see also Figure 8-10). One-letter abbreviations of amino acids are used in this figure. These abbreviations are listed in Appendix E.

control separate from CRH. In many vertebrates, the *pars intermedia* is a discrete structure and contains cells that produce proopiomelanocortin under the positive control of dopamine but not CRH, as in the corticotroph. Whereas the corticotrophic products, after CRH stimulation, are ACTH and β-lipotropin (and some breakdown of β-lipotropin to β-endorphin and γ-lipotropin), the pars intermedia cells, under the influence of dopamine, produce α-MSH and CLIP (the hydrolytic products of ACTH) and β-endorphin. It may be that humans retain these pars intermedia cells, but not in a discrete easily recognized structure.

It can be noted from Figure 3-8 that it is common for releasing hormones to contain unusual constituents at either the N- or C-terminus or at both ends. Pyroglutamate often appears at the N-terminus, and an amidated amino acid frequently appears at the C-terminus. These hormones have short half-lives in blood (ca. 3–7 min), and there are enzymes that cleave N- or C-terminal groupings that appear to be essential for their activity. An example of this metabolism applied to TRH is shown in Figure 3-9.

IV. BIOCHEMISTRY

A. Location of Releasing Hormones in the Central Nervous System

The location of the releasing hormones in various cells of the CNS has been accomplished frequently by the use of fluorescently labeled monospecific antibodies and microscopic detection of fluorescence. The localization of TRH by this method is to the nerve terminals of the hypothalamic nuclei of the dorsomedial nucleus, the paraventricular nucleus, and the perifornical region. There are some other locations for TRH outside of the hypothalamus, suggesting a more diversified role for this peptide. Thus, the TRH neuroendocrine system has its cell bodies in the periventricular region, and these neurons extend to the median eminence. The cell bodies are located in an area described as the thyrotropic area. Thus, TRH is widely distributed in the hypothalamus and also occurs in extrahypothalamic parts of the CNS and in many peripheral tissues and organs. It may coexist with other neurotransmitters and neuromodulators in brain neurons. In addition to releasing TSH, it also releases PRL and norepinephrine.

GnRH is present in neurons located in the external layer of the median eminence closest to the stalk. It is located caudally from this area of the median eminence as well.

Somatostatin (GIH)-containing neurons reside in the median eminence, the periventricular nuclei of the hypothalamus, the peripheral nervous system, and the gut. GIH has also been located in cells of the gastroenteropancreatic endocrine system (see Chapter 7) and in the thyroid, as well as in several areas of the CNS.

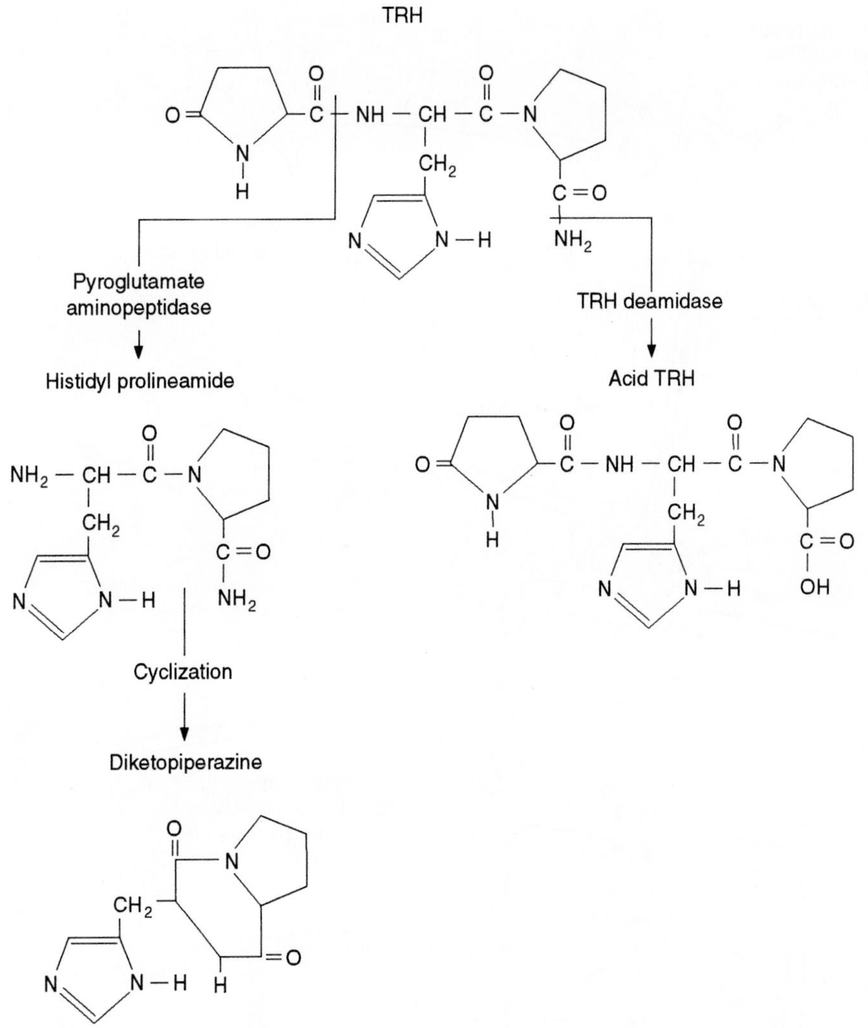

TRH

Pyroglutamate aminopeptidase → Histidyl prolineamide

TRH deamidase → Acid TRH

Cyclization → Diketopiperazine

FIGURE 3-9 Metabolism of thyrotropin-releasing hormone (TRH). Reproduced from Faglia, G., and Persani, L. (1991). Thyrotropin-releasing hormone, basic and clinical aspects. *In* "Brain Endocrinology" (M. Motta, ed.), 2nd ed., pp. 315–350.

GRH has been localized to the arcuate nucleus of the hypothalamus.

CRH also appears to be located elsewhere, in addition to the hypothalamic–hypophyseal axis. It has been found in the central nucleus of the amygdala (of the limbic system) and in polar neurons and dendrites in the cortex, as well as in some cells of the reticular formation. The dorsal motor nucleus has some fibers that may contain CRH. Some CRH-positive fibers terminate in the posterior pituitary and hypothalamic regions in addition to the median eminence.

Polypeptides other than the releasing hormones are produced by neurons. These are substance P (hypothalamus, gut, and various regions of the CNS), enkephalins (CNS, hypothalamic nuclei, including the periventricular nucleus, medial preoptic nucleus, ventromedial nucleus, dorsal and ventral premammillary nuclei, prefornical area, and arcuate nucleus, and 20 other regions outside the hypothalamus, including the adrenal medulla), and gastrin (heterogeneously distributed in the nervous system, including the periventricular, dorso-, and ventromedial nuclei of the hypothalamus, and in the external layer of the median eminence, it is present in the cerebral cortex and the hippocampus and of course has long been known to be located in the stomach). Angiotensin II is located in the CNS in the paraventricular nucleus and prefornical area. In the hypothalamus, angiotensin II-containing nerve terminals are located in the external layer of the median eminence, the dorsomedial nucleus, and the

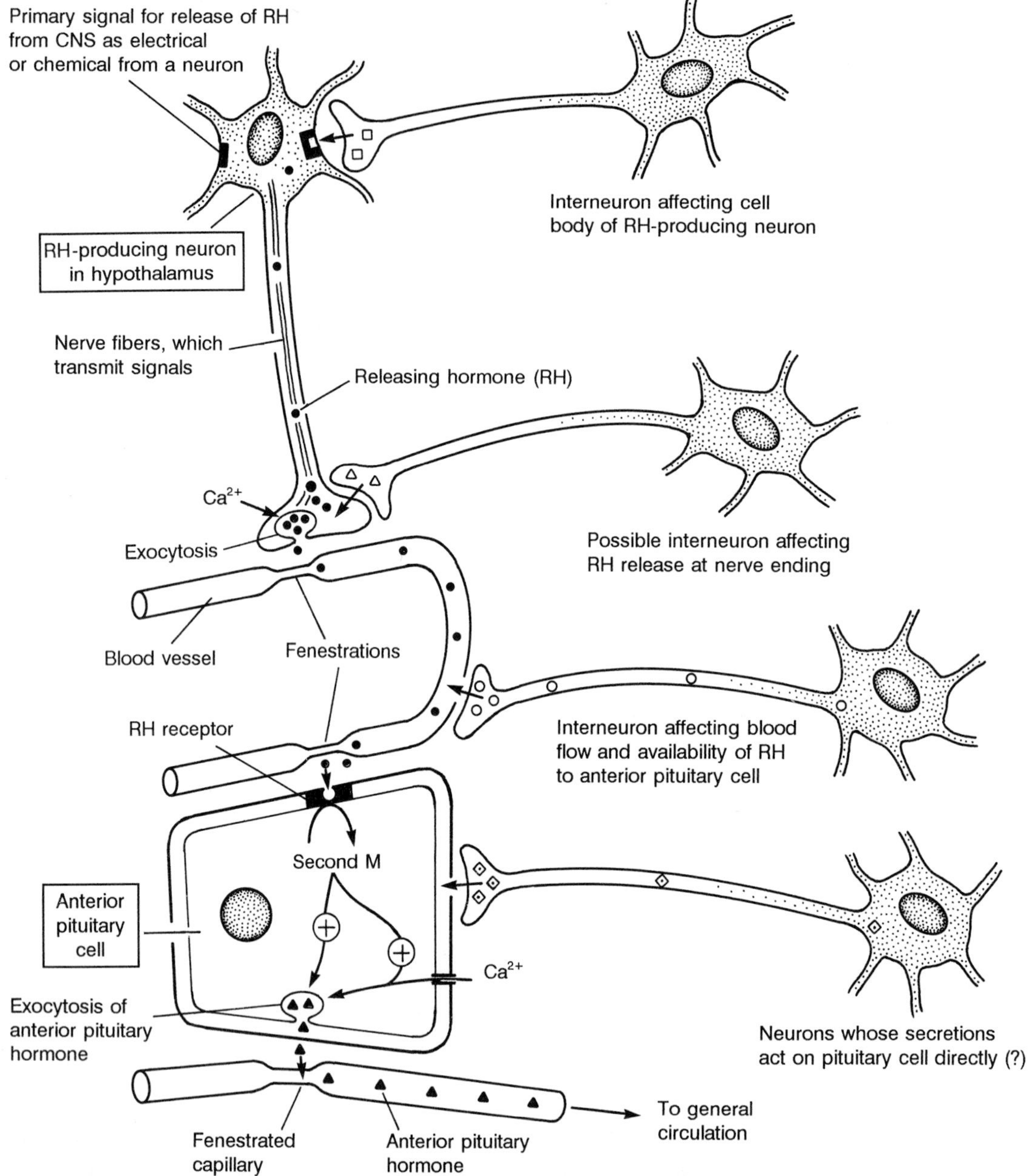

FIGURE 3-10 Neuronal controls of the secretion of releasing hormones. The pathway of synthesis and secretion of releasing hormone (RH) is shown in vertical array at the left. The peptidergic neuron synthesizes RH, which is released in response to electrical or chemical signals delivered by neurons, probably at the level of the cell body. RH is released by exocytosis from the nerve ending, where Ca^{2+} uptake plays a role in the process of exocytosis. RH, released from the nerve ending, enters the total hypothalamic–hypophyseal portal system by way of fenestrations in blood vessels (thin-walled capillaries). RH is transported to the anterior pituitary cell exiting the portal system through fenestrations. RH binds to specific anterior pituitary cell membrane receptor to produce a second messenger (second M), which enhances Ca^{2+} uptake and exocytosis of specific anterior pituitary hormone. Shown on the right are hypothetical interneurons than can act positively or negatively on RH release and transport at a variety of levels of the overall system.

ventral portions. Angiotensin II is also located outside the hypothalamus in other brain areas. There appear to be many other active hormones as well, in addition to the releasing hormones.

B. Secretion of Releasing Hormones and General Actions

Neurons synthesize and package releasing hormone precursors in their cell bodies, and these products are transported down the length of their axons to the nerve terminals (nerve endings), where a signal is awaited for secretion. Since most of the cell bodies of these peptidergic neurons are located in various areas of the hypothalamus, signals for secretion come from higher levels, usually from aminergic or cholinergic neurons in various parts of the brain. The hippocampus (and amygdala) of the limbic system may signal releasing hormone-containing neurons by changes in firing rates of electric signals or by chemical interneuronal contacts (Figure 3-10). The specificity of the overall process is a function of the unique structure of the releasing hormone (Figure 3-8), which interacts with the specific receptor in the target cell membrane. For example, a hypophyseal cell (folliculotroph) producing FSH contains receptors for GnRH just as a cell producing LH (luteotroph) since both anterior pituitary hormones are released by the same releasing hormone, GnRH. In fact, one cell may produce both hormones. TRH interacts with receptors on membranes of the thyrotroph, producing TSH, and the lactotroph, producing PRL. Somatostatin has receptors on somatotrophic cell membranes that also contain GRH receptors since the control of GH secretion is under the influence of both releasing hormones. In addition, there are other regulators of anterior pituitary secretion, such as the aminergic hormones. A particular anterior pituitary cell membrane should contain all of the different receptors necessary for hormonal control.

The aminergic neurons exert similar kinds of control at the level of hypothalamic neurons, which manufacture releasing hormones. Figure 3-10 shows possible mechanisms by which neurons synthesizing releasing hormones are regulated. The primary signal may be to the cell body of the neuron producing a particular releasing hormone. The signal is chemical (e.g., serotonin or other amine; in such cases, synthesis of the amine is likely to occur in the nerve ending rather than in the cell body) or electrical from an interneuron. The secreted releasing hormone passes through fenestrations (thin wall or openings) in the ends of the closed portal system connecting the median eminence with the anterior pituitary cells. Fenestrations overcome the blood–brain barrier, which otherwise might not allow polypeptides to be translocated. The releasing hormones enter the neighborhood of the anterior pituitary cells through fenestrations at the terminus of the portal system (secondary plexus) and then bind to specific plasma membrane receptors of target cells. This interaction produces a second messenger (second M), which directly or indirectly opens the calcium channel so that Ca^{2+} ions accumulate in the cytoplasm (Figure 3-10). The specific anterior pituitary hormone is secreted and enters the general circulation via fenestrations in local capillaries. These hormones then are transported to distant target cells that contain specific membrane receptors. Various interneurons regulate specific processes (Figure 3-10), such as events in the cell body of the releasing hormone-producing cell, secretion of releasing hormone, and blood flow in the closed portal system connecting the hypothalamus (median eminence) with the anterior pituitary.

C. Specific Regulators of Releasing Hormones and Anterior Pituitary Hormone Release

Regulators of releasing hormone release are shown in Table 3-2. Regulators including some of the releasing hormones for pituitary hormone release are listed in Table 3-3. It is obvious that control of secretion is complex. It is becoming more apparent that pituitary hormone release will have both positive and negative controls. In some cases, a regulator has one effect *in vitro* and an opposite effect *in vivo*. This table exemplifies the complexities of regulating the secretion of releasing hormones and anterior pituitary hormones (see Figure 3-10).

The release of pituitary hormones is not governed only by the releasing hormones discussed so far, even

TABLE 3-2 Various Agents Cause Release or Release Inhibition of Releasing Hormones[a]

Agent	CRH	TRH	GHRH	SS[b]
IL-1β	+		(+)[c]	+
IL-2	+			
IL-6	+			
Thyroid hormone			−	
Somatostatin (SS)		−	−	
Galanin			+	
CRH		−		
TNFα	+			
ANF[d]	−			
Serotonin (5-HT)	+			

[a] Data accumulated from the literature 1990–1992.
[b] SS, somatostatin.
[c] Parentheses denote smaller response.
[d] Some authors believe ANF is the physiological CRIF.

TABLE 3-3 **Brain Peptides, Other Agents, and Conditions That May Regulate Pituitary Hormone Release**[a]

Peptide	ACTH	PRL	GH	TSH	LH	FSH	β-Endorphin
Cholecystokinin	+	+	+	−	−		
Motilin			+				
Galanin		+	+				
Substance P		+	+		+		
Neurotensin			+	+			
Opioids	+	+	+	−	−	−	
Vasopressin	+						
Oxytocin	+	+					
CRH	+						+
GHRH			+				
Somatostatin		−	−	−			
GnRH					+	+	
TRH		+	+	+		+	
α-MSH		−			−		
β-Endorphin		+					
GABA		−					
PRF[b]		+					
TNFα		+					
ANF[c]	−						
IL-1β[d]			+				
Retinoic acid deficiency			−				

[a] Assembled in part from summaries presented by McCann, S. M., (1991). Neuroregulatory peptides. *In* "Brain Endocrinology" (M. Motta, ed.), pp. 1–30. 2nd Ed., Raven Press, New York, with permission. (+), causes release, (−), inhibits release.
[b] A peptide of molecular weight less than 5000.
[c] Indirect effect through CRH.
[d] Exerted through a primary effect on GHRH.

though these may represent the major regulators. Table 3-3 summarizes many other brain peptides that can effect the release of pituitary hormones in addition to the releasing hormones themselves. Consequently, the regulation of pituitary hormone release is potentially complicated, and until the prominent regulators are sorted out from the auxiliary or secondary agents the picture will remain complex. Nevertheless, it will suit our purposes to concentrate on the basic system summarized in Table 3-1.

D. TRH and GnRH Receptors

Of the cell membrane receptors for the releasing hormones, the TRH receptor has been cloned and its three-dimensional structure has been determined by X-ray crystallography. It and the GnRH receptor are the only receptors for the releasing hormones whose structural coordinates are registered in the data bases. Consequently, this receptor will be discussed in some detail, trusting that some of the other releasing hormone receptors may have similar characteristics.

Two TRH receptors are encoded by two DNA molecules. In one, there is a 52-basepair deletion to yield a receptor that is shorter by 25 amino acids but contains 12 new amino acid residues at the C-terminal end. This isoform occurs by alternative splicing. Both forms of the receptor are functionally similar in a test system. The second form is of interest because TRH receptors are known to couple to either phospholipase C (to produce diacylglycerol and stimulate the protein kinase C pathway) or adenylate cyclase (to produce cyclic AMP and stimulate the protein kinase A pathway). Importantly, the question remains whether there is one receptor that couples both transducers or two forms of the TRH receptor that couple to each pathway independently.

The secondary structure of TRH receptor is shown in Figure 3-11. It has seven membrane-spanning domains reminiscent of the adrenergic receptor, with three external loops and a side chain on the outside surface of the cell and four loops and an extended side chain from the region of the fourth loop on the cytoplasmic side. The region of the 52-base pair deletion is shown in the figure. The internal side chain has several phosphorylation sites, and these are identified as sites for phosphorylations by protein kinase C, protein kinase A, and casein kinase II.

The human GnRH receptor has been cloned and sequenced. It, like the TRH receptor, has a seven-membrane-spanning domain (Figure 3-12). This receptor has some similarities to the TRH receptor. The extracellular domains are qualitatively similar and the intracellular domains are too, except for the cytoplasmic tail,

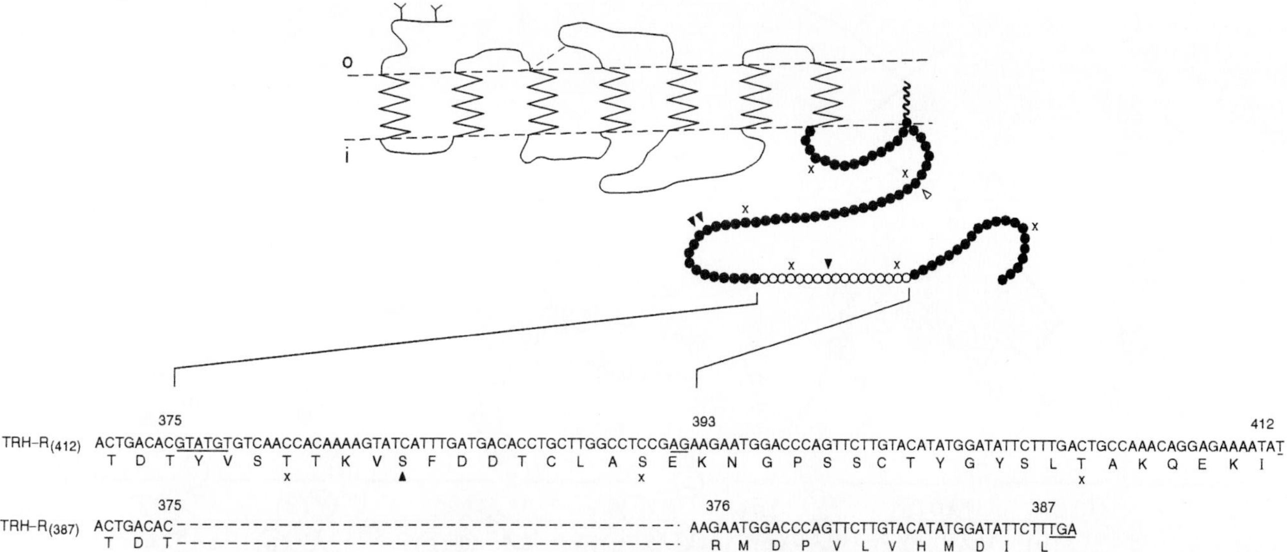

FIGURE 3-11 Seven-transmembrane-spanning model of the TRH receptor showing the situation of the divergent carboxyl-terminal sequence of TRH-$R_{(387)}$. The membrane orientation of the receptor is drawn on the basis of the proposed topography of homologous receptors. The arrangement of the hydrophilic sequences is arbitrary. Putative sizes for phosphorylation by protein kinase C (crosses), protein kinase A (open triangle), and casein kinase II (filled triangles) are indicated in the carboxyl-terminal domain. The nucleotide and amino acid (in single-letter code) sequences of the deleted sequence and its boundaries are shown at the bottom. Numbers indicate the amino acid position in the open reading frame. Reproduced with permission from de la Pena, P., Delgado, L. M., del Camino, D., and Barros, F. (1992). Two isoforms of the thyrotropin-releasing hormone receptor generated by alternative splicing have indistinguishable functional properties. *J. Biol. Chem.* **267**, 25703–25708.

which seems to be lacking in the human GnRH receptor. Possibly other receptors for other releasing hormones may fall into this class of seven-membrane-spanning domain receptors.

E. Molecular Mechanisms of Action

Releasing hormones act at cognate plasma membrane receptor levels either to cause an increase in cyclic AMP or to stimulate the phosphatidylinositol cycle, leading to the stimulation of protein kinase C and an increase in cytoplasmic calcium ion concentration. These actions are summarized in Table 3-4.

1. CRH

The mechanism of CRH action is pictured in Figure 3-13. It operates through a cell membrane receptor and then through a transducing G protein to cause the stimulation of the catalytic subunit of adenylate cyclase (detailed in Chapter 1). Through the stimulation of protein kinase A by the increased level of cyclic AMP (also described in Chapter 1), ACTH is released from the corticotroph of the anterior pituitary. Vasopressin also increases the secretion of ACTH, as does angiotensin II. The main role of vasopressin appears to be one of helping CRH in this activity. Glucocorticoids exert negative feedback (see Chapter 1) on the level of pro-opiomelanocortin message, leading to a decrease in the secretion of ACTH. A three-dimensional structure of human CRH has been modeled from NMR data, as shown in Figure 3-14.

2. GnRH

The molecular mechanism of action of GnRH on the gonadotroph is diagrammed in Figure 3-15. This releasing hormone operates by activating the phosphatidylinositol pathway, which leads to the activation of protein kinase C and calmodulin through increased calcium ion levels in the cytoplasm. These activities lead to the secretion of gonadotropins into general circulation, as well as to an increased level of the β-subunit of the luteinizing hormone by gene activation. Interestingly, the gene encoding GnRH also encodes information for a GnRH-associated protein (GAP). The GAP may be acting as an aid in the correct conformation of GnRH, perhaps analogous in some way to the neurophysins produced by the same gene for either oxytocin or vasopressin.

3. GHRH and Somatostatin

The molecular mechanism of GHRH and somatostatin is outlined in Figure 3-16. These hormones also

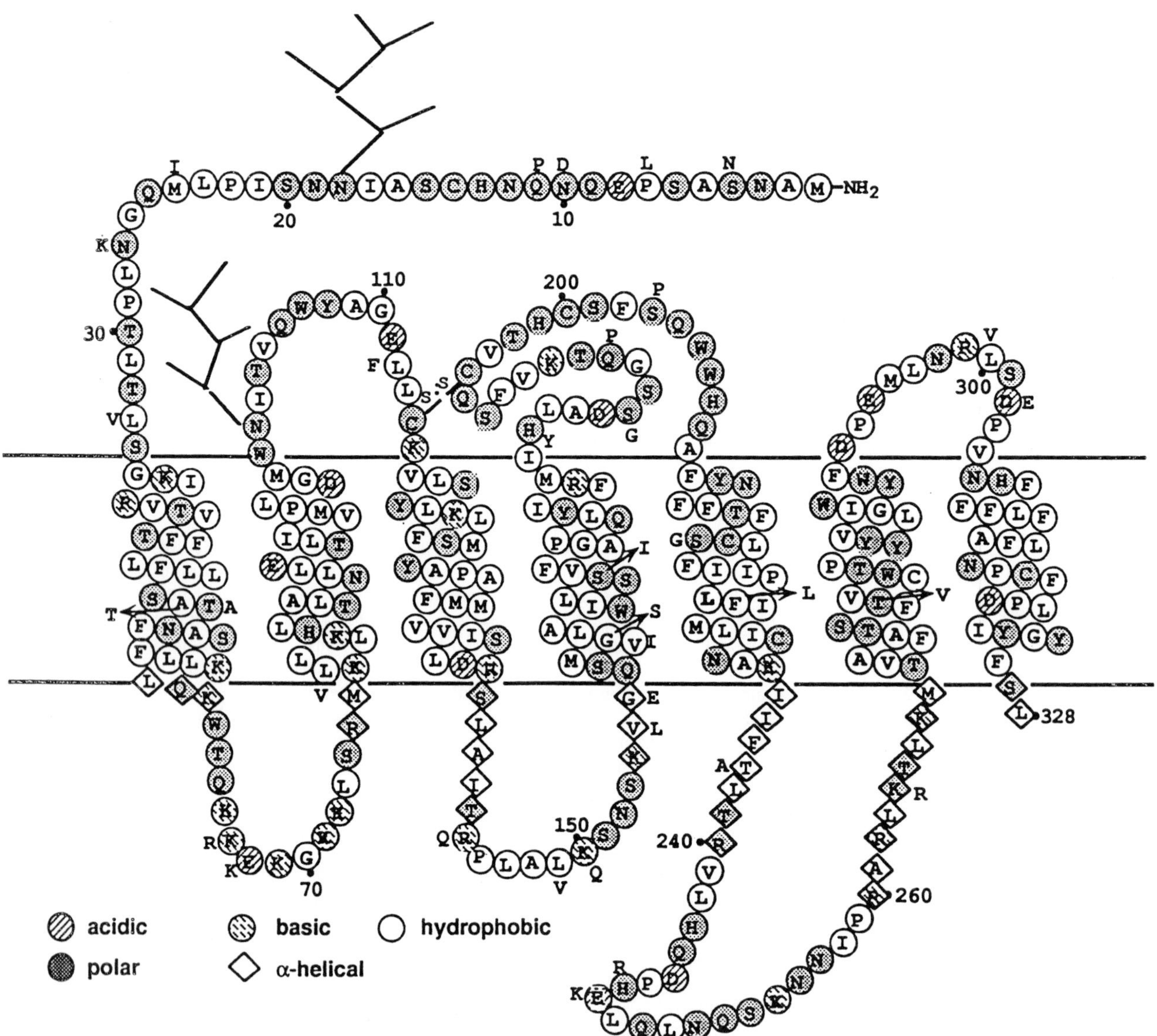

FIGURE 3-12 Schematic diagram of the human GnRHR. Glycosylation sites, possible α-helical extensions of the transmembrane domains, and amino acid class are indicated. The adjacent letters represent the corresponding amino acids, which vary in the mouse GnRHR. K_{191} is not present in the mouse receptor. Reproduced with permission from Chi L., Zhou, W., Prikhozhan, A., Flanagan, C., Davidson, J. S., Golembo, M., IIling, N., Millar, R. P., and Sealfon, S. C., (1993). Cloning and characterization of the human GnRH receptor. *Mol. Cell. Endocrinol.* **91,** R1–R6.

operate through regulation of the catalytic subunit of adenylate cyclase, where GHRH is on the stimulatory pathway and somatostatin is on the inhibitory pathway. These actions culminate in the control of the cytoplasmic level of calcium ions through specific phosphorylation reactions of protein kinase A, and the enhanced level of calcium ions is the direct signal for the secretion of growth hormone. It is not clear whether there may be some stimulation of the phosphatidylinositol cycle and whether this activity could also contribute to the elevated level of calcium ions in the cytoplasm. Moreover, it seems possible (Figure 3-17) that there is direct communication between somatostatin neurons and GHRH-containing arcuate neurons within the hypothalamus. Thus, the secretions of both hormones could be delivered together to the somatomammotroph to regulate the secretion of GH. The other releasing factors would operate in a fashion similar to the two types shown here, as summarized in Table 3-4.

TABLE 3-4 Transduction Pathways of Releasing Hormones

Hormone	Location of action	Pathway	
		PKA[a]	PI[b]
CRH	Corticotrope of anterior pituitary (ACTH)	+	
TRH	Thyrotrope of anterior pituitary (TSH)		+[c]
GnRH	Gonadotrope of anterior pituitary (LH and FSH)		
AVP	Corticotrope of anterior pituitary assists CRH in releasing ACTH		+
GHRH	Somatomammotrope of anterior pituitary (GH)	+	(?)
Somatostatin	Somatomammotrope of anterior pituitary (inhibits GH release)	+[d]	

[a] PKA, protein kinase A.
[b] PI, phosphatidylinositol pathway.
[c] The increase of cytoplasmic calcium concentration may be important in the actions of PI stimulation.
[d] Inhibitory pathway of PKA.

4. *PRF and PIF*

For some time, prolactin release has been known to occur experimentally by the effects of TRH. The negative effector has been considered to be dopamine. However, other work has generated peptides with these activities. Thus, PRF activity appears to be present primarily, if not exclusively, in the intermediate lobe and it is present in cultured posterior pituitary cells. Both PRF and PIF activities can be extracted from bovine neurointermediate lobes. PRF is not oxytocin and is active *in vitro* during dopamine blockade with domperidone and *in vivo* in the presence of dopaminergic tone. PRF appears to be a peptide of molecular weight less than 5000 that rapidly stimulates release of PRL in a dose-dependent and specific manner during dopaminergic inhibition. Oxytocin, arginine vasopressin, and β-endorphin do not contribute significantly to PRF activation in extracts of posterior pituitary.

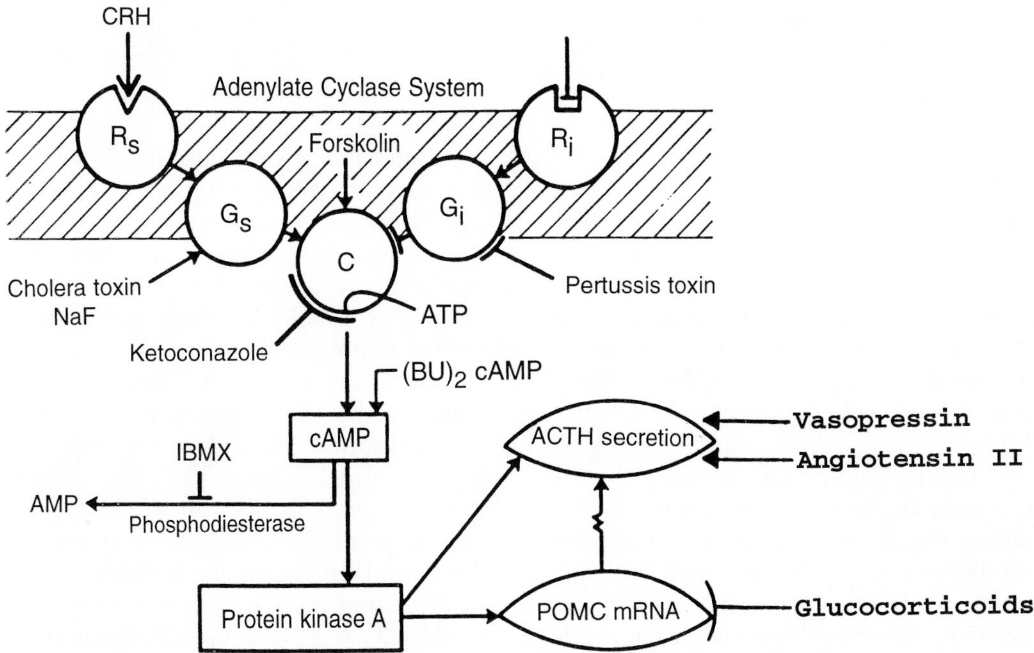

FIGURE 3-13 Model of the interaction between corticotropin-releasing hormone (CRH) and the adenylate cyclase system showing the action of ketoconazole, which is representative of the imidazole derivatives. Abbreviations: AMP, adenosine monophosphate; ATP, adenosine triphosphate; C, catalytic subunit; cAMP, cyclic AMP; G_i, inhibitory guanosine triphosphate (GTP)-dependent protein system; G_s, stimulatory GTP-dependent protein system; IBMX, isobutylmethylxanthine; NaF, sodium fluoride; POMC, proopiomelanocortin; R_i, inhibitory receptor protein; R_s, stimulatory receptor protein; ↓, stimulation; ⊥, inhibition. Reproduced in part with permission from Stalla, G. K., Stalla, J. P., Loeffler, J. P., von Werder, K. and Muller, O. A., Pharmacological modulation of CRH-stimulated ACTH secretion by ketoconazole in Pfeiffer, E. F., Reaven, G. M., Hetzel, W. D., and Hoffman, A. R., eds. (1987). "Corticotropin Releasing Hormone" (O. A. Muller, guest editor), pp. 31–36. Georg Thieme-Verlag, Stuttgart.

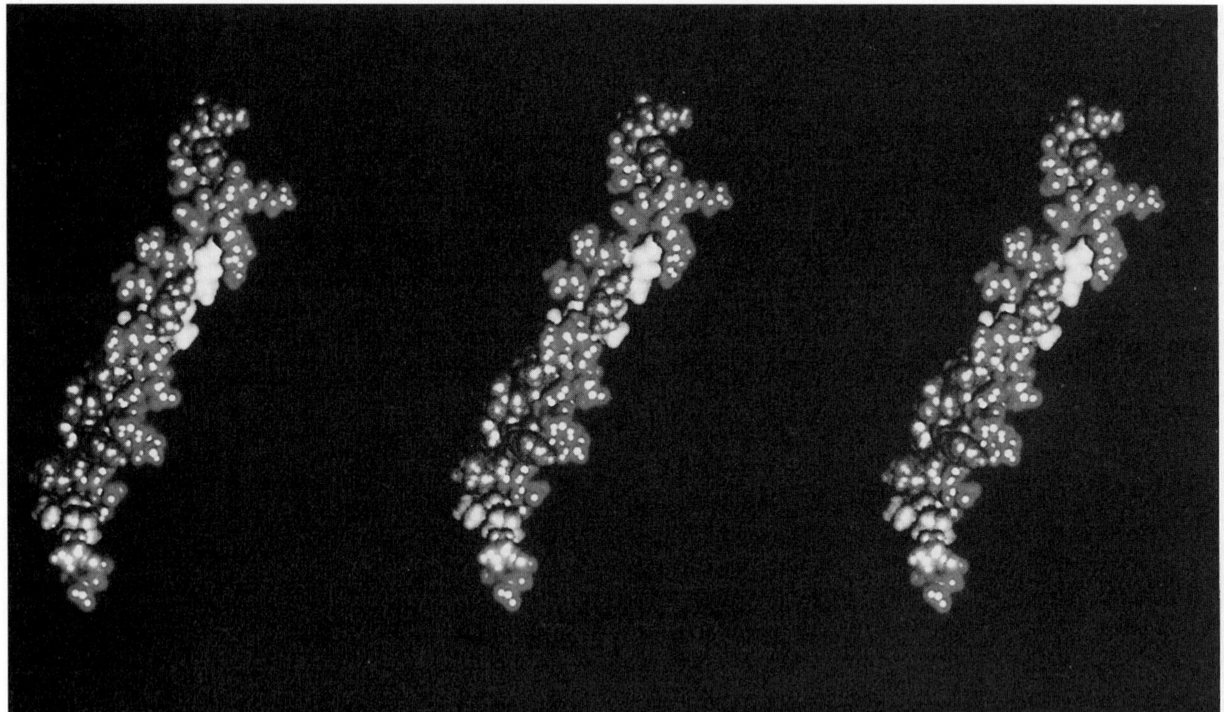

FIGURE 3-14 CPK stereoview of a final solution of hCRF structure according to hydropathy. This photograph was originally published in color and is reproduced here in black and white. Reproduced with permission from Rommier, C., Bernassau, J.-M., Cambillau, C., and Darbon, H. (1993). Solution structure of human corticotropin releasing factor by ^{1}H NMR and distance geometry with restrained molecular dynamics. *Protein Engineer.* **6,** 149–156.

V. CLINICAL ASPECTS

Damage can occur to the fragile structures connecting the hypothalamus to the anterior pituitary, the stalk, and the connecting blood portal system, often by trauma to the head. This can happen in automobile accidents and the like. Tumors, of course, can decrease the transport from the hypothalamus to the pituitary or disconnect it altogether. In any case, such damage has far-reaching effects and these conditions can be collectively listed under panhypopituitarism or hypothalamic hypofunction. Many of the releasing hormones will fail to arrive at their receptors on the plasma membranes of the specific anterior pituitary cell, and the hypothalamic–pituitary–end organ axis will be broken. Diagnosis is now possible through the availability of the releasing hormones, and often patients with these syndromes are treated with the hormone of the terminal gland in the pathway, such as the adrenal hormones, sex hormones, thyroid hormones, etc. The releasing hormones have come into prominence in the clinic. The function of the hypothalamic–pituitary–end organ axis is now possible to define in terms of the affected organ by the availability of releasing hormones that can be injected into the bloodstream. These hormones can gain access to the pituitary gland by this route, and therefore it becomes possible to elicit the function of the pituitary. If the pituitary is malfunctioning, the level of pituitary hormone released in the bloodstream in response to the stimulus of a specific releasing hormone will be subnormal, and it becomes possible to pinpoint the cause of disease in a malfunctioning hypothalamic–pituitary–end organ axis by measuring the level of the blood pituitary hormone, usually with a radioimmunoassay.

Short stature in children requires treatment with growth hormone. Some work also demonstrates that treatment with GHRH instead of growth hormone may be useful in some cases and may also be more economical. However, the regimen for this treatment to obtain the most satisfactory results has not been concluded. Release of GHRH is stimulated by interleukin-1β, while thyroid hormone decreases the secretion of GHRH. Research indicates that GHRH is produced by human lymphocytes, and lymphocytes from rats have also been shown to contain GHRH receptors. These may prove to be useful indicators of the status of GHRH function. The response to GHRH may be greater in women than in men, and GHRH appears to be located in human gonads (corpora lutea) and in Leydig cells in the testes of prepubertal men.

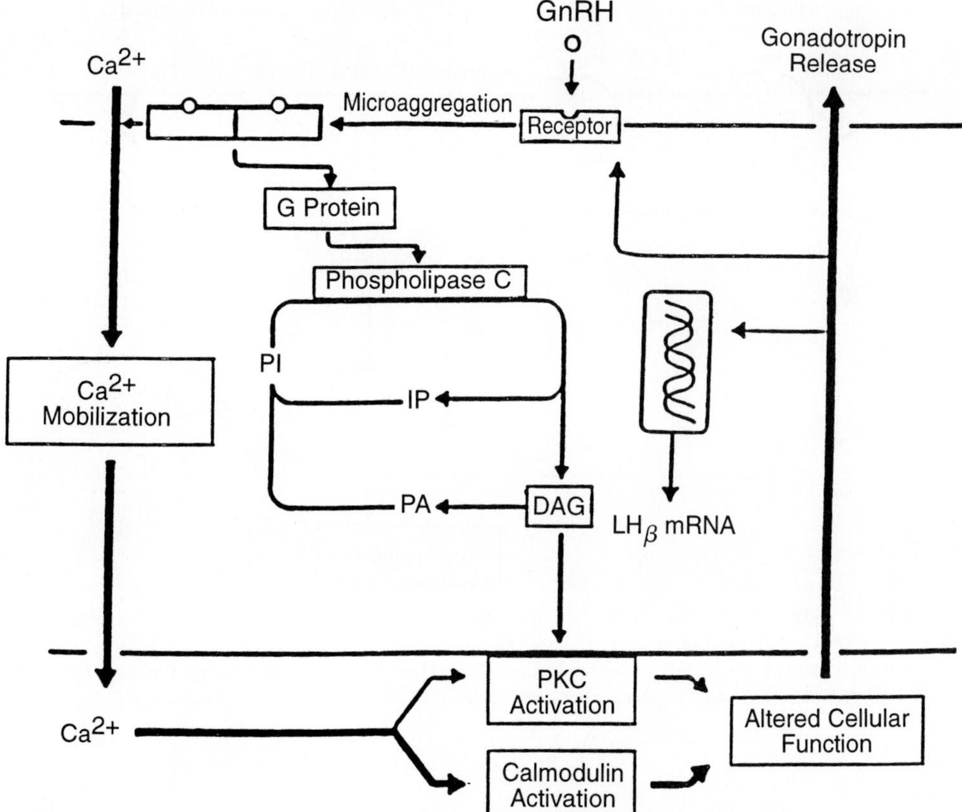

FIGURE 3-15 Mechanism of action of GnRH in the pituitary gonadotroph. GnRH interactions with the plasma membrane receptor. This interaction provokes receptor–receptor interaction (microaggregation, patching, capping, and internalization), activates the mobilization of calcium, and, through a transducing G protein, activates a phospholipase C enzyme. Calcium accumulation activates calmodulin, which appears to mediate the release of gonadotropins. In combination with diacylglycerol (DAG), calcium activates protein kinase C (PKC), which appears to mediate the synthesis of gonadotropins (gene expression). LH_β denotes the β-subunit of luteinizing hormone. Other abbreviations: mRNA, messenger RNA; PI, phosphatidylinositol; IP, inositol phosphate; PA, phosphatidic acid. This figure is reproduced with permission from Conn, P. M., and Crowley, W. F. (1991). *New Engl. J. Med.* **324,** 93–103.

TRH, as well as CRH, has been found to stimulate respiration independent of activation of the pituitary–adrenal axis. CRH acts in the brain to activate the sympathetic nervous system and reduce cellular immune function. Also, the central administration of CRH leads to a reduction in natural killer cell activity in the spleen. Interestingly, there has been a report of the existence of a specific binding protein for human CRH in human plasma. This may indicate that the binding protein can inactivate CRH and may involve as much as one-half of the CRH in the blood. The induction of CRH mRNA by ligands in the PKC mechanism may be blocked more effectively by glucocorticoids than message induction by ligands operating through the PKA pathway. This may explain abnormal CRH production in clinical conditions, such as *anorexia nervosa* and major depression. Stress is known to disturb reproductive function. Stress causes the output of CRH and β-endorphin, which lower the secretion of gonadotropins, oxytocin, and vasopressin. Opioids increase during puberty and fall at menopause. β-Endorphin is elevated by estradiol and progesterone in the luteal phase of the menstrual cycle. Thus, the rapid decline in these steroids and the fall of β-endorphin at menstruation may lead to symptoms of premenstrual syndrome. Conversely, it may be possible to counteract the deleterious effects of stress on reproductive function by the use of opiate antagonists.

Hypothalamic GnRH is secreted in a pulsatile fashion at puberty and not prior to that time. Some evidence suggests that *N*-methyl-D-aspartate (NMDA) receptors, through excitatory amino acids, may facilitate the initiation of puberty. In immature male rats (15 days), there is an inhibitory control of pulsatile GnRH secretion presumably mediated by NMDA receptors, and specific antagonists of NMDA receptors can cause early pubertal development. This inhibitory effect on

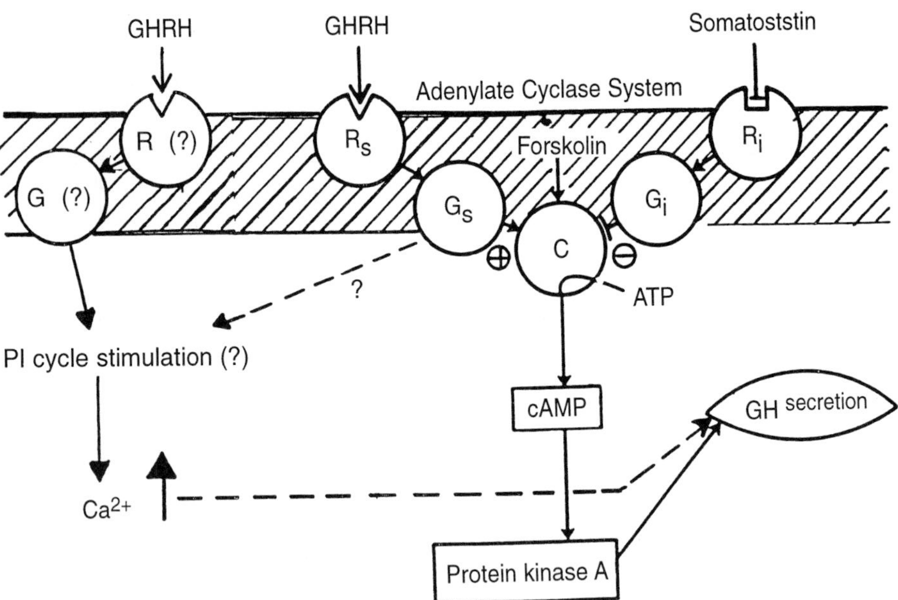

FIGURE 3-16 Speculative scheme for the action of GHRH and somatostatin. Abbreviations: R, receptor; G, GTP-binding protein; C, catalytic subunit of adenylate cyclase; s, stimulatory; i, inhibitory; ? indicates that this phosphatidylinositol (PI) pathway, either through the established GHRH receptor (G_s) or through another receptor and G protein, is highly speculative.

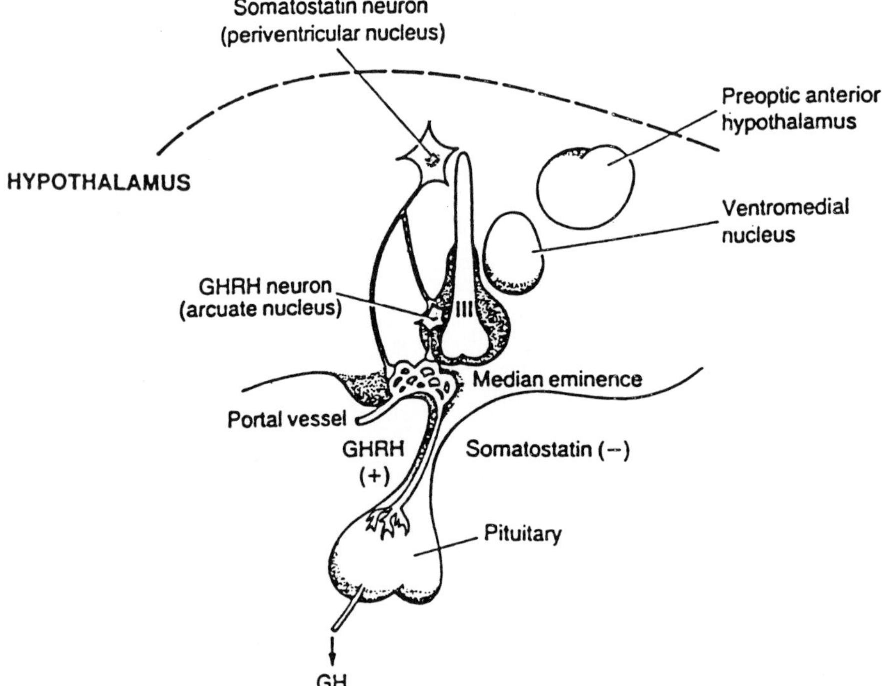

FIGURE 3-17 Hypothalamic peptidergic neurons of the GH neuroendocrine axis. Direct synaptic communication between somatostatin neurons and GHRH-containing arcuate neurons within the hypothalamus is postulated. III indicates the third ventricle. Reproduced from Tannenbaum, G. S. (1991). Neuroendocrine control of growth hormone secretion. *Acta Paediatr. Scand.* **372**[Suppl.], 5–16.

NMDA receptors disappears at puberty (around 25 days), causing an unmasking of the facilitory effects of these same receptors, which leads to the pulsatile release of GnRH.

GnRH assumes an important role in the disease known as Kallman's syndrome, which refers to anosmic (lacking a sense of smell) patients with hypogonadotropic hypogonadism. This probably implies that there is an entire cluster of variations about this general category with some variants lacking some of the clinical manifestations present in the overt syndrome. This syndrome appears to be the result of a deficiency of GnRH, and it may originate from an arrest of the migration of the GnRH neurons at the cribriform plate of the ethmoid sinus during development. By contrast, in the normal fetus, these neurons would have completed their migration to the hypothalamus at this time in development. This migration failure is apparently linked to a deletion on the X chromosome, and the deleted gene (located at Xp22.3) encodes a protein homologous to fibronectin family members, which undoubtedly have functions in neural chemotaxis and cell adhesion and in directing neuronal cell migration during embryonic development. Treatment of individuals with this syndrome usually involves pulsatile administration of GnRH. GnRH or its antagonists can be used in the treatment of sex hormone-dependent cancers in which treatment with the antagonist or a relatively large quantity of GnRH itself will effectively block GnRH receptors, resulting in a diminution of steroid production that would otherwise enhance the growth of certain tumors.

References

A. Books

Ciba Foundation Symposium 172. (1993). "Corticotropin-Releasing Factor." John Wiley & Sons, New York.

Krieger, D. T., and Hughes, J. C., eds. (1980). "Neuroendocrinology." Sinauer Associates, Sunderland, MA.

B. Review Articles

Ben-Jonathan, N., and Liu, J.-W. (1992). Pituitary lactotrophs. Endocrine, paracrine, juxtacrine and autocrine interactions. *Trends Endocrinol. Metab.* **3,** 254–258.

Crowley, W. F., Jr., and Jameson, J. L. (1992). Clinical counterpoint: gonadotropin-releasing hormone deficiency: perspectives from clinical investigation. *Endocrine Rev.* **13,** 635–640.

Dallman, M. F. (1993). Adaptation of the hypothalamic-pituitary adrenal axis to chronic stress. *Trends Endocrinol. Metab.* **4,** 62–69.

Faglia, G., and Persani, L. (1991). Thyrotropin-releasing hormone, basic and clinical aspects. *In* "Brain Endocrinology" (M. Motta, ed.), pp. 315–350. Raven Press, New York.

Fuxe, K., Agnati, L. F., Zoli, M., Biagini, G., Cintra, A., and Eneroth, P. (1991). Regulatory peptides in the neuroendocrine system. *In* "Brain Endocrinology" (M. Motta, ed.), pp. 31–69. Raven Press, New York.

Hyde, J. F., and Ben-Jonathan, N. (1989). The posterior pituitary contains a potent prolactin-releasing factor: in vivo studies. *Endocrinology* **125,** 736–741.

King, J. A., and Millar, R. P. (1992). Evolution of gonadotropin-releasing hormones. *Trends Endocrinol. Metab.* **3,** 339–346.

Krieger, D. T. (1984). Brain peptides. *In* "Vitamins and Hormones" (G. D. Aurbach and D. B. McCormick, eds.), Vol. 41, pp. 1–50. Academic Press, Orlando.

McCann, S. M. (1991). Neuroregulatory peptides. *In "Brain Endocrinology"* (M. Motta, ed.), pp. 1–31. Raven Press, New York.

Orth, D. N. (1992). Corticotropin-releasing hormone in humans. Endocrine Rev. **13,** 164–191.

Owens, M. J., and Nemeroff, C. B. (1991). Physiology and pharmacology of corticotropin-releasing factor. *Pharmacol. Rev.* **43,** 425–473.

Porter, J. C., Sissom, J. F., Arita, J., and Reymond, M. J. (1983). Hypothalamic-hypophyseal vasculature and its relationship to secretory cells of the hypothalamus and pituitary gland. *In "Vitamins and Hormones"* (G. D. Aurbach and D. B. McCormick, eds.), Vol. 40, pp. 145–174. Academic Press, Orlando.

Root, A. W., and Martinez, C. R. (1992). Magnetic resonance imaging in patients with hypopituitarism. *Trends Endocrinol. Metab.* **3,** 283–287.

Schwanzel-Fukuda, M., Jorgenson, K. L., Bergen, H. T., Weesner, G. D., and Pfaff, D. W. (1992). Biology of normal luteinizing hormone releasing hormone neurons during and after their migration from olfactory placode. *Endocrine Rev.* **13,** 623–634.

Steele, M. K. (1992). The role of brain angiotensin II in the regulation of luteinizing hormone and prolactin secretion. *Trends Endocrinol. Metab.* **3,** 295–300.

Tannenbaum, G. S. (1991). Neuroendocrine control of growth hormone secretion. *Acta Paediatr. Scand.* **372** (Suppl.), 5–16.

C. Research Papers

Antoni, F. A., Hunter, E. F., Lowry, P. J., Noble, J. M., and Seckl, J. R. (1992). Atriopeptin: an endogenous corticotropin-release inhibiting hormone. *Endocrinology* **130,** 1753–1755.

Cambronero, J. C., Rivas, F. J., Borrell, J., and Guaza, C. (1992). Adrenalectomy does not change CRF secretion induced by interleukin-1 from rat perifused hypothalami. *Regul. Pept.* **41,** 237–247.

Chi, L., Zhou, W., Prikhozhan, A., Flanagan, C., Davidson, J. S., Golembo, M., Illing, N., Millar, R. P., and Sealfon, S. C. (1993). Cloning and characterization of the human GnRH receptor. *Mol. Cell. Endocrinol.* **91,** R1–R6.

de la Pena, P., Delgado, L. M., del Camino, D., and Barros, F. (1992). Two isoforms of the thyrotropin-releasing hormone receptor generated by alternative splicing have indistinguishable functional properties. *J. Biol. Chem.* **267,** 25703–25708.

Fry, D. C., Madison, V. S., Greeley, D. N., Felix, A. M., Heimer, E. P., Frohman, L., Campbell, R. M., Mowles, T. F., Toome, V., and Wegrzynski (1992). Solution structures of cyclic and dicyclic analogues of growth hormone releasing factor as determined by two-dimensional NMR and CD spectroscopies and constrained molecular dynamics. *Biopolymers* **32,** 649–666.

Gaylinn, B. D., Harrison, J. K., Zysk, J. R., Lyons, C. E., Lynch, K. R., and Thorner, M. O. (1993). Molecular cloning and expression of a human anterior pituitary receptor for growth hormone-releasing hormone. *Mol. Endocrinol.* **7,** 77–84.

Honda, S., Ohashi, S., Morii, H., and Uedaira, H. (1991). Solution structure of human growth hormone-releasing factor fragment

(1–29) by CD: characteristic conformational change on phospholipid membrane. *Biopolymers* **31,** 869–876.

Hyde, J. F., and Ben-Jonathan, N. (1989). The posterior pituitary contains a potent prolactin-releasing factor: *in vivo* studies. *Endocrinology* **125,** 736–741.

Irwin, R., Hauger, R., and Brown, M. (1992). Central corticotropin releasing hormone activates the sympathetic nervous system and reduces immune function: increased responsivity of the aged rat. *Endocrinology* **131,** 1047–1053.

Kay, T. W. H., and Jameson, J. L. (1992). Identification of a gonadotropin-releasing hormone-responsive region in the glycoprotein hormone α-subunit promoter. *Mol. Endocrinol.* **6,** 1767–1773.

Koike, K., Masumoto, N., Kasahara, K., Yamaguchi, M., Tasaka, K., Hirota, K., Miyake, A., and Tanizawa, O. (1991). Tumor necrosis factor-alpha stimulates prolactin release from anterior pituitary cells: a possible involvement of intracellular calcium mobilization. *Endocrinology* **128,** 2785–2790.

Kornreich, W. D., Galyean, R., Hernandez, J. F., Craig, A. G., Donaldson, C. J., Yamamoto, G., Rivier, C., Vale, W., and Rivier, J. (1992). Alanine series of ovine corticotropin-releasing factor (oCRF): a structure-activity relationship study. *J. Med. Chem.* **35,** 1870–1876.

Laatikainen, T. J. (1991). Corticotropin-releasing hormone and opioid peptides in reproduction and stress. *Ann. Med.* **23,** 489–496.

Laudon, M., Grossman, D. A., and Ben-Jonathan, N. (1990). Prolactin releasing factor: cellular origin in the intermediate lobe of the pituitary. *Endocrinology* **126,** 3185–3192.

Murai, I., Reichlin, S., and Ben-Jonathan, N. (1989). The peak phase of the proestrus prolactin surge is blocked by either posterior pituitary lobectomy or antisera to vasoactive intestinal peptide. *Endocrinology* **124,** 1050–1055.

Rosen, L. B., Majzoub, J. A., and Adler, G. K. (1992). Effects of glucocorticoid on corticotropin-releasing hormone gene regulation by second messenger pathways in NPLC and AtT-20 cells. *Endocrinology* **130,** 2237–2244.

Samson, W. K., Fong-Lee, S. H., and Fulton, R. J. (1993). C-type natriuretic peptide mediates the hypothalamic actions of the natriuretic peptides to inhibit luteinizing hormone secretion. *Endocrinology* **132,** 5504–5509.

Samson, W. K., Martin, L., Mogg, R. J., and Fulton, R. J. (1990). A nonoxytocinergic prolactin releasing factor and a nondopaminergic prolactin inhibiting factor in bovine neurointermediate lobe extracts: *in vitro* and *in vivo* studies. *Endocrinology* **126,** 1610–1617.

Saphier, P. W., Faria, M., Grossman, A., Coy, D. H., Besser, G. M., Hodson, B., Parkes, M., Linton, E. A., and Lowry, P. J. (1992). A comparison of the clearance of ovine and human corticotropin-releasing hormone (CRH) in man and sheep: a possible role for CRH-binding protein. *J. Endocrinol.* **133,** 487–495.

CHAPTER

4

Posterior Pituitary Hormones

I. INTRODUCTION

Two important hormones are secreted from the posterior pituitary in both males and females. These are vasopressin (VP), the antidiuretic hormone, and oxytocin (OT), which in females acts as the milk ejection factor. Both are nonapeptides, closely related in structure, and apparently derived from the same ancestral gene.

Although neuronal cells that produce OT and VP share the same locations in the brain, OT neurons predominate in one region and VP neurons predominate in another. Also, OT and VP are the secretions of separate neurons. OT and VP are transported to the nerve ending of a specific neuron, stored there, and released upon appropriate differential stimulation. The nerve endings that store these secretions are located in the posterior pituitary.

The primary recognized function of VP is to stimulate the reabsorption of water from the distal tubular kidney. This is clearly established because in the absence of VP up to as much as 25 liters of diluted urine can be excreted each day. Accordingly, the release of VP is generated by the need to maintain the blood osmolarity of plasma within strict limits (homeostasis), especially with reference to increased Na^+ concentrations in blood produced by the ingestion of NaCl. VP secretion also is increased when blood volume or blood pressure is decreased. The sites of VP synthesis in the hypothalamus appear to be close to the osmoreceptor sites, which sense changes in electrolyte (solute) concentrations in circulation and signal release of the hormone from neuronal terminals in the posterior pituitary. The osmoreceptor is close to the thirst center in the hypothalamus and also interacts with the renin–angiotensin system. Collectively, these systems appear to be the primary elements for regulation of water balance (see also Chapter 15).

The role of OT is less clear. In females, OT plays a major role in milk letdown for nourishing the suckling infant (see also Chapter 14). Obviously there is no corresponding role for OT in males. In addition, oxytocin action on the uterine myometrium may play an important role in inducing the uterine contractions leading to the normal termination of pregnancy. Consequently, it is important to ensure that OT is not released prematurely during pregnancy, and indeed there are safeguards to ensure that this does not occur. Many central functions have been attributed to OT, or it may serve as a precursor for neurotransmitters that signal specific functions. Among these central nervous system activities are maternal behavior, penile erection, lordosis, copulatory behavior, yawning, memory and learning, tolerance and dependence mechanisms, feeding, grooming, and certain aspects of cardiovascular regulation.

Some responses are elicited by both VP and OT. Thus, VP also stimulates uterine contractions in humans and can also stimulate milk ejection to a slight degree. However, the stimuli for the release of each hormone are different, ensuring, to a certain extent, that only one of the hormones is released in response to a specific need. Also, OT and VP are produced in different cells where the hormones are synthesized and may act in a paracrine fashion. Similar biological activities can be expected from homologies in the molecular structures of VP and OT, and the homologous regions may be involved in the responses generated in common. Some functions are affected oppositely by the two hormones. While VP, in pharmacological amounts, elevates blood pressure, OT slightly lowers it, and the same trend is evident in the constriction of coronary arteries by VP and their dilation by OT.

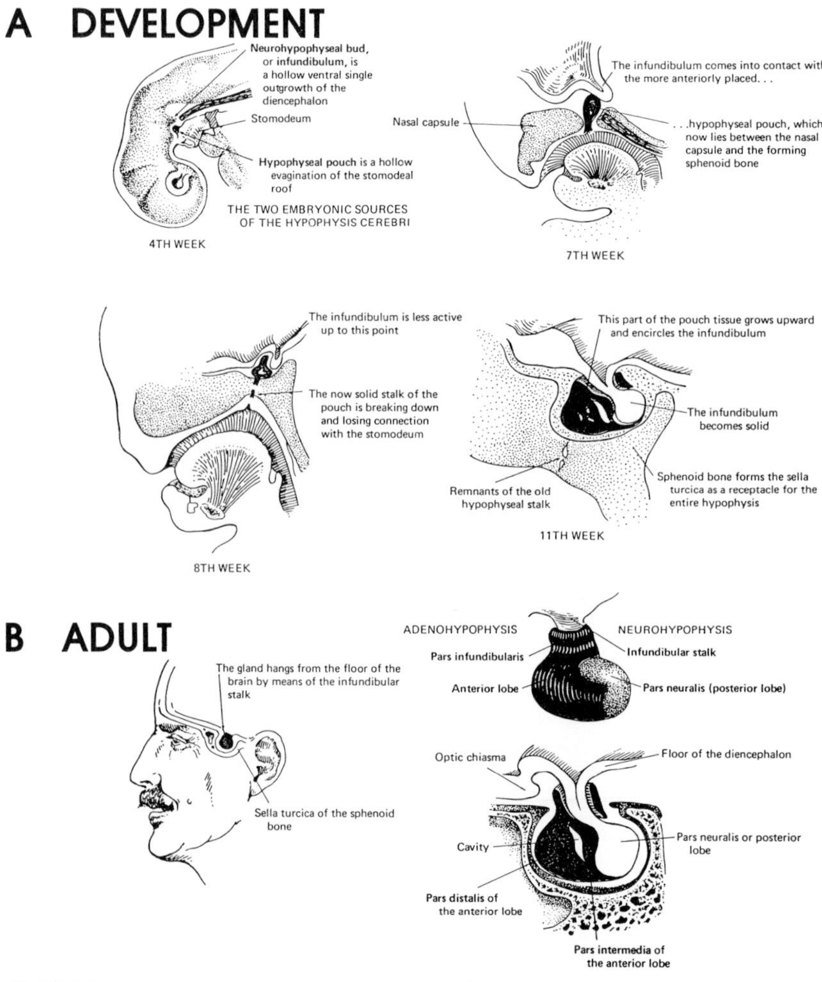

FIGURE 4-1 Development of the pituitary gland. Reproduced by permission from Langebartel, D. A. (1977): "The Anatomical Primer." University Park Press, Baltimore, Maryland.

In its major action, VP inhibits water diuresis, and OT can also produce this effect. Of interest is the possible precursor status of OT for the formation of CNS-active peptides. Since VP is involved in ACTH secretion, and OT produces MIF and MRF in some species (probably not in higher mammals), then there may be regulatory systems to partition the fates of VP and OT between the formation of releasing factors or other biological products active in the CNS and storage in the terminals of the neurohypophysis. The most significant partition occurs through the distinct signals for the release of each hormone and the relative specificity of receptors. As new actions of OT become apparent in tissues common to males and females, the role of this hormone in males may emerge.

II. ANATOMY AND DEVELOPMENT OF THE POSTERIOR PITUITARY

The posterior pituitary or neurohypophysis or *pars neuralis* derives as a hollow ventral outgrowth of the diencephalon of the developing brain. Thus, by the fourth week of human development this outgrowth (neurohypophyseal bud or infundibulum) of the diencephalon becomes apparent (Figure 4-1A). At this stage, a hypophyseal pouch, also hollow, evaginates from the stomodeal roof (Figure 4-1A). These two outgrowths will form the pituitary, while that of the diencephalon will form the posterior pituitary. By the seventh week, both buds have grown steadily and have come into close contact (Figure 4-1A) with the hypophyseal pouch anterior to the infundibulum. By the eighth week the stalk of the pouch has become solid and begins to deteriorate within the sphenoid bone. The precursor of the posterior pituitary, the outpocket of the brain, remains hollow. Some of the tissue of the hypophyseal pouch encircles the infundibular stalk. The sphenoid bone remodels into the *sella turcica,* which serves as a receptacle for the gland, and this process is accomplished by the eleventh week. The surface and cross section of the finished gland are shown in Figure 4-1B, as well as its gross anatomical location in adults.

III. CHEMISTRY

A. Structures of Active and Substituted Posterior Pituitary Hormones

Shown in Figure 4-2 are the structures of active and substituted posterior pituitary hormones. The free amino terminal (amino acid 1) is not important for activity, as deaminooxytocin is more biologically active than OT.

```
    2    1
   ,Tyr -Cys
3 [Ile]    S
   |       |
4 Gln      S
   \       |
   Asn-Cys-Pro- [Leu] -Gly•NH2
    5   6   7     8      9
```

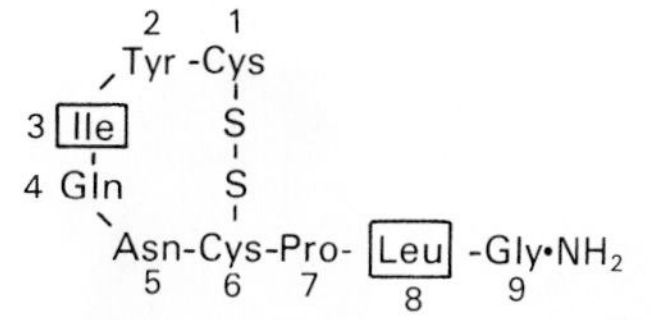

OXYTOCIN (50 mU/mg in supraoptic nucleus or paraventricular nucleus; 1 mg OXYTOCIN= 400-500 units of activity compared to USP posterior pituitary reference standards)

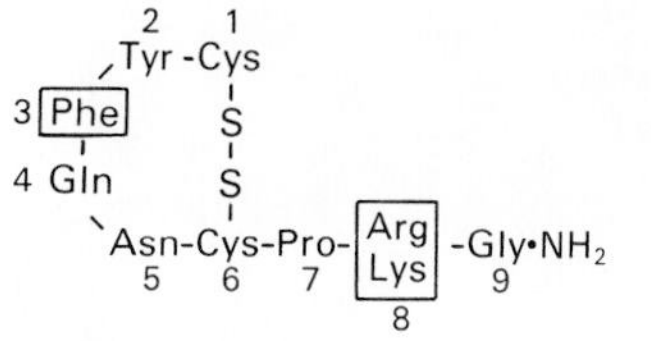

```
    2    1
   ,Tyr -Cys
3 [Phe]    S
   |       |
4 Gln      S
   \       |
   Asn-Cys-Pro- [Arg/Lys] -Gly•NH2
    5   6   7       8        9
```

Arg/Lys **VASOPRESSIN** (~200 mU/mg in supraoptic nucleus; 85 mU/mg in paraventricular nucleus)

```
    2    1
   ,Tyr -Cys
3 [Ile]    S
   |       |
4 Gln      S
   \       |
   Asn-Cys-Pro- [Arg] -Gly•NH2
    5   6   7     8      9
```

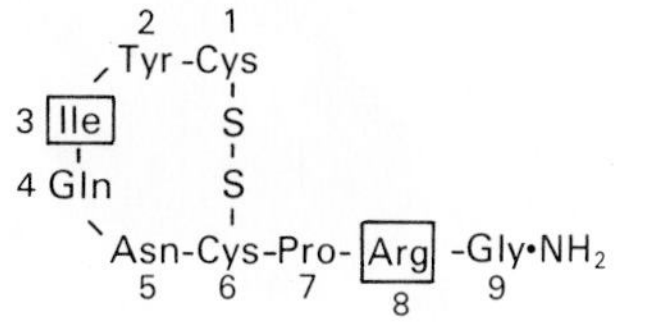

Arg **VASOTOCIN** (ancient structure; questionable whether it is in pineal gland)

FIGURE 4-2 Structures of OT, arginine- or lysine-VP, and arginine vasotocin.

Replacement of Leu8 with Ile reduces activity. If Gln4 is replaced by Glu, the hormone is inactivated. In Arg-vasopressin (AVP), the basic side chain (Arg8) is essential for antidiuretic activity and explains the lack of the crossover effect of OT. Another point of difference is the amino acid residue at position 3. The crystal structure of deaminooxytocin has been solved, and models derived from this analysis are shown in Figure 4-3. Not only are these differences in structure sufficient to guarantee specificity at the receptor level, but oxytocin does not respond to changes in osmolarity that would stimulate vasopressin, and large doses of oxytocin do not generate a natriuretic response. Furthermore, it has been demonstrated that the releases of oxytocin and vasopressin can be initiated separately by estrogen and nicotine. Thus, in spite of structural similarities, these hormones are signaled separately and function distinctly.

The structure of arginine vasotocin (AVT), present in the pineal gland, is also shown in Figure 4-2. AVT reflects the structure of OT at position 3 (Ile) and that of VP at position 8 (Arg). AVT is a very active hormone, especially in the control of developing reproductive functions. More information will appear on this interesting hormone in the chapter on the

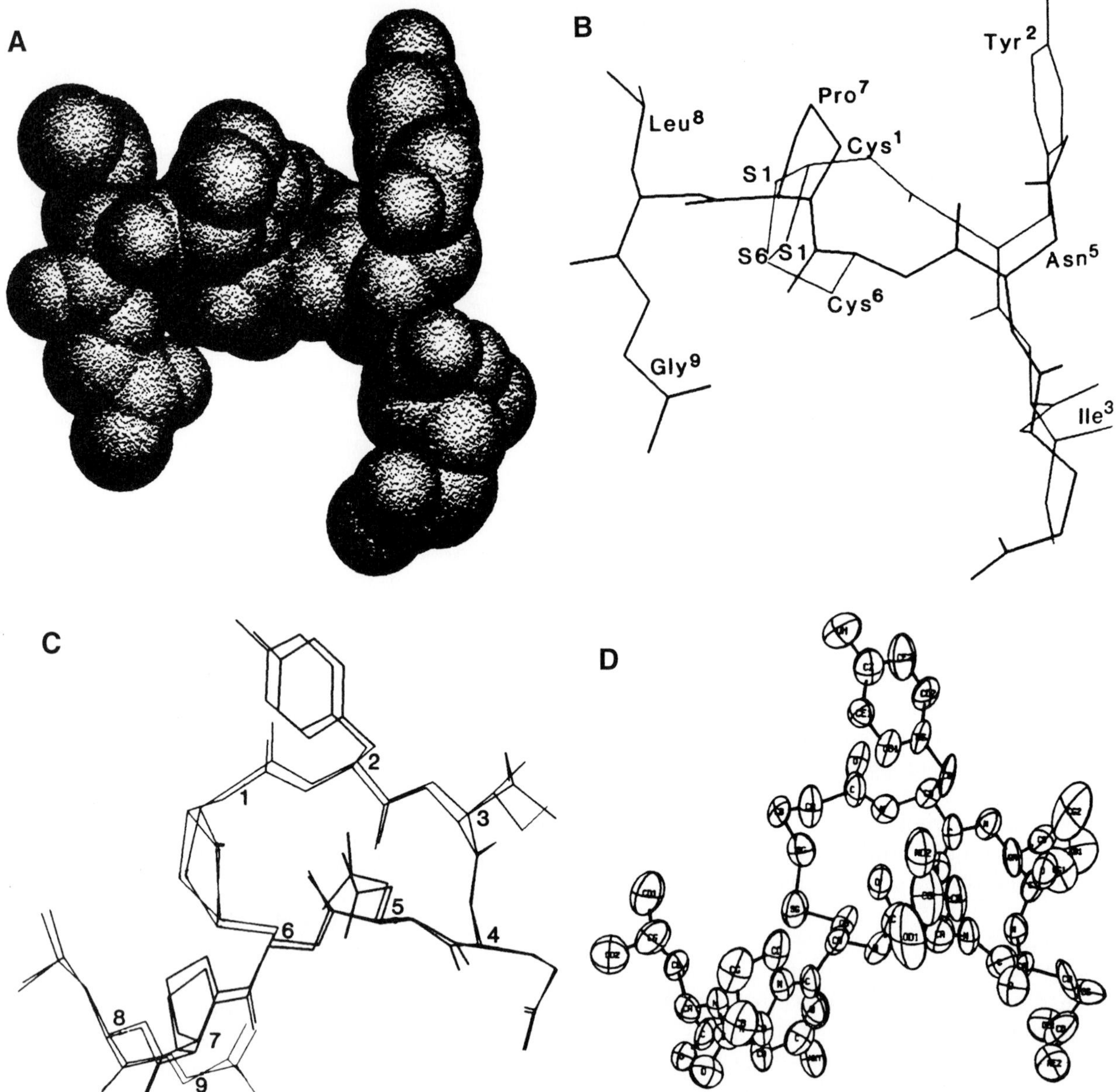

FIGURE 4-3 Structures of deaminooxytocin derived from X-ray diffraction analysis: (A) space-filling model and equivalent stick model (B). This shows the "pleat" of the 20-membered ring and the accentuation of this form by the "tail" peptide and the Gln4 side chain dispositions. In (C), the two molecules of the P2 asymmetric unit are shown superimposed to demonstrate the two conformations in the S1 region, the associated Ile3 rotamers, and other conformational differences. One disulfide chirality is shown in (D). Reproduced from Wood, S. P. *et al*, (1986). *Science* **232,** 633–636.

pineal gland (Chapter 18). AVT appears to be the ancestral precursor of both VP and OT. A scheme showing the evolution of these hormones appears in Figure 4-4.

Two neurophysins (NPs) have been reported in the pineal gland. NPs are proteins derived from the same gene product as either VP or OT in the hypothalamus.

B. Interaction of Posterior Pituitary Hormones with Active Sites of Neurophysins

The posterior pituitary hormones are transported down the long axons of the cell bodies of the neurons responsible for their synthesis. During transport, a preprotein product of a single gene is processed into VP

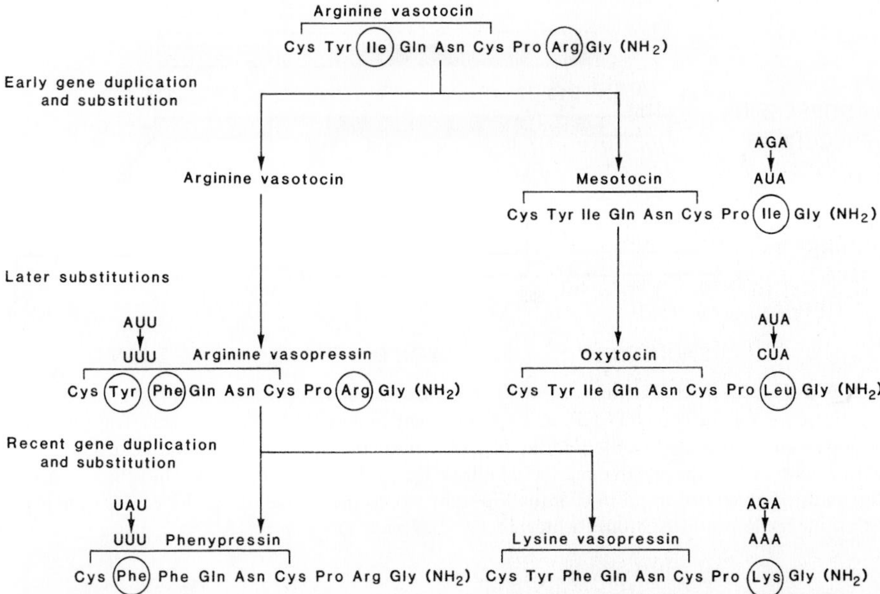

FIGURE 4-4 Evolution of posterior pituitary peptides of mammals. Mammals simultaneously produce two posterior pituitary hormones by gene duplication. Point mutations account for individual amino acid substitutions. Possible changes in nucleotide sequences accounting for point mutations are shown and substituted amino acids are circled. Reproduced from Lincoln, D. W., and Short, R. V., eds. (1984). "Reproduction in Mammals; Book III: Hormonal Control of Reproduction," 2nd ed. Cambridge Univ. Press, London and New York.

or OT (Figures 4-5 and 4-6) and the NP protein with which it complexes. NP I is associated with OT and NP II is associated with VP. The preprotein (Figure 4-7) is of about 20,000 molecular weight and is cleaved by proteases to an NP of about 9500–10,000 molecular weight (Figure 4-5). After removal of the signal peptide, the prohormone is cleaved at a dibasic signal (Gly-Lys-Arg), liberating OT extended by Gly-X-X, where X is a basic amino acid. These enzymes are subtilisin-like serine proteases. Afterward, the carboxy-terminal amino acids are removed by carboxypeptidase E.

The two NPs are very similar in structure, and either OT or VP can form a complex with either NP *in vitro.* In some cases, it has been reported that NPs contain lipids bonded noncovalently, such as cholesterol, phosphatidylcholine, phosphatidylethanolamine, phosphatidylserine, and sphingomyelin. The significance of lipids in these proteins is not clear.

The half-life of VP or OT is increased dramatically from about 3 to about 10–20 min when it is complexed with NP in blood. The hormone–NP complex probably stabilizes the hormone within the neurosecretory granules, but following release from the secretory granule, it is unlikely that the complex survives in the bloodstream.

The first three amino acids of either hormone are active in binding to the NP molecule, as shown for OT in Figure 4-8. Binding of the hormone with the NP probably involves a conformational change in the latter. Cysteine 1 of the hormone incorporates its free amino group in an electrostatic linkage to a charged carboxyl group of the NP-binding site (Figure 4-8). This carboxyl group could be on a glutamate residue. Amino acids 2 (Tyr) and 3 (Phe) of VP (Ile in OT) participate in hydrophobic interactions at the NP-binding site. The peptide bond between amino acid residues 2 and 3 appears to hydrogen bond to NP (Figure 4-8).

The amino acid sequences of NP I and NP II indicate extensive homology between the two. It has been shown that posterior pituitary hormones can be methylated (Arg or Lys residues are most likely), and it is surmised that the modified hormones will have altered biological activities. Apparently, NP serves only a stabilizing role insofar as the activities of the posterior pituitary hormones are concerned. Although NP is released in free form in the blood, a distinct activity of uncomplexed NP has not been found. In spite of this, in pathological states where vasopressin synthesis is impaired, such as in the Brattleboro rat and in *diabetes*

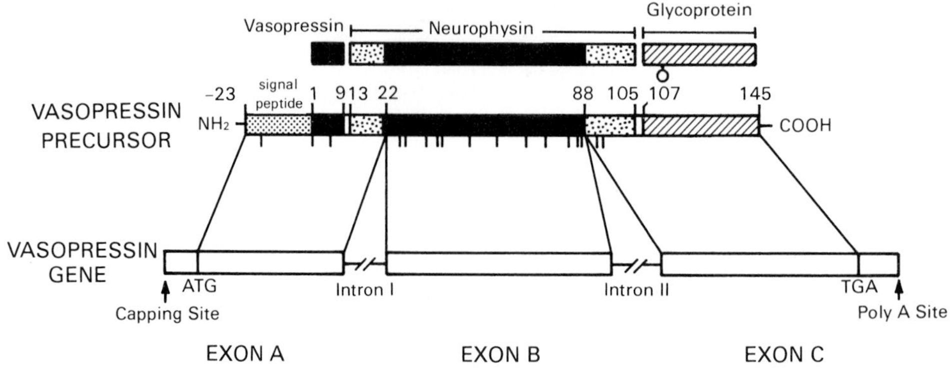

FIGURE 4-5 Proposed structure of the VP gene, its translation product, and the matured peptides. The gene codes for arginine-VP, NP, and glycoprotein and contains two introns. The gene product, provasopressin, is indicated in the middle. Negative numbers indicate the amino acid sequence of the signal sequence, while the positive numbers indicate the prohormone sequence. The attached open circle on the matured glycoprotein product at the top right represents carbohydrate. VP and OT are expressed as polyproteins. Reproduced from Richter, D. (1983). *Trends Biochem. Sci.* **8,** 278–281.

insipidus in humans, the only gene defects encountered are those in the gene sequence encoding VP-associated neurophysin.

The crystal structure of neurophysin II has been determined, and the dimer with two molecules of vasopressin hormone analog is shown in Figure 4-9. The neurophysin dimer is also indicated in Figure 4-7.

IV. BIOCHEMISTRY

A. Formation and Neurosecretion of Posterior Pituitary Hormones

VP and OT are formed in the cell bodies of neurons located in the hypothalamus. Separate neurons synthesize each type of hormone and its accompanying NP. Neuronal cell bodies synthesizing OT are located primarily, but not exclusively, in the paraventricular nucleus, whereas neuronal cell bodies synthesizing VP are located primarily, but not exclusively, in the supraoptic nucleus. Although there have been claims that the same cell can synthesize both types of hormones, this does not appear to be the case in view of the separate interneuronal stimulation of either secretion product. Figure 4-10A shows a concept of the VP-producing neuron and how the neurosecretion may be signaled and controlled by interneurons, including the osmoreceptor, which represents the primary positive signaling mechanism. Electrical signals from the osmoreceptor will be either on or off, depending on the osmolarity of the extracellular fluids. From this figure, we can see that the hormone is synthesized in the cell body and moves down the long axon extending from the hypothalamus to the nerve ending. The processing of the OT precursor occurs during transport from the cell body to the axonal nerve endings in the posterior pituitary. Curiously, it is reported that mRNA for both OT and VP is transported along the axon; however, the significance of this localization is not clear. It is possible that these axonal messages could be *en route* to degradation or storage to be used in replenishment at a later time when new synthesis of hormone is required. VP–NP II secretion is stimulated exclusive of OT–NP I release by using the drug nicotine mediated by the stimulation of a cholinergic receptor. OT–NP I can be released specifically through the action of estradiol.

The VP–NP I complex is released by the osmoreceptor, probably via an interneuron that senses the Na^+ concentration in surrounding fluids. Its own cellular volume shrinks at elevated salt concentrations, and the ionic changes induce receptor deformation that generates an electrical or chemical signal, which is transmitted to the vasopressinergic neuron. This signal, if electrical, may be transmitted down the axon of the vasopressinergic neuron (possibly along axonal nerve fibers), and in the nerve ending the electrical signal depolarizes the membrane, causing exocytosis of VP–NP II (Figure 4-10B). Other interneurons may convey signals from baroreceptors located in the carotid sinus, aortic arch, and left atrium. The signals impinging on the vasopressinergic neuron may be either electrical or chemical, but in Figure 4-10A, they have been pictured as electrical signals for convenience. The VP–NP II complex permeates local fenestrated capillaries in order to reach general circulation.

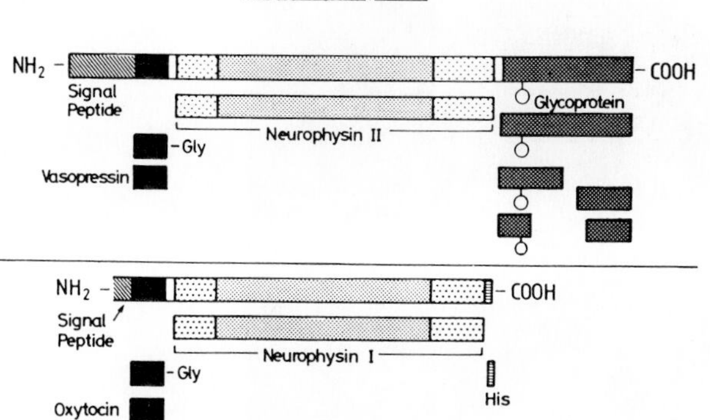

FIGURE 4-6 Preprovasopressin and preprooxytocin. Proteolytic maturation proceeds from top to bottom for each precursor. The organization of the gene translation products is similar in either case, except that a glycoprotein is included in the proprotein of VP in the C-terminal region. Heavily stippled bars of the NPs represent conserved amino acid regions; lightly stippled bars represent variable C- and N-termini. VP and OT are expressed as polyproteins. Reproduced from Richter, D. (1983). *Trends Biochem. Sci.* **8,** 278–281.

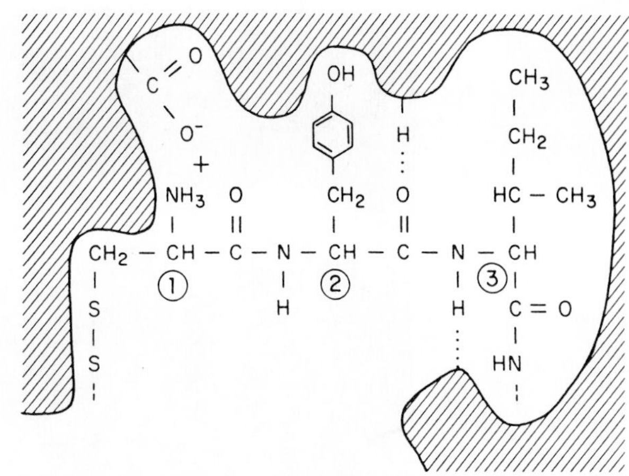

FIGURE 4-8 Interactions of the first three residues of OT with the strong site of NP. The shaded area represents the protein. The first three residues of VP are the same as those of OT, with the exception of phenylalanine in position 3. Residues 4–9 are not shown. Note that the peptide bond between residues 2 and 3 contributes to the hormonal interaction with NP through hydrogen bonding. Reproduced with permission from Breslow, E. (1984), "Cell Biology of the Secretory Process" (M. Cantin, ed.), pp. 276–308. Karger, Basel.

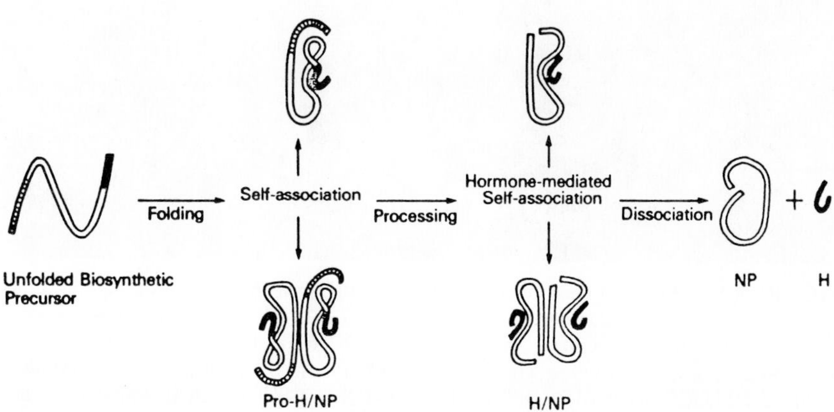

FIGURE 4-7 Relationship of biosynthetic precursor structure to molecular events occurring in neurohypophyseal hormone–NP biosynthesis. The filled and open lines denote hormone and NP segments, respectively. The hatched line represents the COOH-terminal glycopeptide occurring in pro-AVP–NP II. Folding of the precursors is visualized to lead to self-association through the NP domains of the precursors. The NP–NP and H–NP interaction surfaces are retained after enzymatic processing, the latter of which leads to the formation of noncovalent complexes between hormone (H) and NP, as well as to dimers in secretory granules until released by exocytosis. Biosynthesis occurs in the cell body followed by folding of the precursor. These are packaged in granules in the Golgi apparatus, and the granules (neurosecretory vesicles) are transported down the axon (the hormone is bound to NP in the vesicles). Finally, the granules are stored in the nerve terminals of the posterior pituitary until the signal for secretion occurs, after which H and NP may rapidly dissociate, producing the hormone free in the blood. Reproduced from Kanmera, T., and Chaiken, I. M. (1985). Molecular properties of the oxytocin/bovine neurophysin biosynthetic precursor. *J. Biol. Chem.* **260,** 8474–8482. Copyright © 1985 The American Society of Biological Chemists, Inc.

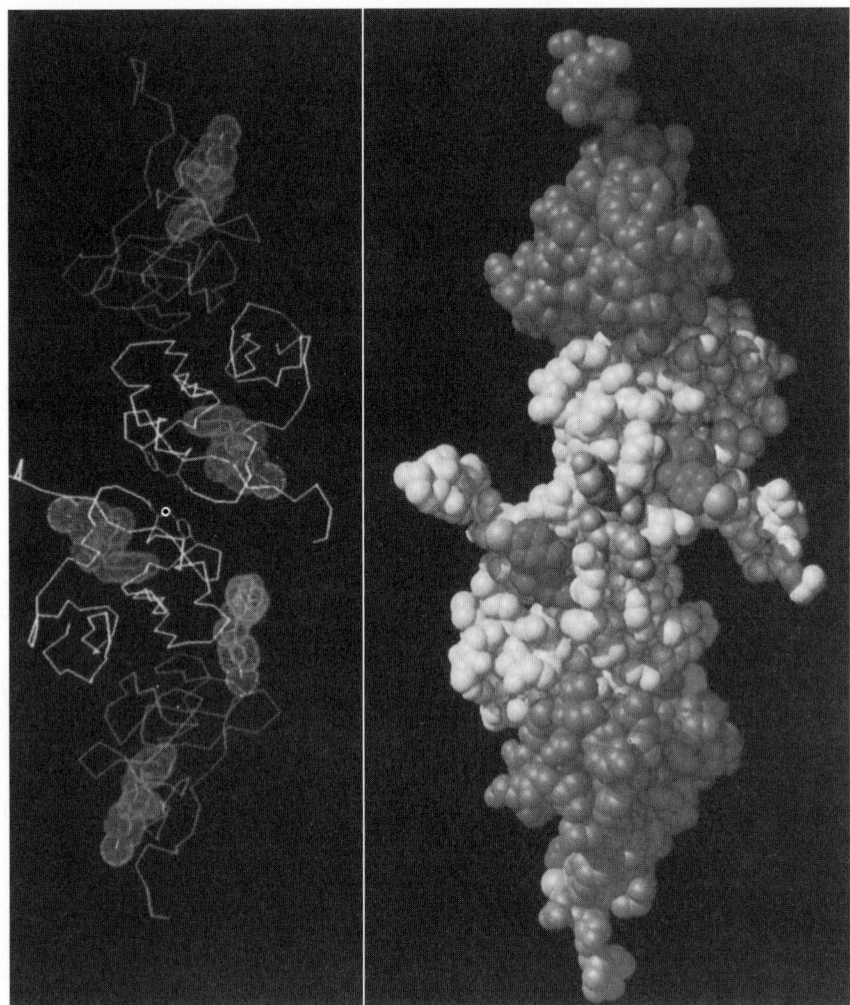

FIGURE 4-9 Models of neurophysin II dimer in contact with two molecules of hormone peptide. NPs are indicated in darker atoms (figure on right), while the hormone corresponds to the lighter atoms in the central part of the model. The left-hand side shows a corresponding stick model. The amino acid sequence of bovine NP II is AMSDLELRQCLPCGPGGKGRCF-GPSICCGDELGCFVGTAEALRCQEENYLPSPCQSGQKPCGSGGRCAAAGICCNDESCVTEP-ECREGVGFPRRV. This figure is a black and white adaptation of a color figure reproduced with permission from Chen, L. *et al.* (1991). *Proc. Nat. Acad. Sci. USA* **88,** 4240–4244.

At some point, the VP–NP complex dissociates and free VP in circulation binds to its membrane receptors. An analogous situation can be pictured for neurons synthesizing OT–NP complexes, except that such neurons are more abundant in paraventricular nuclei than in supraoptic nuclei, although they are present in both locations in the hypothalamus.

An aspect of this overall process of neurosecretion that has not been discussed up to this point is the fact that the membrane of the neurosecretory granule is retrieved by the nerve ending following exocytosis and presumably supplies starting materials for subsequently formed neurosecretory vesicles (endocytosis is coupled to exocytosis). A scheme of this process is presented in Figure 4-11A. The process of exocytosis clearly involves the uptake of Ca^{2+} ions (Figure 4-11C), which, in addition to their ability to aid in depolarization, play an active role in the process of exocytosis. This role is not specifically understood. It may involve derepression of a contractile mechanism when contractile proteins are constituents of the secretory mechanism. Ca^{2+} could bridge, by some action, the neurosecretory granules with the inner nerve ending membrane. A granular ATPase involved in the secre-

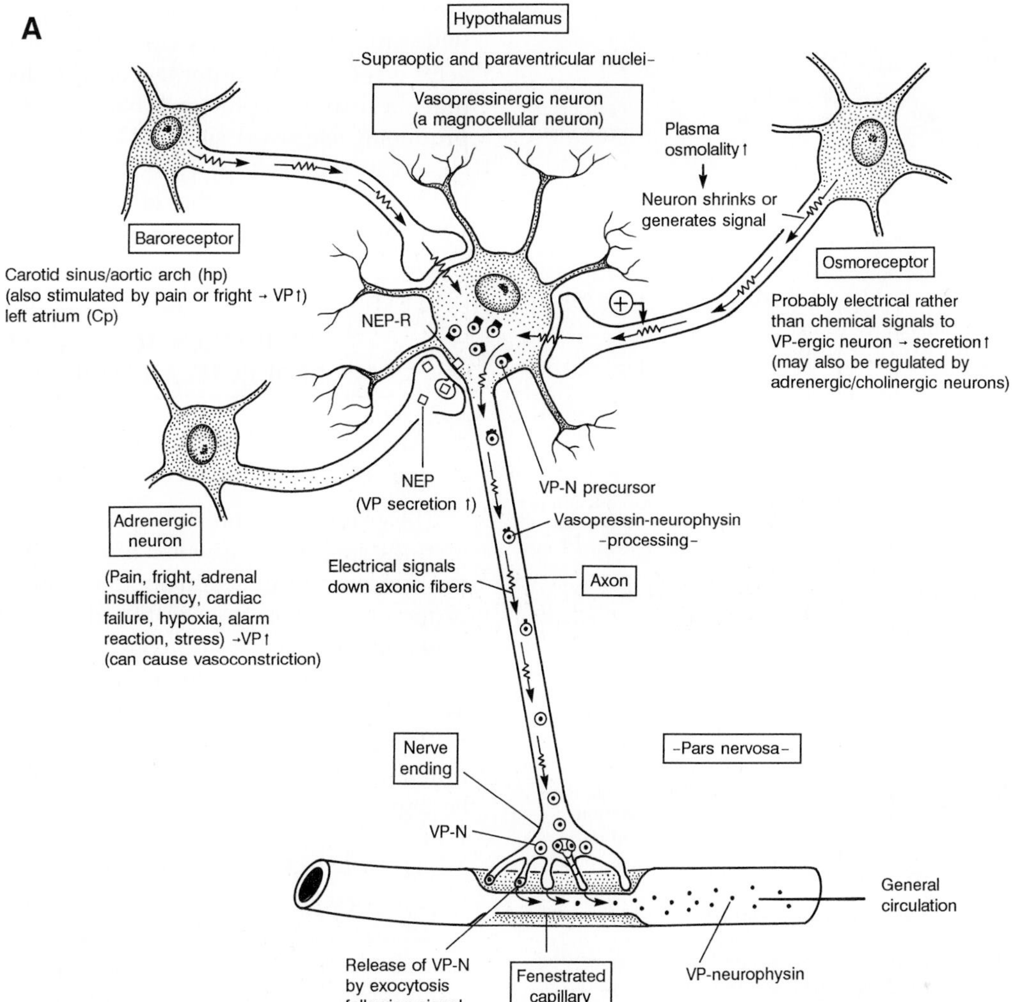

FIGURE 4-10A Vasopressinergic magnocellular neuron and its regulation. Speculative concept of the vasopressinergic neuron and interneurons that provide positive and negative stimuli for the secretion of posterior pituitary hormone. The vasopressinergic neuron in the center has a cell body in the hypothalamus with a long axon extending to the terminal region of the posterior pituitary. The osmoreceptor is the neuron pictured at the right. When plasma osmolality increases, the cell volume of the osmoreceptor neuron shrinks, or ionic changes conformationally alter the receptor to a point where the cell sends an electrical or chemical signal down its axon to the cell body of the vasopressinergic neuron. This signal would generate an action potential or, via a receptor for a chemical signal, a second messenger, whose information would be transmitted down the axon of the vasopressinergic cell to cause the release of the VP–NP complex from the nerve ending by exocytosis. The complex passes through fenestrations in local capillaries and reaches general circulation. Other interneurons that regulate VP secretion are pictured on the left: some aminergic neurons restrict neurotransmitter synthesis to the nerve ending. In the cell body, synthesis of prohormones occurs and the products are packaged into granules. These are moved by fast axonal transport (2 mm/hr) down the axon, and the prohormone is cleaved during this process to the mature products. These are stored in the nerve ending awaiting a signal, as described earlier. After exocytosis, there is retrograde transport of the granule membrane for reuse.

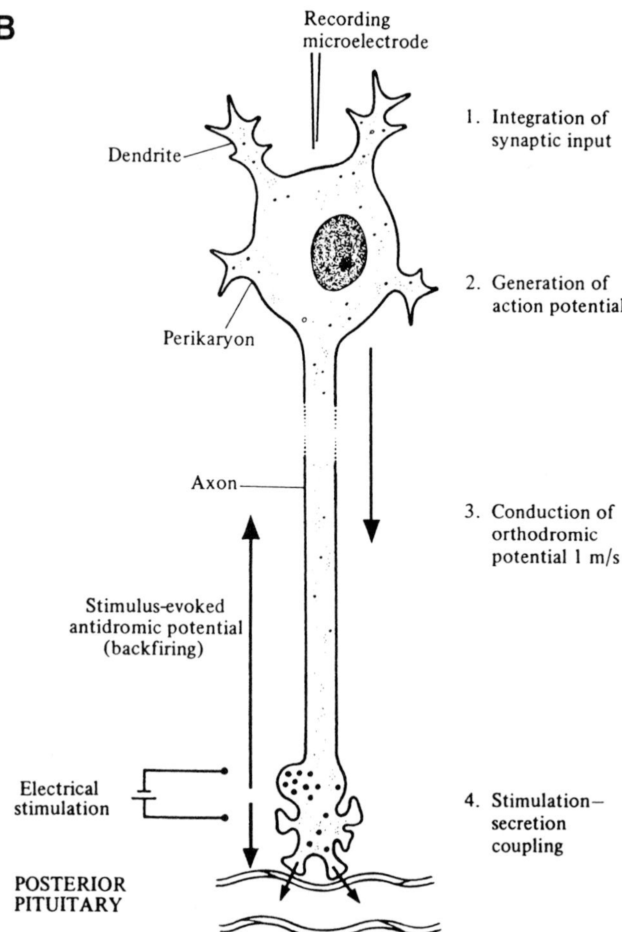

FIGURE 4-10B Electrophysiological properties of the magnocellular neuron. The numbered right-hand column describes the stages in the production of an action potential (the electrical signal) and its conduction along the axon to the nerve ending in the posterior pituitary. This results in depolarization, influx of calcium ions, and hormone release by exocytosis. Hormone release can be induced by electrical stimulation of the posterior pituitary by using an implanted electrode, as indicated at the bottom left. Reproduced from Lincoln, D. W. (1984). The posterior pituitary. *In* "Reproduction in Mammals; Book III: Hormonal Control of Reproduction" (C. R. Austin and R. V. Short, eds.), 2nd ed., pp. 21–50. Cambridge Univ. Press, London and New York. Copyright © 1984 Cambridge University Press.

tion process could be stimulated by Ca^{2+}, and it could interact with calmodulin for the stimulation of enzymes involved in the exocytosis process.

Other regulatory neurons that affect the secretion of posterior pituitary hormones illustrated in Figure 4-10A are those containing norepinephrine (noradrenaline), which are adrenergic neurons, or those containing acetylcholine (cholinergic neurons), which are not shown in Figure 4-10A. These neurons affect the release of posterior pituitary hormones, but it is not clear, as stated before, which modifying neurons are primary for the peptidergic neurons secreting VP or OT and which may be secondary (interneurons modifying the activity of primary interneurons). Indeed, it may be that the osmoreceptor and baroreceptors, instead of producing electrical signals (Figure 4-10A), produce chemical signals of the type shown here. A summary of the neurochemicals that affect the release of OT and VP is presented in Figure 4-12.

V. BIOLOGICAL AND MOLECULAR ACTIONS

A. Role of VP (Antidiuretic Hormone) in the Regulation of Body Fluids

The primary action of arginine vasopressin (AVP) is in the inhibition of water diuresis (loss of water in the urine), and this action operates through the actions of membrane receptors in the kidney. AVP also elevates blood pressure and constricts coronary arteries, and it stimulates intestinal and uterine contractions, as well as slightly stimulating milk ejection. However, the last two activities modestly cross over to the major functions of OT. In most cases, the secretion of OT will be signaled separately from the signaling to release AVP.

Figure 4-13 shows an overview of some of the main features of water balance (additional discussion is presented in Chapter 15). The hypothalamic osmoreceptor detects elevations in serum tonicity. The mechanism is thought to involve the shrinking of the osmoreceptor neuron to the point where a signal (chemical or electrical) is generated and transmits this signal down its axon to its nerve ending. Alternatively, the changed ionic environment produces a conformational alteration in the osmoreceptor, which generates the signal. The signal is then received by the cell body of the vasopressinergic neuron and transmitted to its own nerve ending for the release of VP–NP. Signals to baroreceptors that control the release of AVP–NP are exemplified by emotional states that inhibit the release of AVP–NP and lead to transitory frequency of urination. Nicotine intake and pain may also operate through baroreceptors of the CNS repressive center, which may stimulate AVP–NP release through adrenergic stimulation. The AVP–NP complex is briefly stable in blood, having an experimental half-life of 10–20 min. AVP dissociates from NP and binds to a membrane receptor located in distal tubular cells of the kidney. The second messenger of this interaction is cyclic AMP, which is elevated in the kidney cell and leads to the phosphorylation of proteins. Phosphorylation of proteins causes the admission of water from the lumen (urine side) at

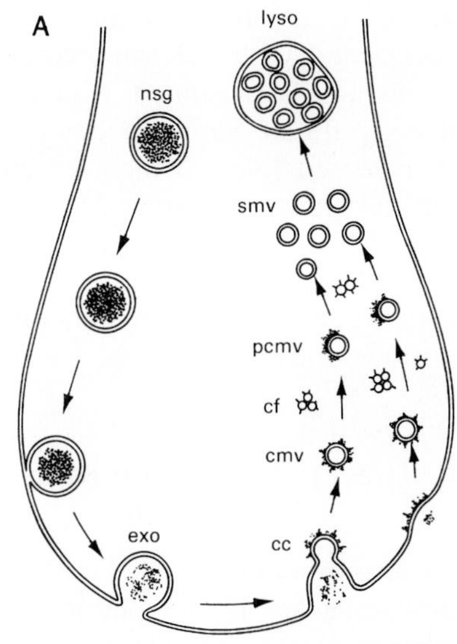

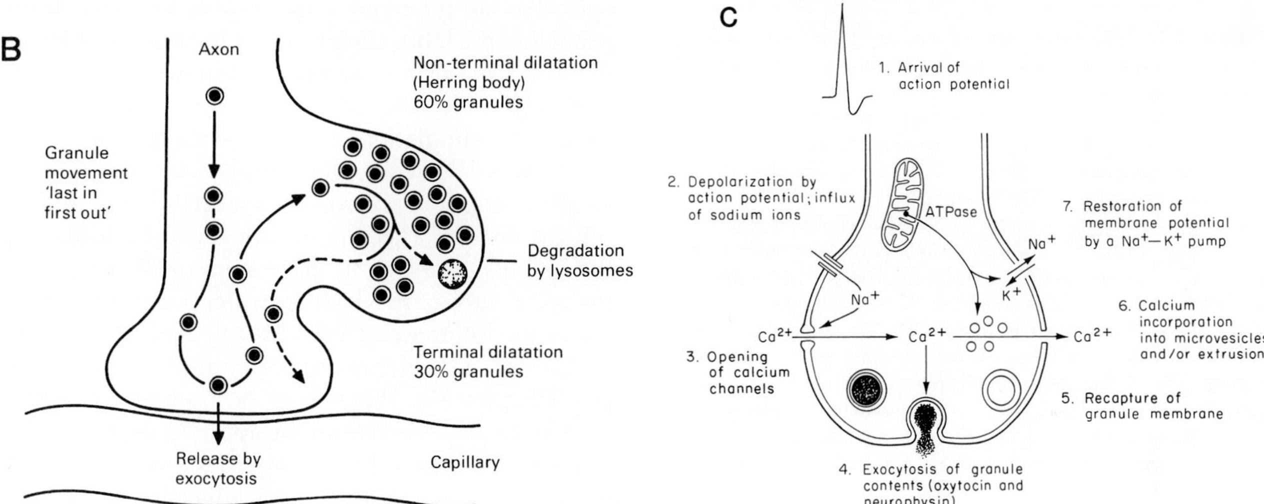

FIGURE 4-11 Mechanism of secretion of posterior pituitary hormones. (A) Scheme showing the exocytosis–vesiculation sequence operating in neurohypophyseal terminals. Contents of the neurosecretory granules (nsg) are emptied into the extracellular space by exocytosis (exo). The granule membrane is reused from the surface of the nerve ending by vesiculation to produce coated caveolae (cc), which pinch off as coated microvesicles (cmv) and then shed coat fragments (cf) to become partially coated microvesicles (pcmv) and then smooth (synaptic) microvesicles (smv). These are subsequently incorporated into lysosomal bodies (lyso). Reproduced from Douglas, W. W. (1974). Mechanism of release of neurohypophysial hormones: Stimulus-secretion coupling. *In* "Handbook of Physiology" (R. O. Greep, E. B. Astwood, E. Knobil, W. H. Sawyer, and S. R. Geiger, eds.), Sect. 7, Vol. IV, Part 1, pp. 191–224. American Physiological Society, Washington, DC. (B) Selectivity of secreted posterior pituitary hormones during exocytosis. The relative distribution of granules between axons, Herring bodies (nonterminal dilatations), and terminal dilatations has been measured by electron microscopy. The movement of granules has been determined by autoradiography. Readily releasable hormone may be about 10% of the gland content. Newly synthesized hormone is released preferentially, suggesting that new neurosecretory granules must first pass through terminals before moving on to less available storage sites. Reproduced from Lincoln, D. W. (1984). The posterior pituitary. *In* "Reproduction in Mammals; Book III: Hormonal Control of Reproduction" (C. R. Austin and R. V. Short, eds.), 2nd ed., pp. 21–50. Cambridge Univ. Press, London and New York. (C) Stages in the process of stimulus–secretion coupling within an axon terminal of the posterior pituitary. Reproduced from Lincoln, D. W. (1984). The posterior pituitary. *In* "Reproduction in Mammals; Book III: Hormonal Control of Reproduction" (C. R. Austin and R. V. Short, eds.), 2nd ed., pp. 21–50. Cambridge Univ. Press, London and New York. (B and C) Reproduced by copyright permission of Cambridge University Press.

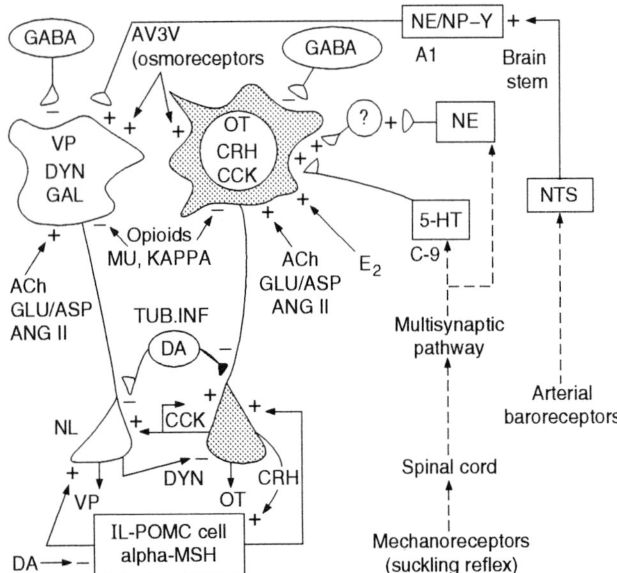

FIGURE 4-12 Schematic view of magnocellular oxytocin (OT) and vasopressin (VP) neurons and the postulated neurochemical influences from local interneurons and distant reflex pathways found *in vivo* that theoretically can influence their gene expression. Arrows and nerve endings end in + or − symbols, which represent excitatory or inhibitory influences, respectively. Coexistent peptides, dynorphin (DYN) and galanin (GAL), are shown in the VP cells, and corticotrophin-releasing hormone (CRH) and cholecystokinin are shown in the OT cells. Their complex interactions at the nerve terminal level in the neural lobe (NL) are also illustrated. Other abbreviations: PVN, paraventricular nucleus; SON, supraoptic nucleus; DA, dopamine; IL, intermediate lobe; POMC, proopiomelanocortin; ACh, acetylcholine; Glu/Asp, excitatory amino acids glutamate/aspartate; ANG II, angiotensin II; NE, norepinephrine; NP-Y, neuropeptide Y; 5-HT, serotonin; NTS, nucleus tractus solitarius; A1, A2, and C-9, brainstem biogenic amine cell groups. Reproduced with permission from Gainer, H. and Wray, S. (1992). *Ann. N.Y. Acad. Sci.* **652,** 14–28.

an increased rate. Presumably, specific phosphorylation results in increasing the transport of proteins called aquaporins (AQPs) that are sequestered in the cell at a constant level. Enhanced transport of AQPs to the apical membrane results in their insertion into the membrane and the creation of new water channels. An increased amount of water is absorbed into the cell by free diffusion and crosses the cell to the basolateral membrane, possibly enclosed in a particle that may be constructed of AQPs.

Water is finally released at the basolateral membrane to augment the fluid of the blood, as shown in Figure 4-15. The net result is an increase in blood pressure and a lowered osmolality. These effects lead to a quenching of the signal to the vasopressinergic neuron (the osmoreceptor takes on water and resumes its unstimulated physical state). Increased serum tonicity also signals the hypothalamic thirst center, which is closely associated with the osmoreceptor. Consequently, more water is consumed, which leads to the dilution of body fluids. Simultaneously, the kidney has sensed increased [Na^+], which inhibits the Na^+ conservation mechanism (Chapter 15) and results in the excretion of Na^+, which in turn tends to reduce blood pressure. Thus, these two systems (and perhaps additional ones) function to maintain homeostatic water balance in the body.

B. The VP Target: The Kidney Distal Tubule

Water is transported in epithelial cells of two major classes. Some cells are highly permeable to water, whereas other types of epithelia, the so-called tight epithelia, are relatively impermeable to water. It appears that the latter class is under the control of VP, so that if these cells are not involved in water absorption normally under the influence of VP, there can be a significant stimulation of water uptake from the urine. In VP-stimulated cells, water appears to move through small aqueous pores across the apical (luminal surface) cell membrane. These pores appear to be formed in the cell membrane as a consequence of VP action. It is the apical surface in which the water pores are inserted under the influence of VP. This cell is also responsive to aldosterone, which stimulates sodium ion resorption (see Chapter 15). Water may be transported within vesicles (aquaporins) from the apical to the tissue side of the cell. Figure 4-14A exemplifies a vesicle capable of carrying water across the cell, which may be a member of the aquaporin family. This particular vesicle operates at the basolateral side of the cell, whereas the VP effect may occur at the apical side, where this type of vesicle (pore) may be formed to move large quantities of water across the cell. Figure 4-14B shows the structure of a single aquaporin molecule.

C. Mechanism of Action of VP in the Kidney Distal Tubule

An overview of the speculated mechanism of action of VP in the distal tubular cell is shown in Figure 4-15. At the top of the figure, the VP–NP complex in the blood is shown to dissociate some time after the release of VP–NP from the posterior pituitary. Free VP reaches the extracellular space and binds to the VP V_2 receptor on the basolateral membrane of the cell. It is required that VP be released from the VP–NP complex prior to associating with its receptor, since the specificity of

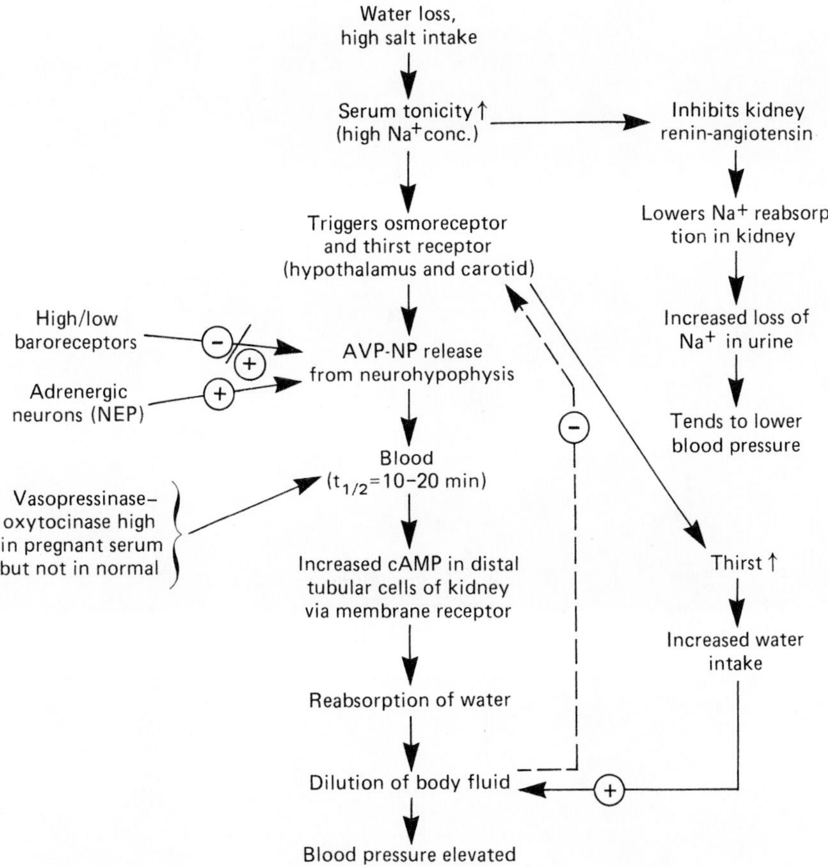

FIGURE 4-13 Scheme describing the regulation of body fluids. Abbreviations: NEP, norepinephrine; AVP–NP, Arg-VP–NP complex.

ligand binding shows that Tyr2 and Asn5 play critical roles in the interaction with receptor. Since Tyr2 is also involved in the interaction with NP, the hormone must be in the free form to interact with the receptor. Through this interaction, there is a stimulation of adenylate cyclase activity on the cytoplasmic side of the membrane. This process involves the transducing guanine nucleotide regulatory protein (Figure 4-16; see Chapter 1 for details of the activities of the G-protein subunits). ATP is converted to cyclic AMP on the cytoplasmic side of the cell membrane by the activated cyclase. Thyroid hormone may affect VP-sensitive adenylate cyclase. Thus, in hypothyroid rats, both the basal activity and the VP-stimulated activity of adenylate cyclase are decreased. Treatment with thyroxine restored the activity to normal levels. The elevated level of cyclic AMP, produced by VP, combines with the regulatory subunits (R) of inactive protein kinase (R_2C_2) to release the catalytically active subunit (C):

$$R_2C_2 + 4cAMP \rightarrow R_2 \cdot 4cAMP + 2C$$

The detailed activation mechanism for PKA is presented in Chapter 1. The active protein kinase (PK_a or C) appears to catalyze the phosphorylation of some specific protein possibly involved in the transport of aquaporins. These insert into the apical membrane to form channels for water uptake, which occurs by a process of free diffusion. Thus, the tight epithelial cell takes up little water in the unstimulated state and takes up a considerable amount of water in the VP-stimulated state due to a profusion of water channels created in the apical membrane. Presumably, water is transported across the cell and released at the basolateral side, involving water molecules transported by aquaporins (Figure 4-14).

Electron micrographs from toad bladder granular cell luminal membrane, untreated or treated with VP, show the aggregation of membrane particles after stimulation with VP. These observations suggest the possibility that water channels (AQP–CD) are being

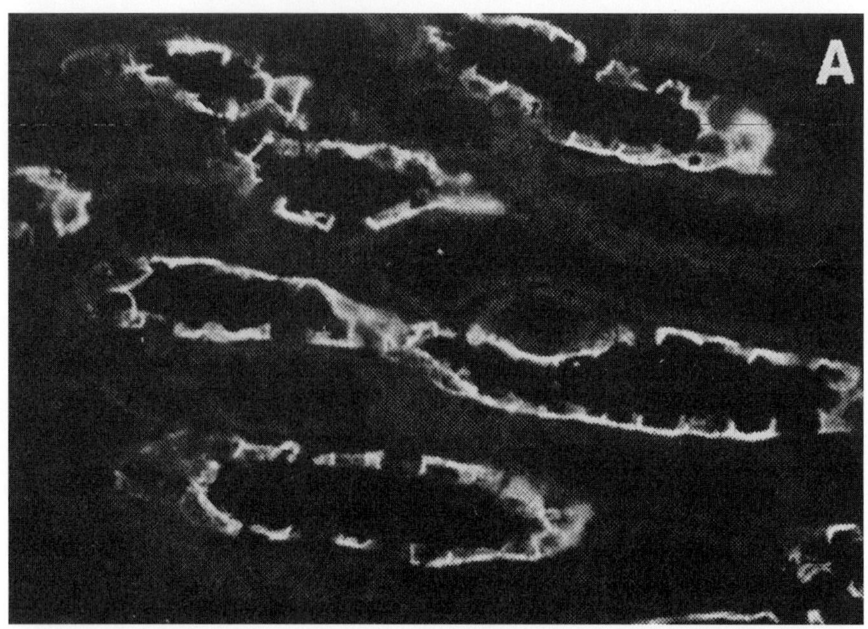

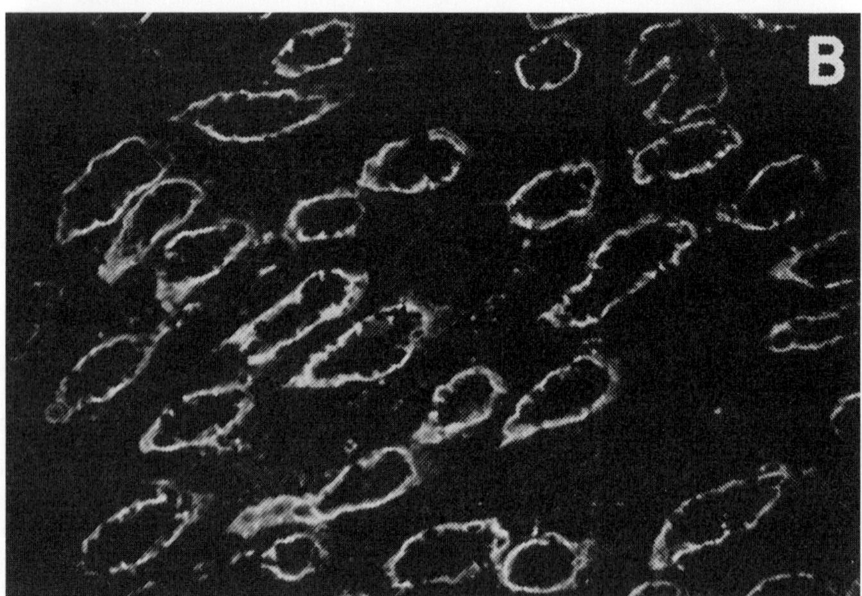

FIGURE 4-14A Visualization of a water-transporting "aquaporin." These vesicles were detected by immunofluorescence in rat kidney cortex (A) and outer medulla (B) with a polyclonal antibody directed against a C-terminal peptide of aquaporin P3. This was detected at the basolateral membrane of collecting duct cells. Reproduced with permission from Ishibashi, K., Sasaki, S., Fushimi, K., Uchida, S., Kuwahara, M., Saito, H., Furukawa, T., Nakajing, K., Yamaguchi, Y., Gojobori, T., and Marumo, F. (1994). *Proc. Nat. Acad. Sci. USA* **91,** 6269–6273. *B,* Structure of aquaporins. Asterisk indicates the site of cysteine in aquaporin (AQP), AQP–CHIP and AQP–CD (collecting duct). AQP–CD is a VP-responsive aquaporin. The cysteine site is not present in MIWC or AQP3. Reproduced with permission from Knepper, M. A., (1994). *Proc. Nat. Acad. Sci. USA* **91,** 6255–6258.

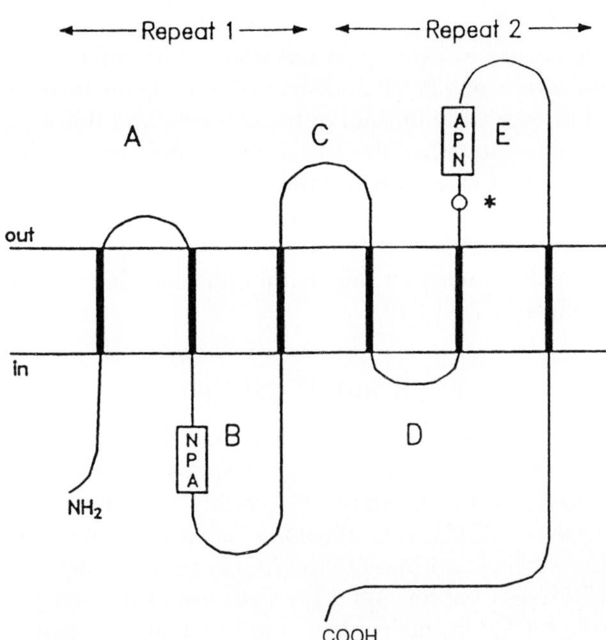

FIGURE 4-14B Structure of aquaporins. Asterisk indicates the site of cysteine in aquaporin (AQP), AQP–CHIP and AQP–CD (collecting duct). AQP–CD is a VP-responsive aquaporin. The cysteine is not present in MIWC or AQP3. Reproduced with permission from Knepper (1994).

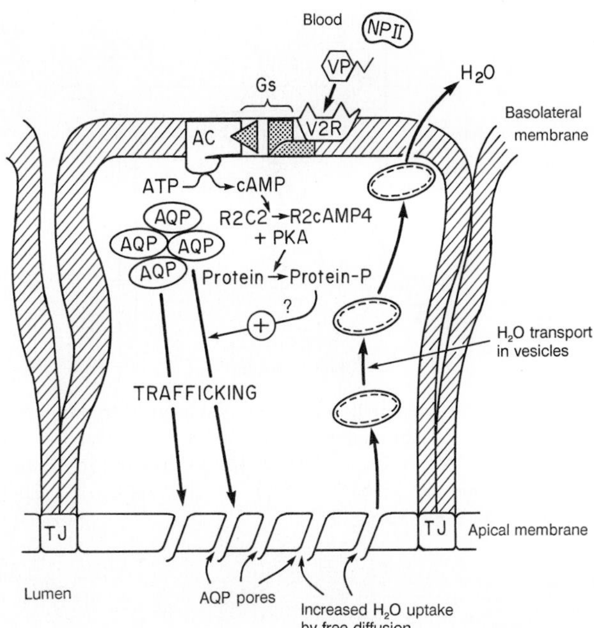

FIGURE 4-15 Speculative model of the mode of action of vasopressin via the V_2 receptor in the collecting duct of the kidney (distal tubule). Starting at the top of the figure, VP is seen in its dissociated form from neurophysin II, and free VP binds to the V_2 receptor. The hormone–receptor complex transmits its signal through the G-protein (G_s)-coupled receptor to adenylate cyclase (AC), which catalyzes the conversion of ATP to cyclic AMP (cAMP). Accumulation of cAMP activates protein kinase A (PKA) by associating with its inactive precursor, R_2C_2, converting it to free regulatory subunit bound with cAMP (R_2cAMP_4) and active kinase (PKA). PKA phosphorylates a presumed specific protein(s), which permits the transport of aquaporin proteins stored in particles (AQP) to the apical membrane, where aquaporin protein (Fig. 4-14B) inserts in the membrane to form a "water pore." This may be stabilized in tetrameric form. With an increase in water pores, more H_2O is taken up by passive diffusion and presumably transported across the cell cytoplasm to the basolateral membrane in vesicles. These vesicles could also involve aquaporin proteins (see Fig. 4-14A). It is unknown whether VP has an effect on AQP mRNA. This is a highly speculative mechanism of enhanced water uptake from urine by the collecting duct cell under the stimulation of VP. TJ: tight junction.

inserted into the membrane under the influence of the hormone.

D. Nature of Kidney VP-Sensitive Adenylate Cyclase

The VP receptor has been studied extensively in beef renal medullary membrane preparations, which contain the receptor and the coupled hormone-sensitive adenylate cyclase. Isolated membrane particles retain nearly all of the specificity of binding and the affinity of the receptor for VP characteristic of the whole cell, enabling the evaluation of ligand binding and consequences for adenylate cyclase activity. In essence, the system can be summarized as shown in Figure 4-16. The binding of the hormone to the membrane receptor results in the activation of adenylate cyclase, so that the intracellular level of cyclic AMP can be raised effectively. The rate-limiting step in the activation of adenylate cyclase is the formation of the hormone–receptor complex. As in other hormone–receptor systems, there appears to be one ligand-binding site per receptor molecule. The transducing action coupling ligand interaction with receptor to activation of adenylate cyclase is a linear function. VP associates with its receptor about 30 times faster than does OT, and the receptor has about 100 times the affinity for VP compared to OT.

GTP accelerates the rate at which VP stimulates adenylate cyclase activity, and this mechanism is elaborated in Chapter 1. The availability of GTP or nucleoside triphosphates, such as ATP, acting similarly is important for the production of this effect in the cell. Under certain conditions, nucleoside trisphosphates can inhibit this process.

E. Effects of VP on the Liver

VP, like α-adrenergic agonists, acts on the hepatocyte to promote glycogenolysis; however, the physiological importance of this action is not well understood. In this effect, VP, like angiotensin II and

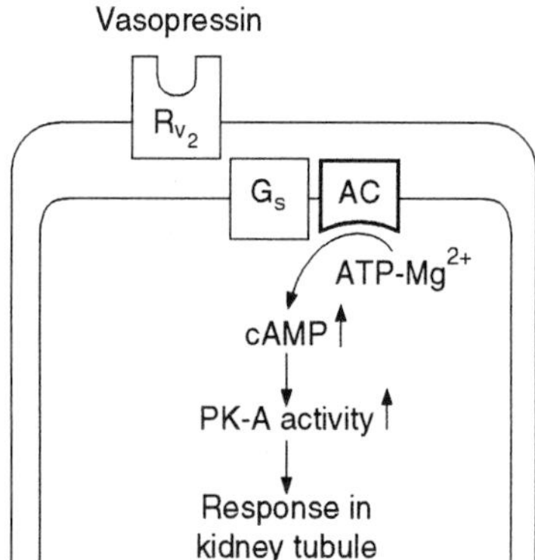

FIGURE 4-16 Vasopressin receptor (V_2 receptor) in the kidney. Abbreviations: G_s, stimulatory G protein; AC, adenylate cyclase; cAMP, cyclic AMP, PKA, protein kinase A. Drawing reproduced from "Hormones" (E.-E. Baulieu and P.A. Kelly, eds.), p. 288. Chapman and Hall, New York (1990).

phenylephrine (see Chapter 11), via an increase in cytoplasmic Ca^{2+}, releases glucose from the liver while promoting the uptake of K^+. Release of glucose is accomplished by stimulating the breakdown of glycogen. These actions of VP can be blocked by phentolamine, an α-blocking agent whose structure is shown in Figure 4-17.

Glycogenolysis is stimulated by VP in liver by mobilizing intracellular stores of Ca^{2+} (mitochondria and/or endoplasmic reticulum), which results in an increase in cytosolic Ca^{2+}. Increased cytosolic Ca^{2+} stimulates phosphorylase *b* kinase and, hence, phosphorylase activation (conversion of phosphorylase *b* to *a*), extrusion of Ca^{2+}, and uptake of K^+. VP interacts with a receptor in the hepatocyte cell membrane, resulting in the generation of a second messenger causing Ca^{2+} mobilization. Ca^{2+} released from the intracellular storage sites (e.g., mitochondrion) is expelled from the cell in part by a specific mechanism in the cell membrane. VP also stimulates the phosphorylation of about a dozen proteins in hepatocyte cytosol, including pyruvate kinase

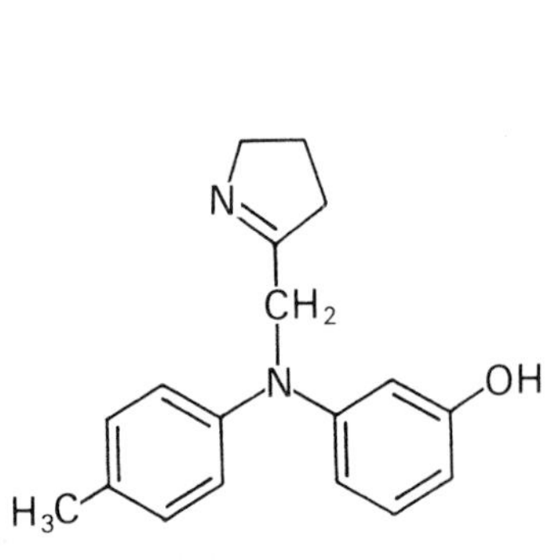

FIGURE 4-17 Structure of phentolamine, an α-blocking agent.

in addition to phosphorylase. Cyclic AMP, a second messenger in β-receptor reactions, is not involved in the effects of VP. VP appears to stimulate the turnover of phosphatidylinositol in hepatocytes, and this effect is linked to either the VP–receptor interaction or the process of extrusion of intracellular Ca^{2+} and uptake of extracellular Ca^{2+}. A model for the V_1 vascular vasopressin receptor action is shown in Figure 4-18. The hepatocyte receptor falls within this general class (see Table 4-1).

F. VP and ACTH Release

It has been known for some time that VP injected at a point where it could gain access rapidly to the hypothalamic–pituitary axis could stimulate the release of ACTH, and indeed, VP is a releasing factor for ACTH in addition to the ACTH-releasing action of CRH (see Chapters 3 and 10). The most likely biological role for VP in this context is to function to potentiate the release of ACTH by CRH, although VP can release some ACTH by itself. CRH has been shown to produce rapid stimulation of cyclic AMP in rat anterior pituitary cells. Although VP agonists cause an increase in CRH-induced ACTH release from rat anterior pituitary cells in culture, VP alone has no effect on cyclic AMP levels in these cells. Its action is mediated by the VP V_{1b} receptor through the G_{plc} protein and is similar to the sequence of events in whole or in part shown in Figure 4-18. However, when combined with CRH, VP causes a twofold increase in the CRH-induced accumulation of cyclic AMP; thus, VP acts by enhancing the effectiveness of CRH. Table 4-1 shows a classification of these receptors.

In addition to its activity in stimulating the direct and indirect (via CRH) release of ACTH from the anterior pituitary, evidence suggests that VP may stimulate cortisol secretion in the nanomolar range directly from the adrenal cortex through the activation of V_1 receptors (Table 4-1). Oxytocin has no similar effect, even at high concentrations. AVP apparently is produced locally in both the adrenal cortex and the medulla together with its neurophysin, so that it acts as a paracrine factor for the stimulation of adrenal steroidogenesis. However, AVP is not present in the human fetal adrenal medulla, but appears to be present in the adult. After stimulation of steroidogenesis by AVP, steroid output decreases, and it appears that AVP desensitizes its receptor. There is evidence that AVP is synthesized in other organs also, such as the ovary, testis, uterus, thymus, and pancreas, where it probably also acts in a paracrine fashion. In addition to its effect in stimulating the release of cortisol from the adrenal cortex, AVP has been reported to enhance the mitotic activity of

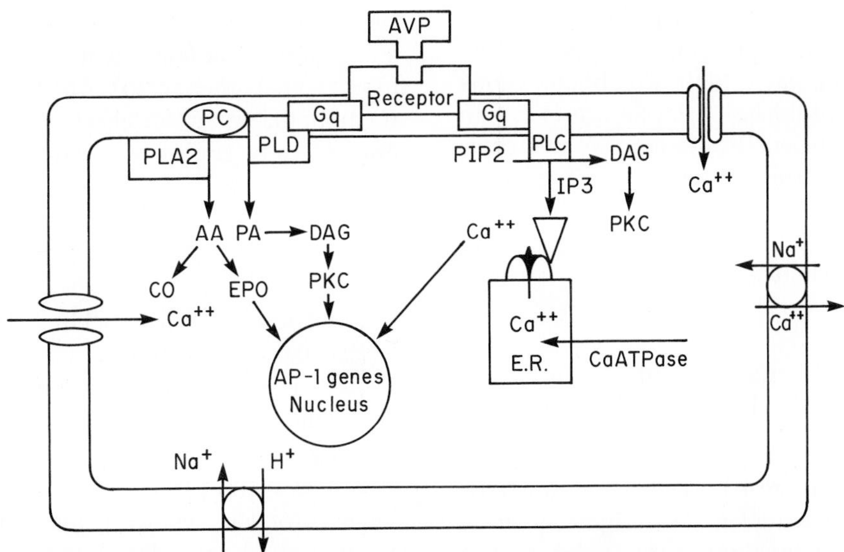

FIGURE 4-18 Cytoplasmic and nuclear signaling pathways of V_1 vascular vasopressin receptors. Abbreviations: PLC, phospholipase C; PLD, phospholipase D; PLA_2, phospholipase A_2; PA, phosphatidic acid; AA, arachidonic acid; PC, phosphatidylcholine; PKC, protein kinase C; ER, endoplasmic reticulum; CO, cyclooxygenase pathway; EPO, epoxygenase pathway; IP_3, 1,4,5-inositol triphosphate; DAG, 1,2-diacylglycerol; Gq, G protein. Reproduced with permission from Thibonnier, M., Bayer, A.L., and Leng, Z. (1993). *Regulatory Peptides* **45**, 79–84, 1993.

rat adrenal cortical cells in culture. Apparently, endogenous AVP may be released together with catecholamines during stress, and it may modulate steroidogenesis in the adrenal gland directly, in addition to the other better known humoral mechanisms leading to cortisol release.

G. Behavioral Effects of VP

Lysine-and arginine-VP have been shown in model systems to positively affect the consolidation of memory and learning. Whereas puromycin induces amnesia, VP attenuates the puromycin effect by a process distinct from that by which puromycin exerts its effects on the cell. A single subcutaneous injection of VP increases resistance to the extinction of a pole-jumping avoidance response in the rat, confirming the positive effects of the hormone on conditioned behavior. So far, the structural requirements for the behavioral effects of VP are less stringent than those for hormonal activities (e.g., stimulation of renal adenylate cyclase activity). This suggests that the VP receptors in kidney and brain are different (Table 4-1). The Brattleboro rat has been used to assess the effects of VP deficiency. These animals are homozygous in their inability to synthesize VP, and they have *diabetes insipidus.* Other work suggests a single base deletion in the VP neurophysin gene that predicts a hormone precursor having a different C-terminus, indicating the possibility of faulty

TABLE 4-1 Classification of VP Receptors and OT Receptor

Receptor/subtype	Coupling system[a]	G protein	Effect	Target
Vasopressin V_{1a} (vasopressor, glycogenolytic)	PLC	G_{plc}	Stimulation	Smooth muscle (blood vessel constriction), liver, CNS
	AC	G_i	Inhibition	Liver
V_{1b} (pituitary)	PLC	G_{plc}	Stimulation	Pituitary
V_2	AC	G_s	Stimulation	Distal and collecting tubule
Oxytocin	PLC	G_{plc}	Stimulation	Uterus, CNS

[a] Abbreviations are the same as those in the legend of Fig. 4-18.

transcription–translation. Norepinephrine levels are elevated in the dorsal septal nucleus and in the supraoptic nucleus compared to those in homozygous nondiabetic rats. Catecholamine levels are lower in the paraventricular nucleus of the diabetic rats compared to those in controls. These results and other studies suggest that VP modulates catecholamine neurotransmission in specific brain regions and may be related to the effects of VP on behavior, learning, and memory.

H. Overview of Regulation of Oxytocin Secretion from the Posterior Pituitary

A summary of the secretion of OT from the posterior pituitary is shown in Figure 4-19. Two major functions are milk ejection in lactating females and participation in signaling uterine contractions at the termination of pregnancy. Some experiments suggest that OT may cause the inhibition of androgen synthesis in the testis, but knowledge of the regulation of secretion of OT in males is not clear. Suckling and related stimuli (auditory, visual) result in milk ejection from the mammary gland of nursing females and are transmitted rapidly over a spinal reflex arc to the paraventricular nucleus of the hypothalamus (see also Chapter 14). This stimulus appears to be cholinergic and is transmitted to nerve endings in the posterior pituitary, resulting in Ca^{2+} uptake, depolarization, and exocytosis of OT–NP into circulation. The OT–NP complex is dissociated after exocytosis.

Transmission of the suckling stimulus to the paraventricular nucleus of the hypothalamus by way of a spinal arc reflex is accomplished in milliseconds. The oxytocinergic neuron is stimulated by acetylcholine and repressed by norepinephrine; thus, it can be envisioned as being positively regulated by a cholinergic interneuron and negatively regulated by an adrenergic interneuron. Visual stimuli (e.g., the sight of a hungry infant or auditory stimuli such at the sound of a baby's cry) can lead to dripping of milk from the breast. These signals are transmitted to the hypothalamus and result in the release of some OT. On the other hand, fear and

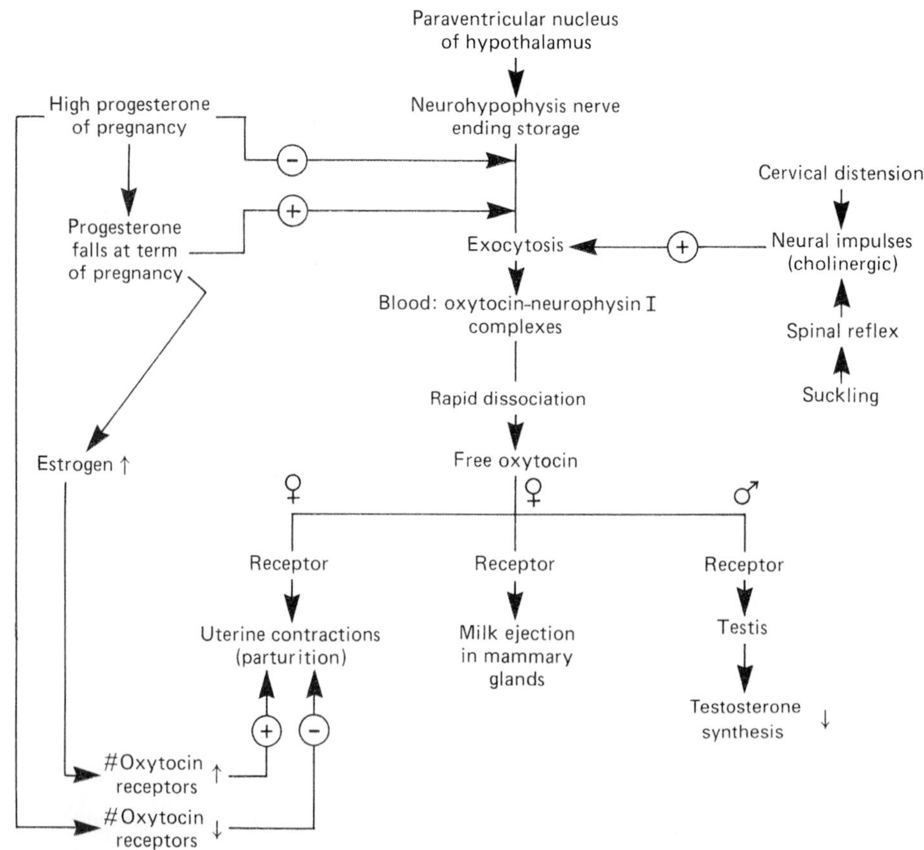

FIGURE 4-19 Secretion of OT from the posterior pituitary. The effect of OT in the testis appears to lower androgen levels and elevate pregnenolone and progesterone.

stress can inhibit OT release. Only a few seconds are required for OT to reach its target in the mammary gland. After dissociation of the NP, OT binds to a specific cell membrane receptor on myoepithelial cells and causes Ca^{2+} uptake in these cells, which derepresses the contractile mechanism. Consequently, ductules and ducts contract and milk is ejected from the gland through the nipple into the infant's mouth. It is not known whether any activities can be ascribed to NP once they have been dissociated from the hormonal complex in the blood.

In the case of terminal uterine contractions at birth, the primary signal for OT release from the posterior pituitary is a marked decline in the level of circulating free progesterone, which was being produced at high levels from the placenta during the latter course of pregnancy. At term, there is a surge of estradiol production with the dramatic decline in free progesterone. Estradiol stimulates the uterine myometrium to become more sensitive to OT by inducing the formation of new OT receptors (Figure 4-20). As a result of this increase in OT receptors in the myometrium, small amounts of oxytocin will stimulate uterine contractions at levels of the hormone that would be ineffective with-

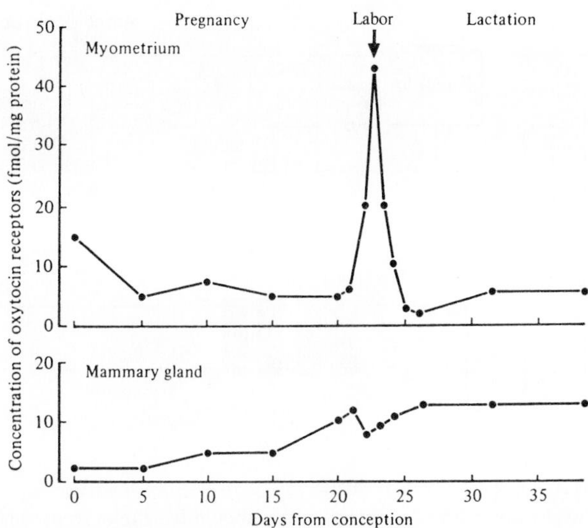

FIGURE 4-20 Concentration of OT receptors in the rat myometrium and mammary gland during pregnancy and lactation. Receptor number is expressed as the specific binding of tritiated OT to particulate fractions (femtomoles/milligram of protein). Reproduced from Lincoln, D. W. (1984). The posterior pituitary. *In* "Reproduction in Mammals; Book III: Hormonal Control of Reproduction" (C. R. Austin and R. V. Short, eds.), 2nd ed., pp. 21–50. Cambridge Univ. Press, London and New York. Reproduced by copyright permission of Cambridge Univ. Press.

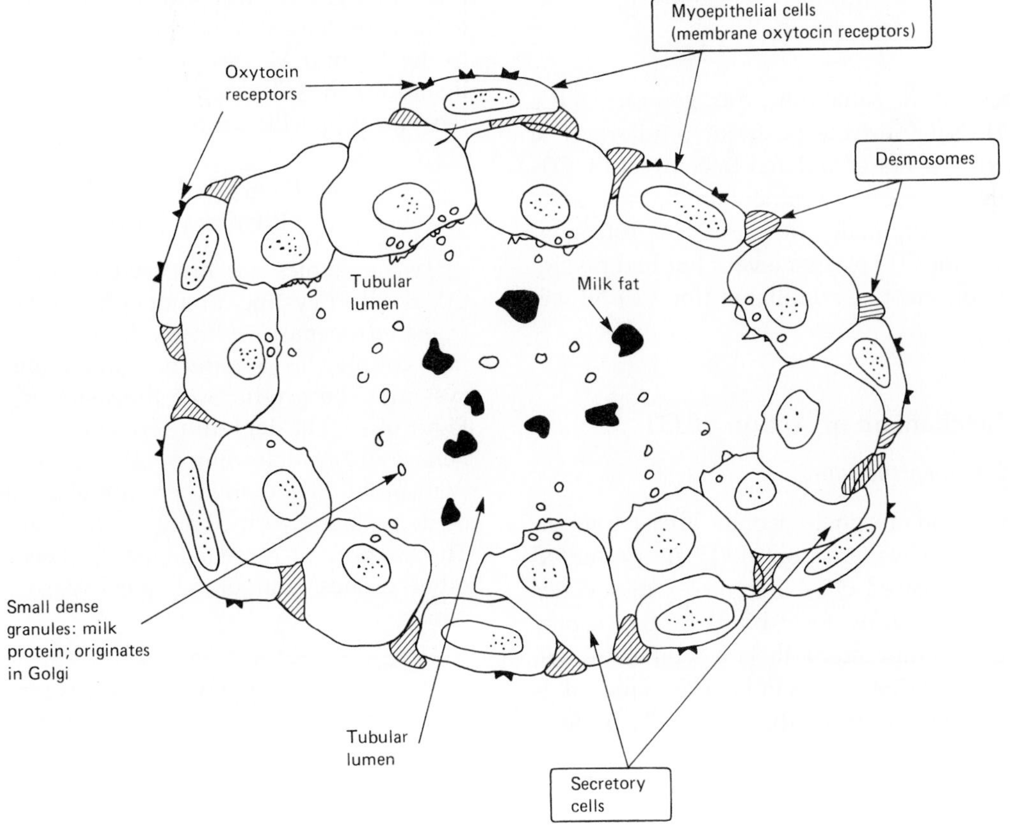

FIGURE 4-21 Drawing of a cross section of a mammary tubular gland.

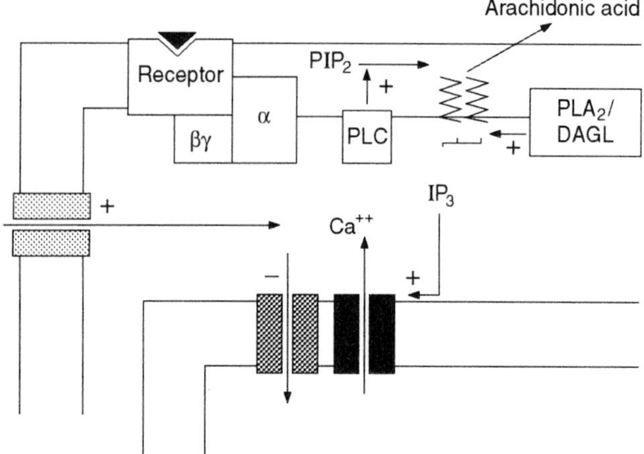

FIGURE 4-22 Model of the oxytocin receptor-mediated signal transduction pathway. This pathway should be similar in myometrium and mammary gland. Binding of OT (black triangle) to its receptor results in G-protein (α- and $\beta\gamma$-subunits are indicated) and phosphoinositol-specific phospholipase C activation, causing IP_3-mediated Ca^{2+} release from internal stores via an IP_3 receptor (black). Additionally, calcium influx via plasma membrane calcium channels (shown in light gray) is enhanced, and sequestration of cytosolic calcium is inhibited through a blockade of the sarcolemmal (Mg^{2+}–Ca^{2+})ATPase (gray). Diacylglycerol is thought to activate protein kinase C and is itself hydrolyzed by phospholipase A_2 or diacylglycerol lipase to generate arachidonic acid, the substrate for prostaglandin synthesis: PIP_2, phosphatidylinositol bisphosphate; DAGL, diacylglycerol lipase; PLA_2, phospholipase A2.

out pregnancy. At the same time, estrogen causes the release of OT–NP from the posterior pituitary. The immature uterus is not stimulated to contract by OT. Experimentally, estrogen treatment in normal male subjects caused dramatic increases in estrogen-stimulated NP (for OT) plasma levels, but had no significant effect on nicotine-stimulated (for VP) NP secretion.

I. Mechanism of Action of OT

1. *Mammary Myoepithelium*

The anatomy and morphology of the breast myoepithelial cells as the sites of action of OT are presented in Chapter 14. Released OT–NP complex dissociates in the blood, and free oxytocin binds to its receptor located on the membrane of the myoepithelial cell. OT regulates increased intracellular Ca^{2+} since it is required for derepression of the contractile system. Contraction occurs, the tubular lumen is narrowed by contraction, and milk is ejected (Figures 4-21 and 4-22).

2. *Uterine Myometrium*

Some reports suggest that OT stimulates the release of $PGF_{2\alpha}$ from the ovine endometrium *in vitro* and from the rat uterus. A direct effect of receptor-bound OT could lead to the regulation of membrane phospholipase to release arachidonic acid, the precursor of $PGF_{2\alpha}$. This prostaglandin seems to be produced by OT in the uterine myometrium as part of its effect on the initiation of labor. How $PGF_{2\alpha}$ would act to promote contractions is unclear. Synthesis of estrogen is stimulated at the termination of pregnancy. This may increase the number of OT-binding sites in uterine myometrium at this time. It could also explain the sensitivity of the near-term uterus to OT coinciding with the burst of OT released from the posterior pituitary as the availability of free progesterone falls dramatically near term. As a result of OT action, the enhanced Ca^{2+} in cytosol is available to derepress the contractile mechanism to generate uterine contractions or contractions of mammary myoepithelial cells, as the case may be.

J. The OT Receptor

The membrane human oxytocin receptor has been cloned and sequenced as shown in Figure 4-23. It is a member of the seven-membrane-spanning domain receptors. The OT receptor thus appears to be similar to the G-protein-coupled rhodopsin-type receptors and to the V_{1a} and V_2 vasopressin receptors. VP is not a ligand for the OT receptor, indicating that the cloned receptor is specific for oxytocin.

K. Degradation of Posterior Pituitary Hormones

Two enzymes are involved in the breakdown of OT and VP: cystine aminopeptidase and glutathione transhydrogenase (with oxidized glutathione reductase coupled as a reduced glutathione-regenerating system). The products of the reaction are shown in Figure 4-24. The degrading enzymes, GSH transhydrogenase and cystine aminopeptidase, can operate in random order as shown. There are also other pathways of degradation, principally in the brain or pituitary. The reactions pictured in Figure 4-24 take place in liver, other tissues, and blood, particularly during pregnancy.

VI. CLINICAL ASPECTS

Many of the clinical applications apply to those considered under releasing factors (Chapter 3), since both hypothalamic releasing factors and posterior pituitary

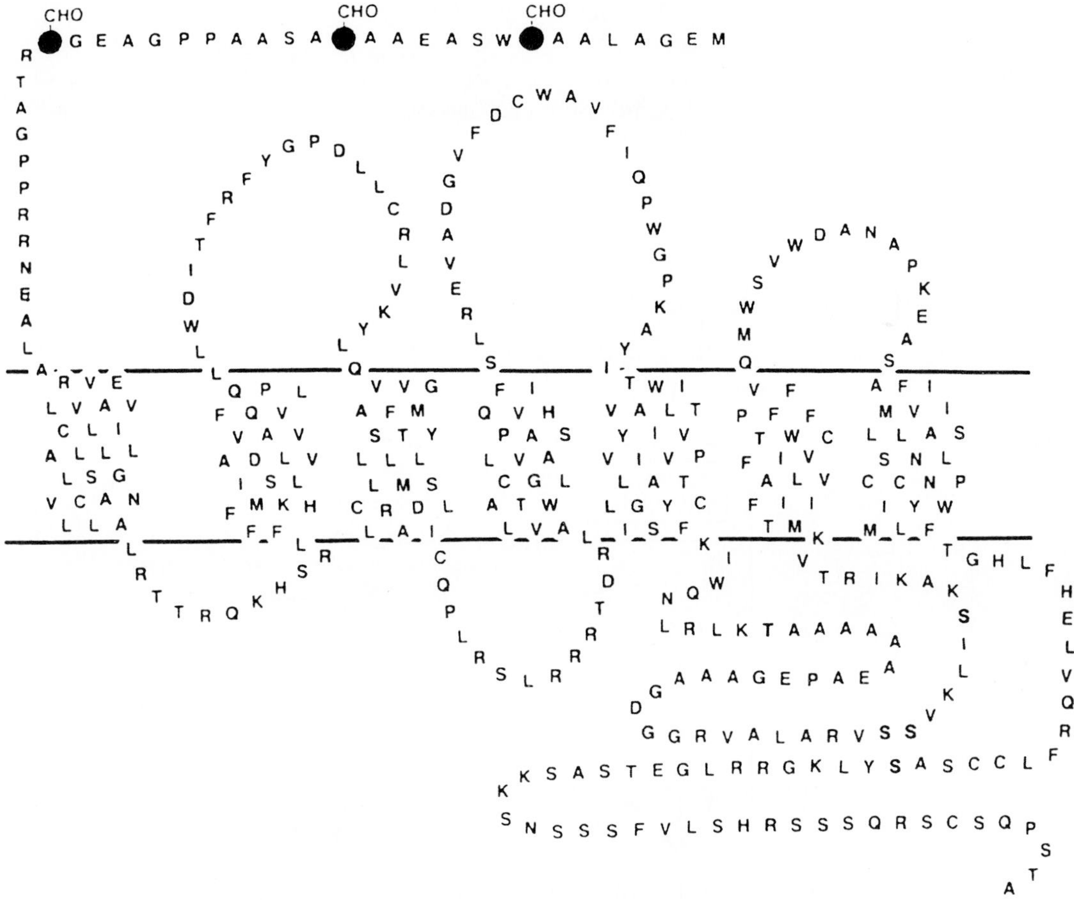

FIGURE 4-23 Amino acid sequence and proposed membrane topography of the human ocytocin receptor. Reproduced with permission from Kimura, T., Azuma, C., Takemura, M., Inoue, T., Kikuchi, T., Kubota, Y., Ogita, K., Saji, F., and Tanizawa, O. (1993). *Regul. Pept.* **45,** 73–77.

hormones travel down the pituitary stalk. Consequently, damage by trauma or tumors will also impede the output of VP and OT. Water balance will be affected, and frequency of urination will be controlled medically by the administration of a hormone or hormone analog to treat diabetes insipidus. Presumably males will have less consequences from OT deficiency. This disease is somewhat rare in occurrence. In principle, a number of conditions can lead to the failure of VP expression from the posterior pituitary or of its effectiveness at the site of action. Damage by tumors or cysts, encephalitis, granulomatous diseases such as tuberculosis, sarcoidosis, syphilis, and meningitis, and arteriosclerosis to the secretory apparatus or the hypothalamohypophyseal portal system may occur. Vascular destruction may also occur to the portal system as a result of degenerative diseases. Brain tumors may be a most important factor. A major cause is trauma. Hereditary diseases may also play a role with abnormal functions at the level of the hypothalamus–pituitary or the kidney. The latter case is presumed to alter the expression or function of VP receptors. Other unsubstantiated possibilities are increased degradation of VP or development of autoimmune antibodies.

OT induces uterine contractions at the start of labor since greatly increased numbers of OT receptors near labor have been induced in the myometrium; however, OT in physiological doses does not induce labor during pregnancy. Hypophysectomized females can have a normal labor, indicating either that the important amounts of OT originate from the fetus or that other factors besides OT are involved. It is still not known whether OT initiates labor. Prior to labor, it has been suggested that relaxin may cause the release of opioids that inhibit the secretion of OT until the time of parturition. These have been discussed in this chapter and in Chapter 14. Excess or deficiency of OT has, so far, not been associated with a disease process.

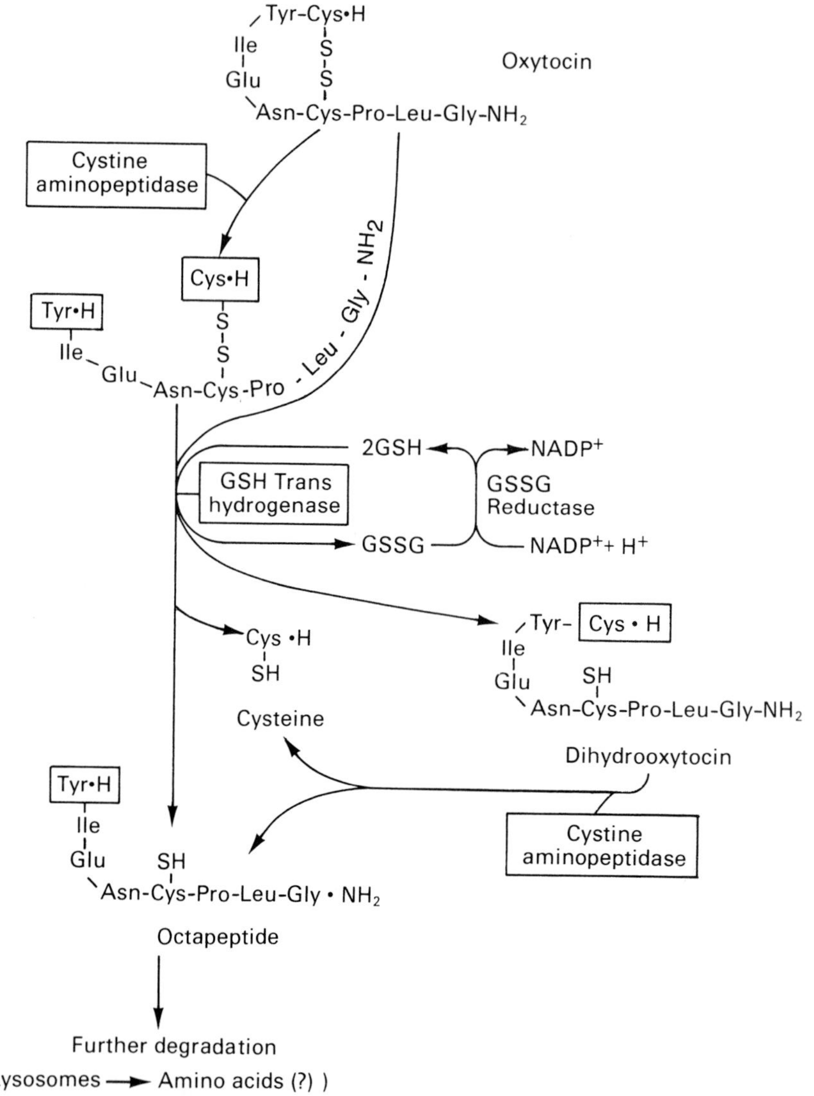

FIGURE 4-24 Degradation of posterior pituitary hormones. OT transhydrogenase (human placenta) is similar to degrading enzymes of other tissues and is also similar to enzymes degrading insulin. These enzymes presumably also degrade VP.

References

A. Books

DeGroot, L. J. (1995). "Endocrinology", 3rd ed. Vol 1. Iyengar, R. and Birnbaumer, L., eds. (1990). "G Proteins." Academic Press, San Diego.

B. Review Articles

Argiolas, A., and Gessa, G. L. (1991). Central functions of oxytocin. *Neurosci. Biobehav. Rev.* **15,** 217–231.

Falke, N. (1991). Modulation of oxytocin and vasopressin release at the level of the neurohypophysis. *Prog. Neurobiol.* **36,** 465–484.

Jenkins, J. S., and Nussey, S. S. (1991). The role of oxytocin: present concepts. *Clin. Endocrinol.* **34,** 515–525.

Knepper, M. A. (1994). The aquaporin family of molecular water channels. *Proc. Natl. Acad. Sci. USA* **91,** 6255–6258.

Mohr, E., and Richter, D. (1994). Vasopressin in the regulation of body functions. *J. Hypertension* **12,** 345–348.

Mohr, E., Myerhof, W., and Richter, D. (1992). The hypothalamic hormone oxytocin: from gene expression to signal transduction. *Rev. Physiol. Biochem. Pharmacol.* **121,** 31–48.

Mohr, E., Meyerhof, W., and Richter, D. (1995). Vasopressin and oxytocin: Molecular biology and evolution of the peptide hormones and their receptors. *Vitamins Hormones* **51,** 235–266.

Vallotton, M. B. (1991). The multiple faces of vasopressin receptors. *Mol. Cell. Endocrinol.* **78,** C73–C76.

C. Research Papers

Breslow, E. (1993). Structure and folding properties of neurophysin and its peptide complexes: biological implications. *Regul. Pept.* **45,** 15–19.

Chen, L., Rose, J. P., Breslow, E., Yang, D., Chang, W.-R., Furey, W. F., Sax, M., and Wang, B.-C. (1991). Crystal structure of a bovine neurophysin II dipeptide complex at 2.8 Å determined from the single-wavelength anomalous scattering signal of an incorporated iodine atom. *Proc. Natl. Acad. Sci. USA* **88,** 4240–4244.

Dyer, R. G. (1988). Oxytocin and parturition—new complications. *J. Endocrinol.* **116,** 167–168.

Kimura, T., Azuma, C., Takemura, M., Inoue, T., Kikichi, T., Kubota, Y., Ogita, K., Saji, F., and Tanizawa, O. (1993). Molecular cloning of a human oxytocin receptor. *Regul. Pept.* **45,** 73–77.

Miller, F. D., Chibbar, R., and Mitchell, B. F. (1993). Synthesis of oxytocin in amnion, chorion and decidua: a potential paracrine role for oxytocin in the onset of human parturition. *Regul. Pept.* **45,** 247–251.

Perraudin, V., Delarue, C., Lefebvre, H., Contesse, V., Kuhn, J.-M., and Vaudry, H. (1993). Vasopressin stimulates cortisol secretion from human adrenocortical tissue through activation of V_1 receptors. *J. Clin. Endocrinol. Metab.* **76,** 1522–1528.

Thibonnier, M., Bayer, A. L., and Leng, Z. (1993). Cytoplasmic and nuclear signaling pathways of V_1-vascular vasopressin receptors. *Regul. Pept.* **45,** 79–84.

Wood, S. P., Tickle, I. J., Treharne, A. M., Pitts, J. E., Mascarenhas, Y., Li, J. Y., Husain, J., Cooper, S., Blundell, T. M., Hruby, V. J., Buku, A., Fischman, A. J., and Wyssbrod, H. R. (1986). Crystal structure analysis of deaminooxytocin: conformational flexibility and receptor binding. *Science* **232,** 633–636.

Yasin, S., Costa, A., Trainer, P., Windle, R., Forsling, M. L., and Grossman, A. (1993). Nitric oxide modulates the release of vasopressin from rat hypothalamic explants. *Endocrinology* **133,** 1466–1469.

CHAPTER

5

Anterior Pituitary Hormones

I. INTRODUCTION

The pituitary gland, as a whole, is made up of three components, the anterior pituitary, the intermediate pituitary (not well-defined in the human), and the posterior pituitary. The anterior pituitary (adenohypophysis), which is the source of 10 polypeptide hormones, is considered to be the "master gland." It comprises the *pars distalis* and the *pars intermedia.* The hormones of the anterior pituitary with their general actions as well as overall regulation are shown in Figure 5-1.

In various cells within the anterior pituitary are expressed the following hormones, which govern a wide variety of important bodily functions: (1) growth hormone (GH, somatotropin, STH), which stimulates the growth of bone, liver, and other tissues and in many cases may act through the agency of somatomedins (insulin-like growth factors, IGFs) produced in the liver and elsewhere; (2) thyrotropic hormone (thyrotropin, TSH), also known as thyroid-stimulating hormone, which stimulates the thyroid gland to release thyroxin (T_4) and triiodothyronine (T_3) into the bloodstream for circulation to the tissues where they are required for cells to burn nutrients as fuels at adequate rates and also for differentiation and development; (3) luteinizing hormone (lutropin, LH) and (4) follicle-stimulating hormone (follitropin, FSH), which play major roles in reproductive activities; (5) adrenocorticotropic hormone (corticotropin, ACTH), which appears to be one

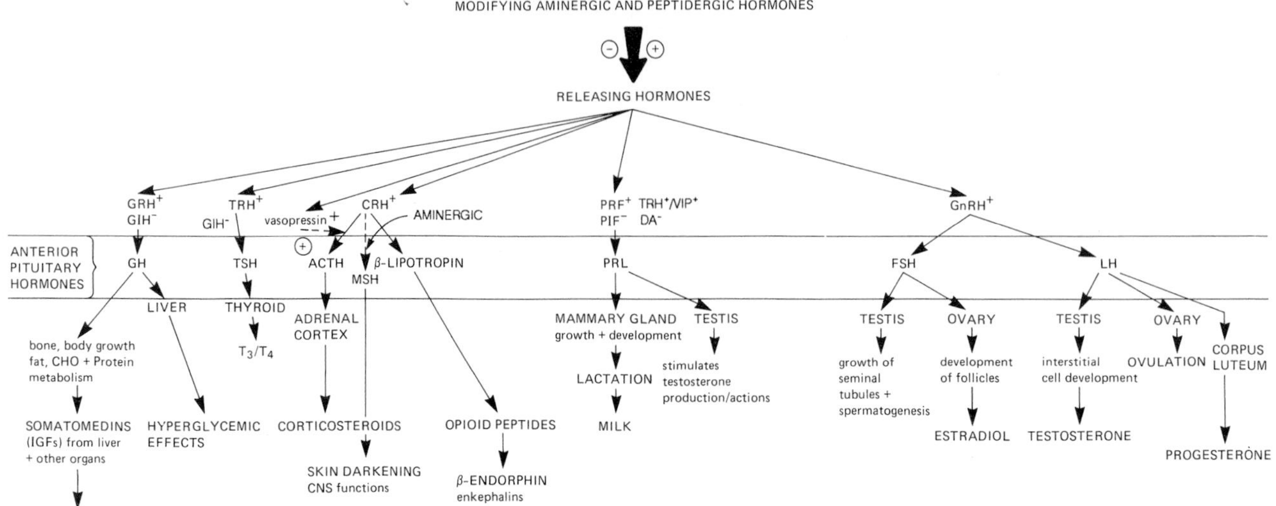

FIGURE 5-1 Overview of the anterior pituitary hormones showing the connections between the aminergic hormones and neurotransmitters of the CNS, the releasing hormones from the hypothalamus, and the anterior pituitary hormones together with the organs upon which they act and their general effects. GRH, Growth hormone releasing hormone, or somatocrinin; GIH, growth hormone release-inhibiting hormone, or somatostatin; TRH, thyroid-stimulating hormone releasing hormone; CRH, corticotropic releasing hormone; PRF, prolactin releasing factor, PIF, prolactin release-inhibiting factor; GnRH, gonadotropic releasing factor; GH, growth hormone; TSH, thyrotropic stimulating hormone; ACTH, adrenocorticotropic hormone; MSH, melanocyte-stimulating hormone; PRL, prolactin; FSH, follicle-stimulating hormone; LH, luteotropic hormone; CHO, carbohydrate; IGFs, insulin-like growth factors; T_3, triiodothyronine; T_4, thyroxine. Superscript plus or minus signs or encircled plus or minus signs refer to positive or negative actions.

of the principal mediators of the stress adaptation mechanism in its tropic stimulation of adrenocortical hormones; (6) prolactin (PRL), which is important for the synthesis of milk constituents during lactation and may have other actions somewhat similar to those of growth hormone as well as being secreted during stress; (7) melanocyte-stimulating hormone (melanotropin, MSH), which plays a role in skin-darkening reactions and influences important CNS functions, such as memory and learning;[1] (8) β-lipotropin (β-LPH), through its proteolytic products (9) β-endorphin and (10) Met-enkephalin, may promote analgesia in stress and act as a neurotransmitter in signaling the release of other hormones or affect ion flux. Thus, through the effects of all of these hormones in the anterior pituitary, the processes of somatic cell growth, metabolic rate, reproductive function, stress adaptation, mammary gland development and function, skin darkening, and CNS reactions are regulated. The importance of these hormones is all too evident in cases of their inadequate production, such as occurs when tumors or other physical pressures reduce the output of one or more of these hormones, or during their overproduction when tumors may secrete them in large quantities ectopically. All of the anterior pituitary hormones are proteins or polypeptides. They are substantially larger (1500–35,000 Da) than many of the hypothalamic releasing hormones recognized earlier, but not larger than CRH and GRH. Anterior pituitary hormones seem to have longer half-lives in the bloodstream than the releasing hormones. As already indicated in Chapter 3, the secretion of anterior pituitary hormones is under the control of the releasing hormones and is sometimes directly or indirectly under neuronal control. The secretion of anterior pituitary hormones also is regulated by an elaborate feedback control by terminal target gland hormones. Thus, for example, cortisol inhibits further output of ACTH (see Table 5.1).

We then focus upon the explanation of the effects of the anterior pituitary hormones. Some mechanisms are better understood than others. The actions of ACTH and TSH are clearer at this point than the actions of the rest; prolactin, growth hormone, and MSH are

[1] MSH and ACTH derive from the same precursor polypeptide and share homologous sequences, but they have distinct receptors. MSH is thought to derive mainly from pars intermedia-like cells under the influence of an aminergic signal. Although there are separate receptors for ACTH and MSH, it has become apparent that ACTH in higher than normal blood concentrations may induce skin darkening, and it may accomplish this by binding to the MSH receptor.

less well understood. Accordingly, more emphasis will be allocated here to prolactin, growth hormone, and β-lipotropin since ACTH actions are covered in Chapter 10, TSH actions are emphasized in Chapter 6, and LH and FSH actions are reported in Chapters 11 and 12.

II. ANATOMICAL, MORPHOLOGICAL, AND PHYSIOLOGICAL RELATIONSHIPS

The location and anatomy of the pituitary are described in Chapter 3 (see Figure 3-2). It is appropriate here to begin with the localization of the anterior pituitary hormones into discrete cell types. The anterior pituitary can be discussed in terms of cells localized to the pars distalis, which occupies the major part of the gland, as shown in Figure 5-2.

TSH is located in the thyrotrope (thyrotropic cells of the pars distalis). This is a basophilic large cell type with very small granules of the hormone. The granule size, measured by electron microscopy, is 120–200 nm. LH and FSH are located in the gonadotrope, also basophilic, containing granules of 200–250 nm. ACTH is located in the corticotropic cell of the pars distalis, a highly staining basophil, with granules 100–200 nm in diameter. This cell also appears to be present in the pars intermedia (Figure 5-2). The pars intermedia is less distinct in the human than in the rat. MSH is located in the melanotrope, a weakly basophilic cell with granules of 200–300 nm. MSH antisera interact with cells of the pars distalis and pars intermedia. GH is in the somatotrope of the pars distalis, an acidophilic cell containing large numbers of granules of 300–400 nm.

Prolactin is in the mammotrope, an acidophilic cell of the pars distalis containing large, dense, and variable secretory granules in the size range of 400–700 nm. β-Lipotropin is also in the pars distalis, presumably in corticotropes, and its processing to β-endorphin can also occur in these cells. MSH is a breakdown product of ACTH, and this breakdown probably does not occur when the corticotrope is stimulated by CRH, resulting in the primary release of ACTH and β-endorphin. The latter is derived from the enzymatic breakdown of β-

TABLE 5-1 Summary of Properties of Secretory Cells of the Anterior Pituitary

Cell	Hormone secreted	Staining type	Diameter of secretory granules (μm)	Distinguishing features
Somatotroph[a]	GH (STH)	Acidophilic (stains with acid dye)	300–350	Typical cell
Mammotroph	PRL	Acidophilic	600–900	Large round, ovoid secretory granules; cells vary in shape; cell number increases in pregnancy and lactation; lysosomes more prominent when secretion is suppressed
Gonadotroph	FSH	Basophilic (stains with basic dye)	~200	Large round cell body; rough endoplasmic reticulum (RER) and Golgi probably play a role in synthesis of carbohydrate moiety of subunits
Luteotroph	LH	Basophilic	~250	Secretory granules slightly larger than FSH-containing cells; granules tend to accumulate in one pole of cell near periphery; Golgi apparatus less extensive than FSH cell
Thyrotroph	TSH	Basophilic	120–150	Smaller than other cell types; irregular in shape; flattened nucleus; smaller granules than in other cell types
Corticotroph	ACTH	Basophilic	~200	Large cell with irregular shape; granules associated with Golgi; gastrin has been reported to occur in these cells
Melanocyte-stimulating hormone cell	MSH	Basophilic		A few layers of cells between pars distalis and pars nervosa; in human cells may extend into neural lobe; cells of pars intermedia are polygonal; morphologically similar to corticotroph; also possible that corticotroph can produce MSH

[a] The suffix troph is used here to connote a growth function. The suffix trope (e.g., gonadotrope) is used to connote a cell containing a substance that produces change rather than growth. There is some confusion in the usage of these terms since at one stage an anterior pituitary hormone may cause a target cell to grow (act as a mitogen), whereas at another stage the same hormone may cause the same cell to alter its metabolism or produce some change other than growth. Most of the information in this table is summarized from description by Lentz, T. L. (1971). "Cell Fine Structure."

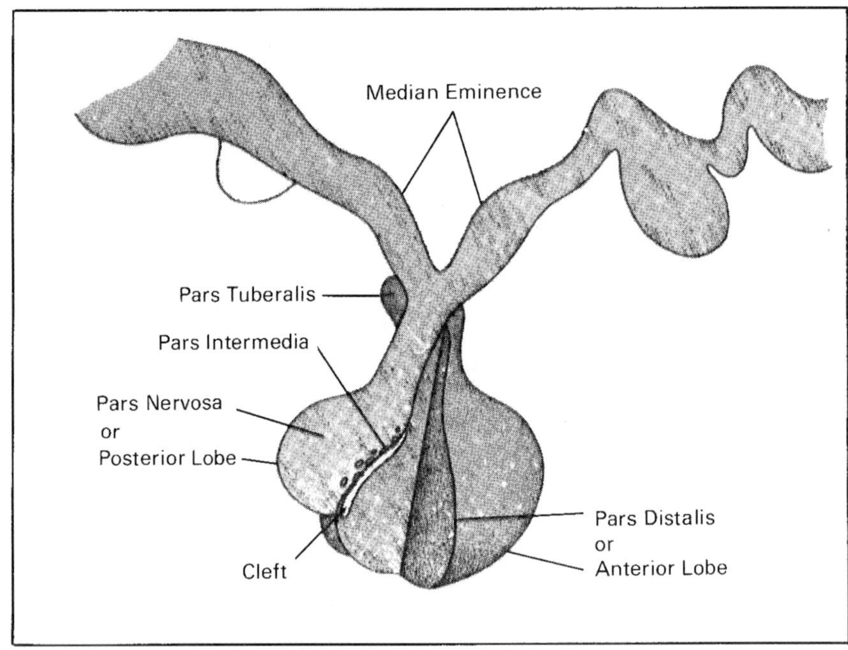

FIGURE 5-2 Pituitary and its structures. Reproduced from Guillemin (1980). Beta lipotropin and endorphins: Implications of current knowledge. *In* "Neuroendocrinology" (D. T. Krieger and J. C. Hughes, eds.), p. 70. Sinauer Associates, Sunderland, Massachusetts, 1980.

lipotropin. α-MSH derives from the proteolytic cleavage of ACTH in pars intermedia-like cells in which the stimulatory signals are different from those of the corticotrope that produces ACTH. Thus, the stimulation of MSH secretion occurs through an aminergic signal to pars intermedia-like cells, while ACTH is elicited by CRH action on corticotropes in the pars distalis.

The cells of the pars distalis are recognized in the light microscope on the basis of their affinity for stains. The two types already mentioned are acidophils (acid stainable) and basophils (base stainable) and are easily distinguished. Cells that are not stained by these procedures usually represent degranulated cells that are undergoing recovery after secretion of granular contents.

The adenohypophysis itself consists of three parts: the pars distalis, pars intermedia, and pars tuberalis, as shown in Figure 5-2. Cells of the pars distalis are arranged in anastomosing cords close to fenestrated capillaries in the secondary capillary plexus of the hypophyseal portal system. The pars intermedia, a narrow region separating the pars distalis from the posterior pituitary (pars nervosa), consists of basophils with MSH and possibly some endorphins, with ACTH virtually absent. The situation in humans is not completely clear in relation to the products of the pars intermedia, which, itself, is not well defined anatomically in comparison with lower forms. Thus, in humans, pars intermedia-like cells seem to be scattered throughout the pars distalis. The pars tuberalis, which is dorsal to the pars distalis and pars intermedia, has no specifically assigned function. The pars distalis has 30–35% acidophils, 15–20 μm in diameter, which contain GH and PRL. The basophils, 15–25 μm in diameter, make up about 10–15% of the cells and contain LH, FSH, TSH, and ACTH.

The release of pars distalis hormones is signaled by the action of releasing hormones, and the general mechanism is shown in Figure 3-10. The synthesis of these hormones occurs on ribosomes that are transported within the rough endoplasmic reticulum cisternae and then concentrated into granules of the Golgi complex. Small granules arise from the Golgi to fuse and form mature secretion granules. At this point the releasing hormones, when triggered, cause the granules to be released by exocytosis.

At the electron microscopic level, there appear to be separate cells for the secretion of each of the six major hormones of the anterior pituitary. However, the total picture is incomplete. One must remember that ACTH and β-LPH derive from the same gene, and both of these peptides can be further degraded even within the anterior pituitary to the MSH hormones as well as β-endorphin. However, there does appear to be some segregation in terms of one or two of these hormones being secreted while the others are not. The corticotroph mainly secretes ACTH, β-lipotropin, and β-endorphin, while the cells in the rat pars intermedia mainly secrete β-endorphin, corticotropin-like intermediate lobe peptide (CLIP), and MSH. Clearly, the derivatization of hormonal peptide precursors and the specific proteolysis taking place are different in the two distinct cell types.

III. CHEMISTRY

A. Growth Hormone (GH)

Table 5-2 summarizes the molecular weights and biochemical properties of the peptide hormones of the anterior pituitary. Human growth hormone is about 21,500 molecular weight containing 191 amino acid residues (Figure 5-3). Various enzymatically modified forms of human growth hormone have been described that may reflect enzymatic processing occurring *in vivo*. A 20,000 molecular weight form (1000–2000 molecular weight less than the normal hormone) appears to lack 15 amino acid residues occurring between positions 32 and 46, but otherwise the hormone is identical to the normal form. The deleted form has normal growth-promoting activity, but lacks the insulin-like activity usually associated with the hormone. It comprises about 15% of the total growth hormone content of the human pituitary. There are three introns (intervening sequences) in the gene encoding growth hormone messenger, and one of these starts at the same point (corresponding to amino acid residues 31 and 32) as the amino acid sequence deletion. Residue 46, corresponding to the end of the deleted amino acid sequence, contains AG in the mRNA, two bases that are usually found at the 3′ end of an intervening sequence in the precursor RNA transcribed from the gene. Thus, processing differences or separate genes account for the appearance of these two forms of human growth hormone.

In addition to the two forms of GH described earlier, larger forms of GH exist. One of these is a GH dimer covalently linked through a disulfide bond. Larger forms than the dimer also exist. Presumably most of the actions of GH are carried out by the smaller forms.

The normal GH gene is expressed primarily in the somatotrope of the anterior pituitary. The transcription factor, Pit-1, permits the expression of the GH gene. Pit-1 is a homeodomain transcription factor that fills two sites followed by other transcription factors, such as the glucocorticoid receptor and the cAMP-induced transcription factor. Other transcriptional activators are USF, Sp1, and NF-1-like.

B. Prolactin (PRL)

Ovine prolactin is about 23,000 molecular weight and comprises 199 amino acids in the sequence. GH and PRL are structurally related and share certain activities. Studies with chemical probes suggest that His27 and His30 near the N-terminus are involved in the interaction of PRL with its receptor. See Section V for further discussion on PRL.

Receptors for GH and PRL are members of the hematopoietic receptor gene family shown in Figure 5-4.

C. Thyroid-Stimulating Hormone (TSH)

Human TSH is 28,300 molecular weight, containing two subunits and 211 amino acid residues. It is structurally related to LH and FSH (and hCG). There is a

TABLE 5-2 Properties of Adenohypophyseal Hormones

Hormone			$t_{1/2}$ in blood (min)		Molecular weight (K = 1000)	Comments on structure
GH		Structurally related	30		21.5K (191 amino acids in human)	Single chain; two S–S bonds
PRL					23K (199 amino acids in ovine)	Single chain; two S–S bonds
TSH		Structurally related	30		28.3K (211 amino acids in human)	Glycoprotein: 16% carbohydrate, two subunits (α–β)[a]
FSH	GTH		240		34K (210 amino acids in human)	16% carbohydrate, two subunits (α–β_2),[a] contains S–S bonds
LH			30		28.5K (204 amino acids in human)	15.5% carbohydrate, two subunits (α–β_3),[a] contains S–S bonds
ACTH		Structurally related	15		4.5K (39 amino acids in human)	Open chain; homology with β-LPH and MSH
MSH				α	(13 amino acids in human)	Linear change; heptapeptide common to α- and β-MSH, ACTH, β-LPH, and γ-LPH
				β	3K (22 amino acids in human)	
β-LPH				β, γ	9.5K (91 amino acids in human)	Open chain; β-LPH is precursor of β-endorphin, enkephalins (ACTH + β-LPH encoded by same mRNA)

[a] α-subunits are similar or identical and can be interchanged experimentally with α-subunits of TSH, FSH, or LH. The β-subunit determines hormonal activity; it is the major immunological determinant and is involved in recognizing a specific binding (receptor) site. α-subunit is involved in penetrating membrane and stimulation of adenylate cyclase. The β-subunit appears to cover most of the surface of the hormones.

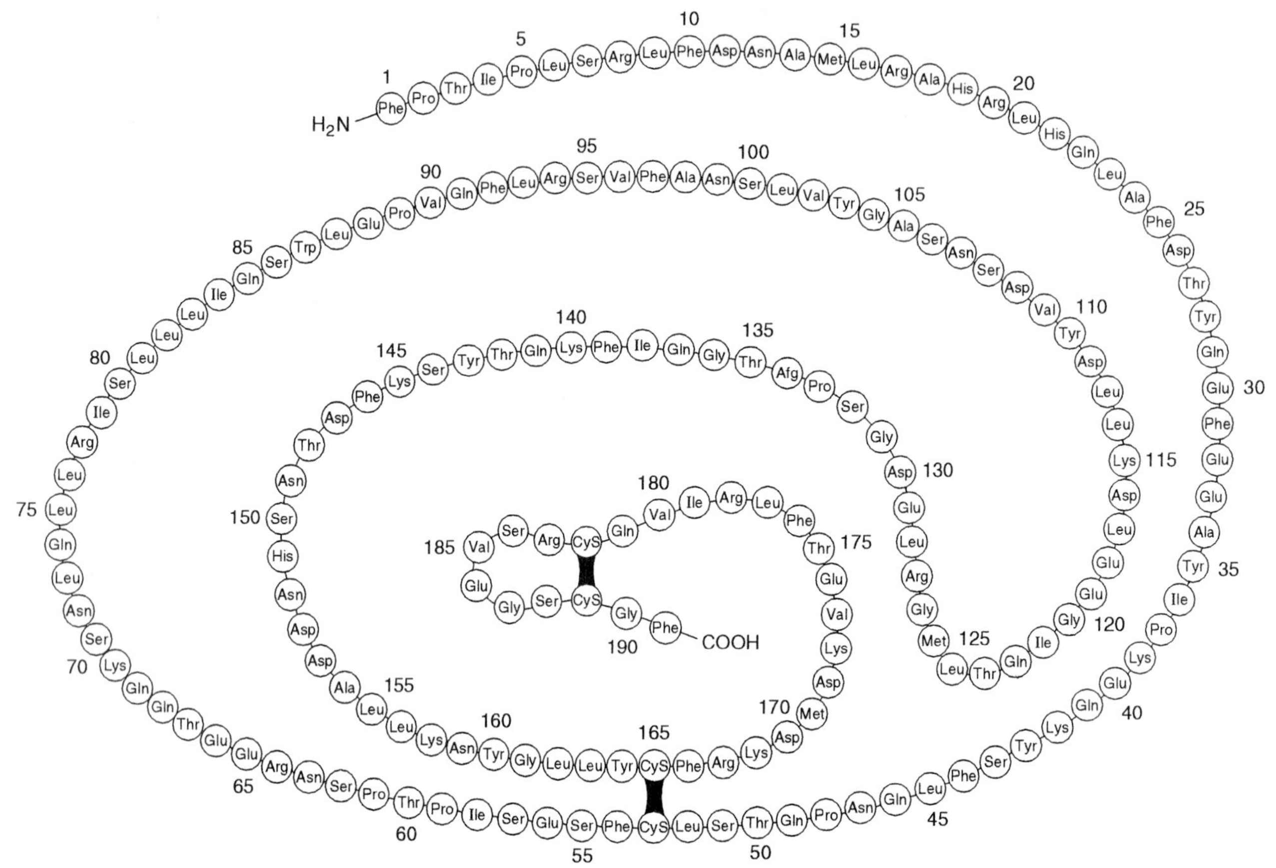

FIGURE 5-3 Covalent structure of hGH. Reprinted with permission from Chawla, R. K., Parks, J. S. and Rudman D. (1983). Structural variants of human growth hormone: Biochemical, genetic and clinical aspects. *Ann. Rev. Med.* **34,** 519–547. Copyright © 1983 by Annual Reviews, Inc. Reprinted with permission from Daughaday, W. H. (1995). Growth hormone, insulin-like growth factors, and acromegaly. *In* "Endocrinology" (L. J. De Groot, ed.), 3rd ed. Vol. 1 pp. 303–329. W. B. Saunders Co., Philadelphia, PA.

high degree of structural homology between the gonadotropic hormones, LH, FSH, and hCG, and one other anterior pituitary hormone, TSH (see Table 13-3 and Chapter 5). They all are composed of two noncovalently linked subunits, designated α and β. The α-subunits of LH, hCG, TSH, and FSH are of identical structure (MW 13,000). The binding activity of each hormone is determined by the β-subunit. Each of these four peptide hormones has a significant number of carbohydrate moieties covalently linked to the peptide chain (Figure 5-5B). These relationships are summarized in Table 5-2.

D. Follicle-Stimulating Hormone (FSH)

Human FSH is a 34,000 molecular weight protein with two subunits consisting of 210 amino acid residues.

The α-subunit is common to all three hormones. Each subunit is encoded by a separate gene, each located on separate chromosomes. The β-subunit gene is drawn in Figure 5-6A.

The single-copy gene is 4.9 kb, located on chromosome 1. It has three exons and two introns. The promoter is shown in Figure 5-6B. The start of transcription is defined by the start of the first exon (Figure 5-6A). The TATA box is 29 bp upstream of the start of transcription (Figure 5-6B). Pit-1 is an important transcription factor that binds to the TSH β-subunit gene between -128 and -58 (Figure 5-6B).

The single-copy gene for the α-subunit is located on chromosome 6 and is 13.5 kb long, with four exons and three introns (Figure 5-7A). The human α-subunit gene promoter (Figure 5-7B) has a consensus TATA box 26 bp upstream of the start of transcription. The α-subunit expression is specific to thyrotropes, gonadotropes, and cells of the placenta, and each cell type

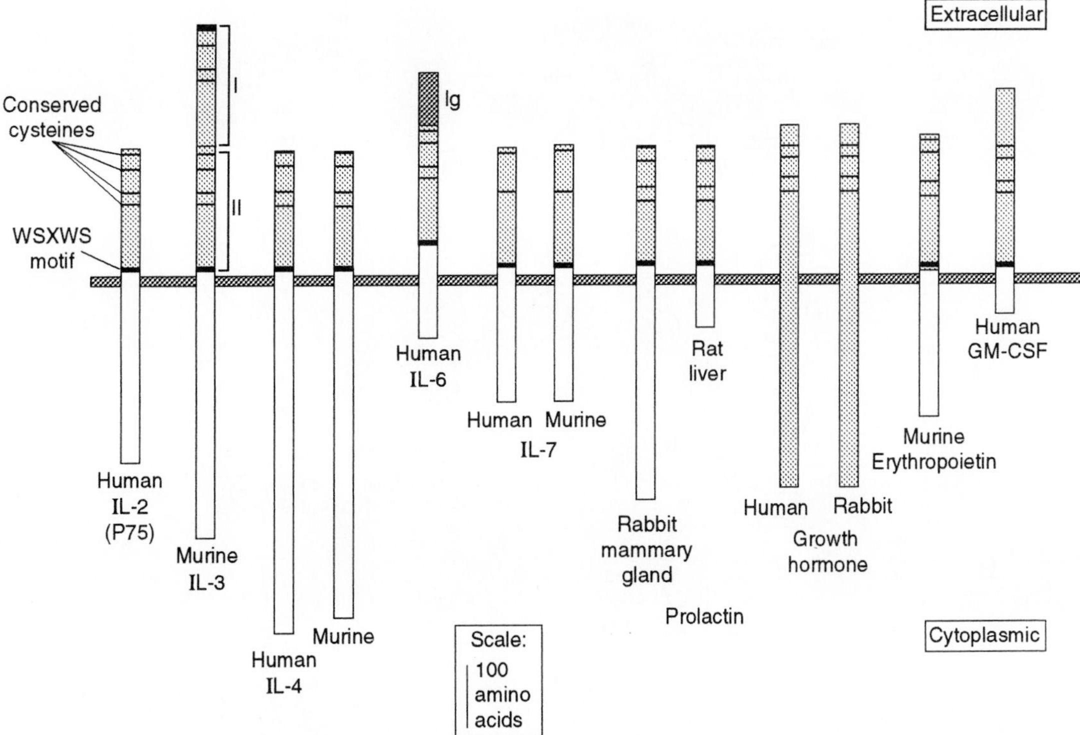

FIGURE 5-4 The hematopoietin receptor superfamily. Schematic representations of the structures of all of the known members of the family are shown. Horizontal bars represent conserved cysteine residues. The black boxes represent the conserved Trp-Ser-X-Trp-Ser (WSXWS) motif. The stippled areas show the stretch of ~210 amino acids within which the receptor homologies are contained. The immunoglobulin-like domain at the N-terminus of the IL-6 receptor is also indicated. Reprinted with permission from Casman, D., Lyman, S. D., Idzerda, R. L., *et al.* (1990). A new cytokine receptor superfamily. *Trends Biochem. Sci.* **15,** 265–270.

has cell-specific regulators of gene expression. In the thyrotrope, expression is dependent on the region −480 to −380, whereas the gonadotropes specify the region −225 to −200 and placental expression depends on the region −200 to about −120. Pit-1 is not as involved in the expression of the α-subunit RNA as it is in the β-subunit, but except for this piece of information, little is known about transactivators of α-subunit gene expression.

TSH is involved in the release of thyroid hormone, as shown in Figure 5-8. In brief, dopamine (DA) drives the release of hypothalamic TRH, which signals the thyrotropic cell of the anterior pituitary to release TSH into the bloodstream. TSH binds to a membrane receptor and affects the release of T_3/T_4 from the thyroid follicle (Chapter 6). T_3/T_4 have a negative feedback effect on TSH production and release, in part by signaling the release of somatostatin (SRIH), which downregulates TSH production. T_3 negatively feeds back on the production of the α-subunit and β-subunit. In hyperthyroidism, the negative effect of T_3 is greater for the TSH β-subunit than for the α-subunit, but in the euthyroid state, both subunits are decreased with a ratio of α/β of about 1/2. In the hypothyroid state, both subunits are affected although the α-subunit can be inhibited slightly more. The negative feedback effect of T_3 is mediated by the T_3 receptor binding to the α-subunit gene at −22 to −7 bp (TRE) (Figure 5-7B) and for the β-subunit gene at about +2 to +10 and about +30 to +37 (Figure 5-6B). Binding of the T_3 receptor to this region may interfere with the functioning of the initiation complex since TRE sites are so close to the start of transcription.

The final subunit protein products are shown in Figure 5-9.

E. Luteinizing Hormone (LH)

Human LH is 28,500 molecular weight, consisting of two subunits and sequences of 204 amino acid residues. Human chorionic gonadotropin (hCG), which has strong homology with LH, is produced during

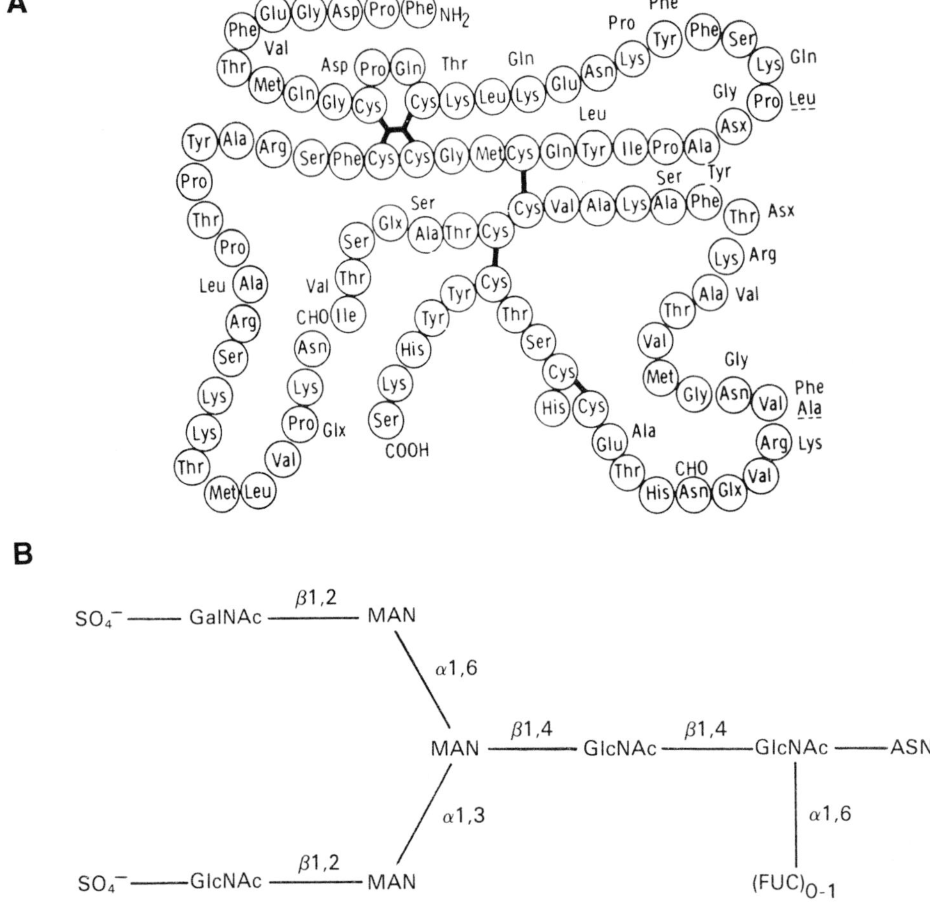

FIGURE 5-5 (A) Amino acid sequence of the α-subunit of ovine–bovine LH. The residues outside the circles show the substitutions reported in human and porcine α-chains. Those underlined are substitutions in porcine LH-α. The LH-α is similar to FSH-α and TSH-α. Reproduced from Pierce, J. G., Faith, M. R., Guidice, L. C., and Reeve, J. R. (1976). *Ciba Found Symp.* (*Excerpta Medica*) **41,** 225–250. (B) General structure of the carbohydrate units of ovine luteotropin (OLH) and bovine TSH. Carbohydrate structures of other anterior pituitary hormones (FSH) are similar. These structures are important since they play a role in the biological function of glycoprotein hormones, although they may not be critical for interaction of the hormone with its receptor. Reprinted by permission from Bedi, G. S., French, W. C., and Bahl, O. P. (1982). Structure of carbohydrate units of ovine luteinizing hormone. *J. Biol Chem.* **257,** 4345–4355. © 1982 by The American Society of Biological Chemists, Inc.

pregnancy by the trophoblast to provide the hormonal stimulus for progesterone production (see Chapter 14). Interestingly, it is rather similar to LH but is distinct from FSH, and its antibody cross-reacts with LH. Consequently, a radioimmunoassay of circulating LH has been developed with the anti-hCG antibody.

F. Adrenocorticotropic Hormone (ACTH)

Human ACTH is 4500 molecular weight in a single chain consisting of 39 amino acids. It is structurally related to MSH and β-lipotropin. All three hormones derive from the same gene product (see Figure 5-10). The sequence of human ACTH is given in Figure 5-11. Note that α-MSH is contained in the first 13 amino acid residues of ACTH.

Human pituitary extracts appear to contain a corticotropin-inhibiting peptide that consists of 32 amino acids with a sequence probably identical to residues 7–38 of ACTH. It cannot stimulate the synthesis of corticosteroids, but inhibits ACTH-stimulated corticosterone production in isolated rat adrenal cells. It has been shown that in genetically obese mice (*ob/ob*) a hormone is located in the neurointermediate lobe of the pituitary (pars intermedia) that stimulates insulin

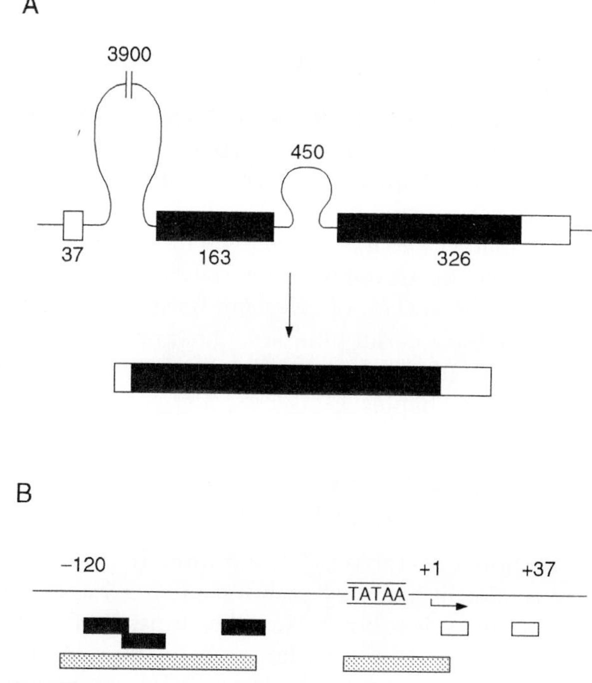

FIGURE 5-6 (A) Schematic representation of the human thyroid-stimulating hormone (TSH) β-subunit gene. The bars indicate the three exons and the lines, which are not drawn to scale, represent intronic DNA. The solid areas represent the coding regions, and the open areas represent the untranslated regions. The intronic DNA is removed by splicing to form mature mRNA. The numbers indicate the size in nucleotides of each exon or intron. (B) Schematic representation of the human thyroid-stimulating hormone (TSH) β-subunit promoter. The transcriptional start site is indicated by the arrow and the TATA box is shown. The numbers above the line denote the position of the nucleotides relative to the transcriptional start site set at +1. The boxes under the line indicate the regions important for T_3 binding (white boxes), Pit-1 binding (black boxes), and thyrotropin-releasing hormone (TRH) response (dotted boxes). Reproduced with permission from Sarapura, V. D., Samuels, M. H., and Ridgway, C. E. (1995). Thyroid-stimulating hormone. *In* "The Pituitary" (S. Melmed, ed.), pp. 187–229. Blackwell Science, Oxford.

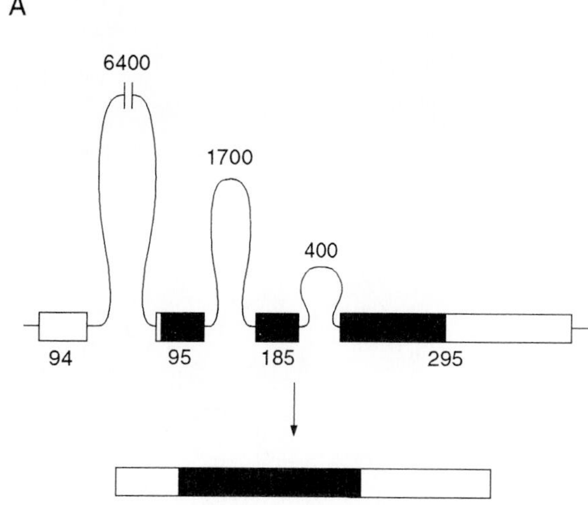

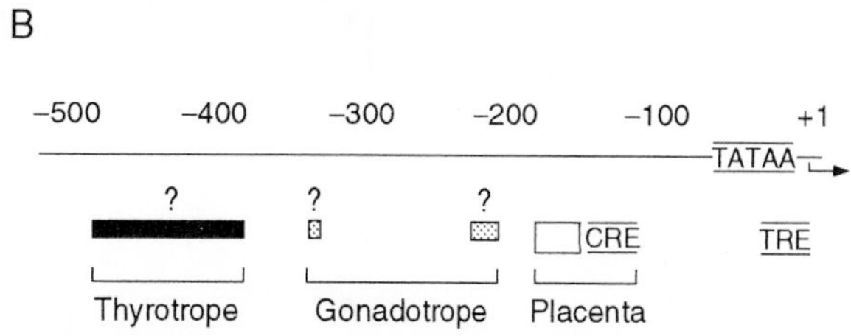

FIGURE 5-7 (A) Schematic representation of the human glycoprotein hormone α-subunit gene. The bars indicate the four exons and the lines, which are not drawn to scale, represent intronic DNA. The solid areas represent the coding regions, and the open areas represent the untranslated regions. The intronic DNA is removed by splicing to form mature mRNA. The numbers indicate the size in nucleotides of each exon or intron. (B) Schematic representation of the human glycoprotein hormone α-subunit promoter. The transcriptional start site is indicated by the arrow and the TATA box is shown. The numbers above the line denote the position of the nucleotides relative to the transcriptional start site set at +1. The boxes under the line indicate the regions important for T_3 binding (TR_3), cAMP binding (CRE), and placental-specific activity (white box) and the putative locations of the gonadotrope-specific (dotted boxes) and thyrotrope-specific (black box) activities. The thyrotrope-specific region and the upstream gonadotrope-specific regions have been described in the mouse α-subunit gene. Reproduced with permission from Sarapura, V. D., Samuels, M. H., and Ridgway, C. E. (1995). Thyroid-stimulating hormone. *In* "The Pituitary" (S. Melmed, ed.), pp. 187–229. Blackwell Science, Oxford.

release. It cross-reacts with an antiserum directed against the C-terminal portion of ACTH, suggesting that it may be related to CLIP, the 18–39 fragment of ACTH (see Figure 5-11). This hormone has been named β-cell tropin and is present in the plasma of *ob/ob* mice. It potentiates glucose-induced insulin secretion. It appears to have properties identical to those of $ACTH_{22-39}$ prepared from ACTH (Figure 5-11).

G. Melanocyte-Stimulating Hormone (MSH)

Human MSH occurs in two major forms, α-MSH and β-MSH. α-MSH is ~1500 molecular weight and consists of 13 amino acid residues. β-MSH is ~2600 molecular weight and consists of 22 amino acid residues. α-MSH is represented in residues 1–13 of the ACTH molecule (Figure 5-10), and β-MSH is derived from cleavage of β-lipotropin (42–134 of translation product; Figure 5-10) to yield γ-lipotropin (42–101) and further cleavage to yield β-MSH (84–101). The sequence of human β-MSH is shown in Figure 5-12.

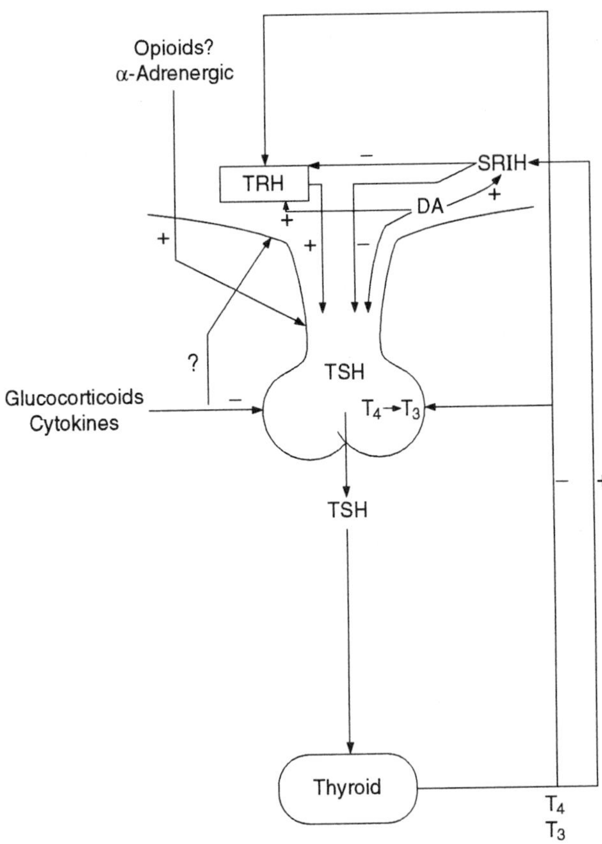

FIGURE 5-8 Neuroendocrine and peripheral control of thyroid-stimulating hormone (TSH) secretion. Abbreviations: T_4, thyroxine; T_3, triiodothyronine; TRH, thyrotropin-releasing hormone; SRIH, somatostatin; DA, dopamine. Reproduced with permission from Sarapura, V. D., Samuels, M. H., and Ridgway, C. E. (1995). Thyroid-stimulating hormone. *In* "The Pituitary" (S. Melmed, ed.), pp. 187–229. Blackwell Science, Oxford.

H. Lipotropin (β-LPH)

There are two forms of lipotropin: β-LPH and γ-LPH. Figure 5-12 shows the primary sequence of β-lipotropin (β-LPH). It contains sequences that become liberated by the actions of endopeptidases, which cleave the chain at basic amino acid residues. These sequences are 1–58, which is γ-LPH, 41–58, which is β-MSH, and 61–91, which is β-endorphin. Thus, α-MSH derives from further cleavage of the ACTH molecule, and β-MSH derives from cleavage of γ-LPH, which, in turn, is broken down from β-LPH. β-LPH is 9500 molecular weight and consists of 91 amino acid residues. It is on the same translation product as ACTH. γ-LPH is a proteolytic cleavage product of β-LPH and consists of 58 amino acid residues. The maturation of these forms is shown in Figure 5-10.

IV. BIOCHEMISTRY

A. Neural Controls

Central controls on the secretion of hormones of the pars distalis comprise the hypothalamic releasing hormones and neurotransmitters. A summary of some of the major controlling factors is presented in Figure 5-13. In many cases it is unclear whether a neurotransmitter is acting through an interneuron to stimulate or inhibit the release of releasing hormone or acting directly at the anterior pituitary. Other possibilities for interaction with the closed portal circulation have been discussed in Chapter 3.

B. Interactions of TSH, FSH, and LH with Receptors

As shown in Table 5-2, these three hormones are very similar in that they each have two subunits and share a similar α-subunit. How then must TSH specifically recognize its cellular target on the thyroid gland cell membrane apart from LH and FSH, which must individually recognize the Leydig cell membrane–ovarian follicle, respectively? Several possibilities are apparent, and they must account for the similar or identical α-subunit common to all three. Thus, Figure 5-14 shows a hypothetical scheme in which this kind of selection can be incorporated and which allows the α-subunit to play a role in facilitating binding while the antigenically unique β-subunit could specify the binding site (see footnote to Table 5-2). Deglycosylated subunits reassociate and interact with receptor normally. Thus, the carbohydrate portions, as demonstrated by the O. P. Bahl laboratory, may not be required for these functions, but could play a role in some other function, such as catabolism or hormonal transport. Interestingly, the β-subunits alone of hCG and LH have steroidogenic activity.

V. PROLACTIN

The primary sequence of the prolactin (PRL) molecule is shown in Figure 5-15. PRL consists of 199 amino acid residues, and the approximate molecular weight of the glycosylated form is about 25,000. Cleaved forms exist and may have physiological significance or may be simply degradation products.

PRL is a member of a gene family that includes GH and CS (chorionic somatomammotropin). A new member of this gene family has been detected in fish, and this form, "somatolactin," is 24% identical to PRL and GH and represents the largest gene (16 kb) family member. Because of this, somatolactin may be the di-

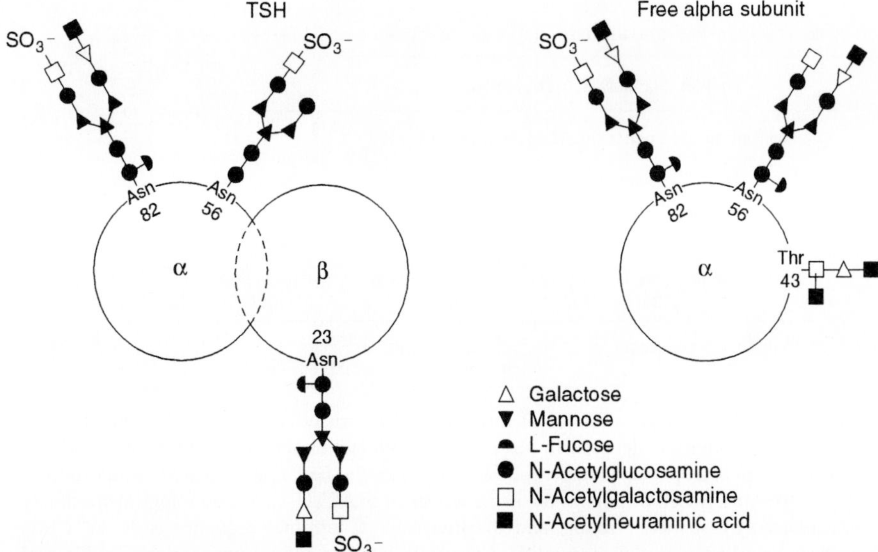

FIGURE 5-9 Oligosaccharide chains of thyroid-stimulating hormone (TSH). Schematic representation of the typical oligosaccharide chains present on the TSH heterodimer and the free α-subunit. Glycosylated asparagine (Asp) and threonine (Thr) residues are indicated. Black and white circles, half-circles, and triangles represent the oligosaccharide chain residues as indicated in the key. Reproduced with permission from Sarapura, V. D., Samuels, M. H., and Ridgway, C. E., (1995). Thyroid-stimulating hormone. *In* "The Pituitary" (S. Melmed, ed.), pp. 187–229. Blackwell Science, Oxford.

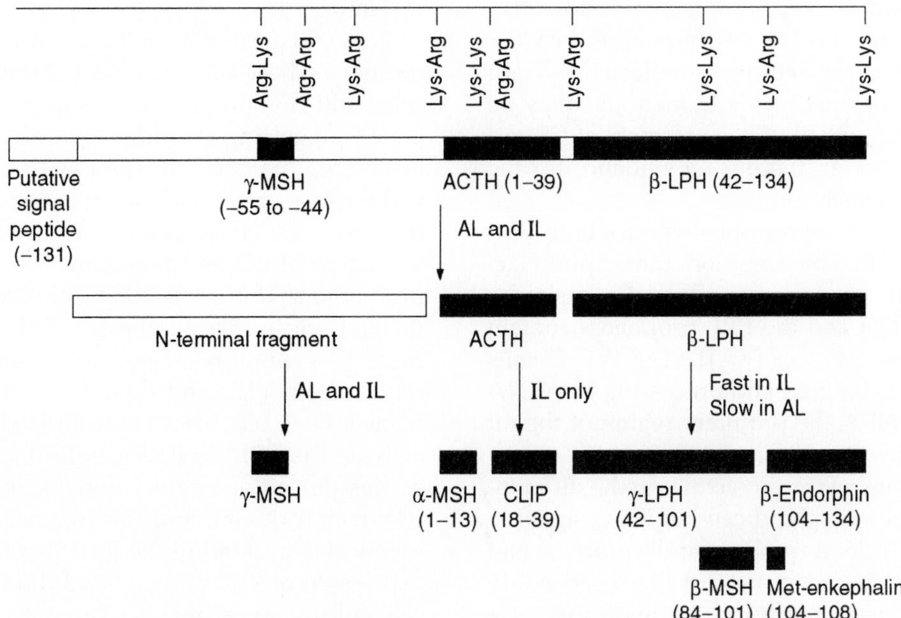

FIGURE 5-10 Schematic representation of the preproopiomelanocortin molecule formed in pituitary cells, neurons, and other tissues. The numbers in parentheses identify the amino acid sequences in each of the polypeptide fragments. For convenience, the amino acid sequences are numbered from the N-terminus of ACTH and read toward the C-terminal portion of the parent molecule. The locations of Lys-Arg and other pairs of basic amino acid residues are also indicated; these are the sites of proteolytic cleavage in the formation of the smaller fragments of the parent molecule. Abbreviations: AL, anterior lobe; IL, intermediate lobe. Reproduced with permission from Ganong, W. E. (1995). "Review of Medical Physiology," 17th ed., p. 368. Appleton and Lange, Norwalk, CT.

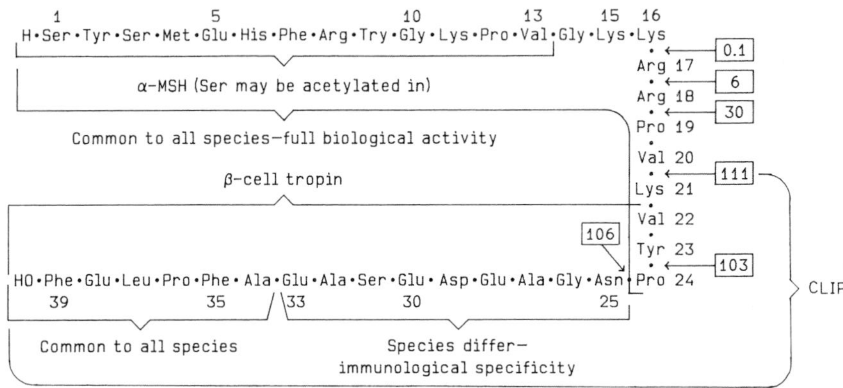

FIGURE 5-11 Linear sequence of human ACTH. Arrows point to hydrolytic cleavage and residual activity is denoted in boxes as a percentage of activity of the native structure after cleavage at a specific linkage. α-MSH comprises the first 13 amino acid sequence from the N-terminus. β-Cell tropin has the same sequence as $ACTH_{22-39}$. Immunological specificity is conferred by sequence 25–33, and the invariable C-terminal sequence is 34–39. CLIP: corticotropin-like intermediary peptide. These data are a summary from a number of different literature sources.

rect precursor of the ancestral gene for both PRL and GH. The human GH gene family contains GH and four placental genes: GH variant and CSA, CSB, and CSL. PRL is a single human gene. The situation of PRL-like genes in the rodent is very complex, and since this complexity is not reflected in the human, the rodent picture will not be discussed. High-molecular-weight forms of CS protein exist in the human and may be explained by the formation of heterodimers of CS and α_2-macroglobulin. So far, there is no evidence for the existence of PRL in such forms.

Because a subset of lactrotrophs secretes both PRL and GH, it appears that the common transcription factor activating both genes may be Pit-1. Two similar sequences of the GH and the PRL promoters contain the AT-rich moiety A(T/A)(T/A)TATNCAT. Details of the human prolactin gene and processing to mRNA and protein, as well as the promoter region of the rat PRL gene, are shown in Figure 5-16A.

The most obvious role of prolactin is in the differentiation of the mammary gland cells and as a signal in the differentiated cells to produce milk proteins and other constituents. Prolactin may act like a secondary growth hormone, particularly on the liver. PRL also can act on the testis to stimulate the production of testosterone. It seems to have some role in stress adaptation as its release from the anterior pituitary is signaled by the release of β-endorphin in response to stress and as a consequence of the action of CRH; β-endorphin then acts, apparently directly, on the mammotrope (lactotrope) to stimulate the release of PRL. It has been known for some time that nursing an infant on demand over a long period of time creates a period in which pregnancy rarely occurs. This is apparently the result of frequent release of PRL from the anterior pituitary in response to suckling, which may have the effect of reducing GnRH release from the hypothalamus.

Prolactin derives from the lactotroph of the anterior pituitary. These cells represent about 20% of pituitary cells, and this proportion is similar in males and females, suggesting a significant role for PRL in the male. PRL receptors occur in the Leydig cells of the testis and other sexual tissues. It appears that PRL maintains the level of LH receptors, which, in turn, facilitates the action of LH and maintains levels of testosterone production. There are other activities of PRL in the male. Lactotrophs synthesize PRL exclusively, but there is a minority of cells that can synthesize and secrete both PRL and GH. A DNA-binding protein named Pit-1 has been characterized that appears to activate the PRL gene. A similar factor called GHF-1 carries out an analogous function in stimulating the production of GH, and the two factors are probably identical. Knockout of the Pit-1 function decreases the expression of PRL and GH and diminishes growth in cell culture experiments. The Pit-1 message is expressed in somatotrophs, lactotrophs, and thyrotrophs. Pit-1, after transfection, will activate PRL gene expression. It is apparent that other cell-specific nuclear factors are required to generate a specific hormonal response. Like several other transcriptional activators, Pit-1 is most efficient in stimulating PRL expression when two molecules of Pit-1 interact at the DNA level. Mutations in the Pit-1 gene have been found in certain genetic strains of dwarf mice, suggesting that Pit-1

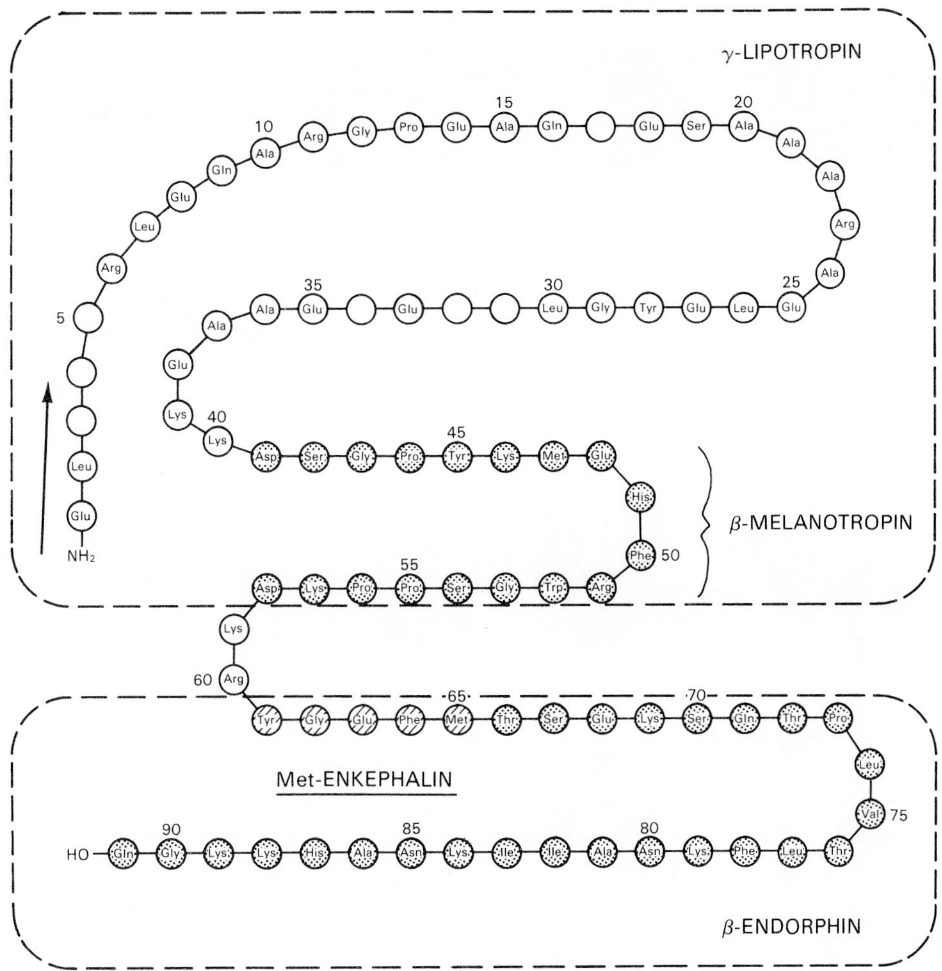

FIGURE 5-12 Structural relationship of γ-LPH, β-MSH, Met-enkephalin, and β-endorphin to the ovine β-LPH structure. Reproduced from Li, C. H. (1982). "Biochemical Actions of Hormones" (G. Litwack, ed.), Vol. 9, pp. 1–41. Academic Press, NY.

function is required for the elaboration of GH (and PRL), and presumably this will have significance for human dwarfism.

Estrogen is known to stimulate the production of PRL. It is likely that the estrogen receptor interacts with Pit-1 since cotransfection of estrogen receptor and Pit-1 results in the stimulation of PRL transcription, whereas transfection of the estrogen receptor alone, in an estrogen receptor-deficient cell line, does not enhance PRL expression. Glucocorticoids, on the other hand, negatively affect the PRL promoter.

A. Regulation of Prolactin Release

There is a discussion of the regulation of PRL release in Chapter 14. Significant positive regulators include estrogen, VIP, and TRH, and the negative regulator is dopamine. A summary of the signal transduction processes for these effectors is shown in Figure 5-17.

PIF, the endogenous inhibitor of PRL release, appears to be dopamine. If dopamine receptor function is blocked by a drug, such as phenothiazine or metoclopramide, a rise in PRL occurs. Dopamine (DA), from the dorsomedial area of the arcuate nucleus and the inferior ventromedial nucleus of the hypothalamus, is released into the median eminence, where DA-carrying axon terminals represent one-third of all the axon terminals in the *zona externa* of the median eminence. This is the tuberoinfundibular dopamine (TIDA) pathway. Certain TIDA axons project into the posterior and intermediate lobes of the pituitary. TI dopamine binds to D_2 dopamine receptors on the lactotroph, resulting in decreased adenyl-

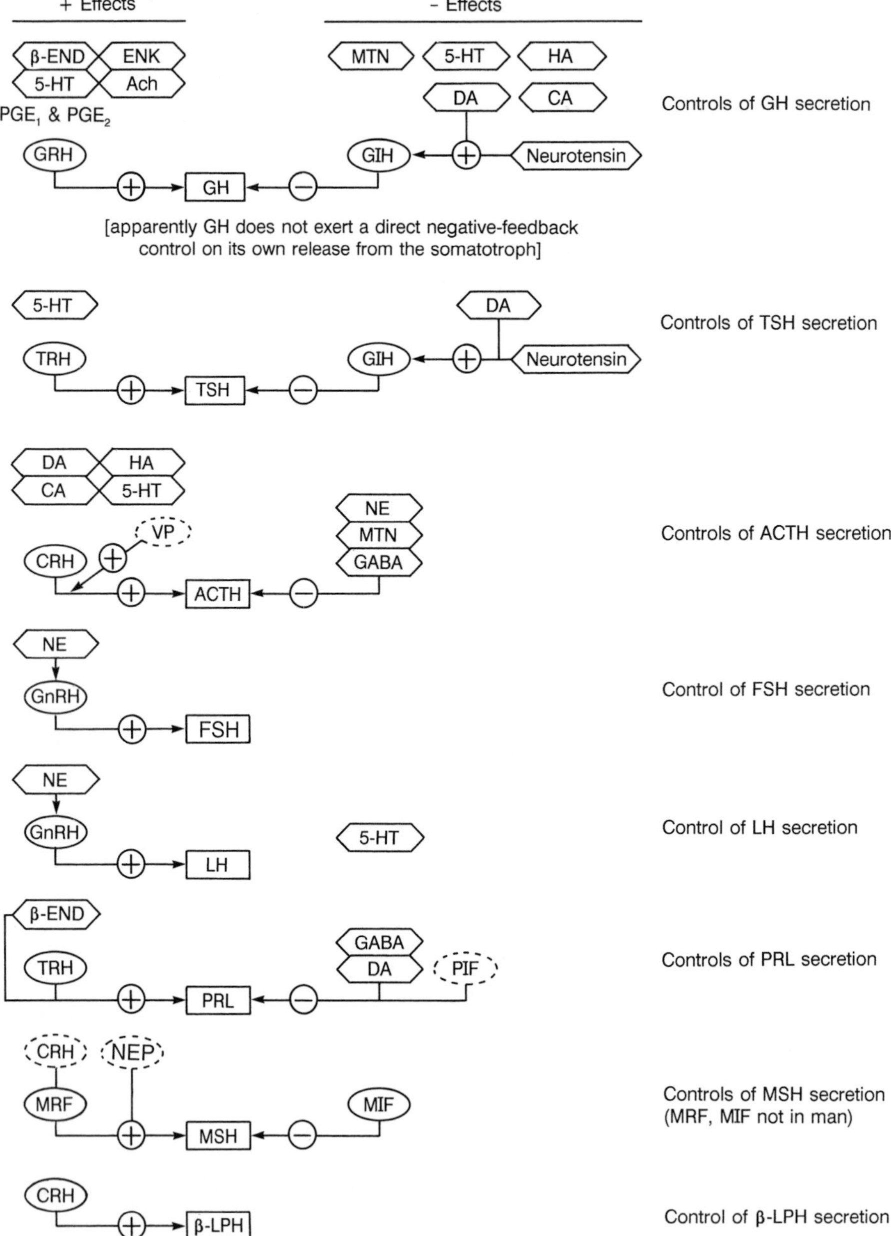

FIGURE 5-13 Summary of some of the major controlling factors on the secretion of hormones from the pars distalis. Factors encircled by ovals are releasing hormones–factors. Factors enclosed by hexagons are neurotransmitters. + effects are those that augment the secretion of the anterior pituitary hormone; − effects are those that decrease the secretion of the anterior pituitary hormone. Abbreviations: ENK, enkephalin; MTN, melatonin; DA, dopamine; 5-HT, serotonin; NE, norepinephrine; MRF, melanocyte hormone-releasing factor; MIF, melanocyte hormone release-inhibiting factor; CA, catecholamine; HA, histamine. See text for other abbreviations. Some of the effects shown here are controversial.

ate cyclase activity, PI cycle activity, and calcium utilization and transport. Multiple D_2 receptors may evolve from alternative splicing events to explain these broad effects. Estrogen blocks DA inhibition by its positive action on the PRL promoter or by other direct effects. The cloned DNA sequence encoding GnRH precursor protein contains an associated peptide called GAP (GnRH-associated peptide). This peptide was found to be an inhibitor of PRL secretion and may be a candidate for PIF.

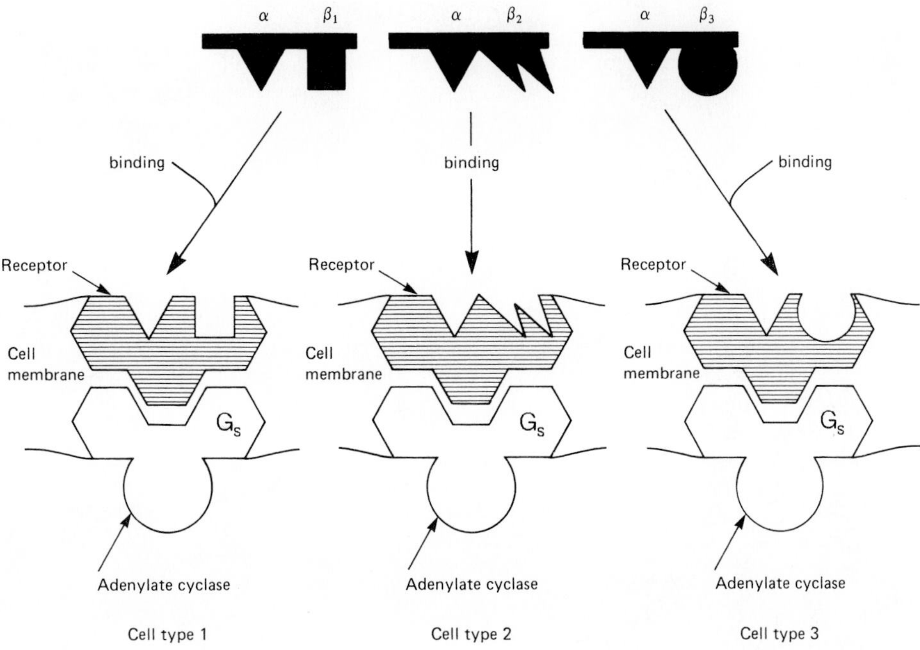

FIGURE 5-14 How anterior pituitary hormones that share a similar subunit (TSH, FSH, LH) specify receptors on different cell types.

B. Stress-Induced Release of Prolactin

During stress, ACTH, β-LPH, and β-endorphin are released simultaneously from the corticotroph (see also Chapter 10). β-Endorphin appears to interact with the μ-receptor, which mediates the release of PRL from the mammotrope. The opiate peptides probably stimulate PRL release by inhibiting DA turnover and release by the TIDA pathway. GH is probably released in addition to PRL as a result of the action of β-endorphin. A serotonergic pathway may cause the stimulation of basal PRL levels in the anterior pituitary as diagnosed by ρ-chlorophenylalanine methyl ester, an inhibitor of serotonin biosynthesis, and by metoclopramide, which stimulates PRL release. The structures of these compounds are shown in Figure 5-18.

Thus, at least two pathways are possible for the regulation of PRL release. A positive pathway operates via a serotonergic mechanism, and a negative pathway operates via a dopaminergic mechanism. A number of factors influence the secretion of PRL and GH. These are summarized in Table 5-3.

C. Prolactin Receptors in Nonmammary and Mammary Tissues

PRL receptor occurs in a wide variety of cells and tissues. PRL receptor is located in the cell membrane, but surprisingly has also been found in endosomes, Golgi apparatus, and lysosomes (degraded receptor has already been discussed). There is also a soluble protein with binding properties similar to those of the membrane receptor, and this form is found in liver, kidney, and mammary cytosols. PRL and GH receptors belong to the hematopoietic–cytokine receptor family shown in Figure 5-19.

The PRL receptor consists of an extracellular domain (210 residues) with five cysteines, a transmembrane domain (24 residues), and a cytoplasmic domain (51 residues). PRL and GH receptors are localized on human chromosome 5. Members of the hemapoietic receptor family are structurally similar in ligand-binding domains but not in cytoplasmic sequences. A fibronectin-like domain occurs in the extracellular domain next to the membrane (Figure 5-19). Mammary gland (rabbit) has a longer PRL receptor than the liver form, and the long form is also present in human. Liver contains the highest number of PRL receptors. The long form of the receptor appears to be involved in the transcription of genes specifying milk proteins. The short form of the receptor may internalize PRL into specific cell types.

Human GH binds to the PRL receptor, and zinc greatly increases this binding. Zinc is chelated by three amino acid residues in hGH and one residue of the

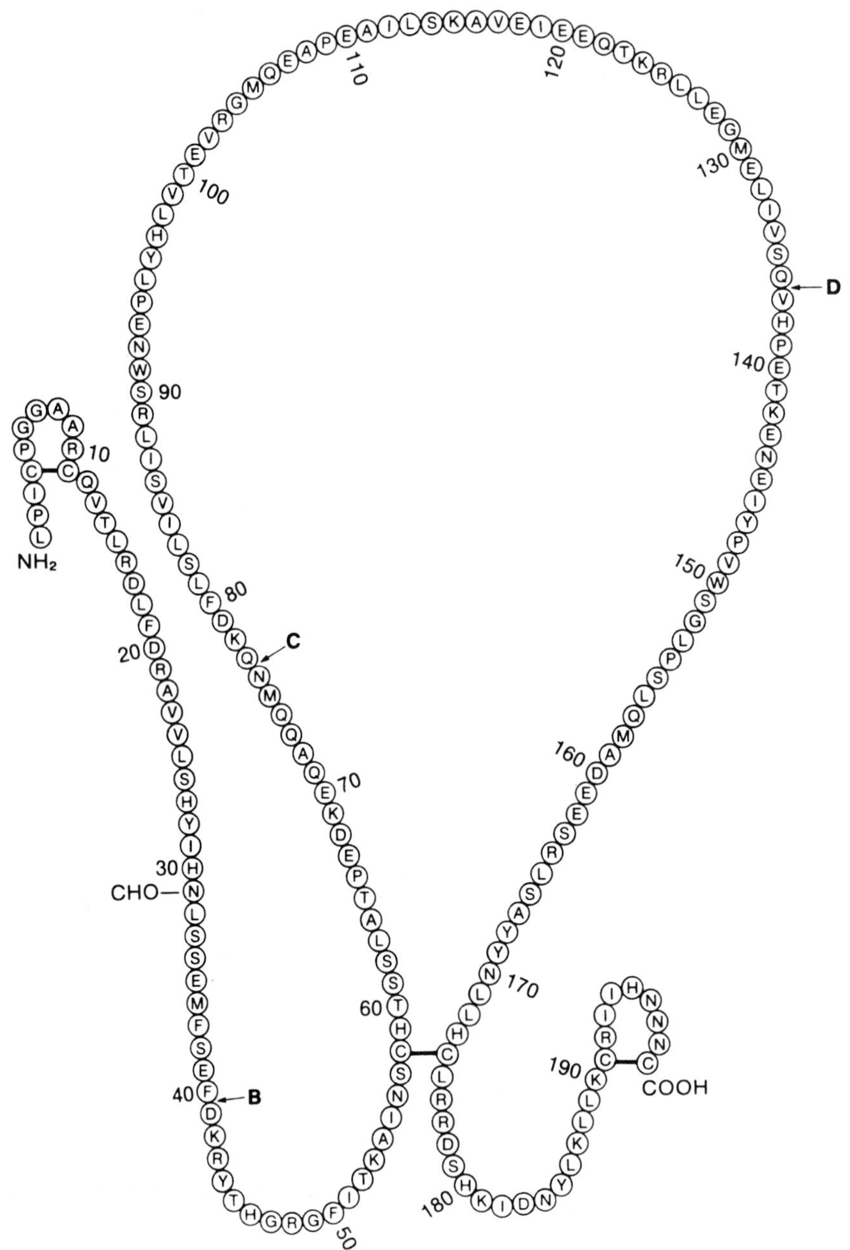

FIGURE 5-15 The human PRL molecule. The human PRL amino acid sequence is displayed in the single-letter amino acid code. The lettered arrows indicate the points at which the coding region of the PRL gene is interrupted by introns. Intron A, located in the signal peptide, is not shown in this diagram, which depicts only the mature protein. The N-linked glycosylation site at residue 31, asparagine, is indicated by CHO. Reproduced with permission from Cooke, N. E. (1995). Prolactin: Basic physiology. *In* "Endocrinology" (L. J. DeGroot, ed.), 3rd ed., Vol. 1, pp. 368–393. W. B. Saunders Co., Philadelphia, PA.

PRL receptor. Zinc is also involved in CS (chorionic somatomammotropin) binding to the PRL receptor, but not in the binding of PRL to the PRL receptor. One GH binds to two molecules of GH receptor, generating the dimeric form of the GH receptor, which is probably the active form for signal transduction to occur. PRL receptor could need to be dimerized also to generate the active form. Although it has been shown that PRL binding to receptor is followed by a rapid stimulation of tyrosine kinase activity, the molecular weights of protein substrates rule out PRL receptor itself as one of the phosphorylated products.

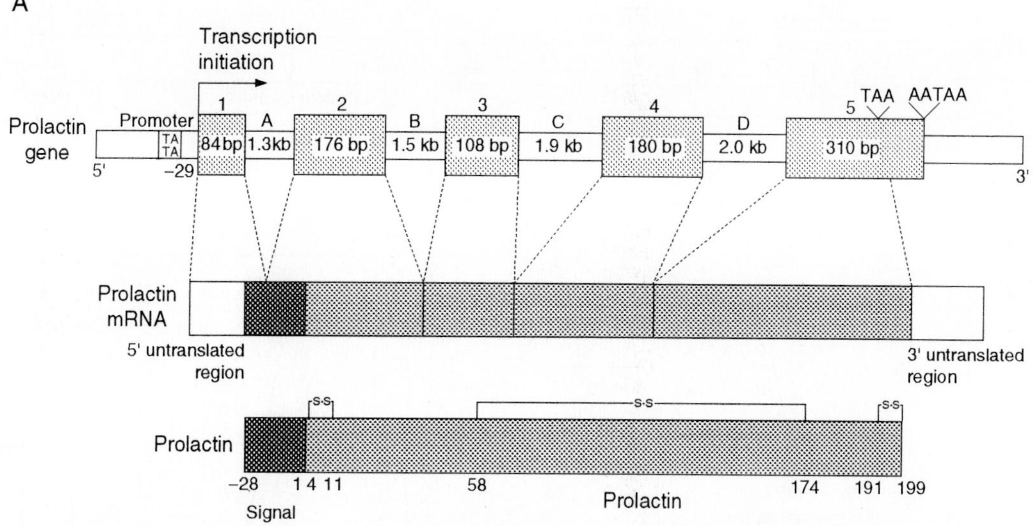

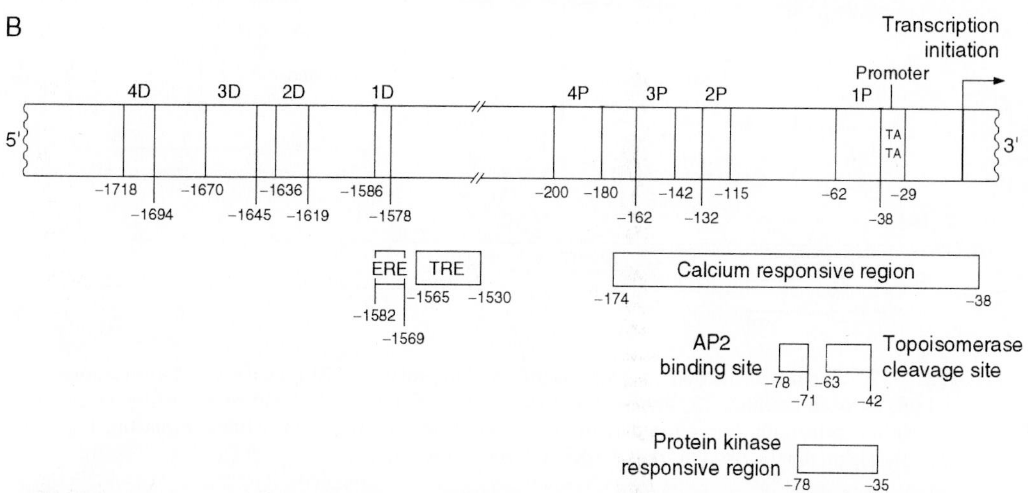

FIGURE 5-16 (A) Processing of the human prolactin gene. The exons are denoted by numbers (1–5) and the introns by letters (A–D). (B) Detailed analysis of the 5′ flanking regions of the rat prolactin gene. Abbreviations: ERE, estrogen response element; TRE, thyroid response element; 1P, 2P, 3P, and 4P refer to proximal binding sites for Pit-1; 1D, 2D, 3D, and 4D refer to distal binding sites for Pit-1. Reproduced with permission from Molitch, M. E. (1995). Prolactin. *In* "The Pituitary" (S. Melmed, ed.), pp. 136–186. Blackwell Science, Oxford.

VI. GROWTH HORMONE

A. Growth Hormone Structure

The sequence of human GH is given in Figure 5-3. Human GH is a peptide containing two disulfide bonds with a molecular weight of 22,000, although a smaller variant of 20,000 accounts for about 10% of GH and lacks amino acids 32–46 through processing of the GH mRNA precursor. A large form of GH is represented by a dimer stabilized by an interchain disulfide bond, and larger forms have been described. The forms larger than 22,000 probably are not physiologically important. A glycosylated form of GH is produced by the placenta that is 25,000. As mentioned previously, GH is related to chorionic somatomammotropin (CS) produced by the human placenta. hCS contains 191 amino acids and is quite similar to hGH. CS could have arisen by a new duplication of the primate GH gene or

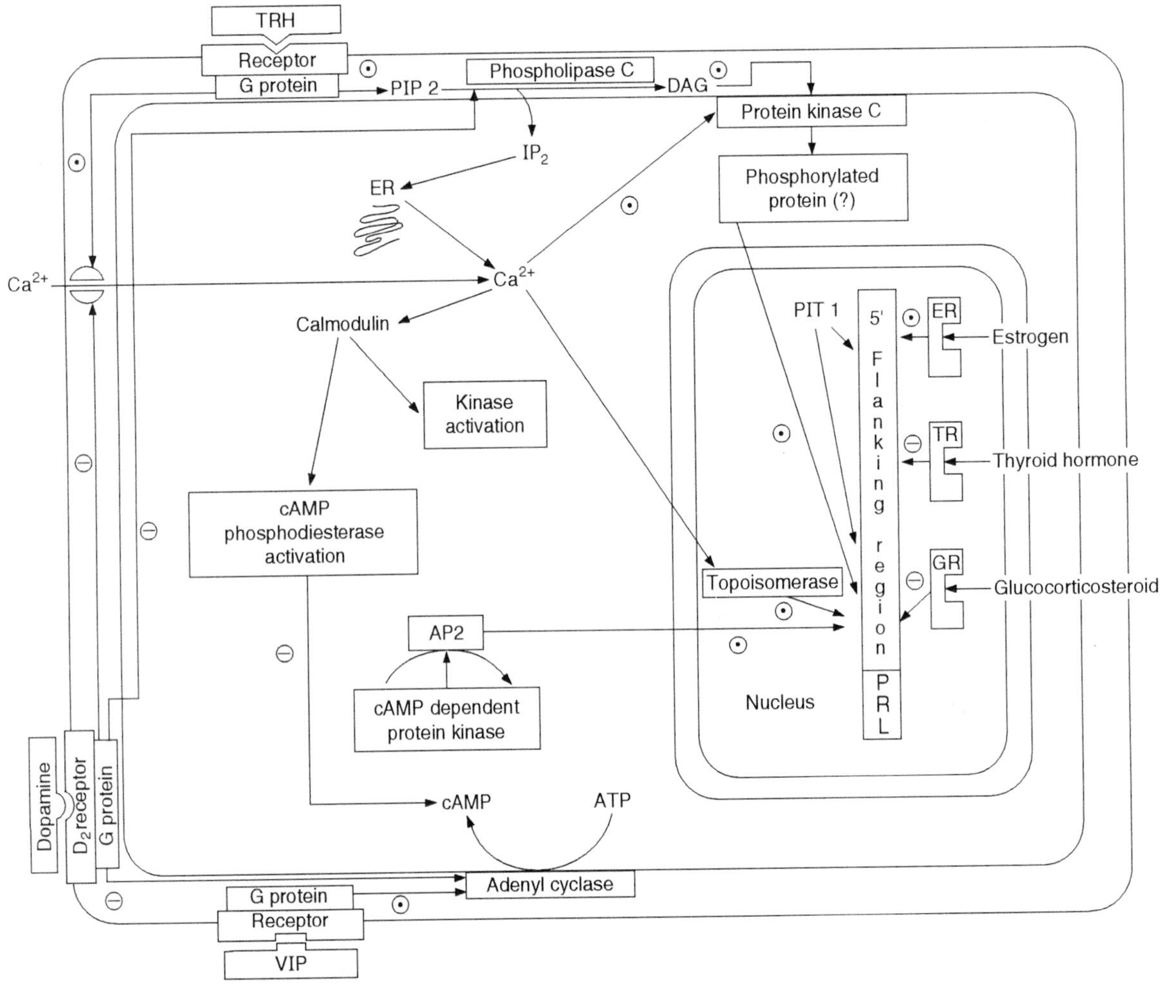

FIGURE 5-17 Signal transduction mechanisms involved in prolactin (PRL) regulation. Abbreviations: (in nucleus) ER, estrogen receptor; TR, thyroid hormone receptor; GR, glucocorticoid receptor; (in cytoplasm) ER, endoplasmic reticulum. Reproduced with permission from Molitch, M. E. (1995). Prolactin. *In* "The Pituitary" (S. Melmed, ed.), pp. 136–186. Blackwell Science, Oxford.

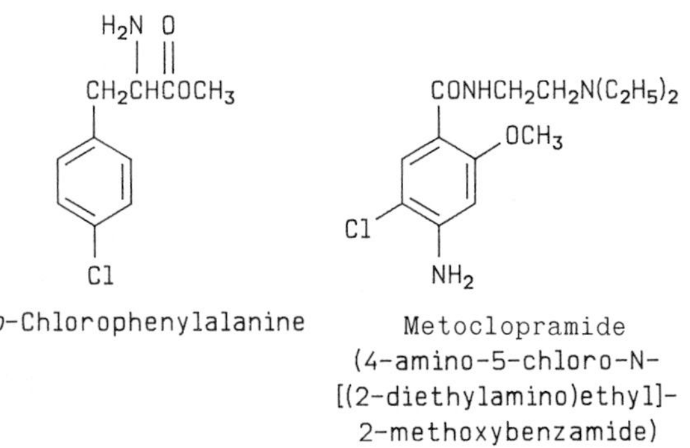

FIGURE 5-18 Structures of inhibitors of serotonin biosynthesis.

extensive interchange of the GH gene with the CS gene. hCS has 0.1% of the growth-promoting activity of hGH. Earlier comparisons were made between the GH–CS gene family and PRL. The gene for hGH is pictured diagrammatically in Figure 5-20.

The gene cluster that gives rise to GH-1, GH-2, CS-1, CS-2, and CS-P is shown in Figure 5-21. The gene cluster consists of five genes, with the gene for GH-1 at the 5′ end. GH-1 or GH-N is the somatotrophic cell pituitary hormone, followed by CS-P, CS-1, GH-2, and CS-2. There is at least 90% base sequence homology among the genes. A 217-amino-acid prehormone is cleaved to yield the mature 191 amino acid hormone with a 22,000 molecular weight. The GH-1 or GH-N gene gives rise to two forms, as shown in Figure 5-20. The two proteins are similar, except that the 20,000

TABLE 5-3 Substances Affecting Prolactin Release[a]

Stimulatory	Inhibitory
Thyrotropin-releasing hormone	Dopamine
Vasoactive intestinal polypeptide	Gonadotropin-associated peptide
Peptide histidine isoleucine	γ-Aminobutyric acid
Serotonin	Somatostatin
Opioid peptides	Acetylcholine
Growth hormone-releasing hormone	
Gonadotropin-releasing hormone	
Oxytocin	
Vasopressin	
Histamine (H_1)	
Bradykinin	
Angiotensin II	
Neurotensin	
Substance P	
Cholecystokinin	
Bombesin	
Secretin	
Gastrin	
Galanin	
Calcitonin	
Calcitonin gene-related peptide	
Thymosin Factor 5	
Melatonin	
Other posterior pituitary factors (?)	
Platelet-activating factor	
Epidermal growth factor	
α-Melanocyte-stimulating hormone	

[a] Reproduced with permission from Molitch, M. E. (1995). Prolactin. *In* "The Pituitary" (S. Melmed, ed.), pp. 136–186. Blackwell Science, Oxford.

form does not express early antilipolytic activity. The lactogenic activity of CS is similar to that of GH and PRL, but its affinity for the GH receptor is one ten-thousandth that of GH.

One gram of somatotropic tissue can produce 2–6 mg of GH-1 (22 kDa) per day. There is also a high level of expression of CS-1 and CS-2 in placenta.

The hGH gene promoter region 500 bp upstream of the transcriptional start is shown in Figure 5-22. Nuclear factors binding to this region are NF-1, USF, AP2, and the glucocorticoid receptor. Pit-1 (GHF-1) is very active, as discussed previously. A silencer is located 200 bp farther upstream, and these sequences will silence pituitary cell expression but not that of placental cells. The protein binding to this region is called pituitary-specific factor 1 (PSF-1), but this factor is absent from placental cells. PSF-1 may interact with Pit-1. Pit-1 contains helix–turn–helix motifs, and the helices are thought to contact the major groove of target DNA and form DNA-dependent homodimers.

Children with no expression of GH-1 have pronounced defective growth patterns. Some of these children raise antibodies against GH when they are treated due to the absence of GH from their systems, although antibody levels seem lower with the use of highly purified biosynthetic human GH preparations than with pituitary extract. Other genes in the GH–CS gene cluster appear to be functional but they do not replace the absence of GH-1. This syndrome seems to entail a loss of 6.7 kb of DNA between two *Hind*III sites normally containing the GH-1 gene. There are several modes of GH-1 gene deletion in this disease, involving unequal recombination and crossing over at meiosis. There is evidence for several independent recombination events. PCR analysis helps to sort out the nature of the deletion, but this approach also has some deficiencies.

B. Regulation of Growth Hormone Secretion

The overall regulation of GH secretion and its actions are summarized in Figure 5-23 and Table 5-3. At the hypothalamic level a number of aminergic neurons can exert influences on GH secretion. These neurons are reviewed generally in Chapter 3. Figure 5-23 shows catecholaminergic neurons, which may stimulate the release of GRH from its neuron, and serotonergic as well as β-endorphinergic neurons, which could stimulate the release of TRH from its nerve ending. Although less certain, melatoninergic neurons or melatonin from the pineal gland could stimulate the release of somatostatin from somatostatinergic neurons, which inhibit the release of GH from the somatotroph. These interneurons are probably linked to various conditions known to operate through the CNS: hypoglycemia (insulin?), exercise, and surgical stress. Once released from the somatotroph, GH circulates in the blood at levels greater than 3 ng/ml, with a total daily output of 1–4 mg.

GH has several major activities. It interacts with GH receptors in adipose cells and brings about an increase in free fatty acids by increasing lipolysis. Phosphorylation of triglyceride lipase possibly is involved in this mechanism. GH interacts with receptors on the cell membranes of liver, kidney, and muscle and causes the release of somatomedins (IGFs) into circulation. Somatomedins are mitogenic (growth stimulatory) for many types of somatic cells and are of great importance in growth. GH also interacts with vascular walls, re-

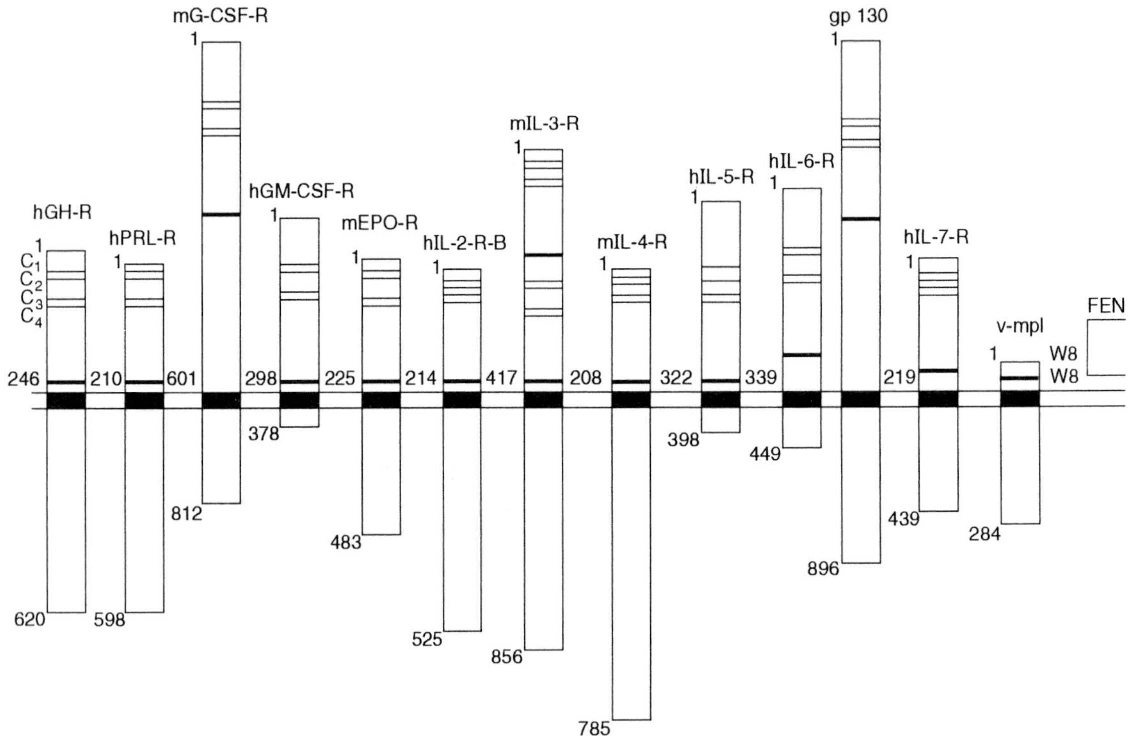

FIGURE 5-19 Members of the hematopoietic–cytokine receptor family that includes the receptors for PRL and GH. Selected amino acids of each receptor are numbered. Homologous regions reside in the extracellular domains, marked at the amino-terminal end by the pairs of cysteines (C1–C4, indicated by thin lines). A common WSXWS motif is indicated by the thick line. The transmembrane regions are in black. Adapted from Kelly, P. A., Djiane, J., Postel-Vinay, M. C., Edery, M. (1991). The prolactin/growth hormone receptor family. *Endocr Rev.* **12,** 235–251. © The Endocrine Society, 1991. Reproduced with permission from Cooke, N. E. (1995). Prolactin: Basic physiology. *In* "Endocrinology" (L. J. DeGroot, ed.), 3rd ed., Vol. 1, pp. 368–393. W. B. Saunders Co., Philadelphia, PA.

sulting in the breakdown of vascular polysaccharides to glucose, which enters circulation.

One form of dwarfism in humans is caused by the defective production of pituitary GH. This condition appears from an autosomal recessive characteristic. Dwarfs in this category completely lack GH but not other pituitary hormones. Treatment of the condition with hGH causes prompt antibody reaction in that the dwarf recognized hGH as a xenobiotic. Through the use of cDNA probes of hGH, there appear to be several hybridizing bands corresponding to this cDNA. Dwarfs lack one of these bands, however, and their immediate family members have this band but in less than normal amounts. The normal GH gene encodes a 22-kDa GH, while the variant encodes, through an alternative splicing event, a 20-kDa GH. A third form of GH also exists that is not very well studied. The 22-kDa form has the growth-stimulating function, but not the hyperglycemic effect that is characteristic of the splice variant form. The ratio of these two forms in the blood is fairly constant in spite of the 22-kDa form being more concentrated in the pituitary. The nearly equal ratio between the two circulating forms may be accounted for by the slower degradation of the 20-kDa splice variant. The fairly constant circulating ratio between the two forms is also independent of age from child to adult, except that GH levels are higher at puberty and decline during senescence. In addition, the responsiveness of the pituitary to releasing factors undoubtedly decreases with aging.

Like many of the other pituitary hormones, the secretion of GH from the pituitary is pulsatile, and peaks of GH secretion can occur throughout a 24-hr period although the magnitude of the peaks can vary with a number of factors, one of which is sleep, during which GH levels are usually higher. The levels of other pituitary hormones also change during sleep but GH is a hallmark. It is not clear what GH is doing in this state, but it could be related to its action on carbohydrate metabolism (Figure 5-23).

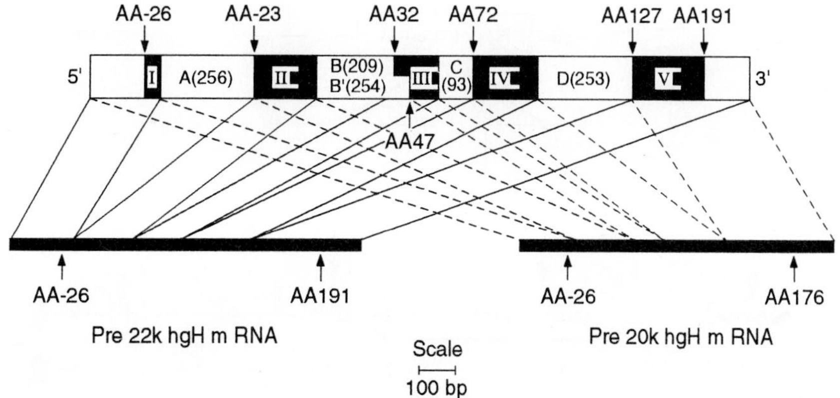

FIGURE 5-20 Theoretical representation of the hGH gene and the mRNA from it. The hGH gene contains three introns: intron A between amino acids (AA) 24 and 25, intron B between AA 31 and 32, and intron C between AA 71 and 72. These are excised when the gene is transcribed into the mRNA for the 22-kDa hGH. Also shown is the alternative splicing transcribed into the mRNA of 20-kDa hGH. Reprinted with permission from Chawla, R. J., Parks, J. H., and Rudman, D. (1983). Structural variants of human growth hormone: Biochemical, genetic and clinical aspects. *Ann. Rev. Med.* **34,** 519–547. Copyright © 1983 by Annual Reviews, Inc. Reprinted with permission from Daughaday, W. H. (1995). Growth hormone, insulin-like growth factors, and acromegaly. *In* "Endocrinology" (L. J. De Groot, ed.). 3rd ed., Vol. 1, pp. 303–329. W. B. Saunders Co., Philadelphia, PA.

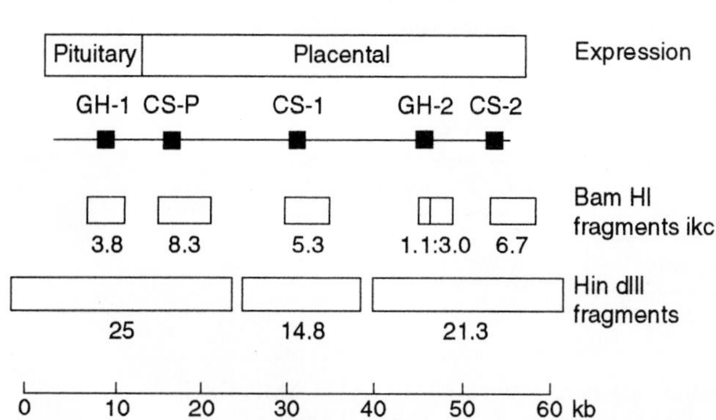

FIGURE 5-21 The GH and CS gene cluster. Five closely related human growth hormone and chorionic somatomammotropin genes occupy a 60-kb region on the long arm of human chromosome 17. The GH-1 gene is expressed in somatotrophic cells of the anterior pituitary, and the other four genes are expressed in syncytiotrophoblastic cells of the fetal placenta. Each gene is under 2 kb in size. Digestion of genomic DNA with *Bam*HI produces six fragments that hybridize to GH or CS cDNA probes. Digestion with *Hind*III produces three hybridizing fragments, with the GH-1 and CS-P genes sharing a 25-kb fragment. Reproduced with permission from Parks, J. S., Pfäffle, R. W., Brown, M. R., Abdul-Latif, H., and Meacham, L. R. (1995). Growth hormone deficiency. *In* "Molecular Endocrinology" (B. D. Weintraub, ed.), pp. 473–490. Raven Press, NY.

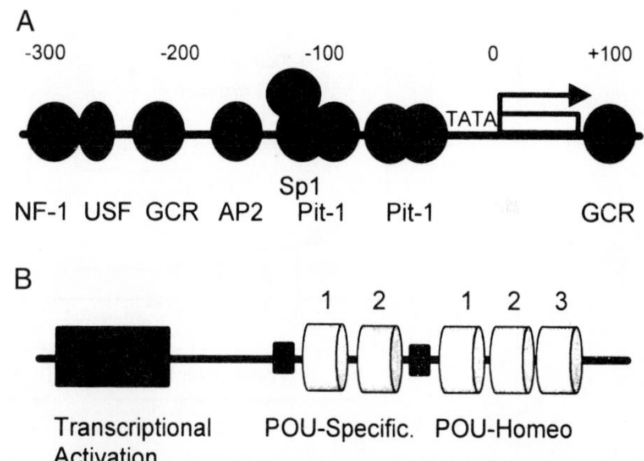

FIGURE 5-22 (A) Cis-active sites in the hGH promoter. Transcription of the GH-1 gene is regulated by proteins that bind to regulatory elements in the 300 bp preceding the start of transcription (arrow). The two Pit-1-binding sites are shown together with sites for NF-1, USF, AP2, Sp1, and the glucocorticoid receptor. (B) Pit-1 protein, helices, and domains. The human Pit-1 gene encodes a 291-amino-acid protein that binds to regulatory elements in the GH, PRL, and Pit-1 promoters. It consists of three recognized domains. They are an amino-terminal transcriptional activation domain, a POU-homeo-specific domain containing two α-helices, and a POU-homeodomain containing three α-helices. Reproduced with permission from Parks, J. S., Pfäffle, R. W., Brown, M. R., Abdul-Latif, H., and Meacham, L. R. (1995). Growth hormone deficiency. *In* "Molecular Endocrinology" (B. D. Weintraub, ed.), pp. 473–490. Raven Press, NY.

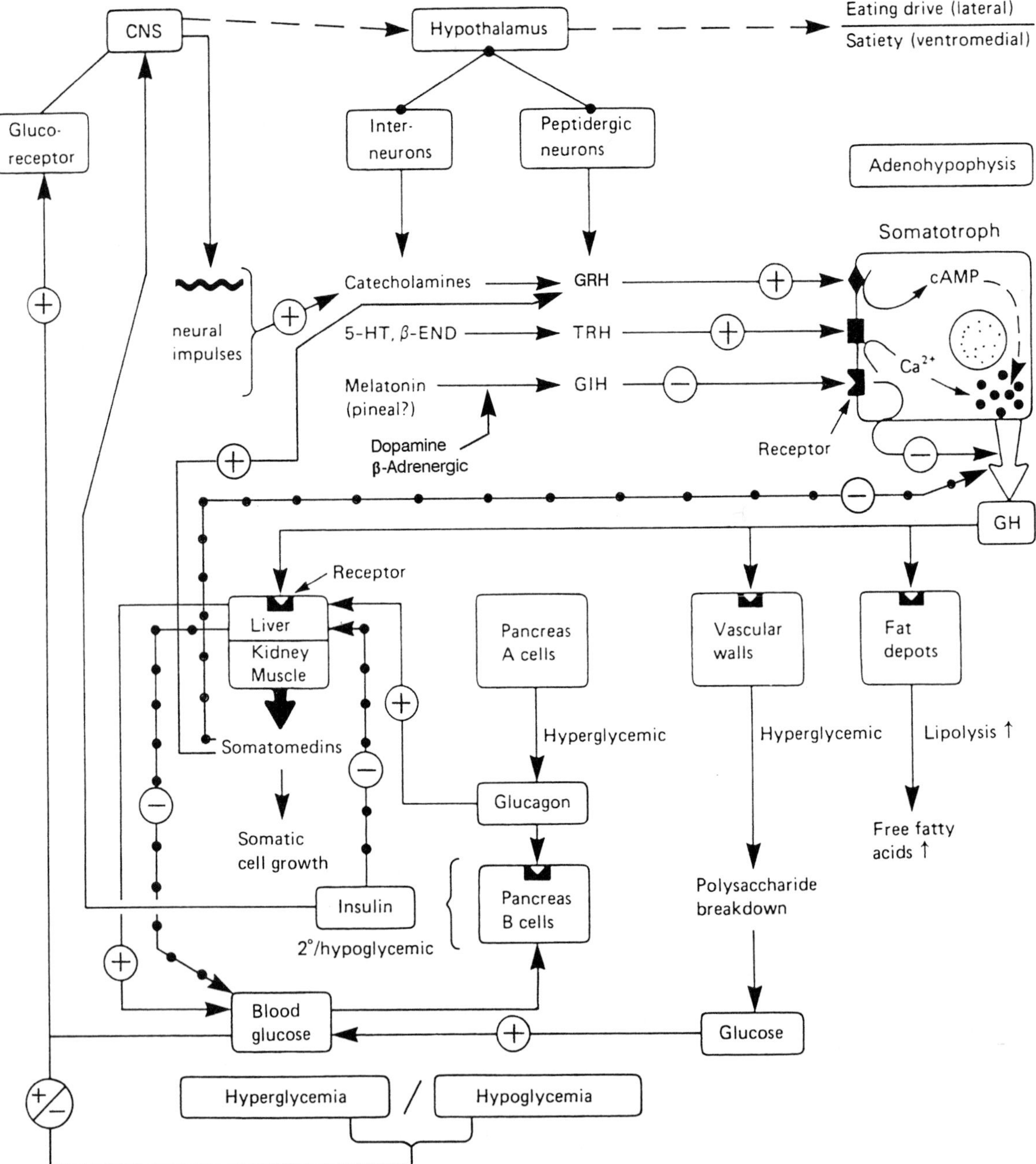

FIGURE 5-23 Control of secretion of growth hormone and its general actions. cAMP, Cyclic AMP; GRH, growth hormone releasing hormone; TRH, thyrotropin releasing hormone; GIH, growth hormone release inhibiting hormone or somatostatin; CNS, central nervous system; 5-HT, 5-hydroxytryptamine or serotonin; β-end, β-endorphin. Solid lines ending in arrows indicate direct actions; beaded lines ending in arrows to feedback effects and encircled sign (+ or −) indicates whether the feedback effect is positive or negative. The same signs are used for direct actions to indicate positive or negative effects. Dashed lines refer to hypothetical stimulatory pathways.

C. The Growth Hormone Receptor (GHR)

A computer model of the GH-binding protein in serum, based on crystallographic coordinates, which is derived from the GH receptor, is shown bound with GH in Figure 5-24. In this model hot spots for mutations are indicated by naming the participant amino acid residues. It appears that mutations in the receptor occur at some locations that are involved in the association with GH, so that these mutations can lead to a GH–GH receptor interaction that is subnormal. Also, there appears to be a lower concentration of GH-binding proteins in the sera of children with idiopathic short stature, suggesting that mutations of the receptor and decreased expression may explain the poorer utilization of GH in these children.

Other forms of short stature include normal immunoreactive GH levels and no other obvious cause. These conditions may be explained by a decreased level of GH receptor, a mutation in the receptor that reduces the effectiveness of normal GH, or a combination. A defective GH may be present that is indistinguishable by radioimmunoassay. Sometimes the administration of hGH in these children may induce growth, but this is a very costly treatment. Work has shown, by use of somatic cell hybridization and following the course of human genomic segregation, that the genes for GH, chorionic somatomammotropin (hCS), also known as placental lactogen, have been localized to chromosome 17 in humans.

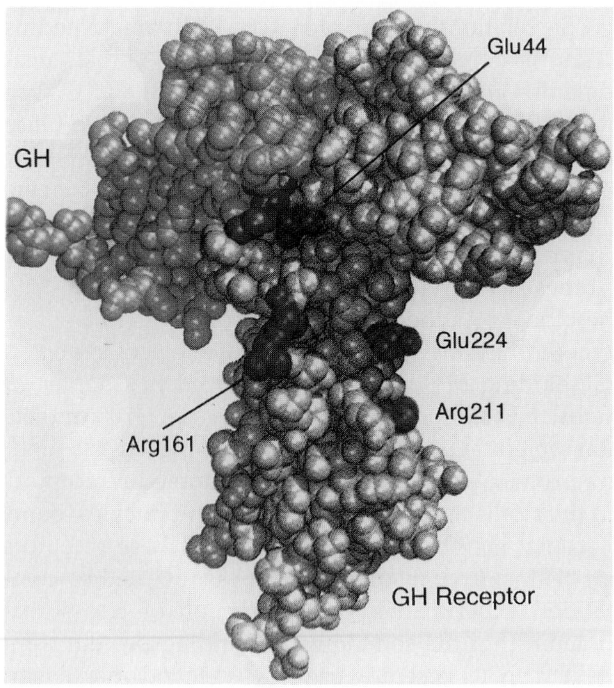

FIGURE 5-24 Structure of the GH receptor–GH complex. The structure of human GH and serum GH-binding protein was derived from crystallographic coordinates. The GH receptor residue Glu44 interacts with Arg64 on the GH molecule. Reproduced with permission from Goddard, A. D., Covello, R., Luoh, S.-M., Clackson, T., Attie, K. M., Gesundheit, N., Rundle, A. C., Wells, J. A., and Carlsson, L. M. S. (1995). Mutations of the growth hormone receptor in children with idiopathic short stature. *New Engl. J. Med.* **333**, 1093–1098.

The GH receptor (GHR) is a member of the hematopoietin receptor gene family already described. The GHR gene is located on chromosome 5 and has 10 exons over a distance of 87 kb. As shown in Figure 5-25, exon 2 encodes the signal peptide, exons 3–7 the extracellular domain, exon 8 the transmembrane domain, and exons 9 and 10 the intracellular domain.

Exon 1A has a TATA box at −31, a CCAAT box at −88, and sites for the binding of several transcription factors involved in liver-specific gene expression, including the glucocorticoid receptor and C/EBP-like proteins. The single gene GHR gives rise to several transcription products, including a 4.2–4.7-kb mRNA encoding the full-length receptor, and a 1.0–1.4-kb transcript encodes a truncated receptor in rat and mouse, but not in humans, rabbits, or pigs. A 2.1–2.6-kb transcript appears in rat liver and adipose. The GHR protein is composed of 638 amino acids, with an extracellular domain of 247 amino acids and a 24-amino acid transmembrane domain. The extracellular domain is extensively glycosylated. Several molecular weight forms of GHR have been observed for complexes with GH ranging from 122,000 to 300,000 for different species. The receptor is probably 100,000 bound with one 22,000 GH molecule. Higher molecular weight forms may involve disulfide bonds between 100,000 subunits. Some GHR in the human is covalently linked to ubiquitin. Like the PRL receptor, GHR is in the transducing active form as a homodimer. Thus, GH is released from the pituitary in homodimer form joined by one Zn^{2+} atom. The dimer dissociates in response to bloodstream dilution into the monomeric form and binds to hGHR. The membrane-bound hGH then complexes with a second hGHR to form the receptor homodimer, which initiates signal transduction. For one molecule of GH to bind two molecules of GHR, GH must possess two interactive domains in its structure. These are suggested in Figure 5-26. These two different sites on GH interact with identical sites on both GHRs.

D. Signal Transduction Process Initiated by GHR

One molecule of GH binds through site 1 of GH to a specific site on the extracellular domain of one GHR, and then the holoreceptor binds a second receptor molecule through site 2 of the GH ligand, as previously

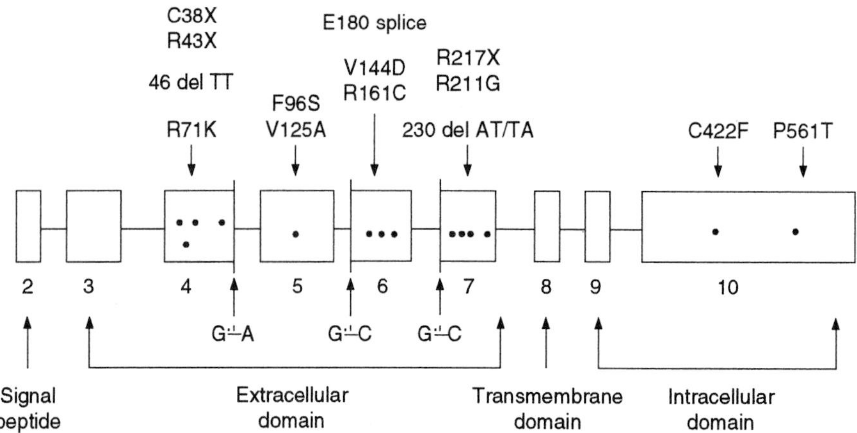

FIGURE 5-25 Structure of the GHR gene and mutations in Laron syndrome. The coding exons (boxes) are shown to approximate relative size; intron sizes are not to scale, and the noncoding part of exon 10 has been omitted. The structural protein domains encoded by the exons are indicated. The three shaded exons are involved in a deletion mutation. The locations (dots) and types of known point mutations within exons 4–7 are summarized above the diagram. The three splice site mutations are indicated by black bars at the exon 4/intron 4, intron 5/exon 6, and intron 6/exon 7 boundaries and are described below the diagram. Reproduced with permission from Harvey, S., and Hull, K. L. (1995). Growth Hormone Action: Growth Hormone Receptors. *In* "Growth Hormone" (S. Harvey, C. G. Scanes, and W. H. Daughaday, eds.), pp. 303–335. CRC Press, Boca Raton, FL.

described, to form the active homodimer through an identical site on the second receptor ($GH_1 \cdot GHR_2$). Homodimer formation undoubtedly produces an altered conformation, so that the intracellular domains now associate with a tyrosine kinase of the Janus family (JAK2). The intracellular receptor domains probably become phosphorylated by JAK2 and other substrates also become phosphorylated, such as MAP kinase and others. The end result is gene activators in the nucleus that may involve a number of transactivating and other factors, such as Spi 2.1, c-*fos*, and c-*jun*, affecting transcriptional activation of genes for IGF-1, P450SC, MUP (major urinary proteins), and others, probably including IGF-binding proteins. The activated transcription of IGF-I is key to the signal transduction mechanism initiated by GH for the growth of tissues. At best our knowledge of this signal transduction mechanism is sketchy. There is also information suggesting that, following GHR homodimer formation, phospholipase C may become activated to produce diacylglycerol and IP_3, which would lead to protein kinase C activation and phosphorylations affecting transcription factors acting in the nucleus. The phosphorylation of GHR in the process of cytoplasmic interaction with JAK2 is different, so far, with our current knowledge of the fate of the PRL receptor homodimer for which there is no strong evidence for its phosphorylation. This series of events leads to the production of IGF-I by the liver and other tissues, which is a key effector (somatomedins) of tissue growth.

E. Growth Hormone and Somatomedins

The relationship between GH and somatomedins (IGFs) is demonstrated in Figure 5-27. Three somatomedins are formed in the human, A, B, and C. One of these (somatomedin C) is identical to IGF-I. Once secreted into the bloodstream, GH binds to a specific receptor on the hepatocyte cell membrane and triggers a second message. In response, the hepatocyte (see Chapter 10 for cell biology of the hepatocyte) either synthesizes or releases preformed somatomedins into the blood, and the hepatocyte also produces a somatomedin-binding protein of ~50,000 molecular weight. IGF-binding proteins are summarized in Chapter 19 (Growth Factors). The somatomedins are ~7000 molecular weight and circulate mostly complexed to a binding protein. There are specific somatomedin receptors on liver cells and cells of adipose, lymphocytes, bone, placental membrane, and others, and these receptors are distinct from insulin receptors as revealed by competition experiments. The interaction of free somatomedin (not the somatomedin-binding protein complex) with its receptor triggers a second messenger, resulting in a mitogenic effect. The second messenger may occur through stimulation of a tyrosine kinase, which is part of the IGF receptor that probably phosphorylates IRS-1 (the insulin receptor substrate) and perhaps other proteins. The sequences of IGF-I, IGF-II, and insulin are shown in Figure 5-28. Because of the rather extensive homology of somatomedins with

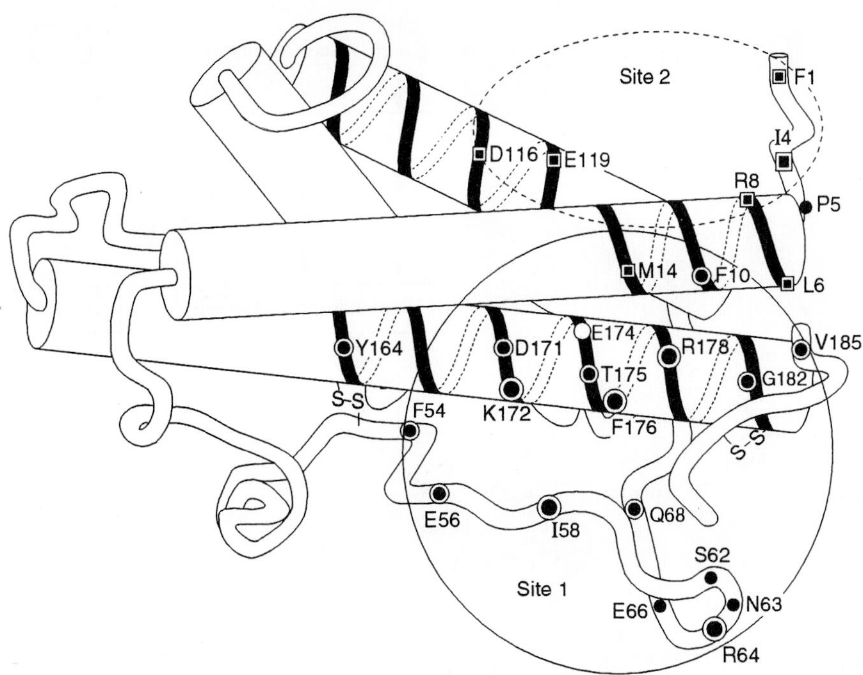

FIGURE 5-26 Probable location of two sites reactive with the GHR projected on a model of the GH structure. The cyclinders represent four predominantly helical domains. Alanine substitutions in hGH that disrupt binding of hGHbp at either site 1 or site 2 are generally delineated by the large shaded circles. Residues for which alanine mutants reduce site 2 binding by 2–4-fold, 4–10-fold, 10–50-fold, and more than 50-fold are shown by increasingly larger squares (■, ■, ■, and ■, respectively). Residues marked by the symbols •, ●, ●, and ○ represent sites where Ala mutations in site 1 of hGH cause reductions of 2–4-fold, 4–10-fold, greater than 10-fold, or a 4-fold increase in binding affinity for the hGHbp, respectively, using the MAb 5 immunoprecipitation assay. Reprinted from Cunningham, B. C., Ultsch, M., de Vos, A. M., *et al.* (1991). Dimerization of the extracellular domain of the human growth hormone receptor: Crystal structure of the complex. *Science* **254,** 821–825, with permission of the American Association for the Advancement of Science). Reprinted with permission from Daughaday, W. H. (1995). Growth hormone, insulin-like growth factors, and acromegaly. *In* "Endocrinology" (L. J. De Groot, ed.), 3rd ed., Vol. 1, pp. 303–329. W. B. Saunders Co., Philadelphia, PA.

insulin, it has been deduced that both hormones derive from the same ancestral gene. The growth-promoting activity of insulin itself is due to its limited ability to bind to IGF-I receptor. To do this, insulin has to be present in high concentrations because the affinity of the IGF-I receptor for insulin is about 1000× less than the affinity of the insulin receptor for insulin. IGF-II can also productively bind to the IGF-I receptor, albeit with much lower affinity. Thus, all three factors are growth-promoting, and IGF-II and insulin become important when IGF-I is low or absent. Presumably, IGFs can cross over to the insulin receptor in the absence of insulin, although this sort of rescue may not be important in insulin-deficient diabetes.

Interestingly, as shown in Figure 5-29, the level of circulating somatomedin increases to a maximum until about 8 years of age when the human adult level is reached. Since growth continues for many years beyond age 8, it is obvious that the amount of circulating somatomedins is not rate-limiting for growth. Thus, some other component of the overall system may limit growth, such as the development of somatomedin receptors or some other factor involved in the utilization of these mitogens. Somatomedins also stimulate sulfation, and this is important to bone growth.

F. Growth Hormone Signal Transduction

The 130-kDa growth hormone receptor contains a single transmembrane domain with a glycosylated extracellular domain (Figure 5-30). While the cytoplasmic domain becomes phosphorylated on tryrosine residues, this domain is different from other growth factor receptors, like those for EGF and PDGF, and has a different tyrosine kinase domain. Interestingly, the human GH receptor is specific for human GH and does not bind GH

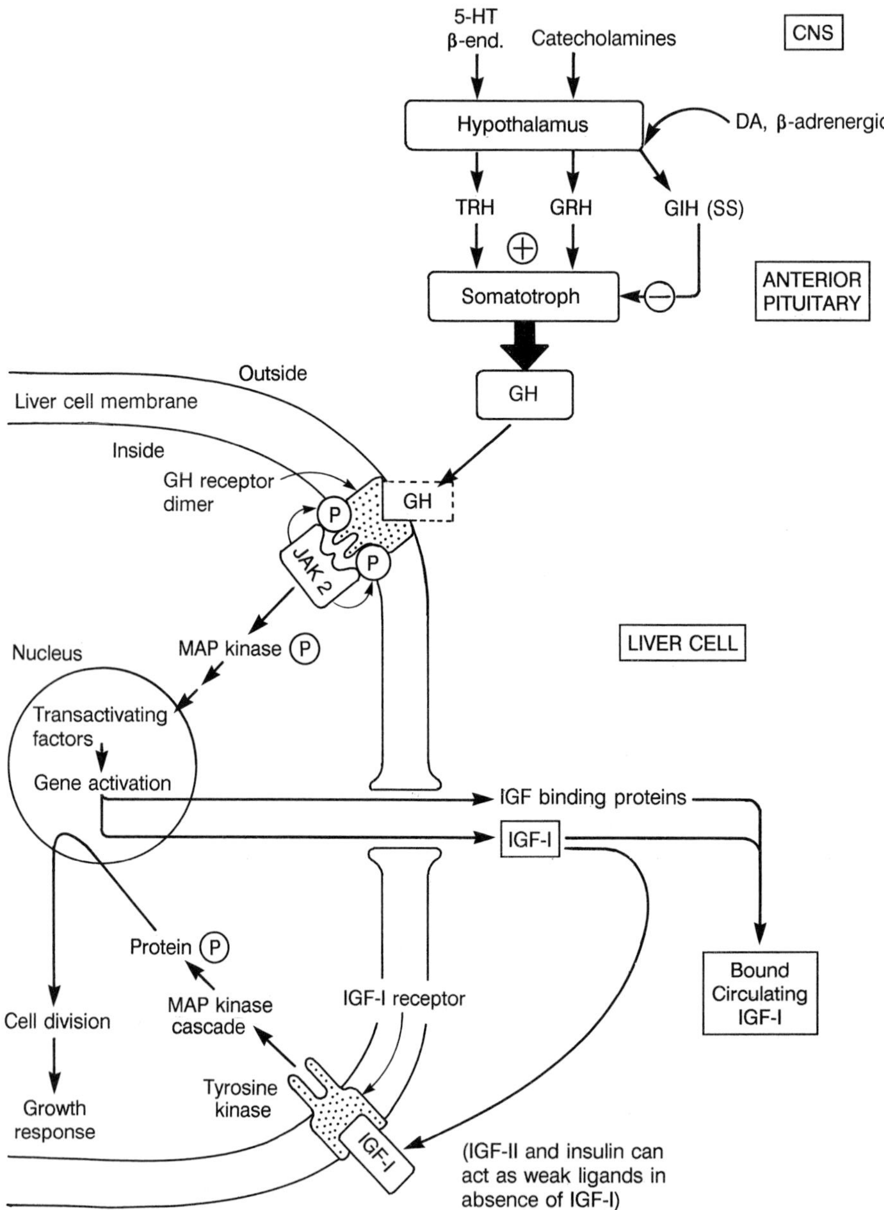

FIGURE 5-27 Somatomedins appear to mediate somatic cell growth. Abbreviations: NSILAs, nonsuppressible insulin-like activities; SM, somatomedin. These factors are now generally referred to as insulin-like growth factors (IGFs). Growth hormone is released from the somatotroph of the anterior pituitary whose regulation is suggested in this figure. After binding to the hepatocyte (in this illustration) GH receptor, reactions follow in the cell that stimulate the release of somatomedins, e.g., SM-C = IGF-I. IGFs bind to their receptors (one shown here) and tyrosine kinase activity is stimulated (this enzyme may be part of the receptor). It is hypothesized that a specific protein may be phosphorylated whose action directly or indirectly results in a growth response.

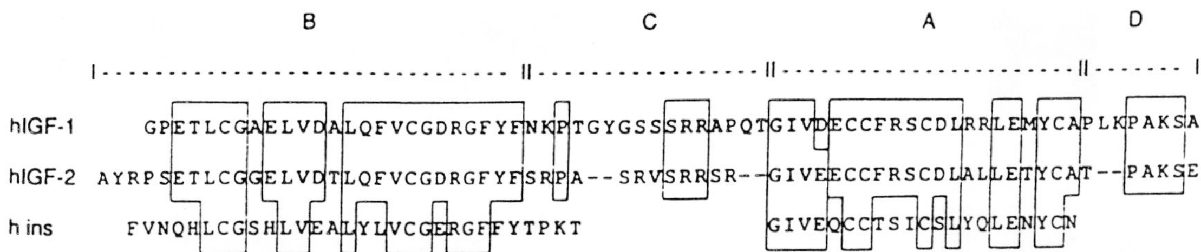

FIGURE 5-28 Structure of human IGF-I, IGF-II, and insulin (ins), with amino acid residues aligned to show the homologies identified by the boxed areas. Single-letter codes are used for amino acid residues. Note that in insulin the C peptide is removed, whereas in IGF-I and IGF-II the A and B chains (domains) remain connected by the C peptide. In addition, there are D domains. Modified and reproduced, with permission, from Sara, Y. R., and Hall, K. (1990). Insulin-like growth factors and their binding proteins. *Physiol. Rev.* **70,** 591. Reproduced with permission from Ganong, W. F. (1995). "Review of Medical Physiology," 17th edition. Appleton and Lange, Norwalk, CT.

of other species, whereas the GH receptor from lower forms is able to bind GH of more than one species. The GH receptor dimerizes in contact with one GH molecule per dimer. Identical binding domains on each receptor monomer are bound by a single GH molecule. Phosphorylation of tyrosine residues on the GH receptor follows by interaction with the cytoplasmic JAK2 kinase, a member of the Janus family of kinases. The GH receptor does not have a tyrosine kinase catalytic center in its cytoplasmic domain. Further downstream events are not clear, except that in certain tissues, such as liver, IGF-I and IGF-binding proteins are expressed. There are reports that GH brings about the activation of protein kinase C, directly or indirectly, and expression of c-*fos* and c-*jun* occurs in some cells. In some cells, diacylglycerol and inositol triphosphate levels are increased after treatment with GH.

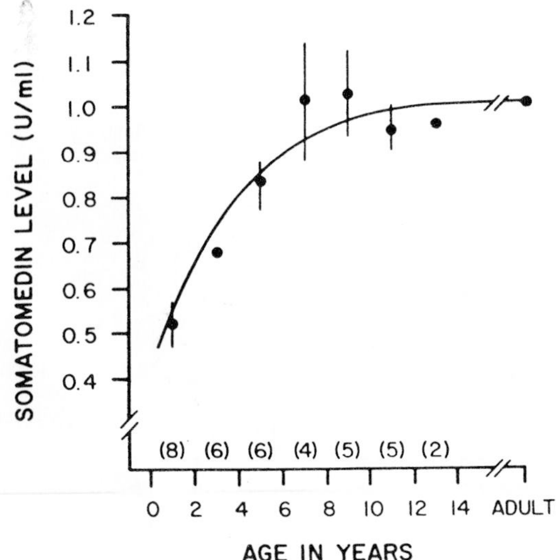

FIGURE 5-29 Levels of somatomedins and somatomedin activity in normal children of different ages. Figures in parentheses denote the number of studies providing data in each age range; in each study, mean levels (±SEM) in normal adults were defined as 1.00 unit/ml and values in children were expressed accordingly. Reproduced by permission from Phillips, L. S., and Vassilopoulou-Sellin, R. (1980). Somatomedins. *New Engl. J. Med.* **302,** 438–446.

VII. β-LIPOTROPIN

β-Lipotropin and ACTH are encoded by the same gene product (see Figure 5-10). ACTH and β-lipotropin are secreted from the anterior pituitary together in response to stress (see Chapter 10). β-Endorphin acts directly at the anterior pituitary level to stimulate the release of PRL. PRL undoubtedly plays some role in stress adaptation and produces hyperglycemic effects in the liver. In purified neurosecretory granules from bovine pituitary glands, proopiomelanocortin-converting enzyme activity (an acid–thiol protease) has been found, making it likely that "maturing" activities operate within the granule itself.

A. β-Endorphin and Stress

β-Endorphin and ACTH were discovered to be secreted simultaneously from the anterior pituitary in response to stress. Long-term adrenalectomy also promoted the secretion of these two peptides, indicating that cortisol might be a feedback inhibitory agent for both and therefore that β-endorphin may be involved, as well as ACTH, in the stress adaptation mechanism. Both of these hormones are secreted in response to corticotropin-releasing hormone (CRH), a further confirmation of their relatedness in terms of mechanism of action and derivation from the same gene product.

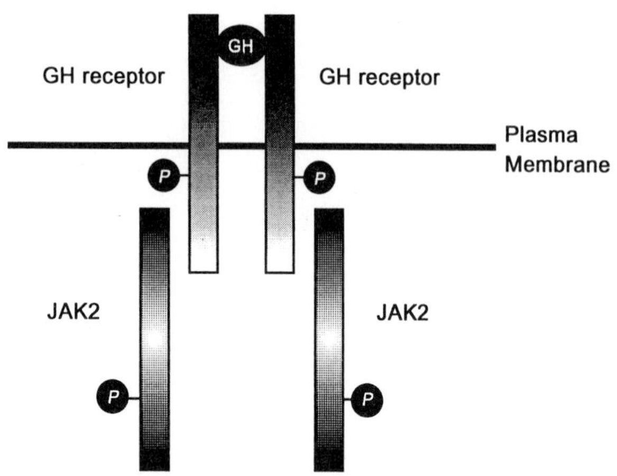

FIGURE 5-30 Structure of single-transmembrane segment receptors that form homologous dimers; the growth hormone receptor. One GH molecule binds to identical binding sites on the extracellular domains of two GH receptors that form a homodimer. The intracellular domains associate with a tyrosine kinase of the Janus family of kinases. Encircled P denotes potential tyrosine phosphorylation sites on the intracellular portion of GH receptor. Reproduced with permission from Gammeltoft. S., and Kahn, C. R. (1995). Hormone signaling via membrane receptors. *In* "Endocrinology" (L. J. DeGroot, ed.), 3rd edition, Vol. 1, pp. 17–65. W. B. Saunders Co., Philadelphia, PA.

Administration of the synthetic glucocorticoid, dexamethasone, inhibited the secretion of both ACTH and β-endorphin. The glucocorticoid functions by reducing the level of translatable mRNA encoding both ACTH and endorphin. Processing of the precursor RNA is unaffected. The exact molecular mechanism by which glucocorticoid receptor operates negatively on the POMC gene is still under study. The receptor may act through negative hormonal response elements in the proopiomelanocortin promoter.

B. β-Endorphin Receptor

Interaction of β-endorphin with receptor involves predominantly the μ-opioid receptor. Enkephalins, morphine, and naloxone, a specific inhibitor of analgesic action (see Figure 5-31), were not as potent competitors as β-endorphin for endorphin-specific binding sites. Interestingly, small amounts of morphine and codeine are produced by the brain. Enkephalins predominantly bind to the δ-opioid receptor, and dynorphins predominantly bind to the κ-receptor. From the point of view of the β-endorphin molecule, there are three binding sites, two of which could explain binding to distinct receptors. Binding to the δ-receptor is located in the Met-enkephalin segment, as shown in Figure 5-32. The μ-receptor, specific for the morphine-binding site, is in the carboxy terminus (Figure 5-32). The middle segment of β-endorphin is the antibody-binding site. A summary of the properties of opioid receptors is presented in Table 5-4.

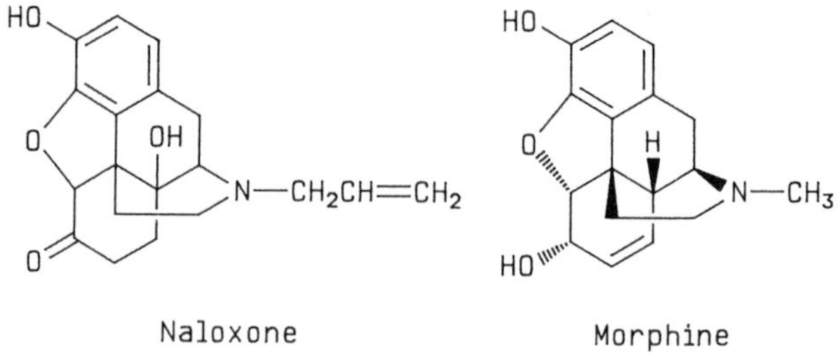

FIGURE 5-31 Structures of the opiate receptor antagonist, naloxone, and the opiate receptor agonist, morphine.

Some general remarks can be made about opioid receptors. μ-Receptors may mediate most opioid analgesic effects, although some negative effects are listed in Table 5-4. κ-Receptors produce analgesia at the spinal level and do not contribute to (drug) dependence. δ-Receptors are important to the periphery but may also contribute to analgesia. Figure 5-33 shows the interactions of opiate peptides in the pain pathway.

C. Processing of β-Endorphin

β-Endorphin is produced as a cleavage product of β-lipotropin, as shown previously (see Figure 5-10). It appears that β-endorphin (1–31) is the first formed, and this molecule is rapidly N-acetylated on its amino-terminal residue and then converted more slowly to

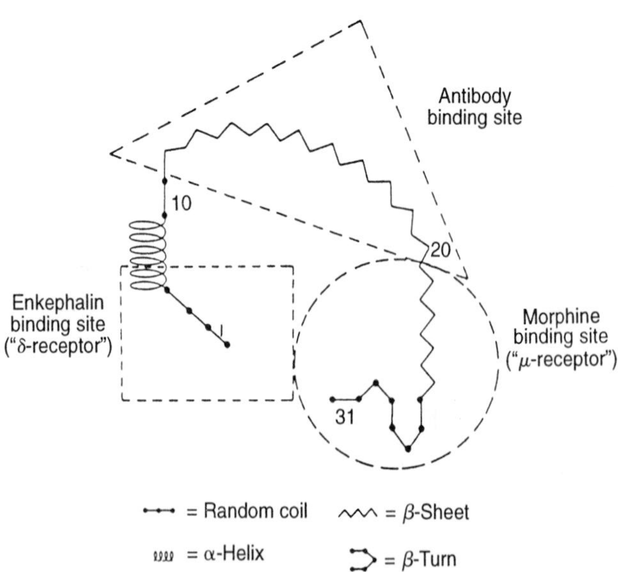

FIGURE 5-32 Potential receptor-binding sites for β-endorphin. Reproduced with permission from Li, C. H. (1982). Beta endorphin. *Cell* **31,** 504–505.

TABLE 5-4 Classification of Opioid Receptors and Their Actions[a]

Receptor	Ligand	Transducer	Molecular action	Physiological effects
μ	β-Endorphin	G_Q (cAMP ↓) G protein coupling to ion channel (?)	K^+ conductance ↑ opening K^+ channels causing hyperpolization	Analgesia, respiratory depression, constipation, euphoria, dependence, meiosis
κ	Dynorphins	G_Q (cAMP ↓) G protein coupling to ion channel (?)	Closure of voltage-gated Ca^{2+} channels inhibiting transmitter release	Analgesia, diuresis, sedation
δ	Enkephalins	G_Q (cAMP ↓) G protein coupling to ion channel (?)	Closure of voltage-gated Ca^{2+} channels inhibiting transmitter release	Analgesia
σ	Not selective	G_Q (cAMP ↓)	Associated with glutamate-activated ion channels; site of action of psychotomimetic drugs, e.g., phencyclidine	Dysphoria

[a] Constructed, in part, from Ganong, W. F. (1995). "Review of Medical Physiology". Appleton and Lange, Norwalk.

α-*N*-acetyl-β-endorphin (1–27). This posttranslational processing is summarized in Figure 5-34.

VIII. THYROTROPIC HORMONE

A. Thyrotropic Receptor

Like other polypeptide hormone receptors, TSH receptor is located in the cellular membrane of thyroid cells. A functional model of the TSH receptor has been hypothesized and is shown in Figure 5-35.

TSHR is a seven-membrane-spanning domain receptor with a large extracellular domain. TSH binding is in the extracellular domain, and trypsinization of cells containing the receptor generates a 15–30-kDa peptide capable of binding TSH. Apparently the loops themselves are not involved in ligand binding and signaling as they are in the adrenergic receptors. The model shown in Fig. 5-35 has been derived from site-directed mutagenesis in terms of functional domains. Antibodies, especially from patients with autoimmune thyroid diseases, have been of great utility in mapping

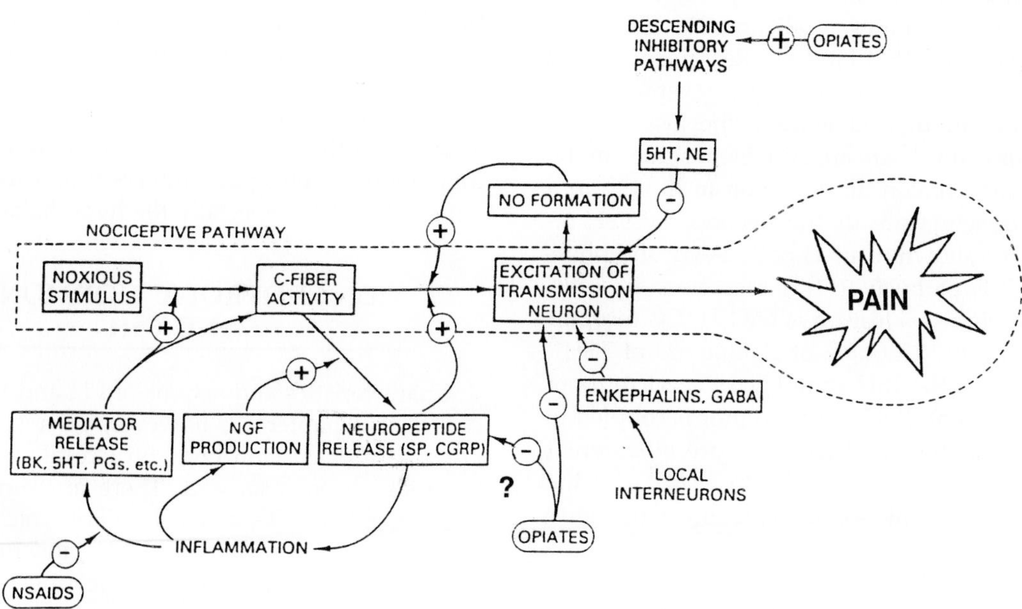

FIGURE 5-33 Summary of modulatory mechanisms in the nociceptive pathway. Abbreviations: CGRP, calcitonin gene-related peptide; 5-HT, serotonin; NE, norepinephrine; BK, bradykinin, PGs, prostaglandins; NGF, nerve growth factor; SP, substance P; GABA, γ-aminobutyric acid; NSAIDS, nonsteroidal anti-inflammatory drugs. Reproduced with permission from Rang, H. P., Dale, M. M., Ritter, J. M., and Gardner, P. (1995). "Pharmacology." Churchill-Livingstone, NY.

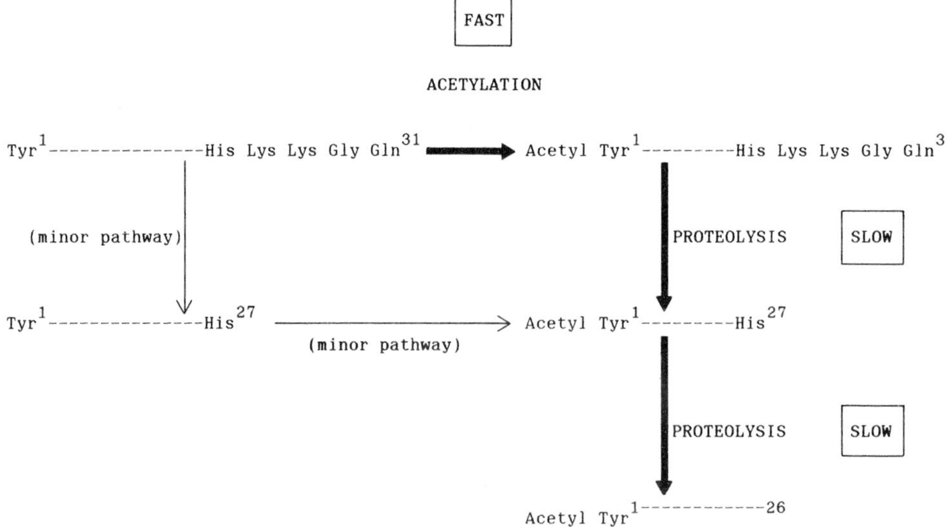

FIGURE 5-34 Posttranslational processing of β-endorphin in rat intermediate pituitary. Reproduced with permission from Eipper, B. A., and Mains, R. E. (1981). *J. Biol. Chem.* **256,** 5689–5695.

the activities of various domains. Thus, antibodies are available that stimulate cAMP levels or block the binding of TSH at different locations on the receptor molecule. Residues on the C-terminal region of the extracellular domain were found to be important for TSH binding. Cysteine residues at 301 and 380 (Figure 5-35) were critical for TSH binding. Residues 303–382 have low homology to gonadotropin receptors, and this region can be deleted without the loss of TSH binding or function. There are cysteine residues in the transmembrane domain that function in linking to a G-protein interaction site. By transfection, the TSH receptor has been shown to affect both cAMP and phosphatidylinositol pathway (PIP_2) responses in cells. Dual coupling is also a feature of the LH/CG receptor in transfected cells. Mutation of alanine 623 of TSHR produces a loss in the TSH-stimulated PI cycle but not cAMP enhancement. Either one receptor is coupled to two pathways or two TSH receptors are in different conformations where one couples to cAMP and the other to the PI cycle. This possibility is shown in Figure 5-35.

Interaction of TSH with TSHR results in a conformational change in the receptor that enables the interaction of receptor and G proteins. This mechanism is suggested in Figure 5-36. Note that the α-subunit, in this model, interacts with the transmembrane domain. The effective portion of the α-subunit, perhaps similar to oxytocin or vasopressin, interacts with the transmembrane domain and G-protein interaction sites.

IX. ACTH

The principal biological action of ACTH is to stimulate corticosteroid production in adrenal cortical cells. This hormone interacts with receptors in membranes primarily of the *zona fasiculata* and *zona reticularis,* although there are also receptors in the *zona glomerulosa.* A further discussion appears in Chapter 10.

An important aspect of ACTH action, as well as for actions of other anterior pituitary hormones, is that they undergo retrograde transport and redistribution within the CNS, especially the hypothalamus.

X. GONADOTROPIC HORMONES LH AND FSH

Characteristics and actions of LH and FSH are discussed in Chapter 13; however, brief space will be given to a discussion here. Structural considerations have also been discussed. There are similarities between LH, FSH, CG, and TSH, all of which contain an identical α-subunit. FSH and LH bind to membrane receptors in their target cells and couple to G proteins through conformational changes in ligand–receptor structures that avail specific sequences in the bound α-subunit similar to AVP (arginine vasopressin) or oxytocin in sequence and also in the receptor for the interaction with the cognate G protein. The result is to activate adenylate cyclase and increase the cellular content of cAMP.

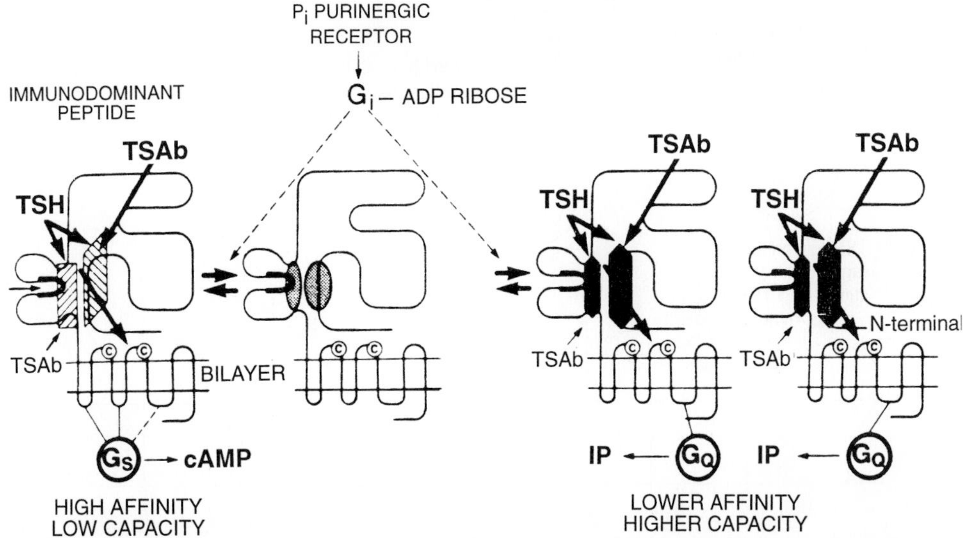

FIGURE 5-35 Speculative model of a single TSHR that binds to different G proteins. The two species are in equilibrium; this equilibrium is assumed to include a receptor molecule with no G protein bound. One species is proposed to have a higher affinity than the other, that coupled to G_s; this species is linked to hormone-increased cAMP levels and is presumed to be lower in number than the TSHR molecules coupled to G_Q. The higher capacity of the G_Q-coupled receptor is signified by a larger number (two) of these forms. The G_Q-coupled receptor has a lower affinity for ligand and is coupled to the inositol phosphate (IP) signal cascade. The equilibrium between the two forms is suggested to be set by the P_1-purinergic receptor via G_1; the regulation of G_1 is further suggested to involve ADP ribosylation. The model suggests that the "precoupling" of the different G proteins induces different conformations in the extracellular domain, as represented by the hatched and dark boxes versus the open circles in TSHR with no G protein bound. The blocking TSHR autoantibody (TSBAb) site is a high-affinity antagonist TSH-binding site that involves tyrosine 385 and residues 295–306 and 387–395, particularly cysteines 301 and 390. The thyroid-stimulating autoantibody (TSAb), which increases cAMP levels, interacts with an agonist site on the extracellular domain; this site involves threonine 40 and residues 30–33, 34–37, 42–45, 52–56, and 58–61. After interacting with these sites, TSH or the TSAb is presumed to transmit its signal to the transmembrane domain; the two exoplasmic cysteines critical for this process are noted. The stimulating autoantibody that increases cAMP levels might not be identical to the stimulating autoantibody activating the PIP_2 signal. The mechanism by which G_1 regulates the equilibrium is unknown. Reproduced with permission from Kohn, L. D., Ban, T., Okajima, F., Shimura, H., Shimura, Y., Hidaka, A., Guiliani, C., Napolitano, G., Kosugi, S., Ikuyama, S., Akamizu, T., Tahara, K., and Saji, M. (1995). Cloning and regulation of glycoprotein hormone receptor genes. *In* "Molecular Endocrinology" (B. D. Weintraub, ed.), pp. 133–153. Raven Press, NY.

The interaction between the α- and β-subunits of the hormones appears to be noncovalent, consisting minimally of hydrogen bonding. Essentially, the free subunits are inactive and must be combined as the α/β-heterodimer for biological activity. Since the α-subunit of all of these hormones is interchangeable, the β-subunit clearly specifies the biological activity of the hormone and accounts for specific binding to the cognate receptor. Accordingly, the β-subunits of LH, FSH, and TSH are encoded by separate genes; however, there are multiple genes that encode β-subunits for CG in the placenta, and some of these arise through alternative splicing. The α- and β-subunits are each stabilized by disulfide bonds, and disruption of these bonds by reducing agents alters the internal configuration and causes dissociation of the heterodimer. Specific domains of the subunits that define the sites interacting with the receptor are not yet known, although specific amino acids, such as lysine 91 of the α-subunit and aspartate 99 of the β-subunit, are required for signal transduction and they could be involved in the interaction of the dimeric hormone and the receptor.

Both of the subunits contain carbohydrate moieties (see Figure 5-9), and both O-linked and N-linked carbohydrates have been found in the α- and β-subunits. N-linked carbohydrate substituents on the β-subunit probably are not important for signal transduction.

Gonadotropin receptors, like the TSH receptor, are members of the seven-membrane-spanning receptor family that couple to G proteins and stimulate adenyl-

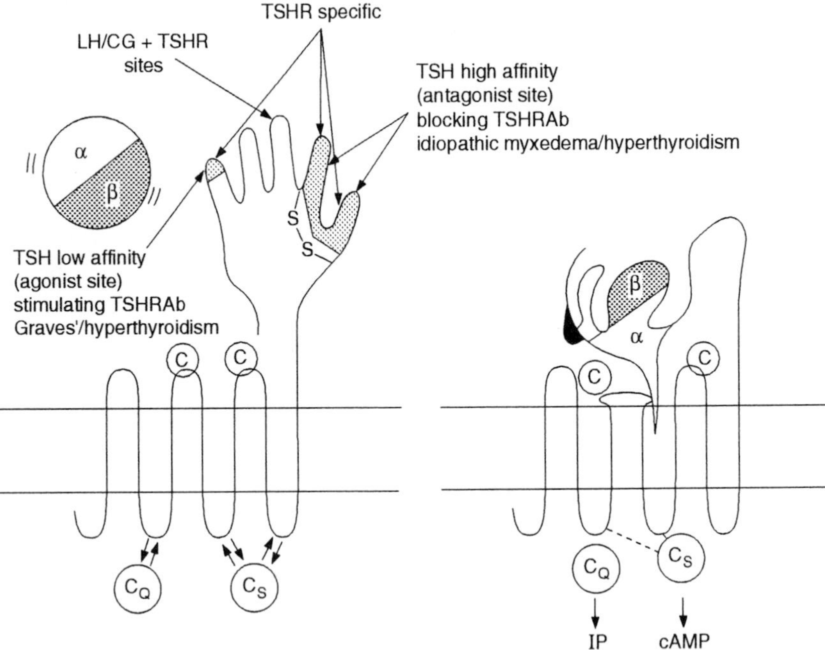

FIGURE 5-36 TSHR expressed in the configuration of a hand with fingers waiting for ligand and attached to the transmembrane domain (left). After interacting with ligand, the TSHR undergoes a conformational change that now pushes the ligand onto the exoplasmic loops of the transmembrane domain. This induces the G proteins on the cytoplasmic surface to activate the transducing signal pathways. The hormone is depicted with two subunits; it is assumed that the β-subunit is the primary surface interacting with the extracellular domain (the hand and fingers) since it generates tissue specificity. The α-subunit is presumed to interact with the transmembrane domain and undergo a conformational change. An important peptide interacting with the transmembrane facet may be one common to all glycoprotein receptors and related to the nonapeptide hormones oxytocin and vasopressin. Reproduced with permission from Kohn, L. D., Ban, T., Okajima, F., Shimura, H., Shimura, Y., Hidaka, A., Guiliani, C., Napolitano, G., Kosugi, S., Ikuyama, S., Akamizu, T., Tahara, K., and Saji, M. (1995). Cloning and regulation of glycoprotein hormone receptor genes. *In* "Molecular Endocrinology" (B. D. Weintraub, ed.), pp. 133–153. Raven Press, NY.

ate cyclase. Receptors in this group have a large extracellular N-terminal domain and a small cytoplasmic C-terminal domain. Because of the similarity of the LH and FSH receptors to those for the catecholamines and their interaction with G proteins and subsequently adenylate cyclase, it is surmised that FSH and LH signal transduction may be similar to that for the catecholamines, about which more information is available. Accordingly, the ligand binds in a pocket generated by the transmembrane domain and then leads to a change in shape as similarly portrayed by the TSH receptor. This conformational change would allow the bound α-subunit (AVP-like sequence) plus domains of the receptor to make contact with a G protein as the first step in the signal transduction process. LH and FSH are probably functional in the signal transduction process as homodimers by analogy to the TSH receptor. Another theory is that glycoprotein hormones have disulfide isomerase activity and that signal transduction could be triggered by the alteration of one or more disulfide bonds in the receptor.

While LH receptors are usually considered to be located in the gonads, they have been detected in the uterus where they could interact with the high level of CG in pregnancy.

XI. ACTH RECEPTOR

The ACTH receptor has now been cloned from several sources, including bovine and human. The bovine receptor has 297 amino acid residues deduced from the DNA and a molecular weight equivalent of about 33,260. It is a seven-membrane-spanning domain receptor. It is reported to have glycosylations in the N-terminus. The receptor appears to have phosphorylation sites for protein kinases A and C, and the consequent participation of these kinases in ACTH

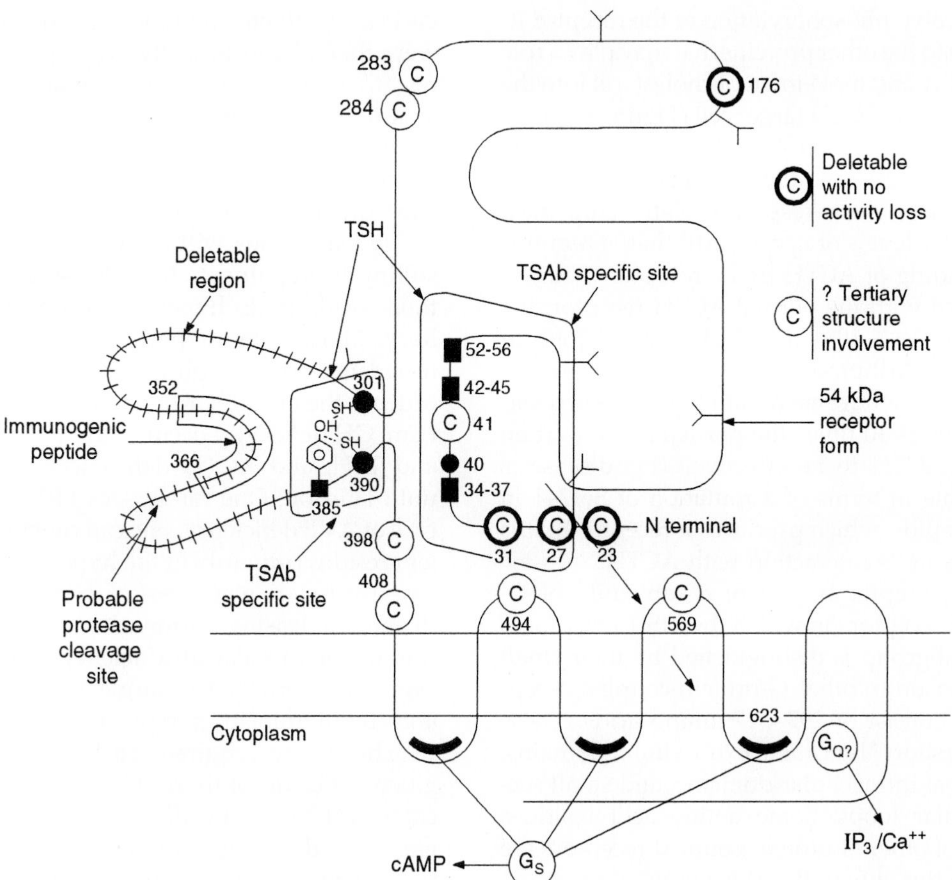

FIGURE 5-37 Model of TSHR with the TSH site and the TSHRAb epitopes defined in a three-dimensional array. Determinants for blocking TSHRAbs (TSBAb) and stimulating TSHRAbs (TSAb) are approximated to compose the TSHR site. The former are implicated in the expression of disease by patients with idiopathic myxedema and hypothyroidism; the latter are implicated in patients with Graves' disease and hyperthyroidism. They are presumed to comprise the high-affinity TSH-binding antagonist site and the agonist site for TSH, respectively, on the basis of the competitive antagonism or agonism of TSH with monoclonal inhibiting and stimulating receptor antibodies. The loop between residues 303 and 392 is X-marked and separated from the remainder of the external domain since residues within it can be deleted with no loss in receptor function. This loop includes residues 352–366, which comprise the immunogenic peptide used to produce a specific antibody to the receptor. The immunogenic peptide is approximated near the blocking TSHRAb site because immunization with it or an adjacent peptide can produce blocking TSHRAbs that do not inhibit TSH binding and because they can, with time, produce antibodies reactive with peptides containing residues identified as blocking TSHRAb and high-affinity TSH binding. The model does not preclude the possibility that other sites contribute to the TSHR site, i.e., residues that react with peptides common to TSH binding using TSHR peptides. These are presumed to lie within residues 200–285 in particular. We identify cysteines that are (open circles) or are not (bold circles) likely to affect tertiary structure. The two cysteines on the exoplasmic loops of the transmembrane domain are assumed to carry the conformational signal of the receptor–ligand interaction with the extracellular domain of the receptor; it is possible that these cysteines are also involved in ligand binding as proposed in adrenergic receptors. The third intracytoplasmic loop, particularly alanine 623 but also its five N- and C-terminal residues, is identified as the critical link for hormone and TSHRAb coupling to a G protein important for signal-transducing activity, leading to the phosphatidylinositol biphosphate cascade. It also interacts with G_S, but only regulates constitutive or basal cAMP levels. Intracytoplasmic loops 1 and 2 are presumed to be coupled to G_S. The figure further delineates possible mechanisms for coupling to G proteins. Reproduced with permission from Kohn, L. D., Ban, T., Okajima, F., Shimura, H., Shimura, Y., Hidaka, A., Guiliani, C., Napolitano, G., Kosugi, S., Ikuyama, S., Akamizu, T., Tahara, K., and Saji, M. (1995). Cloning and regulation of glycoprotein hormone receptor genes. *In* "Molecular Endocrinology" (B. D. Weintraub, ed.), pp. 133–153. Raven Press, NY.

action may involve phosphorylation of the receptor itself in addition to the other proteins that may play a role in the availability and movement of cholesterol into the mitochondria of the ACTH target cell (Figure 5-37).

The gene for the human ACTH receptor has been expressed in COS cells, and these cells become responsive to ACTH in a dose–response relationship measured by cellular levels of cyclic AMP that is reminiscent of the binding of ACTH to the receptor. Thus, in cells transfected with the normal ACTH receptor, the cells respond to 10^{-9}–10^{-8} M ACTH by a 5–6-fold increase in cyclic AMP production.

Interestingly, a syndrome of familial glucocorticoid deficiency requires 10 times the normal concentration of circulating ACTH to be effective. This disease is now explainable in terms of a mutation of Ser741 in the ACTH receptor, which produces a receptor that is less efficacious in its interaction with ACTH.

The ACTH receptor is part of a subfamily of G-protein-linked receptors in which the MSH receptor is a part. This subgroup is distinguished by their small size in comparison to other G-protein-coupled receptors and they consist of 297–317 amino acids. These receptors have short N-terminal extracellular domains, short C-terminal intracellular domains, and small second extracellular loops. Some amino acid residues characteristic of other G-protein-coupled receptors are also lacking in this subfamily. This family of receptors is shown in Figure 5-38.

XII. CLINICAL ASPECTS

Abnormalities of the anterior pituitary stem from a number of causes. Trauma to the anterior pituitary can occur often by mechanical pressure from tumors. The anterior pituitary may decline in function because there has been trauma to the delicate stalk connecting the pituitary to the hypothalamus and the releasing hormone concentrations in the pituitary become inadequate. Tumors of individual cell types may arise and produce high amounts of one or two anterior pituitary hormones, or hormones that act like them, ectopically. Detection of primary abnormal functioning at the pituitary level is easier now due to the availability of many of the releasing hormones that make evocator tests possible (see Chapter 3). In cases of inadequate availability of a pituitary hormone, the target gland hormone can be supplied (e.g., cortisol for inadequate ACTH supply).

Autoimmune diseases are an important consideration. In Graves' disease, an antibody is produced, apparently directed against the lipid component of the TSH receptor. In binding to the receptor, adenylate cyclase is activated and the thyroid responds as if TSH were there. Unfortunately, thyroid hormone does not negatively feed back on the autoantibody production as it does with TSH production and hyperthyroidism develops. Treatment usually involves ablating the function of the thyroid gland and maintaining normal subsequent hormone levels by replacement therapy.

Two situations will be discussed briefly: those resulting in hypofunction of the anterior pituitary and those resulting in hyperfunction of the anterior pituitary. Many anterior pituitary hormones may be affected in "panhypopituitarism." There is usually an order to the loss of hormonal secretions in hypopituitarism. GH deficiency occurs early followed by LH, FSH, and TSH, and ACTH deficiences are not visualized until later on. One rarely sees PRL deficiency. In the case of AVP deficiency, one can conclude that the problem resides primarily in the hypothalamus. In this disease, the function or access of the pituitary to the hypothalamic releasing hormones is destroyed by trauma, tumors, or vascular insufficiency. Such conditions are eventually corrected by surgical intervention if a tumor and by administration of the terminal hormones, whichever are required, such as the sex hormones, glucocorticoids, or thyroid hormone. Fortunately, evocator tests with purified releasing hormones (e.g., GnRH) and anterior pituitary hormones are available to localize the center of the disturbance as being in the hypothalamus, pituitary, or terminal gland. A particular disease important to development is dwarfism, which can occur because of a deficiency of GH, often beginning at infancy. Other anterior pituitary hormones may be affected in this disease. Primary hypothyroidism, for example, can produce substantial inhibition of the release of GH, resulting in dwarfism. Human growth hormone can be used to treat GH hyposecretion, but it is presently available in limited quantity. Presumably the cloned gene product will help in the treatment of dwarfism. Other conditions are associated with hypofunctions in ACTH and TSH, and in some cases the results are not severe. ACTH deficiency is substituted therapeutically. There has become an increasing availability of releasing hormones for clinical testing and diagnosis. This advance makes it possible to specifically localize malfunctions in the hypothalamus–pituitary–end organ axis.

Overproduction of pituitary hormones can occur when tumors ectopically produce a pituitary hormone(s). Disease states occur with excessive prolactin and growth hormone overproduction, leading to gigantism during growth (epiphyseal cartilages are open) or to acromegaly if epiphyseal cartilages are fused. Overproduction of prolactin leads to galactorrhea and amenorrhea in the absence of acromegaly

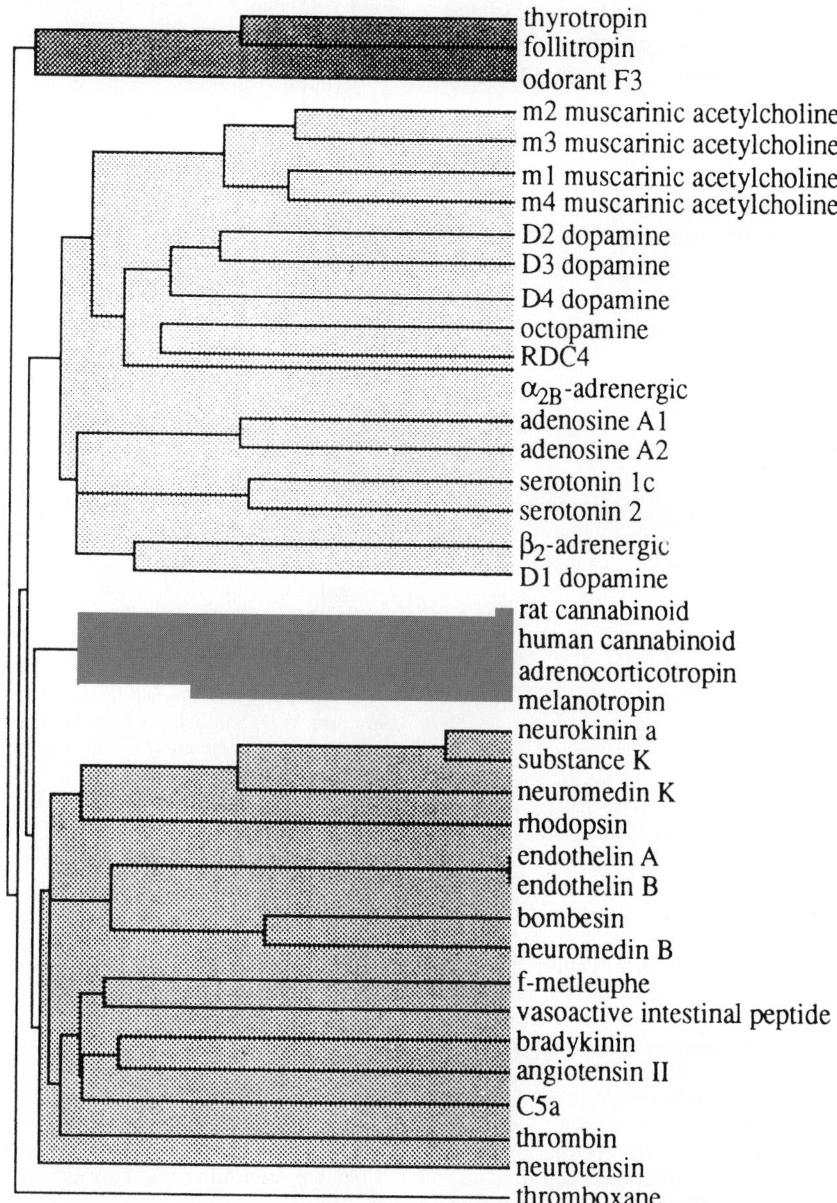

FIGURE 5-38 Multiple sequence alignment of melanocortin receptors and other representative members of the G-protein-coupled receptor family. Amino acid sequences were obtained from the Protein Identification Resource (PIR) database and aligned with the melanocortin receptors with the Pileup program, version 7.0, from the Genetics Computer Group analysis software (Madison, WI.) The branch length is proportional to the similarity between sequences, which suggests that the melanocortin and cannabinoid receptors are members of a distinct subfamily of the G-protein-coupled receptors. Reproduced with permission from Mountjoy, K. G., Robbins, L. S., Mortrud, M. T., and Cone, R. D. (1992). The cloning of a family of genes that encode the melanocortin receptors. *Science* **257**, 1248–1251.

and pregnancy. Galactorrhea is often controlled by reducing prolactin secretion using a drug (bromocryptine) that inhibits prolactin formation and release.

Hyperproduction of growth hormone can result from a tumor of the somatotroph (acidophil pituitary adenoma) during growth and result in gigantism. The process of growth is accelerated under this condition, leading to giants. This form of GH overproduction is relatively rare. If left untreated, the tumor destroys the functional gland, impairing the other pituitary hormones and resulting in death. This disease usually involves enlargement of the *sella,* but is principally manifested by unusually high circulating levels of GH. Therapy involves removal of the acidophil tumor, but sometimes high levels of GH persist even after surgical intervention.

Acromegaly is also relatively uncommon. There are two metabolic types of acromegaly: those in which circulating GH levels fall after glucose administration and those whose GH levels are unaffected by glucose. The precise explanation of acromegaly is unknown, and some views infer that humans can revert to a more primitive phase and resemble the Neanderthal man.

References

A. Books

DeGroot, L. J. (1995). "Endocrinology," 3rd ed., Vol. 1. W. B. Saunders Co., Philadelphia, PA.

Ganong, W. F. (1995). "Review of Medical Physiology," 17th ed. Appleton and Lange, Norwalk, CT.

Harvey, S., Scanes, C. G., and Daughaday, W. H. (1995). "Growth Hormone." CRC Press, Boca Raton, FL.

Melmed, S. (1995). "The Pituitary." Blackwell Science.

Rang, H. P., Dale, M. M., Ritter, J. M., and Gardner, P. (1995). "Pharmacology." Churchill, Livingstone, NY.

B. Review Articles

Casman, D., Lyman, S. D., Idzerda, R. L., *et al.* (1990). A new cytokine receptor superfamily. *Trends Biochem Sci.* 15, 265–270.

Chawla, R. K., Parks, J. S., and Rudman, D. (1983). Structural variants of human growth hormone: Biochemical, genetic and clinical aspects. *Annu. Rev. Med.* **34,** 519–547.

Cooke, N. E. (1995). Prolactin: Basic Physiology. *In* "Endocrinology" (L. J. DeGroot, ed.), 3rd ed., Vol. 1, pp. 368–393. W. B. Saunders Co., Philadelphia, PA.

Daughaday, W. H. (1995). Growth hormone, insulin-like growth factors, and acromegaly. *In* "Endocrinology" (L. J. De Groot, ed.), 3rd ed., Vol. 1, pp. 303–329. W. B. Saunders Co., Philadelphia, PA.

Gammeltoft, S., and Kahn, C. R. (1995). Hormone signaling via membrane receptors. *In* "Endocrinology" (L. J. DeGroot, ed.), 3rd ed., Vol. 1, pp. 17–65. W. B. Saunders Co., Philadelphia, PA.

Guillemin, R. (1980). Beta-lipotropin and endorphins: Implications of current knowledge. *In* "Neuroendocrinology" (D. T. Krieger and J. C. Hughes, eds.), p. 70. Sinauer Associates, Sunderland, MA.

Harvey, S., and Hull, K. L. (1995). Growth Hormone Action: Growth Hormone Receptors. *In* "Growth Hormone" (S. Harvey, C. G. Scanes, and W. H. Daughaday, eds.), pp. 303–335. CRC Press, Boca Raton, FL.

Imura, H. (1995). Adrenocorticotropic hormone. *In* "Endocrinology" (L. J. De Groot, ed.), 3rd ed., Vol. 1, pp. 355–367. W. B. Saunders Co., Philadelphia, PA.

Kohn, L. D., Ban, T., Okajima, F., Shimura, H., Shimura, Y., Hidaka, A., Guiliani, C., Napolitano, G., Kosugi, S., Ikuyama, S., Akamizu, T., Tahara, K., and Saji, M. (1995). Cloning and regulation of glycoprotein hormone receptor genes. *In* "Molecular Endocrinology" (B. D. Weintraub, ed.), pp. 133–153. Raven Press, New York.

Li, C. H. (1982). "Biochemical Actions of Hormones" (G. Litwack, ed.), Vol. 9, pp. 1–41. Academic Press, New York.

Miller, E. E. (1995). Role of neurotransmitters and neuromodulators in the control of anterior pituitary hormone secretion. *In* "Endocrinology" (L. J. De Groot, ed.), 3rd ed., Vol. 1, pp. 178–191. W. B. Saunders Co., Philadelphia, PA.

Molitch, M. E. (1995). Prolactin. *In* "The Pituitary" (S. Melmed, ed.), pp. 136–186. Blackwell Science.

Moyle, W. R., and Campbell, R. K. (1995). Gonadotropins. *In* "Endocrinology" (L. J. De Groot, ed.), 3rd ed., Vol. 1, pp. 230–241. W. B. Saunders Co., Philadelphia, PA.

Parks, J. S., Pfäffle, R. W., Brown, M. R., Abdul-Latif, H., and Meacham, L. R. (1995). Growth hormone deficiency. *In* "Molecular Endocrinology" (B. D. Weintraub, ed.), pp. 473–490. Raven Press, New York.

Phillips, L. S., and Vassilopoulou-Sellin, R. (1980). Somatomedins. *New Engl. J. Med.* **302,** 438–446.

Pierce, J. G., Faith, M. R., Guidice, L. C., and Reeve, J. R. (1976). Structure and structure–function relationships in glycoprotein hormones. *Ciba Found. Symp. Excerpta Med.* **41,** 225–250.

Sara, V. R., and Hall, K. (1990). Insulin-like growth factors and their binding proteins. *Physiol. Rev.* **70,** 591.

Sarapura, V. D., Samuels, M. H., and Ridgway, C. E. (1995). Thyroid-stimulating hormone. *In* "The Pituitary" (S. Melmed, ed.), pp. 187–229. Blackwell Science.

Wass, J. A. H., and Besser, M. (1995). Tests of pituitary function. *In* "Endocrinology" (L. J. De Groot, ed.), 3rd ed., Vol. 1, pp. 487–496. W. B. Saunders Co., Philadelphia, PA.

C. Research Papers

Bedi, G. S., French, W. C., and Bahl, O. P. (1982). Structure of carbohydrate units of ovine luteinizing hormone. J. Biol Chem. **257,** 4345–4355.

???

Eipper, B. A., and Mains, R. E. (1981). *J. Biol. Chem.* **256,** 5689–5695.

Goddard, A. D., Covello, R., Luoh, S.-M., Clackson, T., Attie, K. M., Gesundheit, N., Rundle, A. C., Wells, J. A., and Carlsson, L. M. S. (1995). Mutations of the growth hormone receptor in children with idiopathic short stature. *New Engl. J. Med.* **333,** 1093–1098.

Kelly, P. A., Djiane, J. Postel-Vinay, M. C., and Edery, M. (1991). The prolactin/growth hormone receptor family. *Endocr. Rev.* **12,** 235–251.

Li, C. H. (1982). Beta-endorphin. *Cell* **31,** 504–505.

Mountjoy, K. G., Robbins, L. S., Mortrud, M. T., and Cone, R. D. (1992). The cloning of a family of genes that encode the melanocortin receptors. *Science* **257,** 1248–1251.

Raikhinstein, M., Zohan, M., and Hanukoglu, I. (1994). cDNA cloning and sequence analysis of the bovine adrenocorticotropic hormone (ACTH) receptor. *Biochim. Biophys. Acta* **1220,** 329–332.

CHAPTER

6

Thyroid Hormones

I. INTRODUCTION

A. Background

The thyroid gland and its hormonal products mediate a myriad of physiological effects that cause alterations in virtually all metabolic pathways and organs. Thyroid hormone modulates oxygen consumption, basal metabolic rate (BMR), and lipid, carbohydrate, and protein metabolism. In addition, it also changes the synthesis and degradation rates of a wide variety of other growth factors and hormones, such that many of its effects emerge as secondary influences on other endocrine pathways.

The effects of thyroid hormone can be ascribed to two categories of biological responses: (a) effects on cellular differentiation and development, and (b) effects on metabolic pathways. These two actions are interconnected in that changes in development and growth are a consequence of hormonal modulation of metabolism. In addition, changes in cellular differentiation will alter patterns of gene expression, which, in turn, result in modulation of metabolic pathways. An understanding of thyroid hormone can best be appreciated via merging of selected events at the level of gene transcription, with effects on cellular metabolism that are then reflected at the whole animal physiological level.

The principal target tissues of action of the thyroid hormones with respect to the stimulation of oxygen consumption are skeletal muscle, cardiac muscle, liver, gastrointestinal tissues, and kidney. The basic metabolic rate of an intact organism is a measure of the energy expended after completion of the intestinal absorption of food; it is proportional to both the oxygen consumption and the surface area (body volume) of the organism. The BMR is traditionally expressed in kilocalories per square meter per hour (or day). The normal BMR for adult euthyroid males is 35–40 kcal/m^2 body surface/hr; for euthyroid women of the same age the BMR is 6–10% lower. In the hormonally deficient (hypothyroid) state, the BMR can fall to 20–25 kcal/m^2/hr, whereas in instances of hormone excess (thyrotoxicosis or hyperthyroidism) it may rise as high as 60–65 kcal/m^2/hr.

B. Iodine Metabolism

The principal hormones produced by the thyroid are thyroxine (T_4) and triiodothyronine (T_3), which contain, respectively, four and three atoms of organically bound iodine (see Section III). Accordingly, the normal functioning of the thyroid is dependent upon adequate and regular dietary intake of iodine.

Iodine is a rare element; although it is present in ocean water, it is distributed unevenly in the soils of the various land masses of the world. For adults the recommended daily allowance (RDA) of iodine established by the U.S. National Research Council is 150 μg/day. In the United States, the average daily intake of iodine by adults is in the range of 250–700 μg/day, largely as iodide. The main dietary sources of iodine are iodized salt and iodate, which is used as a bread preservative. Present-day iodized salt contains 100 mg of KI/kg of salt (0.01% KI).

In the absence of adequate dietary access to iodine, an individual will adaptively develop iodine deficiency or endemic goiter. The definition of endemic goiter can be described in statistical terms and exists when 10–20% or more of preadolescent children in a given geographic or population grouping have enlarged, mildly hypertrophied thyroid glands. In most adults with endemic goiter, iodine intake will be below 50–75 μg/day and urinary excretion can fall to 20-50 μg/day. Endemic goiter does not exist in the United States. However, endemic goiter can be found in areas of the world in which there is a suboptimal level of iodine in the soil and in food crops grown thereon, e.g., large areas of central Africa, central Asia, the Andes of South America, and Indonesia. It has been clearly demonstrated that endemic goiter can be prevented by the oral administration of potassium iodide at 6-month intervals.

C. Metabolic Effects of Thyroid Hormones

The thyroid hormones, depending upon whether they are present at elevated or reduced levels, have a wide range of effects in humans that are evident at many different levels of organization; these effects range from behavioral changes, growth effects, changes in cardiac output, gastrointestinal function, and tissue oxygen consumption, muscle myopathy (weakness), to perturbations of the immune mechanism.

In instances of thyroid hormone excess, the subject has an accelerated BMR and an elevated cardiac output associated with high body temperature, warm skin, and inappropriate sweating for the ambient temperature. Because the muscle mass of the body constitutes 50% of body weight, and because muscle tissue has a high rate of oxygen consumption, the increased muscular metabolic activity substantially contributes to the higher BMR. Although hyperthyroid subjects often increase their food intake, there is usually a concomitant weight loss related to the greater gastrointestinal activity and associated diarrhea. Such an individual is often hyperactive, with rapid movements and exaggerated reflexes, and often exhibits short attention spans. The muscle myopathy is attributable to both abnormal protein catabolism with negative nitrogen balance and abnormal neuromuscular transmission. A prominent external physical feature of hyperthyroidism is exophthalmos or bulging of the eyeball.

In many respects, the symptoms of a hypothyroid subject are the inverse of these described earlier for thyroid hormone excess. The BMR is reduced, with an associated enlarged heart that has reduced cardiac output. Accordingly, the body temperature is lowered, the skin is "cool," and sweating is reduced in relation to the ambient temperature. Also, the hypothyroid's appetite is poor and there is reduced gastrointestinal activity. Although the skeletal muscles are somewhat enlarged, there is an obvious myopathy. The principal external clinical features of hypothyroidism include a "myexdemic appearance." This describes the accumulation under the skin of mucoproteins and fluids, which result in the subject having a "puffy" appearance as a consequence of alterations in electrolyte and water balances.

II. ANATOMICAL AND MORPHOLOGICAL RELATIONSHIPS

A. Thyroid Gland

1. Gross Level

Figure 6-1 shows the location and structure in humans of the thyroid gland. In adults the thyroid gland weighs 10–20 g. It consists of brownish-red right and left pear-shaped lobes. The two lobes are connected by an isthmus and lie on the lower part of the larynx and the upper part of the trachea. From the inner capsule of the thyroid, septa extend into the gland, dividing it into lobules of varying shape and size. The separate lobules are composed of follicles, which are the

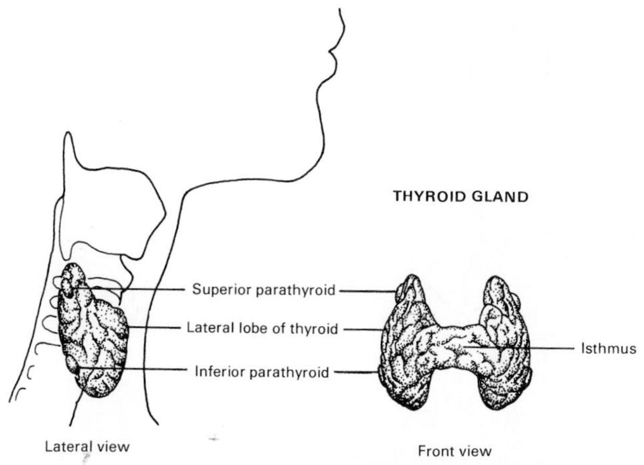

FIGURE 6-1 Location and structure of thyroid gland in humans.

main structural and functional units of the thyroid parenchyma. In the human thyroid gland there are 20–30 million follicles. The thyroid gland has a rich blood supply, with each lobe being serviced by three arteries. The blood flow to the thyroid gland is ~5 ml/g/min.

2. *Ultrastructure*

The thyroid tissues comprise two principal cell types—the follicle cell (see the following paragraph) and parafollicular cells or C cells. The C cells are found in the follicular wall or the interfollicular spaces and are responsible for the production of the polypeptide hormone, calcitonin (see Chapter 9).

A follicle consists of a single layer of epithelial cells enclosing a cavity termed the follicle lumen. This lumen is filled with a viscous proteinaceous solution, the colloid. Each follicle is encased by a thin basement membrane (see Figures 6-2 and 6-7). The fresh colloid is clear in appearance and composed of homogeneous granules that are secretory products of the follicle cell, principally thyroglobulin. The colloid represents the storage form of the thyroid hormones.

The follicle cell contains a visible rough endoplasmic reticulum (RER), which occupies most of the space in the basal and paranuclear portions of the cell. The cisternae of the RER are of varying size and shape; most are dedicated to thyroglobulin synthesis. The Golgi apparatus in the follicle cells shows a composition similar to that in many other secretory cells. It is normally located apically to the nucleus. Also visible in the follicle cell are the exocytotic vesicles containing the thyroglobulin destined for the colloid.

A prerequisite for thyroid hormone release from the colloid is the degradation of the thyroglobulin. It is now generally accepted that the degradation of thyroglobulin occurs after reentry of the colloid into the follicular cell followed by release of the thyroid hormones; these processes are discussed later in this chapter.

III. CHEMISTRY

A. Endogenous Hormones

The mammalian thyroid gland biosynthesizes, stores, and secretes two molecular species of thyroid hormones. They are thyroxine (T_4) or L-3,5,3′,5′-tetraiodothyronine and triiodothyronine (T_3) or L-3,5,3′-triiodothyronine (see Figure 6-3). The parent compound for the iodinated series of thyroid-active hormones is thyronine. Thyronine is 4-(4′-hydroxyphenoxy)-L-phenylalanine; the conventional numbering system for the carbon atoms of the thyronine molecule is also indicated in Figure 6-3.

B. Thyroid Hormone Analogs

A detailed study of the structural requirements for thyromimetic action suggests that thyroid hormone "active" compounds must have, at a minimum, a central lipophilic core containing bulky 3, 5, and 3′ substituents (not necessarily iodine atoms) and two anion groups at each distal end of the molecule. Thus, extensive structural modification of the molecule is possible without a major loss of biological activity. Methyl groups, bromine, fluorine, and nitrate are tolerated with decreasing activity in the 3, 5, 3′, and 5′ positions, and the alanine side chain can be replaced by formate, acetate, propionate, or pyruvate groups with some decrease in, but not loss of, activity. Also, the ether link between the two phenolic rings can be replaced by a sulfur or methylene linkage. The biological properties of some thyroxine analogs are summarized in Table 6-1.

Thyroid-active compounds lacking a substitution in the 5′ position have higher biological activity than those retaining a group at this position. Thus, triiodothyronine (T_3) is some 5–8 times more active than tetraiodothyronine (T_4) (see Table 6-1). Compounds that are di- or trisubstituted but that lack two substituents in the first ring [e.g., 3,3′,5′-triiodothyronine (reverse T_3 or rT_3), 3,3′-diiodothyronine, or 3′,5′-diiodothyronine] usually have low or undetectable thyromimetic action.

Reverse T_3 or 3,3′,5′-triiodothyronine is present in low concentrations in the thyroid and blood. It is largely produced by peripheral monodeiodination of thyroxine (T_4). Reverse T_3 has only 10% of the thyromimetic activity of T_4 and under some circumstances can antagonize the calorigenic effects of thyroxine.

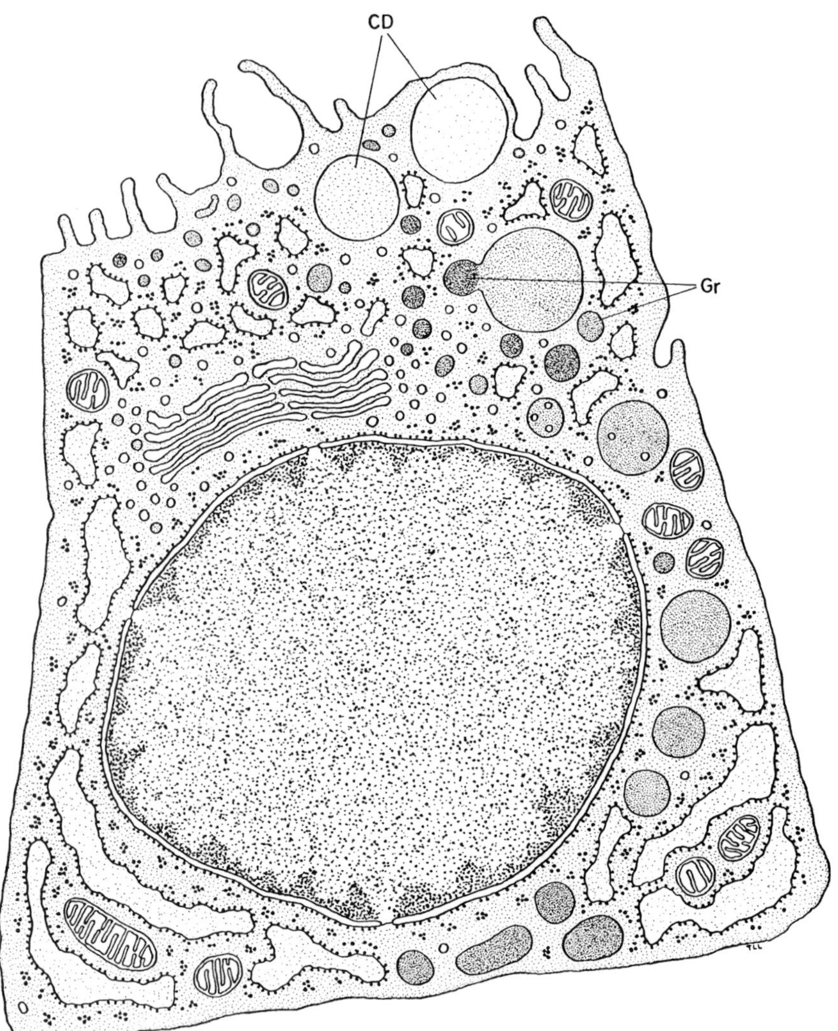

FIGURE 6-2 Thyroid follicular epithelial cell and its adjacent follicle lumen (Fl). The cells have a well-developed Golgi apparatus, which is associated with the synthesis of thyroglobulin and its transport into the follicle lumen. Also, membrane-bound dense granules (Gr) and colloid droplets (CD) are present in the cytoplasm. [Reproduced with permission from Lentz, T. L. (1971). "Cell Fine Structure." W. B. Saunders, Philadelphia, PA.]

Although D-thyroxine is not produced endogenously, the administration of this analog will result in 5–20% of the calorigenic response of L-thyroxine. Surprisingly, D-T_3 and D-T_4 have been found to bind *in vitro* to the nuclear thyroid receptor protein with the same high affinity as the comparable L compounds. Thus, the differences in activity of the D compounds between *in vivo* and *in vitro* systems may relate to differences in blood transport or tissue metabolism.

The acetic acid analogs derived from T_3 and T_4, namely, tri- or tetraiodothyroacetic acid (TRIAC or TETRAC) (see Figure 6-3), are physiologically important since they are known to be produced endogenously by deamination and decarboxylation of the parent compound.

IV. BIOCHEMISTRY

A. Thyroglobulin

1. General Comments

The biosynthesis of thyroid hormone is dependent upon the continuous dietary availability of iodine. Approximately 90–95% of all iodide in the body is present in the thyroid. The dietary iodine is absorbed by the

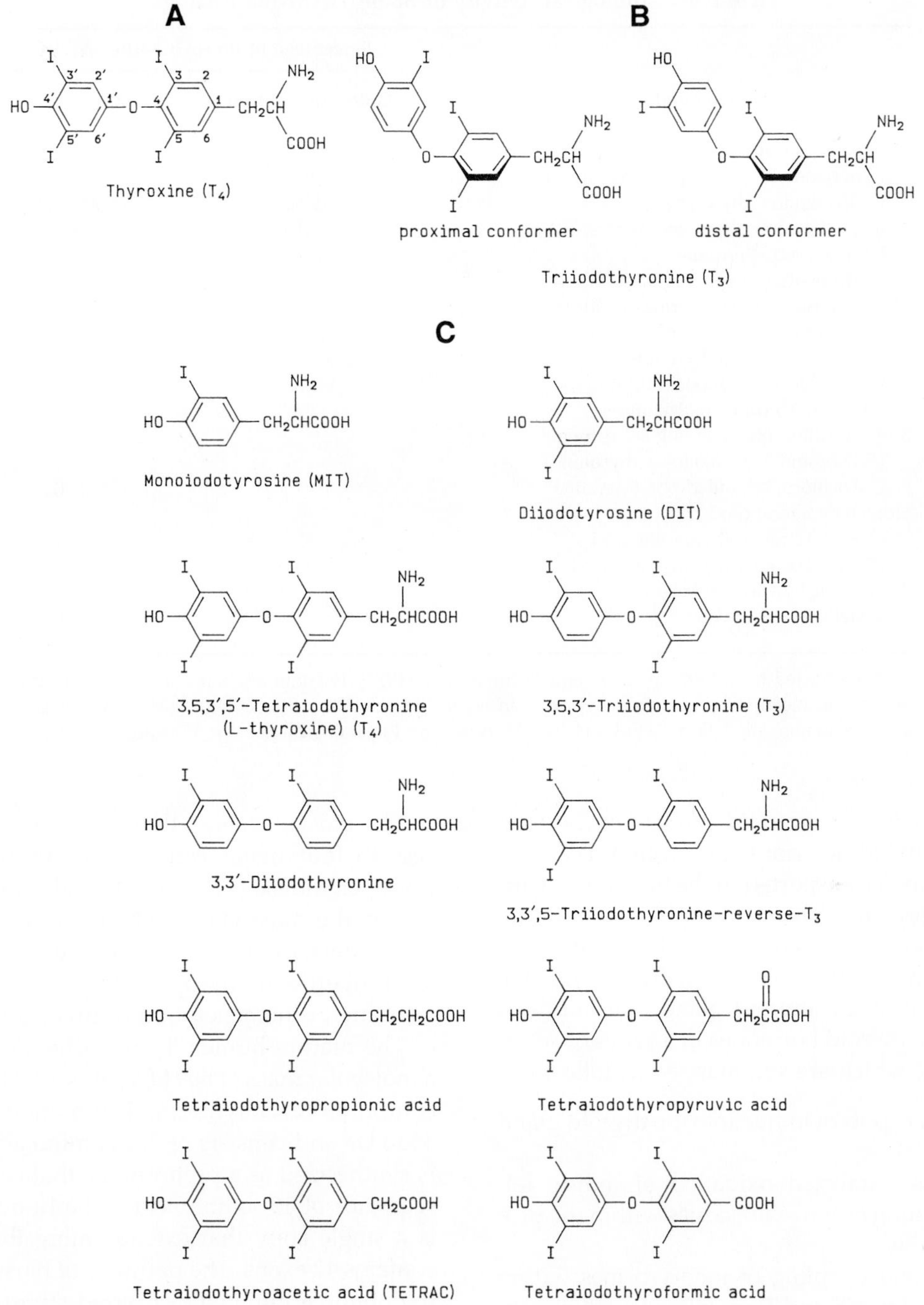

FIGURE 6-3 Structure of thyroxine, triiodothyronine, precursors, and analogs. (A) The numbering system of the carbons for the two aromatic rings of thyroxine or 3,5,3′,5′-tetraiodothyronine (T_4) is indicated. (B) Schematic representation of the conformation of triiodothyronine (T_3), illustrating that the phenyl rings exist in planes that are normal to one another and intersect at an angle of ~120°. Two conformations are possible for T_3 analogs that are monosubstituted in the ortho position relative to the 4′-OH group: (1) the proximal conformer with the single outer ring substituent oriented toward the inner ring; (2) the distal conformer with the same substituent directed away from the inner ring. The distal conformation is believed to correspond to the active conformation of the molecule. (C) Structure of other thyroxine analogs.

TABLE 6-1 Biological Activity of Some Thyroxine Analogs[a]

Compound	Percentage of thyroxine-like activity: Goiter prevention	Calorigenic
Iodinated thyronines		
L-Thyroxine (3,5,3′5′-tetraiodo-L-thyronine)	100	100
3,5,3′-Triiodo-L-thyronine	500–800	300–500
3,3′,5′-Triiodo-L-thyronine (reverse T_3)	<1	<1
3,3′ Diiodo-DL-thyronine	<1	<3
3-lodo-DL-thyronine	<2	<2
3′ or 5′ position phenolic ring substituents		
3′,5′-Dibromo-3,5-diiodothyronine	7–10	
3′-Bromo-3,5-diiodothyronine	130–200	
3′,5′-Dichloro-3,5-diiodo-L-thyronine	15–27	
3′-Chloro-3,5-diiodo-L-thyronine	27	
3 or 5 position phenolic ring substituents		
3,5-Dibromo-3′,5′-diiodo-DL-thyronine		12
3,5-Dichloro-3′,5′-diiodo-DL-thyronine		0.2
Side chain alterations		
3,5,3′,5′-Tetraiodothyroacetic acid	57	10–15
3,5,3′,5′-Tetraiodothyropyruvic acid	75	10–20
4′-Phenolic hydroxyl substituents		
O-Methyl-DL-thyroxine		50

[a] Abstracted from Pittman, C. S., and Pittman, J. A. (1974). Relation of chemical structure to the action and metabolism of thyroactive substances. *In* "Handbook of Physiology" (M. A. Greer and D. H. Solomon, eds.), Sect. 7, Vol. 111, p. 233. American Physiological Society, Washington, DC.

intestine after reduction to iodide; iodate in the food is also reduced to iodide prior to absorption. The inorganic iodide is then transported in the blood by a variety of plasma proteins. The concentration of total iodide in the blood plasma is 8–15 μg/100 ml, while the protein-bound concentration is normally 6–8 μg/100 ml. After arrival at the thyroid gland, the iodide is converted to the thyroid hormones in a series of seven metabolic steps, which are summarized as follows:

(a) Active transport of iodide into the thyroid gland follicular cells

(b) Peroxidase-catalyzed oxidation of iodide followed by iodination of tyrosyl residues within the protein thyroglobulin

(c) Transfer and coupling of iodotyrosines within thyroglobulin to form T_4 and T_3 (mediated by thyroperoxidase)

(d) Storage of thyroglobulin in the lumen of the thyroid follicle as colloid

(e) Endocytosis of colloid

(f) Proteolysis of thyroglobulin with concomitant release of T_4 and T_3 as well as free iodotyrosines and iodothyronines

(g) Secretion of iodothyronines into the blood

(h) Scavenger deiodination of iodotyrosines within the thyroid follicular cells for reutilization of the liberated iodine

The biosynthesis of thyroid hormones requires access to four major components: thyroglobulin (Tg), thyroperoxidase (TPO), H_2O_2, and iodide. Thyroglobulin is the most abundant protein of thyroid tissue. An understanding of the biochemistry and biological properties of thyroglobulin is integral to an understanding of the generation of thyroid hormone.

The mature human Tg is a dimeric molecule with a molecular mass of 669 kDa; about 10% of the Tg mass comprises carbohydrates. The monomeric subunit is 330 kDa and consists of 2748 amino acid residues. Tg is synthesized as a prehormone that contains a leader sequence of 19 amino acids. The human gene for Tg is a single copy that extends more than 300 kb and contains 42 exons. The pathway of biosynthesis of thyroglobulin is initiated by thyroid-stimulating hormone through interaction with its membrane receptor (see Figure 6-4). After its biosynthesis, Tg is subject to extensive glycosylation; the carbohydrate units are linked to 20 potential glycosylation sites on the protein.

As shown in Figure 6-5, the primary structure of human Tg is divided into four domains. Bovine Tg has an almost identical organization. Interestingly, Tg contains 134 tyrosine residues, only 25-30 of which have been shown to become iodinated; only 4 of these iodotyrosines are actively utilized to form iodothyronines. Four principal hormonogenic sites have been

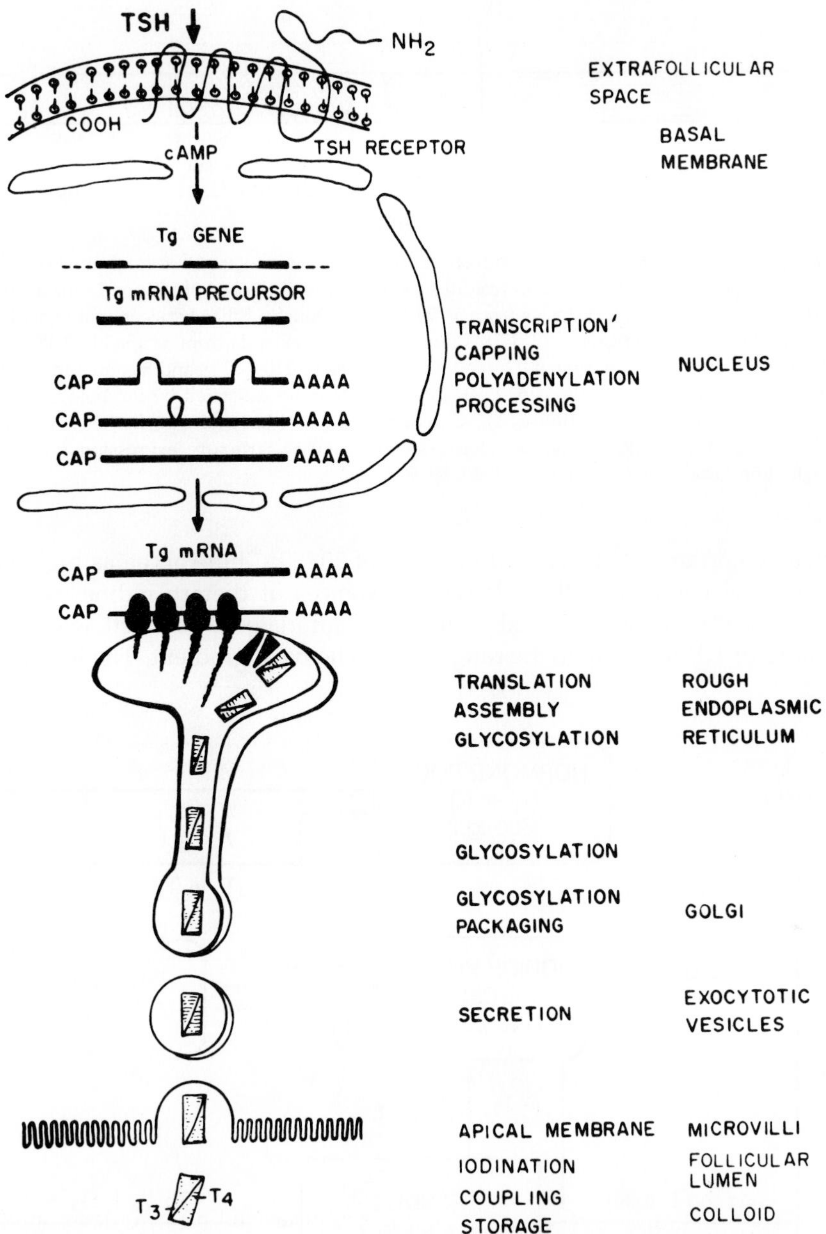

FIGURE 6-4 Biosynthetic pathway for the production of thyroglobulin. TSH initiates the process via stimulation of the cAMP pathway. The mature mRNA is translated into the 2748-amino acid subunit of 330 kDa. The mature Tg is a dimer of two subunits. Before Tg is secreted, extensive glycosylation occurs. [Modified with permission from Medeiros-Neto, G., Targovnik, H. M., and Vassart, G. (1993). Defective thyroglobulin synthesis and secretion causing goiter and hypothyroidism. *Endocr. Rev.* **14,** 165–183.]

identified in Tg at amino acid residues 5, 2553, 2567, and 2746. By definition, a hormonogenic site contains a tyrosine that ultimately becomes iodinated and is converted to a thyroid hormone (see later discussion).

B. Iodide Transport

A summary of iodine physiology and metabolism in adults is given in Figure 6-6. The transfer of iodide from the blood across the basal lateral membrane

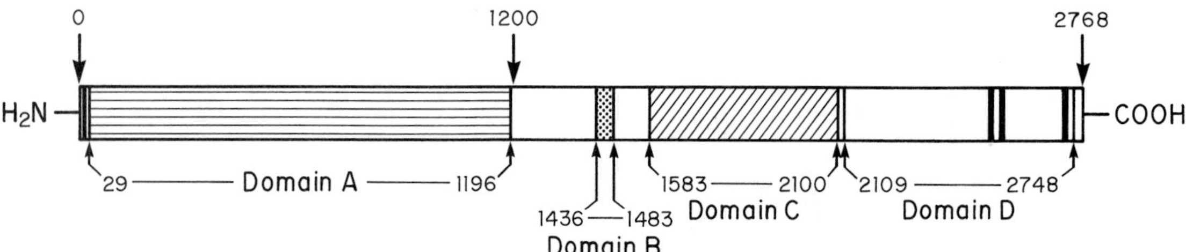

FIGURE 6-5 Schematic diagram of the primary amino acid organization of thyroglobulin (Tg). The mature human Tg is composed of 2748 amino acid residues. Domain A is composed of 10 type 1 repeats of 28 highly conserved residues, including 6 cysteines from aa 29 to 1196. Domain B lies between amino acids 1436 and 1483 and contains 3 type 2 repeats of 12–17 residues. Domain C extends from aa 1583 to 2100 and consists of a type 3 motif that is repeated 5 times. Domain D comprises aa 2109–2748 and has no internal homology. The hormonogenic sites at aa positions 5, 2553, 2567, and 2746 are indicated by the bold lines. [Modified with permission from Bacolla, A., Brocas, H., Christophe, D., de Martynoff, G., Leriche, A., Mercken, L., Parmaj, J., Pohl, V., Targovnik, H., and van Heuverswyn, V. (1985). Structure, expression and regulation of the thyroglobulin gene. *Mol. Cell. Endocrinol.* **40,** 89–97.]

of the thyroid follicular cell occurs by an active transport process. Under normal conditions, the relative amounts of the halide in the thyroid as compared with the serum or plasma (T : S or T:P ratio) are in the range of 20–30:1, but can be as high as 300:1 under circumstances of dietary iodine deprivation. The iodine is accumulated by the cell against both an electrical and a chemical gradient. The iodide transport is inhibited

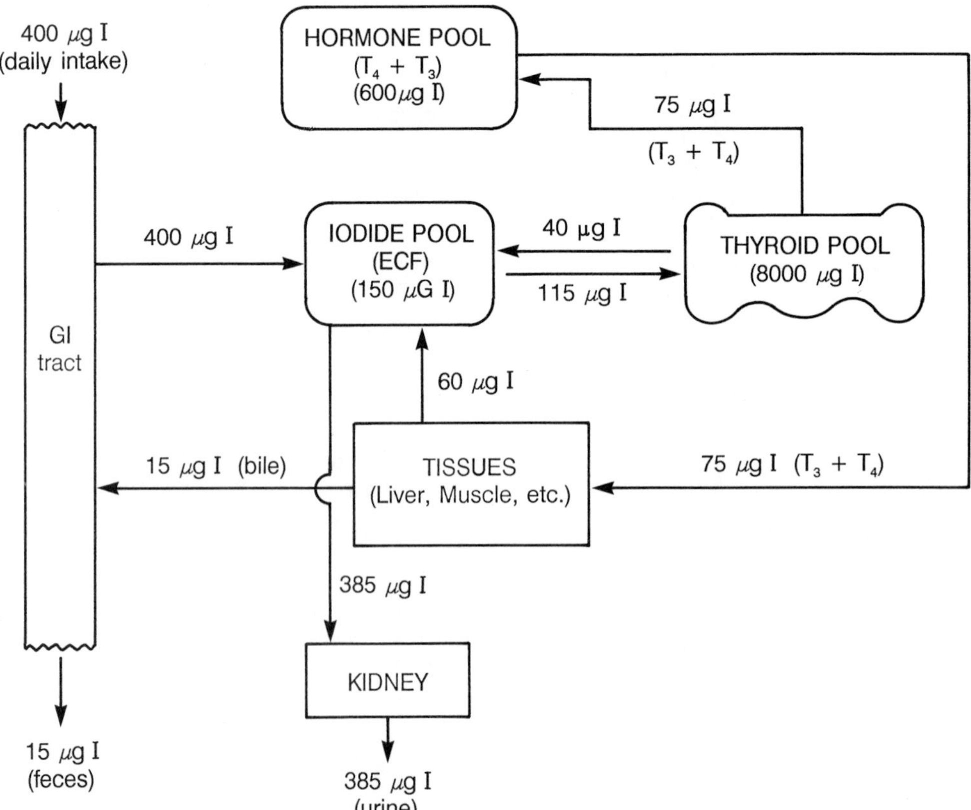

FIGURE 6-6 Iodine physiology and metabolism. The values indicated are typical for a healthy adult with a dietary intake of 400 μg of iodine (I) per day. ECF: extracellular fluid. Since the ECF volume is 25 liters, the ECF iodide concentration is 0.6 μg/100 ml. Approximately 115 μg of I is accumulated per day across the thyroid membrane by an active transport process; of this amount, 75 μg is converted to hormone-bound iodine, with the remaining 40 μg/day leaking back to the ECF pool.

by ouabain, suggesting the involvement of a Na^+,K^+ ATPase. The K_m for iodide transport into the follicular cell is $\sim 3 \times 10^{-5}$ *M;* the maximum concentration of iodide attainable by the rat thyroid is 1 m*M*.

The rate of iodide transport is proportional to the extracellular Na^+ concentration, which suggests the existence of a Na^+ I^- cotransport system. In this model, the putative iodide carrier must bind Na^+ ions on the external surface of the cells (which then move up a concentration gradient as they enter the cell) in order to transport I^- ions into the cell against the electrochemical gradient.

Certain monovalent anions such as ClO_4^-, SCN^-, and pertechnetate (TcO_4^-) are able to effectively compete with I^- for access to the transport process on the follicular cell. It has been proposed that the charged group of the iodide carrier may be either the guanidinium group of nitrogen or the quaternary nitrogen of phosphatidylcholine.

A similar transport system for iodide exists in the mammary gland, salivary glands, parietal cells of the gastric mucosa, choroid plexus, and ciliary body. These tissues, however, do not organify or store the iodide.

Thyrotropin or TSH is the most important physiological factor affecting iodide uptake by the thyroid. After TSH occupies its membrane receptors on the thyroid follicular cell, there is a stimulation of cAMP and an elevation in $[Ca^{2+}]_i$. Further molecular details are not yet available on how these changes are coupled to the iodide uptake process. Once the iodide has entered the follicle cell, it is believed that a significant proportion moves across the cell and is stored in the lumen with the colloid. As yet, no specific iodide-binding protein has been identified in the colloid compartment.

The iodide transport system of the follicle is controlled in an autoregulatory fashion by the gland. Under conditions of a chronic dietary excess of iodine, the activity of the iodide transport system is down-regulated. A small number of patients with familial goiter have been studied in which the disease is attributable to a defective thyroid gland iodide transport system.

C. Iodide Organification and Thyroid Peroxidase

1. *Introduction*

The process of incorporation of 3–4 iodide atoms on the aromatic rings of a tyrosine followed by conversion to a thyroid hormone form is complex; it involves three general steps:

(a) Oxidation of iodide to iodine
(b) Addition of iodine to a limited number of specific tyrosine acceptor residues present in thyroglobulin, particularly at the four hormonogenic sites (see Figure 6-5)
(c) Coupling of two iodotyrosyl residues to generate equimolar amounts of thyroid hormone and dehydroalanine residues within the thyroglobulin molecule

The release of the thyroid hormone coupled to the thyroglobulin occurs in a separate process involving reabsorption of the Tg from the colloid compartment of the follicle followed by selective proteolysis; this process will be discussed separately. Figure 6-7 presents a general overview of the steps of thyroid hormone biosynthesis and secretion.

2. *Thyroid Peroxidase*

After the entry of iodide into the follicular cell, it must first be oxidized to a higher oxidation state before organification. The redox potential for the couple is

$$2I^- \rightarrow I_2 + 2e^-, E^\circ = +0.535 \text{ V}$$

This is relatively high in relation to the E° values for most biological oxidants. Only O_2 and H_2O_2 are sufficiently good electron acceptors at pH 7 to be able to receive the electrons derived from the oxidation of iodide to iodine.

Much evidence has accumulated indicating that the enzyme thyroid peroxidase (TPO) is responsible for both the oxidation of iodide to iodine and the iodination of tyrosine residues on thyroglobulin (Tg). In rat thyroid cells, TPO is widely distributed in the membranes of the rough endoplasmic reticulum, in the Golgi apparatus, and along the apical plasma membrane facing the plasma surface. In contrast, TPO is not present in the membranes of the pseudopods extending into the follicular lumen.

Human TPO is a heme-containing enzyme comprising 933 amino acid residues (~101 kDa). The deduced amino acid sequence contains both a putative signal peptide at the NH_2-terminus and 25 hydrophobic amino acids at the COOH-terminus, which probably create a transmembrane domain that anchors the TPO to the plasma membrane of the thyroid follicular cells.

3. *Catalytic Properties of TPO*

TPO has the capability to catalyze two separate reactions utilizing iodide as a substrate:

$$\text{(a) Tg–tyrosine} + I^- \xrightarrow{\text{TPO}} \text{Tg–iodinated tyrosine}$$

$$\text{(b) } I^- \xrightarrow{\text{TPO}} I_2$$

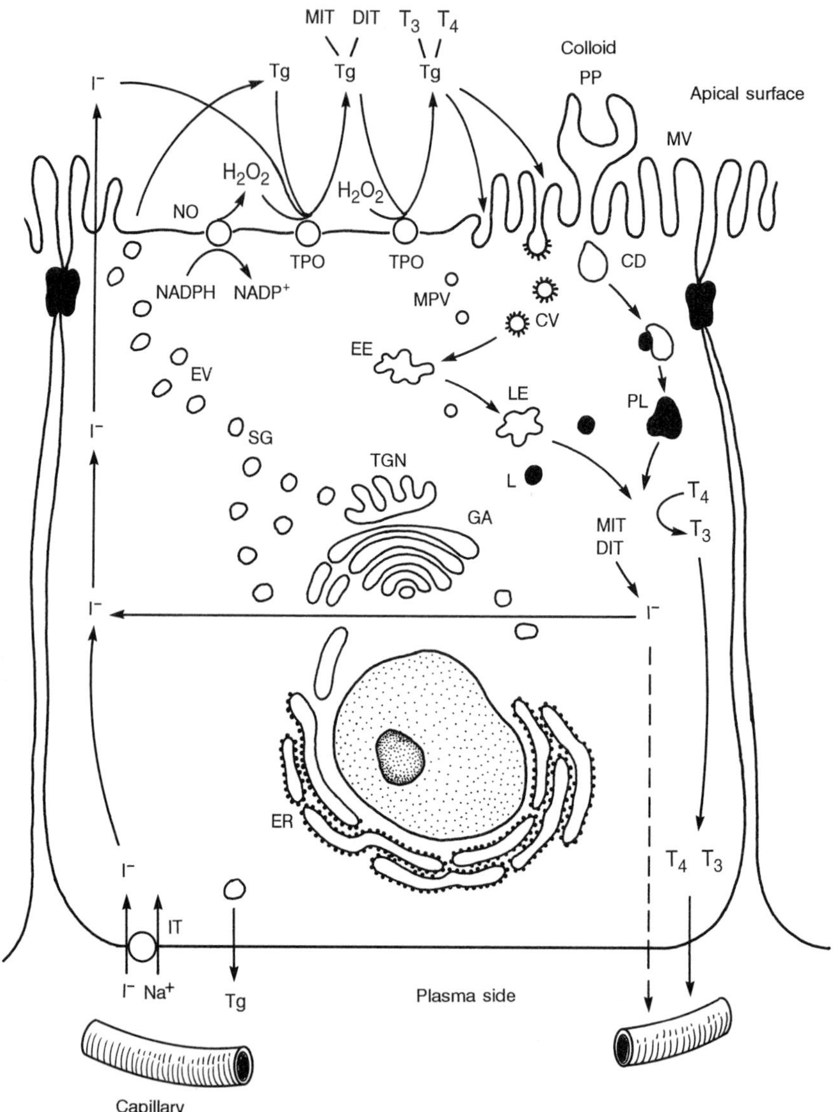

FIGURE 6-7 Process of thyroid hormone biosynthesis and secretion by a thyroid follicular cell. Abbreviations: IT, iodide transporter; ER, endoplasmic reticulum; GA, Golgi apparatus; SG, secretory granules; EV, exocytotic vesicles; NO, NADPH oxidase; TPO, thyroid peroxidase; MV, microvilli (which are present all over the apical surface but have been omitted for clarity); MPV, micropinocytotic vesicles (involved in Tg reabsorption); PP, pseudopods; CD, colloid droplets; EE, early endosomes; LE, late endosomes. The deiodination of iodotyrosine, iodide recycling, Tg conversion to T_3, and $T_4 + T_3$ secretion are also indicated. [Modified with permission from Gentile, F., DiLauro, R., and Salvatore, G. (1995). Biosynthesis and secretion of thyroid hormones. *In* "Endocrinology" (L. J. DeGroot *et al.*, eds.), Vol. 1, Chapt. 34, pp. 517–542. W. B. Saunders Philadelphia, PA.]

Although free iodine is capable of chemically iodinating tyrosine residues, reaction b is not believed to be a significant pathway since the K_m for iodide conversion to I_2 is ~100× greater than the K_m of iodide for iodination of a tyrosine residue (reaction a).

In addition, TPO catalyzes a third reaction, the coupling of Tg-bound iodotyrosines into Tg-bound thyronines.

$$\text{c) Tg–iodinated tyrosine} \xrightarrow{\text{TPO}} \text{Tg–thyronines}$$

4. The H_2O_2 Generation System

Peroxide (H_2O_2) is an obligatory electron receptor for the reactions catalyzed by TPO. In fact, the rate of generation of H_2O_2 controls the relative rates of tyrosyl iodination and coupling. As shown in Figure 6-8, two mechanisms of generation of H_2O_2 have been proposed; as yet there is no agreement on which pathway is favored. The required NADPH is derived from the pentose phosphate pathway.

5. Molecular Mechanisms of Tyrosine Iodination and Iodotyrosine Coupling

The enzymatic production of iodotyrosines on thyroglobulin (phase A) and their subsequent coupling to generate thryoglobulin-bound thyronines (phase B) are both mediated by the enzyme thyroid peroxidase (TPO). The precise details of this series of complex enzymatic transformations are not yet available. However, Figure 6-9 presents a schematic pathway that invokes the involvement of three participants designated compounds I, II, and III; compounds I and II are believed to be conjugates with TPO (see Figure 6-9), but their precise structures are not yet known.

In phase B, TPO mediates a concerted succession of three steps, which results in the formation of equimolar amounts of thyronine hormone and dehydroalanine within the polypeptide chain of thyroglobulin (see Figure 6-10). The locations of the preferred hormonogenic sites on Tg for the generation of thyronine are indicated in Figure 6-5.

One of the major physiological controls of the formation of monoiodotyrosine formation is the dietary iodide supply. In the absence of ample dietary iodine, the iodotyrosine:iodothyronine and MIT:DIT ratios increase due to poor iodination of thyroglobulin.

It has also been observed that the iodination of thyroglobulin is inhibited by excess iodide; this is known

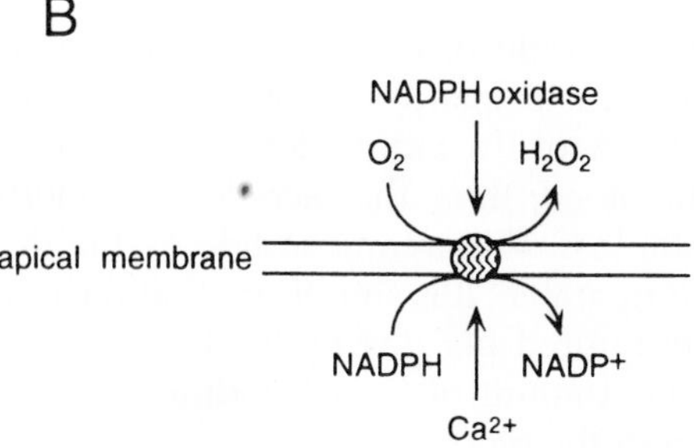

FIGURE 6-8 Proposed mechanisms for production of H_2O_2. (A) Superoxide anion, O_2^-, is first produced from O_2 by the NADPH oxidase enzyme and then converted by dismutation, mediated by the enzyme superoxide dismutase (SOD), on the inside of the apical membrane (see Fig. 6-7); the H_2O_2 is then released to the outside. (B) H_2O_2 is generated directly from O_2 by the NADPH oxidase outside the cell. As indicated, changes in intracellular [Ca^{2+}] have been shown to stimulate the production of H_2O_2. [Modified with permission from Gentile, F., DiLauro, R., and Salvatore, G. (1995). Biosynthesis and secretion of thyroid hormones. *In* "Endocrinology" (L. J. DeGroot *et al.*, eds.), Vol. 1, Chapt. 34, pp. 517–542. W. B. Saunders, Philadelphia, PA.]

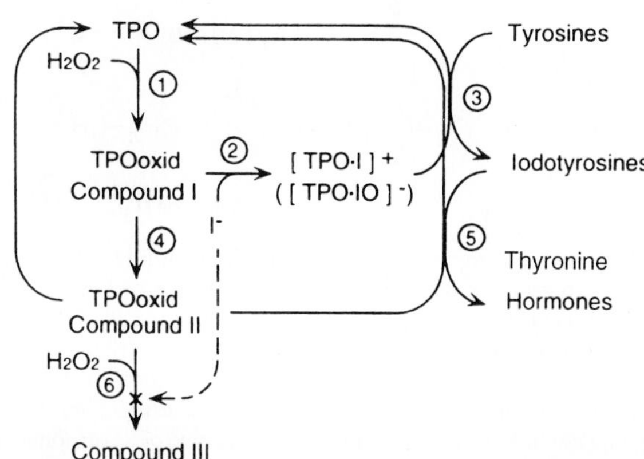

FIGURE 6-9 Summary of reactions catalyzed by thyroid peroxidase (TPO) to generate iodotyrosines and thyroid hormones. Five separate reactions have been postulated, which can be separated into two phases. Phase A utilizes reactions 1–3 to generate iodotyrosine, while phase B utilizes reactions 4–6 to effect coupling of iodotyrosines to generate thyronine hormones. In phase A, reaction 1 utilizes H_2O_2, which forms a complex with TPO, designated as compound I. Compound I can participate as a substrate in both reactions 2 and 3. Reaction 2 utilizes compound I to oxidize I so as to generate an active iodinating agent (shown here as either an iodinium ion or hypoiodate bound to TPO). Reaction 3 then catalyzes the iodination of tyrosine and regenerates TPO. In phase B, in the absence of iodide, reaction 4 effects the conversion of compound I into compound II. Next, reaction 5 catalyzes the coupling of two iodinated thyronines, which generates a thyronine hormone and regenerates free TPO (see Fig. 6-10). In the presence of excess H_2O_2, reaction 6 converts compound II into compound III, which is inactive; reaction 6 can be prevented by I. The chemical nature of compounds I, II, and III is not yet known. [Modified with permission from Gentile, F., DiLauro, R., and Salvatore, G. (1995). Biosynthesis and secretion of thyroid hormones. *In* "Endocrinology" (L. J. DeGroot *et al.* eds.), Vol. 1, Chapt. 34, pp. 517–542. W. B. Saunders, Philadelphia, PA.]

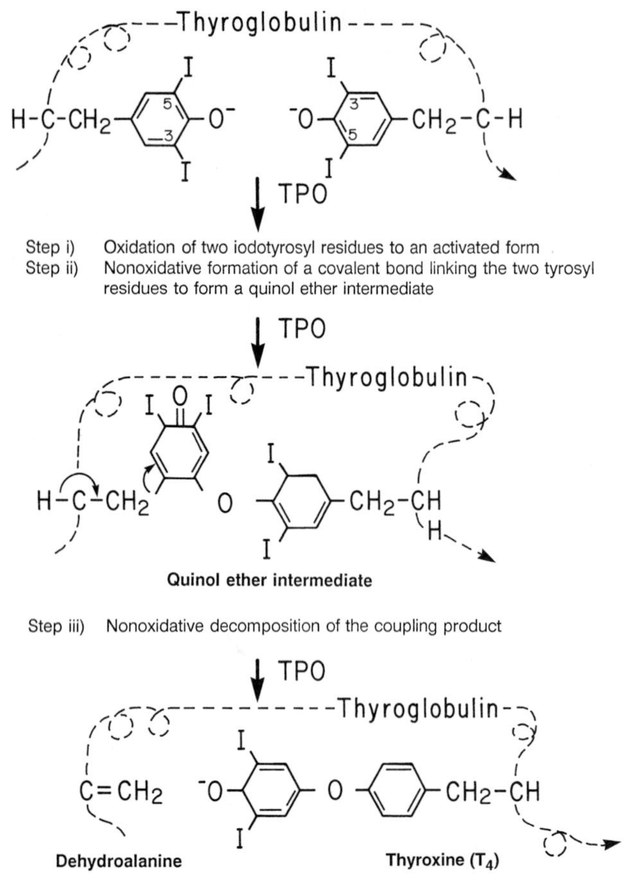

Step i) Oxidation of two iodotyrosyl residues to an activated form
Step ii) Nonoxidative formation of a covalent bond linking the two tyrosyl residues to form a quinol ether intermediate

Quinol ether intermediate

Step iii) Nonoxidative decomposition of the coupling product

Dehydroalanine

Thyroxine (T_4)

FIGURE 6-10 Proposed mechanism for the coupling of two iodotyrosines to generate tyrosines. The entire coupling reaction is catalyzed by TPO as reaction 5 of phase B (see Fig. 6-9). The three concerted reactions are indicated. It is not clear whether TPO-mediated coupling occurs via an ionic or a free radical mechanism. [Modified with permission from Gentile, F., DiLauro, R., and Salvatore, G. (1995). Biosynthesis and secretion of thyroid hormones. *In* "Endocrinology" (L. J. DeGroot *et al.*, eds.), Chapt. 34, Vol. 3, pp. 517–542. W. B. Saunders, Philadelphia, PA.]

as the Wolff–Chaikoff effect. This inhibitory effect in intact animals is only transient, probably because the iodide transport system in the follicular cells undergoes some type of adaptation that results in a reduction in intracellular iodide levels. The adaptation to the circumstances of excess iodide appears to be due to a decrease in the unidirectional influx of iodide caused by an increase in the K_m for the pump.

D. Secretion Mechanisms of Thyroid Hormones

In order for thyroid hormone (T_4 and T_3) to be produced, it is necessary for thyroglobulin, with its conjugate thyronine residues, to be reabsorbed from the follicular lumen and degraded. The main control over T_4 and T_3 secretion is mediated in a complex fashion by TSH. Basically, interaction of TSH with its membrane receptor on the follicular cell results in pinocytosis of Tg from the colloid, followed by Tg digestion and subsequent hormone release.

At the apical cell membrane–colloid interface, colloid (Tg) is engulfed into a colloid vesicle by the processes of both macropinocytosis and micropinocytosis, so that Tg is delivered inside the follicular cell. The rate of pinocytosis is proportional to the magnitude of the TSH stimulus. Then lysosomes (containing proteolytic enzymes, endopeptidases, cathepsin B, phosphatases) fuse with the colloid vesicles, and hydrolysis of thyroglobulin ensues. This results in the release of T_1, T_3, DIT, MIT, and an assortment of peptide fragments and free amino acids.

1. *Thyroid hormone secretion*

The predominant thyroid hormone product released from Tg is T_4, with lesser amounts of T_3, rT_3, and monoiodinated tyrosines. Under normal physiological circumstances, T_4 is the chief thyroid hormone secreted by the thyroid; *in toto,* 80–90 μg of T_4 is released per day, while only 8 μg of T_3 and 1 μg of rT_3 are released per day. The mechanism by which T_4 and T_3 are released into circulation is unknown. Not only is TSH a stimulator of T_4 and T_3 release, but vasointestinal peptide (VIP) and β-adrenergic agonists, both acting through an adenylcyclase-cAMP pathway, and forskolin, acting through a PKC pathway, have been shown to stimulate the release of T_3 and T_4. Approximately 80% of the T_4 secreted is deiodinated further in peripheral tissues to generate additional T_3.

a. Iodine Reutilization by the Thyroid

Normally the ratio of iodotyrosines to iodothyronines in the thyroid gland is approximately 4:1. Approximately 25% of the iodide derived from tyrosine is lost to the blood; thus, the secretion of iodothyronines and the loss of inorganic iodide by the thyroid gland are comparable. It is significant that the amount of iodide recaptured per day by the thyroid gland is 2–3 times the amount of "new" iodide captured by the gland from the blood.

The monoiodotyrosines released from the thyroglobulin that do not leave the gland are rapidly deiodinated inside the gland by an NADPH-dependent microsomal flavoprotein deiodinase. This enzyme has little affinity for iodothyronine compounds.

2. *Feedback Actions of T_4/T_3 at the Pituitary–Hypothalamus*

The physiological regulation of the secretion of the thyroid hormone is a complex system that involves, in addition to the thyroid gland, participation by the hypothalamus, pituitary, and neural activity. In addition to the thyroid–piuitary–hypothalamus feedback control loop, a neural control of thyroid metabolism has been proposed.

The activity of the thyroid–pituitary loop is a balance between the peripheral consumption/catabolism of T_4/T_3 and the combined hypothalamic and feedback actions of $T_4 + T_3$ on the pituitary secretion of TSH. Each day the body catabolizes ~100 μg of T_4 and 30 μg of T_3. These lost hormones are replenished by the secretion of new $T_4 + T_3$ by the thyroid; as the blood levels of $T_4 + T_3$ fall, this is sensed by the hypothalamus with a resultant increase in the secretion of TSH by the pituitary. The thyroid hormones, $T_4 + T_3$, do not appear to have negative feedback influence on the hypothalamus. The role of the hypothalamus is to stimulate the pituitary to biosynthesize and secrete TSH through the action of thyrotropin-releasing hormone (TRH) (see Chapter 5).

The set point for the control of the system lies largely in the thyrotropes of the anterior pituitary. Thyroid hormones, through interaction with nuclear receptor, suppress at the transcriptional level the synthesis of prepro-TSH in thyrotropes in the anterior pituitary (see Chapter 5). This represents the classic negative feedback loop of the pituitary–thyroid endocrine axis. In addition, some negative feedback actions of T_3 are believed to be at the hypothalamus. TRH release from the hypothalamus, which travels to the pituitary via hypophyseal portal vessels, interacts with the pituicyte membrane at specific receptor sites, leading to an activation of adenyl cyclase and an increase in intracellular cAMP (see Chapter 5). This causes increased synthesis and secretion of TSH, which in its initial phases is not dependent upon protein synthesis.

a. Long-Acting Thyroid Stimulator

The blood of many hyperthyroid subjects contains a substance known as "long-acting thyroid stimulator" or LATS. Other terminology for LATS reclassifies them into mouse thyroid-stimulating autoantibodies (MTS) or human thyroid-stimulating autoantibodies (HTS). Physiologically, LATS is capable of mimicking, over a somewhat slower time interval, all of the actions of TSH (10–12 hr instead of 2 hr for TSH) and the stimulation of thyroid follicular cell adenylate cyclase and the consequent cascade of metabolic events, which lead to increased secretion of $T_4 + T_3$.

Biochemically, LATS has been identified as a 7 S globulin of the IgG class of immunoglobulins; thus, it is an antibody. The formation of this antibody is believed to result from the inappropriate release of an antigen from the thyroid follicular cells and the subsequent production of autoantibodies by the thymus-dependent lymphocytes. The molecular basis of the biological action(s) of LATS is that the antibody is postulated to be complementary to the TSH receptor protein in the basal membrane of the thyroid follicular cell.

An important difference between LATS and TSH is that the production of LATS is not suppressed by high blood levels of the thyroid hormones. Thus, LATS production is believed to be the pathophysiological mechanism leading to the generation of Graves' disease or hyperthyroidism.

E. Systemic Transport of Thyroid Hormone

1. Plasma Concentrations of Thyroid Hormones

Table 6-2 lists the blood concentrations and kinetics in humans of the principal iodine-containing compounds. The main secretory product of the thyroid is T_4. Thyroxine is present in the blood at a concentration of 40–100 times that of T_3. The concentrations of DIT and TETRAC are comparable to that of T_3 (i.e., ~110–140 ng/ml). The blood level of reverse T_3 is ~30–50% of that of T_3 and 1% of that of T_4.

2. Thyroid Hormone-Binding Proteins

In most vertebrates, the bulk of the thyroid hormone circulates in the blood bound to proteins. The three blood proteins responsible for systemic transport of the thyroid hormones are thyroxine-binding globulin (TBG), thyroxine-binding prealbumin (TBPA; also

TABLE 6-2 Kinetics of T_4, T_3, and rT_3 Metabolism in Normal Human Subjects[a]

Parameter	Hormone		
	T_4	T_3	rT_3
Serum concentration			
Total, μg/100 ml (nmol/ liter)	8.0 (103)	0.12 (1.8)	0.04 (0.5)
Free, ng/100 ml (pmol/liter)	2.1 (27)	0.3 (4)	0.2 (3.4)
Body pool, μg (nmol)	810 (1000)	46 (71)	40 (61)
Volume of distribution (liter)	10	38	98
Metabolic clearance rate (liters/day)	1.1	22	90
Disposal rate, μg/day (nmol/day)	82 (100)	26 (34)	36 (46)
Half-life in plasma (days)	7	1	0.2

[a] Modified from Pittman, C. S. (1979). Hormone metabolism in endocrinology. *In* "Endocrinology" (L. J. DeGroot, G. F. Cahill, L. Martini, D. H. Nelson, W. D. Odell, J. T. Potts, E. Steinberger, and A. I. Winegrad, eds.), Vol. 1, pp. 365–372. Grune & Stratton, New York.

known as transthyretin), and albumin (ALB) (see Table 6-3). The normal distribution of T_4 among these proteins is 70–75% bound to TBG, 15–20% bound to TBPA, and 5–10% bound to ALB. In contrast, T_3 is bound 70–75% to TBG and 25–30% to ALB, and T_3 has little or no affinity for TBPA.

TBG is a single 54-kDa polypeptide chain, which is synthesized in the liver. TBG has four carbohydrate chains that comprise 23% of the molecule by weight; in addition, there are 8–10 sialic acid residues per molecule. Each molecule of TBG has a single ligand-binding site. Although only approximately 0.04% of T_4 and 0.4% of T_3 are "free," it is the free fraction that initiates thyroid hormone biological responses.

Several instances of inherited abnormalities in blood TBG levels have been reported. Absent or low TBG is more common than excess TBG; the incidence rate of abnormalities may approach 1 in 2000. No genetic abnormalities of TBPA have been found in humans.

Gestation causes a marked increase in the thyroid physiological status of the mother. As estrogen secretion becomes elevated in the first weeks of pregnancy, there is a doubling of TBG concentration. This in turn stimulates T_4 secretion, and shortly thereafter a new equilibrium is established wherein the blood level of total T_4 is double that of the nonpregnant state. This occurs without any alteration in the "free" T_4 levels.

TBPA is composed of four identical polypeptide subunits, each containing 127 amino acids. It has a molecular mass of 55 kDa and a single hormone-binding site. TBPA binds T_3 only 1% as avidly as T_4. In plasma, TBPA binds 11% of the total T_4. Interestingly, TBPA also contains a separate specific binding site for the retinol-binding protein. Retinol-binding protein binds vitamin A precursors.

The crystallographic structure of TBPA is known (see Figure 6-11).

TABLE 6-3 Some Properties and Metabolic Parameters of the Principal Thyroid Hormone-Binding Proteins in Serum[a]

Parameter	TBG[a]	TBPA[b]	Albumin
Molecular weight	63,000	55,000	69,000
Structure	Monomer	Tetramer	Predominantly monomeric
Carbohydrate content (%)	30	1	
Number of binding sites for T_4 and T_3	1	1	1
Association constant, K_a (M^{-1})			
For T_4	1×10^{10}	1×10^{8}	1.5×10^{6}
For T_3	5×10^{8}	1×10^{6}	1×10^{7}
Concentration in serum (mg/100 ml)	2.0	25	3500
Occupancy, or saturation (%)			
For T_4	30	0.5	0.0015
For T_3	0.6	0.001	0.001
Half-life (hr)	5	2	15
Degradation rate (mg/day)	15	650	17,000

[a] Modified from Refetoff, S. (1979). Thyroid hormone transport in endocrinology. *In* "Endocrinology" (L. J. DeGroot, G. F. Cahill, L. Martini, D. H. Nelson, W. D. Odell, J. T. Potts, E. Steinberger, and A. I. Winegrad, eds.), Vol. 1, pp. 347–356. Grune & Stratton, New York.

[b] TBG, thyroid-binding globulin; TBPA, thyroxine-binding prealbumin.

F. Thyroid Hormone Metabolism and Catabolism

1. Production of Thyroid Hormones

The blood concentration of the thyroid hormones for each substance is dependent upon not only the amount of hormone biosynthesized and eventually se-

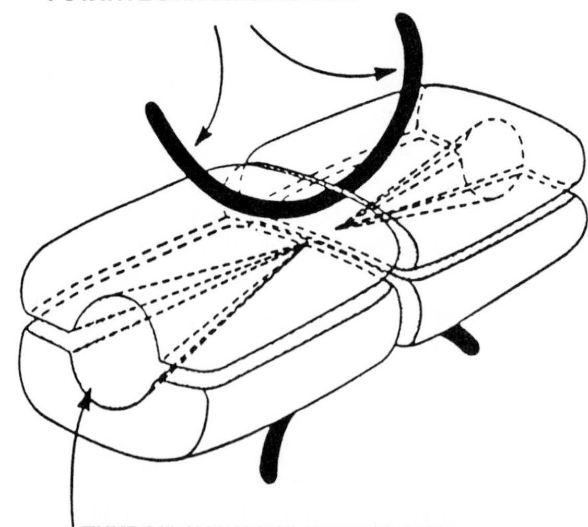

FIGURE 6-11 Model of thyroxine-binding prealbumin (TBPA). The schematic model is derived from X-ray crystallographic studies of prealbumin, illustrating the interaction of four identical subunits forming a channel through the interior of the molecule, thereby forming two identical thyroid hormone-binding sites. The channel narrows at the center of the molecule. Although the binding sites are identical, T_4 binds cooperatively; occupation of T_4 at one site presumably alters the conformation of the second site, which results in a lower affinity for the second T_4-binding interaction. Depicted on the upper and lower sides of the molecule is a symmetric β-pleated sheet structure that is relatively rich in amino acids with ionic side chains and that also contains tryptophan molecules. On the basic of computer graphic modeling studies, this site has been proposed to be a DNA-binding site. [Reproduced with permission from Eberhardt, N. L., Apriletti, J. W., and Baxter, J. D. (1980). The molecular biology of thyroid hormone action. *In* "Biochemical Actions of Hormones" (G. Litwack, ed.), Vol. 7, pp. 311–375. Academic Press, New York, © 1980 Macmillan Journals Limited.]

creted but also the hormone's affinity for carrier proteins, affinity for target tissues, rate of catabolism, and finally rate of clearance. It is apparent that the physiologically mediated changing relationships between the interconversions of $T_4 \rightarrow T_3$ and $T_4 \rightarrow rT_3$ provide insight into many aspects of disease states related to the thyroid hormones.

Figure 6-12 summarizes the daily production of T_4, T_3, and rT_3 in humans. T_4 also is converted in the peripheral tissues to T_3 and rT_3. The conversion of T_4 to T_3 has clearly been shown to occur in many principal tissues of humans, rats, and sheep. Because T_3 is 3–5 times more active than T_4, and because ~30% of daily T_4 production is diverted to produce ~80% of the daily T_3 turnover, the key role of T_3 in meeting the energy requirements of humans is obvious. The hormone T_4 is largely restricted to the vascular pool with a slow turnover, while T_3 has a small vascular pool and a high turnover rate and is found in higher concentrations in the target tissues (see Table 6-2). From these considerations, the view has emerged that T_4 may have little intrinsic biological activity.

In a normal human, ~40% of the T_4 is deiodinated by a pathway that yields reverse T_3 (rT_3). The biological role of rT_3 is not known. It has been suggested that a physiological concentration of rT_3 may regulate the conversion of $T_4 \rightarrow T_3$. The plasma concentration of rT_3 is known to vary from that of T_3 in a number of clinical conditions, including patients suffering from hepatic and renal diseases, malnutrition, or treatment with dexamethasone.

2. *Metabolic Transformations of Thyroid Hormones*

Figure 6-13 depicts the metabolic pathways for the further metabolism of thyroxine. Only very small amounts of unaltered T_4 are excreted in the urine or feces. The liver and kidney are the major sites of metabolism. In the liver, stepwise deiodination is the prime pathway for inactivation (Figure 6-13A). Other transformations that have been detected include β-glucuronide and sulfate conjugation with the phenol functionality and transamination or deamination of the alanine side chain (Figure 6-13B).

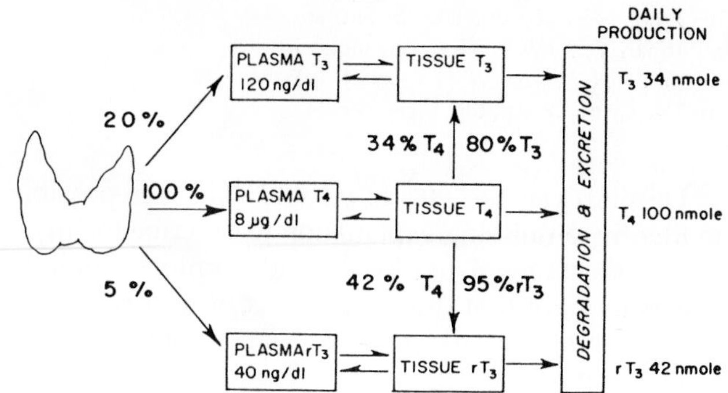

FIGURE 6-12 Daily production of T_3, T_4, and rT_3 in normal human subjects. [Modified from Pittman, C. S. (1979). Hormone metabolism. *In* "Endocrinology" (L. J. DeGroot, G. F. Cahill, L. Martini, D. H. Nelson, W. D. Odell, J. T. Potts, E. Steinberger, and A. I. Winegrad, eds.), Vol. 1, p. 365. Grune & Stratton, New York.]

G. Antithyroid Drugs

Specific compounds that interfere with the thyroid follicular peroxidase are used both in research and in certain clinical circumstances. In general, small doses of goitrogens preferentially block the coupling reactions in the thyroid, while large doses block the iodination of the tyrosyl residues in thyroglobulin. The most potent antithyroid compounds contain the thionamide group (see Figure 6-14). The simplest member of the group is thiourea; the compounds thiouracil and 6-*n*-propylthiouracil are significantly more potent. The structures of several antithyroid drugs are given in Figure 6-15.

V. BIOLOGICAL AND MOLECULAR ACTIONS

A. Introductory Comments

Thyroid hormones produce a wide array of physiological effects in virtually all of the body's organs and metabolic pathways; these are summarized in Table 6-4. Principal effects include the modulation of oxygen consumption, as well as carbohydrate, protein, and fat metabolism. In addition, thyroid hormones modulate the degradation and synthesis rates of many other growth factors and hormones, which results in many secondary effects. The biological effects of thyroid hormone are divisible into two general categories: (1) effects on cellular differentiation and development; (2) effects on metabolic pathways.

One of the hallmarks of thyroid hormone action occurs during the process of metamorphosis in amphibians (tadpoles → frogs) and in supporting fetal development in animal models. In humans, the requirement for thyroid hormone is emphasized by the occurrence of cretinism in the absence of adequate amounts of T_3. Many of the developmental effects of thyroid hormone cannot be reversed by later application of thyroid hormone, which suggests that the hormone is only able to act in a developmentally predetermined window of time.

Another of the hallmarks of thyroid hormone action has been its ability to alter oxygen consumption (see Figure 6-16). Oxygen consumption is significantly stimulated by thyroid hormone in heart, diaphragm, liver, kidney, and pancreas.

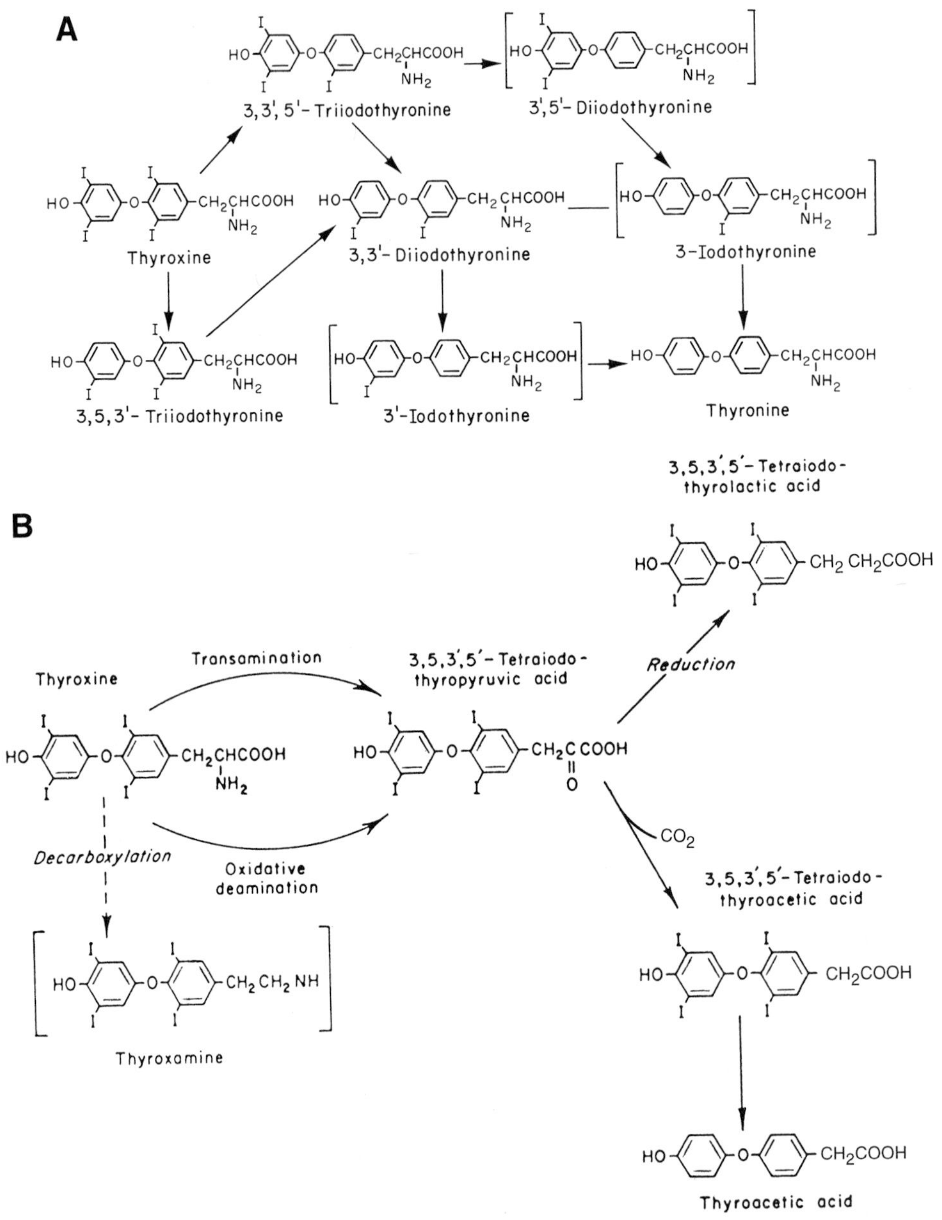

FIGURE 6-13 (A) Deiodination of thyroxine and its further metabolism (in brackets). (B) Metabolism of the thyroxine side chain. The metabolites have not been identified by *in vivo* studies. [Modified with permission from Figs. 28-2 and 28-3 in Pittman, C. S. (1979). Hormone metabolism. *In* "Endocrinology" (L. J. DeGroot, G. F. Cahill, L. Martini, D. H. Nelson, W. D. Odell, J. T. Potts, Jr., E. Steinberger, and A. I. Winegrad, eds.), Vol. 1, p. 367. Grune & Stratton, New York].

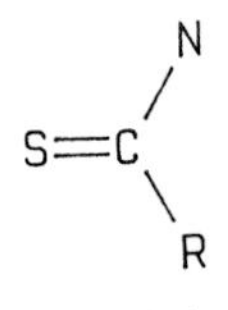

FIGURE 6-14 General structure of antithyroid compounds. The thionamide group R may be either S, O, or N.

Despite extensive efforts, it has not yet been possible to identify a unifying explanation to describe the myriad of effects mediated by thyroid hormone. Initially it was thought that the major effect of thyroxine was to uncouple oxidative phosphorylation; however, this model is no longer favored. Subsequent descriptions of thyroid action focused on models describing an increase in energy expenditure via stimulation of Na^+, K^+ ATPase activity. While these effects of thyroxine are clearly demonstrable, they are unable to rationalize

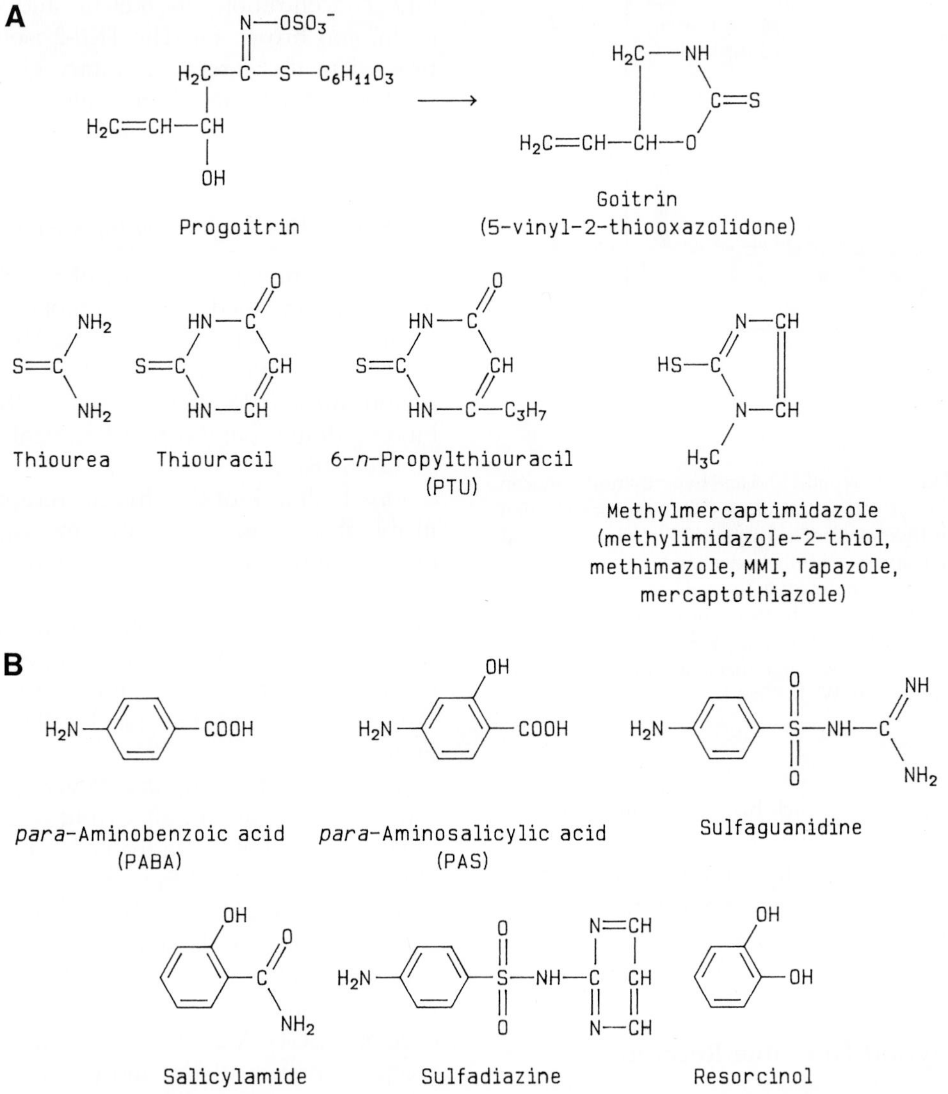

A

Progoitrin → Goitrin (5-vinyl-2-thiooxazolidone)

Thiourea

Thiouracil

6-*n*-Propylthiouracil (PTU)

Methylmercaptimidazole (methylimidazole-2-thiol, methimazole, MMI, Tapazole, mercaptothiazole)

B

para-Aminobenzoic acid (PABA)

para-Aminosalicylic acid (PAS)

Sulfaguanidine

Salicylamide

Sulfadiazine

Resorcinol

FIGURE 6-15 (A) Antithyroid compounds with the thionamide group. Progoitrin is present in a number of vegetables, including cabbage, kale, and cauliflower. (B) Aromatic compounds with antithyroid activity.

TABLE 6-4 Physiological Effects of Thyroid Hormones

Response	Comment(s)
Fetal development	The thyroid and anterior pituitary TSH systems begin to function in the human fetus at ≈11 weeks. The fetus does not receive much T_3 + T_4 from the mother.
Oxygen consumption, heat production	T_3 increases O_2 consumption and heat production by stimulation of Na^+,K^+ ATPase in all tissues except brain, spleen, testis, and ovary.
Cardiovascular effects	T_3 increases gene transcription of myosin heavy chain α and Ca^{2+} ATPase and mediates inotropic and chronotropic effects on the heart, which both increase cardiac output and heart rate in hyperthyroidism.
Sympathetic effects	T_3 increases the number of β-adrenergic receptors in heart muscle, skeletal muscle, adipose tissue, and lymphocytes.
Skeletal actions	Stimulate increased bone turnover (resorption > formation).
Carbohydrate and lipid metabolism	Hyperthyroidism increases hepatic gluconeogenesis and glycogenolysis.
Gastrointestinal effects	T_3 stimulates gut motility.

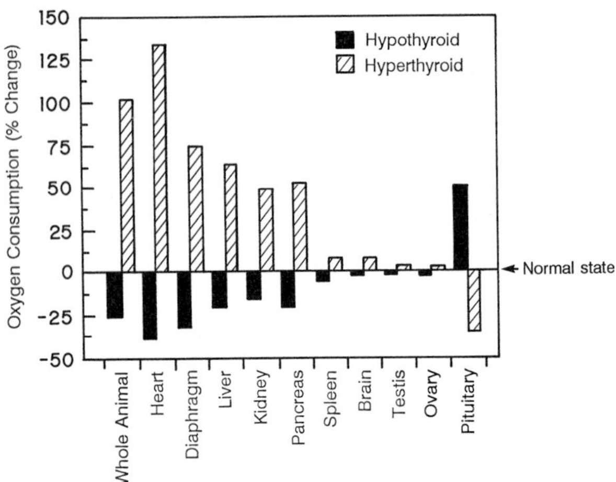

FIGURE 6-16 Effects of hypothyroid and hyperthyroid status on oxygen consumption in various rat tissues. Oxygen consumption was determined in thyroidectomized rats before (solid bars) and after (striped bars) treatment with thyroxine for 4–6 days. Note the paradoxical response of the pituitary to thyroid hormone in relation to the other tissues of the body; the basis for this response is not presently understood. [Modified from Barker, S. B. (1964). Physiological activity of thyroid hormones and analogues. In "The Thyroid Gland", (R. Pitt-Rivers and W. R. Trotter, eds.), p. 200. Butterworths London.]

the many effects of thyroid hormone on gene expression.

The majority of effects of thyroid hormones are now believed to be mediated through interactions of thyroxine with its nuclear receptors, so as to bring about changes in gene expression.

B. Thyroid Hormone Receptor

1. *Introduction*

In analogy to the steroid hormones, retinoic acid, and $1\alpha,25(OH)_2D_3$, thyroid hormones are known to bind to a nuclear receptor that interacts with transcription factors to either up-regulate or down-regulate the transcription of a wide array of genes. The nuclear receptor for thyroid hormones belongs to the steroid receptor supergene family (see Figure 1-27). Since the landmark discovery in 1986 that the nuclear protein encoded by the *c-erb*-A gene (a cellular homologue of the *v-erb*-A oncogene) was structurally identified to be the thyroid nuclear protein, multiple isoforms of the thyroid receptor have been discovered. As summarized in Figure 6-17, five isoforms of the thyroid receptor are now known. Interestingly, two of these isoforms, TRα-2 and TRα-3, do not bind a ligand. There is marked tissue-specific expression of the isoforms of the thyroid receptor: TRβ-1 is expressed in most tissues, while TRα-1 and TRα-2 are present in relatively high concentrations in skeletal and cardiac muscle, brain, and brown fat. The TRβ-2 isoform is only expressed in the anterior pituitary gland and selected regions of the brain. There is also emerging evidence that the TRα gene mRNA may be alternatively spliced, so as to generate a number of distinct protein products.

2. *Thyroid Hormone Receptor Structural Properties*

The functional properties of the thyroid receptor have been mapped via the preparation of scanning deletion mutants (see, for example, Figure 1-44); thus, it has been possible to precisely define the DNA-binding region, nuclear localization signal, thyroxine ligand-binding domain, and heterodimerization domains (see Figure 6-18).

One hallmark of the thyroid receptor is that, even in the absence of ligand, the unoccupied receptor is always found to be associated with the nucleus and its chromatin; this is because the receptor is not cytoplasmically anchored to heat shock proteins, and thus the nuclear localization signal is operative. This is dictated by the sequence of amino acids in the region of 176–186, the so-called nuclear localization domain (See Figure 6-18).

The ability to recognize target genes is a critical aspect of the action of all steroid receptors, including the thyroid receptor. On the basis of structural homology within the DNA-binding domain and other members of the steroid superfamily (see Table 1-10), the thyroid receptor is classified as a member of the estrogen–vitamin D receptor–thyroid subfamily. These receptors all share the same consensus sequence preference of AGGTCA in the DNA-binding domain. These receptors preferentially bind to this so-called half-site when it is present as a single or multiple copy in the promoters of genes that are subject to regulation.

3. *Transcriptional Regulation by Thyroid Receptor*

The three-dimensional structure of the rat α, thyroid receptor ligand binding domain with a thyroid hormone agonist has been determined at a 2.2 Å resolution via X-ray crystallographic procedures. The receptor secondary structure was found to be composed of twelve α-helices and four β-strand domains (see Figure 6-19).

The front cover of this book presents a schematic representation of the secondary structure α-helices (red for helix$_1$ >>> royal blue for helix$_{12}$) and β-strand elements (yellow color) and indicates the orientation of the ligand. The hormone (purple color) is depicted as a space-filling model. There are two notable features of this structure: (a) the ligand is completely buried within the ligand binding domain as part of the hydrophobic core; (b) the carboxy-terminal heterodimeriza-

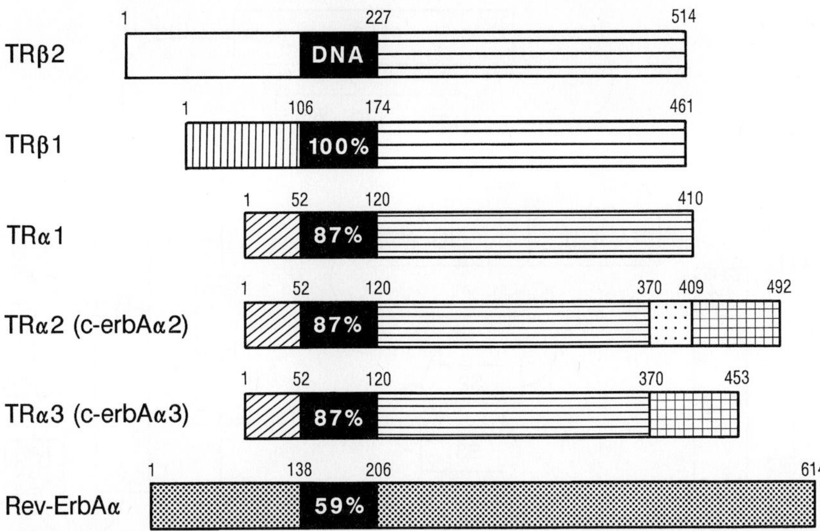

FIGURE 6-17 Isoforms of the thyroid nuclear receptor isolated from the rat. The amino acid sequences have been deduced from the cDNA. The numbering of the amino acid residues is indicated above each form. The percent homology of the DNA-binding domain of TRβ-1 (black region) is also indicated. Identical patterns indicate 100% aa sequence identity. Similar patterns (horizontal lines) indicate highly similar sequences. The TRα-2 and TRα-3 isoforms contain unique carboxyl-terminal sequences that eliminate ligand binding. [Modified with permission from Lazar, M. A. (1993). Thyroid hormone receptors: Multiple forms, multiple possibilities. *Endocr. Rev.* **14,** 184–193.]

tion domain (see Figures 6-18 and 6-19) forms an amphipathic helix with its hydrophobic face constituting part of the ligand hormone binding cavity. These two observations suggest a structural role for the ligand in determining the functionally active conformation for the receptor related to the receptor's role in the regulation of gene expression.

The functional stoichiometry of the thyroid receptor as it interacts with the promoters of regulated genes is variable. Both monomers and homodimers as well as heterodimers of the thyroid hormone receptor are known to be biologically active in the regulation of gene transcription (see Figures 6-20 and 1-45). The usual heterodimer partner is either the RAR or RXR receptor isoform, but under some circumstances the thyroid receptor can also form a heterodimer with the receptor for $1\alpha,25(OH)_2D_3$ (VDR). In addition, the ligand-binding domains of these receptors can be either occupied or unoccupied, depending upon the levels of free hormones available. Thus, there is a wide array of receptor species (both with and without ligand) that are able to regulate the transcription of target genes.

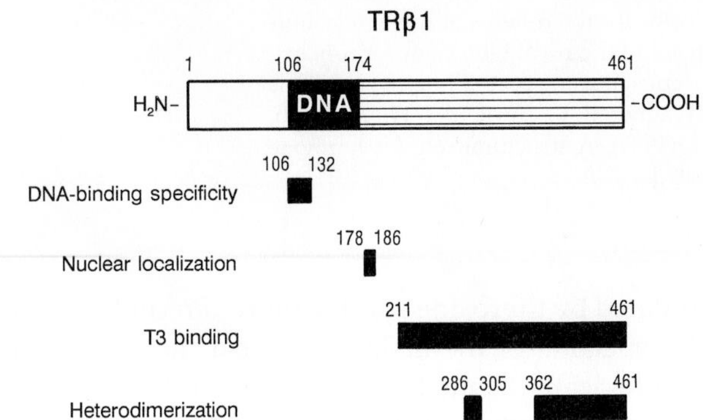

FIGURE 6-18 Mapping of the functional domains of the thyroid receptor isoform TRβ-1. [Modified with permission from Lazar, M.A. (1993). Thyroid hormone receptors: Multiple forms, multiple possibilities. *Endocr. Rev.* **14,** 184–193.]

The unliganded thyroid receptor is functionally active and can repress the transcription of genes activated by T_3. T_3 binding to the thyroid receptor also can either activate or repress gene transcription (see Table 6-5). It is not yet known with certainty which form of the thyroid receptor (monomer, homodimer, or heterodimer) participates in negative or positive regulation; this is an area of research under intense investigation.

C. Production of Biological Responses

As already emphasized, the thyroid hormones are able to mediate a wide array of biological responses in many tissues of the body; Table 6-4 enumerates many of these responses. Evidence indicates that most of these responses are the consequence of the action of

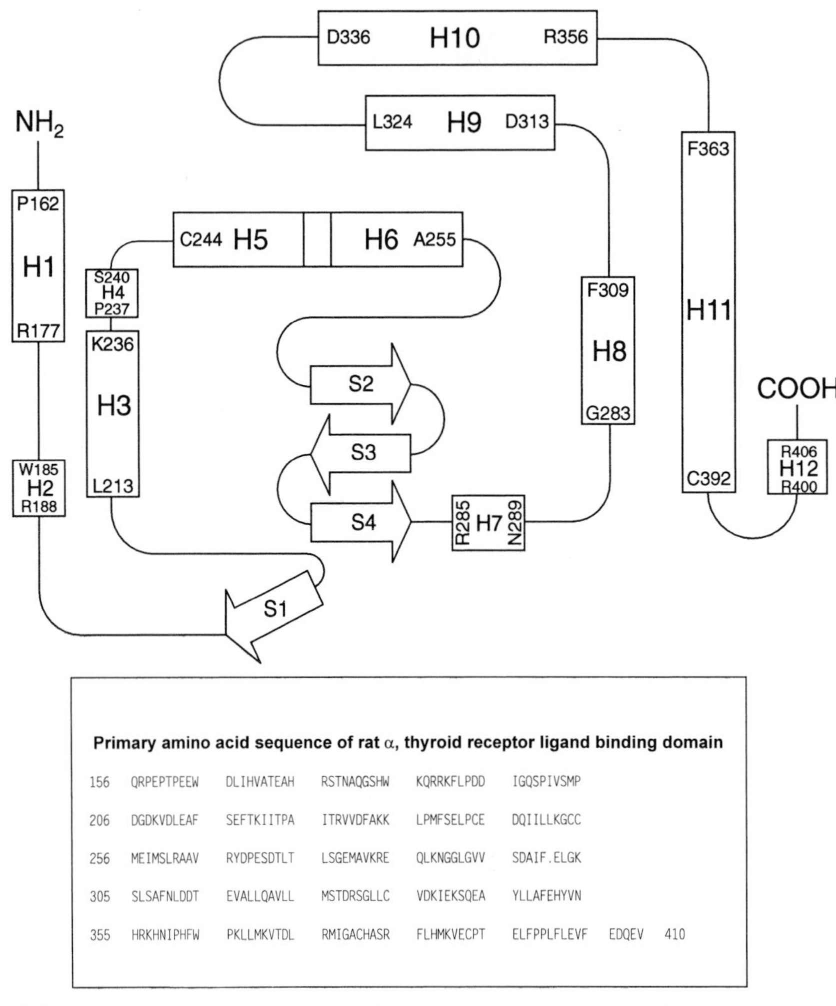

FIGURE 6-19 Schematic model of the structural domains of the thyroid receptor as determined via X-ray crystallographic techniques. The ligand binding domain of the thyroid receptor is composed of 255 amino acids which represent residues 156–410 in the intact thyroid receptor. This figure illustrates the 12 α-helices, designated H_1 to H_{12} and the 4 β-strands, designated S_1 to S_4 which comprise the secondary structure. The amino acid residues at the beginning and at the end of each α-helix are indicated using the one-letter code (see Appendix E). The inset box provides the primary amino acid sequence of the thyroid α_1 receptor ligand binding domain. The actual three-dimensional ribbon structure of the rat α_1 thyroid receptor ligand binding domain with a thyroid hormone agonist is shown on the front cover of this book; it begins with the red color for the α-helix$_1$ at the N-terminus and progresses down the colors of the visible spectrum to the royal blue color for α-helix$_{12}$ at the carboxy terminus. The hormone (purple) is depicted as a space-filling model. [This figure was modified with permission from Fig. 1 of Wagner, R. L., Apriletti, J. W., McGrath, M. E., West, B. L., Baxter, J. D., and Fletterick, R. J. (1995). A structural role for hormone in the thyroid hormone receptor. *Nature* **378,** 690–697.]

the family of thyroid receptors. This family can utilize various thyroid receptor isoforms (Figure 6-17) as well as unliganded or liganded forms of homo- and heterodimers to achieve specificity with regard to the selection of the genes to be regulated.

Table 6-6 presents a selected summary of the many genes known to be either up-regulated or down-regulated by thyroxine and the thyroid receptor. Careful inspection of the table will reveal the wide array of biological systems that are subject to the influence of thyroid hormonal status. Some selective examples include the following: intermediary metabolism enzymes (malic enzyme, phosphoenolpyruvate carboxy kinase, 6-phosphogluconate dehydrogenase), choles-

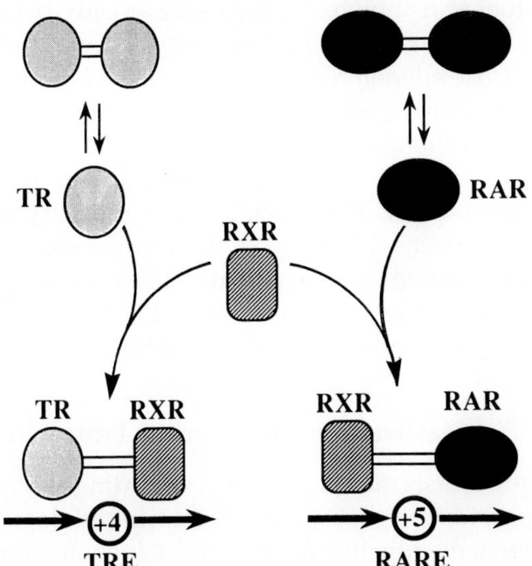

FIGURE 6-20 Model of functional forms of the thyroid receptor. Thyroid receptors (TR) can bind to DNA as monomers, homodimers, or heterodimers. The heterodimer consists of one subunit of TR + one subunit of the retinoic acid receptor, either the RXR or RAR isoform. Another level of complexity involves whether the receptor ligand-binding domain is occupied with hormone. Thus, thyroxine binding to TR and 9-*cis*-retinoic acid binding to RXR can shift the equilibrium between homodimers and heterodimers, which can provide another level of regulation of the receptor interactions with specific response elements. [Modified with permission from Jameson, J. L., and DeGroot, L. J. (1995). Mechanisms of thyroid hormone action. *In* "Endocrinology" (L. J. DeGroot *et al.*, eds.). Volume 1, Chapt. 37, pp. 583–601. W. B. Saunders, Philadelphia, PA.]

terol biosynthesis (HMG-CoA reductase), hormone systems (growth hormone, oxytocin, nerve growth factor), oxygen metabolism (cytochrome C oxidase), ion transport, and energy metabolism (Ca^{2+} ATPase, Na^+, K^+ ATPase).

VI. CLINICAL ASPECTS

A. Hormone Deficiency

Thyroid hormone deficiency may result in a wide variety of clinical and physiological disturbances in virtually every organ system. The classic disease state is myxedema; the accumulation of mucinous edema results in facial periorbital and peripheral edema involving both the hands and feet. The periorbital edema contributes to the patient's haggard appearance and includes a "droop" of the upper eyelid.

Table 6-6 lists the major causes of hypothyroidism. Hashimoto's disease is an autoimmune thyroid disease in which the subject has circulating thyroid antibodies. It is not known whether these antibodies functionally contribute to the state of hypothyroidism. The typical patient with Hashimoto's disease is a middle-aged woman with an enlarged, but firm nodular goiter who is euthyroid or hypothyroid, that is, presenting an appearance of mild myxedemas. The thyroglobulin antibodies usually are of the IgG type. They are highly species-specific. Complexes of antibody and thyroglobulin frequently are deposited in the thyroid. Hypothyroidism in elderly patients is usually considered to be an inactive phase of Hashimoto's disease and the major cause of idiopathic hypothyroidism.

Without question, in many sections of the world the prime cause of hypothyroidism is dietary insufficiency

TABLE 6-5 Examples of Genes Up-Regulated or Down-Regulated by Thyroid Hormones[a]

Positively regulated genes	Negatively regulated genes
Rat growth hormone*	Human growth hormone*
Oxytocin*	TSHα*, TSHβ* (thyroid-stimulating hormones)
Nerve growth factor (NGF)	EGF receptor*
Myosin heavy chain α*	Myosin heavy chain β
Myelin basic protein*	
Phosphoenolpyruvate carboxy kinase*	
Ca^{2+} ATPase	
Na^+,K^+ ATPase	
Cytochrome oxidase	
6-Phosphogluconate dehydrogenase	
HMG-CoA reductase	

[a] In most cases, the effects of T_3 + the thyroid receptor were assessed by determinations of steady-state mRNA levels. For those genes with an asterisk (*), the presence of an HRE or hormone response element has been identified. [This information was abstracted from Tables 37-3 and 37-4 of Jameson, J. L. and DeGroot, L. J. (1995). Mechanisms of thyroid hormone action. *in* "Endocrinology" (L. J. DeGroot *et al.*, eds.), Vol. 1, Chapt. 37, pp. 583–601. W. B. Saunders, Philadelphia, PA.]

TABLE 6-6 Major Causes of Hypothyroidism

Loss of functional thyroid tissue
Chronic autoimmune thyroiditis (Hashimoto's disease)
Idiopathic hypothyroidism
Postradioactive iodine treatment
Postthyroidectomy
Biosynthetic defects in thyroid hormonogenesis
Inherited defects
Iodine deficiency
Antithyroid agents
Hypothyrotropic hypothyroidism
TSH deficiency
TRH deficiency
Peripheral resistance to thyroid hormones

of iodine. This results in the adaptive appearance of an endemic goiter. In large part, the adaptive responses are triggered by increased TSH secretion, which attempts to stimulate the iodide-trapping mechanisms and subsequent steps in the intrathyroidal metabolism of iodine. The morphological consequence of this sequela is the development of a goiter.

Endemic cretinism is the most severe version of goiter. In countries such as Zaire and Ecuador and regions of the Himalayas, 5–12% of whole communities can be affected. Goiter is quite common, hypothyroidism frequent, but cretinism rare. The dominant characteristics of the disease are mental retardation and an associated impairment of the central nervous system.

There are a variety of hereditary diseases in which there are defects in one or more of the many steps required for the biosynthesis of thyroid hormone. These include defects in (a) iodide concentration, (b) iodide organification, (c) thyroglobulin biosynthesis, (d) iodotyrosine dehalogenase, and (e) thyroid gland insensitivity to TSH. Another cause of hypothyroidism relates to perturbation in the thyroid–pituitary axis, such that there may be a deficiency or total lack of pituitary-derived TSH or thyrotropin-releasing hormone (TRF) (see Chapter 5).

B. Hormone Excess

There are two principal types of hyperthyroidism that result in thyrotoxicosis. These are Graves' disease and toxic adenoma. Clinically, Graves' disease is characterized by a distinct bulging of the eyeball (exophthalmos). This results from an accumulation of fluid and a connective tissue ground substance. Also characteristic is goiter and pretibial myxedema. Graves' disease is also an autoimmune disease that results from the presence of an IgG globulin—a long-acting thyroid stimulator (LATS)—in plasma, which causes responses in the thyroid gland that are qualitatively similar to these of TSH, but with a more persistent effect. LATS is believed to be produced by thymus-dependent lymphocytes in individuals with a genetic predisposition. It has been suggested that the disorder is inherited by a simple autosomal recessive mechanism, with relative sex limitation to the female and reduced penetrance in homozygotes. The incidence of Graves' disease in England and probably the United States is estimated to be 23 in 100,000 overall population, with an established sex ratio of 4–5:1 for females over males.

The presence of an adenoma in the thyroid gland (toxic adenoma) can lead to the overproduction of thyroid hormones. An understanding of toxic adenoma is best found by approaching the pathology of the disease as being due to a benign hyperplasia. There is a gradual transition from inordinate sensitivity to TSH to autonomous function. Although the hyperthyroidism of toxic adenoma can be controlled by antithyroid drugs, definitive resolution is normally achieved by ablation of the neoplasm(s).

There are also a variety of other minor types of hyperthyroidism. They include (1) a pituitary tumor secreting excess TSH, (2) thyrotoxicosis factitia wherein a patient chronically ingests an excess of T_4 or T_3, usually in an effort to lose weight, (3) a functioning metastatic carcinoma, and (4) a multinodular goiter.

C. Measurement of Thyroid Function

Thyroid tests may be classified according to whether they provide insight into the anatomical, etiological, or functional dysfunction of the gland. Given the complex cellular and molecular biology of the many steps associated with the production of thyroid hormones and the obvious multitude of thyroid-related diseases, a wide battery of tests, assays, and manipulations are now available to the clinic which will assist in the determination of the problem. A detailed discussion of these procedures is beyond the scope of this text.

References

A. Books

Burrow, G. N., Oppenheimer, J. H., and Volpe, R. (1989). "Thyroid Function and Disease," pp. 1–335. W. B. Saunders, Philadelphia, PA.

Dunn, J. T., and Van der Haar, F. (1990). "A Practical Guide to the Correction of Iodine Deficiency." International Council for Control of Iodine Deficiency Disorders.

Greer, M. A., ed. (1990). "The Thyroid Gland." Raven Press, New York.

B. Review Articles

Brent, G. A. (1994). The molecular basis of thyroid hormone action. *New Engl. J. Med.* **331,** 847–853.

Chin, W. W., Carr, F. E., Burnside, J., and Darling, D. S. (1993). Thyroid hormone regulation of thyrotropin gene expression. *Recent Prog. Hormone Res.* **48,** 393–414.

Christophe, D., and Vassart, G. (1990). The thyroglobulin gene: evolutionary and regulatory issues. *Trends Endocrinol. Metab.* **1,** 351–356.

Freedman, L. P. (1992). Anatomy of the steroid receptor zinc finger region. *Endocr. Rev.* **13,** 129–145.

Gentile, F., Di Lauro, R., and Salvatore, G. (1995). Biosynthesis and secretion of thyroid hormones. *In* "Endocrinology" (L. J. DeGroot, M. Besser, H. G. Burger, J. L. Jameson, D. L. Loriaux, J. C. Marshall, W. D. Odell, J. T. Potts, Jr., and A. H. Rubenstein, eds.), 3rd ed., Vol. 1, Chapter 34, pp. 517–542. W. B. Saunders Company, Philadelphia.

Jameson, J. L., and DeGroot, L. J. (1995). Mechanisms of thyroid hormone action. *In* "Endocrinology" (L. J. DeGroot, M. Besser, H. G. Burger, J. L. Jameson, D. L. Loriaux, J. C. Marshall, W. D.

Odell, J. T. Potts, Jr., and A. H. Rubenstein eds.), 3rd ed., Vol. 1, Chapter 37, pp. 583–601. W. B. Saunders, Philadelphia.

Lazar, M. A. (1993). Thyroid hormone receptors: multiple forms, multiple possibilities. *Endocr. Rev.* **14,** 184–193.

Medeiros-Neto, G., Targovnik, H. M., and Vassart, G. (1993). Defective thyroglobulin synthesis and secretion causing goiter and hypothyroidism. *Endocr. Rev.* **14,** 165–183.

Wagner, R. L., Apriletti, J. W., McGrath, M. E., West, B. L., Baxter, J. D., and Fletterick, R. J. (1996). A structural role for hormone in the thyroid hormone receptor. *Nature* **378,** 690–697.

Weiss, R. E., and Refetoff, S. (1992). Thyroid hormone resistance. *Annu. Rev Med.* **43,** 363–375.

C. Research Papers

Bernier-Valentin, F., Kostrouch, Z., Rabilloud, R., and Rousset, B. (1991). Analysis of the thyroglobulin internalization process using *in vitro* reconstituted thyroid follicles: evidence for a coated vesicle-dependent endocytic pathway. *Endocrinology* **129,** 2194–2201.

Chazenbalk, G. D., Nagayama, Y., Russo, D., Wadsworth, H. L., and Rapoport, B. (1990). Functional analysis of the cytoplasmic domains of the human thyrotropin receptor by site-directed mutagenesis. *J. Biol. Chem.* **265,** 20970–20975.

Corvilain, B., Van Sande, J., Laurent, E., and Dumont, J. (1991). The H_2O_2-generating system modulates protein iodination and the activity of the pentose phosphate pathway in dog thyroid. *Endocrinology* **128,** 779–785.

Dali, G., Levy, O., and Carrasco, N. (1996). Cloning and characterization of the thyroid iodide transporter. *Nature* **379,** 458–460.

Datta, S., Magge, S. N., Madison, L. D., and Jameson, J. L. (1992). Thyroid hormone receptor mediates transcriptional activation and repression of different promoters *in vitro*. *Mol. Endocrinol.* **6,** 815–825.

Dumont, J. E., and Vassart, G. (1995). Thyroid regulation. *In* "Endocrinology" (L. J. DeGroot, M. Besser, H. G. Burger, J. L. Jameson, D. L. Loriaux, J. C. Marshall, W. Odell, J. T. Potts, Jr., and A. H. Rubenstein, eds.), 3rd ed., Vol. 1, Chapter 35, pp. 543–559. W. B. Saunders, Philadelphia.

Dupuy, C., Virion, A., Ohayon, R., Kaniewski, J., Deme, D., and Pommier, J. (1991). Mechanism of hydrogen peroxide formation catalyzed by NADPH oxidase in thyroid plasma membrane. *J. Biol. Chem.* **266,** 3739–3743.

Eggo, M. C., Lippes, H., and Burrow, G. N. (1992). Control of thyroid secretion: effects of stimulators of protein kinase C, thyrotropin, and calcium mobilization on secretion of iodinated compounds from sheep thyroid cells. *Endocrinology* **130,** 2274–2283.

Kliewer, S. A., Umesono, K., Mangelsdorf, D. J., and Evans, R. M. (1992). Retinoid X receptor interacts with nuclear receptors in retinoic acid, thyroid hormone and vitamin D_3 signalling. *Nature* **355,** 446–449.

Loosfelt, H., Pichon, C., Jolivet, A., Misrahi, M., Caillou, B., Jamous, M., Vannier, B., and Milgrom, E. (1992). Two-subunit structure of the human thyrotropin receptor. *Proc. Natl Acad. Sci. USA* **89,** 3765–3769.

Magnusson, R. P., Chazenbalk, G. D., Gestautas, J., Seto, P., Filetti, S., DeGroot, L. J., and Rapoport, B. (1987). Molecular cloning of the complementary deoxyribonucleic acid for human thyroid peroxidase. *Mol. Endocrinol.* **1,** 856–861.

McGrath, M. E., Wagner, R. L., Apriletti, J. W., West, B. L., Ramalingam, V., Baxter, J. D., and Fletterick, R. J. (1994). Preliminary crystallographic studies of the ligand-binding domain of the thyroid hormone receptor complexed with triiodothyronine. *J. Mol. Biol.* **237,** 236–239.

Musti, A. M., Avvedimento, E. V., Polistana, C., Ursini, Y., Obici, S., Nitsch, L., Cocozza, S., and Di Lauro, R. (1986). The complete structure of the rat thyroglobulin gene. *Proc. Natl. Acad. Sci. USA* **83,** 323–327.

Palumbo, G., Gentile, F., Condorelli, G. L., and Salvatore, G. (1990). The earliest site of iodination of thyroglobulin is residue number 5. *J. Biol. Chem.* **265,** 8887–8892.

Parmentier, M., Libert, F., Maenhaut, C., Lefort, A., Gerard, C., Perret, J., Van Sande, J., Dumont, J. E., and Vassart, G. (1989). Molecular cloning of the thyrotropin receptor. *Science* **246,** 1620–1622.

Raby, C., Lagorce, J.-F., Jambut-Absil, A.-C., Buxeraud, J., and Catanzano, G. (1990). The mechanism of action of synthetic antithyroid drugs: iodine complexation during oxidation of iodide. *Endocrinology* **126,** 1683–1691.

Sakurai, A., Miyamoto, T., Refetoff, S., and DeGroot, L. J. (1990). Dominant negative transcriptional regulation by a mutant thyroid hormone receptor-beta in a family with generalized resistance to thyroid hormone. *Mol. Endocrinol.* **4,** 1988–1994.

Sugawara, A., Yen, P. M., Apriletti, J. W., Ribeiro, R. C. J., Sacks, D. B., Baxter, J. D., and Chin, W. W. (1994). Phosphorylation selectively increases triiodothyronine receptor homodimer binding to DNA. *J. Biol. Chem.* **269,** 433–437.

Yen, P. M., Darling, D. S., Carter, R. L., Forgione, M., Umeda, P. K., and Chin, W. W. (1992). Triiodothyronine (T_3) decreases binding to DNA by T_3-receptor homodimers but not receptor-auxiliary protein heterodimers. *J. Biol. Chem.* **267,** 3565–3568.

Zhang, X. K., Hoffmann, B., Tran, P. B., Graupner, G., and Pfahl, M. (1992). Retinoid X receptor is an auxiliary protein for thyroid hormone and retinoic acid receptors. *Nature* **355,** 441–446.

CHAPTER

7

Pancreatic Hormones: Insulin and Glucagon

I. INTRODUCTION

A. Background Information

A feature essential for life of higher vertebrates is their ability to maintain a relatively constant blood glucose concentration. In the higher animals, glucose is essential as an energy source for all cells. Although some cells can utilize alternate "fuel metabolites," such as amino acids or fatty acids, the brain and its neurons are dependent upon a continuous supply of blood-delivered glucose. Superimposed on the requirement for the maintenance of a constant blood glucose level are the perturbations in blood glucose that may occur naturally as a consequence of ongoing physiological and metabolic events. These can include (1) the intestinal absorption and concomitant systemic transport to storage depots (liver, muscle, and/or adipose tissue) of foodstuffs (e.g., carbohydrates, proteins, fat); (2) muscular activity; (3) thermogenesis and response to environmental extremes of heat and cold; (4) starvation; (5) pregnancy/lactation; and (6) disease or injury states. The endocrine gland largely responsible for the maintenance of blood glucose levels and the proper cellular uptake and exchange of "fuel metabolites" is the pancreas. The principal hormones secreted by the pancreas are insulin and glucagon. In addition, the pancreas secretes a number of other peptide hormones (see Table 7-6).

Insulin is a pluripotent hormone in that it has a wide sphere of influence; directly or indirectly it affects virtually every organ and tissue in the body. The main function of insulin is to stimulate anabolic reactions for carbohydrates, proteins, and fats, all of which will have the metabolic consequence of producing a lowered blood glucose level. Glucagon can be thought of as an indirect antagonist of insulin. Glucagon stimulates catabolic reactions that ultimately lead to an elevation in blood glucose levels. Thus, the pancreas is continuously adjusting the relative amounts of glucagon and insulin secreted, in response to the continuous perturbations of blood glucose and other fuel metabolites occurring as a consequence of changes in anabolism and catabolism in the various tissues.

B. Regulation of Blood Glucose

The blood concentration of glucose normally lies within the range of 80–110 mg/100 ml (4.4–6.1 m*M*). Reduction in blood glucose levels below 45–55 mg/100 ml for a continued interval of time will lead to an impairment of brain function, tremors, and convulsions due to activation of the sympathetic nervous system and, ultimately, death. Conversely, prolonged hyperglycemia (a relative lack of insulin) leads to a devastating wasting of metabolic energy, osmotic diuresis, and metabolic acidosis. Because glucose is a "small molecule," all blood glucose is completely filtered by the glomerulus of the kidney (see Figure 15-3). Under normal physiological circumstances, virtually all of this filtered glucose is reabsorbed; however, since this reabsorptive process is saturable and can only accommodate a finite throughput of glucose, in instances of excessive hyperglycemia, a significant fraction of the filtered blood glucose will be lost in the urine. If there is not a balanced increase in the dietary intake of glucose, there will be, of necessity, a compensatory breakdown or catabolism of the storage forms of glucose, namely, glycogen; over an extended interval of time this can result in a significant loss of "stored" metabolic energy.

Related to prolonged hyperglycemia and its concomitant glucosuria is an associated increase in the loss of body water from the general circulatory system; this is due to the passive loss of water across the kidney tubule, which occurs as a consequence of the high osmotic pressure of the sugar-rich urine. In extreme cases of chronic glucosuria there will be cellular dehydration, leading to a loss of potassium and a reduction in blood volume, which can ultimately lead to a marked lowering of systemic blood pressure (see Chapter 15). In a normal (70 kg) man, the 24-hr urine volume ranges from 600 to 2500 ml, with a glucose concentration of 10–20 mg/100 ml of urine. In a typical untreated diabetic (without renal complications), the 24-hr urine volume may exceed 3000–3500 ml, with a glucose concentration of 500–5000 mg/100 ml of urine.

However, the adverse consequences of a relative lack of insulin (hyperglycemia) are not restricted to a derangement of carbohydrate metabolism. In the absence of insulin, triglycerides and fatty acids will be mobilized from adipose tissue and amino acids from muscle tissue (discussed in detail later). These substances then proceed to the liver, where the fatty acids and branched chain amino acids are converted into ketone bodies. The ketone bodies include β-hydroxybutyrate, acetoacetate, and acetone. Eventually, this results in progressive ketonemia, and when the renal threshold for acetoacetate and β-hydroxybutyrate is exceeded, both substances will appear in the urine. Because they are excreted as the sodium salts, this has the consequence, when extended over a period of time, of depleting the body of base. This in turn elevates the $[H_2CO_3]/[NaHCO_3]$ ratio and leads to the condition of metabolic acidosis (see footnote 1). Ultimately this will trigger a rapid and deep respiratory rate, which is diagnostic of diabetic acidosis. Tabulated in Table 7-1 is a resume of the changes in several blood and urine components that may result as a consequence of altered blood glucose levels.[1]

In addition to the dominant partnership of insulin and glucagon in maintaining glucose homeostasis, there is an extensive contribution of other physiological factors and hormones to the regulation of blood glucose. These are summarized in Table 7-2.

C. Nutritional and Metabolic Interrelationships

1. *Introduction*

Any detailed understanding of the integrated actions of glucagon and insulin to effect blood glucose homeostasis cannot be achieved by limiting the assessment of their actions to those on carbohydrate metabolism alone. Due to the ready metabolic interchanges that occur between carbohydrate, protein, and fat constituents, it is essential to have a clear understanding of the intermediary metabolism of all of these substances. In addition, some appreciation of the principles of dietary nutrition is required because many key enzymes of intermediary metabolism of higher animals are adaptively regulated to reflect the current dietary intake of carbohydrate, protein, and lipid. The general design is such that the ingested components are diverted to storage sites during periods of feeding and are later reutilized by the metabolic processes of glycogenolysis, gluconeogenesis, and ketogenesis during intervals of food deprivation.

[1] It should be recalled that both acetoacetic acid and β-hydroxybutyric acid have a primary carboxyl group with a pK $\approx$ 4.2; thus, at the physiological pH of 7.2–7.4, both acids are ionized so as to generate acetoacetate, β-hydroxybutyrate, *and* protons (H^+). These protons will be neutralized by the physiological base bicarbonate (HCO_3^-). Over time, if large quantities of acetoacetate and β-hydroxybutyrate are produced (as in diabetes) and are excreted, there is a significant reduction in the available HCO_3^-, thus elevating the ratio of $[H_2CO_3]/[HCO_3^-]$, which ultimately leads to the condition of metabolic acidosis.

TABLE 7-1 Effects of Altered Blood Glucose Levels on Several Constituents of the Blood and Urine

Parameter	"Normal" state	Chronic elevated blood glucose level (diabetes)	Chronic lowered blood glucose level (starvation)
Blood insulin (μU/ml)	50–100	8–16[a]	5–10
Glucagon (pg/ml)	<50	200–300	150
Glucose (mg/100 ml)	80–110	>130	50–60
Free fatty acids (mM)	0.4–0.7	2.0	4.0
Ketones[b] (mM)	1–2	20	10–20
Urine			
Volume (ml/24 hr)	600–2500	3000–3500	600–1200
Glucose (mg/100 ml)	6–8	500–5000	5–10

[a] For type II or insulin-dependent diabetes.
[b] The ketones consist of β-hydroxybutyrate, acetocetate, and acetone.

2. *Substrate Stores*

Table 7-3 lists for a normal (70 kg) man the magnitudes of body pools of carbohydrate, fat, and protein and their theoretical caloric yields (in kilocalories) if they were completely metabolized to CO_2 and water.

Virtually the complete body pool of protein is in a continuous state of flux. In the steady state there is a balance between biosynthesis and degradation. The amino acid efflux from human muscle is 0.5–1.0 g/kg/day. Physiological amounts of insulin are known to markedly reduce amino acid efflux from muscle. The released amino acids are further catabolized principally by the liver. Although a normal 70-kg man has a total of 10 kg of protein, only ~60% of this is theoretically metabolically mobilizable; the remaining 40% represents structural proteins. In reality, no more than 2 kg of body protein can be metabolically mobilized before there is marked muscular hypertrophy and myopathy (weakness). In cases of extreme starvation, death ensues not from hypoglycemia but from loss of respiratory muscle function, which usually leads to terminal pneumonia.

Fat is the body's principal form of stored energy; 77% of the calories derived from the substrate stores tabulated in Table 7-3 comes from metabolic oxidation of triglycerides and free fatty acids. Fat or adipose tissue represents the most efficient body storage form of energy. This results not only from the higher caloric yield per gram mass of adipose tissue (see Table 7-4) but also from the fact that the biological "packing" of

TABLE 7-2 Factors Contributing to Glucose Homeostasis

Factors resulting in a reduction in blood glucose level	Factors resulting in an elevation in blood glucose level
Insulin	Glucagon actions in the liver to stimulate glycogenolysis
	Epinephrine stimulation of glycogenolysis
Glucose uptake by peripheral tissues	Cortisol (gluconeogenesis)
Glucosuria	Insulin antagonists; growth hormone; cortisol
Exercise	Dietary intake of carbohydrates and proteins; mobilization from storage sites (glycogenolysis)
Stimulation of glucagon catabolism	Stimulation of insulin catabolism

TABLE 7-3 Body Pools of Carbohydrate, Fat, and Protein in a Normal (70 kg) Man[a]

Class	Amount (g)	Caloric yield (kcal)
Carbohydrate		
Muscle glycogen	350	1450
Liver glycogen	85	350
Extracellular glucose	20	80
Fat		
Muscle triglycerides	300	2800
Adipose triglycerides	15,000	140,000
Plasma free fatty acids	0.4	4
Plasma triglycerides	4.0	40
Protein		
Muscle	10,000	41,000
Total		185,600

[a] Adapted with the permission of the authors and publisher from Wahren, J. (1979). Metabolic adaptation to physical exercise in man. *In* "Endocrinology" (L. J. DeGroot *et al.*, eds.), Vol. 3., p. 1912. Grune & Stratton, New York.

TABLE 7-4 Caloric Yield of Food Substances

	Respiratory quotient[a]	Energy yield (kcal/g)[b]
Carbohydrates	1.00	4.0
Proteins	0.80	4.1–4.2
Fat (triglycerides)	0.70	9.0

[a] Respiratory quotient is defined as the molar ratio of CO_2 released to O_2 consumed during the complete oxidation of a food substance to CO_2 and H_2O.

[b] The energy yield is expressed as the derived kilocalories that are generated from completely oxidizing 1.00 g of the indicated food component to CO_2 and H_2O. The energy yield is identical irrespective of whether the oxidation occurs in a bomb calorimeter or in a living organism.

triglycerides is more condensed than that of protein or glycogen due to the absence of water molecules in the triglyceride structures. It is the fat depots that allow humans to tolerate prolonged intervals of dietary calorie deprivation. The energy stored in the body fat depots is sufficient to permit the average individual to survive 2–3 months of total food deprivation. Lipolysis or hydrolysis of adipose tissue triglycerides leads to the release of glycerol and free fatty acids into the bloodstream. During intervals of exercise, the sympathetic neurotransmitter, norepinephrine, is a potent stimulator of triglyceride mobilization. The free fatty acids are then further catabolized by both muscle and liver tissue.

Approximately 80% of the storage form of carbohydrate, namely, glycogen, is found in muscle; the concentration is 9–16 g/kg of wet muscle. The muscle glycogen levels are depleted rapidly during vigorous exercise, but reduced only slowly during prolonged fasting. Because muscle, in contrast to the liver, does not contain the enzyme glucose 6-phosphatase, and because normally there is no hydrolysis of the phosphate esters of the various glycolytic pathway intermediates, it is not possible for significant amounts of glycogen (glucose) to leave the muscle. Thus, muscle glycogen can only serve as a metabolic fuel in the cell in which it is localized.

The liver glycogen pool plays a key role in facilitating the metabolic adjustment by the body to varying energy requirements. The concentration of hepatic glycogen is 50 g/kg of wet tissue; it represents 20% of the total body glycogen pool. Liver glycogen may be depleted gradually during periods of fasting (within 10–12 hr of fasting, the hepatic glycogen is mobilized at a rate of 50 mg/min/kg of liver) or very rapidly in response to exercise. During 40 min of severe exercise, the liver glycogen pool may be depleted by 18–20 g; that is, glycogen is mobilized at a rate of 330 mg/min/kg wet weight. Replenishment of the hepatic levels of glucose occurs relatively slowly; 12–36 hr are required, depending on the diet and extent of physical activity.

The body pool of extracellular glucose is quite modest by comparison with the muscle and liver stores; only ~20 g of free glucose is present in the extracellular water and the intracellular pool of the liver. The principal function of the blood pool of glucose is to provide a continuous supply of glucose to all of the glucose-dependent tissues. In the basal state the brain is the chief glucose-consuming tissue. During a 24-hr interval the brain requires 125 g of glucose.

The exclusive source of replenishment of the blood glucose pool is the liver. Under usual circumstances, 70% of the hepatic output of glucose is derived from liver glycogenolysis and 30% from gluconeogenesis. During intervals of prolonged exercise (40 min), ~50% of the oxidative metabolism of skeletal muscle is derived from glucose uptake from blood.

The ketones or "ketone bodies" accumulate in the blood under normal circumstances, where they function to ensure the survival of the organism, and under pathological conditions, where they can cause coma and death. The ketone bodies are produced almost exclusively by the liver from fatty acids and other carbon fragments derived from amino acids. Acetone is removed from the body via the lungs. The metabolic role of the ketones varies depending upon the nutritional state. Their principal function is to serve under conditions of fasting as a primary substrate for energy metabolism by muscle, heart, and particularly the brain. Although muscle and heart have the capacity for cellular uptake and oxidation of free fatty acids (derived from lipolysis of adipose tissue), the brain does not have this capacity. Thus, under conditions of fasting when the liver glycogen and eventually blood glucose pools become depleted, the physiological significance of the ketone bodies assumes a far greater importance. Under conditions of adequate feeding, the blood levels of the ketones are relatively low (see Table 7-1).

Figure 7-1 summarizes the balance of glucose production and utilization in humans that ensues over a 24-hr interval after dietary intake and in the absence of excessive exercise. Maintenance of a constant blood glucose level is governed by the liver and the regulatory actions of the pancreatic hormones. In the absence of hepatic glucose production, the blood level of glucose would fall by 50% in 40–60 min.

3. Dietary Nutritional Requirements

The daily caloric requirements can be met by adequate dietary intake of protein, carbohydrate, or fat. The current recommendation is that 15, 30, and 55%

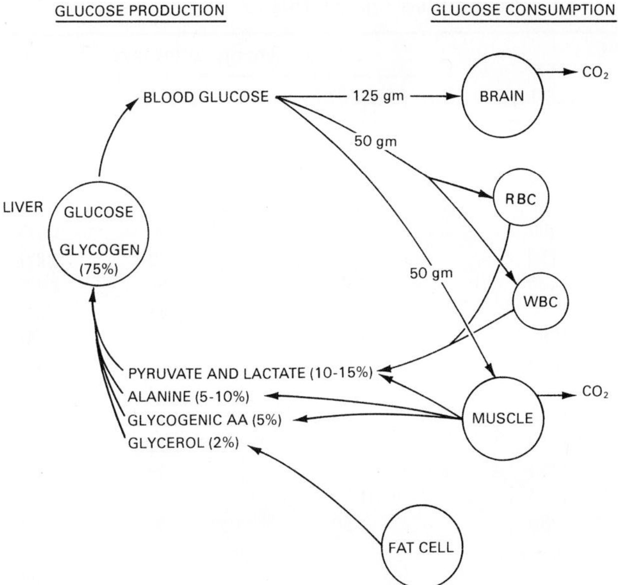

FIGURE 7-1 Schematic diagram describing the utilization and production of glucose by a normal man in a postabsorptive state (i.e., not actively ingesting a meal). The values for glucose uptake represent the amounts in grams consumed per day; in the case of muscle this refers to the resting state. Glucose output from the liver derives from glycogenolysis; thus, approximately 25% of the liver glycogen is released as glucose and 75% is retained by the liver. As starvation extends beyond 12 hr, glycogen stores are depleted and the contribution from gluconeogenesis increases. Abbreviations: RBC, red blood cell; WBC, white blood cell; AA, amino acid. [Reproduced with permission from the author and publisher of Felig, P. (1979). Starvation. *In* "Endocrinology" (L. J. DeGroot *et al.*, eds.), Vol. 3, pp. 1927–1940. Grune & Stratton, New York.]

of the daily energy intake should be derived from protein, fat, and carbohydrate, respectively. Each food substance has its own characteristic energy yield (see Table 7-4), which reflects the relative extent of reduction of the carbon atoms of that food component.

The unit of metabolic energy historically utilized by nutritionists is the kilocalorie (1000 cal); 1.0 kcal represents the energy required to raise the temperature of 1000 g of water from 15.5 to 16.5°C. However, increasingly nutritionists are expressing energy in terms of joules; 1.0 kcal is equivalent to 4.185 kJ.

In addition to kilocalories or kilojoules, the diet must supply, on a regular basis, adequate amounts of the various essential dietary constituents; these include water and fat-soluble vitamins, the essential unsaturated fatty acids, the essential amino acids, and both bulk and trace minerals.

The basal metabolic rate (BMR) is defined as the energy expenditure required to maintain those cellular processes that are essential for the continuing activities of the organism; these include the metabolic activity of the heart, lungs, kidneys, and other vital organs. The BMR of an adult 70-kg male is 35–40 kcal/m^2 body surface/hr. The comparable value for a woman of the same age is 29–36 kcal/m^2 body surface.

Not all of the caloric content of dietary constituents is available to support the daily kilocalorie requirements. When food is eaten there is a rise in total body oxygen consumption; this is known as the dietary-induced thermogenesis (DIT) of foodstuffs. Although a detailed molecular explanation of the process of DIT has not been given, in essence it represents energy that is spent in the metabolic processing of food, which is therefore lost or not available to support other bodily energy requirements. The approximate DIT is ~30% for protein, 0% for carbohydrate, and 4% for lipid, each in terms of the theoretical energy value of the food component ingested.

The dietary energy requirements necessary for normal bodily functions vary with age and sex; the recommended values are tabulated in Table 7-5. Of course, the energy intake must be increased appropriately during intervals of prolonged exercise.

II. ANATOMICAL, MORPHOLOGICAL, AND PHYSIOLOGICAL RELATIONSHIPS

A. Introduction

The hepatopancreatic complex, along with the gall bladder, is responsible for the integrated digestion and subsequent processing of most dietary nutrients. The complex series of steps and the components of the digestive system are summarized in Table 8-3. The pancreas is both an exocrine and an endocrine gland. The exocrine pancreas biosynthesizes and secretes the major digestive enzymes, for example, the proteases (chymotrypsin and trypsin), amylase, and lipase, as well as bicarbonate, while the endocrine tissue of the pancreas produces the peptide hormones insulin, glucagon, pancreatic polypeptide, somatostatin, and small amounts of other peptide hormones (see Table 7-6).

Figure 7-2 illustrates the gross anatomical features of the human pancreas. Anatomically, the pancreas has distinct dorsal and ventral lobes. The pancreas can be divided into lobules that contain the exocrine acinar glands and the endocrine islets of Langerhans. The islet cells are separated from the surrounding acinar cells by a thin layer of reticular tissue. The endocrine protein of the pancreas is only 1–2% of the weight of the gland. Figure 7-3 schematically illustrates the relationship between the number and distribution of the insulin- and the glucagon-secreting cells in typical rat and human pancreatic islets.

TABLE 7-5 Median Heights and Weights and Recommended Energy Intake[a]

Category	Age (years) or condition	Weight (kg)	Height (cm)	REE[b] (kcal/day)	Average energy allowance: Multiples of REE (kcal)	kcal/kg	kcal/day
Infants	0.0–0.5	6	60	320		108	650
	0.5–1.0	9	71	500		98	850
Children	1–3	13	90	740		102	1300
	4–6	20	112	950		90	1800
	7–10	28	132	1130		70	2000
Males	11–14	45	157	1440	1.70	55	2500
	15–18	66	176	1760	1.67	45	3000
	19–24	72	177	1780	1.67	40	2900
	25–50	79	176	1800	1.60	37	2900
	51+	77	173	1530	1.50	30	2300
Females	11–14	46	157	1310	1.67	47	2200
	15–18	55	163	1370	1.60	40	2200
	19–24	58	164	1350	1.60	38	2200
	25–50	63	163	1380	1.55	36	2200
	51+	65	160	1280	1.50	30	1900
Pregnant	1st trimester						+0
	2nd trimester						+300
	3rd trimester						+300
Lactating	1st 6 months						+500
	2nd 6 months						+500

[a] This table represents recommendations on energy intake that were reported in Recommended Dietary Allowances (1989), prepared by the Food and Nutrition Board of the Commission on Life Sciences, National Research Council, 10th ed., National Academy Press, Washington, DC.

[b] REE is the resting energy expenditure in kilocalories/day; the values were derived from basal metabolic rate (BMR) measurements.

B. Development and Embryologic Origins

In invertebrates, the functional cells of the exocrine pancreas and liver are combined and there is no clear-cut presence of the endocrine pancreas cells. In vertebrates, there is increasing structural organization and differentiation of the exocrine and endocrine cells in relation to the liver as one proceeds from fish (endocrine pancreas is a separate organ; exocrine pancreas is partially incorporated into the liver) up through amphibia (endocrine cells scattered in exocrine tissue or islets) to birds and mammals where the pancreas is a separate organ (endocrine pancreas found as vascularized islets embedded in the exocrine tissue).

In mammals and birds, the adult pancreas develops embryologically from two diverticula of the duode-

TABLE 7-6 Hormones and Proteins Secreted by the Pancreas

Pancreatic cell of origin	Hormone/protein	No. of amino acid residues (or kDa)	Function
α (A)	Glucagon	29	Elevation of blood glucose
β[c] (B)	Insulin	51	Reduction of blood glucose
	Islet amyloid polypeptide (IAPP) or "amylin" (see footnote *a*)	37	Inhibit glucose uptake by muscle; insulin antagonistic effect
	Chromogranin A (see footone *b*)	49 kDa	Precursor of (i) pancreastatin (4.9 kDa), which inhibits insulin secretion and (ii) β-granin (24 kDa)
	Parathyroid-related protein (PTHrP)	141	Paracrine factor (see Chapter 9)
δ (D)	Somatostatin	14	Inhibits secretion of insulin, glucagon, and PP.
(F)	Pancreatic polypeptide (PP)	36	Stimulates secretion of pepsin and HCl by the stomach

[a] IAPP is structurally homologous to calcitonin gene-related peptide (CGRP); see Figure 9-4.

[b] The chromogranins A, B, and C are a family of closely related acidic proteins that are produced by many neuroendocrine cells. Chromogranins A are produced by pancreatic α, β, and F cells of the pancreatic islet.

[c] The B cells also secrete small quantities of procathepsin B, carboxypeptidase H, endopeptidases I and II, and peptidylglycan α-amidating monooxygenase.

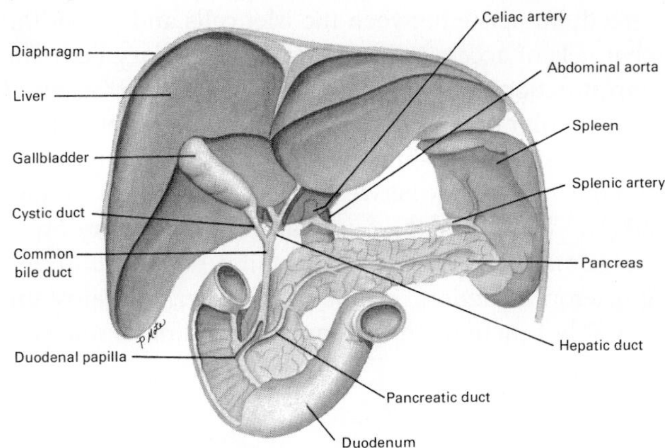

FIGURE 7-2 Gross anatomical features of the pancreas, liver, and biliary system.

num. It has been generally accepted that the endocrine cells of the pancreas develop from pancreatic ducts originally of endodermal origin. In the developing rat embryo, insulin biosynthesis can be detected by day 16.

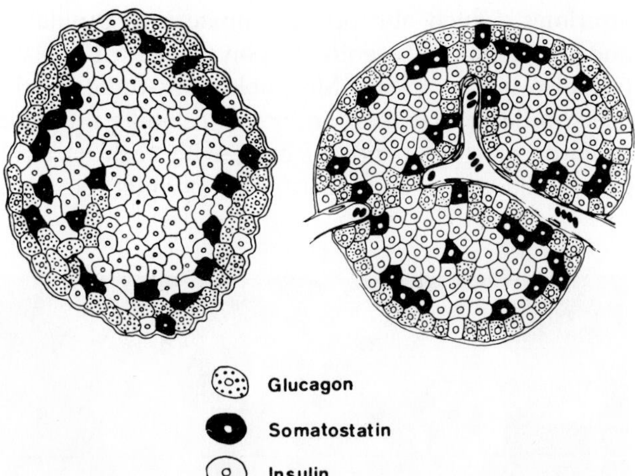

FIGURE 7-3 Schematic diagram of a pancreatic islet. (Left) Schematic representation of the number and distribution of insulin-, glucagon-, and somatostatin-containing cells in the normal rat islet. Note the characteristic position of most glucagon- and somatostatin-containing cells at the periphery of the islet, surrounding the centrally located insulin-containing cells. Cell types in the islet for which a characteristic function and/or morphology are not defined are intentionally omitted. (Right) Schematic representation of the number and distribution of insulin-, glucagon-, and somatostatin-containing cells in the normal human islet. Large vascular channels penetrate the islet and are surrounded by glucagon- and somatostatin-containing cells. This pattern divides the total islet mass into small subunits, each of which contains a center formed mainly of insulin-containing cells and surrounded by glucagon- and somatostatin-containing cells. Cell types for which definite functions and/or morphologies have not yet been determined are intentionally omitted. [Reproduced with permission from Orci, L. (1977). *In* "Insulin and Metabolism" (J. S. Bajaj, ed.), p. 262. Excerpta Medica, Amsterdam.]

C. Ultrastructure of Pancreatic Islets

In both humans and rats, pancreatic islets are composed of at least three major cell types: the α or A (glucagon-secreting) cells, the β or B (insulin-secreting) cells, and the δ or D cells. By histological staining with neutral red it has been estimated that there are 13,500 or 890,000 islets, respectively, in the rat or human pancreas. The islets in the rat pancreas range in diameter from 50 to 400 μm. The distribution of cells in a typical rat islet is 15–18% α, 75–80% β, and 2–10% δ. The islets are surrounded by a basement membrane that encloses all three cell types.

The pancreatic β-cell also produces and secretes a neuropeptide-like molecule known as islet amyloid polypeptide (IAPP). IAPP was first described as a major protein component of the amyloid deposits that are found in the islets of elderly type I diabetics. The physiological function of IAPP is not yet known.

The pancreas of a number of species also secretes a substance known as pancreatic polypeptide (PP). Avian PP is a potent gastric secretagogue; it stimulates the release of both pepsin and HCl. In the chicken pancreas, immunofluorescent studies have indicated that PP is present in cells scattered throughout the exocrine pancreas. A peptide homologous to avian PP has been isolated from the human pancreas; immunofluorescent studies have indicated human PP present in cells that are localized on the periphery of the human islet. Low concentrations of cells that secrete gastrin are also dispersed throughout the endocrine portions of the pancreas.

As enumerated in Table 7-6, there are a number of other hormones and proteins that are produced and secreted by the endocrine pancreas. In most instances their precise physiological role remains to be clarified.

The α-cells can be differentiated from other islet cells (see Figure 7-4) on the basis of the ultrastructural appearance of their secretory granules. Usually the center of the granule is exceedingly electron-dense. The β-cells can be specifically identified histochemically by staining with aldehyde fuchsin or aldehyde thionin, which apparently reacts with sulfhydryl groups present in insulin. A remarkable feature of most β-cells is the visible presence of a crystalline matrix or array; in the β-cells of the human, bat, or dog, there is a repeating periodicity of ~50 Å. This may be related to the property of insulin to form dimers and hexamers (discussed later).

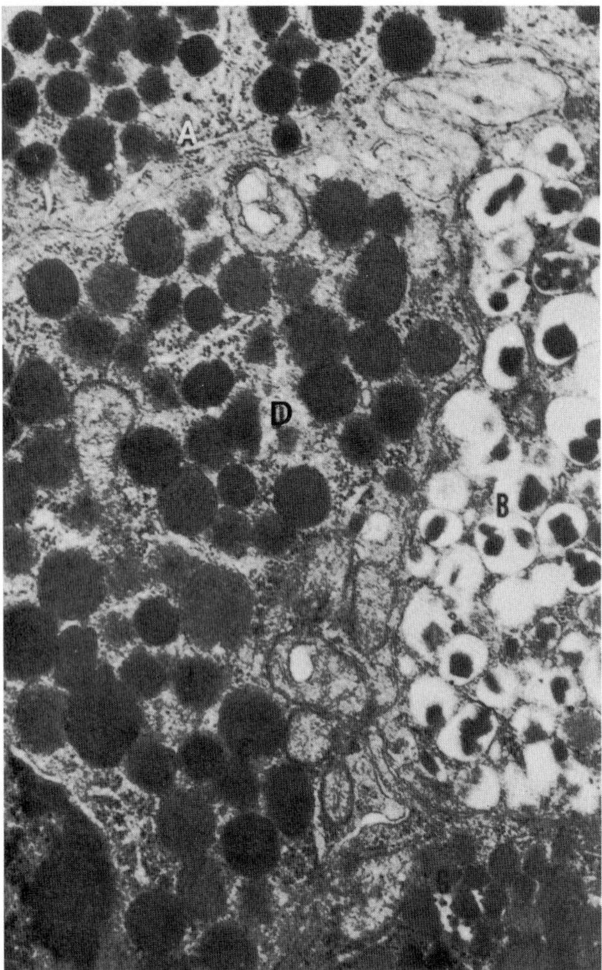

FIGURE 7-4 Electron micrograph of normal α (A), β (B), and δ (D) cells of a human pancreatic islet (×32,000). [Reproduced with permission from Lacy, P. E., and Greider, M. H. (1979). Anatomy and ultrastructural organization of pancreatic islets. *In* "Endocrinology" (L. J. DeGroot *et al.*, eds.), Vol. 2, p. 909. Grune & Stratton, New York.]

It has been proposed by R. Unger, L. Orci and others that the α-, β-, and possibly δ-cells function together as a hormonal secretory unit; this unit releases suitable proportions of glucagon and insulin necessary to regulate the minute-to-minute blood glucose levels and also modulates metabolism either toward anabolism or catabolism in accordance with physiological needs. The integrated secretion of glucagon and insulin by the α- and β-cells may well have a functional basis due to the close anatomic intimacy of these two cells. Orci has identified, by electron microscopic, freeze–fracture methods, a preponderance of both gap junctions and tight junctions between adjacent α- and β-cells. It is believed that the tight junctions may form dynamically to trap insulin released extracellularly from secretory granules, thus effectively compartmentalizing the intercellular space between the islet cells and providing channels of access to the pancreatic capillary bed. The gap junctions also could well serve as pathways of intercellular communication between adjacent cells. In other systems, gap junctions have been shown to permit the interchange between cells of molecules of under 500 Da. The role of the δ-cells and their secretory product somatostatin, which can inhibit both insulin and glucagon secretion, is not yet known. Somatostatin possibly functions in the pancreas as a paracrine substance.

D. Vascularization and Innervation of Pancreatic Islets

Figure 7-3 illustrates the gross anatomical features of the pancreas in relation to the biliary system and liver.

The arterial supply of the pancreas arises from the splenic, hepatic, and mesenteric arteries, and venous drainage is into the splenic and mesenteric veins. As shown in Figure 7-5, the individual islets of the pancreas are vascularized by an extensive labyrinth of capillaries. Each islet is normally vascularized by 1–3 arterioles, which abruptly terminate into capillaries and 1-6 veinules, depending upon the size of the islet (see also Figure 15-6B). Morphological studies of the islets indicate that their endocrine cells are arranged in cords or short bands of cells, with each islet cell being adjacent to a capillary. This permits the rapid

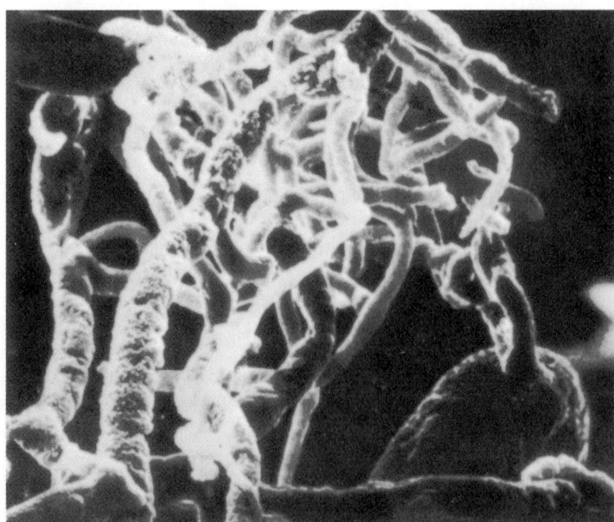

FIGURE 7-5 Scanning electron micrograph of the capillary network in a pancreatic islet. [Reproduced with permission from Lacy, P. E., and Greider, M. H. (1979). Anatomy and ultrastructural organization of pancreatic islets. *In* "Endocrinology" (L. J. DeGroot *et al.*, eds.), Vol. 3, p. 911. Grune & Stratton, New York.]

transfer of the secreted hormones into the general vascular system.

The capillaries present in the pancreatic islets comprise endothelial cells that are fenestrated to permit the rapid uptake of the peptide hormones (see Figure 2-34 for an example of capillary wall fenestration). It has been estimated from an infusion of horseradish peroxidase (40 kDa) that the fenestrated islet capillary is 4–7 times more permeable than a nonfenestrated capillary. The hormonal products secreted by islet cells into the surrounding extracellular fluid must traverse the basement membrane of the endothelium before entering the bloodstream.

Figure 7-6 schematically illustrates the innervation of the pancreatic islets by both parasympathetic cholinergic neurons and sympathetic adrenergic neurons. There are no specialized morphological membrane structures in the islet cells characteristic of neuromuscular synapses; the nerve terminals end abruptly within the islet, underneath the basal lamina that surrounds the islet cell.

Both adrenergic and cholinergic nerve fibers are present in both the acinar and the islet regions of the pancreas. Stimulation of the parasympathetic nervous system leads to insulin secretion and inhibition of glucagon secretion, irrespective of whether the stimuli occur at the lateral hypothalamic nuclei, the motor nuclei of the vagus, or the mixed pancreatic nerves. Stimulation of the sympathetic nervous system or application of epinephrine likewise can stimulate glucagon production and inhibit insulin secretion. The hypothalamus appears to play the major integrating role in the balance between sympathetic and parasympathetic regulation of the islet.

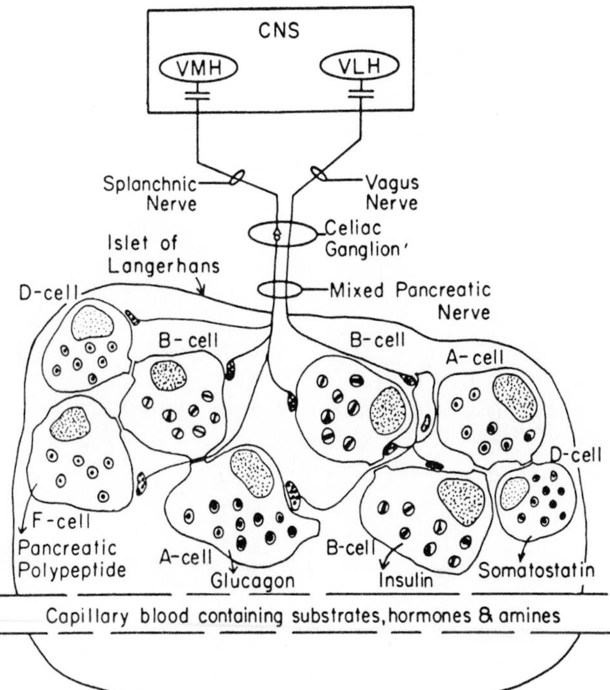

FIGURE 7-6 Schematic representation of the autonomic innervation of the pancreatic islets. Abbreviations: VMH, ventromedial hypothalamus; VLH, ventrolateral hypothalamus; CNS, central nervous system [Reproduced with permission from Woods, S. C. *et al.* (1981). The role of the nervous system in metabolic regulation and its effect on diabetes and obesity. *In* "Handbook of Diabetes Mellitus" (M. Brownlee, ed.), p. 213. Garland Publishing Co., New York.]

III. CHEMISTRY

Table 7-6 lists the peptide hormones and other proteins produced by the pancreas. The chemistry of each of the major hormones will be discussed separately.

A. Insulin

The history of insulin occupies a unique place in the evolution of our understanding of peptide hormones. Table 7-7 summarizes the notable series of "firsts" associated with the development of the chemistry and physiology of this important hormone. Due to its relatively modest size, insulin was an ideal molecule for the pioneering developments of peptide chemistry, amino acid sequencing techniques, and recombinant DNA cloning of the insulin genome.

Insulin is a molecule of molecular weight 5700–6100, which falls on the borderline between large polypeptide and small protein. It is composed of two separate peptide chains, designated the A chain and the B chain. These two chains are joined together by two disulfide

TABLE 7-7 History of Insulin

Year	Event	Investigator(s)
1869	Discovery of pancreatic islets	P. Langerhans
1921	Discovery and isolation of insulin	F. G. Banting[a] and C. Best
1955	Determination of primary amino acid sequence	F. Sanger[a]
1967	Discovery and structure determination of proinsulin	D. F. Steiner and R. Chance
1969	Determination of three-dimensional structure	D. Crowfoot-Hodgkin[a]
1979	Cloning of the insulin gene	W. Rutter and H. Goodman

[a] Received Nobel Prize.

bridges. The A chain consists of 21 amino acid residues and the B chain of 30 amino acid residues. The primary amino acid structure of human and hagfish insulin is detailed in Figure 7-7.

The primary amino acid sequence of some 72 vertebrate insulins has been determined. Amino acid substitution apparently can occur at many positions without the loss of biological activity; indeed, as many as 34 of the 51 positions can be replaced (see Figure 7-7).

From careful analysis of structure–function relationships of both naturally occurring and chemically modified insulins, it is apparent that certain structural features of insulin have been conserved throughout evolution. These include (i) the precise positions of the three disulfide bonds; (ii) the N- and C-terminal regions of the A chain; and (iii) the hydrophobic residues of the C-terminal region of the B chain (residues 23–26). From biological studies, it has been shown that insulin binds to receptors from tissues of differing species in a way that supports the view that there has been little or no evolutionary drift in the binding site specificity of the insulin receptor. The insulin receptor specifically recognizes an extensive region of ~11 residues around residues A-21 and B-23–B-27 and is dependent upon an intact tertiary structure (see the following).

Since the amino acid sequence of the hagfish insulin differs from that of human at only 19 of 52 residues (37%), it can be calculated that amino acid substitutions) have occurred in insulin at a rate of $\sim 1 \times 10^{-9}$ site/year; this is a typical figure for a highly conserved protein. These findings also imply that the membrane receptor for insulin has not undergone significant changes throughout the course of vertebrate evolution.

D. Crowfoot-Hodgkin, T. Blundell, and associates have determined via X-ray crystallographic techniques the three-dimensional structure of porcine insulin initially at 2.8-Å resolution and more recently at 1.5 Å (see Figure 7-8). Å highly similar structure of insulin in solution has been achieved via nuclear magnetic resonance (NMR) spectroscopy.

Insulin can exist as a monomer, dimer, tetramer or hexamer. The insulin dimers in the crystalline state are held together by hydrogen bonds between the peptide

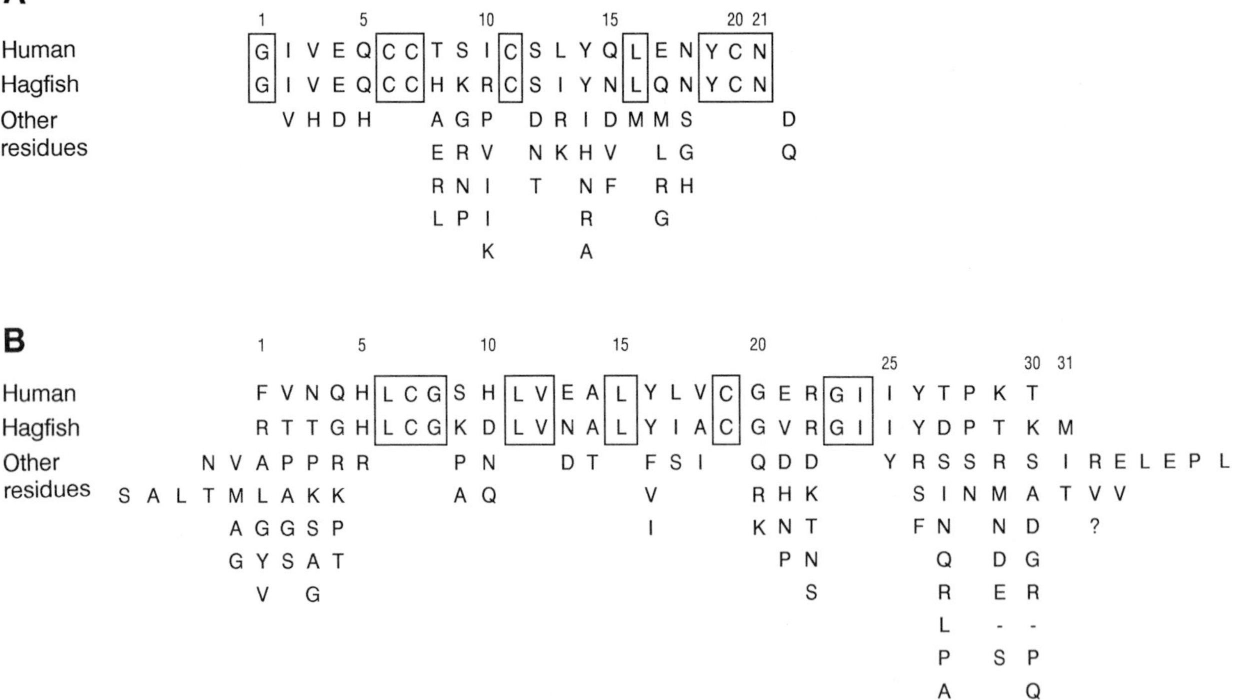

FIGURE 7-7 Primary amino acid sequences of human and hagfish insulins. In addition, the amino acid sequences of 70 other vertebrate insulins are known, and the substitutions that occur at each position (indicated as "other") are shown for comparison. The invariant residues are enclosed in boxes. [Reproduced with permission from Steiner, D. F. Bell, G. I., Tager, H. S. and Rubenstein, A. H. (1995). Chemistry and biosynthesis of the islet hormones: Insulin, islet amyloid polypeptide (amylin), glucagon, somatostatin, and pancreatic polypeptide. *In* "Endocrinology" (L. J. DeGroot, M. Besser, H. G. Burger, J. L. Jameson, D. L. Loriaux, J. C. Marshall, W. D. Odell, J. T. Potts, Jr., and A. H. Rubenstein, eds.), 3rd edition, Vol. 2, Chapter 76, pp. 1296–1328. W. B. Saunders, Philadelphia, PA.]

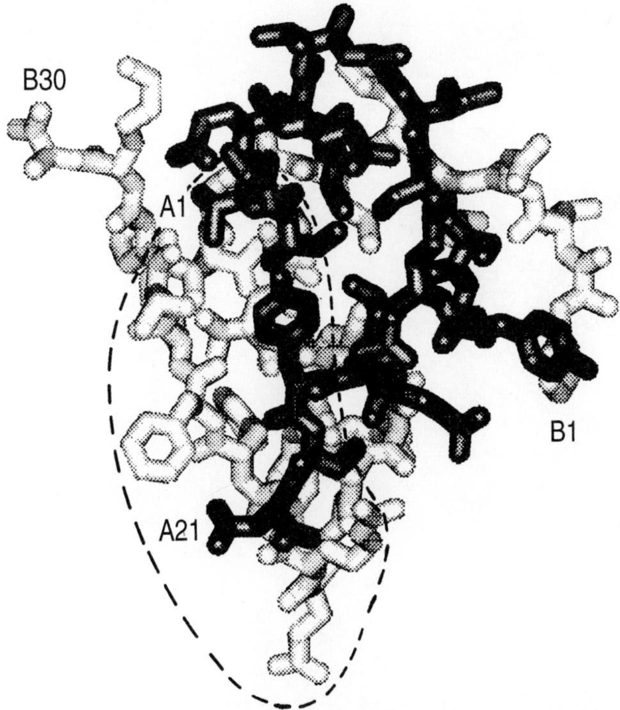

FIGURE 7-8 Three-dimensional structure of the monomeric form of insulin oriented perpendicular to the 3-fold axis. The side chains of all of the amino acids are shown; the A chain is dark gray and the B chain in light gray. The dashes indicate an important region on the surface of the insulin monomer that is believed to contact the receptor. [Reproduced with permission of B. Xiao, University of York, York, UK.]

groups of residues B-24 and B-26, which form an antiparallel pleated sheet structure (see Figure 7-9). The hexameric structure of crystalline insulin consists of three dimers ordered around a major 3-fold axis containing two zinc atoms, each coordinated at the imidazole groups of three B-10 histidine residues (see Figure 7-9). Indeed, the propensity for insulin to crystallize into hexamers may be related to the regular arrays seen in the electron microscopic evaluation of the storage granules of the β-cell.

While for many years it was thought that insulin was not a member of any hormone family, either in a structural sense (e.g., oxytocin and vasopressin are structural analogs of one another) or in a protein-processing sense (proopiomelanocortin produces β-lipotropin, ACTH, endorphins, etc.), it now appears that there is a hormone family of homologous growth factors that includes proteins with regions of amino acid sequences identical to that of insulin. These include the following: (i) relaxin, a polypeptide hormone from the *corpus luteum* that is responsible for the dilation of the *symphysis pubis* prior to parturition; (ii) insulin-like growth factors (IGFs) I or II (formerly known as NSILA or nonsuppressible insulin-like activity); (iii) somatomedins (particularly somatomedin C); and (iv) nerve growth factor (NGF). Some of these factors are discussed in further detail in Chapter 19.

B. Glucagon

Glucagon, which is secreted by the α-cells of the pancreas, has a molecular weight of 3450. It is the most potent hepatic glycogenolytic agent known. The amino acid sequence of glucagon is presented in Figure 7-10. In contrast to insulin, glucagon is composed of a single amino acid chain of 29 residues and is devoid of disulfide linkages. The amino acid sequences of rat, rabbit, human, porcine, bovine, turkey, chicken, and duck glucagons have been determined. There is a high degree of structural conservation in all of these peptides.

The secondary and tertiary structures of glucagon have been studied in solution by optical rotary dispersion–circular dichroism and in the crystalline state by X-ray crystallography (3-Å resolution). In dilute solutions, the glucagon peptide, due to its relatively short chain length, is flexible, with many different conformations and some indication of a stable interaction between valine-23 and tryptophan-25. In more concentrated solutions and in the crystalline state, the molecules self-associated into α-helical trimers. The helical region extends from residues 10 to 25 and results in the formation of two hydrophobic "sticky" patches that are involved in the self-association process. Thus, the glucagon molecule is unique among medium-sized polypeptides. Although it has a clearly definable secondary structure (α helix) and quaternary structure (self-association), it has no definable tertiary structure.

From biological studies, it has been determined that virtually the entire glucagon peptide chain is required for optimal interaction with the glucagon-sensitive adenylate cyclase of hepatocyte membranes. The evidence suggests that glucagon binds to its receptor as a consequence of entropic considerations, resulting in the selection of the optimal solution helical conformer, which has an available hydrophobic region for stereoselective interaction with the membrane receptor.

Several potent antagonists of glucagon have been synthesized by site-directed mutagenesis; thus deletion of the N-terminal histidine, coupled with substitution of Asp9 by Glu and α-carboxyamidation of the C-terminal Thr, generates a peptide that binds to the glucagon receptor with high affinity but that does not stimulate the production of cAMP.

Glucagon is also structurally homologous with a family of peptide hormones found in the gastrointestinal tract (see Chapter 8); these include secretin, vaso-

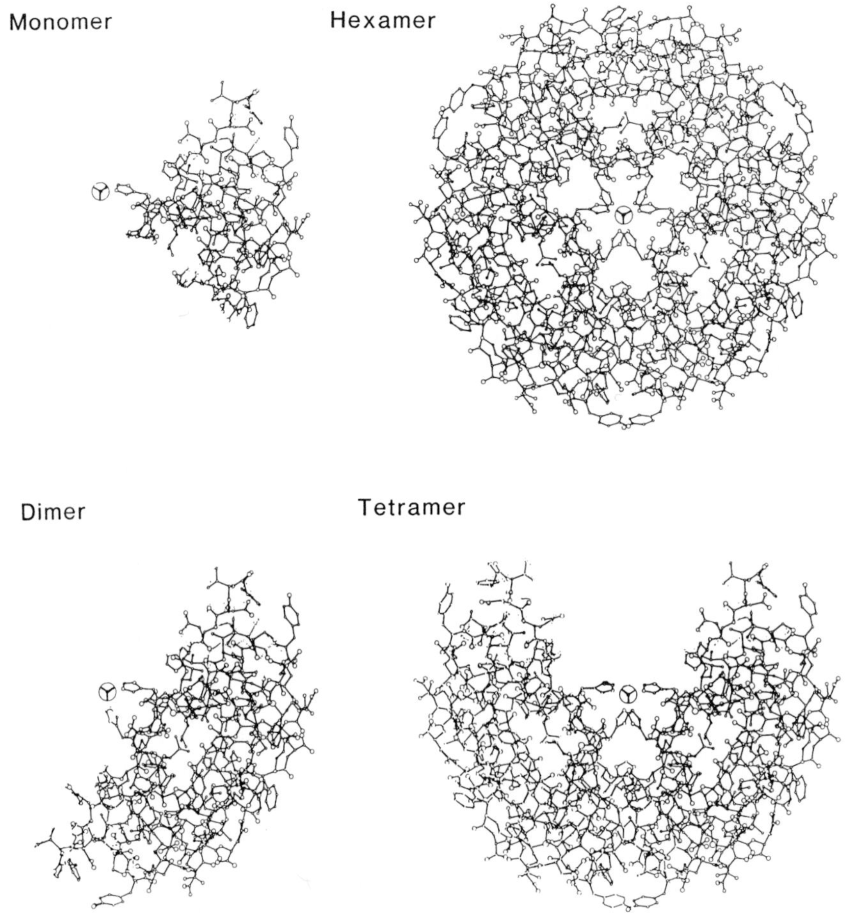

FIGURE 7-9 Hexameric, dimeric, and monomeric forms of insulin, showing the development of dimers from monomers and their organization into the hexamer. [The insulin structures were obtained courtesy of G. G. Dodson (York University, York, UK) and redrawn by F. D. Coffman and M. F. Dunn (University of California, Riverside).]

active intestinal polypeptide (VIP), gastric inhibitory peptide (GIP), enteroglucagon (a 69-amino-acid peptide that has all 34 amino acids of glucagon at its N-terminus), and glicentin (a 100-amino-acid peptide isolated from the intestine). Interestingly, glicentin has many of the immunodeterminants of glucagon and can account for much of the glucagon-like activity of the intestine.

C. Pancreatic Polypeptide

Pancreatic polypeptide (PP) is a 36-amino-acid peptide that appears to stimulate the gastric secretion of HCl and pepsin; it may also act as a satiety factor. PP is known to be released after a protein meal.

The amino acid sequences of several PP are given in Figure 7-11. All sequenced PP have amidated carboxy-termini. With the exception of the avian PP, which has 20 out of 36 residues different from those of human PP, there is strict conservation of sequences (only 2–3 amino acid differences) among the other four sequenced PP species.

Like insulin, PP molecules can self-associate to dimers and, in the presence of zinc ions, can form higher oligomers. Because a homologous polypeptide has been isolated from pig intestine, it has been proposed that PP may be a member of a larger family of pancreatic and gastroenteric hormones.

PP belongs to a family of structurally related carboxyamidated neuroendocrine peptides; these include

```
                1           5           10          15          20          25        29
Human   H2N- H  S  Q  G  T  F  T  S  D  Y  S  K  Y  L  D  S  R  R  A  Q  O  F  V  Q  W  L  M  N  T  -COOH
```

FIGURE 7-10 Primary amino acid sequence of human glucagon.

	1 5 10 15 20 25 30 35 36
Human	A P L E P Q Y P G D D A T P E Q M A Q Y A A D L K K Y I N M L T R P R Y -COOH
Porcine	A P L E P V Y P G D D A T P E Q M A Q Y A A E L K K Y I N M L T R P R Y -COOH
Avian	G P S Q P T Y P G D D A P V E A L I R E Y D N L Q Q Y L N V V T R H R Y -COOH

FIGURE 7-11 Primary amino acid sequences of human, porcine, and avian pancreatic polypeptides (PP). Amino acids underlined are different from those of human glucagon. [Data are derived from Kimmel, J. R., Pollock, H. G., Chance, R. E., Johnson, M. G., Reeve, J. R., Jr., Taylor, I. L., Miller, C., and Shivery, J. E. (1984). "Pancreatic polypeptide from rat pancreas." *Endocrinology* **114**, 1725–1731.]

neuropeptide Y (NPY) and peptide YY (PYY), which are found in many neurons of the central and peripheral nervous systems (see Table 8-1).

D. Somatostatin

Somatostatin is the smallest of the pancreatic hormones; it contains 14 amino acids and one disulfide bridge. Somatostatin is derived from a 116-amino-acid preprosomatostatin (see Figure 7-12). Although somatostatin was originally discovered in the hypothalamus, it is also known to be produced by the δ-cells of the endocrine pancreas and dispersed cells in the gastrointestinal tract. Somatostatin, when produced by the hypothalamus, is frequently termed growth hormone release-inhibiting hormone (GIF).

The physiological role of circulating somatostatin is not clear; it may function in a paracrine fashion to regulate pancreatic islet functions.

E. Islet Amyloid Polypeptide

A second peptide hormone that is produced by the pancreatic β-cells is islet amyloid polypeptide (IAPP) or amyloid (see Figure 7-13). IAPP is a 37-amino-acid peptide that is structurally homologous to a neuropeptide known as calcitonin gene-related peptide (CGRP) (CGRP is discussed in Figure 9-4).

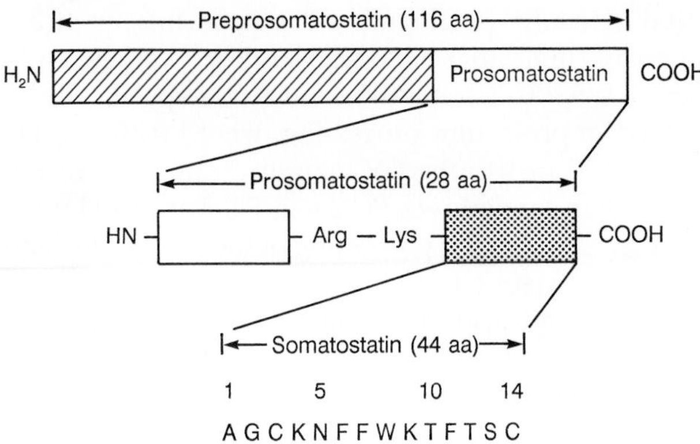

FIGURE 7-12 Primary amino acid sequence of human somatostatin and its relationship to preprosomatostatin. Native somatostatin has a disulfide bond connecting amino acids 3 and 14.

Amyloid is a highly insoluble protein that is deposited in the pancreas of elderly type I diabetics; it is readily visualized at the light microscopic level. It is not known by what mechanism amyloid accumulation occurs in diabetics. Although amyloid-like material has been known to be associated with the pancreas since 1901, amyloid was not solubilized until 1986. Knowledge of the amino acid sequence of IAPP has permitted, through recombinant techniques, the characterization of both the cDNA and a prepro-IAPP, as well as the gene.

Pro-IAPP is packaged along with proinsulin into the secretory granules of the β-cell; and then mature forms of both peptides are cosecreted. The biological function of IAPP is believed to be related to its ability to antagonize insulin's stimulation of glucose uptake by muscle cells.

F. Leptin

The maintenance of body weight is a complex process involving homeostasis of nutrient intake (carbohydrate, protein, and fat) as well as the metabolic processes of gluconeogenesis and glycolysis in muscle and liver cells and particularly in adipose cells. Normally higher vertebrates maintain the body's energy reserve of fat (triglycerides) with high precision; thus, a typical nonobese, healthy adult would have a body mass containing 15–20% fat. However, in some vertebrates (including humans), there is a loss of regulation of the storage of triglycerides in adipose cells, such that the body mass of fat can increase to 30%.

Classical parabiotic[2] studies conducted in the early 1970s suggested that genetically obese mice lacked a

[2] Parabiosis: In a parabiotic experiment, the circulatory system of one animal is connected to a second animal. Thus, Coleman [Diabetologia 9, 294 (1973)] indirectly connected the circulatory system of adult *ob/ob* (obese) mice to that of normal mice, and the *ob/ob* mice lost weight; this suggested the existence of a weight-regulatory substance in the serum of the normal mice, i.e., a hormone associated with the maintenance of a normal dietary caloric intake.

```
          1       5          10          15          20          25          30          35
IAPP      K C N T A T C A T Q R L A N F L V H S S N N F G A I L S S T N V G S N TY-COOH
CGRP-I    A C D T A T C V T H R L A G L L S R S G G V V K N N F V P T N V G S K AF-COOH
CGRP-II   A C N T A T C V T H R L A G L L S R S G G M V K S N F V P T N V G S K AF-COOH
```

FIGURE 7-13 Comparison of the amino acid sequences of human islet amyloid polypeptide (IAPP) and calcitonin gene-related peptide (CGRP). The underlining indicates substitution of different amino acids in CGRP-I and CGRP-II as compared with IAPP. [Abstracted from Nishi, M., Sanke, T., Nagamatsu, S., Bell, G. I., and Steiner, D. F. (1990). Islet amyloid polypeptide—A new β cell secretory product related to islet amyloid deposits. *J. Biol. Chem.* **265,** 4173–4176.]

blood-borne factor (i.e., a hormone) that regulated, via appetite control, the stimulation of glycolysis and thus the extent of body fat deposition. Also, it is known that mutations in the *obese* gene in the C57BL/6 mouse result in the onset of obesity, which includes increased deposition of body fat, hyperglycemia, and hyperinsulinemia (similar to type II diabetes).

These observations led in 1995, via molecular biology techniques, to the biochemical and biological characterization of a new hormone, termed leptin.[3] Leptin is a 16-kDa protein that is the plasma form of the gene product encoded by the *ob* gene. Daily administration of recombinant leptin to *obese* mice reduced their body weight by 30% within 2 weeks; in addition, leptin reduced the food intake and increased the *ob/ob* mouse expenditure of metabolic energy. A full understanding of the endocrinology of leptin and its potential application to the regulation of body weight will be an area of intense research activity in the near future.

IV. BIOCHEMISTRY

A. Biosynthesis of Hormones

1. *Biosynthesis of Insulin*

a. Identification of Proinsulin

For years one of the intriguing problems relating to insulin biosynthesis was that of describing the independent synthesis of the separate A and B chains followed by the correct formation of the three disulfide linkages. It was not at all clear whether these formed spontaneously on the basis of peptide chain folding, conformation, and entropic considerations or whether there were specialized "biological mechanisms" that mediated this precise process. The discovery of proinsulin by Steiner and colleagues in 1967 provided a resolution to this knotty problem.

Proinsulin is a precursor form of insulin that exists as a single-chain polypeptide of ~9000 Da; this chain contains, within its sequence, the 21 amino acid residues of the A chain and the 30 residues of the B chain of insulin. In addition, it contains a connecting sequence, known as the C peptide, which falls between the N-terminus of the A chain and the C-terminus of the B chain (see Figure 7-14). It is now clear that proinsulin represents an intermediate species that appears on the biosynthetic pathway, leading from the immediate polysomal translation product to the final granule stored form of insulin. In the case of proinsulin, the excised C peptide is stored in the secretory granule along with mature insulin.

Mammalian proinsulins contain from 76 (hagfish) to 89 (angler fish) amino acid residues; these variations in amino acid units reflect only differences in the length of the connecting C peptide between the A and B chains. Theoretically it is not necessary to have a connecting sequence of 25–38 amino acid residues; from model-building exercises related to the tertiary structure of insulin (Figure 7-9), it appears that only 5–8 residues would be minimally required. The additional residues of the C peptide do not completely "mask" the biological activity of the molecule, since proinsulin contains 3–5% of the biological activity of native insulin.

Both the prospective N- and C-termini of the C peptide of all known mammalian proinsulins end in a pair of basic amino acid residues (either lysine or arginine). These represent the recognition signals for proteolytic cleavage that are necessary for the conversion of proinsulin to insulin (see Figure 7-15). As a consequence of extensive molecular biological studies, two new endopeptidases PC-2 and PC-3, which are specifically involved in proinsulin processing, were identified. PC-2 and PC-3 are members of a family of subtilisin-related proprotein convertases. PC-2 and PC-3 are also known to be involved in the processing of the proopiomelanocortin precursor (see Figure 1-9B).

The amino acid composition of the C peptide is known for fifteen mammalian and one avian species. It is apparent that there is a 15× higher rate of mutation in the C peptide domain than in the A or B chains of the related insulins; this is consistent with the hypothesis that the C peptide does not have any extrapancreatic hormonal function. The C peptide is secreted in equi-

[3] Leptin is derived from the Greek root *leptós*, meaning thin.

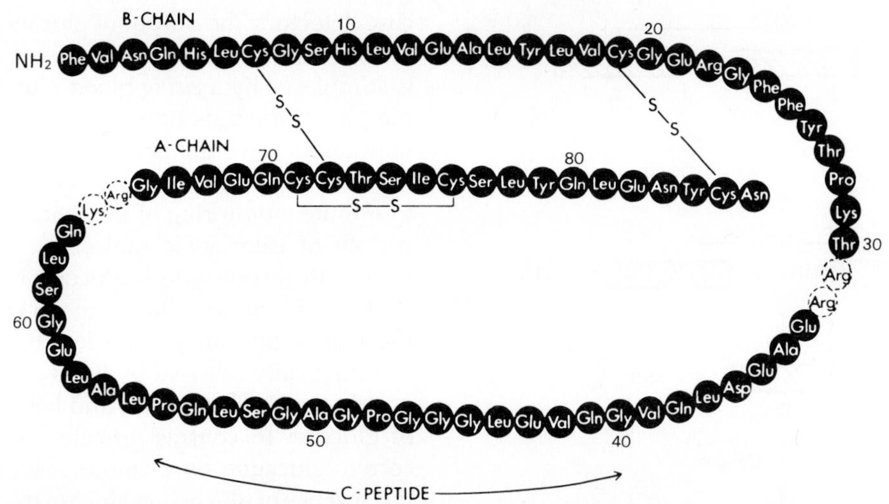

FIGURE 7-14 Amino acid sequence of human proinsulin. Dashed circles indicate sites of peptidase cleavage so as to generate the mature insulin and free C peptide. [Modified with permission from Oyer, P. E., Cho, S., Peterson, J. D., and Steiner, D. F. (1971). Studies on human proinsulin: Isolation and amino acid sequence 4 of the human pancreatic C-peptide. *J. Biol. Chem.* **246,** 1375–1386.]

molar amounts with insulin and is known to circulate in the blood; radioimmunoassays are now available to measure this peptide.

b. Biosynthetic Pathway for Insulin

The structure of the gene is now known for seven species (see Figure 7-16). The insulin genes belong to an insulin superfamily of hormones and growth factors that are concerned with metabolism and growth; the other family members include the insulin-related growth factors IGF-I and IGF-II, as well as the ovarian hormone relaxin (IGF; see Chapter 19).

In contrast to IGF-I and IGF-II, which are biosynthesized by most tissues, insulin is only produced in the β-cells of the pancreatic islet. A summary of the regulation of expression of the insulin gene is given in Figure 7-17. Glucose has been shown to modulate both the transcription and translation processes at several distinct steps. This level of control by glucose is consistent with the observation that an elevation in glucose levels can result in a 20-fold increase in insulin biosynthesis.

As is the case with other secreted proteins (e.g., parathyroid hormone, honeybee mellitin, myeloma L chains, and serum albumin), in insulin–proinsulin the immediate translation product released from the polyribosomes has an N-terminal extension or "leader sequence" of ~23 amino acid residues (see Figure 7-16). This product released from the polysomes is designated preproinsulin [by analogy, see preproparathyroid hormone (Figure 1-11)].

The conversion of proinsulin to insulin occurs in the rough endoplasmic reticulum (RER). The details of the synthesis and secretion of insulin are described in Figure 7-18. As the insulin is released from the proinsulin, it tends to crystallize in the secretory granules as a consequence of the relatively high levels of zinc ions that are present. Morphological studies have shown that the repeating units characteristic of mature pancreatic granules are quite similar to those found in the *in vitro* crystallized hexameric zinc insulin.

2. *Biosynthesis of Glucagon*

The biosynthesis and secretion of the 29-amino-acid glucagon by the pancreas α-cell are under the regulatory control of nutrients. In addition, an intestinal form of glucagon, termed glicentin, is produced that consists of 69 amino acid residues; this includes an 8-amino acid C-terminal extension added to glucagon and a 32-residue N-terminal extension. In contrast to glucagon, the secretion of glicentin from the intestinal L cells is not regulated by nutrients. Glicentin is released from the small intestine during the absorption of glucose, triglycerides, and amino acids; its physiological function remains to be clarified.

Both glucagon and glicentin are encoded by the same gene. As detailed in Figure 7-18, differential processing of the initial gene transcript by the pancreatic α-cell or the intestinal L cells results in the biosynthesis of mature glucagon or the peptides glicentin (GLP-I), IVP-II, and GLP-II, respectively (see Figure 7-19). Intestinal L cells presumably lack the endopeptidases required to complete the processing of proglucagon to glucagon.

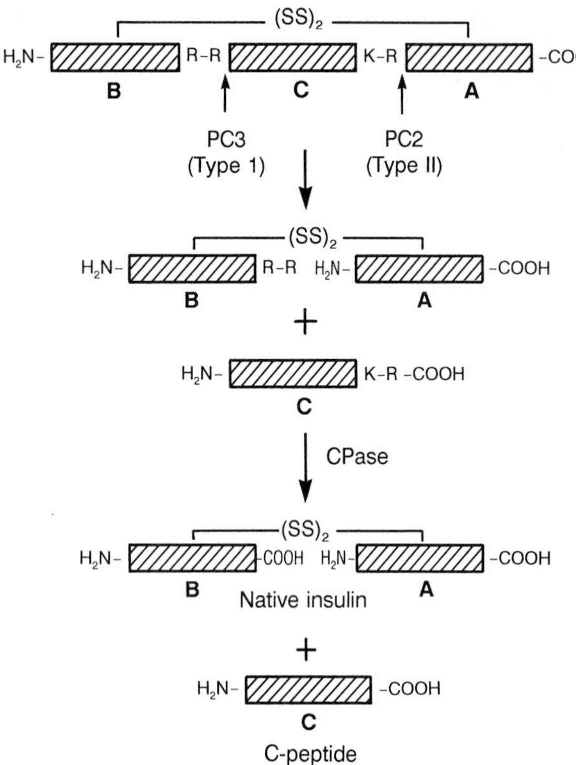

FIGURE 7-15 Cleavage of preproinsulin by specific endopeptidases to generate mature insulin and the C peptide. In the preproinsulin, the two pairs of dibasic signal residues (RR and KR) are indicated. Two specific endopeptidases, PC-2 and PC-3, cleave on the C-terminal side of each of the dibasic signal residues, releasing mature insulin and a C peptide with two additional amino acids (K and R) on the C-terminus. These two residues are removed by carboxypeptidase H (CPase) to generate the mature C peptide. [Modified with permission from Steiner, D. F., Bell, G. I., Tager, H. S., and Rubenstein, A.H. (1995). Chemistry and biosynthesis of the islet hormones: Insulin, islet amyloid polypeptide (amylin), glucagon, somatostatin and pancreatic polypeptide. *In* "Endocrinology" (L. DeGroot, M. Besser, H. G. Burger, J. L. Jameson, D. L. Loriaux, J. C. Marshall, W. D. Odell, J. T. Potts, Jr., and A. H. Rubenstein, eds.), 3rd ed., Vol. 2, Chapter 76, PP. 1296–1328. W. B. Saunders, Philadelphia, PA.]

3. *Biosynthesis of Pancreatic Polypeptide*

The pathway of biosynthesis of PP is shown in Figure 7-20; the general features are similar to those employed by most peptide hormones (see Figure 1-10). The propancreatic peptide, which is 9–10 kDa, serves to ultimately generate the mature PP and a second secretory peptide, known as icosapeptide. The icosapeptide has no known biological activity.

B. Secretion of Pancreatic Hormones

1. *Background*

The process of the regulated and integrated secretion of insulin and glucagon is very complex. Independently, the α- and β-cells are exquisitely sensitive glucose detectors; the output of glucagon is stimulated by a *falling* blood glucose level, while the output of insulin is stimulated by a *rising* blood glucose level. Together the pancreatic cells function as a "fuel molecule" homeostat for the organism. The β-cell can be envisioned as a fuel receptor, which is responsible for the minute-to-minute monitoring of the changes in the organism's supply of calorigenic molecules. The β-cell secretes insulin in response to D-glucose, L-amino acids, fatty acids, and ketones. There are no known inhibitors of β-cell function among the fuel molecules. However, insulin is only released by the β-cell in response to the L-amino acids, fatty acids, and ketones in the presence of glucose. In contrast, α-cells can be stimulated to secrete glucagon by L-amino acids and fatty acids in the absence of glucose; D-glucose is also a potent inhibitor of glucagon release. These relationships are summarized in Table 7-8.

The typical response of both α- and β-cells to a maximal physiological signal for secretion is a biphasic output of their characteristic hormones (see Figure 7-21). Thus, in the isolated pancreas, infusion of 20 m*M* D-glucose will elicit within 1 min a rapid burst of insulin secretion; this will be followed by a resting interval of 5–7 min and then a slowly rising "second phase" of insulin secretion, which will last as long as the glucose is present. A similar biphasic release pattern is followed when the α-cells sense the presence of 1–5 m*M* amino acids in the absence of glucose.

2. *Secretion of Insulin*

Insulin is secreted from the β-cells of the pancreatic islets by the process of emiocytosis (see Figure 7-22). In this process, the granules inside the cell are believed to migrate to the peripheral cell membrane down or along a microtubular network. In the region of the cell membrane, the microtubules coalesce into a microfilamentous network that is immediately adjacent to the cell membrane. This microfilamentous system, which contains actin, has fibers that are 50–70 Å in diameter. Once the β-granules encounter the cell membrane there is a fusion of the cell and granule membranes (see Figure 7-22), which results in the dissolution of the membrane at the point of contact, with concomitant extrusion of the granule insulin contents out of the cell.

The most important physiological secretagogue for insulin release is glucose. There are two general models to describe how changes in the blood level of glucose both initiate and modulate the secretion of insulin from the pancreatic β-cell; these include (i) the membrane glucose receptor model of stimulus recognition and (ii) the fuel metabolism model of stimulus recognition.

The operation of a β-cell membrane receptor for glucose, which is coupled to the opening of an outward

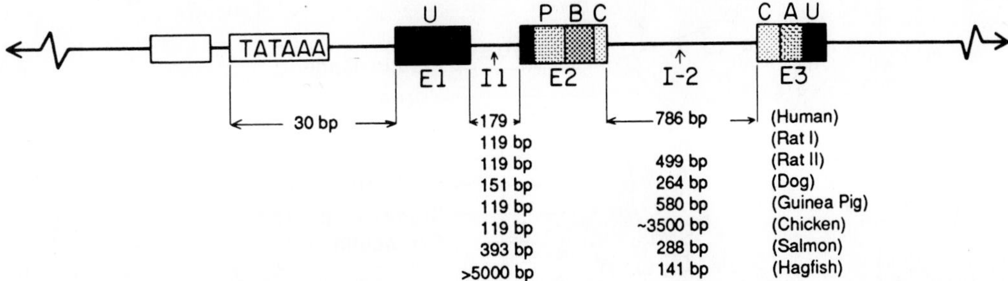

FIGURE 7-16 Schematic representation of the vertebrate insulin gene. Exons (E) that appear in mature preproinsulin mRNA are shown as bars, whereas the sizes of the two introns (i) are not on the same scale, but are indicated below. Abbreviations: U, untranslated region; P, prepeptide coding region; B, B chain coding region; C, C peptide coding region; A, A chain coding region. A typical TATA box, which signals the initiation of transcription, is shown 30 bp upstream; it is preceded by a promoter region (the unfilled box). [Modified with permission from Steiner, D. F., Bell, G. I., Tager, H. S., and Rubenstein, A. H. (1995). Chemistry and biosynthesis of the islet hormones: Insulin, islet amyloid polypeptide (amylin), glucagon, somatostatin and pancreatic polypeptide. *In* "Endocrinology" (L. DeGroot, M. Besser, H. G. Burger, J. L. Jameson, D. L. Loriaux, J. C. Marshall, W. D. Odell, J. T. Potts, Jr., and A. H. Rubenstein, eds.), 3rd ed. Vol. 2, Chapter 76, pp. 1296–1328. W.B. Saunders, Philadelphia, PA.]

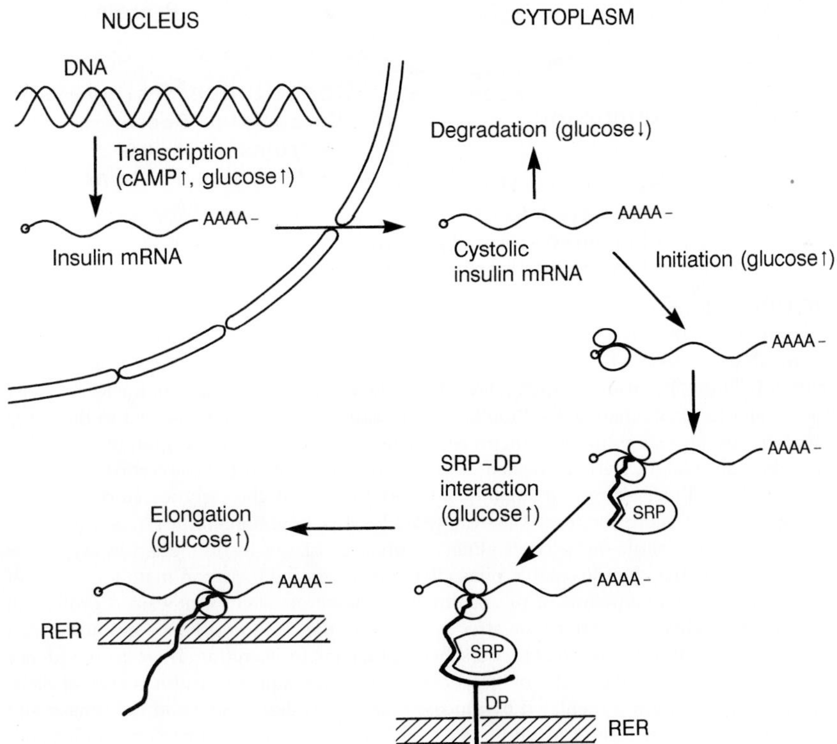

FIGURE 7-17 Synopsis of the regulation of expression of the insulin gene. The primary transcription of the insulin gene is selectively stimulated by elevated cAMP and/or glucose. In addition, insulin mRNA levels are stabilized at high glucose levels, which also selectively stimulates the translation of insulin mRNA at the levels of initiation, elongation, and interaction of signal recognition particles (SRP) with a docking protein (DP) [Modified with permission from Steiner, D. F., Bell, G. I., Tager, H. S., and Rubenstein, A. H. (1995). Chemistry and biosynthesis of the islet hormones: Insulin, islet amyloid polypeptide (amylin), glucagon, somatostatin and pancreatic polypeptide. *In* "Endocrinology" (L. DeGroot, M. Besser, H. G. Burger, J. L. Jameson, D. L. Loriaux, J. C. Marshall, W. D. Odell, J. T. Potts, Jr., and A. H. Rubenstein, eds.), 3rd ed., Vol. 2, Chapter 76, pp. 1296–1328. W.B. Saunders, Philadelphia, PA.]

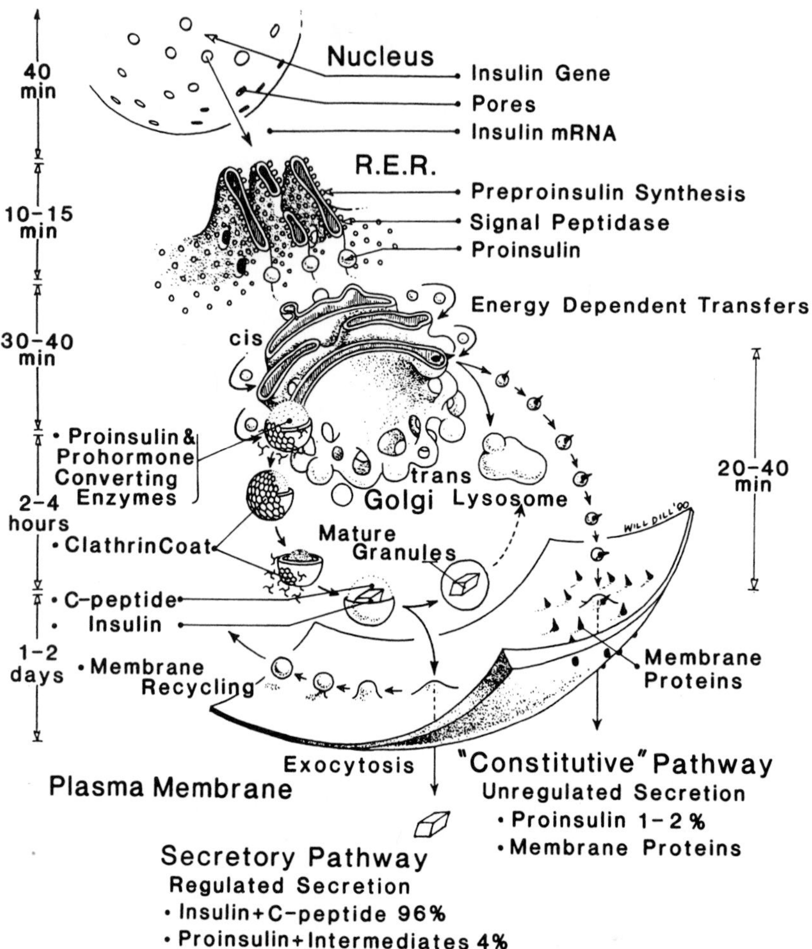

FIGURE 7-18 Schematic diagram describing the synthesis and secretion of insulin. After the insulin mRNA leaves the nucleus, the translation process of preproinsulin occurs on the rough endoplasmic reticulum (RER). This is followed by rapid cleavage to proinsulin (within 1–2 min). Proinsulin is then released into the intracisternal spaces of the RER, where it folds and forms the native disulfide bonds of insulin. It is then transported to the Golgi apparatus by an energy-dependent process. The clathrin-coated early granules budding from the trans-Golgi cisternae are rich in proinsulin and contain the converting proteases, PC-2 and PC-3. Processing occurs mainly, if not exclusively, in the early secretory granules, giving rise to the more condensed mature granules. Fractionation studies have confirmed that the mature granule-dense cores almost entirely consist of insulin, often in crystalline arrays, whereas the granule-soluble phase that surrounds the inclusion mainly consists of C peptide and small amounts of proinsulin. The release of newly synthesized proinsulin and insulin begins only about 1 hr after synthesis in the RER, and hence granules must undergo a maturation process that renders them competent for secretion. There is no evidence for significant nongranular routes of secretion of either proinsulin or insulin in normal islets. Exocytosis of granules is regulated by glucose and many other factors and in humans and dogs results in the release of insulin and C peptide in equimolar proportions under both basal and stimulated conditions. [Modified with permission from Steiner, D. F., Bell, G. I., Tager, H. S., and Rubenstein, A. H. (1995). Chemistry and biosynthesis of the isltet hormones: Insulin, islet amyloid polypeptide (amylin), glucagon, somatostatin and pancreatic polypeptide. *In* "Endocrinology" (L. DeGroot, M. Besser, H. G. Burger, J. L. Jameson, D. L. Loriaux, J. C. Marshall, W. D. Odell, J. T. Potts, Jr., and A. H. Rubenstein, eds.), 3rd ed., Vol. 2, Chapter 76, pp. 1296–1328. W. B. Saunders Company, Philadelphia, PA.]

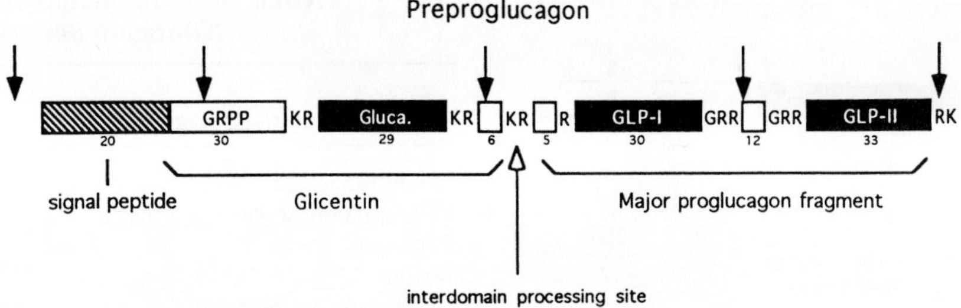

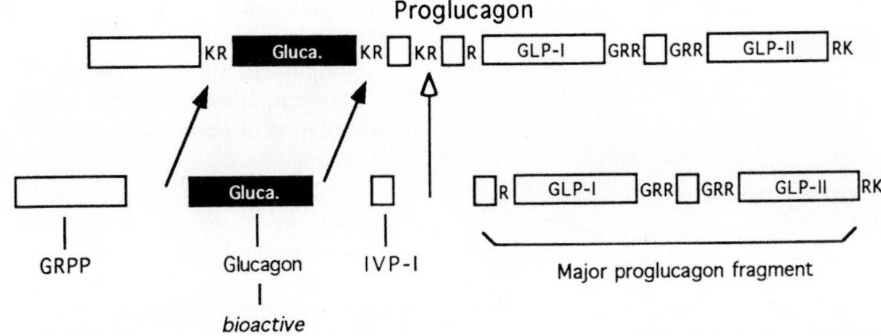

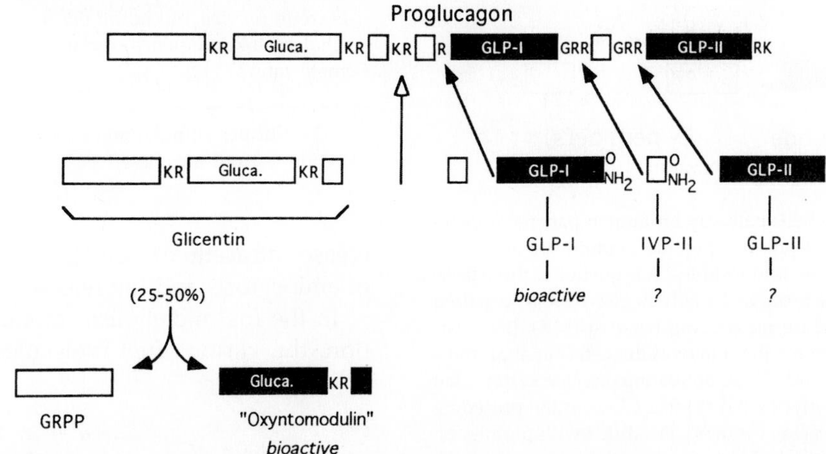

FIGURE 7-19 Differential processing of preproglucagon by the pancreas and large intestine. The processing sites that must be cleaved to generate glucagon and the glucagon-like peptides are shown between indicated segments. The C-terminus of glucagon-like peptide 1 is amidated; the sequence Gly-Arg-Arg directs cleavage of the precursor and provides a substrate for the polypeptide-amidating activity enzyme complex (PAM). The tissue-specific processing of this precursor is indicated. In α-cells, processing predominates at the more N-terminally located Lys-Arg pairs to produce glucagon. In gut cells, residues 1–69 are released *en bloc* as glicentin or gut glucagon, whereas the C-terminal major proglucagon fragment (residues 70–160) is efficiently processed to release GLP-1 and -2 peptides. The arrows above the preproglucagon sequence at the top indicate the positions of introns in the human gene; each exon of the preproglucagon gene encodes a domain of the messenger RNA or of preproglucagon. The human glucagon gene, abbreviated CCG, is located on the long arm of chromosome 2 in the region q36–q37. [Modified with permission from Steiner, D. F., Bell, G. I., Tager, H. S., and Rubenstein, A. H. (1995). Chemistry and biosynthesis of the islet hormones: Insulin, islet amyloid polypeptide (amylin), glucagon, somatostatin and pancreatic polypeptide. *In* "Endocrinology" (L. J. DeGroot, M. Besser, H. G. Burger, J. L. Jameson, D. L. Loriaux, J. C. Marshall, W. D. Odell, J. T. Potts, Jr., and A. H. Rubenstein, eds.), 3rd ed., Vol. 2, Chapter 76, PP. 1296–1328. W. B. Saunders, Philadelphia, PA.]

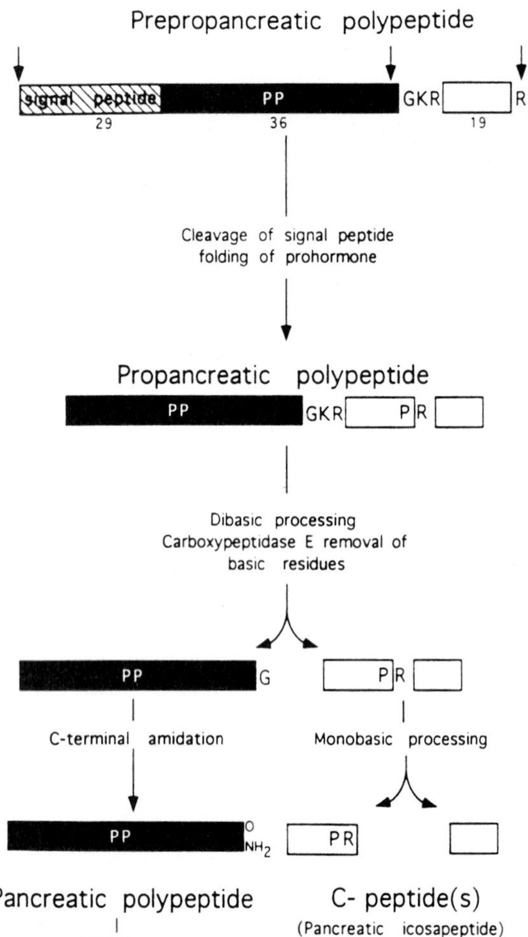

FIGURE 7-20 Biosynthetic pathway for human pancreatic polypeptide. The prepropancreatic polypeptide peptide has 66 amino acids preceded by 29 amino acid residues. A sequence of three basic amino acids (G, K, R) is processed by three enzymes: (i) a paired basic residue endopeptidase recognizing Lys-Arg (K,R), (ii) a carboxypeptidase B-like enzyme that removes the C-terminal arginine and lysine successively, and (iii) an amidating enzyme system that oxidatively removes the glycine (G) to yield CO_2 and the preceding carboxyamidated amino acid (Tyr or Y). [Modified with permission from Steiner, D. F., Bell, G. I., Tager, H. S., and Rubenstein, A. H. (1995). Chemistry and biosynthesis of the islet hormones: Insulin, islet amyloid polypeptide (amylin), glucagon, somatostatin and pancreatic polypeptide. *In* "Endocrinology" (L. DeGroot, M. Besser, H. G. Burger, J. L. Jameson, D. L. Loriaux, J. C. Marshall, W. D. Odell, J. T. Potts, Jr., and A. H. Rubenstein, eds.), 3rd edi., Vol. 2, Chapter 76, pp. 1296–1328. W. B. Saunders Co., Philadelphia, PA.]

K^+ channel and inward Ca^{2+} channel, is described in Figure 7-23. In this model, the intracellular enzyme glucokinase, which converts glucose to glucose 6-phosphate, is believed to play a key role in determining the rate of metabolism of glucose, which will ultimately lead to changes in the intracellular ratio of [ATP]/[ADP]. When this ratio is elevated, the K^+ channel is inhibited, resulting in membrane depolarization, with concomitant opening of the Ca^{2+} channel so as to increase intracellular [Ca^{2+}], leading to the stimulation of emiocytosis and the release of insulin by the β-cell.

In the fuel metabolism model of stimulus recognition, the various fuel molecules are recognized as a

TABLE 7-8 Modulators of Insulin and Glucagon Secretion

Modulator/secretagogue	Insulin secretion[a]	Glucagon secretion
Rising blood glucose	↑	↓
Falling blood glucose	↓	↑
Glucagon	↑	↓
Insulin	→	↓
Glucogenic amino acids		
Alanine, arginine	→	↑
Lysine	↑	↑
Leucine	↑	→
Growth hormone	↑	→
Nervous stimulation		
Sympathetic (norepinephrine)	↓	↑
Parasympathetic (acetyl choline)	↑	↓
Stimulators of adenyl cyclase		
Glucagon	↑	→
ACTH	↑	→
TSH	↑	→
Isoproterenol	↑	→
cAMP	↑	→
Oral hypoglycemic agents		
Tolbutamide	↑	→
Chloropromamide	↑	→
Gastrointestinal peptides		
Enteroglucagon	↑	→
Secretin (*in vivo*, but not *in vitro*)	↑	↓
Cholecystokinin–pancreozymin	↑	↑
Somatostatin	↓	↓

[a] ↑ indicates stimulation of secretion; ↓ indicates inhibition of secretion; → indicates no effect.

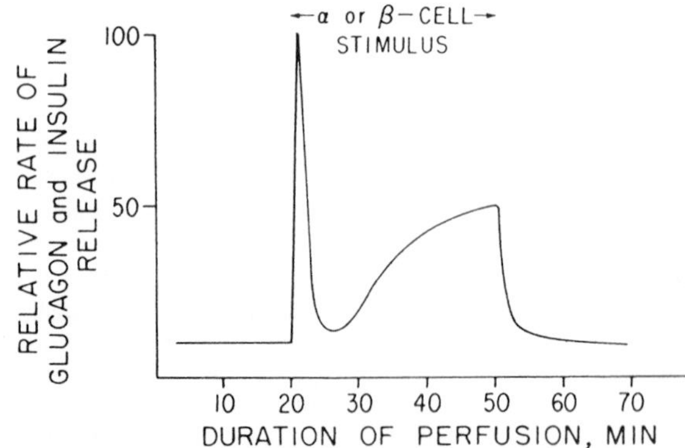

FIGURE 7-21 Typical biphasic hormone-releasing profile as seen with the isolated perfused pancreas from different species as a result of α- or β-cell stimulation. [Reproduced with permission from Matschinsky F. M., Pagliara, A. A., Zawalich, W. S., and Trus, M. D. (1979). Metabolism of pancreatic islets and regulation of insulin and glucagon secretion. *In* "Endocrinology" (L. J. DeGroot *et al.*, eds.), Vol. 2, p. 936. Grune & Stratton, New York.]

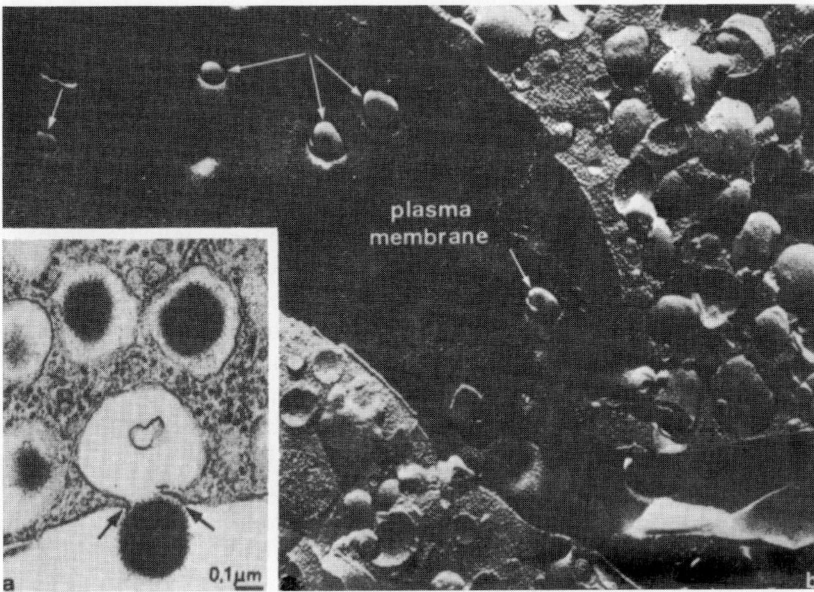

FIGURE 7-22 Electron micrograph and freeze–fracture study of insulin secretion. Isolated islets were exposed to a high (3.0 mg/ml) glucose concentration. (a) Thin section of part of a β-cell. A granule core is being discharged into the extracellular space through the opening resulting from the coalescence of the membrane limiting the granule with the plasma membrane (arrows) (×56,000). (b) Freeze–fracture replica. On the exposed face of the plasma membrane there are several exocytotic (emiocytotic) stomata (thin arrows). The channel extending in the cytoplasm (large arrow) is possibly the result of the fusion of the membranes of several granules whose cores could be discharged through a single aperture in the plasma membrane (×31,000). Courtesy of Dr. Lelio Orci, Geneva. [Reproduced with permission from Lacy P. E., and Greider, M. H. (1979). Anatomy and ultrastructural organization of pancreatic islets. *In* "Endocrinology" (L. J. DeGroot *et al.*, eds.), Vol. 2, p. 913. Grune & Stratton, New York.]

consequence of their uptake into the β-cell and subsequent metabolism via intermediary metabolism. The recognition process might consist of the identification of the presence of a certain family of key metabolites or cofactors, such as NADH/NAD, ATP, and intracellular pH. This then could lead to activation of the stimulus–secretion mechanism, resulting in insulin secretion.

A third plausible model would be a combination of the two separate models described. Figure 7-24 presents a schematic working model proposed by F. Matschinsky to describe the coupling of the "fuel receptor" to the stimulus–secretion mechanism, which results in insulin secretion.

In terms of the biphasic secretion of insulin (Figure 7-21) then, the first acute phase of insulin secretion represents the release of insulin already stored in the granules of the β-cell. The slower second phase of insulin release is likely supported by an initiation of *de novo* insulin biosynthesis.

3. Secretion of Glucagon

The principal physiological function of the pancreatic α-cell is to prevent hypoglycemia by appropriately regulating the secretion of glucagon. Glucagon secretion by the α-cells particularly is dedicated to safeguarding the delivery of fuel (usually glucose) to the cerebellum. As documented in Table 7-8, there is an array of substances that are potent modulators of glucagon secretion. In addition, there is evidence of both adrenergic and cholinergic control of glucagon secretion, which comes into play during insulin-induced hypoglycemia; this redundant neural mechanism apparently exists to combat the life-threatening situation of hypoglycemia. Glucagon release is clearly stimulated by epinephrine and blocked by the β-adrenergic blocking agent propanolol, which indicates the presence of a β-adrenergic receptor on the D cell membrane. Also, somatostatin can block glucagon release from the D cell. Electron micrographic studies suggest that the basic mechanism for release of the glucagon-containing granules is emiocytosis. At the present time it is not clear whether there is an involvement of a microtubule–microfilament system or whether there is a glucose receptor in the plasma membrane.

4. Integrated Secretion of Insulin and Glucagon

The overriding responsibility of insulin and glucagon is to maintain blood glucose within normal limits. Shown in Figure 7-25 are the dose–response curves of

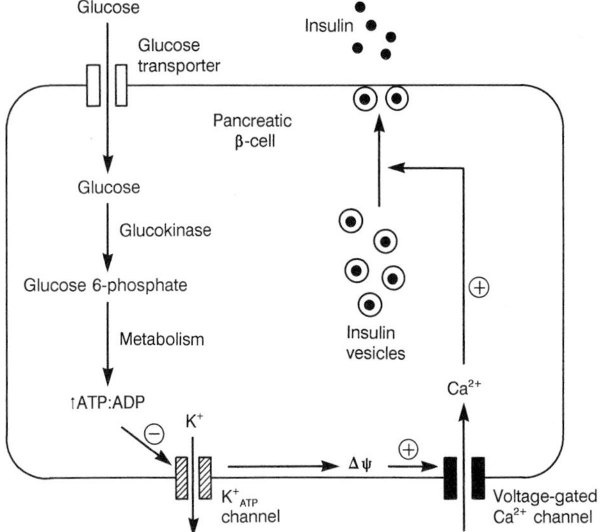

FIGURE 7-23 Model describing cellular events linking occupancy of a membrane glucose receptor to secretion of insulin by the pancreatic β-cell. The membrane glucose receptor is proposed to be the high K_m passive glucose transporter, GLUT-2 (see Table 7-9). The K_m of the transporter for glucose is $\approx$17 mM; thus, the influx of glucose into the β-cell is proportional to the blood glucose level up to 10–12 mM. The glucose that enters the β-cell is phosphorylated by glucokinase to yield glucose 6-phosphate. The increased concentration of glucose 6-phosphate stimulates glycolysis and oxidative metabolism, thereby increasing the ratio of [ATP]/[ADP]. The increased [ATP]/[ADP] ratio is proposed to inhibit the ATP-sensitive K^+ channels; the resultant depolarization of the plasma membrane triggers the opening of the voltage-gated Ca^{2+} channels so that there is an influx of Ca^{2+}. The resulting increase in cytosolic [Ca^{2+}] acts to stimulate the fusion of the insulin-containing granules with the β-cell membrane. [Modified with permission from Efrat, S., Tal, M., and Lodish, H. F. (1994). The pancreatic β-cell glucose sensor. *Trends Biochem. Sci.* **19,** 535–538.]

glucose suppression of glucagon release stimulated by amino acids and of glucose stimulation of insulin secretion. The chemistatic properties of the α- and β-cells toward the prevailing blood glucose concentration are nicely interdependent. The β-cell threshold for glucose is ~5 mM, while the α-cell threshold for an amino acid mixture is 10 mM, provided that glucose is already present. The steep portions of the insulin release curve occur at blood concentrations of amino acids and glucose that occur postprandially. Conversely, the half-maximal glucose-mediated suppression of glucagon secretion occurs at a glucose concentration of 3 mM. These two dose–response curves describe the responses of the α- and β-cells as they work to effect a stable blood glucose concentration.

Given the possibility of metabolic interconversion of amino acids, fatty acids, and glycogen into glucose in liver, muscle, or adipose tissue, and given the reality of a changing dietary intake of these same nutrients, it is to be anticipated that these fuel metabolites themselves would coordinately modulate glucagon and insulin secretion. These relationships are incorporated into the model of a fuel receptor in a pancreatic β-cell receptor illustrated in Figure 7-24 and are further illustrated schematically in Figure 7-26.

Figure 7-26 schematically describes the secretion of insulin and glucagon under the conditions of (i) a normal basal metabolic state, (ii) exercise, and (iii) after ingestion of a carbohydrate meal. A crucial function of the integrated operation of the α–β-cell unit is to permit the appropriate regulation of the homeostasis of nutrients, such as amino acids, by allowing increases in the secretion of insulin without the danger of concomitant hypoglycemia. Thus, the aminogenic actions of insulin, which are manifest after the ingestion of a "protein-rich" meal, do not lead to hypoglycemia because there is a coupled stimulation of glucagon release; the glucagon then initiates hepatic glycogenolysis and increased release by the liver of as much glucose as leaves the extracellular space as a consequence of the insulin-mediated cellular uptake of glucose. Similarly, this same coupling of α- and β-cells makes possi-

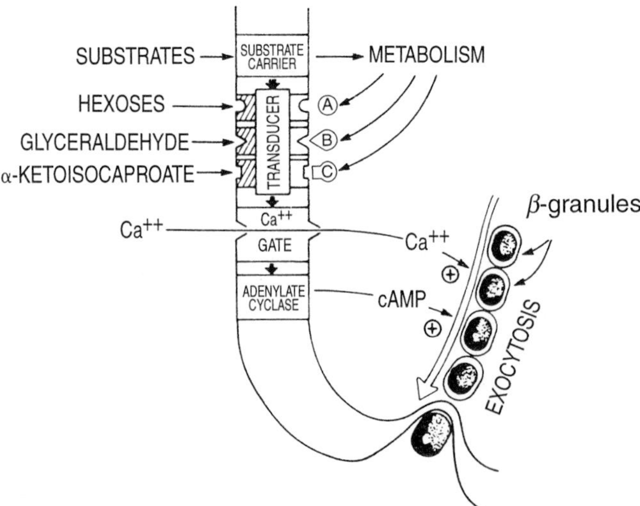

FIGURE 7-24 Model of a hypothetical fuel receptor in a pancreatic β-cell. The cell membrane of the islet cell is postulated to contain five coupled systems: (1) the substrate carriers, which may or may not function as fuel receptors; (2) a receptor–transducer complex with fuel receptors (e.g., for hexoses, glyceraldehyde, and α-ketoisocaproate) on the outside and sites for various fuel-derived metabolites and cofactors (e.g., A, B, C) on the inside of the membrane; (3) a Ca^{2+} gate that controls Ca^{2+} entry; (4) the adenylate cyclase system; and (5) the secretory complex comprising microtubule and secretory granules involved in the process of exocytosis driven by, Ca^{2+} and cAMP. [Modified from Matschinsky, F. M., Pagliara, A. A., Zawalich, W. S., and Trus M. D. (1979). Metabolism of pancreatic islets and regulation of insulin and glucagon secretion. *In* "Endocrinology" (L. J. DeGroot *et al.*, eds.), Vol. 2, p. 946. Grune & Stratton, New York.]

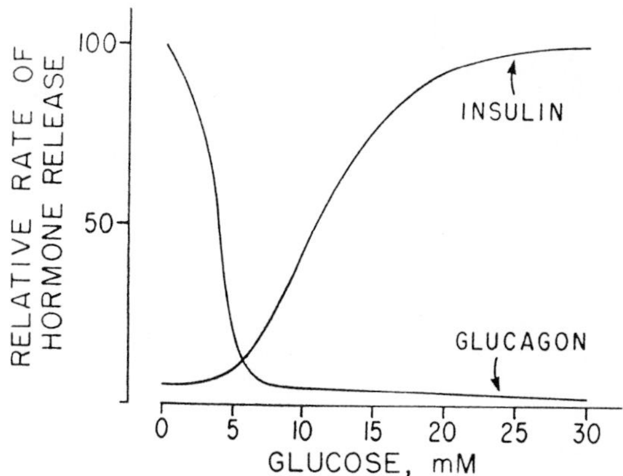

FIGURE 7-25 Dose–response curves of glucose suppression of glucagon release stimulated by amino acids and of glucose stimulation of insulin secretion as obtained by studies *in vitro* with the isolated perfused pancreas of the rat. [Modified from Matschinsky, F. M., Pagliara, A. A., Zawalich, W. S., and Trus, M. D. (1979). Metabolism of pancreatic islets and regulation of insulin and glucagon. *In* "Endocrinology" (L. J. DeGroot *et al.*, eds.), Vol. 2, p. 937. Grune & Stratton, New York.]

ble the inhibition of lipolysis without an associated fall in glucose. A key unresolved issue concerns the detailed role of somatostatin in pancreatic islet functioning. Although somatostatin is known to inhibit the secretion of both insulin and glucagon, the molecular basis and physiological rationale for this effect are not clear.

C. Pharmacological Agents Related to the Pancreas

Diabetes can be induced chemically by the acute administration of the drug alloxan 7 or the antibiotic streptozotocin (Figure 7-27). Each of these agents will selectively destroy the pancreatic β-cells, leaving the glucagon-secreting α-cells relatively functional. The biological actions of alloxan apparently relate to its affinity for zinc; hence, it is selectively accumulated by the zinc-rich β-cells. The seven-carbon sugar mannoheptulose also is known to disrupt the secretory response of the β-cells. These agents are frequently used to produce diabetes in experimental animal systems.

There are two groups of drugs that when given orally, are capable of effecting a sharp drop in the elevated blood glucose levels associated with diabetes (see Figure 7-27). These are the sulfonylurea derivatives (tolbutamide, chlorpropamide, glibenclamide, tolazamide, and glipizide) and the biguanide derivative (phenformin). When these oral drugs were first introduced in the early 1960s, there was much hope that they would provide a long-term solution to the management of diabetes. However, their general use has fallen off since the report in 1971 that persons treated with tolbutamide for 8 years had a higher death rate from cardiovascular disease than did control subjects.

The biological mechanism of action of the sulfonylurea agents is believed to be dependent upon their

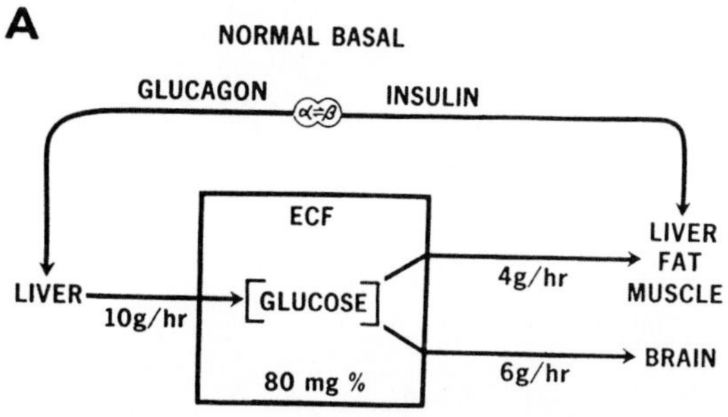

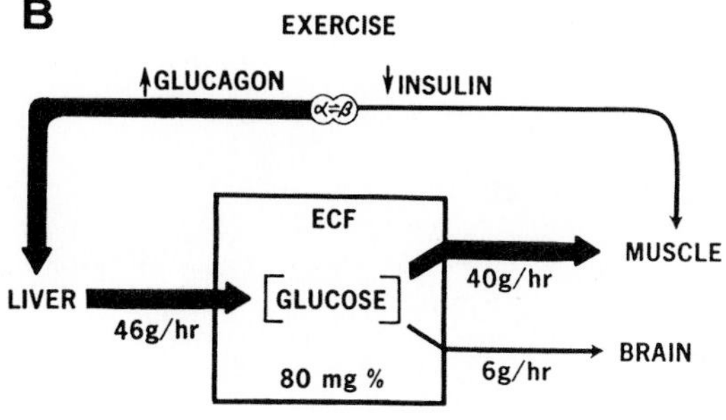

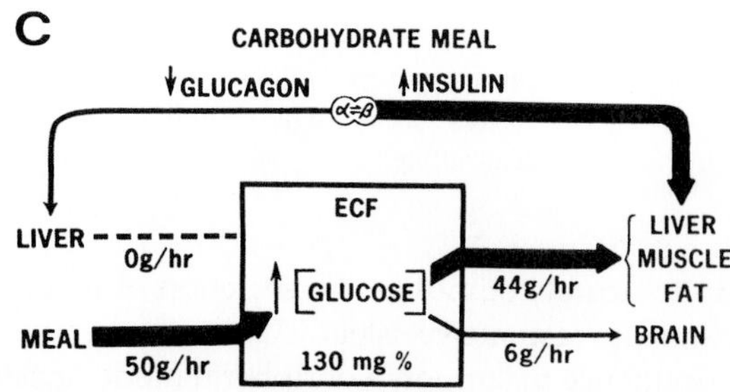

FIGURE 7-26 Comparison of the relative contributions of insulin and glucagon to the maintenance of blood glucose homeostasis in (A) normal basal state; (B) an instance of exercise; and (C) after ingestion of a carbohydrate-rich meal. (A) The role of glucagon in maintaining glucose influx at a rate equal to that of glucose efflux into insulin-independent tissues so as to maintain the extracellular fluid (ECF) glucose concentration above a hypoglycemic range is emphasized. (B) The role of glucagon in maintaining glucose influx equal to markedly increased glucose efflux into exercising muscle so as to maintain the ECF glucose concentration above a hypoglycemic range and maintain cerebral glucose delivery is emphasized. (C) The role of insulin in increasing glucose efflux during the rapid influx of ingested glucose so as to prevent an abnormal rise in ECF glucose is emphasized. Reduced glucagon secretion may facilitate this by helping to diminish hepatic glucose production. [Reproduced with permission from Unger, R. H., and Orci, L. (1979). Glucagon and diabetes. *In* "Endocrinology" (L. J. DeGroot *et al.*, eds.), Vol. 2, p. 962. Grune & Stratton, New York.]

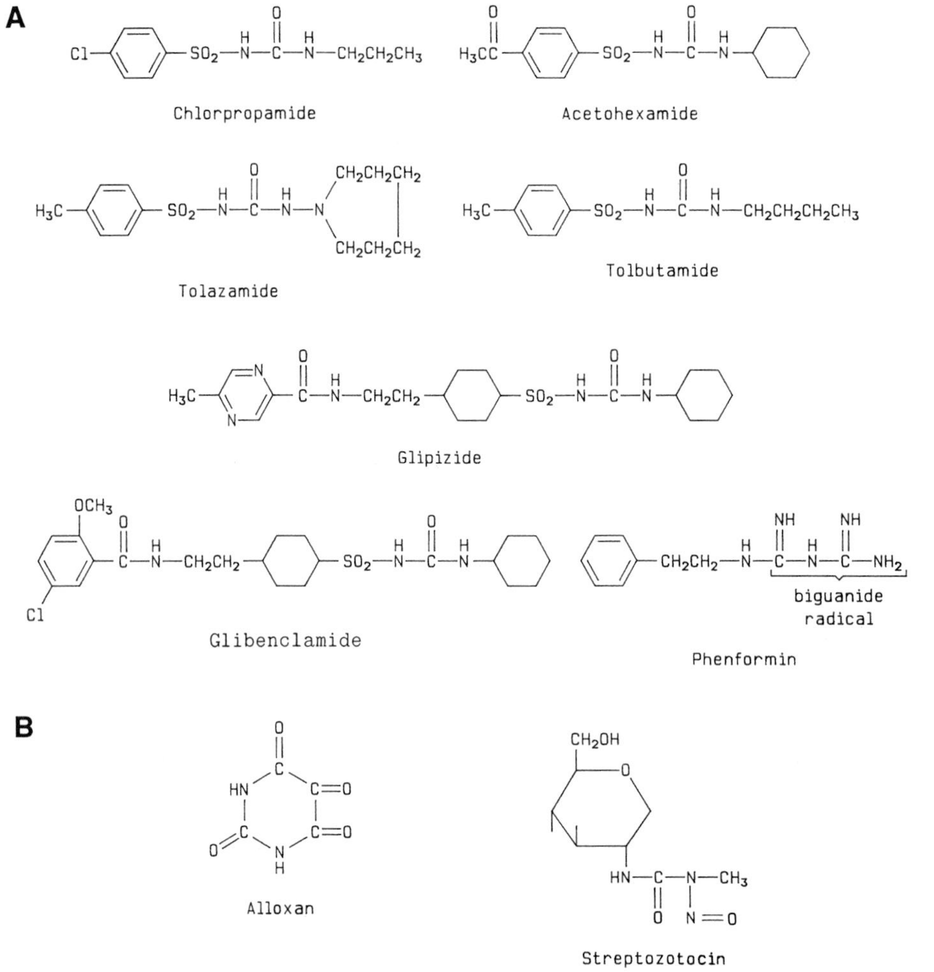

A

Chlorpropamide

Acetohexamide

Tolazamide

Tolbutamide

Glipizide

Glibenclamide

Phenformin (biguanide radical)

B

Alloxan

Streptozotocin

FIGURE 7-27 (A) Structures of the most frequently employed oral antidiabetic compounds; the "active" portion of the molecule is the sulfonylurea radical. (B) Structures of alloxan and streptozotocin, 2-substituted D-glucose derivatives that are employed to selectively destroy pancreatic β-cells.

ability to acutely stimulate the secretion of insulin by functionally responsive islets. Alternatively, there is some evidence to support a peripheral mode of action, wherein the sulfonylurea agents increase the sensitivity of tissues such as the liver to the prevailing levels of insulin.

The biguanides have long been known to be potent hypoglycemic agents; regrettably, they are also very toxic to many tissues. Although a detailed mode of action is not available, from *in vivo* studies it appears that these agents decrease the rate of intestinal absorption of glucose and increase the peripheral tissue uptake of blood glucose.

The principal agents that have been employed for the treatment of the some 6–10 million diabetic individuals in the United States who require drug-assisted management of their blood glucose levels are various preparations of insulin. These preparations have all been purified from extracts of either bovine or porcine pancreas. Because both bovine and porcine insulins have amino acid sequences significantly different from that of human insulin, a complicating problem has been the generation of antibodies by the patient to the heterologous insulin drug. Thus, a major success of the new recombinant DNA technology has been to clone the gene for human insulin and then mass-produce human insulin in quantities that have made it possible to provide human insulin as a drug for the diabetic population.

D. Metabolism of Insulin and Glucagon

The plasma half-life of insulin is 3–5 min. The principal pathway for the lowering of plasma insulin levels is receptor-mediated endocytosis of the insulin–receptor complex; a major site of this activity is in the liver, with the kidney as the second most important site. Mutant insulins that bind weakly to the insulin receptor are degraded 6–10 times more slowly.

The plasma half-life of glucagon is 6–7 min. Glucagon is largely inactivated by the liver and kidney. A glucagon-degrading enzyme has been purified from the liver; it specifically cleaves off the N-terminal tripeptide, H_2N-His-Ser-Gln-COOH.

V. BIOLOGICAL AND MOLECULAR ACTIONS

A. Interactions with Target Tissues

1. Glucagon

Glucagon secretion is stimulated by a fall in the blood glucose level or a rise in the blood levels of free fatty acids or certain amino acids (see Table 7-8). Most of the biological consequences of glucagon lead to an increase in the blood level of glucose. Virtually all of the glucagon secreted by the pancreatic α-cell is sequestered by the hepatocyte cells of the liver; thus, the extrahepatic blood levels of glucagon are quite low.

Glucagon elicits its plasma biological responses through interaction with a membrane receptor, which leads to an increase in intracellular [Ca^{2+}] and an elevation of cAMP. Figure 1-6 describes the cascade mechanism linking the occupancy of the glucagon receptor to the stimulation of glycogenolysis. The second messenger, cyclic AMP, then is the intracellular initiator of specific responses. Glucagon has a high affinity for its liver cell membrane receptor; the K_d is 4×10^{-9} *M*.

2. Insulin

Insulin is the chief hormone controlling intermediary metabolism. It affects virtually every tissue in the body, but principally liver, muscle, and adipose tissue. Its short-term effects are to reduce blood glucose and to conserve body fuel supplies. There are also a number of effects of insulin on the regulation of gene transcription and cell replication. Thus, an understanding of the mode of action of insulin is complex; Figure 7-28 presents an introductory schematic diagram of the signal transduction events that link the formation of an insulin–receptor complex to the generation of its many biological responses.

The number of insulin receptors in insulin target cells ranges from less than 100 to more than 200,000 per cell; the highest concentrations of the insulin receptor are on hepatocytes and adipocytes. The insulin receptor is a heterotetrameric transmembrane glycoprotein of 350–440 kDa. It is composed of two α-subunits (135 kDa) linked by disulfide bonds to two β-subunits (95 kDa). There is uncertainty whether the insulin receptor binds one or two insulin molecules. As shown in Figure 7-29, each α-subunit appears to contain an insulin-binding domain; however, when one site is occupied, it induces negative cooperativity for the second insulin-binding site.

Interestingly, like insulin, which consists of A and B chains derived from a common precursor (see Figure 7-13), both subunits of the insulin receptor are derived from a single amino acid chain proreceptor. Also, a variety of amino acid substitution mutations have been identified in both the α- and β-subunits of the insulin receptor present in patients with severe insulin resistance. Occupancy of the receptor initiates a complex series of response cascades that involve over 50 proteins–enzymes. The response cascades can be classified at three levels (see Figure 7-30). Level I describes the immediate consequences of occupancy of the α_1-subunit of the insulin receptor and includes activation of the receptor's tyrosine kinase, which catalyzes the phosphorylation of the cytoplasmic protein *I*nsulin *R*eceptor *S*ubstrate-1; this protein then interacts with several other proteins, including PI 3-kinase and SH2-GRB2. Level II involves a series of serine and threonine kinases and phosphatases, particularly the enzymes MAP kinase[4] and RAF kinase. These latter two kinases then initiate a cascade of phosphorylation and dephosphorylation steps that are coupled to both the cytoplasmic and nuclear responses described in level III. In level III, the principal cytoplasmic responses are the regulation of glycogen synthesis via glycogen synthase and the regulation of glucose transporter translocation. The nuclear responses include stimulation of gene expression, cell growth, and protein synthesis.

The time scale of response to insulin ranges from immediate (occurring within seconds to minutes after occupancy of the insulin receptor), to intermediate (minutes to hours), to long-term (many hours to days).

As emphasized in Figure 7-28, the first step of signal transduction initiated by the occupied insulin receptor is the activation of its tyrosine kinase activity when the α-subunit-binding domain is occupied by insulin. The key substrate for the tyrosine kinase activity has been identified as being a 131-kDa intracellular protein termed IRS-1. IRS-1 is essential for some, if not all, of insulin's biological actions. The most noteworthy feature of IRS-1 is that it possesses 21 potential tyrosine phosphorylation sites; 6 sites

[4] The serine–threonine kinase was originally named because it phosphorylates *m*icrotubule-*a*ssociated *p*rotein (MAP). The enzyme is now known as MAP kinase, because it is stimulated by many mitogens, or as extracellular signal-related kinase (ERK), because it is regulated by a variety of extracellular ligands.

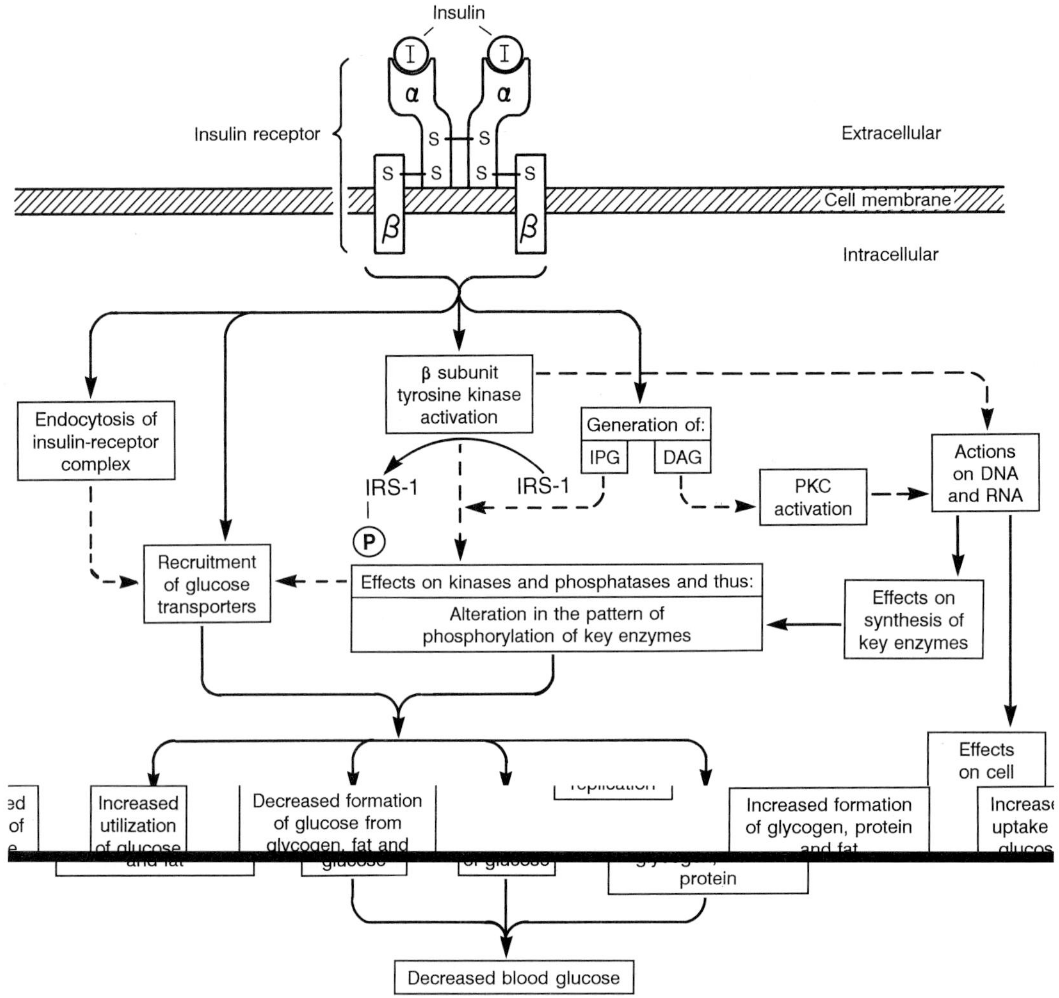

FIGURE 7-28 Schematic diagram of the linking of the insulin receptor to the activation of signal transduction pathways that generate biological responses. Abbreviations: I, insulin; IRS-1, insulin receptor substrate; IPG, inositol phosphoglycan; DAG, diacylglycerol; PKC, protein kinase C. [Modified with permission from Rang, H. P., Dale, M. M., Ritter, J. M., and Gardner P. (1995). "Pharmacology," p. 406. Churchill-Livingstone, New York.]

use the Tyr-Met-X-Met motif, 3 use the Tyr-X-X-Tyr motif, and 12 utilize other hydrophobic motifs. IRS-1 also contains over 30 potential serine–threonine phosphorylation motifs. At least eight of these IRS-1 tyrosine sites become phosphorylated by the activated insulin receptor.

After multiple phosphorylation of IRS-1, it binds several proteins with SH2 domains, including phosphatidylinositol-3′-kinase (PI 3′-kinase) and the GRB2 protein comprising two subunits, an 85-kDa regulatory subunit (which has two SH2 domains) and a 110-kDa catalytic subunit. Activation of PI 3-kinase results in the activation within minutes of a nonclassical phosphatidylinositol pathway (for comparison see Figure 1-39), which results in the phosphorylation of phosphatidylinositol (PI), PI-4P, and PI-4,5P_2 on the 3′ position, yielding, respectively, PI-3P, PI-3,4P_2, and PI-3,4,5P_3. None of the PI products phosphorylated by the PI 3′-kinase can be hydrolyzed by any known phospholipase. Thus, it remains to be elucidated how the PI 3′-kinase generates subsequent biological responses. Also, the precise details of the other interaction of IRS with SH2 proteins (including GRB2 and *ras*, including the *ras*-related proteins that lead to modulation of gene expression) remains to be elucidated. The biochemical properties of *ras* proteins closely resemble those of G proteins involved in the signal transduction of transmembrane receptors. The

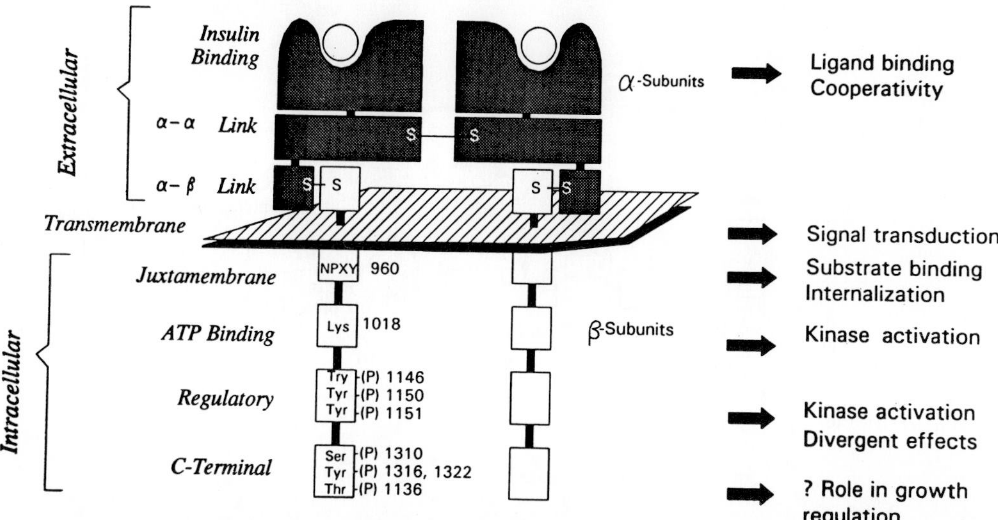

FIGURE 7-29 Schematic model of the insulin receptor tetramer. The α-subunits, which are entirely extracellular, consist of 735 amino acids and define the insulin-binding site. The two α-subunits are joined to each other by a disulfide bond; each α-subunit is also linked to a β-subunit by a disulfide bond. The β-subunits span the membrane; they have 193 extracellular amino acids, 23 transmembrane amino acids organized in an α-helix, and an intracellular domain of 402 amino acids. The approximate locations of the various domains are indicated. [Modified with permission from Kahn, C. R., and White, M. F. (1995). Molecular mechanism of insulin action. *In* "Endocrinology" (L. J. DeGroot, M. Besser, H. G. Burger, J. L. Jameson, D. L. Loriaux, J. C. Marshall, W. D. Odell, J. T. Potts, Jr., and A. H. Rubenstein, eds), 3rd edi., Vol. 2, 1373–1388. W. B. Saunders Co., Philadelphia. PA.]

ras proteins are known to bind GTP in analogy with the G proteins.

As indicated in Figure 7-28, one of the principal cytoplasmic effects of the activated insulin receptor is to stimulate glucose uptake in muscle and adipose tissue. The principal pathway for glucose entry into the cell is via a Na^+-independent glucose transporter, which facilitates the movement of glucose down a concentration gradient (outside to inside the cell).

There is a family of glucose transporter isoforms (see Table 7-9). From the perspective of insulin action, the most important transporter is GLUT-4; it is expressed in the prime target organs of insulin action, e.g., skeletal and cardiac muscles and adipose tissue. GLUT-4 is a 520-amino acid protein that can span the cell membrane 12 times (see Figure 7-31). It is believed that the response of insulin to stimulate the cellular uptake of glucose is mediated via energy-dependent recycling of the GLUT-4 (where it is catalytically inactive) from intracellular vesicles to the cellular membrane, where it catalyzes glucose uptake into the cell. Thus, when the insulin receptor is occupied, GLUT-4 moves to the cell membrane, and following removal of insulin the GLUT-4 returns to its intracellular pool. Exactly how insulin receptor kinase activation is coupled to glucose transporter migration remains to be elucidated.

B. Biological Actions

1. Introduction

The principal biological actions of insulin and glucose are complex and interdependent upon the presence or absence of the partner hormone and the delicate balance of anabolism and catabolism that is occurs throughout the body in response to changes in caloric intake, caloric composition, and the degree of physical activity. In most situations the roles of insulin and glucagon are antagonistic. Tables 7-10 and 7-11, respectively, summarize the actions of insulin and glucagon in their principal target tissues, the liver, muscle, and adipose tissue. Also, it should be appreciated that the fundamental intermediary metabolism capabilities of these three tissues and their "normal" oxidative substrates vary depending upon the prevailing metabolic condition. The actions of insulin and glucagon in the liver, muscle, and adipose tissue will be discussed separately. Figure 7-32 compares the major metabolic effects that are found in the absence and presence of insulin for liver, muscle, and adipose tissue.

2. Liver

The liver performs an indispensable role in maintaining an adequate blood level of glucose. It possesses the enzymatic capability either to generate glucose

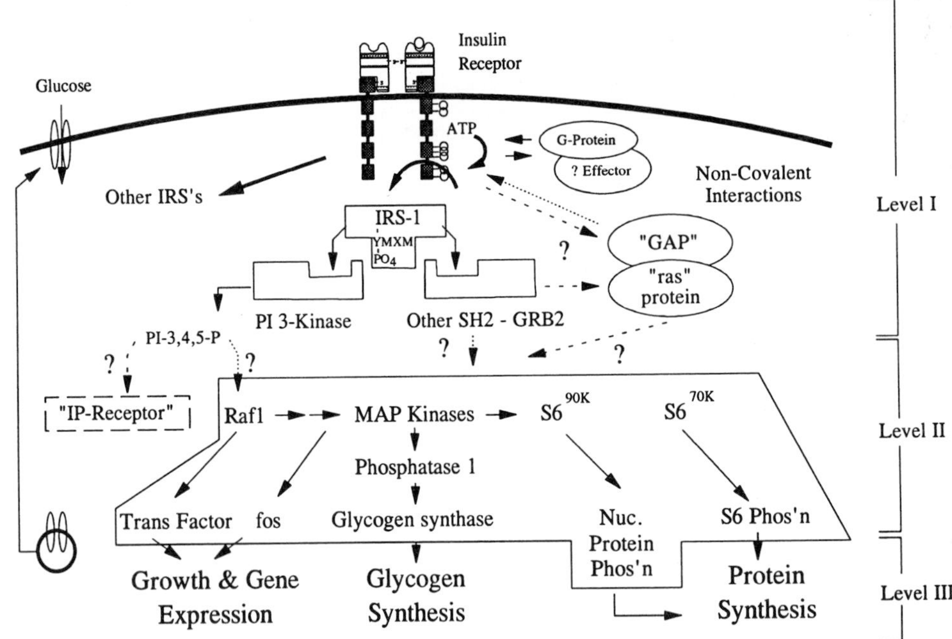

FIGURE 7-30 Model of insulin actions in a generic target cell. The responses to occupancy of the insulin receptor occur at three levels. Other details of the insulin receptor and signal transduction are presented in Figures 7-25 and 7-26. Abbreviations: IRS-1, insulin receptor substrate-1; PI 3′-kinase, phosphatidylinositide-3′-kinase; *ras* protein, a GTP-binding protein; MAP kinase, see text for discussion; PI-3,4,5-P, phosphatidylinositide-3,4,5-P_3; Raf[1]. [Modified with permission from Kahn, C. R., and White, M. F. (1995). Molecular mechanism of insulin action. *In* "Endocrinology" (L. J. DeGroot, M. Besser, H. G. Burger, J. L. Jameson, D. L. Loriaux, J. C. Marshall, W. D. Odell, J. T. Potts, Jr., and A. H. Rubenstein, eds), 3rd edi., Vol. 2, pp. 1373–1388. W. B. Saunders Co., Philadelphia. PA.]

from stored glycogen or to store excess glucose as glycogen. In addition, the liver contains an active gluconeogenesis capability that will permit the production of glucose from three- and four-carbon fragments derived from amino acids. Also, the liver can oxidatively metabolize glucose and other smaller metabolites to H_2O and CO_2, which generates ATP, or convert two-carbon fragments to larger free fatty acids, which then are incorporated into triglycerides and phospholipids.

Figure 7-33 diagrams the enzyme reactions involved in the storage of glucose and mobilization of glycogen. Several key enzymes in this scheme are markedly af-

TABLE 7-9 Family of Glucose Transporters

	Characteristics of known human glucose transporters			
Code	**Number of amino acids**	**Tissue specificity**	**Proposed function**	**Insulin stimulation**[a]
		Facilitative glucose transporters		
GLUT1	492	Blood–brain barrier, erythrocyte, fibroblast	Basal glucose uptake into cells	+
GLUT2	524	Liver, β-cell, intestine	Low-affinity glucose transport	−
GLUT3	496	Brain, fibroblast	Neuronal glucose uptake	?
GLUT4	509	Fat, skeletal muscle, heart	Insulin-stimulated glucose uptake	+++
GLUT5	501	Small intestine, sperm	Fructose transport	?
		Active glucose transporters		
SGLT1	664	Intestine, kidney	Intestinal absorption, renal reabsorption	−

[a] Indicates the extent of stimulation of insulin secretion: +, modest; +++, vigorous; −, no effect; ?, effect not known.
[b] Modified with permission from Stephens, J. M., and Pilch, P. F. (1995). The metabolic regulation and vesicular transport of GLUT4, the major insulin-responsive glucose transporter. *Endoc. Rev.* **16,** 529–546.

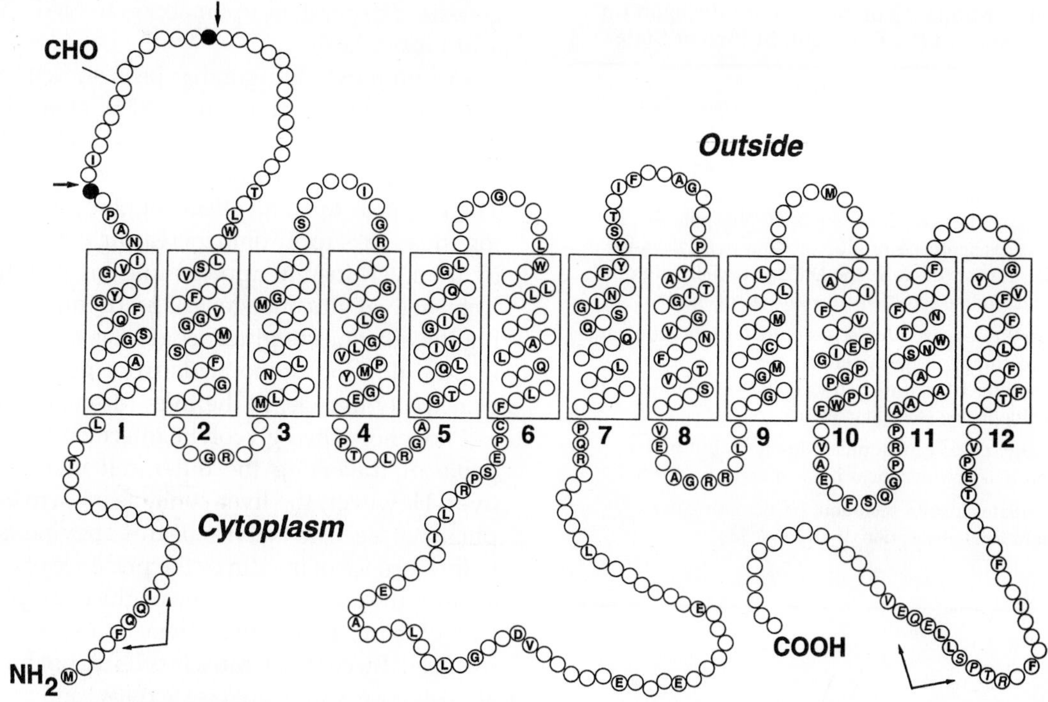

FIGURE 7-31 Structural model of the major insulin-responsive glucose transporter, GLUT4 (see also Table 7-9). The transmembrane positioning of the GLUT4 sequence is based upon the predicted topography of GLUT1; the 12 membrane-spanning helices are numbered and boxed. Those amino acids that are conserved among GLUT 1 → 4 are indicated by a boldface single-letter code. Sites of glycosylation are indicated by CHO. [Modified with permission from Stephens J. M. and Pilch, P. F. (1995). The metabolic regulation and vesicular transport of GLUT4, the major insulin-responsive glucose transporter. *Endocr. Rev.* **16**, 529–546.]

fected by the presence or absence of glucagon or insulin. Activation of the glycogenolytic system involves the phosphorylation of key enzymes, while inactivation necessitates dephosphorylation. On the one hand glycogen is phosphorylyzed (broken down) by the cascade of cAMP-governed reactions that convert the relatively inactive phosphorylase *b* into the more active phosphorylase *a*. On the other hand, under conditions where it is appropriate to convert glucose into glycogen, there is activation of a phosphatase that converts the phosphorylated D or *dependent* form of glycogen synthetase into the dephosphorylated I or *independent*

TABLE 7-10 Summary of Actions of Insulin on Several Tissues

Tissue	Insulin-sufficient state	Insulin-deficient state
Liver	No effect on glucose uptake Stimulates biosynthesis of hexokinase IV and activates glycogen synthetase I Promotes glycolysis and formation of ATP	Uptake of free fatty acids and conversion to ketones
Muscle	Stimulation of glucose uptake Stimulates biosynthesis of hexokinase II and pyruvate kinase Stimulates glycolysis and formation of ATP Increases muscle glycogen levels and creatine phosphate	Impaired blood glucose
Adipose	Stimulation of glucose uptake	Decreases triglyceride synthesis due to a lack of glycerol phosphate
	Enchances glycolysis, which makes available glycerol phosphate, which, in turn, enhances triglyceride synthesis Inhibits lipase activity	Stimulation of lipolysis and release of FFA into the bloodstream
Brain	No direct actions of insulin; brain dependent upon blood glucose	None

TABLE 7-11 Summary of Actions of Glucagon on Several Tissues in the Glucagon-Sufficient State

Tissue	Effect
Liver	Inactivates glycogen synthetase and activates phosphorylase α, which leads to an activation of glycogenolysis Increases the activity of glucose 6-phosphatase Enhances synthesis of glucose from pyruvate and lactate as well as amino acids, especially arginine and alanine (i.e., activates gluconeogenesis)
Muscle	Muscle does not contain receptors for glucagon, and accordingly it has no effect Major effects on muscle metabolism are mediated by epinephrine (see Chapter 11)
Adipose	In large doses can stimulate lipolysis, but under normal circumstances it has little or no effect
Pancreas	Stimulates insulin secretion, particularly after intestinal absorption of amino acids
Brain	None

form. It will be noted that a key feature of this system is the unique role played by the protein phosphorylase *b* kinase-kinase, which has two distinct catalytic activities. When it is phosphorylated (as a consequence of the presence of glucagon), it initiates the cascade that ultimately leads to the conversion of phosphorylase *b* into phosphorylase *a* and the production of glucose-1-P. Alternatively, when the phosphorylase *b* kinase-kinase is not phosphorylated, it acquires the activity of glycogen-I-synthetase-kinase. Thus, the presence of glucagon favors the generation of an active phosphorylase *b* kinase-kinase, while the presence of insulin activates a phosphatase that leads to glycogen formation.

In addition to the glucagon–insulin modulation of glycogen storage and mobilization in the liver, these two peptide hormones also modulate the balance between gluconeogenesis and lipogenesis. Under circumstances of glucose demand, gluconeogenesis will predominate to convert into glucose those carbon skeletons derived from amino acids or glycolytic intermediates produced prior to pyruvate. Under conditions of glucose excess, glycolytic intermediates and free fatty acids can be directed either to storage as triglycerides, to oxidation by the TCA cycle, or to production of ketone bodies. It is not yet clear which detailed cellular or hormonal signals determine whether a given free fatty acid is dedicated to storage as triglyceride or to conversion to CO_2 and / or ketones. These relationships are diagrammed in Figure 7-34. Reversal of the glycolytic pathway requires access to NADH, whereas lipogenesis is dependent upon access to NADPH for fatty acid biosynthesis.

In summary, glucagon has been shown to stimulate the conversion of pyruvate, lactate, alanine, and glycerol into glucose. This is accomplished largely by modulating key enzymes of the gluconeogenic pathway. There appear to be no effects of glucagon on stimulating substrate supply or in increasing amino acid uptake by the liver. Insulin appears to exert its inhibitory effects on liver gluconeogenesis by (i) inhibiting or slowing the enzymes of gluconeogenesis and (ii) diminishing the flow of amino acids from peripheral tissues, principally muscle, to the liver (discussed later).

Phosphorylated glycolytic intermediates are not capable of traversing the outer cell membrane of the liver. However, the liver contains an active glucose-6-phosphatase. The activity of this enzyme is increased in the absence of insulin or the presence of cortisol; this ensures the ready conversion of glucose 6-phosphate to free glucose, which may then be exported from the liver cell. By contrast, muscle cells do not have measurable glucose-6-phosphatase activity, and thus the muscle glycogen stores cannot be mobilized for the maintenance of blood glucose levels.

3. *Muscle*

In the absence of insulin there is a stimulation of net protein catabolism in muscle. The resulting free amino acids are released into the bloodstream and delivered to the liver where they are oxidatively deaminated. The resulting carbon fragments are then committed to gluconeogenesis or catabolism to yield ketone bodies and / or CO_2. The resulting increase in nitrogen is converted to urea and excreted in the urine.

Muscle tissue is known to contain insulin but not glucagon receptors. Insulin occupancy of these receptors leads to an increased uptake of both glucose and amino aids; there is also an associated stimulation of protein synthesis. The biochemical bases for these effects are not yet clearly understood, in that they cannot be explained by changes in cellular cAMP levels. Also, actinomycin D does not block insulin-mediated stimulation of protein synthesis. It has been proposed that the 40 S large subunit of ribosomes obtained from the muscle of diabetic animals is defective.

After an overnight fast and in the absence of physical activity, muscle tissue is largely dependent upon the oxidation of free fatty acids to meet its energy demands. Under these conditions there would be a significant pool of muscle glycogen. With the initiation of mild-to-moderate exercise, the muscle tissue successively oxidizes its own glycogen, then blood glucose,

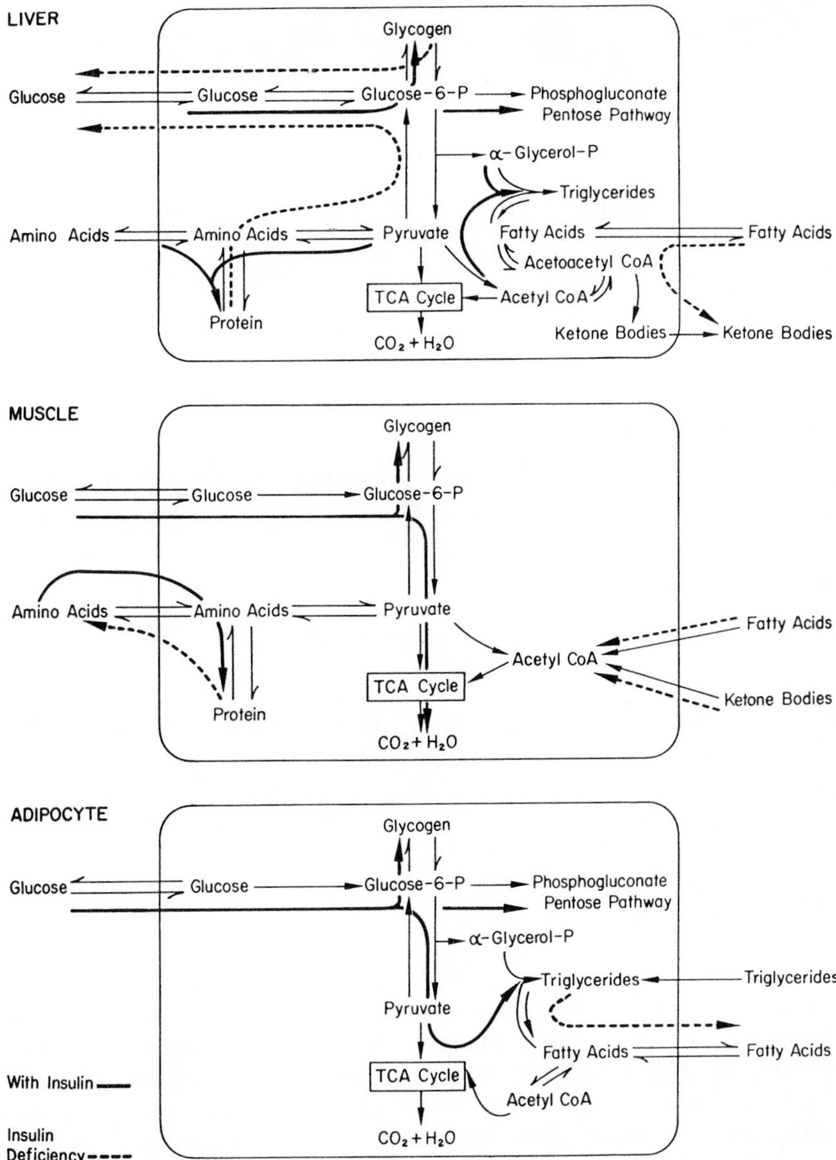

FIGURE 7-32 Comparison for liver, muscle, and adipose tissue of the major metabolic effects that occur in the absence and presence of insulin. Thick solid arrows show the pathways that are favored in the presence of insulin, while thick broken arrows depict those that predominate when the action of the hormone is insufficient. For an explanation, see the text. [Modified with permission from Gilman, A. G., Goodman, L. S., and Gilman, A. (1980). "The Pharmacological Basis of Therapeutics," 6th ed. Macmillan, New York.]

and finally blood-delivered free fatty acids derived from adipose and hepatic stores.

4. *Adipose Tissue*

Adipose tissue is one of the principal target organs for insulin action. The extensive amount of triglycerides stored in adipocytes serves as an important fuel source under conditions of dietary caloric restriction or prolonged exercise. The mobilized free fatty acids are systematically delivered to a number of key organs (heart and kidney), which can directly utilize them as a substrate for oxidative metabolism. If the free fatty acids were taken up by the liver, they could be esterified into triglycerides or phospholipids, oxidized to CO_2, or more likely converted into ketones, particularly β-hydroxybutyrate and acetoacetate. These ketones in turn can also serve as fuels for most extrahepatic tis-

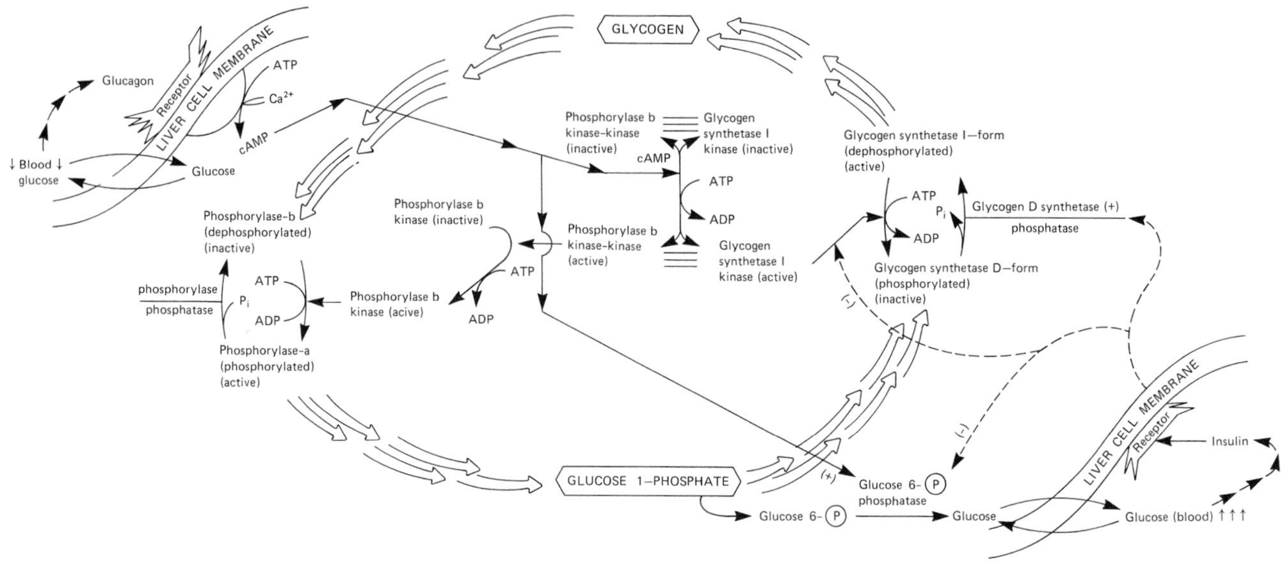

FIGURE 7-33 Summary of the biochemical consequences of insulin and/or glucagon actions in the storage of glucose or the mobilization of glycogen in a liver cell. Events depicted occur in the cell cytoplasm after binding of insulin or glucagon to their respective membrane receptors.

sues, particularly the brain, under conditions of very low blood glucose levels.

Insulin can suppress ketogenesis via two general mechanisms: (i) In the adipocyte, insulin inhibits lipolysis, which generates free fatty acids. (ii) In the liver, insulin also blocks or reduces the extent of liver oxidation and enhances the storage of the free fatty acids as triglycerides.

The rate-limiting step in the biosynthesis of triglycerides by both the liver and adipose tissue is the availability of glycerol phosphate. This three carbon sugar is not transported through the blood; hence, it must be generated *in situ* in the cell where triglyceride synthesis is occurring. This in turn implies that there should be adequate supplies of substrates to the glycolytic pathway to ensure the availability of the glycerol phosphate in sufficient amounts.

VI. CLINICAL ASPECTS

A. Insulin

Diabetes mellitus, or more usually diabetes, is the most common disease associated with the pancreatic islets; it may affect 1–5% of the total population. In the United States it is estimated that more than 12 million people have some form of diabetes; annually some 200,000 individuals are newly diagnosed as diabetic. Diabetes is the eighth leading health-related cause of death in the United States. More than 50% of diabetics die because of coronary disease; individuals with juvenile-onset diabetes usually succumb because of renal failure. Diabetes is the second highest causative factor of blindness.

The term diabetes means "running through"; characteristic features are polydipsia and polyuria. *Mellitus* is Latin for honey (i.e., sweet). In a classic sense, diabetes mellitus is not a single disease because it has no single definable and distinct etiology.

The major pathophysiological abnormalities generated by diabetes mellitus include the following: (i) glucose intolerance or inappropriately high blood glucose levels in the presence of elevated or normal insulin concentrations; (ii) acidosis and ketosis; (iii) reduction of the enzymes for gluconeogenesis; (iv) reduced growth; (v) a microangiopathy, including a thickened basement membrane of capillaries in muscle and retinal blood vessels and of the capillaries of the renal glomerulus; and (vi) a neuropathy, including peripheral sensory and motor defects.

Clinically there are two general classifications of diabetes: type I or insulin-dependent diabetes mellitus, formerly designated juvenile-onset diabetes, and type II or insulin-independent mellitus, formerly designated as maturity-onset diabetes. It is not yet clear for either form of diabetes, in molecular terms, the specific defects that likely are present in the complex biosynthetic and secretory steps required for normal insulin secretion by the β-cell. What is clear is that in type I diabetes the β-cells in the islet are partially or totally destroyed, and consequently there is no endogenous insulin secretion. This is determined by the absence of plasma levels of C peptide and an absolute requirement of insulin for the maintenance of life. Individuals

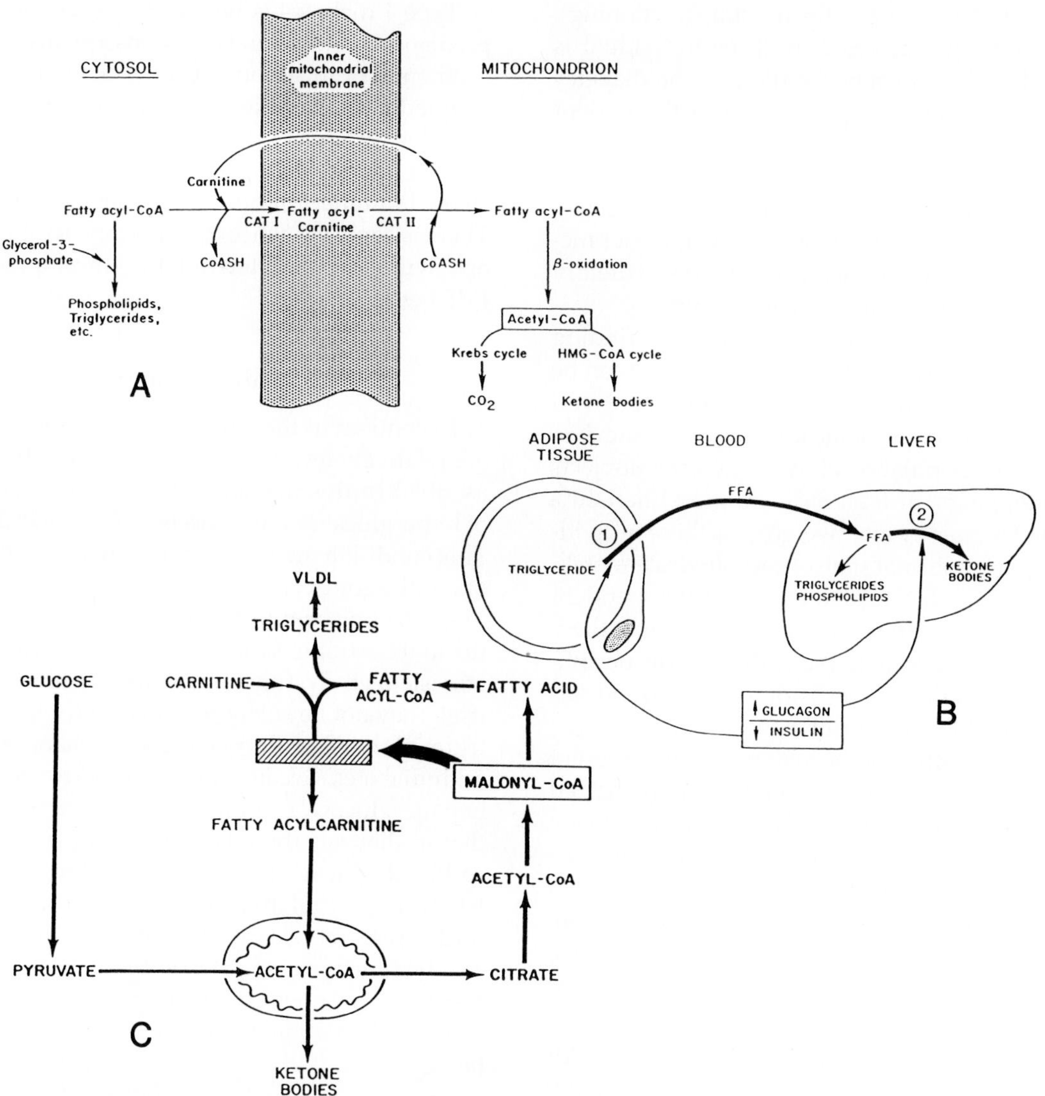

FIGURE 7-34 (A) Major pathways of fatty acid metabolism in the liver, where CAT I and CAT II refer to carnitine acyltransferases located on the outer and inner surfaces of the inner mitochondrial membrane. (B) Summary of the effects of insulin and/or glucagon on hepatic ketone production. (C) Interrelationships between the pathways of fatty acid synthesis and oxidation in the liver. [Modified from McGarry, J. D. (1979). New perspectives in the regulation of ketogenesis. *Diabetes* **28**, 517–523, and reproduced with permission from the American Diabetes Association, Inc.]

with type I diabetes are highly prone to develop metabolic ketosis, and in the untreated state there are the acute symptoms of polyuria, polydipsia, weight loss, and fatigue, all of which are secondary to the absence of insulin. Patients with juvenile-onset diabetes require mandatory daily treatment with some form of drug insulin preparation.

Diabetic ketoacidosis is a clinical condition characterized by hyperglycemia, hyperketonemia, and concomitant metabolic acidosis that largely results from elevated blood levels of β-hydroxybutyrate and acetoacetate. A propensity to diabetic ketoacidosis is the primary basis for clinical classification of a subject with juvenile-onset diabetes. The clinical symptoms arise from a low-to-undetectable blood level of insulin and result in the adverse consequence of persistent hyperglycemia. In the extreme, an afflicted subject will become unconscious due to "diabetic coma." Diabetic ketoacidosis is an acute life-threatening emergency and if left untreated leads to hypokalemia and cardiovascular failure. Diabetic ketoacidosis accounts for ~65% of all diabetic hospital admissions in the age group 1–20 years and 40% in the 21–34-years age group.

Type II or maturity-onset diabetes is a form of diabetes mellitus in which there is only a relative lack of insulin, but the secretion is not correctly "geared" for glucose signals (i.e., the β-cells are still functioning). It is frequently not diagnosed until the individual is middle-aged. In this form of the disease, the diabetes is stated to be insulin-independent since the patient will not develop ketosis without insulin therapy. Type II diabetics may require only intermittent adjunctive insulin therapy, usually to manage fasting hyperglycemia. In many subjects, adequate glucose and fuel metabolite homeostasis can be achieved by careful dietary management of nutritional and caloric intake.

The most commonly associated physical finding with maturity-onset diabetes is obesity. Obesity can be defined as an enlargement of the adipose depot and is usually associated with some kind of pancreatic dysfunction. For the population of hyperglycemic subjects older than 20 years, at least 90% are more than 10% overweight. Frequently after weight loss these individuals may regain a normal state of carbohydrate metabolism. A significant fact related to insulin-independent diabetes associated with or without obesity is that there is an apparent decrease in the sensitivity of the peripheral tissues to insulin. As a result, obese individuals have an elevated level of insulin secretion.

From the perspective of the insulin receptor present in the various target tissues, it should be emphasized that in the obese form of diabetes, characterized by glucose intolerance, hyperinsulinemia, and resistance to both endogenous and exogenous insulin, there is no evidence for a "nonfunctional form" of the insulin. The circulating ratio of proinsulin to insulin is normal; both cross-react normally in radioimmunoassays for insulin. The defect appears to be both in the number of receptors present in the target cells and in postreceptor events.

The pathogenesis of diabetes mellitus, besides being related to environmental and dietary factors, also clearly contains a genetic or familial component. Although the precise statistics supporting this statement vary with their source, it is estimated that between 25 and 50% of diabetic patients have "family histories" consistent with a genetic component. The reported incidence of diabetes is much higher in children derived from parents who both have diabetes; also, monozygotic (identical) twins have a significantly higher incidence of diabetes than dizygotic (fraternal) twins. Current evidence favors the position that the inheritance is polygenic or multifactorial; in some specific instances there is clear evidence of a recessive gene for diabetes. The Puma Indians have an incidence rate of diabetes that approaches 50% or higher. Also, there are animal models of diabetes (the *ob/ob* and *db/db* diabetic mouse and the Zucker rat) that clearly reflect genetic transmission.

Type I diabetes is believed to occur in genetically predisposed individuals as a consequence of toxic environmental or infectious insult to the pancreatic β-cells coupled with the actions of an aggressive immune system that gradually, but persistently, destroys the β-cells. It has been proposed that certain β-cell proteins share immune epitopes with certain viral peptides. There is some evidence that immune suppression therapy will slow or interrupt the development of type I diabetes.

B. Glucagon

In contrast to the situation relating to insulin, there are relatively few disease states that can be specifically ascribed to dysfunctions of glucagon. The syndrome of hyperglucagonemia was only first described in 1966. This condition arises as a consequence of the presence of a glucagon-secreting tumor of the pancreatic α-cell.

There is also mounting evidence that diabetes mellitus in its extreme form is a double disorder of pancreatic islet cell dysfunction. Thus, insulin deficiency by itself may not be solely responsible for the broad spectrum of intermediary metabolic changes associated with diabetes. Because insulin is a known inhibitor of hepatic glucose production and secretion, it may be that in diabetes there is an inappropriately high level of blood glucagon for the prevailing blood glucose levels (i.e., a "relative" excess of glucagon). If indeed future research clearly documents that there is an α-cell dysfunction associated with diabetes mellitus, then a substantial revision of the pathophysiological concepts and treatment modalities for this disease will be necessary.

References

A. Books

Ganong, W. F. (1995). "Review of Medical Physiology," 17th ed. Appleton & Lange, Norwalk, CT.

Rang, H. P., Dale, M. M., Ritter, J. N., and Garder, P. (1995) "Pharmacology." Churchill-Livingston, New York

B. Review Articles

Bell, G. I., and Reisine, T. (1993). Molecular biology of somatostatin receptors. *Trends Neurosci.* **16,** 34–38.

Campbell, J. S., Seger, R., Graves, J. D., Graves, L. M., Jensen, A. M., and Krebs, E. G. (1995). The MAP Kinase Cascade. *In* "Recent Progress in Hormone Research" (C. W. Bardin, ed.), Vol. 50, pp. 131–159. Academic Press, Inc., San Diego.

Cooper, G. J. S. (1994). Amylin compared with calcitonin gene-related peptide: Structure, biology, and relevance to metabolic disease. *Endocr. Rev.* **15,** 163–201.

Efrat, S., Tal, M., and Lodish, H. F. (1994). The pancratic β-cell glucose sensor. *Trends Biochem. Sci.* **19,** 535–538.

Halaas, J. L., Gajiwala, K. S., Maffei, M., Cohen, S. L., Chait, B. T., Rabinowitz, D., Lallone, R. L., Burley, S. K., and Friedman, J. M. (1995). Weight-reducing effects of the plasma protein encoded by the *obese* gene. *Science* **269,** 543–546.

Kahn, C. R., and White, M. F. (1995). *In* "Endocrinology" (L. J. DeGroot, *et al.*, eds.), 3rd ed., Vol. 2, pp. 1373–1388. W. B. Saunders Company, Philadelphia, PA.

Kahn, C. R., White, M. F., Shoelson, S. E., Backer, J. M., Araki, E., Cheatham, B., Csermely, P., Folli, F., Goldstein, B. J., Huertas, P., Rothenberg, P. L., Saad, M. J. A., Siddle, K., Sun, X., Wilden, P. A., Yamada, K., and Kahn, S. A. (1993). The insulin receptor and its substrate: molecular determinants of early events in insulin action. *In* "Recent Progress in Hormone Research" (C. W. Bardin, ed.), Vol. 48, pp. 291–339. Academic Press, Inc., San Diego.

Malaisse, W. J. (1995). *In* "Endocrinology" (L. J. DeGroot, *et al.*, eds.), 3rd ed., Vol. 2, pp. 1329–1336. W. B. Saunders Company, Philadelphia, PA.

Mommsen, T. P., and Plisetskaya, E. M. (1991). Insulin in fishes and agnathans: history, structure, and metabolic regulation. *Rev. Aquat. Sci.* **4,** 225–259.

Pelleymounter, M. A., Cullen, M. J., Baker, M. B., Hecht, R., Winters, D., Boone, T., and Collins, R. (1995). Effects of the *obese* gene product on body weight regulation in *ob*/*ob* mice. *Science* **269,** 540–543.

Philippe, J. (1991). Structure and pancreatic expression of the insulin and glucagon genes. *Endocr. Rev.* **12,** 252–271.

Stephens, J. M., and Pilch, P. F. (1995). The metabolic regulation and vesicular transport of GLUT4, the major insulin-responsive glucose transporter. *Endocr. Rev.* **16,** 529–546.

Steiner, D. F., Chan, S. J., Welsh, J. M., and Kwok, S. D. M. (1985). Structure and evolution of the insulin gene. *Annu. Rev. Genet.* **9,** 463–484.

Taylor, S. I., Cama, A., Accili, D., Barbetti, F., Quon, M. J., Sierra, M. L., Suzuki, Y., Koller, E., Levy-Toledano, R., Wertheimer, E., Moncada, V. Y., Kadowaki, H., and Kadowaki, T. (1992). Mutations in the insulin receptor gene. *Endocr. Rev.* **13,** 566–595.

Unger, R. H. (1991). Diabetic hyperglycemia: Link to impaired glucose transport in pancreatic β cells. *Science* **251,** 1200–1205.

Unger, R. H., and Orci, L. (1995). *In* "Endocrinology" (L. J. DeGroot, *et al.*, eds.), 3rd ed., Vol. 2, pp. 1337–1353. W. B. Saunders Company, Philadelphia, PA.

White, M. F., and Kahn, C. R. (1994). The insulin signaling system. *J. Biol. Chem.* **269,** 1–4.

C. Research Papers

Baker, E. N., Blundell, T. L., Cutfield, J. F., Cutfield, S. M., Dodson, E. J., Dodson, G. G., Hodgkin, D. M. C., Hubbard, R. E., Isaacs, N. W., Reynolds, C. D., Sakabe, K., Sakabe, N., and Vijayan, N. M. (1988). The structure of 2 Zn pig insulin crystals at 1.5 Å resolution. *Philos. Trans. R. Soc. Lond on (Biol.)* **319,** 369–456.

Blackstone, C. D., Seino, S., Takeuchi, T., Yamada, T., and Steiner, D. F. (1988). Novel organization and processing of the guinea pig pancreatic polypeptide precursor. *J. Biol. Chem.* **263,** 2911–2916.

Brimble, K. S., and Ananthanarayanan, V. S. (1993). Calcium binding and translocation properties of glucagon and its fragments. *Biochemistry* **32,** 1632–1640.

Considine, R. V., Sinha, M. K., Heiman, M. L., Kriauciunas, A., Stephens, T. W., Nyce, M. R., Ohannesian, J. P., Marco, C. C., McKee, L. J., Bauer, T. L., and Caro, J. F. (1996). Serum immunoreactive-leptin concentrations in normal-weight and obese humans. *New Engl. J. Med.* **334,** 292–295.

Cooper, G. J. S., Leighton, B., Dimitriadis, G. D., Parry-Billings, M., Kowalchuk, J. M., Howland, K., Rothbard, J. B., Willis, A. C., and Reid, K. B. M. (1988). Amylin found in amyloid deposits in human type 2 diabetes mellitus may be a hormone that regulates glycogen metabolism in skeletal muscle. *Proc. Natl. Acad. Sci. USA* **85,** 7763–7766.

Cooper, G. J. S., Willis, A. C., Clark, A., Turner, R. C., Sim, R. B., and Reid, K. B. M. (1987). Purification and characterization of a peptide from amyloid-rich pancreases of Type 2 diabetic patients. *Proc. Natl. Acad. Sci. USA* **84,** 8628–8632.

Eck, M. J., Shoelson, S. E., and Harrison, S. C. (1993). *Nature* **362,** 87–91.

Fischbarg, J., Cheung, M., Czegledy, F., Li, J., Iserovich, P., Kuang, K., Hubbard, J., Garner, M., Rosen, O. M., Golde, D. W., and Vera, J. C. (1993). Evidence that facilitative glucose transporters may fold as β-barrels. *Proc. Natl. Acad. Sci. USA* **90,** 11658–11662.

Halaas, J. L., Gajiwala, K. S., Maffei, M., Cohen, S. L., Chait, B. T., Rabinowitz, D., Lallone, R. L., Burley, S. K., and Friedman, J. M. (1995). Weight-reducing effects of the plasma protein encoded by the *obese* gene. *Science* **269,** 543–546.

Jelinek, L. J., Lok, S., Rosenberg, G. B., Smith, R. A., Grant, F. J., Biggs, S., Bensch, P. A., Kuijper, J. L., Sheppard, P. O., Sprecher, C. A., O'Hara, P. J., Foster, D., Walker, K. M., Chen, L. H. J., McKernan, P. A., and Kindsvogel, W. (1993). Expression cloning and signaling properties of the rat glucagon receptor. *Science* **259,** 1614–1616.

Johnson, J. H., Ogawa, A., Chen, L., Orci, L., Newgard, C. B., Alam, T., and Unger, R. H. (1990). Underexpression of β-cell high K_m glucose transporters in noninsulin-dependent diabetes. *Science* **250,** 546–549.

Nagamatsu, S., Nishi, M., and Steiner, D. (1991). Biosynthesis of islet amyloid polypeptide. *J. Biol. Chem.* **266,** 13737–13741.

Nishi, M., Sanke, T., Nagamatsu, S., Bell, G. I., and Steiner, D. F. (1990). Islet amyloid polypeptide—a new β cell secretory product related to islet amyloid deposit. *J. Biol. Chem.* **265,** 4173–4176.

Rohner-Jeanrenaud, F., and Jeanrenaud, B. (1996). Obesity, leptin, and the brain. *New Engl. J. Med.* **334,** 324–325.

Rouille, Y., Martin, S. K., and Steiner, D. F. (1995). Differential processing of proglucagon by the subtilisin-like prohormone convertases, PC2 and PC3 to generate glucagon or glucagon-like peptides, *J. Biol. Chem.* **270,** 26488–26496. *USA.*

Sasaki, K., Dockerill, S., Adamiak, D. A., Tickle, I.J., and Blundell, T. (1975). X-ray analysis of glucagon and its relationship to receptor binding. *Nature* **257,** 751–757.

Steiner, D. F., Smeekens, S. P., Ohagi, S., and Chan, S. J. (1992). The new enzymology of precursor processing endoproteases. *J. Biol. Chem.* **267,** 23435–23438.

Unson, C. G., MacDonald, D., Ray, K., Durrah, T. L., and Merrifield, R. B. (1991). Position 9 replacement analogs of glucagon uncouple biological activity and receptor binding. J. Biol. Chem. **266,** 2763–2766.

Vionnet, N., Stoffel, M., Takeda, J., Yasuda, K., Bell, G.I., Zouali, H., Lesage, S., Velho, G., Iris, F., Passa, Ph., Froguel, Ph., and Cohen, D. (1992). Nonsense mutation in the glucokinase gene causes early-onset non-insulin-dependent diabetes mellitus. *Nature* **356,** 721–722.

Weiss, M. A., Frank, B. H., Khait, I., Pekar, A., Heiney, R., Shoelson, S. E., and Neuringer, L. J. (1990). NMR and photo-CIDNP studies of human proinsulin and prohormone processing intermediates with application to endopeptidase recognition. *Biochemistry* **29,** 8389–8401.

Yamada, Y., Post, S. R., Wang, K., Tager, H. S., Bell, G. I., and Seino, S. (1992). Cloning and functional characterization of a family of human and mouse somatostatin receptors expressed in brain, gastrointestinal tract, and kidney. *Proc. Natl. Acad. Sci. USA* **89,** 251–255.

CHAPTER

8

Gastrointestinal Hormones

I. INTRODUCTION

A. Background

The stomach, intestines (small and large), liver, gallbladder, and pancreas operate as a physiological unit to effect the digestion and absorption of bodily nutrients. A discussion of the principal pancreatic hormones, insulin and glucagon, is presented in Chapter 7. This chapter will focus only on the gastrointestinal hormones.

The gastrointestinal hormones are a family of polypeptides produced by specialized endocrine cells present in the stomach and small and large intestines, which function both as traditional hormonal messengers and as neurotransmitters. These hormones mediate a variety of specific biological responses by the stomach, small intestine, both the endocrine and exocrine pancreas, liver, and gallbladder, which, when integrated, optimize the physiological conditions necessary to permit the efficient digestion and absorption of protein, carbohydrates, and fat from the lumen of the intestine.

Gastrointestinal function is modulated by a complex series of hormonal and neural interrelationships. There is also increasing evidence of neuroendocrine communication between the brain and gut that provides an additional level of integration and complexity to these digestive processes.

The gastrointestinal tract holds a special place in the history of endocrinology; it was the site of discovery, by Bayliss and Starling in 1902, of secretin, which was the first hormone to be described and which permitted the formulation of the concept that hormones are chemical messengers.

B. Resume of the Gastrointestinal Hormones

Table 8-1 tabulates the 11 gastrointestinal (GI) hormones, including the 3 major GI hormones, gastrin, secretin, and cholecystokinin (CCK), and summarizes their sites of production and biological actions; Table 8-2 summarizes the neurotransmitter hormones produced by the gastrointestinal tract. As will be reviewed in Table 8-3, many of the gastrointestinal hormones play key roles in the plethora of steps and processes associated with the digestion and absorption of food.

C. Problems of Food Processing and Digestion

All animals require continued and regular intake of food so that they can meet their bodily nutrient requirements. Briefly, these food components include the following: (a) Adequate caloric intake of carbohydrate, fat, and protein to provide the spectrum of substrates required for the generation of metabolic energy, which is indispensable for the biosynthesis, modification, and repair of tissue components, and the generation of mechanical (muscle contraction) and electrical energy (nerve impulse); (b) adequate intake of essential substances not capable of being biosynthesized by the animal in question (i.e., essential amino acids, essential fatty acids, and all vitamins); and (c) adequate intake of macro and trace minerals. Thus, the term food is a general label to include the variety of chemical substances, including proteins, carbohydrates, fats, minerals, and vitamins, required to provide nourishment (i.e., maintenance of life).

However, the simple process of swallowing food does not mean that these food substances are available to sustain life. It is only after they have left the lumen of the intestine and appeared in the blood or lymph that they can be said to be "in the body." Although some food substances, such as minerals and free amino acids, can be absorbed with little or no change in the chemical form in which they are ingested, other food components must be subjected to extensive physical and chemical modification prior to their entry into the lymph or bloodstream. Thus, proteins, polysaccharides (both macromolecules), fats (triglycerides), and phospholipids must all be degraded to their constituent

TABLE 8-1 Summary of Gastrointestinal Hormones[a]

Hormone	Stimulatory agent(s)	Site of production	Major action(s)
Major GI Hormones			
CCK[b]	Chyme, fats, fatty acids	Duodenal mucosa	Stimulates pancreatic amylase, trypsin, and chymotrypsin secretion and gallbladder contraction
Gastin	Distension of stomach	Gastric mucosa of pylorus	Stimulates acid and pepsin secretion
Secretin	Acidity of intestinal contents	Duodenal mucosa	Pancreatic bicarbonate and H_2O secretion
Other GI Hormones			
Enkephalins			Opiate-like action; reduce intestinal motility
Enteroglucagon	Ileum and colon		Tropic effects on enterocytes
GIP[c]	Oral glucose and amino acids	Duodenal mucosa	Inhibits gastric acid secretion; enhances glucose-mediated insulin secretion
Motilin	Alkaline conditions	Duodenal mucosa	Stimulates intestinal motility
Neurotensin		Ileum	Stimulates pancreatic HCO_3^- secretion; stimulates colonic motility and inhibits gastric and small intestinal motility
Pancreatic Polypeptide (PP)[d]		Endocrine pancreas	Inhibits pancreatic HCO_3^- and protein secretion
PYY[d]			Inhibits gastric motor motility and acid secretion
Somatostatin	Gastrin	Nerve cells in intestine	Exerts inhibitory actions on release of CCK, VIP,[e] GIP

[a] Table 8-2 lists neurotransmitters produced by the gastrointestinal tract.
[b] CCK is the abbreviation for the peptide hormone cholecystokinin–pancreozymin; it is usually designated simply as cholecystokinin.
[c] GIP is known as both glucose-dependent insulinotropic polypeptide and also gastrin-inhibiting polypeptide.
[d] The pancreatic polypeptide family consists of three peptides, pancreatic polypeptide (PP), peptide YY (PYY), and neuropeptide Y (NPY; see Table 8-2). All three are amino acid peptides.
[e] See Table 8-2.

TABLE 8-2 Neurotransmitters Produced by the Gastrointestinal Tract[a]

Neurotransmitter	Actions
Bombesin[b] (gastrin releasing peptide GRP)	Stimulates gastrin release
Calcitonin gene-related peptide (CGRP)	Stimulates somatostatin release and smooth muscle contraction
Endothelin	Potent vasoconstrictor; acts locally in the GI tract; function not known (see Chapter 15)
Galanin	Not definitely known; may inhibit postprandial release of somatostatin, neurotensin, and PP
Neuropeptide Y (NPY)	Potent vasoconstrictor; inhibits acetylcholine release
Neuromedin B	Stimulates gastrin release; inhibits hypothalamus TSH release
Neuromedin U	Regulates anterior pituitary function
Substance P and the tachykinins	Smooth muscle contractions
Vasoactive intestinal peptide (VIP)	Smooth muscle relaxation; stimulates pancreatic HCO_3^- secretion

[a] Table 8-1 lists hormones produced by the gastrointestinal tract.
[b] Bombesin is a member of the bombesin-like family of peptides found in the skin, gut, and brain of frogs; it shares biological activity with the mammalian gastrin-releasing peptide (GRP), the tachykinins, and neuromedin B.

components—amino acids, mono- and disaccharides, fatty acids, and glycerol, respectively—before they can be efficiently absorbed.

The overall physical and chemical properties of the various foodstuffs (e.g., water solubility, net charge), as well as the nature of their structural units and bonding (e.g., amide, glycoside, ester), pose diverse biochemical problems to be resolved by the digestive system. The system of digestion, then, is the processing of the ingested bulk food into molecular forms capable of entering the intestinal absorptive process.

Table 8-3 summarizes the steps and processes associated with the digestive system. The digestive process involves the complex integration of voluntary and involuntary muscle contractions, parasympathetic and sympathetic neural actions, release of gastrointestinal and other hormones, and biosynthesis and release of a host of digestive enzymes (e.g., pepsin, trypsin, chymotrypsin, amylase) and digestive reagents (e.g., HCl and HCO_3^-), along with digestive detergents (e.g., bile acids). These substances all collectively work together to process the food and present it to the intestinal absorptive machinery in an optimal form for absorption. The intestinal absorptive machinery includes the integrated functioning of the mouth, stomach, small and large intestines, pancreas, liver, and gallbladder.

As a consequence of evolutionary pressures, a number of hormonal systems have emerged to participate in the regulation of digestion and absorption of key dietary nutrients. These include the gastrointestinal hormones (this chapter), insulin and glucagon (see Chapter 7), and vitamin D metabolites (one of the calcium-regulating hormones—see Chapter 9). In addition, many other hormones are known to exert effects on either the digestion process or the gastrointestinal tract to modulate the absorption of various nutrients.

II. ANATOMICAL, MORPHOLOGICAL, AND PHYSIOLOGICAL RELATIONSHIPS

A. Gastroenteropancreatic System

The gross anatomical organization of the digestive tract, the stomach, the small intestine, and the large intestine of humans is shown in Figure 8-1. The anatomical relationships between the liver, pancreas, and gallbladder are given in Figure 7-2.

The digestive tube or alimentary canal in humans is a 9-m-long muscular tube extending from the lips to the anus. Indicated in Figure 8-1 are the individual components of the alimentary canal; each part makes an essential contribution to the overall process of making food available to the cells of the body. The operations within the digestive tube are supplemented by several accessory organs, including the teeth, tongue, salivary glands, pancreas, and liver.

B. Stomach

Figure 8-2 presents a cross-sectional view of the stomach. The gastric secretions of hydrochloric acid and pepsin are dependent upon the functional capacity and mass of the various secretory elements in the gastric mucosa, as well as being a consequence of the action of the several hormonal and neural stimulants and inhibitors.

The gastric mucosa is divided into three anatomical regions: cardiac, oxyntic, and pyloric. The total surface area of the gastric mucosa is ~800 cm^2. Anatomically, the cardiac portion encompasses the first few millimeters below the gastroesophageal junction and includes the cardiac mucus glands; it secretes alkaline mucus and a small amount of electrolytes. The oxyntic or fundic portion comprises 75–80% of the entire stomach. The oxyntic glands, which number ~35 million

TABLE 8-3 Steps and Components of the Digestive Process

Process	Anatomical location	Purpose
Food entry		
Consumption of food/mastication	Mouth	Support bodily nutrient requirements
Salivary secretion (stimulated by the nervous system): 1200 ml is secreted per day.	Salivary glands	Conversion of starches into dextrins and maltose
Voluntary swallowing of food	Tongue/pharnyx/esophagus	Entry of food into digestive tract
Involuntary peristaltic muscular contraction	Esophagus	To move food along the digestive tract
Digestion in the stomach		
Secretion of gastric juice (HCl + pepsin) as a consequence of neural, mechanical, and hormonal (gastrin) stimuli (2000 ml secreted/day)	Stomach	Conversion of proteins to polypeptides
Involuntary mechanical contractions	Stomach	Mixing of intestinal contents and when pylorus opens transfer of stomach contents (chyme) to the small intestine
Digestion in the small intestine		
Neural and hormonal (secretion of pancreozymes); mediated secretion of pancreatic juice (1200 ml secreted/day)	Exocine or acinar pancreas secretions are ducted into the duodenum	
Bicarbonate secretion		HCO_3^- neutralizes HCl from stomach and creates a favorable duodenal pH (6.0–7.0) so that pancreatic enzymes may operate efficiently
Amylase, trypsinogen, and chymotrypsinogen secretion and conversion to "active" enzymes		Hydrolysis of amide, glycoside, and ester bonds of peptides, carbohydrates, and lipids
Neural stimulation and duodenal muscular distension and stimulation of intestinal juice (4000–5000 ml secreted/day)	Duodenum	
Enterokinase (activates trypsinogen conversion into trypsin)		Further processing of proteins → peptides
Proteases (chymotrypsinogens, trypsinogen, lipases, amylases)		Further processing of proteins → peptides
Sucrase, maltase, lactase; specific transport systems (all bound to brush border membranes)		Further processing of dextrins and other partially hydrolyzed substances
Synthesis and secretion of bile acids (by the liver) followed by storage in the gallbladder; hormonally stimulated secretion (by cholecystokinin and secretin) of the gallbladder to release bile acids (600–800 ml secreted/day)	Liver (site of production), gallbladder (storage), transport (through bile duct) into duodenum	Emulsify fats to facilitate absorption of fatty acids
Involuntary mechanical activity		
Segmentation	Duodenum, jejunum, and ileum	Squeezing and mixing of intestinal contents (8–10 contractions/min in duodenum)
Peristalsis	Duodenum, jejunum, and ileum	Onward movement of intestinal contents
Chemical activity	Duodenum, jejunum, and ileum	Continued saponification of fat and hydrolysis of proteins → peptides to amino acids and carbohydrates–dextrins to mono- and disaccharides followed by cellular absorption and transport to the blood or lymph compartments
Digestion in the large intestine Involuntary mechanical activity	Large intestine	Mixing movements that promote reabsorption of H_2O and compaction of intestinal contents (7000–8000 ml of H_2O reabsorbed/day)
Bacterial action	Large intestine	Acts on undigested residues to effect fermentation of carbohydrate and further degradation of proteins as well as biosynthesis of vitamin K

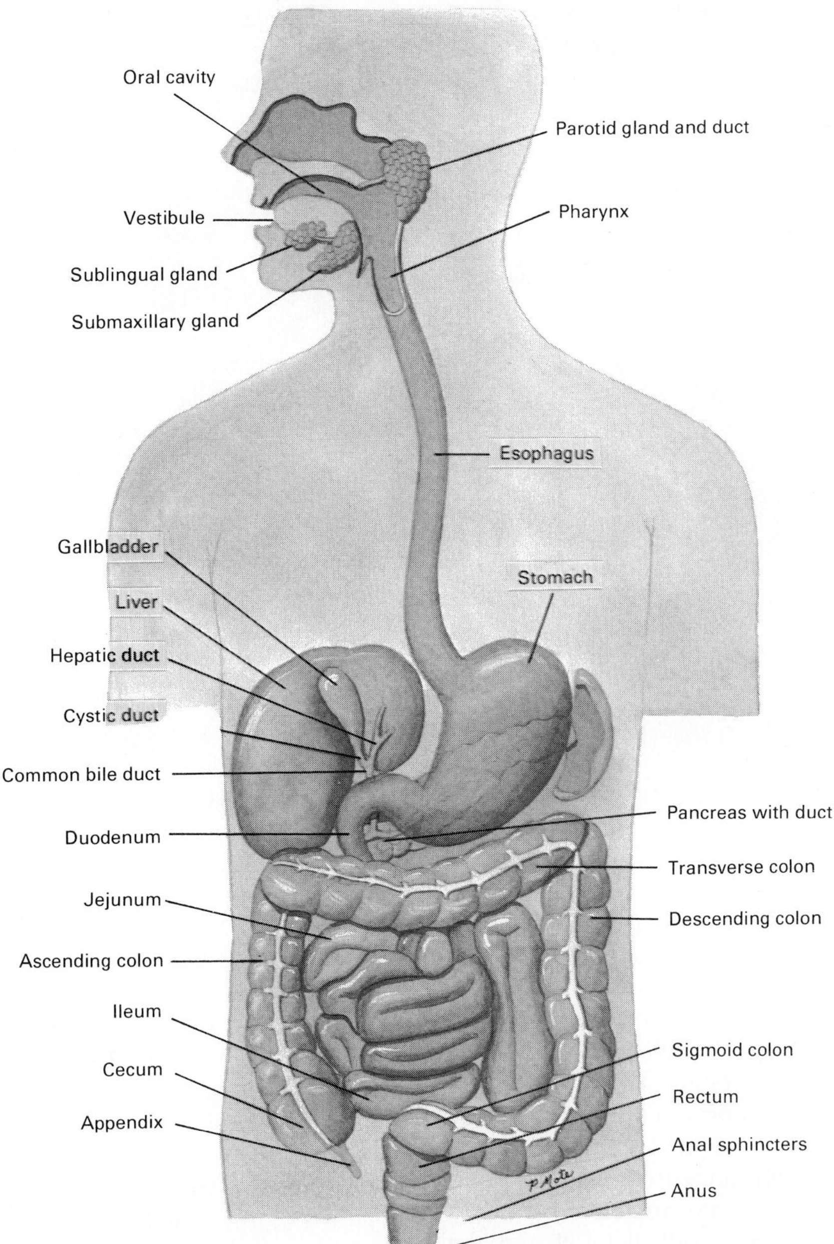

FIGURE 8-1 Diagram of the human digestive system.

in the stomach of humans, are straight or slightly coiled tubules and are largely composed of parietal cells and mucus-secreting cells. The parietal cells, constituting 75% of the mucosal volume, have large numbers of mitochondria, which are required for the metabolic energy demands involved in the production of concentrated hydrochloric acid. The parietal or oxyntic cells undergo a dramatic morphological transformation during stimulation, so that the internal tubulovesicles coalesce and become transformed into microvillar membranes. The chief cells are the primary site of the secretion of pepsin. The pyloric gland area constitutes the lower 20% of the gastric mucosa and contains pyloric glands, which are largely composed of mucus-secreting cells.

C. Small and Large Intestines

The small intestine of humans is divided anatomically into the duodenum (~25 cm), which is connected to the stomach via the pylorus, the jejunum (~275 cm), and the ileum (~425 cm), which connects to the large

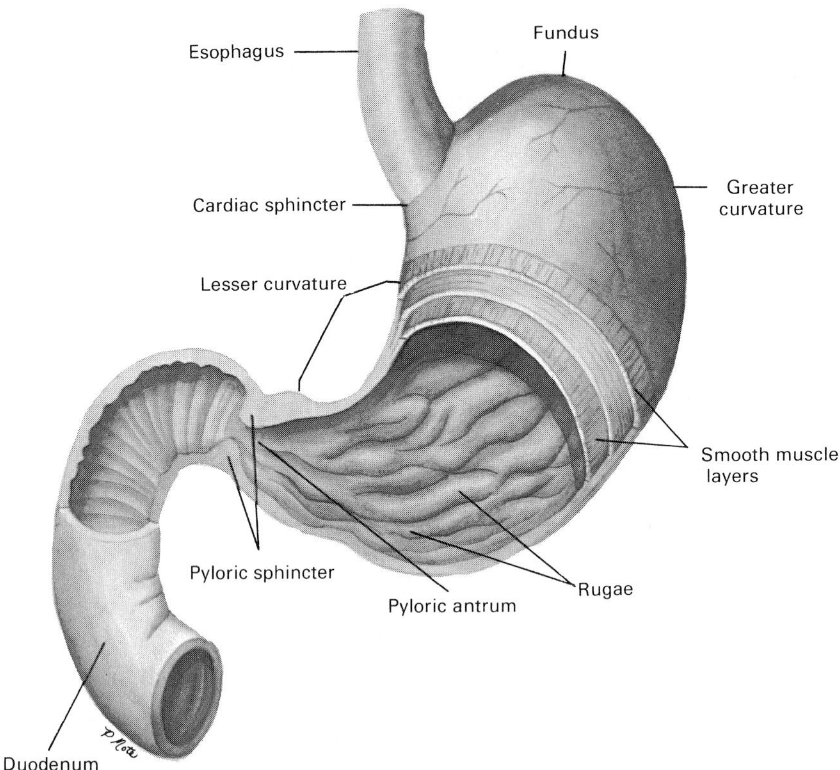

FIGURE 8-2 Longitudinal section of the human stomach and upper duodenum.

intestine. The large intestine (~180 cm) is sequentially divided into the colon, appendix, rectum, and anal canal.

The morphology and cellular organization of the intestine are superbly adapted to efficiently effect the absorption of dietary constituents. The lumen of the small intestine is composed of a multitude of finger-like projections, which are termed villi (see Figure 8-3A). Typically these villi range from 0.5 to 1.5 mm in length. They are composed principally of two types of cells: the columnar epithelial cell and the goblet or mucus-secreting cell. The cells on the villus are rapidly renewed. The mucosal cells originate in a progenitor population existent at the base of the villus in the crypt of Lieberkuhn, and ultimately, after migration up the villus, the cells are sloughed off into the lumen of the intestine. Only after epithelial cells enter the villus do they differentiate into fully functional absorptive cells. This differentiation process is characterized by a change in the RNA/protein ratio and a change in the activity of various enzymes as a function of the position of the epithelial cell on the villus. Intestinal epithelial cell turnover times of 70–100 hr have been reported in mice, rats, and humans. The rate of epithelial cell renewal increases when there is an increased demand for new cells, as in the recovery from radiation injury, response to parasite infestation, or adaptation to partial intestinal resection.

The apical surface of these cells, which is exposed to the lumen, is referred to as the brush border or the microvillar membrane (see Figure 8-3B). These microvilli are ~1 μm long and ~0.1 μm wide. As a consequence, the surface area of the mucosa is increased by 600-fold compared to the total area of a hypothetical cell without microvilli. A variety of carbohydrate hydrolases and an alkaline phosphatase are specifically associated with this brush border region of the intestinal mucosal cell. This is believed to be the site of localization of permeases that are postulated to be involved in the absorption of food materials into the cell. Beneath the columnar epithelial cells lie the lymphatic and blood vascular systems, which project into each villus and thus provide an efficient mechanism for the translocation of the various substrates once they exit from the epithelial cells.

The outermost surface of the columnar epithelial cell, that is, the microvilli or brush border region, is covered with a filamentous, mucopolysaccharide coat termed the glycocalyx or "fuzzy coat." The function of the glycocalyx is not specifically known.

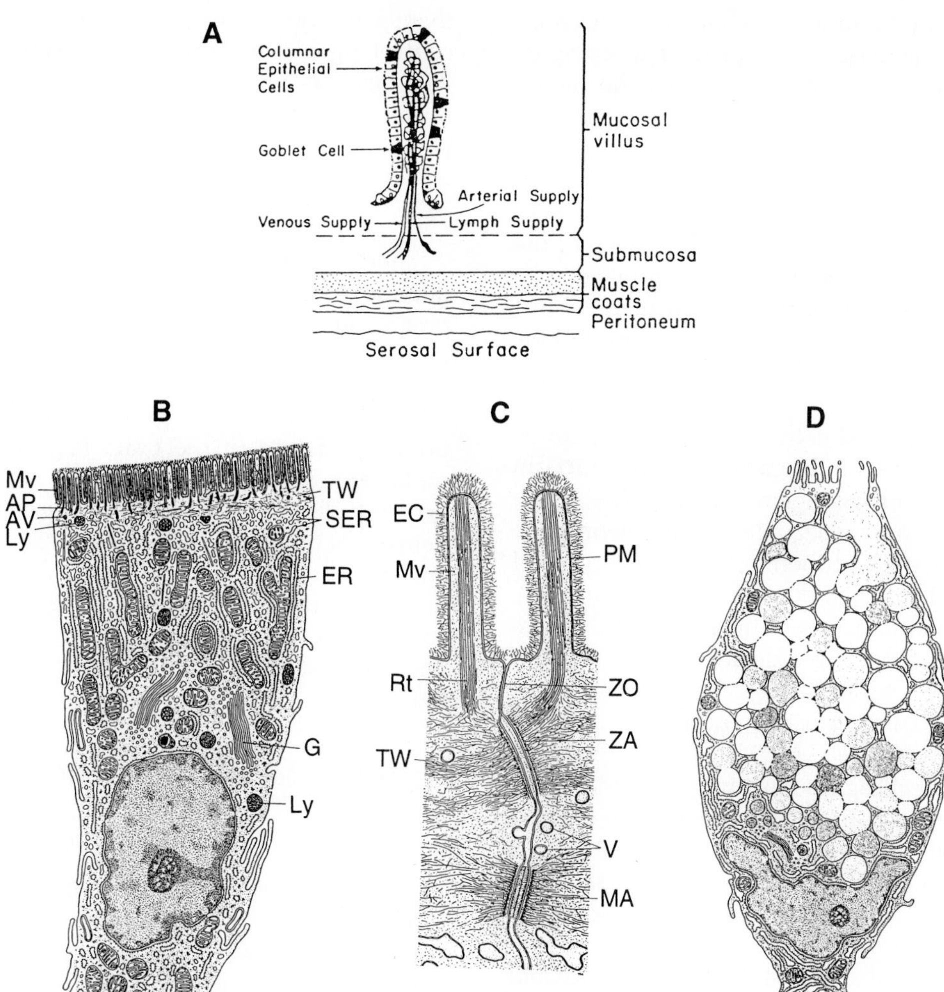

FIGURE 8-3 Organization of the intestinal mucosa. (A) Diagrammatic representation of the intestinal villus at the macro level. (B) Representation of an intestinal columnar epithelial cell, which lines the intestinal mucosa. Microvilli (Mv) extend from the free surface and form the striated edge. The surfaces of the microvilli are coated with a filamentous mucopolysaccharide material termed the glycocalyx. The terminal web (TW) lies immediately under the microvilli. Abbreviations: G, Golgi; SER, smooth endoplasmic reticulum; Ly, lysosomes. (C) The junctional complex of an intestinal epithelial cell. The junction comprises three elements: the zonula occludens (ZO or tight junction), zonula adherens (ZA or intermediate junction), and macula adherens (MA or desmosome). Abbreviations: PM, plasma membrane; MA, macula adherens; Rt, filamentous rootlets; EC, glycocalyx or filamentous coat. All three elements appear to be involved in cell-to-cell attachment. (D) Goblet cells occur singly in the intestinal epithelium and large intestine. The mucous secretion of the goblet cells lubricates the lumen of the gastrointestinal tract. [Reproduced with permission from Lentz, T. L. (1971). "Cell Fine Structure" pp. 179, 181, 183. W. B. Saunders, Philadelphia, PA.]

D. Hormone-Secreting Cells: Their Distribution in the Gastroenteropancreatic Complex

A variety of endocrine cells are dispersed throughout the gastric mucosa, small intestine, and colon. Two basic kinds of cells are identifiable by classical histological procedures of staining with silver and chrome: (a) enterochromaffin cells and (b) argophyl cells. The enterochromaffin cells are sparsely present in the gastric mucosa, but are fairly prevalent in the small and large intestines. Either both cell types are located between the bases of other gastrointestinal gland cells, or they extend from the basal lamina to the lumen of the intestine. In humans there are approximately twice as many argophyl cells as enterochromaffin cells in the intestine.

By using immunofluorescent techniques, it has been shown that in the gastroenteropancreatic complex there may be as many as 14 different types of hormone-secreting cells. It has been documented that these cells are capable of secreting gastrin, motilin, secretin, cholecystokinin, somatostatin, GIP, VIP, enkephalins, insulin, and pancreatic polypeptide. In many cases, these cells have microvillar borders facing the lumen of the stomach or the intestine, so that they can respond to chemicals in the stomach or intestinal contents. Most of the cells appear to produce a single peptide hormone, but in some instances they also produce biogenic amines such as histamine or 5-hydroxytryptamine.

There are four general pathways by which peptide hormone-secreting cells of the gastrointestinal tract can make their chemical messengers available; these pathways include (a) paracrine delivery; (b) neurocrine delivery, (c) "open" endocrine cell delivery; and (d) "closed" endocrine cell delivery (see Figure 8-4). Examples of GI hormones that function via these pathways are as follows: paracrine fashion (PP, VIP), neurocrine (enkephalins, NPY, CGRP), open endocrine delivery (gastrin), and closed endocrine delivery (secretin, motilin, CCK).

A. Pearse and colleagues have proposed that all of the gastroenteropancreatic endocrine cells are derived from a common embryological ancestor, the APUD cell, which is originally derived from the neural crest. The acronym APUD stands for amine precursor uptake and decarboxylation, which reflects the common biochemical properties of all of these cells to produce and metabolize amines. Pearse noted that both endocrine cells and neurons that produce peptide hormones share similar ultrastructural, cytochemical, and biochemical properties.

E. Coordination of Gastroenteropancreatic Hormone Release

A number of gastrointestinal hormones can only be secreted by their cells of origin as a consequence of the active digestion and absorption of nutrients in the stomach and duodenum. Thus, the stomach has been recognized to be a major control center to coordinate digestion. The ingestion of nutrients is postulated to be the initiating signal for gastroenteropancreatic hormone release. Carbohydrate ingestion leads to the secretion of GIP and enteroglucagon, while fat ingestion stimulates CCK, GIP, neurotensin, and possibly motilin release. Because these hormonal responses occur within 15 min of ingestion, it seems probable that cellular absorption of the dietary component cannot have played a significant role in stimulating the prompt release of the GI hormones. Instead, it is now known that gastric hormones or releasing factors function as control signals for hormone release from jejunal and ileal cells, as well as from the pancreas. A gastrin-releasing protein has been isolated from the fundic portion of the stomach.

F. Brain–Gut Axis

There is now good neurophysiological and neuroendocrine evidence describing the brain–gut axis. A surprising number of gut hormones have also been found in the brain (e.g., CCK, gastrin, secretin, VIP, glucagon, PP, PYY, NPY). Further, some peptide hormones that clearly function in the gastrointestinal system were originally found in the brain (e.g., enkephalins, somatostatin, substance P, CGRP). It is also known that neurons of the central nervous system interact with neurons of the enteric nervous system via both afferent and efferent pathways to influence digestive processes. Finally, evidence is now emerging that GI hormones and peptides provide a feedback signal to the central nervous system to regulate satiety and thermoregulation.

III. CHEMISTRY AND BIOCHEMISTRY

A. General Relationships

All of the gastrointestinal hormones are polypeptides, and they are each produced in special endocrine cells that are dispersed throughout the stomach and small and large intestines. The gastrointestinal hormones were originally discovered during the search for physiological agents that regulated and stimulated the exocrine secretions and activities of the gut. Application of molecular biology techniques has resulted in significant advances with respect to the elucidation of the gene, cDNA, and, therefore, amino acid sequence of many GI hormones; also, for many of these hormones there is an emerging understanding of their membrane receptors and mode of signal transduction. As a consequence of these studies, it has been possible to identify structural homologies between the GI hormones and group them into families (see Table 8-4).

The discussion of the chemistry, biochemistry and biological actions of the GI hormones and neurotransmitters in the following sections follows their order of presentation in Tables 8-2 and 8-3.

B. Cholecystokinin (CCK)

1. Chemistry

A 115-amino-acid prepro form of cholecystokinin is cleaved to generate biologically active peptides of several sizes, including 59-, 38-, 33-, and 8-amino-acid

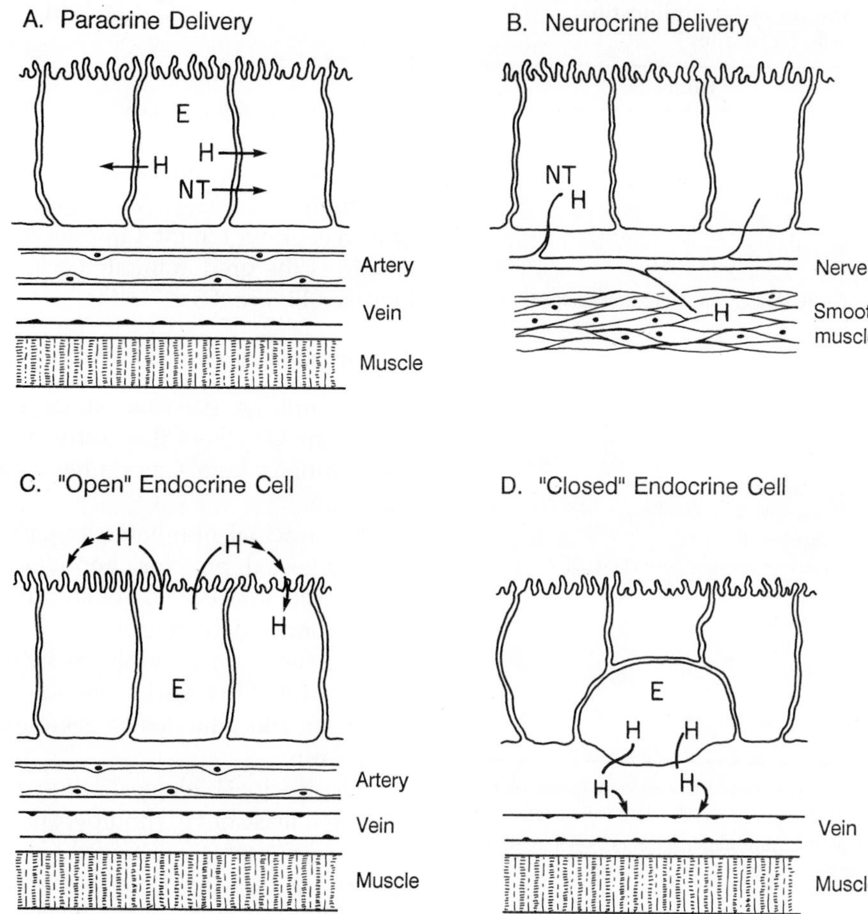

FIGURE 8-4 Schematic models describing pathways by which hormone-producing cells (E) make their chemical messengers available: (A) paracrine delivery, where the hormone (H) or neurotransmitter (NT) moves from an endocrine cell to immediately adjacent cells; (B) neurocrine delivery, where there is release of NT by the nerve to the local environment, which includes mucosa, other endocrine cells, or smooth muscle; (C) release of H from an endocrine cell that is "open" to the intestinal lumen; (D) release of H to the blood from an endocrine cell that is "closed" and does not have an opening to the intestinal lumen.

species. The amino acid sequences of CCK and gastrin are presented in Figure 8-5. All of the biological activity of CCK is contained in the octopeptide, and it is likely that the other larger peptides represent intermediates in processing. In order to achieve full biological activity of CCK, the tyrosine at position 7 must be sulfated.

The first five C-terminal amino acids of CCK and gastrin are identical. Interestingly, in gastrin the tyrosine that is at position 6 need not be sulfated for full biological activity.

2. *Biochemistry and Biology*

CCK, as determined by immunohistochemical techniques, is found principally in the duodenum and jejunum and also in the central nervous system (CNS), as well as in the nerves of the myenteric plexus of the intestine and the nerves of the urinary bladder and uterus.

The two major biological responses affected by CCK are stimulation of the acinar cells of the pancreas to secrete pancreatic enzymes and stimulation of gallbladder contraction. The release of CCK from the endocrine cells in the intestinal mucosa is stimulated by the breakdown products of food in the duodenum, particularly 10–18-carbon fatty acids and L-amino acids.

The biological actions of CCK result as the consequence of interactions with two classes of receptors; the CCK-A receptor is predominantly present in the pancreas and gallbladder, while the CCK-B receptor is found in the central nervous system (see Table 8-5). There is some evidence that the occupancy of CCK receptors by CCK may contribute to a meal termination signal, i.e., to function as a satiety factor.

TABLE 8-4 Families of Gastrointestinal Peptide Hormones

Peptide family[a]	Hormones
Secretin	Enteroglucagon
	Glicentin
	Glucagon
	GIP
	Secretin
	VIP
	Growth hormone-releasing factor (GRF)
Gastrin	CCK
	Gastrin
Pancreatic polypeptide	PP
	NPY
	PYY
Tachykinin–bombesin	Neurokinins
	Neuromedins
	Substance K
	Substance P
	Gastrin-releasing peptide (GRP)
Opioids	Enkephalin
	β-Endorphin
	ACTH
	α-Melanocyte-stimulating hormone (α-MSH)
	Dynorphins

[a] The grouping into families is based upon comparison of the hormones' primary amino acid sequences. [The data were abstracted from Walsh, J. H. (1989). Gastrointestinal peptides. *In* "Gastrointestinal Diseases" (L. R. Johnson, ed.), pp. 131–142. Raven Press, New York.]

Both the CCK-A and CCK-B receptors are highly glycosylated. The CCK-B receptor has an amino acid sequence that is 48% identical to that of the CCK-A receptor. Both receptors contain seven transmembrane domains characteristic of G protein-coupled receptors. In the pancreatic acinar cell, CCK stimulates amylase release via a complex signal transduction system involving IP_3, nitric oxides, cGMP, and Ca^{2+} channels (see Figure 8-15).

	14				10				5				1	
CCK								D	Y#	M	G	W	M	D F -COOH
Gastrin	W	L	E	E	E	E	E	A	Y*	G	W	M	D	F -COOH

FIGURE 8-5 Amino acid sequences of CCK and gastrin. For CCK, the minimum peptide with full biological activity has 8 amino acids; however, peptides of 33, 38, and 59 amino acids are known to be present in circulation. Sulfation of the tyrosine at position 7 (indicated by #) is mandatory for biological activity. For gastrin there are three principal circulating forms, G34 (big gastrin), G17, and G14 (minigastrin); the minimum peptide with full biological activity has 14 amino acids. Sulfation of the tyrosine at position 6 (indicated by *) is known to occur, but is not considered mandatory for full biological activity.

C. Gastrin

1. *Chemistry*

A pro form of gastrin of 104 amino acids is cleaved to generate biologically active, variable-length peptides; they are known as big gastrin (G_{34}), small gastrin (G_{17}), and minigastrin (G_{14}). All of the biological activity of gastrin is present in the 14-amino-acid minigastrin. G_{34} gastrin is the predominant species present in circulation (see Figure 8-5).

2. *Biochemistry and Biology*

The bulk of gastrointestinal gastrin is localized within the G cells of the gastric antrum; the G_{17} form predominates here. Gastrin has also been localized in the brain.

The principal stimulants for gastrin release are partially digested proteins, peptides, and amino acids. Gastrin secretion can also be stimulated by vagus nerve stimulation and the condition of insulin hypoglycemia. Gastrin secretion is inhibited by antral acidification to pH ≈1.0–2.5. Gastrin release is also stimulated by bombesin and inhibited by secretin, GIP, VIP, and somatostatin.

The principal biological action of gastrin is to stimulate the release of HCl and pepsin by the parietal cells. These responses are achieved via direct interaction of gastrin with a cell-surface membrane receptor and via indirect stimulation of the release of histamine from enterochromaffin-like cells present in the gastric mucosa (see Figures 8-13 and 8-14).

D. Secretin

1. *Chemistry*

Unlike most other peptide hormones, only one form of secretin is known; this single species of secretin has 27 amino acids, with the amino acid sequence given in Figure 8-6. The entire molecule of 27 amino acids is required for full biological activity.

Secretin has a structural homology with glucagon, GIP, and VIP (see Table 8-4 and Figure 8-6). Like other peptide hormones, secretin is biosynthesized from a prepro form consisting of 104 amino acids (see Figure 8-7). The mRNA for secretin contains a typical signal peptide and flanking N-terminal and C-terminal peptides.

2. *Biochemistry and Biology*

Within the GI tract, secretin is found in the S cells of the duodenum and jejunum. Secretin secretion is stimulated by unknown mechanisms when the acidity of the duodenal contents falls below pH ≈4.5.

TABLE 8-5 Properties of CCK Receptors[a]

	Receptor class	
	CCK-A	CCK-B
Amino acids (number)	444	452
Molecular weight	49,600	48,900
Glycosylation sites, N-linked (number)	4	4
Glycosylated molecular weight	85,000	95,000
Fold enhancement in K_D for sulfated CCK peptide analog over nonsulfated CCK peptide analogs	500–1000	3–10
Principal location	GI tract	Brain and spinal cord (CNS)
Biological responses	To stimulate (i) gallbladder contraction; (ii) secretion of enzymes from ancinar pancreas	Modulate anxiety Neuroleptic activity

[a] The information in this table describes properties of cloned rat CCK-A and CCK-B receptors. [Abstracted from Wank, S. A., Harkins, R., Jensen, R. T., *et al.* (1992). Purification, molecular cloning, and functional expression of the cholecystokinin receptor from rat pancreas. *Proc. Natl. Acad. Sci. USA* **89,** p 3125–3129; and from Wank, S. A., Pisegna, J. R., and DeWeerth, A. (1992). Brain and gastrointestinal cholecystokinin receptor family: Structure and functional expression. *Proc. Natl. Acad. Sci. USA* **89,** 8691–8695.]

The principal biological response of secretin is to stimulate the secretion of bicarbonate and water by the acinar cells of the pancreas. This response is stimulated by ligand occupancy of the secretin receptor, which is located on the outer cell membrane of the pancreatic acinar cells. The human secretin receptor has been cloned and found to be a 419-amino acid peptide (47,900 molecular weight) that belongs to the family of G protein-coupled receptors with seven transmembrane domains. Other members of the receptor family include VIP, glucagon-like peptide (GLP-1), parathyroid hormone (PTH), and calcitonin (CT). The mature receptor protein contains 5 potential sites for asparagine-linked glycosylation as well as 10 cysteines for disulfide formation. The human secretin receptor has 48% homology with the human VIP receptor.

The biological effects of secretin are believed to occur via receptor-mediated stimulation of adenylate cyclase, which leads to the secretion of bicarbonate and H_2O by the acinar cells.

E. Enkephalins

There is direct evidence of the biosynthesis of both Met-enkephalin and Leu-enkephalin within the myenteric neurons of the gut. As summarized in Figure 1-9, the enkephalins are produced as a consequence of processing of the proopiomelanocortin mRNA. In addition, a second gene has been found in the gut that codes for precursor molecules of Leu-enkephalin and Met-enkephalin. Each of the enkephalins consists of five amino acids (Tyr-Gly-Gly-Phe-X), and they differ only at the COOH residue where X may be either Leu or Met. The biological actions of the enkephalins in the gut are not understood in detail. Their principal effects are to inhibit gut motility and mucosal secretions while stimulating sphincter tone.

F. Enteroglucagon

The topic of enteroglucagon is covered in Chapter 7. Enteroglucagon, also known as glicentin, represents a larger form of the pancreatic 29-amino-acid glucagon. The structures are contained within the 160-amino-acid proglucagon sequence: enteroglucagon consists of 69 amino acids (see Figure 8-6). Enteroglucagon is biosynthesized in the mucosal L cells. Enteroglucagon has been postulated to be a trophic factor for the intestinal mucosa.

G. GIP

GIP is an acronym for both gastrin-inhibitory peptide and glucose-dependent insulinotropic polypeptide; both terms apply to the same 42-amino-acid pep-

```
          1       5         10        15        20        25        30        35        40  42
Secretin  H S D G T F T S E L S R L R E G A R L Q R L L Q G L V
Glucagon  H S Q G T F T S D Y S K Y L D S R R A Q D F V Q W L M N T
VIP       H S D A V F T D N Y T R L R K Q M A V K K Y L N S I L N
GIP       Y A E G T F I S D Y S I A M D K I H Q Q D F V N W L L A E K G K K N D W K H N I T Q -COOH
```

FIGURE 8-6 Amino acid sequences of the secretin peptide hormone family, including secretin, glucagon, vasoactive intestinal peptide (VIP), and gastrointestinal-inhibitory peptide (GIP).

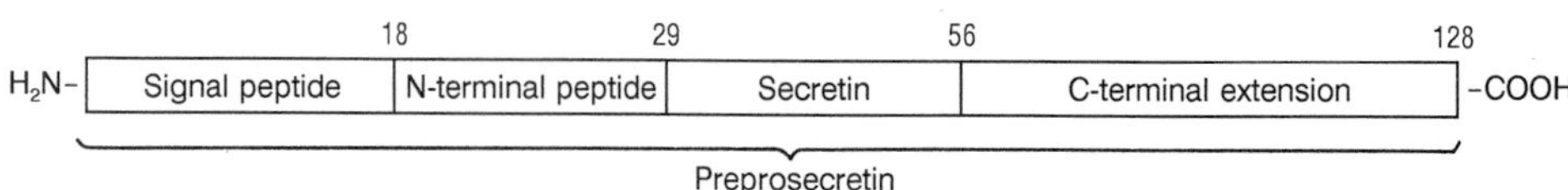

FIGURE 8-7 Organization of preprosecretin. [The information was abstracted from Kopin, A. S., Wheeler, M. B., and Leiter, A. B. (1990). Secretin: Structure of the precursor and tissue distribution. *Proc. Nat. Acad. Sci. USA* **87,** 2299–2303.]

tide. GIP, which is a member of the secretin peptide family (Table 8-4), shares a high sequence homology with both glucagon (15 amino acids identical) and secretin (9 amino acids identical) (see Figure 8-6). GIP is biosynthesized largely by K cells present in the small intestine. Receptors for GIP have been identified in selected regions of the GI tract as well as the brain.

Little information is available concerning the mechanisms by which GIP elicits biological responses. Two categories of biological responses are known: (i) inhibition of gastric acid secretion and gastric motility and (ii) potentiation of pancreatic insulin secretion after oral as opposed to intravenous glucose administration.

H. Motilin

Motilin is a 22-amino-acid linear peptide (see Figure 8-8); the biological activity is believed to be associated with the nine N-terminal residues. The amino acid sequence is entirely different from that of all other known gastrointestinal hormones. The solution structure of motilin has been determined in detail via nuclear magnetic resonance (NMR) studies.

A

1 5 10 15 20 22
Motilin F V P I F T T Y E L Q R M E E K E R N R G Q -COOH

B

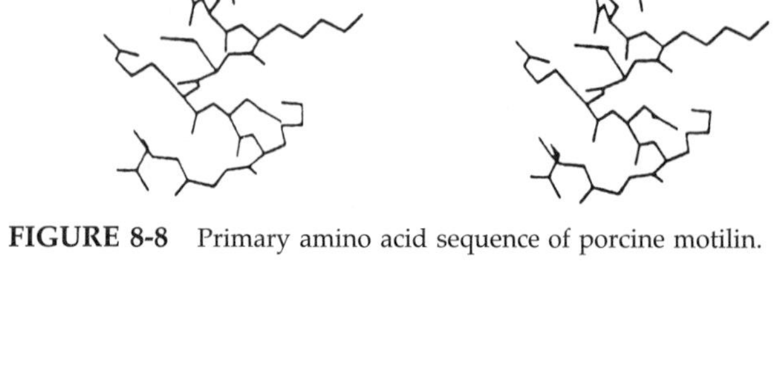

FIGURE 8-8 Primary amino acid sequence of porcine motilin.

Motilin is believed to play a role in the stimulation of small intestinal motility during fasting. Its release from the endocrine EC cells of the duodenum is cyclical, occurring approximately every 2 hr. Significantly, most of motilin's actions do not coincide with dietary intake, but with fasting. It has been suggested that it enhances the movement of digested food particles in preparation for the next meal. The physiological role of motilin remains to be clarified.

I. Neurotensin

Neurotensin is a 13-amino acid polypeptide (see Figure 8-9) found in the N cells of the ileum; it is also present extensively in nerve cells of the CNS. The secretion of neurotensin into circulation is effected by the ingestion of a meal. Neurotensin was named because of its potency to cause hypotension and vasodilation; in the gut, neurotensin mediates inhibition of gastric motility and stimulation of the secretion of PP and glucagon, but inhibits insulin release.

J. Pancreatic Polypeptide Family (PP)

The pancreatic polypeptide family comprises three 36-amino-acid peptides; they include pancreatic peptide (PP), peptide YY (PYY), and neuropeptide (NPY). Table 8-6 provides a comparison of the biological properties of this peptide family.

There is an emerging area of research linking leptin (see Chapter 7) and neuropeptide Y. Leptin is a hormone produced by adipocytes that may signal to the hypothalamus to inhibit the secretion of appetite-stimulating (orexigenic) NPY. High concentrations of NPY are known to stimulate food intake, while low concentrations have the opposite effect.

K. Somatostatin

Somatostatin is a cyclic peptide with two naturally occurring forms: either a 28 (SS-28)-or a 14 (SS-14)-amino-acid peptide (see Figure 8-10). Somatostatin was

1 5 10 13
Neurotensin E L Y E N K P R R P Y I L -COOH

FIGURE 8-9 Amino acid sequence of neurotensin.

TABLE 8-6 Pancreatic Polypeptide Family[a]

	Pancreatic polypeptide (PP)	Peptide YY (PYY)	Neuropeptide (NPY)
Tissue location	Pancreas	Ileum, colon, rectum	Brain, enteric neurons of the mucosa, submucosa, and smooth muscle
Release stimulated by	Protein or cholinergic stimulation	Intestinal fat	Nerve transmission; absence of leptin (?)
Proposed function(s)	Inhibit gastric acid secretion, pancreatic exocrine secretins, and gallbladder contraction	Inhibit gastric acid secretion, small intestine motility, pancreatic enzyme secretion	Vasoconstrictor; stimulation of appetite (orexigenic)
Possible modes of action	?	Via pertussis toxin-sensitive G_i protein	Via a pertussis toxin-sensitive G_i protein

[a] The information in this table was abstracted from Wilding, J. P. H., Ghatei, M. A., and Bloom, S. R. (1995). Hormones of the Gastrointestinal tract. *In* "Endocrinology" (L. J. DeGroot, M. Besser, H. G. Burger, J. L. Jameson, D. L. Loriaux, J. C. Marshall, W. D. Odell, J. T. Potts, Jr., and A. H., Rubenstein, eds.), Vol. 3, Chapter 153, pp. 2870–2894. W. B. Saunders, Philadelphia, PA.

first isolated from the hypothalamus, where it functions to inhibit the secretion of growth hormone (see Figure 3-8).

While the principal form of somatostatin is the SS-14 species, both SS-14 and SS-28 display overlapping but distinct biological activities and act through different receptors (see Table 8-7). It is known, however, that the 14 amino acids at the N-terminus of SS-28 have no intrinsic biological activity. Aspects of the actions of somatostatin are also covered in Chapters 3 and 7.

L. Bombesin and Tachykinin Peptide Families

The bombesins include structurally related peptides that were first isolated from amphibian skin and that were found to be biologically active in the GI tract. Later, additional members of the bombesin family were isolated directly from the gut, e.g., gastrin-releasing peptide (GRP) and neuromedin B; they are tabulated in Table 8-8 and their amino acid sequences are presented in Figure 8-11. In amphibians, three subfamilies of the bombesin family have been identified; these include the bombesins, the ranatensins, and the phylloitorins. Each subfamily is characterized by a common amino acid near its carboxy-terminus.

The tachykinins comprise three related mammalian-derived peptides, substance P, neurokinin A, and neurokinin B; each of these hormones also displays some structural homology to portions of the bombesins (see also Table 8-8 and Figure 8-11). Substance P, which was discovered in 1931, was the first peptide to be found in both the brain and the GI tract.

The tachykinins (fast acting), which mediate rapid contractions of smooth muscle, are so named to distinguish them from bradykinins (see Chapter 15), which have a much slower action on smooth muscle. The tachykinin-related peptides are characterized by a COOH-terminal sequence—Phe-X-Gly-Leu-Met-NH_2 (see Figure 8-11).

The pattern of biological responses generated by the tachykinins is complex and includes smooth muscle contraction, hyperalgesia, activation of most cells (see Chapter 15), release of histamine from salivary cells, and involvement with asthma, hay fever, migraine headache, and inflammatory bowel disease. Three distinct tachykinin receptors have been cloned and found to belong to the seven transmembrane proteins that are coupled to G proteins (see Figure 1-23).

M. Calcitonin Gene-Related Peptide

Calcitonin gene-regulated peptide (CGRP) was the first neuropeptide to be discovered solely by molecular biology techniques. CGRP represents an alternative splicing of the mRNA transcribed from the calcitonin gene (see Figure 9-4); the mature hormone has 37 amino acids.

CGRP is widely distributed throughout the CNS and selected regions of the GI tract, particularly in sensory neurons. The GI-related biological effects of CGRP include stimulation of somatostatin secretion, smooth muscle contraction, and increases in mucosal blood flow.

N. Endothelin

Endothelin is a 21-amino-acid peptide that is a highly potent vasoconstrictor. The endothelins are discussed in detail in Chapter 15 (see Figures 15-13, 15-

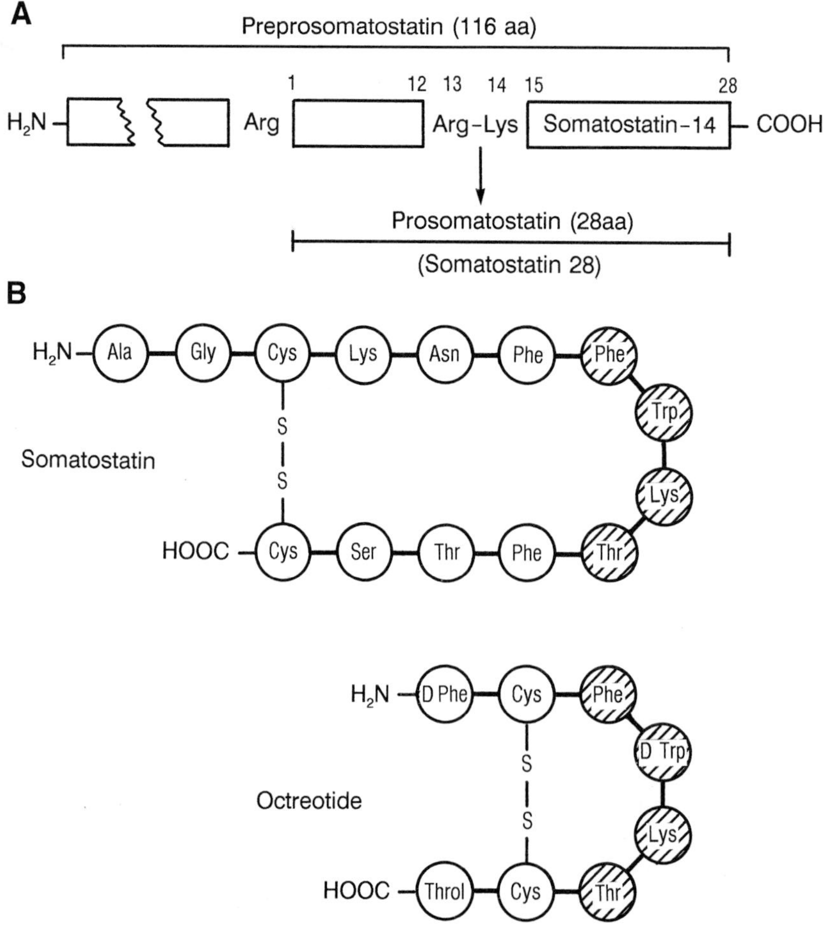

FIGURE 8-10 (A) Structures of preprosomatostatin, somatostatin 28 (SS-28), and somatostatin 14 (SS-14). The preprosomatostatin has 116 amino acids; it is cleaved to SS-28 and then, under some circumstances, to SS-14. However, both SS-14 and SS-28 are biologically active. B illustrates the disulfide structure of SS-14 and the synthetic, long-acting SS-14 analog octreotide; this drug is used to suppress SS-14 secretion in acromegaly and in gut endocrine tumors. [Information abstracted from Weiss, J. A. H. (1995). Somatostatin. *In* "Endocrinology" (L. J. DeGroot, M. Besser, H. G. Burger, J. L. Jameson, D. L. Loriaux, J. C. Marshall, W. D. Odell, J. T. Potts, Jr., and A. H. Rubenstein, eds.), Vol 1, Chapter 17, pp. 266–279. W. B. Saunders, Philadelphia PA.]

14, and 15-15). Endothelins are produced by a variety of cells in the GI tract, including plexus nerves and mucosal cells. The precise biological role of the endothelins in the GI tract remains to be elucidated.

O. Galanin

Human galanin is a 30-amino-acid peptide (see Figure 8-12) derived from a prepro form of 201 amino acids. Galanin does not have any structural homology to other GI hormones. Galanin is found in neurons associated with the submucosa. As yet, the physiological role of galanin is not clearly defined. There is some evidence that it can inhibit the release of somatostatin, insulin, pancreatic polypeptide, and neurotensin.

P. Vasoactive Intestinal Peptide (VIP)

Vasoactive intestinal peptide (VIP) is a straight chain 28-amino-acid intestinal peptide. VIP has extensive structural homology with both glucagon and secretin, as well as some homology with GIP (see Figure 8-6). VIP occurs widely in both the intestinal and nervous systems of many mammals and lower animals.

TABLE 8-7 Properties of Somatostatin 28 (SS-28) and Somatostatin 14 (SS-14)[a]

	Comments	Relative amounts (%)	
		SS-28	SS-14
Tissue location	Hypothalamus (see Chapter 3) Pancreatic D cells (see Chapter 7)	Hypothalamus: ~5–10	 ~95–90
	Stomach, duodenum, colon	Pancreas ~10	 ~90
		GI tract: ~75	 ~25
Release stimulated by	Pancreas: glucose, arginine Hypothalamus: inhibit growth hormone release GI tract: after ingestion of a mixed meal, but SS-28 $\gg$ SS-14		
Proposed function	Hypothalamus: inhibit release of growth hormone Pancreas: inhibit insulin and glucagon secretion GI tract: paracrine agent to inhibit gastrin release and acid secretion		
Possible mode of action	Interaction with specific membrane receptors, coupled to Gs proteins and adenylate cylase; there is evidence for separate receptors for SS-28 and SS-14		

[a] The information in this table was abstracted from Wilding, J. P. H., Ghatei, M. A., and Bloom, S. R. (1995). Hormones of the Gastrointestinal tract. *In* "Endocrinology" (L. J. DeGroot, M. Besser, H. G. Burger, J. L. Jameson, D. L. Loriaux, J. C. Marshall, W. D. Odell, J. T. Potts, Jr., and A. H. Rubenstein, eds.), Vol. 3, Chapter 153, pp. 2870–2894. W. B. Saunders, Philadelphia, PA.

VIP is present in a large number of neurons within the gut, CNS, and urogenital tract. The biological actions of VIP, which are largely mediated in a paracrine fashion, are summarized in Table 8-9.

The biological actions of VIP are mediated via interaction with an outer cell membrane receptor that has seven transmembrane domains and significant homology with the secretin receptors. The VIP receptor is believed to be coupled to adenylate cyclase.

IV. BIOLOGICAL AND MOLECULAR ACTIONS

A. Gastric Acid Secretion

The parietal cells of the stomach secrete a solution of 0.15 *M* HCl and 7 m*M* KCl, with only a trace of other electrolytes. The concentration of hydrogen ions is 1 million times higher (pH ~1) than that of plasma. The secretion of HCl by the parietal cells of the stomach

TABLE 8-8 Bombesin–Tachykinin Family of Neuropeptide Hormones[a]

Hormone	No. of amino acids	Source	Biological action(s)
Bombesin-like			
Bombesin	14	Skin of frog *Bombina bombina*	Stimulate gastrin release and gallbladder contraction
Litorin	9	Amphibian skin	Similar to bombesin
Ranatensin	11	Amphibian skin	Similar to bombesin
Alytensin	14	Amphibian skin	Similar to bombesin
Gastrin-releasing peptide	27	Mammalian gut and brain	Stimulation of gastrin release from antral G cells
Neuromedin B	10	GI tract, pancreas, CNS	Stimulates gastrin release; in hypothalamus inhibits TS-11 release
Phylcaerulein	9	Amphibian skin	
Caerulein	10	Amphibian skin	Mimics CCK actions to stimulate gallbladder contractions
Tachykinin-like			
Substance P	11	Hypothalamus	Nociception; pain pathway vasodilation
Neurokinin A	10	Brain	Smooth muscle contraction, vasodilation
Neurokinin B	10		Smooth muscle contraction

[a] The information in this table was abstracted from Wilding, J. P. H., Ghatei, M. A., and Bloom, S. R. (1995). Hormones of the Gastrointestinal tract. *In* "Endocrinology" (L. J. DeGroot, M. Besser, H. G. Burger, J. L. Jameson, D. L. Loriaux, J. C. Marshall, W. D. Odell, J. T. Potts, Jr., and A. H. Rubenstein, eds.), Vol. 3, Chapter 153, pp. 2870–2894. W. B. Saunders, Philadelphia, PA; and from (1995) Peptides as mediators. *In* "Pharmacology" (H. P. Rang, M. M. Dale, J. M. Ritter, and P. Gardner, eds.), Chapter 9, pp. 191–202. Churchill-Livingstone, New York.

Bombesins

	1				5					10				14
Bombesin	E**	Q	R	L	G	N	Q	W	A	V	G	H	L	M*
Alytensin	E**	G	R	L	G	T	Q	W	A	V	G	H	L	M*
Ranatensin				G**	V	P	Q	W	A	V	G	H	L	M*
Neuromedin B	R	G	N	L	W	A	T	G	H	F	M	G**		
Litorin GIP						G**	Q	W	A	V	G	H	L	M*

Tachykinins

	1				5					10	11
Substance P	R	P	K	P	Q	Q	F	F	G	L	M*
Neurokinin A		H	R	T	D	S	F	V	G	L	M*
Neurokinin B		D	M	H	D	F	F	V	G	L	M*

FIGURE 8-11 Amino acid sequences of bombesins and tachykinins. Neuromedin B is the mammalian homolog of the amphibian bombesin peptide ranatensin. * denotes the amide of the carboxyl of the C-terminal amino acid. ** denotes pyroglutamyl; see Figure 3-8. [The bulk of the information was abstracted with permission from Edmondson, S., Khan, N., Shriver, J., Zdunek, J., and Gräslund, A. (1991). The solution structure of motilin from NMR distance constraints, distance geometry, molecular dynamics, and an iterative full relaxation matrix refinement. *Biochemistry* **30,** 11271–11279. The sequence of the human neuromedin B is abstracted from Krane, I. M., Naylor, S. L., Heilin-Davis, D., Chen, W. W., and Spindel, E. R. (1988). "Molecular cloning of cDNAs encoding the human bombesin-like peptide neuromedin B". *J. Biol. Chem.* **263,** 13317–13323.]

is effected by endocrine (gastrin), neuroendocrine (acetylcholine), and paracrine (histamine) pathways.

The parietal cell is known to have separate receptors for gastrin, histamine, and acetylcholine. Figure 8-13 presents a working model of the signal transduction pathways that are utilized to stimulate the secretion of acid. Figure 8-14 presents a model describing the production of HCl by the parietal cell.

Pepsin is the principal proteolytic enzyme present in the gastric secretions. Pepsin is stored in an inactive form as pepsinogen in secretory granules of the chief cells of the oxyntic mucosa of the stomach. Smaller amounts of pepsinogen are also found in the cardiac and pyloric cells of the stomach and upper duodenal mucosa. The process of pepsin secretion is probably analogous to that of secretion of the proenzymes trypsinogen and chymotrypsinogen from the exocrine portion of the pancreas.

Pepsinogen secretion is stimulated by gastrin and related peptides, as well as by a vagal cholinergic stimulation that may be induced by feeding. Also, secretin

TABLE 8-9 Biological Actions of Vasoactive Intestinal Peptide (VIP)[a]

Location	Action
Esophagus	Relaxation of lower sphincter
Stomach	Relaxation of gastric fundus; inhibition of gastric acid secretion
Small intestine	Modulation of gut motility; contraction of longitudinal muscle and relaxation of the circular muscle; stimulation of intestinal secretion
Anus	Relaxation of anal sphincter
Pancreas	VIP-ergic neurons innervate the islets of Langerhans
Penis	Increased blood flow

[a] The information in this table was abstracted from Wilding, J. P. H., Gahatei, M. A., and Bloom, S. R. (1995). Hormones of the gastrointestinal tract. *In* "Endocrinology" (L. J. DeGroot, M. Besser, H. G. Burger, J. L. Jameson, D. L. Loriaux, J. C. Marshall, W. D. Odell, J. T. Potts, Jr., and A. H. Rubenstein, eds.), Vol. 3, Chapter 153, pp. 2870–2894. W. B. Saunders, Philadelphia, PA.

is a strong stimulator of pepsinogen secretion. The enzymatically inactive pepsinogen (40.4 kDa) is, in the presence of acid, autocatalytically converted to the active form, pepsin, by the cleavage of a 42-amino-acid residue from the N-terminal portion of the peptide chain. Pepsin is active at acidic pH (below 3.5) and is irreversibly inactivated at neutral or slightly alkaline pH. The active form of pepsin is 32 kDa and has a proetolytic specificity for Trp, Phe, Tyr, Met, and Leu amino acid residues in a protein substrate.

B. Pancreatic, Biliary, and Intestinal Secretions

1. *Exocrine Pancreas Secretion*

After the ingestion of a meal, and associated with the appearance of the acidified chyme from the stomach in the duodenum, there is a coupled secretion by the exocrine pancreas of the pancreatic juice, which is a rich source of many enzymes; the composition is given in Table 8-10. Many of the potent enzymes are secreted in inactive pro forms and must be converted separately to active enzymes in the lumen of the intestine. The pancreatic secretion is mainly dependent upon the actions of CCK; in addition, the stimulation of appro-

1 5 10 15 20 25 30

Galanin G W T L N S A G Y L L G P H A V G N H R S F S D K N G L T S -COOH

FIGURE 8-12 Amino acid sequence of human galanin. The mature galanin is derived from a pro precursor consisting of 201 amino acids. [The information was abstracted from McKnight, G. L., Karlsen, A. E., Kowalyk, S. *et al.* (1992). Sequence of human galanin and its inhibition of glucose-stimulated insulin secretion from RIN cells. *Diabetes* **41**, 82–87.]

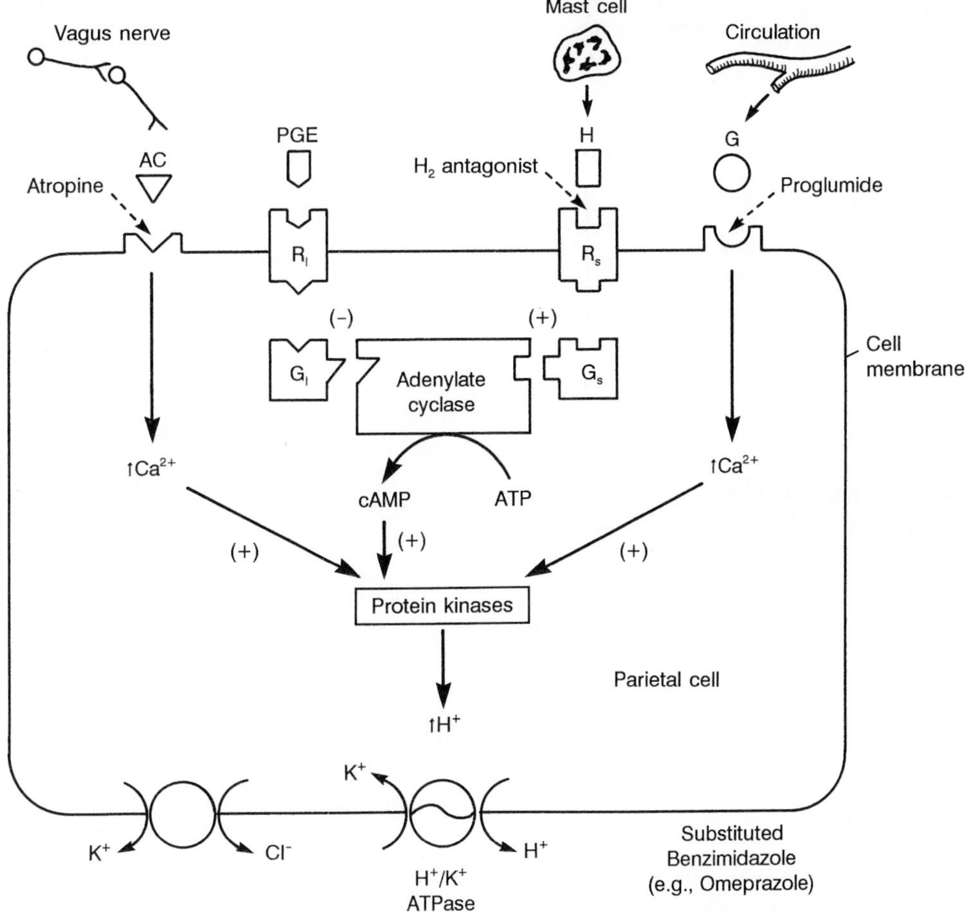

FIGURE 8-13 Model of agonist stimulation of a parietal cell leading to the secretion of acid. The parietal cell (outer cell membrane) has receptors for gastrin (G), acetylcholine (AC), and histamine (H), as well as for prostaglandins (PGE). The main stimulants of HCl acid secretion are G, AC, and H; collectively they are believed to activate, via a G protein, adenylate cyclase, which eventually leads to an increase in intracellular [Ca^{2+}], and protein kinases, which initiate by as yet unknown mechanisms the secretion of acid (see Figure 8-14). The secretion of HCl can be blocked by PGE, proglumide, which blocks the gastrin receptor, atropine, which blocks the AC receptor, and certain drugs such as omeprazole which interacts with the H^+,K^+ ATPase. [This figure was adapted from Wolfe, M. M., and Soll, A. H. (1988). The physiology of gastric acid secretion, *New Engl. J. Med.* **319**, 1707–1715.]

priate cholinergic or peptidergic nerves leading to the pancreas can stimulate exocrine pancreatic secretions (see Figure 8-15).

Pancreatic secretions of electrolytes and H_2O from the ductular and centroacinar cells are primarily under the influence of secretin. Secretin is known to stimulate intracellular cAMP in these cells, which activates unknown factors necessary to alter the secretory cell membrane permeability to sodium and hydrogen ions. Thus, there is an increased exchange of external Na^+ ions for internal H^+ ions. The increased concentration of extracellular H^+ lowers the local pH, which then increases the production of CO_2 from circulating bicarbonate. This CO_2 then diffuses into the cells, where it combines with water to form carbonic acid (mediated by carbonic anhydrase) and results in the production of HCO_3^-, which can be utilized for secretion.

The pancreatic secretion of digestive enzymes occurs in the acinar cells of the exocrine pancreas under the influence of CCK. In the pancreatic acinar cells, CCK interacts with a receptor present on the outer cell membrane. This activates a G protein-coupled signal transduction pathway (see Figure 8-15) to ultimately stimulate the secretion of amylase, trypsinogen, and chymotrypsinogen stored in secretory granules.

2. *Biliary Secretion*

Hormonal regulation of bile secretion, primarily by CCK and secretin, occurs by action on the ductules and ducts of the gallbladder. Table 8-11 presents the

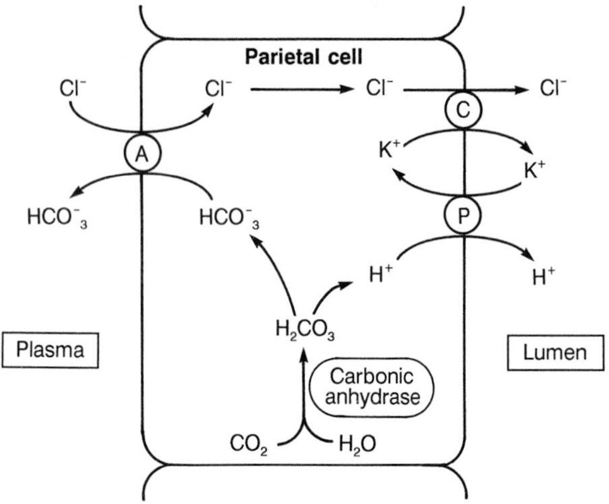

FIGURE 8-14 Schematic model describing the production of hydrochloric acid by the parietal cell. The H^+ ions that are secreted are generated inside the parietal cell by the action of carbonic anhydrase-mediated conversion of $CO_2 + H_2O$ into $H_2CO_3 \rightarrow H^+ + HCO_3^-$. The resulting bicarbonate is exported into the plasma compartment in a 1:1 exchange for Cl^-. The H^+ and Cl^- are then secreted into the lumen of the cell via the integrated actions of a Na^+,H^+ ATPase and a symport-combined transfer of K^+Cl^- to the lumen. Thus, for every H^+Cl^- secreted, a K^+ ion makes a round trip from the parietal cell → lumen → parietal cell. [Modified with permission from Rang, H. P., Dale, M. M., Ritter, J. M., and Gardener, P. (1995). "Pharmacology," pp. 386. Churchill-Livingstone, New York.]

major components of human bile. These constituents are secreted into the intestinal lumen in response to dietary fat and they act as detergents to disrupt and

TABLE 8-10 Composition of Human Pancreatic Juice[a]

Components		
Cations	Na^+, K^+, Ca^{2+}, Mg^{2+}, pH ≈8.0	
Anions	HCO_3^-, Cl^-, SO_4^{2-}, HPO_4^{2-}	
Digestive enzymes	**Activator**	**Substrate**
Trypsinogen	Enteropeptides	Proteins, polypeptides
Chymotrypsinogen	Trypsin	Proteins, polypeptides
Prolactases	Trypsin	Elastin, other proteins
Procarboxypeptidase A	Trypsin	Proteins, polypeptides
Procarboxypeptidase B	Trypsin	Proteins, polypeptides
Prolipase	Trypsin	Fat droplets
Pancreatic lipase	Trypsin	Triglycerides
Cholesterol ester hydrolase	Trypsin	Cholesterol esters
α-Amylase	Trypsin	Starch
Ribonuclease	Trypsin	RNA
Deoxyribonuclease	Trypsin	DNA
Prophospholipase A_2	Trypsin	Phospholipids

[a] Information in this table was abstracted from Ganong, W. F. (1995). "Review of Medical Physiology," Chapters 25–26. Appleton-Lange, New York.

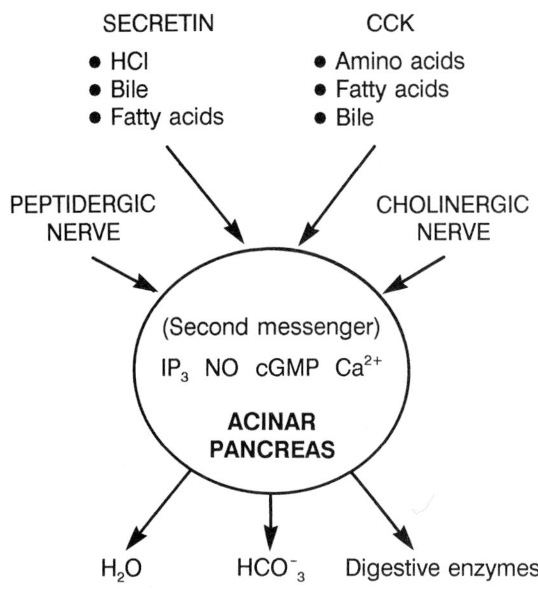

FIGURE 8-15 Biological actions of secretin and CCK on the acinar pancreas, which result in the secretion of H_2O, HCO_3^-, and digestive enzyme.

disperse the oil droplets. As yet there is no detailed mechanism describing how CCK mediates gallbladder emptying. The CCK-A receptor (see Table 8-5) is present in the gallbladder, and occupancy of this receptor by a CCK antagonist specifically blocks gallbladder emptying.

3. *Intestinal Secretion*

Several gut hormones, including VIP, GIP, secretin, CCK, and glucagon, and prostaglandins E_1, E_2, and $F_{2\alpha}$ have been shown to affect the small intestine and colon by either inhibiting the active absorption of electrolytes and H_2O or stimulating the secretion of H_2O and electrolytes. Both VIP and the prostaglandins are potent stimulators of the adenylate cyclase–cAMP system,

TABLE 8-11 Major Organic Components of Human Bile[a]

Component	Concentration (mg/ml bile)
Dihydroxycholanic acids	49
Cholic acid	32
Phospholipid (mostly lecithin)	27
Protein	5
Cholesterol	3
Bilirubin	3
Hexose and hexosamine	1

[a] Abstracted from Nakayma, F., and Miyake, H. (1966). Species differences in cholesterol-complexing macromolecular fractions in bile in relation to gallstone formation. *J. Lab. Clin. Med.* **67**, 78–86.

and it is probable that this is the basis for their actions on the intestinal secretion of H_2O and electrolytes. The cellular mechanisms of the other gut hormones in this system are not known.

C. Motor Functions of the Intestinal Tract

The gastrointestinal hormones play an important physiological role in the regulation of the motor activity of the gastrointestinal tract. This includes effects on the stomach, small intestine, colon, gallbladder, and bile ducts. The hormones may have indirect effects (neurally mediated) or direct actions (muscular) on smooth muscle motor activity. The peptides that primarily stimulate motor activity are gastrin, CCK, and motilin, while the peptides that inhibit motor activity are secretin, VIP, glucagon, and enteroglucagon. Gastrin exerts its effects partly via direct interaction with intestinal or stomach muscle cell receptors and partly via postganglionic cholinergic fibers; it is effective at very low concentrations [e.g., $(1–100) \times 10^{-12}$ *M*].

The actions of CCK on gallbladder contraction are receptor-mediated. Emptying of bile is the result of a number of successive events: (a) First, there is a progressive increase in the tension of the gallbladder wall; (b) then an intermittent opening of the sphincter-like cholecystocystic junction occurs; (c) next, there are phasic contractions along the common axis of the bile duct; and (d) finally, sequential opening and closing of the choledochoduodenal sphincter occur, resulting in the periodic discharge of bile into the duodenum. No molecular mechanisms of hormone action are available to describe these complex and critical physiological processes.

V. CLINICAL ASPECTS

A. Peptic Ulcer Disease

The term peptic ulcer refers to an open sore on the esophagus, stomach, or duodenum. The reason why peptic ulcers develop is not well understood. It has been postulated that there is a shift in the balance of mucosa-protecting (secretion of HCO_3^- and mucus) and mucosa-damaging (secretion of acid and pepsin) mechanisms. Peptic ulcer disease can account for as many as 4,000,000 hospital days per year in the United States. Although individuals with duodenal ulcer disease clearly have increased rates of gastric acid secretion, and although gastrin is a potent secretagogue of HCl secretion, it has not been possible to directly implicate gastrin in the etiology of duodenal ulcer disease.

The drugs cimetidine and ranitidine have a proven value in the treatment of duodenal ulcer disease through their capacity to inhibit the secretion of gastric acid in hypersecretory states, particularly those involving peptic ulceration.

B. Zollinger–Ellison Syndrome

Zollinger–Ellison syndrome is characterized by an advanced ulcer disease of the upper gastrointestinal tract, hypersecretion of gastric acid, and tumors of the pancreas islets. If is estimated that ~1% of operative cases of peptic ulcer disease are individuals with Zollinger–Ellison syndrome. The syndrome results from the release of large quantities of gastrin by the pancreatic islet tumors, probably D cells. This then results in increased secretion of HCl by the parietal cells of the stomach.

C. Carcinoid Syndrome

The term carcinoid or argentaffinosis is employed to describe a group of intestinal tumors that grow more slowly than the more common intestinal carcinoma. The carcinoid tumor is derived from the enterochromaffin cells, which are cytochemically indistinguishable from many of the gastrointestinal hormone-secreting cells of the intestinal tract. The most frequent sites of their location are the ileum and the appendix. The chief clinical signs of carcinoid syndrome are flushing and diarrhea.

The chief biochemical lesion is an elevated production of 5-hydroxytryptamine, which results from the conversion of tryptophan to 5-hydroxytryptophan followed by decarboxylation to yield 5-hydroxytryptamine. Elevated blood levels of 5-hydroxytryptamine and an increased urinary excretion of 5-hydroxyindoleacetic acid are often diagnostic of the carcinoid syndrome. Also, the tumor contains the enzyme kallikrein (see Chapter 15), which leads to increased plasma levels of bradykinin.

Both pharmacological and surgical treatments are employed to treat carcinoid syndrome.

References

A. Books

Rang, H. P., Dale, M. M., Ritter, J. M., and Gardner, P. (1995). "Pharmacology," 3rd ed., pp. 1–855. Churchill Livingstone, New York.

B. Review Articles

Blundell, J. (1991). Pharmacological approaches to appetite suppression. *Trends Pharmacol. Sci.* **12,** 147–157.

Green, G. M. (1994). Feedback inhibition of cholecystokinin secretion by bile acids and pancreatic proteases. *Ann. NY Acad. Sci.* **713,** 167–179.

Kreil, G., and Wechselberger, C. (1994). Peptides related to cholecystokinin in nonmammalian vertebrates. *Ann. NY Acad. Sci.* **713,** 32–38.

Pandol, S. J., Gukovskaya, A., Bahnson, T. D., and Dionne, V. E. (1994). Cellular mechanisms mediating agonist-stimulated calcium influx in the pancreatic acinar cell. *Ann. NY Acad. Sci.* **713,** 41–48.

Reeve, J. R., Jr., Eysselein, V. E., Ho, F. J., Chew, P., Vigna, S. R., Liddle, R. A., and Evans, C. (1994). Natural and Synthetic CCK-58. Novel reagents for studying cholecystokinin physiology. *Ann. NY Acad. Sci.* **713,** 11–21.

C. Research Papers

Black, J. W., and Shankley, N. P. (1987). How does gastrin act to stimulate oxyntic cell secretion? *Trends Pharmacol. Sci.* **8,** 486–489.

DeGiorgio, R., Sternini, C., Anderson, K., Brecha, N. C., and Go, V. L. W. (1992). Tissue distribution and innervation pattern of peptide immunoreactivities in the rat pancreas. *Peptides* **13,** 91–98.

DelValle, J., and Yamada, T. (1990). The gut as an endocrine organ. *Annu. Rev. Med.* **41,** 447–455.

Dourish, C. T., Rycroft, W., and Iversen, S. D. (1989). Postponement of satiety by blockade of brain cholecystokinin (CCK-B) receptors. *Science* **245,** 1509–1511.

Drewe, J., Gadient, A., Rovati, L. C., and Beglinger, C. (1992). Role of circulating cholecystokinin in control of fat-induced inhibition of food intake in humans. *Gastroenterology* **102,** 1654–1659.

Edmondson, S., Khan, N., Shriver, J., Zdunek, J., and Graslund, A. (1991). The solution structure of motilin from NMR distance constraints, distance geometry, molecular dynamics, and an iterative full relaxation matrix refinement. *Biochemistry* **30,** 11271–11279.

Evers, B. M., Izukura, M., Chung, D. H., Parekh, D., Yoshinaga, K., Greeley, G. H., Jr., Uchida, T., Townsend, C. M., Jr., and Thompson, J. C. (1992). Neurotensin stimulates growth of colonic mucosa in young and aged rats. *Gastroenterology* **103,** 86–91.

Eysselein, V. E. (1992). Regulation of gastric acid secretion by gastrin in duodenal ulcer patients and healthy subjects. *Gastroenterology* **102,** 1142–1148.

Fatatis, A., Holtzclaw, L. A., Avidor, R., Brenneman, D. E., and Russell, J. T. (1994). Vasoactive intestinal peptide increases intracellular calcium in astroglia: synergism with α-adrenergic receptors. *Proc. Natl. Acad. Sci. USA* **91,** 2036–2040.

Fried, M., Erlacher, U., Schwizer, W., Lochner, C., Koerfer, J., Beglinger, C., Jansen, J. B., Lamers, C., Harder, F., Bischof-Delaloye, A., Stalder, A., and Rovati, L. (1991). Role of cholecystokinin in the regulation of gastric emptying and pancreatic enzyme secretion in humans. Studies with the cholecystokinin receptor antagonist loxiglumide. *Gastroenterology* **101,** 503–511.

Hearn, S. C., Jones, P. M., Ghatei, M. A., Byrne, J., Hill, S. F., and Bloom, S. R. (1992). The presence, characterization and synthesis of neuromedin B in the human pituitary gland. *Neuroendocrinology* **56,** 729–735.

Ishihara, T., Shigemoto, R., Mori, K., Takahashi, K., and Nagata, S. (1992). Functional expression and tissue distribution of a novel receptor for vasoactive intestinal polypeptide. *Neuron* **8,** 811–819.

Ishihara, T., Nakamura, S., Kaziro, Y., Takahashi, T., Takahashi, K., and Nagata, S. (1991). Molecular cloning and expression of a cDNA encoding the secretin receptor. *EMBO J.* **10,** 1635–1641.

Jaffe, B. M. (1992). Current issues in the management of Zollinger–Ellison syndrome. *Surgery* **111,** 241–250.

Jia, X., Brown, J. C., Ma, P., Pederson, R. A., and McIntosh, C. H. S. (1995). Effects of glucose-dependent insulinotropic polypeptide and glucagon-like peptide-I-(7–36) on insulin secretion. *Am. J. Physiol. (Endocrin. Metab. 31)* **268,** E645–E651.

Jiang, S., and Ulrich, C. (1995). Molecular cloning and functional expression of a human pancreatic secretin receptor. *Biochem. Biophys. Res. Commun.* **207,** 883–890.

Kopin, A. S., Lee, Y. M., McBride, E. W., Miller, L. J., Lu, M., Lin, H. Y., Kolakowski, L. F., and Beinborn, M. (1992). Expression cloning and characterization of the canine parietal cell gastrin receptor. *Proc. Natl. Acad. Sci. USA* **89,** 3605–3609.

Kopin, A. S., Wheeler, M. B., Nishitani, J., McBride, E. W., Chang, T., Chey, W. Y., and Leiter, A. B. (1991). The secretin gene: Evolutionary history, alternative splicing, and developmental regulation. *Proc. Natl. Acad. Sci. USA* **88,** 5335–5339.

Kopin, A. S., Wheeler, M. B., and Leiter, A. B. (1990). Secretin: Structure of the precursor and tissue distribution of the mRNA. *Proc. Natl. Acad. Sci. USA* **87,** 2299–2303.

Krane, I. M., Naylor, S. L., Helin-Davis, D., Chin, W. W., and Spindel, E. R. (1988). Molecular cloning of cDNAs encoding the human bombesin-like peptide neuromedin B. *J. Biol. Chem.* **263,** 13317–13323.

Kroog, G. S., Sainz, E., Worland, P. J., Akeson, M. A., Benya, R. V., Jensen, R. T., and Battey, J. F. (1995). The gastrin-releasing peptide receptor is rapidly phosphorylated by a kinase other than protein kinase C after exposure to agonist. *J. Biol. Chem.* **270,** 8217–8224.

Lacroix, A., Bolte, E., Tremblay, J., Dupre, J., Poitras, P., Fournier, H., Garon, J., Garrel, D., Bayard, F., Taillefer, R., Flanagan, R. J., and Hamet, P. (1992). Gastric inhibitory polypeptide-dependent cortisol hypersecretion: a new cause of Cushing's syndrome. *New Engl. J. Med.* **327,** 974–979.

Liddle, R. A. (1994). Regulation of cholecystokinin gene expression in rat intestine. *Ann. NY Acad. Sci.* **713,** 22–31.

Maton, P. N. (1988). The carcinoid syndrome. *J. Am. Med. Assoc.* **260,** 1602–1605.

McKnight, G. L., Karlsen, A. E., Kowalyk, S., Mathewes, S. L., Sheppard, P. O., O'Hara, P. J., and Taborsky, G. J., Jr. (1992). Sequence of human galanin and its inhibition of glucose-stimulated insulin secretion from RIN cells. *Diabetes* **41,** 82–87.

Miralles, P., Peiero, E., Degano, P., Silvestre, R. A., and Marco, J. (1990). Inhibition of insulin and somatostatin secretion and stimulation of glucagon secretion by homologous galanin in perfused rat pancreas. *Diabetes* **39,** 996–1001.

Nagalla, S. R., Gibson, B. W., Tang, D., Reeve, J. R., and Spindel, E. R. (1992). Gastrin-releasing peptide (GRP) is not mammalian bombesin: identification and molecular cloning of a true amphibian GRP distinct from amphibian bombesin in *Bombina orientalis*. *J. Biol. Chem.* **267,** 6916–6922.

Niederau, C., Luthen, R., and Heintges, T. (1994). Effects of CCK on pancreatic function and morphology. *Ann. NY Acad. Sci.* **713,** 180–198.

Rang, H. P., Dale, M. M., Ritter, J. M., and Gardner, P. (1995). The gastrointestinal tract. *In* "Pharmacology," 3rd ed., Chapter 19, pp. 385–402. Churchill Livingstone, New York.

Rattan, S. (1991). Role of galanin in the gut. *Gastroenterology* **100,** 1762–1768.

Rohner-Jeanrenaud, F., and Jeanrenaud, B. (1996). Obesity, leptin, and the brain. *New Engl. J. Med.* **334,** 324–325.

Rothstein, R. D., Johnson, E., and Ouyang, A. (1991). Distribution and density of substance P receptors in the feline gastrointestinal tract using autoradiography. *Gastroenterology* **100,** 1576–1581.

Sandvik, A. K., and Waldum, H. L. (1991). CCK-B (gastrin) receptor regulates gastric histamine release and acid secretion. *Am. J. Physiol.* **260,** G925–G928.

Schjoldager, B. T. G. (1994). Role of CCK in gallbladder function. *Ann. NY Acad. Sci.* **713,** 207–213.

Shriver, J., and Edmondson, S. (1993). Defining the precision with which a protein structure is determined by NMR. Application to motilin. *Biochemistry* **32,** 1610–1617.

Spindel, E. R., Zilberberg, M. D., and Chin, W. W. (1987). Analysis of the gene and multiple mRNAs encoding human gastrin releasing peptide (GRP): Alternate RNA splicing occurs in neural and endocrine tissue. *Mol. Endocrinol.* **1,** 224–232.

Sreedharan, S. P., Patel, D. R., Xia, M., Ichikawa, S., and Goetzl, E. J. (1994). Human vasoactive intestinal peptide$_1$ receptors expressed by stable transfectants couple to two distinct signaling pathways. *Biochem. Biophys. Res. Commun.* **203,** 141–148.

Symes, A., Lewis, S., Corpus, L., Rajan, P., Hyman, S. E., and Fink, J. S. (1994). STAT proteins participate in the regulation of the vasoactive intestinal peptide gene by the ciliary neurotrophic factor family of cytokines. *Mol. Endocrinol.* **8,** 1750–1763.

Tanaka, K., Masu, M., and Nakanishi, S. (1990). Structure and functional expression of the cloned rat neurotensin receptor. *Neuron* **4,** 847–854.

Thor, K., and Rosell, S. (1986). Neurotensin increases colonic motility. *Gastroenterology* **90,** 27–31.

Wada, E., Way, J., Lebacq-Verhaden, A. M., and Batley, J. F. (1990). Neuromedin B and gastrin-releasing peptide mRNAs differentially distributed in the rat nervous system. *Neuroscience* **19,** 2917–2930.

Wank, S. A., Harkins, R., Jensen, R., Shapira, H., deWeerth, A., and Slattery, T. (1992a). Purification, molecular cloning, and functional expression of the cholecystokinin receptor from rat pancreas. *Proc. Natl. Acad. Sci. USA* **89,** 3125–3129.

Wank, S. A., Pisegna, J. R., and de Weerth, A. (1992b). Brain and gastrointestinal cholecystokinin receptor family: structure and functional expression. *Proc. Natl. Acad. Sci. USA* **89,** 8691–8695.

Wilding, J. P. H., Ghatei, M. A., and Bloom, S. R. (1995). Hormones of the gastrointestinal tract. *In* "Endocrinology" (L. J. DeGroot, M. Besser, H. G. Burger, J. L. Jameson, D. L. Loriaux, J. C. Marshall, W. D. Odell, J. T. Potts, Jr., and A. H. Rubenstein, eds.), 3rd ed., Vol. 3, Chapter 153, pp. 2870–2894. W. B. Saunders Company, Philadelphia, PA.

Wiley, J. W., Lu, Y., and Owyang, C. (1991). Mechanism of action of peptide YY to inhibit gastric motility. *Gastroenterology* **100,** 865–872.

Wolfe, M. M., and Soll, A. H. (1988). The physiology of gastric acid secretion. *New Engl. J. Med.* **319,** 1707–1715.

Yamada, Y., Post, S. R., Wang, K., Tager, H. S., Bell, G. I., and Seino, S. (1992). Cloning and functional characterization of a family of human and mouse somatostatin receptors expressed in brain, gastrointestinal tract, and kidney. *Proc. Natl. Acad. Sci. USA* **89,** 251–255.

Zachary, I., Gil, J., Lehmann, W., Sinnett-Smith, J., and Rozengurt, E. (1991). Bombesin, vasopressin and endothelin rapidly stimulate tyrosine phosphorylation in intact Swiss 3T3 cells. *Proc. Natl. Acad. Sci. USA* **88,** 4577–4581.

C H A P T E R

9

Calcium-Regulating Hormones: Vitamin D, Parathyroid Hormone, Calcitonin

I. INTRODUCTION

A. Background Information

Calcium and phosphorus are the most abundant of the inorganic elements in humans; together they are the key structural minerals of the human body. A 70-kg man contains about 1200 g of calcium and 770 mg of phosphorus present as phosphate. Calcium and phosphorus are also essential to a great number of cellular processes (see Table 9-1).

For optimal growth and function, living organisms require an adequate supply of calcium and phosphate to meet their metabolic and structural needs. Higher organisms must be dietarily supplied with calcium and phosphorus on a continual basis. The plasma concentrations of these substances are maintained within a surprisingly narrow limit by an endogenous control mechanism, often of great subtlety and elegant precision.

The three primary target tissues involved in calcium and phosphorus homeostasis in humans and higher animals are the intestine, where these ions enter into the physiological milieu, bone, where they are stored and made available for minute-by-minute regulation of the serum levels of these ions, and kidney, where their extent of excretion can be monitored and regulated. The maintenance of calcium and phosphorus homeostasis thus involves the delicate and coordinated

TABLE 9-1 **Biological Calcium and Phosphorus**

Calcium	Phosphorus
Body content	Body content
70-kg man has 1200 g of Ca^{2+}	70-kg man has 770 g of P
Utilization	Utilization
Structural: bone has 95% of body Ca^{2+}	Structural: bone has 90% of body P_i
Plasma [Ca^{2+}] is 9.0–10.2 mg/100 ml (2.2–2.5 m*M*)	Plasma [P_i] is 2.1–4.0 mg/100 ml (0.68–1.1 m*M*)
Muscle contraction	Intermediary metabolism (phosphorylated intermediates)
Nerve impulse transmission	Genetic information (DNA and RNA)
Blood clotting	Phospholipids
Membrane structure	Enzyme–protein components (phosphohistidine, phosphoserine)
Enzyme cofactors (amylase, trypsinogen, lipases, ATPases)	Membrane structure
Egg shell (birds)	
Daily requirements (70-kg man)	Daily requirements (70-kg man)
Dietary intake: 600–1600[a]	Dietary intake: 600–2000[a]
Fecal excretion: 300–600[a]	Fecal excretion: 200–600[a,b]
Urinary excretion: 100–300[a,b]	Urinary excretion: 400–1400[a,b]
Sweat: 100–200[a,b]	

[a] Values in milligrams per day for an adult.
[b] Based on the indicated level of dietary intake.

TABLE 9-2 **Distribution of Calcium and Phosphate in Normal Human Plasma**[a]

State	Plasma concentration		Percentage of total
	mmol/liter	mg/100 ml[b]	
Phosphorus			
Free HPO_4^{2-}	0.50	1.55	44
Free $H_2PO_4^-$	0.11	0.34	10
Protein-bound	0.14	0.43	12
$NaHPO_4^-$	0.33	1.02	28
$CaHPO_4$	0.04	0.12	3
$MgHPO_4$	0.03	0.10	3
Total	1.15	3.56	
Calcium			
Free Ca^{2+}	1.18	4.72	48
Protein-bound	1.14	4.56	46
Complexed	0.08	0.32	3
Unidentified	0.08	0.32	3
Total	2.48	9.92	3

[a] Adapted from Walser, M. (1961). Ion association. VI. Interaction between calcium, magnesium, inorganic phosphate, citrate and protein in normal human plasma. *J. Clin. Invest.* **40,** 723.
[b] For plasma inorganic phosphate ($H_2PO_4^-$, HPO_4^{2-}, and PO_4^{3-}), the values are expressed as milligrams of elemental phosphorus or P per 100 ml of plasma, in accordance with long-standing clinical practice.

interrelationships of absorption by the intestine, accretion and reabsorption by bone tissue, and renal tubular reabsorption and urinary excretion by the kidney.

The three principal endocrine regulators of calcium and phosphorus metabolism are the two peptide hormones parathyroid hormone and calcitonin and vitamin D and its metabolites. The interdependent actions of these three hormones of calcium and phosphorus homeostasis reflect the crucial roles of both calcium and phosphate in the biological processes of higher animals. It is vital that the extracellular calcium ion concentration is maintained within narrow limits; accordingly, the higher the phylogenetic order, the greater the complexity of endocrinological interrelationships and mechanisms developed to ensure ion homeostasis.

The total concentration of calcium in plasma is normally 9.0–10.2 mg/100 ml or 2.2–2.5 m*M*; consistent deviations outside this narrow "range" of concentrations are reflective of significant perturbations of the calcium-regulating hormones.

Calcium in the plasma exists in three forms: free (ionized), complexed (chelated) by organic ions such as citrate, and protein bound (Table 9-2). Under usual circumstances, ~45% of the total plasma calcium is bound to protein, primarily to albumin, and the remaining 55% is ultrafilterable. There are two basic mechanisms for changing the proportion of ionized calcium in the plasma: that which can be mediated by endocrine mechanisms and that which can occur by changing the concentration of plasma protein. All evidence indicates that the biologically functional and therefore regulated form of calcium is the ionized species.

The plasma concentrations of the several ionic species of phosphate are not regulated so stringently. Under normal circumstances, phosphate is present in the plasma at a level of 2.1–4.0 mg of phosphate per 100 ml of plasma or 0.68–1.1 m*M*. (In the United States, plasma concentrations of phosphate are formally expressed as elemental phosphorus contained in the inorganic phosphate.) Approximately 10% of the plasma phosphate is protein bound, and the remainder exists as free phosphate, as either HPO_4^{2-} or $H_2PO_4^-$. The relative proportion of these two species is dependent upon the plasma pH. There are many biological functions of phosphate, extending from its role as an inorganic buffer ion to multiple involvements as organic phosphates in various enzymatic and structural proteins, phospholipids, and nucleic acids.

B. Calcium and Phosphorus Homeostasis

The principal organs of the body involved in the maintenance of calcium and phosphate homeostasis are the intestine, kidney, and bone; it is here that the three calcium-regulating hormones initiate an integrated set of biological responses that results in calcium and phosphorus homeostasis. Vitamin D is known to play a dominant role in increasing the intestinal absorption of calcium and phosphorus; this uptake process is regulated according to the nutritional or physiological needs of the animal and is dictated by certain physiological signals. Once calcium and phosphorus are made available in the plasma, a delicate balance occurs in the bone between accretion and mobilization and in the kidney between urinary excretion and reabsorption. In the event that the dietary availability of calcium and phosphorus is diminished (or increased), the balance is tipped in favor of increased bone mobilization (or increased urinary excretion) to meet the stringent requisite of a constant serum calcium level. Thus, serum calcium may become elevated by stimulation of intestinal calcium absorption, bone calcium mobilization, or parathyroid hormone stimulation of the tubular reabsorption of calcium at the kidney. It is clear that bone is the central organ in calcium metabolism, acting as a reservoir for calcium and phosphorus. There are a variety of endocrine-mediated events that govern this organ's involvement in calcium and phosphorus metabolism; these will be discussed later in the chapter.

Figure 9-1 is a schematic diagram illustrating the 24-hr "metabolic balance" of calcium and phosphorus metabolism in a normal adult male. Calcium and phosphorus are both absorbed into the body primarily in the duodenum and jejunum regions of the "intestine."

The Recommended Daily Allowance (RDA) of calcium for an adult male or female is 800 mg; the RDA increases to 1200 mg for a pregnant or lactating female. In addition to the calcium ingested in the diet, approximately 600 mg is added to the intestinal contents by intestinal secretions. Of the approximately 1500–1800 mg of calcium present in the lumen of the intestine, $\approx$900 mg is absorbed by intestinal epithelial cells and transported to the blood compartment, leaving the remaining 600–900 mg to be excreted in the feces.

After calcium has entered the extracellular pool, it is in constant exchange with the calcium already present in the extra- and intracellular fluids of the body and in certain compartments of the bone and glomerular filtrate. The entire extracellular pool of calcium turns over between 40 and 50 times per day. The glomerulus of the kidney filters some 10,000 mg of calcium per day, but the renal tubular reabsorption of this ion is so efficient that under normal circumstances only between 100 and 300 mg of Ca appears in the urine. In the event of hypercalcemia, the urinary excretion of calcium rises in a compensatory fashion; however, it rarely exceeds a value of 400–600 mg/day. The renal tubular reabsorption of calcium is stimulated by parathyroid hormone and possibly by $1\alpha,25(OH)_2D_3$. The increased urinary excretion of calcium is stimulated by phosphate deprivation, acidosis, adrenal steroids, and saline diuresis. Also, it should be noted that an additional 50–200 mg of calcium may be lost per day through the skin via sweating.

The dynamics of phosphate metabolism are not particularly different from that of calcium. Under normal circumstances, $\approx$70% of phosphate in the diet is absorbed. Absorption of phosphate is interrelated in a complex fashion with the presence of calcium and can be stimulated by a low-calcium diet and also by $1\alpha,25(OH)_2D_3$. Intestinal absorption of phosphate is inhibited by high dietary calcium levels, aluminum hydroxide ingestion, and beryllium poisoning. Phosphate in the body is also partitioned among three major pools: the kidney ultrafiltrate, the readily exchangeable fraction of bone, and the intracellular compartments in the various soft tissues.

The major excretory route for phosphate (Figure 9-1) is through the kidney. The handling of phosphate by the kidney is determined by the rates of glomerular filtration, tubular reabsorption, and possibly tubular secretion. Every day the kidney glomerulus filters some 6000–10,000 mg of phosphorus. A normal man, given a diet containing 900 mg of phosphorus, excretes $\approx$600 mg/day in the urine.

II. ANATOMICAL, MORPHOLOGICAL, AND PHYSIOLOGICAL RELATIONSHIPS

A. Parathyroid Gland

Parathyroid glands have been identified in all vertebrate species higher than fishes. In humans there are normally four parathyroid glands. Anatomically they are localized on the surface of each side of the thyroid (see Figure 6-1); in total there is $\approx$120 mg of tissue. The parathyroid glands are derived from the endoderm of the third and fourth pharyngeal pouches.

The two main classes of epithelial cells of the parathryoid gland are the more abundant chief cells and the less abundant oxyphil cells. The chief cell is responsible for the biosynthesis of parathyroid hormone (PTH). This cell undergoes a series of cyclical changes, which can be observed histologically and are associated with the synthesis, packaging, and secretion of PTH, followed by cellular involution. Each individual chief cell appears to undergo its cycle independent of adjacent chief cells.

The oxyphil cells appear in the normal human parathyroid gland at the time of puberty. After puberty the number of oxyphil cells increases throughout life. The

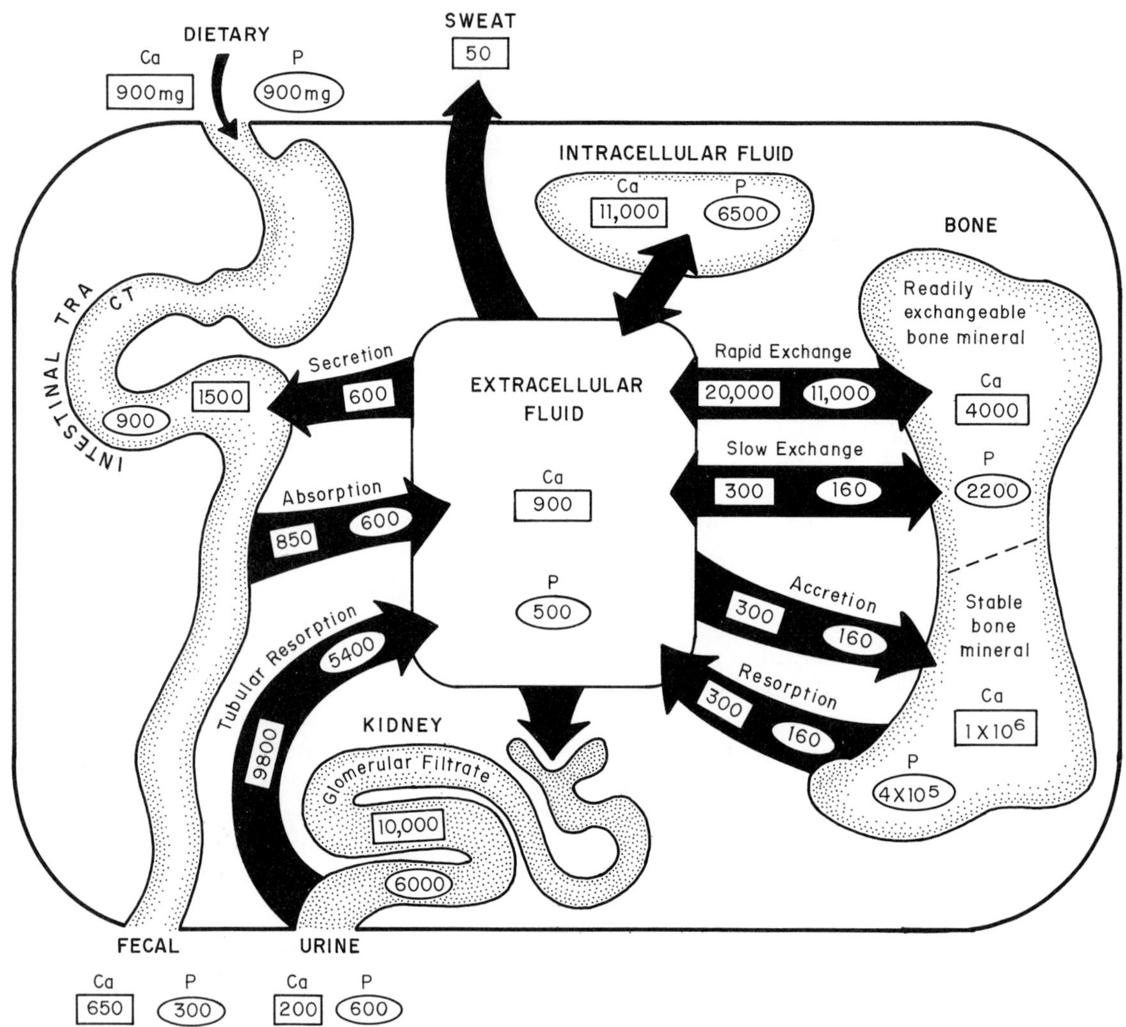

FIGURE 9-1 Schematic model of calcium and phosphorus metabolism in an adult man having a calcium intake of 900 mg/day and a phosphorus intake of 900 mg/day. All numerical values are milligrams per day. All entries relating to phosphate are calculated as phosphorus, and are enclosed in ovals. Entries related to calcium are enclosed in rectangles. Modified by permission from Norman, A. W. (1979) "Vitamin D: The Calcium Homeostatic Steroid Hormone," p. 278. Academic Press, New York.

biological function of oxyphil cells is not known. Under some circumstances of the pathological dysfunction hyperparathyroidism (e.g., parathyroid adenomas and chief cell hyperplasia), the oxyphil cells are believed to secrete PTH.

B. Calcitonin-Secreting Cells

Calcitonin (CT) is secreted by the parafollicular or C cells, which are located in the thyroid glands of higher animals. The anatomical relationship of the C cells to the thyroid follicles varies with the species. The thyroid C cells have been established as being embryologically derived from the neural crest.

In lower vertebrates, including the teleost fish, elasmobranchs, anurans, urodeles, and aves, the calcitonin-secreting C cells are localized in the anatomically distinct ultimobranchial body. The ultimobranchial body persists as a separate gland in all jawed vertebrates except mammals.

C. Intestine

The morphology and cellular organization of the intestine are superbly adapted to efficiently effect the absorption of dietary constituents, including calcium and phosphorus. The anatomical organization of the intestinal mucosa, which optimizes the surface/volume ratio of the cell and thereby facilitates the intestinal absorptive processes, is discussed in Chapter 8 (see Figure 8-3).

D. Kidney

The kidney is responsible not only for indispensable homeostatic actions with regard to the electrolytes of the body and filtration and removal of nitrogenous wastes, but also as an endocrine gland for several classes of hormones, including the hormonally active forms of vitamin D. The anatomical organization of the kidney is presented in Chapter 15. Shown in Figure 15-4 is the principal site of reabsorption of calcium and phosphorus, as well as the putative sites of action of parathyroid hormone and of vitamin D and / or its metabolites.

The concentration of calcium in the blood, particularly its ionized form, is regulated stringently through complex interactions among (1) the movement of calcium in and out of the bone, (2) its absorption by the intestine, and (3) the renal tubular reabsorption of calcium lost into the renal tubule by glomerular filtration. The kidneys of a typical adult male transfer some 11,000 mg (275 mmol) of calcium per day from the plasma into the glomerular filtrate. However, only 0.5–1.0% of this large amount of filtered calcium is lost into the urine; this is a reflection of the remarkably effective tubular reabsorption mechanism(s) for calcium. Humans normally lose 100–200 mg of calcium daily in the urine. It is known that several nonhormonal factors, as well as parathyroid hormone, vitamin D, and its metabolites, may alter the renal handling and excretory rate of calcium.

From consideration of the fluid dynamics of the kidney, it is possible to postulate that an increase in the urinary excretion of calcium may arise from either decreased tubular reabsorption, increased filtered load of calcium, or both. An increase in the filtered load produced by an elevation in the rate of glomerular filtration normally has only a small effect on calcium excretion. However, an increase in the filtered load produced by hypercalcemia usually will result in a more marked reduction in the tubular reabsorption of this cation.

The concentration of phosphorus in the blood is not regulated as stringently as that of calcium. In the case of phosphate, 40–60% of that available in the diet is absorbed by the intestine. The inorganic phosphate in the blood is also very efficiently filtered at the glomerulus, so that in a typical day ≈6000–10,000 mg is transferred from the blood to the glomerular filtrate. Of this amount, the renal tubule reabsorbs 80–90%. Humans normally lose 700–1500 mg of phosphate daily in the urine. Thus, in comparison to calcium, a significantly larger amount of phosphate is excreted on a daily basis (see Figure 9-1). As will become evident later, both parathyroid hormone and vitamin D metabolites play important roles in modulating the amount of phosphate actually excreted in the urine. Under circumstances of dietary phosphate restriction, the kidney becomes very efficient in its tubular reabsorption of phosphate from the glomerular filtrate, thus conserving this important anion for the body. In contrast, in situations of dietary excess, the plasma phosphorus concentrations may become elevated, ultimately resulting in a significant increase in the amount of phosphorus excreted in the urine.

E. Bone

Bone is a complex tissue made up of cells and extracellular material; it is composed of organic and mineral components. The cells are of a wide variety of morphological and functional types, but all have a common origin in the mesenchymal tissue. The principal cell types are the (1) chondrocytes or cartilage cells, which secrete the collagen matrix of the cartilage region, (2) osteoblasts or bone-forming cells, (3) osteoclasts, which are multinucleated giant cells responsible for both bone resorption and bone remodeling at specific sites, and (4) osteocytes, which represent osteoblasts trapped in a mineralized matrix. On a dry weight basis, bone consists of ≈65–70% inorganic crystals of the salt hydroxyapatite and 30–35% organic matrix known as osteoid. Of the osteoid, 40% is composed of the extracellular protein collagen.

Figure 9-2 shows a schematic representation of bone organization at both the macro (A) and cellular (B) levels. There are two major categories of bone. Cortical bone is composed of densely packed columns of mineralized collagen laid down in layers and is the major component of tubular bones. Trabecular or cancellous bone is spongy in appearance, providing both strength and elasticity, and is present in the axial (vertebrae) skeleton.

The formation of functional bone tissue can be divided into two phases: (1) that concerned with the production and secretion of the extracellular collagen bone matrix and (2) the deposition of the mineral calcium hydroxyapatite crystals in the matrix. It should be emphasized that bone is a dynamic tissue and that both of these processes occur continuously throughout the life of the skeletal system.

Neither the collagen matrix, the extracellular mineral crystals, nor the several different cell types associated with bone exclusively determine the behavior of the bone tissue. It is a unique combination of the organic and inorganic phases, as well as the particular biochemical properties of these various cell types, that collectively confers on bone both its unusual mechanical properties (to support the weight of the soft tissues

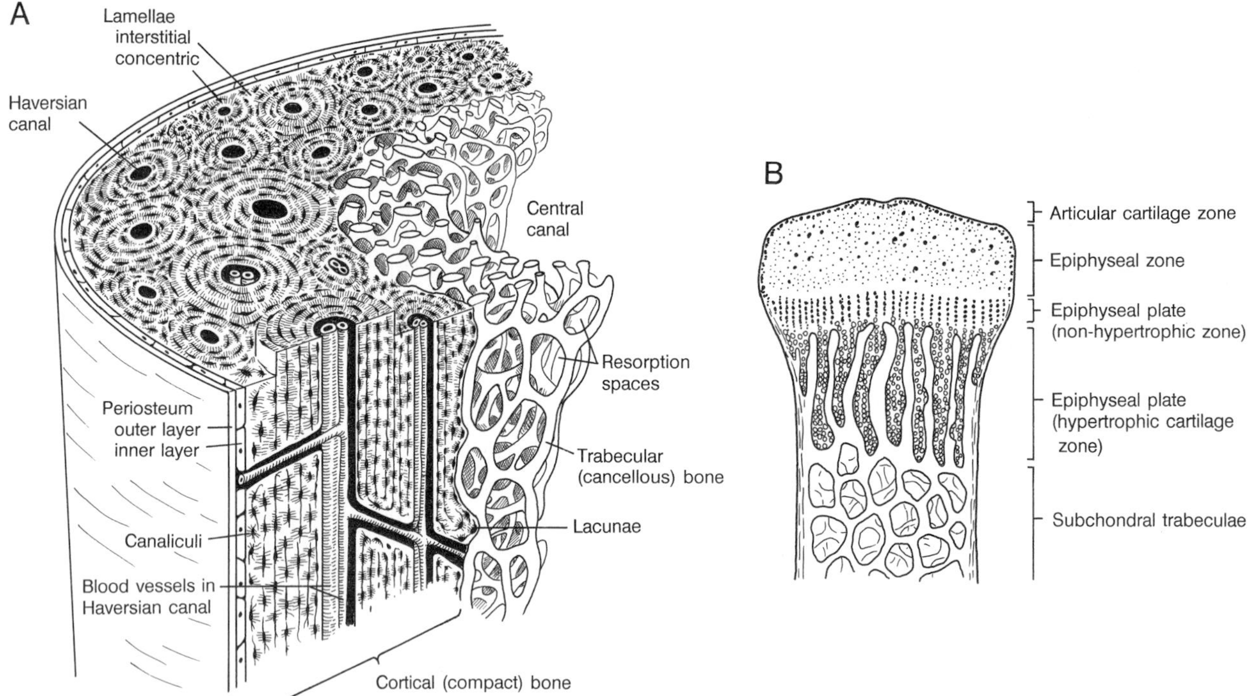

FIGURE 9-2 Schematic representation of bone at the macro and cellular levels of organization. (A) Diagram of some of the main features of the structure of bone seen in both the transverse (top) and longitudinal sections. Areas of cortical (compact) and trabecular (cancellous) bone are indicated. The central area in the transverse section simulates a microradiograph, with the variations in density reflecting variations in mineralization (see also Figure 9-18). Note the distribution of the osteocytic lacunae, Haversian canals, and resorption spaces and the different views of the structural basis of bone lamellation. Reproduced with permission from "Gray's Anatomy" (R. Warwick and P. L. Williams, eds.), 35th edition, Longman, Philadelphia, 1973. (B) Schematic sagittal section of the proximal portion of a typical long bone. Cartilage cell (chondrocytes) proliferation occurs in the nonhypertrophic zone of the eplphyseal plate. In the bottom portion of the hypertrophic zone, the chondrocytes undergo hypertrophy and degeneration (apoptosis) at the same time as the region becomes vascularized. Calcification first occurs after the invasion of bone-forming osteoblasts, which in turn produces the typical osseous tissue within the template created by the hypertrophic cartilage cells. Reproduced with permission from Norman, A. W., and Hurwitz, S. (1993). *J. Nutr.* **123,** 310–316.

of the body) and its ability to serve as a dynamic reservoir for the calcium and phosphate ions needed for mineral homeostasis in the whole organism. Table 9-3 presents an outline of the elements of bone structure and metabolism.

Bone matrix is biosynthesized, secreted, organized, mineralized, and finally destroyed by reabsorption, all in accordance with the physiological and hormonal signals operative at any particular time. The production of organic matrix by osteoblast cells first involves the intracellular synthesis of protocollagen molecules by the ribosomal system through conventional protein biosynthetic pathways. The protocollagen is then polymerized to yield a three-stranded triple helix, each strand of which has a separate unique amino acid sequence. Next, approximately half of the proline residues and a small number of lysine residues of the protocollagen helix are enzymatically hydroxylated, and these new hydroxyamino acid residues are then glycosylated to yield the macromolecule known as procollagen. The procollagen is secreted into the extracellular space around the osteoblast cells, where it undergoes further conversion to yield tropocollagen via cleavage of an N-terminal fragment of 20,000 Da from each of the three individual peptide genes of the macromolecule. After proteolysis, the tropocollagen molecule is some 15 Å in diameter. It immediately undergoes polymerization to yield microfibrils several centimeters in length and consisting of five tropocollagen molecules across the diameter. These microfibril pentamers further polymerize to yield the final form of collagen fibers, which range in diameter from 150 to 1300 Å. The polymerization process, although spontaneous, is highly ordered and principally in-

TABLE 9-3 Outline of the Elements of Bone Structure and Metabolism[a]

I. Structure of bone	
A. Macroscopic level	
1. Spongiosa or cancellous	Occurs in metaphyseal region of long bone and is composed of spicules known as trabeculae; is the most metabolically active region
2. Cortical bone	Occurs in cortex and diaphysis of long bone
B. Microscopic level	
1. Collagen fiber distribution	
a. Woven bone	Composed of loosely and randomly arranged collagen bundles; predominant in young bones
b. Lamellar bone	Composed of highly ordered bundles of collagen fibers arranged in layers interspersed between osteocytes; is characteristic of adult bone
2. Three-dimensional collagen fiber network	Principally characteristic of adult bone
a. Trabecular	Synonymous with cancellous bone
b. Haversian	Occurs as a consequence of remodeling of cortical bone
C. Types of bone cells	
1. Osteoprogenitor	Synonymous with mesenchymal cells
2. Osteoblasts	Cells that synthesize and secrete bone matrix
3. Osteoclasts	Cells that resorb bone; are multinucleate
4. Osteocytes	Precise function not known
5. Chondrocytes	Cartilage cells; they synthesize and release collagen
II. Inorganic phase of bone	On a dry-weight basis bone is 65–70% inorganic material
A. Components	
1. Crystalline calcium hydroxyapatite	Constitutes 55–60% of bone mineral; molecular formula $[(Ca^{2+})_{10-y}(H_3O)^{+}_{2+y}(PO_4^{3-})_6(OH^-)_2]$
2. Amorphous calcium phosphate	Constitutes 40–45% of bone mineral; may be formed before crystalline calcium hydroxyapatite, which is then produced with time
B. Ionic composition	Other ions include K^+, Na^+, Mg^{2+}, Sr^{2+}, Cl^-, $CO_3^=$, F^-, and citrate^{3-} and many trace constituents
III. Organic phase of bone	On a dry-weight basis bone is 30–35% organic material, often referred to as bone matrix or osteoid
A. Collagen	
1. Biosynthetic steps	
a. Intracellular phase	
i. Protocollagen biosynthesis	Occurs on ribosome; composed of 20% proline and 33% glycine
ii. Protocollagen hydroxylation	Specific hydroxylation of 50% of proline and a small number of lysines
iii. Glycosylation of protocollagen to yield procollagen	Procollagen molecular weight is 360,000; is secreted into extracellular space
b. Extracellular phase	
i. Conversion of procollagen to tropocollagen	An N-terminal fragment of procollagen of 20,000 Da is removed; tropocollagen molecule is 3000 × 15 Å
ii. Tropocollagen polymerization into microfibrils	44-Å diameter × several centimeters length
iii. Polymerization of microfibrils into collagen fibrils	150–1300-Å diameter × several centimeters length
iv. Cross-linking of strands of collagen fibers	Formation of Schiff bases, followed by aldol condensation
B. Ground substance	5% bone matrix is ground substance
1. Mucopolysaccharide	
2. Mucoproteins	
C. Bone matrix proteins	
1. Osteocalcin	Major extracellular cellular protein of bone (the sixth most abundant protein in the body); produced by osteoblasts; molecular weight = 13,000; contains γ-carboxyglutamic residues (Gla)
2. Matrix Gla protein (MGP)	A Gla-containing protein; molecular weight = 15,000
3. Osteopontin	Molecular weight = 41,000; contains N- and O-linked oligosaccharides
4. Bone sialoprotein (BSP)	Protein = 33 kDa; contains 50% carbohydrate
5. Osteonectin	An acidic glycoprotein; molecular weight = 44,000

[a] Modified with permission from Norman, A. W. (1979). "Vitamin D: The Calcium Homeostatic Hormone." Academic Press, New York.

volves hydrophobic interactions between adjacent peptide chains. The final phase of formation of mature collagen molecules is the generation of covalent chemical bonds, either by an aldol condensation or by a Schiff base conjugation.

In the normal process of human bone formation, there is usually a delay of 5–10 days between the synthesis of the extracellular organic matrix and its ultimate mineralization. It is believed that the normal coupling time between matrix formation and subsequent mineralization depends on the presence of vitamin D metabolites. After the final secretion of protocollagen by the osteoblasts and the ultimate formation of mature collagen, the osteoblasts differentiate and are incorporated into the bone matrix as osteocytes.

As yet a detailed molecular description of the total calcification process cannot be provided. However, in the final mineralization steps, an extracellular accumulation of calcium and phosphate occurs, which ultimately results in the appearance of amorphous calcium phosphate. This in turn is crystallized into the formal hydroxyapatite structure of bone within the matrix of the mature collagen fibers.

In understanding the hormonal regulation of calcium and phosphorus homeostasis, it is important to appreciate that bone tissue is dynamic, in terms both of the metabolic activity of the various cell populations present and of the kinetics of mineral flux into and out of the various bone compartments. As summarized in Figure 9-1, there are three "compartments" of bone mineral; they represent the various fractions of the total calcium present in bone that can be identified on the basis of radioactive tracer studies (utilizing either radioactive $^{45}Ca^{2+}$ or $^{47}Ca^{2+}$). These compartments include the readily exchangeable bone mineral compartment, which has both a rapid (20,000 mg/day) and a slow (300 mg/day) component of calcium. The term rapid implies the capability of exchange on a minute-by-minute basis. The third compartment is stable bone mineral and in adult men is composed of approximately 1 kg of calcium. However, the stable bone mineral compartment is not biologically inert, and it is that pool of calcium that participates in the bone remodeling processes associated with bone mineralization (accretion, as in growth) and/or resorption. Both the accretion and resorption processes are those that are affected by the hormonal regulators of calcium metabolism, whereas the rapid and slow exchange pools largely represent chemical equilibria between the bone and the bathing extracellular fluids.

In a normal adult, a remodeling of the stable bone mineral is occurring continuously, such that, on any given day, approximately 300 mg of calcium is resorbed and another 300 mg of calcium is replaced by accretion. Thus, in the normal state there is a balance in the extent of daily bone resorption and bone accretion, leading to no net change in total skeletal calcium content. In contrast, in certain disease states, such as osteoporosis, there can be an imbalance in the daily bone remodeling rates so that there is a small net daily loss of calcium from the skeleton. For example, if bone resorption exceeds bone accretion by only 10 mg/day, when extended for 20 years (e.g., from menopause at age ≈50 to age ≈70) there will be 10 mg/day × 365 days × 20 years = 73,000 mg of calcium lost from the skeleton or about 7.3% of the total skeletal calcium content.

III. CHEMISTRY AND BIOCHEMISTRY

A. Parathyroid Hormone

Parathyroid hormone (PTH) is an 84-amino acid-containing protein that is secreted by the chief cells of the parathyroid gland. PTH is a single-chain polypeptide of 84 amino acids (molecular weight = 9300) with no cysteine residues and hence no disulfide bridges. There is a preponderance of basic residues (arginine + lysine), conferring an overall positive charge on the molecule. A distinctly hydrophobic region is found in the central portion of the sequence (residues 31–43), including the single tyrosine residue, which is the site of iodine-125 labeling for radioimmunoassay procedures. A fragment of the intact hormone consisting of the first 28–34 amino acids at the N-terminal region of the molecule is sufficient for the peptide to exert its entire spectrum of characteristic biological effects.

Because PTH represents a secreted protein, it is biosynthesized as a larger precursor by the parathyroid gland (see Figure 1-11). Two precursor species have been identified, including a prepro-PTH with 31 additional amino acids added onto the N-terminal region of PTH, representing the primary translation product of the mRNA in the ribosomal fraction of the parathyroid gland. This prepro-PTH then is secreted into the cisterna of the rough endoplasmic reticulum, where it is processed within seconds into a form known as pro-PTH. This occurs by the removal of the NH_2-terminal methionyl residue and the next 24 amino acids (i.e., residues −30 through −7) after biosynthesis. By 20 min after synthesis, pro-PTH reaches the Golgi region where it is stored in vesicles and converted into PTH by removal of the NH_2-terminal hexapeptide. PTH is stored in the secretory granule until it is released into circulation in response to a fall in the blood concentration of calcium. These steps are summarized schematically in Figure 1-11. Through the application of recombinant DNA techniques, a cDNA probe to the prepro-PTH messenger RNA has been prepared and used to evaluate cloned genomic DNA sequences containing the gene for PTH.

The introduction of radioimmunoassay techniques has permitted the detection of circulating levels of PTH and its changing concentration in disease states. Three predominant peptide species have been identified in human plasma. The first appears to be identical to the intact or native form of the PTH hormone as isolated from parathyroid tissue; this has 84 amino acids and a molecular weight of ~9500. There are two other populations of PTH fragments, one with a molecular weight of ~7000 and one that is smaller, having a molecular weight in the range of 4500. Conflicting views are held on the importance of these three peripheral forms of circulating PTH. One view is that only the 9500 molecular weight material is biologically active, with the smaller species representing degradation products. The other view is that one of the smaller species is the major biologically active form of the hormone. There is evidence that the liver Kupfer cells play an important role in the metabolism of the intact PTH molecule. The half-life of the intact PTH molecule in normal human plasma is only 20 min.

B. Parathyroid Hormone-Related Protein

Parathyroid hormone-related protein (PTHrP) was originally purified and an N-terminal amino sequence was obtained from human tumors associated with humoral hypercalcemia of malignancy (HHM); this information permitted its molecular cloning and study of the total amino acid sequence. PTHrP is a 141-amino-acid protein; a comparison of the structures of PTH and PTHrP is presented in Figure 9-3. Of the first 13 amino acids, 8 are identical and 3 more represent conservative amino acid substitutions. The close structural homology of PTHrP with PTH over the first 13 amino acids allows PTHrP to mimic many of the biological actions of PTH (see later discussion).

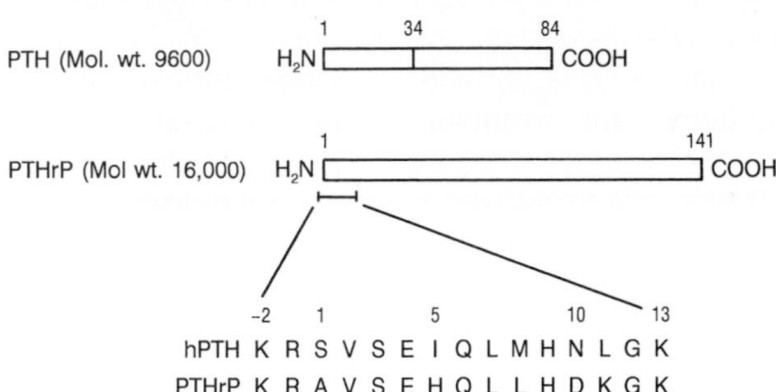

FIGURE 9-3 Comparison of the size and amino acid sequence of parathyroid hormone (PTH) and parathyroid hormone-related protein (PTHrP). Reproduced with permission from Broadus, A. E., Mangin, M., Ikeda, K., Insogna, K. L., Weir, E. C., Burtis, W. J., and Stewart, A. F. (1988). *New. Engl. J. Med.* **319**, 556–563.

C. Calcitonin

Calcitonin (CT) is a small polypeptide hormone secreted by the ultimobranchial gland in fish, amphibians, reptiles, and birds; in mammals it is secreted by the specialized C cells that are found primarily in the thyroid gland, but also at a small number of extrathyroidal sites. The amino acid sequences of nine calcitonins have been determined; these include five mammalian species—porcine, bovine, ovine, human, and rat—as well as four nonmammalian species—salmon I, II, and III and eel. All have a similar structure consisting of a straight chain peptide of 32 amino acids, with a seven-membered disulfide ring at the N-terminus and a prolinamide residue at the C-terminus. Compared to the mammalian calcitonins, the nonmammalian hormones are much more stable and have a potency 10–50 times greater.

Studies of the molecular biology of the biosynthesis of calcitonin mRNA have demonstrated that the calcitonin gene can support the production of multiple protein products from a single transcription unit (see Figure 9-4). Thus, the same gene produces, as a consequence of alternative processing of RNA transcripts, the hormone calcitonin in the thyroid and a protein termed the calcitonin gene-related peptide (CGRP) in the hypothalamus. CGRP thus represents a putative hormone that was discovered through the application of recombinant DNA methodology without prior knowledge of its structure or biological function. As yet the biological function of CGRP has not been clearly defined; it is distributed selectively in neural tissues and has been postulated to be involved with nociception (the ability to perceive noxious stimuli), ingestive behavior, and modulation of the autonomic and nervous systems.

The discovery that alternative mRNA processing can produce multiple protein products may provide insight into the mechanisms by which endocrine genes increase the diversity and specificity of their biological responses. Calcitonin may be measured via bioassay or radioimmunoassay. The bioassay of calcitonin is based on the hypocalcemic actions of the peptide. Several radioimmunoassays for calcitonin have been developed; the radioimmunoassay for human calcitonin is important in the detection and treatment of medullary carcinoma of the thyroid.

D. Vitamin D

The molecular structure of vitamin D is closely allied to that of classical steroid hormones (see Figure 9-5). Technically, vitamin D is a secosteroid. Secosteroids

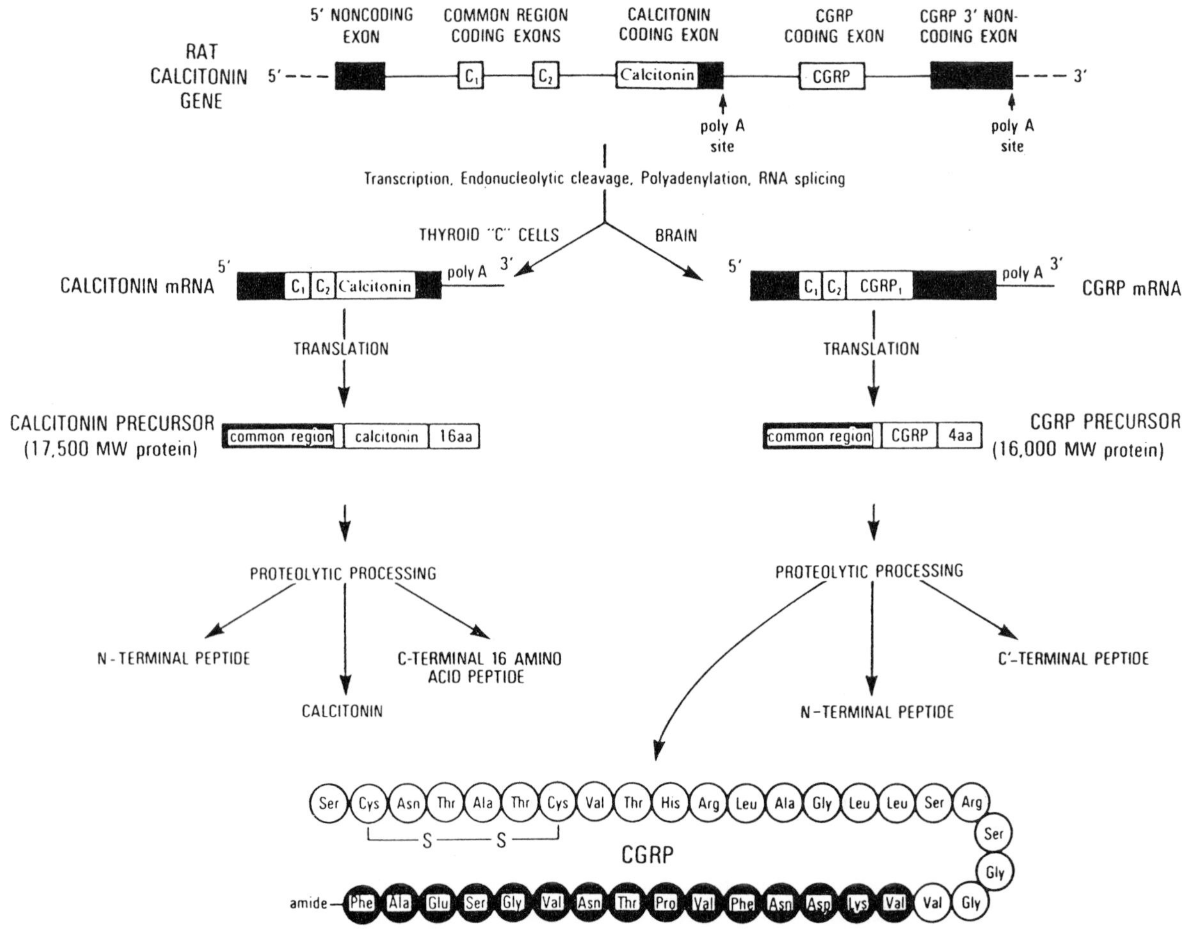

FIGURE 9-4 Model for the production of calcitonin (CT) and calcitonin gene-related peptide (CGRP) via the alternative RNA processing pathways utilized in the expression of the calcitonin gene. The calcitonin gene supports the production of CT in the thyroid and of CGRP in the hypothalamus. Mature CT, which has 32 amino acid residues, and mature CGRP, which has 37 amino acid residues, are derived from different precursor proteins. However, these two precursor proteins have an identical "common region" of 76 amino acid residues, which is derived from the C_1 and C_2 exons of the gene. Reproduced by permission of Rosenfeld, M. G. (1983). *Nature (London)* **304,** 129–135. Copyright © 1983 Macmillan Journals Limited.

are those in which one of the rings of the cyclopentanoperhydrophenanthrene ring structure of the classic steroids (see Figure 2-2) has undergone fission by breakage of a carbon–carbon bond; in the instance of vitamin D this is the 9, 10-carbon bond of ring B.

There is a family of vitamin D-related steroids depending upon the precise structure of the side chain attached to carbon-17. The vitamin D_2 family is derived from ergosterol, while the vitamin D_3 family is derived from 7-dehydrocholesterol. The naturally occurring form of vitamin D is that which has the side chain structure identical to that of cholesterol; this is known as vitamin D_3 or cholecalciferol. Vitamin D_2 or ergocalciferol is not a naturally occurring form of the vitamin. Collectively, vitamin D_3 and vitamin D_2 can be termed the calciferols or simply vitamin D.

Vitamin D_3 is normally produced by exposure to sunlight of the precursor, 7-dehydrocholesterol, present in the skin. Vitamin D_2 is produced synthetically via ultraviolet irradiation of the sterol ergosterol. One International Unit (IU) of vitamin D_3 is equivalent to 25 ng or 65 pmol. In the United States, the principal form of vitamin D supplementation for food was vitamin D_2 in the interval 1930–1965; since 1965 the predominant form of vitamin D supplementation has been vitamin D_3. The Recommended Dietary Allowance (RDA) of vitamin D (either D_2 or D_3) for an adult (ages 25–50) is 200 IU (10 mg), and that for pregnant or

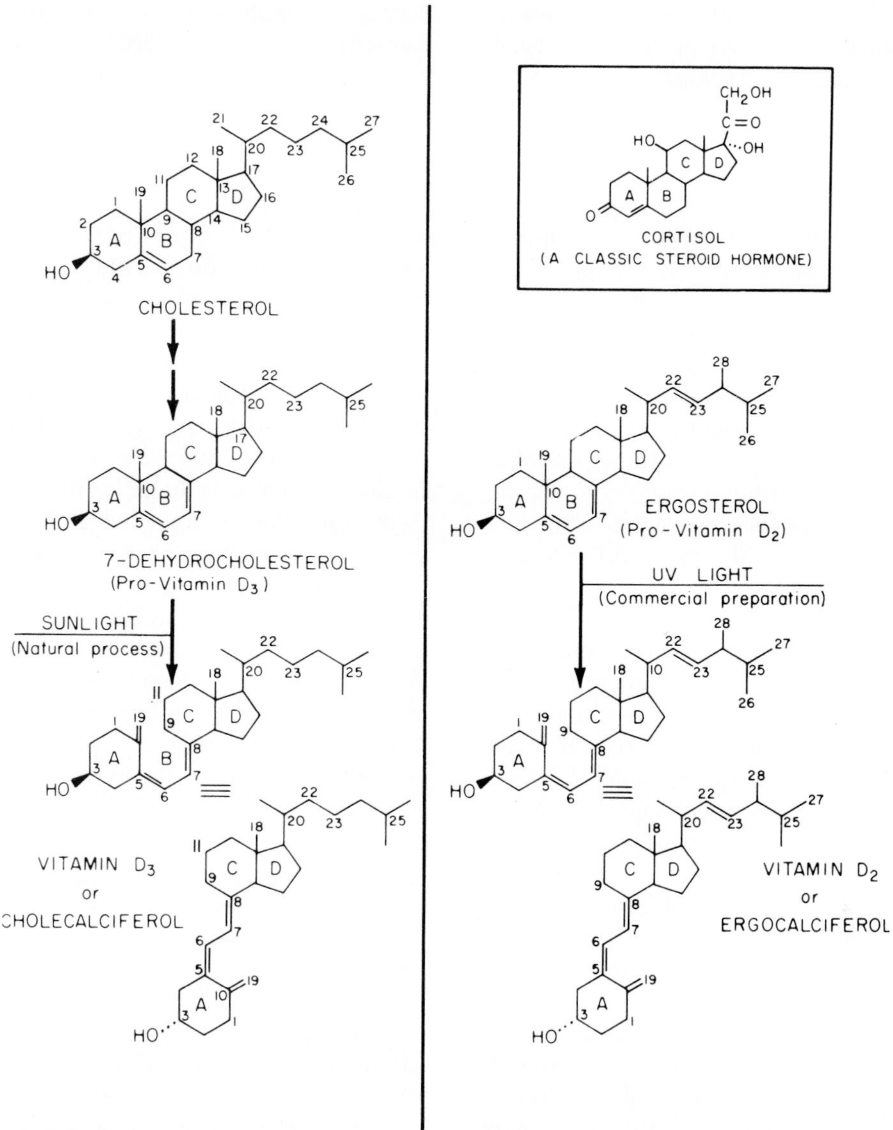

FIGURE 9-5 Structural relationship of vitamin D_3 (cholecalciferol) and vitamin D_2 (ergocalciferol) with their respective provitamins, cholesterol, and a classic steroid hormone, cortisol (see inset box). The two structural representations presented at the bottom for both vitamin D_3 and vitamin D_2 are equivalent, these are simply different ways of drawing the same molecule. It is to be emphasized that vitamin D_3 is the naturally occurring form of the vitamin; it is produced from 7-dehydrocholesterol, which is present in the skin, by the action of sunlight (see Figure 9-6). Vitamin D_2 (which is equivalently potent to D_3 in humans and many mammals, but not birds) is produced commercially by the irradiation of the plant sterol ergosterol with ultraviolet light.

lactating women or children less than 4 years of age is 400 IU.

A formal definition of a vitamin is that it is a trace dietary constituent required to effect the normal function of a physiological process. The emphasis here is on *trace* and the fact that the vitamin *must* be supplied in the diet; this implies that the body is unable to synthesize it. Thus, cholecalciferol is only a vitamin when the animal does not have access to sunlight or ultraviolet light. Under normal physiological circumstances, all mammals, including humans, can generate via ultraviolet photolysis adequate quantities of vitamin D to meet the (RDA). It is largely through historical accident that calciferol has been classified as a vitamin rather than as a steroid hormone. Chemists had certainly appreciated the strong structural similarity between vitamin D and other steroids (see Figure 2-3), but this correlation was never widely acknowledged in the biological, clinical, or nutritional sciences until 1965–1970.

The chief structural prerequisite of a sterol to be classified as a provitamin D is its ability to be converted, upon ultraviolet irradiation, to a vitamin D; thus, it is mandatory that it have in its B ring a $\Delta^{5,7}$-conjugated double-bond system. A summary of the photochemical pathway involved in the production of vitamin D is presented in Figure 9-6. In the skin, the principal ultraviolet irradiation product is previtamin D_3. The conversion of previtamin D to vitamin D involves an intramolecular hydrogen transfer from C-19 to C-9 (see Figure 9-6 for details); this transformation can occur in the absence of further ultraviolet exposure. The resulting vitamin D_3 is then transported into the general circulatory system by the vitamin D-binding protein.

Interest has focused on the effects of latitude (diminished UV-B intensity), skin pigmentation (concentration of melanin), and photochemical regulation upon the efficiency of conversion of 7-dehydrocholesterol present in the skin to epidermal previtamin D_3. In order of importance, the significant determinants limiting the rate of cutaneous production of previtamin D_3 are (i) photochemical regulation, (ii) pigmentation, and (iii) latitude.

Although the chemical structure of vitamin D was determined in the 1930s, it was not until the era 1975–1995 that the truly unique structural aspects of the molecule became apparent. In contrast to other steroid hormones (see Figure 2-3), the vitamin D molecule has three structural features that contribute to the extreme conformational flexibility of this secosteroid molecule; these include the presence of (i) an eight-carbon side chain and (ii) the broken B ring, which "unlocks" the A ring so that it (iii) can undergo chair–chair interchange many times per second. See Figure 9-7 for a detailed consideration of the conformational flexibility of vitamin D molecules.

A totally new era in the field of vitamin D has opened since 1964 with the discovery of the metabolism of vitamin D into 37 daughter metabolites (see Figure 2-25). It is now recognized that there is an endocrine system for processing the prohormone, vitamin D, into its hormonally active daughter metabolite(s) (see Figure 9-8). The biologically active form of vitamin D, particularly in the intestine, is the steroid 1α,25-dihydroxyvitamin D_3 [$1\alpha,25(OH)_2D_3$]. The endocrine gland producing the biologically active form(s) of vitamin D is the kidney. After metabolic conversion of vitamin D_3 into 25-hydroxyvitamin D_3 by a liver microsomal enzyme, this circulating form of the secosteroid serves as a substrate for either the renal 25(OH)D-1α-hydroxylase or the 25(OH)D-24-hydroxylase. Both enzymes are located in the mitochondrial fraction of the kidney cortex. The 1α-hydroxylase is localized in the kidneys of members of every vertebrate class from teleosts, through amphibians, reptiles, and aves, to mammals, including primates. It has been shown that the 25(OH)D-1α-hydroxylase enzyme system is a classical mixed-function steroid hydroxylase similar to the steroid hormone hydroxylases found in the adrenal cortex mitochondria (see Figure 2-27A). The 1α-hydroxylase is a cytochrome P450-containing enzyme that involves an adrenodoxin component incorporating molecular oxygen into the 1α-hydroxyl functionality of $25(OH)D_3$. The $1\alpha,25(OH)_2D_3$ form of the vitamin is also known to induce the renal 24-hydroxylase to permit the coproduction of $24,25(OH)_2D_3$. Emerging evidence suggests an important role for $24,25(OH)_2D_3$ in mediating normal bone development and parathyroid function. Thus, $1\alpha,25(OH)_2D_3$ and $24,25(OH)_2D_3$ are the two principal forms of the parent vitamin D_3 that mediate the biological responses characteristic of this vitamin. The other metabolites shown in Figure 2-25 are believed to represent pathways for inactivation of the biologically active metabolites of vitamin D.

IV. BIOLOGY AND PHYSIOLOGICAL SIGNIFICANCE

A. Parathyroid Hormone

The major target organs for PTH actions are kidney and bone. PTH may also act upon the intestine, but its effects here are probably indirect and are achieved by virtue of the tropic actions of PTH in stimulating the renal biosynthesis of $1\alpha,25(OH)_2D_3$.

The secretion of PTH is stimulated in response to a lowered blood concentration of calcium (see Figure 9-9). In terms of the several forms of blood calcium summarized in Table 9-2, it is known that the biosynthesis and secretion of PTH are only responsive to the ionic and not to the protein-bound forms of calcium. The secretion of PTH can be (1) stimulated when hypocalcemia is induced by the infusion of calcitonin or the divalent metal-chelating agent, EDTA (ethylenediaminetetraacetic acid), or (2) decreased when hypercalcemia is induced by the infusion of calcium. Thus, there is an inverse correlation between serum calcium concentration and PTH concentration in the range of 4-10 mg of calcium/100 ml (see Figure 9-9). The most stringent control of serum calcium concentration is achieved in the range of 9–10.5 mg of Ca^{2+}/100 ml of serum, which is considered to be the normal physio-

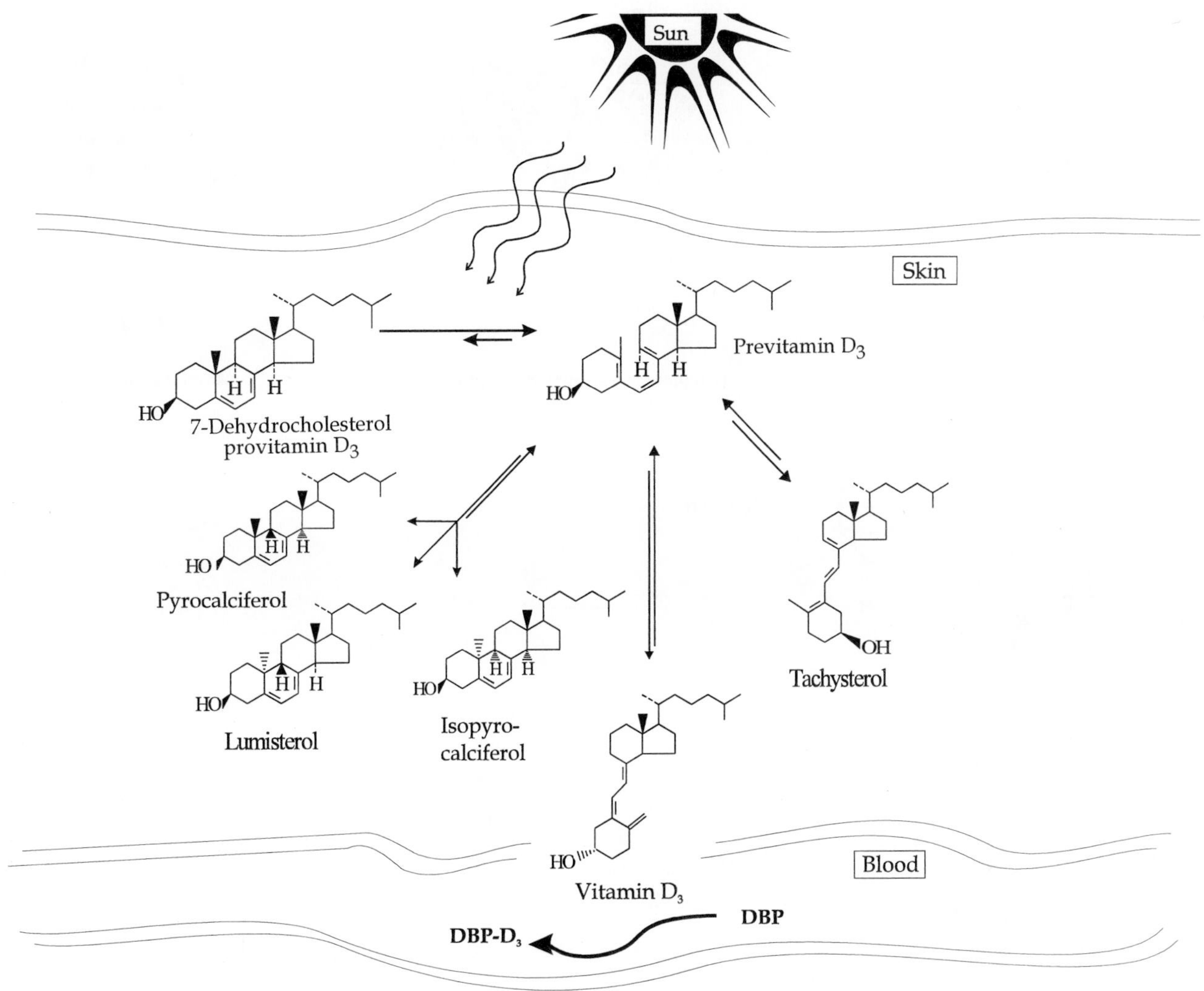

FIGURE 9-6 Photochemical pathway of production of vitamin D_3 (calciferol) from 7-dehydrocholesterol. The starting point is the irradiation of a provitamin D, which contains the mandatory $\Delta^{5,7}$-conjugated double bonds; in the skin this is 7-dehydrocholesterol. After absorption of a quantum of light from sunlight (UV-B), the activated molecule can return to the ground state and generate at least six distinct products. The four steroids that do not have a broken 9, 10-carbon bond (provitamin D, lumisterol, pyrocalciferol, and isopyrocalciferol) represent the four diastereomers with either an α- or a β-orientation of the methyl group on carbon-10 and the hydrogen on carbon-9. The three secosteroid products, vitamin D_3, previtamin D and tachysterol, have differing positions of the three conjugated double bonds. In the skin the principal product is previtamin D, which then undergoes a 1,7-sigmatropic hydrogen transfer from C-19 to C-9, yielding the final vitamin D. Vitamin D can be drawn as either a 6-*s-trans* representation (this figure) or a 6-*s-cis* representation (see Figure 9-7), depending upon the state of rotation about the 6,7-single bond. The resulting vitamin D_3, which is formed in the skin, is removed by binding to the plasma transport protein, the vitamin D-binding protein (DBP), present in the capillary bed of the dermis. The DBP–D_3 then enters the general circulatory system.

logical range of this divalent cation. Serum concentrations of calcium that fall above or below this range are likely to be due to the presence of disease states that perturb the integrated calcium–phosphorus homeostatic mechanisms.

A key signal transduction question relates to the identification of the mechanism(s) by which the surface of a parathyroid cell can sense a fall in the ambient extracellular ionized calcium concentrations and signal the secretion of PTH. A Ca^{2+}-sensing receptor from bovine parathyroid tissue has been cloned and sequenced. This 120-kDa protein (1085 amino acids) features a large extracellular domain (600 amino acid) with clusters of acidic amino acid residues possibly

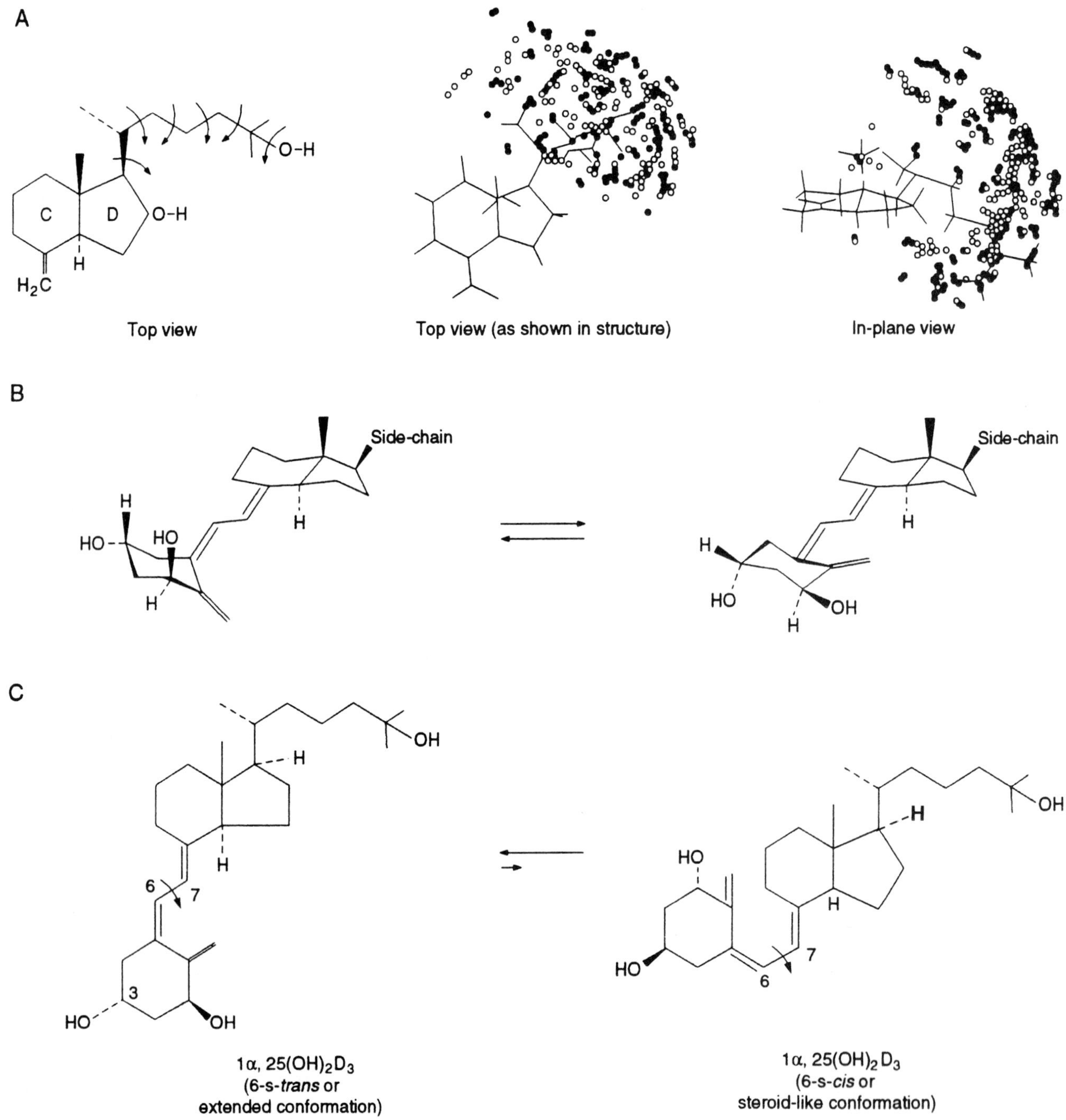

FIGURE 9-7 Conformational flexibility of vitamin D molecules using 1α,25(OH)$_2$-vitamin D$_3$ [1α,25(OH)$_2$D$_3$] as an example. (A) The dynamic single bond rotation of the cholesterol-like side chain of 1α,25(OH)$_2$D$_3$ (i.e., 360° rotations about five single carbon bonds and the oxygen, as indicated by the curved arrows). The dots indicate the position in three-dimensional space of the 25-hydroxyl group for some 394 readily identifiable side chain conformations which have been determined from energy minimization calculations. Two orientations of the C/D/side chain are presented: a "top" view and an "in-plane" view. (B) The rapid (thousands of times per second) chair–chair interconversion of the A-ring of the secosteroid which effectively equilibrates the 1α-hydroxyl between the axial and equatorial orientations. (C) The 360° rotational freedom about the 6,7 carbon–carbon bond of the seco β-ring which generates conformations ranging from the more steroid-like (6-s-*cis*) conformation, to the open and extended (6-s-*trans*) conformation of 1α,25(OH)$_2$D$_3$.

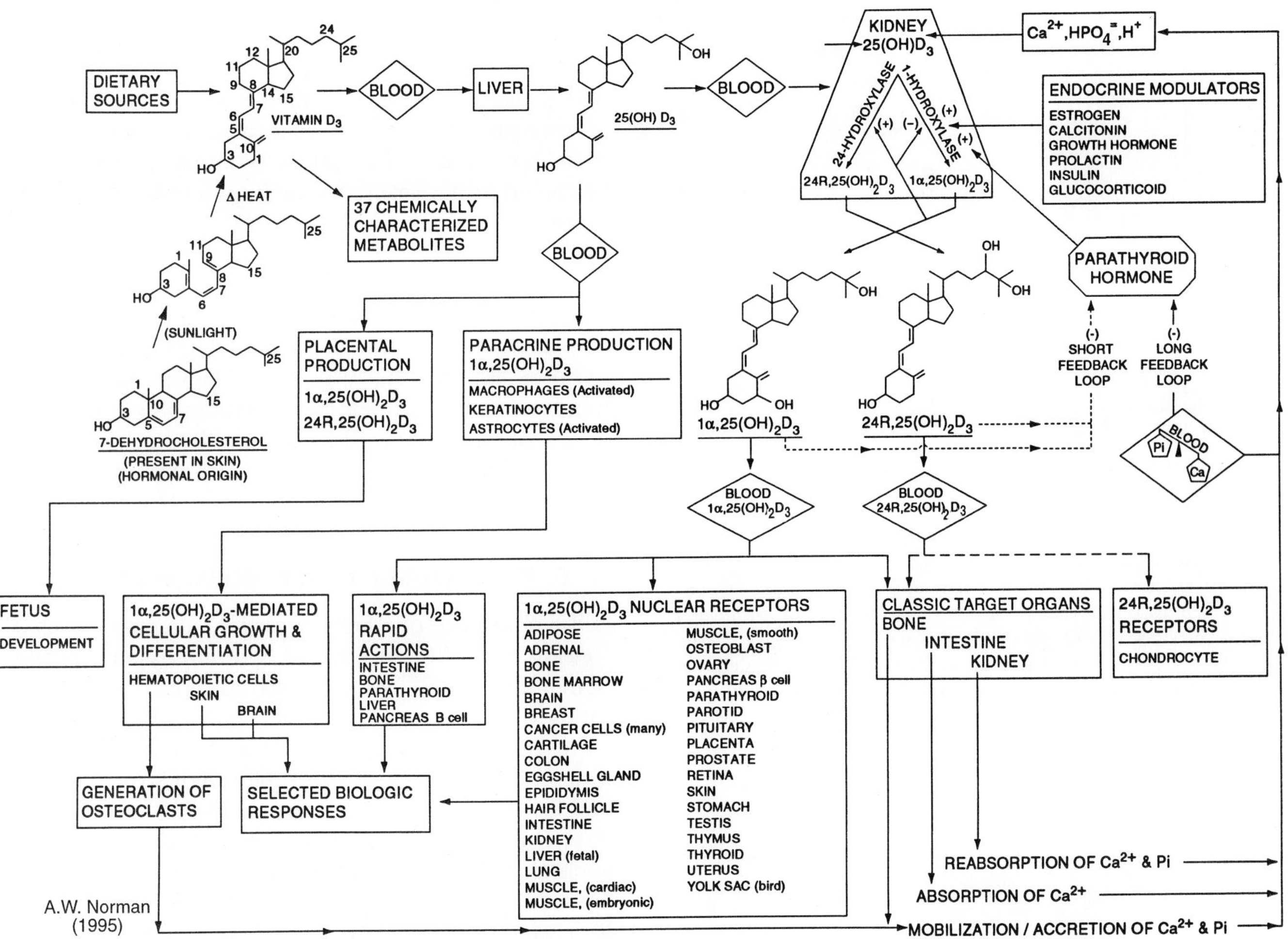

FIGURE 9-8 Summary of the vitamin D endocrine system. In addition to production of 1α,25(OH)$_2$D$_3$ and 24R,25(OH)$_2$D$_3$ by the endocrine gland function of the kidney, small amounts of 1α,25(OH)$_2$D$_3$ are also produced in a paracrine fashion and by the placenta during pregnancy. Target organs and cells for 1α,25(OH)$_2$D$_3$ by definition contain nuclear receptors for 1α,25(OH)$_2$D$_3$ (nVDR). Also, 1α,25(OH)$_2$D$_3$ generates biological effects involving rapid signal transduction pathways utilizing a putative membrane receptor. The precise biological roles of 24,25(OH)$_2$D$_3$ are not yet defined although it is believed to function in bone and cartilage.

involved in Ca^{2+} binding (sensing?) coupled to a seven-membrane-spanning domain like those in the G-protein-coupled receptor superfamily.

The most important biological actions of PTH are the following: (1) to increase the rate of conversion of 25(OH)D$_3$ to 1α,25(OH)$_2$D$_3$ in kidney tissue and, thus, increase the serum concentration of 1α,25(OH)$_2$D$_3$; (2) to increase the plasma calcium concentrations; (3) to increase the extent of osteoclastic and osteocytic osteolysis in bone (bone resorption and remodeling); and (4) to increase the urinary excretion of phosphate and hydroxyproline-containing peptides and to decrease the urinary excretion of calcium.

The opossum kidney cell PTH receptor has been cloned, sequenced, and found to be a member of a distinct family of G-protein-coupled receptors with seven-transmembrane-spanning domains (see Figures 1-19C and 1-20). The mature receptor protein contains 585 amino acids. Both PTH and PTHrP bind with equal affinity to the cloned and expressed receptor, and both ligands equivalently stimulate adenyl cyclase.

Intriguingly, there is close structural homology between the PTH receptor, the PTHrP receptor, and the calcitonin receptor, as well as with the secretin, vasoactive intestinal polypeptide (VIP), and glucagon receptors. All of these receptors couple receptor occu-

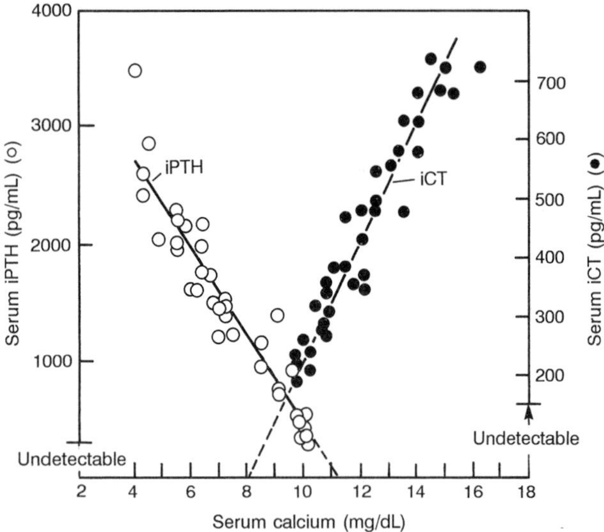

FIGURE 9-9 Changes in plasma levels of immunoreactive parathyroid hormone (iPTH) and calcitonin (iCT) as a function of plasma total calcium. The data were obtained in pigs given EDTA to decrease plasma calcium or calcium infusions to increase plasma calcium. Note that, as serum calcium increases, iPTH falls and serum iCT increases; as serum calcium decreases the reverse occurs. Reproduced with permission of Arnaud, C. D. *et al.* (1970). *In* "Calcitonin: Proceedings of the Second International Symposium" (S. Taylor, ed.). Heinemann, London, p. 236.

pancy to activation of adenyl cyclase and, in some instances, increases in the concentration of intracellular Ca^{2+}.

The PTH receptor has been found to be expressed in both kidney and osteoblast cells, which are prime targets for PTH action, and also in aorta, brain, heart, ileum, liver, placenta, skin, uterus, and testes. The functional significance of these apparent new target organs for PTH remains to be established.

The actions of PTH on bone are complex and are an area of intense investigation. The response of bone to PTH is biphasic; the immediate action is largely that of bone mineral mobilization (i.e., an elevation of the blood levels of both calcium and phosphorus). These effects may be seen within minutes following hormone administration and have, in fact, been used in parathyroidectomized animals as a bioassay for PTH.

A second or slower action of PTH is its effect upon bone cell activity. [PTH is believed to increase the size and number of the bone-resorbing cells, namely, the osteoclasts. This, then, upsets the balance between osteoblasts (bone-forming cells) and osteoclasts and can lead to an increased remodeling rate of bone.] The net effect is normally that resorption is somewhat greater than accretion, so that a net negative skeletal balance of Ca^{2+} and P_i is observed. Although PTH is a potent bone-resorbing agent, receptors for PTH are not found on osteoclasts and are only found on osteoblasts (see later discussion on bone remodeling). Also associated with prolonged bone resorption is an increased release of lysosomal enzymes, so that there is a mobilization of the bone matrix that results in both increased blood levels and urinary excretion of the amino acid hydroxyproline.

The principal actions of PTH on the kidney are to stimulate the renal excretion of phosphate and to enhance the renal tubular reabsorption of calcium. The site of action of PTH in the kidney has been localized to be in the proximal tubule (see Figure 15-4). Also, there are well documented physiologically important effects of PTH that lead to increased excretion of potassium, bicarbonate, sodium, and amino acids and decreased excretion of ammonia and calcium.

B. Parathyroid Hormone-Related Protein

The existence of PTHrP was first encountered as a consequence of the disease of humoral hypercalcemia of malignancy (HHM). In HHM, patients have significantly elevated serum Ca^{2+} levels and a tumor that secretes PTHrP. Although PTH and PTHrP are quite different proteins, because of their high structural homology over the first 13 amino acids (see Figure 9-3) PTHrP is able to interact with the PTH receptor and generate PTH-like biological responses in bone and kidney. As discussed earlier, the N-terminus of PTH is crucial for interaction with its membrane receptor and activation of adenylate cyclase. It has not yet been unequivocally established whether there is a separate and distinct membrane receptor for PTHrP. PTHrP is known to be an effective agonist for the PTH receptor.

The PTHrP is known to be synthesized in several tissues (growth plate of bone, placenta, uterus, skin) with specific temporal and spatial profiles appropriate for its location; it is believed that PTHrP has largely paracrine functions.

Current research into the biological actions of PTHrP suggests that it is involved in regulating the rate of cartilage differentiation. During vertebrate embryogenesis, the first element of the skeleton to appear is a template comprised of cartilage cells (chondrocytes). These chondrocytes are ultimately replaced by mineralized bone in a process termed endochondral ossification. PTHrP has been shown to be a key regulator of the rate of chondrocyte differentiation. It has been postulated that if chondrocyte differentiation proceeds too slowly or too quickly, abnormalities in the length of the long bones will appear, which may explain some forms of dwarfism.

In other target tissues, PTHrP has been shown to increase placental calcium transport and inhibit contraction of uterine muscle and, as well, has been suggested to play a role in the onset of labor.

C. Calcitonin

The dominant biological action of calcitonin is to mediate a lowering of serum calcium levels. Calcitonin is secreted in response to elevated blood levels of ionized calcium; the rate of calcitonin secretion is a direct function of the plasma calcium concentration (Figure 9-9). Extensive studies have demonstrated that one or more hormones of the digestive tract have the ability to elicit increased calcitonin secretion. The most widely tested hormones in this regard have been gastrin and its synthetic analog pentagastrin. These results are consistent with the observation that the secretion of calcitonin is stimulated shortly after the ingestion of high dietary levels of calcium.

In addition to the onset of hypocalcemia, there is also normally an accompanying hypophosphatemia after the administration of calcitonin. Also, in experimental animals, the blood level of calcitonin is elevated in pregnancy and lactation.

The biological effects of calcitonin are mediated as a consequence of the interaction of CT with a receptor present in the outer membrane of target cells of both skeletal and renal tissues. The human calcitonin receptor has been cloned and sequenced. Like the PTH receptor, the calcitonin receptor belongs to a distinct family of G-protein-coupled receptors with seven-transmembrane-spanning domains. The mature receptor comprises 490 amino acids. When the cloned CT receptor was transfected into COS cells, the resulting cells bound both salmon CT and human CT and displayed dose-dependent production of cAMP. Also, calcitonin, but not calcitonin gene-related peptide, can generate a receptor-mediated increase in intracellular Ca^{2+} and IP_3, suggesting activation of the phospholipase C signal transduction pathway as well as the adenyl cyclase pathway. As yet no detailed description of the mechanism of action of calcitonin is available.

Evidence exists for multiple isoforms of the CT receptor, and the CT receptor is known to be expressed in porcine kidney cells, human osteoblasts, mouse brain cells, and human ovary cells. The functional significance of these various isoforms of the calcitonin receptor remains to be established.

The hypocalcemic and hypophosphatemic effects of calcitonin are believed to be due to an inhibition of PTH-mediated calcium resorption from bone, mediated by cyclic AMP. In addition, calcitonin has independent actions in the kidney, where it can stimulate calcium and phosphate excretion.

At least three important biological functions for calcitonin have been proposed: (1) protection of the young animal or newborn against postprandial hypercalcemia; (2) blocking of the actions of PTH in mobilizing bone calcium and phosphorus; and (3) stimulation of the urinary excretion of both calcium and phosphate in the kidney. The net effect of these three actions is to mediate a reduction in serum calcium levels.

D. Vitamin D and Its Metabolites

The vitamin D molecule itself has no intrinsic biological activity. All biological responses attributed to vitamin D are now known to arise only as a consequence of the metabolism of this secosteroid into its biologically active daughter metabolites, namely, $1\alpha,25(OH)_2D_3$ and $24,25(OH)_2D_3$.

Figure 9-8 summarizes the scope of the vitamin D endocrine system. The steroid hormone $1\alpha,25(OH)_2D_3$ is produced only in accord with strict physiological signals dictated by the calcium "demand" of the organism; a bimodal mode of regulation has been suggested. On a time scale of minutes, changes in the ionic environment of the kidney mitochondria resulting from the accumulation and release of calcium and/or inorganic phosphate may alter the enzymatic activity of the 1-hydroxylase. In addition, parathyroid hormone has been shown, on a time scale of hours, to be capable of stimulating the production of $1\alpha,25(OH)_2D_3$, possibly by stimulating the biosynthesis of the 1-hydroxylase. It is also relevant that $1\alpha,25(OH)_2D_3$ is a stimulant for the renal production of $24,25(OH)D_3$. Thus, under normal physiological circumstances, both renal dihydroxylated metabolites are secreted and are circulating in the plasma. There is evidence of a "short feedback loop" for both of these metabolites to modulate and/or reduce the secretion of PTH. There is also some evidence that other endocrine modulators such as estrogens, androgens, growth hormone, prolactin, and insulin may affect the renal production of $1\alpha,25(OH)_2D_3$ (see Figure 9-8). Thus, the kidney is clearly an endocrine gland, in the classic sense, that is capable of producing, in a physiologically regulated manner, appropriate amounts of $1\alpha,25(OH)_2D_3$.

The plasma compartment contains a specific protein, termed the vitamin D-binding protein (DBP), that is utilized to transport vitamin D secosterols. DBP is similar in function to the corticosteroid-binding globulin (CBG), which carries glucocorticoids (see Chapter 10), and the steroid hormone-binding globulin (SHBG), which transports estrogens or androgens (see Chapter 12). DBP is a slightly acidic (pI = 5.2) monomeric glycoprotein of 53,000 Da, which is synthesized and secreted by the liver as a major plasma constituent. From analysis of the cloned cDNA, it has been determined that DBP is structurally homologous to albumin and α-fetoprotein; these three plasma proteins are members of the same multigene family, which likely is derived from the duplication of a common ancestral gene. DBP, originally called group-specific component (Gc), was initially studied electrophoretically as a polymorphic marker in the α-globulin region of human serum.

DBP appears to be a multifunctional protein. It possesses one ligand-binding site for secosterols of the vitamin D family; thus, it may carry the parent, D_3, or the daughter metabolites: 25(OH)D_3, 1α,25$(OH)_2D_3$, and 24,25(OH)D_3. DBP is also able to bind with high affinity to monomers of actin, thereby preventing their polymerization in the blood compartment. The suggestion has been made that DBP may also function as a precursor of a macrophage-activating factor (MAF). Preliminary evidence has been presented that MAF is a cytokine that increases the production of osteoclasts. Since the total plasma concentration of vitamin D sterols is only $\approx 0.2\ \mu M$, while DBP circulates at 9–13 μM, under normal circumstances only 12% of the sterol-binding sites on DBP are occupied.

The principal mode of action of the vitamin D metabolites is believed to occur by a steroid hormone-like mechanism. Definitive biochemical evidence supports

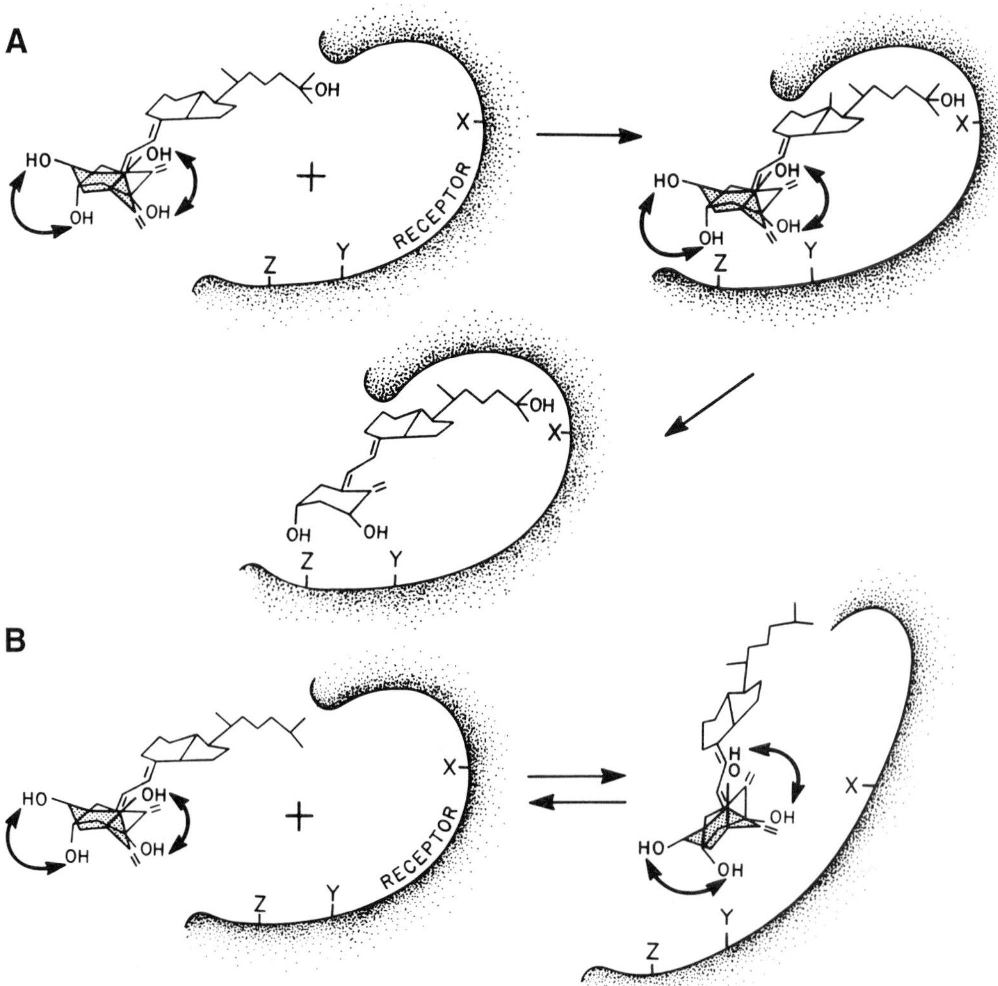

FIGURE 9-10 Schematic model of how the nuclear 1α,25$(OH)_2D_3$ receptor (VDR) "captures" the conformationally mobile A ring to form a stable receptor–ligand complex. Panel A illustrates the proposed steering effects of the 25-hydroxyl group to permit capture of the conformationally active A ring, while panel B illustrates the proposed consequences of the absence of the 25-hydroxyl group on vitamin D ligand capture by the VDR. X, Y, and Z indicate postulated binding domains on the VDR. [Reproduced with permission from A. W. Norman, A. W., and Henry, H. L. (1979). Vitamin D to 1,25-dihydroxycholecalciferol: Evolution of a steroid hormone. *Trends Biochem. Sci.* **4,** 14–18.]

the existence of nuclear receptors for 1α,25(OH)$_2$D$_3$ (nVDR) in at least ≈30 different tissues (see Figure 9-8). While it is not surprising to find 1α,25(OH)$_2$D$_3$ nVDR in the classic target organs of intestine, kidney, and bone, the presence of the nVDR in such diverse tissues as the pancreas, eggshell gland, parathyroid gland, pituitary, reticuloendothelial system cells, and cerebellum emphasizes the diversity of the vitamin D endocrine system and the pleiotropic actions of 1α,25(OH)$_2$D$_3$.

Evidence has been presented that the steroid hormone is able to generate biological responses via signal transduction pathways, which involve its interaction with a nuclear 1α,25(OH)$_2$D$_3$ receptor (nVDR) to regulate gene transcription, or via a putative membrane receptor (mVDR), which is believed to be coupled to the opening of voltage-gated Ca^{2+} channels. These topics will be discussed separately.

The nuclear 1α,25(OH)$_2$D$_3$ receptor is a protein with a molecular weight of 50,000 (see Table 1-5). Extensive evidence exists that supports the view that the unoccupied VDR is largely associated with the nuclear chromatin fraction. The VDR binds 1α,25(OH)$_2$D$_3$ tightly ($K_d = (1\text{–}5) \times 10^{-11}$ M) and with great ligand specificity. Table 9-4 summarizes the structural features of the ligand that optimize its tight binding to the VDR. Figure 9-10 presents a schematic model that emphasizes the importance of the simultaneous presence of the 1α-, 3β-, and 25-hydroxyl groups on the ligand for stabilization of the conformationally mobile 1α,25(OH)$_2$D$_3$. Absence of the 1α- or 3β-hydroxyl group dramatically reduces interaction with the VDR by >99.9% Also, when the eight-carbon side chain is lengthened by two carbons or shortened by one carbon, the interaction with the VDR is reduced by 76%. The data of Table 9-4 also emphasize that neither the parent vitamin D$_3$ nor the intermediate 25(OH)D$_3$, both of which are present in the blood at vastly higher concentrations than 1α,25(OH)$_2$D$_3$, is able to significantly interact with the nVDR.

Preliminary evidence has suggested the existence of protein receptors for 24,25(OH)D$_3$ in chondrocytes (bone cartilage cells). As yet no specific function for the 24,25(OH)D$_3$ receptor has been identified, although there is a possibility of its involvement in some aspects of mineralization and in bone fracture healing.

As can be appreciated from an examination of the scope of the vitamin D endocrine system (Figure 9-8), the biological actions of 1α,25(OH)$_2$D$_3$ are known to extend far beyond the classical actions of vitamin D in the intestine, kidney, and bone. The nuclear VDR is present in at least 32 target tissues. In each of these tissues, the 1α,25(OH)$_2$D$_3$–nuclear receptor complex is involved in selective regulation of gene transcription; this includes genes associated with mineral homeostasis, vitamin D metabolism, extracellular matrix proteins, cell differentiation and proliferation, oncogenes, chromosomal proteins, growth factors, signal transduction proteins, peptide hormones, and energy metabolism. The list of genes shown to be specifically regulated by the VDR now exceeds 70. Furthermore, in some cells the presence of a second putative membrane receptor for 1α,25(OH)$_2$D$_3$ is operative, generating biological responses via other signal transduction pathways.

Three distinct vitamin D-induced calcium-binding proteins have been isolated and biochemically characterized. These proteins are all produced *de novo* in target tissues as a consequence of the 1,25(OH)$_2$D$_3$ receptor complex selective genome activation. The two proteins with a K_d for Ca^{2+} of 10^{-7} M are termed calbindins and belong to a family of tight calcium-binding proteins; other members of this family include the muscle proteins troponin C and parvalbumin and the widely distributed calmodulin. All of these proteins are believed to have a high degree of structural similarity in their calcium-binding sites. The third vitamin D-induced calcium-binding protein is an extracellular protein found associated with bone; it is termed osteocalcin or the bone Gla protein (BGP), since it has many γ-carboxyglutamic acid residues and binds Ca^{2+} with a $K_d \sim 10^{-3}$ M.

A detailed biochemical function for the vitamin D-dependent calbindins (CaBP) has not yet been elucidated. Both the 9000- and the 28,000-Da CaBP species

TABLE 9-4 Ligand Specificity of the Nuclear 1α,25(OH)$_2$D$_3$ Receptor

Ligand	Structural modification	RCI[a]
1α,25(OH)$_2$D$_3$		100
1α,25(OH)$_2$-24-nor-D$_3$	Shorten side chain by one carbon	67
1α,25(OH)$_2$-3-epi-D$_3$	Orientation of 3β-OH altered	24
1α,25(OH)$_2$-24a-dihomo-D$_3$	Lengthen side chain by two carbons	24
1β,25(OH)$_2$D$_3$	Orientation of 1α-OH changed	0.8
1α(OH)D$_3$	Lacks 25-OH	0.15
25(OH)D$_3$	Lacks 1α-OH	0.15
1α,25(OH)$_2$-7-dehydrocholesterol	Lacks a broken B ring; is not a secosteroid	0.10
Vitamin D$_3$	Lacks 1α, and 25-OH groups	0.0001

[a] The relative competitive index (RCI) is a measure of the ability of a nonradioactive ligand to compete, under *in vitro* conditions, with radioactive 1α,25(OH)$_2$D$_3$ for binding to the nuclear 1α,25(OH)$_2$D$_3$ receptor (nVDR). [Data taken from Bouillon, R., Okamura, W. H., and Norman, A. W. (1995). Structure–function relationships in the vitamin D endocrine system. *Endocr. Rev.* **16**, 200–257.]

are soluble proteins found exclusively inside the intestinal and kidney cells, and they constitute 1–3% of the soluble cellular proteins. There is some evidence to implicate these CaBPs in some of the steps involved in the translocation of calcium ions across the cell. Normally, in all cells the intracellular concentration of "free" Ca^{2+} is maintained at 10^{-7} *M* or lower. It has been suggested that calbindin may function in the intestinal and kidney cells, which are actively involved in calcium translocation, as well as in the pancreatic B cell as an intracellular calcium buffer to prevent the adverse effects of a high free Ca^{2+} concentration.

The principal classical biological effects of vitamin D (mediated by its daughter metabolites) are as follows: at the intestine to stimulate the absorption of dietary calcium and phosphorus; at the skeleton to promote both the mineralization of bone matrix and to stimulate bone resorption; and at the kidney to reduce the urinary excretion of phosphate and calcium (i.e., to stimulate the renal tubular reabsorption of these two ions). Of these three systems, the most thoroughly studied biochemically is the intestine.

The mode of action of $1\alpha,25(OH)_2D_3$ in the target organ, intestinal mucosa, in stimulating intestinal calcium and phosphorus absorption is believed to involve the integrated actions of the nuclear and membrane receptors for $1\alpha,25(OH)_2D_3$.

The process of intestinal calcium translocation and the effects of $1\alpha,25(OH)_2D_3$ therein are complex. At least three steps are involved: (1) uptake of the Ca^{2+} ions across the brush border membrane and packaging into lysosomal-like vesicles; (2) translocation of the Ca^{2+}-containing vesicles across the cell; and (3) efflux of the Ca^{2+} out of the cell across the basal lateral membrane. A vesicular model for the action of $1\alpha,25(OH)_2D_3$ in the intestine is given in Figure 9-11.

$1\alpha,25(OH)_2D_3$ increases the permeability of the brush border membrane to Ca^{2+} by altering the membrane protein composition and topology, which may include the induction of a membrane-bound ATPase.

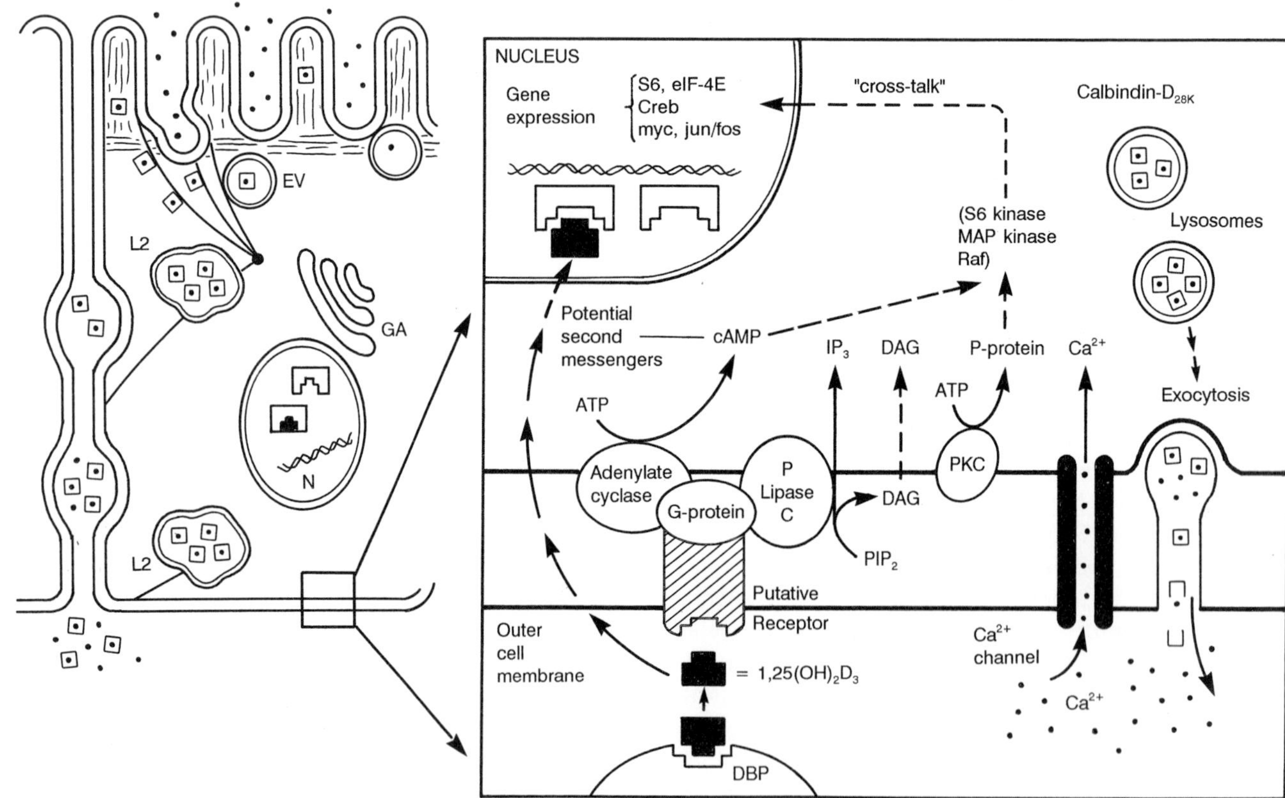

FIGURE 9-11 Schematic model describing the actions of $1\alpha,25(OH)_2D_3$ in the intestine in stimulating intestinal calcium transport. In this model, the nuclear VDR in the intestine in the presence of $1\alpha,25(OH)_2D_3$ is responsible for up-regulating the genes for proteins involved with the intestinal Ca^{2+} transport process, including brush border alkaline phosphatase and the cytosolic calbindin D. Occupancy of the putative membrane receptor by $1\alpha,25(OH)_2D_3$ results in the initiation of a series of signal transduction events involving PKC, DAG, and the opening of voltage-gated Ca^{2+} channels, which collectively results in the exocytotic event that releases Ca^{2+} outside the cell and adjacent to the capillary system. EV, endocytic vesicle; GA, Golgi apparatus; L2 = lysozomal vesicle; N, nucleus; (·) Ca^{2+}; (□) calbindin$_{28K}$.

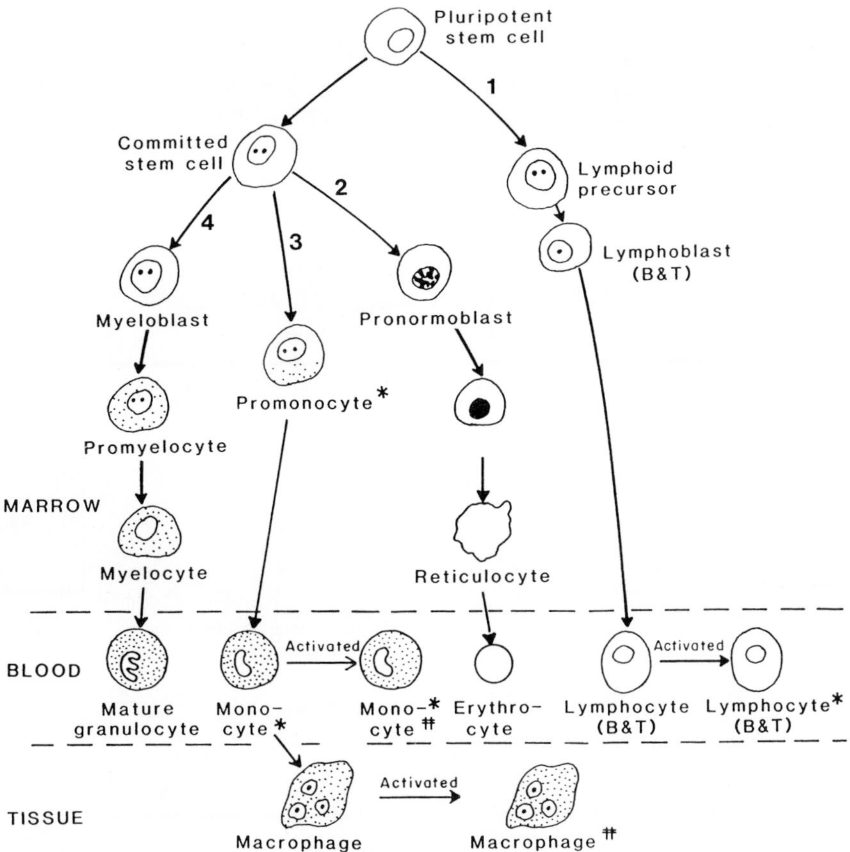

FIGURE 9-12 Flowchart of stem cell differentiation. A pluripotent stem cell can differentiate along one of four lineages toward (1) B and T lymphocytes, (2) erythrocytes, (3) monocytes and macrophages, and (4) granulocytes. An asterisk indicates cells that possess $1\alpha,25(OH)_2D_3$ receptors; # indicates cells that can produce small amounts of $1\alpha,25(OH)_2D_3$. See also Figure 15-21.

It is believed that the absorption of Ca^{2+} from the lumen of the intestine occurs via endocytic internalization of the Ca^{2+} across the brush border membrane. This is followed by fusion of the resulting vesicles with lysosomes, which also contain the Ca^{2+}-binding protein, calbindin; these vesicles then move across the cell to the basal lateral membrane. In a complex process involving the $1\alpha,25(OH)_2D_3$ putative membrane receptor-mediated opening of voltage-gated Ca^{2+} channels, the exocytosis of the Ca^{2+}-bearing membrane vesicles is stimulated. A further detailed description remains to be provided.

A major new action of $1\alpha,25(OH)_2D_3$ has been described within the hematopoietic system (see Figure 9-12). Nuclear receptors for $1\alpha,25(OH)_2D_3$ exist in promonocytes, monocytes, and activated B and T lymphocytes. The principal biological effect of $1\alpha,25(OH)_2D_3$ is to promote cell differentiation along the lineage from the committed stem cell to promono-

TABLE 9-5 Modulators of Bone Cell Resorptive Activity

Stimulators of bone resorption	Inhibitors of bone resorption
PTH	
PTHrP	Calcitonin
$1\alpha,25(OH)_2D_3$	Glucocorticoids
Prostaglandins	Estrogens
PGE_2	Androgens
Interleukin-1 (IL-1)	Insulin-like growth factor
Tumor Necrosis Factors	TGF-1
TNF-α	Fluoride (F^-)
TNF-β	
Thyroxine	
Retinol	
Growth factors	
EGF (epidermal growth factor)	
FGF (fetal growth factor)	

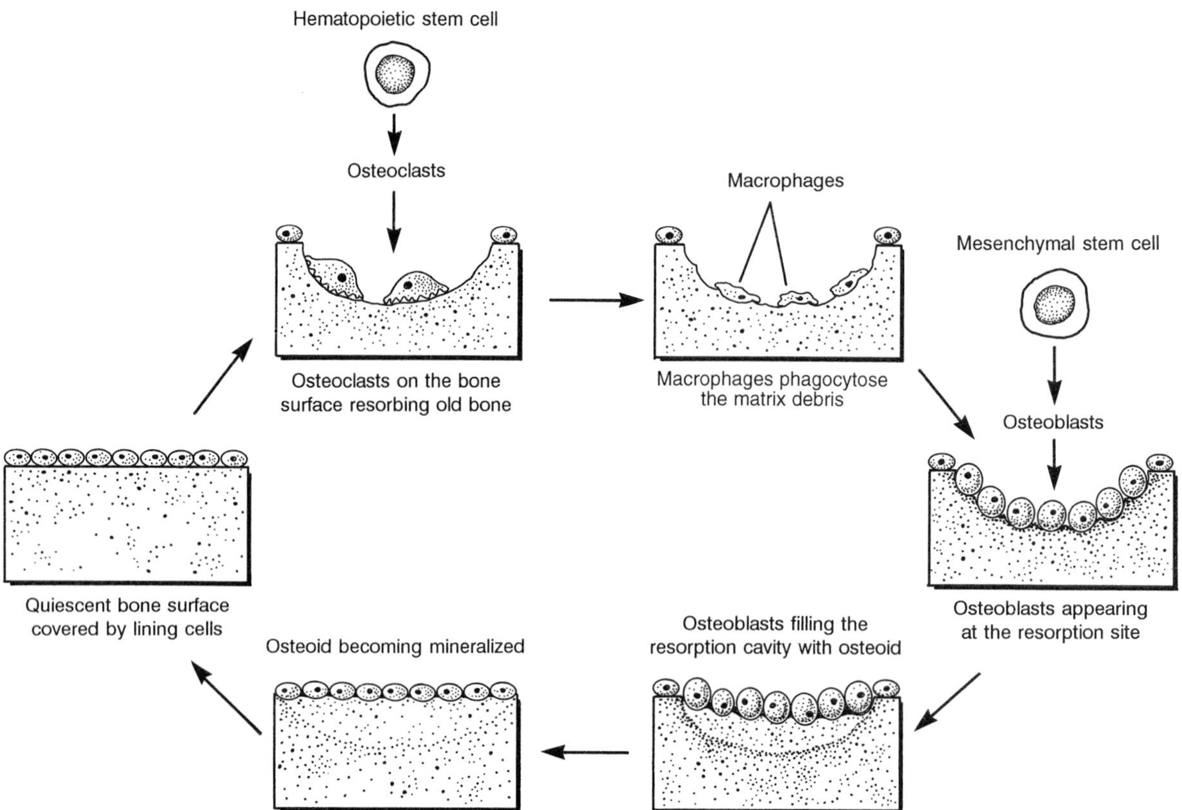

FIGURE 9-13 Schematic model of a bone remodeling unit; see also Table 9-6. In this model, the initiation of the remodeling event occurs through activation of the osteoclast. The osteoclast is believed to attach its ruffled border to the bone surface in order to carry out its biological actions in a defined local environment. The osteoclast then actively resorbs bone locally as a consequence of local acidification; the resulting mobilized Ca^{2+} is then directed to the blood compartment. Also, the osteoclast secretes proteases that locally digest the bone matrix proteins (collagen and osteocalcin). Next macrophages phagocytose the matrix debris; the net result is the creation of a local bone pit that can be viewed under the microscope. Then in a reversal phase, the osteoblasts move to the resorbed area and, under suitable hormonal activation, biosynthesize new extracellular matrix proteins followed by the orderly importation of Ca^{2+} and Pi to create the mineral phase. Thus, the resorption cavity has become filled with new hydroxyapatite. The process culminates in the conversion of the active osteoblasts to lining cells, which cover the quiescent bone surface. Linked to this remodeling event, of necessity, is the orderly generation via cell differentiation of the necessary numbers of osteoclasts and osteoblasts. Osteoblasts are ultimately derived from mesenchymal stem cells and are more immediately derived from osteoprogenitor cells. Osteoclasts are ultimately derived from a hematopoietic stem cell, which, under the influence of $1\alpha,25(OH)_2D_3$ and other cytokines, differentiates to yield functional osteoclasts (see Figure 9-12) Adapted by permission of Manolagas, S. C., and Jilka, R. L. (1975). *New Engl. J. Med.* **332,** 305–311.

cyte ≫ monocyte ≫ macrophage. Macrophages are believed to be precursors of osteoclasts, the cells responsible for bone resorption. Further, γ-interferon derived from activated T lymphocytes will induce bone marrow macrophages to biosynthesize small amounts of $1\alpha,25(OH)_2D_3$. Thus, there is a paracrine system operative for $1\alpha,25(OH)_2D_3$, such that "locally" produced $1\alpha,25(OH)_2D_3$ in the bone marrow acts on neighboring cells to promote differentiation toward macrophages and osteoclasts, thereby increasing the bone calcium resorption capability.

E. Biology of Bone Remodeling

Bone is a metabolically active organ, undergoing throughout life a continual turnover and remodeling process involving bone resorption and accretion. The balance between the rates of bone resorption and bone formation will determine, both at a local level or globally (the entire skeleton), whether there is a negative, neutral, or positive calcium balance. The biological activities of bone cells are subject to the actions of a multitude of hormones, cytokines, and other physiological regulators (see Table 9-5).

TABLE 9-6 Steps of Bone Remodeling

	Cell or cellular event				
	Activation →	Resorption →	Reversal →	Formation →	Mineralization
Control factors	Differentiation of hematopoietic stem cell → osteoclast	Modulation of osteoclast	Differentiation of mesenchymal stem cell → osteoblast	Modulation of osteoblast	Active osteoblast → quiescent osteocyte
Positive	$1\alpha,25(OH)_2D_3$ PTH Thyroxine Ca^{2+}	PTH Thyroxine Ca^{2+}	CT Estrogen Phosphate IGF-1	$1\alpha,25(OH)_2D_3$ Fluoride Phosphate	$1\alpha,25(OH)_2D_3$ $24,25(OH)_2D_3$ (?) Phosphate
Negative	CT Estrogen (?)	CT Fluoride	Cortisol Immobilization Age	Cortisol Immobilization Age	Fluoride

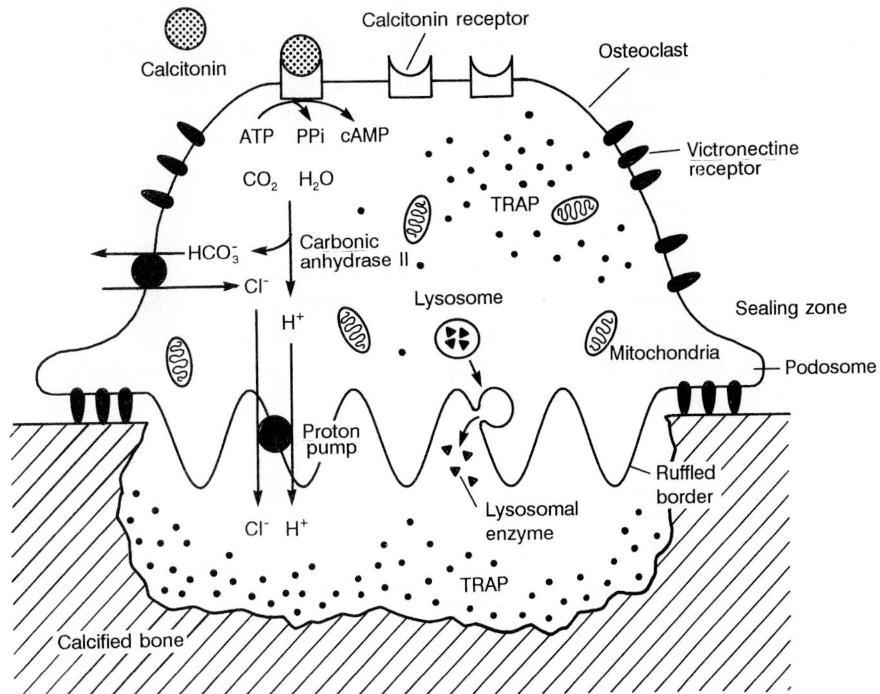

FIGURE 9-14 Functional morphology of an osteoclast. This simplified diagram indicates some of the functional elements of the osteoclast. Numerous receptors for calcitonin and vitronectin are expressed in the plasma membrane. The vitronectin receptors appear to function to facilitate the binding of osteoclasts to bone matrix; the receptor utilizes the recognition sequence of Arg-Gly-Asp on the bone matrix protein. Osteoclasts possess a significant level of tartrate-resistant acid phosphatase (TRAP), which has been used as a specific histochemical marker of osteoclasts. The basolateral membrane has ATP-dependent ion transport systems to extrude calcium and maintain intracellular potassium, similar to other cells. The osteoclast is enriched in mitochondria to provide the large amount of ATP necessary for these pumps as well as for an electrogenic proton pump, which raises the hydrogen ion concentration at the ruffled border area. Lysosomal enzymes are synthesized and taken up in Golgi vesicles that contain mannose 6-phosphate receptors. These vesicles presumably circulate through the cell, discharging their enzyme contents into the ruffled border area. The sealing zone, with its actin filaments and specialized attachment structures (podosomes), surrounds the ruffled border area and separates it from the extracellular fluid. Carbonic acid is generated from the CO_2 produced by mitochondrial metabolism and provides the source for hydrogen ions. The excess HCO_3^- is removed by exchange with chloride. [Reproduced with permission from Suda, T., Takahashi, N., and Martin, T. J. (1992). Modulation of osteoclast differentiation. *Endocr. Rev.* **1,** 66–80.]

One model of bone resorption and formation describes a "bone remodeling unit," which comprises the integrated and sequential actions of osteoclasts and osteoblasts. The detailed functioning of this model through the phases of activation, resorption, reversal, formation, and mineralization is described in Figure 9-13 and Table 9-6. Also, a model for the functional operation of the osteoclast is presented in Figure 9-14.

Although the dominant activators of bone resorption and osteoclast activation are PTH and $1\alpha,25(OH)_2D_3$, intriguingly, the osteoclast does not have a receptor for either of these hormones. Thus, the activation process involving PTH is believed to occur as a consequence of PTH interacting with its receptor on the osteoblast. Then the activated osteoblast releases paracrine agents, whose nature has not been precisely defined, which then achieve activation of the osteoclast, so that the bone-resorptive event is initiated. In contrast, the actions of $1\alpha,25(OH)_2D_3$ to increase bone resorption are believed to be mediated more distally by increasing a hematopoietic cell differentiation process (see Figure 9-12) to produce increased numbers of osteoclasts. The osteoclast, however, has receptors for calcitonin, and occupancy of this receptor leads to a reduction in bone resorption.

Bone formation is achieved through the activation of the osteoblast. $1\alpha,25(OH)_2D_3$, through the actions of its nVDR in the osteoblast, is a major stimulator of transcription, leading to increased biosynthesis of collagen, osteocalcin, and other matrix proteins. The detailed steps of osteoblast-mediated Ca^{2+} transport and bone mineralization are not yet known.

F. Integrated Actions of Parathyroid Hormone, Calcitonin, and Vitamin D Metabolites

The maintenance of calcium and phosphorus homeostasis involves the integration of absorption by the intestine, accretion and reabsorption by bone tissue, and urinary excretion of these two ions by the kidney. Table 9-7 summarizes the physiological effects of calcitonin, parathyroid hormone, and vitamin D metabolites in these three organs. The steroid $1\alpha,25(OH)_2D_3$ plays a dominant role in increasing the intestinal absorption of calcium and phosphorus; however, this uptake process is regulated carefully according to the needs of the animal and is dictated by certain physiological signals, which are dependent in large part on the plasma levels of calcium and possibly phosphorus. Once calcium and phosphate enter into the plasma, a delicate balancing operation occurs between accretion and mobilization in the bone and between excretion and reabsorption by the kidney. In the event that the dietary intake or availability of calcium and phosphorus is diminished or increased, it is possible to tip the balance in favor of increased bone mobilization or

TABLE 9-7 Physiological Effects of Calcitonin, Parathyroid Hormone, and Vitamin D (Metabolites) Related to Mineral Metabolism[a]

	Calcitonin	Parathyroid hormone	$1\alpha,25(OH)_2D_3$
Intestinal			
Calcium absorption	↓ ?	↑ (indirect)	↑
Phosphate absorption	?	?	↑
Renal			
Phosphate excretion	↑	↑	↓
Calcium excretion	↑	↓	↓
Adenyl cyclase activity	↑	↑	?
Skeletal			
Calcium mobilization	↓	↑	↑
Mineralization of bone matrix	—	—	↑
Other			
Plasma levels of calcium	↓	↑	↑
Plasma levels of phosphate	↓	↓	—

[a] The important role of bone as a central organ in calcium and phosphorus metabolism, acting both as a source of and a reservoir for these two ions, is discussed in the text. It is apparent that bone remodeling processes may contribute to both short- and long-term events necessary for calcium and phosphorus homeostasis. The relative actions of bone formation and resorption are known to be modulated by various endocrine regulators during times of skeletal growth and lactation and in birds during the process of egg laying. Also, it is not surprising that bone is involved in a wide variety of disease states that reflect perturbations in calcium and phosphorus homeostasis.

increased urinary excretion, respectively, to meet the stringent requirement of a constant serum calcium level. Thus, serum calcium may become elevated by PTH–1α,25$(OH)_2$D-mediated stimulation of bone calcium mobilization or by parathyroid hormone stimulation of the tubular reabsorption of calcium at the kidney. Concomitantly, PTH also stimulates urinary phosphate excretion, so that as the serum calcium concentration increases, there is usually an associated reduction in the plasma phosphate level. This then prevents the inappropriate precipitation of calcium phosphate in soft tissues, which would be the logical consequence of exceeding the solubility product, K_{sp}, for [calcium] × [phosphate].

When serum calcium levels become too elevated, the action of calcitonin may come into play. Simply speaking, calcitonin is believed to block many of the actions of PTH at the skeletal level, thereby preventing further elevation of serum calcium levels. Secretion of calcitonin may be stimulated during intervals of gastrointestinal absorption of calcium to prevent short-term intervals of hypercalcemia. Also associated with the increased secretion of calcitonin is an increase in calcium excretion by the kidney, which can contribute to a reduction in the circulating levels of serum calcium and prevent soft tissue calcification.

V. CLINICAL ASPECTS

A. Parathyroid Hormone

The common disorders of the parathyroid gland fall into two main categories: (1) those associated with hypofunction (e.g., hypoparathyroidism) and (2) those associated with hyperfunction (e.g., hyperparathyroidism).

Hyperparathyroidism is the most common disorder involving the parathyroid glands. Primary hyperparathyroidism is a disorder of mineral metabolism characterized by a defect in the normal feedback control of PTH secretion by the plasma calcium concentration. Secondary hyperparathyroidism is a disorder characterized by primary disruption of mineral homeostasis, such as impaired production of 1α,25$(OH)_2D_3$ or low dietary intake of calcium, which leads to a low blood calcium level and a compensatory increase in parathyroid gland function, size, and secretion of PTH. Hyperparathyroidism can result from a single adenoma, multiple adenomas, hyperplasia, usually of the chief cells, and carcinoma.

The principal clinical features of hyperparathyroidism are a markedly elevated plasma level of serum calcium (11–13 mg/100 ml) and, when present for prolonged intervals of time, extensive resorption of the skeleton. At the kidney level, the presence of renal kidney stones or calculi is a consequence. Deposition of calcium in the collecting tubules of the kidney produces nephrocalcinosis. The hyperactive gland may be removed surgically, resulting in a fall in serum calcium levels and often minimization of the urolithiasis. The principal deleterious consequence of prolonged hyperparathyroidism is the greatly decreased mineral content of the bone.

The most common cause of hypoparathyroidism is damage to or removal of the parathyroid glands in the course of an operation on the thyroid gland. A much less common cause of hypoparathyroidism is an idiopathic lack of function. Most cases are found in children; approximately twice as many females as males are affected. The most prominent clinical feature of hypoparathyroidism is hypocalcemia, and in extreme instances tetany is observed.

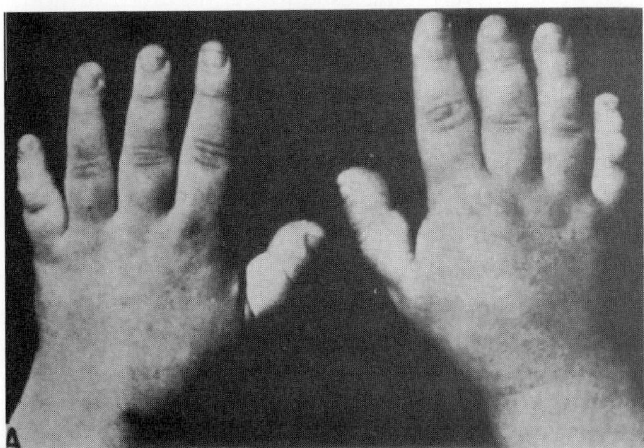

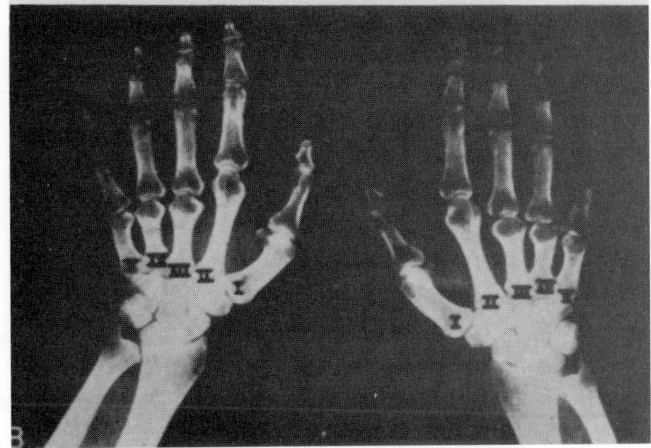

FIGURE 9-15 Photograph and X-ray film of hands of a patient with pseudohypoparathyroidism. Note that all digits, except the thumbs and index fingers, are shorter than normal. The shortness results mainly from short metacarpals. Modified from Albright, F., and Reifenstein, E. C. (1948). "The Parathyroid Glands and Metabolic Bone Disease." Williams & Wilkins, Baltimore.

Pseudohypoparathyroidism is a rare genetic disorder involving bone and mineral metabolism. It is probably inherited as an X-linked dominant trait with variable penetrance and is characterized by signs and symptoms similar to those of hypoparathyroidism. However, in this disease state there is a peripheral resistance to the biological actions of parathyroid hormone and an elevated secretion of PTH. Characteristic clinical features include a round face, short stature, brachydactylia, especially of the metacarpal and metatarsal bones as a result of early epiphyseal closure (see Figure 9-15), and in some instances ectopic soft tissue calcifications.

B. Parathyroid Hormone-Related Protein

Disorders of mineral metabolism are frequently found in patients with cancer. While primary bone tumors rarely lead to systemic disorders of calcium homeostasis, there can be a malignancy that does not alter the skeleton directly, but does alter the balance of one or more of the calcium-regulating hormones. Such an example is humoral hypercalcemia of malignancy (HHM). In this instance, a tumor (e.g., renal or squamous carcinoma or breast cancer) secretes parathyroid hormone-related protein, which because of its structural homology to PTH acts as an agonist on the PTH receptor in osteoblast cells and leads to inappropriate activation of bone resorption and the onset of hypercalcemia.

C. Calcitonin

The only disease state known to be related to calcitonin is that resulting from a disorder of C-cell function; this is described as a medullary carcinoma of the thyroid, which results in the hypersecretion of calcitonin. The etiology of this condition is unknown. Over 50% of the cases exhibit a familial incidence, and the pattern of inheritance indicates an autosomal dominant mode of transmission. The disease is rare, only occurring in one of every several thousand patients seeking medical attention. It is one of the few forms of cancer that can be detected by radioimmunoassay of the plasma level of calcitonin. This is carried out by measurement of its concentration before and after a standard provocative infusion with either calcium or pentagastrin. Ectopic calcitonin production is commonly seen in some tissues, especially small-cell carcinoma of the lung.

D. Vitamin D

Conceptually, human clinical disorders related to vitamin D can be considered to arise because of one of the following: (1) altered availability of the parent vitamin D; (2) altered conversion of vitamin D to its

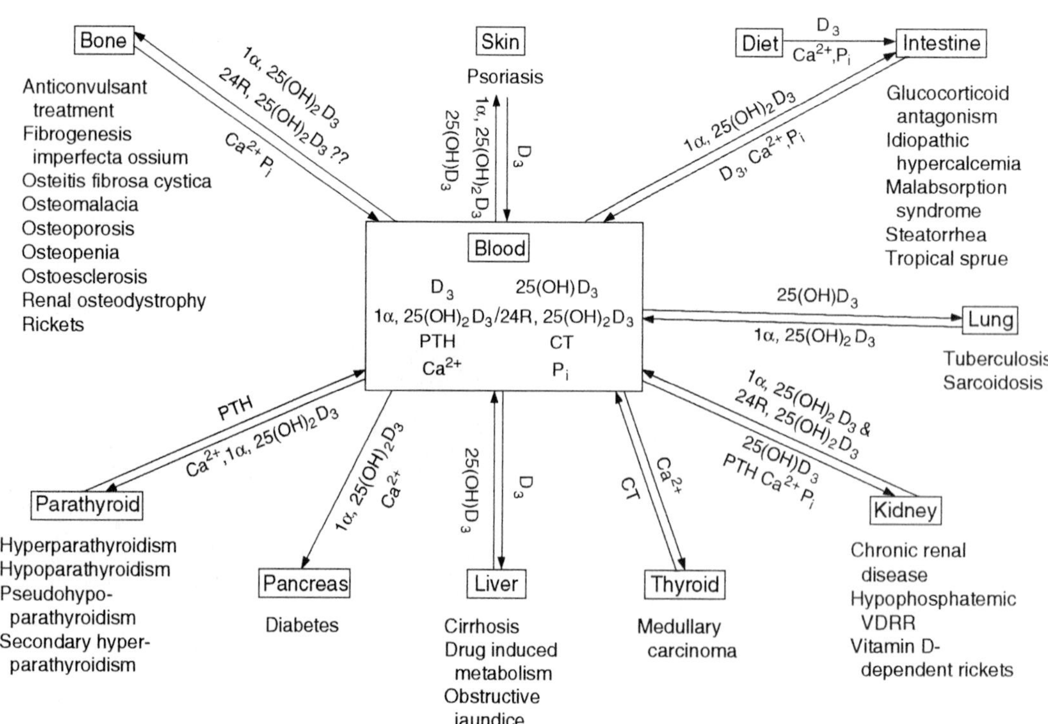

FIGURE 9-16 Human disease states related to vitamin D.

principal daughter metabolites $1\alpha,25(OH)_2D_3$ and $24,25(OH)_2D_3$; (3) conditions that may be due to variations in organ responsiveness to these dihydroxylated metabolites; and (4) perturbations in the integrated interactions of these metabolites with PTH and calcitonin. Figure 9-16 presents a schematic diagram of the relationship between human disease states related to vitamin D and the metabolic processing of vitamin D via its endocrine system. A wide variety of organs have diseases related to vitamin D: the intestine (malabsorption), the parathyroid gland (hyper- and hypoparathyroidism), the bone (osteomalacia, osteoporosis, rickets), or the kidney, such as chronic renal failure. All of these, in their own way, reflect a disturbance in or a malfunction of the body's normal endocrine processing of vitamin D and its interaction with the other calcemic hormones.

The classic deficiency state resulting from a dietary absence of vitamin D or lack of ultraviolet (sunlight) exposure is the bone disease called rickets in children or osteomalacia in adults. Historically, the identification of rickets as a disease state allowed G. Mellanby in 1920 to induce rickets in puppies experimentally by nutritionally withholding vitamin D. This work demonstrated that, in the absence of sunlight, vitamin D was truly a trace essential dietary constituent (e.g., a vitamin).

The clinical features of rickets and osteomalacia depend upon the age of onset. The classical skeletal disorders of rickets include deformity of the bones, especially in the knees, wrists, and ankles (see Figure 9-17), as well as associated changes in the costochondrial joint junctions, which have been termed by some as the rachitic rosary. If rickets develops in the first 6 mo.

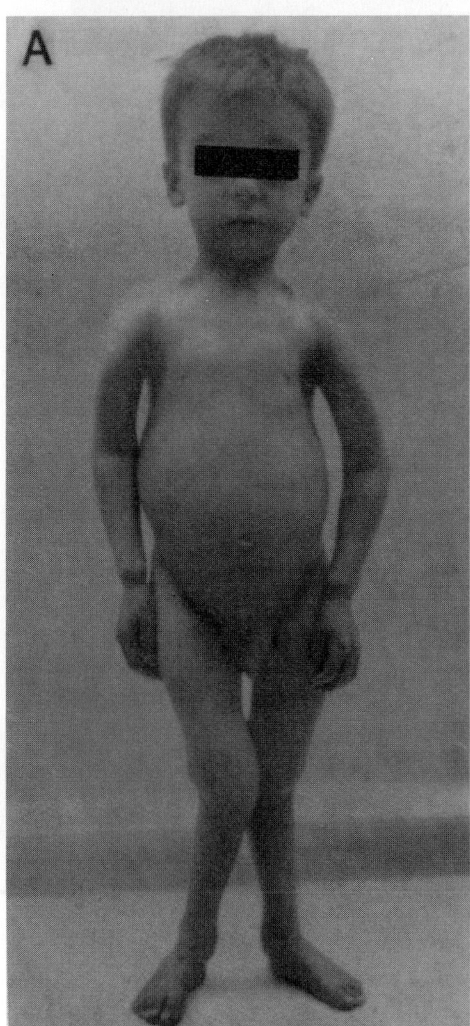

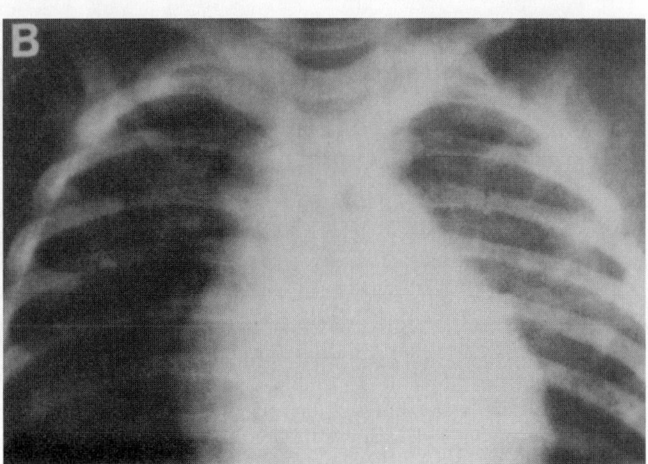

FIGURE 9-17 Classic appearance of rickets in a child. Reproduced by permission from Norman, A. W. (1979). "Vitamin D: The Calcium Homeostatic Steroid Hormone," p. 442. Academic Press, New York.

of life, infants may suffer from convulsions or develop tetany due to a low blood calcium level (usually <7 mg/100 m), but may have only minor skeletal changes. After 6 mo., bone pain as well as tetany is likely to be present. Since osteomalacia occurs after growth and development of the skeleton are complete

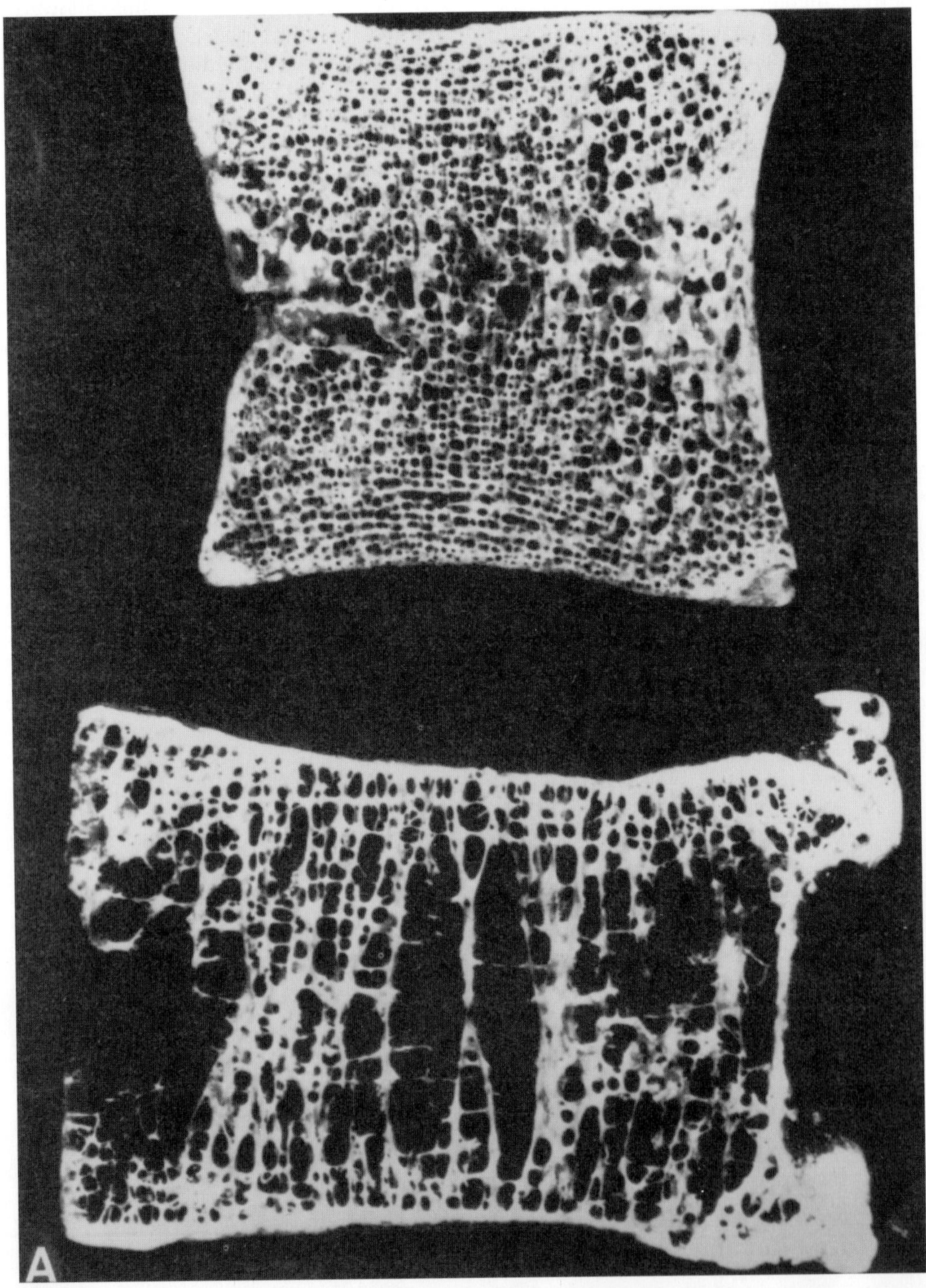

FIGURE 9-18 Effect of long-standing osteoporosis on the mineral content of lumbar vertebrae. Shown are the microradiographs of a lumbar vertebra from a young (upper panel) and an elderly (lower panel) female. [Reproduced from Avioli, L. V., and Krane, S. M. (1977). "Metabolic Bone Disease," Vol. 1, p. 320. Academic Press, New York.]

(i.e., the adult stage of life), its main symptoms are muscular weakness and bone pain, with little bone deformity.

A characteristic feature of bone osteomalacia and rickets is the failure of the organic matrix of bone (osteoid) to calcify. This leads to the appearance of excessive quantities of uncalcified bone matrix, termed the osteoid. In addition, there is often a high serum level of alkaline phosphatase, a fact that is often used to assist in the clinical diagnosis of the presence of osteomalacia.

E. Osteoporosis

Osteoporosis is the most common generalized disorder of bone; it is estimated that in the United States over 25 million people, 80% of whom are female, have some form of osteoporosis. It is most simply characterized as a state of insufficiently calcified bone and is the end result of a number of metabolic abnormalities that affect the skeleton. Some of the causes of osteoporosis that have been recognized or proposed include (1) adrenal cortical hyperfunction, (2) prolonged dietary calcium deficiency, (3) reduced levels of estrogen, as in postmenopausal women, (4) prolonged skeletal immobilization, and (5) chronic vitamin D deficiency. As a consequence of these "changes," an imbalance results in the remodeling rate of bone, such that there is either an increase in the relative rate of bone resorption or a decrease in the rate of bone formation.

The chief clinical sign of osteoporosis is thinning of the bones with a concomitant increase in the number of fractures (see Figure 9-18). Each year, this disease leads to more than 1.3 million fractures, mostly of the hip, spine, and wrist; more than 250,000 osteoporosis-related hip fractures occur annually. In the 1990s it is estimated that ≈$5 billion will be expended annually to provide health care for hip fractures attributable to postmenopausal osteoporosis.

References

A. Books

Bilezikian, J. P., Raisz, L. G., and Rodan, G. A., eds (1996). "Principles of Bone Biology." Academic Press, San Diego.

Coe, F. L., and Favus, M. J., eds. (1992). "Disorders of Bone and Mineral Metabolism." Raven Press, New York.

B. Review Articles

Bouillon, R., Okamura, W. H., and Norman, A. W. (1995). Structure-function relationships in the vitamin D endocrine system *Endocrine Rev.* **16,** 200–257.

Hannah, S. S., and Norman, A. W. (1994). $1\alpha,25(OH)_2$ Vitamin D_3 regulated expression of the eukaryotic genome. *Nutr. Rev.* **52** (11), 376–381.

Haussler, M. R., Mangelsdorf, D. J., Komm, B. S., *et al.* (1988). Molecular biology of the vitamin D hormone. *In* "Recent Progress in Hormone Research" (R. P. Greep, ed.), pp. 263–305. Academic Press, New York.

Pike, J. W. (1991). Vitamin D_3 receptors: structure and function in transcription. *Annu. Rev. Nutr.* **11,** 189–216.

Suda, T., Shinki, T., and Takahashi, N. (1990). The role of vitamin D in bone and intestinal cell differentiation. *Annu. Rev. Nutr.* **10,** 195–211.

Suda, T., Takahashi, N., and Martin, T. J. (1992). Modulation of osteoclasat differentiation. *Encocrine Rev.* **13,** 66–80.

Vortkamp, A., Lee, K., Lanske, B., Segre, G. V., Kronenberg, H. M. and Tabin, C. J. (1996). Regulation of rate of cartilage differentiation by indian hedgehog and PTH-related protein. *Science* **273,** 613–622.

Walters, M. R. (1992). Newly identified actions of the vitamin D endocrine system. *Endocrine Rev.* **13,** 719–764.

C. Research Papers

Bringhurst, F. R., Juppner, H., Gus, J., Potts, J. T., Jr., Kronenberg, H. M., Abou-Samra, A. B., and Segre, G. V. (1993). Clonal, stably expressed PTH/PTHrP receptors activate multiple messenger signals and biological responses in LLC-PKI kidney cells. *Endocrinology* **132,** 2090–2098.

Ferguson, J. E., Gorman, J. V., Bruns, D. E., Weir, E. C., Burtis, W. J., Martin, T. J., and Bruns, M. E. (1992). Abundant expression of PTHrP in human amnion and its association with labor. *Proc. Natl. Acad. Sci. USA* **89,** 8384–8385.

Frost, H. M. (1990). Skeletal structural adaptations to mechanical usage—redefining Wolff's law: the bone remodeling problem. *Anat. Rec.* **226,** 403–413 and 414–422.

Holick, M. F., MacLaughlin, J. A., and Doppelt, S. H. (1981). Regulation of cutaneous previtamin D photosynthesis in man: skin pigment is not an essential regulator. *Science* **211,** 590–593.

Juppner, H., Abou-Samra, A. B., Freeman, M., Kong, X. F., Schipani, E., Richards, J., Kolakowski, L. F., Jr., Hock, J., Potts, J. T., Jr., Kronenberg, H. M., and Segre, G. V. (1991). A G protein-linked receptor for parathyroid hormone and parathyroid hormone-related peptide. *Science* **254,** 1024–1026.

Lanske, B., Karaplis, A. C., Fee, K. et al. (1996). PTH/PTHrP receptor in early development and indian hedgehog-related bone growth *Science* **273,** 663–666.

Nemere, I., Dormanen, M. C., Hammond, M. W., Okamura, W. H., and Norman, A. W. (1994). Identification of a specific binding protein for 1α,25-dihydroxyvitamin D_3 in basal lateral membranes of chick intestinal epithelium and relationship to transcaltachia. *J. Biol. Chem.* **269** (38), 23750–23756.

Norman, A. W., Bouillon, R., Farach-Carson, M. C., Bishop, J. E., Zhou, L.-X., Nemere, I., Zhao, J., Muralidharan, K. R., and Okamura, W. H. (1993a). Demonstration that 1β,25-dihydroxyvitamin D_3 is an antagonist of the nongenomic but not genomic biological responses and biological profile of the three A-ring diastereoisomers of 1α,25-dihydroxyvitamin D_3. *J. Biol. Chem.* **268** (27), 20022–20030.

Norman, A. W., Okamura, W. H., Farach-Carson, M. C., Allewaert, K., Branisteanu, D., Nemere, I., Muralidharan, K. R., and Bouillon, B. (1993b). Structure-Function Studies of 1,25-dihydroxyvitamin D_3 and the vitamin D endocrine system: 1,25-dihydroxy-penta-deuterio-previtamin D_3 (as a 6-s-cis analog) stimulates nongenomic but not genomic biological responses *J. Biol. Chem.* **268** (19), 13811–13819.

Rhoten, W. G., and Sergeev, I. N. (1994). Calbindin-D appears to buffer intracellular Ca^{2+} in a butyrate-treated rat insulinoma cell line. *Endocrine* **2,** 989–995.

Tian, X. Q., Chen, T. C., Matsuoka, L. Y., Wortsman, J., and Holick, M. F. (1993). Kinetic and thermodynamic studies of the conversion of previtamin D_3 to vitamin D_3 in human skin. *J. Biol. Chem.* **268,** 14888–14892.

Yamin, M., Gorn, A. H., Flannery, M. R., Jenkins, N. A., Gilbert, D. J., Copeland, N. G., Tapp, D., Krane, S. M., and Goldring, S. R. (1994). Cloning and characterization of a mouse brain calcitonin receptor cDNA and mapping of the calcitonin receptor gene. *Endocrinology* **135,** 2635–2643.

CHAPTER

10

Adrenal Corticoids

I. INTRODUCTION

The hormones produced by the cells of the human adrenal cortex in response to appropriate stimuli are steroid hormones: cortisol, the major substance secreted in response to stress as well as a biorhythmic secretion, and aldosterone, the steroid hormone acting on the kidney and other lumenal epithelia to conserve sodium ion from being excreted in the urine. Aldosterone is primarily secreted during stress and is often referred to as a stress hormone, whereas glucocorticoids, such as cortisol, are secreted by a basal rhythmic mechanism probably controlled by a serotonergic neuron, in addition to during stress. The rhythmic mechanism reaches a high level of secretion early in the morning. Aldosterone, consequently, plays an important role in water balance and the regulation of blood pressure. Secretion of these two hormones is controlled by different systems. A third major hormone, dehydroepiandrosterone (DHEA), and its sulfatide–sulfate derivatives are secreted by the innermost layer of cells of the adrenal cortex in considerable amounts. Little is known about the physiological activity of this hormone, except that it is a weak androgen and can be converted in various cells by the aromatase enzyme system into an active estrogen. This becomes an important consideration in mammary cancer where these cells are proliferated by estrogen.

Glucocorticoids are essential to life, and they act on different cells in different ways. Without the secretion of cortisol during stress, a human could not survive. Even in patients who lack the ability to release cortisol from the adrenal in the face of stress, the oral administration of the hormone can be of concern when extra amounts are needed and may not be anticipated when acute traumatic stress occurs. Glucocorticoids can induce proteins or repress the expression of certain proteins by transcriptional actions. A general result of glucocorticoid action is the direct effects of increased glycogen, especially in the liver, and increased level of circulating glucose. The latter effect may occur through direct stimulation of certain rate-limiting enzymes in the gluconeogenesis pathway and through the negative effects of the hormone on peripheral cells, which cause them to cease taking up nutrients (glucose) from the blood. Prolonged high levels of glucocorticoids can lead to the death of susceptible cells, which accounts for muscle waste and immunodeficiency. However, early effects of glucocorticoids probably do not include immunodeficiency and β-endorphin, which is secreted from the pituitary together with ACTH in the stress response, but may actually stimulate the action of certain immune cells. Glucocorticoids are known to exert permissive effects on cells, without which many other hormones could not induce particular cellular proteins. The permissive action of glucocorticoids may require mediation through the transcriptional level. Some permissive actions of glucocorticoids are related to their ability to induce cyclic AMP-dependent protein kinase or to transcriptionally interact with this pathway. As powerful anti-inflammatory agents, glucocorticoids act on many cells through the normal receptor mechanism to induce a protein(s) known as lipocortin or one of the annexins. Lipocortin is an inhibitor of phospholipase A_2 and thus prevents the release of arachidonic acid (and other fatty acids) from phospholipids in the cell membrane. These fatty acids are precursors of prostaglandins, prostacyclin, thromboxanes, leukotrienes, and lipoxins, some of which mediate processes of inflammation and pain. Their production is depressed by the glucocorticoid-induced lipocortin, whose action may represent one pathway by which glucocorticoids exert their anti-inflammatory effects. Another effect on reducing inflammation involves the glucocorticoid suppression of the inducible cyclooxygenase, an enzyme involved in the production of inflammatory prostaglandins and prostaglandin relatives. Glucocorticoids have been shown to stimulate the uptake of Na^+ into tubular epithelial cells of the large intestine and kidney by a process (presumably involving the Na^+–H^+ antiporter) that is discrete from that regulated by mineralocorticoids (Na^+ conductance channel), so that Na^+ uptake into tissues appears to be a specific response to glucocorticoids as well as to mineralocorticoids. Water balance would also be affected by this activity.

All of these effects, i.e., cell-specific metabolic changes, specific cell death in certain cases, permissive effects in a wide variety of cell types, profound anti-inflammatory effects exerted through the induction of lipocortin (annexin I), suppression of cyclooxygenase, and effects upon Na^+ retention, contribute to the systemic effects of this hormone. In addition, some actions of glucocorticoids occur so rapidly that one must consider a nontranscriptional route of signal transduction. A case in point is the rapid action of glucocorticoids in the treatment of acute asthmatic attack, a palliation that can sometimes occur in minutes in the airway. In this connection, it has been established that some nerve endings contain a nonclassical membrane glucocorticoid receptor that may mediate ion transport or some related activity by a nontranscriptional signaling mechanism.

Mineralocorticoids promote sodium and potassium transport, usually followed by changes in water balance. This function is essential to life, but in some cells can be produced by glucocorticoids. Glucocorticoids and mineralocorticoids generally act through their own specific receptors, but in certain situations they can bind to the same hormone receptor. Thus, each can be expected to exert effects typical of the other, especially when large amounts are administered. In fact, both hormones seem to be involved in Na^+ ion uptake in tubular epithelial cells. Mineralocorticoids act on the kidney to maintain water balance through the control of sodium ion uptake and mesh in function with the vasopressin system through the control of water reabsorption. Potassium ion transport is also affected by mineralocorticoids. Because of the overlapping specificities of mineralocorticoids and glucocorticoids, much of the molecular interactions of these two hormones remains to be sorted out. A critical feature of aldosterone is that it is secreted in effective amounts only during stress. In certain cells such as mononuclear cells, there appears to be a small number of membrane-associated aldosterone receptors, which would mediate direct effects on ion transport. An important aspect of the crossover effect of glucocorticoids acting on the mineralocorticoid receptor, especially in the kidney, is the ability or inability to inactivate cortisol as a potential ligand for the mineralocorticoid receptor through the 11β-hydroxysteroid reductase enzyme. This enzyme catalyzes the removal of a proton from the 11β-hydroxy group of cortisol to render it ineffective as a ligand for the mineralocorticoid receptor in the kidney and thus inactivates its potential for increasing Na^+ reabsorption and hypertension. There are some known genetic defects in this enzyme that could be the cause of hypertension.

Dehydroepiandrosterone circulates at very high levels in the bloodstream, mainly as the sulfate derivative. A receptor for this hormone had not been discovered for a long time but now may have been found. It is probably the main source for androgens (via testosterone) in females and can be converted to estrogen as well by the aromatase enzyme system. It is apparently important in fetal development in supplying cells with a precursor for estrogen synthesis.

II. ANATOMY, DEVELOPMENT, AND CELLULAR FINE STRUCTURE OF THE ADRENAL CORTEX

The adrenal glands are small organs embedded in fat and enclosed by coverings located above the kidneys. For this reason, they are often referred to as the suprarenal glands. The glands on each side of the body have different shapes; the right one is shaped like a pyramid, whereas the left one is crescent shaped. Nevertheless, they are easily recognized even in small experimental animals. In men, the glands weigh about 5 g each, and their general locations are indicated in Figure 10-1. These glands are well supplied with blood vessels, including a branch from the aorta and a branch from the renal artery (Figure 10-2). The blood entering the adrenal cortex is drained out through the adrenal medulla. This is significant in view of the regulation of an important enzyme of the medulla by cortisol from the adrenal cortex. The adrenal blood supply empties into a major single central vein, which merges with the renal vein (Figure 10-2). The adrenal gland consists of two major parts, an outer cortex and a central medulla. The cortex is the part responsible for the formation of the steroid hormones, aldosterone, cortisol (hydrocortisone), and dehydroepiandrosterone (DHEA), as well as amounts of analogous hormones that are metabolites in the pathways of syntheses of these hormones.

The adrenal medulla is made up of modified nervous cells and secretes the catecholamines, principally epinephrine (and norepinephrine) (see Chapter 11). The cortex is involved in short- and long-term stress reactions, whereas the medulla takes part in sudden or alarm reactions and is under nervous system control. Catecholamines and enkephalins are released from the medulla instantaneously following a signal transmitted by the nervous system. This signal probably originates in the hypothalamus and arrives in the medulla by way of splanchnic nerves from the spinal cord.

As expected from their greatly differing functions, the cortex and the medulla arise from different precursor tissues during development. The adrenal cortex is mesodermal in origin, formed by ingrowth of the peritoneum. The medulla arises from the neural crest and is ectodermal in origin. Essentially, the medulla may be thought of as specialized nervous tissue designed for secretion. By the fifth week of development a suprarenal groove appears above and on each side of the dorsal mesentary (Figure 10-3A). The cells lining the groove (mesothelial cells) begin to proliferate (Figure 10-3B), and a "primary cortex" is derived from this proliferation (Figure 10-3C) by the seventh week of development. While this process has been going on, neural crest cells have moved downward toward the cortex. By 7 weeks the mesothelial layer becomes active again (Figure 10-3C) and produces a new group of cells that are laid down upon the cells of the primary cortex and become the "secondary cortex" (Figure 10-3D). Eventually this will be covered by a connective tissue capsule. In the meantime, the cells of the sympathetic ganglia (neural crest cells) have invaded the primary cortex to form the forerunner of the medulla (Figures 10-3E,F).

In fetal life the human adrenal is large. At term and for the first postnatal year or longer, the primary cortex begins to atrophy while the cells of the secondary cortex actively divide. The primary cortex is often referred to as the "fetal cortex" because of its transient life. It functions like the testis and is uncharacteristic of the mature adrenal (see Chapter 14). As the fetal cortex disappears, the *zona glomerulosa* appears immediately adjacent to the capsule. It is present at the termination of pregnancy. With continued proliferation, the next inward zone, the *zona fasciculata,* arises and within a few months after birth the most inward layer of cells, the *zona reticularis,* is formed.

The zones of adrenal cortex cells are shown in Figure 10-4. The zona glomerulosa is associated with the production of aldosterone, the salt-retaining hormone. Its cells are tall, rather unpigmented, and arranged in clusters. The zona fasciculata, the main producer of cortisol, which is the major glucocorticoid in humans, comprises cells with many sides (polygonal). This is the largest of the three zones in the cortex. The zona reticularis is the innermost layer of cells adjacent to the medulla. The fasciculata–reticularis zona probably function as a unit to produce cortisol and DHEA. ACTH converts fasciculata cells to reticularis cells, which are responsible for most of the steroid production. Capillaries are found in this cellular network (Figure 10-2). These cells are smaller and more pigmented (golden brown) than fasciculata cells. The pigment is lipofuscin. Lipofuscin granules are derived from lysosomes and are called secondary lysosomes. Although the cells of the zona reticularis are able to secrete glucocorticoids, they primarily produce and secrete the weak

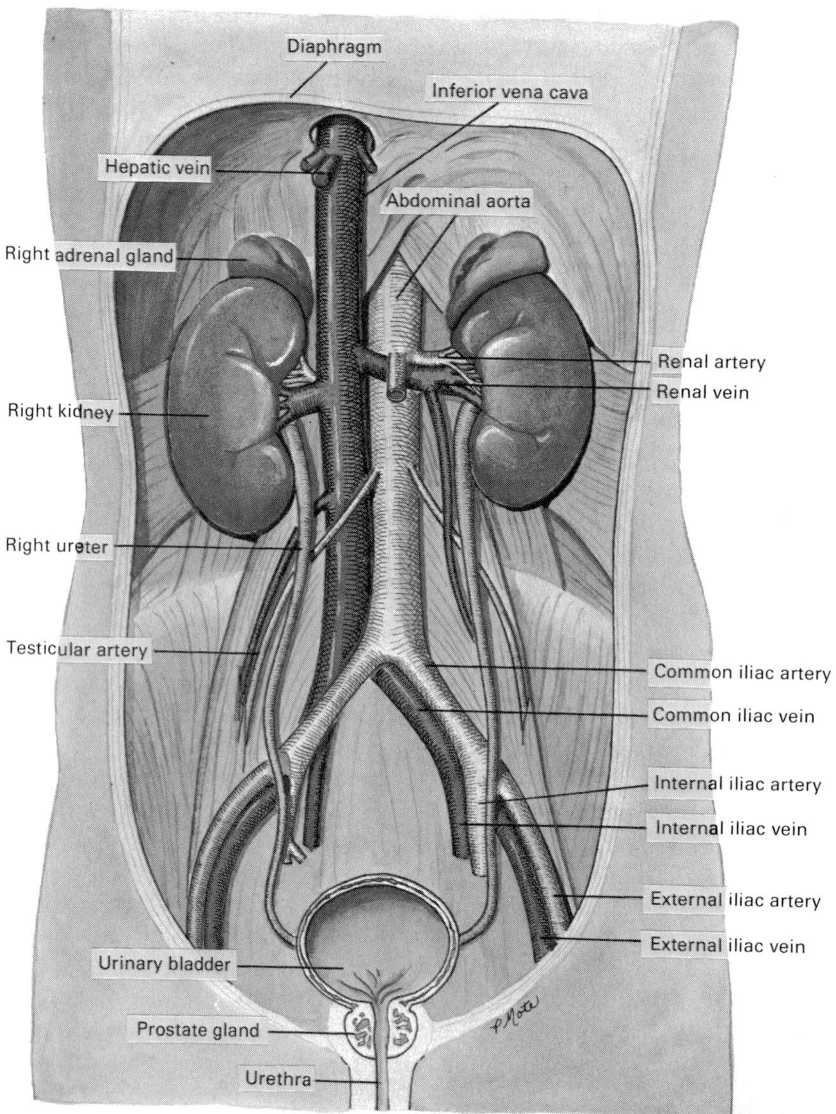

FIGURE 10-1 Summary of the major anatomical components of the urine-forming system. Reproduced with permission from Chaffee, E. E., and Lytle, I. M. (1980). "Basic Physiology and Anatomy," 4th ed., p. 483. Lippincott, Philadelphia. PA.

androgen dehydroepiandrosterone (DHEA) in rather large quantities (15–25 mg/day). In addition to DHEA, a small amount of testosterone is formed in the reticularis layer.

The fine structure of the cells of the zona glomerulosa contains relatively long mitochondria with lamellar cristae, and these differ from the mitochondria in the cells of the other adrenal cortical layers. The zona fasciculata is the largest of the three zones and is made up of columnar cells that are polygonal and form cords that radiate from the zone above. The mitochondria of these cells vary in size, but are usually larger than the mitochondria in cells of the other zones. The zona reticularis contains networks of interconnecting cells. Mitochondria in these cells resemble those in the cells of the zona fasciculata.

A. Glucocorticoid Targets

Although the glucocorticoids act on many cells of the body, the liver is an important target organ because of the relatively large concentration of glucocorticoid receptors in the hepatocyte. Figure 10-5 shows the various levels of organization of the liver. A basic subdivision is the liver lobule shown in the upper right. Cords of hepatocytes are clustered together, with portal canals at each apex. The portal triad consists of a bile duct, hepatic artery, and portal vein, as shown in the

A

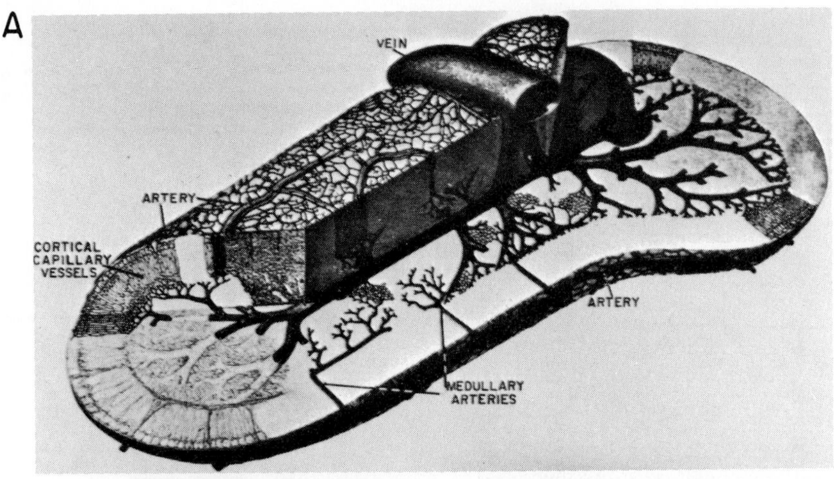

B

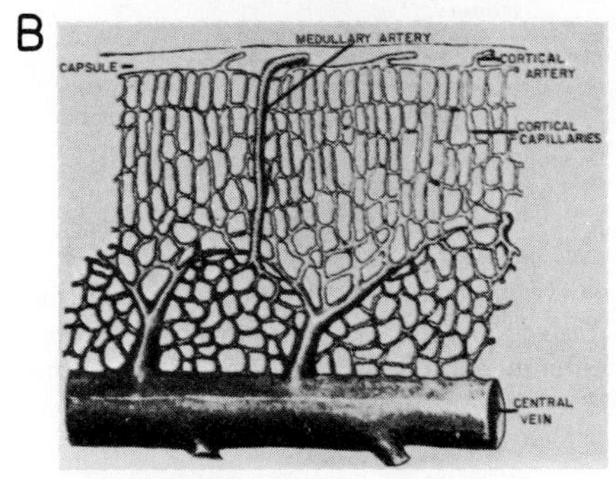

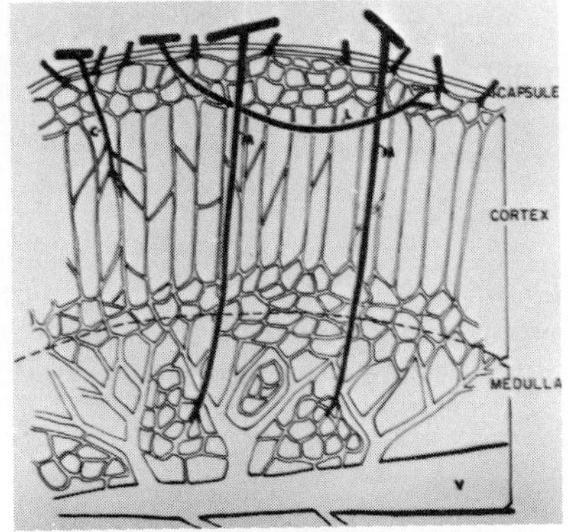

FIGURE 10-2 (A) Arterial blood supply and venous drainage of the adrenal gland in the dog. (B) Blood supply of the adrenal cortex and medulla in the dog. Reproduced from Coupland, R. E. (1975). "Handbook of Physiology" (R. O. Greep and E. B. Astwood, eds.), Section 7, Vol. VI, pp. 282–294. American Physiological Society, Washington, D.C.

upper center. From this drawing it is clear how the circulation efficiently bathes the hepatocytes. Also clear from this figure is the fact that the large majority of cells in the liver are hepatocytes, with bile duct cells and Kupffer (scavenging) cells being in the minority. At least 65% of the liver cells are probably hepatocytes.

Another major target of glucocorticoid hormones is the thymus. It is located above the upper end of the heart. The thymus has a relatively large size in newborns and is invested with immunological functions. Its size decreases with age until adult proportions are attained largely due to the death of thymus cortex glucocorticoid-sensitive cells as the adrenal zona fasciculata becomes more active in secreting cortisol. The thymus cells killed off by apoptosis in this way during development probably contain large numbers of glucocorticoid receptors per cell. The histology of the thymus is shown in Chapter 17. Besides a capsule, there are two main layers of cells: the cortex and the medulla. Thymocytes are similar to the lymphocytes of other lymphoid organs and in circulation. There are many free ribosomes and a large nucleus containing chromatin condensed in large masses in close proximity to the inside nuclear envelope.

III. CHEMISTRY AND BIOCHEMISTRY

The structures of the principal steroid hormones of the adrenal cortex are shown in Figure 10-6. The chemistry and structural relationships of the glucocorticoids compared to the other steroid hormones are reviewed in Chapter 2. The metabolic pathway for the production of glucocorticoids is presented in Figure 2-22.

A. Glucocorticoids and Stress

Glucocorticoids are secreted in large amounts in humans, up to 25 mg or more per day, and represent a major chemical response of the body to stress. A person undergoing long-term stress will have higher

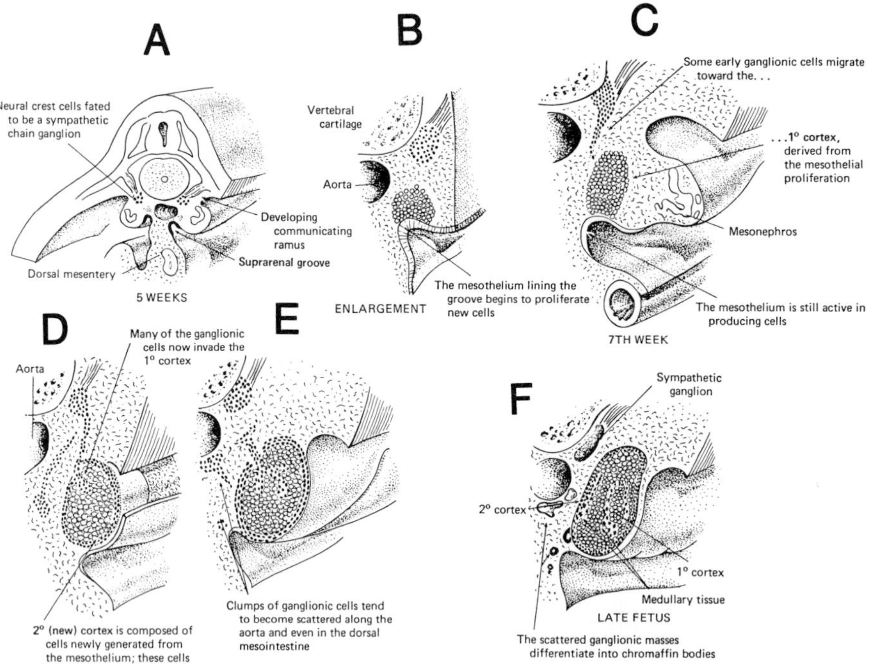

FIGURE 10-3 Development of the adrenal glands: (A) at 5 weeks development of the human fetus; (B) enlargement of the mesothelium; (C) seventh week of development; (D) secondary cortex; (E) appearance of ganglionic cells; (F) differentiation into chromaffin bodies in the late fetus. Reproduced from Langebartel, D. A. (1977). "The Anatomical Primer," p. 443. University Park Press, Baltimore, Maryland.

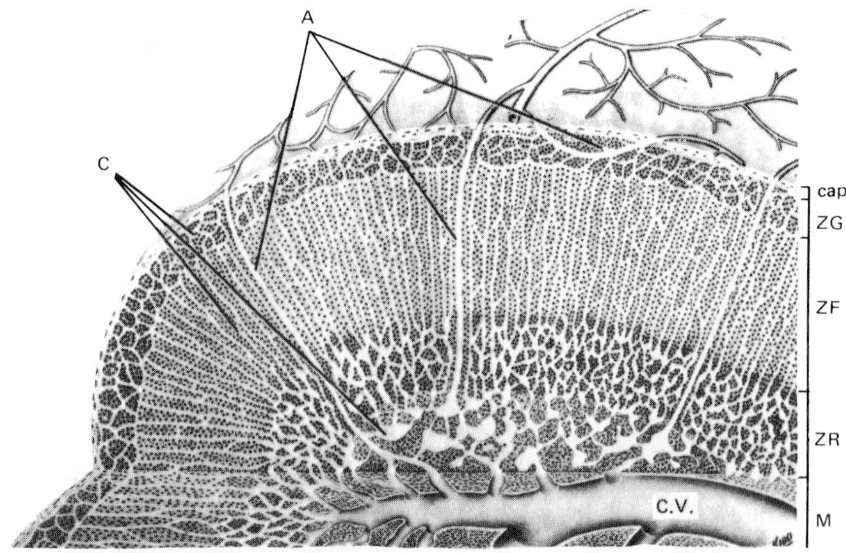

FIGURE 10-4 Schematic view of the circulation in a mammalian adrenal gland. The arteries and arterioles are shown in the capsule (cap), the outer layer of which has been removed. Other vessels shown are the radially arranged capillaries (C), a few arterioles (A) in the cortex, and "arterial" and "venous" capillaries in the medulla. The zones of the cortex are identified as zona glomerulosa (ZG), zona fasciculata (ZF), and zona reticularis (ZR). M denotes the medulla and the CV central vein. Reproduced from Harrison, R. G. (1959). "A Textbook of Human Embryology," Vol. 1, p. 48. Blackwell Scientific, London.

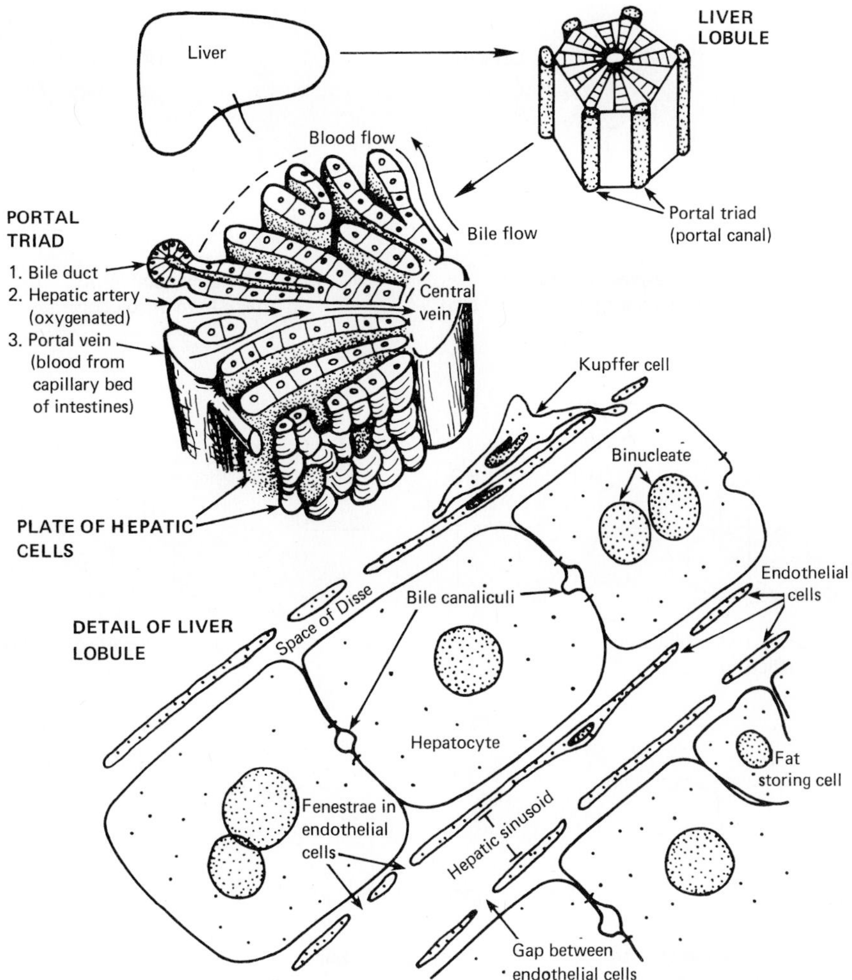

FIGURE 10-5 Anatomical components of the liver and hepatocyte, a major target cell for glucocorticoids. This drawing was made originally by Dr. Laurie Paavola, Department of Anatomy, Temple Medical School, and is redrawn here with her permission. Other work with a polyclonal antibody to the glucocorticoid receptor indicates that the hepatocyte is the cell in the liver having a large number of receptor antigens, while the bile and Kupffer cells have little or no antigenic activity.

amounts of cortisol circulating in the bloodstream than the unstressed person. The steroid acts on many of the tissues of the body to an extent determined by the number of glucocorticoid receptors present in the cell of a particular tissue. The liver, which contains about 65,000 receptor molecules per cell in an experimental animal (rat), predominates as a major target of corticosterone, the principal glucocorticoid in rat. Other important targets are the lymphoid cells, thymus gland, and kidney. Many other tissues seem to have enough receptor molecules to provide a response to stress, especially if it is long-term. In fact, most tissues of animals or of cells in culture seem to contain measurable amounts of receptor, making it theoretically possible for nearly all tissues of the body to be affected by stress.

Studies using antibodies to the receptor indicate that rat biliary and Kupffer cells of the liver contain small amounts of receptor, and cells of the *pars intermedia* have small amounts or lack receptor altogether. Long-term stress may be distinguished from short-term stress, often referred to as "alarm" or "fright." In short-term stress, the minute-to-minute changes in metabolism are under the control of catecholamine hormones, primarily epinephrine, secreted by the adrenal medulla. The secretion of epinephrine is in turn controlled by the autonomic nervous system. Both stresses, long- or short-term, lead to the release of glucocorticoids. Figure 10-7 delineates the overall pathways by which the two types of stress call into play the secretion of glucocorticoid hormones or epinephrine. Glucocorti-

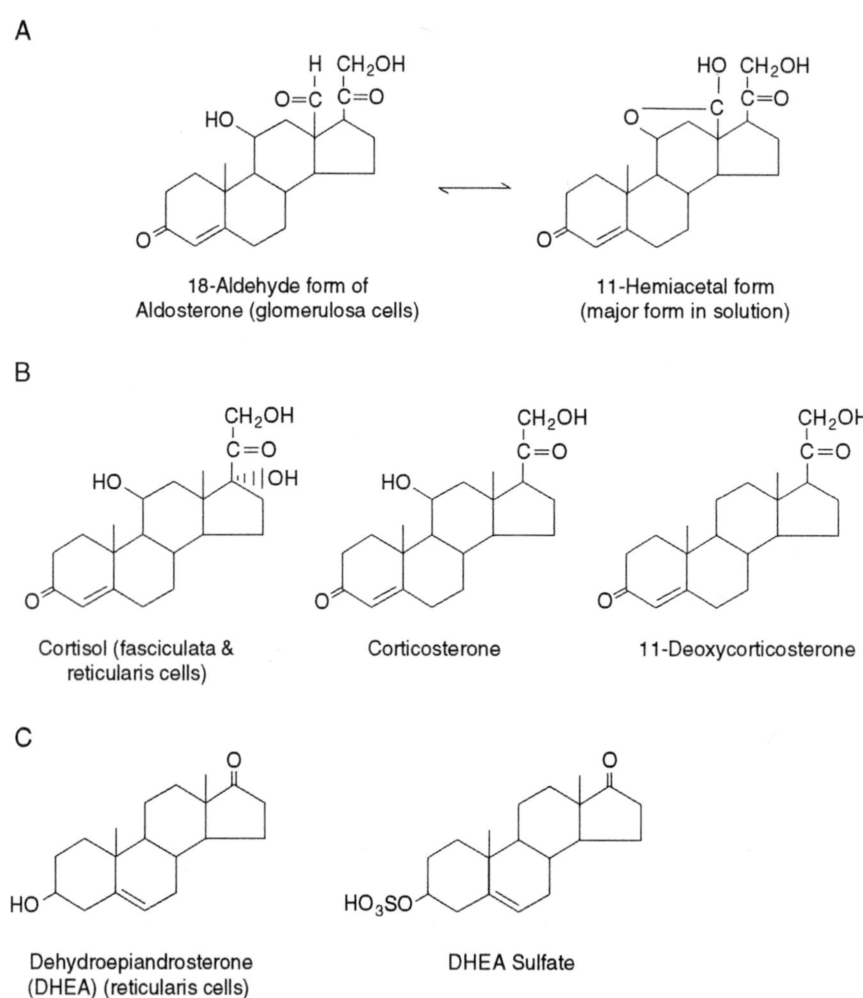

FIGURE 10-6 Structures of major hormones of the adrenal cortex. Aldosterone is secreted from the outer layer of the adrenal cortex, the zona glomerulosa. Two forms are in equilibrium, as shown in (A). (B) The structures of cortisol, the major human product of the zona fasciculata, together with corticosterone and deoxycorticosterone. (C) The structures of dehydroepiandrosterone (DHEA) and its sulfate derivative secreted from the inner layer of cells of the cortex, the zona reticularis.

coids are so named because they influence the storage of carbohydrate macromolecules in the form of energy supplies (glycogen), which can be tapped by the conversion of glycogen to glucose on a minute-to-minute basis by epinephrine for instantly usable energy in "flight or fight" situations or for the expenditure of "nervous energy." Many other hormones are involved in stress, including glucagon, growth hormone, prolactin, β-endorphin, vasopressin, angiotensin II, and prostaglandins.

Thus, we can classify two types of stressors: those that are relatively long-term and evoke glucocorticoids from the adrenal cortex, such as intensive cold, prolonged loud noise, serious injury, burns, surgery, and changes in the environment necessitating adaptation by the organism with serious consequences if adaptation does not occur. The alarm emergency or shorter term stressor would be an event of a surprise nature, such as fright induced by a specific happening. This type of stressor would primarily evoke the secretion of epinephrine (and norepinephrine) from the adrenal medulla as well as cortisol from the adrenal cortex. Both types usually occur simultaneously; one chain of events (Figure 10-7) in operation does not preclude the utilization of the other pathway. There are, however, certain conditions that can cause the pathways to operate separately. The sympathetic system generating epinephrine and norepinephrine is activated when the organism attempts to escape from or deal with the environmental challenge, the flight or flight response.

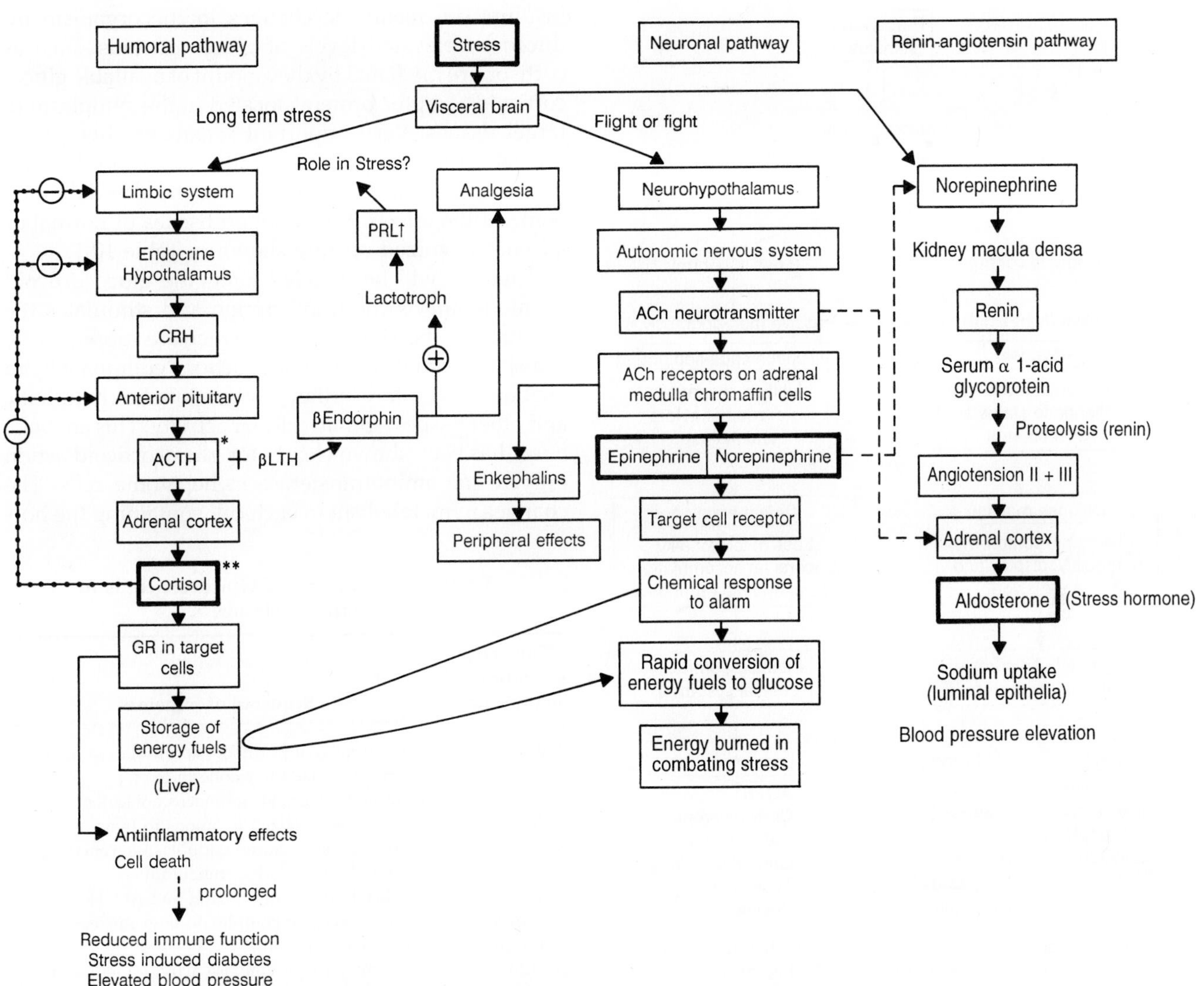

FIGURE 10-7 Pathways activated by stress. Abbreviations: +, stimulates; −, inhibits; CRH, corticotropic releasing hormone; ACTH, adrenocorticotropic hormone; β-LTH, β-lipotropin; GR, glucocorticoid receptor; PRL, prolactin; ACh, acetylcholine; *, ACTH is a growth factor for cells of the adrenal cortex, especially the zona fasciculata cells; **, cortisol is also released through a rhythmic pathway under the control of a serotonergic neuron. In this 24-hr biorhythm, cortisol secretion is highest in the morning; thus, cortisol is secreted by an endogenous pathway that does not rely on stress. The secretion of cortisol can be increased greatly by stress, as shown here through the humoral pathway, whereas the secretion of aldosterone is largely a stress response.

When the organism becomes immobile or passive, the adrenocortical axis is preferentially activated. This is discussed in detail by Henry (1993).

The catecholamines will be considered in-depth in Chapter 11, but are presented in outline form here because stress reactions involve both glucocorticoids and catecholamines. Figure 10-7 shows three pathways, including a dual pathway responding to stress: the neural pathway and the humoral pathway generating cortisol. It is important to emphasize that these two pathways are sometimes operating together, but they can also be dissociated especially when the individual is confronted with an emotionally stressful situation and copes on different levels. This concept is elaborated in Figure 10-8. The sympathetic adrenomedullary pathway is activated when a fight or flight response is issued to a specific challenge. On the other hand, the humoral pathway, ending in cortisol release from the adrenal gland, is operative when the individual becomes immobile, passive, and depressed

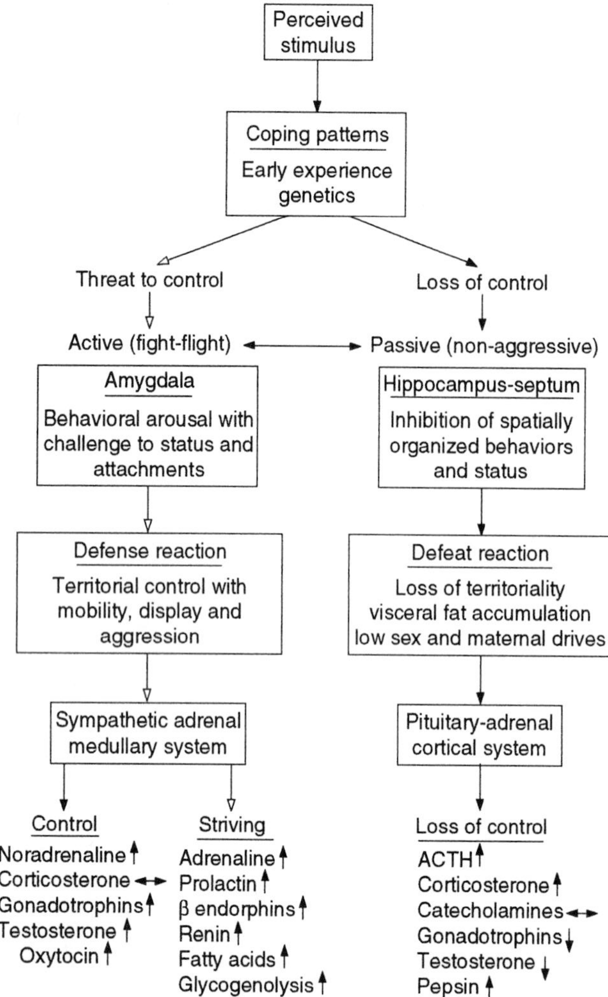

FIGURE 10-8 Defense reaction is activated when the organism is challenged but remains in control. With the loss of control there is activation of the hypothalamopituitary–adrenal axis, and the gonadotrophic species preservative system shuts down. Visceral fat accumulates with a Cushingoid distribution, and there is a shift from active defense to a passive nonaggressive coping style. Reproduced with permission from J. P. Henry, (1993). Biological basis of the stress response. *NIPS* 8:69–73.

as loss of control is perceived. A chronic emotional reaction of passivity and defeat to a stressful situation can produce dire consequences as the adrenal hypertrophies and levels of cortisol continue to increase. This can generate a Cushingoid-like bodily reaction in which visceral fat accumulates (Figure 10-8), blood pressure becomes elevated, and arteriosclerosis and type II diabetes eventually develop. Sequential episodes of elevated glucocorticoids cause sufficient repression of glucose uptake in peripheral cells to involve insulin release from the β-cells of the pancreas. Eventually, chronic repetitions of this scenario could lead to exhaustion of the β-cells' ability to produce and secrete insulin. The metabolic changes in the organism induced by elevated levels of glucocorticoids, such as cortisol, are mediated by the amount of available glucocorticoid receptor proteins located in the cytoplasm of target tissues. Very important organs are liver, lymphocytes (including thymus cells), adipose cells, kidney, anterior pituitary, and various parts of the brain. Some of the effects of cortisol on tissues of normal or adrenalectomized rats are shown in Table 10-1.

Thus, as will be detailed later, the glucocorticoid hormone affects the transcription and stimulates the production of certain mRNAs, which are subsequently translated into proteins whose actions culminate in an anabolic (liver and kidney) or a catabolic (lymphatic and other tissues) shift in cellular activity. This anabolic effect has been shown clearly for glucocorticoid action on tyrosine aminotransferase in hepatoma cells. The changes in metabolism in each cell containing the hor-

TABLE 10-1 Effects of Glucocorticoids in Various Tissues[a]

Time after glucocorticoid administration	Sequence of responses
2 min	Liver uptake of hormone, accumulation of unmetabolized hormone in liver
<5 min	Binding to liver cytosol macromolecules
~15 min	Cortisol metabolized to anions in liver, conversion of rough endoplasmic reticulum to smooth endoplasmic reticulum
<30 min	Feedback inhibition of CRH and ACTH
<30 min	Steroid–receptor complex in liver nucleus
~90 min	Glycogen deposition
30–120 min	Nuclear RNA polymerase activity increased
2–4 hr	Increased RNA synthesis, increased fatty acid release from adipose, decreased glucose utilization in many sensitive tissues peripheral to liver
	Protein breakdown in peripheral tissues: decreased glucose utilization, nucleic acid synthesis and protein synthesis decreased, ornithine decarboxylase activity increased in liver, tyrosine aminotransferase activity peaks in liver, tryptophan oxygenase activity peaks in liver, polysome aggregation
8–20 hr	Increased general liver protein synthesis, decreased glucose utilization, increased hepatic gluconeogenesis, lympholysis
4 hr to days	Small increase in threonine dehydrase activity (liver), alanine aminotransferase activity increases (liver), many glycolytic enzyme activities increase (liver, kidney), urea cycle enzymes increase: arginine synthetase system; arginine succinase; arginase

[a] Some of the data of this table are reproduced from Shulster, D., Burstein, S., and Cooke, B. A. (1976). "Molecular Endocrinology of the Steroid Hormones," p. 275. Wiley, New York.

mone receptor are the culmination of the hormonal effect at the level of the cell. The summation of metabolic changes in different tissues constitutes the adaptation of the organism to stress at this level. Antiinflammatory and antipain effects and Na^+ movements contribute as major alterations produced by glucocorticoids. Not all of these biochemical changes are understood. For example, the stress signals mediated by the brain and the feedback effects of cortisol on elements of the brain remain to be elucidated clearly. Behavioral modifications, if any, to stress may be clarified by future research in this direction. Elaboration of β-endorphin by the anterior pituitary in response to stress is bound to have important stress adaptation effects mediated by the central nervous system.

B. Corticotropin-Releasing Hormone (CRH)

The 41-amino acid primary sequence of CRH peptide is given in Figure 3-8. Figure 10-7 shows that the response of the hypothalamus to signals from the limbic system is the secretion and subsequent resynthesis of the CRH. CRH is released from specific cells in the hypothalamus. CRH activity appears to be present in other tissues, such as cerebral cortex and liver. It appears to be produced by the placenta and fetal membranes throughout gestation, and a myometrial CRH receptor whose affinity increases during pregnancy has been reported. CRH appears in several tissues as a result of arthritic inflammation.

C. Mode of Action of Releasing Hormones (e.g., CRH)

Following signals from the limbic system to the hypothalamus (Figure 10-7), CRH is released into a closed portal circulation intimately connected with the anterior pituitary (Figure 3-2). Detailed information on anatomical considerations and modes of action of hypothalamic releasing hormones may be found in Chapters 3 and 5.

D. Mode of Action of ACTH

ACTH is the peptide hormone secreted from the anterior pituitary in response to CRH. ACTH is a 39-amino acid peptide; the amino acid sequence is presented in Figure 5-11. The common N-terminal sequence of residues 1–13 of ACTH and residues 1–13 of α-MSH constitutes an important homologous structure. Also, sequence 11–17 of β-MSH and residues 4–10 of ACTH are homologous. These homologies in amino acid sequence are suggestive of a close relationship or of a common precursor between ACTH and MSH. However, in humans, only ACTH, β-lipotropin, and β-endorphin appear to be secreted by corticotrophic cells of the anterior pituitary. Normally, α-MSH derives from the intermediate pituitary cells. Thus, in corticotropic cells of the anterior pituitary, the precursor protein proopiomelanocortin is cleaved by proteases, ultimately to β-lipotropin with some further cleavage to ACTH, to γ-LPH and β-endorphin together with a 16,000 fragment (big γ-MSH) and a smaller fragment. Thus, the main products in corticotropes are ACTH, β-LTH, γ-LPH, and β-endorphin. In neurointermediary cells (and in certain extra pituitary tissues), proopiomelanocortin is cleaved to the same products mentioned for corticotropic cells, but the processing of these products is continued. ACTH is further broken down into α-MSH and CLIP (corticotropic-like intermediary peptide); γ-LPH is further cleaved to β-MSH and β-endorphin could be cleaved further (into enkephalins?). Also, big γ-MSH is further cleaved into two or three smaller products.

In the anterior pituitary corticotroph, endopeptidase cleavages occur at Lys-Arg and Arg-Arg sites; however, Arg-Lys, Lys-Lys, and Lys-Lys-Arg-Arg sites are not cleaved. The specificity apparently resides in the enzymes because the resistant sites in the corticotroph are susceptible sites in the melanotroph.

In the intermediary pituitary cells, ACTH is processed extensively into α-MSH and CLIP. This occurs by endopeptidase and exopeptidase cleavage together with acetylation and amidation. The major products of these cells are α-MSH, CLIP, and *N*-acetyl-β-endorphin. In diseases characterized by altered secretion of ACTH, skin pigmentation (theoretically due to MSH action) can occur. It is not completely clear how MSH is secreted in this case; possibly in certain diseases ACTH is degraded to α-MSH. Thus, hyperpigmentation occurs in diseases of negative feedback by reduced cortisol levels, and the level of ACTH rises. However, other work suggests that ACTH may be more important than MSH in regulating melanocyte pigmentation. Consequently, the elevated levels of circulating ACTH in Addison's disease may be directly related to changes in skin pigmentation and MSH may not be a critical factor.

It is known that ACTH, MSH, and β-lipotropin are all coded for by a single gene. The gene gives rise to an mRNA that is translated to a preprotein called the opiocortin precursor or proopiomelanocortin. This is represented diagrammatically in Figure 5-10. Not all of the information in the preprotein sequence is known, as depicted by the blank spaces in the parent peptide chain.

Following the secretion of ACTH into the blood circulation after stimulation by CRH from the hypothalamus, ACTH molecules bind to a specific receptor on the outer cell membranes of all three layers of cells of the adrenal cortex, the *zona glomerulosa,* the *zona fasciculata,* and the *zona reticularis.* The results of many experiments appear in Figure 10-9 in the form of a speculative mechanism of action.

Molecules of ACTH bind to an outer cell membrane receptor. Formation of this complex activates adenylate cyclase of the inner membrane and increases the affinity of the enzyme for ATP. This could be accomplished by a mechanism described in Chapter 1. The cytoplasmic level of cyclic AMP is increased and inactive protein kinases are converted to active catalytic subunits and cyclic AMP-bound regulatory subunits (see Chapter 1). The active catalytic subunits phosphorylate proteins, which increase the rates of hydrolysis of cholesteryl esters to free cholesterol. The rate of transport of free cholesterol into the mitochondria is increased (possibly by the sterol carrier protein, StAR protein), and, perhaps most importantly, the cholesterol side chain cleavage reaction in the mitochondrion, which appears to be catalyzed by a single protein, is stimulated. A novel adrenal peptide of low molecular weight, about 3000, has been shown to appear after ACTH stimulation. It could arise from a precursor protein or by synthesis *de novo.* Conversion of a precursor to the active peptide could be under the control of ACTH, operating perhaps by way of cyclic AMP, which would control some step in the system, for example, proteolytic processing. Generation of this peptide in cells of the adrenal cortex stimulates, in some way, the side chain cleavage system of enzymes in the mitochondrion, converting cholesterol to Δ^5-pregnenolone. This system is reminiscent of the hypothetical insulin mediator peptide that is claimed to regulate phosphoprotein phosphatases (see Chapter 7). The StAR (steroid acutely regulated) protein may account for many of the changes required for the increased synthesis of cortisol (see Figure 2-33).

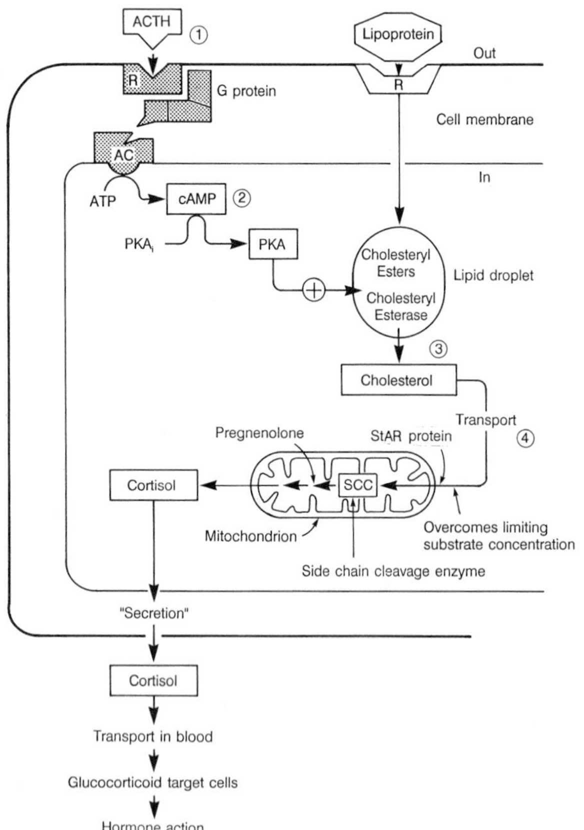

FIGURE 10-9 Action of ACTH on an adrenal fasciculata cell to enhance the production and secretion of cortisol. Abbreviations: AC, adenylate cyclase; cAMP, cyclic AMP; PKA, protein kinase A; SCC, side chain cleavage system of enzymes. StAR (steroid acutely regulated) protein is a cholesterol transporter functioning between the outer and inner mitochondrial membranes (see Figure 2-33).

This peptide of about 3000 molecular weight could be the "steroidogenesis activator polypeptide," which appears to be generated in the adrenal cortex in response to ACTH and which stimulates cholesterol side chain cleavage in adrenal mitochondria obtained from rats treated with cycloheximide. The level of this activator increases through stimulation by ACTH or cyclic AMP, and its appearance is inhibited by cycloheximide. There are two other candidates for this stimulatory factor. However, other work indicates the rapidly turning over protein to be the StAR protein, which is a steroid transporter between the outer and inner mitochondrial membranes. Thus, the central effect of ACTH is to activate PKA, which in turn phosphorylates inactive cholesteryl esterase to produce the active enzyme. This causes cholesteryl esters in the droplet to be broken down to free cholesterol, which then enters the mitochondrion. By this means, the substrate-limited side chain cleavage enzyme can now act with greater velocity. In addition, there is some mechanism that is responsive to ACTH, perhaps through the StAR protein, that accounts for a stimulation in the activity of the side chain cleavage event. These effects generate higher levels of cortisol in the zona fasciculata cells.

Since the side chain cleavage enzyme is rate limiting in the mitochondria at the beginning of cortisol synthesis, and since this step seems to be substrate limited, the role of ACTH is to produce biochemical changes that increase the substrate cholesterol level in the mitochondrion (Figure 10-9). The intracellular level of Ca^{2+} may also increase, and Ca^{2+} has been shown to stimulate the activity of 11 β-hydroxylase in adrenal mitochondria.

There are longer acting effects of ACTH on adrenal cells than the effects described for increasing the availability of cholesterol to the mitochondria and enhancing the rate of cholesterol side chain cleavage. These relate to increasing the synthetic rate of steroids by ACTH-driven enhancement in the levels of mRNAs of steroid hydroxylases. With bovine adrenocorticol cells in primary culture, ACTH was demonstrated to enhance the levels of message for P450scc, P45011β, adrenoxin, P45017α, P450C21, and adrenoxin reductase. This probably occurs by increased transcription. The spatial arrangement of some of these proteins in relation to the mitochondrial uptake of cholesterol is shown in Figure 10-10.

E. Sources of Cholesterol and Production of Glucocorticoids

The immediate substrate for steroid hormone biosynthesis is cholesterol (derived from circulating lipoproteins) from intracellular stores of cholesteryl esters or from free cholesterol (Figure 10-9). For a discussion of cholesterol biosynthesis, see Chapter 2, Figures 2-13–2-15.

Cholesterol is synthesized chiefly by the liver and the intestine and to a smaller extent in other tissues. Cholesterol can be transported to peripheral cells as a component of circulating lipoproteins. Lipoproteins are absorbed by many cells, such as the adrenal cortical cells by first interacting with a membrane receptor (Figure 10-9), and degraded within the cell to avail free cholesterol, which is stored as fatty acid esters in the "lipid droplet." Activation of cholesteryl esterase results from cellular stimulation by ACTH, which elevates levels of cyclic AMP, and the esterase is activated through phosphorylation by protein kinase A.

Cortisol is the main product in the zona fasciculata of human adrenal cortex. Aldosterone is the major steroid of the outer zona glomerulosa. Dehydroepiandrosterone and its sulfate are the principal products of the zona reticularis. These biochemical conversions are shown in Figure 2-21. It is important to realize that the nature of the steroid hormone product is governed by the specificities of the converting enzymes produced in a given cell type. In essence, this specialization constitutes the cell's phenotypic function. For example, genes are expressed in the fasciculata cell encoding information for cytosolic 17α-hydroxylase, 21-hydroxylase, and mitochondrial 11β-hydroxylase, whose reactions together with other cellular enzymes produce cortisol. A zona glomerulosa cell expresses genes for cytosolic 21-hydroxylase, but 17α-hydroxylase is not expressed. Mitochondrial 11β-hydroxylase and 18-hydroxylase are also expressed, resulting in the

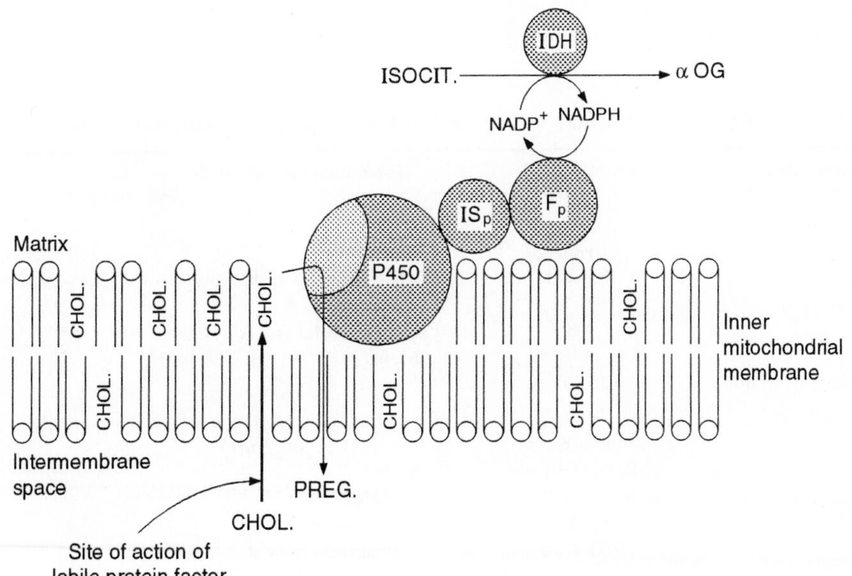

FIGURE 10-10 Arrangements of the proteins involved in cholesterol side chain cleavage in the inner mitochondrial membrane, showing the postulated site of action of the labile protein factor. Abbreviations. IS_p (probably the StAR protein), adrenodoxin; F_p, adrenodoxin reductase; IDH, isocitrate dehydrogenase; ISOCIT, isocitrate; αOG, α-oxoglutarate. Reproduced with permission from Simpson and Waterman (1995). Steroid hormone biosynthesis in the adrenal cortex and its regulation by adrenocorticotropin in Endocrinology; DeGroot, L. J. (ed.) (1995). W.B. Saunders, Philadelphia, PA. 3rd ed., ed. Vol. 2, pp. 1630–1641.

production of aldosterone rather than cortisol. Note that the side chain cleavage of cholesterol, the rate-limiting step in steroid hormone biosynthesis, occurs in the mitochondria. This is not a simple one-step reaction, but apparently the total reaction is catalyzed by a single protein (Figure 2-21). Specific signals to the cells of the adrenal cortex determine which steroid hormone will be synthesized at an increased rate compared to the unstimulated state. Thus, ACTH stimulates fasciculata and reticularis cells to produce cortisol and dehydroepiandrosterone. Glomerulosa cells also have an ACTH receptor that plays a role in aldosterone synthesis through the stimulation of protein kinase A. The regulation of aldosterone synthesis will be discussed later.

F. Mechanism of Secretion of Glucocorticoids

Secretion can be said to occur at two levels in the processes of synthesis and release of steroid hormones. Biosynthetic intermediates move between cytoplasmic and mitochondrial compartments and finally to the cytoplasm and cell exterior in the case of cortisol as the end product. Thus, the removal of Δ^5-pregnenolone from the mitochondria to the cytoplasm for action by the microsomal 3β-hydroxy-Δ^5-steroid dehydrogenase to produce progesterone is a transportation event. 17α-Hydroxylase is also located on the microsomes in the cytoplasm, as is 21-hydroxylase. The resulting 17α-hydroxydeoxycorticosterone must be transported back to the mitochondria, perhaps by way of a transporting protein, but 11β-hydroxylation to form cortisol occurs on the inner mitochondrial membrane (see Table 2-5). Cortisol comes out of the mitochondrion and moves out of the cell into the extracellular space and into the bloodstream. It is difficult to find evidence of "packaging" of newly synthesized cortisol, and little is known about its exit from the cell. Consequently, it seems possible that the newly formed steroid simply diffuses into the cell membrane and out into the extracellular space.

Free cholesterol is the substrate for the mitochondrial synthesis of cortisol. Electron microscopy shows that lipid droplets reside near mitochondria. Cholesteryl ester hydrolase and cholesterol-transporting protein could be activated by enhanced levels of cyclic AMP produced by ACTH action (if they are active as phosphorylated forms). Values for the normal ranges of corticosteroids produced in a day, plasma levels, protein binding, and urinary excretion are presented in Table 10-2.

G. Transport of Glucocorticoids in the Blood (CBG)

A corticosteroid-binding globulin (CBG) is present in blood. This protein is also referred to as transcortin. It is synthesized in the liver and exported to the

TABLE 10-2 Normal Values in the Investigation of Adrenal Cortex[a]

Hormones and steroids	Normal range	Hormones and steroids	Normal range
Production rates of major hormones		Urine excretion	
Cortisol (mg/day)	10–20	Free cortisol	150 μg/day
Aldosterone (μg/day)	60–200	Cortisol in certain disease states	Several hundred thousand μg/day
Testosterone (mg/day)			
Males	5–8	Estrogens, total (μg/day)	
Females	0.3–0.8	Males	4–25
Dehydroepiandrosterone (mg/day)	25–30	Females	4–60
Plasma levels (μg/100 ml)		Pregnanetriol (mg/day)	0–4
Cortisol	5–20 (at 8:00 a.m.)	Pregnanediol (mg/day)	
Aldosterone	0.002–0.01	Males	0–1.5
Testosterone		Females	
Males	0.5–1.0	Follicular	0.5–1.5
Females	0.03–0.06	Postmenopausal	0.2–1.0
Protein binding (%)		Aldosterone (μg/day)	4–17
Cortisol	80	Free cortisol (μg/day)	0–200
Aldosterone	70		
Testosterone			
Males	90		
Females	95		

[a] Reproduced from Ezrin, C., Godden, J. O., Volpe, R., and Wilson, R. (eds.) (1973). "Systematic Endocrinology," p. 170. Harper & Row, New York.

circulation. This protein binds cortisol with relatively high affinity (binding constant $\cong 10^8\ M^{-1}$; the dissociation constant for the reaction, CBG + cortisol $\leftrightarrow$ CBG–cortisol, is about $10^{-8}\ M$ cortisol). Because of the affinity of the protein for cortisol, most of the hormone circulates in the bound form as reflected in the equilibrium, which favors the complex: CBG + cortisol $\leftrightarrow$ CBG–cortisol; there is only a small amount of the free hormone. Nevertheless, at the target cell it is the free steroid that enters the cell, probably by a free diffusion process. The mechanism of entry of steroid hormones from the blood across the target cell membrane into the cytoplasm has been an unsolved question for many years. Although thermodynamic measurements of the uptake process equate to the energy of free diffusion, some investigators have postulated that carrier proteins are present to facilitate the uptake process. Over the years there have been reports of membrane steroid receptors, and these might fill the role suggested earlier. These questions possibly will be resolved by the newer approaches of molecular genetics. The driving force behind the movement of free hormone into the target cell appears, in part, to be proportionate to the number of specific hormone receptor molecules in the target cell cytoplasm that have unfilled ligand-binding sites (unoccupied receptors). The affinity of this receptor for cortisol is similar (20 n*M*) to that of the circulating CBG–cortisol complex. Steroid hormones will cycle into and out of a target cell, and retained molecules will be determined by the number of unoccupied receptors.

H. Feedback Effects of Glucocorticoids

When cortisol is produced in response to ACTH, it has negative feedback effects on various elements of the hormonal cascade system (Figure 10-7). Thus, the negative long feedback loop refers to the effect of elevated cortisol on the limbic system, the hypothalamus, and the anterior pituitary. Feedback inhibition of ACTH is very fast. Feedback on the hippocampus may shut off further electrical activity responsible for the release of CRH and the subsequent release of ACTH. These actions are mediated by glucocorticoid receptors located in these cells and that operate transcriptionally. A glucocorticoid-like receptor has been reported to be located in the membranes of certain nerve endings that do not contain nuclei. Consequently, whatever process this receptor serves is probably carried out by a nongenomic mechanism. Either it serves as part of the negative feedback process, or it is somehow involved in ion transport.

IV. BIOLOGICAL AND MOLECULAR ACTIONS

A. Molecular Actions of Glucocorticoids

1. Immunosuppression and Apoptosis Induced by Glucocorticoids

Among the cell types that involute and ultimately die under prolonged elevated levels of glucocorticoids are the cells involved in the production of immunoglobulin. Generally speaking, antibody production is only marginally affected by glucocorticoids in humans following relatively minor momentary incidents of stress. The effects of increased glucocorticoids in the circulation by stress are exemplified by *in vitro* experiments with immunoglobulin-producing plasma cells (derived from B cells).

Diminished immunoglobulin responses occur after treatment with relatively large amounts of glucocorticoid and probably when stress is severe and extended over a long duration. Suppressor T cells seem to be very sensitive to steroid treatment, but helper T cells are relatively unaffected. B-cell-derived plasma cells, the primary immunoglobulin producers, are depressed in their ability to produce all classes of immunoglobulins. Immunosuppression by glucocorticoids results from the hormonal suppression of various lymphokines and monokines, which are required for the function of antibody production. This is part of the catabolic response of certain cell types to glucocorticoids, as discussed in the following (see Figure 10-11).

While clinical treatment with pharmacological doses of glucocorticoids represents one end of the spectrum, the physiological rise in circulating levels of cortisol following stress represents the other. Some workers believe that there is a close link between stress, elevated glucocorticoids resulting from it, and injury to the immunological apparatus, which may leave the individual vulnerable to the action of latent oncogenic viruses, newly transformed cancer cells, or other disease processes that are normally held in control by the immunological apparatus. On the other hand, some believe that, in stress, elevated levels of glucocorticoids serve to suppress the body's normal defenses against stress, preventing them from overshooting and causing damage to the organism. A more detailed discussion of the immune system and the effects of various factors on its function is found in Chapter 17 on thymus hormones. Thus, the negative effects of glucocortiocids are to be weighed against their positive effects of anti-inflammation, etc. This could also be a mechanism whereby individuals who cannot adapt to stressful changes are selected out of the population. In fact, little is known about the overall effects of stress adaptation and the role of the glucocorticoid hormone for the

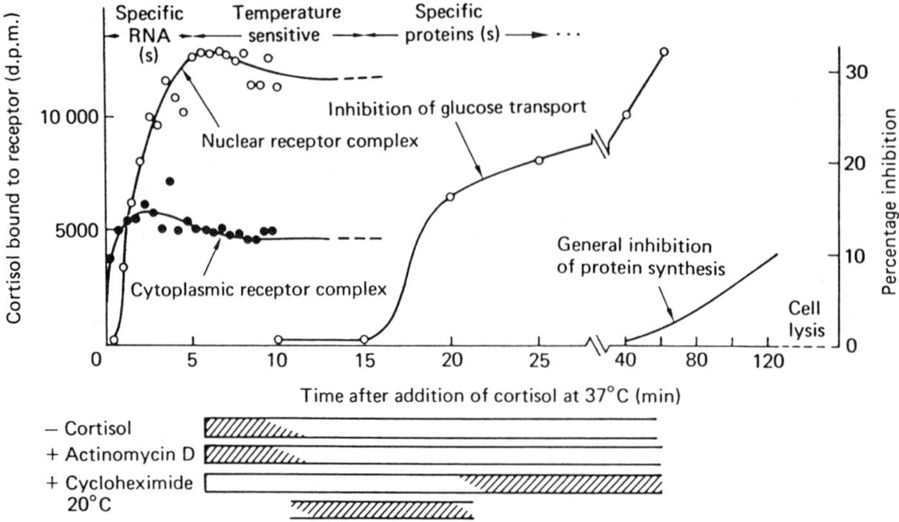

FIGURE 10-11 Time course in rat thymus cell suspensions at 37°C of cortisol–receptor complex formation, cortisol-induced inhibition of glucose transport, and inhibition of protein synthesis. Kinetics of receptor complex formation was determined with [^{3}H]cortisol at about 0.1 μM. Inhibitory metabolic effects were produced with about 1 μM cortisol. Glucose transport was measured with 2-min pulses of radioactive hexose initiated at the times indicated. Shaded segments of the horizontal bars in the lower part of the figure roughly indicate the time intervals during which emergence of the cortisol effect on glucose metabolism can be blocked by treatment with cortexolone (which displaces cortisol from the glucocorticoid receptors), actinomycin D, and cycloheximide and can be delayed by lowering the temperature. Open bars indicate periods during which these treatments have no effect. At the top of the figure is the sequence of steps by which it is hypothesized that the cortisol–receptor complex leads to the synthesis of a specific protein that inhibits glucose transport. Reproduced from A. Munck and G. R. Crabtree, Glucocorticoid-induced lymphocyte death. Bower, I. D., and Lockshin, R. A. (eds.), (1981). In "Cell Death in Biology and Pathology, pp. 329–357. Chapman and Hall, London and New York.

whole organism. The effects of this hormone are extremely broad, but clearly, it is key to survival and humans cannot live without it.

The mechanism involved in cell death produced by glucocorticoids is intriguing. There have been many advances in understanding cell death by apoptosis, a genetic mechanism that is set in motion by cells susceptible to glucocorticoids, in this case including T and B cells. Other work on the apoptotic mechanism in nematodes has identified critical genes, *ced*-3, *ced*-4, and *ced*-9, in this process. *ced*-3 encodes a protease that is homologous to the mammalian interleukin-1β-converting enzyme (ICE), which catalyzes the conversion of the interleukin-1β precursor into the active inflammatory cytokine. There are other ICE homologs in mammalian cells, such as ICH (ICE homolog), *nedd*-3 (mouse homolog), CPP32 (cysteine protease, 32,000 Da, and *mch* (mammalian *ced*-3 homologue) to name a few. The three-dimensional structure of ICE has been solved (Figure 10-12). This cysteine protease, which cleaves an aspartyl–X bond, is synthesized in the cell as a proprotein containing two subunits, a p20 and a p10, that are released in free form through the cleavage of aspartyl–X bonds on either end of each subunit. Initital inactive precursors are either activated by a distinct protease, as yet unknown, or autocatalyzed by a small number of endogenously active ICE-like molecules. Because some cells contain respectable amounts of endogenous ICE, autoactivation seems less likely, so that its activity may be controlled by another process, possibly by a highly regulated protease. In fact, ICEε, a splice variant giving rise to only a p10-like protein, can combine with a p20 subunit to form an inactive p30 enzyme, or it could prevent dimerization $[(p20)_2–(P10)_2]$. So far, it is unlikely that ICE is on the pathway of glucocorticoid-induced apoptosis, but it seems likely that one of the other homologs could be a key member of that pathway. A speculative scheme of glucocorticoid-induced apoptosis in susceptible T or B cells can be contemplated (Figure 10-13).

That the glucocorticoid receptor is a key element in this process is emphasized by the demonstration that a negative regulator of the receptor, called the "modulator," which maintains the receptor in its nonactive oligomeric form, is capable of preventing apoptosis of leukemia cells in the presence of glucocorticoids.

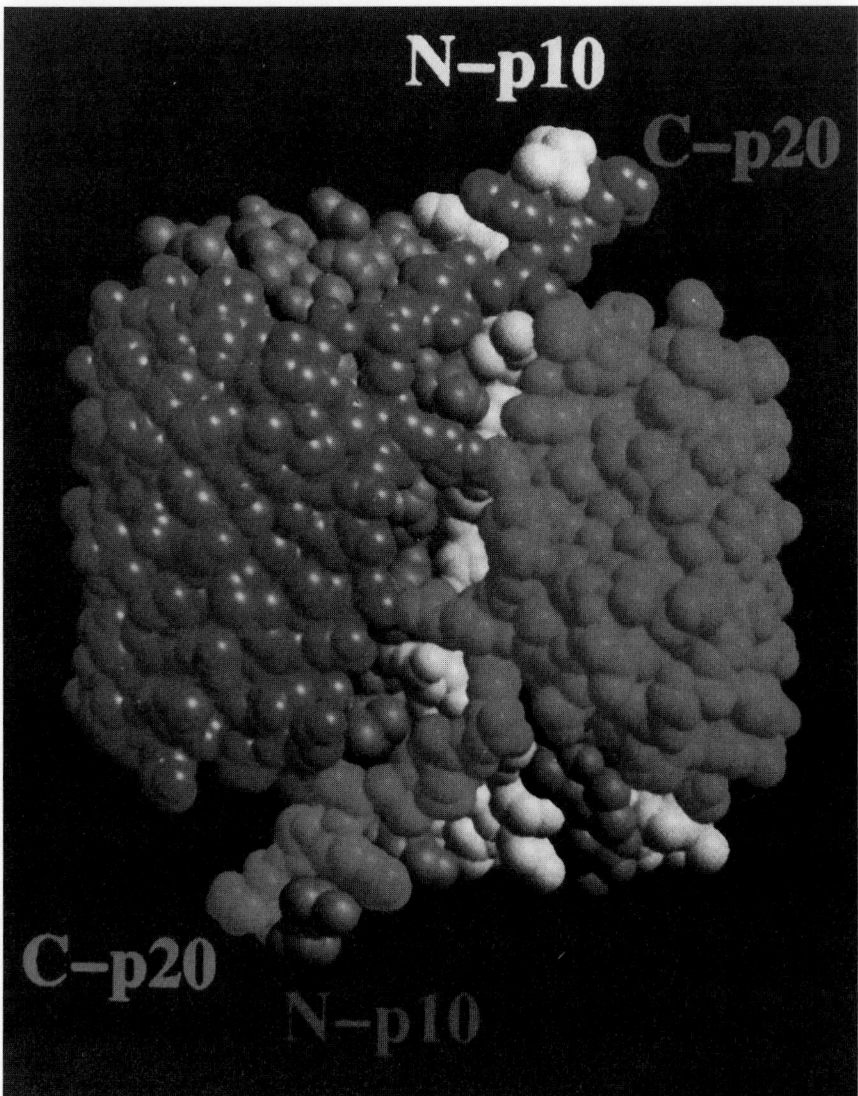

FIGURE 10-12 CPK model of the $(p20)_2$–$(p10)_2$ tetramer of ICE. Two p20 subunits surround two adjacent p10 subunits. The tetrapeptide aldehyde inhibitor is represented by the lighter atoms. The crystallographic 2-fold axis is roughly perpendicular to the plane of drawing and runs through the small hole at the center of the interface between the two p10 subunits. β-Strands 5 and 6 protrude from the top and, by 2-fold symmetry, the bottom of the model. The N- and C-terminal ends of each subunit are labeled. Dynamics simulations suggest that the formation of a tetramer would stabilize the active site of ICE by reducing the mobility of the catalytic Cys285 and other residues near the p20 C-terminus. This is an adaptation of a colored photograph. Reproduced with permission from Wilson, K. P., Black, J. A. F., Thomson, J. A., Kim, E. E., Griffith, J. P., Navia, M. D., Murcko, M. A., Chambers, S. P., Aldape, R. A., Raybuck, S. A., and Livingston, D. J. (1994). "Structure and mechanism of interleukin-1β converting enzyme". *Nature* **370**;270–275.

2. *Actions of Glucocorticoids on Cells*

The cellular mode of action of glucocorticoid hormone is summarized in Figure 10-14, where the target gland is the liver cell, representing the predominant tissue whose overall response is anabolic. Later, the actions of the hormone on the thymus cell, which responds to the hormone catabolically, will be described.

In this model (Figure 10-14), the free hormone is shown permeating the liver cell membrane, as previously discussed. Once inside the cytoplasm of the cell, it can combine with the specific, high-affinity receptor to form a steroid–receptor complex. This com-

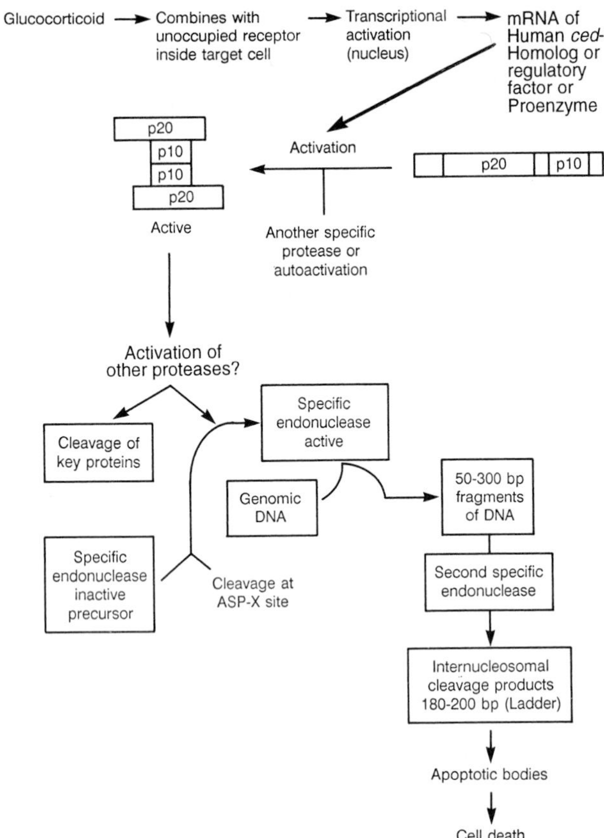

FIGURE 10-13 Speculative scheme of the induction of apoptosis in susceptible T or B cells by glucocorticoids that includes a key role for an ICE/*ced*3-like protease (*mch*). There is as yet no direct evidence that an ICE/*ced*3-like protein is on the glucocorticoid pathway. If not, it means that there may be more than one principal path to apoptotic cell death. Glucocorticoid enters the cell and binds to the cyctoplasmic unoccupied receptor. The receptor complex is activated in the cytoplasm by disaggregation of the oligomeric non-DNA-binding form, releasing the monomeric active receptor. The receptor dimerizes and is translocated to the nucleus through the nucleopore. Once inside the nucleus, it is presumed to bind to a hormone-responsive element upstream of a gene encoding a specific ICE/*ced*-like enzyme (*mch*). There is no evidence for this point. This would result in the transcription of mRNA for *mch* or some regulatory factor, levating its active concentration in the cytoplasm after translation. This *mch* in a proprotein form would be cleaved to p20 and p10 subunits, which could combine into a p20–p10 dimer and become active. It is possible that the dimer could dimerize into a tetramer $(p20)_2$–$(p10)_2$. There is no direct evidence of what enzyme splits out p20 and p10 subunits from the proenzyme form. It could be accomplished by the *mch* itself if there were a small amount of the active form available, or there may be another Asp-X protease that accomplishes the activation. Once activated the *mch* could bind and cleave a substrate, which could be a specific pronuclease (no data on this point), as well as other important proteins in the cell. The nuclease would be activated by this protease cleavage, and it could translocate to the nucleus (again no data here) and attack genomic DNA, generating fragments of 50–300 base pairs. The larger pieces might serve as substrates for another nuclease to generate internucleosome-sized DNA (180–200 base pairs). Nuclear condensation would occur and apoptotic bodies would form, followed by cell death. The dead apoptotic cell would be quickly recognized by macrophages and removed so that an inflammatory process would not be generated.

plex is the non-DNA-binding oligomer of the glucocorticoid receptor, in which the receptor is complexed with other proteins in a manner in which the zinc finger DNA-binding region of the receptor is occluded, presumably by hsp 90 dimer. In this form, when the receptor is associated with these other proteins, it is in a conformation favorable for productive binding by hormone, as compared to the activated or free monomeric form. When steroidal ligand enters the cell, presumably by free diffusion, it binds to the unoccupied receptor with high affinity ($K_d \cong nM$). This complex can then be activated or transformed (cytoplasmic activation) by dissociation of the oligomeric complex to yield the free receptor subunit in the DNA-binding form. This activated receptor forms a homodimer and is translocated to the nucleus through the nucleopore with the participation of the nucleoporins (see Chapter 1), possibly involving hsp 70. Inside the nucleus the receptor can "search" DNA for the hormone-responsive element, to which it binds with high affinity. Binding to the major groove by the upstream zinc fingers strengthens the homodimer interaction that involves the downstream Zn^{2+} fingers and other sites on the molecule, in the steroid-binding domain, as well. Other transactivating proteins bind to the pretranscriptional complex until the complex can support the binding and function of RNA polymerase II (see the following discussion on transcription).

RNA polymerase II initiates the transcriptional process, which results in the generation of mRNA encoded by the hormone-sensitive gene. This mRNA is transported to the cytoplasm for translation into protein. Genes activated in this manner generate proteins in the cell whose actions culminate in the cellular response to the hormone. Not all receptor–DNA interactions lead to gene activation, and some can result in suppression. In this case, the negative hormone-responsive element is slightly altered in the DNA sequence, and suppression may take the form of the receptor interacting with the negative responsive element and competing with a required transcription factor (for activation) whose responsive site on DNA overlaps the negative hormone-responsive element. Other explanations are also possible, and less is known about negative regulation than about positive regulation.

The glucocorticoid receptor has been purified from rat liver, and polyclonal as well as monoclonal antibodies have been prepared to it. The full-length receptor and truncated forms have been overexpressed in the baculovirus system. In rat liver the properties of the activated receptor include the following. It appears to be a single polypeptide chain of molecular weight about 94,000. It has one steroid-binding site per molecule and a DNA-binding site. The DNA-binding site

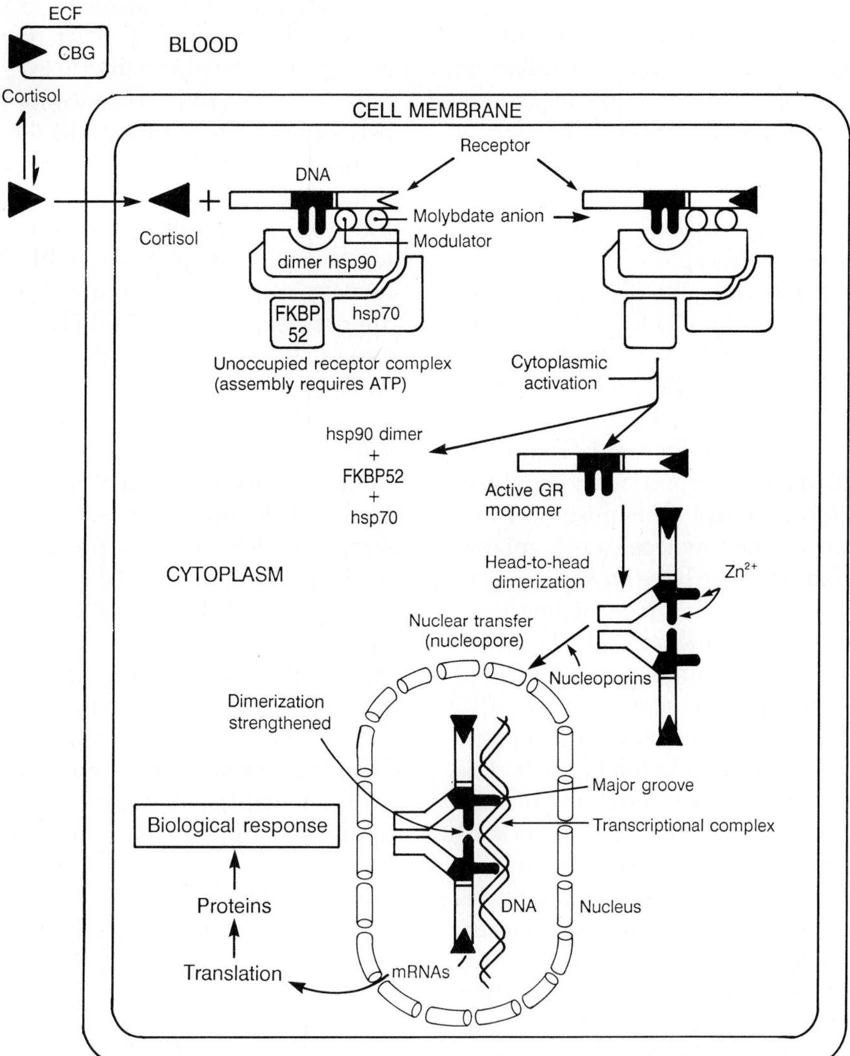

FIGURE 10-14 Action of the glucocorticoid receptor on a target cell.

is unavailable at the surface of the unactivated form of the larger receptor complex (Figure 10-14). The properties listed here pertain to the activation or DNA-binding (or nuclear binding) form of the steroid–receptor complex.

A large number of *in vitro* experiments describe the conditions under which the activation–transformation process occurs (see Figure 10-14). The steroid–receptor complex undergoes a further conformational change that enables it to translocate to the cell nucleus, where it brings about an increase in the rate of synthesis of certain mRNAs coding for the translation of specific enzymes in the cytoplasm. The activated–transformed state of the steroid–receptor complex can be determined by its ability to bind to DNA. Another view of activation, which is not incompatible with the conformational change idea, is that a low-molecular-weight inhibitor is bound to the receptor in the unactivated form, thereby holding receptor subunits together in a way that obscures its DNA-binding sites (Figure 10-14). Upon activation, this inhibitor dissociates from the surface of the receptor, exposing the Zn^{2+} fingers of the DNA-binding domain after other proteins have dissociated. Considerable work has now been accomplished on the mode of entry of activated–transformed cytoplasmic receptors into the nucleus. The receptors apparently are phosphorylated and associate with specific proteins (nucleoporins) that ferry the receptors through the nucleopore. In addition, hsp 70 may bind to this transportation complex and provide the energy required for transport via its ATPase activity.

The receptor for nuclear localization signals is a sequence of seven or eight amino acids that are presumed to be the binding site for nucleoporins or other specific proteins that affect attachment to the nucleopore and subsequent transport. Nucleoporin proteins are found in both the cytoplasmic and nucleoplasmic compartments, suggesting their carrier role in the process of translocation. A nuclear translocation signal was first demonstrated for the SV40 T antigen and has the sequence 126-PKKKRKV, indicating the importance of basic amino acids. Nuclear localization sequences for various steroid receptors are as follows: human glucocorticoid receptor, 491-RKTKKKIK; human mineralocorticoid receptor, 673-RKSKKLGK; human androgen receptor, 628-RKLKKLGN; human progesterone receptor, 637-RKFKKFNK; human estrogen receptor, 256-RKDRRGGR; human vitamin D_3 receptor, 102-RKREMILL; and human thyroid hormone receptor, 179-KRLAKRKL. The second amino acid, lysine, is required and is present in this position in all receptors cited. It is likely that molecules recognizing the nuclear localization sequence have an acidic amino acid sequence. In support of this idea, it has been shown that antibodies to the sequence DDDED, which could be considered the acidic counterpart of the SV40 T antigen nuclear localization sequence KKKRK, blocked all of the nuclear-targeted protein receptors. It is likely that a cytoplasmic receptor binds to this sequence and assists in transport through the nucleopore. It is also possible that the "receptor" is dissociated from the protein being transported to the nucleoplasmic side, generating concentrations of molecules on both sides of the nucleopore. In addition, as mentioned before, the transport process may involve the energy of ATP, and possibly hsp 70, a known ATPase, may be involved in the transport process. Once in the nucleoplasm, the receptor is released from its carriers and is able to interact with chromatin. The nuclear localization sequence (NLS) receptors may involve at least two cytoplasmic proteins of about 60,000 and 76,000 molecular weight. It has even been hypothesized that the hsp 56, which has been shown to be an immunophilin, could be the NLS-binding protein for the cytoplasmic glucocorticoid receptor. Also, this molecular weight is close to the 60,000 mentioned earlier. This protein has an acidic sequence (rabbit) 140-EDLTDDED-147 that conceivably could interact with the rat glucocorticoid receptor NLS, 510-RKTKKKIK-517. These proteins bind directly to the NLS sequence of the steroid receptor and probably act as transporting proteins to and through the nucleopore.

In order to appreciate the structure of the nucleopore, a schematic representation is shown in Chapter 1. Its dimensions are about 80 nm in length and 100 nm in diameter. The protein molecular weight of the complex is estimated at 125 million by scanning electron microscopy. The aperture in the center can change from 10 nm in the closed state to 40 nm in the open state. There is evidence for fibrillar connections between a pore complex and the nucleoskeleton, the cytoskeleton, and other pore complexes. Presumably, solid-state transport could occur through this structure, which is involved in the nuclear import of proteins and export of RNAs with the assistance of the nucleoporins. The locations of nucleoporin residence are not clear, but they could be attached to the spokes. The nucleoporins are a family of proteins that contain carbohydrate structures and are sensitive to the action of wheat germ agglutinin. The nucleoporins could be different from the nuclear localization signal receptors in the cytoplasm, but they could play a structural role in the nucleopore-transporting process. Figure 10-15 shows models of the structural organization of the nucleoporin family. Once in the nucleus, the exposed Zn^{2+} fingers of the activated receptor complex permit interaction with the major groove of DNA. A second site on the steroid–receptor complex in the steroid-binding domain assists in the specific binding to DNA.

With the use of cell culture, several types of cells have been isolated: cells with normal function of the receptor that respond to glucocorticoid and cells that apparently have a defective receptor. Other types are

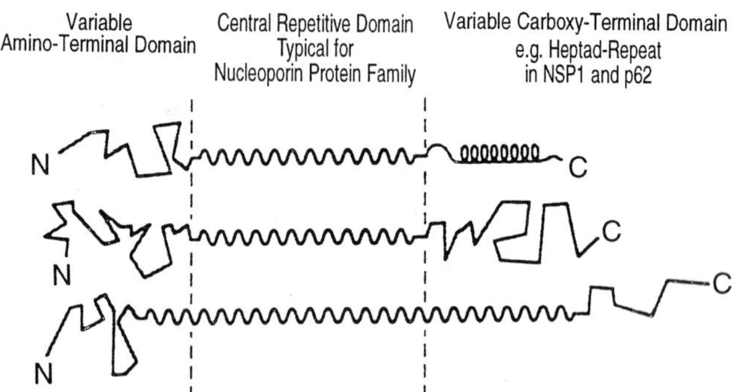

FIGURE 10-15 Model of the structural organization of the nucleoporin protein family. The common structural motif diagnostic for nucleoporins is suggested to be a central repetitive domain consisting of many degenerate repeat sequences (which can vary in length) able to fold into short amphipathic β-sheet structures (wavy line). Apart from this common motif, various members of the nucleoporin protein family are not homologous in the flanking carboxy- and amino-termini, which are involved in their specific function (e.g., NSP1 and NUP1). In the case of yeast NSP1 and human p62, which appear to be functional homologs, the carboxy-terminal domains show a heptad repeat organization and, thus, may form coiled-coil structures (α-helical line). Reproduced with permission from Carmo-Fonseca, M., and Hurt, E. C. (1991). Across the nuclear pores with the help of nucleoporins. *Chromosoma* **101**, 199–205.

cells that have a "receptor" that does not translocate to the nucleus [it probably has an altered DNA-binding site (see the following)] and cells that have a receptor and translocate to the nucleus, but are unresponsive to the hormone. In this last case, which resembles the condition of the late rat fetus, the receptor complex can bind to its acceptor on chromatin, but there is a block in the steps of the receptor mechanism beyond the interaction of receptor and acceptor. The glucocorticoid receptor has a developmental program of its own, and its expression in cells corresponds to the time during development when glucocorticoid responsiveness first appears. This also intersects with the availability of the steroidal ligand secreted from the developing adrenal gland. A molecular analysis of these mutant cells, which have blocks at various points in the receptor mechanism, will help to explain receptor functions. In addition to the types described here, there are also mutations that lead to the rapid degradation of receptor once it has been activated.

The liver enzymes synthesized in response to glucocorticoid action are involved in the metabolism of amino acids or in the synthesis of glycogen and to a limited extent with the synthesis of glucose (Figure 10-16). Many of these enzymes are listed in Table 10-1.

B. Glucocorticoids, Analogs, and Antiglucocorticoid Compounds

So far, we have discussed the actions of the major natural glucocorticoids; cortisol in humans and corticosterone in the rat. Other compounds have been synthesized in the laboratory, and some of these have greater potency than the natural hormones. Steroids may be divided into classes, depending on (1) ability to compete for binding to the ligand-binding site of the glucocorticoid receptor and act like glucocorticoids (permit binding to DNA or the cell nucleus compared to the native hormone); (2) ability to bind to the ligand-binding site and permit binding of the ligand–receptor complex to DNA, but to a considerably smaller extent than the natural glucocorticoids; and (3) ability to bind to the glucocorticoid receptor and prevent its binding to DNA in the cell nucleus under activating conditions. Steroids have been classified according to their potency in inducing specific enzymes in cell culture. The enzy-

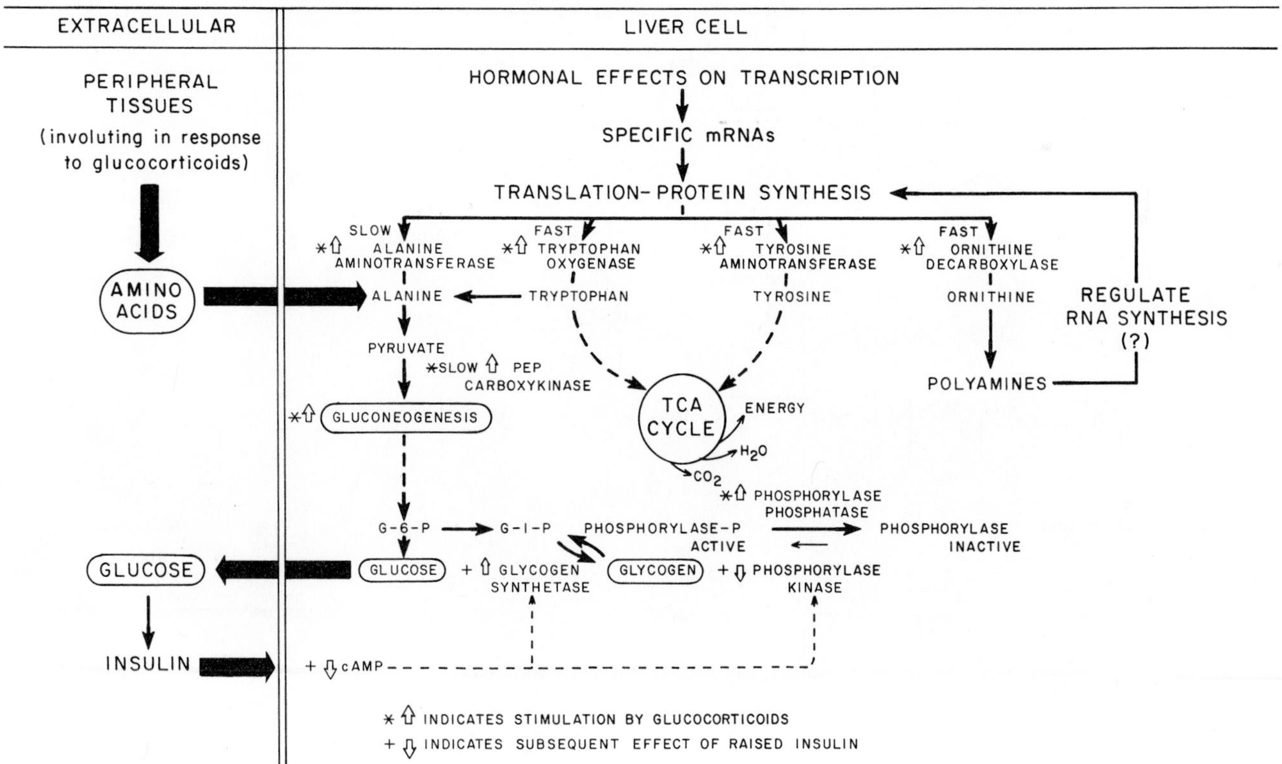

FIGURE 10-16 Overview of the actions of glucocorticoids on metabolism and on specific enzyme levels in the rat hepatocyte. Reproduced from Cake, M. H., and Litwack, G. (1975). *In* "Biochemical Actions of Hormones" (G. Litwack, ed.), Vol 3, pp. 317–190. Academic Press, New York.

TABLE 10-3 Potency of Natural and Synthetic Glucocorticoids

Steroid	Structure	Binding to ligand-binding site of glucocorticoid receptor (affinity)	Ability to induce TAT in hepatoma cell culture	Classification (N, natural steroid; S, synthetic steroid)
Cortisol	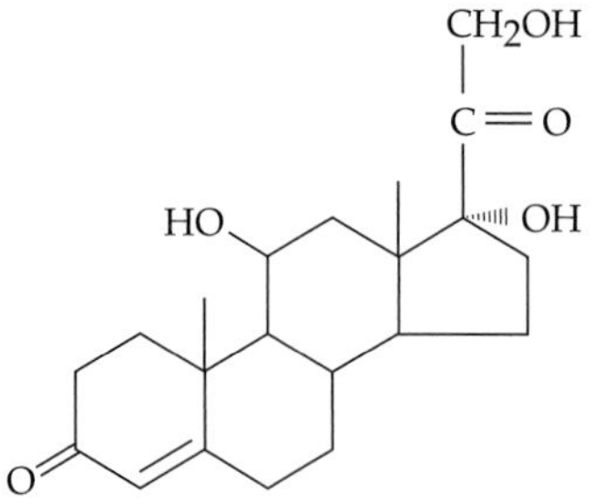 CH_2OH, C=O, HO, OH, O	Strong binding	++++	(N) Optimal inducer
Corticosterone	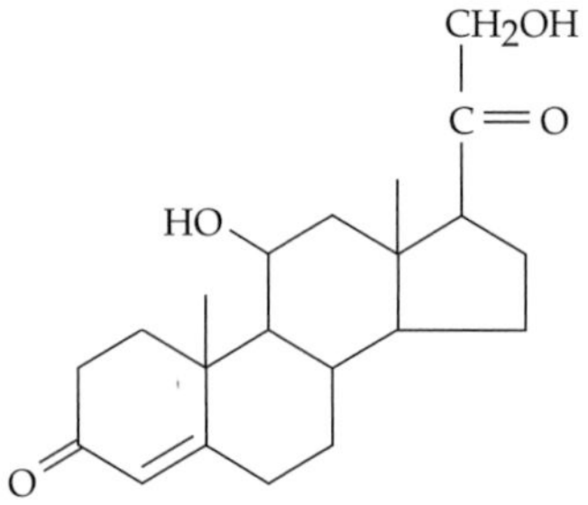 CH_2OH, C=O, HO, O	Strong binding	++++	(N) Optimal inducer
Dexamethasone	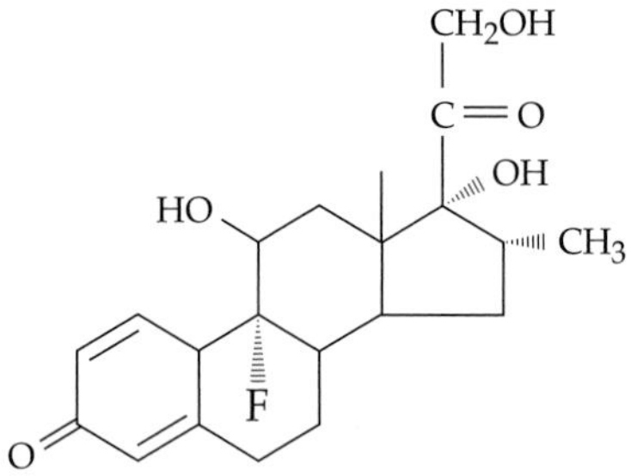 CH_2OH, C=O, OH, HO, CH_3, F, O	Strong binding	++++	(S) Optimal inducer
Triamcinolone acetonide	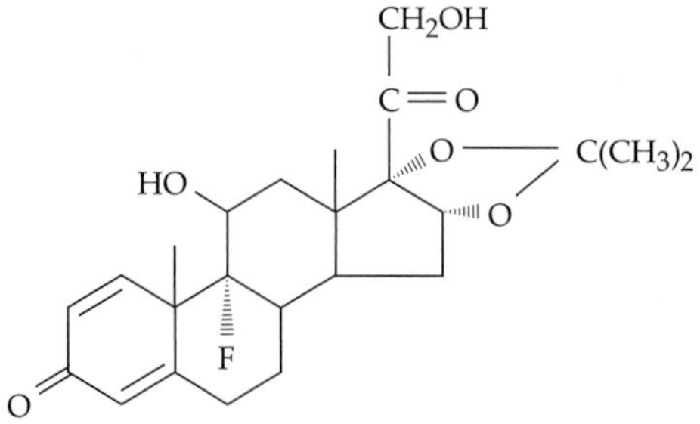 CH_2OH, C=O, O, $C(CH_3)_2$, HO, O, F, O	Strong binding	++++	(S) Optimal inducer
11β-Hydroxyprogesterone	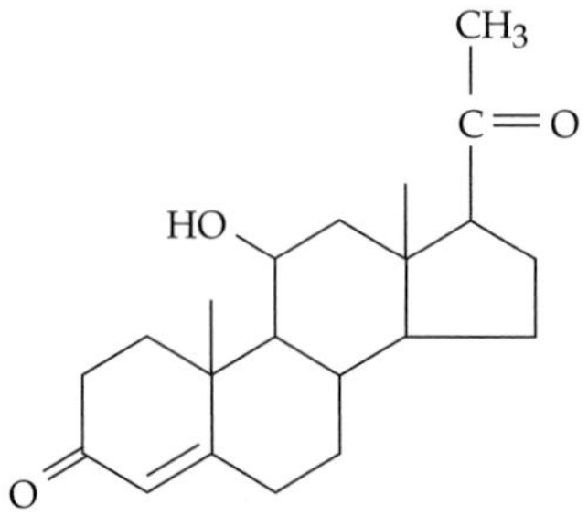 CH_3, C=O, HO, O	Strong binding	++	(N) Suboptimal inducer

(continues)

TABLE 10-3 — *Continued*

Steroid	Structure	Binding to ligand-binding site of glucocorticoid receptor (affinity)	Ability to induce TAT in hepatoma cell culture	Classification (N, natural steroid; S, synthetic steroid)
5α-Dihydrocortisol	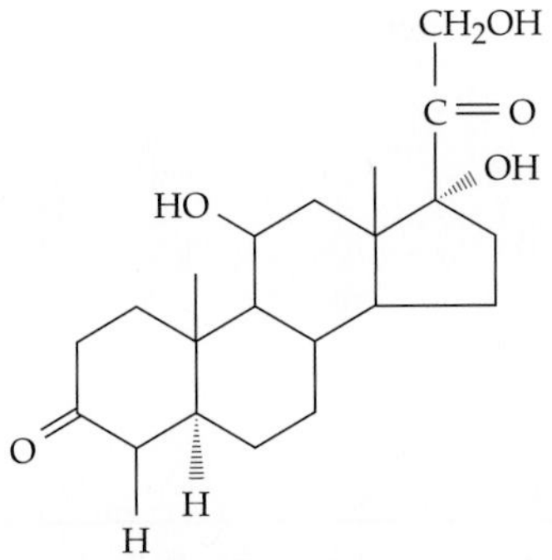 CH_2OH, C=O, OH, HO, O, H, H	Strong binding	++	(N) Suboptimal inducer
Dexamethasone mesylate	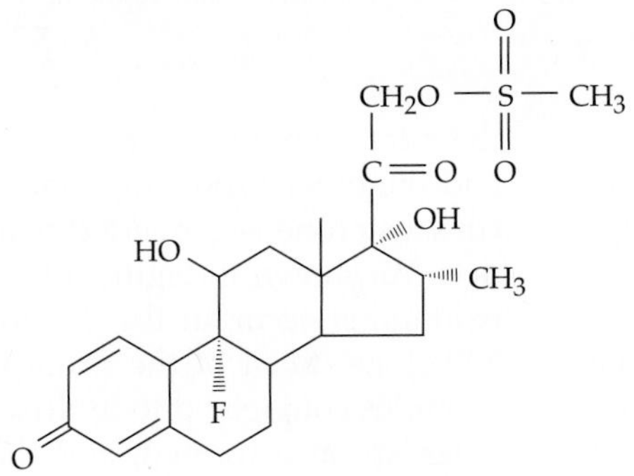 O, CH_2O—S—CH_3, O, C=O, OH, HO, CH_3, F, O	Moderately strong binding	++	(S) Optimal (?) inducer
Deacylcortivasol	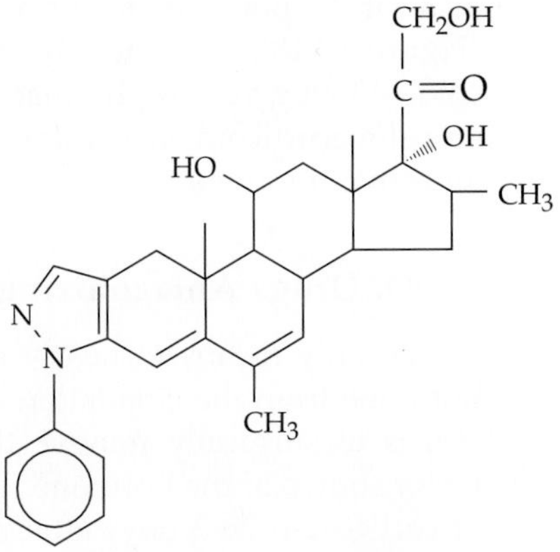 CH_2OH, C=O, OH, HO, CH_3, N, N, CH_3	Very strong binding	++++	(S) Optimal inducer
Progesterone	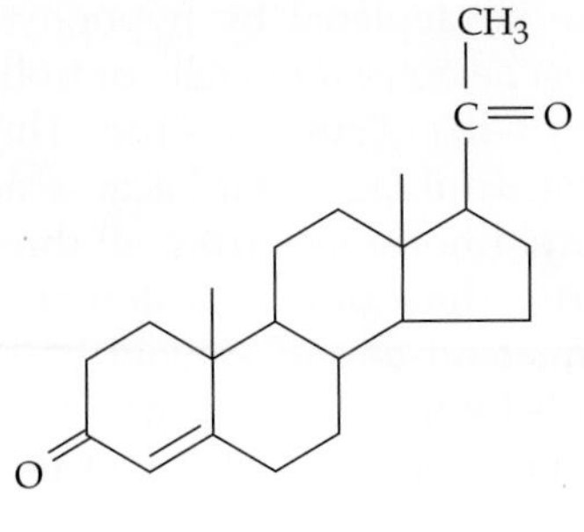 CH_3, C=O, O	Weak binding	0	(N) Noninducer or anti-inducer[a]

(*continues*)

TABLE 10-3 — *Continued*

Steroid	Structure	Binding to ligand-binding site of glucocorticoid receptor (affinity)	Ability to induce TAT in hepatoma cell culture	Classification (N, natural steroid; S, synthetic steroid)
17α-Hydroxyprogesterone	CH_3, C=O, OH	Strong binding	0/+	(N) Anti-inducer[a]

[a] An anti-inducer binds to receptor, but will not induce TAT. Therefore, it does not translocate to the nucleus, or if it does it cannot act productively in the nucleus.

matic activity that has been used is that of tyrosine aminotransferase (TAT) in hepatoma cells in culture. These cells are immortal in culture and retain the property of TAT induction by glucocorticoids after transformation. The three groups of agents mentioned reflect glucocorticoid potency in the induction of this enzyme *in vitro*. The groups of compounds cluster into those described for ligand binding to the glucocorticoid receptor and ability to induce TAT. They are called (1) agonists, (2) weak agonists, and (3) antagonists (Table 10-3). A new series of very high-affinity glucocorticoid agonists–antagonists has been developed by Roussel-UCLAF in France. The structures of these compounds tell us something about the structures of ligands and the ligand-binding domain of the receptor. Some of these structures are shown in Figure 10-17. For high affinity to the glucocorticoid receptor, the C-11 position needs to be substituted. Substituents around the D ring also increase affinity, but the key seems to be the substitution of the C-11 and also the pattern of double bonds in the A and B rings. It is not essential to have a C-19 methyl if there is a bulky substituent at C-11.

C. Catabolism of Glucocorticoids

The major sites of inactivation of glucocorticoids are the liver and to some extent the kidney. Enzymes are present that reduce the double bond at C-4,5 of the unsaturated 3-ketosteroids. The product contains either a 5α- or a 5β-proton, and reduced steroids generally are much less active. The principal routes of metabolism of cortisol and corticosterone are shown in Figure 10-18A. These are NADPH requiring enzymes, one in the microsomal fraction that produces the 5α-product and the other producing mainly the 5β-isomer, while corticosterone is converted to a mixture of both products. As shown in Figure 10-18, further reductions or oxidations occur in the 3-ketone, 11β-hydroxyl, and 20-ketone. Most of the urinary metabolites are glucuronides conjugated to hydroxyl groups. Sulfate derivates are also formed, and these often occur in bile as 3- or 21-monosulfates or 3,21-disulfates. Cortisone (Figure 10-18B) is relatively inactive as a glucocorticoid. When given in pharmacological amounts, it exerts glucocorticoid effects after its reduction to cortisol, an active hormone.

D. Drugs Antagonizing Glucocorticoids

The only means to nearly completely remove the hormone from the circulation of an experimental animal is to surgically remove the adrenal glands, the major source of the hormone. The rat is most depleted of corticosterone 3 days after adrenalectomy and can also be depleted by hypophysectomy. Such animals must be housed in well-controlled environments, since they will not survive stress. They must be given salt in their drinking water because adrenalectomy removes steroid hormones from all three layers of the adrenal cortex, including aldosterone. There is no means to render an animal devoid of endogenous glucocorticoids by the use of drugs. Two agents are worth mentioning because they lower the amount of glucocorticoids in the circulation. Drugs of this type are metyrapone and aminoglutethimide, shown in Figure 10-19. Metyrapone inhibits 11β-hydroxylation in the adrenal cortex, leading to decreased cortisol produc-

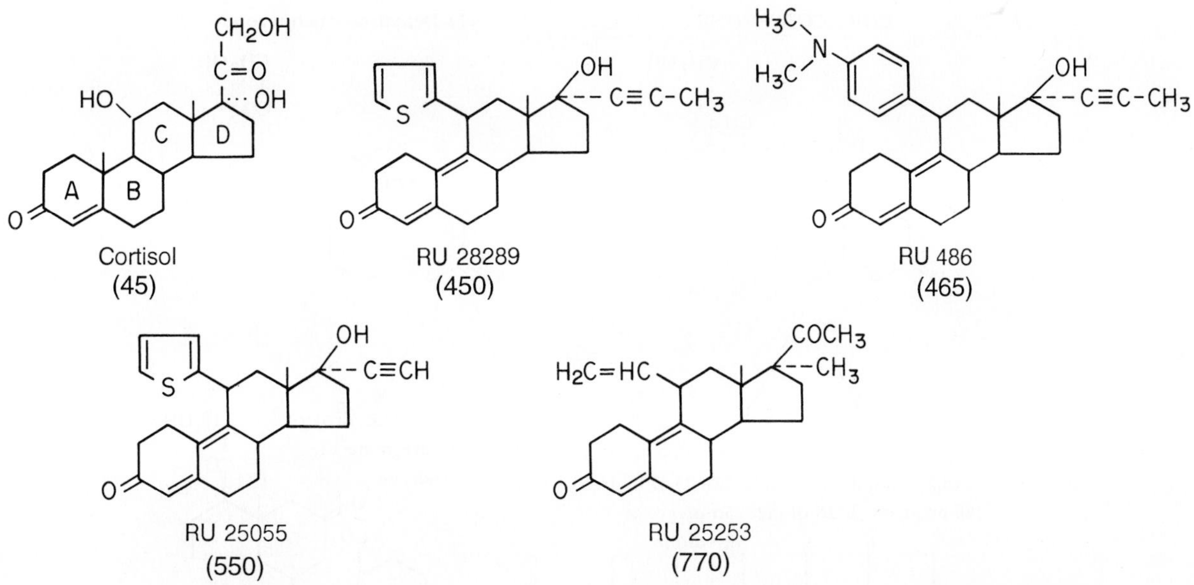

Cortisol (45)

RU 28289 (450)

RU 486 (465)

RU 25055 (550)

RU 25253 (770)

FIGURE 10-17 Structures of agonists–antagonists of the glucocorticoid receptor developed by Roussel-UCLAF. Cortisol is presented for comparison. The number below the structure in parentheses indicates the relative affinity for the glucocorticoid receptor. Thus, RU28289 has about 10 times the affinity of cortisol. These data were derived from Ojasoo, T., Raynaud, J. P., and Doré, J. C (1994). Affiliations among steroid receptors as revealed by multivariate analysis of steroid binding data. *J. Steroid Biochem. Molec. Biol.* **48**, 31–6.

tion, and it may inhibit the side chain cleavage reaction. The 11β-hydroxylase is a mitochondrial enzyme and would be expected to cause the accumulation of inactive precursors of cortisol, such as 11-deoxycorticosterone or 11-deoxycortisol. As a consequence of decreased cortisol, the long-loop negative feedback by cortisol on the anterior pituitary, hypothalamus, and limbic system (Figure 10-7) is much reduced, with the result that there is increased output of ACTH by the corticotrophic cells of the anterior pituitary. 11-Deoxycortisol can bind weakly to the glucocorticoid receptor and has minimal glucocorticoid activity. Measurement of elevated plasma concentrations of ACTH, usually by radioimmunoassay, is diagnostic of hypofunction of the adrenal cortex (Addison's disease). Not only is ACTH the key hormone of the stress humoral pathway, but it is also a growth factor for cells of the adrenal cortex. Consequently, when glucocorticoid therapy is used, the hypothalamic–pituitary–adrenal axis is interfered with due to the function of the negative feedback mechanism (Figure 10-7). Such therapy is usually given on alternate days to allow for recovery of the axis.

E. Anti-inflammatory Actions

Experiments, especially with cell culture systems and lung tissue, demonstrate that cortisol and other glucocorticoids inhibit the appearance of prostaglandins. The glucocorticoids, via a receptor mechanism, appear to cause the synthesis of a protein called annexin I (lipocortin) of about 40,000 molecular weight. This polypeptide is an inhibitor of cell membrane phospholipase A_2 that is responsible for the release of fatty acid precursors of prostaglandin synthetase, such as arachidonic acid (Figure 10-20; refer also to Chapter 16). This mechanism may represent part of the anti-inflammatory action of the glucocorticoids. The gene for the 40,000-molecular-weight lipocortin has been cloned, and glucocorticoids also inhibit inducible cyclooxygenase, which is responsible in large part for the formation of prostaglandins and relatives such as leukotrienes, which are potent inflammatory agents. Two forms of cyclooxygenase exist: COX 1 and COX 2. COX 1 is a constitutive enzyme that probably functions in vascular responses and generates required levels of circulating hormones under noninflammatory conditions. COX 2 is an inducible enzyme in inflammatory cells and becomes induced as part of the inflammatory response. COX 2 is located in the endoplasmic reticulum and catalyzes the oxidation of arachidonic acid to the peroxy form (cyclic endoperoxide). From this intermediate (PGG2), a variety of prostaglandins, thromboxanes, and prostacyclin can be synthesized. The first two groups are generally inflammatory, so that suppression of COX 2 synthesis by glucocorticoids accounts for a major part of its anti-inflammatory

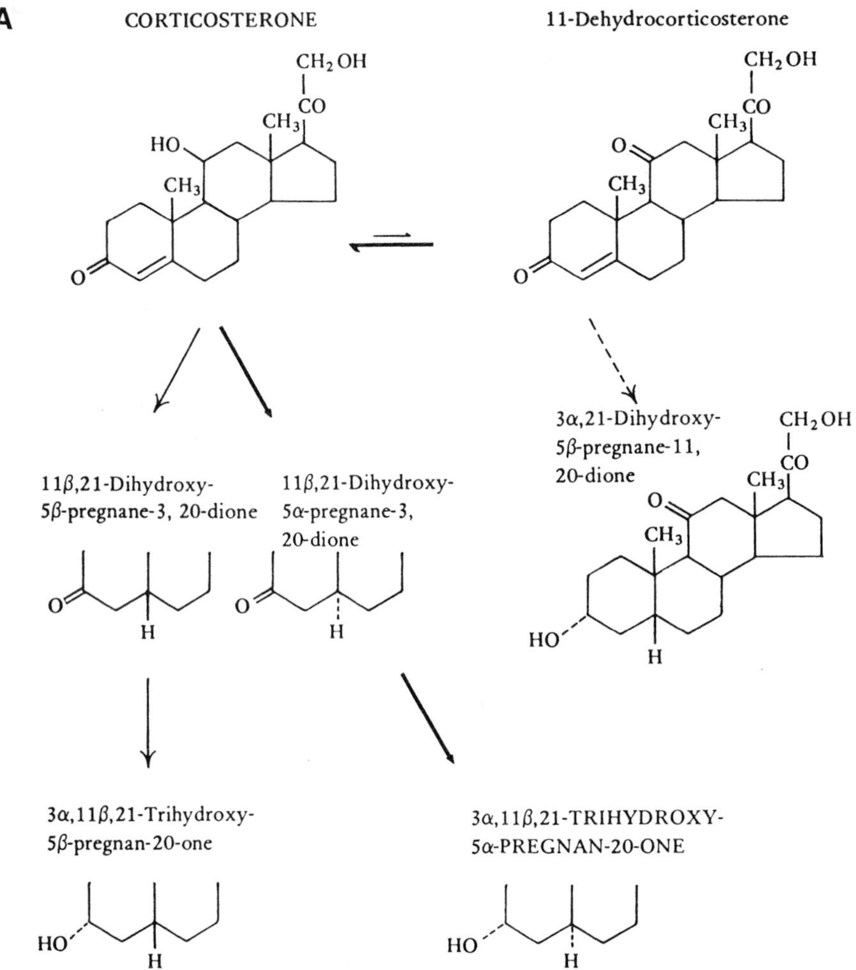

FIGURE 10-18 Metabolism of corticosterone (A) and cortisol (B). Reproduced from Schulster, D., Burstein, S., and Cooke, B. A., (1976). "Molecular Endocrinology of the Steroid Hormones," pp. 156–157. Wiley, New York, by copyright permission.

action (see Chapter 16 for further information). The glucocorticoids have also been shown to exert anti-inflammatory action at the level of the IL-6 promoter. Since IL-6 is an inflammatory cytokine, its inhibition would reduce the inflammatory process (Figure 10-21).

V. GLUCOCORTICOID RECEPTOR

The glucocorticoid receptor is a member of the nuclear steroid–thyroid receptor gene gamily (Chapter 1). Like the other receptors in this superfamily, the glucocorticoid receptor can be divided into functional domains (Figure 1-26). The major functional regions consist of the ligand-binding domain (C-terminal), the DNA-binding domain (central), and the antigenic domain (N-terminal). Bars in the figure locate additional functions, including transactivation and dimerization. The glucocorticoid receptor superfamily consists of the glucocorticoid receptor, the progesterone receptor, the mineralocorticoid receptor, and the androgen receptor. These receptors can bind to the palindromic hormone response element shown in Table 1-10, which differs from the response elements for the estradiol receptor–thyroid hormone receptor superfamily. These elements differ by only a few bases. In the case of the thyroid-responsive element, the sequence is the same as that for the estradiol receptor, except that three nucleotides do not separate the elements. With the vitamin D_3 receptor the responsive elements are in tandem, requiring that the DNA-binding domains of the homodimeric receptor are in a head-to-tail orientation rather than a head-to-head, which would be the case with the glucocorticoid receptor (see Figure 10-22). The homodimeric structure of the glucocorticoid receptor, for example, is stabilized by DNA binding in which several amino acid residues of the second Zn^{2+} finger interact with DNA, while other residues interact with the second

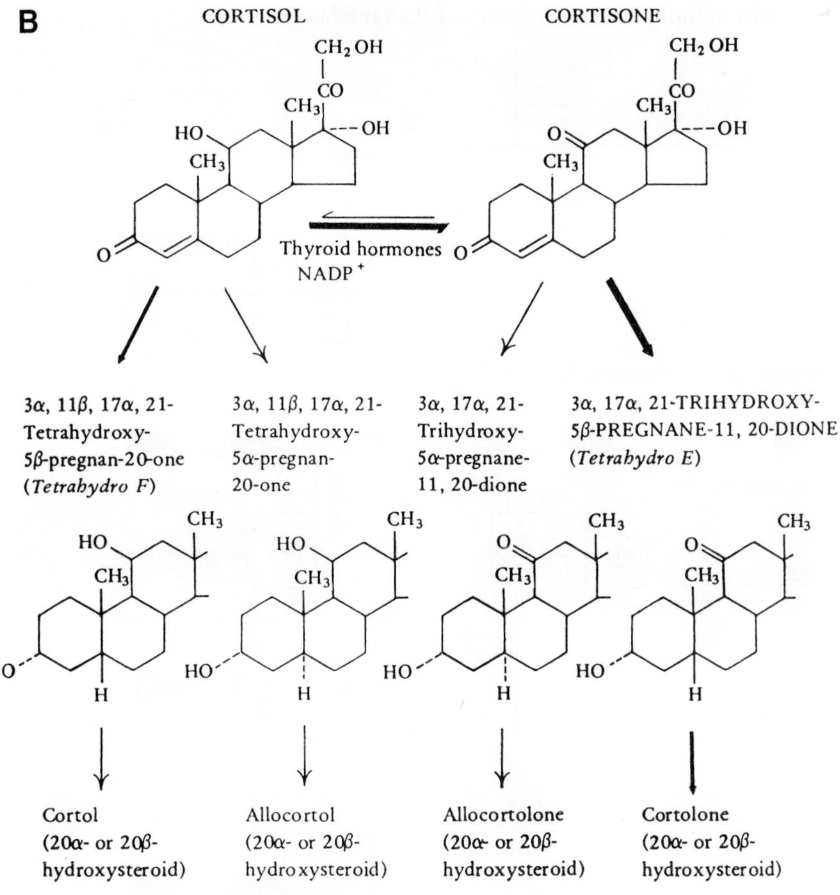

FIGURE 10-18 *(Continued)*

zinc finger of the the other receptor partner to produce this stabilized configuration. The interaction of the glucocorticoid homodimer with the double standard hormone-response element, based on X-ray diffraction data, is shown in Chapter 1.

The DNA-binding domain of the glucocorticoid receptor is shown in Figure 10-23. It consists of two Zn^{2+} fingers. The left (N-terminal) is the specific finger and contains the P (proximal) box of five amino acids that specify the DNA sequence to which the receptor will bind, while the right finger contains the D (distal) box of amino acids that specifies the spacing between the half-sites of the responsive element in terms of the receptor conformation that enables it to interact with the elements that are spaced appropriately. Other func-

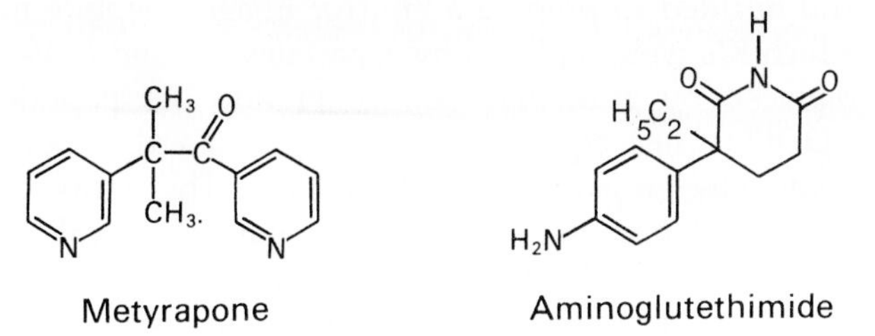

Metyrapone

Aminoglutethimide

FIGURE 10-19 Structures of metyrapone and aminoglutethimide.

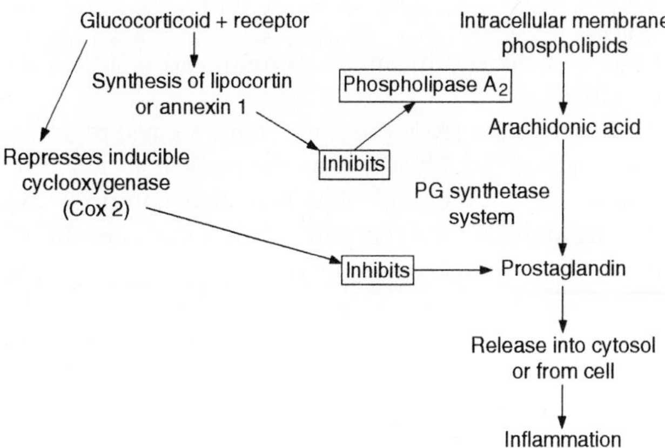

FIGURE 10-20 Action of glucocorticoids on the prostaglandin (PG) system.

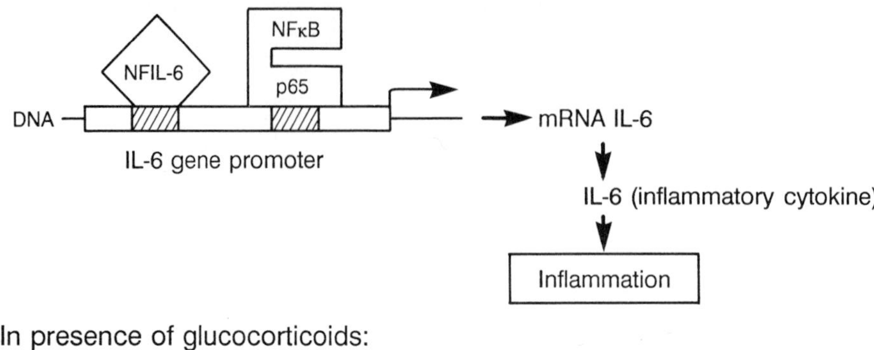

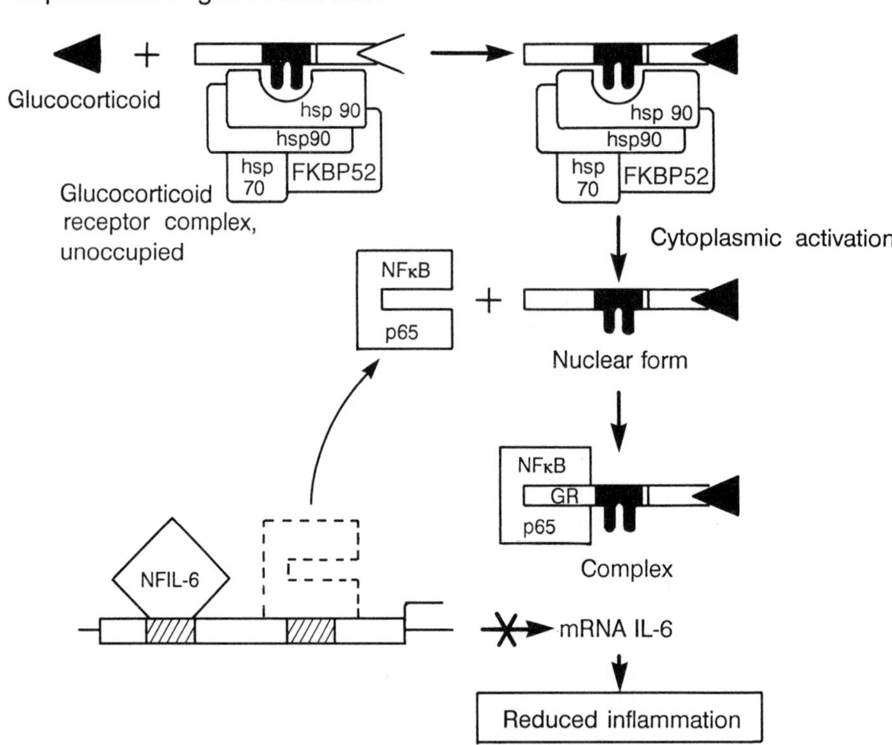

FIGURE 10-21 Anti-inflammatory action of glucocorticoid mediated by the direct interaction of NFκB p65 and the glucocorticoid receptor.

tions and the significance of amino acid residues are specified.

Downstream from the Zn^{2+} finger a basic region is encountered, which serves as the nuclear localization signal. The amino acid sequences that comprise this signal among several receptors of this family are shown in Figure 10-24. These sequences are bound by specific cytoplasmic proteins, called the nucleoporin family, which are presumed to dock the cytoplasmic receptor within the structure of the nucleopore (Figure 10-15) for transport into the nucleus. Other proteins may be involved in the transport process, and hsp 70 and α-G protein (RAN) have been suggested to participate. hsp 70 could contribute energy to the process by its ATPase activity. The transporting family of cytoplasmic proteins was discovered in an interesting way. Some Japanese researchers took the SV40 T antigen sequence of the nuclear localization signal, KKKRK, and decided to prepare a "mirror image" peptide in which K's were replaced by aspartates (D) and R was replaced by glutamate (E), generating the peptide DDDED. This sequence was synthesized and an antibody was prepared against DDDED. The antibody cross-reacted with the SV40 T antigen nuclear localization signal and also with a 69-kDa protein, which was found to bind to the nuclear localization signal. These cytoplasmic proteins constitute a family of proteins that are involved in nuclear transport.

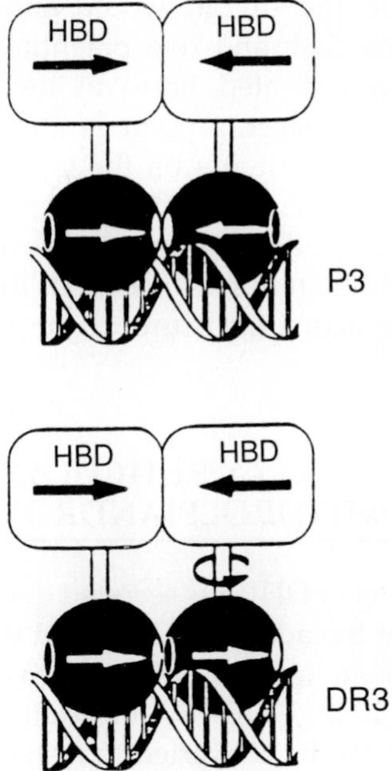

FIGURE 10-22 Mode of binding of steroid receptor to inverted repeats (P3) versus the binding of the vitamin D receptor to direct repeats (DR3). The orientation of the recognition element in the major groove is implied by the arrows of the DBD. The symmetrical dimerization region of the steroid receptor DBDs is indicated by white ovals, whereas the heterologous dimerization region of vitamin D is indicated by black and white ovals. Like the vitamin D receptor, other nuclear receptors that bind to direct repeats may form analogous heterologous associations. The hormone-binding domain, which has a stronger dimerization function, may form a symmetrical dimerization interface. The two domains may be linked by a flexible tether. Reproduced with permission from Luisi, B. F., Schwabe, J. W. R., and Freedman, L. P. (1994). The steroid/nuclear receptors: From three-dimensional structure to complex formation''. *Vitam. Horm.* **49**, 1–48.

SV40 T Antigen	P	K	K	K	R	K	V	
GR	491R	K	T	K	K	K	I	K
MR	673R	K	S	K	K	L	G	K
AR	628R	K	L	K	K	L	G	N
PR	637R	K	F	K	K	F	N	K
ER	256R	K	D	R	R	G	G	R
T3RB	179K	R	L	A	K	R	K	L
RAR	162R	K	A	H	Q	E	T	F
VDR	102R	K	R	E	M	I	L	L

FIGURE 10-24 Amino acid sequences from the literature that comprise the nuclear translocation signal.

Both the glucocorticoid and mineralocorticoid receptors are cytoplasmic in their unoccupied (by ligand) forms. The glucocorticoid receptor is complexed with a dimer of hsp 90 and a monomer of immunophilin p52. hsp 70 is also found in this complex, although not always as a stoichiometric partner, and is thought to play a role in the assembly of the oligomeric form shown in Figure 10-14. Interestingly, the oligomeric form of the glucocorticoid receptor generates a conformation that productively binds the steroidal ligand in cellular concentrations, whereas the free receptor monomer does not rebind the steroid well once the ligand has dissociated. In addition to generating the steroid-binding form of the receptor, the oligomer probably serves to anchor the receptor to a cytoplasmic skeletal element, such as tubulin or actin. Once the

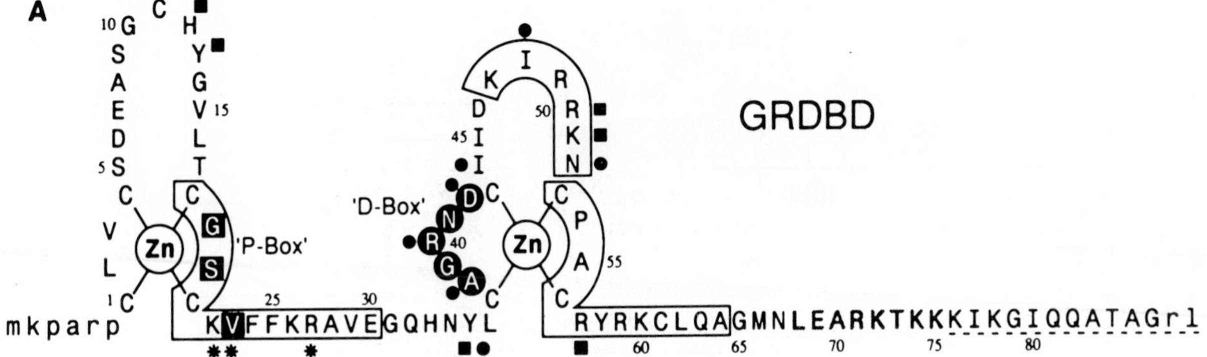

FIGURE 10-23 Two Zn^{2+} fingers of the glucocorticoid receptor. Single-letter abbreviations of the amino acids are used. Functional aspects of amino acid residues are indicated. AA 457–462 represent critical amino acids in the P box that generate binding specificity to the response element. AA 477–481 represent the D box involved in the recognition of spacing between half-sites of the response element and dimerization. Reproduced with permission from Luisi, B. F., Schwabe, J. W. R., and Freedman, L. P. (1994). The steroid/nuclear receptors: from three-dimensional structure to complex function. *Vitam. Horm.*, **49:1–48.**

steroidal ligand binds to the receptor in this form, the complex is triggered to dissociate, so that free receptor monomer is formed. This process has been referred to as cytoplasmic activation or transformation. The receptor becomes hyperphosphorylated. The "activated" receptor can enter the nucleopore as a monomer or as a homodimer and be transported to the nucleoplasm, where it searches DNA until the hormone-responsive element is encountered. There it binds tightly to DNA as a homodimer and probably binds DNA into a configuration that is preferred for the entry of the other transactivating proteins that comprise the transcriptional complex. In the case of mouse mammary tumor virus LTR, the transactivating proteins of the transcriptional complex are shown in Figure 10-25. A cartoon of the retinoic acid receptor–retinoic acid X receptor heterodimer facilitates a view of how these proteins could assemble in a transcriptional complex (Figure 10-26). Once racheted in place, RNA polymerase II is able to start transcribing the open reading frame of the activated gene. Eventually, the steroid–receptor complex dissociates from the DNA and is transported back into the cytoplasm after a dephosphorylation event takes place. In some tissues, for example, the liver, the dissociated steroid is quickly metabolized to an inactive form. The liberated receptor becomes phosphorylated and reassembled into the inactive oligomeric state. Some of the receptor may be degraded rather than entering the reassembly process. The reassembled oligomer is now poised for another event in which glucocorticoid ligand is released from the adrenal gland.

A hypothetical model of the steroid receptor-binding domain has been developed by analogy with the structures inherent in certain protease enzymes. This model is shown in Figure 10-27. For the glucocorticoid receptor, the steroid would enter the binding pocket A ring first, and two potential conflicts with the A ring are indicated. Part way toward the mouth of the binding pocket, a point of attachment is seen with the C-11 substituent on the C ring. Then closer to the mouth of the pocket, two interactions with the D ring are indicated. Although not proven, models like these will stimulate further research that will help to arrive at the actual structure of the steroid-binding domain.

VI. ZONA RETICULARIS AND DEHYDROEPIANDROSTERONE

Up to 30 mg of DHEA is secreted daily by the innermost zone of the adrenal cortex, and the blood levels of this hormone and its sulfate derivative are fairly high (Table 10-2). DHEA is considered to be a weak androgen. It can be converted to testosterone and androstenediol, as shown in Figure 2-23. When the aromatase system of enzymes is present, DHEA can be converted to estrogens, as summarized in Figure 14-11. Surprisingly, little is known concerning the direct actions of DHEA other than its serving as a substrate for the production of sex hormones, although there have been reports of the discovery of a receptor for DHEA.

VII. MINERALOCORTICOID HORMONE

A. Introduction

The mechanisms of response to stress involve the glucocorticoid hormone secretion activated by a cascade system consisting of the limbic system of the

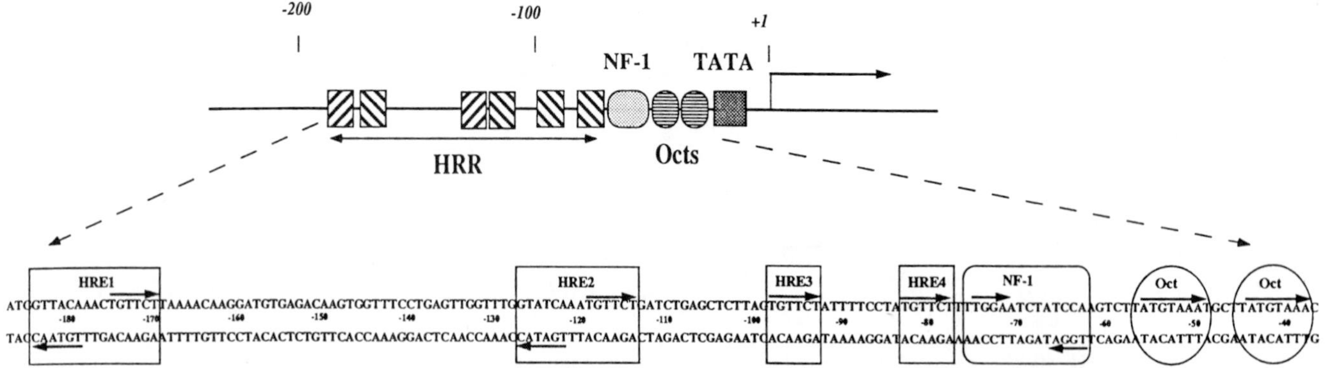

FIGURE 10-25 HRR of the MMTV promoter. Schematic representation of the MMTV regulatory region, including the hormone regulatory region (HRR) with the receptor-binding sites (HRE 1–4), the palindromic binding site for CTF/NFI (NF-1), and the two octamer motifs (Octs) upstream of the TATA box (TFIID). The nucleotide sequence of the region between -187 and -37 is shown. Note the varied spacing of the half-palindromes in HRE1 and HRE2. Reproduced with permission from Truss, M., and Beato, M. (1993). Steroid hormone receptors: Interaction with deoxyribonucleic acid and transcription factors. *Endocr. Rev.* **14,** 459–479.

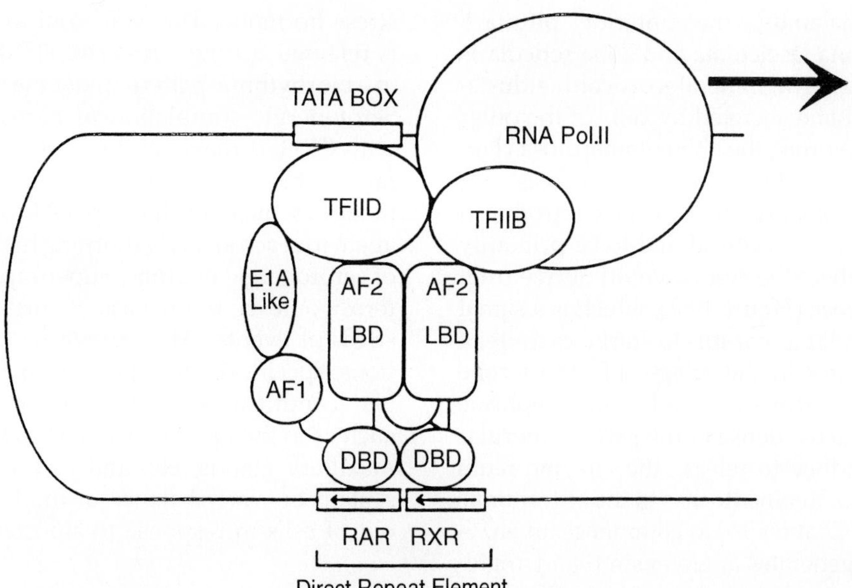

FIGURE 10-26 Model for the stabilization of the preinitiation complex by an RXR–RAR heterodimer. Reproduced with permission from Pfahl, M., Apfel, R., Bendik, I., Fanjul, A., Graupner, G., Lee, M.-O., La-Vista, N., Lu, X-P., Piedrafita, J., Ortiz, M. A., Salbert, G., and Zhang, X.-K. (1993). Nuclear retinoid receptors and their mechanism of action. *Vitam. Horm.*, **49,** 327–382.

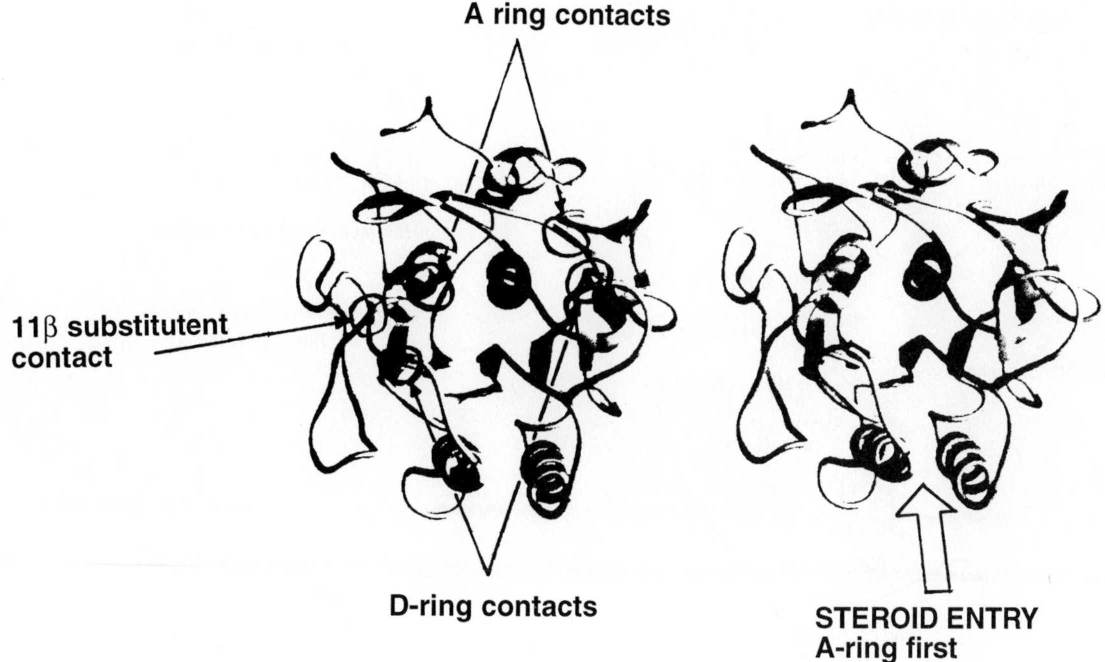

FIGURE 10-27 Completed predicted structure of the GR HBD family, based on an alignment with thermitase, in crossed stereo. The original structure is in color, therefore some details are not obvious. Locations of contacts and point of entry of the steroid molecule are indicated by the current authors of HORMONES. The model is reproduced by permission from Goldstein, R. A., Katzenellenbogen, J. A., Luthey-Schulten, Z. A., Seielstad, D. A., and Wolynes, P. G. (1993). Three-dimensional model for the hormone binding domains of steroid receptors, *Proc. Nat. Acad. Sci. USA* **90**, 9949–9953.

brain, the hypothalamus, the anterior pituitary (ACTH), and the zona fasciculata and zona reticularis of the adrenal cortex. The mineralocorticoid, aldosterone, is synthesized and secreted by cells of the outer layer of the adrenal cortex, the zona glomerulosa (Figure 10-4). The secretion of this hormone involves a set of signals different from those that we have introduced for cortisol. Aldosterone is considered to be primarily a stress hormone where the signal would derive from norepinephrine release (Figure 10-7), which is a signal for the juxtaglomerular apparatus to start a cascade of events that culminates in the release of aldosterone (Chapter 15). Thus, stress-released norepinephrine would signal the macula densa of the juxtaglomerular apparatus of the kidney to release the enzyme renin into the bloodstream. Renin attaches to the N-terminus of a blood protein (Chapter 15) to commence an enzymatic cascade that generates angiotensin II and angiotensin III, which are hormones. Angiotensins II and III bind to a membrane receptor on cells of the adrenal zona glomerulosa. This effect, together with ACTH, which is also generated by stress through the humoral pathway (Figure 10-7), increases the synthesis and release of aldosterone. As this seems to be the key mechanism for the release of aldosterone, it is considered a stress hormone. This is in contrast to cortisol, which is released during stress (ACTH) but is also released in a biorhythmic pattern under the control of serotonin. Serotonergic stimulation of cortisol release, starting with CRH, is maximal during the morning hours and falls off beforehand and afterwards until the end of 24 hr, after which time the cortisol level begins to increase again to reach an early morning high due to the activity of serotonergic neurons. Superimposed upon this pattern would be the release of cortisol generating from stressful events. Aldosterone has salt conserving actions, specifically for Na^+, which is reabsorbed by the Na^+ conductance channel in lumenal epithelial cells such as in the gastrointestinal tract, sweat glands, submaxillary glands, etc. and therefore is a part of the system of water balance in the body. K^+ is pumped out of cells in response to aldosterone. However, the

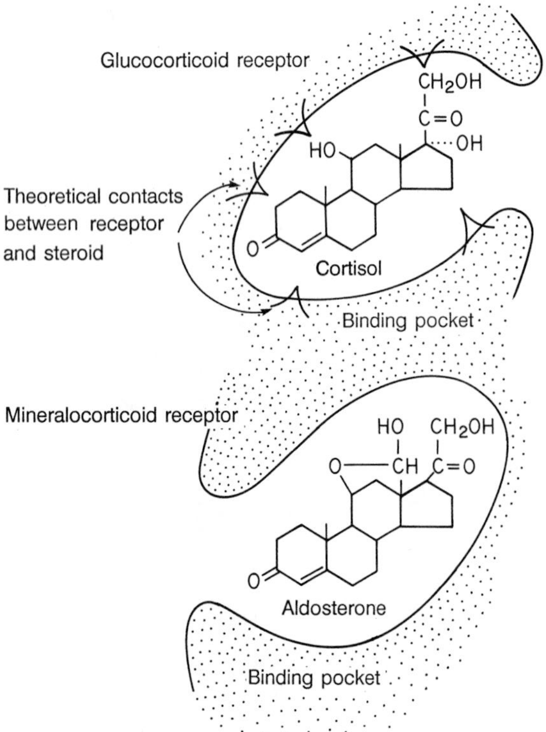

FIGURE 10-28 Speculative views of the binding of cortisol by the glucocorticoid receptor and of aldosterone by the mineralocorticoid receptor.

Free Aldehyde form

Hemiacetal with C-21

Hemiacetal with C-11 oxygen

Major form of aldosterone binding to the mineralocorticoid receptor

Structure of the antagonist spironolactone

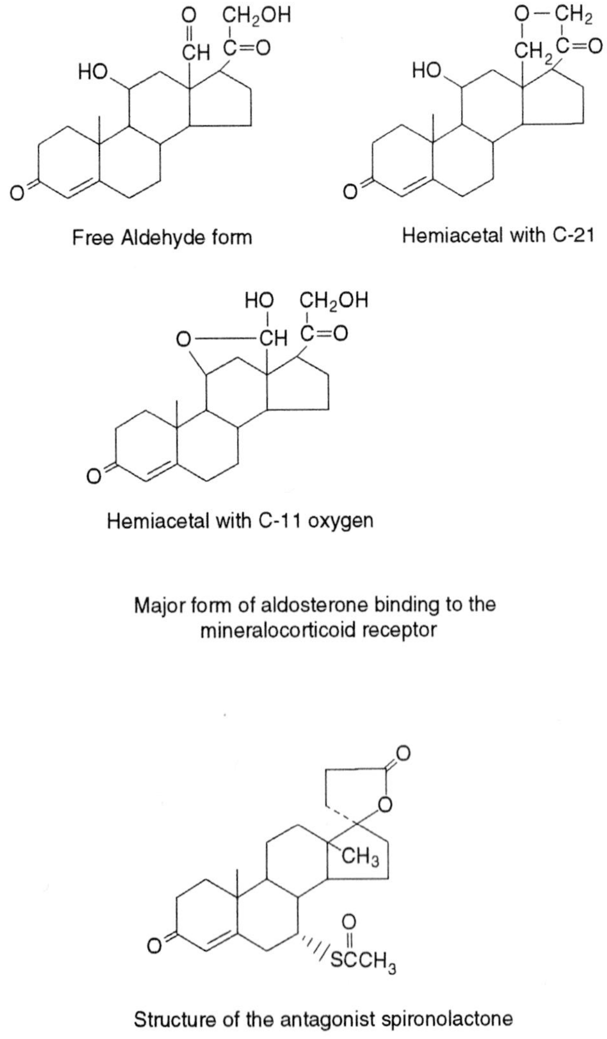

FIGURE 10-29 Potential structures of soluble aldosterone and structure of the antagonist, spironolactone.

aldosterone-secreting system is not the only system controlling water balance. The topic of water homeostasis is covered in Chapter 15, where the mode of action of aldosterone is reviewed, and in Chapter 4, where the actions of vasopressin are presented.

B. Aldosterone-Producing System

This system is diagrammed in Figure 15-7. The initial signal is a fall in blood volume, a decrease in Na^+ concentration in blood, a fall in blood pressure or release of catecholamines (stress) at the juxtaglomerular apparatus of the kidney, or the effect of angiotensin II. Any of these stimuli causes the release of renin, an enzyme with proteolytic activity, from juxtaglomerular cells. Renin acts on a plasma α_2-globulin, releasing an N-terminal decapeptide by hydrolyzing, at a hydrophobic sequence (amino acids 10–13), a leucyl–leucyl peptide bond. The resulting decapeptide is angiotensin I, an intermediate with no hormonal activity. Angiotensin I is then attacked by a converting enzyme, which has a broad distribution in the vascular epithelium, lung, liver, adrenal cortex, pancreas, kidney, spleen, and neurohypophysis. This enzyme is a carboxypeptidase that cleaves off the carboxy-terminal dipeptide, His-Leu, at a peptide bond connecting Phe and His. The resulting dipeptide is inactive and the octapeptide is angiotensin II, a hormone. This hormone has a short half-life of about 1 min because its activity is destroyed by angiotensinase, a Ca^{2+}-activated enzyme found in the peripheral vascular beds. This hormone, in addition to its effects on zona glomerulosa cells of the adrenal cortex, sensitizes vascular smooth muscles to the contractile effects of norepinephrine. There are specific outer cell membrane receptors for angiotensin II in the zona glomerulosa of the adrenal cortex. Binding of the hormone to the receptor causes enhanced PI cycle turnover, with the activation of phosphatidylinositol protein kinase. In addition, ACTH binds to its receptor on these cells and elevates cAMP and PKA. Both pathways work in concert to stimulate aldosterone synthesis. Apppreciable levels of aldosterone are generated, primarily during stress.

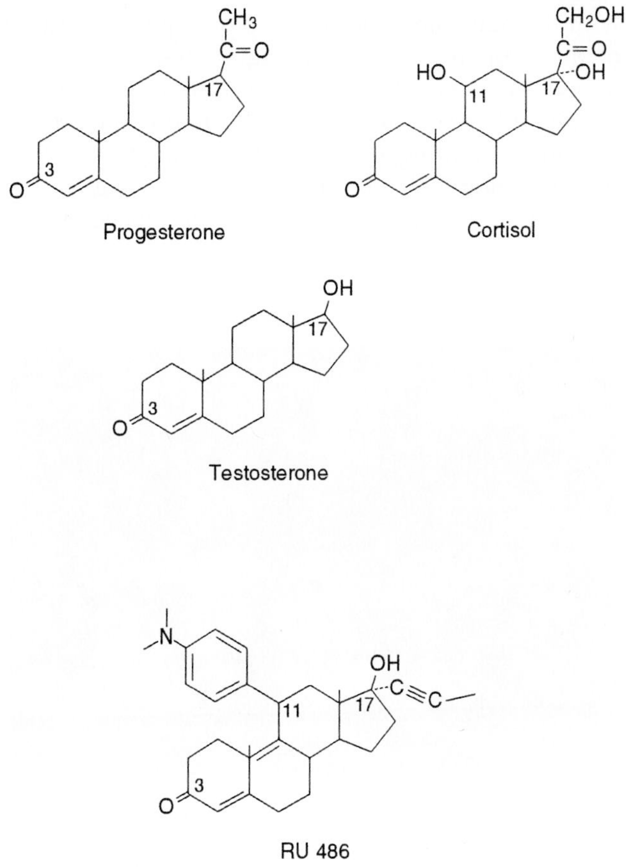

Progesterone

Cortisol

Testosterone

RU 486

FIGURE 10-30 Structure of RU-486 compared to progesterone, cortisol and testosterone whose receptors are affected by RU-486.

TABLE 10-4 Effects of Overproduction of Cortisol (Cushing's Syndrome)[a,b]

Protein metabolism
Thinning of skin
Reddish striae
Loss of bone matrix and demineralization
Poor wound healing
Muscle wasting and weakness
Capillary fragility and bruising
Impaired growth (children)
Carbohydrate metabolism
Abnormal glucose tolerance curve
Overt diabetes mellitus
Lipid metabolism
Centripetal fat distribution
Moon face
Electrolyte balance
Sodium retention, potassium loss
Hypertension
Hypervolemia
Hematopoietic effects
Eosinopenia, lymphopenia
Polymorphonuclear leukocytosis
Erythrocytosis
General effects
Hypercalciuria and renal calculi
Gastric ulceration
Psychosis
Impaired immunological tolerance

[a] Cushing's syndrome is most often caused iatrogenically by administration of large amounts of glucocorticoids.

[b] Reproduced from Ezrin, C., Godden, J. O., Volpe, R., and Wilson, R. (eds.) (1973). "Systematic Endocrinology," p. 171. Harper & Row, New York.

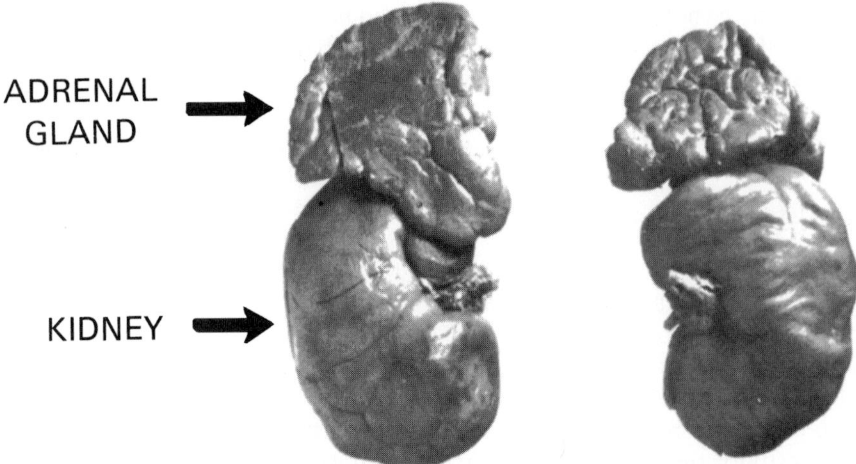

FIGURE 10-31 Bilateral diffuse adrenocortical hyperplasia with moderate folding and segmentation of the surface in a 5-year-old girl with congenital adrenogenital syndrome. The bean-shaped organ below is the kidney and the arrow points to the hyperplastic adrenal glands. Reproduced from Labhart, A. (ed.,) (1976). "Clinical Endocrinology," p. 362. Springer-Verlag, Berlin and New York.

C. Aldosterone Receptor

Like the glucocorticoid receptor, the mineralocorticoid receptor resides in the cytoplasm in the unoccupied (by ligand) form and also appears to contain hsp 90 in a heteromeric structure. Different from the glucocorticoid receptor, it appears that aldosterone binds in the binding pocket of the mineralocorticoid receptor D ring first, as contrasted with A ring first in the case of the glucocorticoid receptor (Figure 10-28). In addition, there seem to be three possible configurations of aldosterone, as shown in Figure 10-29. There may be a preponderance of the C-11 hemiacetal form, and this may be the major form binding to the mineralocorticoid receptor. Presumably the mineralocorticoid receptor undergoes activation–transformation in the cytoplasm and is transported into the nucleus by mechanisms similar to those described for the glucocorticoid receptor. An important issue is the crossover of ligand binding, and the mineralocorticoid receptor is an effective binder of cortisol. Thus, when cortisol is not inactivated by kidney 3β-hydroxysteroid dehydrogenase, to produce the inactive hormone, cortisone, cortisol can bind to the kidney mineralocorticoid receptor and enhance Na^+ uptake by stimulation of the sodium ion channel, leading to increased blood pressure. On the other hand, aldosterone is not an effective ligand for the glucocorticoid receptor.

RU-486 is an antagonist of the progesterone receptor, the glucocorticoid receptor, and to a lesser degree the androgen receptor. However, it does not interact with the mineralocorticoid receptor. Upon inspection of the structure of RU-486 (Figure 10-30), it can be hypothesized that the environment around the D ring is quite different from that of aldosterone (Figure 10-29) thus preventing RU-486 from entering the binding pocket D ring first (Figure 10-28.) The A ring, on the other hand, closely resembles that of the glucocorticoid

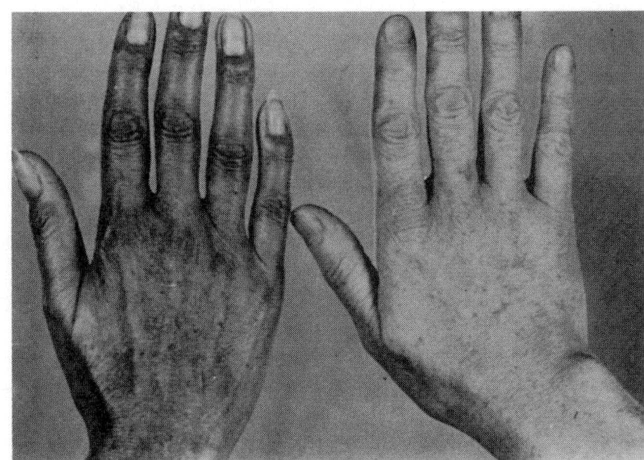

FIGURE 10-32 Idiopathic Addison's disease. The pigmentation of the hand on the left is contrasted with that of a normal hand. Reproduced from Ezrin, C., Godden, J. O., Volpe, R., and Wilson, R. (eds.) (1973). "Systematic Endocrinology," p. 178. Harper & Row, New York.

receptor, suggesting that it could enter the binding pocket A ring first, and the bulky substituent of C-11 may account, at least in part, for its inhibitory properties.

Other work describes membrane mineralocorticoid receptors. These receptors may be linked to the Na^+–H^+ antiporter located in nonepithelial cells such as lymphocytes and smooth muscle cells and are apparently insensitive to cortisol and spironolactone, a mineralocorticoid receptor antagonist whose structure is shown in Figure 10-29.

The mineralocorticoid receptor should activate the transcription of specific targets such as the α-subunit of the Na^+,K^+ ATPase of the kidney. However, there appear to be cases where the mineralocorticoid receptor, the glucocorticoid receptor, and both ligands are present. From experiments, it has been learned that the glucocorticoid receptor and the mineralocorticoid receptor can form a heterodimer that is less active as a transcriptional activator than either homodimer. This level of control undoubtedly will command future investigation.

D. Aldosterone Antagonists

Synthetic antagonists of aldosterone action are available and are known as spironolactones. The structure of spironolactone is given in Figure 10-29. This substance has no Na^+-conserving action on its own, and its action appears to be through competition for aldosterone binding at the ligand-binding site of the aldosterone receptor. Although some other compounds have indirect antimineralocorticoid effects, spironolactone is the compound used most frequently.

E. Aldosterone Metabolism

More than a dozen metabolites of radioactive aldosterone have been characterized in the urine. Nearly all of the hormone is metabolized within 48 hr. The major urinary form is a glucuronide derivative obtained by substitution of the reduced C-3 position. Conjugation occurs in the liver and kidney.

VIII. CLINICAL ASPECTS

A. Glucocorticoids

Major clinical problems associated with the adrenal cortex center around hyper- or hypoproduction of adrenal steroids. The causes of these conditions can be quite complicated since there are several organs involved in the hormonal cascade. In the case of the glucocorticoids, the limbic system of the brain, the hypothalamus, the blood portal system connecting the hypothalamus with the anterior pituitary, and the adrenal gland are parts of the system. The normal values of steroidal hormones produced by the adrenal are presented in Table 10-2. When cortisol is overproduced, often by a pituitary tumor causing high levels of circulating ACTH, the resulting disease is known as Cushing's disease. When cortisol is underproduced, the resulting disease is known as Addison's disease, which is most frequently the result of adrenal destruction. The effects of Cushing's disease are summarized in Table 10-4. Hyperproduction of cortisol from the adrenal gland would be expected to occur if there is a tumor of the adrenal producing abnormal amounts of the steroid or if there is a pituitary tumor producing high levels of ACTH (also if there is overproduction of CRH at the hypothalamic level, which apparently has not been documented so far). Adrenalectomy is often a treatment of the primary disease, with maintenance levels of glucocorticoids taken orally. Adrenal tumors are more common in women. Figure 10-31 shows a case of adrenocortical hyperplasia in a 5-year-old girl with congenital adrenogenital syndrome. The more malignant forms of adrenal tumors result in excess androgen production.

Addison's disease or hypoproduction of cortisol is usually the result of atrophy of the adrenal cortex by unexplained causes (idiopathic). This can be the result of failure at other levels of the cascade from the limbic system downward. The results of Addison's disease are sodium loss (since aldosterone usually is also deficient), excess potassium in the blood, low blood pressure, hypoglycemia, high levels of ACTH and MSH (loss of negative feedback), and pigmentation of the skin due to the high levels of MSH and/or ACTH (Figure 10-32). The acute onset of Addison's disease, referred to as Addisonian shock, is shock associated with adrenal insufficiency and most commonly occurs when an individual with long-standing adrenal insufficiency is exposed to a stressful stimulus. It is treated with large intravenous doses of cortisol, fluid replacement, and antibiotics.

B. Mineralocorticoids and Androgens

Mineralocorticoid excess can be a feature of Cushing's syndrome, as mentioned earlier. It can be the cause of reduced potassium levels in blood, resulting in weakness, sodium retention, and hypertension.

An inborn error of metabolism can result in a deficiency of the adrenal cortical enzyme, C-21 hydroxylase, whose function is required for the synthesis of cortisol. This deficiency results in lowered cortisol lev-

TABLE 10-5 Forms of Adrenal Hyperplasia[a]

Deficiency	Syndrome	Ambiguous genitalia	Postnatal Virilization	Salt metabolism	Steroids increased	Steroids decreased	Enzyme	Chromosomal location	Frequency	Gene cloned
	Lipoid hyperplasia	Males	No	Salt wasting	None	All	$P450_{cc}$	15	Rare	Yes
3β-Hydroxysteroid dehydrogenase	Classic	Males Females: mildly virilized	Yes	Salt wasting	DHEA, 17-hydroxypregnenolone	Aldo, T, cortisol	3β-Hydroxysteroid dehydrogenase type II	1q	Rare	Yes
	Nonclassic	No	Yes	Normal	DHEA, 17-hydroxypregnenolone	—	3β Hydroxydehydrogenase	1q	?Frequent	—
17α-Hydroxylase	—	Males	No	Hypertension	DOC, corticosterone	Cortisol, T	$P450_{c17}$	10	Rare	Yes
17,20-Lyase	—	Males	No	Normal	—	DHEA, T, Δ^4-A	$P450_{c17}$	10	Rare	Yes
21-Hydroxylase	Salt wasting	Females	Yes	Salt wasting	17-OHP, Δ^4-A	Aldo, cortisol	$P450_{c21}$	6p	1/14,000	Yes
									75% cases	Yes
	Simple	Females: virilized	Yes	Normal	17-OHP, Δ^4-A	Cortisol	$P450_{c21}$	6p	25% cases	Yes
	Nonclassic	No	Yes	Normal	17-OHP, Δ^4-A	—	$P450_{c21}$	6p	0.1–1% (3% in European Jews)	Yes
11-Hydroxylase	Classic	Females	Yes	Hypertension	DOC, 11-deoxycortisol (S)	Cortisol, ±aldo	$P450_{c11}$	8q	1/100,000	Yes
	Nonclassic	No	Yes	Normal	11-Deoxycortisol ±DOC	—	$P450_{c11}$	8q	?Frequent	Yes
Corticosterone methyl oxidase type II	Salt wasting	No	No	Salt wasting in infancy	18-Hydroxycorticosterone	Aldo	$P450_{cmo}$	8q	Rare (except in Iranian Jews)	Yes

[a] Abbreviations: Aldo, aldosterone; T, testosterone; Δ^4-A, Δ^4-androstenedione; DOC, 11-deoxycorticosterone; 17-OHP, 17α-hydroxyprogesterone. This table is reproduced from New, M. I. (1995). "Congenital adrenal hyperplasia. In "Endocrinology" (L. J. DeGroot, ed.), 3rd ed. p. 1814, W. B. Saunders, CO., Philadelphia, PA.

els and enlarged output of ACTH by the anterior pituitary due to reduced negative feedback. High ACTH results in adrenal hyperplasia and hypersecretion, especially in the products of the zona reticularis (DHEA), which, together with adrenal testosterone, can lead to masculinization of female babies. Precocious puberty in males can also result from this condition. Treatment is centered on cortisol replacement therapy, which restores the negative feedback mechanism.

C. Congenital Adrenal Hyperplasia and Others

Under this heading is a group of diseases caused by an enzymatic defect that results in the reduced synthesis of cortisol. This impairment engenders an oversecretion of ACTH, a growth factor for adrenal cells, leading to adrenal hyperplasia. A large number of disturbances arise that are treatable by glucocorticoids and salt-retaining steroids. Prenatal diagnosis is especially important in this disease. Table 10-5 lists the various forms of adrenal hyperplasia.

Glucocorticoid resistance is rare and may be based on point mutations in the glucocorticoid receptor that reduce the effectiveness of negative feedback by cortisol. As a consequence, there will be elevated levels of circulating cortisol. Clinical problems can be mild to excessive due to high levels of circulating cortisol or steroids.

Other variations generating high or low levels of glucocorticoids or mineralocorticoids can be found in the clinical literature.

References

A. Books

Cake, M. H., and Litwack, G. (1975). *In* "Biochemical Actions of Hormones" (G. Litwack, ed.), Vol. 3, pp. 317–390. Academic Press, New York.

Chaffee, E. E., and Lytle, I. M. (1980). "Basic Physiology and Anatomy," 4th ed., p. 483. Lippincott, Philadelphia.

Coupland, R. E. (1975). *In* "Handbook of Physiology" (R. O. Greep and E. B. Astwood, eds.) Sect. 7, Vol. VI, pp. 282–294. Am. Physiol. Soc., Washington, DC.

DeGroot, L. J., ed. (1995). "Endocrinology," 3rd ed., Vol. 1. W. B. Saunders, Philadelphia.

Ezrin, C., Godden, J. O., Volga, R., and Wilson, R., eds. (1973). "Systematic Endocrinology." Harper & Row, New York.

Felig, P., Baxter, J. D., and Frohman, L. A., eds. (1995). "Endocrinology and Metabolism," 3rd ed. McGraw-Hill, Inc., New York.

Harrison, R. G. (1959). "A Textbook of Human Embryology," Vol. 1, p. 48. Blackwell Scientific, London.

Labhart, A., ed. (1976). "Clinical Endocrinology," p. 362. Springer-Verlag, Berlin and New York.

Langebartel, D. A. (1977). "The Anatomical Primer," p. 443. University Park Press, Baltimore, MD.

Munck, A., and Crabtree, G. R. (1981). Glucocorticoid-induced lymphocyte death. *In* "Cell Death in Biology and Pathology" (I. D. Bower and R. A. Lockshin, eds.), pp. 329–357. Chapman and Hall, London and New York.

Schulster, D., Burstein, S., and Cooke, B. A. (1976). "Molecular Endocrinology of the Steroid Hormones," pp. 156–157. Wiles, New York.

B. Review Articles

Agutter, P. S., and Prochnow, D. (1994). Nucleocytoplasmic transport. *Biochem. J.* **300,** 609–618.

Baniahmad, A., and Tsai, M. J. (1993). Mechanisms of transcription activation by steroid hormone receptors. *J. Cell Biochem.* **51,** 151–156.

Carmo-Fonseca, M., and Hurt, E. C. (1991). Across the nucleopores with the help of nucleoporins. *Chromosoma* **101,** 199–205.

Carson-Jurica, M. A., Schrader, W. T., and O'Malley, B. W. (1990). Steroid receptor family. *Endocrine Rev.* **11,** 201–220.

Edwards, C. R. W. (1995). Primary mineralocorticoid excess syndromes. *In* "Endocrinology" (L. J. DeGroot, ed.), 3rd ed., Vol. 2, pp. 1775–1803. W. B. Saunders Co., Philadelphia.

Henry, J. P. (1993). Biological basis of the stress response. *NIPS* **8,** 69–73.

Imura, H. (1995). Adrenocorticotropic hormone. *In* "Endocrinology" (L. J. DeGroot, ed.), 3rd ed., Vol. 1, pp. 355–366. W. B. Saunders, Philadelphia.

Ing, N. H., and O'Malley, B. W. (1995). The steroid receptor superfamily: molecular mechanisms of action. *In* "Molecular Endocrinology" (B. D. Weintraub, ed.), pp. 195–215. Raven Press, New York.

Loriaux, D. L., and McDonald, W. J. (1995). Adrenal insufficiency. *In* "Endocrinology" (L. J. DeGroot, ed.), 3rd ed., Vol. 2, pp. 1731–1740 W. B. Saunders Co., Philadelphia.

Malchoff, D. M., and Malchoff, C. D. (1995). Glucocorticoid resistance. *In* "Endocrinology" (L. J. DeGroot, ed.), 3rd ed., Vol. 2, pp. 1770–1774. W. B. Saunders Co., Philadelphia.

Meisfeld, R. L. (1995). Overview of glucocorticoid action at the molecular level. *In* "Endocrinology" (L. J. DeGroot, ed.), 3rd ed., Vol. 2, pp. 1656–1667. W. B. Saunders Co., Philadelphia.

New, M. I. (1995). Congenital adrenal hyperplasia. *In* "Endocrinology" (L. J. DeGroot, ed.), 3rd ed., Vol. 2, pp. 1813–1835. W. B. Saunders Co., Philadelphia.

Orth, D. N. (1995). Cushing's syndrome. *New Engl. J. Med.* **332,** 791–803.

Pratt, W. B. (1993). The role of heat shock proteins in regulating the function, folding and trafficking of the glucocorticoid receptor. *J. Biol. Chem.* **268,** 1455–21458.

Pratt, W. B., Czar, M. J., Stancato, L. F., and Owens, J. K. (1993). The hsp56 immunophilin component of steroid receptor heterocomplexes: could this be the elusive nuclear localization signal-binding protein? *J. Steroid Biochem. Mol. Biol.* **46,** 269–279.

Simpson, E. R., and Waterman, M. R. (1995). Steroid hormone biosynthesis in the adrenal cortex and its regulation by adrenocorticotropin. *In* "Endocrinology" (L. J. DeGroot, ed.), 3rd ed., Vol. 2, pp. 1630–1641. W. B. Saunders, Philadelphia.

Truss, M., and Beato, M. (1993). Steroid hormone receptors: interaction with deoxyribonucleic acid and transcription factors. *Endocrine Rev.* **14,** 451–479.

White, P. C. (1994). Disorders of aldosterone biosynthesis and action. *New Engl. J. Med.* **331,** 250–258.

C. Research Papers

Goldstein, R. A., Katzenellenbogen, J. A., Luthey-Schulten, Z. A., Seielstad, D. A., and Wolynes, P. G. (1993). Three-dimensional model for the hormone binding domains of steroid receptors. *Proc. Natl. Acad. Sci. USA* **90,** 9949–9953.

Imamoto-Sonebe, N., Matsuoka, Y., Semba, T., Okada, Y., Uchida, T., and Yoneda, Y. (1990). A protein recognized by antibodies to Asp-Asp-Asp-Glu-Asp shows specific binding activity to heterogeneous nuclear transport signals. *J. Biol. Chem.* **265,** 16504–16508.

Kliewer, S. A., Umesono, K., Mangelsdorf, D. J., and Evans, R. M. (1992). Retinoid X receptor interacts with nuclear receptors in retinoic acid, thyroid hormone and vitamin D_3 signalling. *Nature* **355,** 446–449.

LaCasse, E. C., Lochnan, H. A., Walker, P., and Lefebvre, Y. A. (1993). Identification of binding proteins for nuclear localization signals of the glucocorticoid and thyroid hormone receptors. *Endocrinology* **132,** 1017–1025.

Luisi, B. F., Schwabe, J. W. R., and Freedman, L. P. (1994). The steroid/nuclear receptors: from three-dimensional structure to complex formation. *Vitamins Hormones* **49,** 1–48.

Madan, A. P., and DeFranco, D. B. (1993). Bidirectional transport of glucocorticoid receptors across the nuclear envelope. *Proc. Natl. Acad. Sci. USA* **90,** 3588–3592.

Moore, M. S., and Blobel, G. (1993). The GTP-binding protein Ran/TC4 is required for protein import into the nucleus. *Nature* **365,** 661–663.

Newcomer, M. E., Pappas, R. S., and Ong, D. E. (1993). X-ray crystallographic identification of a protein binding site for both *all-trans-* and 9-*cis*-retinoic acid. *Proc. Natl. Acad. Sci. USA* **90,** 9223–9227.

Ojasoo, T., Raynaud, J. P., and Dore, J. C. (1994). Affiliations among steroid receptors as revealed by multivariate analysis of steroid binding data. *J. Steroid Biochem. Mol. Biol.* **48,** 31–46.

Pfahl, M., Apfel, R., Bendik, I., Fanjul, A., Graupner, G., Lee, M. O., La Vista, N., Lu, X.-P., Piedrafita, J., Ortiz, M. A., Salbert, G., and Zhang, X.-K. (1993). Nuclear retinoid receptors and their mechanism of action. *Vitamins Hormones* **49,** 327–382.

Pierce, D. (1994). A mechanistic basis for distinct mineralocorticoid and glucocorticoid receptor transcriptional specificities. *Steroids* **59,** 153–159.

Ray, A., and Prefontaine, K. E. (1994). Physical association and functional antagonism between the p65 subunit of transcription factor NF-1cB and the glucocorticoid receptor. *Biochemistry* **33,** 752–756.

Rupprecht, R., Arriza, J. L., Spengler, D., Reul, J. M., Evans, R. M., Holsboer, F., and Damm, K. (1993). Transactivation and synergistic properties of the mineralocorticoid receptor: relationship to the glucocorticoid receptor. *Mol. Endocrinol.* **7,** 597–603.

Wehling, M. (1994). Novel aldosterone receptors: specificity-conferring mechanism at the level of the cell membrane. *Steroids* **59,** 160–163.

Wilson, K. P., Black, J.-A. F., Thomson, J. A., Kim, E. E., Griffith, J. P., Navia, M. A., Murcko, M. A., Chambers, S. P., Aldape, R. A., Raybuck, S. A., and Livingston, D. J. (1994). Structure and mechanism of interleukin-1β converting enzyme. *Nature* **370,** 270–275.

CHAPTER

11

Hormones of the Adrenal Medulla

I. INTRODUCTION

Catecholamines are synthesized and released from adrenergic neurons of the nervous system, and together with cholinergic neurons they represent two of the major vehicles for chemical neural communication. The adrenergic neurons of the nervous system synthesize and secrete norepinephrine (noradrenaline), and the cholinergic neurons synthesize and secrete acetylcholine. There are a variety of other neurons that synthesize additional neurotransmitters: serotonergic neurons, dopaminergic neurons, melatonergic neurons, enkephalinergic neurons, etc. Even some polypeptides such as insulin, now known to be present in the brain as well as in the pancreas, may act as neurotransmitters. Consequently, the demarcation between the central nervous system and the endocrine system becomes blurred at some point.

For the moment, however, the emphasis on catecholamines as hormones is a subject focusing on the adrenal medulla rather than on the central nervous system. Whereas in the nervous system the adrenergic neurotransmitter is invariably norepinephrine (noradrenaline), the product of the adrenal medulla is uniquely epinephrine (adrenaline), in most species but not all, in addition to some norepinephrine. In this case epinephrine, released as a result of sudden environmental stress (e.g., fright), is a hormone.

The adrenal medulla is made up of modified neurons no longer possessing axons or nerve endings and so are essentially cell bodies that have been adapted to secretory function. Thus, in the endocrine sense, the trigger to the adrenal medulla is a unique environmental signal. This is transmitted by way of a single cholinergic neuron from the sympathetic nervous system terminating in a synapse with the chromaffin cell of the adrenal medulla.

The catecholamine hormone released from the chromaffin cell binds to a receptor on the hepatocyte and elsewhere and generates the breakdown of glycogen to glucose, which is passed into circulation for use as a ready fuel during stress. Other physiological changes attributable to the adrenal catecholamines include changes in blood pressure and heart function, which also occur through adrenergic receptors.

The overall system is not essential to life in the same sense as the requirement for the adrenal cortex. The adrenal medulla can be removed (sympathectomy), and, clinically, this is done sometimes without endangering survival in cases of oversecretion of catecholamines. However, when the adrenal cortex is removed, life can be sustained only in a totally nonstressful environment, with salt provided to the experimental animal unless glucocorticoid therapy is given. It is clear that when the adrenal is not secreting cortisol (in the human), the hormone must be supplied orally for continued survival.

II. ANATOMICAL, MORPHOLOGICAL, AND PHYSIOLOGICAL RELATIONSHIPS

A. Introduction

The adrenal glands are embedded and enclosed by coverings located (Figure 10-1) in fat above the kidneys. The blood circulation of the adrenal is described in Figure 10-2. The important detail to remember is that blood drains from the adrenal cortex, carrying its secretions (e.g., cortisol, aldosterone) into the medulla, and then joins larger vessels that ultimately merge with the bloodstream. Thus, the intimacy of the cortical steroids with the medulla is apparent. Since the glucocorticoids induce the enzyme converting norepinephrine to epinephrine in the chromaffin cell, and since glucocorticoids as well as epinephrine are increased by stress, the capacity for the response of the specific secretion of the adrenal medulla is guaranteed by the events occurring in the cortex.

B. Development of the Adrenal Medulla

The adrenal medulla arises from the neural crest and is ectodermal in origin. As described in Chapter 10, the development of the primary cortex is signaled by the appearance of a suprarenal groove above the dorsal mesentary on each side (Figure 10-3). The cells lining the groove proliferate into the primary cortex. Neural crest cells, which give rise to sympathetic neuroblasts, move down toward the cortex and invade it to form the precursor of the medulla. The cells of the medulla, postnatally, display a segmental arrangement, which may relate to the structures of the nerve fibers from the sixth thoracic to the first lumbar segments, probably representing the sources of the migrating nerve cells forming the medulla. Ultimately, the medulla becomes invested with cells that secrete norepinephrine, in addition to those secreting epinephrine. The persistence of the former suggests their relationship with the postganglionic sympathetic neurons.

C. Autonomic Nervous System and the Adrenal Medulla

It is becoming clear that the same structures are involved, at the central nervous system level, for triggering the adrenal cortex and the medulla. Thus, the limbic system is a candidate for early stimulation followed by the hypothalamus. It has been shown experimentally that electrical stimulation of the dorsomedial nuclei and posterior hypothalamic areas leads to increased secretion of epinephrine and norepinephrine from the adrenal medulla. One possibility is that separate hypothalamic centers exist that control the secretion of norepinephrine and epinephrine, since there are situations when one type of medullary catecholamine is secreted predominantly. Signals are transmitted from the hypothalamus to the sympathetic nervous system, specifically to a neuron located in the lateral horn gray matter of the thoracolumbar spinal cord (Figure 11-1). This neuron sends out a long axon that travels in splanchnic nerves, passing through the celiac ganglion and continuing to the adrenal medulla. This first neuron releases acetylcholine from its nerve ending. The acetylcholine-containing secretor vesicles are about 40–65 nm in diameter. The acetylcholine released binds to a receptor in the synaptic junction to the cell body of a second neuron in this chain, which contains the acetylcholine receptor. In this case, the second neuron of the sympathetic autonomic nervous system is the specialized chromaffin cell.

The chromaffin cell itself may be considered to be a modified postganglionic sympathetic neuron. It is analogous to the second neuron of the sympathetic nervous system and releases catecholamines. Whereas the catecholamines of the usual postganglionic second neuron synapse with an effector organ or tissue, chromaffin cells release catecholamines directly into the blood circulation. The nature of the precise signals at the beginning of the overall network is unclear. It would not be surprising to learn that the limbic system stimulates the hypothalamus electrically by alteration in the firing rate of the electrical signal. Presumably, further transmission from the hypothalamus to the sympathetic system is neuronal.

D. Histology of the Adrenal Medulla

The cells of the medulla are polyhedral, epithelioid, and arranged in connecting cords. Many capillaries and venules are present (Figure 11-1). As previously

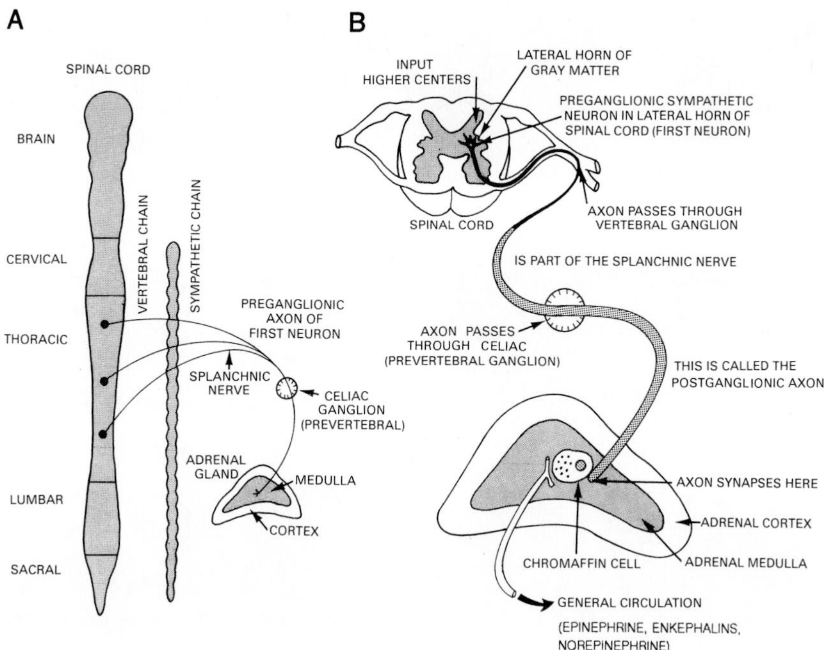

FIGURE 11-1 Autonomic nervous system and the innervation of chromaffin cells of the adrenal medulla. (A) Connections between the central nervous system and the adrenal medulla. Signals emanate from the hypothalamus down the spinal cord by way of descending autonomic pathways to the appropriate level in the spinal cord (segment 7 of the thoracic spinal cord). The signal is transported further from segment 7 over the long preganglionic axon, which passes through the celiac ganglion and innervates the adrenal medulla by releasing acetylcholine at the level of the chromaffin cell. (B) Details of the pathway from the spinal column described in A. Redrawn from notes compiled by Dr. Laurie Paavola, Department of Anatomy, Temple University School of Medicine, with permission.

described, the neuron of the autonomic sympathetic nervous system delivers acetylcholine to the chromaffin cell and makes terminal contact with it. In most mammals, the epinephrine-containing chromaffin cell is more abundant than the norepinephrine chromaffin cell.

Chromaffin cells are of two types and they share many common features. The cell is polyhedral, with a large round nucleus containing one or more nucleoli. A large number of catecholamine-containing granules 1000–3000 Å in diameter are obvious in the cytoplasm. Chromaffin cells are also found in the kidney, ovary, testis, liver, heart, and gastrointestinal tract. Similar cell types are found in the carotid and aortic bodies, which function as chemoreceptors (see Chapter 4).

A target tissue of interest with respect to epinephrine is the hepatocyte. A discussion of the biology of this cell and the liver was presented in Chapter 10. Important target cells for the action of epinephrine are pericytes and vascular smooth muscle cells. The membranes of these cells contain α-receptors that, when occupied by ligand, lead to movements of ions in and out of the cell and activation of contractile elements of the cell. The pericyte is found outside the endothelial lining of arterioles, capillaries, and sinuses. Although the function of these cells is not clearly understood, they have been suggested to regulate the size of the vessel lumen in that they possess contractile elements. There are tight junctions between pericytes and endothelial cells suggestive of ion transfer. Smooth muscle cells are elongated with a contractile apparatus probably similar to the myosin system. Actin is undoubtedly present in these cells, but the scarcity of thick filaments characteristic of myosin makes it difficult to explain contractile behavior on a basis similar to that occurring in striated muscle.

III. CHEMISTRY

The structures of norepinephrine, epinephrine, and a number of other compounds are shown in Figure 11-2. Various epinephrine-like compounds are compared for their ability to elevate blood pressure or their ability to cause the contraction of an aortal strip *in vitro.* The phenylethylamine structure is basic to this activity.

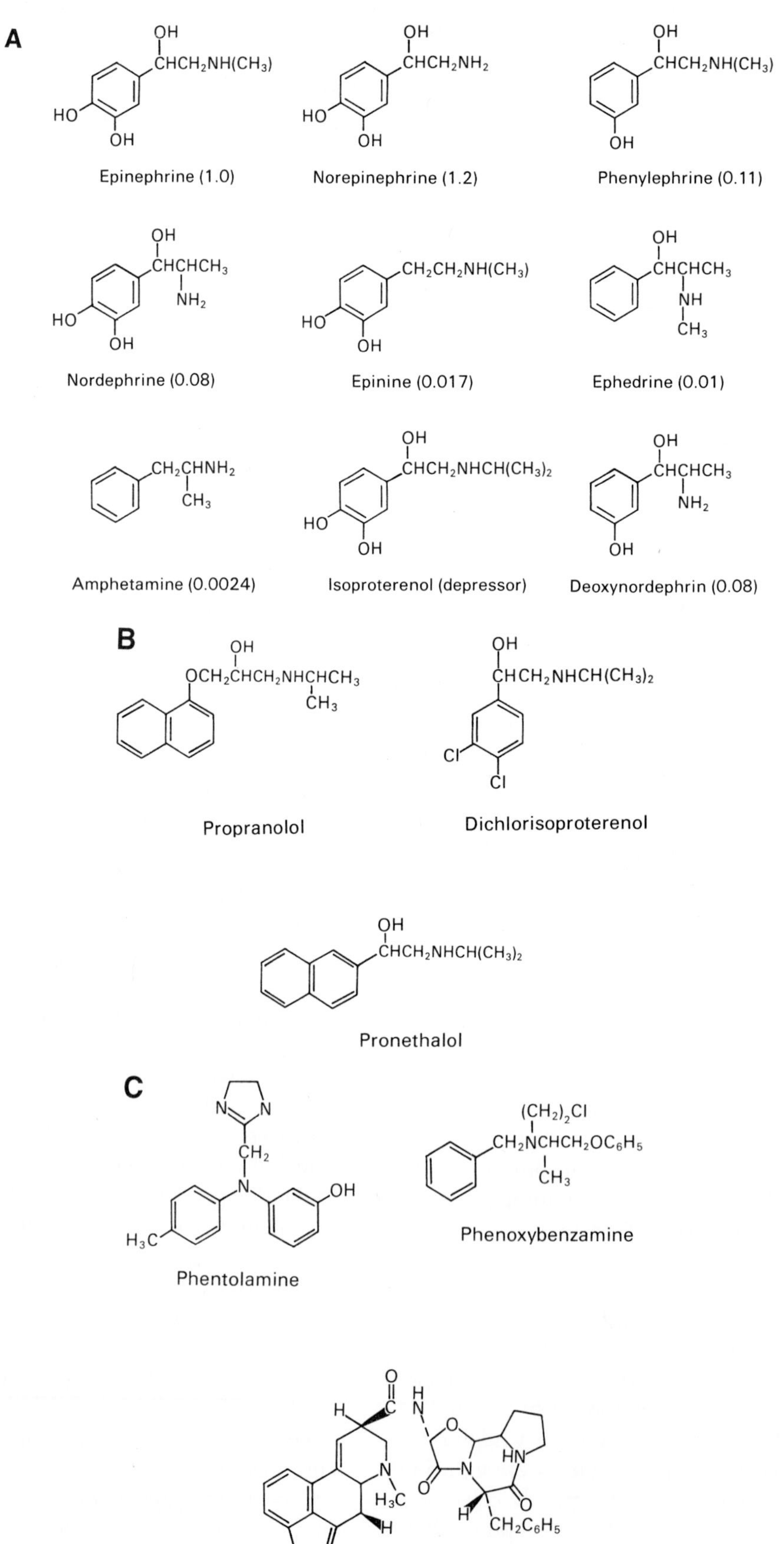

FIGURE 11-2 (A) Compounds with epinephrine-like (pressor) activity. The numbers in parentheses represent the relative pressor activity (epinephrine = 1.0). Reproduced from Frieden, E. H. (1976). "Chemical Endocrinology," p. 46. Academic Press, New York. (B) Specific antagonists of β-adrenergic receptors. (C) Specific antagonists of α-adrenergic receptors.

FIGURE 11-3 Structure of 2-benzylimidazoline.

Hydroxylation of the ring amplifies pressor activity. Some substituted imidazolines, such as benzylimidazoline (see Figure 11-3), have adrenergic activity. This type of structure can be viewed as a phenylethylamine in which the ethylamine portion has been cyclized. Further development of structure–function and antagonism occurs with the later discussion of specific receptors of catecholamines.

IV. HORMONE ACTION AND BIOCHEMISTRY

A. Biochemistry of the Chromaffin Cell of the Adrenal Medulla

The relationship of the adrenal cortex to the medulla is shown in Figure 11-4. In Figure 11-4A, the vascular relationships are evident. The secretions from the cortex bathe the medulla on their way to the general circulation. Since unique environmental signals similar to those that cause epinephrine secretion from the medulla also stimulate glucocorticoid synthesis in the cortex, a ready supply of cortisol is available to the medulla for enhancement in the rate of conversion of norepinephrine to epinephrine. The availability of glucocorticoid is demonstrated in Figure 11-4B. Here it is shown that the initial signals resulting in ACTH stimulation of the *zona fasciculata* of the cortex also stimulate the release of neurosecretory granules by way of the limbic system, hypothalamus, and autonomic nervous system via a cholinergic neuron, which is shown to release acetylcholine synapsing with a receptor on the membrane of a chromaffin cell. Some rearrangement of macromolecules in the cell membrane occurs, probably involving phospholipids, and a "calcium channel" is constituted, permitting influx of this ion from the extracellular space. In some way the Ca^{2+} facilitates secretion of the neurosecretory granules. This may occur by the binding of Ca^{2+} to the granules and also to the inner plasma membrane, acting as a bridge, facilitating the process of exocytosis, and pouring the contents of the granule into the venous sinus and then into local arterioles through fenestrations. Catecholamine secretion is stimulated by both nicotinic and muscarinic agonists. Recent work suggests that both types of agents may be binding to the same acetyl-

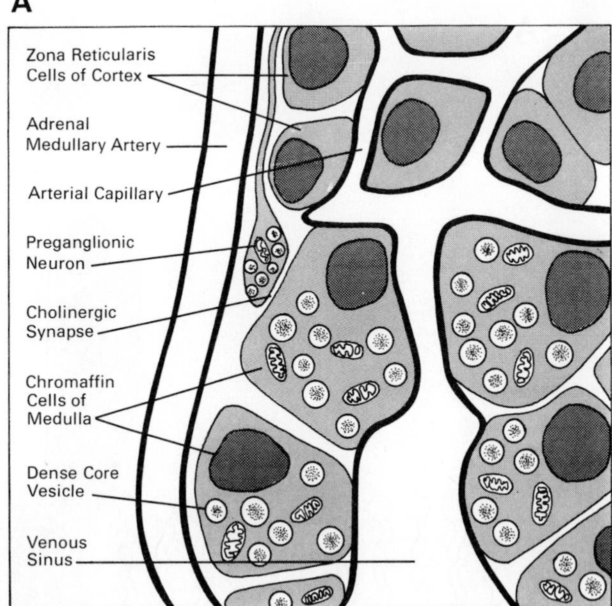

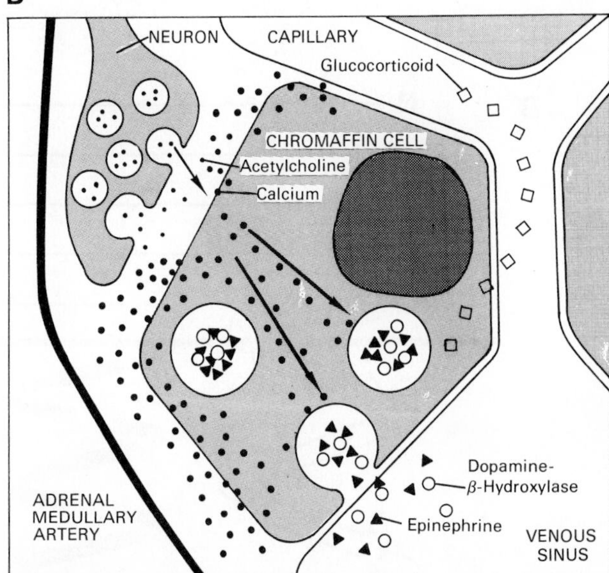

FIGURE 11-4 Diagrams showing the relationship of adrenal medulla chromaffin cells to preganglionic neuron innervation and the structural elements involved in the synthesis of epinephrine and the discharge of catecholamines in response to acetylcholine. (A) Functional relationship between cortex and medulla for control of the synthesis of adrenal catecholamines. Glucocorticoids that stimulate enzymes catalyzing the conversion of norepinephrine to epinephrine reach the chromaffin cells from capillaries shown in B. (B) Discharge of catecholamines from storage granules in chromaffin cells after nerve fiber stimulation, resulting in the release of acetylcholine. Calcium enters the cells as a result, causing the fusion of granular membranes with the plasma membrane and exocytosis of the contents. Reproduced with permission from Krieger D. T. and Hughes, J. C. (eds.) (1980). "Neuroendocrinology." Sinauer Associates, Sunderland, Massachusetts.

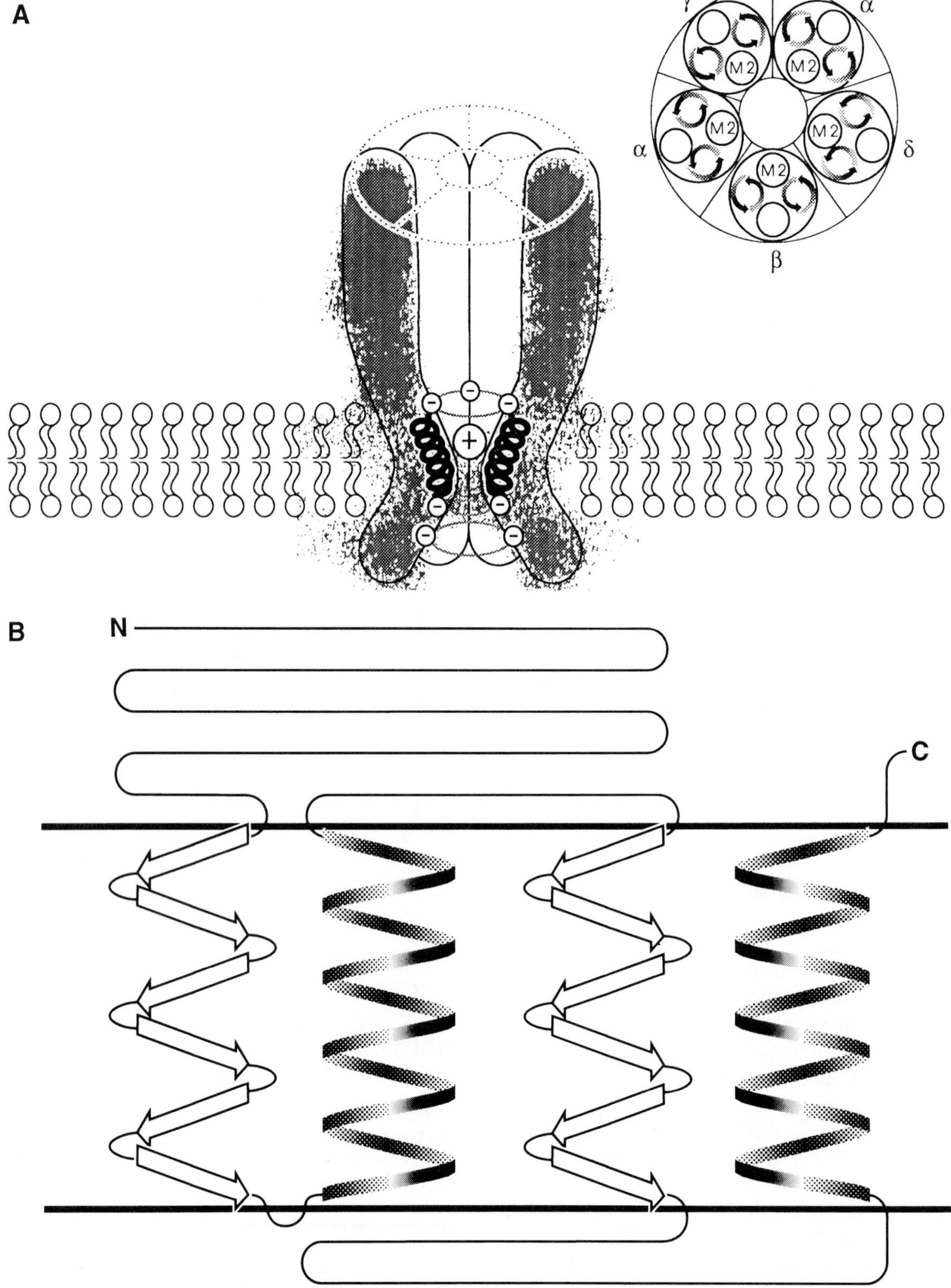

FIGURE 11-5 (A) Model of the human acetylcholine receptor, longitudinal section (center), and cross section at the plane of the membrane surface (upper right). The model combines results from electron microscopy (shaded part showing the dimensions of the receptor protein and its position in the membrane), molecular biology (three rings of negative charge shown to be important for the properties of the receptor's channel), and biochemistry (solid lines, depicting the quaternary structure and the channel wall formed by the α-helical M2

choline receptor at two different sites. This receptor may be unusual in that it has both nicotinic and muscarinic features. Recent molecular modeling work is available on the structure of the nicotinic acetylcholine receptor as shown in Figures 11-5A,B. The four-transmembrane helix structure of this receptor comprises an ion channel formed by homologous structural elements in the subunits of the receptor (Figure 11-5A). Although the information for this model comes from indirect experiments and modeling, it provides a workable view of the receptor and its inherent ion channel.

In the chromaffin cell the process of catecholamine synthesis is ongoing, as shown in Figure 11-6. Tyrosine (or phenylalanine) enters the cell via the blood circulation and extracellular fluid. In the cytosol, tyrosine is converted by tyrosine hydroxylase to dihydroxyphenylalanine (dopa) (Figure 11-7). Tyrosine hydroxylase is the rate-limiting enzyme in the synthesis of catecholamines in all adrenergic neurons. L-dopa is converted to dopamine by an L-amino acid decarboxylase located in the cytosol. Dopamine must then enter the neurosecretory granule since the enzyme catalyzing the subsequent catalytic biosynthetic pathway is localized there. Norepinephrine formed within the granule must perfuse out to be converted to epinephrine by the final enzyme of the pathway, phenylethanolamine *N*-methyltransferase (PNMT), which is located in the cytosol. Norepinephrine, as an intermediate in this overall process, would be expected to occur in lower concentrations in the adrenal medulla, and this is true for the human adrenal medulla where it may comprise only about 20% of the total catecholamines. Obviously, the epinephrine in the cytosol must move back into the neurosecretory granule and is stored there, awaiting secretion. The re-uptake process is linked to an ATPase of the granule membrane. The scenario of movement of intermediates in a biosynthetic pathway between intracellular compartments is not new, and we have seen similar movements during the synthesis of the steroid hormones.

segments of the receptor subunits). This model is also in accordance with recent computer modeling studies. Reproduced from Hucho F., Görne-Tschelnokow, U., and Strecker, A. (1994). β-structure in the membrane spanning part of the nicotinic acetylcholine receptor. *Trends Biochem. Sci.* **19**, 383–387. (B) Schematic representation of the transmembrane folding of nAChR subunits. The model takes into account the facts that there are approximately equal amounts of α- and β-structure and that M2 and probably M4 are α-helices. The β-structure (arrows), assigned to M1 and M3, is depicted pointing in different directions. FTIR measurements (linear dichroism) have shown that the intramembrane β-structures are not oriented uniaxially. The intramembrane peptides have not yet been identified. Therefore, the number of membrane passings is hypothetical in this model. Good experimental evidence exists for both the amino and the carboxyl termini being on the extracellular side of the membrane. Reproduced from Hucho, F., Görne-Tschelnokow, U., and Strecker, A. (1994). β-structure in the membrane spanning part of the nicotinic acetylcholine receptor. *Trends Biochem. Sci.* **19**, 383–387.

B. Regulation of Biosynthesis

Tyrosine hydroxylase is the rate-limiting enzyme in the overall biosynthesis of epinephrine (Figure 11-6). It is allosterically inhibited (heterotrophy) by norepinephrine. Presumably the initial enzyme contains an allosteric site with a K_i value that from time to time is exceeded by the amount of norepinephrine in the cytosol. This is undoubtedly of physiological significance because norepinephrine cannot be metabolized further without first entering the neurosecretory granule. Although perhaps not rate-limiting for the overall pathway, its concentration in cytosol could easily become effective. Also, the tyrosine hydroxylase rarely functions at maximal velocity and therefore, at least to a certain extent, is probably substrate-controlled. At suitable concentrations, norepinephrine can bind to the allosteric site on tyrosine hydroxylase, producing a change in conformation and a rise in the K_m^{Tyr}.

Interestingly, ACTH, which so far we have considered in the context of the *zona glomerulosa, zona fasciculata,* and *zona reticularis* of the cortex in the stimulation of steroid biosynthesis, also apparently binds to chromaffin cell membrane receptors and stimulates the intracellular level of cyclic AMP. Elevated cyclic AMP levels activate protein kinases, which may phosphorylate some proteins in the cell. The activated protein kinase, alternatively, may translocate to the nucleus where reactions are stimulated, leading to an increased rate of synthesis of tyrosine hydroxylase.

Finally, the levels of tyrosine hydroxylase are also under neuronal control. The dopa decarboxylase enzyme does not appear to be under specific hormonal or neural regulation. However, dopamine β-hydroxylase is a regulated enzyme. Like tyrosine hydroxylase, its synthesis is under neuronal control. Its level in the cell, and consequently within the neurosecretory granules, is also affected by ACTH by way of elevation of the intracellular level of cyclic AMP. The final enzyme in the pathway, PNMT, can be induced by cortisol mediated by a cytoplasmic glucocorticoid receptor, presumably at the transcriptional level (see Figure 10-14). The extent of the increase in PNMT by cortisol is only about 50% with long-term stress. Under similar conditions, tyrosine hydroxylase, through the indicated regulators, can be increased 3-fold.

Tyrosine hydroxylase activity appears to have a half-life of about 3 days, as judged from its exponential

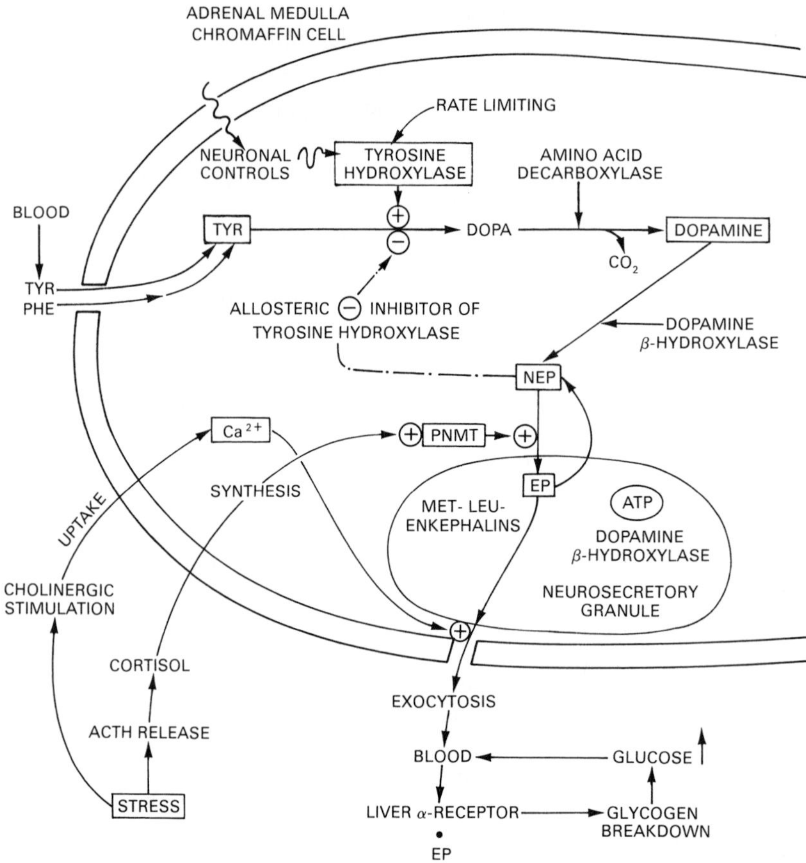

FIGURE 11-6 Biosynthesis, packaging, and release of epinephrine in the adrenal medulla chromaffin cell. Abbreviations: PNMT, phenylethanolamine *N*-methyltransferase; EP, epinephrine; NEP, norepinephrine; SAM, *S*-adenosylmethionine. Arrows pointing upward following the name of a compound refer to an increase in the concentration of that substance. Neurosecretory granules contain epinephrine, dopamine β-hydroxylase, ATP, and Met- or Leu-enkephalin, as well as larger enkephalin-containing peptides or norepinephrine in place of epinephrine. Epinephrine and norepinephrine are contained in different cells. Enkephalins could also be contained in separate cells, although that is not completely clear. Presumably epinephrine, once secreted into the bloodstream, not only affects α-receptors of hepatocytes to ultimately increase blood glucose levels, as shown here, but also interacts with α-receptors on vascular smooth muscle cells and on pericytes to cause cellular contraction and increase blood pressure.

rate of decline following the removal of long-term stress. Denervation of the adrenal gland by severing the splanchnic nerve prevents elevations of tyrosine hydroxylase and dopamine β-hydroxylase in stressed animals, whereas the activity of PNMT can still be induced by prolonged stress. Thus, nerve impulses, as indicated in Figure 11-6, are required for the stimulation of tyrosine hydroxylase and dopamine β-hydroxylase. In repeated or long-term stress in which epinephrine is secreted from the adrenal medulla in appreciable amounts, there is an elevation of the capacity for further catecholamine synthesis. This is accomplished by increases in the enzymes of the biosynthetic pathway discussed earlier.

C. Content of Neurosecretory Granules and Secretion

The neurosecretory granules have been found to contain epinephrine, ATP, dopamine β-hydroxylase, enkephalins, and enkephalin-containing peptides. The granules have an acidic interior, a condition that is known to stabilize epinephrine. The loss of dopamine β-hydroxylase from chromaffin when the content of granules has been exhausted provides a rationale for the interval required to produce epinephrine again in high levels. This is related to the time involved in the resynthesis of dopamine β-hydroxylase. The epinephrine released causes the contraction of the vascular system already discussed and leads to glycogen breakdown in the liver. Both of these processes probably proceed by the mediation of α-receptors.

Enkephalins exert an analgesic effect in the central nervous system. The fate of these polypeptide hormones released into the bloodstream from the adrenal medulla is not clear. Obviously, there could be some problem in transporting them into the brain in view of the blood–brain barrier, and they do not function there.

The release of enkephalins from the adrenal medulla secretory granule is proportional to the release of epinephrine. Enkephalin is stored in the granule together with epinephrine. Epinephrine, as indicated before, must enter the chromaffin granule after its conversion in the cytosol by PNMT. This uptake into the neurosecretory granule is stimulated by Mg^{2+} and ATP coupled to an H^+ electrochemical gradient. The proton pumping Mg ATPase and an anion transport site apparently exist as parts of a macromolecular complex in the membrane of the chromaffin granule. The internal acidic milieu (pH 5.7) of the granule coincides with a pH optimum for dopamine β-hydroxylase of pH 5.5 and explains the need for compartmentation of this enzyme.

Some new information on stimulators and inhibitors of catecholamine secretion is summarized in Table 11-1.

The chromaffin cells, maintained in cell culture conditions, are able to synthesize Met- and Leu-enkephalin *de novo* from radioactive amino acids. Reserpine (Figure 11-8), a drug that has catecholamine-depleting activity, causes nearly complete removal of catecholamines from the chromaffin cells, but the synthesis of enkephalins is enhanced under these conditions, particularly that of Met-enkephalin. The adrenal medulla usually contains Met-enkephalin at several times the level of Leu-enkephalin. Free enkephalin pentapeptides are about 5% of the total enkephalin-containing peptides in the adrenal medulla.

The details of the mechanism of the intracellular events preceding exocytosis of the neurosecretory granules are unclear. Once again, Ca^{2+} (see Chapter 1) is involved and is elevated in the cytosolic compartment following the release of acetylcholine from the sympathetic autonomic cholinergic neuron and the postsynaptic binding of the transmitter to the chromaffin cell membrane receptor. The exocytotic event appears to be an all-or-none process in which the total contents of the granule are released into the extracellular space and into local blood circulation by entrance through *fenestrated* capillaries, which overcome diffusion barriers in the normal arteriolar wall. Even though epinephrine and enkephalins may be contained in the same chromaffin cell, there is evidence that their syntheses may be under different controls. The empty granule is probably reabsorbed by the chromaffin cell (see Chapter 4) or recycled from its fused state with the chromaffin membrane, because empty granules can be found in the medulla cells after secretion has occurred. During exocytosis, all of the products of the processing of proenkephalin are released, including the largest and even a small amount of proenkephalin itself. Three of the enkephalin-containing peptides are more active in isolated systems than the pentapeptides themselves, including peptide E, the heptapeptide, and the octapeptide. Peptide E is a 25-amino acid residue chain that contains one copy of Tyr-Gly-Gly-Phe-Met at the N-terminus and one copy of Tyr-Gly-Gly-Phe-Leu at the C-terminus. These may not be processed to free enkephalin. Enkephalin receptors are present in many tissues where these hormones may act, but it appears that they do not act in the brain. Some activities of enkephalins are reviewed in Table 11-2. Some of these actions may be in the realm of functions for the adrenal medulla enkephalins, although not much is known about how and where they act. It has been shown that Met-enkephalin changes membrane fluidity and modulates the production of nitric oxide and GMP in rat brain synaptosomes, and Met-enkephalin can potentiate immune reactivity when injected in low doses into the cerebral cavity. Both Met- and Leu-enkephalins appear to be agonists for the delta opioid receptor, and in cardiac myocytes Leu-enkephalin stimulates the levels of IP_3 and intracellular Ca^{2+}. The delta opioid receptor has been cloned, and a three-dimensional model of the ligand-bound receptor has been generated.

Prostaglandin E_2 (PGE_2) is bound specifically by particulate fractions of bovine, ovine, and human adrenal medulla cells. The binding is sensitive to treatment with phospholipases, glycosidases, trypsin, and sulfhydryl reagents. The dissociation constant of 2 n*M* PGE_2 indicates tight binding. The receptor site is quite specific for PGE_2 compared to other prostaglandins, and prostacyclin (PGI_2) has a lower affinity than PGE_2. The prostaglandins bind to this receptor site in the same order that they demonstrate potency to release epinephrine from chromaffin granules. Thus, PGE may play a role in controlling the activity of adrenal medulla secretions.

D. Catabolism of Catecholamines

As with most other hormones, epinephrine is active at its receptor site in the unmetabolized form. The routes of metabolic inactivation have been well-established. These pathways of metabolism are shown

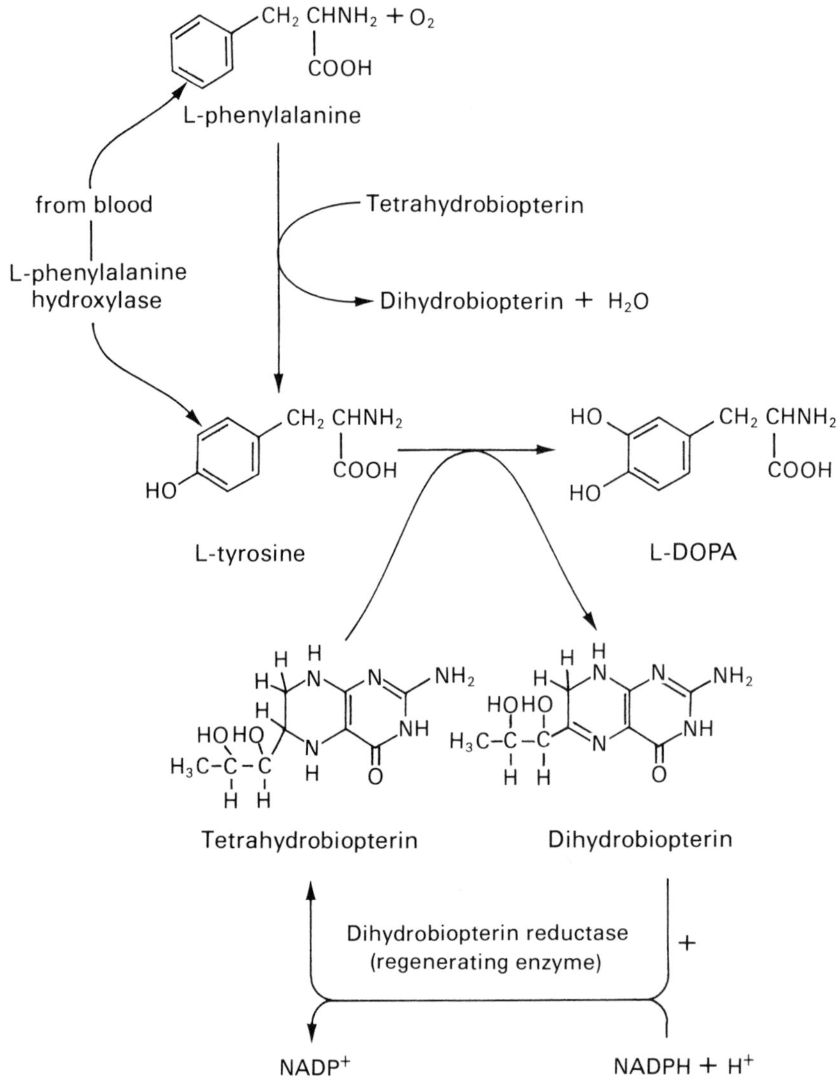

FIGURE 11-7 Biochemical steps in the synthesis of epinephrine by the chromaffin cell of the adrenal medulla.

in Figure 11-9. The major reactions involve oxidation by monoamine oxidase (MAO), a mitochondrial enzyme in liver, kidneys, brain, and adrenergic nerve endings. The second major catabolic reaction is methylation. This is catalyzed by catechol *O*-methyltransferase (COMT), which is widely distributed, especially in liver and kidneys, but it is absent from nerve endings. Its location may be in the cytosol and possibly in the cell membrane. The aldehydes produced are converted to acids by aldehyde dehydrogenase. The aldehydes can also be reduced to form alcohols. COMT methylates catecholamines at the C-3 hydroxyl of the ring and, to a vastly smaller extent, the *p*-hydroxyl group. There are variable amounts of MAO and COMT in different tissues, but the catecholamines eventually reach the liver and kidneys, which contain these enzymes. Consequently, the urinary products appear in a fairly consistent pattern (Figure 11-9). The majority of these excretion products are conjugated as glucuronides or sulfates, the latter being the more abundant form.

V. ACTIONS OF EPINEPHRINE

A. General Effects of Catecholamines

Catecholamines act through pairs of outer cell membrane receptors that produce opposite actions. A similar phenomenon is characteristic of the prostaglandins (Chapter 16). A partial listing of the opposite actions of catecholamines in various tissues is shown in Table 11-3. The opposing responses are generated by ligand

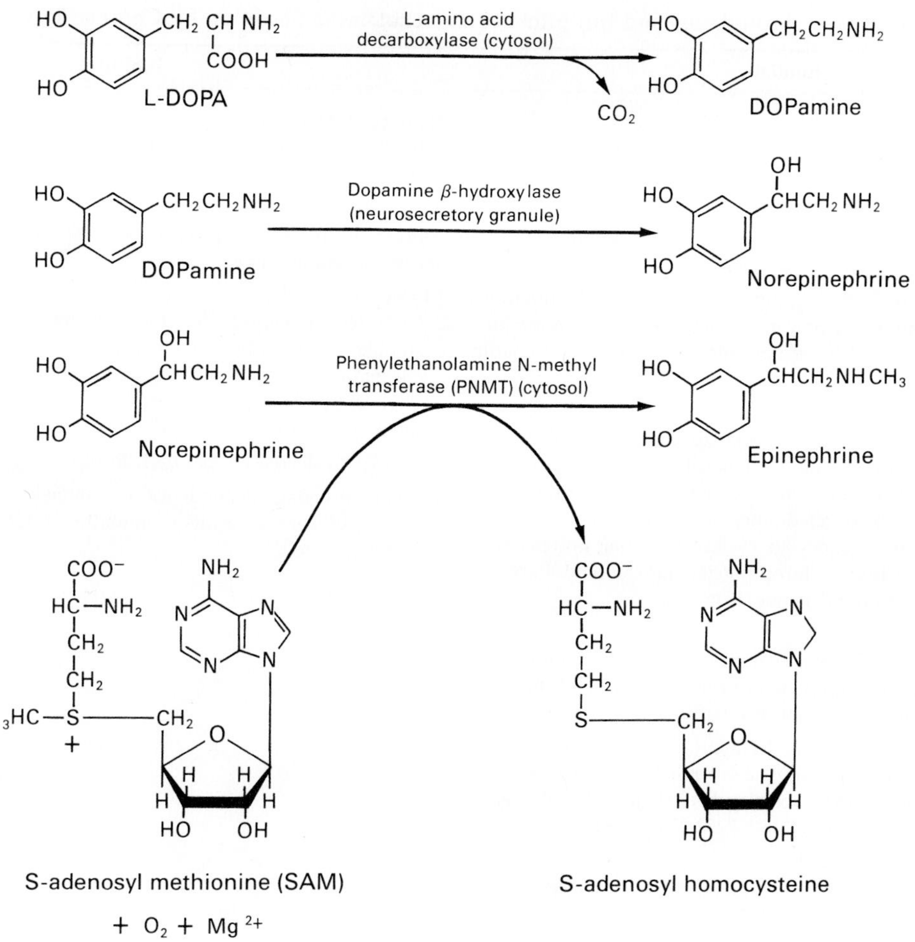

L-DOPA → DOPamine (CO_2)
L-amino acid decarboxylase (cytosol)

DOPamine → Norepinephrine
Dopamine β-hydroxylase (neurosecretory granule)

Norepinephrine → Epinephrine
Phenylethanolamine N-methyl transferase (PNMT) (cytosol)

S-adenosyl methionine (SAM) + O_2 + Mg^{2+} → S-adenosyl homocysteine

FIGURE 11-7—*Continued*

binding to different receptors located on the same or different cell membrane, called α- and β-adrenergic receptors. α-Receptors act by altering the pattern of ion flow into the cell, particularly with regard to Ca^{2+}, whereas the β-receptor operates by the stimulation of adenylate cyclase and the intracellular level of cyclic AMP. The two receptor types are somewhat difficult to distinguish because usually either one can bind the same ligand (e.g., epinephrine). Many of the receptors have subclasses wherein each class may operate through a different mechanism. This is true for the α-receptors. Two principal classes are α_1-and α_2-receptors. α_1-Receptors control Ca^{2+} ion flux, whereas α_2-receptors inhibit adenylate cyclase. α-and β-receptors can be resolved by a specific order of affinity for different compounds and especially by the use of specific inhibitors of each receptor type. α-Adrenergic receptors bind ligands (agonists) in the following order (starting with highest affinity), epinephrine > norepinephrine > phenylephrine > isoproterenol, whereas the order of agonist binding for β-adrenergic receptors is isoproterenol > epinephrine > norepinephrine > phenylephrine. The structures of these agonists are shown in Figure 11-2. Thus, epinephrine can bind to both receptors, and the actual physiological effect achieved may be a direct consequence of the available amount of catecholamine in a particular location since the binding constants for the various types of the adrenergic receptors may be quite different. Table 11-4 lists the adrenergic receptor types with agonist preferences and antagonists.

There are specific inhibitors or blockers of α- and β-adrenergic receptors (Figure 11-2, Table 11-4). α-Adrenergic antagonists are represented by phentolamine, phenoxybenzamine, and ergot alkaloids. Specific antagonists of β-adrenergic receptors are propranolol, dichlorisoproterenol, and pronethalol. Catecholamines, especially epinephrine, represent an emergency stimulus for alterations in carbohydrate metabolism to make available an energy source that can be burned quickly for "flight-or-fight" responses. Thus, the initial responses in various tissues are as

TABLE 11-1 Stimulators and Inhibitors of Catecholamine Secretion by Chromaffin Cells

Stimulate	Inhibit
Acetylcholine Activates both nicotinic and muscarinic receptors to trigger catecholamine secretion in rat adrenal medulla.[a]	Neuropeptide Y Inhibits nicotinic receptor-induced ^{3}H-NEP secretion from bovine chromaffin cells.[j]
Vasoactive intestinal polypeptide (VIP) Able to evoke catecholamine secretion from a chromaffin cell.[b]	Nitric oxide From chromaffin cell-potential autoinhibitory process of catecholamine release.[k]
Protein kinase C activators Those such as phorbol esters induce secretion and redistribution of filamentous actin which prevents contact between chromaffin granules and the cortical surface plasma membrane of chromaffin cells.[c]	Chromostatin (A 20-residue peptide derived from chromogranin A, the major soluble component of secretory granules in adrenal medullary chromaffin cells) stimulates protein phosphatase 2-A dephosphorylating a target protein and lowering its activity in the secretory process.[l]
K^+ channel blockers (Such as tetraethylammonium) depolarize chromaffin cell membrane and induce elevation in cytosolic Ca^{2+} and dose-dependent catecholamine secretion. Action dependent on Ca^{2+} and blocked by $MgCl_2$/or nifedipine.[d] Also, PACAP (pituitary adenylate cyclase-activating polypeptide) stimulates secretion from cultured porcine adrenal medullary chromaffin cells in a dose-dependent manner.[e]	IL-1α/TNFα decreased met ENK levels[g]
Histamine Greater stimulation in EP cells than in NEP cells. Histamine acts through chromaffin H1 receptors to increase PI metabolism and stimulates secretion of CA from chromaffin cells[f]	Blockers of voltage-dependent Ca^{2+} channels e.g., by idebenone results in inhibition of CA secretion in adrenal chromaffin cell.[m]
IL-1α elevates VIP TNFα[g] (Ref. A (Eskay, R. L., and Eiden, L. E., IL-1α and TNFα differentially regulate enkephalin, vasoactive intestinal polypeptide, neurotensin and substance P biosynthesis in chromaffin cells. Endocrinol. 130, 2252-2258, 1992).	
Depolarizing concentrations of K+ Cause secretion of high and low molecular weight enkephalin-like peptides from chromaffin cells.[h]	
Nicotine and muscarinic agonists (e.g., oxotreniorine-M) promote CA secretion. Apparently both types of agonists bind at 2 different sites on the same receptor which is coupled to secretion representing an unusual cholinergic receptor that has both nicotine and muscarinic features.[i]	

[a,b] Chowdhury, P. S., Guo, X., Wakade, T. D., Przywara, D. A., and Wakade, A. R. (1994). Exocytosis from a single rat chromaffin cell by cholinergic and peptidergic neurotransmitters. *Neuroscience* **59,** 1–5.

[c] Trifaro, J. M., Vitale, M. L., and Rodriguez Del Castello, A. (1993). Scinderin and chromaffin cell actin network dynamics during neurotransmitter release. *J. Physiol. (Paris)* **87,** 89–106.

[d] Gonzalez-Garcia, C., Cena, V., Keiser, H. R., and Rojas, E. (1993). Catecholamine secretion induced by tetraethylammonium from cultured bovine adrenal chromaffin cells. *Biochim. Biophys. Acta* **1177,** 99–105.

[e] Isobe, K., Nakai, T., and Takura, Y. (1993). Ca^{2+}-dependent stimulatory effect of pituitary adenylate cyclase-activating polypeptide on catecholamine secretion from cultured porcine adrenal medullary chromaffin cells. *Endocrinology* **132,** 1757–1765.

[f] Choi, A. Y., Cahill, A. L., Perry, B. D., and Perlman, R. L. (1993). Histamine evokes greater increases in PI metabolism and CA secretion in EP-containing than in NEP-containing chromaffin cells. *J. Neuro. Chem.* **61,** 541–549.

[g] Eskay, R. L., and Eiden, L. E., (1992). IL-1α and TNFα differentially regulate enkephalin, vasoactive intestinal polypeptide, neurotensin and substance P biosynthesis in chromaffin cells. *Endocrinology* **130,** 2252–2258.

[h] Cherdchu, C., and Hexum, T. D. (1991). Differential secretion of enkephalin like peptides from bovine adrenal chromaffin cells. *Neuropeptides* **19,** 237–242.

[i] Shirvan, M. H., Pollard, H. B., and Heldman, E. (1991). Mixed nicotinic and muscarinic features of cholinergic receptor coupled to secretion in bovine chromaffin cells. *Proc. Natl. Acad. Sci. USA* **88,** 4860–4864.

[j] Hexum, T. D., Zheng, J., and Zhu, Jr. (1994). Neuropeptide Y inhibition of nicotinic receptor-mediated chromaffin cell secretion. *J. Pharm. Exp. Therap.* **271,** 61–66.

[k] Torres, M., Ceballos, G., and Rubio, R. (1994). Possible role of nitric oxide in catecholamine secretion by chromaffin cells in the presence and absence of cultured endothelial cells. *J. Neurochem.* **63,** 988–996.

[l] Galindo, E., Zwiller, J., Bader, M. F., and Aumis, D. (1992). Chromostatin inhibits catecholamine secretion in adrenal chromaffin cells by activating a protein phosphatase. *Proc. Natl. Acad. Sci. USA* **89,** 7398–7402.

[m] Houchi, H., Azuma, M., Oka, M., and Moritaki, A. (1991). *Jpn J. Pharmacol.* **57,** 553–558.

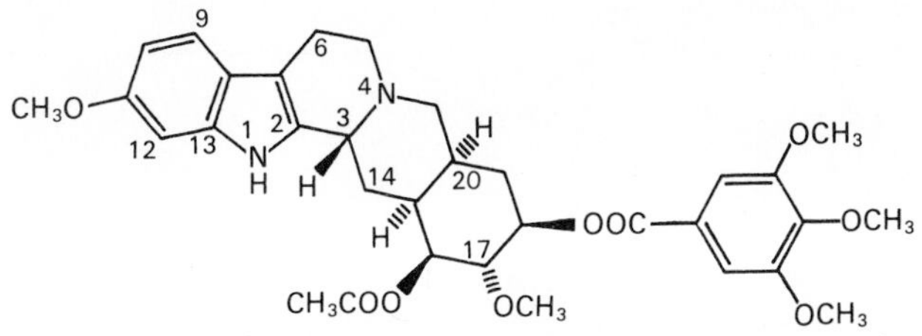

FIGURE 11-8 Structure of reserpine.

follows: increases in glycogenolysis and gluconeogenesis in the liver, increased gluconeogenesis in skeletal muscle, increased inotropic effect (affecting the force of muscular contraction in the heart), increased amylase secretion by the salivary gland, relaxation of the uterine musculature, conversion of triglycerides to free fatty acids and glycerol in adipose, elevations in blood pressure, heart rate, and cardiac output (depending on amount of catecholamine), and dilation of bronchial musculature. Although norepinephrine is a pressor, its activity in this regard is much smaller than that of epinephrine. The structure of the β_2-adrenergic receptor is shown in Figure 11-10.

B. Effects of Epinephrine on Glycogenolysis in the Liver

The breakdown of glycogen to glucose is accomplished by epinephrine operating at an α-adrenergic receptor on the hepatocyte membrane through a mechanism involving calcium ion fluxes. A scheme for the operation of the α-receptor in hepatocyte glycogenolysis is shown in Figure 11-11. As the result of a stressor

TABLE 11-2 Suggested Functions of Enkephalins[a]

Function	Explanatory remarks
Circulatory homeostatic mechanisms (chm)	These peptides are released during intense stimulation and modify chm. Primarily endorphins and enkephalins influence chm responses to stress at behavioral, endocrinological and neural levels.[b]
Enhancement of cholecystokinin (CCK) release in brain	CCK and enkephalins colocalized in several brain structures and they interact physiologically. The delta-opioid receptors may be involved in CCK release.[c]
Antagonistic actions on auditory potentials	Many opioid peptides are expressed within the olivocochlear efferent terminals involved in auditory function. Neurotransmitters of the olivocochlear efferents, such as enkephalins, dymorphins, acetylcholine may have antagonistic actions on auditory potentials. It is important to know whether adrenal chromaffin enkephalins can gain access to these nerves.[d]
Anti-stressing function in cases of heart failure	Heart failure is a stressor releasing opioids, endorphin, met-enkephalin, leu-enkephalin, dynorphin, casomorphin and others split out from precursor proteins. These play a role in improving hemodynamics but may be reduced in chronic heart failure through exhaustion of the opioid system reflected in reduced plasma levels of endorphin and lipotropin. Blood-brain barrier important here.[e]
Low doses of enkephalins potentiate immune responses	Met-enk is more efficient when applied intracerebroventricularly but also active when given intrathecally. Met-enk is more potent than leu-enk. High doses of met-enk suppressed immune inflammatory reactions and skin reactions to protein antigen and other conditions.[f]
Activation of endogenous enkephalinergic system by alcohol intake involved in continuance of alcohol drinking	Alcohol induces the endogenous enkephalinergic system and subsequent occupation of delta opioid receptors which is involved in continued alcohol drinking.[g]

[a] An important consideration is whether adrenal medulla enkephalins can be transported to the sites of action listed in this Table or whether only locally produced enkephalins are involved. Blood–brain barrier function in the case of circulatory enkephalins has to be considered.

[b] McCulbin, J. A. (1993). Stress and endogenous opioids: Behavioral and circulatory interactions. *Biological Psychiat.* **35,** 91–122.

[c] Ruiz-Gayo, M., Durieux, C., Fournie-Zaluski, M. C., and Roques, B. P. (1992). Stimulation of delta-opioid receptors reduces the *in vivo* binding of the cholecystokinin (CCK)-B selection agonist [^{3}H]pBC 264: Evidence for a physiological regulation of CCKergic systems by endogenous enkephalins. *J. Neurochem.* **59,** 1805–1811.

[d] Sakley, T. L., Kalish, R. B., Musiek, F. E., and Hoffman, D. W. (1991). Effects of opioid drugs on auditory evoked potentials suggest a role of lateral olivocochlear dymorphins in auditory function. *Hearing Res.* **55,** 133–142.

[e] Lowe, H. (1991). Role of endogenous opioids in heart failure. *Zeitschr. Kardiol.* **8,** 47–51.

[f] Jankovic, B. C. (1991). Enkephalins and immune inflammatory reactions. *Acta Neurol.* **13,** 433–441.

[g] Froehlich, J. C., Zweifel, M., Harts, J., Lumeng, L., and Li, T. K. (1991). Importance of delta opioid receptors in maintaining high alcohol drinking. *Psychopharmacology* **103,** 467–472.

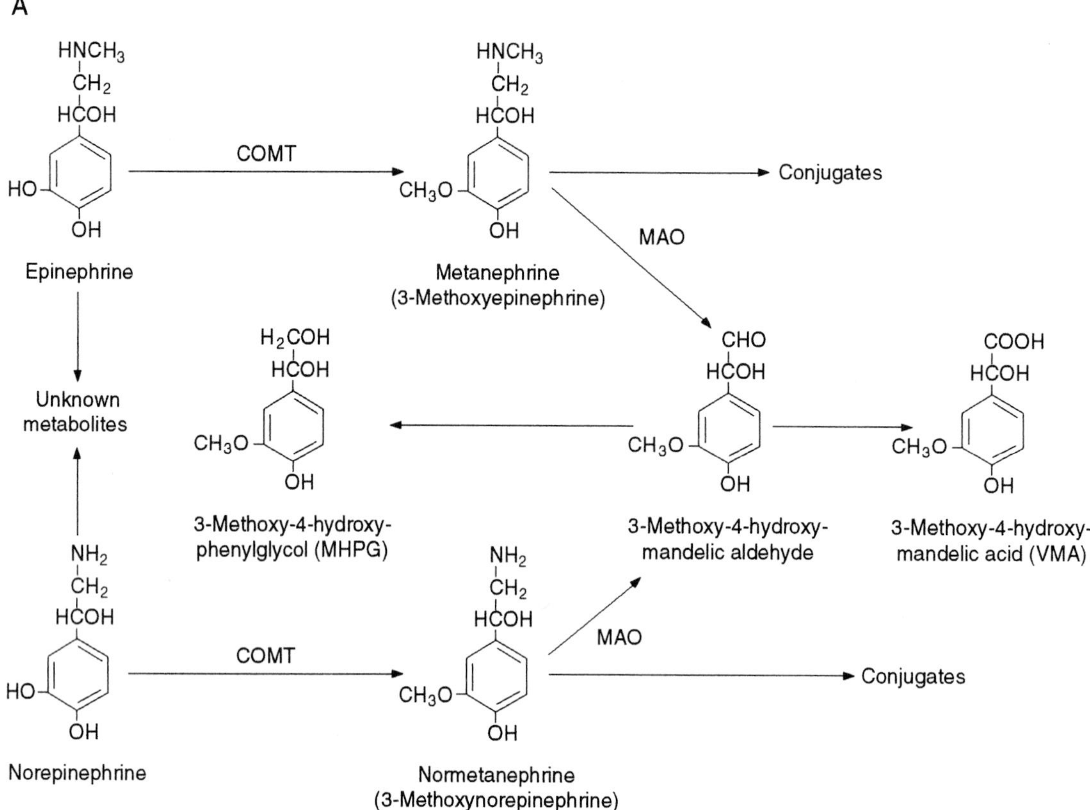

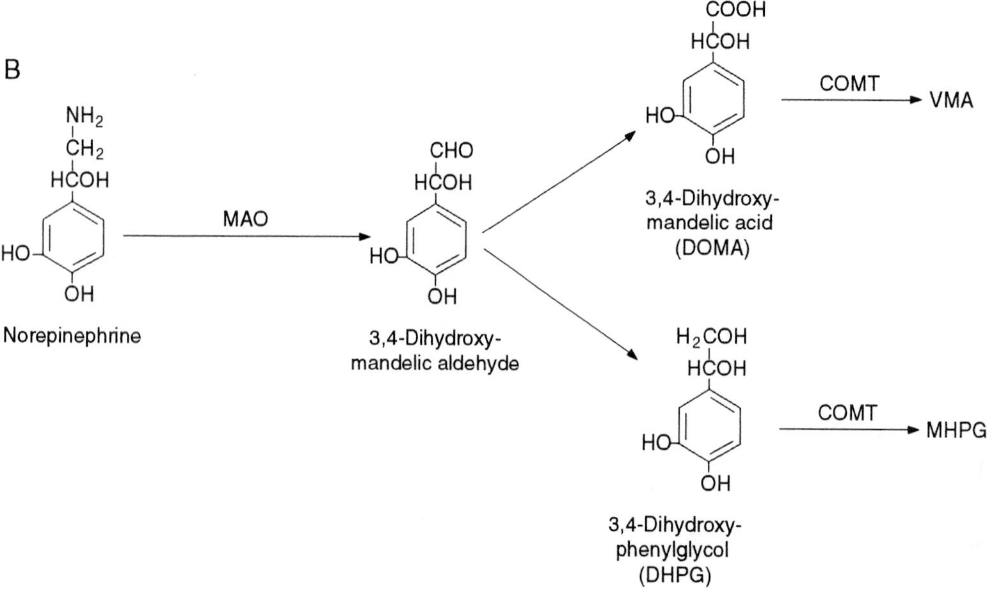

FIGURE 11-9 Catabolism of catecholamines in the blood and of norepinephrine in nerve endings. (A) The main site of catabolism of circulating epinephrine and norepinephrine is the liver. The conjugates formed there are mostly glucuronides and sulfates. (B) Catabolism of norepinephrine in adrenergic nerve endings. The acid and glycol enter circulation and may be subsequently O-methylated in VMA and in 3-methoxy-4-hydroxyphenylglycol. Abbreviations: MAO, monoamine oxidase; COMT, catechol *O*-methyltransferase; MHPG, monohydroxyphenylglycol. Reproduced from Ganong W. F. (1995). "Review of Medical Physiology," 17th ed., p. 92. Lange Medical Publications, Los Altos, California.

TABLE 11-3 Responses of Effector Organs to Autonomic Nerve Impulses and Circulating Catecholamines[a]

Effector organs	Cholinergic impulse response	Noradrenergic impulses	
		Receptor type[b]	Response
Eyes			
Radial muscle of iris		α_1	Contraction (mydriasis)
Sphincter muscle of iris	Contraction (miosis)		
Ciliary muscle	Contraction for near vision	β_2	Relaxation for far vision
Heart			
S-A node	Decrease in heart rate, vagal arrest	β_1	Increase in heart rate
Atria	Decrease in contractility and (usually) increase in conduction velocity	β_1	Increase in contractility and conduction velocity
A-V node	Decrease in conduction velocity	β_1	Increase in conduction velocity
His-Purkinje system	Decrease in conduction velocity	β_1	Increase in conduction velocity
Ventricles	Decrease in conduction velocity	β_1	Increase in contractility
Arterioles			
Coronary	Constriction	α_1, α_2	Constriction
		β_2	Dilation
Skin and mucosa	Dilation	α_1, α_2	Constriction
Skeletal muscle	Dilation	α_1	Constriction
		β_2	Dilation
Cerebral	Dilation	α_1	Constriction
Pulmonary	Dilation	α_1	Constriction
		β_2	Dilation
Abdominal viscera		α_1	Constriction
		β_2	Dilation
Salivary glands	Dilation	α_1, α_2	Constriction
Renal		α_1, α_2	Constriction
		β_1, β_2	Dilation
Systemic veins		α_1	Constriction
		β_2	Dilation
Lungs			
Bronchial muscle	Contraction	β_2	Relaxation
Bronchial glands	Stimulation	α_1	Inhibition
		β_2	Stimulation
Stomach			
Motility and tone	Increase	α_1, α_2, β_2	Decrease (usually)
Sphincters	Relaxation (usually)	α_1	Contraction (usually)
Secretion	Stimulation		Inhibition
Intestine			
Motility and tone	Increase	α_1, α_2, β_2	Decrease (usually)
Sphincters	Relaxation (usually)	α_1	Contraction (usually)
Secretion	Stimulation	α_2	Inhibition
Gallbladder and ducts	Contraction	β_2	Relaxation
Urinary bladder			
Detrusor	Contraction	β_2	Relaxation (usually)
Trigone and sphincter	Relaxation	α_1	Contraction
Ureters			
Motility and tone	Increase (?)	α_1	Increase (usually)
Uterus	Variable[c]	α_1, β_2	Variable[c]
Male sex organs	Erection	α_1	Ejaculation
Skin			
Pilomotor muscles		α_1	Contraction
Sweat glands	Generalized secretion	α_1	Slight, localized secretion[d]
Spleen capsule		α_1	Contraction
		β_2	Relaxation
Adrenal medulla	Secretion of epinephrine and norepinephrine		

(Continued)

TABLE 11-3—*Continued*

Effector organs	Cholinergic impulse response	Noradrenergic impulses: Receptor type[b]	Noradrenergic impulses: Response
Liver		α_1, β_2	Glycogenolysis
Pancreas			
Acini	Increased secretion	α	Decreased secretion
Islets	Increased insulin and glucagon secretion	α_2	Decreased insulin and glucagon secretion
		β_2	Increased insulin and glucagon secretion
Salivary glands	Profuse, watery secretion	α_1	Thick, viscous secretion
		β_2	Amylase secretion
Lacrimal glands	Secretion	α	Secretion
Nasopharyngeal glands	Secretion		
Adipose tissue		β_1, β_3	Lipolysis
Juxtaglomerular cells		β_1	Increased renin secretion
Pineal gland		β	Increased melatonin synthesis and secretion

[a] Modified from Gilman, A. G., *et al.* (eds.) (1990). "Goodman and Gilman's The Pharmacological Basis of Therapeutics," 8th ed. Pergamon Press, New York.
[b] Where a receptor subtype is not specific, data are as yet inadequate for characterization.
[c] Depends on stage of menstrual cycle, amounts of circulating estrogen and progesterone, pregnancy, and other factors.
[d] On palms of hands and in some other locations ("adrenergic sweating"). Reproduced with permission from Ganong, W. F. (1995). Review of Medical Physiology," 17th ed. Appleton & Lange, Norwalk, CN.

epinephrine binds to the α_1-receptor, but changes in the turnover of phospholipids are involved. This action is thought to lead to an increase in cytosolic Ca^{2+} levels, as a result generating inositol triphosphate and mobilizing calcium from the endoplasmic reticulum (or from mitochondria presumably deposited there as calcium phosphate). The elevated cytosolic Ca^{2+} level stimulates phosphorylase kinase, which catalyzes the conversion of phosphorylase *b* to phosphorylase *a*, the active form, and glucose 1-phosphate is converted to glucose and secreted, raising the blood level for use as an emergency fuel. The synthesis of glycogen is also blocked under these conditions. In most species, particularly in humans, β_2-adrenergic receptors are also important in epinephrine actions on the liver. Thus, epinephrine-induced glycogen breakdown may be thought of as being mediated by both α_1- and β_2-receptors for epinephrine.

TABLE 11-4 Cloned Adrenergic Receptors and Their Characteristics[a]

Mammalian subtype	Peptide length	Amino acid identity (MSD %)	Introns	Chromosomal location (human)	Agonist potency	Antagonist potency	Tissue distribution (rat)	Major G protein	Effector system
β_1	477	71 (β_2)	0	10 q24–q26	ISO > NE ≥ E	BETAX > > ICI-118551	Heart, pineal		↑ Adenylyl cyclase
β_2	413		0	5 q31–q33	ISO > E > NE	ICI-118551 > BETAX	Lung, prostate	G_g	↑ Ca^{2+} channel
β_3	402	63 (β_2)	0		NE > ISO > E	ICI-118551 >> ALP	Adipose tissue		
α_{2A}	450	42 (β_2)	0	10 q24–q26	OXY > > E ≥ NE	YOH > > > PRAZ	Aorta, brain		↓ Adenylyl cyclase
α_{2B}	450	36 (β_2), 74 (α_{2A})	0	2	OXY = E ≥ NE	YOH > > PRAZ	Liver, kidney	G_i	↑ K^+, ↓ Ca^{2+} channel
α_{2C}	461	39 (β_2), 75 (α_{2A})		4	OXY = E ≥ NE	YOH > PRAZ	Brain		↑ Na+/H+ exchange ↑ Phospholipase C, A_2
α_{1A}					OXY > > E ≥ NE	PRAZ = WB-4101	Vas deferens, brain		↑ Phospholipase C
α_{1B}	515	42 (β_2)	1	5 q31–q32	OXY > E ≥ NE	PRAZ > WB-4101	Liver, brain		↑ Ca^{2+} channel
α_{1C}	466	42 (β_2), 72 (α_{1B})	1	8	OXY > > E ≥ NE	PRAZ = WB-4101	Olfactory bulb	G_q	↑ Phospholipase A_2
α_{1D}	560	43 (β_2), 76 (α_{1B})	—	20	NE ≥ E > OXY	PRAZ ≥ WB-4101	Vas deferens, brain		↑ Phospholipase D

[a] Abbreviations: BETAX, betaxolol; NE, norepinephrine; PRAZ, prazosin; E, epinephrine; OXY, oxymetazoline; YOH, Yohimbine. Reproduced from Lomasney, J. W., Allen, L. F., and Lefkowitz, R. J. (1995). Cloning and regulation of catecholamine receptor genes. *In* "Molecular Endocrinology" (B. D. Weintraub, ed.), pp. 115–131. Raven, New York.

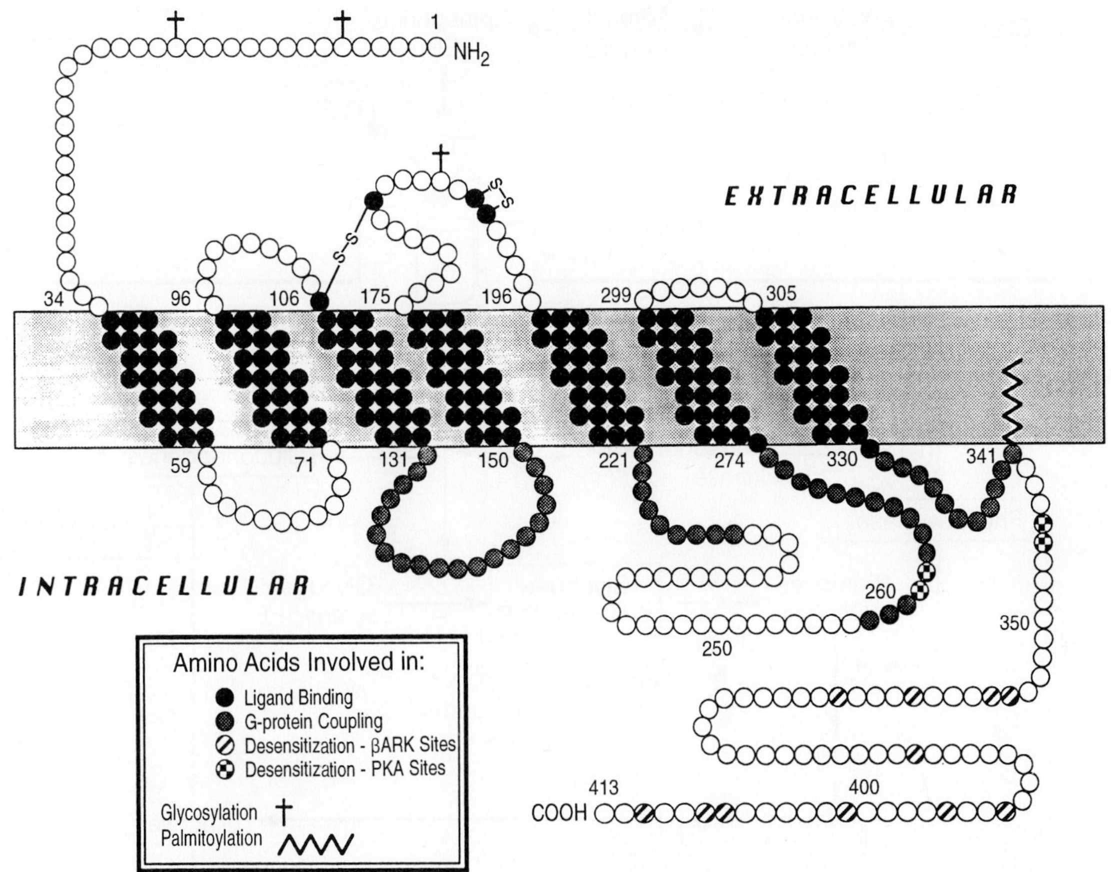

FIGURE 11-10 Topology–functional domains of the β_2-adrenergic receptor. Structural features of the β_2-adrenergic receptor as a model for the functional domains of G-protein-coupled receptors. Regions depicted are involved in ligand binding, G-protein coupling, and regulation by kinases. Abbreviations: βARK, β-adrenergic receptor kinase; PKA, protein kinase A. Reproduced from Lomasney, J. W., Allen, L. F., and Lefkowitz, R. J. (1995). Cloning and regulation of catecholamine receptor genes. In "Molecular Endocrinology" (B. D. Weintraub, ed.), pp. 115–13). Raven Press, New York.

C. Ligand Detection of the Adrenergic Receptor Subclasses

The β-receptor is a membrane site that acts to increase the rate of adenylate cyclase activity on the cytosolic side, producing stimulation in the level of intracellular cyclic AMP (see Chapter 1). The β-receptors of the heart have been classified as β_1-adrenergic receptors, and the other receptors are classified as β_2-adrenergic receptors. Epinephrine and norepinephrine bind equally well to the β_1-receptor, whereas epinephrine is about a 10-fold better ligand than norepinephrine for the β_2-receptor.

The α-receptors are also subdivided into α_1- and α_2-adrenergic receptors. These classes of receptors may be distinguished through the use of prazosin and yohimbine (Figure 11-12). The action of the receptors leads to a lowering of cyclic AMP (α_2-receptor; see Figure 11-13) and operates by altering the flux of Ca^{2+} within the cell. The α_1-receptors refer to postsynaptic receptors or those located on effector cells and have a great affinity for the antagonist prazosin (Table 11-4). The α_2-receptors refer to the presynaptic receptors, although apparently it is still unclear whether α_2-receptors actually are located on nerve terminals. They may also refer to postsynaptic effects that are coupled to adenylate cyclase in an inhibitory way, resulting in a lowering of intracellular cyclic AMP levels.

D. Regulation of Adrenergic Receptors

Adrenergic receptors are regulated by thyroid hormone and by glucocorticoids. In the hyperthyroid state, the heart β_1-adrenergic receptor is elevated in concentration. In the absence of glucocorticoids, there

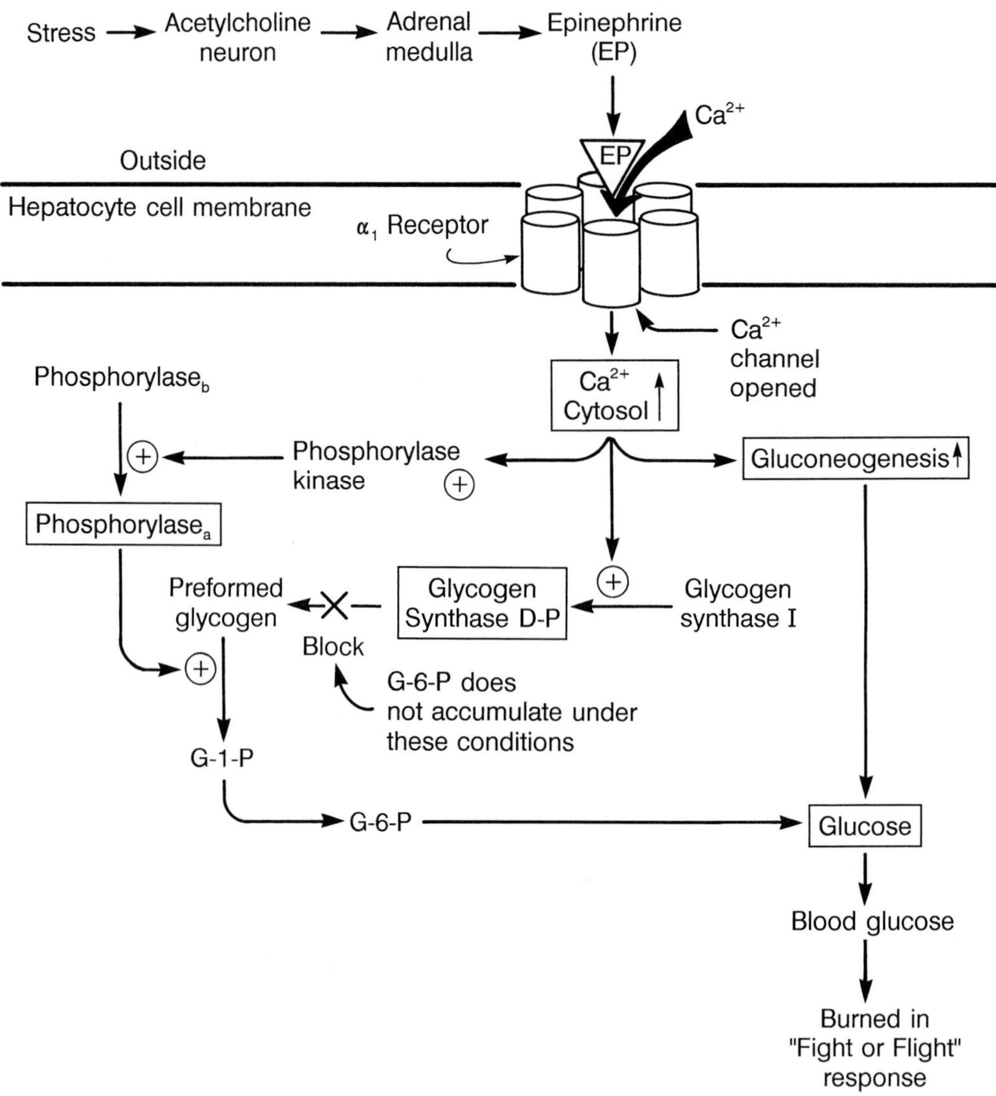

FIGURE 11-11 Possible scenario for the mode of action of epinephrine on the hepatocyte to generate glucose from increased glycogenolysis and gluconeogenesis. Epinephrine is released through the stress pathway by way of a two-neuron system, ending on an acetylcholinergic neuron innervating the adrenal medulla. The acetylcholine binds to membrane receptors on the chromaffin cell, causing an increased cellular uptake of Ca^{2+}. This results in the release of epinephrine (and other components) from granules through exocytosis. The epinephrine released binds to α-receptors on hepatocytes (and on cells that bring about contraction and increased blood pressure), causing an elevation in the cytosolic Ca^{2+} concentration. This may be brought about by activating the α_1-receptor-linked Ca^{2+} channel. Elevated Ca^{2+} stimulates phosphorylase kinase activity, glycogen synthase I (the effect of Ca^{2+} may be indirect here), and gluconeogenesis, all of which lead to the generation of glucose and an increase in circulating glucose levels for use in fight-or-flight responses in adapting to the original stress.

is an increase in the number of β-receptors without a detectable change in the number of α-receptors.

Desensitization is another regulatory aspect of adrenergic receptors. Prolonged exposure to an agonist such as epinephrine leads to decreased responsiveness to a given target tissue. This desensitization occurs with both α- and β-receptors. The mechanisms of these responses are summarized in Figure 11-14. Certain cells in culture contain adenylate cyclase but lack β-adrenergic receptors. The membranes have been isolated from such cells and combined with extracts from other cells containing the receptor. The entire system can be reconstituted in the formerly receptor-deficient membrane. The reconstituted system reproduces the properties of a normal cell membrane in terms of the effects of agonists and antagonists in stimulating adenylate cyclase. Such experiments tend to strengthen our conception of the receptor system.

FIGURE 11-12 Structures of prazosin (A), an α_1-receptor antagonist, and yohimbine (B), an α_2-receptor antagonist.

VI. ENKEPHALINS

β-Lipotropin and β-endorphin are produced in response to stress together with ACTH and secreted from the corticotroph of the anterior pituitary. Although enkephalins are smaller components that are contained in the endorphin structure, they appear to be the functional products of genes separate from that transcribing the proopiomelanocortin message in the corticotroph (see Chapter 5). Presumably they can be produced in the CNS by further breakdown of lipotropin and endorphin. The major source of circulating enkephalins appears to be the adrenal medulla. These are secreted, like epinephrine, by the exocytosis of granules following cholinergic stimulation and calcium influx.

A precursor protein exists in the adrenal medulla that contains multiple copies of the enkephalins. The individual enkephalins are released by the actions of trypsin-like enzymes. Proenkephalin, the precursor of these several individual enkephalins of the adrenal medulla, contains one Leu- and six Met-enkephalin sequences. The structure of human adrenal proenkephalin mRNA and protein is shown in Figure 11-15. This

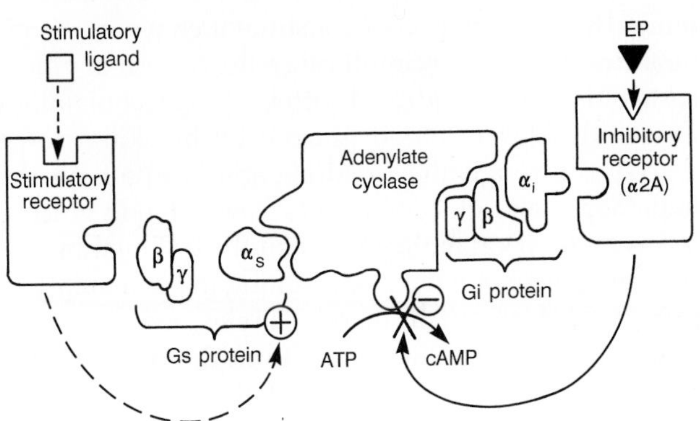

FIGURE 11-13 Outline of the function of an α_2-receptor for epinephrine. The hormone interacts with the α_2-receptor in the cell membrane and exerts inhibition of adenylate cyclase activity. This requires the binding of GTP to the guanine nucleotide-transducing agent G_i, whose concerted action with the ligand receptor results in the inhibition of the enzymatic reaction.

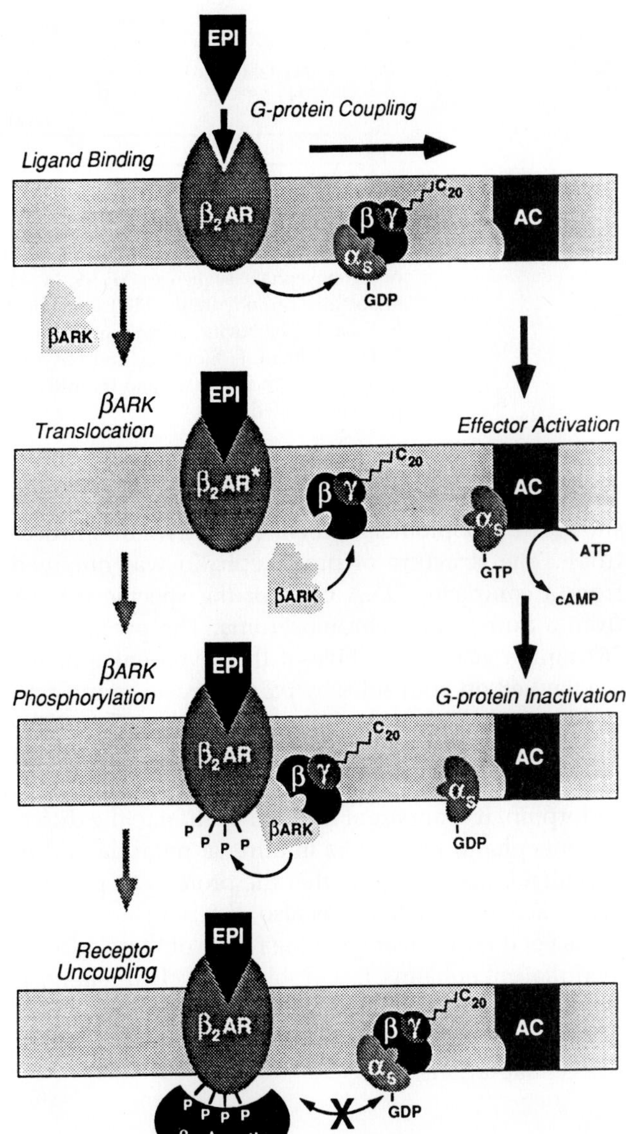

FIGURE 11-14 Adrenergic receptor activation–inactivation. Agonist binding to the β_2-adrenergic receptor initiates two concurrent processes, one of which results in effector activation, while the other activates pathways that modulate receptor function and result in receptor desensitization. One important mechanism of desensitization is effected by a receptor-specific kinase (i.e., βARK) that phosphorylates the β_2-adrenergic receptor and causes uncoupling from the G protein. Abbreviations: βARK, β-adrenergic receptor kinase; AC, adenylyl cyclase; EPI, epinephrine; α_s, β, γ, G-protein subunits. Reproduced from Lomasney, J. W., Allen, L. F., and Lefkowitz, R. J. (1995). Cloning and regulation of catecholamine receptor genes. In "Molecular Endocrinology" (B. D., Weintraub, ed.), pp. 115–131. Raven Press, New York.

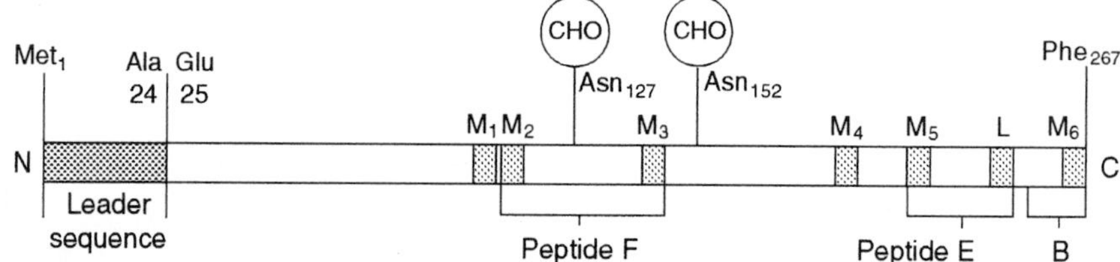

FIGURE 11-15 Low-resolution model of enkephalin precursor showing the distribution of Met-enkephalin sequences (M_1, M_2, M_3, M_4, M_5, and M_6) and Leu-enkephalin (L). The position of potential carbohydrate (CHO) attachment sites and the sequences corresponding to peptides F, E, and B of bovine adrenal medulla are shown. Reproduced from Comb, M., Seeburg, P. H., Adelman, J., Eiden, L., and Herbert, E. (1982). Primary structure of human Met- and Leu-enkephalin precursor and its mRNA. *Nature (London)* **295**, 663–666. © 1982 Macmillan Journals Limited.

figure should be compared with Figure 5-10, which shows the proopiomelanocortin (POMC) of the corticotroph. The structure of proenkephalin was obtained from a complete cDNA copy of the specific mRNA from a human pheochromocytoma. The precursor is 267 amino acids long. Five of the seven enkephalins are flanked on both sides by pairs of basic amino acid residues, which probably serve as cutting sites for proteases. The precursor does not contain sequences of the opioid peptides, dynorphin, α-neoendorphin, or β-endorphin. It is apparent that, in addition to the different enkephalin precursors in adrenal medulla and in hypothalamus, it is likely that the proteolytic processing to active enkephalins is also different.

As yet it is not clear what the role is of these adrenal medulla enkephalins (see Table 11-2). Ultimately, it can be considered that since the release of these hormones is triggered by stress, they likely play a role in stress adaptation.

VII. CLINICAL ASPECTS

A tumor of the adrenal medulla chromaffin cell, known as a pheochromocytoma, is characterized by the sporadic hypersecretion of epinephrine. Intermittent-to-permanent hypertension as well as death may result. Usually such tumors can be diagnosed and surgically removed. There are no known diseases associated with hypofunction of the chromaffin cells of the adrenal medulla. Pheochromocytoma is involved for a small percentage (<0.5%) of hypertensive patients, yet autopsy evidence suggests that some pheochromocytomas may go undetected and may occur more frequently than generally believed. For diagnosis, urinary levels of epinephrine, norepinephrine, and their metabolites (metanephrine, normetanephrine, dihydroxyphenylglycol, and vanillylmandelic acid) should be analyzed. An epinephrine-secreting tumor requires the measurement of urinary metabolites, and such a tumor cannot be dismissed by the measurement of norepinephrine alone. Tumors secrete various catecholamines singly or in combination, and secretion products are inconsistent from one neoplasm to the next. After the measurement of plasma catecholamines, following urinary measurements, imaging can be carried out. Sometimes pheochromocytomas occur outside of the adrenal and may be found at various locations along the neural crest. Blood can be sampled at various locations along the aorta and analyzed for catecholamines. The locations at which their levels are the greatest may indicate the general location of the extra adrenal tumor and could guide surgical removal. These tumors usually do not express phenylethanolamine *N*-methyltransferase (PNMT) and, accordingly, overproduce norepinephrine and its metabolites, but not epinephrine. However, some of these tumors were found to produce epinephrine. Tumors outside the adrenal are more likely to be malignant.

Repeated stress causes many episodes of epinephrine release from the adrenal medulla and potentiates the ability of the medulla to produce more catecholamines by the induction of constituent enzymes of epinephrine synthesis. Stimuli cause the release of glucocorticoids from the adrenal cortex. The catecholamines constantly elevate blood glucose by breaking down glycogen through the α_1-adrenergic receptor on the hepatocyte, and glucocorticoids elevate blood glucose levels by their catabolic effects on the cell membranes of many peripheral tissues, resulting in the inability to extract glucose from the blood. The elevated circulating glucose evokes insulin release from the β-cells of the pancreas. Constant repetition of this scenario could lead to the exhaustion of the β-cells and the precipitation of "stress-induced" diabetes. A hallmark of this diabetes is a deficiency of tissue stores of carbohydrates. In addition, the effects on blood pressure by catecholamines have been described and may lead to permanent hypertension. In the rat the magnitude of the response to catecholamines appears to be under genetic control, and spontaneously hypertensive rats show the greatest increase in plasma catecholamines during exposure to specific stresses.

β-Adrenergic receptors have been postulated to play a crucial role in asthma. A defect in the β_2-receptor may be associated with the cause of this disease; however, leukotrienes (see Chapter 16) also may be of considerable importance.

Reports indicate that the number of β-adrenergic receptor sites is decreased on the chronic lymphocytic leukemia lymphocyte surface. It is postulated that these membrane alterations interfere with basic mechanisms of communication, which may underlie the disease process characterized by the loss of regulation of cell division.

References

A. Books

Ganong, W. F. (1995). "Review of Medical Physiology," 17th ed. Appleton and Lange, Norwalk, CT.

B. Review Articles

Collins, S. (1993). Recent perspectives on the molecular structure and regulation of the β_2 adrenoreceptor. *Life Sci.* **552,** 2083–2091.

Cryer, P. E. (1995). Diseases of the sympathochromaffin system. *In* "Endocrinology and Metabolism" (P. Felig, J. D. Baxter, and L. A. Frohman, eds.), 3rd ed. McGraw-Hill, New York.

Jankovic, B. D., and Radulovic, J. (1992). Enkephalins, brain and immunity: modulation of immune responses by methionine-enkephalin injected into the cerebral cavity. *Int. J. Neurosci.* **67,** 241–270.

Knapp, R. J., Malatynska, E., Collins, N., Fang, L., Wang, J. Y., Hruby, V. J., Roeski, W. R., and Yamamura, H. I. (1995). Molecular biology and pharmacology of cloned opioid receptors. *FASEB J.* **9,** 516–525.

Lomasney, J. W., Allen, L. F., and Lefkowitz, R. J. (1995). Cloning and regulation of catecholamine receptor genes. *In* "Molecular Endocrinology" (B. D. Weintraub, ed.), pp. 115–131. Raven Press, New York.

Trifaro, J. M., Vitale, M. L., and Rodriguez Del Castillo, A. (1993). Scinderin and chromaffin cell actin network dynamics during neurotransmitter release. *J. Physiol. (Paris)* **87,** 89–106.

C. Research Papers

Bryant, S. D., Attila, M., Salvadori, S., Guerrini, R., and Lazarus, L. H. (1994). Molecular dynamics conformations of deltorphin analogues advocate γ opioid binding site models. *Pept. Res.* **7,** 175–184.

Callewaert, G., Johnson, R. G., and Morad, M. (1991). Regulation of the secretory response in bovine chromaffin cells. *Am. J. Physiol.* **260,** C851–C860.

Cherdchu, C., and Hexum, T. D. (1991). Differential secretion of enkephalin-like peptides from bovine adrenal chromaffin cells. *Neuropeptides* **19,** 237–242.

Choi, A. Y., Cahill, A. L., Perry, B. D., and Perlman, R. L. (1993). Histamine evokes greater increases in phosphatidylinositol metabolism and catecholamine secretion in epinephrine-containing than in norepinephrine-containing chromaffin cells. *J. Neurochem.* **61,** 541–549.

Chowdhury, P. S., Guo, X., Wakade, T. D., Przywara, D. A., and Wakade, A. R. (1994). Exocytosis from a single rat chromaffin cell by cholinergic and peptidergic neurotransmitters. *Neuroscience* **59,** 1–5.

Eskay, R. L., and Eiden, L. E. (1992). Interleukin-1 alpha and tumor necrosis factor-alpha differentially regulate enkephalin vasoactive intestinal polypeptide, neurotensin and substance P biosynthesis in chromaffin cells. *Endocrinology* **130,** 2252–2258.

Galindo, E., Zwiller, J., Bader, M. F., and Aunis, D. (1992). Chromostatin inhibits catecholamine secretion in adrenal chromaffin cells by activating a protein phosphatase. *Proc. Natl. Acad. Sci. USA* **89,** 7398–7402.

Gonzalez-Garcia, C., Cena, V., Keiser, H. R., and Rojas, E. (1993). Catecholamine secretion induced by tetraethylammonium from cultured bovine adrenal chromaffin cells. *Biochim. Biophys. Acta* **1177,** 99–105.

Hexum, T. D., Zheng, J., and Zhu, J. (1994). Neuropeptide Y inhibition of nicotinic receptor-mediated chromaffin cell secretion. *J. Pharm. Exp. Ther.* **271,** 61–66.

MacArthur, L., Iacangelo, A. L., Hsu, C. M., and Eiden, L. E. (1992). Enkephalin biosynthesis is coupled to secretory activity via transcription of the of the proenkephalin A gene. *J. Physiol. (Paris)* **86,** 89–98.

Parker, T. L., Kesse, W. K., Mohamed, A. A., and Afework, M. (1993). The innervation of the mammalian adrenal gland. *J. Anat.* **183,** 265–276.

Shirvan, M. H., Pollard, H. B., and Heldman, E. (1991). Mixed nicotinic and muscarinic features of cholinergic receptor coupled to secretion in bovine chromaffin cells. *Proc. Natl. Acad. Sci. USA* **88,** 4860–4864.

Tischler, A. S., McClain, R. M., Childers, J., and Downing, J. (1991). Neurogenic signals regulate chromaffin cell proliferation and mediate the mitogenic effect of reserpine in the adult rat adrenal medulla. *Lab. Invest.* **65,** 374–376.

Torres, M., Ceballos, G., and Rubio, R. (1994). Possible role of nitric oxide in catecholamine secretion by chromaffin cells in the presence and absence of cultured endothelial cells. *J. Neurochem.* **63,** 988–996.

Vitale, M. L., Rodriguez Del Castillo, A., and Trifaro, J. M. (1992). Protein kinase C activation by phorbol esters induces chromaffin cell cortical filamentous actin disassembly and increases the initial rate of exocytosis in response to nicotinic receptor stimulation. *Neuroscience* **51,** 467–474.

Wilson, S. P. (1992). Vasoactive intestinal peptide is a secretagogue in bovine chromaffin cells pretreated with pertussis toxin. *Neuropeptides* **23,** 187–192.

CHAPTER

12

Androgens

I. INTRODUCTION

A. General Comments

The endocrine physiology of the male and the interplay of the several hormones associated with male sex determination, fetal development, and, following birth, growth and sexual maturation are still another example of the effectiveness of a highly differentiated endocrine system. Its integrated operation is dependent upon the interaction of signals—both hormonal and neural—between the central nervous system, hypothalamus, pituitary, and testes. The two major functions of the testes are steroid hormone production and gametogenesis.

The hormones that are responsible for the development and maintenance of the male phenotype comprise the gonadotropins, luteinizing hormone (LH) and follicle-stimulating hormone (FSH) produced by the anterior pituitary, the androgenic steroid hormones, testosterone, androstenedione, dehydroepiandrosterone, and 5α-dihydrotestosterone produced by the gonads, and inhibin, a protein hormone also produced by the gonads. The female steroid hormones, estradiol and estrone, also play an important role in the male in certain select circumstances.

This chapter discusses the biology and biochemistry of the androgens and gonadotropins in the male; the concepts of this chapter should be compared and contrasted with those of Chapter 13, which provides similar information for the female. In addition, a portion of Chapter 14 is devoted to the hormonal aspects of fertilization and sex determination.

B. Characteristics of a Male

As discussed in Chapter 14, the gonads of both males and females in the early embryonic state are morphologically identical. It is only after the onset of sex differentiation (during the fifth and sixth weeks of fetal development) that the inevitable consequences of expression of the genetic information resident in the XY (male) or XX (female) chromosomes normally manifest themselves to convert the "indifferent" gonad into the male testes or female ovaries.

The male testes have the dual function of being responsible for both the production and release of the germ cell, the spermatozoan, as well as the biosynthesis and secretion of the key androgenic steroid hormone, testosterone. This steroid hormone and others produced from it play a dominant role in the differentiation, growth, and maintenance of the sexual reproductive tissues necessary for continuation of the species, the growth and maintenance of secondary sex characteristics, and the anabolic effects of skeletal muscle growth and skeletal growth.

Sexually the male can be classified by six characteristics: (1) chromosomal composition and structure; (2) gonads that are functionally and structurally testes; (3) tonic androgen production in adequate amounts; (4) external and internal genitalia that are appropriate for a male; (5) rearing as a male; and (6) self-acceptance of a male role. Thus, the male sexual identity is the summation of the four genetically determined organic characteristics as well as the two psychological characteristics of gender role and sex of rearing. Recent studies by B. McEwen indicate the key role of both androgens and estrogens early in fetal and postnatal development for the sexual development of the brain. In certain male species androgens also play a key role in mediating coital behavior patterns.

II. ANATOMICAL AND MORPHOLOGICAL RELATIONSHIPS OF THE MALE REPRODUCTIVE SYSTEM

A. Introduction

The male reproductive system comprises the gonads (two testes), excretory ducts (epididymis, vas deferens, and ejaculatory duct), and several accessory structures (prostate, seminal vesicles, bulbourethral glands, and penis). These are illustrated in Figure 12-1.

The key steps of organogenesis of the sexually "indifferent," yet chromosomally determined male gonad are discussed in Chapter 14. The presentation of this chapter will begin with the generation of the morphologically identifiable testes.

B. Testes

The male gonads consist of two testes contained in the scrotum or pouch, which is suspended outside the abdominal cavity between the thighs. The scrotum, besides housing the testes, plays an important role in temperature regulation necessary for normal testicular function. The formation of the male sex cells optimally occurs only at the scrotal temperature of 32.5°C. A key developmental step in the male, prior to birth, is the translocation of the gonads from the abdominal cavity into the scrotum; this process is androgen dependent. Failure of the testes to descend into the scrotum is termed cryptorchidism and results in sterility.

Each adult human testis is 4–5 cm long and is an ellipsoid of 35–45 g. Structurally each testis consists of the parenchyma or seminiferous tubules, which are surrounded by a capsule composed of three layers (see Figure 12-2).

The tunica albuginea divides the parenchymal seminiferous tubules into discrete lobules. The seminiferous tubules are the site of production of the male sex cells, the spermatozoa, and have a total length of ~800 m (one-half mile). The seminiferous tubules are organized in a highly convoluted irregular pattern, with some blind pouches, ultimately terminating in the tubulus recti, which in turn empties into the rete testis and the epididymal duct.

The seminiferous tubule walls are composed of germinal epithelial cells and Sertoli cells, which are the site of the spermatogenesis process (discussed in Section II.E). The newly formed spermatozoa are transported through the lumen of the seminiferous tubules, where they empty into a highly convoluted network of ducts known as the rete testis. Next, the spermatozoa are transported, as a consequence of ciliary motion and testicular fluid pressure, through the efferent ductules to the epididymis, where they are stored.

The walls of the seminiferous tubules have a complex structure in that there is an array of many layers of epithelial germ cells interspersed with the Sertoli or "nurse" cells. The specialized Sertoli cells (see Figure 12-3) line the basement membrane of the seminiferous tubules; their cytoplasm is in close contact with the innermost layer of the basement membrane, which in a convoluted fashion frequently surrounds the germ cells as they undergo their maturation process.

The Sertoli–Sertoli cell tight junctions (see Figure 12-3) tend to form a blood–testis barrier (analogous to the blood–brain barrier) that divides the germinal epithelium into an adluminal and a basal compartment. As shown in Figure 12-3, the basal compartment contains the immature spermatozoa or spermatogonia, and the adluminal compartment contains the remainder of the germ cells. An important function of the blood–testis barrier is to prevent the back-diffusion of autoantigens from the interior of the tubule to the bloodstream. The spermatozoa which carry surface antigens can stimulate an antibody response even in the host male. Thus, the Sertoli–Sertoli membrane barrier is an important protective mechanism. While the detailed functions of the Sertoli cells in the adult male are not known, it is believed that they perform both a nutritive and a phagocytic role, being responsible for the clearing of damaged germ cells from the seminifer-

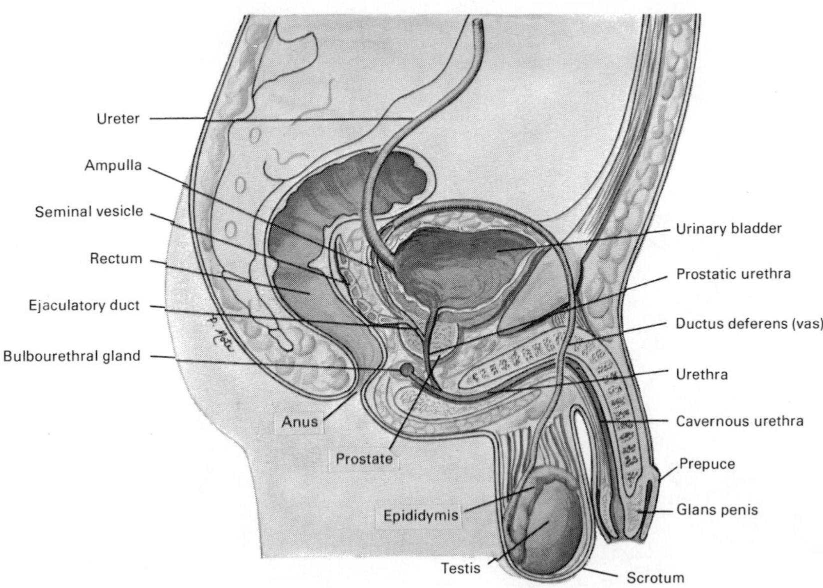

FIGURE 12-1 Male reproductive organs (midsagittal view).

ous epithelium. It has been demonstrated that FSH can stimulate the Sertoli cells to secrete an androgen-binding protein (ABP) into the seminiferous lumen.

As discussed in Chapter 14, in the male fetus the Sertoli cell secretes the Müllerian Inhibitory Factor (MIF), which acts to induce involution of the Müllerian ducts and also to block the development of the uterus and fallopian tubes. Thus, the Sertoli cell has two major categories of biological action, depending upon whether it is operative in a fetal (production of MIF) or adult testis (gamete production).

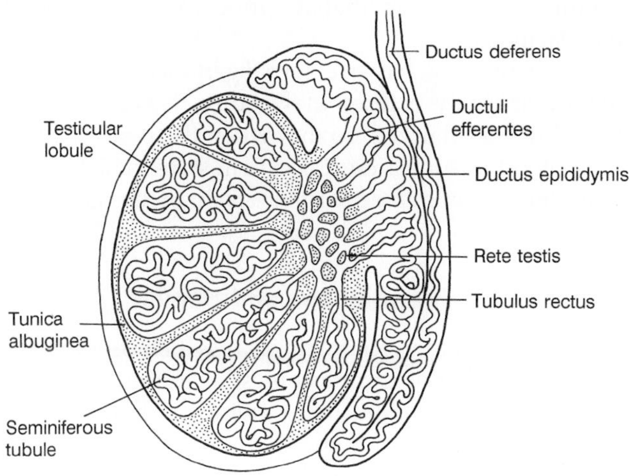

FIGURE 12-2 Cross-sectional schematic diagram of the testis and epididymis. Note the communication between the testicular lobules and the convoluted seminferous tubules inside the testicular lobules. Reproduced with permission from Junqueira, L. C. Carneiro, J. and Kelley, R. O. (1991). "Basic Histology" 7th ed., p. 392. Appleton & Lange, Norwalk, CT.

The interstitial tissue (see Figure 12-3), which makes up 5% of the total testicular volume, contains the primary endocrine cells of the testes, the Leydig cells. Leydig cells under the action of LH are the major site of production of testosterone. The Leydig cells are located in clusters that lie in the connecting tissue stroma between the seminiferous tubules. The histology of a Leydig cell is comparable to that of other steroid-secreting cells, particularly the lutein cells of the corpus luteum (see Figure 13-3).

C. Duct System

The duct system connects each testis to the urethra and functions to transport the mature spermatozoa during ejaculation. It is composed of the epididymis, ductus deferens, and ejaculatory duct.

The epididymis is a small narrow structure attached to the surface of the testis. Approximately 12 days are required for the transit of the spermatozoa through the epididymis; during this interval they undergo the morphological and functional changes necessary to provide them with the capacity to fertilize an ovum. The mature spermatozoa then enter the vas deferens.

The vas deferens is a 7- to 8-cm tubule connecting the epididymis with the ejaculatory duct (see Figure 12-1). The passage of sperm through the vas deferens is accomplished by peristaltic contractions of smooth muscle in the duct wall. Vasectomy, which is the bilateral ligation of the vas deferens, has been established as an effective and safe male contraceptive procedure. However, the current success rate for the reversal of vasectomy (as judged by the subsequent production

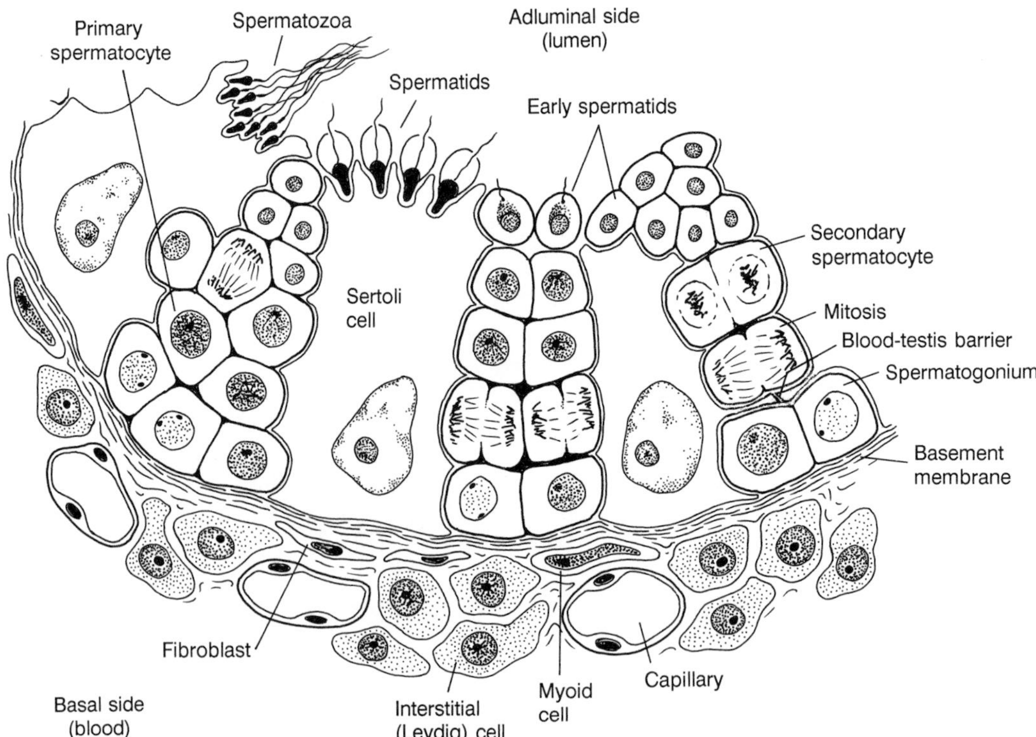

FIGURE 12-3 Schematic diagram of the wall of a seminiferous tubule and adjacent interstitial tissue. The tight junctions between the walls of adjacent Sertoli cells form a blood–testis barrier that prevents large molecules from passing into the lumen of the seminiferous tubules. Note that the germ cells are surrounded by two adjacent Sertoli cells, and that as the germ cells mature, they move from the basement membrane side (spermatogonium) to the adluminal side (early spermatids). The interstitial or Leydig cells are the source of testosterone. Reproduced with permission from Junqueira, L. C., Carneiro, J., and Kelley, R. O. (1991). "Basic Histology", 7th ed., p. 392. Appleton & Lange, Norwalk, CT.

of pregnancy) is only 50%. Also, approximately half of the men who elect to undergo a vasectomy develop antibodies against their own spermatozoa.

The ejaculatory duct is located at the merger of the vas deferens and the seminal vesicles and extends through the prostate to the urethra.

D. Accessory Structures

The accessory structures include the seminal vesicles, prostate gland, bulbourethral glands, and penis.

In adult males, the two seminal vesicles are 10–20 cm in length and lie distal to the ejaculatory duct. Under the actions of androgens they produce the mucoid secretions that constitute the major fluid volume of the ejaculate.

The prostate gland is a muscular organ containing 30–50 tubuloalveolar glands that encompass the junction of the ejaculatory duct with the urethra. The prostatic secretions, which are highly responsive to androgens, are rich in the enzyme acid phosphatase, which is a constituent of the ejaculate.

The two bulbourethral glands, sometimes referred to as Cowper's glands, are 2–3 mm in diameter and secrete prostaglandins and an alkaline mucoid solution into the urethra to neutralize the acidity and coat the urethral lining immediately prior to the arrival of the ejaculate.

The penis, which is an external genital organ, carries the urethra to the exterior of the body. Its erectile tissue is highly vascularized, so that when the penis is filled with blood it is in an erect and firm state, which facilitates deposition of the spermatozoa in the female vaginal canal.

E. Spermatozoa and Semen

A mature human spermatozoon, which contains the haploid number of chromosomes, including either the male, Y, or female, X, determining chromosome, is ~50–60 μm long. Anatomically, the spermatozoon is composed of the head and tail (see Figure 12-4). Under normal circumstances, in the adult human ~30 million germ cells are produced per day and available for maturation over an 8- to 10-week interval into spermatozoa.

The ejaculate or seminal fluid, which normally consists of 2.5–3.5 ml, contains an average of 100 million

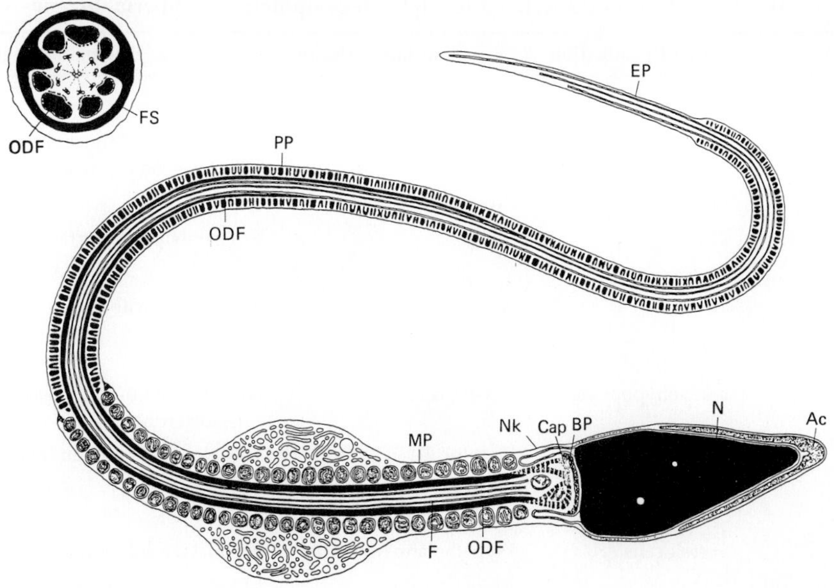

FIGURE 12-4 Schematic diagram of a spermatozoon. The spermatozoon consists of two principal regions: the head, containing the nucleus, and the tail, which can be subdivided into four regions: the neck (Nk), middle piece (MP), principal piece (PP), and end piece (EP). The head contains the nucleus (N) with its condensed chromatin enclosed by a nuclear envelope; the anterior portion of the nucleus is enclosed by the acrosomal cap (Ac) or acrosome. The middle piece contains elongated mitochondria wound helically around the flagellar fibers (F). As shown in the inset, the axial filament of the sperm flagellum is composed of two central single fibers surrounded by nine double fibrils (ODF). Reproduced with permission from Lentz, T. L. (1971). "Cell Fine Structure," p. 247. W. B. Saunders Co., Philadelphia, PA.

spermatozoa and a mixture of secretions from the various accessory genital glands. It is rich in fructose, ascorbic acid, prostaglandins, carnitine, and a variety of enzymes, including acid phosphatase, aminotransferases, muramidase, and dehydrogenases. Even though fertilization of an ovum requires only one sperm, 50% of men with sperm counts of 20–40 million/ml and all men with sperm counts under 20 million/ml are sterile.

III. CHEMISTRY, BIOCHEMISTRY, AND BIOLOGICAL RESPONSES

Table 12-1 tabulates the 9–10 hormones pertinent to male development and reproduction.

A. Steroid Hormones

1. Androgens

Androgens are steroid hormones that induce the differentiation and maturation of the male reproductive organs, the development of male secondary sex characteristics, and the behavioral manifestations consistent with the male role in reproduction. The two most important steroid hormones of the adult male are testosterone and 5α-dihydrotestosterone. The structures of these compounds are presented in Figure 12-5.

As reviewed in Chapter 2 (see Figures 2-3, 2-22, and 12-5), the naturally occurring androgens typically are 19-carbon steroids. Testosterone is the principal male androgen produced and secreted by the testes. In addition, a number of other androgen intermediates are released in low concentrations, particularly androstenedione and androstene-3β, 17-diol. Table 12-2 summarizes the secretion rates, plasma levels, and metabolic clearance rates of the principal sex steroids in the male.

The biochemical pathway for the production of androgens is discussed in Chapter 2. The side chain cleavage of cholesterol is confined to the mitochondria of the Leydig cells (see Figure 2-21). The subsequent conversion of pregnenolone into testosterone requires four enzymatic reactions, which are divided into two parallel pathways: one proceeding via 17-hydroxypregnenolone, known as the Δ^5-pathway, and the other proceeding via 17-hydroxyprogesterone, known as the Δ^4-pathway (see Figure 2-22). The relative activities of the Δ^5- versus Δ^4-pathway vary among various mammalian species; in the rodent the Δ^4-pathway is predominant, while in the human testes the Δ^5-

TABLE 12-1 Hormones Related to Male Development and Spermatogenesis

Hormone	Site of production	Principal target tissue	Principal biological function
		Steroid Hormones	
Testosterone	Leydig cells of testes	Many	Maintenance of functional male reproductive system and secondary male sex characteristics
5α-Dihydrotestosterone (DHT)	Prostate	Prostate	See Table 12-3
Androstenediol	Testes	Many	Not known with certainty
Dehydroepiandrosterone	Testes		
Estradiol	Testes		Not known with certainty
		Peptide Hormones	
LH	Adenohypophysis	Leydig cells	Stimulate steroidogenesis and production of testosterone
FSH	Adenohypophysis	Sertoli cells	Secretion of androgen-binding protein
Gonadotropin-releasing hormone (GnRH)	Hypothalamus		
Inhibin	Sertoli cells	Hypothalamus–pituitary	Feedback inhibition of FSH secretion
Müllerian inhibitory factor (MIF)	Sertoli cells	Müllerian duct	In very early male fetus, MIF induces Müllerian duct regression (see Chapter 14)
Prolactin	Adenohypophysis	Leydig cells	Potentiates the actions of LH

pathway is apparently dominant. All of these reactions occur in the microsomal fraction of the cell.

In the liver, testosterone is converted to the two 17-keto compounds, androsterone and etiocholanolone (see Figure 2-35), which are in turn conjugated to either glucuronic acid or sulfate to yield a water-soluble form amenable to urinary excretion.

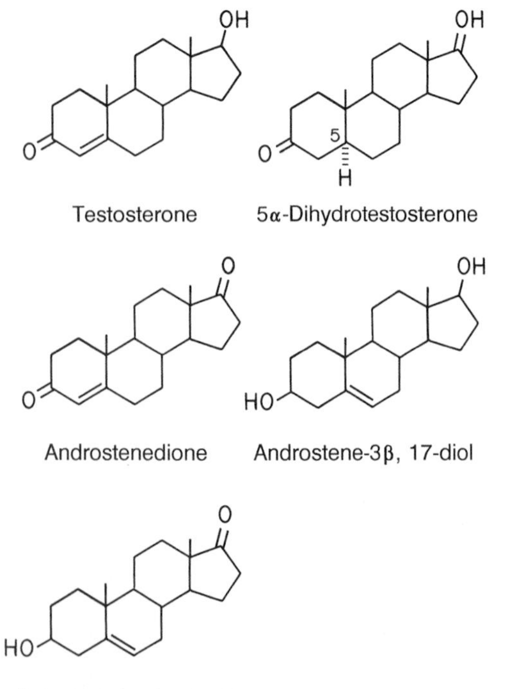

FIGURE 12-5 Structure of the important steroid hormones for the male (see also Figure 2-22).

The two principal androgens present in target cells are testosterone (T) and 5α-dihydrotestosterone (DHT). While both T and DHT are produced by the testes and circulate in the plasma, there is also significant target organ metabolism of T to DHT. Both T and DHT are potent activators of the androgen nuclear receptor. It is not yet clear whether T or DHT is the predominant initiator of gene transcriptors, although both bind very effectively to the androgen nuclear receptor. Thus, the mechanism by which two different steroid hormones bind to the same receptor but achieve different biological functions is not yet known.

a. Testosterone 5α-Reductase

The major site of nontesticular production of DHT is in the prostate, but there is also evidence of its production in other target tissues. The enzyme responsible for the conversion of T into DHT is a Δ^3-ketosteroid-5α-oxidoreductase that requires NADPH as a cofactor.

It is now known from molecular biological studies that there are two genes for both the rat and the human 5α-reductase. Both genes code for 24.9-kDa proteins that share a 50% amino acid homology, but have somewhat different kinetic properties.

The tissue distribution of the mRNA for the two 5α-reductase proteins is different. The type I enzyme, which has been postulated to be involved in the catabo-

TABLE 12-2 Production Rates, Metabolic Clearance, and Plasma Levels of Sex Steroids in the Human Male

Steroid	Plasma concentration (ng/100 ml)	Testes secretion rate (μg/day)	Metabolic clearance rate[a] (liters/day)
Testosterone	300–1100	5000	980
Dihydrotestosterone (DHT)	25–75	500–100[b]	500
Androstenedione	50–200	2500	2300
Androstane-3α,17β-diol	130	200	>1200
Estradiol	2–3	10–15[c]	1700

[a] Metabolic clearance rate (MCR) is an estimation of the rate at which the steroid is irreversibly removed from the plasma by inactivation (further metabolism). The plasma flow through the liver is ~1500 liters/day. Abstracted from Makin, H. T. J. (ed.) (1975). "Biochemistry of Steroid Hormones." William Clowes & Sons, Ltd., London, UK.

[b] Approximately 300–400 μg of DHT is estimated to be synthesized outside the testes.

[c] Most blood estradiol in the male is derived from peripheral aromatization of secreted testosterone.

lism (i.e., inactivation) of testosterone, is found in highest concentrations in the liver, kidney, intestine, brain, adrenal, and lung of the rat. In contrast, the type II 5α-reductase, which is believed to be involved in the generation of "active androgen," is found in typical androgen target tissues such as the prostate, seminal vesicles, epididymis, and liver. There is also evidence that the type I 5α-reductase is associated with microsomal membranes of the cell, while the type II enzyme is associated with the cell nucleus.

Because of the importance of the 5α-reductase in determining the response of a tissue to dihydrotestosterone, there has been a search to identify synthetic inhibitors of this enzyme. As a consequence of these efforts, several drug candidates have been identified; these include the orally active finasteride and 17β-(*N,N*-diethylcarbamoyl)-4-aza-5α-androstan-3-one (4-MA). The hope is that these drugs may allow the selective treatment of DHT-dependent clinical disorders, such as benign prostatic hyperplasia, alopecia, hirsutism, and acne.

An extensive series of studies has documented that mutations in the 5α-reductase gene result in the appearance of several clinical disorders, including some cases of male pseudohermaphroditism (see Section V). Not only do these studies provide a molecular description of several clinical diseases, but they also emphasize that dihydrotestosterone may be the "active androgen."

The biological responses of the androgens are summarized in Table 12-3. They may be divided into four categories: (1) a growth-promoting or androgenic effect on the male reproductive tract; (2) a stimulatory or anabolic effect on body weight (skeletal muscle) and nitrogen balance; (3) development of male secondary sex characteristics; and (4) actions in the central nervous system and brain.

While it is clear that both T and DHT are the predominant androgens, it has not been possible to unequivocally determine which steroid is the primary or sole initiator of biological responses in which tissues; however, Table 12-3 includes some proposed assignments. Certainly it is not true that DHT is the only active androgen. Whether DHT or T is the primary androgenic initiator presumably is a reflection of the tissue distribution of the 5α-reductase.

TABLE 12-3 Biological Responses of the Androgens

Response	Proposed androgenic mediator[a]
I. Androgenic actions on male reproductive tract	
Differentiation and growth of male reproductive tract: epididymus, prostate, seminal vesicles, vas deferens, bulbourethral glands	DHT
II. Androgenic stimulation of male secondary sex characteristics	
Growth of external genitalia (penis, scrotum)	
Deepening of voice through enlargement of larnyx and thickening of vocal cords	T
Hair, both growth and bodily distribution	
III. Anabolic actions	
Skeletal growth	
Skeletal muscle growth	T
Subcutaneous fat distribution	
Growth of accessory organs	
Prostate	DHT
Seminal vesicle	T, DHT
IV. Actions in the CNS and brain	
Differentiation of selected regions (hypothalamus, preoptic area, brain cortex)	Metabolism of T to E
Development of libido	T

[a] T, testosterone; DHT, dihydrotestosterone; E, estradiol. The proposed mediator assignments were made on the basis of the ligand selectivity of the receptor–binding protein present in the tissue in question.

The effects of androgens on the brain and central nervous system are complex and believed to occur as a consequence of the metabolism of testosterone into both DHT as well as estradiol. The 5α-reductase is present in the hypothalamus, midbrain, amygdala, hippocampus, cerebellum, and cerebral cortex. As yet the specific function of the localized DHT is not known; conceivably it may play a role in the development of the brain and the initiation of puberty. In addition, T, but not DHT, can be converted into estradiol in specific neurons by aromatization.

2. *Estrogens*

As documented in Table 12-2, there are finite levels of estradiol and estrone present in the male. Approximately 10–20% of these are generated by the testes; the remainder is produced in a variety of nonendocrine tissues including brain, liver, fat, and skin, all of which have low levels of the cytochrome P450 aromatase necessary to transform androgens into estrogens (see Figure 2-23). With the exception of the effects of testosterone-derived estradiol in the male brain, the biological role of estrogens in the male is not well delineated.

B. Steroid Hormone-Binding Globulin

Following their secretion from the tissue of origin, all steroid hormones are bound to one or more plasma proteins. For the sex steroids there is one plasma β-globulin protein that serves to transport both selected androgens as well as estrogens. This protein, termed steroid hormone-binding globulin (SHBG), has been isolated and purified and is a dimeric glycosylated protein with a molecular mass of 84 kDa (see Chapter 2, Table 2-7). SHBG has a preference for steroids with a 17β-hydroxyl [$K_d \approx (1\text{–}5) \times 10^{-10}\ M$]; accordingly, it binds T, DHT, and estradiol, but not progesterone or cortisol, with high affinity. SHBG is synthesized by the liver, and its plasma levels, which are 2-fold greater in normal women than in men, are increased in pregnancy and hyperthyroidism.

The functions of SHBG are not clear. Since its ligands are reasonably water soluble, it cannot function simply as a means of steroid solubilization. SHBG also is not believed to participate directly in the mode of action of either androgens or estrogens. It has been suggested that an important function of SHBG is to provide a "reservoir" of bound hormone that could effectively dampen oscillations in the free concentrations.

C. Peptide Hormones

1. *Gonadotropins*

The chemistry of luteinizing hormone (LH), which was formerly designated in the male as interstitial cell-stimulating hormone (ICSH), and follicle-stimulating hormone (FSH) is discussed in detail in Chapter 5. Both FSH and LH are secreted by the adenohypophysis; their release is governed in a complex fashion by gonadotropin-releasing hormone (GnRH), the blood level of steroid hormones, and possibly other as yet uncharacterized factors.

a. Luteinizing Hormone

The production and secretion of testosterone by the Leydig cells is under the control of LH in the adult male and by chorionic gonadotropin (hCG) in the developing male fetus. The secretion of LH is reciprocally related to the blood levels of testosterone and estradiol.

The actions of LH on the Leydig cells to stimulate testosterone production are produced as a consequence of its interaction with a membrane receptor that stimulates cAMP production; this in turn activates a cholesterol side chain cleavage pathway. A similar mechanism for LH is operative in the female corpus luteum and in both the male and female in the adrenal cortex, where ACTH stimulates the production of the glucocorticoids (see Figure 2-20) (see later discussion).

b. Follicle-Stimulating Hormone

FSH in the male, in conjunction with testosterone, acts on the Sertoli cells of the seminiferous tubule at the time of puberty to initiate sperm production. In the rat once the germinal epithelium differentiative process is established, testosterone alone can maintain viable sperm production; it is not yet certain whether this is also true in the primate. FSH interacts on the Sertoli cell with a membrane receptor, which results in a concomitant increase in cAMP. This in turn stimulates additional metabolic processes related to spermatogenesis. The biological actions of FSH in the male are summarized in Table 12-4.

c. GnRH

The secretion by the pituitary adenohypophysis of the gonadotropins FSH and LH is governed by the central nervous system hypothalamus-mediated re-

TABLE 12-4 Biological Actions of Follicle-Stimulating Hormone (FSH) in the Male

Cell	Action
Mature Sertoli cell	Initiate spermatogenesis Initiate synthesis of inhibin Stimulate production of new proteins, including androgen-binding protein (ABP), GnrH-like peptide, Müllerian inhibiting factor (MIF), Plasminogen activator, transferrin
Immature Sertoli cell	Stimulate mitotic division

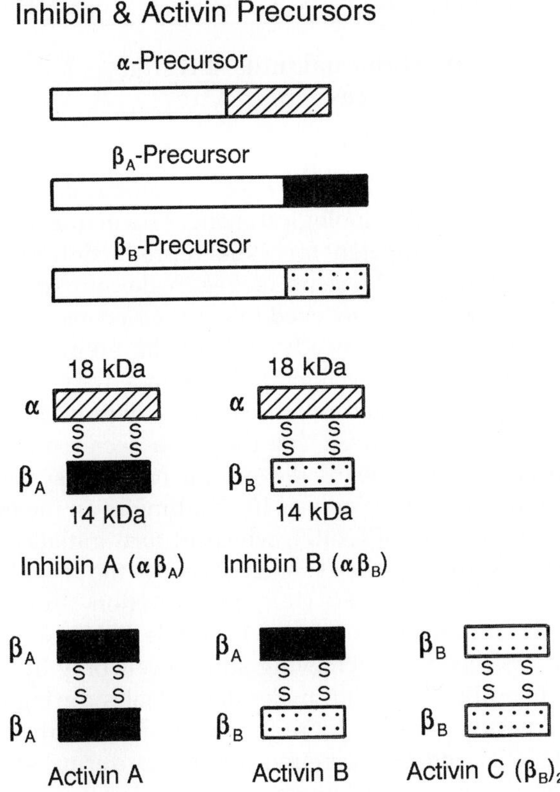

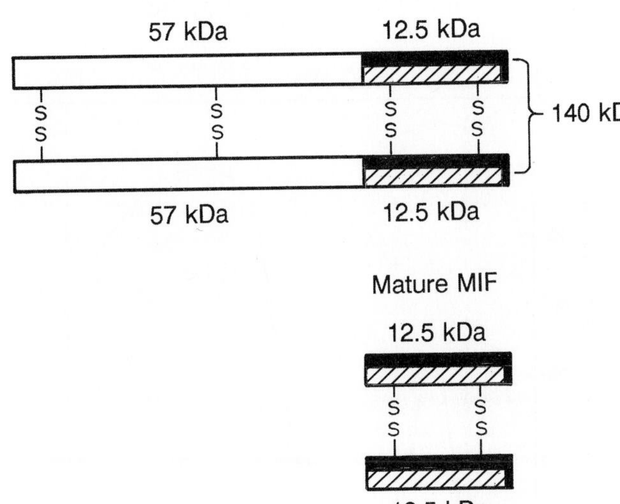

FIGURE 12-6 Structural relationships of the inhibins, activins, and Müllerian inhibiting factor (MIF). All three protein hormones belong to the transforming growth factor-β gene family. The inhibins and activins are derived from the same precursor protein; there is one precursor for the α-subunit and two precursors for the β-subunit. The inhibins are disulfide-linked heterodimers of $\alpha\beta_A$ or $\alpha\beta_\beta$. The activins are disulfide-linked hetero- or homodimers of the β-subunits ($\beta_A\beta_A$, $\beta_B\beta_B$, β_A β_B). The inhibins inhibit FSH secretion, while the activins stimulate FSH secretion from the pituitary. The precursor protein of MIF is a heavily glycosylated disulfide-linked dimer of 140 kDa, which undergoes proteolytic cleavage of both chains to yield a 24-kDa C-terminal fragment that has full MIF activity. The MIF shows structural homology to the β-chains of inhibin and activin. Modified with permission from Russell, W. E., and VanWyk, J. J. (1995). Role of hormones and aberrant growth. *In* "Endocrinology" (L. J. DeGroot, M. Besser, H. G. Burger, J. L. Jameson, D. L. Loriaux, J. C. Marshall, W. D. Odell, J. T. Potts, Jr., and A. H. Rubenstein, eds.), Vol.3, 3rd ed., pp. 2590–2623. W. B. Saunders Company, Philadelphia, PA.

lease of gonadotropin-releasing hormone, GnRH. GnRH is a decapeptide (see Figure 3-8) with a C-terminal glycinamide and an N-terminal pyroglutamyl residue. GnRH release by the hypothalamus stimulates the secretion and release of LH and/or FSH from the adenohypophysis.

2. *Inhibin*

The regulation of secretion of FSH from the adenohypophysis is only partly dependent on gonadal steroids; in addition, the protein hormone inhibin acts principally at the level of the pituitary to diminish the secretion of FSH. Inhibin is a hormone secreted in males by the Sertoli cells and in females by the ovarian follicles.

Inhibin is a heterodimeric protein; it has a common α-subunit and one of two β-subunits. Thus, there are two isoforms of inhibins, either inhibin A ($\alpha\beta_A$) or inhibin B ($\alpha\beta_B$) (see Figure 12-6). The β-subunit is 18 kDa and the α-subunit is 14 kDa. Structurally the inhibins belong to the transforming growth factor (TGF-$_\beta$) multigene family (see Chapter 19). The α- and β-subunits are cross-linked by one or more disulfide bridges. As yet, it is not known whether the detailed structures of male- and female-derived inhibins are identical.

The structure of the inhibins is also related to that of another peptide hormone, the activins (see the following). Figure 12-6 compares the structures of inhibins and activins.

Both subunits of mature inhibin are derived from precursor prohormone species. Thus, the α-subunit's primary transcript contains 364 amino acids, while the β-subunit's primary transcript contains 424 amino acids. Both the mature α- and β-subunits lie at the extreme carboxyl end of their respective prohormone species. Each mature subunit is preceded by either two α-subunit or five β-subunit arginines, which constitute the cleavage sites for proteolytic release of the mature subunits. It is known that the α- and β-subunits of biologically active inhibin are linked by disulfide bridges, which suggests that their structure is more similar to that of immunoglobulins than that of the dimeric glycoprotein hormones TSH, FSH, or LSH (see Figure 12-6).

3. *Activin*

Activins are a class of protein hormones that are either heterodimers or homodimers of the two β-subunits of inhibin, e.g., either $\beta_A\beta_B$, $\beta_A\beta_A$, or $\beta_B\beta_B$ (see Figure 12-6). The activins have been shown under experimental conditions to stimulate FSH release from the pituitary. The physiological function of the activins in male reproduction is not yet settled.

4. *Müllerian Inhibitory Factor*

Müllerian inhibitory factor (MIF) is a gonadal peptide hormone that causes the regression of the Müllerian ducts[1] during male embriogenesis. MIF is a 14-kDa disulfide-linked homodimer that is proteolytically cleaved to generate biologically active peptide fragments from the C-terminus (see Figure 12-6). MIF is a member of the transforming growth factor-β (TGF-β) multigene family of glycoproteins involved in growth and regulation (see Chapter 19).

5. *Prolactin*

The blood levels of the adenohypophyseal hormone prolactin (PRL) in the male are only slightly lower than those in the female. The chemistry of PRL is described in Chapter 5. The precise biological functions and responsibility of PRL in the male are not yet known. However, the secretion of PRL is reduced under circumstances of androgen deficiency. It is known that there are PRL receptors on the plasma membrane of Leydig cells and that PRL can augment the stimulatory actions of LH on steroidogenesis. Some evidence also suggests that PRL has direct actions on the male reproductive tract, particularly the prostate and seminal vesicles, to increase the concentration of androgen receptors.

There are several reports of the clinical consequences of hyperprolactinemia in the male (usually resulting from a pituitary tumor); the common observation is testicular atrophy, a reduction in plasma testosterone levels, and a high incidence of impotence. All of these symptoms can be reversed by removal of the tumor.

[1] The Müllerian ducts represent an immature precursor of the female internal reproductive structures. This topic is discussed in more detail in Chapter 14 (see Figure 14-15).

IV. PHYSIOLOGICAL RELATIONSHIPS

A. Hypothalamus–Pituitary–Testicular Axis

1. *General Comments*

Puberty, or the integral series of anatomical, physiological, and endocrinological changes occurring to produce a male competent for sexual reproduction, occurs between 10 and 17 years of age in the human. The onset of puberty is believed to occur as a consequence of a change in the steady state of the prepubescent pituitary–gonadal system, resulting in a dramatic rise in the plasma testosterone level (see Figure 12-7). It has been hypothesized that there is a decrease in the feedback sensitivity of the central nervous system–pituitary axis over years 6–10, resulting in an increase in the secretion of GnRH, which in turn initiates the increased secretion of LH and FSH above the prepubescent low basal levels. Thus, LH secretion rises and reaches adult levels by age 15, while FSH secretion rises more slowly, achieving adult levels only by age 17. In addition, in both males and females during the interval of puberty, there are intermittent bursts of both FSH and LH secretion that occur during sleep; their etiology is not known.

2. *Integrated functioning of the Hypothalamus–Pituitary–Testicular Axis*

The hypothalamus–pituitary–testicular axis is diagrammed schematically in Figure 12-8. The production and secretion of LH is governed by the medial–basal

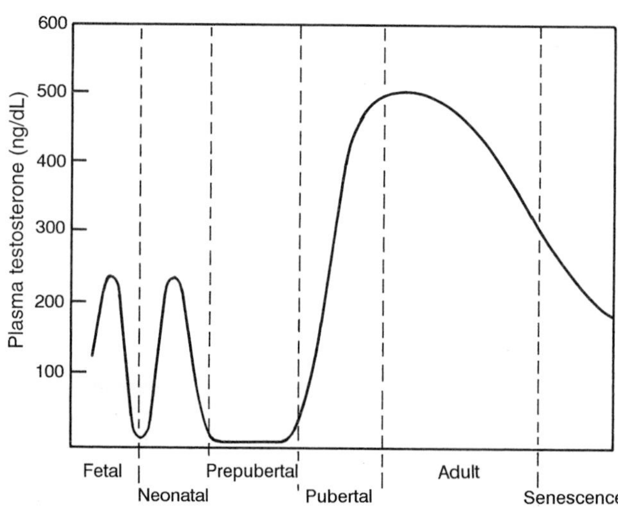

FIGURE 12-7 Changes in human male plasma testosterone levels as a function of age. Modified with permission from Ganong, W. F. (1995). "Review of Medical Physiology," 17th ed. Appleton & Lange, Norwalk, CT.

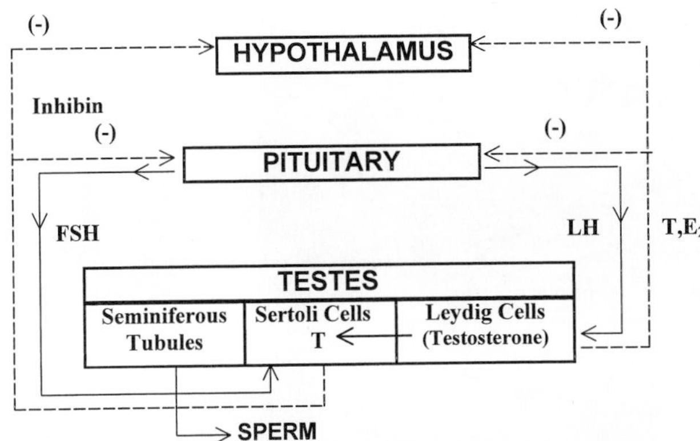

FIGURE 12-8 Schematic diagram of the male hypothalamic–pituitary–testicular axis. Abbreviations: LH, luteinizing hormone; GnRH, gonadatropin-releasing hormone; FSH, follicle-stimulating hormone; T, testosterone; E_2, estradiol. Dashed lines indicate a negative influence, and solid lines indicate a positive influence. This figure should be compared with Figure 13-6, which diagrams the female hypothalamic–pituitary–ovary axis.

region of the hypothalamus. Destruction of the arcuate nucleus of the brain leads to decreased secretion of both LH and testosterone. Neurons having their origins in the central nervous system impinge upon hypothalamic cells and locally secrete catecholamines, endorphins, and/or dopamine, which results in the episodic production and release of GnRH into the hypophyseal portal system. The presence of GnRH on specific adenohypophyseal cell membrane receptors then results in the release of LH. The LH is transported systemically to the Leydig cells of the testes.

In the human, the Leydig cells differentiate and begin secretion of testosterone during the seventh week of fetal life; during the same time interval there is also activation of the fetal pituitary secretion of LH. Then, following birth, the Leydig cells revert to a relatively undifferentiated state until they are again activated at puberty (see also Figure 12-7).

B. Leydig Cell Control and Function

Figure 12-9 illustrates the complexity of the local anatomical organization of the Leydig cells and seminiferous tubules in the testes. Although it is pedagogically useful to consider the physiological relationships of the testis Leydig cells and Sertoli cells as separate topics, it is quite clear that their activities are intimately related. In this regard, it is helpful to understand the cellular distribution of the male hormone receptors in this system. Although it is known that LH is able to produce biological responses in both Leydig and Sertoli cells, receptors for LH are found only on the Leydig cells. This implies that the LH actions in Sertoli cells are mediated through paracrine processes. Receptors for FSH are present principally on Sertoli cells and to some extent on spermatogonia. In addition, functional androgen nuclear receptors have been demonstrated in Leydig, Sertoli, and peritubular cells.

Leydig cells serve two principal functions: (a) they are the site of production of testosterone, producing approximately 7 mg daily for systemic transport to distal target tissues; and (b) they initiate paracrine interactions with the immediately adjacent seminiferous tubules and Sertoli cells to influence the process of spermatogenesis (see Figures 12-3 and 12-8).

LH-mediated stimulation of testosterone synthesis and secretion is initiated by the binding of LH to hormone-specific receptors on the outer membranes of the Leydig cell, which results in the concomitant production of cAMP inside the cell. During puberty, as a consequence of the increased secretion of LH, an increase in the secretion of testosterone by the Leydig cells results. In some circumstances, prolactin, through binding to its Leydig cell membrane, is also known to potentiate the actions of LH on testosterone production.

The rate of testosterone biosynthesis and secretion is positively correlated with the blood levels of LH. The secretion of the gonadotropin can be diminished by increasing blood concentrations of sex steroids (both testosterone and estradiol), which facilitate their binding to steroid receptors in both the hypothalamus and pituitary; this is termed "suppressive negative feedback." In circumstances of a fall in the blood levels of testosterone and estradiol, LH levels can increase; this is termed the "recovery phase of negative feedback." The precise details of the feedback mechanisms are not yet clear. Since both androgens and estrogens potentially are subject to further metabolism in selected regions of the hypothalamus, it is possible that the initiating signal for the reduction of LH secretion is a metabolite of the sex steroid (T or E_2) rather than the parent steroid. It is believed that the feedback effects on LH secretion are mediated both by influencing the amount of GnRH released by the hypothalamus and by changing the sensitivity of the adenohypophyseal LH-secreting cells to GnRH.

C. Seminiferous Tubule Function

During puberty, due to the increased secretion of hypothalamic GnRH and FSH, there is a maturation of the Sertoli cells in terms of both their biochemical capability as well as their cellular anatomical development. Thus, the blood–testis barrier (see Figure 12-3) forms at puberty, and the Sertoli cells initiate five

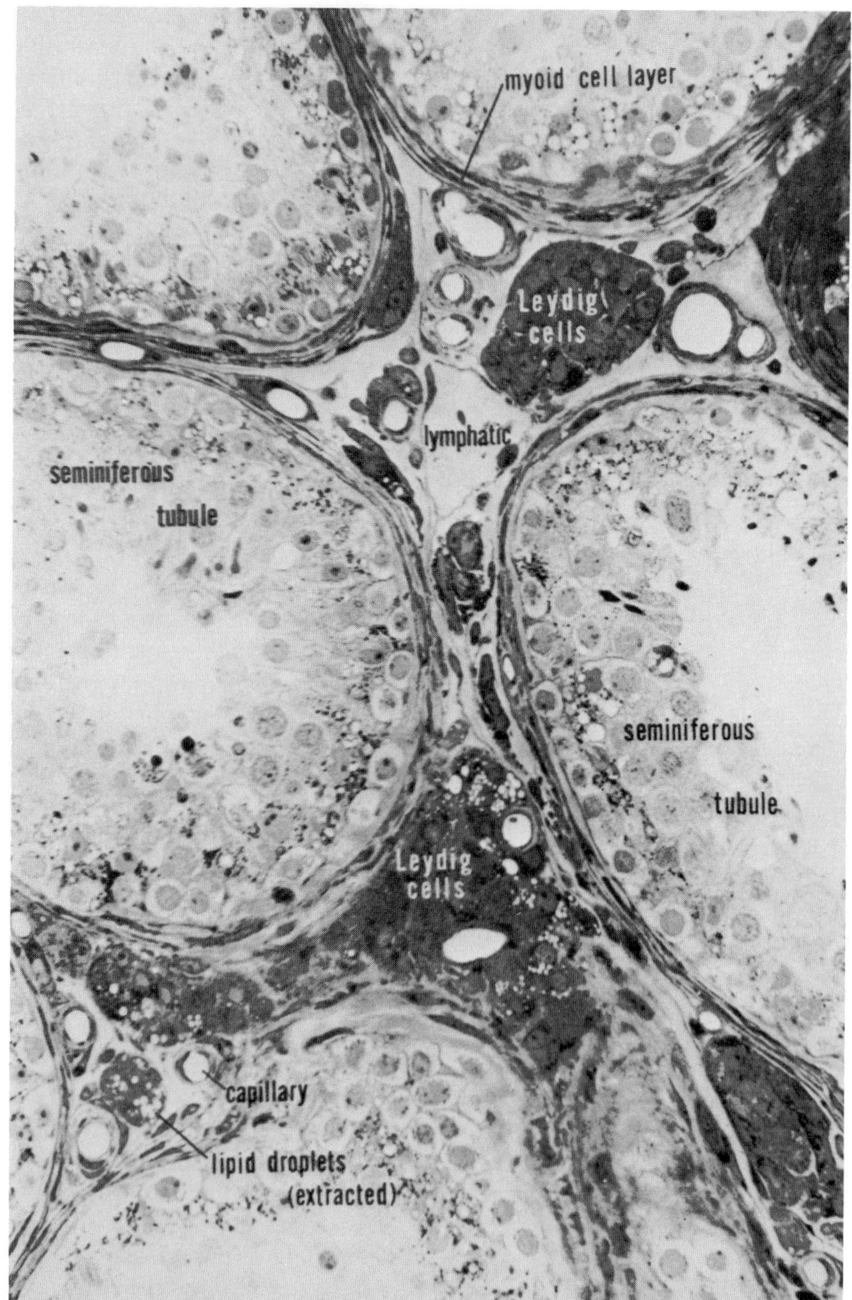

FIGURE 12-9 Cellular organization of cells of the seminiferous tubules of a human testis. Leydig cells occur in clusters surrounded by several seminiferous tubules as well as capillaries. Reproduced with permission from Christensen, A. K. (1975). Leydig cells. *In* "Handbook of Physiology" (D. W. Hamilton and R. O. Greep, eds.), Section 7, Vol. V, pp. 57–94. Williams & Wilkins, Baltimore, MD.

important functions: (1) production of unique proteins, including androgen-binding protein (ABP; see later); (2) nourishment of developing spermatozoa; (3) phagocytosis of damaged spermatozoa; (4) production of a bicarbonate- and potassium-rich fluid used for transport of the mature sperm; and (5) production of estradiol from testosterone.

FSH actions on the seminiferous tubules are initiated by binding to specific receptors on the external plasma membrane of the Sertoli cells. The FSH receptor is a member of the G-protein-coupled seven-transmembrane protein receptor superfamily (see

Figure 1-23). Occupancy of the FSH receptor leads to an increase in cAMP, which in turn stimulates the phosphorylation of a number of unknown cellular proteins, as well as to an increase in intracellular Ca^{2+}. Collectively these responses mediate FSH stimulation of mRNA and protein synthesis by the Sertoli cells.

Some of the proteins whose production is stimulated by FSH include the androgen-binding protein (ABP; see later discussion), transferrin, P450, GnRH-like peptides, ceruloplasmin, Müllerian duct-inhibiting substance (MIS; see Chapter 14), and inhibin (see Table 12-4).

A major action of FSH on the Sertoli cells is to stimulate the production and secretion of inhibin. Inhibin, along with testosterone and estradiol, is a negative feedback regulator of the pituitary secretion of FSH (see Figure 12-8). The chief evidence supporting the existence of inhibin is that, following orchiectomy (removal of the testes), FSH secretion rises. Further elevation in FSH cannot be blocked by only the administration of testosterone or estradiol. A further discussion of the actions of inhibin is given in Chapter 13.

FSH also stimulates the production of androgen-binding protein (ABP) by the Sertoli cell. The production of ABP in the Sertoli cell is stimulated by testosterone as well. In this regard, ABP is unique in that its biosynthesis is stimulated by both a peptide hormone and a steroid hormone. ABP is a protein of molecular mass 90 kDa, which binds T and DHT with high affinity ($K_d = 10^{-9}$ *M*). The ABP is secreted by the Sertoli cells into the tubular lumen (see Figure 12-3) so as to ensure that the androgen concentration of the intraluminal fluid is maintained at a high level. The detailed role of ABP is not known, although it may function to deliver testosterone to the seminiferous tubule and epididymis. ABP normally does not circulate in the blood.

In addition, FSH stimulates mitosis in immature Sertoli cells. The complement of Sertoli cells in the adult human testes is known to be largely dependent upon FSH-mediated amplification in the late phases of fetal development.

D. Spermatogenesis

The process of gametogenesis in the male is termed spermatogenesis. In contrast to the comparable process in the female (oogenesis), which occurs exclusively in the embryonic phase, the process of spermatogenesis occurs in the male from puberty throughout the bulk of adult life. Some of the other fundamental differences between the processes of male spermatogenesis and female oogenesis are summarized in Table 12-5.

The overall process of the production of mature spermatozoa by the process of gametogenesis is dependent on both a specialized cellular anatomical relationship between the developing germ cells and the surrounding cells and the presence of the gonadotropins, FSH and LH. Figure 12-9 illustrates the complexity of the local cellular organization in a section of the testis. At least five cell types are involved in the overall process of spermatogenesis: (1) Sertoli cells, (2) Leydig cells, (3) developing germ cells, (4) myoepithelial cells, and (5) epithelial cells of the duct system.

The Sertoli cell is unusual in that it possesses receptors for both the steroid hormone, testosterone, as well as the peptide hormone, FSH. While both hormones are critical for the process of spermatogenesis, FSH is only mandatory for the maturation and testosterone-sensitizing process of the Sertoli cell that occurs during puberty. After puberty, if FSH is removed by hypophysectomy, spermatogenesis can be sustained in the rat by the immediate administration of large doses of testosterone. In man, however, there is a continuing requirement for FSH along with testosterone or LH to attain spermatogenesis.

The entire process of spermatogenesis takes place while the developing germ cell is completely embedded in the seminiferous tubule wall. The three main phases of spermatogenesis, which requires 70 days in man, are as follows: (i) proliferation and differentiation of the germ cell, known as a spermatogonium; (ii) meiotic maturation to yield primary spermatocytes, then secondary spermatocytes, and finally spermatids; and (iii) transformation of spermatids into mature spermatozoa. These stages are illustrated in Figure 12-10A,B.

When embryonic gonocytes become committed to the future production of sperm cells, they are known as spermatogonia; they remain in this state until puberty. After puberty, selected spermatogonia are converted into a primary spermatocyte, which in turn can yield, after meiosis, two secondary spermatocytes. Secondary spermatocytes then divide again to generate two haploid spermatids. The cellular structures of a spermatocyte and a spermatid are shown in Figure 12-11.

As emphasized in Figure 12-9, there is exquisite local cellular architecture describing the relationship of the cells of the seminiferous tubules. Here the seminiferous epithelium is composed of a very tight packing of the spermatogonia in intimate contact and supported by adjacent tall columnar Sertoli cells. The Sertoli cells, which rest upon the basement membrane (outer side of tubule), extend apically toward the lumen of the tubule. Thus, as the spermatogonia divide and mature to spermatids, they migrate from a peripheral position adjacent to the Sertoli cell(s) outer membrane to a position inward on the Sertoli cells closer to the lumen of the seminiferous tubules. In the very final stages of maturation, the mature spermatozoon

TABLE 12-5 Comparison of Male Spermatogenesis and Female Oogenesis

Topic	Male spermatogenesis	Female oogenesis
Period of process	After puberty, throughout adulthood	Embryonic life
Number of functional germ cells produced in a lifetime	Many trillions over a lifetime or about 30×10^6/day	7,000,000 (20th week of fetus), 1–2 million at birth, ~40,000 ova available at puberty
Time required for production of a mature germ cell	Approximately 60–65 days for sperm production followed by 10–14 days for epididymal transport	12–50 years for a mature ovum; ~4000 ova used from puberty to menopause.
Type of cell division	"Even" division of cytoplasm	"Reductive" division yields production of one potential ovum and three polar bodies
Structural organization of the mature germ cell	Has a specialized anatomical structure (see Figure 12-4)	Relatively uncomplicated anatomical structure (see Figure 13-4C)

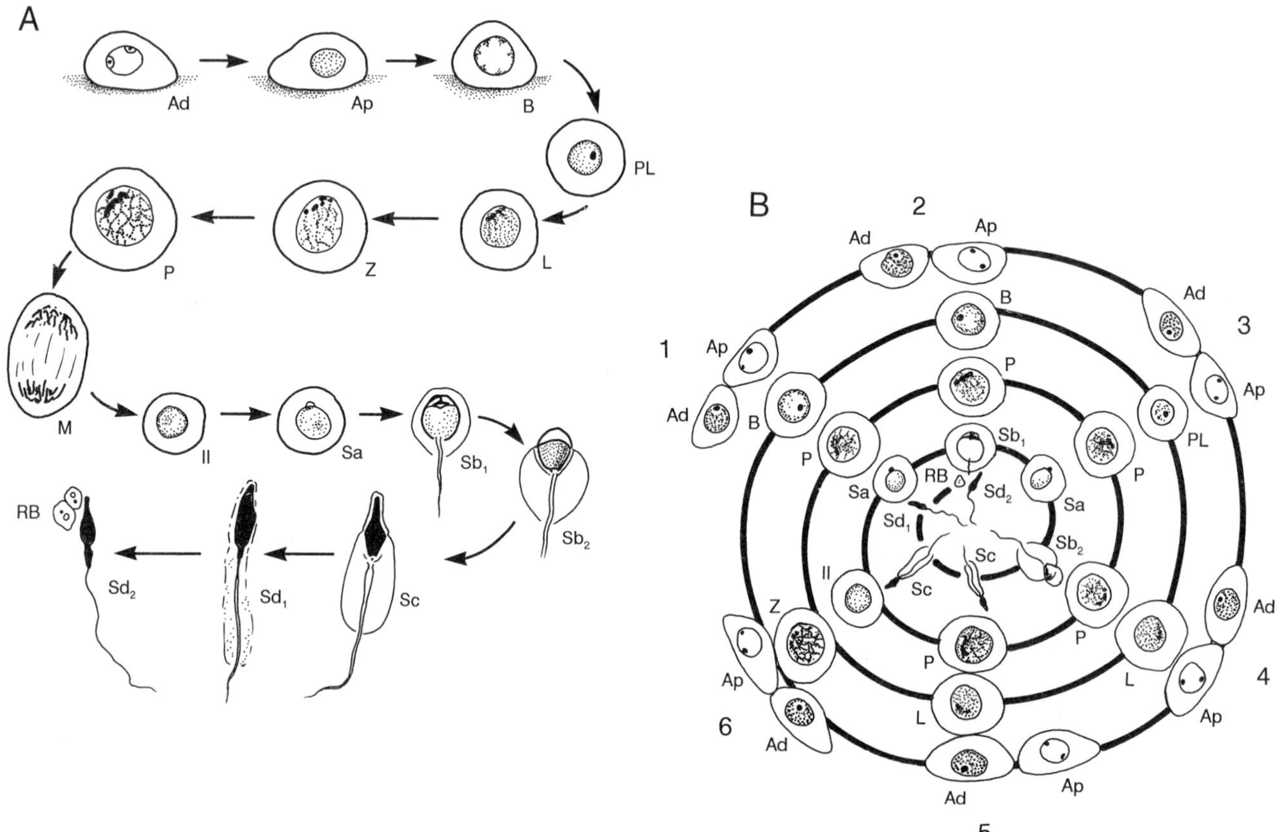

FIGURE 12-10 Summary of the spermatogenic cycle occurring over 70 days in the seminiferous tubules of man. (A) Sequence of steps in the transformation of a spermatogonium into a spermatozoon. (B) Representation of the six stages of the spermatogenic cycle, with their complement of germ cells arranged to show how their sequence of development in collaboration with the Sertoli cells corresponds to the geometry of spiral pathways leading conically toward the lumen of the seminiferous tubule. Definition of the symbols for both panels A and B: Ad, Ap, and B, spermatogonial cells, dark type A, pale type A, and type B; primary spermatocytes, PL, preleptotene, L, leptotene, Z, zygotene, P, pachytene; M, meiotic division; II, secondary spermatocyte; Sa, Sb, Sc, and Sd, spermatids; RB, excess spermatid cytoplasm or residual body. A was modified with permission from DeKretser, D. M., Risbridger, G. P. and Kerr, J. B. (1995). Basic endocrinology of the testis. *In* "Endocrinology" (L. J. DeGroot, M. Besser, H. G. Burger, J. L. Jameson, D. L. Loriaus, J. C. Marshall, W. D. Odell, J. T. Potts, Jr., and A. H. Rubenstein, eds.), 3rd ed., Vol. 3, pp. 2307–2335. W. B. Saunders Co. Philadelphia, PA. B was modified from Kerr, J. B. (1992). Functional cytology of the human testis. *In* "Bailliere's Clinical Endocrinology and Metabolism" (D. M. de Kretser, ed.), Vol. 6, pp. 235–250. Bailliere Tindall, Philadelphia.

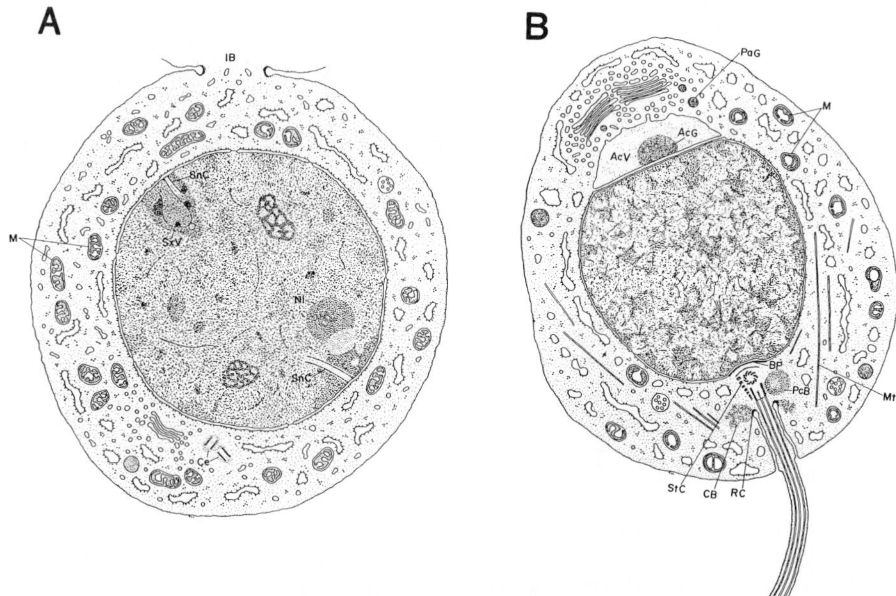

FIGURE 12-11 (A) Schematic diagram of a spermatocyte. The cytoplasmic structure of a spermatocyte is unspecialized. The process of division of spermatozoa (giving rise to spermatocytes) and spermatocytes is incomplete, so that the daughter calls remain connected to intercellular bridges (IB). Abbreviations: M, mitochondria; SxV, sex vesicle; Ce, centriole. Meiotic division of a spermatocyte yields two secondary spermatocytes, which then divide again to generate two haploid spermatids. (B) Schematic diagram of a spermatid. Spermatogenesis is the process of differentiation of the spermatid into the mature sperm. As the spermatid develops, proacrosomal granules (AcG) spread to cover the tip of the mature spermatozoon (see Figure 12-4). The cell centrioles migrate from the Golgi zone distally to give rise to the flagellum. Abbreviations: Mt, mitochondria; AcV, acrosomal vesicle; StC, stritiated columns; CB, chromatoid body; RC, ring centriole; PcB, paracentriolar body. Reproduced with permission from Lentz, T. L. (1971). "Cell Fine Structure," pp. 243–245. W. B. Saunders Company, Philadelphia, PA.

separates from the Sertoli cell outer membrane and is released into the lumen of the seminiferous tubule. There is significant evidence to support the two-way paracrine effects of Sertoli cells on developing spermatids and vice versa.

V. MOLECULAR ACTIONS

A. Production of Steroid Hormones

1. Steroid Biosynthesis—Actions of LH

The molecular actions of LH to stimulate steroidogenesis and the production of testosterone by the Leydig cell are analogous to those discussed in Chapter 10 for the action of ACTH (see Figure 10-9). The mechanism involves the production of cAMP as a consequence of the binding of LH to the Leydig cell plasma membrane. The cAMP then activates protein kinases, which phosphorylate as yet unidentified proteins, causing increased protein synthesis and ultimately increased rates of hydrolysis of cholesterol esters into cholesterol. Cholesterol is then subjected to side chain cleavage in the mitochondria to yield pregnenolone, which is the rate-limiting step in androgen biosynthesis (see Figure 2-20). The subsequent steps in the generation of androgens are reviewed in Figure 2-22.

B. Cellular Mechanisms of Action(s) of Androgens

1. Androgen Receptor

The biological effects mediated by the androgenic steroids in the male reproductive system, as well as in those tissues associated with the secondary sex characteristics (see Table 12-3), are all believed to occur as a consequence of the association of the appropriate androgen with a cytoplasmic receptor in a given target tissue. Table 12-6 tabulates the tissue distribution of androgen receptors–binding proteins. As discussed earlier in some tissues dihydrotestosterone has been

TABLE 12-6 Tissue Distribution of Androgen Receptor[a]

Category of response	Tissue
Androgenic effects	
Male reproductive tract	Testes, prostate, seminal vesicle, epididymus
Secondary sex characteristics	Skin, hair follicle, cockerel comb, and wattles
Brain	Hypothalamus, pituitary preoptic area, cortex
Anabolic effects	*Levator ani* muscle, thigh muscle
Other responses	Sebaceous and preputial gland, androgen-sensitive tumor, kidney, uterus, liver

[a] Most of the entries were derived from experimental studies in either the rodent or adult chicken, where either the androgen receptor mRNA or protein was quantitated. Information for this table was abstracted from Quigley, C. A., DeBellis, A., Marschke, K. B., El-Awady, M. K., Wilson, E. M., and French, F. S. (1995). Androgen receptor defects: Historical clinical and molecular perspectives. *Endocr. Rev.* **16,** 271–321.

shown to be the initiating signal, while in other tissues it is believed to be testosterone.

The human androgen receptor (AR) gene and cDNA have been cloned and evaluated (see Figure 12-12). The AR gene codes for a protein of 918 amino acids with a molecular mass of 110-114 kDa. Surprisingly, the AR gene is located on the X chromosome.

The AR has been shown to function, as do all steroid receptors, as a transcriptional regulatory protein that is essential for male sexual differentiation and development. As shown in Figure 12-12, the AR has many functional domains that participate in its activation of gene transcription. The key region is the DNA-binding domain, which interacts with the hormone response elements (HRE) of the promoters of regulated genes. The consensus sequence of the HRE for the AR is similar to the HRE for the glucocorticoid and progesterone receptors (see Table 1-10). For these three receptors, the HRE is AGAACAnnnTGTTCT. Although the AR has a partial palindrome-like HRE, which is compatible with the formation of a functional dimer as has been shown for the retinoid X, $1\alpha,25(OH)_2D_3$ and thyroid receptors, no evidence of a dimer of the AR has been reported.

The ligand-binding domain of the androgen receptor, which represents 30% of the protein, has a high specificity for its ligand (see Table 12-7). Both testosterone and dihydrotestosterone display the highest affinity for the AR, which is consistent with their designation as "active" androgens.

The AR receptor is localized in the cytoplasmic–nuclear portion of the target cell and, when occupied with ligand (usually testosterone or dihydrotestoster-

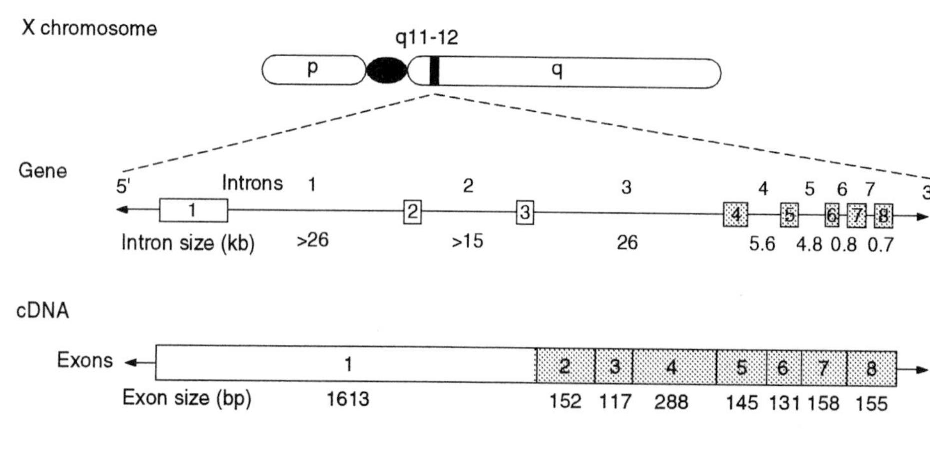

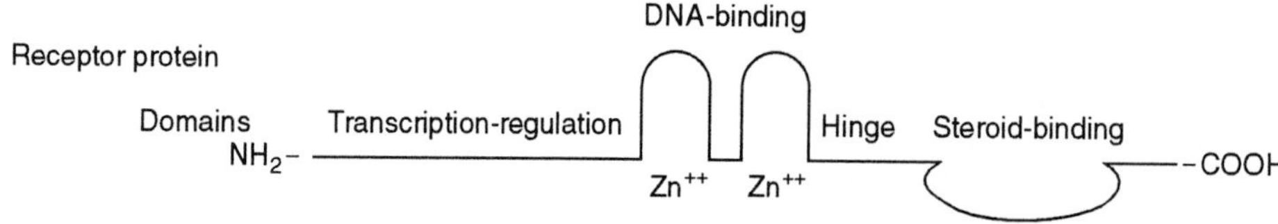

FIGURE 12-12 Schematic diagram of the relationships of the androgen receptor, chromosomal location, gene, cDNA, and protein. For the receptor protein, the regions of functional domains are indicated. The androgen receptor is known to be phosphorylated on serine 650. Modified with permission from Quigley, C. A. DeBellis, A., Marschke, K. B., El-Awady, M. K., Wilson, E. M., and French, F. S. (1995). Androgen receptor defects: Historical, clinical and molecular perspectives. *Endocr. Rev.* **16,** 271–321.

TABLE 12-7 Ligand Specificity of the Androgen Receptor

Steroid	Relative competitive index[a,b] (%)
Dihydrotestosterone	100
Dimethylnortestosterone	85
Testosterone	44
Cyproterone acetate	21
Progesterone	10
Estradiol	10
Cortisol	<0.0001
Aldosterone	<0.0001

[a] The relative competitive index (RCI) is a measure of the ability of a given ligand to compete with [^{3}H]dihydrotestosterone (DHT). The RCI for DHT is set to 100%. Data were abstracted from Hiipakka, R. A., and Liao, S. (1995). Androgen physiology. *In* "Endocrinology" (L. J. DeGroot, M. Besser, H. G. Burger, J. L. Jameson, D. L. Loriaux, J. C. Marshall, W. D. Odell, J. T. Potts, Jr., and A. H. Rubenstein, eds.), 3rd ed., Vol. 3, pp. 2336–2350. W. B. Saunders Company, Philadelphia, PA.

one) in its target cells, translocates to the nucleus of the cell where it functions as a ligand-activated transcriptional regulatory protein. A general discussion of this model for steroid hormone action is presented in Figure 1-41.

A wide variety of genes have been shown to be regulated by androgens and the AR; this is consistent with the phenotypic differences of females and males. Table 12-6 tabulates the long list of target organs that possess the AR.

TABLE 12-8 Summary of Androgen Receptor Mutations Associated with Androgen-Insensitivity Syndrome (AIS)[a]

Structural defects (~8%)	
Complete gene deletions	2
Partial gene deletions	8
1–4 bp deletions–insertions	6
Total	16
Single base mutations (~92%)	
Premature termination codons	13
Splice junction abnormalities	4
Amino acid substitutions, complete AIS	87
Amino acid substitutions, partial AIS	69
Total	173
Grand total	189

[a] The data from 180 patients were abstracted for this table from Quigley, C. A., DeBellis, A., Marschke, K. B., El-Awady, M. K., Wilson, E. M., and French, F. S. (1995). Androgen receptor defects: Historical clinical and molecular perspectives. *Endocr. Rev.* **16,** 271–321.

An extensive evaluation of mutations in the androgen receptor has been carried out (see Table 12-8). More than 50 different naturally occurring mutations have been identified and documented as to the change in the nucleotide sequence. The presence of a mutation in the AR gene can result in a dysfunction at any level in the receptor-mediated steps of androgen action, including androgen production and secretion, androgen binding to its receptor, nuclear localization of the AR, and binding of the AR to promoters of regulated genes. Collectively these defects lead to clinical disorders categorized as the androgen-insensitivity syndrome (AIS), which are all related to the condition of male pseudohermaphroditism.[2] At the functional level, two categories of AIS have been linked to the AR: (a) abnormalities of binding of androgen to the AR and (b) abnormalities of binding of the AR to the HRE on gene promoters. Clearly either of these defects will disrupt the normal transcriptional activity of the AR and lead to the onset of the clinical condition.

The detailed evaluation of the AR and its correlation with AIS represent a modern triumph of molecular biology application to clinical endocrinology.

2. *Antiandrogen Compounds*

Some synthetic analogs of testosterone compete for binding to the AR, but do not generate androgenic effects; such analogs are termed antiandrogens. The principal antiandrogen compounds available include cyproterone acetate, α,α,α-trifluoro-2-methyl-4′-nitro-*m*-propionotoluide (flutamide), and 6α-bromo-17α-methyl-17β-hydroxy-4-oxa-5α-androstan-3-one (BOMT) (see Figure 12-13). The biological actions of these steroids result from their blocking active androgens from interacting with their target organ intracellular receptors.

Estrogens are also capable of generating antiandrogenic responses; these effects are largely mediated either by (1) inhibition of testicular androgen secretion via blocking the secretion of LH or (2) direct suppression of testosterone synthesis by Leydig cells.

3. *Anabolic Steroids*

Anabolic steroids are analogs of testosterone that mediate an array of responses in the skin, skeleton, and muscle, including nitrogen, potassium, and inorganic phosphorus retention, as well as increased skeletal muscle mass. Chemically it has been possible to produce compounds that maximize the anabolic activity and minimize androgenicity. The examples of nandrolone decanoate, oxandrolone, and stanozolol are given in Figure 12-13. The biochemical basis for their actions in muscle and skeleton is not known. Although these

[2] A pseudohermaphrodite is an individual with the genetic composition and gonads of one sex and the genitalia of the other. Thus, male pseudohermaphroditism results in the appearance in the male of female external genitalia in genetic 46 XY males.

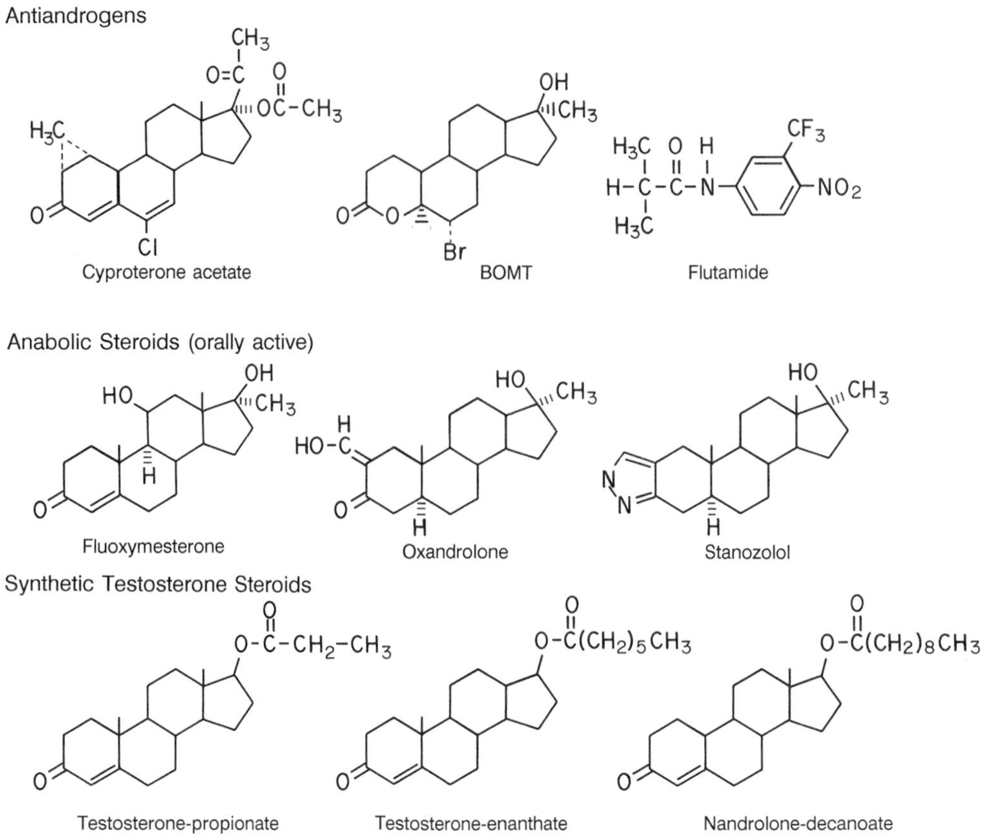

Antiandrogens

Cyproterone acetate BOMT Flutamide

Anabolic Steroids (orally active)

Fluoxymesterone Oxandrolone Stanozolol

Synthetic Testosterone Steroids

Testosterone-propionate Testosterone-enanthate Nandrolone-decanoate

FIGURE 12-13 Structures of three classes of representative androgens: (a) antiandrogens, inhibitors of the steroid 5α-reductase; (b) orally active anabolic steroids; and (c) synthetic testosterone steroids.

tissues do contain receptors for endogenous androgens, it has also been suggested that the anabolic steroids may function by competing with endogenous glucocorticoids for their receptors in these tissues.

VI. CLINICAL ASPECTS

A. Prostatic Cancer

Cancer of the prostate is the second most frequently occurring cancer in males in the United States. The pioneering work of the Nobel Laureate C. Huggins established the role of androgens in prostatic functions and suggested that a potentially effective management of the disease was the surgical removal of the source of the testosterone (i.e., orchiectomy). This procedure was widely adopted starting in the period 1940–1956. Subsequently this procedure was replaced by treatment of the malignancy with large doses of the estrogen diethylstilbestrol. The logic was that if the growth of the tumor was dependent upon the presence of circulating androgens, this could be eliminated by estrogen feedback actions on pituitary gonadotropin secretions. More recent studies indicate that the use of an antiandrogen in combination with either small doses of estrogen or LHRH may be more efficacious in the treatment of prostate cancer, since adrenal as well as testicular androgens are blocked by this combination treatment.

B. Steroid 5α-Reductase Deficiency

It has been established that mutations in the gene for the steroid 5α-reductase, type II, can result in a deficiency in the conversion of T to DHT; thus, this class of mutations can explain the appearance of one form of hereditary male pseudohermaphroditism. This is an extremely rare genetic disorder. Affected individuals are characterized as 46 XY males who have an autosomal recessive disorder characterized by the presence at birth of an external female phenotype, but with normally virilized wolffian structures that terminate in a vagina. The clinical management of individuals is complex and frequently necessitates reconsideration of the sexual identity.

C. Male Hypogonadism

Male hypogonadism is used to describe a spectrum of disorders of testicular function associated with inadequate sperm production and/or androgen secretion. The clinically observable end result may also occur as a consequence of a hypothalamic disorder, pituitary failure, or testicular disease.

Under circumstances of low gonadotropin levels, particularly LH, a wide spectrum of disorders including pituitary dwarfism, hypogonadotropic eunuchoidism (failure of appropriate gonadotropin secretion before and at puberty), and Prader–Willi syndrome (often associated with cryptorchidism) may be encountered. Under the circumstances of elevated gonadotropin production of Klinefelter's syndrome (a genetic disorder frequently with a chromosomal karotype of 47 XXY), Noonan's syndrome or male Turner's syndrome (a genetic disorder) and Sertoli cell-only syndrome (the seminiferous tubules contain only Sertoli cells) can be encountered.

An example of a clinical syndrome due to absent LHRH is Kallman's syndrome, which is associated with hypogonadotropic hypogonadism and the absence of smell.

Finally, types IV, V, and VI of congenital adrenal hyperplasia (see Table 10-5) represent, respectively, a deficiency of the 3β-steroid dehydrogenase-Δ^5, Δ^4-isomerase (type IV), 17α-hydroxylase (type V), and cholesterol side chain cleavage enzymes (type VI). In each instance there is a generalized inability of the steroidogenic tissue to biosynthesize androgens, and accordingly there is no feedback suppression of gonadotropin secretion.

D. Cryptorchidism

Cryptorchidism, or the condition of one or two undescended testes, occurs at birth in 10% of males. Normally it is a self-correcting situation, such that during the first year the testes will descend into the scrotum in all but 2–3% of males, and after puberty only 0.3–0.4% of males will have either unilateral or bilateral undescended testes. If the cryptorchid state is not corrected by puberty, the exposure of the germinal epithelium to the internal body temperature will result in impaired spermatogenesis.

E. Gynecomastia

Gynecomastia is defined as the inappropriate development of the male mammary glands. It occurs predominantly during puberty and adolescence, as well as in neonatal life, and has been estimated to have an incidence rate of ~8 per 100,000. The pathophysiology of gynecomastia is complex, and although estrogen is important for breast development (see Chapter 14), the disease is not simply explained by excess estrogen. A further discussion of this disease state is beyond the scope of this book.

F. Anabolic Steroids

Anabolic androgenic steroids represent a class of steroid analogs of testosterone that can stimulate both masculinization and protein anabolism. The structures of several orally active anabolic steroids are given in Figure 12-13. The misuse of androgens can involve either the inappropriate use of prescriptions or, in the absence of a clinical indication, anabolic steroid abuse by athletes and bodybuilders. The expectation of these users–abusers is that they will increase their body's muscle mass and enhance their athletic performance. In males who are not androgen deficient or in females, the application of exogenous androgens for prolonged intervals of time can lead to the onset of dangerous clinical side effects, including hepatotoxicity, hirsutism in the female, fluid retention, and gynecomastia.

G. Infertility

Infertility is defined as a failure of conception by a couple who have been having regular intercourse for 1 year in the absence of contraception. Male-related factors are responsible for approximately 40% of the cases of infertility.

H. Male Contraception

Although the classical methods of male contraception, which include the use of condoms, interrupted coitus, and vasectomy (surgical removal of a small segment of the vas deferens), are reasonably effective, there has been a renewed interest in developing new endocrinological methods of male contraception. In principal, endocrinological approaches could have the potential of interrupting the male reproductive system in a more "natural" fashion to effectively produce a nonfertile, but otherwise functionally competent state. While an effective and safe male endocrinologically based drug equivalent to the female "pill" has yet to be developed, it is anticipated that this will be an active area of future clinical investigations.

References

A. Books

Braunstein, G. D. (1994). Testes. *In* "Basic & Clinical Endocrinology" (F. S. Greenspan, and J. D. Baxter, eds.), Chapter 9, pp. 391–418. Appleton & Lange, Norwalk, CT.

B. Review Articles

Catlin, D. H. (1995). Anabolic steroids. *In* "Endocrinology" (L. J. DeGroot J., Besser, H. G. Burger, J. L. Jameson, D. L. Loriaux, J. C. Marshall, W. D. Odell, J. T. Potts, Jr., and A. H. Rubenstein, eds.), 3rd ed., Vol. 3, Chapter 132, pp. 2362–2376. W. B. Saunders Company, Philadelphia, PA.

Chang, C., Saltzman, A., Yeh, S., Young, W., Keller, E., Lee, H., Wang, C., and Mizokami, A. (1995). Androgen receptor: an overview. *Crit. Rev. Eukaryot Gene Express.* **5,** 97–125.

De Kretser, D. M., Risbridger, G. P., and Kerr, J. B. (1995). Basic endocrinology of the testis. *In* "Endocrinology" (L. J. DeGroot, M. Besser, H. G. Burger, J. L. Jameson, D. L. Loriaux, J. C. Marshall, W. D. Odell, J. T. Potts, Jr., and A. H. Rubenstein, eds.), 3rd ed., Vol. 3, Chapter 130, pp. 2307–2335. W. B. Saunders Company, Philadelphia, PA.

Griffin, J. E., and Wilson, J. D. (1992). Disorders of sexual differentiation. *In* "Campbell's Urology" (P. C. Walsh, A. B. Retik, T. A. Stamey, and E. D. Vaughan, Jr., eds.), 6th ed., pp. 1509–1542. W. B Saunders Company, Philadelphia, PA.

Handelsman, D. J. (1995). Testosterone and other androgens: physiology, pharmacology, and therapeutic use. *In* "Endocrinology" (L. J. DeGroot, M. Besser, H. G. Burger, J. L. Jameson, D. L. Loriaux, J. C. Marshall, W. D. Odell, J. T. Potts, Jr., and A. H. Rubenstein, eds.), 3rd ed., Vol. 3, Chapter 131, pp. 2351–2361. W. B. Saunders Company, Philadelphia, PA.

Herdt, G. (1990). Mistaken gender: 5α-reductase hermaphroditism and biological reductionism in sexual identity reconsidered. *J. Am. Anthropol. Assoc.* **92,** 433–446.

Hiipakka, R. A., and Liao, S. (1995). Androgen physiology. *In* "Endocrinology" (L. J. DeGroot, M. Besser, H. G. Burger, J. L. Jameson, D. L. Loriaux, J. C. Marshall, W. D. Odell, J. T. Potts, Jr., and A. H. Rubenstein, eds.), 3rd ed., Vol. 3, Chapter 131, pp. 2336–2351. W. B. Saunders Company, Philadelphia, PA.

Jegou, B. (1991). Spermatids are regulators of Sertoli cell function. *In* "The Male Germ Cell Spermatozonium to Fertilization" (B. Robaire, ed.), Chapter 637, pp. 340–353. Annals of the New York Academy of Science, New York.

Jegou, B. (1992). The Sertoli cell. *Baillieres Clin. Endocrinol. Metab.* **6,** 273–311.

Kierszenbaum, A. L. (1994). Mammalian spermatogenesis *in vivo* and *in vitro:* a partnership of spermatogenic and somatic cell lineages. *Endocr. Rev.* **15,** 116–134.

Klaij, I., Timmerman, M. A., Blok, L. J., *et al.* (1992). Regulation of inhibin β_B subunit mRNA expression in rat Sertoli cells: consequences for the production of bioactive and immunoactive inhibin. *Mol. Cell. Endocrinol.* **85,** 237–246.

Lee, M. M., and Donahoe, P. K. (1993). Müllerian inhibiting substance: a gonadal hormone with multiple functions. *Endocr. Rev.* **14,** 152–164.

MacLaughlin, D. T., Hudson, P. L., Graciano, A. L., *et al.* (1992). Müllerian duct regression and antiproliferative bioactivities of Müllerian inhibiting substance reside in the carboxyterminal. *Endocrinology* **131,** 291–296.

McEwen, B. S. (1991). Steroid hormones and thyroid hormones modulate a changing brain. *J. Steroid Biochem. Mol. Biol.* **40,** 1–14.

Pfeiffer, D. C., and Vogl, A. W. (1991). Evidence that vinculin is codistributed with actin bundles in ectoplasmic ("junctional") specializations of mammalian Sertoli cells. *Anat. Rec.* **231,** 89–100.

Quigley, C. A., De Bellis, A., Marschke, K. B., El-Awady, M. K., Wilson, E. M., and French, F. S. (1995). Androgen receptor defects: historical, clinical, and molecular perspectives. *Endocr. Rev.* **16,** 271–321.

Russell, D. W., Berman, D. M., Bryant, J. T., Cala, K. M., Davis, D. L., Landrum, C. P., Prihoda, J. S., Silver, R. I., Thigpen, A. E., and Wigley, W. C. (1994). The molecular genetics of steroid 5α-reductases. *Recent Prog. Hormone Res.* **49,** 275–284.

Skinner, M. K. (1991). Cell-cell interactions in the testis. *Endocr. Rev.* **12,** 45–77.

Zhou, Z., Wong, C., Sar, M., and Wilson, E. M. (1994). The androgen receptor: an overview. *Recent Prog. Hormone Res.* **49,** 249–274.

C. Research Papers

Andersson, S., and Russell, D. W. (1990). Structural and biochemical properties of cloned and expressed human and rat steroid 5α-reductases. *Proc. Natl. Acad. Sci. USA* **76,** 3640–3644.

George, F. W., Russell, D. W., and Wilson, J. D. (1991). Feed-forward control of prostate growth: dihydrotestosterone induces expression of its own biosynthetic enzyme, steroid 5α-reductase. *Proc. Natl. Acad. Sci. USA* **88,** 8044–8047.

Hodgkin, J. (1991). Sex determination and the generation of sexually dimorphic nervous systems. *Neuron* **6,** 177–185.

LeVay, S. (1991). A difference in hypothalamic structure between heterosexual and homosexual men. *Science* **253,** 1034–1037.

Normington, K., and Russell, D. W. (1992). Tissue distribution and kinetic characteristics of rat steroid 5α-reductase isozymes: evidence for distinct physiological functions. *J. Biol. Chem.* **267,** 19548–19554.

Petra, P. H. (1991). The plasma sex steroid binding protein (SBP or SHBG): a critical review of recent developments on the structure, molecular biology and function. *J. Steroid Biochem. Mol. Biol.* **40,** 735–753.

CHAPTER

13

Estrogens and Progestins

I. INTRODUCTION

A. General Comments

The endocrine physiology of the female and the interplay of the many hormones associated with female sex determination, conception, fetal development, birth, growth, puberty, the reproductive years (lactation), and finally menopause beautifully illustrate the complexity and responsivity of this highly differentiated endocrine system. Its integrated operation is dependent upon the interaction of signals—both hormonal and neural—between the central nervous system (CNS), the pituitary, and the ovaries.

The hormones that are responsible for the development and maintenance of the female phenotype comprise the gonadotropin-releasing hormone (GnRH), the gonadotropins, luteinizing hormones (LH) and follicle-stimulating hormones (FSH), both produced by the anterior pituitary, and the female steroid hormones, the estrogens and progestins.

The gonadotropin-releasing hormone and the gonadotropins are not believed to have any direct actions on bodily functions except through their specific actions on the pituitary and ovaries. In contrast to this limited sphere of action, the female steroid hormones have a wide range of actions in many tissues. Finally, specialized hormones such as relaxin, placental lactogen, and human chorionic gonadotropin are utilized at certain key intervals to achieve essential endocrinological responses.

This chapter will discuss the biology and biochemistry of the estrogens and progestins in the nonpregnant female; the concepts of this chapter should be compared and contrasted with these of Chapter 12, which provides similar information for the male. Chapter 14 will cover the hormonal relationships of pregnancy, lactation, and development, as well as the hormonal aspects of fertilization and sex determination.

B. Characteristics of a Female

As discussed in Chapter 14, the gonads of both males and females in the early embryonic stage are morphologically identical. It is only after the onset of sex differentiation (during the fifth and sixth weeks of fetal development) that the inevitable consequences of expression of the genetic information resident in the XX (female) or XY (male) chromosomes normally manifest themselves to convert the "indifferent" gonads into the fetal female ovaries or male testes.

The female ovary has the dual function of being responsible for both the production and release of the germ cell or ovum as well as the biosynthesis and secretion of the key steroid hormones, progesterone and estrogen. These steroid hormones play a dominant role in the differentiation, growth, and maintenance of the sexual reproductive tissues necessary for continuation of the species.

Sexually the female can be classified by six characteristics: (1) chromosomal composition and structure; (2) gonads that are functionally and structurally ovaries; (3) female sex hormone production (cyclically in the adult); (4) external and internal genitalia that are morphologically appropriate for a female; (5) rearing as a female; and (6) self-acceptance of a female role.

Thus, the female sexual identity is the summation of the four genetically determined organic characteristics as well as the two psychological characteristics of gender role and sex of rearing. Studies by B. McEwen indicate the key role of the sex steroid hormones (both estrogens and androgens) early in fetal and postnatal development for the sexual development of the brain.

II. ANATOMICAL AND MORPHOLOGICAL RELATIONSHIPS OF THE FEMALE REPRODUCTIVE SYSTEM

A. Introduction

The human female reproductive system comprises the gonads (two ovaries) and uterine tubes, a single uterus, a vagina, external genitalia, and the mammary glands. These are diagrammed in Figure 13-1 with the exception of the mammary glands, which are diagrammed in Figure 14-5.

The key steps of organogenesis of the sexually indifferent, yet chromosomally determined female gonad are discussed in Chapter 14. The presentation in this chapter will begin with the generation of a morphologically identifiable ovary.

B. Ovaries

The adult human ovary is 4–5 cm in length and is almond-shaped. Structurally each ovary consists of the cortex (outer) and medulla (inner) zones (see Figure 13-2). Just below the surface layer of connective tissue is the germinal epithelium, which is the site of generation of the follicles. Each primary follicle contains one central germ cell or oogonium, which is surrounded by a layer of epithelial cells. Surrounding the follicular cells, but separated by a basement membrane, are the *theca externa* and *theca interna* cells (see insert box of Figure 13-3). The medulla comprises the central region of the ovary, which is devoid of follicles. The normally functioning ovary undergoes, over a 4–5 decade interval, a series of profound changes that are essential for female development, puberty, reproduction, and menopause.

The ovarian follicle consists of a large round oocyte surrounded by follicular cells. After ovulation, the walls of the follicle collapse and cells from the thecal layer invade the granulosa layer to form the *corpus luteum.* If pregnancy does not occur, the corpus luteum lasts for only 14 days when it becomes replaced with fibrous tissue to form the *corpus albicans.* Prior to ovulation, the theca interna cells and follicular cells collaborate to biosynthesize estrogens.

The process of gametogenesis in the female is termed oogenesis. In contrast to the comparable process in the male (spermatogenesis), which is initiated in puberty and continues throughout the bulk of adult life, the process of germ cell production in the female occurs exclusively in embryonic life (see also Table 12-5 for other female–male comparisons).

During embryogenesis, the oogonia or primordial germ cells available for a lifetime of ovulation are produced. The process of oogenesis is initiated in weeks 8–20 of intrauterine life and continues in defined stages through puberty until the final stage of generation of a mature ovum occurs in a menstrual cycle (see Figure 13-3). By the 20th week of gestation there are ~7 million mitotic germ cells in each ovary. These germ cells then cease to divide by mitosis; some become atretic while others mature to the oogonium stage and proceed by meiotic division to the leptotene or arrested prophase stage. These individual oogonia become surrounded by mesenchymal cells to form the primary follicles (see Figure 13-2). By birth the number of follicles in an ovary number only 1–2 million; this number then decreases

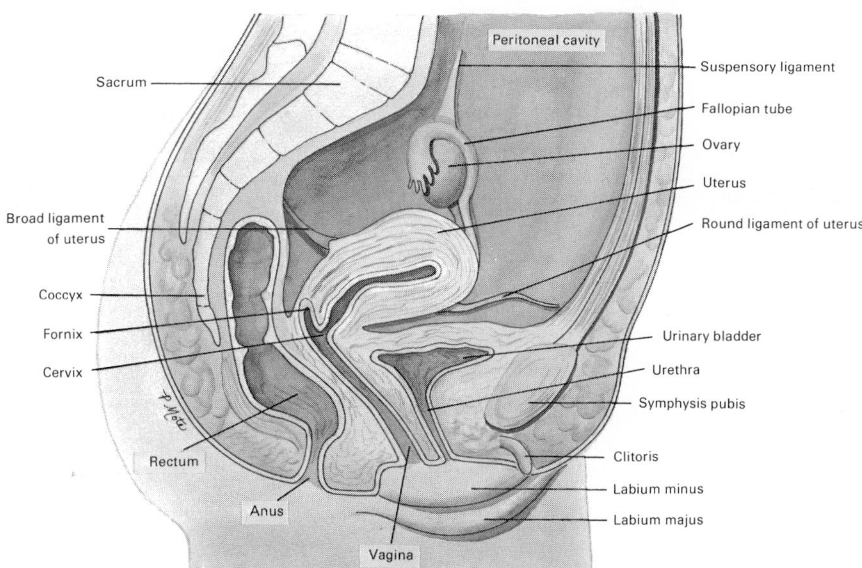

FIGURE 13-1 Female reproductive organs (midsagittal view).

by further atresia and cell death, so that by puberty there are only ≈ 400,000 oogonia in follicles to support the monthly ovulatory cycles over the next 35–40 years. These oogonia continue in the leptotene (arrested prophase) state until a limited number are selected by an as yet poorly understood endocrine process for further maturation, which leads to ovulation and the release of one mature ovum per month.

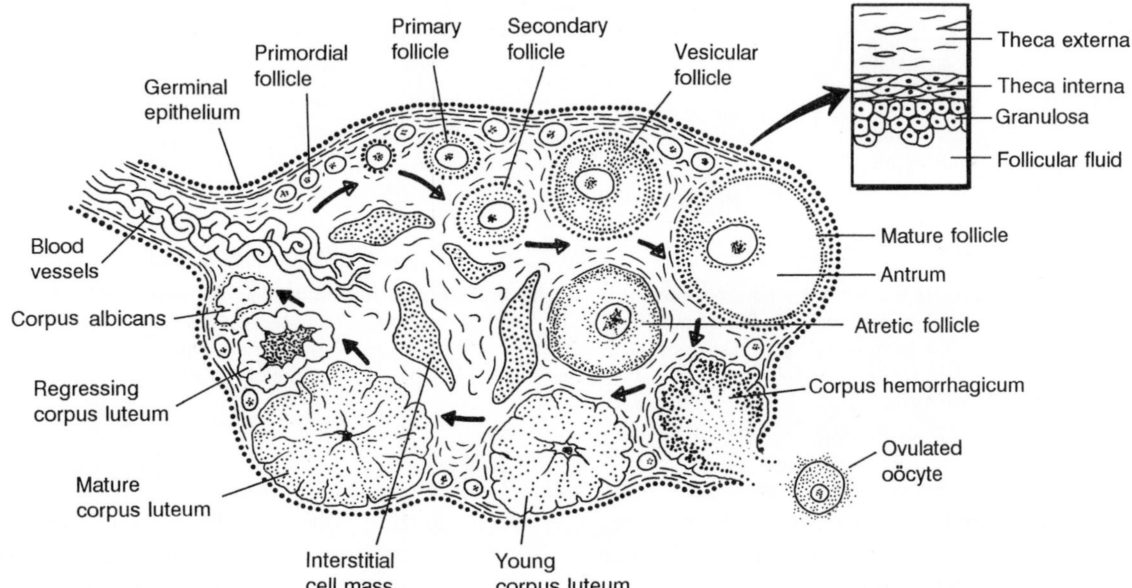

FIGURE 13-2 Schematic diagram of a mammalian ovary. The sequence of events necessary for the origin, growth, and rupture of a typical ovarian–Graafian follicle and the concomitant formation and subsequent regression of the corpus luteum is presented. The sequence should be followed clockwise around the ovary beginning with the germinal epithelium at 10 o'clock. The inset box illustrates a section of the wall of a mature follicle. Both the thecal and granulosa cells acquire the capability for the biosynthesis of estradiol, while the corpus luteum is the principal site of production of progesterone (see also Figure 13-12). Modified with permission from Katzund, B. G. (ed) (1984). "Basic and Clinical Endocrinology," 2nd ed. Lange Medical Publications, Los Altos, CA.

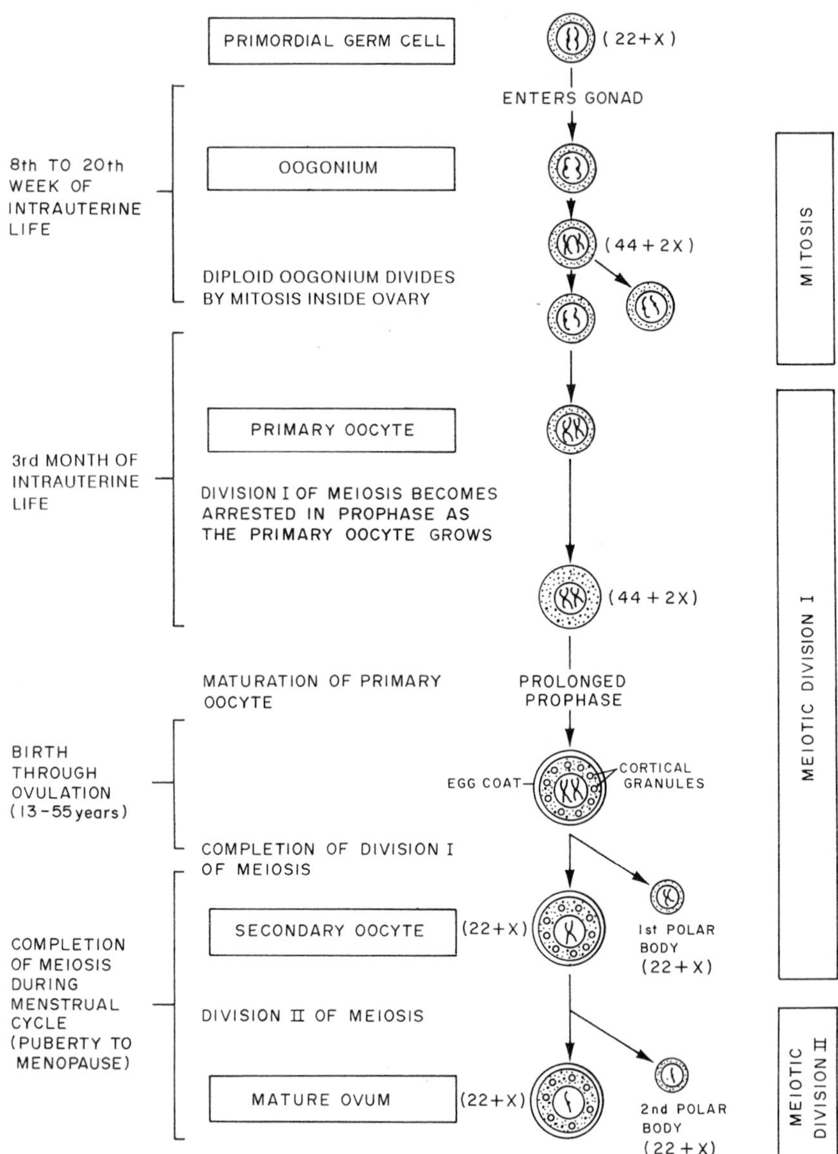

FIGURE 13-3 Meiotic processses in germ cells in women as related to their age and sexual development. Modified from Odell, W. D. (1979). The reproductive system in women. *In* "Endocrinology" (L. J. DeGroot, G. F. Cahill, L. Martini, D. H. Nelson, W. D. Odell, J. T. Potts, Jr., E. Steinberger, and A. I. Winegard, eds.), Vol. 3, p. 1385. Grune & Stratton, New York.

Ovulation of a single ovum each month requires only about a total of 400 oogonia over the 30–40 year period of reproductive fertility. At the time of menopause the number of viable oogonia in the ovary is virtually zero. These events are summarized in Figure 13-3. It is interesting that the ovum that is ultimately fertilized is the product of an intense selection procedure; there is a ratio of atresia to selection of ~1000:1. Figure 13-4 presents a schematic diagram of a single oocyte. During follicular development just prior to ovulation, the primary oocyte undergoes a meiotic or reductive division so that the number of chromosomes is reduced by one-half (see Figure 13-3). This yields the secondary oocyte, which contains 22 autosomes, one X sex chromosome, and the bulk of the cell cytosol; in addition, a polar body or polycyte is formed that also has 22 autosomes, one sex chromosome, and only very little of the cell cytosol.

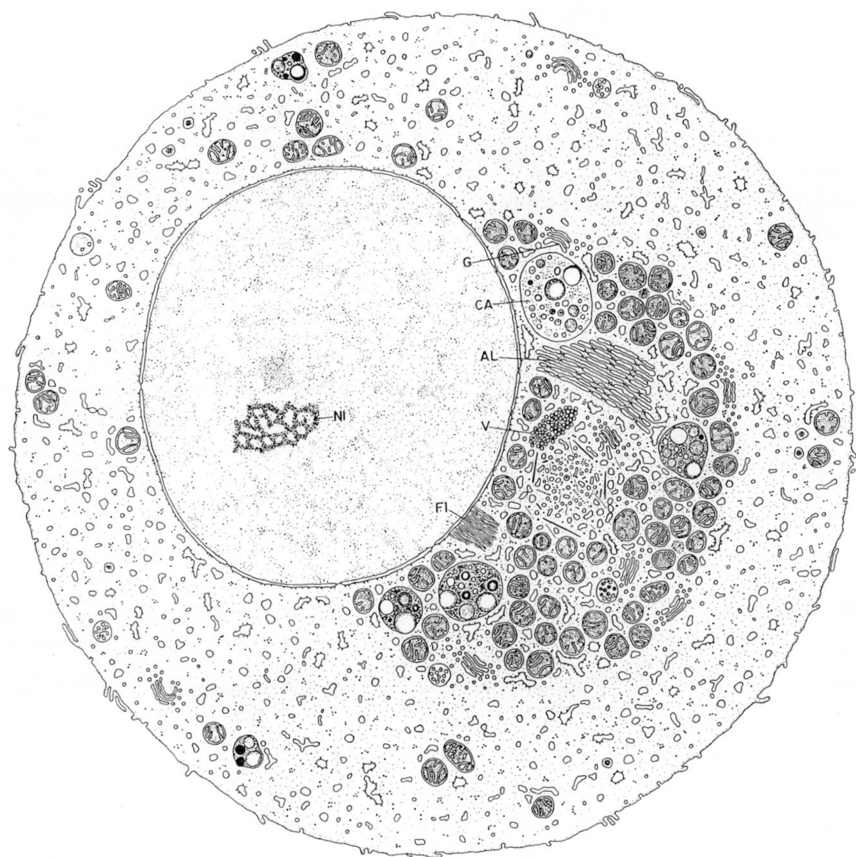

FIGURE 13-4 Schematic diagram of a mature primary oocyte. Abbreviations: G, multiple Golgi complexes; CA, compound aggregates; AL, annulate lamellae; V, vesicles; FI, wavy filaments; NI, nucleolus. Reproduced with permission from Lentz, T. L. (1971). "Cell Fine Structure," p. 269. Saunders, Philadelphia, PA.

C. Corpus Luteum

At the site of ovulation where the mature follicle ruptures and releases the ovum, the cells comprising the follicle, under the actions of LH, enlarge and differentiate into lutein cells; as a consequence a yellowish substance (lutein) accumulates in the cytoplasm. The lutein cells are derived by maturation from both the theca interna and follicular cells. These cells are the principal sites of production of progesterone and estrogen after ovulation. Subsequently, capillaries grow into these cells, giving rise to the corpus luteum, which is a typical endocrine organ. In the human, the corpus luteum may range in size from 1.5 to 4 cm.

If the released oocyte is not fertilized within 1–2 days, then the corpus luteum will continue to increase in size for 10–12 days. This is then followed by virtual total regression of the gland to produce a small white ovarian scar known as the corpus albicans and concomitant cessation of progesterone and estrogen secretion (see Figure 13-3).

If the released oocyte is fertilized, the corpus luteum continues to grow and function for the first 3 months of pregnancy. Then it slowly regresses, leaving a white scar on the ovary. The progesterone from the corpus luteum is essential for maintenance of the first 2 months of pregnancy; after this time the production of progesterone by the placenta is adequate for the continued maintenance of pregnancy.

D. Fallopian Tubes

The fallopian tubes are also variously referred to as uterine tubes or oviducts. Each tube is 11–12 cm long and extends from the trumpet-shaped end, which is in contact with the ovary, down to the proximal end, which penetrates the uterine wall (see Figure 13-1). During the ovulatory process, the many small finger-like projections or *fimbriae* of the ciliated cells on the inner surface of the fallopian tubes actively massage or undulate to aid in the translocation of the released mature oocyte down to the uterus. During the tubular passage of the oocyte, if it is not fertilized, it will eventually disintegrate and disappear; the biochemical–endocrine basis of these changes is not yet known.

If the oocyte becomes fertilized, it promptly completes its mitotic division and becomes a zygote. Usually the zygote will reach the uterus in 4–5 days, where it will implant and continue to develop during pregnancy. If the zygote fails to reach the uterus, the woman is said to be subject to an ectopic pregnancy.

E. Uterus

In the nonpregnant adult woman the uterus is a muscular, thick-walled, pear-shaped organ 5 cm wide × 7.5 cm long × 2.5 cm thick. The cavity of the uterus communicates below with the vagina and above with the fallopian tubes. The uterine wall comprises three chief layers: (1) the outer or perimetrium; (2) the middle or myometrium; and (3) the inner or endometrium.

The endometrium is a mucous membrane composed of two layers: the thin, basal layer and the outer, functional layer. The functional layer changes dramatically under hormonal influence during the menstrual cycle and is almost completely lost during the process of menstruation. In the secretory phase of the menstrual cycle (14–28 days), the cells are quite tall and columnar with numerous microvilli.

The volume of the uterine cavity changes in capacity from 2–5 ml before pregnancy to 5000–7000 ml at the term of pregnancy; the increase in uterine mass is from 60 g at puberty to 1000 g at the end of pregnancy, which is a 16-fold increase. The primary stimulus for the growth of the myometrium during pregnancy is estrogen produced by the placenta.

TABLE 13–1 Hormones Related to Female Development, Reproduction, and Lactation

Hormone	Site of production	Principal target tissue	Principal biological function
Steroid hormones			
17β-Estradiol	Ovary and follicle	Uterine endometrium	Cell proliferation
Estrone	Placenta		
Dehydroepiandrosterone sulfate (DHEA)	Fetal adrenal		
Estriol (from DHEA sulfate)	Placenta		
Progesterone	Corpus luteum	Uterine endometrium, mammary gland	Prepare for implantation of the blastocyte and development of mammary alveolar system
Peptide hormones (nonpregnant state)			
Gonadotropin-releasing hormone (GnRH)	Hypothalamus	Adenohypophysis	Stimulation of release of FSH and LH
Follicle-stimulating hormone (FSH)	Adenohypophysis	Ovarian granulosa and thecal cells	Maturation of ovarian follicle and stimulation of estrogen production
Luteinizing hormone (LH)	Adenohypophysis	Corpus luteum	Stimulate progesterone production
Inhibin	Follicle granulosa cell	Hypothalamus–pituitary	Participate in feedback inhibition of FSH secretion
Activin	Follicle granulosa cell	Hypothalamus	Increase secretion of FSH
Peptide hormones (pregnant/lactating state)[a]			
Prolactin (of the pregnant/lactating states)	Adenohypophysis	Mammary tissue	Stimulate milk production
Human chorionic gonadotropin	Trophoblast and placenta	Maternal corpus luteum	Stimulate progesterone production
Human placental lactogen (HPL) [also known as human chorionic somatomammotropin (HCS)]	Trophoblast and placenta	Maternal tissue	Produce peripheral insulin resistance in the mother
Relaxin	Ovary	Uterine cervix	Softening of uterine cervix
Oxytocin	Neurohypophysis	Uterus and mammary tissue	Milk release
Other			
Prostaglandins	Fetus	Uterus	

[a] All of the peptide hormones of the nonpregnant state are also functional during pregnancy and lactation.

F. Vagina

The vagina is a 9-cm membranous tube that extends from the lower portion or cervix of the uterus downward and forward to the external opening in the vestibule (see Figure 13-2). The wall of the vagina is composed of fibroelastic tissue and smooth muscle, which is lined with a mucous membrane formed of squamous epithelial cells. These cells are quite responsive to estrogen, and their physical appearance under the light microscope changes corresponding to the phase of the ovulatory cycle.

The Papanicolaou test, or Pap smear, which consists of a light microscopic examination of cells obtained painlessly from the vagina and cervix of the uterus, is an important diagnostic test that is of considerable clinical importance in the early diagnosis of cancer of the uterus. The foreign cancerous cells can easily be distinguished from the regular uniform appearance of the epithelial cells.

A description of the anatomical organization of the mammary glands and the placenta is presented in Chapter 14.

III. CHEMISTRY, BIOCHEMISTRY, AND BIOLOGICAL RESPONSES

Table 13-1 tabulates the 13–15 hormones pertinent to female development, reproduction, and lactation.

A. Steroid Hormones

1. *Structural and Metabolic Relationships*

The two most important steroid hormones of the adult female are estradiol and progesterone. In addition, estrone, estriol, and dehydroepiandrosterone play important roles in pregnancy. The structures of these compounds are presented in Figure 13-5.

As reviewed in Chapter 2, the naturally occurring estrogens are typically 18-carbon steroids that have an aromatic A ring with a phenolic hydroxyl. The ovarian follicular cell in the nonpregnant female is the cellular site of production of estradiol. Also, substantial quantities of estrone may be secreted by the ovarian follicular cell, as well as smaller quantities of 17α-estradiol, 16α-estriol (commonly referred to as estriol), and 6α-hydroxy-17β-estradiol. Figure 2-23 reviews the pathways of biosynthesis of the principal estrogens in the nonpregnant female

In the pregnant female the principal estrogen is estriol; it has a biological activity approximately equivalent to that of 17β-estradiol. Estriol is synthesized in the placenta from the precursor dehydroepiandrosterone

Progesterone

17β-Estradiol

Estrone

16α-Estriol

Dehydroepiandrosterone (DHEA)

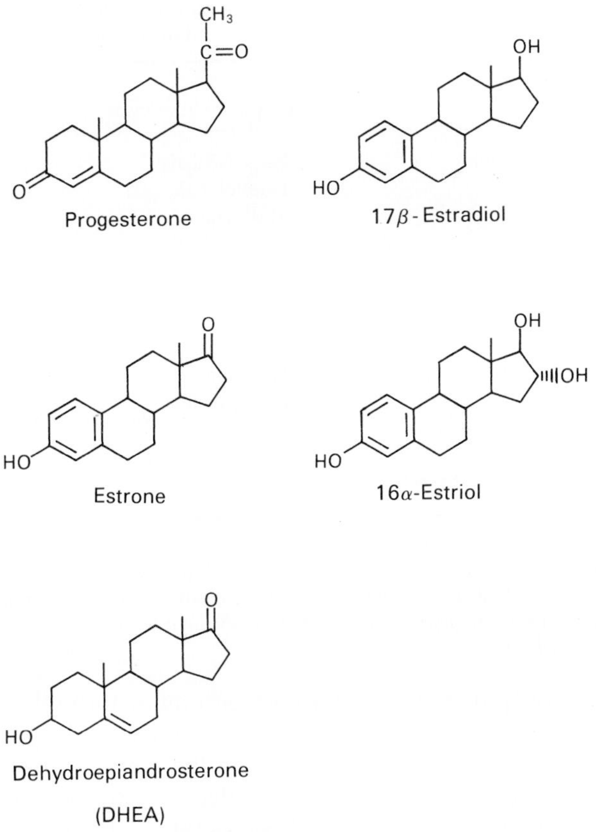

FIGURE 13-5 Structures of the important steroid hormones for the female (see also Figures. 2-23 and 2-30).

sulfate, which is provided by the adrenal cortex of the fetus (see Figure 14-10).

As reviewed in Chapter 2, the naturally occurring progesterones have 21 carbons, with an oxo functionality on both C-3 and C-20. The principal progestin produced by the corpus luteum is progesterone. Small quantities of 20β-hydroxyprogesterone, 10α-hydroxyprogesterone, and 17α-hydroxyprogesterone are also secreted. Figure 2-20 reviews the pathways of biosynthesis of the principal progestins in the nonpregnant female. Progesterone plays an indispensable role in the maintenance of pregnancy (see Figure 14-9).

2. *Secretion Rates*

The secretion rates of the various steroids produced by the ovaries and corpus luteum have been estimated either by careful determination of their concentration in the periphery and ovarian vein coupled with a measurement of the ovarian blood flow or by use of radioisotopes.

Table 13-2 tabulates the changes that occur in the secretion rate, plasma concentration, and metabolic clearance rates of the principal steroids produced by

TABLE 13-2 **Plasma Levels, Production Rates, and Metabolic Clearance of Steroid Hormones Produced by the Human Ovaries**[a]

	Phase of menstrual cycle	Ovarian secretion rate (mg/day)	Metabolic clearance rate[b] (liters/day)	Total plasma concentration (ng/100 ml)
17β-Estradiol	Early follicular	0.07		6
	Late follicular	0.4–0.8	1300	30–60
	Midluteal	0.3		20
Estrone	Early follicular	0.08		4
	Late follicular	0.3–0.6	2200	15–30
	Midluteal	0.2		10
Progesterone	Follicular	1.5	2200	50–100
	Midluteal	24.00		1000–1500
17-Hydroxyprogesterone	Early follicular	0.2	2000	3
	Late follicular	3–4		200
Androstenedione		0.8–1.6	2000	130–160
Testosterone			700	35
Dehydroepiandrosterone		0.3–3.0	1000	400–500

[a] Abstracted from Tagatz, G. E., and Gurpide, E. (1973). Hormone secretion by the normal human ovary. *In* "Handbook of Physiology" (R. O. Greep and E. B. Astwood, eds.), Vol. II, Part I, Section 7, pp. 603–613. American Physiological Society, Bethesda, MD.

[b] Metabolic clearance rate (MCR) is an estimation of the rate at which the steroid is irreversibly removed from the plasma by inactivation (further metabolism). The plasma flow through the liver (a frequent site of steroid catabolism) is ~1500 liters/day.

the ovarian follicles and corpus luteum throughout the various phases of the human menstrual cycle. It is apparent that the actions of the gonadotropins, FSH and LH, have dramatic effects on the steroid-metabolizing enzymes of these tissues. In the plasma compartment, the estrogens are transported by a specific plasma protein, the steroid hormone-binding globulin (SHBG), and the progestins are transported by the plasma protein termed the corticosteroid-binding globulin (CBG). The biochemistry of CBG is described in Chapter 10, while SHBG is reviewed in Chapter 12. Both proteins effectively reduce the "free" concentration of both classes of steroids to on the order of 1–2%.

B. Peptide Hormones

As summarized in Table 13-1, a family of peptide hormones is associated with female reproduction. Those strictly associated with pregnancy and lactation, for example, human chorionic gonadotropin, prolactin, human placental lactogen, oxytocin, and relaxin, will be discussed in Chapter 14.

1. *Gonadotropins*

The chemistry of follicle-stimulating hormone (FSH) and luteinizing hormone (LH) is discussed in detail in Chapter 5. Both FSH and LH are secreted by the adenohypophysis; their release is governed in a complex fashion by gonadotropin-releasing hormone (GnRH), the blood level of circulating steroid hormones, and the interrelated actions of the gonadal regulatory peptides (inhibin, activin, and follistatin). The regulation of the secretion of FSH and LH as well as a description of their biological responses are given in Section IV.B.

The secretory functions of the pituitary gonadotropes are controlled by the interaction of neural (CNS), local, and peripheral signals (see Figure 13-6 and Chapters 3 and 5). The decapeptide GnRH is secreted from the hypothalamus in a pulsatile fashion with short latency and is the primary stimulator of both LH and FSH. In addition, the steroid hormones, progesterone and estradiol, function as negative feedback signals at both the central and pituitary levels to diminish GnRH and gonadotropin release. The system is further modulated by three additional peptide hormones, inhibin, activin, and follistatin.

2. *GnRH*

The secretion of the gonadotropins FSH and LH by pituitary adenohypophysis is governed by the central nervous system (CNS) hypothalamus-mediated release of gonadotropin-releasing hormone (GnRH). GnRH is a decapeptide (see Figure 3-8) with an N-terminal pyroglutamyl residue and a C-terminal glycinamide. Various chemically synthesized analogues of GnRH have been found to be useful in the clinical management of problems of female infertility and both female and male contraception.

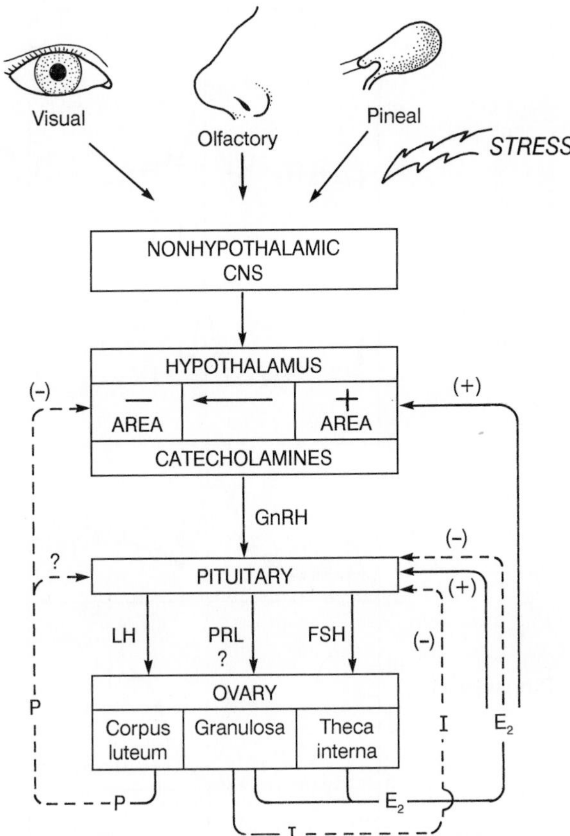

FIGURE 13-6 Schematic diagram of the female hypothalamic–pituitary–ovarian axis. The model emphasizes both the neural and hormonal feedback controls. Depending upon hormone concentration, time course, and prior hormonal status, E_2 can either inhibit or stimulate LH secretion through effects at either the hypothalamus or pituitary. Inhibin produced by the granulosa cells inhibits FSH secretion. This figure should be compared with Figure 12-8, which diagrams the male hypothalamic–pituitary–testicular axis. Abbreviations: GnRH, gonadotropin-releasing hormone; FSH, follicle-stimulating hormone; I, Inhibin; LH, luteinizing hormone; PRL, prolactin hormone; E_2, estradiol; P, progesterone. A dashed line indicates a negative influence, and a solid line indicates a positive influence.

3. *Inhibin*

Inhibin is a heterodimeric protein; it has a common α-subunit and one of two β-subunits. Thus, there are two inhibin isoforms, either inhibin A ($\alpha\beta_A$) or inhibin B ($\alpha\beta_B$); see Figure 12-6. The β-subunit is 18 kDa and the α-subunit is 14 kDa. The inhibins structurally belong to the transforming growth factor (TGF_β) multigene family (see Chapter 19). The α- and β-subunits are cross-linked by one or more disulfide bridges. As yet, it is not known whether the detailed structures of male- and female-derived inhibins are identical.

The structure of the inhibins is also related to another group of peptide hormones, the activins (see the following). Figure 12-6 compares the structures of inhibins and activins.

4. *Activins*

Activins are a class of protein hormones that are either heterodimers or homodimers of the two β-subunits of inhibin, e.g., either $\beta_A\beta_B$, $\beta_A\beta_A$, or $\beta_B\beta_B$ (see Figure 12-6).

Activins have been detected in gonadal fluids (particularly follicular fluid), plasma, and the pituitary and other tissues. Activins are known to have a wide range of biological actions, including inhibition of ACTH and growth hormone production, stimulation of erythroid and megakaryocyte cell differentiation (see Chapter 15), and importantly, in the context of this chapter, stimulation of FSH production and release from the pituitary. The activins may also play a role in embryonic differentiation, particularly the induction of mesoderm.

5. *Follistatin*

Follistatin is a monomeric single-chain protein of 288 amino acids present in follicular fluids and pituitary gonadotroph. Several isoforms (31–45 kDa) exist as a consequence of differential splicing of the mRNA and as a consequence of differential glycosylation. Follistatin has some structural homology with epidermal growth factor (EGF) and human pancreatic trypsin inhibitor.

Follistatin will bind to both activin and inhibin through interaction with their β-subunits. Thus, activin, a $\beta\beta$-dimer, will bind two molecules of follistatin, while inhibin, an $\alpha\beta$-dimer, binds only one molecule of follistatin. Follistatin is believed to function biologically as a local modulator of activin function, so as to reduce or block activin-mediated stimulation of FSH secretion.

IV. PHYSIOLOGICAL RELATIONSHIPS

A. Puberty and Sexual Development

Puberty, or the developmental stage at which anatomical, physiological, and endocrinological changes occur to develop a female competent for sexual reproduction, occurs over the time of 9–14 years of age in the human. Puberty in the female is initiated endocrinologically at age 9–10 by the gradually increasing CNS-mediated secretion of gonadotrophins; the menarche or first menstrual period normally occurs between ages 12 and 16. Previously the juvenile years before puberty were postulated to be endocrinologically relatively inert, but now due to increasing sensitivity of radioimmunoassay techniques for quantitating very low levels of various hormones it is clear that significant changes in many hormones do occur. However, as yet there are no unifying hypotheses to explain these changes.

Puberty, then, is initiated by increased pituitary output of the gonadotrophins, FSH and LH. It is not precisely clear what factors initiate this change. Possibly, throughout childhood the low levels of estrogen secreted by the ovaries of the prepubescent female are sufficient at the CNS–hypothalamus level to block the release of GnRH (see Figure 13-6). Then, as a consequence of a changing CNS–hypothalamus sensitivity to the prevailing blood estrogen levels, GnRH release is gradually increased, which results in an increased release of FSH and LH. In both females and males during the interval of puberty, there are bursts of both FSH and LH secretion that occur during sleep; their etiology is not known. Associated with these changes is an increased sensitivity of the ovaries to FSH and LH, which results in a gradually increasing metabolic production of estrogens and androgens. This in turn accelerates the growth of the uterus, vagina, accessory sex glands, genitalia, pelvis, breasts, and axillary and public hair. These changes then culminate after the menarche in the monthly cyclical processes of ovulation and, in the absence of fertilization, menstruation.

B. The Female Reproductive Cycle

1. *Background*

Figure 13-7 details the cyclical changes in FSH, LH, progesterone, estradiol, and inhibin, while Figure 13-8 depicts the morphological changes in the uterine endometrium that occur during a normal adult female menstrual cycle. The cycle can be conveniently divided into two phases: pre- and postovulatory. The key events of the preovulatory phase are growth and maturation of the ovarian follicle and initiation of the maturation of the uterine endometrium. The key events of the postovulatory phase are the growth, development, and involution of the corpus luteum and, in the absence of initiation of pregnancy, the shedding of the uterine endometrium. These changes are all orchestrated by CNS–hypothalamus regulation of pituitary release of FSH and LH (see Figure 13-6).

A 28-day cycle, as shown in Figure 13-8, is generally regarded as the mean length of normal cycles, but the range for different women extends from 25 to 35 days. The preovulatory phase is usually much more variable than the postovulatory or luteal phase; the latter is remarkably constant at 13 ± 1 days.

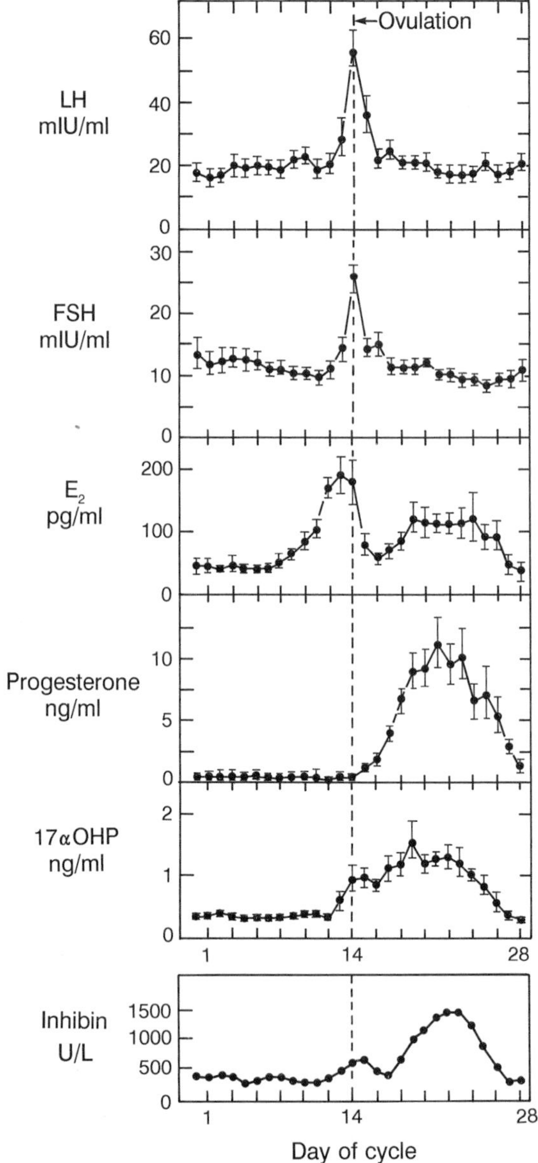

FIGURE 13-7 Presentation of the cyclical hormonal changes throughout the human menstrual cycle for luteinizing hormone (LH), follicle-stimulating hormone (FSH), progesterone, 17-hydroxyprogesterone (17α-OHP), and estradiol (E_2). The cyclic changes in hormones diagrammed in this figure should be correlated with morphological changes of the follicle–corpus luteum and endometrium illustrated in Figure 13-8. The inhibin data were abstracted from McLachlan, R. I., Robertson, D. M., Healy, D. L., Burger, H. G., and de Kretser, D. M. Circulating immunoreactive inhibin levels during the normal human menstrual cycle *J. Clin. Endocrinol. Metab.* **65**, 954–961. The top five panels were modified from Thorneycroft, I. H., Mishell, D. R., Jr., and Stone, S. C. (1971) (1987). *Am. J. Obstet. Gynecol.* **111**, 947–951.

Table 13-3 summarizes the ovarian uterine and hormonal changes that occur throughout the menstrual

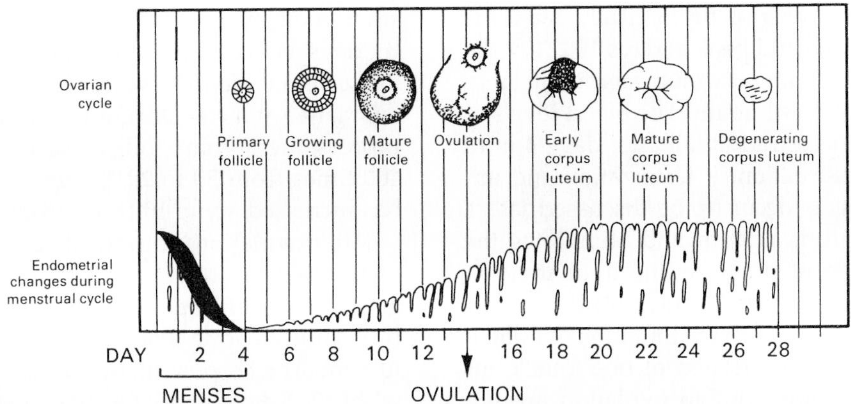

FIGURE 13-8 The sequential changes throughout the human menstrual cycle of the follicle–corpus luteum are shown in relation to the changes in the thickness of the endometrium. These morphological changes should be compared to the cyclical hormone changes presented in Figure 13-7.

cycle in relation to the morphological changes occurring in the ovaries and uterine endometrium.

2. *Regulation of Secretion of FSH and LH*

The multiple factors controlling the release of GnRH by the CNS–hypothalamus throughout the menstrual cycle are not clearly understood. As diagrammed in Figure 13-7, throughout the menstrual cycle there are significant changes in the ratio of FSH to LH. From days 1 to 10 of the cycle, FSH blood levels and LH levels are approximately equivalent. Then at days 12–14 LH rapidly becomes elevated to a very large peak or surge on the day of ovulation; blood levels of FSH are also moderately increased at ovulation, but not to the extent achieved by LH. Throughout the luteal phase the levels of FSH are low, while LH levels are relatively high on days 14–18 and then fall to low levels in the absence of fertilization and implantation by day 28.

3. *Hormonal Events in the Ovaries and Corpus Luteum*

The dominant hormonal changes of the menstrual cycle, particularly in the uterine endometrium, are mediated by the steroid hormones secreted by the ovaries and corpus luteum (see Table 13-3). Associated with the cyclical changes in the gonadotropins described in the preceding section are related changes in the blood levels of estrogen and progesterone. In the early preovulatory or follicular phase, estradiol levels remain low until ~7–8 days before the LH surge. Then estradiol increases and reaches a peak 1 day before the LH surge; next, there is a drop in estradiol at days 14–16, followed by a rise to a second peak at days 20–26 (see Figure 13-7).

Progesterone secretion by the ovaries in the preovulatory period is very low, and accordingly the blood levels are low. The blood levels of progesterone and

TABLE 13-3 Summary of Ovarian and Uterine Morphological Changes with Hormonal Activities in the Human Menstrual Cycle

Menstrual cycle					Production of		
Days	Phase	Uterine endometrium	Ovaries	Estrogen	progesterone	FSH	LH
1–4	Menstrual	Shedding of outer layers	Initiation of follicular development	Low	Very low	Increasing	Low
5–12	Follicular	Reorganization and proliferation	Maturation of follicle	Increasing	Very low	High	Low
13–15	Ovulation	Further growth	Ovulation	High	Low	Low	High
16–25	Luteal	Highly vascularized and active secretion	Functional corpus luteum	First declining, then a secondary rise	Increasing	Low	High
26–28	Premenstrual	Initiation of degeneration	Regression of corpus luteum	Decreasing	Decreasing	Increasing	Decreasing

17α-hydroxyprogesterone rise dramatically after the LH surge and ovulation and peak at days 18–24 of the cycle; this is coincident with maximum steroid metabolism activity of the corpus luteum. The blood levels of both progesterone and estradiol fall after day 24 until the initiation of menstruation. If fertilization and implantation of the zygote occur (to be discussed later), hCG rescues the corpus luteum and stimulates the continued production of progesterone until the placenta becomes functional (see Chapter 14).

The mechanisms that govern the initiation of follicular growth as well as the selection of one follicle for the necessary maturation prior to ovulation are not clearly understood. This involves the sequence of maturation diagrammed in Figure 13-2. Important factors are the amount of LH and the ratio of FSH:LH; also, the availability of estrogen and possibly androgens is critical. In the immature female hypophysectomized rat model, estrogens clearly promote ovarian growth, reduce follicular atresia, induce granulosa cell hyperplasia, and increase ovarian response to FSH. In contrast, androgens in the same model promote follicular atresia. Thus, if LH were to stimulate androgen (androstenedione) production in certain follicles, or if there were differences in different secondary follicles of the ratio of estradiol to androgen combined with a favorable cellular receptor concentration for FSH, then one "selected" follicle would be stimulated to develop into a Graafian follicle. The Graafian follicle, in turn, would be subject to the influence of the gonadotropins and eventually ovulate.

Shown in Figure 13-2 is a pictorial representation of the developmental stages of follicular maturation and corpus luteum development in the ovary. In the selected secondary follicle on days 1–3 of the cycle, the oocyte becomes progressively larger and the surrounding granulosa cells proliferate further. The granulosa cells produce principally only estradiol. The cells adjacent to the follicle become enlarged and arranged in concentric circles; this cellular array is termed the thecal. The thecal cells are active in steroid metabolism and can produce both estradiol and androstenedione. As the follicle cell develops (days 6–10), the granulosa cells produce, as a consequence of the trophic actions of FSH, increasing amounts of estradiol. Blood levels of estradiol increase slowly (days 3–7) and then more rapidly to reach an apex (days 11–13) just prior to the LH surges. This has the effect of establishing the high levels of estradiol required for positive feedback at the CNS–hypothalamus and pituitary levels. Simultaneously, receptors for LH are appearing in increasing concentrations on both the thecal and granulosa cells in preparation for ovulation. Very late (days 11–13) in the follicular phase, the thecal cells are stimulated by LH to begin producing progesterone. By days 11–12 of the cycle the general maturation of the secondary follicle is complete, and it is referred to as a Graafian follicle. During this 12-day developmental process, the primary follicle has increased in size some 400 times from 50 to 29,000 μm, while the oogonium has increased some 10 times from 15 to 150 μm.

After ovulation, both the thecal and granulosa cells of the follicle undergo rapid mitosis. Capillaries are generated from the theca that invade the granulosa cells, thus creating a new endocrine organ, the corpus luteum. The corpus luteum reaches its maximum size within 2–8 days, and if fertilization and zygote implantation do not occur by 8–9 days, then it undergoes regression and degeneration, ultimately leading to the production of the corpus albicans. The corpus luteum, under the stimulus of LH, actively produces progesterone.

4. *Changes in the Uterine Endometrium*

The endometrium (see Figure 13-8) and vaginal epithelium undergo very striking morphological changes throughout the 28 days of the menstrual cycle. During the first half of each menstrual cycle, which is often denoted the follicular or preovulatory phase, the estrogen concentrations gradually increase to reach a maximum 24 hr before ovulation, while the progesterone concentrations are relatively low. These changes in estrogen, with a low progesterone value, stimulate the endometrium to increase in thickness from 1 to 3–5 mm. Then, 36 hr after ovulation, which signals the initiation of the luteal or postovulatory phase, the progesterone concentration rises sharply due to its secretion from the corpus luteum (see Figure 13-7). Simultaneously estrogen levels are maintained at levels two-thirds of their previous maximum. This balance of progesterone and estrogen induces further specific morphological changes in the endometrium. Glycogen-containing granules appear in the glandular cells; additionally, the glands of the endometrium become increasingly tortuous. This phase of the cycle is also termed the secretory phase since the lumens of the endometrial glands are filled with secretions. The endometrium also becomes highly vascularized with the ingrowth of new spiral arteries from the arcuate vessels of the myometrium. The growth of these spiral arteries is stimulated by estrogen.

In the event that fertilization and implantation do not occur, the stromal edema decreases, the glandular secretions diminish, and the endometrium is invaded by lymphocytes. The concentrations of estrogen and progesterone both fall sharply at days 27–28 of the cycle as a consequence of the involution of the corpus luteum. At the time of the rapid fall in estrogen and progesterone at the end of the luteal phase, the spiral arteries become constricted. Accordingly, blood flow through the capillaries supported by these arteries diminishes. Eventually blood stasis and stromal degeneration follow, and the endometrial tissue is sloughed off.

C. Menopause

Menopause is defined as the cessation of ovulation by the ovaries. This occurs in women between ages 40 and 50 years due to the utilization of the fixed number of follicles that were established in the fetal ovaries. Each month of ovulation, several follicles disappear by the process of atresia, and normally only one matures to contribute an ovum. Thus, at some interval in the fourth decade there are no more follicles available to support this cyclical process. The endocrinological consequences of cessation of ovulation are varied, and a variety of medical problems may develop, depending upon the exact nature of the reproductive hormonal balance achieved

In the postmenopausal woman, the average blood concentrations of FSH and LH remain elevated for the remainder of life. The postmenopausal secretion of immunoreactive LH is 7-fold higher and that of FSH is 3-fold higher than in a premenopausal woman. While the postmenopausal ovary is not totally devoid of the ability to secrete steroids, the production level is markedly reduced in comparison to a younger, ovulating woman. In menopausal women the major circulating estrogen is estrone; this is formed almost exclusively by aromatization of androstenedione by the adrenals as well as by adipose tissue and muscle. The control of LH and FSH secretion in postmenopausal women is mediated via a short feedback loop wherein FSH and LH themselves act on the hypothalamus in conjunction with steroids that are largely derived from the adrenals.

V. BIOLOGICAL AND MOLECULAR ACTIONS

A. Cellular Mechanisms of Action of the Reproductive Hormones

1. GnRH

Elucidation of all of the factors governing the release of GnRH as well as a biochemical description of their actions are only now emerging. The problem has not been readily approachable through the classical biochemical techniques of fractionation, purification, and so on due to the complex, but critically important, anatomical relationships that exist between the CNS, the hypothalamus, which secretes GnRH, and the adenohypophysis of the pituitary, which is the site of action of GnRH.

The release of GnRH from the hypothalamus is determined by an integration of visual, olfactory, pineal, and stress signals, as well as endocrine factors; thus, the hypothalamus acts as a clearinghouse to determine whether GnRH should be released (see Figure 13-6). The release of FSH and LH is a highly regulated process determined by (a) negative feedback, (b) positive feedback, and (c) neural components.

The decapeptide GnRH is released from the hypothalamus in a pulsatile fashion with short latency and initiates the concomitant secretion of both LH and FSH (see Figure 13-9). The steroid hormones progesterone and estradiol can also function as both negative and positive feedback signals at both the CNS and pituitary levels to diminish GnRH and gonadotropin release.

In addition, the system is modulated further in a complex fashion by the combined actions of inhibin, activin, and follistatin. Inhibin and follistatin effectively reduce GnRH-stimulated release of FSH by interfering with the actions of activin (see Figure 13-10). In addition, the neuropeptide Y (NPY) and the intestinal and hypothalamic peptide galanin (see Chapter 8) have

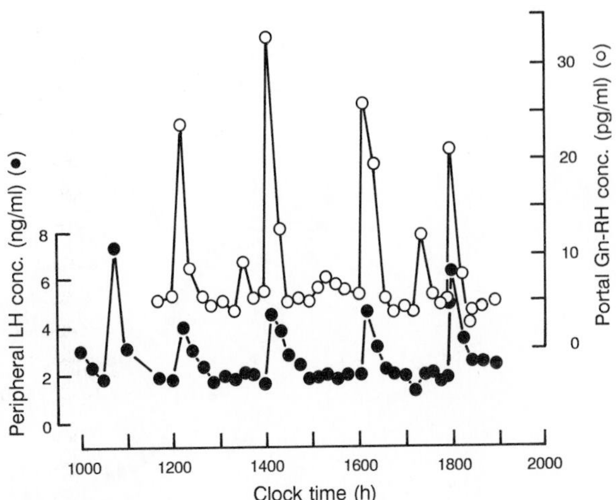

FIGURE 13-9 Pulsatile secretion of gonadotropin-releasing hormone (GnRH) in hypothalamic–portal blood and luteinizing hormone (LH) in jugular veinous blood in ovariectomized ewes. Reproduced with permission from Clark, I. J. and Cummins, T. J. (1982). The temporal relationship between gonadotropin releasing hormone (GnRH) and luteinizing hormone (LH) section in ovariectomized ewes." *Endocrinology* **11**, 1737–1739.

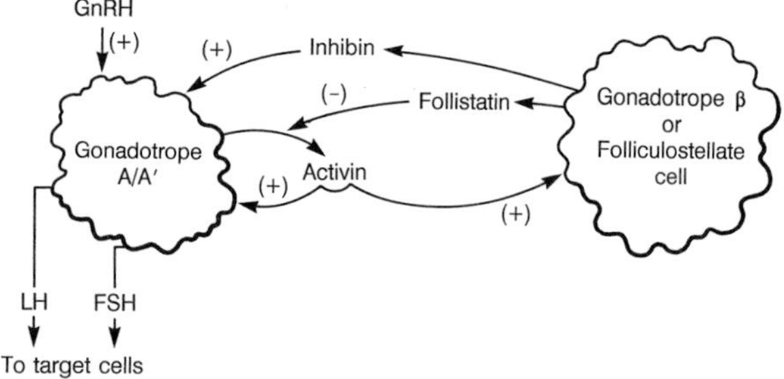

FIGURE 13-10 Schematic model of paracrine and autocrine regulation of the pituitary gonadotrope by activin, follistatin, and inhibin. The secretion of LH and FSH occurs from separate gonadotropes, either A or A', both of which are stimulated principally by GnRH. However, a fine-tuning of the secretion of LH and FSH is effected by the combined actions of inhibin, activin, and follistatin. In this model, activin, which is produced by gonadotrope A, has its actions on both gonadotropes A and B diminished by follistatin Modified from Bilezikjian, L. M., Corrigan, A. Z., and Vale, W. W. (1993). Activin-β and follistatin as autocrine/paracrine factors of the rat anterior pituitary. From 11th International Symposium on Inhibin and Inhibin-Related Proteins. Siena, Italy.

both been shown to be capable of stimulating GnRH release.

The primary action of GnRH is to stimulate both the acute release and the biosynthesis of LH and FSH. The first step is the binding of GnRH to its receptor on the plasma membrane of the gonadotroph (see Figure 13-11). This initiates a complex series of signal transduction events (involving PKC, intracellular Ca^{2+}, IP_3, and DAG), which result in a prompt (within minutes) secretory burst of LH or FSH. The mechanism(s) by which GnRH stimulates the expression of mRNA for LH and FSH is not yet known.

2. *Gonadotropins (FSH and LH)*

The principal biochemical actions of FSH and LH are to induce the production of estrogen (by the ovarian thecal cell) and progesterone (by the corpus luteum), respectively. Their generally accepted mechanism of action is not dissimilar from the actions of the peptide hormone ACTH in stimulating steroidogenesis in the adrenal cortex (see Chapter 10).

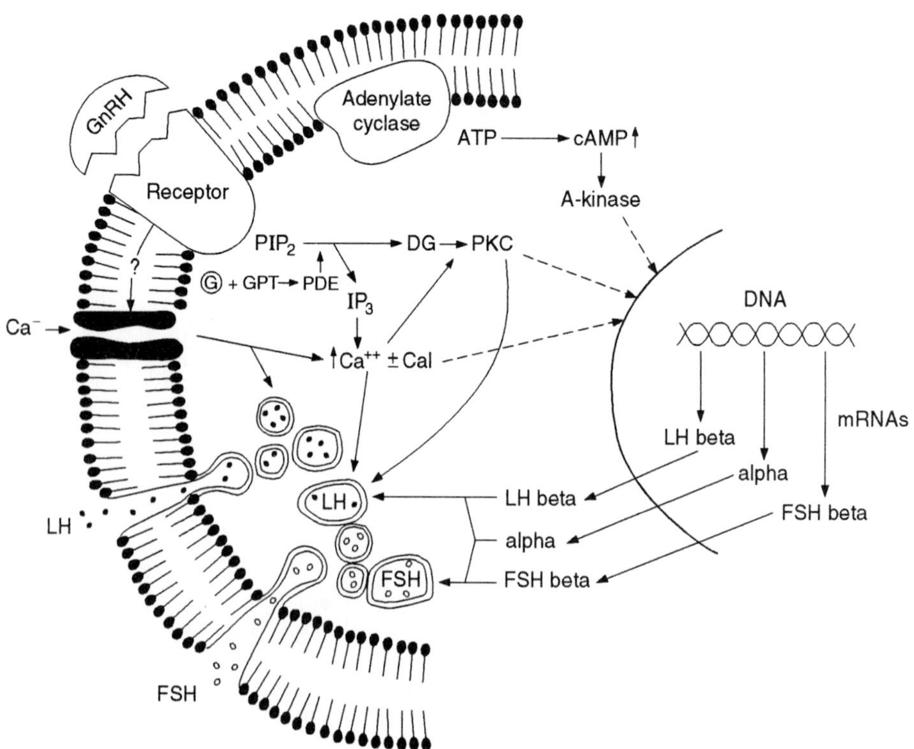

FIGURE 13-11 Schematic model of GnRH-mediated secretion of the gonadotropins LH and FSH. Dashed lines indicate pathways where the mechanisms are unknown. Abbreviations: PIP_2, phosphatidylinositol 4,5-phosphate; DG, diaclyglycerol; PKC, protein kinase C; IP_3, inositol triphosphate; G, G protein; PDE, phosphodiesterase (phospholipase C); cal, calmodulin. Modified with permission from Marshall, J. C. (1995). Regulation of gonadotropin sectetion. *In* "Endocrinology" (L. J. DeGroot, H. G. Burger, J. L. Jameson, D. L. Loriaux, J. C. Marshall, W. D. Odell, J. T. Potts, Jr., and A. H. Rubenstein, eds.), Vol. 3, pp. 1993–2007. W. B. Saunders, Philadelphia, PA.

Receptors for both FSH and LH are present in the cell surface of their respective target cells. Both FSH and LH intracellularly stimulate the production of cAMP by adenyl cyclase. Ovarian cells also contain receptors for prostaglandins and catecholamines, both of which can mimic some of the actions of LH and cAMP. The cAMP then, as shown in Figure 13-12, activates a protein kinase that stimulates cholesterol side chain cleavage, eventually increasing the production of estradiol and progesterone. It has been demonstrated that there is a good correlation between receptor occupancy by LH, stimulation of cAMP, activation of protein kinase, and progesterone production.

LH mediates a large increase in the activity of the cholesterol side chain cleavage enzyme, which is the rate-limiting step in the ovarian steroidogenesis pathways. LH also stimulates the gene expression of 3β-steroid dehydrogenase, 17α-hydroxylase, and 17,20-lyase (see Table 2-6).

Further work is required to describe, at the biochemical level, the complex series of steps associated with maturation of the ovarian follicle, luteinization of the granulosa cells, and formation of a functional corpus luteum. It has been proposed that the cyclical nature of the menstrual cycle is largely governed by temporal changes in the cell surface receptors of the ovarian follicles for FSH and LH. These changes in turn drive, in a cyclical fashion, the biosynthetic and mitotic activities of the follicular granulosa cells and the steroid-producing thecal cells.

In the early phases of the cycle of granulosa cell development (day 26 of the cycle), the granulosa cell is quite dependent upon FSH. At this point, it has largely FSH receptors and only a limited number of LH receptors; it is rapidly dividing, growing, and acquiring the enzymatic capacity to aromatize androgens to estradiol. At this point the production of estradiol is a collaboration between the surrounding thecal cells, which convert cholesterol to androgens, and the granulosa cell, which converts the androgen to estradiol. This granulosa-produced estrogen then collaborates with FSH to speed up the replication rate of the granulosa cells. Thus, although there is now (days 8–10 of the cycle) a falling plasma concentration of FSH, there is an increased capacity to respond to FSH (due to the increased concentration of FSH receptors) and to produce estradiol. Concurrently there is an increased concentration of LH receptor on the thecal and granulosa cells in preparation for ovulation and formation of the luteal phase of the cycle. Inhibin, which is secreted by the granulosa cells, has been shown to effect the inhibition of FSH production and release by the pituitary. Then as the late follicular phase ensues (days 11–13), the aromatization capacity of the granulosa cells becomes maximal and the blood level of estrogen rises dramatically (see Figure 13-7), which leads to a positive feedback loop at the hypothalamus–pituitary axis and initiates the midcycle surge in LH secretion (days 13–14). Figure 13-13 summarizes the pathways of steroidogenesis that are associated with the ovarian follicle and the corpus luteum.

After ovulation, the thecal and granulosa cells are transformed into the corpus luteum, which is the site of production and secretion of large quantities of progesterone and moderate quantities of estradiol.

3. Estrogen

a. Background

The dominant actions of estrogen occur in the female reproductive tract, although there are also significant biological actions mediated in the hypothalamus and brain as well as in a variety of other visceral organs. Table 13-4 tabulates the tissue distribution of the estrogen receptor. Summarized in Table 13-5 are the temporal events initiated in the immature or ovariectomized rat uterus as a consequence of estradiol administration. It is not clear whether all of the actions reported in Table 13-5 are mediated as a consequence of estrogen interaction with target organ-specific nuclear receptors for the steroid hormone. The presence of the estrogen receptor in a tissue, however, is presumptive of the biological actions of estrogen at that location. There is also emerging evidence that some biological responses to estrogen occur too rapidly to be mediated via the nuclear receptor.

b. Estrogen Receptors

The human estrogen receptor (ER) gene and cDNA have been cloned and evaluated (see Figure 13-14). As with all classical steroid receptors, the ER gene has eight coding exons. Although the transcription of the ER gene is initiated from two promoters, which generate two mRNA species with alternate 5′ ends, only one species of ER is generated after translation. The human ER is a polypeptide chain of 595 amino acids (66.2-kDa molecular mass).

The ER has been shown to function as a transcriptional regulatory protein that is essential for female sexual differentiation and development. As shown in Figure 13-14, the ER has many functional domains that participate in its activation of gene transcription. The key region is the DNA-binding domain, which interacts with the hormone response elements (HRE) of the promoters of regulated genes. The consensus sequence of the HRE for the ER is shown here; it represents an inverted repeat with three spacer nucleotides (see also Table 1-10): AGGTCAnnnTGACCT.

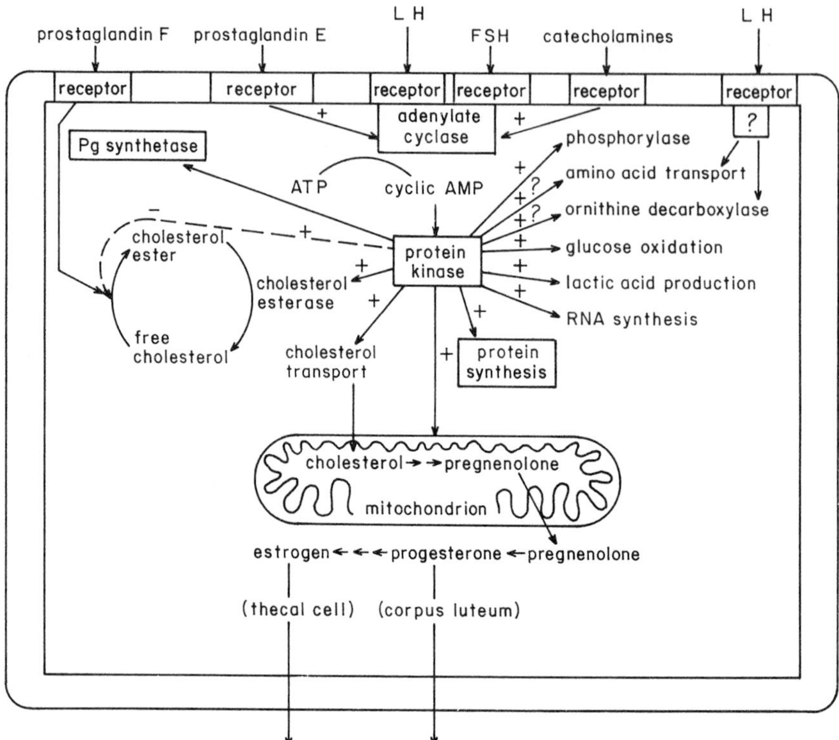

FIGURE 13-12 Model describing the steroidogenic actions of FSH and LH on the ovary. Modified from Siiteri, P. K., and Febres, F. (1979). Ovarian hormone synthesis, circulation, and mechanisms of action. *In* "Endocrinology" (L. J. DeGroot, G. F. Cahill, L. Martini, D. H. Nelson, W. D. Odell, J. T. Potts, Jr., E. Steinberger, and A. J. Winegrad, eds.), Vol. 3, p. 1405. Grune & Stratton, New York.

The ER is predominantly localized in the nucleus of the cell, irrespective of whether it has a bound ligand, e.g., estradiol; a comparison of the subcellular locations of other unoccupied receptors is given in Table 1-5. As noted in Figure 13-14, the ER has two nuclear localization signals inherent in its structure.

Prior to binding ligand, the ER is present in an inactive complex with the heat shock protein, hsp 90. As

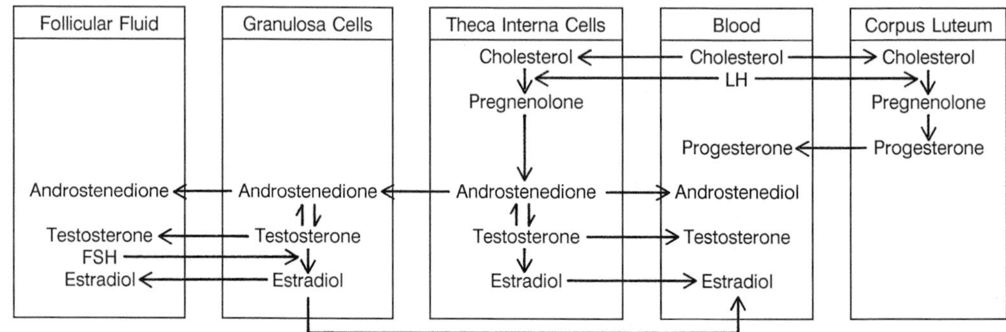

FIGURE 13-13 Summary of pathways of steroidogenesis of estradiol, progesterone, and testosterone (limited quantities) by the ovarian follicle (theca interna cells and granulosa cells) and the corpus luteum. Figure 13-2 describes the morphological changes occurring in the mammalian ovary that are necessary for the formation of a mature follicle (major source of estradiol) and after ovulation the generation of the corpus luteum (major source of progesterone).

TABLE 13-4 Tissue Distribution of Estrogen Receptors

Reproductive tract
Uterus (mammals)
Vagina (mammals)
Placenta (mammals)
Oviduct (chicken)
Ovary
Corpus luteum
Mammary tissue (mammals)
Neuroendocrine system
Pituitary
Hypothalamus
Brain (preoptic area, septum, amygdala, cortex)
Testis
Epididymis
Prostate gland
Seminal vesicles
Müllerian duct (chick embryo)
Visceral organs
Liver
Kidney
Lung

described in Figure 13-15, the binding of estradiol to the ER facilitates the formation of an ER homodimer, which then binds to the hormone response elements (HRE) located on the promoter of the gene so as to generate a functional transcriptional complex.

TABLE 13-5 Temporal Actions of Estrogens in the Rat Uterus[a]

Time after estradiol administration	Sequence of responses
30 min	Stimulation of RNA polymerase II
1 hr	Stimulation of glucose and phospholipid metabolism
1.5 hr	Stimulation of RNA polymerase I
2–5 hr	Stimulation of general protein synthesis
5–7 hr	Acceleration of water imbition
10–20 hr	Stimulation of net protein, RNA, and DNA synthesis
20–30 hr	Acceleration of cell division

[a] Modified from Fig. 1 of Katzenellenbogen, B. S., and Gorski, J. (1975). Estrogen action on syntheses of macromolecules in target cells. *In* "Biochemical Actions of Hormones" (G. Litwack, ed.), Vol. 3, p. 187. Academic Press, New York.

4. *Progesterone*

a. Background

In contrast to estradiol, the biological actions of progesterone are largely restricted to the female reproductive tract and mammary tissue. Table 13-6 tabulates many of the known biological actions of progesterone. Receptors for progesterone have been found in the uterus, mammary tissue, placenta, and anterior pituitary.

b. Progesterone Receptors

The human progesterone receptor (PR) gene and cDNA have been cloned and evaluated. While there is only one gene, two distinct mRNA species generate two functionally different protein receptor forms, A and B (see Figure 13-16). The two mRNA species are transcribed from the single gene under the control of separate A and B promoters, both of which are known to be activated by estradiol and the ER.

The human PR_B protein consists of 933 amino acids (115-kDa molecular mass), while the PR_A protein consists of 768 amino acids (82 kDa). The two PR proteins are identical except that PR_A lacks the first 165 amino acids present on the amino terminus of PR_B.

Both the A and B forms of the progesterone receptor are functional, like all steroid receptors, as a transcriptional regulatory protein. The general model of the ER regulation of gene transcription shown in Figure 13-15 is also applicable to the PR. The consensus sequence of the hormone response element (HRE) for the PR is similar to that of the glucocorticoid HRE (see Table 1-10) and is shown here: AGAACAnnnTGTTCT.

There is emerging evidence that, although both the PR_A and PR_B are able to bind ligands and interact with the HRE, they are functionally different. On the basis of transfection studies, the two receptor forms were found to have both promoter-specific and cell-specific differences. This suggests that the cellular responsiveness to progesterone may be modulated via alteration in the ratio of tissue expression of PR_A and PR_B in target cells.

The presence of PR in breast tumors is generally indicative of a possible responsiveness to endocrine treatments. Intriguingly, it has been found that a significant proportion of PR-positive breast tumors expressed atypically low amounts of PR_B; the significance of this observation remains to be elucidated.

B. Mechanisms of Progesterone and Estrogen Receptor Action (As Exemplified by the Chick Oviduct System)

One system that has contributed extensively to our current understanding of the mode of action of steroid hormones is the chick oviduct. W. Schrader, B. O'Mal-

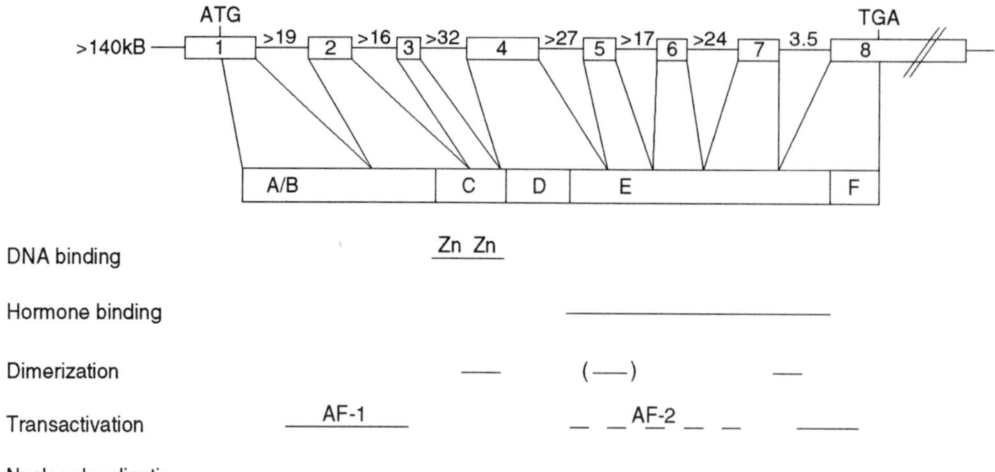

FIGURE 13-14 Schematic diagram of the genomic organization, domain structure, and functional characteristics of the estrogen receptor. The ER gene spans over 140 kb of DNA. Note that the exon structure does not correlate with the domains of the ER (see Figure 1-26). The DNA-binding domain C is composed of two zinc fingers (see Figure 1-42). Two dimerization domains on the ER have been detected in domains C and E. The hinge domain D codes for a nuclear localization signal. Two transactivation domains in the A/B and E/F regions have also been described; they act independently and confer on the ER both promoter and target organ specificity. Modified with permission from Pfeffer, E. U., Fecarotta, E., Arena, G., Forlani, A., and Vidali, G. (1996). Alternative splicing of the estrogen receptor primary transcript normally occurs in estrogen receptor positive tissues and cell lines. *J. Steroid Biochem. Mol. Biol.* **56**, 99–105.

ley, and colleagues as well as R. Palmiter and R. Schimke have carried out detailed biochemical analyses of estrogen- and progesterone-mediated induction of ovalbumin, conalbumin, and other egg white proteins in the chick oviduct. Figure 13-17 presents a schematic diagram of the hormone-mediated events in the immature chick oviduct. As shown here, estrogen plays a key role in priming the immature oviduct to be responsive to subsequent treatment with either estrogen or progesterone. Then, secondary stimulation with either progesterone or estrogen leads to rapid biosynthesis of ovalbumin and other egg white proteins.

The synthesis of DNA complementary to ovalbumin messenger RNA (cDNA probes) has permitted a detailed study of gene expression in the chick oviduct system. By employing this cDNA probe to analyze the consequences of primary estrogen treatment or secondary estrogen or progesterone treatment, extensive data have been accumulated that strongly support the primary hypothesis that steroid hormones act by gene activation.

C. Hormonal Effects on Ovulation and the Menstrual Cycle

1. *Steroid Receptors*

The mammalian uterus structure and function are determined in large part as a consequence of its exquisite sensitivity to progesterone and estrogen. In general, progesterone and estradiol exert opposing effects on the estrogen and progesterone receptor systems. The presence of estradiol increases, via protein synthesis pathways, the concentration of receptors for both progesterone and estradiol, while progesterone generally leads to the down-regulation of receptor-binding sites for these steroids. An additional regulatory component relates to the blood levels of the steroid hormones.

In the golden hamster there is ample evidence of a positive relationship between the increase in estrogen secretion occurring during the follicular phase of the estrous cycle and the elevation of the uterine estrogen and progesterone receptors (see Figure 13-18). The details of the biochemical changes contributing to receptor changes during ovulation are not clear. The ovulatory surge of gonadotropin probably stimulates a transitory rise in serum estradiol; however, the occupied estrogen receptor fails to accumulate due to sup-

TABLE 13-6 Biological Actions of Progesterone

Thermogenesis (in woman)
Regulation of egg movement through fallopian tubes
Preparation of the uterus to receive the blastocyst
Alteration of electrical activity in the brain
Control of uterine contraction (at parturition)
Generation of the secretory system of breasts (during pregnancy)

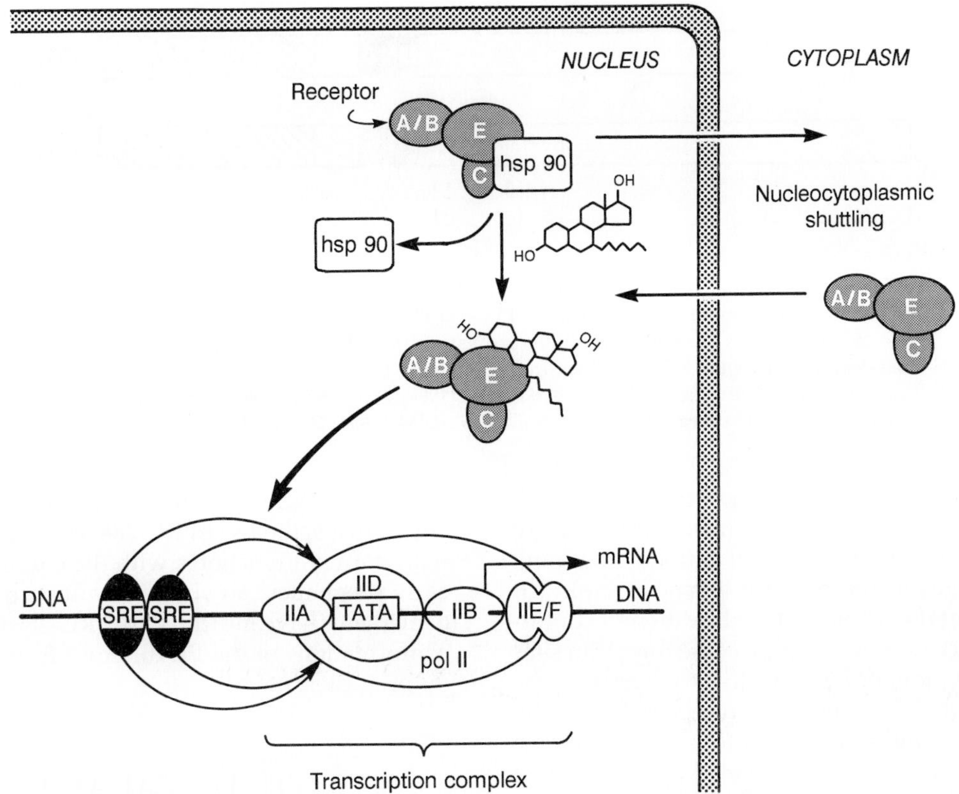

FIGURE 13-15 Schematic model of estrogen receptor action. Although the ER is predominantly found in the cell nucleus, it constantly shuttles between the nucleus and cytoplasmic compartments. After the ER binds to the ligand, it then dissociates from the inactive hsp 90–ER complex and forms a homodimer, $(ER)_2$. The liganded ER homodimer then seeks out estrogen-specific hormone response elements, SRE, and forms a DNA–receptor complex that, in some fashion, stabilizes the formation of a transcription complex (TC). The TC comprises RNA polymerase II (pol II), transcription factors IIA, IIB, IID, and IIE/F, and a TATA box, all of which are located on the promoter of the gene that will be transcribed. The A/B, C, and E shown on the receptor (top portion of figure) refer to the domains of the ER (see Figures 1-26 and 13-14). [For a detailed discussion, see Tsai, M.-J., and O'Malley, B. W. (1994). Molecular mechanisms of actions of steroid/thyroid receptor superfamily members. *Ann. Rev. Biochem.* **63,** 451–486.]

pression by the preovulatory levels of progesterone. Then, when the serum concentrations of estradiol fall, the nuclear levels of the estrogen receptor fall precipitously as a consequence of unopposed progesterone action. Thus, changes in the levels of the estrogen receptor are a balance between, first, the combined actions of the simultaneous presence of progesterone and estradiol and, second, the unopposed action of progesterone during estrogen withdrawal.

D. Estrogen and Progestin Antagonists

As a consequence of the emergence of a detailed understanding of how steroid hormones interact with receptors to regulate gene transcription (see Figures 1-41, 10-14, 10-26, and 13-15), a new field of research has emerged concerning the design of steroid hormone antagonists. Steroid antagonists can be either steroidal analogues or nonsteroidal compounds that selectively function as inhibitors of steroid hormone action. Table 13-7 summarizes some proposed mechanisms of action of antagonists of steroid hormones. Table 13-8 tabulates some examples of steroid antagonists for each class of classical steroid hormones. Two models, based on extensive studies of the progesterone and estrogen receptors, which provide a conceptual basis for the mode of action of steroid antagonists, are presented in Figure 13-20.

In the first model (Figure 13-20A), the antihormone is envisioned to interfere with the normal process of cycling of the unoccupied receptor–heat shock protein 90 (hsp 90) complex between the nuclear and cytoplasmic compartments. The consequence of the presence of the antihormone is that the receptor accumulates in the cytoplasm where it can become degraded.

The second model has as its basis the fact that a major consequence of hormone binding to the progesterone receptor is the induction of a conformational

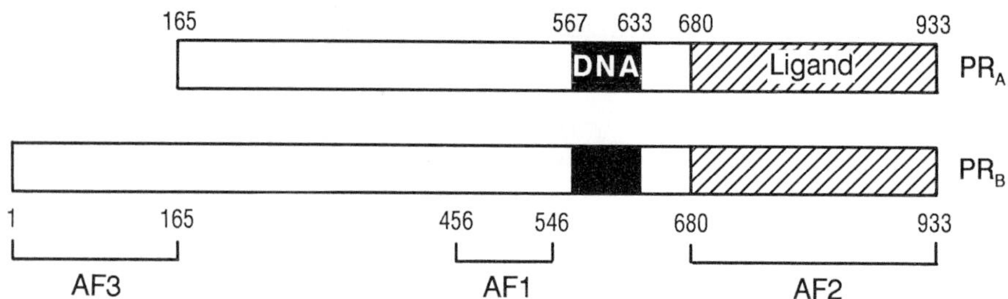

FIGURE 13-16 Schematic representation of the two forms of the progesterone receptor protein, PR_A and PR_B. The amino acid sequences of PR_A and PR_B are identical except that PR_A lacks the first 165 amino acids present on the amino terminus of PR_B. The domain regions are similar to that described in Figure 1-26. Modified from Graham, J. D., Yeates, C., Balleine, R. L., Harvey, S. S., Milliken, J. S., Bilous, A. M., and Clarke, C. L. (1996). Progesterone receptor A and B protein expression in human breast cancer. *J. Steroid Biochem. Mol. Biol.* **56,** 93–98.

change in the receptor protein. The experimental evidence supporting this conclusion included the use of protease digestion and monoclonal antibody mapping to detect changes in the receptor–ligand complex. It was argued that if a ligand induced a different conformation in the receptor, various protease digestion sites would become more or less sensitive to digestion, depending upon their location in the receptor. The consequence of the ligand-induced conformational change is that the receptor becomes activated and competent to form a functional transcription complex on the promoter of a gene. In contrast (see Figure 13-20B), when an antagonist binds to a receptor, either no conformational change or a different conformational change occurs, which does not yield a normally activated receptor; accordingly, the receptor–ligand complex cannot generate a functional transcription complex.

One example of a clinically useful progestational antagonist is the use of RU-486 as an agent to interfere with normal PR actions, with the consequence that, in early pregnancy, an abortion will occur. Several lines of evidence have indicated that RU-486 induces unique conformations of the PR different from that mediated by progesterone.

VI. CLINICAL ASPECTS

A. Female Contraception

A major clinical topic related to female endocrinology, both on a personal level as well as on a societal level, concerns the ability of a woman to control her periods of fertility. Table 13-9 summarizes the effec-

HORMONAL STATE	OVIDUCT GROWTH	OVIDUCT WEIGHT	STATE OF DIFFERENTIATION	HORMONE-INDUCED PROTEINS: ESTROGEN	HORMONE-INDUCED PROTEINS: PROGESTIN
UNSTIMULATED		0.01g	UNDIFFERENTIATED CELLS	NONE	NONE
	P ↓ NO GROWTH; E ↓				
PRIMARY STIMULATION		2g	TUBULAR GLANDS; GOBLET CELLS; — LUMEN	OVALBUMIN, OTHERS	NONE; – – – AVIDIN
	DISCONTINUE ESTROGEN				
WITHDRAWAL		0.25g	REGRESSED STRUCTURE	NONE	NONE
	P ↓ OR E ↓				
SECONDARY STIMULATION		0.5g	TUBULAR GLANDS; GOBLET CELLS; — LUMEN	OVALBUMIN, OTHERS	OVALBUMIN, OTHERS; – – – AVIDIN

FIGURE 13-17 Schematic model describing the mode of action of estrogen and progesterone in the chick oviduct. Modified from Vedekis, V. W., Schrader, W. T., and O'Malley, (1978). The chick oviduct progesterone receptor. *In* "Biochemical Actions of Hormones" (G. L. Litwack, ed.), Vol. 5, pg. 359. Academic Press, New York.

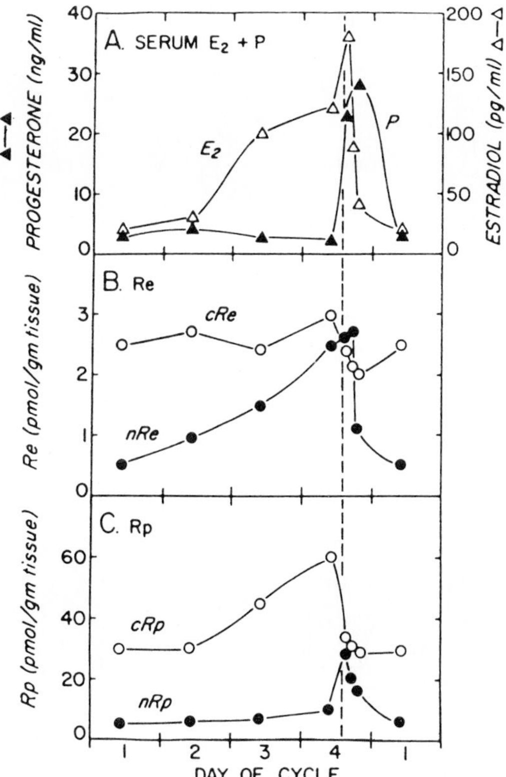

FIGURE 13-18 Serum steroid relationships during the 4-day hamster estrous cycle. The dashed vertical line indicates the time of the critical period for the ovulatory surge of the gonadotropins. Abbreviations: c, cytosol; E_2, 17β-estradiol; n, nuclear; P, progesterone; Re, estrogen receptor; Rp, progesterone receptor. Modified with permission from Leavitt, W.W. (1983). Hormonal regulation of estrogens and progesterone receptor system. *In* "Biochemical Actions of Hormones" (G. Litwack, ed.), Vol. 10, p. 329. Academic Press, New York.

tiveness of various contraceptive methods. Of the various methods available for contraception, only the use of oral steroids and the "rhythm" method are strictly based on endocrinological principles.

The rhythm method of contraception is predicated on the following facts: (1) the human ovum can be fertilized for only about 24 hr after ovulation; (2) spermatozoa are competent to fertilize an ovum only up to 48 hr after coitus; and (3) ovulation usually occurs 12–16 days before the onset of menstruation. Therefore, if a woman has accurately established the length of her menstrual cycles, she can calculate her personal fertile period by subtracting 18 days from her previous shortest cycle and 11 days from her previous longest cycle. Then, in each subsequent cycle, the couple should abstain from coitus during this calculated fertile interval. The failure rate in one study assessing the effectiveness of the rhythm method was reported to be 15 pregnancies per 100 woman years. The principal reason for failure is the irregularity of the menstrual cycle even in women with previously regular cycles. Since progesterone causes an increase in basal body temperature, the effectiveness of the rhythm method can be increased by abstinence for 48 hr after the rise in body temperature. Under these circumstances, the failure rate was only 6.6 pregnancies per 100 woman years (Marshall, 1968).

The steroid contraceptive pill is the most widely used method of contraception; worldwide some 50 million women take some form of oral contraceptive. Figure 13-21 presents the structure of some of the presently employed synthetic oral contraceptives. There are two types of oral contraceptives: (a) a formulation containing combinations of estrogen with a progestin (the "combined" pill) and (b) a formulation with only a progestin.

In the combined pill, the most frequently used form of estrogen is ethinylestradiol or mestranol, while the progestin is present as either norethindrone, levonorgestrel, ethynodiol, or medroxyprogesterone. The combined pill is normally given continuously for 3 weeks, followed by a "blank" pill for the fourth week to permit the process of menstruation to occur and to establish the regularity of taking the daily pill. The combination pill prevents the development of the ovarian follicle via the estrogen inhibiting the release of FSH and interfering with GnRH release. In addition, the progestin inhibits the release of LH, so as to prevent ovulation.

In the progestin-only pill, the drugs employed are either ethynodiol, levonorgestrel, megestrol, norgestrol, or norethindrone. The pill is taken continuously. The progestin-only pill primarily acts upon the cervical mucus to make the environment inhospitable to sperm and also to block implantation.

The antiprogestin, mifepristone (RU-486), is used in some countries to provide a medical alternative to surgical termination of pregnancy (see Figure 13-21). When mifepristone is administered orally within 50 days of the last menstrual period, followed 48 hr later by a prostaglandin (given as an intravaginal pessary), a complete abortion results in 95% of the cases.

B. Anovulation

The inability of a woman to become pregnant may, in some instances, be due to a lack of ovulation. There are four chief ways to induce ovulation; these include the administration of clomiphene (a nonsteroidal compound with weak estrogen activity), gonadotropins (LH and or FSH), or bromoergocriptine (see Figure 14-23) or the surgical procedure of ovarian

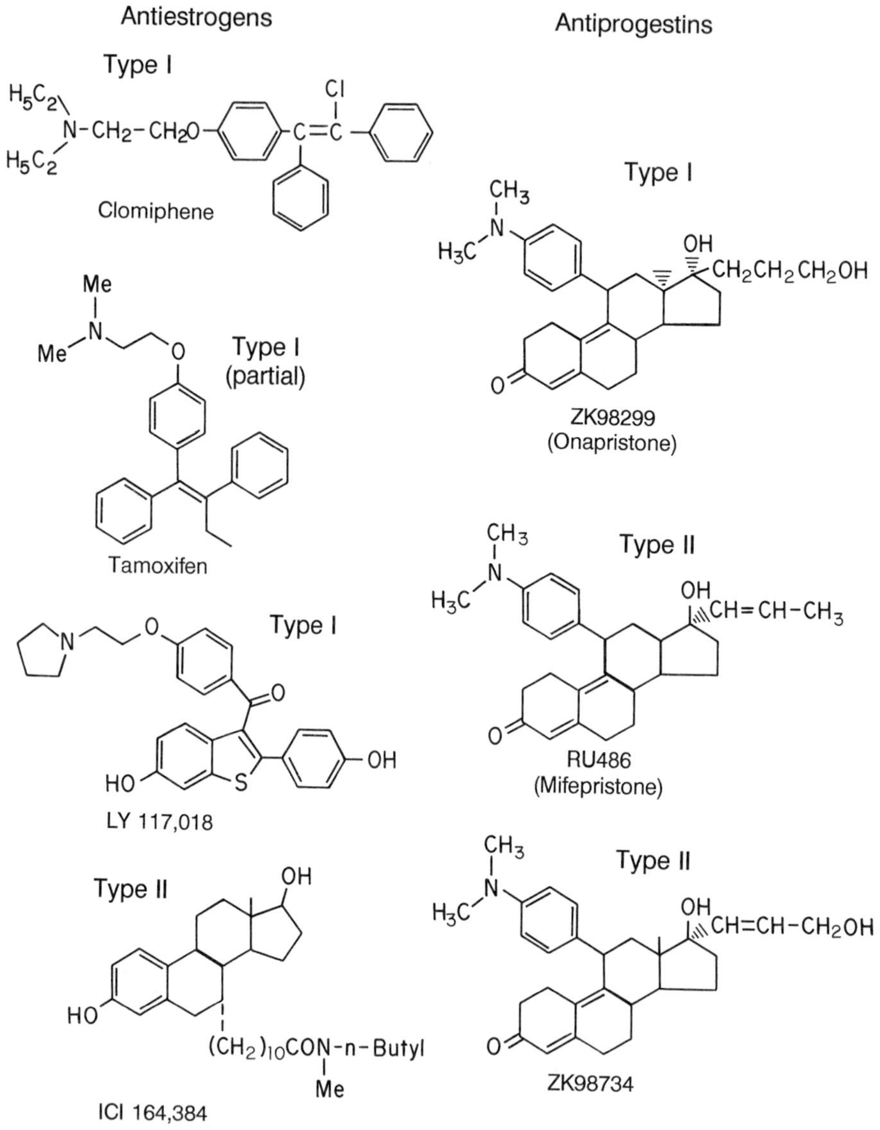

FIGURE 13-19 Examples of antiestrogenic and antiprogestinal compounds. Each category of antagonist can be divided into class I (partial antagonist) or class II (pure antagonist).

TABLE 13-7 Proposed Steps in Steroid Hormone Action for Site of Action of Steroid Antagonists (Antihormones)

- Transport through blood on serum protein
- Entry into target cell
- Receptor binding by ligand
  - Competition with natural hormone
  - Allosteric modification of natural hormone binding
  - Generation of a new conformational form of the receptor
- Receptor–ligand actions
  - Interference with receptor dimer formation (either homo- or heterodimers)
  - Reduction or inhibition of transcription-activating function(s)

TABLE 13-8 Examples of Steroid Antagonists

Class	Examples	Figure in this edition
Antiglucocorticoids	RU-486	13–19
Antimineralocorticoids	Spironolactone	10–29
Antiprogestins	RU-486 (an abortifacient)	13–19
Antiestrogens	Tamoxifen	13–19
Antiandrogens	Flutamide	12–13
	Cyproterone acetate	

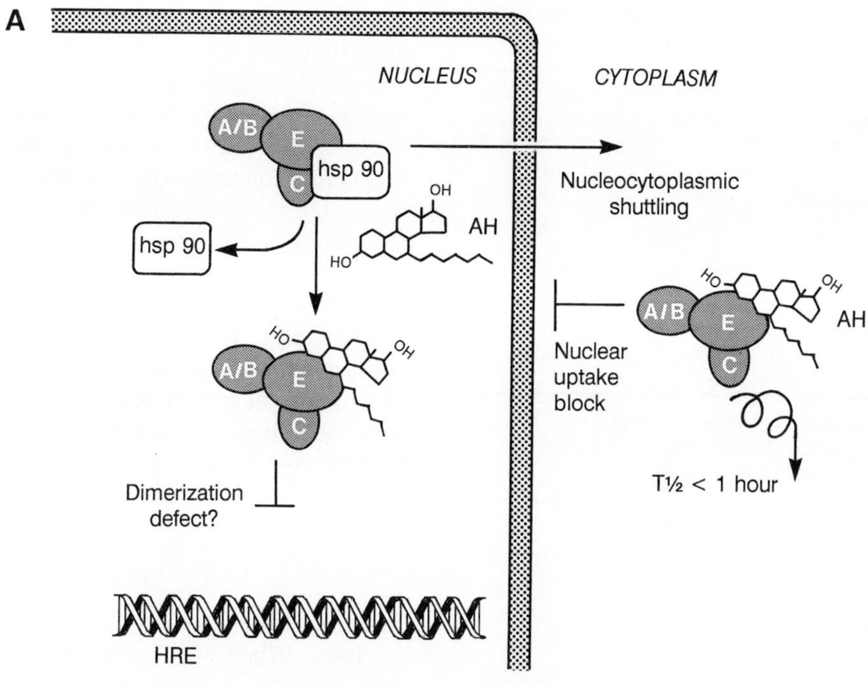

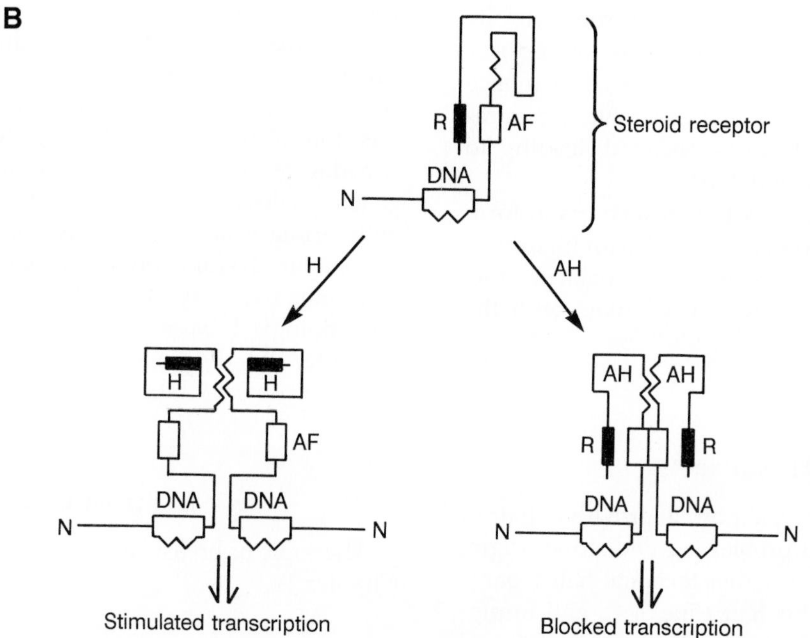

FIGURE 13-20 Schematic models illustrating how steroid antagonists (antihormones) might block the formation of a competent ligand–receptor-mediated transcription complex. (A) In this model, a pure antagonist interferes with the nuclear–cytoplasmic shuttling process, with the result that the receptor accumulates in the cytoplasm and is degraded. Abbreviations: HRE, hormone response element; hsp 90, heat shock protein 90; AH, antihormone. (B) In this model, the key point is that an antihormone (AH) does not induce a receptor conformational change that removes a surface repressor (R) function. Abbreviations: H, hormone; AH, antihormone; AF, activation function; R, repressor function; DNA, DNA-binding domain; ⩔ dimerization domain; N, amino terminus. Modified with permission from Tsai, M.-J., and O'Malley, B. W. (1994). Molecular mechanisms of actions of steroid/thyroid receptor superfamily members. *Ann. Rev. Biochem.* **63,** 451–486.

TABLE 13-9 Relative Effectiveness of Various Contraceptive Methods

Contraceptive method	Number[b] (millions)	Failures/100 Women Years[a]
Vasectomy of the male	4.1	0.02
Tubal ligation of the female	9.6	0.13
Oral contraceptives (E_2 + P)	10.7	0.30
Intrauterine devices (IUD)	0.7	1.2
Diaphragm	2.0	1.9
Condom	5.1	3.6
Withdrawal (during coitus)	0.8	6.7
Rhythm[c]	0.8	15.5

[a] Data were abstracted from Vessey *et al.* (1982).

[b] The data concerning the number of women in the United States that employ the indicated contraceptive procedure were abstracted from Mosher, W. D., and Pratt, W. F. (1990). Contraceptive use in the United States. *Patient Education Council* **16,** 163.

[c] Periodic abstinance from coital activity during periods of presumed fertility.

wedge resection. Clomiphene is an antiestrogen compound that blocks the binding of estrogen to its receptor. Clomiphene's effectiveness as an inducer of ovulation is believed to occur as a consequence of it functioning as an analogue of circulating estradiol. This permits GnRH to be secreted, leading to pituitary release of LH and FSH.

Polycystic ovary disease is the most common form of anovulation and is often associated with hirsutism. The hirsutism results from excessive androgen production that is in some unknown way linked with the inability to ovulate.

C. Hirsutism

Hirsutism or the heavy, abnormal growth of hair in the female is a medical problem of endocrine origin. Hair may be classified as either terminal hair (coarse and pigmented) or vellus hair (fine, soft, and unpigmented). The hair follicle may produce either terminal or vellus hair in response to appropriate stimuli. In hirsutism there is an androgen-dependent transition to the production of terminal hair. Androgens neither increase the number of hair follicles nor increase the mitotic rate of the cells of the connective tissue of the papillae of the hair follicle. Instead, androgens mediate an increase in the number and size of the cells of the papilla, which leads to an increase in hair growth rate and diameter.

Most commonly hirsutism results from inappropriately high plasma levels of testosterone and its precursors. If the elevated plasma level of testosterone is extensive over a prolonged time interval (e.g., years), virilization (masculinization) may also occur. The most common causes of elevated levels of testosterone in the female are congenital adrenal hyperplasia (see Chapter 10) or the presence of a tumor in the adrenal cortex, hyperplasia of the androgen-producing cells (the stromal and thecal cells) of the ovaries or an ovarian tumor, and occasionally polycystic ovary disease. Hirsutism may result from the inappropriate production of testosterone or the testosterone precursors dehydroepiandrosterone, 3α-androstanediol, or androst-4-ene-3,17-dione (see Figure 2-22).

D. Primary Amenorrhea

Amenorrhea is the absence of expected menstrual periods. When the menarche has not occurred by age 16, it is appropriate to seek an explanation. If there has been no change in the growth of breasts, appearance of pubic hair, or changes in vaginal smears, this suggests a tentative diagnosis of primary amenorrhea. The disorder can be due to either extragonadal or gonadal problems. One retrospective analysis found that gonadal dysgeneses and primary gonadal failure in phenotypic females who were genetic males accounted for about 40% of the patients with primary amenorrhea. An additional 20% have a gonadal dysfunction (e.g., polycystic ovaries) or "resistant ovary syndrome." In about 40% of cases, the etiological basis for the amenorrhea could be attributed to dysplasia of the Müllerian ducts or hypogonadotropic states.

E. Breast Cancer

The topic of breast cancer is reviewed at the end of Chapter 14.

F. Menopause

Cessation of reproductive activity is associated with a number of bodily changes, resulting from changes in the circulating levels of estrogen, progesterone, FSH, and LH. A major problem can be the development of the bone disease osteoporosis. Osteoporosis is a thinning of the skeleton that, if it occurs over a long enough time interval, incurs the definite risk of increased accidental or spontaneous fracture. In the view of many,

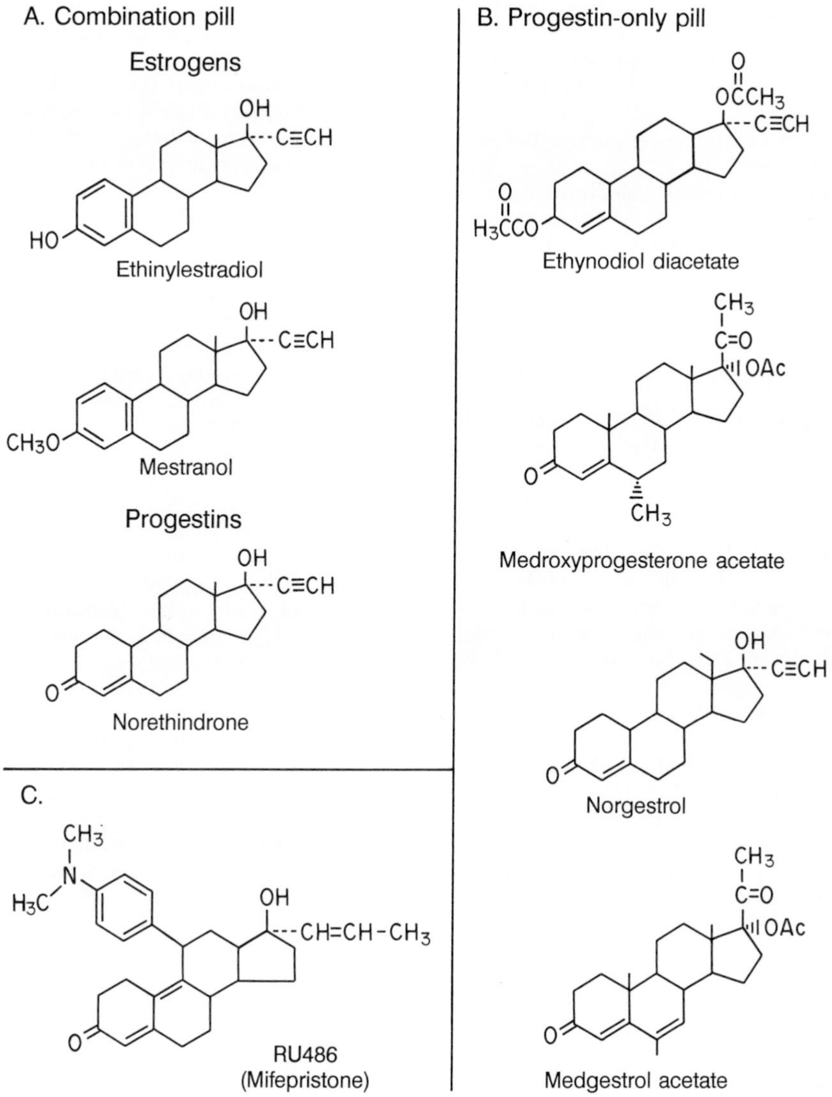

A. Combination pill

Estrogens

Ethinylestradiol

Mestranol

Progestins

Norethindrone

B. Progestin-only pill

Ethynodiol diacetate

Medroxyprogesterone acetate

Norgestrol

Medgestrol acetate

C.

RU486 (Mifepristone)

FIGURE 13-21 Structures of steroids used as oral contraceptives: (A) combination pills containing an estrogen + progestin; (B) the progestin-only pill; (C) RU-486 or mifepristone, which can be utilized as an abortifacient.

the onset of osteoporosis is initiated by the absence of estradiol (see Chapter 9).

References

A. Books

Knobil, E., and Neill J. D. (1994). "The Physiology of Reproduction," 2nd ed., Vol. 2. Raven Press, New York.

B. Review Articles

Burger, H. G. (1992). Inhibin. *Reprod. Med. Rev.* **1,** 1–20.

Djerassi, C. (1995). The mother of the pill. *In* "Recent Progress in Hormone Research" (C. W. Bardin, ed.), Vol. 50, pp. 1–17. Academic Press, San Diego.

Gaddy-Kurten, D., Tsuchida, K., and Vale, W. (1995). Activins and the receptor serine kinase superfamily. *In* "Recent Progress in Hormone Research" (C. W. Bardin, ed.), Vol. 50, pp. 109–129. Academic Press, San Diego.

Gorski, J., Furlow, J. D., Murdock, F. E., Fritsch, M., Kaneko, K., Ying, C., and Maleeyer, J. R. (1993). Perturbations in the model of the estrogen receptor regulation of gene expression. *Biol. Reprod.* **48,** 8–14.

Hall, J. E., and Crowley, W. F., Jr. (1995). Gonadotropins and the gonad: normal physiology and their disturbances in clinical endocrine diseases. *In* "Endocrinology" (L. J. DeGroot, M. Besser, H. G. Burger, J. L. Jameson, D. L. Loriaux, J. C. Marshall, W. D. Odell, J. T. Potts, Jr., and A. H. Rubenstein, eds.), 3rd ed., Vol. 1, Chapter 15, pp. 242–258. W. B. Saunders Company, Philadelphia, PA.

Mathews, L. S. (1994). Activin receptors and cellular signaling by the receptor serine kinase family. *Endocr. Rev.* **15,** 310–325.

McEwen, B. S. (1991). Steroid hormones and thyroid hormones modulate a changing brain. *J. Steroid Biochem. Mol. Biol.* **40,** 1–14.

Odell, W. D. (1979). The reproductive system in women. *In* "Endocrinology" (L. J. DeGroot, *et al.*, eds.), Vol. 3, p. 1385. Glunt & Stratton New York.

Parker, M. G. (1995). Structure and function of estrogen receptors. *Vitamins Hormones* **51,** 267–287.

Richards, J. S. (1994). Hormonal control of gene expression in the ovary. *Endocr. Rev.* **15,** 725–751.

Tsai, J.-J., and O'Malley, B. W. (1994). Molecular mechanisms of action of steroid/thyroid receptor superfamily members. *Annu. Rev. Biochem.* **63,** 451–486.

C. Research Papers

Billig, H., Furuta, I., and Hsueh, A. J. W. (1993). Estrogens inhibit and androgens enhance ovarian granulosa cell apoptosis. *Endocrinology* **133,** 2204–2212.

Chen, C. L. (1993). Inhibin and activin as paracrine/autocrine factors. *Endocrinology* **132,** 4–5.

Conn, M. P., and Crowley, W. F. (1994). Gonadotropin releasing hormone and its analogs. *Annu. Rev. Med.* **45,** 391–405.

DePaolo, L. V., Bicsak, T. A., Erickson, G. F., Shimasaki, S., and Ling, N. (1991). Follistatin and activin: a potential intrinsic regulatory system within diverse tissues. *Proc. Soc. Exp. Biol. Med.* **198,** 500–512.

Fitzpatrick, S. L., and Richards, J. S. (1994). Identification of a cyclic 3′,5′-monophosphate response element in the rat aromatase promoter that is required for transcriptional activation in rat granulosa cells and R2C Leydig cells. *Mol. Endocrinol.* **8,** 1309–1319.

Hashimoto, M., Nakamura, T., Inoue, S., Kondo, T., Yamada, R., Eto, Y., Sugino, H., and Muramatsu, M. (1992). Follistatin is a developmentally regulated cytokine in neural differentiation. *J. Biol. Chem.* **267,** 7303–7306.

Heckert, L. L., Daley, I. J., and Griswold, M. D. (1992). Structural organization of the follicle stimulating receptor gene. *Mol. Endocrinol.* **6,** 70–80.

Hill, N. C. W., Furguson, J., and MacKenzie, I. Z. (1990). The efficacy of oral mifepristone (RU 38,486) with a prostaglandin E_1 analog vaginal pessary for the termination of early pregnancy: Complications and patient acceptability. *Am. J. Obstet. Gynecol.* **162,** 414–417.

Inouye, S., Guo, Y., DePaolo, I., Shimonaka, M., Ling, N., and Shimasaki, S. (1991). Recombinant expression of human follistatin with 315 and 288 amino acids: Chemical and biological comparison with native porcine follistatin. *Endocrinology* **129,** 815–822.

Marshall, J. A. (1968). A field trial of the basal body temperature method of regulating births. *Lancet* **2,** 8–10.

Michel, U., Farnworth, P., and Findlay, J. K. (1993). Follistatins: more than follicle-stimulating hormone suppressing proteins. *Mol. Cell. Endocrinol.* **91,** 1–11.

Nakamura, T., Takio, K., Eto, Y., Shibai, H., Titani, K., and Sugino, H. (1989). Activin-binding protein from rat ovary is follistatin. *Science* **247,** 836–838.

Robertson, D. M., Foulds, L. M., Prisk, M., and Hedger, M. P. (1992). Inhibin/activin b subunit monomer: Isolation and characterization. *Endocrinology* **130,** 1680–1687.

Shimonaka, M., Inouye, S., Shimasaki, S., and Ling, N. (1991). Follistatin binds to both activin and inhibin through the common beta-subunit. *Endocrinology* **128,** 3313–3315.

Thomsen, G., Woolf, T., Whitman, M., Sokol, S., Vaughan, J., Vale, W., and Melton, D. A. (1990). Activins are expressed early in *Xeropus* embryogenesis and can induce axial mesoderm and anterior structures. *Cell* **63,** 485–493.

Ueno, N., Ling, N., Ying, S. Y., Esch, F., Shimasaki, S., and Guillemin, R. (1987). Isolation and partial characterization of follistatin: A single chain M_r 35,000 monomeric protein that inhibits the release of follicle stimulatory hormone. *Proc. Natl. Acad. Sci. USA* **84,** 8282–8286.

CHAPTER

14

Hormones of Pregnancy and Lactation

I. INTRODUCTION

A. General Comments

The ultimate purpose of reproduction is to produce a new female or male in order to maintain or prolong the species. As described in Chapters 12 and 13 for the male and female, respectively, there is extensive anatomical and endocrinological specialization that, besides defining the individual's fundamental male or female status, collectively emphasizes his or her functional complementarity. It is the successful union both in an anatomical as well as in an endocrinological sense that is essential for procreation (i.e., the meeting of a competent sperm with the ovum). Thus, the fundamental unit of reproductive biology is not the individual, but the triad of the mother, father, and offspring. This chapter describes the human endocrinology that is unique to the processes of pregnancy and lactation. It is beyond the scope of this chapter to describe the cell biology and endocrinology involving the many growth factors and paracrine substances of growth and development of the fetus and placenta.

B. Sequence of Events in Reproduction

The process of conception or fertilization of an ovum by a spermatozoon initiates a complex series of developmental and metabolic events for both the fetus and the mother, which are controlled by an array of endocrine factors. In this regard, a primordial endocrine event occurring shortly after the merging of the spermatozoon with the ovum to yield the zygote is the determination of the sex or gender of the offspring. This next leads to the implantation process, which is followed by development of the placenta as a mechanism for delivering nutrients to the developing fetus. Eventually, through further endocrine intervention, parturition or birth occurs, which then initiates the process of lactation.

II. ANATOMICAL AND MORPHOLOGICAL RELATIONSHIPS

A. Implantation

After fertilization, the zygote immediately initiates cell division by mitosis and passes through the 2-, 4-, and 8-cell stages until a tiny cluster of cells known as a morula is formed. After 3–4 days the morula reaches the uterus where it continues to grow. The morula now differentiates to create an outer shell, designated as the trophoblast, and an inner mass of cells that will ultimately become the embryo. This fluid-filled ball or blastocyst adheres to the surface of the uterine endometrium and begins the implantation process by the secretion of proteolytic enzymes, which digest a cavity in the endometrium (see Figure 14-1).

The sequelae of events leading from implantation of the blastocyst to formation of the fetal placental–decidual unit with its supporting fetal–maternal circulatory systems are summarized in Figures 14-2, 14-3, and 14-4A,B, respectively.

B. Fetal Placental–Decidual Unit

1. Background

The fetal placental unit collectively serves three important functions: (i) as a source of protein and steroid hormones, which are delivered to the maternal circulation; (ii) as a selective barrier that determines the nature of the communication and interaction between the maternal and fetal systems; and (iii) as a participant in the control of fetal growth, development, and endocrine function.

The subsequent stages of maturation of the blastocyst into the early fetus (56 days) are illustrated in Figure 14-2. The maternal–fetal circulatory system is established with the creation of an elaborate labyrinth network. As diagrammed in Figure 14-3, the fetus and maternal placenta have entirely separate circulatory systems; there is no direct communication between these two tissues.

Following implantation, the trophoblast of the blastocyst proliferates new villi rapidly. These villi infiltrate the endometrial decidua and generate the syncyti-

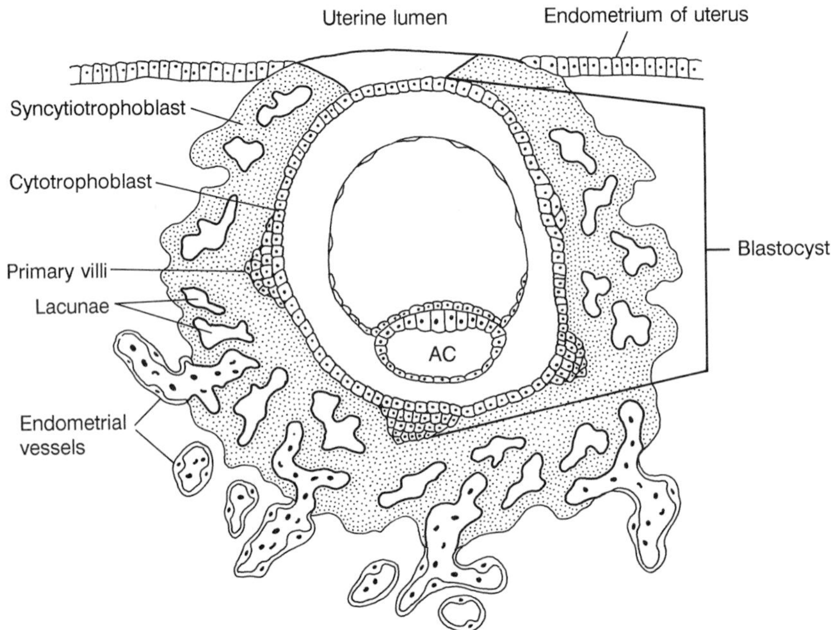

FIGURE 14-1 Illustration of the blastocyst implantation site in the uterine endometrium at 12 days of gestation. The syncytiotrophoblast (stippled) layer of cells of the blastocyst invades and surrounds some endometrial capillaries. Columns of cytotrophoblast have begun to grow outwardly in a few areas to create primary villi (see Figure 14-3). AC is the amniotic cavity. Figure 14-8 also illustrates the close anatomical relationship of the cytotrophoblast and the syncytiotrophoblast as they collaborate to produce a variety of hormones. Modified with permission from Strauss III, J. F., Gafrels, M., and King, B. F. (1995). Placental hormones. *In* "Endocrinology" (L. J. DeGroot, M. Besser, H. G. Burger, J. L. Jameson, D. L. Loriaux, J. C. Marshall, W. D. Odell, J. T. Potts, Jr., and A. H. Rubenstein, eds.), 3rd ed., Vol. 3, Chapt. 124, pp. 2171–2206. W. B. Saunders Co., Philadelphia, PA.

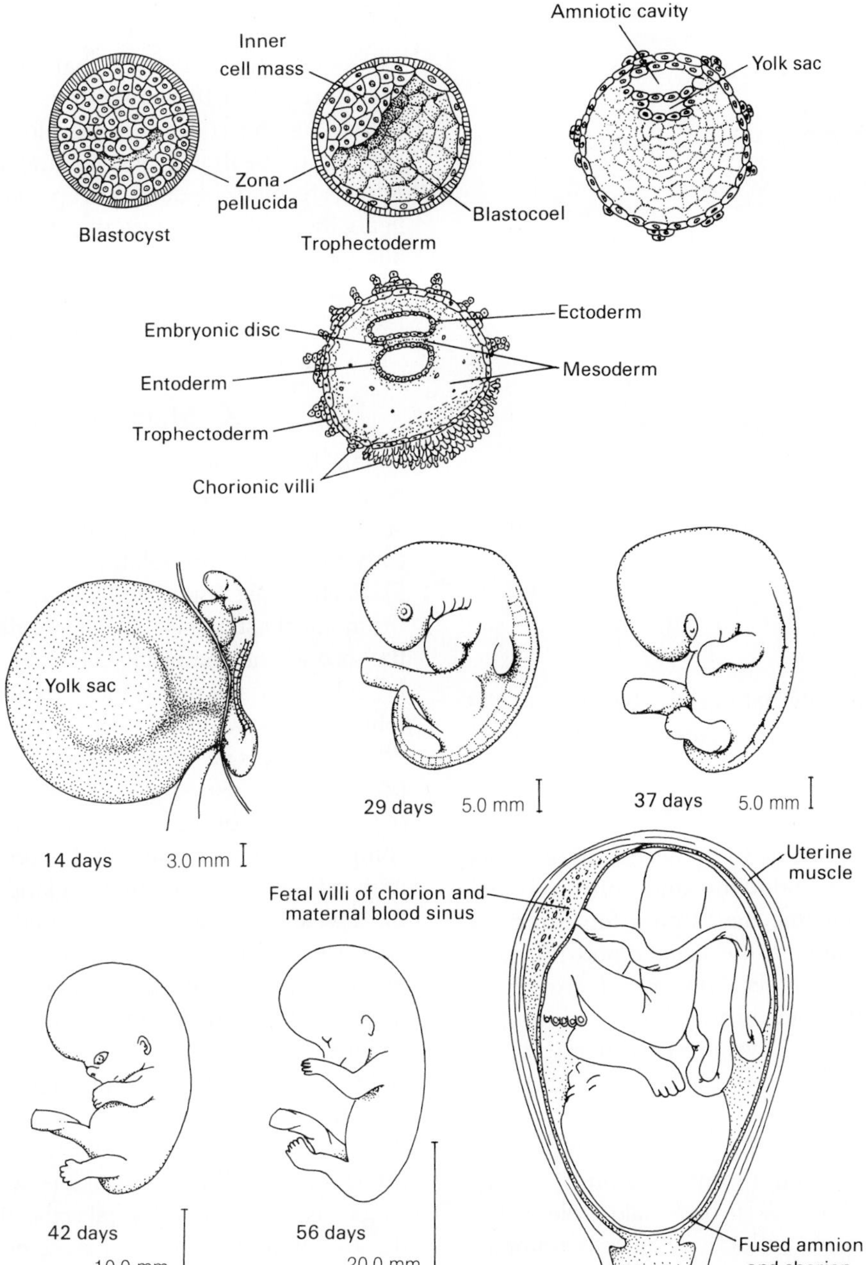

FIGURE 14-2 Representation of the various stages of development from blastocyst to early fetus. The blastocyst (upper left corner), which has resulted from the union of the spermatozoon and ovum, at ~7 days of age adheres to the surface of the maternal endometrium and initiates the implantation process, which ultimately leads to the development of the fetus (lower right).

otrophoblast cells, which are in direct contact with the maternal blood, and the cytotrophoblast cells, which lie directly underneath (see Figure 14-3). Simultaneously the embryonic blood vessels develop in the trophoblast, and as soon as it is vascularized, it is termed the chorion (see Figure 14-3). Subsequently, another fetal membrane, the amnion, develops around the growing embryo; the amnion is connected to the chorion by the body stalk. The villi in this region enlarge to form the chorion frondosum, which ultimately develops into the fetal portion of the placenta. The smooth outer surface of the chorion is termed the chorion laeve. The body stalk then ultimately develops into the umbilical cord. Thus, the fully developed pla-

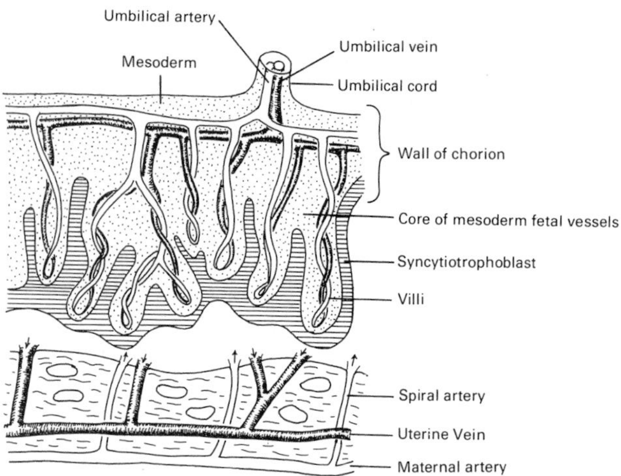

FIGURE 14-3 Diagram of the maternal and fetal circulating circulatory systems that emphasizes the entirely separate nature of the two systems.

centa consists of the maternal decidua basalis and the fetal chorion frondosum (see Figure 14-4A,B).

2. *Placenta*

The primary function of the placenta is to effect communication (both endocrine and nutrition) between the mother and the developing fetus and, at the same time, to maintain the genetic and immune integrity of both individuals. The rate of growth of the placenta is most rapid during the first trimester; by full term the placenta weighs between 500 and 600 g and measures 2–3 cm in thickness and 15–20 cm in diameter.

One obvious function of the placenta is to provide the fetus with oxygen and all nutrients and to carry away excretory products. A second major role of the placenta is to serve as an endocrine gland to maintain pregnancy. In this respect, the placenta takes over the endocrine function of the ovary and the anterior pituitary.

Cholesterol from the mother is metabolized by the placenta into progesterone. In the interim, until placental production of progesterone is firmly established, the trophoblast secretes human chorionic gonadotropin, or hCG, which maintains the production of progesterone by the maternal corpus luteum. To form the other steroid hormones, placentally derived progesterone is transported to the fetal liver and fetal adrenal and then returned to the placenta for transformation into estrogens and a group of androgens, including testosterone.

3. *Decidua*

The endometrial stoma that surrounds the conceptus shortly after implantation is referred to as the decidua. As the conceptus develops into a fetus, the decidua tissue that grows over the invaded blastocyst–fetus is termed the decidua capsularis, while the tissue beneath the blastocyst–fetus is termed the decidua basalis (see Figure 14-4A). Decidual cells are known to have some endocrine capabilities and can produce prolactin, relaxin, and, during labor, prostaglandins.

C. Mammary Glands

The breasts or mammary glands are functionally related to the female reproductive system in that they secrete milk for nourishment of the newborn; anatomically the breasts are related to the skin (see Figure 14-5). Each breast comprises 15–20 lobes, which overlie the major chest muscle, the pectoralis. They lie between the second and sixth ribs. Each lobe consists of glandular, lobular tissue that is individually drained by intralobular ducts, which in turn empty into the main lactiferous ducts. The breast tissue is responsive during pregnancy to estrogen, progesterone, prolactin, and oxytocin. The process of lactation is one of the most complex endocrine-mediated events in the body. The maturation of the mammary glands is initiated when the developing zygote is implanted in the uterus. Estrogens are responsible for the growth of the duct system, while progesterone mediates generation of the secretory system; then prolactin stimulates the production of milk.

In the first trimester of pregnancy, initial rapid growth and branching from the terminal portions of the gland are followed by the appearance of true glandular acini. The hollow alveolus becomes lined with a single layer of myoepithelial cells, ultimately enclosing the glandular alveolus in a loose network surrounded by capillaries so that the lumen of the alveolus is connected to an interlobular duct (see Figure 14-6). In the second trimester of pregnancy, the alveolar secretion begins. In the third trimester, particularly in the ninth month, breast enlargement occurs as a consequence of the hypertrophy of parenchymal cells and distention of the alveoli with colostrum.

In the human, the rate of milk production is normally, by 1 week postpartum, 500 ml/day; after 3 weeks milk production increases to 800–1000 ml/day. Table 14-1 tabulates the composition of human and bovine milk. The colostrum or milk obtained in the several days immediately following parturition differs from milk in its biological and physical properties. The

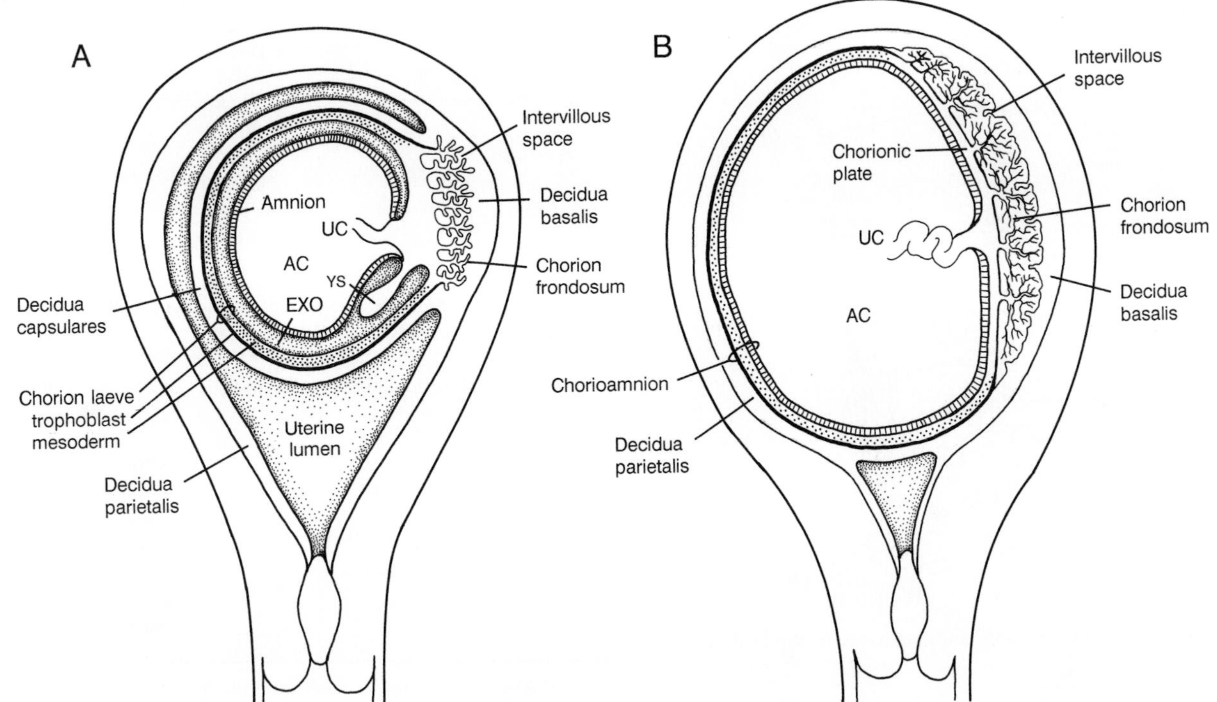

FIGURE 14-4 Midsagittal section of a uterus with developing embryo, illustrating the placental and fetal membranes (panel A) at 1 month of gestation and (panel B) at 9 months of gestation. In A the villi adjacent to the decidua basalis are well-developed (chorion frondosum), whereas elsewhere the villi have regressed (chorion laeve). The chorion laeve is made up of the trophoblast (black) and mesoderm (stippled); at this stage it is covered by the decidua capsularis, but has not yet fused with the decidua parietalis. In B the chorion frondosum and decidua basalis have the same relation as in A; however, the chorion laeve and amnion have fused (chorioamnion) and the decidua capsularis has degenerated, causing the chorioamnion to become opposed to the decidua parietalis, which obliterates the uterine lumen. Abbreviations: AC, amniotic cavity; UC, umbilical cord; YS, yolk sac; EXO, exocoelom. Modified with permission from Strauss III, J. F., Gavels, M., and King, B. F. (1995). Placental hormones. *In* "Endocrinology" (L. J. DeGroot, M. Besser, H. G. Burger, J. L. Jameson, D. L. Loriaux, J. C. Marshall, W. D. Odell, J. T. Potts, Jr., and A. H. Rubenstein, eds.), 3rd ed., Vol. 3, pp. 2171–2206. W.B. Saunders, Philadelphia, PA.

principal difference is a higher protein content (up to 8 and 20% in the human and bovine, respectively).

III. CHEMISTRY, BIOCHEMISTRY, AND BIOLOGICAL RESPONSES

A. Pregnancy

1. *Introduction*

The endocrine changes accompanying pregnancy are remarkable. A pregnant woman in the late phase of the third trimester produces, on a daily basis, some 250–300 mg of progesterone, 15-20 mg of 17β-estradiol, 50–100 mg of estriol, 75–100 mg of cortisol, 3–8 mg of deoxycorticosterone (DOC), and 1–2 mg of aldosterone. In addition, there is a massive production of human chorionic somatomammotropin (in excess of 1 g/day), human chorionic gonadotropin, human chorionic thyrotropin, and chorionic ACTH, as well as increased plasma levels of the angiotensins and renin. Figure 14-7 summarizes the temporal changes of many of these hormones throughout gestation.

2. *Peptide Hormones*

a. Background

Each of the morphological variants of the trophoblast cells of the chorionic villi (see Figure 14-3) has a distinctive capability to produce a wide variety of steroid, peptide, and protein hormones (see Figure 14-8). The syncytiotrophoblast cells are in direct contact with the maternal blood where chorionic gonadotropin (CG), chorionic somatomammotropin (CS), chorionic proopiomelanocortin, and progesterone are present. However, these hormones do not readily gain access to the fetus due to the several barriers, which include the cytotrophoblast, the mesenchymal stroma of the

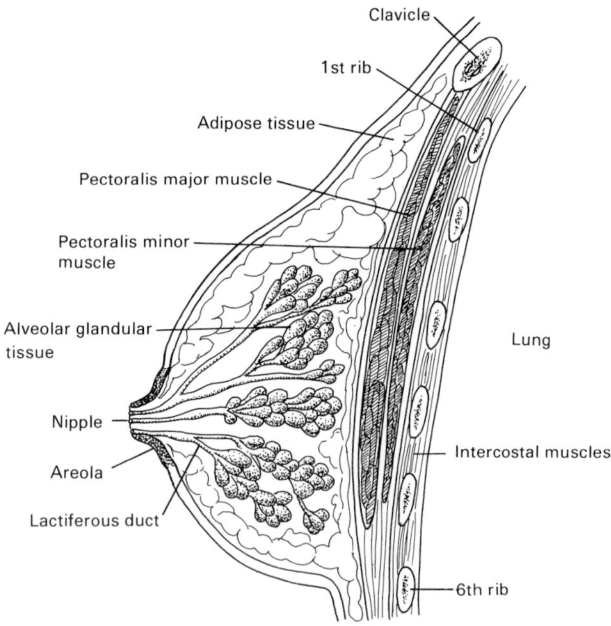

FIGURE 14-5 Ducts and glandular tissue of the human mammary gland.

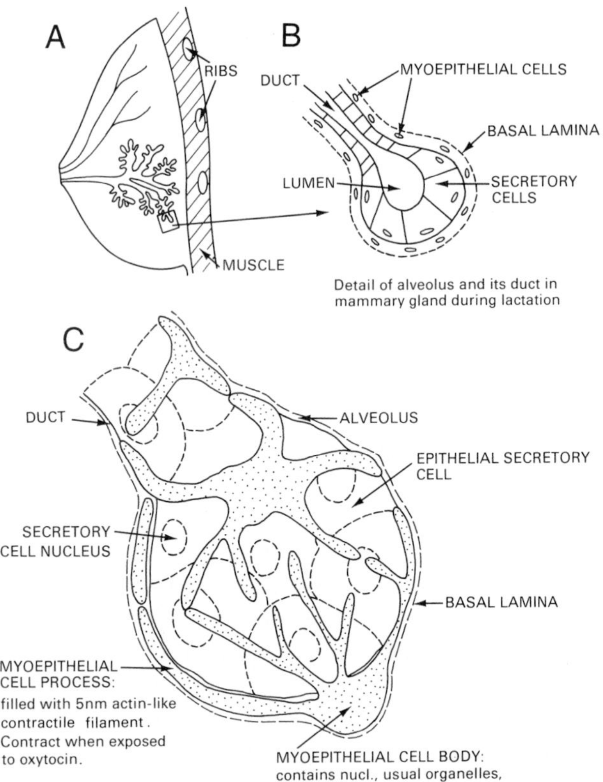

FIGURE 14-6 Schematic diagram of a mammary gland alveolus and its duct: (A) mammary gland; (B) alveolus and its duct; (C) cellular organization of alveolar duct. (Drawn by Dr. L. Paavola, Department of Anatomy, Temple University Medical School.)

TABLE 14-1 Composition of Milk

Component	Bovine (%)	Human (%)
Water	87	87.5
Total solids	13	12.5
Carbohydrate	4.5–5.0	7.0–7.5
Lipid	3.5–5.0	3.0–4.0
Protein	3.0–4.0	1.0–1.5
Ash	0.75	0.2

Specific components of human milk[a]

Ash	Lipids	Carbohydrates	Proteins
Ca^{2+}	Triglycerides with fatty acids	Lactose	Caseins (80% of protein nitrogen)
HPO_4^{2-}	Oleic (32%)		β-Lactoglobulin (50% of whey protein)
K^+	Palmitic (15%)		Immunoglobulins (IgA)
Na^+	Myristic (20%)		Albumin
Mg^{2+}	Stearic (15%)		α-Lactalbumin
Cl^-	Other (6%)		Lactoperoxidase
	Cholesterol (4%)		Xanthine oxidase
	Phospholipids (8%)		Alkaline phosphatase

[a] Abstracted from Vorherr, H. (1972). To breast-feed or not to breast-feed. *Postgrad. Med. J.* **51,** 127–139.

villus core, and the fetal capillary endothelium (see Figures 14-4A,B).

b. Chorionic Gonadotropin

Human chorionic gonadotropin, hCG, is a heterodimeric glycoprotein of 57 kDa consisting of a noncovalently bound α (92 amino acids) and a distinctive β (134 amino acids) subunit. There is one gene for the α-subunit located on chromosome 6q21.1-23; this α-subunit is also very similar to the TSH, LH, and FSH α-subunits. Chromosome 19 has a cluster of six genes for the hCG β-subunit and one gene for the LH β-subunit.

Table 14-2 summarizes the structural homology of several of the released pituitary and placental hormones. The β-CG has an 80% homology to the 121-amino acid subunit of the LH β-subunit.

hCG is produced by the syncytiotrophoblast and thus, technically, hCG is produced by the fetus. The rate of production of hCG is maximal by the 10th week of pregnancy, and then it falls slowly to a nadir at 17 weeks and remains at a low but readily measurable level for the duration of the pregnancy.

The major biological function of CG is to function as a luteotropin, that is, to stimulate the production of progesterone by the corpus luteum. This ensures a continual supply of ovarian progesterone until the placenta has developed to the point (usually at 6–8 weeks) when it can generate adequate quantities of progesterone; at this time the corpus luteum atrophies. By anal-

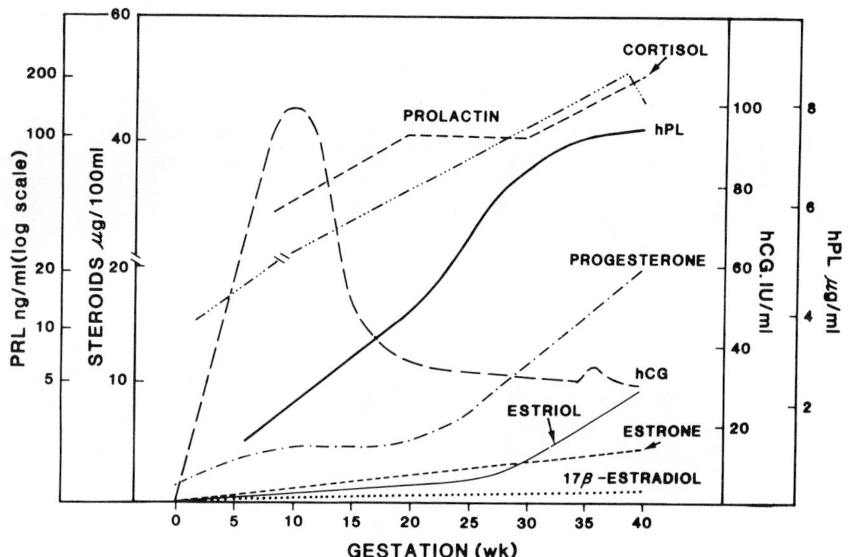

FIGURE 14-7 Temporal changes during human pregnancy in the serum concentrations of several hormones. Abbreviations: hPL, human placental lactogen; hCG, human chorionic gonadotropin; PRL, prolactin. Modified from Aragona, C., and Friesen, H. G. (1979). Lactation and galactorrhea. *In* "Endocrinology" (L. J. DeGroot, G. F. Cahill, L. Martin, D. H. Nelson, W. D. Odell, J. T. Potts, E. Steinberger, and A. I. Winegrad, eds.), Vol. 3, pp. 1614–1617. Grune & Stratton, New York.

ogy with LH (Chapter 13), the β-subunit of hCG is believed to interact specifically with a membrane receptor to stimulate the production of progesterone from cholesterol.

Human CG is the first hormone to be produced in significant quantities by the conceptus; accordingly, a radioimmunoassay for hCG has been developed as a means of diagnosing pregnancy even before the menstrual period has been missed. The hCG is believed to be produced by the syncytiotrophoblast cells of the villi (see Figures 14-1 and 14-3).

c. Chorionic Somatomammotropin

Human chorionic somatomammotropin (hCS), which was formerly referred to as human placental lactogen (hPL), is a 190-amino acid polypeptide (single

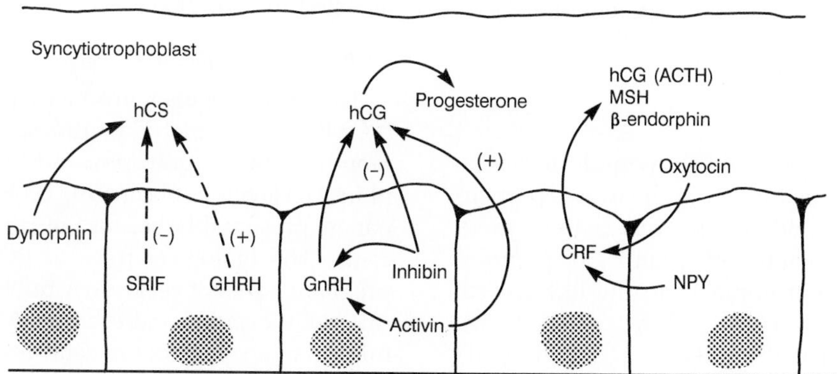

FIGURE 14-8 Model for hormone production by paracrine interaction of the cytotrophoblast and syncytiotrophoblast cells. (See also Figures 14-1 and 14-3 for anatomical details.) The syncytiotrophoblasts produce chorionic somatomammotropin (CS), chorionic gonadotropin (CG), and proopiomelanocortin (POMC) polypeptide hormones. Abbreviations: NPY, neuropeptide Y; ACTH, adrenocorticotropic hormone; GHRH, growth hormone-releasing hormone; SRIF, somatostatin; GnRH, gonadotropin-releasing hormone; MSH, melanocyte-stimulating hormone. Modified with permission from Jaffe, R. B. (1991). "Reproductive Endocrinology" (S. S. C. Yen, and R. B. Jaffe, eds.), 3rd ed. W.B. Saunders, Philadelphia, PA.

TABLE 14-2 Structural Homology of Pituitary and Placental Hormones

Hormone	Molecular mass (kDa)	Number of amino acids in the chains		Carbohydrate containing	Produced by	
		α	β		Pituitary	Placenta
Glycoproteins						
FSH (follicle-stimulating hormone)	29	89	115	+	+	
LH (luteinizing hormone)	29	89	121	+	+	
hCG (human chorionic gonadotropin)	57	92	139	+		+
TSH (thyroid-stimulating hormone)	29	89	112	+	+	
PRL (prolactin)	22	(198)[a]		−	+	
GH (growth hormone)	21.5	(191)[a]		−	+	
hCS (human chorionic somatomammotropin)[b]	22	(191)[a]		−		+

[a] Has only one amino acid chain.
[b] hCS was formerly known as human placental lactogen.

chain) of molecular mass 21.5 kDa. hCS has significant structural homology with growth hormone. Little information is available concerning the production and secretion of hCS by the trophoblasts and placenta.

The major biological roles of hCS are not well-described; its effects on the mobilization and metabolism of maternal fat stores have been described. hCS is also an insulin antagonist and in this role is postulated to be involved in the regulation of maternal blood glucose levels to ensure optimal availability of blood glucose to meet the caloric requirements of the fetus. In addition, hCS has been suggested as being one of the contributory agents to the development of diabetic ketoacidosis in pregnant women who have no prior history of diabetes.

d. Relaxin

Relaxin (RLX) is a small 6-kDa peptide hormone produced by the corpus luteum of the ovary, particularly during the first and second trimesters of pregnancy. Relaxin is a member of a family of peptides structurally related to insulin and insulin-like growth factors. The primary structure of RLX consists of two peptide chains (designated α and β) of 22 and 31 amino acid residues, respectively. The α- and β-chains are covalently linked by two disulfide bonds, with an additional intradisulfide link in the α-chain.

In some species (pig, rat), parturition is preceded by a significant increase in relaxin. In these species, relaxin has been shown to play a role in the softening of the fibrous connective tissue at the cervix. However, in human pregnancy, the role of RLX is not clear; relaxin concentrations are lowest in the third trimester and unchanged during parturition.

e. Oxytocin

Oxytocin is a nonapeptide secreted by the neurohypophysis. An extensive discussion of its chemistry and mode of action is presented in Chapter 4. Milk letdown and milk ejection from the mammary tissue are the chief biological responses. There is also some preliminary evidence to support the action of oxytocin on the uterine endometrium at parturition to stimulate its contraction. High concentrations of oxytocin can be detected in fetal blood.

f. Chorionic Proopiomelanocortin

The placenta also produces proopiomelanocortin (POMC; see Chapter 5) peptides; these include ACTH, β-endorphin, β-lipotropin, and three forms of dynorphin. The levels of plasma ACTH increase during pregnancy, presumably because of production by the placenta. The biological roles of POMC and ACTH in pregnancy are not yet known, but it has been suggested that β-endorphins and other POMC-derived peptides might have some role in fetal response to stress, hypoxia, and parturition.

g. Neuropeptide Y

The 36-amino acid neuropeptide Y (NPY; see Chapter 8) is found in the cytotrophoblast. The levels of NPY rise in the first trimester and remain elevated through parturition; the levels of NPY fall dramatically after delivery. The biological function of NPY in the placenta is not known, although NPH receptors are known to be present in the placenta.

h. Other Peptide Hormones

Evidence, both chemical and immunochemical, has been presented to indicate that human placental tissue produces and secretes human chorionic thyrotropin (an analog of TSH), chorionic ACTH, GnRH- and TRH-releasing hormones, and proopiomelanocortin (see Figure 14-8).

3. *Steroid Hormones*

a. Introduction

The steroid hormones generated specifically as a consequence of the circumstances of pregnancy are produced in a number of tissues; these include the maternal–fetal placenta, the maternal and fetal adrenals, and the maternal ovaries and liver. The steroids produced at these sites are tabulated in Table 14-3. Figures 2-20, 2-21, and 2-22 and Table 2-6 should also be consulted for additional details concerning steroid metabolism.

There is a changing pattern of production of steroid hormones throughout the trimesters of pregnancy, so that as the fetus differentiates and develops, the fetal adrenals and liver have an increasing responsibility for steroid metabolism. Immediately following conception through weeks 12–13 of the pregnancy, the principal source of progesterone and estrogen is the corpus luteum. By week 7 and until parturition, the placenta produces significant quantities of both estrogens (in the forms of 17β-estradiol, estriol, estrone, and estetrol) and progesterone. Both the placental and corpus luteum production of progesterone is stimulated by hCG.

Since the placenta does not have the complete array of steroid-metabolizing enzymes to convert cholesterol into estradiol, progesterone, or other steroids, during the second and third trimesters both the maternal adrenal cortex and the fetal adrenal cortex serve as major sources of precursors for placental steroidogenesis. In this regard, dehydroepiandrosterone sulfate from the maternal adrenals is converted by the placenta to estrogens. The fetal adrenal cortex becomes enzymatically competent to produce steroids by day 50 of the pregnancy. In this fetal compartment, many of the steroids are sulfate conjugates. The fetal adrenal also assumes major responsibility for the production of C-19 androgens and sulfoconjugated Δ^5 C-21 steroids such as pregnenolone sulfate in the second and third trimesters.

b. Production of Progesterone

Throughout the course of pregnancy there are three principal forms of the C-21 progestins present: progesterone, 16α-hydroxyprogesterone, and 17α-hydroxyprogesterone. Figure 14-9 summarizes their separate pathways of biosynthesis.

Progesterone is produced largely by the corpus luteum up to weeks 5–6 of gestation; by week 12 of gestation the placenta becomes the dominant site of biosynthesis. Thus, the plasma level of progesterone rises from 1–2 ng/ml at conception to over 100 ng/ml by parturition. The placenta contains all of the enzymes required for the conversion of maternally derived cholesterol into progesterone.

17α-Hydroxyprogesterone levels in the plasma rise from 0.5 ng/ml at conception to a plateau of 50–

TABLE 14-3 Steroid Hormones of Pregnancy and Their Sites of Production

Mother	Placenta	Fetus
	Progesterone (after 12 weeks) Estriol Estradiol, estetrol Δ^5-Pregnenolone	
Adrenal cortex Cholesterol	Dehydroepiandrosterone sulfate Cortisol	Adrenal cortex Pregnenolone sulfate 17α-Hydroxypregnenolone sulfate Dehydroepiandrosterone sulfate 16α-Dehydroepiandrosterone sulfate Δ^5-Pregnenolone sulfate
Ovaries (corpus luteum) Progesterone	Estrone Estradiol	
Liver	Estriol	Liver Cholesterol (fetal source) 16α-Hydroxyprogesterone 17α-Hydroxyprogesterone Estetrol

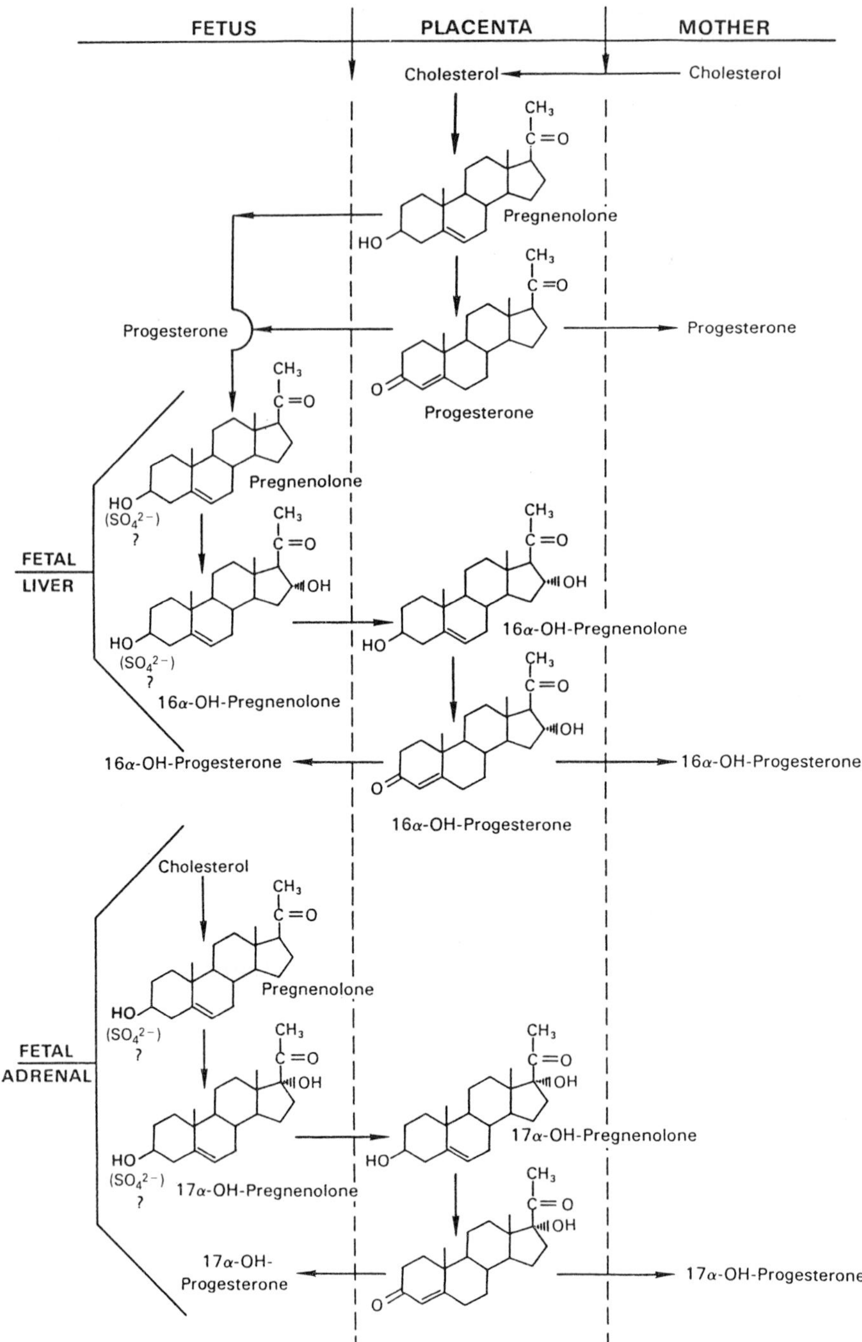

FIGURE 14-9 Pathways and anatomical compartmentalization of the production of the C-21 progestins, particularly progesterone, 17α-hydroxyprogesterone, and 16α-hydroxyprogesterone in the human fetal placental unit. This figure was derived in part from Buster, J. E., and Marshall, J. R. (1979). Conception, gamete and ovum transport, implantation, fetal placental hormones, hormonal preparation for parturition and parturition control. *In* "Endocrinology" (L. J. DeGroot, G. F. Cahill, L. Martini, D. H. Nelson, W. D. Odell, J. T. Potts, E. Steinberger, and A. I. Winegrad, eds.), Vol. 3, pp. 1595–1612. Grune & Stratton, New York.

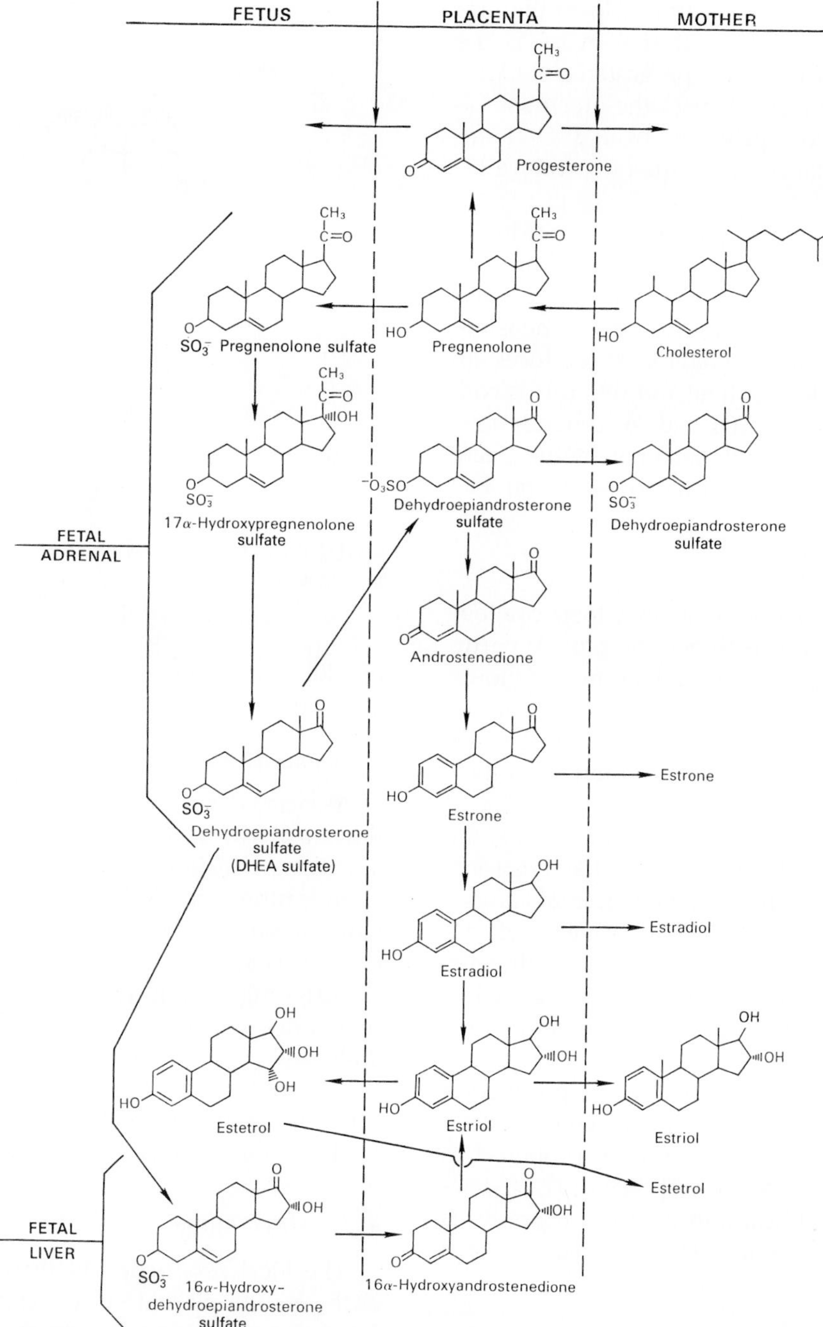

FIGURE 14-10 Pathways and anatomical compartmentalization of the production of the C-18 estrogens (estradiol, estrone, estriol, and estetrol) in the human fetal placenta unit. This figure was derived from Buster, J. E., and Marshall, J. R. (1979). Conception, gamete and ovum transport, implantation, fetal placental hormones, hormonal preparation for parturition and partition control. *In* "Endocrinology" (L. J. DeGroot, G. F. Cahill, L. Martini, D. H. Nelson, W. D. Odell, J. T. Potts, Jr., E. Steinberger, and A. I. Winegrad, eds.), Vol. 3, pp. 1595–1612. Grune & Stratton, New York.

60 ng/ml at weeks 6–36 of gestation. Up to the first 8–12 weeks of gestation the maternal ovary is the principal site of 17α-hydroxyprogesterone. After the first trimester, the placenta uses the precursor 17-hydroxy-Δ^5-pregnenolone produced from Δ^5 C-21 sulfoconjugates in the fetal adrenal cortex to produce 17-hydroxyprogesterone.

16α-Hydroxyprogesterone plasma levels gradually rise from 0.5 ng/ml at conception to a level of 120–140 ng/ml by 32 weeks of gestation. The precise biosynthetic pathway of 16α-hydroxyprogesterone is not known; it is believed that the fetal liver produces 16-hydroxy-Δ^5-pregnenolone sulfate and that this is converted by the placenta into 16α-hydroxyprogesterone, which is then available to both mother and fetus. There is no known specific biological response attributable to 16α-hydroxyprogesterone.

c. Production of Estrogens

Throughout the course of pregnancy there are four principal forms of the C-18 estrogens present. At parturition the relative serum concentrations are as follows: estradiol, 10–30 ng/ml; estriol 5–10 ng/ml; estrone, 5–8 ng/ml; and estetrol, 2–4 ng/ml. Estetrol is 15, 16-dihydroxyestradiol. Figure 14-10 summarizes their pathways of biosynthesis.

By the end of the first trimester the placenta is the principal site of biosynthesis of estradiol and estrone. Estriol is produced largely from the placental conversion of 16-hydroxydehydroepiandrosterone sulfate derived successively from the fetal liver and adrenals. Finally, estetrol is believed to be largely produced in the fetal compartment from placentally generated estriol.

Since the fetus plays a key role in the production of estetrol and estriol, the measurement of the maternal blood levels of these steroids has been proposed to provide some insight into fetal well-being. Thus, deteriorating fetoplacental health in the third trimester is often associated with falling unconjugated maternal serum estriol or estetrol concentrations.

d. Production of Androgens

The principal C-19 androgen in the pregnant female is dehydroepiandrosterone sulfate; prior to conception the maternal serum level is 1600 ng/ml, and this decreases throughout gestation to a value of 800 ng/ml at parturition. As described in Figure 14-11, the principal source of androgens in the female is from the maternal adrenal cortex; here cholesterol is converted to pregnenolone and then finally to dehydroepiandrosterone sulfate. The fall in blood levels of dehydroepiandrosterone sulfate is believed to reflect an increased metabolic clearance rate due to a significant uptake of this steroid by the placenta for conversion into estrogens.

Cholesterol

Maternal adrenal cortex

CH_3

C=O

HO

Pregnenolone

O

O

SO_3^-

Dehydroepiandrosterone sulfate (DHEA)

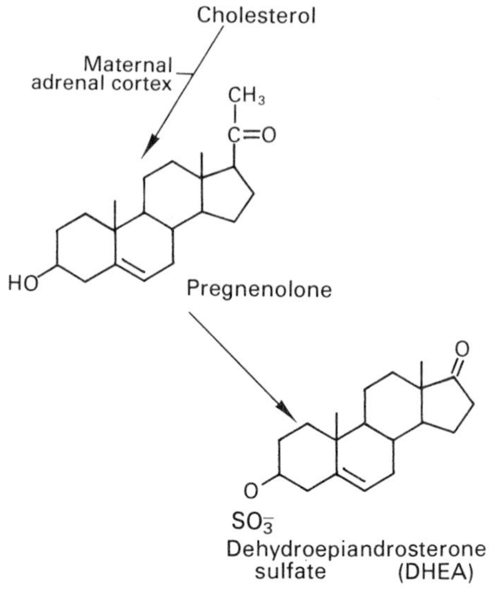

FIGURE 14-11 Pathways for the production of androgens related to pregnancy.

e. Vitamin D Steroids

Associated with fetal growth and development in the third trimester is the requirement of significant amounts of calcium for skeletal development. This calcium is ultimately derived from either the dietary calcium or skeletal calcium reserves of the mother. The calcium is actively transported across the placenta. It is intriguing, therefore, that the placenta and also the fetal kidney have the enzymatic capability to convert 25-hydroxyvitamin D_3 into 1,25-dihydroxyvitamin D_3; this latter secosteroid is a hormonally active form of vitamin D_3 (see Chapter 9) that is known to be required for both intestinal as well as bone calcium processes.

4. Prostaglandins

The local production of prostaglandins PGE_2 and $PGF_{2\alpha}$ by the placenta is associated with the onset of labor and parturition. The structures of PGE_2 and $PGF_{2\alpha}$ are given in Figure 16-3. The enzymatic capability for the production of these prostaglandins resides in the decidua. The released prostaglandins then act on the myometrium to stimulate an adenyl cyclase.

B. Parturition

In human pregnancy, a normal birth can occur after 34–36 weeks of gestation. The process of labor and birth represents the culmination of a complex series of endocrinological events in the mother and fetus.

After 34 weeks of gestation there is a large increase in fetal exposure to glucocorticoids, particularly cortisol. Contributing to the rise in cortisol is a sharp decrease in corticosteroid-binding globulin (CBG), also called transcortin, in fetal serum. This protein normally has bound to it the greatest proportion of circulating cortisol. The resulting increase in "free" glucocorticoids triggers major biochemical changes in the placenta, as shown in Figure 14-12.

Throughout pregnancy, the placenta secretes large amounts of progesterone, which has a profound effect on mammary gland development and blockage of the action of prolactin on differentiated mammary gland cells, so that they are not producing milk proteins. Although still controversial, some evidence suggests that the large amount of available progesterone binds to the glucocorticoid receptor present in the mammary gland instead of cortisol and prevents the activity of the glucocorticoid receptor, which may be required, in addition to prolactin, for the expression of milk proteins. Also blocked by progesterone is the release of oxytocin from the neurohypophysis. This is an important control since oxytocin mediates uterine contractions.

When free cortisol levels rise near term, the large output of progesterone by the placenta is depressed by a fall in the secretion of hCG from the trophoblast. The sharp fall in progesterone levels relieves the brake on the maternal neurohypophysis exerted throughout pregnancy and results in the release of oxytocin. Cortisol also acts on the placenta to stimulate estrogen production. Concomitantly the amount of $PGF_{2\alpha}$ is increased; the uterus may be the primary source of this prostaglandin. Both estrogen and $PGF_{2\alpha}$ sensitize the uterus to the action of oxytocin. Thus, the oxytocin released from the neurohypophysis causes smooth muscular contractions of the uterus, which culminates in the expulsion of the fetus. Cortisol appears to be the main trigger in initiating the process.

Cortisol may also play a role in the induction of the fetal pulmonary surfactant, a substance essential to lung alveolar stability, as well as in the storage of glycogen in fetal skeletal and cardiac muscles and in the liver. This latter event may assist the fetus to avoid the stresses of hypoxia during labor.

However, for the parturition process to proceed normally, it is essential that connective tissue changes have occurred so that the cervix is converted from a firm semirigid structure to a soft compliant ring that is capable of stretching sufficiently to allow passage of the fetus. The "ripening" of the cervix occurs over a time frame of hours (in multigravid[1] women) to weeks (in primagravid women); biochemically this results in a splitting of collagen fibers as well as a 50% reduction in glycosaminoglycan sulfate and collagen. The initiator of these changes has not yet been elucidated.

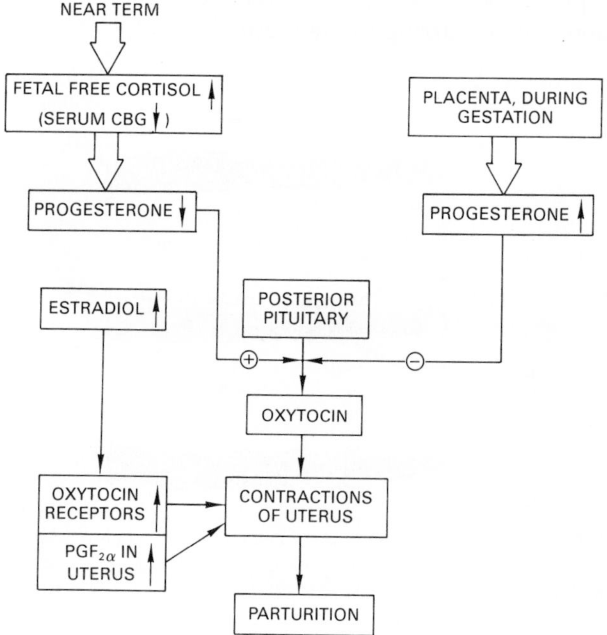

FIGURE 14-12 Probable roles of glucocorticoid hormone in the normal termination of pregnancy. + indicates stimulation; − indicates inhibition.

C. Lactation

1. Introduction

The endocrine changes associated with lactation are no less remarkable than those encountered in pregnancy. The process of lactation can be divided into three stages: (i) mammogenesis, which is the growth and development of the tissues; (ii) lactogenesis, the process of milk formation and secretion; and (iii) milk ejection, which occurs during suckling. The process of lactation normally imposes adaptive changes upon the mother. The metabolic cost of generating more than 1000 ml of milk/day usually requires a substantial change in dietary intake to balance the loss of calories, protein, minerals, and fluids (see Table 14-1) while maintaining an ideal body weight.

For lactation to ultimately occur, it is necessary during pregnancy to effect duct, lobule, and alveolar growth. It has been concluded from experiments using combinations of oophorectomized, hypophysectomized, and adrenalectomized rats that no less than six pituitary hormones, as well as chorionic somatomammotropin, estrogen, progesterone, glucocorticoids, thyroxine, and insulin, are all involved in some aspect

[1] Gravid is distended by pregnancy.

of mammary growth and development. However, it is beyond the scope of this chapter to consider in detail these developmental changes.

Summarized in the following paragraphs is an enumeration of the effects of hormones necessary for the actual process of lactation.

2. *Peptide Hormones*

a. Prolactin

The principal peptide hormone of lactation is the pituitary hormone prolactin. Prolactin is a 22.5-kDa protein composed of a single polypeptide chain of 198 amino acids (see Chapter 5). There is strong structural homology between prolactin and growth hormone.

In females the biological action of prolactin is to mediate the proliferation and differentiation of the breast and, thus, to permit, after appropriate stimuli, the secretion of milk (discussed later). The physiological role of prolactin in the male is not known. However, newborn infants of both sexes have plasma levels of prolactin that exceed the highest levels of the mother during either pregnancy or lactation.

b. Other Peptide Hormones

Insulin (see Chapter 7) has been shown to be required for proper functioning of the mammary gland during lactation. Although the biochemical bases of its actions have not been elucidated, it seems likely that insulin will stimulate glucose uptake by mammary cells, which will facilitate lipogenesis.

Parathyroid hormone (PTH) (see Chapter 9) is also necessary for optimal lactation. Parathyroidectomy of lactating animals depresses lactation; however, it is not known whether the reduction in lactation is due to perturbation in the actions of PTH (i) directly on the mammary tissue, (ii) on the mobilization of bone calcium to provide calcium for milk secretion, or (iii) in stimulating the renal production of 1,25-dihydroxyvitamin D_3 [$1,25(OH)_2D_3$]. The $1,25(OH)_2D_3$ in turn would mediate the intestinal absorption of adequate amounts of calcium to support its secretion in milk (see Table 14-1) and avoid the deleterious consequences of extensive bone calcium mobilization (see chapter 9).

c. Other Hormones

The thyroid hormone, thyroxine (see Chapter 6), is believed to be important in humans for proper milk secretion. Hypothyroid subjects are sometimes associated with galactorrhea, which is the secretion of milk from the breast under nonphysiological circumstances.

With respect to the involvement of steroid hormones in lactation, surprisingly neither estrogen nor progesterone is essential for established milk secretion; oophorectomy does not abolish milk production. In the mouse, rat, and goat, adrenal steroids have been shown to be essential for the induction and maintenance of milk secretion.

IV. CELL BIOLOGY AND MOLECULAR ACTIONS

A. Hormonal Aspects of Fertilization

1. *Capacitation*

Within 5–15 min after sperm are deposited in the vagina, motile spermatozoa are present in the fallopian tubes. Estrogen, secreted late in the follicular phase, causes the endocervical glands to secrete a mucus that favors sperm transport and motility. The process of fertilization is initiated by the union of a competent spermatozoon with an ovum of the same species.

Capacitation of a spermatozoon is required before a successful union with an ovum can be achieved. Capacitation involves the removal of an outer coating from the surface of the spermatozoon; this is followed by the removal of the acrosome or tip of the spermatozoon (see Figure 14-13 and also Figure 12-4). This latter process, which has been termed the acrosome reaction, allows the acrosomal hyaluronidase enzymes to come into contact with the substances surrounding the ovum, so that the spermatozoon may penetrate and unite. While estrogens do not have an effect on the capacitation process, progesterone can inhibit capacitation from occurring in the uterus.

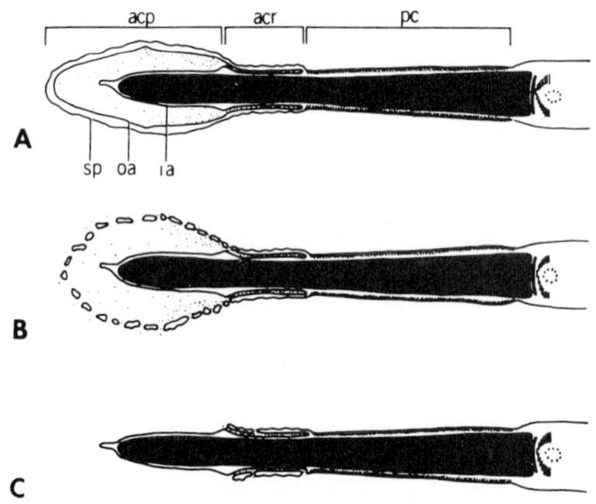

FIGURE 14-13 Capacitation of a spermatozoon before (A), during (B), and after (C) acrosomal reaction. Abbreviations: acp, acrosomal cap region; acr, acrosomal collar region; ia, inner acrosomal membrane; oa, outer acrosomal membrane; pc, postnuclear cap region; sp, plasma membrane of the spermatozoon. Modified from Yanagimachi, R., and Noda, Y. (1970). Electron microscope studies of sperm incorporation into the golden hamster egg. *Am. J. Anat.* **128,** 429–462.

TABLE 14-4 Chromosomal Karotypes Related to Human Disorders

Clinical condition	Total number of chromosomes	Sex chromosome	Number of autosomal chromosomes
Normal male	46	XY	44
Normal female	46	XX	44
Turner's syndrome	45	X	44
Klinefelter's syndrome	47	XXY	44
Superfemale	47	XXX	44
Mongolism (female)	47	XX	45
Mongolism (male)	47	XY	45

2. *Conception*

Of the 200 × 10^6 sperm that are deposited in the vagina during coitus, only approximately 100–200 sperm reach the distal one-third of the fallopian tubes (see Figure 13-1) where fertilization normally occurs. The species specificity of the fertilization process is determined by a receptor-like mechanism present in the vitelline membrane of the ovum. W. Lennarz and associates have isolated a glycoprotein from sea urchin egg membranes that will only cross-react with a species-specific protein, termed bindin, present on the surface of a homologous sea urchin spermatozoon.

The attachment of a single spermatozoon to any region of the zona pellucida on a single ovum is the formal process of fertilization; the entire spermatozoon then enters the ovum, so that there is both a nuclear and a cytoplasmic contribution to the zygote. After sperm penetration, the fertilized ovum forms the second polar body (see also Figure 13-3). This is followed by the formation of female and male pronuclei; fusion of these haploid nuclei then creates, for the first time, the diploid nuclei of a totally new and unique individual. Some 30 hr later the first mitotic cleavage occurs, yielding a zygote with two cells; then within another 10 hr mitotic cleavage occurs again, producing a 4-cell state. Within 50–60 hr after fertilization the morula arises, followed by the blastocyst at 3–4 days (see Figures 14-1 and 14-2).

B. Sex Determination

1. *Definition of Sex*

Union of the spermatozoon with the ovum not only restores the diploid number of chromosomes (46 in humans) but also determines the genetic sex of the new individual. Since the oocyte always has an X chromosome, the sex of the offspring is determined by the fertilizing spermatozoon; if it is an X-bearing spermatozoon, the sex will be female (XX), while if it is a Y-bearing spermatozoon, the sex will be male (XY).

Under normal circumstances of the union of the human spermatozoon with the secondary oocyte, there is the creation of a zygote with 46 chromosomes. However, in some instances there are chromosomal abnormalities that may result in the offspring having an incorrect number of either autosomal chromosomes or sex chromosomes. Some of these abnormalities are summarized in Table 14-4. It is beyond the scope of this chapter to consider in detail the wide array of endocrine and nonendocrine disorders resulting from perturbations of genetic mechanisms of sexual development. The interested reader is referred to the review by Grumbach *et al.* (1992).

There are several other parameters besides the genetic sex that are utilized in describing the total definition of the sex of an individual. These are summarized in Table 14-5. Gonadal sex is a reflection of the morphology of the gonads and the hormones that they elaborate and is of dominant practical importance in defining the sex of an individual. Phenotypic sex is a reflection of the appearance of the external genitalia (i.e., the penis and vulva) as well as the secondary sex characteristics, including the beard and breasts, while somatic sex reflects the difference in the structures of the internal sex organs. Finally, psychological sex is a reflection of the social environment and behavioral aspects in which an individual may be raised.

TABLE 14-5 Parameters Involved with the Definition of Sex

Parameter	Description
Physiological sex	
Genetic or gender	Interaction of sex chromosomes XY = ♂, XX = ♀
Gonadal	Comparison of ovaries with testes
Somatic	Comparison of the internal sex organs in the male and female
Phenotypic	Comparison of the external genitalia and secondary sex characteristics: male, penis, beard; female, vulva, breasts
Birth certificate	Legal document usually certified by the physician present at birth, but may not correctly state the gonadal or genetic sex of the individual
Nonphysiological sex	
Psychological	Reflection of the social dimensions around the individual as she/he is reared, which may also reflect behavioral attitudes of key individuals in the environment

In the instance of individuals who are transsexuals, transvestites, and homosexuals, their physiological sex may not agree with their psychological sex. Thus, when an individual experiences normal development and sexual differentiation and is exposed to a normal psychosocial environment, then her/his genetic, gonadal, somatic, phenotypic, and psychological sexes will match. However, disturbances in any one of these parameters can result in aberrant gender identification and pose serious clinical, endocrinological, and psychological problems.

2. *Sex Differentiation*

The ultimate purpose of sex differentiation is to produce a female or male of the species. While the genetic sex of the new offspring is determined at fertilization, no morphological differences in the reproductive system of the embryo can be detected until the initiation of sexual differentiation and appearance of characteristic gonadal sex attributes at approximately the fourth week of development.

The gonadal anlage or gonadal precursor cells can be identified in the 4-week-old embryo and are preformed from the coelomic epithelium and from condensation with its underlying mesenchyme. At this stage, since it is impossible to distinguish whether the gonadal anlage will differentiate into ovaries or testes, it is designated as an "indifferent gonadal anlage." The indifferent anlage has two putative genital ducts: the Müllerian ducts, which can ultimately become the internal reproductive tract of the adult female, and the Wolffian ducts and mesonephros, which can become the adult male internal reproductive system. If the embryo is destined to become a male, the Müllerian ducts degenerate, while if the embryo is destined to become a female, the Wolffian ducts degenerate.

There is an inherent tendency for the indifferent gonadal anlage to undergo differentiation to yield ovaries provided that the correct X chromosome is present. The production of the male gonads, the testes, is believed to be related to the *SRY* gene (for sex-determining region Y), which is located on the Y chromosome (see the following). The chronology of these relationships is diagrammed schematically in Figure 14-14. The indifferent gonads begin to develop in the male at 43–50 days and in the female slightly later at 50–60 days.

Figure 14-15 diagrams the steps of human sex determination that lead to the conversion of the indifferent gonad anlage into either a testis or an ovary. This process of gonad differentiation is determined by the individual's genetic sex, and it is the subsequent male or female gonadal secretions that determine the somatic and phenotypic sexes. Figure 14-16 diagrams the reciprocal changes occurring in the male and female with respect to either Wolffian or Müllerian duct development.

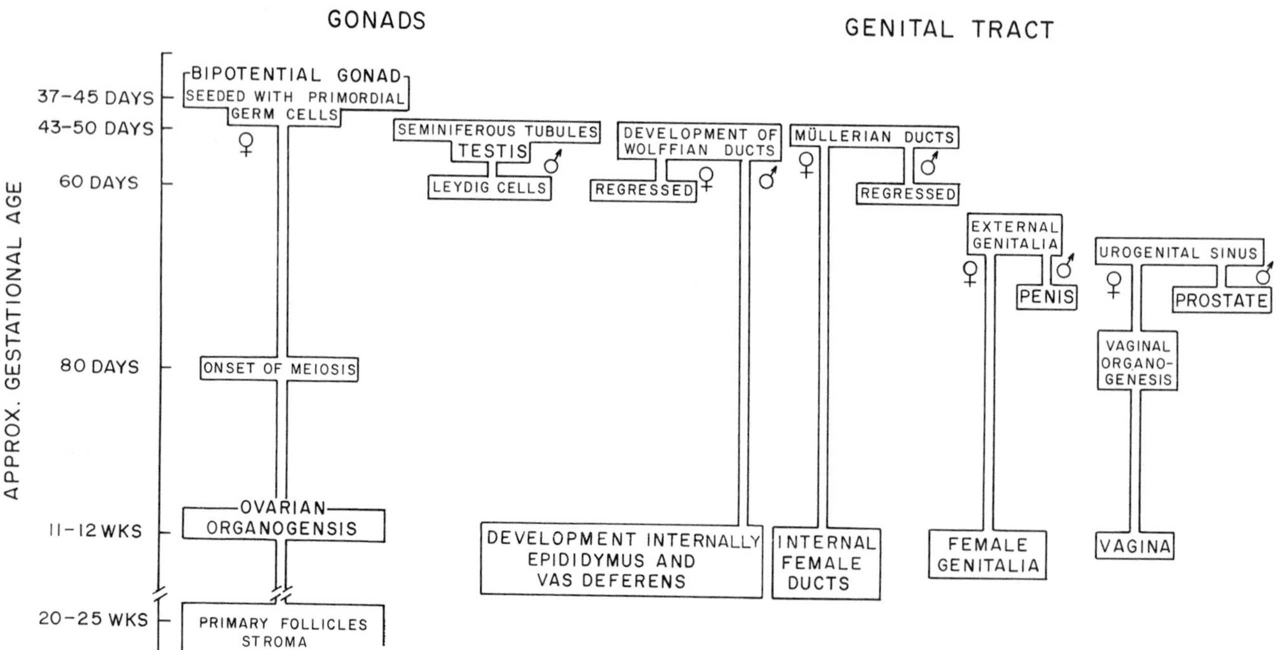

FIGURE 14-14 Chronology of the development of the external and internal male and female sex organs. Modified from Conte, F. A., and Grumbach, M. M. (1979). Pathogenesis, classification, diagnosis, and treatment of anomalies of sex. *In* "Endocrinology" (L. J. DeGroot, G. F. Cahill, L. Martini, D. H. Nelson, W. D. Odell, J. T. Potts, E. Steinberger, and A. I. Winegrad, eds.), Vol. 3, p. 1318. Grune & Stratton, New York.

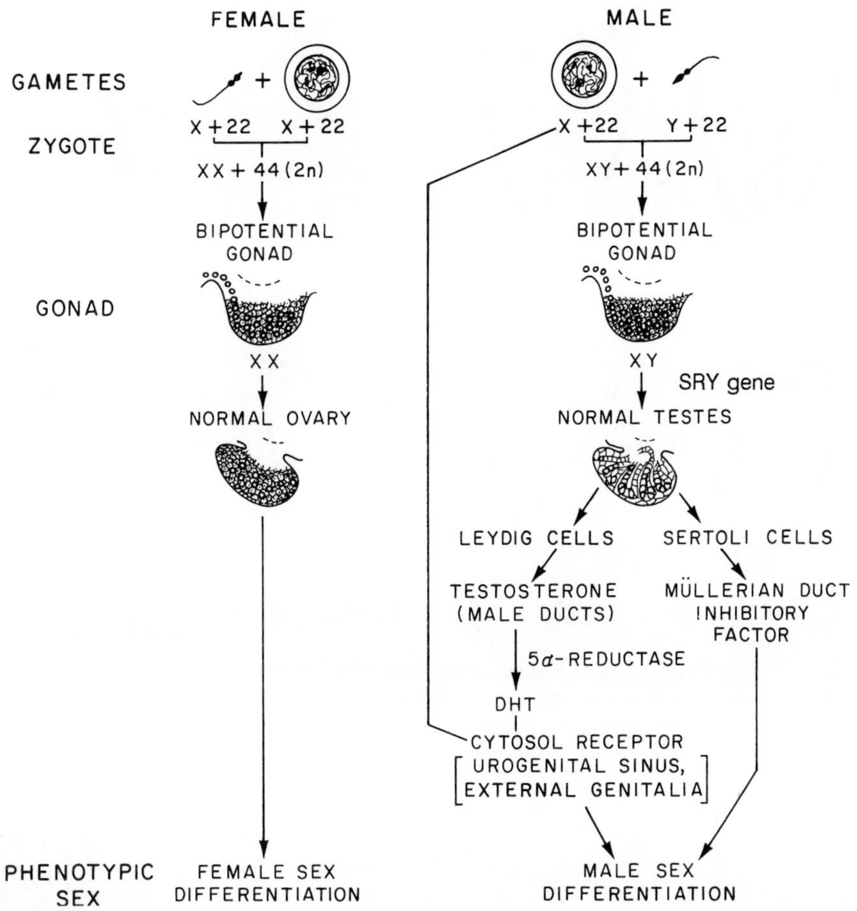

FIGURE 14-15 Schematic diagram of male and female sex determination and differentiation. The key contribution of the *SRY* gene or testis-determining factor is emphasized on the right for the male; see also Figures 14-17 and 14-18 for additional details. Modified from Grumbach, M. M. (1967). "Biologic Basis of Pediatric Practice" (R. E. Cooke, ed.), p. 1060. McGraw-Hill, New York.

In the absence of the *SRY* gene, female sex differentiation will lead to conversion of the indifferent gonad anlage into ovaries and to the regression of the wolffian ducts. Thus, the development of the female reproductive tract requires no hormonal stimulation from either ovaries or testes.

3. *SRY or Testis-Determining Factor*

The recent discovery of the *SRY* gene which codes for a protein that mediates testicular organogenesis has provided insight into one of the most challenging problems of biology; namely, the molecular basis of sex determination. The *SRY* gene is located on the Y chromosome. The *SRY* human gene codes for a DNA-binding protein that has an 80-amino-acid domain similar to that present in "high-mobility group" (HMG) proteins. This DNA-binding domain interacts specifically with high affinity to the DNA sequence AACAAG and causes bends in the DNA.

Importantly, the *SRY* gene-derived protein interacts with the promoter of two sex-specific genes: (a) the cytochrome P_{450} aromatase which mediates the conversion of testosterone to estradiol; and (b) Müllerian inhibitory factor (MIF) which mediates Müllerian duct regression, effectively blocking the development of female sex organs. These concepts are illustrated in Figures 14-17 and 14-18. Thus, the combined actions of the SRY protein ensures that the fetus has access to adequate quantities of the male steroid hormone testosterone and, at the same time, promote the regression of the Müllerian ducts which are the precursors of the female sex organs. In this context it is important to appreciate that a fetus (male or female) in its uterine-placental environment is continuously exposed to the female estrogen (see Figure 14-10).

This model was validated by introduction of a 14-kb DNA fragment containing mouse *SRY* into mouse embryos by microinjection into fertilized eggs. When an XX female embryo was injected with the

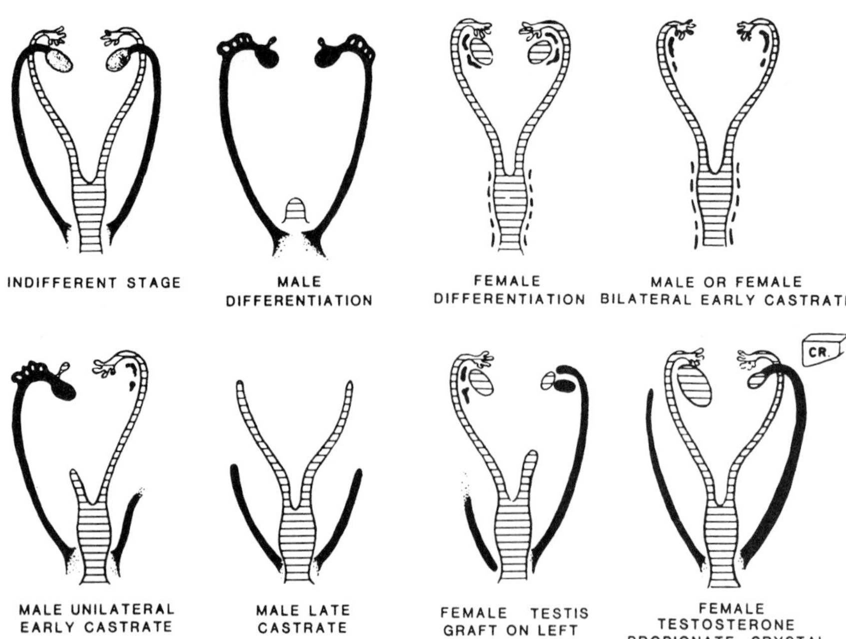

FIGURE 14-16 Schematic diagram of male (Wolffian duct) and female (Müllerian duct) differentiation. Testosterone stimulates Wolffian duct development, but has no effect on Müllerian duct inhibitory factor (MIF). Modified from Grumbach, M. M. Styne, D. M., Wilson, J. D., and Foster, D. W. (1992). "Williams Textbook of Endocrinology" Philadelphia, PA.

SRY–DNA, the phenotype was an XX male with functional testes that were morphologically indistinguishable from those of XY male mice.

Differentiation of the male indifferent gonad anlage occurs as a consequence of the presence of the *SRY* and produces functional testes. With the appearance of the fetal testes (by week 7 of embryo life), two hormones are secreted:-anti-Müllerian hormone and testosterone. The existence of a Müllerian duct-inhibiting factor (MIF) (or the anti-Müllerian hormone) had been inferred from studies where castrated male fetuses or normal female fetuses treated with androgens did not result in regression of the Müllerian ducts. This observation led to the suggestion that there must be two testes-derived hormones required for complete masculinization of the internal reproductive tract: testosterone and a factor designated anti-Müllerian hormone. This suggestion is supported both by clinical studies of the syndrome of testicular feminization as well as by studies employing antiandrogens (see Figure 12-13), where it has been observed that these compounds blocked the ability of the testes to induce the differentiation of the Wolffian duct, but did not block the ability of the testes to induce the regression of the Müllerian ducts.

4. *Müllerian Inhibitory Factor (MIF)*

Müllerian inhibitory factor (MIF) is the gonadal hormone that mediates regression of the Müllerian ducts during male embryogenesis (see Figure 14-14). MIF is a 140-kDa disulfide-linked homodimer protein that is a member of the transforming growth factor (TGF-β) multigene family of glycoproteins that includes the inhibins and activins (see Figures 12-6 and 14-18). The proteins of this gene family are all biosynthesized as dimeric precursors that require posttranslational processing for activation, as well as cleavage and dissociation to release functional C-terminal fragments.

MIF is expressed in the Sertoli cells of the testis. Human MIF is synthesized as a highly conserved protein of 536 amino acids with a 24-amino acid leader sequence. This protein is "activated" by cleavage at arginine-427 to yield 57.5- and 12.5-kDa fragments. The 12.5-kDa fragment of MIF is the biologically active protein.

The Leydig cells of the testes (see Chapter 12) are the principal site of steroidogenesis, leading to the production of testosterone. Testosterone, after further metabolism in target cells by the 5α-reductase enzyme to dihydrotestosterone, functions as a steroid hormone to cause differentiation of the Wolffian ducts into the

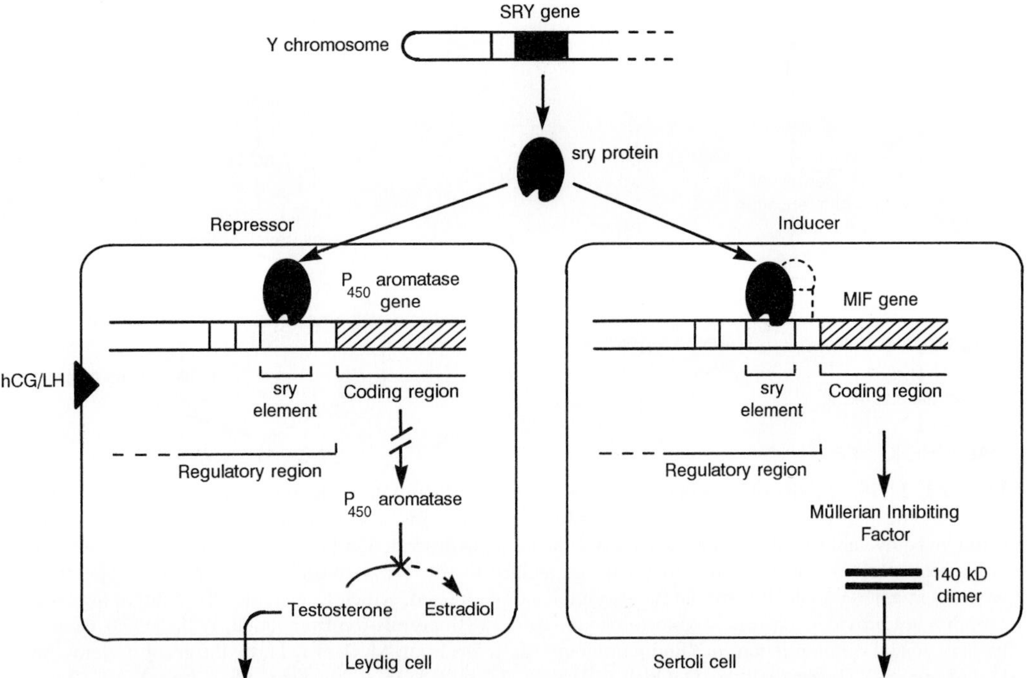

FIGURE 14-17 Schematic model describing the biological actions of the SRY protein on inducing the expression of the Müllerian inhibiting factor (MIF) and inhibiting the activation of the cytochrome P450 aromatase gene. SRY protein, a transcription factor, binds to the sry element in the regulatory region of the *MIS* gene and induces the expression of the 140-kDa MIF dimer in fetal Sertoli cells, causing regression of the paramesonephric (Müllerian) ducts. SRY protein binds to the sry element of the P450 aromatase gene and down-regulates its expression in fetal Leydig cells stimulated by hCG/LH, preventing the conversion of testosterone to estradiol. Testosterone is required for the development of the mesonephric (Wolffian) ducts into the male phenotype. Modified with permission from Kierszenbaum, A. L. (1994). "Mammalian spermatogenesis *in vivo* and *in vitro*: A partnership of spermatogenic and somatic cell lineages". *Endocr. Rev.* **15,** 116–134.

epididymis, seminal vesicles, and vas deferens (see Figures 14-14 and 14-15).

Interestingly, the gene for the androgen receptor protein, which is present in all androgen-dependent tissues, has been shown to be present on the X chromosome. Thus, an X-linked gene controls the androgen response of somatic cell types by generating the messenger RNA coding for the nuclear cytosol receptor protein.

C. Pregnancy

1. Placental Transfer of Hormones

While the maternal placental unit provides all of the nutrients, electrolytes, water, vitamins, and thermogenic, respiratory, and excretory functions required for fetal growth, an important endocrinological problem has been to elucidate the relative independence–dependence of the fetus from the mother with respect to the plethora of peptide and steroid hormones. Since there is no direct circulatory connection between the maternal placenta and fetal placenta (see Figure 14-3), transfer of hormones cannot occur by "direct" pathways. It has been determined that the placenta is impermeable to all polypeptide and thyroid hormones. However, steroid hormones and the catecholamines epinephrine and norepinephrine are capable of transfer across the placenta. Thus, as described in Figures 14-9 and 14-10, the fetus, mother, and placenta work cooperatively to produce and transfer the required amounts of progesterone and estrogen to the maternal placental unit to maintain pregnancy. Presumably the transfer of the steroid hormone occurs by a simple diffusional process.

2. Fetal Endocrinology

In view of the general lack of hormone transfer to the fetus from the mother, it is clear that the growing fetus must develop its endocrine system in coordination with its own growth and development. It is essential by the time of parturition that all endocrine systems be operative and competent to assume responsibility for their respective domains. However, growth of the

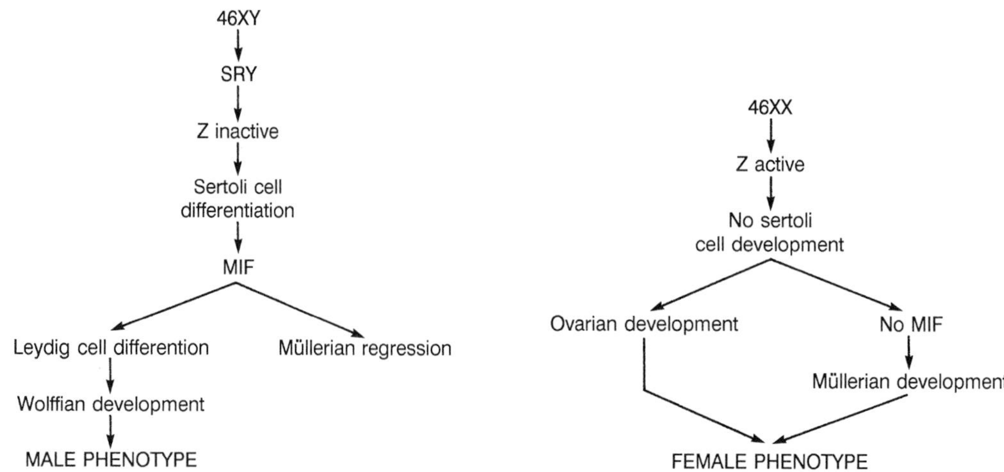

FIGURE 14-18 Model of the genetic bases for sexual differentiation illustrating the primordial role of the *SRY* gene and its protein product, which is functional only in the 46XY male. In this model, the *Z* gene could be equivalent to the cytochrome P450 aromatase gene described in Figure 14-17. In the absence of the SRY protein (a 46XX female), the enzymatic capability to convert testosterone to estradiol is operative, whereas in a 46XY male, the aromatase enzyme is not expressed, which thus denies the embryo access to estradiol but provides it access to testosterone. Modified with permission from Odell, W. D. (1995). Genetic basis of sexual differentiation. *In* "Endocrinology" (L. J. DeGroot, M. Besser, H. G. Burger, J. L. Jameson, D. L. Loriaux, J. C. Marshall, W. D. Odell, J. T. Potts, Jr., and A. H. Rubenstein, eds.), 3rd ed., Vol. 2, pp. 1881–1887. W. B. Saunders, Philadelphia, PA.

fetus appears to be largely controlled by inherent tissue factors and genetics rather than by placental hormones. Selected aspects of fetal metabolism are affected to a degree by the fetal pituitary and placental hormones.

With the exception of the catecholamines, the fetus is capable of functioning autonomously. The basal secretion rates of all of the anterior pituitary hormones are quite high by the end of the second trimester; by the third trimester their secretion is somewhat reduced, probably due to inhibitory feedback control. The production of cortisol is known to occur in the human fetus by the 10th week of gestation; both cortisol and dehydroepiandrosterone sulfate have been shown to increase after ACTH stimulation. Both glucagon and insulin secretion are reduced in the fetus; this may be a reflection of the relative constancy of the fetal blood sugar. Finally, the relatively high fetal serum calcium levels effectively suppress parathyroid hormone secretion and promote calcitonin secretion. The fetal kidney also is believed to be able to produce the dihydroxylated metabolites of vitamin D_3, namely, $1\alpha,25(OH)_2D_3$ and $24(R),25(OH)_2D_3$.

D. Breasts and Lactation

1. *Hormonal Regulation of Breast Development*

There have been extensive studies dealing with hormonal control of mammary growth and function. The three principal hormones required for mammary gland maturation are prolactin, insulin, and cortisol. There are two distinct phases of mammary secretory cell development: the proliferative phase and the differentiative phase.

In the proliferative phase, stem cells present in the mammary gland divide into precursors or secretory cells. Proliferation is regulated by insulin, epidermal growth factor (EGF), and growth hormone, all of which act as mitogenic stimuli. Prolactin (PRL) appears to sensitize the stem cell to the action of insulin and may act like a mitogen itself.

In the differentiative phase of mammary gland secretory cell development, a number of hormones are involved. Glucocorticoids (cortisol) are important and appear to play a role in this process. Prolactin and insulin are also critical. The secretory cells can appear in the absence of prolactin (in organ culture), but they will not produce milk constituents. Apparently the mRNAs for the milk proteins are under the control of PRL, but cortisol probably plays a role here. Cortisol may be required for PRL to elevate the mRNAs for milk proteins.

Both estrogen and progestin also stimulate the development of the mammary gland and the estradiol-stimulating duct, as well as causing a lowering of prolactin release inhibitory factor (PIF) to generate PRL release, progesterone-stimulating alveolar development, and a lowering of the level of PIF (Figure 14-19). In addition, estradiol may stimulate the release of PRL at the anterior pituitary through a mechanism involving the corticomedial amygdala of the limbic system. An important point is that, although PRL is being secreted in pregnancy and the mammary gland is developing, milk is not being synthesized due to the high levels of progesterone, which may block the glucocorticoid receptor. At term, free progesterone levels fall and the mammary cells become functional, as shown in Figure 14-19.

2. *Hormonal Regulation of Lactation*

a. Effects on Milk Proteins

The principal target organ for prolactin in the female is the mammary gland. Figure 14-20 shows a postulated mechanism of action for prolactin to describe its biological actions in stimulating the synthesis of milk constituents in the mammary gland. Eventually, after PRL interacts with its membrane receptor, nuclear activation occurs and the messenger RNA levels increase for the milk proteins casein, α-lactalbumin, β-lactoglobulin, and so on (see Table 14-1).

b. Prolactin Receptor

The prolactin receptor, which was first cloned from liver, consists of 285 amino acid residues; 210 aa are in the extracellular domain, 24 aa are in a hydrophobic transmembrane domain, and 54 aa are in a cytoplasmic domain. The prolactin receptor, as does the growth

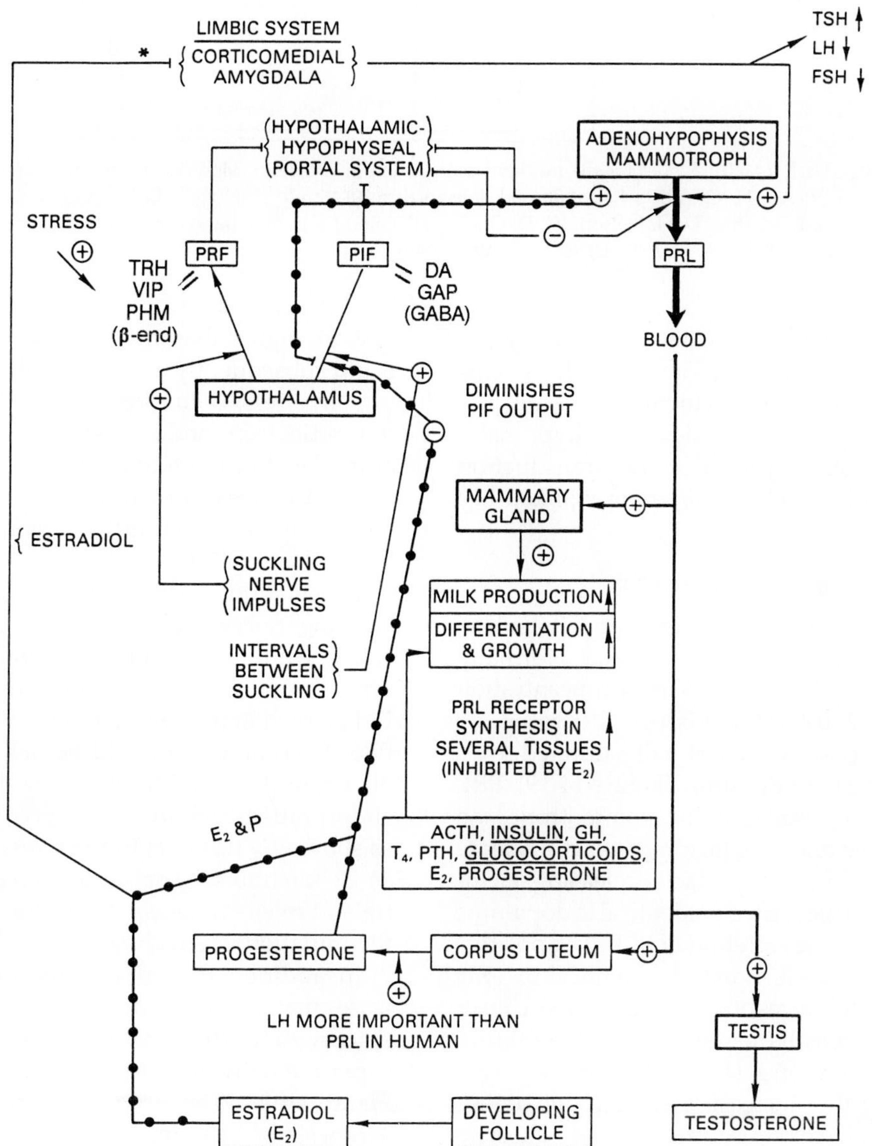

FIGURE 14-19 Schematic diagram of the regulated secretion of prolactin from the pituitary and summary of its biological actions. + indicates stimulation; − indicates inhibition. Abbreviations: VIP, vasointestinal peptide; PHM, peptide histamine methionine; GAP, gonadotropin-associated peptide.

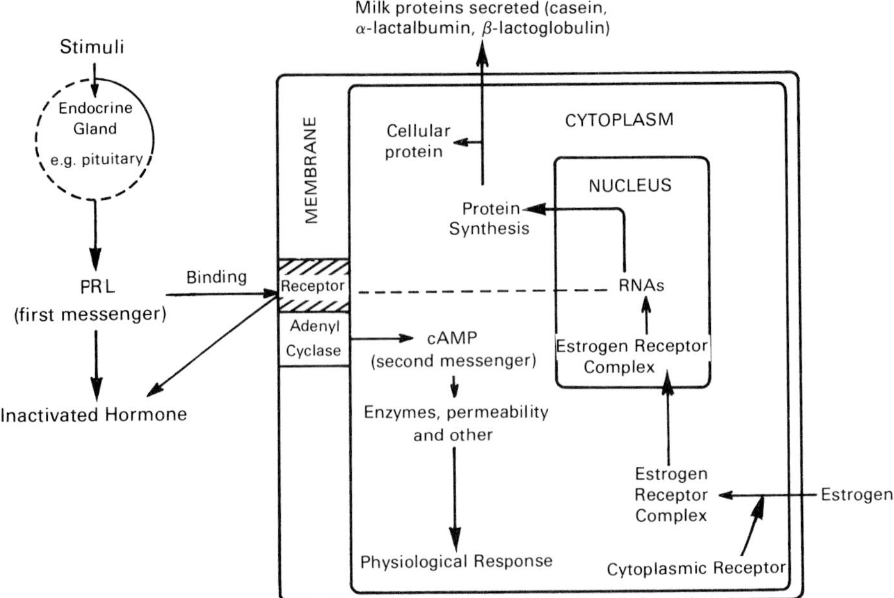

FIGURE 14-20 Model for the mechanism of action of prolactin. Modified from Aragona, C., and Friesen, H. G. (1979). Lactation and galactorrhea. *In* "Endocrinology" (L. J. DeGroot, G. F. Cahill, L. Martini, D. H. Nelson, W. D. Odell, J. T. Potts, Jr., E. Steinberger, and A. I. Winegrad, eds.), Vol. 3, pp. 1613–1627. Grune & Stratton, New York.

hormone receptor, belongs to the hematopoietic cytokine superfamily of receptors (see Figure 15-22). Specific prolactin receptors are located in mammary tissue, placenta, testis, prostate, seminal vesicles, and hypothalamus, to name a few. As yet, the signal transduction pathway to which the prolactin receptor is linked has not been identified.

c. Effects of Suckling on Prolactin Secretion

The regulation of prolactin release and its subsequent actions is diagrammed in Figure 14-21. Normally in the male and female the PRL serum concentration is 5–10 ng/ml or 0.2–0.4 n*M*; at this level PRL receptors in the mammary tissue are about half-saturated.

Starting with the hypothalamus (Figure 14-19), PRL is postulated to be released from the mammotrope by a pair of presumptive releasing factors, prolactin release inhibitory factor (PIF) and prolactin-releasing factor (PRF). PIF appears to be related or identical to dopamine (DA), which inhibits the secretion of PRL. The positive releasing factor, PRF, which may be identical to TRH (see Chapter 5 for structures of releasing factors), causes PRL to be released from the mammotrope of the anterior pituitary. Progesterone and estrogen inhibit the release of PIF (dopamine), thus increasing the secretion of PRL by diminishing a negative constraint. Estradiol acts indirectly on the mammotrope to effect the release. Apparently this is accomplished by the direct action of estradiol on the corticomedial amygdala of the limbic system, higher than the hypothalamus. Such an effect could subsequently act on hypothalamic releasing factors. In stress, β-endorphin is produced via β-LPH in the corticotrope and later stimulates the release of PRL at the level of the mammotrope. One of the end hormones in stress, cortisol from the adrenal gland, also negatively feeds back on the mammotrope to inhibit further release of PRL.

In lactation, a dominant overriding signal is the suckling nerve impulse, which travels through the spinal column, reaching the brain in a millisecond time frame, and stimulates the prompt release of PRF, which in turn acts on the mammotrope to release PRL. Such an effect could be delivered by serotonergic neurons and additionally by the stimulation of β-endorphinergic neurons, which could also release PRL as shown by its effect in the stress mechanism. Figure 14-22 illustrates the episodic changes in plasma prolactin and oxytocin levels that occur as a consequence of the initiation of suckling.

In addition to endogenous inhibitors of PRL release, that is, PIF or dopamine, the drug 2-bromo-α-ergocryptine (bromocriptine) is an inhibitor used experimentally and clinically. Its structure is given in Figure 14-23. The drug appears to cause decreased synthesis of PRL and an increase in the degradation rate of PRL; thus, the drug finds use in terminating milk production at weaning. If it is administered abruptly, its use may prevent the development of mas-

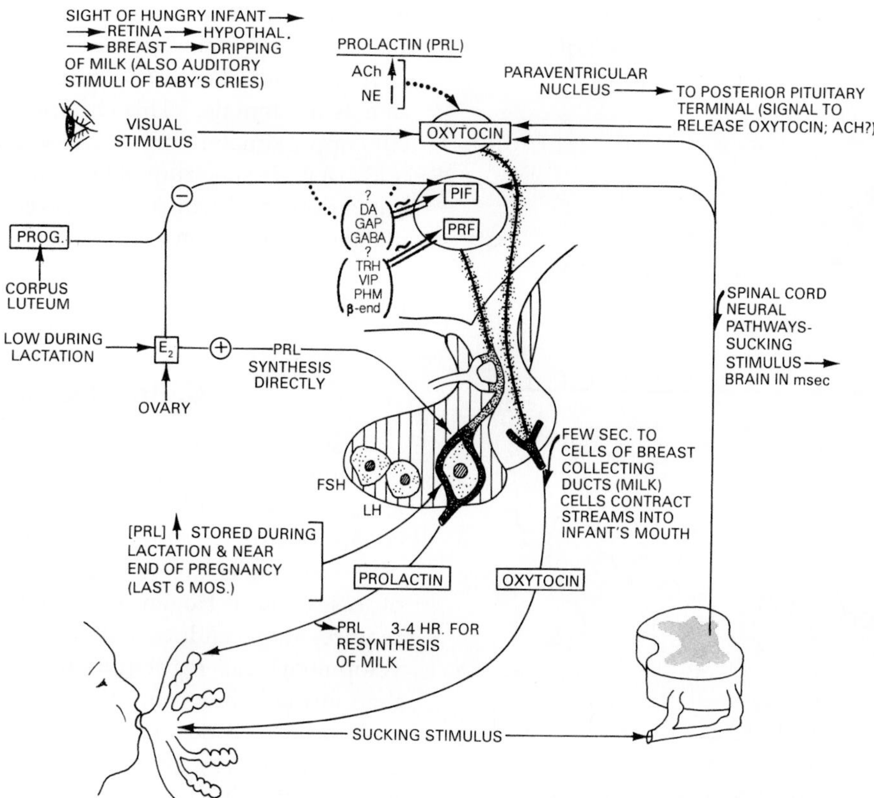

FIGURE 14-21 Integrated model to describe the neuroendocrine control of suckling and lactation. Abbreviations: GAP, gonadotropin-associated peptide; PHM, peptide histidine methionine; VIP, vasointestinal peptide. This figure is redrawn in part from Ezrin, C., Godden, J. O., Volpe, R., and Wilson, R. (1973). "Systematic Endocrinology." Harper & Row, Hagerstown, MD.

titis and infection in the mammary gland due to the stasis of milk. It has been used experimentally to determine whether mammary tumors, whose growth might be stimulated by PRL, could be regressed.

V. CLINICAL ASPECTS

A. Pregnancy–Gestational Neoplasms

Gestational neoplasms are the group of related tumors that can occur as a consequence of pregnancy. Morphologically, gestational tumors can be categorized as (i) a hydatidiform mole occurring after a molar pregnancy and (ii) a choriocarcinoma, which can occur following any kind of gestation.

A hydatidiform mole is a gestational neoplasm that occurs as the trophoblast initiates the implantation process in the uterine endometrium. Normal fetal vascularity is lost, and the trophoblastic cells surrounding the villi show varying degrees of proliferation. Choriocarcinoma represents the more malignant histological type of gestational neoplasm.

The incidence of gestational neoplasms is low: 1 in 2000 to 1 in 15,000. Further, of 900 patients studied at the New England Trophoblastic Disease Center, 71% had a molar pregnancy, and in the remaining 29% there was no evidence of metastases; an overall cure rate of 93% was achieved.

In instances of molar pregnancy and choriocarcinoma there is frequently massive production of human chorionic gonadotropin (hCG). Thus, plasma determination of hCG is quite useful for diagnosis as well as for evaluation of the success of management of the disease.

B. Anomalies of Sex Determinants

There is an extremely wide spectrum of endocrine disorders that may arise from perturbations in the many processes contributing to sexual organization. Thus, there are clinical problems present at each level of sex definition, that is, genetic sex, gonadal sex, somatic sex, and phenotypic sex (see Table 14-4). It is beyond the scope of this chapter to consider other than a few of these disorders

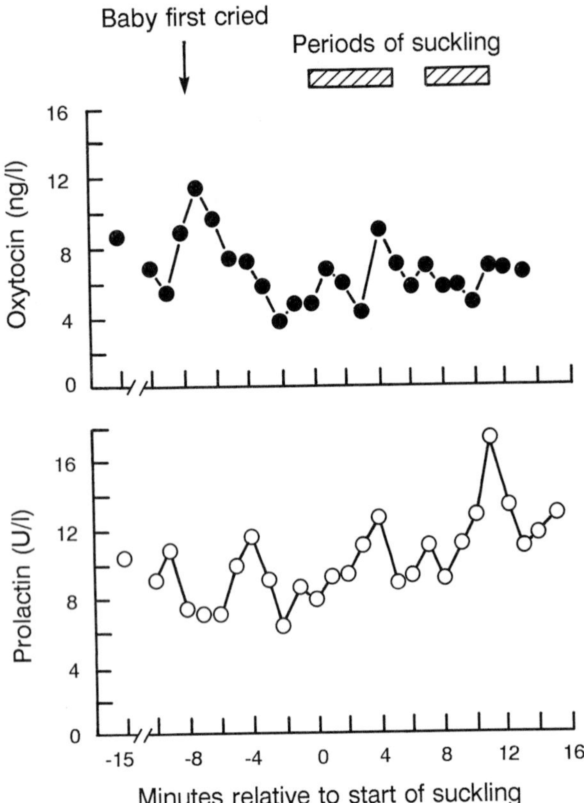

FIGURE 14-22 Changes in plasma oxytocin (●) and prolactin (○) concentrations before and during suckling. Modified with permission from McNeilly, A. S., I. C., Robinson, M. J., Houston, and P. W. Howie, (1983). Release of oxytocin and prolactin in response to suckling. *British Medical Journal,* **286,** 257–259.

Klinefelter's syndrome or seminiferous tubule dysgenesis is the most common cause of male hypogonadism; it results from an extra X chromosome to give an XXY genotype. Such individuals are characterized by the presence of small testes, gynecomastia (partial breast development), azoospermia, and elevated gonadotropins.

Subjects with Turner's syndrome are female and have a single X chromosome; they are usually small in stature and exhibit undeveloped secondary sex characteristics and amenorrhea.

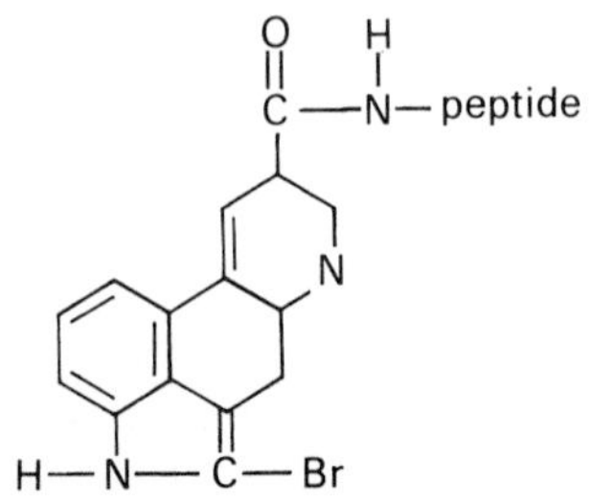

An individual with XXX chromosomes is designated as a "superfemale." This chromosomal aberration occurs approximately once in 1600 female births. There is also a wide spectrum of hermaphrodites that reflects the various problems that can arise from incorrect gonadal differentiation. A true hermaphrodite is an individual who possesses both testicular and ovarian tissues.

O H
C—N—peptide
N
H—N—C—Br

FIGURE 14-23 Structure of bromocriptine (or 2-bromo-α-ergocryptine), an inhibitor of pituitary prolactin secretion.

C. Breast Cancer

Breast cancer is the most prevalent malignancy in the United States, with 175,000 new cases occurring annually. Women have approximately a 1:9 lifetime chance of experiencing the disease. While the causes of breast cancer are not known, approximately 35% of cases are estrogen-dependent. Mammary tissue requires more endocrine factors for its structural development and regulation of its lactation function than virtually any other tissue; this includes estrogen and progesterone, prolactin, vasopressin, insulin, growth hormone, glucocorticoids, thyroxine, 1,25-dihydroxyvitamin D_3, and oxytocin. While it is conceivable that any or all of these hormones could mediate cell growth of a breast tumor, to date only estrogen, progesterone, and prolactin have been linked to malignant breast growth. The dramatic demonstration by Beatson in 1896 that mammary cancer regressed after bilateral oophorectomy in premenopausal women initiated the view that mammary tumors could be controlled by manipulating the endocrine environment. Jensen has shown a good correlation between the presence of an estrogen receptor in breast tumors and the likelihood of response to endocrine therapy. Approximately 66% of all breast tumors have significant levels of estrogen receptor. Such information is now utilized as part of the diagnostic evaluation of women with breast cancer; those premenopausal individuals who are estrogen receptor-positive become candidates for castration. Postmenopausal women who are estrogen receptor-positive are candidates for treatment with antiestrogens such as tamoxifen. Tamoxifen (see Figure 13-19) is a nonsteroidal competitive inhibitor of estradiol binding to the estrogen receptor. Tamoxifen is utilized as an oral drug for the treatment of postmenopausal women with advanced breast cancer. Regrettably in the human, endocrine therapy rarely achieves total extinction of the tumor and other management procedures, including chemotherapy or surgical procedures, must often be instituted.

D. Galactorrhea

Galactorrhea is defined as a persistent discharge from the breast of a milklike fluid that is not associated with parturition and lactation. Galactorrhea may occur as a consequence of a wide variety of endocrine and nonendocrine disorders; these endocrine-related instances include pituitary tumors, which produce prolactin, the Chiari–Frommel syndrome (which is defined as galactorrhea and amenorrhea persisting more than 6 months postpartum) in the absence of nursing, oral contraceptive usage, and primary hypothyroidism.

Serum levels of prolactin are usually elevated in galactorrhea. If the cause of the elevated plasma level of prolactin is an inappropriate secretion by the adenohypophysis, then the possibility of a pituitary tumor must be considered. In this instance, surgical removal or chemotherapy with drugs such as L-dopamine or the ergot alkaloid bromocriptine may be initiated. These two compounds interfere with prolactin secretion by the pituitary. The structure of bromocriptine is given in Figure 14-23.

References

A. Review Articles

Bryant-Greenwood, G. D., and Schwabe, C. (1994). Human relaxins: chemistry and biology. *Endocr. Rev.* **15,** 5–26.

Cooke, N. E. (1995). Prolactin: basic physiology. *In* "Endocrinology" (M. Besser, H. G. Burger, J. L. Jameson, D. L. Loriaux, J. C. Marshall, W. D. Odell, J. T. Potts, Jr., and A. H. Rubenstein, eds.), 3rd ed., Vol. 1, Chapter 23, pp. 368–393. W. B. Saunders Company, Philadelphia, PA.

Ezrin, C., Godden, J. O., Volge, R., and Wilson, R. (1973). "Systematic Endocrinology." Harper & Row, Hagerstown, MD.

Gaddy-Kurten, D., Tsuchida, K., and Vale, W. (1995). Activins and the receptor serine kinase superfamily. *In* "Recent Progress in Hormone Research" C. W. Bardin, ed.), Vol. 50, pp. 109–129.

Grumbach, M. M., Styne, D. M., Wilson, J. D., and Foster, D. W. (1992). "Williams' Textbook of Endocrinology," 8th ed., pp. 1–1712. W. B. Saunders Company, Philadelphia, PA.

Kierszenbaum, A. L. (1994). Mammalian spermatogenesis *in vivo* and *in vitro*: a partnership of spermatogenic and somatic cell lineages. *Endocr. Rev.* **15,** 116–134.

Lee, M. M., and Donahoe, P. K. (1993). Müllerian inhibiting substance: a gonadal hormone with multiple functions. *Endocr. Rev.* **14,** 152–164.

Odell, W. D. (1995). Genetic basis of sexual differentiation. *In* "Endocrinology" M. Besser, H. G. Burger, J. L. Jameson, D. L. Loriaux, J. C. Marshall, W. D. Odell, J. T. Potts, Jr., and A. H. Rubenstein, eds.), Vol. 2, Chapter 107, pp. 1881–1887. W. B. Saunders Company, Philadelphia, PA.

Qu, J., and Thomas, K. (1995). Inhibin and activin production in human placenta. *Endocr. Rev.* **16,** 485–507.

Strauss, J. F., III, Gafvels, M., and King, B. F. (1995). Placental Hormones. *In* "Endocrinology" L. J. DeGroot, M. Besser, H. G. Burger, J. L. Jameson, D. L. Loriaux, J. C. Marshall, W. D. Odell, J. T. Potts, Jr., and A. H. Rubenstein, eds.), 3rd ed., Vol. 3, Chapter 124, pp. 2171–2206. W. B. Saunders Company, Philadelphia, PA.

Walker, W. H., Fitzpatrick, S. L., Barrera-Saldana, H. A., Resendez-Perez, D., and Saunders, G. F. (1991). The human placental lactogen genes: Structure, function, evolution, and transcriptional regulation. *Endocr. Rev.* **12,** 316–328.

B. Research Articles

Beatson, G. T. (1896). On the treatment of inoperable cases of carcinoma of the mamma: Suggestions for a new method of treatment of inoperable cases of carcinoma of the mamma. Lancet **2,** 162–164.

Berkovitz, G. D., Fechner, P. Y., Marcantonio, S. M., Bland, G., Stetten, G., and Goodfellow, P. N. (1992). The role of the sex-determining region of the Y chromosome (SRY) in the etiology of 46,XX true hermaphroditism. *Human Genet.* **88,** 411–416.

Berta, P., Hawkins, J. R., Sinclair, A. H., Taylor, A., Griffiths, B. L., Goodfellow, P. N., and Fellous, M. (1990). Genetic evidence equating SRY and the testis-determining factor. *Nature* **348,** 448–450.

Haqq, C. M., King, C. Y., Donahoe, P. K., and Weiss, M. A. (1993). SRY recognizes conserved DNA sites in sex-specific promoters. *Proc. Natl. Acad. Sci. USA* **90,** 1097–1101.

Hercz, P., Siklos, P., and Ungar, L. (1989). Quantitative comparison of serum steroid and peptide hormone concentrations in male and female fetuses in the maternal–fetoplacental system during the 28th–40th weeks of pregnancy. *Eur. J. Obstet. Gynecol. Reprod. Biol.* **30,** 201–204.

Jahn, G. A., Edery, M., Ali, S., Belair, L., Kelly, P. A., and Djiane, J. (1991). Prolactin receptor gene expression in rat mammary gland and liver during pregnancy and lactation. *Endocrinology* **128,** 2976–2984.

Koopman, P., Gubbay, J., Vivian, N., Goodfellow, P. N., and Lovell-Badge, R. (1991). Male development of chromosomally female mice transgenic for SRY. *Nature* **351,** 117–121.

McElreavey, K., Vilain, E., Abbas, N., Herskovitz, I., and Fellous, M. (1993). A regulatory cascade hypothesis for mammalian sex determination: SRY represses a negative regulator of male development. *Proc. Natl. Acad. Sci. USA* **90,** 3368–3372.

Page, D. C., Mosher, R., Simpson, E. M., Fisher, E. M., Mardon, G., Pollack, J., McGillivray, B., Chapelle, A. D. L., and Brown, L. G. (1987). The sex-determining region of the human Y chromosome encodes a finger protein. *Cell* **51,** 1091–1104.

Palmer, M. S., Berta, P., Sinclair, A. H., Pym, B., and Goodfellow, P. N. (1990). Comparison of human ZFY and ZFX transcripts. *Proc. Natl. Acad. Sci. USA* **87,** 1681–1685.

Rozakis-Adcock, M., and Kelly, P. A. (1990). Mutational analysis of the ligand binding domain of the prolactin receptor. *J. Biol. Chem.* **266,** 16472.

CHAPTER

15

Hormones Related to the Kidney and Cardiovascular System

I. INTRODUCTION

A. Background

The kidney, heart, and circulatory system together play an indispensable role in the maintenance of life in higher organisms, not only from the perspective of maintaining the constancy of many components of the extracellular fluid, including filtering out of nitrogenous wastes and maintenance of normal blood pressure, but also as major endocrine organs.

The kidney as an endocrine gland is the site of production of renin and the following hormones: (1) erythropoietin, which is a peptide hormone essential for the process of erythropoiesis or red blood cell formation by the bone marrow; (2) 1α,25-dihydroxyvitamin D_3, the hormonally active form of vitamin D, which is essential for calcium homeostasis (see Chapter 9); and (3) the kallikreins, a group of serine proteases that act on blood proteins to produce bradykinin, a potent vasodilator. Renin is an enzyme with proteolytic activity that acts on a plasma protein, α_2-globulin, to produce the hormonal angiotensins, which in turn act at the adrenal cortex to stimulate the biosynthesis and secretion of the mineralocorticoid aldosterone. In addition, the kidney serves as an endocrine target organ for a number of hormones. Table 15-1 summarizes the endocrine aspects of the kidney both as an endocrine secretory gland and as an endocrine target organ.

An important physiological function of the kidney was observed by J. Peters in 1835, "The kidneys appear to serve as the ultimate guardians of the constitution of the internal environment." In this regard, it is clear that the kidney occupies a unique position within the

TABLE 15-1 Hormones That Are Produced or Have Major Actions in the Kidney

Hormones	Major target organ(s)	Function(s)
Produced by the kidney		
Erythropoietin	Bone marrow	Stimulate red blood cell formation
1α,25-Dihydroxyvitamin D_3 (+other vitamin D metabolites)	Intestine, bone, kidney	Maintenance of calcium homeostasis and regulation of gene transcription in many target organs
Renin (an enzyme)[a]	Blood to mediate production of the hormonal angiotensins	To mediate production of aldosterone (by the adrenal cortex)
Prekallikreins	Serum protein α_2-globulins	To produce kinins (e.g., bradykinin) that are potent vasodilators
Prostaglandins	Kidney	To antagonize hypertension by inhibiting Na^+ reabsorption
Acting on the kidney		
Aldosterone		Stimulate Na^+ reabsorption
Atrial natriuretic factor (ANF)		Increase glomerular filtration rate
1α,25-Dihydroxyvitamin D_3		Stimulate Ca^{2+} reabsorption
Endothelins		(Secondary actions)
Vasopressin		Promote H_2O reabsorption
Prostaglandins		
Catecholamines (epinephrine, norepinephrine)		
Cortisol		Diverse
Insulin		Diverse
Glucagon		Diverse
Thyroxine		Diverse

[a] Technically renin, an enzyme, does not fit the classical or modern definition of a hormone (see Chapter 1).

physiological network of the living organism, in that it functions as the ultimate organ for the monitoring and conservation of body water and all its electrolyte constituents, as well as many small organic molecules. In addition, the kidney plays an important physiological role in maintaining correct acid–base balance. It is thus perhaps not surprising to see concentrated in the kidney such an array of hormone production sites, as well as sites of hormone action. It seems likely that the exquisite anatomical organization of the kidney (discussed later) offers significant advantages to the organism in terms of the regulation of production of key hormones, as well as the regulation of the biological responses to various hormones that affect the homeostasis of key bodily electrolytes.

Many aspects of these processes are also closely linked to the process of blood pressure regulation. Accordingly, a secondary objective of this chapter is to consider hormonal aspects of the cardiovascular system. Table 15-2 summarizes the endocrine aspects of the cardiovascular system.

B. Scope of This Chapter

It is not pedagogically feasible to present a description of the physiological participation of the kidney in the formation of urine, namely, glomerular filtration and determination of electrolyte and acid–base balances. Also, it is beyond the scope of this book to present a detailed consideration of the heart and cardiac physiology and function. Thus, after a brief review of the anatomy of the kidney and the cardiovascular system, this chapter will present separately the renin–angiotensin–aldosterone–atrial natriuretic factor–Na^+ reabsorption system, the endothelin system, the kallikrein–kinin system, and the nitric oxide system. Finally, this chapter also provides a brief introduction to the topics of the hormones related to blood cell production and the hormones associated with the inflammation process.

A portion of the renin–angiotensin–aldosterone system, particularly that involving the biosynthesis and secretion of aldosterone, is included in Chapter 10. The roles of the kidney in calcium and phosphorus homeostasis, including its production of key vitamin D metabolites and its action as a target organ for parathyroid hormone, calcitonin, and 1α,25-dihydroxyvitamin D_3, are covered in Chapter 9.

II. ANATOMICAL, MORPHOLOGICAL, AND PHYSIOLOGICAL RELATIONSHIPS

A. The Kidney

1. Introduction

The kidneys, along with the ureters, urinary bladder, and male or female urethra, comprise the anatomical units responsible for the multiplicity of endocrine,

TABLE 15-2 Hormones That Are Produced by or Have a Major Action on the Cardiovascular System[a,b]

Hormones	Major target organ(s)	Function(s)
Produced by cardiovascular system		
Atrial natriuretic hormone (ANF)	Kidney glomerulus	Increase filtration rate
Endothelins (ET-1, ET-2, ET-3)	Endothelial tissue Brain	Vasoconstriction
Nitric oxide (NO)	Vascular smooth muscle	Vasorelaxation
Acting on the cardiovascular peripheral system		
Endothelins		Vasoconstriction
Angiotensin II		Vasoconstriction
Vasopressin		Vasoconstriction
Prostaglandins (thromboxanes)		Vasoconstriction
Epinephrine, norepinephrine		Vasoconstriction
Bradykinin		Vasorelaxation
Nitric oxide		Vasorelaxation
Atrial natriuretic factor		Vasorelaxation

[a] The cardiovascular system comprises the heart and the arterial and venous circulatory systems; see Figure 15-6.
[b] This table excludes hormones that are produced in the blood (e.g., angiotension) and hormones that use blood components as substrates (e.g., renin).

metabolic, and filtering actions of the kidney. Figure 10-1 presents a gross anatomical overview of the urinary system. With the completion of the metabolic and filtration actions of the kidneys, the urine, containing a wide variety of bodily wastes, is conveyed via the two ureters (one per kidney) to the urinary bladder for temporary storage. Then, at the time of micturition or emptying of the urinary bladder, the urine is voided through the urethra. The normal output of urine per day in adults may range from 600 to 2500 ml. Table 15-3 lists the principal electrolytes of the intracellular and extracellular fluids and urine.

2. *Kidney — Gross Structure*

The two kidneys in humans are located in a posterior position in the abdominal cavity; they present a ''bean-like'' appearance and are enclosed in a fibrous capsule. Diagrammed in Figure 15-1 is the cross-sectional appearance of the kidney. In the central portion of the concave border is the hilus, which contains all of the blood vessels and nerves for the kidney as well as the ureter. There are three general regions of the kidney: the renal pelvis, the renal cortex, and the renal medulla. The renal pelvis, which is attached to the upper end of the ureter, receives or collects the urine from all regions of the kidney. The renal medulla, which is the inner portion of the kidney, is composed of cone-shaped renal masses termed renal pyramids. The renal cortex constitutes the outer portion of the kidney.

3. *Kidney Microscopic Structure*

The fundamental operational unit of the kidney is the nephron (see Figure 15-2). Each human kidney contains $\approx$1 million nephrons. The nephron is the anatomical structure responsible for the formation of the urine as a consequence of the filtration of the blood and concomitant associated selective absorption and secretion of constituents in the newly created urine.

Each nephron consists of a renal tubule (largely in the medulla) and a renal corpuscle (located in the cortex). The renal corpuscle (see Figure 15-3), in turn, is composed of a glomerulus enclosed by a capsule, termed Bowman's capsule. The glomerulus is an exquisite anatomical structure containing a group of capillaries interposed between two arterioles. The vessel transporting blood to the glomerulus is known as the afferent arteriole; it divides into 40–60 capillary loops, which then ultimately reunite to form the exit pathway for the blood known as the efferent arteriole. Associated with the cells of the efferent arteriole (see Figure 15-3) is a cluster of cells termed the juxtaglomerular or JG cells. The JG cells are the site of production of the enzyme renin. As noted in Figure 15-2, the efferent arterioles subsequently arborize to form a network of capillaries surrounding the renal tubule.

Bowman's capsule constitutes the beginning of the renal tubule. Specialized cells, termed podocytes, in Bowman's capsule form slits or pores of such molecular dimensions that restrict and effectively prevent the passage of macromolecules in the blood into the top of the nephron; thus, only small molecules pass into the filtrate.

The renal tubule (see Figure 15-2) is anatomically structured to permit countercurrent distributive processes between its proximal and distal regions and

TABLE 15-3 Principal Constituents of Major Body Fluids in the Normal Adult

Constituent	Intracellular compartment (meq/liter)	Extracellular compartment (meq/liter)[a]	Urine[b] (g)[c]	Urine ratio [plasma]
Cations				
Sodium	11	142	2–4	1.0
Potassium	162	4	1.5–2	10–15
Calcium	2	5	0.1–0.3	
Magnesium	28	2	0.1–0.2	
Anions				
Chloride	3	101	17.1	1.5
Bicarbonate	10	27	0.15–2.5	1.5
Phosphate (HPO_4^{2-})	102	2	0.6–1.8	25
Sulfate (SO_4^{2-})	20	1		50
Other				
Protein	65	16		
Urea			30.0	35
Creatinine			1.0	70
Organic acids	3	6		
Glucose (mg/100 ml)		80–110	6–8	

[a] The values reported for the extracellular compartment are also applicable to plasma.
[b] Calculated by assuming a daily urinary output of 2 liters.
[c] Amount excreted in a 24-hr interval by a normal man consuming an average mixed diet.

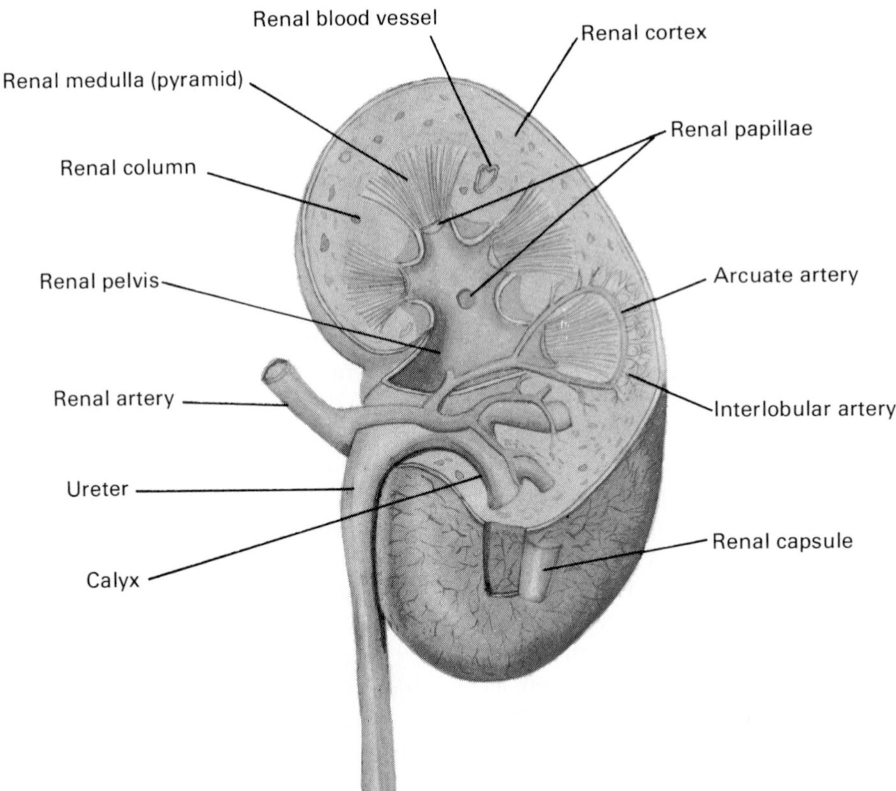

FIGURE 15-1 Cross section of a human kidney.

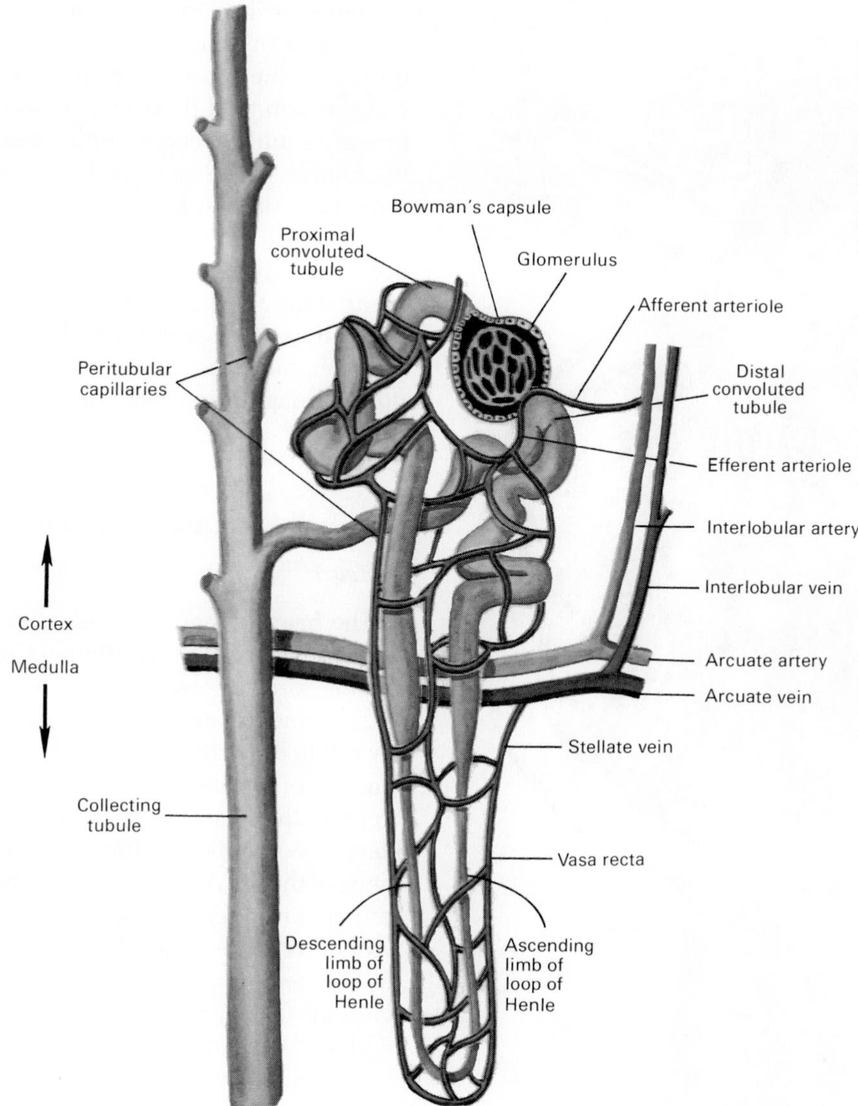

FIGURE 15-2 Schematic diagram of a nephron with its associated vascular system.

the loop of Henle as well as the arborized network of capillaries.

An important distinction to note is that the far distal region of each renal tubule is anatomically connected to the afferent arteriole of its own glomerulus. The point of attachment of each distal tubule to its glomerulus is the *macula densa;* the cells of the macula densa interact with the juxtaglomerular cells. Thus, this anatomical specialization allows for physiological, metabolic, and hormonal communication to occur between the distal tubule (containing the exiting urine) and the afferent arteriole (containing the entering blood, which will be filtered).

The distal convoluted tubule empties into a branch of the collecting ducts that ultimately leads to a main collecting duct, in turn connecting to a network of renal papillae coursing through the renal pyramids to the renal pelvis.

4. *Physiological Processes*

The kidney is the principal organ responsible for the homeostasis of a wide spectrum of electrolytes, as well as the conservation of body water. The kidney carries out its homeostatic actions sequentially by selective glomerular filtration (mediated by a high blood pressure in the glomerulus), tubular secretion, and tubular reabsorption; these are all processes that collectively regulate the concentration of the metabolic end products, osmotic pressure, ionic composition, and

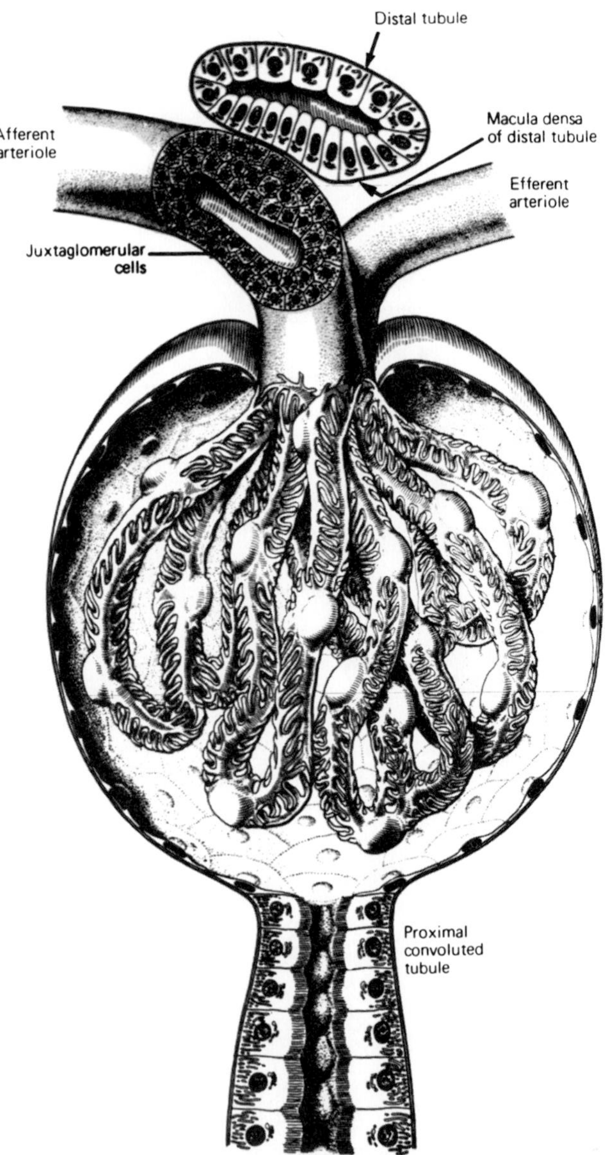

FIGURE 15-3 Cross section of a renal corpuscle. The upper part shows the vascular pole with afferent and efferent arterioles and the macula densa. Note the juxtaglomerular cells in the wall of the afferent arteriole. Podocytes cover the glomerular capillaries. Their nuclei protrude on the cell surface. Podocyte processes can be seen. The cells of the parietal layer of Bowman's capsule are shown. The lower part of the drawing shows the urinary pole and the proximal convoluted tubule. Modified from Junquiera, L. C. and Carneiro, J. (1983). "Basic Histology," 4th ed., p. 400. Lange Medical Publications, Los Altos, CA.

volume of the internal environment. Figure 15-4 is a schematic diagram of a typical kidney nephron, indicating the sites of reabsorption of the various ionic substances. The key to the achievement of homeostasis is the process of countercurrent distribution. Countercurrent distribution occurs as a consequence of the anatomical organization of the nephron and is supported by the processes of passive diffusion, tubular reabsorption, and tubular secretion. Both of the latter processes utilize energy-dependent active transport mechanisms. The end result of these activities is the formation of the residual urine containing bodily wastes (both ionic, organic, and nitrogenous); however, in the process of forming the urine from the glomerular filtrate, the tubule has returned many essential nutrients and electrolytes to the blood. A comparison of the extracellular concentrations to urine concentrations of major urine constituents is presented in Table 15-2.

B. Cardiovascular System

1. Heart

The heart, although weighing less than 450 g, is a remarkable organ. Its continuous beating provides the driving force to ensure the delivery of oxygen, nutrients, hormones, and other regulatory molecules to every cell in the body, as well as effecting the continuous removal of metabolic waste products. Over an average lifetime, the heart pumps approximately 300 million liters ($\approx$80 million gallons) of blood. The dynamic range of the heart pumping capacity is also impressive; cardiac output can range from a low of 5 to an upper limit of 35 liters of blood per minute. The heart is also the source of an important hormone, atrial natriuretic hormone (ANF), which is involved with the regulation of blood pressure.

Figure 15-5 presents a cross-sectional view of the heart. The human heart is divided longitudinally into right and left halves, each of which consists of two chambers, the atrium and the ventricle. The function of the heart is to pump the blood into the arterial system that has been delivered to the heart from the venous system. The right side of the heart receives from the *venae cavae* the oxygen-depleted blood from the peripheral tissues; this is then pumped into the pulmonary artery for circulation to the lungs. The left side of the heart receives the oxygen-enriched blood back from the lungs via the pulmonary veins; this blood is then pumped into the aorta for disbursement to all of the peripheral tissues.

The pressure necessary for these tasks is created by the strong contractions of the various musculatures of the heart. However, the force and rate of these heart contractions, as well as the volume of blood that may exit the heart with each heartbeat, are determined via a complex set of physiological and hormonal signals

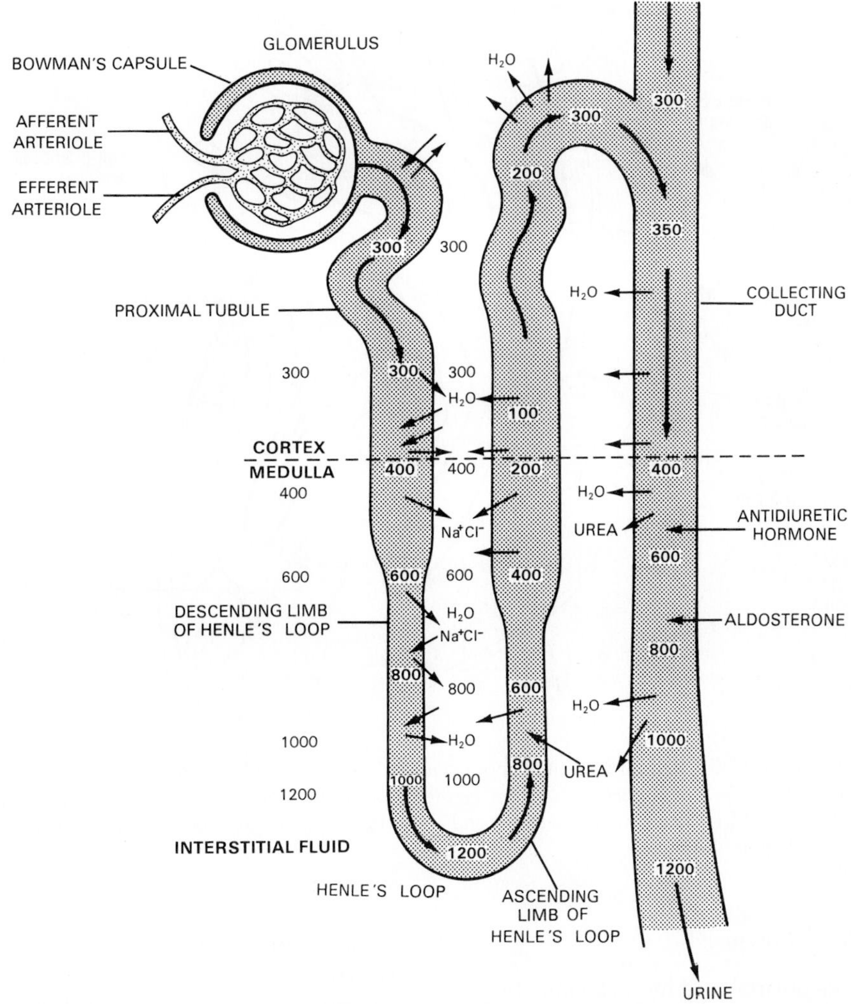

FIGURE 15-4 Schematic diagram of a kidney nephron illustrating the process of countercurrent exchange. The numbers indicate the gradients in osmolarity that result as a consequence of simple diffusion and transport or exchange. The filtrate, as it enters the loop of Henle, becomes further concentrated by diffusion of water into the hypotome interstitium. In the ascending limb of the loop of Henle, Na^+ is actively transported out of the filtrate (thereby diluting it) into the interstitium, where it is concentrated. The osmolarity of the interstitial fluid can increase from 300 to 1100 mOsm/kg Hg_2 as the papilla is approached. Then, as the distal tubule passes back past the glomerulus, additional Na^+ may be reabsorbed in exchange for H^+ or K^+ ions; eventually water is reabsorbed in the distal tubule and collecting duct. Also indicated are the sites of reabsorption of various constituents. Aldosterone and antidiuretic hormone act principally on the collecting duct. Modified from Pitts, R. F. (1974). "Physiology of the Kidney and Body Fluids," 3rd ed., p. 134. Year Book Medical Publishers, Inc., Chicago.

operative in the circulatory system. A detailed discussion of these topics is beyond the scope of this text.

2. *Cardiovascular System*

The circulatory system or cardiovascular system comprises the heart and blood vessels, or the body's "plumbing system;" see Figure 15-6A. The heart is the pumping organ that ejects blood into the arterial network; the blood is then returned to the heart via the venous system. The capillaries represent microscopic vesicles that interconnect between the small arteries (arterioles) and small veins (venules); see Figure 15-6B. The walls of the capillaries are only one endothelial cell in thickness. Tiny openings, or fenestrations, facilitate the delivery and exchange between the circulating blood in the capillary and the interstitial fluid that surrounds and bathes the neighboring cells.

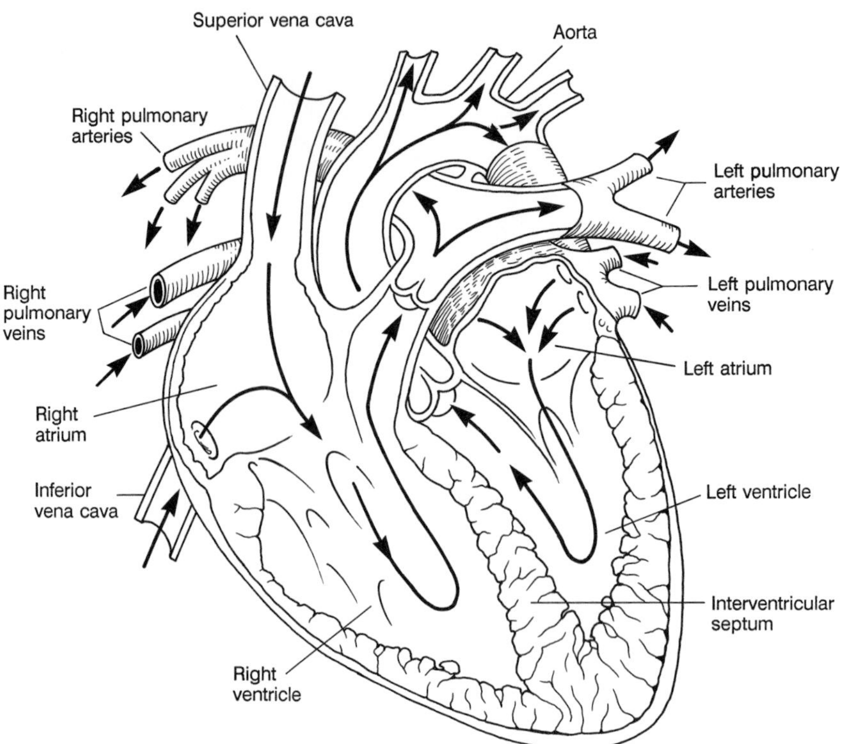

FIGURE 15-5 Interior view of the human heart. The arrows indicate the direction of blood flow. [Modified with permission from Figure 20-6, of Solomon, E. P., Schmidt, R. R., and Adragna, P. J. (1990). "Human Anatomy and Physiology," p. 676. Saunders College Publishing, Philadelphia, PA.

3. *Regulation of Blood Pressure*

Table 15-4 classifies normal and hypertensive blood pressures. The principal determinants of mean arterial blood pressure are the volume of cardiac output (stroke volume and heart rate), systemic blood flow, and resistance to blood flow in the various perfused organs. These functions are regulated by a variety of hormones, neurotransmitters, and local paracrine factors, as well as by the health and vigor of the heart.

III. HOMEOSTASIS OF FLUID, ELECTROLYTES, AND BLOOD PRESSURE

A. Introduction

Hormones are intimately involved in the regulation of both cardiovascular and renal functions. The maintenance of salt homeostasis, circulatory volume, and blood pressure requires the integrated actions of the renin–angiotensin–aldosterone system, the adrenergic nervous system, vasopressin, atrial natriuretic hormone, kinins, endothelins, prostaglandins, and the nitric oxide system.

Some of these hormones can affect heart rate and contractility directly or effect vasoconstriction or vasodilation of the arteries and veins. In addition, the growth factor properties of hormones are able to influence cardiovascular development and muscular hyperplasia or hypertrophy, as well as being involved with pathological changes that can be associated with atherosclerosis, cardiac hypertrophy, and heart failure. These hormone systems are also capable of mediating important biological effects in the kidney.

The volume of the extracellular fluid (ECF) is governed by its Na^+ concentration. The Na^+ concentration of the ECF is determined by regulating the extent of excretion of Na^+ in the urine, which is mediated by aldosterone and ANF. The principal, although not exclusive, factors governing the excretion of Na^+ are the steroid hormone aldosterone and the glomerular filtration rate. The glomerular filtration of the kidney can be markedly increased by the actions of ANP, thus increasing the extraction of blood Na^+. Also, ANP acts on the smooth muscle present in large arteries and vascular beds to effect relaxation and, thus, achieve a reduction in blood pressure. The function of aldosterone is to directly stimulate the absorption of Na^+ by the renal tubules; this has the consequence of increasing the extracellular fluid volume. Thus, in the renin–angiotensin system, the rate of secretion of aldosterone by the adrenal cortex is ultimately regulated by the extracellular fluid volume.

As discussed in Chapter 10, bilateral adrenalectomy is fatal; this results from the absence of aldosterone, which, in turn, leads to an increased loss of Na^+ in the urine, a concomitant retention of K^+ in the extracellular fluid, and loss of water from both the extracellular and intracellular compartments. If this process continues, death inevitably follows.

B. Renin–Angiotensin II

Renin is an enzyme of 347 amino acids of molecular mass ≈42 kDa. Renin may be isolated from both kidney and mouse submaxillary glands. The original isolation of the pure protein required a 3 million-fold purification; the primary sequence of human renin has been deduced through cloning and sequence analysis of cDNA prepared from human kidney mRNA. Renin as an enzyme belongs to the class of aspartyl proteinases.

Renin is biosynthesized as a preprorenin consisting of 406 amino acids. The prepro form is converted to prorenin by the removal of 20 amino acids. Then the mature renin is generated by specific cleavage of a 43-amino-acid pro sequence by a prorenin processing enzyme (PPE). The PPE has been highly purified, but not fully characterized. On the basis of both immunohistochemical and biochemical studies, the suggestion has been made that PPE is identical to the protease cathepsin B; both PPE and cathepsin antibodies stained proximal renal tubules as well as renal juxtaglomerular cells. Approximately 50% of the prorenin in the juxtaglomerular cells is constitutively secreted continuously without proteolytic processing, while the other 50% of the prorenin is sorted to secretory granules, where it is proteolytically processed and stored as renin. The release of this renin is regulated by cAMP. Renin in the circulatory system is variably glycosylated and circulates as 4–5 isoenzymes.

Human renin has, at the amino acid level, 63% homology with mouse submaxillary gland renin and 34% homology with human pepsinogen (see Chapter 8). In renin the aspartyl residues at positions 38 and 226 are believed to be important for catalytic activity. The X-ray crystallographic structure of human renin has been determined to 2.5 Å. The general shape of renin is bilobal with a long deep cleft that contains the aspartyl 38 and 226 residues.

Changes in the secretion of renin by the renal juxtaglomerular cells can occur in response to changes in renal arterial pressure, via sympathetic nervous system signals and also via changes in the status of a number of hormones (see Table 15-5). The secretion of renin is governed at the level of the renal glomerulus by both long and short feedback loops. The dominant negative long feedback loop, which diminishes renin secretion, involves increased renal arterial pressure, which is sensed by stretch receptors present in the glomerular afferent arteriolar wall so as to result in a reduction in renin release and, ultimately, a diminution in aldosterone production. A negative short feedback loop is mediated by angiotensin II, which directly inhibits renin release.

C. Angiotensins I and II

The natural substrate for renin is the plasma protein α_2-globulin, which is termed angiotensinogen. Angiotensinogen is a glycoprotein of 57 kDa that is synthesized and secreted into the bloodstream by the liver. The biosynthesis of angiotensinogen is increased by glucocorticoids, estrogens, and some oral contraceptives. The details of the conversion of angiotensinogen into angiotensin I (a decapeptide) and angiotensin II (an octapeptide) are summarized in Figure 15-7. In the circulatory system, renin hydrolyzes the Leu–Leu bond of angiotensinogen at residues 10 and 11 to generate the biologically inactive decapeptide angiotensin I. Angiotensin I is then converted by a converting enzyme (molecular mass ~ 200 kDa) that removes the carboxyl-terminal His-Leu dipeptide to yield the octapeptide hormone angiotensin II.

Angiotensin-converting enzyme (ACE) is a zinc-containing protein. Its most important actions are (a) to convert angiotensin I → II and (b) to inactivate bradykinin (a very potent vasodilator). The principal site of conversion of angiotensin I to II is in the vascular epithelium of the lung; however, ACE activity is also present in the kidney, vascular epithelium, heart, brain, and testis. A potent orally active synthetic inhibitor of the converting enzyme is the drug captopril or *l*-(D-3-mercapto-3-methylpropanoyl)-2-proline. The converting enzyme is also called kininase II because of its action on bradykinin (see Figure 15-16). Angiotensin III, a nonapeptide, is produced by the action of an N-terminal peptidase on angiotensin I. Table 15-6 summarizes the biological actions of the angiotensins. Angiotensin II is the most potent vasoconstrictive agent known. All of the components of the renin–angiotensin system necessary to produce angiotensin II are also present in the brain, where the latter substance may function as a neurotransmitter.

The two principal biological actions of angiotensin II are (a) to stimulate the production of aldosterone in the adrenal zona glomerulosa and (b) to function as a highly potent vasoconstrictor in the circulatory system,

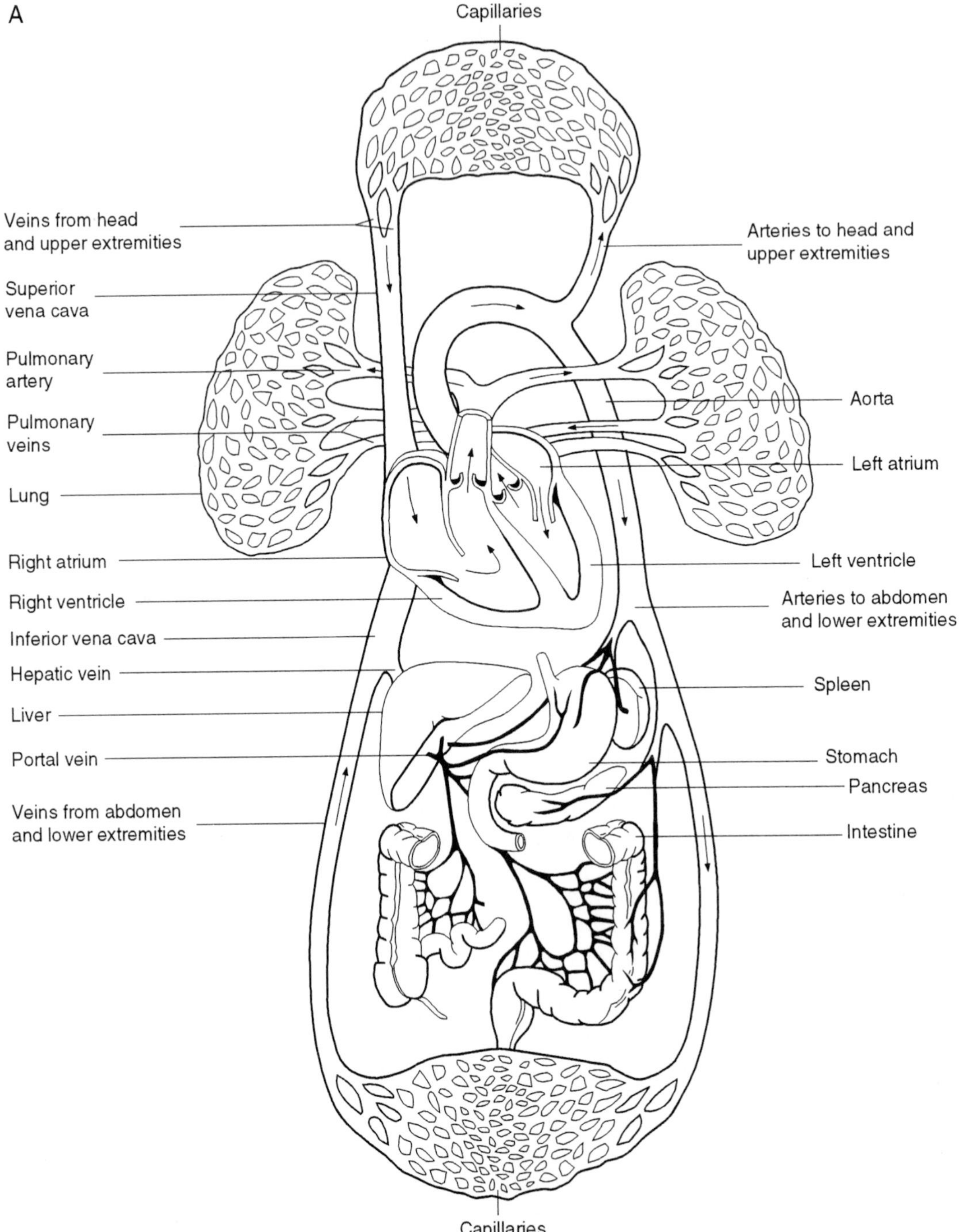

FIGURE 15-6 Diagram of the human circulatory system. (A) Illustration of the pulmonary and systemic circulatory sysems. The pulmonary circulation includes the pulmonary arteries, capillaries, and veins. The systemic circulation includes all of the other arteries, capillaries, and veins of the body. (B) Diagram of a capillary bed. Precapillary sphincters are relaxed, thus permitting the flow of blood through the capillary network. A greatly magnified portion of capillary wall is shown in the inset at the upper left of B. Modified with permission from Figure 16-4 of Chaffee, E. E. and Lytle, I. M. (1980). ''Basic Physiology and Anatomy,'' 4th ed., p. 336. J. B. Lippincott Co., Philadelphia, PA.

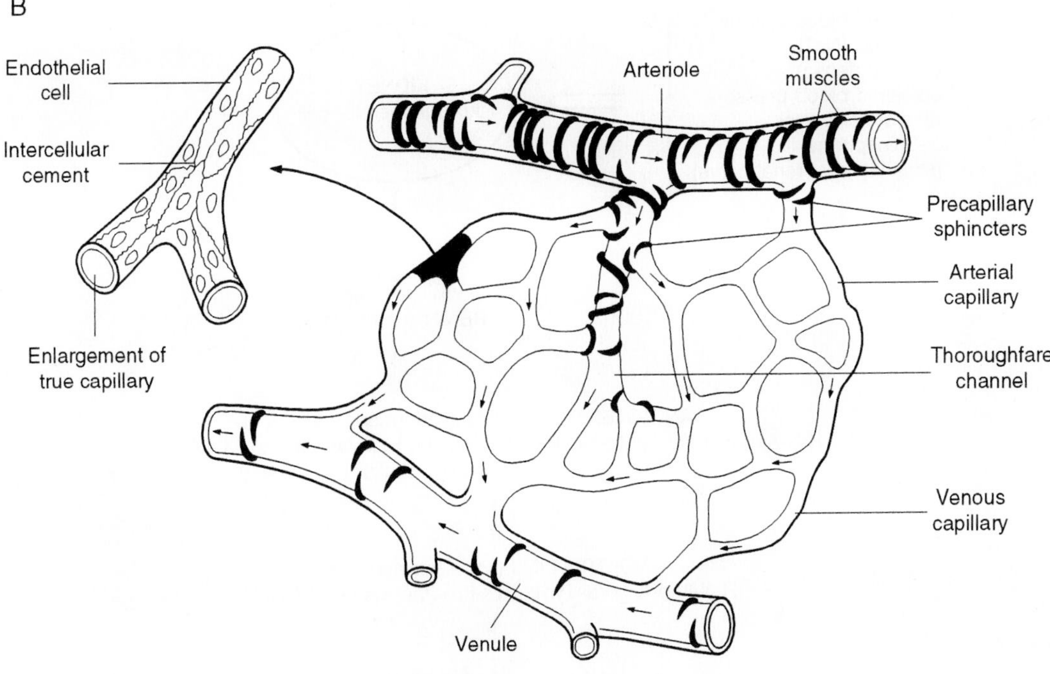

FIGURE 15-6—*Continued*

where it acts within seconds. In both locations, the biological response generated by angiotensin II is mediated as a consequence of its binding to high-affinity membrane receptors; these receptors have been cloned and found to have a seven transmembrane domain that is linked to a G protein (see Chapter 1). The immediate postangiotensin II receptor-binding signal transduction events include the elevation of both IP_3 and intracellular Ca^{2+} concentrations.

1. *Aldosterone Biosynthesis*

The site of production of the mineralocorticoid aldosterone is the adrenal zona glomerulosa cell, which employs a variety of signal transduction mechanisms that include receptors for angiotensin II, ACTH, dopamine, and atrial natriuretic peptide as well as a K^+-modulated voltage-gated Ca^{2+} channel. The dominant agonist is angiotensin II, but under appropriate physiological circumstances the other agonists are also effective. Within seconds of exposure to angiotensin II, IP_3, diacylglycerol (DAG), and intracellular Ca^{2+} concentrations all increase, followed shortly thereafter by an elevated biosynthesis of aldosterone; the details are summarized in Figure 15-8.

D. Aldosterone Actions in Renal Tubular Reabsorption

In humans, the principal biologically active mineralocorticoids are aldosterone, cortisol, and, to some limited extent, deoxycorticosterone; their structures are given in Figure 2-20. The mineralocorticoids mediate

TABLE 15-4 Classification of Blood Pressure Levels

Classification[a]	Systolic (mm Hg)	Diastolic (mm Hg)
Normal	<130	<85
High normal	130–139	85–89
Hypertension		
Mild	140–159	90–99
Moderate	161–179	100–109
Severe	180–209	110–119

[a] The classifications are for age >18 years. The data are obtained from the Joint National Committee on Detection, Evaluation, and Treatment of High Blood Pressure.

TABLE 15-5 Factors Regulating Renin Release

Inhibitory factors	Stimulatory factors
Increased renal arterial pressure	Decreased renal arterial pressure
Increased extracellular fluid volume	Decreased extracellular fluid volume
Angiotensin II	
Atrial natriuretic factor (ANF)	PGI_2
Vasopressin	ACTH
α-Adrenergic stimulation (dopamine)	β-Adrenergic stimulation (catecholamines)

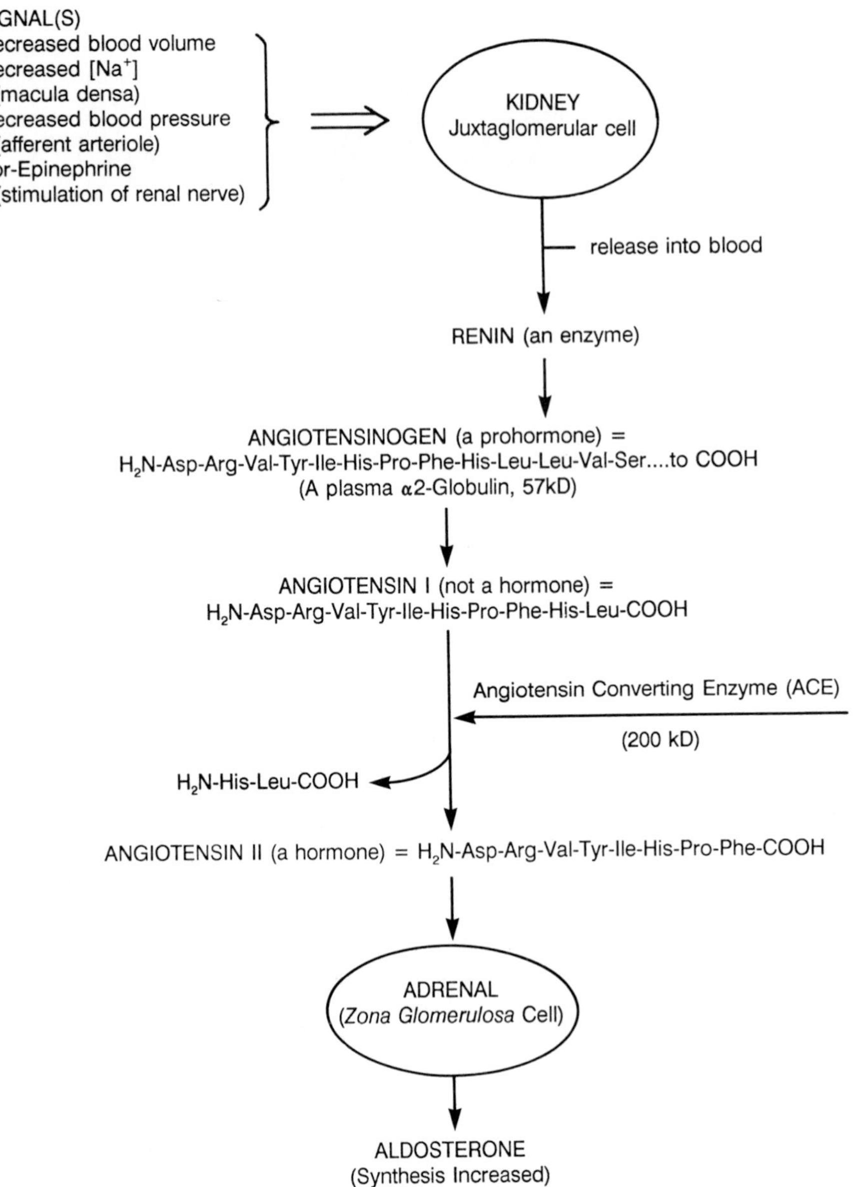

FIGURE 15-7 Schematic description of the role of renin in the production of angiotensin I (not a hormone) and angiotensin II (a hormone).

their actions on ion balance principally in the kidney, but also to some limited extent in the salivary glands, gut, and sweat glands. In the kidney, the actions of the mineralocorticoids result in increased cortical collecting tubule reabsorption of Na^+ with a concomitant secretion of K^+ and H^+ (as ammonium). Only a fraction of the Na^+ filtered by the glomerulus is actually reabsorbed as a consequence of aldosterone action; however, this fraction can effect significant consequences on electrolyte balance.

After aldosterone is secreted from the zona glomerulosa of the adrenals, it enters the circulation and reaches the kidney mucosal cells which concentrate aldosterone. Figure 15-9 summarizes the series of aldosterone-mediated responses which occur in the kidney collecting duct cells. This includes induction by the aldosterone-receptor of at least three proteins; these include (a) a Na^+ ion permease enzyme which increases the number of atoms of Na^+ entering the mucosal cell from the urine, (b) an increase in the amount of citrate synthase in the mitochondrial matrix; and (c) an increase in the levels of the Na^+,K^+ ATPase which is responsible for pumping Na^+ from the cell cytoplasm on the peritubular side of the cell into the peritubular fluid.

Finally, there is some emerging evidence supporting the existence of a membrane receptor for aldosterone that is linked to the Na^+ permease, so as to facilitate

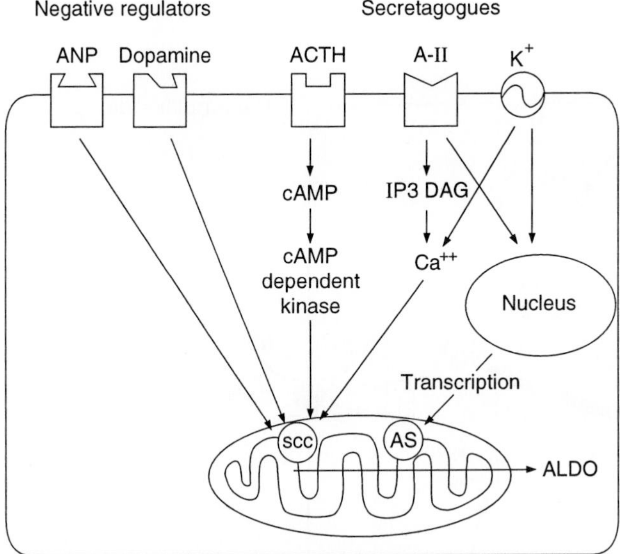

FIGURE 15-8 Receptor model stimulation of aldosterone biosynthesis. Receptor-mediated effects leading to aldosterone production by the adrenal zona glomerulosa cell. Illustrated is the acute stimulation by agonists to increase the first step in aldosterone biosynthesis, which is likely mediated by the StAR cholesterol transport protein (see Chapter 2) that delivers substrate to the P450scc enzyme. Abbreviations: A-II, angiotensin II; ANP, atrial natriuretic protein; AS, enzyme associated with synthesis of aldosterone (see Table 2-6); K^+, K^+-modulated voltage-gated Ca^{2+} channels; SCC, side chain cleavage of cholesterol. Modified from Mortensen, R. M., and Williams, G. H. (1995). Aldosterone action: Physiology. *In* "Endocrinology" (L. J. DeGroot *et al*, eds.) 3rd ed. Vol. 3, pp. 1668–1676. W. B. Saunders, Philadelphia, PA.

Na^+ uptake into the renal tubular cell; this represents a nongenomic action of the steroid.

E. Atrial Natriuretic Protein System

1. Introduction

In the interval of 1970–1985, a new family of peptide–protein hormones was discovered that are produced by the heart and brain; this is known as the atrial natriuretic peptide (ANP) system. Cardiac ANP is secreted by atrial cells following volume expansion or in circumstances of elevated blood pressure. Table 15-7 summarizes the physiological actions of ANP.

TABLE 15-6 Biological Actions of the Angiotensins

Activity	Angiotensin
Stimulation of aldosterone biosynthesis and secretion by the adrenal	II = III >>> I
Elevation of blood pressure via vasoconstriction	II > III
Stimulation of release of catecholamines by adrenal medulla	II
Stimulation of thirst by action on central nervous system	II > III

2. Chemistry, Biosynthesis, and Secretion of ANP

There are three structural forms of the natriuretic peptides (NP), designated as atrial NP (ANP), brain NP (BNP), and C-type NP (CNP); the structural relationships are illustrated in Figure 15-10. Outside of the heart and brain (hypothalamus), some evidence of ANP gene expression has also been found in the lung, aortic arch, kidney, pituitary, and adrenal medulla. BNP was originally isolated from the brain, but now it is known that cardiac atrial cells secrete both ANP and BNP. CNP is believed to be produced only in the brain.

All three natriuretic peptides are characterized by a 17-member amino acid ring formed by a disulfide bridge between two cysteine residues. Each of the three natriuretic peptides is the product of a separate gene transcript.

The genes that code for ANP produce a prepro-ANP of 151 amino acids. Loss of the 25-amino-acid hydrophobic leader sequence generates a 126-amino-acid pro-ANP, which is stored in granules of the atrial myocytes. ANP is generated at the time of secretion by cleavage of the pro-ANP at Arg98-Ser99 to yield the mature peptide of 28 amino acid residues (residues 98–126 of prepro-ANP) and the N-terminal peptide (residues 1–98 of prepro-ANP). The N-terminal fragment is also secreted into the blood, but does not have any known biological activity. As yet the specific peptidases that process pro-ANP into ANP have not been characterized. The amino acid sequence of secreted ANP has been determined in many species and is identical in all except for position 110, where isoleucine replaces methionine in rat, mouse, and rabbit.

The secretion of ANF is stimulated by any of the following: (a) atrial stretch caused by volume expansion; (b) elevated blood pressure; (c) atrial tachycardia; and (d) high-salt diet.

3. Biological Actions of ANP

The several biological effects of ANP (see Table 15-7) include inhibition of angiotensin II stimulation of aldosterone biosynthesis, stimulation of the renal glomerulosa filtration rate (GFR) so as to promote Na^+ natriuresis, and vasorelaxation; these are summarized in Figure 15-13. All of these biological effects are mediated by ANP binding to a specific outer cell membrane receptor. As shown in Figure 15-10 B, three distinct

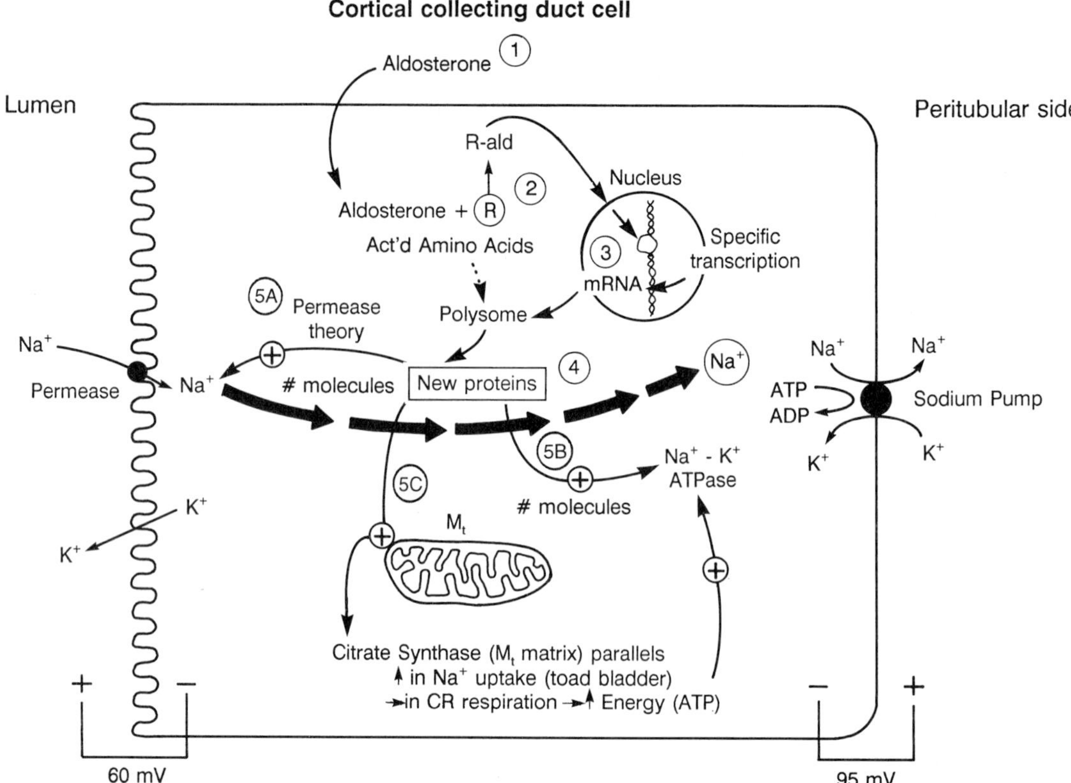

FIGURE 15-9 Cellular model of the actions of aldosterone on Na^+ and K^+ transport processes in the mucosal cells of the collecting duct of the kidney. In the absence of aldosterone, there is only basal update of Na^+ from the lumen by the Na^+ permease and a basal rate of translocation of the Na^+ across the cell so as to exit the peritubular side of the cell via the action of a Na^+/K^+ ATPase. The figure illustrates the five steps which are believed to be operative in the presence of aldosterone. Step ①, aldosterone enters the cell and in ② binds to a high-affinity receptor by a reaction similar to that described for cortisol in the liver cell (see Chapter 10). The complex of aldosterone–receptor presumably is activated and translocated to the nucleus (③) where it stimulates the rate of transcription of certain mRNAs which generate several aldosterone-induced proteins indicated as new proteins (④). These include the α- and β-subunits of the Na^+ permease pump (step 4A), which is believed to be the rate-limiting step in Na^+ reabsorption. The Na^+/K^+ ATPase (step 4B) is responsible for pumping Na^+ from the cell cytoplasm on the peritubular side of the cell into the peritubular fluid which increases the removal of Na^+ from the cell on its way to the blood circulation and would positively influence the reactions of Na^+ leading up to this step by mass action. The citrate lyase (step 4C) is postulated to increase the rate of synthesis of citrate which, after traversing the mitochondrial oxidative pathway, would give rise to correspondingly more ATP which could be available as a substrate for the Na^+/K^+ ATPase. Modified from O'Neill, R. G. (1990). Aldosterone regulation of Na^+ and K^+ transport in the cortical collecting duct. *Semin. Nephrol.* **10,** 371.

ANP receptors have been identified through recombinant DNA techniques.

The atrial natriuretic peptide receptor (ANPR) A and B forms have both extracellular and intracellular domains coupled by a linear transmembrane domain; the intracellular domain has a region that is structurally homologous to a protein kinase, which functions as a regulatory element, and a second region that is a guanylylcyclase catalytic domain. The third receptor, ANPR-C, has a truncated cytoplasmic domain and, thus, is devoid of both kinase and guanylylcyclase actions.

The ANPR-A shows its highest concentration in the kidney glomerulus, adrenal zona glomerulosa, brain, and cardiac aorta, which is in accord with the sites of physiological actions of ANP. The ANPR-B is found almost exclusively in tissues of neural origin, in accordance with the view that CNP acts mostly within the central nervous system. The function of the ANPR-C is not clear. It is widely distributed in the vascular system, but since the ANPR-C has no intracellular domain, it is likely that it is biologically inert.

The most striking renal effect of ANF is the prompt and sustained increase in the glomerular filtration rate.

TABLE 15-7 **Physiological Actions of Atrial Natriuretic Peptide**

Site of action	Effect produced
Kidney	(a) Increased glomerular filtration rate (GFR) leading to increased Na^+ excretion and correlated reduction in plasma volume.
	(b) Suppression of renin secretion
Smooth muscle of arteries, capillaries, and veins	Muscle relaxation with reduction in blood pressure
Adrenal zona glomerulosa cell	Antagonism of angiotensin II stimulation of aldosterone biosynthesis
Brain	Complex set of neuroendocrine and neurotransmitter effects

This effect frequently occurs in the presence of decreased arterial pressure and, thus, is likely mediated via selective constriction of the efferent arterioles. This leads to an increased urine volume and natriuresis. ANF also suppresses renal renin secretion, which will ultimately reduce the delivery of aldosterone to the renal tubule. This action of ANF then effectively reduces the extent of aldosterone-mediated renal tubular Na^+ reabsorption, thus supporting the natriuresis process. ANF also blocks adrenal aldosterone secretion (see Figure 15-9), as well as the vasoconstrictive actions of angiotensin II.

The second major biological action of ANF is its powerful relaxation of preconstricted renal vasculature and large arteries and other vascular beds. This action has been shown to result from ANF interation with the ANPR-A, which results in an increase in cyclic guanosine monophosphate (cGMP) in the vascular smooth muscle. Infusion of ANF into rats or dogs results in an immediate drop in blood pressure, which persists for the duration of the infusion. This response has also been attributed to an induced fall in cardiac output or venous return, but the mechanism of this reponse is not known.

F. Water Balance

Osmoreceptors (Figure 4-10A) located in the hypothalamus (and also in the carotid artery) are capable of sensing increased concentrations of solutes, particularly Na^+. This leads to a secretion of vasopressin, which stimulates renal tubular water reabsorption. At the same time, the thirst center in the hypothalamus, which is linked to the osmoreceptor, is stimulated with a concomitant increase in water consumption.

Dilution of body fluids results from activation of the osmoreceptor and the thirst center and inactivation of the aldosterone-producing mechanism. The response of the system is a fall in the concentration of Na^+, so that the original signals from the osmoreceptors and the thirst centers are quenched.

Evidence supports the existence of a second system in the neurohypophysis, involving the formation of angiotensin II, a hormone that in that location could elevate cyclic AMP levels and cause additional release of vasopressin. In addition, there is new evidence of a polypeptide hormone called neurotensin found in the hypothalamus, which acts as a hypotensive agent (its action would be similar to the effect of lowering the circulating concentration of Na^+).

1. Renal Relationships of K^+ and Na^+

Although Na^+ is the principal extracellular cation and K^+ is the principal intracellular cation, the metabolism and homeostasis of these monovalent cations are closely interrelated. It is essential to life that serum K^+ levels be maintained within normal limits (3.8–5.4 meq/liter), so that the normal K^+ concentration of 150–160 meq/liter of intracellular water can be supported.

The homeostatic control of K^+ is not regulated as stringently as that of Na^+. An important endocrine contribution to K^+ homeostasis is the major stimulatory effects of K^+ on aldosterone secretion. Aldosterone in turn acts on the renal tubules to restore plasma K^+ to normal by enhancing the renal tubular excretion of K^+. Also, the falling levels of plasma K^+ are known to stimulate renin secretion, which in turn will ultimately lead to the normal action of aldosterone upon increasing Na^+ reabsorption. It is to be emphasized that there is no stoichiometric relationship between Na^+ reabsorption and K^+ excretion; thus, K^+ excretion may increase prior to an effect on Na^+, and normally the aldosterone effect on K^+ diminishes before its action on Na^+ terminates. Finally, K^+ can stimulate the secretion of insulin. This in turn leads to the promotion by insulin of glucose and K^+ entry into cells, which also has the consequence of lowering blood K^+.

G. Endothelins

1. Introduction

The vascular endothelium, which provides the barrier between the blood and the vascular wall (see Figure 15-6), is the site of production of three hormones important to the regulation of blood pressure; these are prostacyclin, nitric oxide, and the endothelins. Endothelin (ET) is a 21-amino-acid peptide that is a highly potent vasoconstrictor.

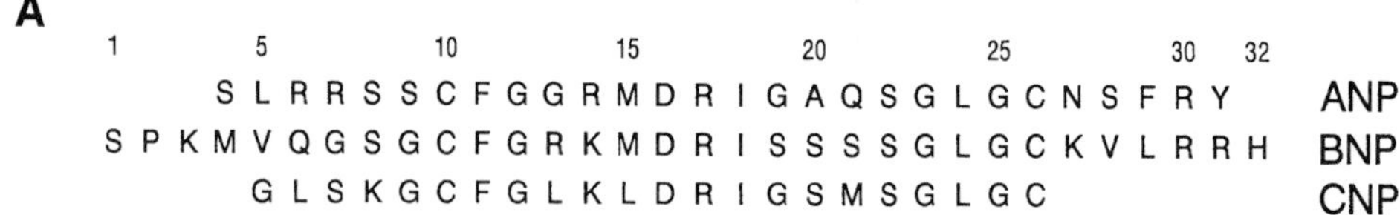

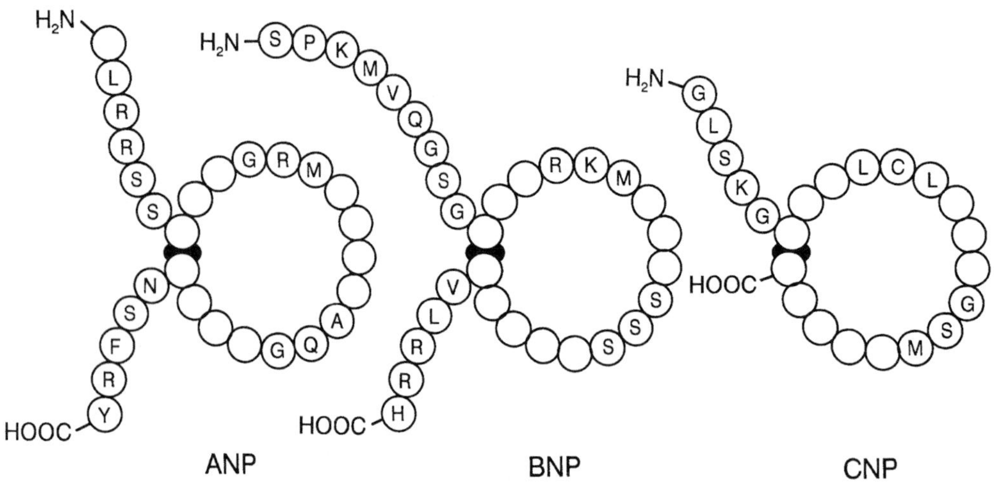

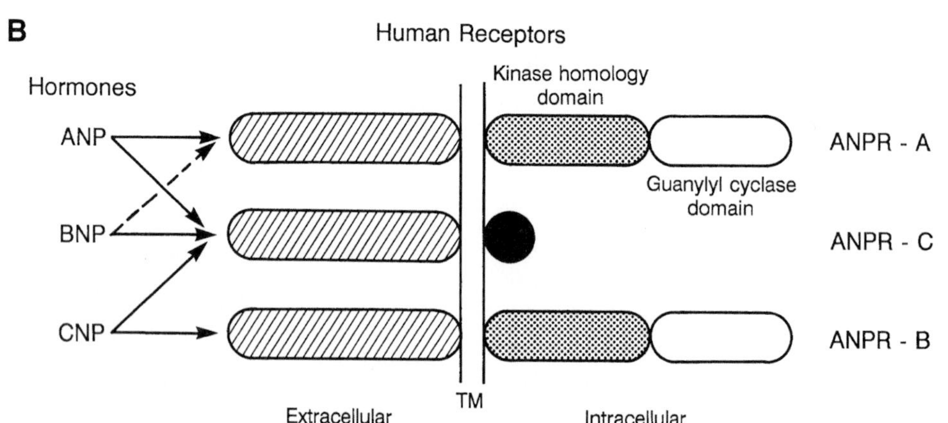

FIGURE 15-10 Structural organization of atrial natriuretic peptides and their receptors. (A) Structural homology of the A, B, and C types of human ANP. In the 17-membered ring, amino acids that are identical are shown as open circles. Abbreviations: ANP, atrial natriuretic peptide; BNP, brain natriuretic peptide; CNP, C-type natriuretic peptide. Modified with permission from Sudoh *et al.* (1990). (B) Schematic representation of receptor classes for human ANP. Note: The terminology for designation of the receptors ANPR-A, ANPR-B, and ANPR-C does not coincide with the terminology for their preferred ligands ANP, BNP, and CNP. The solid lines connect the respective natriuretic peptide with its preferred receptor(s). TM: transmembrane region of the receptor. Modified with permission from Koller, J. K., Lowe, D. G., Bennett G. L. *et al.* (1991). Selective activation of the B-natriuretic peptide receptor by c-type natriuretic peptide (CNP). *Science* **252**. 120–123.

2. *Chemistry, Biosynthesis, and Secretion*

Three isoforms of endothelin have been identified; they are termed endothelins 1, 2, and 3 (ET-1, ET-2, and ET-3). The structures of the endothelins and their corresponding proendothelins are presented in Figure 15-12. Three separate ET genes have been identified; each codes for a separate and distinct endothelin. After transcription of the ET-1 gene, the mature mRNA codes for a prepro-ET-1 of 203 amino acids. Removal of the amino-terminal 20-amino-acid leader sequence generates a 183-amino acid protein that is an intermediate

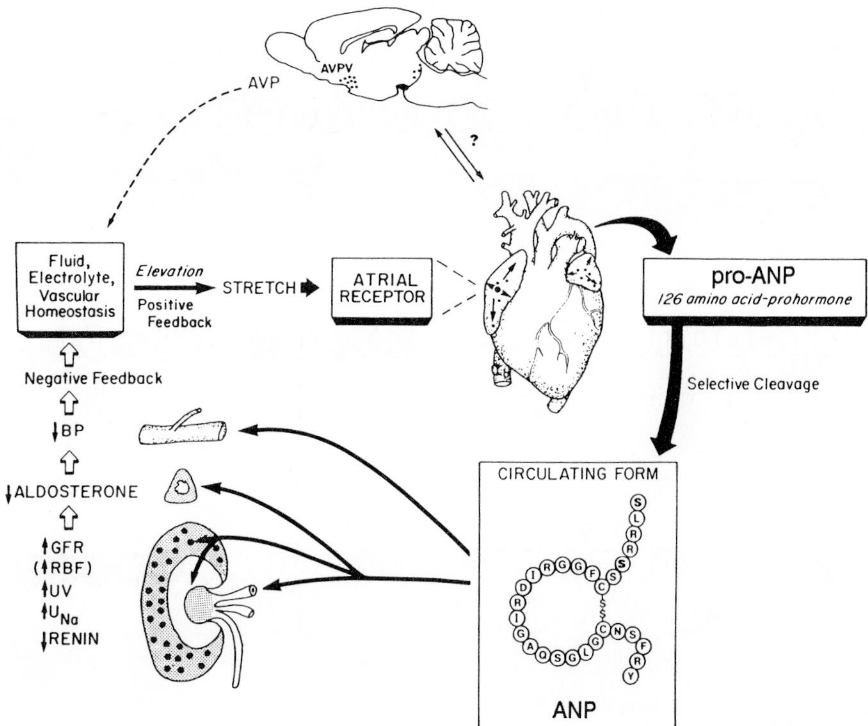

FIGURE 15-11 Schematic diagram of the ANP hormonal system. An elevated vascular volume results in cleavage and release of atriopeptin, which acts on the kidney (glomeruli and papilla) to increase the glomerular filtration rate (GFR) so as to increase renal blood flow (RBF), increase urine volume (UV) and Na^+ excretion (U_{Na}), and decrease plasma renin activity. Natriuresis and diuresis are also enhanced by the suppression of aldosterone production and its actions and by the release from the posterior pituitary of arginine vasopressin (AVP). Diminution of vascular volume provides a negative feedback signal that suppresses circulating levels of atriopeptin. Modified from Needleman, P., and Greenwald, J. E. (1986). Atriopeptin: A cardiac hormone intimately involved in fluid, electrolyte, and blood pressure homeostasis. *New Engl. J. Med.* **314,** 828–834.

form of ET-1. Processing of this "intermediate" ET-1 by a pair of dibasic specific endopeptidases (at Lys51-Arg52 and Arg91-Arg92) generates the 38-amino-acid polypeptide pro-ET-1, which is sometimes referred to as "big" ET-1. Pro-ET-1 represents the secreted form of endothelin; further processing occurs outside the endothelin cell by the action of a putative endothelin-converting enzyme (ECE). The sites of ECE cleavage for the proendothelins are shown in Figure 15-12. Pro-ET-1 has only 1% of the biological activity of mature ET-1.

3. *Biological Actions of Endothelins*

In endothelial cells ET-1 is predominant, while ET-2 and ET-3 are virtually undetectable. The mRNAs for all three endothelins are present in human kidney and jejunum, while in the nervous system the mRNA for ET-1 is the major species. The variable but diverse distribution of the ET isoforms suggests that ET has multiple functions both inside and outside the vascular system.

Radioligand-binding studies with the isoforms of ET have indicated the presence of two classes of receptors. Type I receptors are involved with vasoconstriction, bronchoconstriction, and stimulation of aldosterone biosynthesis. Type II receptors are linked to the inhibition of platelet aggregation and vasorelaxation. Through molecular cloning of ET receptors, the deduced amino acid sequence of seven ET receptors has been elucidated; they can be divided into two classes designated ET_A and ET_B (see Table 15-8). Both classes of endothelin receptors belong to the seven-transmembrane, G-protein-coupled superfamily (see Figure 1-23).

The current view is that ET_A endothelin receptors mediate the paracrine (vasoconstrictor) actions of ET-1, while the ET_B receptors mediate the more "nonselective" actions of endothelins, including autocrine actions related to endothelin-relaxing factor (EDRF) and vasodilation (see Figure 15-13).

Occupancy of the endothelin receptors is coupled to the activation of phospholipase C via a pertussis

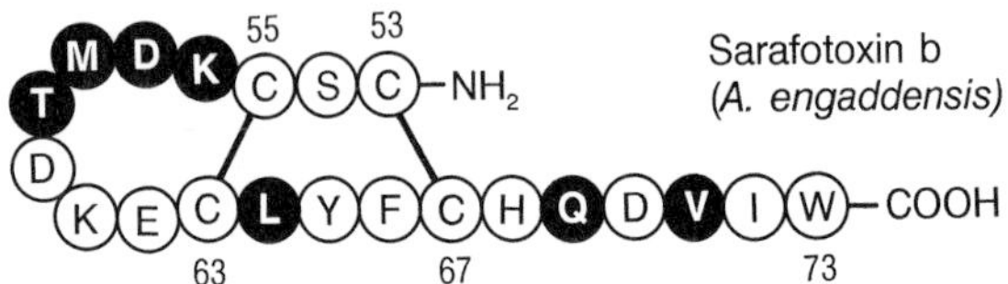

FIGURE 15-12 Endothelins. Amino sequence of human endothelin isoforms and their respective proendothelins. The filled circles for ET-2 and ET-3 represent amino acid differences compared to ET-1 and its pro form. The arrow indicates the site of proteolytic cleavage by the endothelin-converting enzyme (ECE), which generates the mature, biologically active endothelins. The two solid lines indicate disulfide linkages. Also shown is the structure of the structurally homologous sarafotoxin b, which is present in the venom of the Egyptian asp *Atractaspesis engaddensis*. Adapted with permission from Rubanyi, G. M., and Bothelo, L. H. (1991). Endothelins. *FASEB* J. **5,** 2713–2720.

toxin-insensitive G protein in all tissues expressing the ET_A or ET_B forms. This results in the production of IP_3 and the mobilization of intracellular free Ca^{2+}, leading to a transient increase in $[Ca^{2+}]_i$. Associated with these changes are the elevation of diaclyglycerol (DAG), activation of protein kinase C (PKC), and opening of a Ca^{2+} channel. Then, as diagrammed in Figure 15-14, depending upon the target cell, a chloride (outward movement), K^+ (outward movement), or Na^+ (outward movement) channel is opened. Thus, a wide variety of the pleiotropic biological effects of endothelins can be explained by these many signal transduction events. However, even though endothelins are highly potent vasoconstrictors, much additional detail is required to achieve an in-depth understanding of the contributions of these hormones to their minute-by-minute contribution to the regulation of blood pressure.

TABLE 15-8 Properties of Endothelin Receptors

Receptor class	Affinity	Biological response
ET_A	ET-1 = ET-2 > ET-3	Vasoconstriction Bronchoconstriction Stimulation of aldosterone secretion
ET_B	ET-1 = ET-2 = ET-3	Vasodilation *Ex vivo* platelet aggregation

There are emerging reports of a separate sphere of actions of endothelins associated with the reproductive system. Plasma levels of ET-1 increase during pregnancy, and estradiol increases the ET receptor present in the endometrium. ET-1 is also present in ovarian follicular fluid. In addition, in the hypothalamus–hypophyseal system, ET is a local modulator of the secretion of gonadotropins, prolactin, growth hormone, and TSH. The basis for these diverse actions of ET is not yet known.

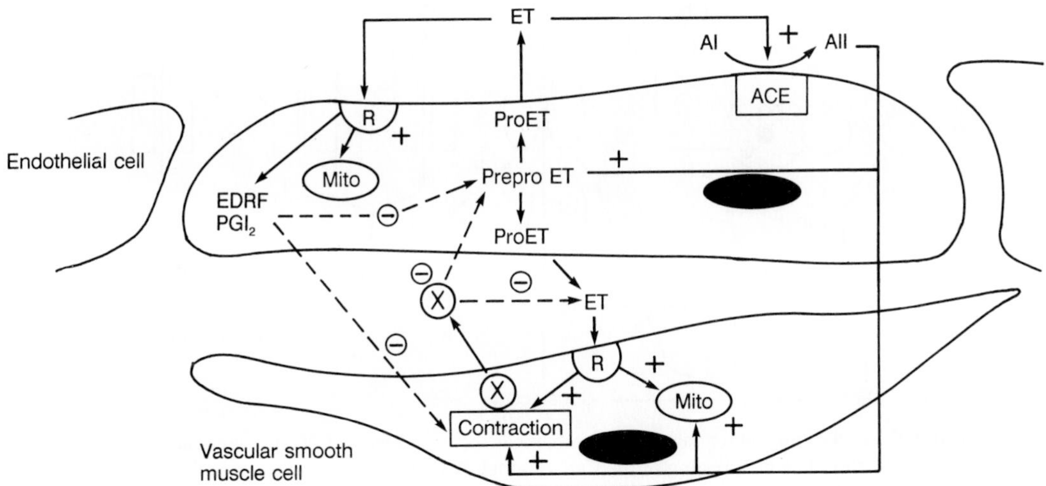

FIGURE 15-13 Paracrine and autocrine actions of endothelins in the vascular wall. ET, formed from big ET outside the endothelial cell, can act in an autocrine fashion by stimulating the production of endothelium-derived relaxing factors (EDRF) and PGI_2, activating the angiotensin-converting enzyme (ACE), and promoting gene expression and mitogenesis (Mito). In a paracrine fashion, ET binds to the underlying vascular smooth muscle (VSM) cell, triggers vasoconstriction, and may facilitate mitogenesis. The release of prostacyclin (PGI_2) and EDRF by ET from endothelial cells can mediate vasodilation. Facilitated production of angiotensin II (AII), on the other hand, acts synergistically with ET. The biosynthesis of ET is inhibited by EDRF and by unknown factors (X) released from VSM cells. Proteolytic enzymes at or near the VSM cell surface can degrade ET-1, reducing its bioactivity. Continuous lines and arrows with plus (+) signs indicate facilitation; interrupted lines with minus (−) signs indicate inhibition. Modified from Rubanyi, G. M., and Parker-Botelho, L. H. (1991). Endothelins. *FASEB J.* **5**, 2713–2720.

H. Kallikreins and Kinins

1. Introduction

The kallikreins are a group of serine proteases known as kininogens, which act on plasma α_2-globulins to release kinins such as bradykinin. Bradykinin is the most potent vasodilator substance known. The kallikrein–kinin system is linked in the endothelial cell to stimulate the production of another vasodilator, the prostacyclins.

2. Biochemistry and Physiology

Two classes of kallikreins have been identified: (a) those present in plasma and (b) those present in organs, including the kidney, salivary glands, and pancreas. The plasma kallikreins have molecular weights of 110,000 and 70,000. All of the kallikreins are serine proteases.

The plasma kallikrein is normally found as a proenzyme termed prekallikrein. Prekallikrein is activated to kallikrein by one of the blood-clotting factors, factor XII, or Hageman factor. In turn, Hageman factor is activated by the plasma kallikrein (see Figure 15-15). The kallikreins utilize their protease activity to release peptide kinins from their precursor substrate forms.

The principal kinin is a nonapeptide bradykinin. The bradykinin receptor, which is principally associated with endothelial cells, has been cloned and found to belong to the seven-transmembrane G-protein-coupled receptor superfamily. The potent vasodilation properties of bradykinin are largely due to its stimulation in the endothelium of prostaglandins and nitric oxide (see later discussion).

It is noteworthy that the principal mode of inactivation of bradykinin is the removal from the COOH-terminal of a dipeptide by an enzyme termed kininase II. The enzymatic activities of kininase II and the enzymatic activities of angiotensin I converting enzyme (ACE) have shown them to be the same protein. Thus, ACE, which generates the potent vasoconstrictor, angiotensin II, inactivates the vasodilator bradykinin.

I. Prostaglandins

1. Renal Prostaglandins

Prostaglandins and other eicosanoids acting in a paracrine and/or autocrine fashion play extensive roles in the regulation of blood pressure in both the kidney and the circulatory system. These topics will be discussed separately.

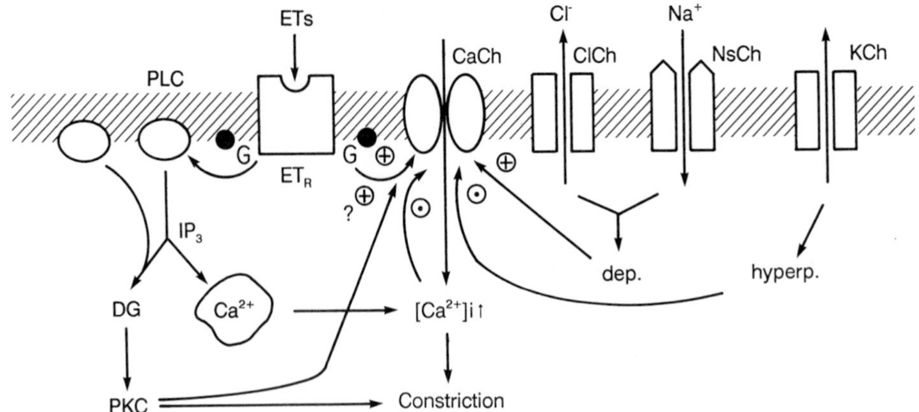

FIGURE 15-14 Intracellular signaling mechanisms activated by endothelins. Activation of PKC is known to stimulate the opening of the Ca^{2+} channel. Activation of phospholipase C is central to the generation of cellular responses to ET. An increase in the concentration of intracellular free Ca^{2+} is indispensable to eliciting responses to the endothelins. ET also stimulates a plasma membrane Ca^{2+} channel and induces an increase in the influx of extracellular Ca^{2+}. Since endothelins have been shown to open a nonselective cation channel or chloride channel and to induce depolarization (dep), it is likely that the Ca^{2+} channel may also be opened by depolarization. Endothelins also open a K^+ channel and induce hyperpolarization (hyperp). This inhibits the Ca^{2+} channel. An increase in the concentration of $[Ca^{2+}]_i$ inhibits Ca^{2+} influx. Activation of PKC is known to stimulate the opening of the Ca^{2+} channel. Abbreviations: ET_R, ET receptor; G, G protein; PLC, phospholipase C; PLD, phospholipase D; IP_3, inositol triphosphate; DG, diaclyglycerol; PKC, protein kinase C; $[Ca^{2+}]_i$, concentration of intracellular free Ca^{2+}; CaCh, calcium channel; ClCh, chloride channel; NsCh, nonselective cation channel; KCh, K^+ channel. Inositol trisphosphate stimulates the intracellular calcium pool to release Ca^{2+}. Modifed with permission from Masaki, T. (1993). Endothelins: Homeostatic and compensatory actions in the circulatory and endocrine systems. *Endocri. Rev.* **14**, 256–268.

The complex series of steps that ultimately lead to the synthesis of prostaglandins in the kidney are called into play after the renin–angiotensin system has been in operation. Certain signals that set the system in motion can occur via the autonomic nervous system; examples of this kind of signal would be intellectual

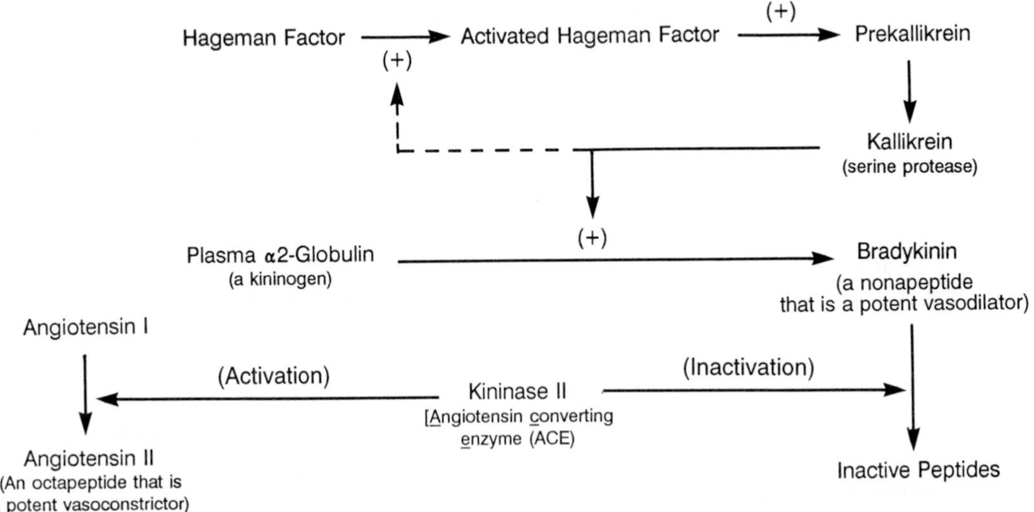

FIGURE 15-15 Schematic diagram of the Kallikrein–kinin pathway. Hageman's factor (factor XII) also participates in the blood-clotting sequence. Kallikrein is a plasma serine protease that both activates Hageman factor and releases bradykinin from α_2-globulin. Bradykinin, a vasodilator, is inactivated by kininase II, also known as angiotensin-converting enzyme (ACE) (see Figure 15-7), which also mediates the production of the vasoconstrictor, angiotensin II.

activity or fright, either of which is a hypertensive stimulus. As a consequence, increased adrenergic activity results (norepinephrine release), and stimulation of the secretion of renin by the juxtaglomerular apparatus of the kidney follows. This leads to an elevation in blood pressure through the angiotensin system and aldosterone release. Ultimately, the increase in blood pressure follows from Na^+ reabsorption and the subsequent increase in fluid volume of the circulatory system. The increased blood pressure is sensed by the kidney medulla, resulting in increased blood flow. The increased blood flow through the kidney medulla sets the events of Figure 15-16 into play, which results in the release of PGA_2 (PGE_2 can be released also) by the interstitial cells. PGA_2 is circulated to the kidney cortex where it antagonizes the hypertension produced by the reabsorption of Na^+.

2. *Endothelial Prostaglandins*

A variety of prostaglandins are also produced in the endothelial cells of the circulatory system. PGI_2 is a potent vasodilator, while thromboxane A_2 has potent vasoconstrictor actions. PGI_2 release from endothelial cells is stimulated by endothelin 1; PGI_2 then acts in a paracrine fashion in collaboration with nitric oxide (see next section) to effect vasorelaxation.

J. Nitric Oxide System

1. *Introduction*

A relatively surprising addition to the family of chemical messengers is nitric oxide (NO). NO is a free radical gas of limited solubility in water. Since NO is noncharged, it can rapidly diffuse across cell membranes and into cells; it has been shown to act as both an intracellular and an intercellular (paracrine) messenger to elicit a wide spectrum of biological responses. A physiological function for NO was first established in the vascular system when the endothelin-derived relaxing factor (EDRF) could be quantitatively explained by the formation of NO. NO is now known to be an integral participant in the signal transduction processes associated with the vascular, immune, and neural systems.

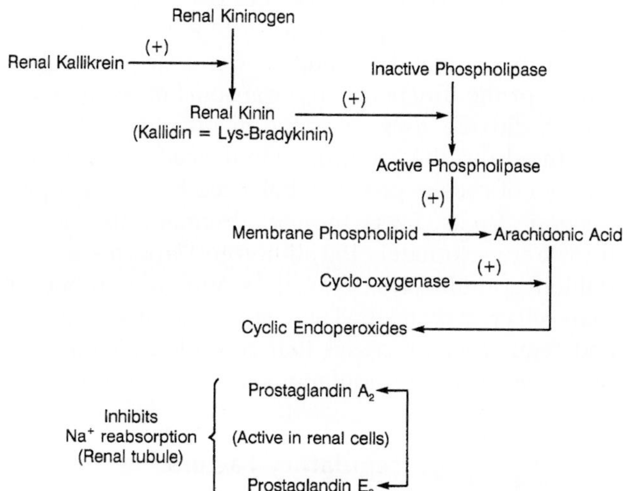

FIGURE 15-16 Pathway for the generation of renal prostaglandins. See also Figure 15-16 for additional details of the kallikrein–kinin system and Chapter 16 for a detailed discussion of prostaglandin synthesis.

2. *Chemistry, Biosynthesis, and Secretion*

The formation of NO is an enzyme-mediated reaction (see Figure 15-17); the nitrogen donor is the amino acid L-arginine and the oxygen donor is molecular oxygen. The reaction is catalyzed by an NADPH requiring nitric oxide synthase (NOS). NOS enzymes are structurally related to cytochrome P450 reductase (see Chapter 2) and range in size from 130 to 160 kDa.

NOS exists both as a constitutive enzyme, which is regulated by Ca^{2+} and the calcium-binding protein, calmodulin, and as an inducible enzyme, which is not regulated by calmodulin. The regulatable forms of NOS are induced by interferon-α (INF-α), tumor necrosis factor-α (TNF-α), and interleukin-1β (IL-1β) and inhibited by glucocorticoids. The inducible forms of NOS are associated with components of the host defense immune system.

3. *Biological Actions of NO*

The wide spectrum of biological actions of NO in the cardiovascular, nervous, and host defense systems is summarized in Table 15-9. Only the actions of NO in the cardiovascular system will be discussed in this chapter.

After the generation of NO in the endothelial cell by the NOS enzyme, the NO diffuses to an adjacent smooth muscle where it acts as an agonist to initiate biological responses (see Figure 15-18). Virtually all of the known biological actions of NO on the cardiovascular system are mediated by the activation of a soluble guanylate cyclase. The guanylate cyclase has heme as a prosthetic group, and NO binds tightly to this moiety. The resulting activation of the guanylate cyclase results in the production of cGMP, which is then postulated to have actions on protein kinases, nucleotide-sensitive phosphodiesterases, ion channels, or other unknown cellular proteins that are linked to the generation of the smooth muscle response of vasodilation and also to the inhibition of platelet adhesion and aggregation.

NO is a short-lived agonist in the cellular environment. NO is inactivated either by its chemical linkage

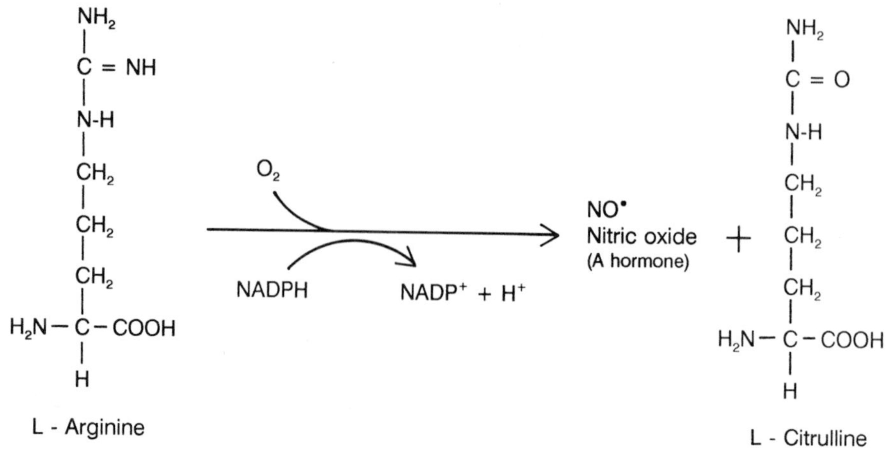

L - Arginine + O_2 + NADPH → $NO^\bullet$ Nitric oxide (A hormone) + L - Citrulline + $NADP^+ + H^+$

FIGURE 15-17 Enzymatic reaction catalyzed by nitric oxide synthase (NOS).

to proteins (nitrosylation) or by oxidation to nitrite (NO_2^-) and then nitrate (NO_3^-).

K. Summary

In the normal physiological state in humans, the distribution, composition, and volume of body fluids are held within relatively narrow limits, despite wide variations in the intake of water and Na^+. Such homeostasis or stability of the internal environment requires a multifactoral collaboration of the hormones and other physiological processes that affect electrolyte and water metabolism. Thus, as has been reviewed in the preceding sections of this chapter, a diverse array of hormones (aldosterone, angiotensin II, renin, ANP, NO, endothelin, vasopressin, prostaglandins, and kinins) and signal transduction systems, each responding to different stimuli, is integrated to provide the kidney and cardiovascular system with coherent messages to effect the collective regulation of blood pressure, electrolyte concentration, and water volume. These relationships are summarized in Figure 15-19.

TABLE 15-9 Biological Actions of Nitric Oxide

System	Response
Cardiovascular	
Smooth muscle	Initiate vasorelaxation; control of regional blood flow and blood pressure
Platelets	Limitation of aggregation and adhesion
Nervous	
Peripheral	Neurotransmission (penile erection, gastric emptying)
Central	Neurotransmission; long-term potentiation (appetite control, nociception)
Host defense	
Macrophages, Leukocytes, Monocytes	Defense against bacteria, fungi, protozoans, parasites, and viruses

IV. Hormones and Blood Cell Production

A. Introduction

The various cellular elements of blood are produced by the hematopoietic system (see Table 15-10). The main components of the hematopoietic system are bone marrow and blood; in addition, the liver and spleen are important accessory organs. The spleen is a blood reservoir, and it can also degrade time-expired red blood cells. The liver stores vitamin B_{12}, which is essential for erythrocyte production as well as the hemoglobin biosynthesis.

The term "erythron" describes the circulating cells and their precursors present in the blood compartment. The erythron may be thought of as a disperse organ whose prime function is the transport of oxygen and carbon dioxide, as well as the maintenance of blood pH. In adults, the erythron is in a steady state where the loss of cells is precisely balanced by the new production of cells. Every minute a human must biosynthesize approximately 150 million erythrocytes and 120 million granulocytes, as well as numerous mononuclear cells and platelets. There are numerous hormones and regulatory processes that control crucial aspects of the various cell populations of the erythron.

B. Regulatory Factors

The regulatory factors for hematopoiesis (see Table 15-11) comprise classical hormones, growth factors, and cytokines, which control and direct the division and maturation of the hematopoietic process down

A

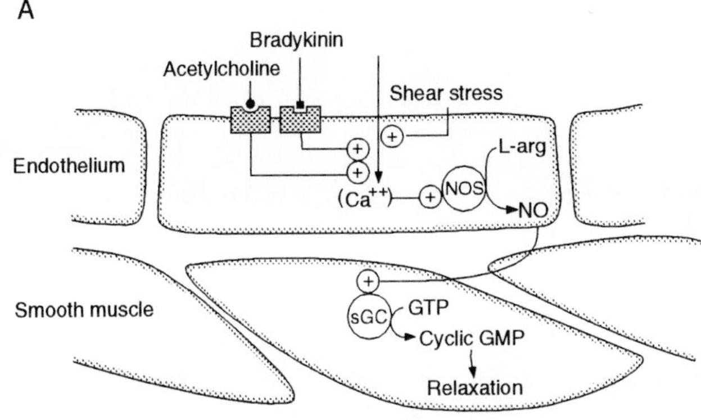

B

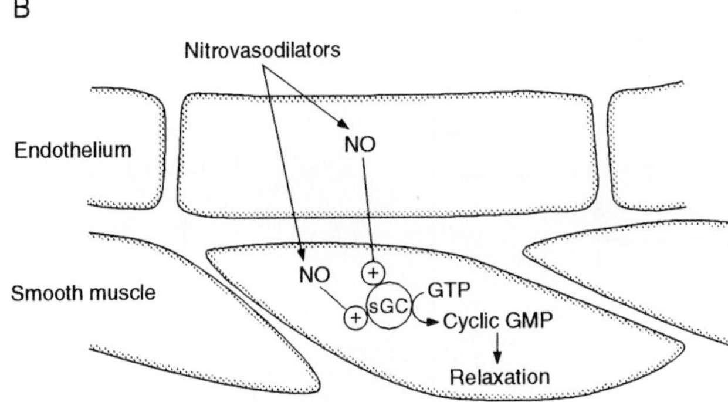

FIGURE 15-18 Model for the synthesis of NO in endothelial cells and its paracrine actions in smooth muscle cells. In (A), shear stress or receptor activation of vascular endothelium by bradykinin or acetylcholine results in an influx of calcium. The consequent increase in intracellular calcium stimulates the constitutive nitric oxide synthase (NOS). The nitric oxide (NO) formed from L-arginine (L-Arg) by this enzyme diffuses to nearby smooth muscle cells, in which it stimulates the soluble guanylate cyclase (sGC), resulting in enhanced synthesis of cyclic GMP from guanosine triphosphate (GTP). This increase in cyclic GMP in the smooth muscle cells leads to their relaxation. In (B), nitrovasodilators such as sodium nitroprusside and nitroglycerin release nitric oxide spontaneously or through an enzymatic reaction. The liberated nitric oxide stimulates the soluble guanylate cyclase in the vascular smooth muscle cell, resulting in relaxation. Adapted from Moncada, C., and Higgs, A. (1993). The L-arginine-nitric oxide pathway. *New Engl. J. Med.* **329**, 2002–2012.

eight possible paths of cell development (see Figure 15-20).

1. *Cytokines*

Cytokines are peptide hormones produced mainly by macrophages and lymphocytes, but also by leucocytes and fibroblasts present in the erythron. Most cytokines act in a paracrine or autocrine fashion. They function in the complex cell differentiation process of hematopoiesis (see Table 15-11) and are also involved in the processes of inflammation and immune reactions; the latter topic is not discussed in this chapter.

The regulatory actions of each of the cytokines upon hematopoiesis are mediated by unique receptors present in the cell membranes of the responding cells. Typically these are transmembrane glycoproteins; collectively they belong to a superfamily of cytokine growth factor receptors that also includes the receptors for growth hormone and prolactin (see Figure 15-21).

C. Cell Differentiation

Table 15-11 presents a menu of regulatory factors that are involved in the process of hematopoiesis. A simplified scheme of hematopoietic cell differentiation is presented in Figure 15-21, including an indication of the site of action of the hormones and cytokines. The process of hematopoiesis has its origin in a single small population of self-renewing pluripotent stem cells that can be stimulated to differentiate along one of four general pathways; these include (a) the production of the antibody-generating cells the B and T lymphocytes, which are stimulated by interleukins 1–7; (b) the production of the white blood cell line, including monocytes, granulocytes, and macrophages whose collective production is largely stimulated by the family of colony-stimulating factors; (c) the production of platelets, which is promoted by several cytokines and hormones including erythropoietin; and (d) the production of the red blood cells, which is largely stimulated by erythropoietin.

The stem cells appear to require exposure to multiple cytokines to initiate the cell-differentiative process to create committed stem cells. It should be appreciated that although Figure 15-22 and Table 15-11 imply a linear cell differentiation process along precise pathways with precise locations for effects of the cytokines and growth factors, the *in vivo* reality is believed to be considerably more complex; e.g., there are no examples where one cytokine or regulatory action is confined to cells of a single hematopoietic lineage.

A detailed consideration of the complexities of the hematopoietic process is beyond the scope of this presentation; however, a brief discussion of the various classes of growth factors and hormones is presented in the following.

D. Hematopoietic Regulating Factors

As detailed in Table 15-11, 21 hematopoietic regulatory factors have been identified, characterized, cloned, and expressed in a biologically active form. The biological effects of these hematopoietic regulators are mediated by their interaction with their specific cognate receptor present in the outer cell membrane of their responding cells. These receptors belong to the cytokine superfamily of transmembrane receptors (see Figure 15-21).

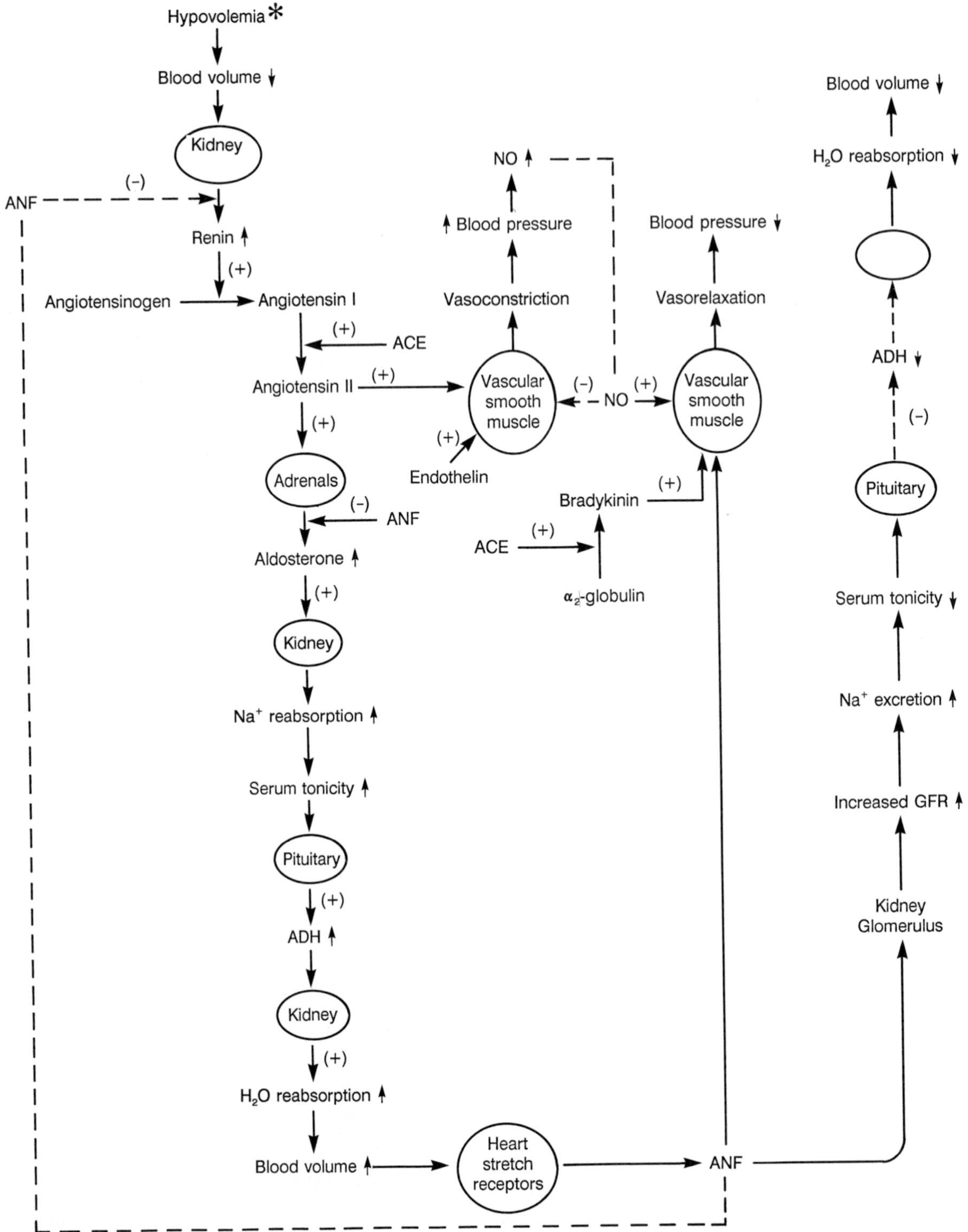

FIGURE 15-19 Simplified diagram of the integrated actions of hormones in the kidney and cardiovascular system to achieve homeostasis of blood pressure, blood volume (H_20), and electrolyte (Na^+) concentration. The reader should evaluate the figure starting at the * symbol.

TABLE 15-10 Cellular Elements of Blood

Cell type	Normal range	Function
Red blood cells (RBC) (erythrocytes)	Male: 5.6×10^6 Female: 4.8×10^6	Oxygen transport
Platelets (thrombocytes)	$(130–370) \times 10^3$	Essential for clotting
White blood cells (WBC)	$(5–10) \times 10^3$	
Nongranular leukocytes		
Lymphocytes	25–30% of WBC	
B cells		Antibody production
T cells		Participate in cell-mediated immune responses
Monocytes	6% of WBC	Precursors of macrophages
Macrophages		Phagocytosis
Granular leukocytes		
Neutrophils	60% of WBC	Phagocytosis
Eosinophils	1–3% of WBC	Participate in allergic responses
Basophils	1% of WBC	Prevention of inappropriate clotting

TABLE 15-11 Hematopoietic Hormones and Cytokines[a]

Regulator	Abbreviation	Responding hematopoietic cells	Responses
Erythropoietin	Epo		
Granulocyte macrophage stimulatory factor	GM-CSF	Committed progenitor cells	See text and Figure 5-21
Granulocyte colony stimulating factor	G-CSF		
Macrophage colony stimulatory factor	M-CSF		
Interleukin-1	IL-1	PSC(*b*)	*b*
			c
			d
			Promotes proliferation of B cells
			Promotes proliferation of B cells
			e
Interleukin-2	IL-2		
Interleukin-3	IL-3 (multi-CSF)	PSC	
Interleukin-4	IL-4	B, T, G, MC	
Interleukin-5	IL-5	E, B	
Interleukin-6	IL-6	B, G, PSC, Meg	
Interleukin-7	IL-7	B, T	Promotes proliferation of B-cell progenitors
Interleukin-8	IL-8	M	Activates macrophages
Interleukin-9	IL-9	T, Meg, MC	Enhances most cell growth
Interleukin-10	IL-10	T	Inhibits clonal expansion of T cells
Interleukin-11	IL-11	Meg, B	
Interleukin-12	IL-12	NK	
Interleukin-13	Meg CSF		
Stem cell factor	SCF		
Leukemia inhibitory factor	LIF		
Tumor necrosis factor (α and β)	TNFα, TNFβ		
Interferon-γ	IFNγ	M	

[a] The information in this table was adapted from two sources: Metcalf, D., Nicola, N., and Gough, N. M. (1995). Hormone and blood cell production. *In* "Endocrinology" (L. J. DeGroot *et al.*, eds.), Vol. 3, pp. 2943–2963. W.B. Saunders, Philadelphia, PA; and Rang, H. P., Dale, M. M., Ritter, J. M., and Gardner, P. (1995). "Pharmacology," p. 242. Churchill-Livingston, New York. Abbreviations employed are as follows: PSC, pluripotent stem cells; G, granulocyte; M, macrophage; E, eosinophils; Ey, erythroid cells; Meg, megakarocytes; MC, mast cells; T, T lymphocytes; B, B lymphocytes; NK, natural killer cells.

[b] IL-1 is produced as a consequence of infection or antigenic challenge. It produces a wide array of effects in the immune responses and induces other cytokines.

[c] IL-2 stimulates proliferation and maturation of $T_{1/2}$ cells and production by $T_{1/2}$ cells of cell cytokines that control the differentiation and proliferation of B cells.

[d] IL-3 acts on the pluripotent hematopoietic stem cell to promote the development of G, M, F, Ey, Meg, and MC.

[e] IL-6 activates stromal bone marrow cells to produce colony-stimulating factors and acts as a pyrogen (i.e., raises body temperature; promotes proliferation of B and T cells).

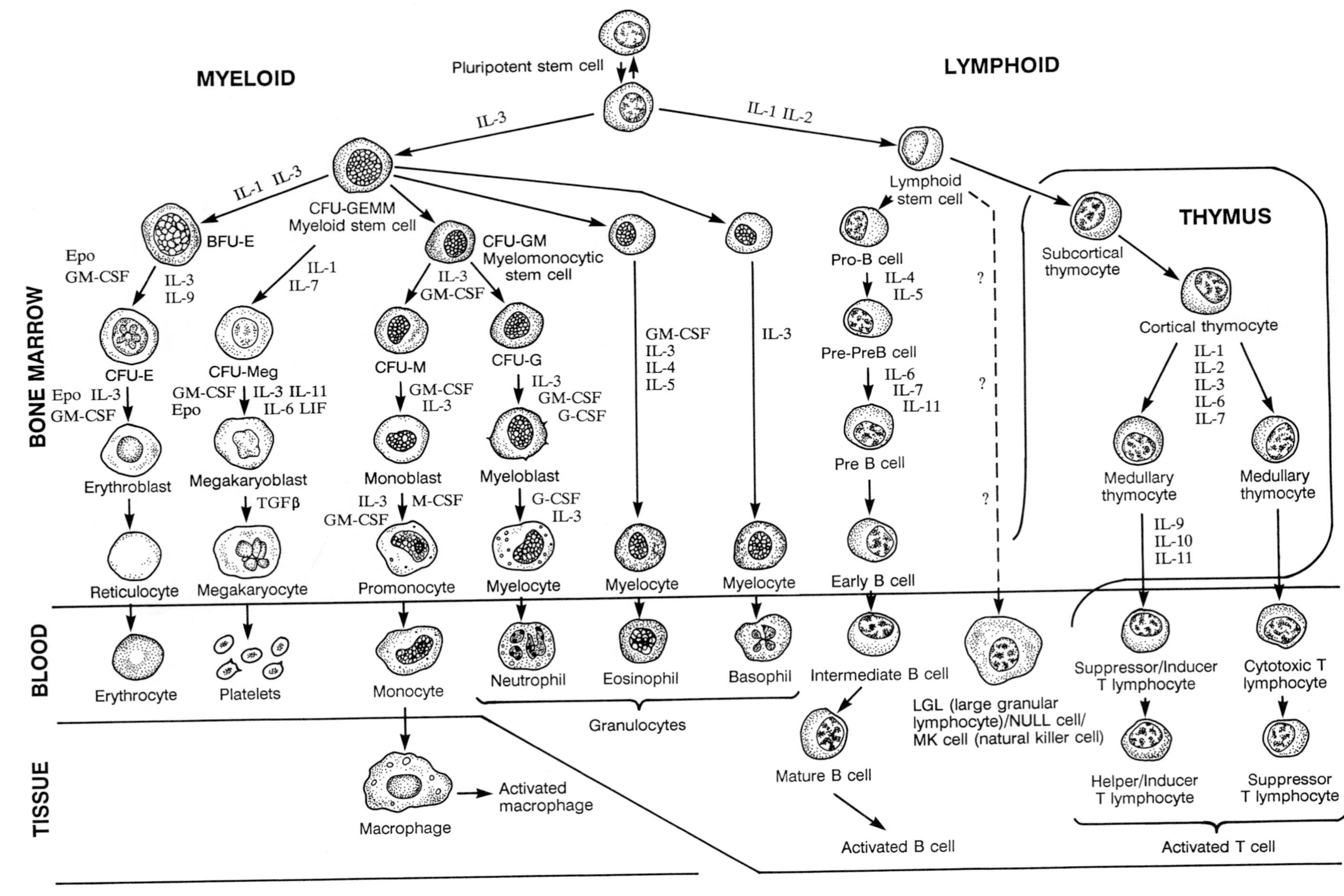

MYELOID
LYMPHOID
Pluripotent stem cell
IL-3
IL-1 IL-2
IL-1 IL-3
CFU-GEMM
Myeloid stem cell
Lymphoid
stem cell
THYMUS
BFU-E
Epo
GM-CSF
IL-3
IL-9
IL-1
IL-7
CFU-GM
Myelomonocytic
stem cell
IL-3
GM-CSF
GM-CSF
IL-3
IL-4
IL-5
IL-3
Pro-B cell
IL-4
IL-5
Subcortical
thymocyte
Cortical thymocyte
IL-1
IL-2
IL-3
IL-6
IL-7
BONE MARROW
CFU-E
Epo IL-3
GM-CSF
CFU-Meg
GM-CSF
Epo
IL-3 IL-11
IL-6 LIF
CFU-M
GM-CSF
IL-3
CFU-G
IL-3
GM-CSF
G-CSF
Pre-PreB cell
IL-6
IL-7
IL-11
?
?
?
Erythroblast
Megakaryoblast
TGFβ
Monoblast
IL-3
GM-CSF
M-CSF
Myeloblast
G-CSF
IL-3
Pre B cell
Medullary
thymocyte
IL-9
IL-10
IL-11
Medullary
thymocyte
Reticulocyte
Megakaryocyte
Promonocyte
Myelocyte
Myelocyte
Myelocyte
Early B cell
BLOOD
Erythrocyte
Platelets
Monocyte
Neutrophil
Eosinophil
Basophil
Granulocytes
Intermediate B cell
LGL (large granular
lymphocyte)/NULL cell/
MK cell (natural killer cell)
Suppressor/Inducer
T lymphocyte
Cytotoxic T
lymphocyte
TISSUE
Macrophage
Activated
macrophage
Mature B cell
Activated B cell
Helper/Inducer
T lymphocyte
Suppressor
T lymphocyte
Activated T cell

1. *Colony-Stimulating Factors*

The three colony-stimulating factors (GM-CSF, G-CSF, and M-CSF) and IL-3 are the major paracrine–autocrine regulators of hematopoiesis. A summary of their biochemical properties is given in Table 15-12. Their major biological action is to stimulate the proliferation (i.e., amplify the number of cells) of selected points in the erythroid and myeloid lineages.

2. *Erythropoietin*

Erythropoietin is a hematopoietic hormone secreted by the kidney that functions as a major stimulator of the production of erythrocytes. Erythropoietin is a secreted glycoprotein of 165 amino acids, but with a mature molecular mass of 30–34 kDa due to the presence of a significant amount of covalently linked carbohydrate. Hypoxia (a low-oxygen environment) can stimulate the synthesis of erythropoietin mRNA in the renal capillary tubular cells by an unknown mechanism(s). The erythropoietin receptor belongs to the cytokine receptor superfamily (see Figure 15-21).

In the adult human, erythropoietin is a classic hormone, secreted by the kidney and transported systemically to its site of action in the erythron. However, in the fetus, it is a paracrine–autocrine agent since the liver is the major site of production of both erythropoietin and erythropoietic cells.

FIGURE 15-20 Schematic diagram of the general scheme of blood cell differentiation, including an indication of the role of hematopoietic hormones and cytokines. All of the cells of the hematopoietic cell system are derived from a pluripotent stem cell; this cell can differentiate to generate the daughter stem cells that support the production of the lymphoid and myeloid lineages. The lymphoid stem cell supports the production of the B and T lymphocytes. The myeloid stem cell supports the production of the reticulocyte–erythrocyte, the megakarocyte, and the monocyte and granulocyte lineages (see Table 15-10). The different progenitor cells that are identified in the *in vitro* culture systems are CFU-GEMM (colony-forming unit, granulocyte–erythrocyte–monocyte–megakaryocyte), CFU-Meg (CFU, megakaryocyte), CFU-GM (CFU, granulocyte–monocyte), CFU-E (CFU, erythroid), and BFU-E (burst-forming unit, erythroid). The abbreviations of the regulatory cytokine factors are defined in Table 15-11. Note: The proposed sites of action of the cytokines, which are likely to change, are an area of intense investigation. The assignments shown here were derived from Metcalf, D., Nicola, N. A., and Gough, N. M. (1995). Hormones and blood cell production. *In* "Endocrinology" (L. J. DeGroot *et al.*, eds.), Vol. 3, p. 2943. W. B. Saunders and Co. New York. This figure was modified from the original by Clark, S. C., and Kamen, R. (1987). The human hematopoietic colony-stimulating factors. *Science* **236**, 1229–1237; the updated information was extrapolated from both a "Hematopoetry" poster (1993) prepared by the Genzyme Corporation, Cambridge, MA and a poster (1993) prepared by Harlan Bioproducts for Science, Inc., Indianapolis, IN.

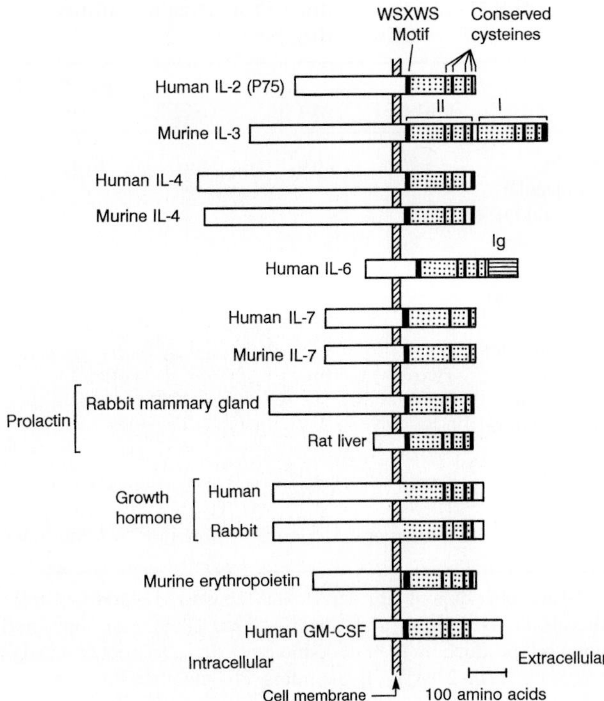

FIGURE 15-21 Diagram of selected members of the family of membrane receptors that participate in hematopoiesis. Schematic representations of the structures of all the known members of the family are shown. Horizontal bars represent conserved cysteine residues. The black boxes represent the conserved Trp-Ser-X-Trp-Ser (WSXWS) motif. The stippled areas shown the stretch of ~210 amino acids within which the receptor homologies are contained. The immunoglobulin-like domain at the N-terminus of the IL-6 receptor is also indicated. Modified from Casman, D., Lyman, S. D., Idzerda, R. L., *et al.* (1990). A new cytokine receptor superfamily. *Trends Biochem. Sci.* **15**, 265–270.

V. HORMONES AND INFLAMMATION

A. Introduction

The body's protection against infection, invading pathogens, or injury is a complex, overlapping network that includes the immune, inflammatory, and allergic systems. It is beyond the scope of this presentation to provide an in-depth discussion of these topics. What is presented is a brief overview that illustrates the extensive involvement of classical hormones, paracrine agents, lymphokines, cytokines, and growth factors generated by cells present in the circulatory system, which function in the inflammation and immune responses.

B. Inflammation Response

The acute inflammatory response comprises innate reactions (reactions not involving immunological components) that occur locally at the site of injury or infec-

TABLE 15-12 Biochemical Properties of Colony-Stimulating Factors[a]

	GM-CSF	M-CSF	G-CSF	Multi-CSF (IL-3)
Protein molecular mass (kDa)	14.4	26.0	18.6	15.4
Glycosylated molecular mass (kDa)	15–35	15–30	20	15–30
Number of amino acids	127	158	174	133
Cysteine disulfide	Two	Three	Two	One
Glycosylation sites				
N	Two	Two		Two
O	Three	?	Two	?
Active form	Monomer	Monomer	Monomer	Homodimer

[a] The abbreviations are defined in Table 15-11. Modified from Metcalf, D., Nicola, N., and Gough, N. M. (1995). Hormones and blood cell production. *In* "Endocrinology" (L. J. DeGroot *et al.*, eds.) Vol. 3, pp. 2943–2963. W. B. Saunders, Philadelphia, PA.

tion and that can be divided into cellular and vascular events. The vascular events involve initial dilations of the small arterioles, which result in increased local blood flow (mediated by histamines and prostaglandins) and exudation of fluid. The fluid exudate contains protein components of four enzyme cascades: (i) the kinin system, (ii) the blood coagulation system, (iii) the complement system, and (iv) the fibrinolytic system (see Figure 15-22). As a consequence of local processing by these enzyme cascades (which results in inflammation; see the following), the products of the infective agents are carried by the lymphatic system to nearby lymph glands so as to initiate an immune response.

The cells associated with the inflammation process are of two classes: (a) those cells normally present in tissues (e.g., tissue macrophages, mast cells, and endothelial cells) and (b) leukocytes derived from blood (e.g., polymorphonuclear cells, also known as granulocytes, which include the neutrophils, eosinophils, and basophils, and the monocytes).

The mast cells, granulocytes, and monocytes–macrophages are all derived from the hematopoietic cell differentiation process (see Figure 15-21). The granulocytes are the first cells to enter a site of inflammation; they first adhere to the vascular endothelial cells, migrate by chemotaxis to the site of the invading pathogen, and then engulf, kill, and digest the invading organism. The monocytes–macrophages arrive at a later stage in the inflammation process; the monocytes further differentiate into macrophages, which then become activated. The macrophages have many properties; they can aggressively phagocytose (engulf microorganisms) and secrete cytokines, paracrine hormones (NO, prostaglandins), and digestive enzymes. The mast cells are capable of secreting many local paracrine substances, including histamine, leukotrienes, platelet-activing factor, and interleukins; in addition the mast cell membranes have receptors for the IgE class of antibodies.

The immune response is mediated by specialized cells, the B and T lymphocytes, which are also generated by the hematopoiesis cell differentiation system. The B cells are responsible for antibody production (IgG, IgE, IgM, IgA, and IgD). The T cells are responsible for so-called cell-mediated reactions, which include a complex array of responses, including the killing of virally infected cells and the secretion of lymphokines (e.g., INF_{α}, which activates macrophages), which are complementary to the production of antibodies by the B cells.

VI. CLINICAL ASPECTS

A. Hypertension

Hypertension is a disorder that may occur throughout life; it is defined as the continued elevation of blood pressure beyond defined limits (see Table 15-4). In adults the criterion of moderate hypertension is where the systolic and diastolic pressures exceed 160 and 100 mm Hg, respectively. As many as 50 million Americans have elevated blood pressure. Hypertension has a worldwide incidence of 9–18% in the adult population. In the presence of continuing uncontrolled hypertension, there is frequently a high incidence of heart disorders (congestive failure, coronary heart disease), stroke, kidney disease, and/or aneurysms of the aorta.

There are three general causative factors that may lead to the development of hypertension: (a) primary aldosteronism, which results in the unrestrained production of aldosterone by the adrenals (see Chapter 10); (b) a pheochromocytoma, which is a tumor that produces catecholamines (see Chapter 11); and (c) renovascular hypertension, which results in a lowered blood supply to one kidney with normal blood flow to the second kidney.

An important component of the diagnosis of the etiology of hypertension is a determination of the level of Na^+ excretion in relation to the plasma levels of renin and also the daily amount of aldosterone excreted in the urine. Such an analysis may provide the formulation of a rational basis of treatment, which can include the use of inhibitors of the renin–angiotensin system such as the antagonist saralasin, use of β-blocking

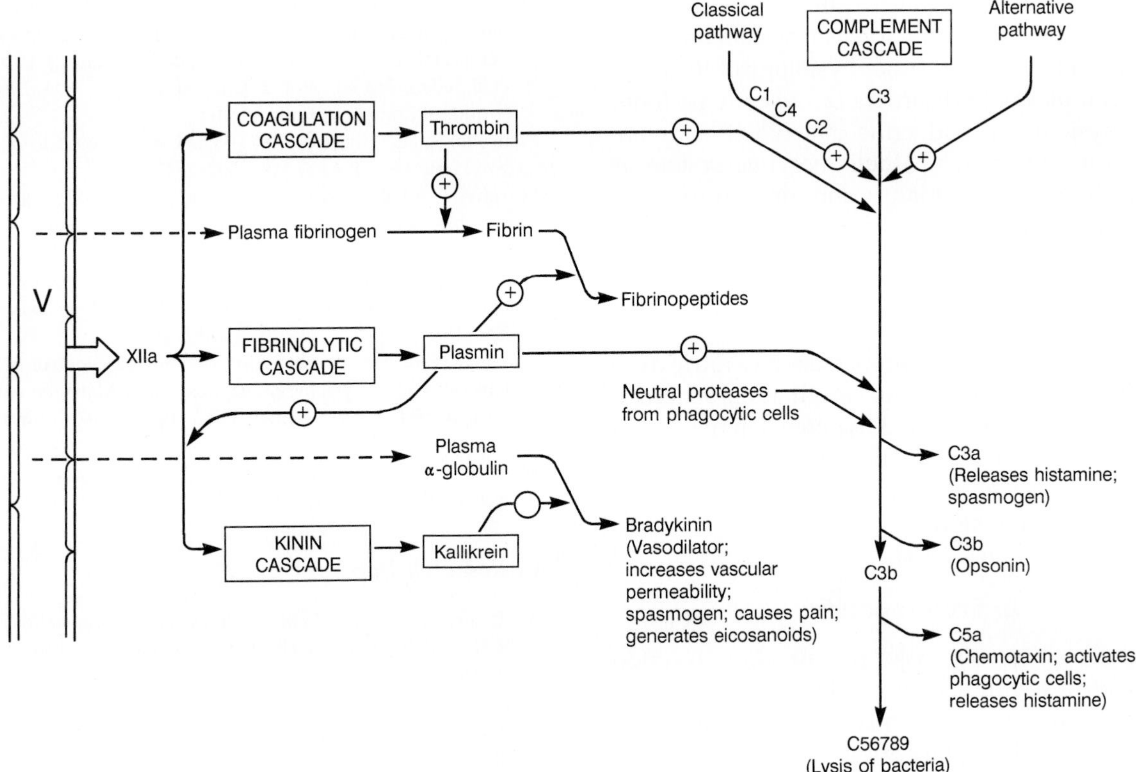

FIGURE 15-22 Enzyme cascades associated with the inflammation response. The processes are initiated at the time of local injury when plasma leaks into the tissues as a consequence of increased vascular permeability. The complement components are indicated by C1, C2, etc. V indicates a postcapillary venule. The coagulation cascade results, at the site of injury, in the conversion of fluid blood into a solid gel or clot of blood. The principal event is the conversion of soluble fibrinogen into insoluble fibrin; the blood coagulation cascade involves at least 13 tissue factors. When the coagulation system is activated, the fibrinolytic cascade or clot-dissolving process also becomes operative. This involves the action of plasminogen activators. Plasminogen activators are serine proteases that cleave an Arg–Val bond in plasminogen to release the enzyme plasmin. Plasmin is a trypsin-like enzyme that cleaves Arg–Lys bonds in the local fibrin clot. (Plasminogen, a β-globulin of 143 kDa, is a constitutive blood protein that forms a complex with the fibrin strands of the blood clot.) The kinin cascade results in the production of bradykinin (see Figure 15-16), which causes vasodilation and increased vascular permeability. The complement cascade is an enzyme system with nine major components; the system is activated by substances derived from microorganisms. The ultimate objective of the complement system is to release chemotaxins (which attract phagocytic cells), histamines, and opsonin (which attaches to the surface of microorganisms and facilitates their ingestion by macrophages and granulocytes). Adapted with permission from M. M., Dale, J. C., Foreman, and T.-P., Fan (1994). "Textbook of Immunopharmacology," 3rd ed. Blackwell Scientific Publications, Oxford.

drugs such as propanolol, or dietary restriction of salt intake. It is beyond the scope of this presentation to consider this topic in detail.

B. Aldosterone

Defects in aldosterone biosynthesis and action have marked effects on blood pressure and electrolyte metabolism. Addison's disease or chronic aldosterone deficiency is discussed in Chapter 10. Mineralocorticoid deficiency, as it results in several of the congenital adrenal hyperplasias resulting from defects in steroid metabolism (see Chapter 10), causes hyponatremia, hypovolemia, and hyperkalemia due to inadequate renal reabsorption of water and Na^+ and decreased excretion of K^+.

Primary aldosteronism can result from adrenal tumors. Hypertension can result in patients with hyperaldosteronism or hypertensive forms of congenital adrenal hyperplasia. Fortunately these disorders are rare.

C. Edematous Disorders

Congestive heart failure, nephrotic syndrome, and hepatic cirrhosis are a series of disorders characterized by excessive retention by the body of Na^+ and water, which leads to edema. Frequently there is an associated stimulation of the renin–angiotensin system and aldosterone production. The pathophysiology of these diseases is complex, and disorders can usually be found in the physiological endocrine matrix of the renin–angiotensin system–sympathetic nervous system activity, the humoral vasodilatory and/or natriuretic factors, and the vasopressin system.

D. Nitric Oxide

The topic of clinical disorders related to nitric oxide is relatively new. The potent vasodilation actions of excess NO and the associated hypotension may be associated with septic shock. There are also suggestions that a relative lack of NO may be associated with certain forms of impotence.

E. Erythropoietin

There are no known major disease states or endocrine disorders with a high incidence attributable to erythropoietin. There can, in some circumstances, be a tumor that results in the ectopic production of erythropoietin. The use of recombinant erythropoietin is helpful to treat the anemia associated with chronic renal failure.

References

A. Books

Fisher, J. W., ed. (1977). "Kidney Hormones," Vol. 2. Academic Press, New York.

B. Review Articles

Baxter, J. D., Duncan, K., Chu, W., James, M. N. G., Russell, R. B., Haidar, M. A., DeNoto, F. M., Hsueh, W., *et al.* (1991). Molecular biology of human renin and its gene. *Recent Prog. Hormone Res.* **47,** 211–251.

Brenner, B. M., Ballermann, B. J., Gunning, M. E., and Zeidel, M. L. (1990). Diverse biological actions of atrial natriuretic peptide. *Physiol. Rev.* **70,** 665–699.

Clark, S. C., and Kamen, R. (1987). The human hematopoietic colony-stimulating factors. *Science* **236,** 1229–1237.

Koller, K. J., Lowe, D. G., Bennett, G. L., *et al.* (1991). Selective activation of the B-natriuretic peptide receptor by c-type natriuretic peptide (CNP). *Science* **252,** 120–123.

Krantz, S. B. (1991). Erythropoietin. *Blood* **77,** 419–434.

Lowenstein, C. J., and Snyder, S. H. (1992). Nitric oxide, a novel biological messenger. *Cell* **70,** 705–707.

Marletta, M. A. (1993). Nitric oxide synthase structure and mechanism. *J. Biol. Chem.* **268,** 12231–12234.

Masaki, T. (1993). Endothelins: homeostatic and compensatory actions in the circulatory and endocrine systems. *Endocrine Rev.* **14,** 256–268.

Metcalf D., Nicola N. A., and Gough N. M. (1995). Hormones and blood cell production. *In* "Endocrinology" (L. J., DeGroot, M. Besser, H. G. Burger, J. L. Jameson, D. L. Loriaux, J. C. Marshall, *et al.*, eds.), 3rd ed., Vol. 2, Chapter 157, pp. 2943–2963. W. B. Saunders Company, Philadelphia.

Moncada, S., and Higgs, A. (1993). The L-arginine-nitric oxide pathway. *New Engl. J. Med.* **329,** 2002–2012.

Murphy, T., Alexander, R., Griendling, K., *et al.* (1991). Isolation of a cDNA encoding the vascular type-1 angiotensin II receptor. *Nature* **351,** 233–236.

Rosenzweig, A., and Seidman, C. E. (1991). Atrial natriuretic factor and related peptide hormones. *Annu. Rev. Biochem.* **60,** 229–255.

Said, S. I. (1990). Neuropeptides in blood pressure control. *In* "Hypertension, Pathophysiology, Diagnosis, and Management" (J. H. Laragh and B. M. Brenner, eds.), pp. 791–804. Raven Press, New York.

White, P. C. (1994). Disorders of aldosterone biosynthesis and action. *New Engl. J. Med.* **331,** 250–258.

C. Research Papers

Bradt, D. S., Hwang, P. M., Glatt, C. E., Lowenstein, C., Reed, R. R., and Snyder, S. H. (1991). Cloned and expressed nitric oxide synthase resembles cytochrome P450-reductase. *Nature* **351,** 714–718.

Dhanaraj, V., Dealwis, C. G., Frazao, C., *et al.* (1992). X-ray analyses of peptide-inhibitor complexes define the structural basis of specificity for human and mouse renins. *Nature (London)* **357,** 466–472.

Hausdorff, W. P., Sekura, R. D., Aguilera, G., and Catt, K. J. (1987). Control of aldosterone production by angiotensin II is mediated by two genuine nucleotide regulatory proteins. *Endocrinology* **120,** 1668–1678.

Ikeda, U., Hyman, R., Smith, T. W., and Medford, R. M. (1991). Aldosterone-mediated regulation of Na^+-K^+-ATPase gene expression in adult and neonatal rat cardiocytes. *J. Biol. Chem.* **266,** 12058–12066.

Inoue, A., Yanagisawa, M., Kimura, S., Kasuya, Y., Miyauchi, T., Goto, K., and Masaki, T. (1989). The human endothelin family: three structurally and pharmacologically distinct isopeptides predicted by three separate genes. *Proc. Natl. Acad. Sci. USA* **86,** 2863–2867.

Janes, R. W., Peapus, D. H., and Wallace, B. A. (1994). The crystal structure of human endothelin. *Struct. Biol.* **1,** 311–319.

Kambayashi, Y., Bardhan, S., Takahashi, K., Tsuzuki, S., Inui, H., Hamakubo, T., and Inagami, T. (1993). Molecular cloning of a novel angiotensin II receptor isoform involved in phosphotyrosine phosphatase inhibition. *J. Biol. Chem.* **268,** 24543–24546.

Levin, E. R. (1995). Mechanisms of Disease. *New Engl. J. Med.* **333,** 356–363.

Longmore, G. D., and Lodish, H. F. (1991). An activating mutation in the murine erythropoietin receptor induces erythroleukemia in mice: A cytokine receptor superfamily oncogene. *Cell* **67,** 1089–1102.

Mukoyama, M., Nakajima, M., Horiuchi, M., Sasamura, H., Pratt, R. E., and Dzau, V. J. (1993). Expression cloning of type 2 angiotensin II receptor reveals a unique class of seven-transmembrane receptors. *J. Biol. Chem.* **268,** 24539–24542.

Needleman, P., and Greenwald, J. E. (1986). Atriopeptin: A cardiac hormone involved intimately in fluid, electrolyte, and blood pressure homeostasis. *New Engl. J. Med.* **314,** 828–834.

Rubanyi, G. M., and Bothelo, L. H. (1991). Endothelins. *FASEB* **5,** 2713–2720.

Sielecki, A. R., Hayakawa, K., Fujinaga, M., Murphy, M. E. P., Fraser, M., Muir, A. K., Carilli, C. T., Lewicki, J. A., *et al.* (1989). Structure of recombinant human renin, a target for cardiovascular-active drugs, at 2.5 Å resolution. *Science* **243,** 1346–1351.

Sudoh, A. M., Minamino, N., Kangawa, K., and Matsuo, H. (1990). A new member of natriuretic peptide family identified in porcine brain. *Biochem. Biophys. Rev. Commun.* **168,** 863–870.

Verrey, F., Kraehenbuhl, J. P., and Rossier, B. C. (1989). Aldosterone induces a rapid increase in the rate of Na, K-ATPase gene transcription in cultured kidney cells. *Mol. Endocrinol.* **3,** 1369–1376.

CHAPTER

16

Prostaglandins

I. INTRODUCTION

A. Background

The prostaglandins (PG) represent a class of substances produced in a wide variety of cells. These act on the cells that produce them, on neighboring cells, or usually over short distances and can be classified as autocrine hormones. Some PGs, such as prostacyclin (PGI_2), may be regarded as acting like the more traditional endocrine hormones in the sense that they are synthesized in blood vessel cells, survive in the bloodstream for a period of time, and can exert effects somewhat distant from their sites of synthesis. However, we usually think of PGs and their relatives as being potent local hormones (autocrine and paracrine) acting over a short lifetime. Many of the prostaglandins and relatives are important mediators of inflammatory reactions. Prostaglandins are misnamed for an early activity of semen, presumably originating in the prostate, which contracted uterine smooth muscle.

PGs and their relatives, prostacyclin (PGI_2), thromboxanes (TX), leukotrienes (LT), and lipoxins (LP), derive from fatty acids stored in cellular membranes as phospholipids or triglycerides. Fatty acid precursors, typically arachidonic acid, are released by a phospholipase or by a lipase in the cell membranes following a stimulatory event. This event will usually be a signal to activate the enzyme liberating arachidonic acid from lipids in the membrane. A series of synthetic reactions catalyzed by enzymes take place in the membrane, culminating in the release of the PG product from the membrane into the cellular cytoplasm. The released PG may bind to its receptor located within the plasma membrane or other internal membrane, or it may be released to the cell exterior and ultimately produce an effect by binding to a receptor in the cell membrane of a neighboring cell. There is little information available on the mechanism by which PGs are secreted from cells.

PGs are produced by many different cells in the body. It is not yet clear whether all cells are capable of producing them, but this seems a distinct possibility. Ultimately, the release of PGs from cells may be a product of the action of other hormones or neurotransmitters (i.e., the signal to the cell to synthesize PGs).

PGs exert a wide variety of effects on different target tissues. They affect behavior through direct actions on individual neurons and substructures of the brain, such as cerebellar and reticular formation (the latter is responsible for screening various kinds of environmental signals), and they act on the hypothalamus and the pituitary. Vasomotor and temperature-regulatory centers are affected by PGs. Autonomic and neuromuscular junctions are also affected. As is evident in other chapters, PGs act on anterior pituitary trophic hormone target tissues such as thyroid, adrenal, ovary, and testis, on exocrine hormone targets such as pancreas and gastric mucosa, and on endocrine target tissues such as renal tubules, bone, and adipocytes. They act on the smooth muscles of the reproductive, alimentary, and respiratory tracts and on cardiovascular smooth muscles. Some of these effects will be elaborated in this chapter. PGs act on red blood cells, leukocytes, and platelets, the last of which will be described here. The role of PGs in pain will be mentioned, consistent with the inflammatory actions of certain PGs (see Chapter 10). The roles of prostaglandins and prostaglandin relatives in asthma will be discussed.

B. Structural Classes of Prostaglandin (PG) Relatives

Although there is a complexity of structures under the heading of PGs and their relatives, it will be helpful to describe briefly the differences between the major classes. All of these compounds are derived from fatty acids, usually arachidonic acid, with an open chain, 20-carbon structure. The PGs, including TX and PGI_2, all arise by mediation through a cyclic endoperoxide. On the other hand, LTs utilizing cyclooxygenase are formed directly from a fatty acid metabolite via lipoxygenases and do not involve a cyclic endoperoxide intermediate.

PGs resemble hairpins with a five-membered ring and two chains extending from the ring. Substituents on the five-membered ring determine the subclass and activity of the PGs. TXs have a six-membered ring with two chains extending from the ring. The six-membered ring has one or more oxygens associated with it. PGI_2 has two adjacent rings, one of which contains an oxygen with one chain extending from each ring. Often TX and PGI_2 are found to have opposing biological activities. LTs are essentially modified open chain fatty acids that may be conjugated to glutathione or glutathione degradation products. Lipoxins are open chain fatty acids. There is a wide array of related structures with characteristic activities or inactivity if such structures result from metabolism. The characteristics of the

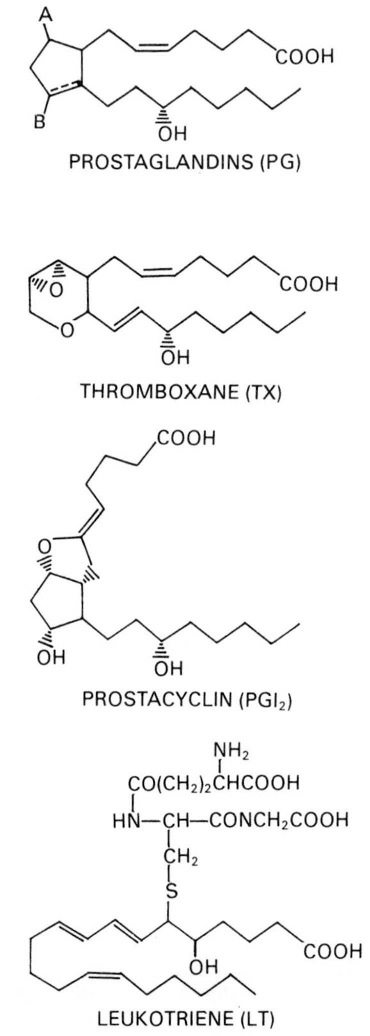

A

B

COOH

OH

PROSTAGLANDINS (PG)

O

O

COOH

OH

THROMBOXANE (TX)

COOH

O

OH

OH

PROSTACYCLIN (PGI_2)

NH_2

$CO(CH_2)_2CHCOOH$

$HN—CH—CONCH_2COOH$

CH_2

S

COOH

OH

LEUKOTRIENE (LT)

FIGURE 16-1 General features of major classes of PG relatives. A and B represent hydroxyl or ketone substituents. LT and LP are fatty acid derivatives. LT contains glutathione.

general structures mentioned here are shown in Figure 16-1.

C. Classification of PGs

Although the many structures in this group of compounds may seem bewildering at first (Figure 16-2), there are some simple generalizations that can be set forth to allow a facile understanding of structure–function relationships. PGs arise from a cyclic endoperoxide generated by the enzyme system, PG synthetase. This is a complex of enzymes, including cyclooxygenase, required to produce the key intermediate, the cyclic endoperoxide derivative of arachidonic acid or other fatty acids. There is a constitutive (COX1) and an inducible cyclooxygenase (COX2). COX2 is induced under conditions of inflammation, and the induction of this enzyme is inhibited by glucocorticoids representing one of its pathways of anti-inflammatory action. The cyclic endoperoxide intermediate is acted on by various isomerases to produce PG subclasses. The cyclic endoperoxide is also a precursor of PGI_2 and TX. On the other hand, other groups of compounds in this class, LT and lipoxins (LP), are derived directly from arachidonic acid without the mediation of a cyclic endoperoxide (Figure 16-2). Figure 16-3 repeats Figure 16-2 but shows individual structures. The one-double-bond members of some of the classes, such as PGE_1 or $PGF_{1\alpha}$, are not shown here, but are illustrated elsewhere in the chapter. Of the many groups of compounds, the PGs are composed of only three abundant groups: PGF, PGE, and PGA. F stands for phosphate buffer-soluble (phosphate begins with an F in Swedish) structures, which are the most hydrophilic or polar of the PGs. The F series has the general structure shown in Table 16-1. All of the PGs have the five-membered ring shown here. The F compounds (most polar) have two hydroxyls in the C-9 and C-11 positions, constituting the most hydrophilic of the classical PGs. The E series (PGE) is intermediate in solubility (is ether-soluble, thus, E) and has one hydroxyl and one ketone substituted on the five-membered ring (Table 16-1). The A

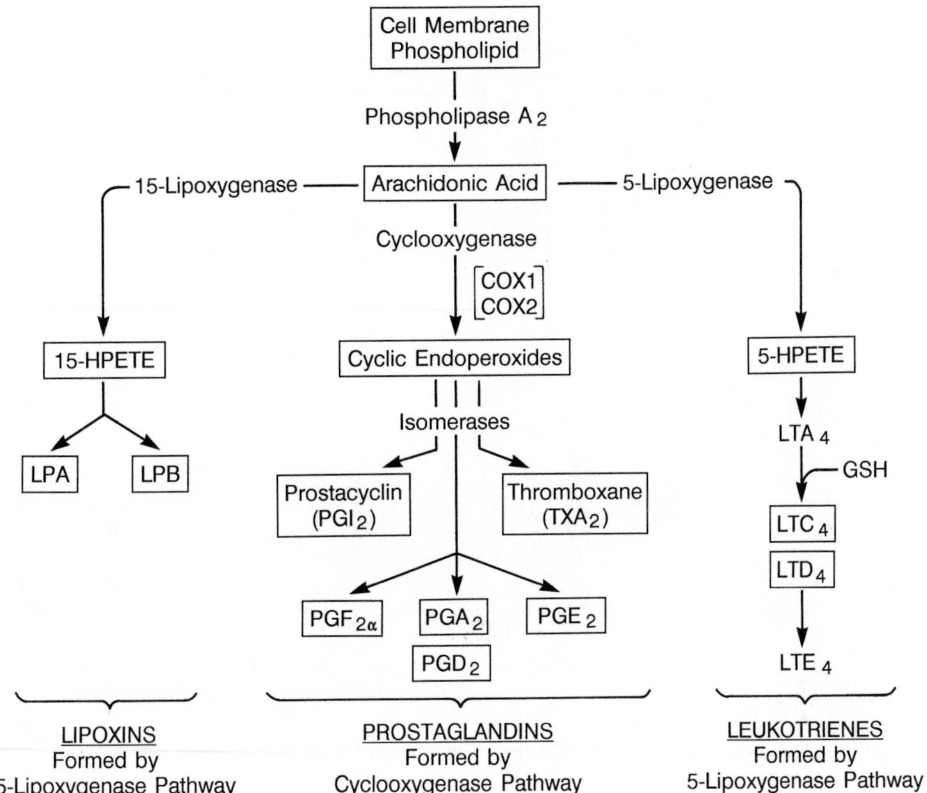

FIGURE 16-2 Overview of enzymatic pathways giving rise to major prostaglandin relatives. Subscript number refers to the number of double bonds in the structure (most have equivalent structures with one double bond, not shown here). COX1 is constitutive cyclooxygenase; the conjugated leukotrienes are LTC_4 and LTD_4. α, in $PGF_{2\alpha}$, refers to the steric position of the C-9 substituent.

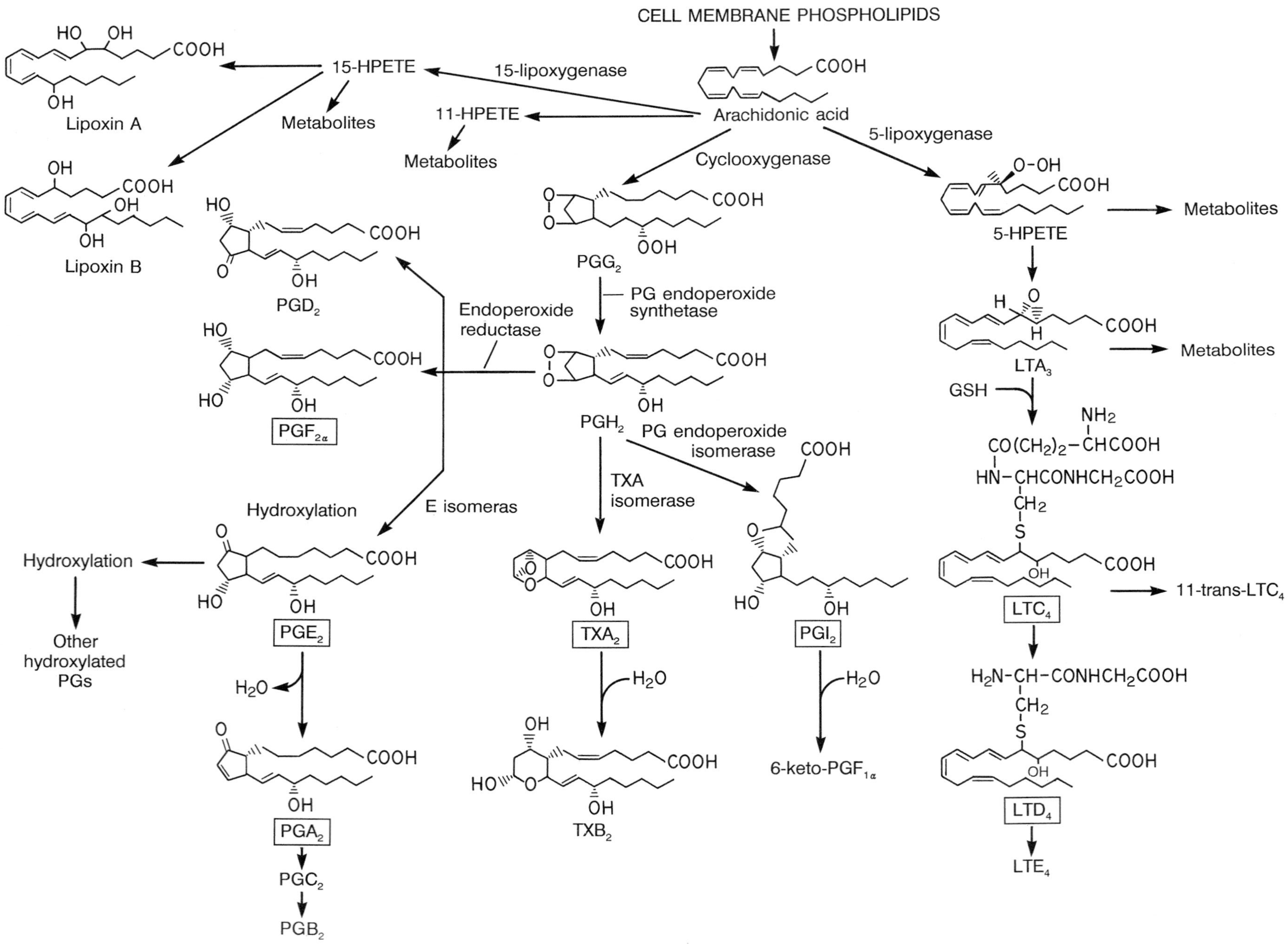

CELL MEMBRANE PHOSPHOLIPIDS

Arachidonic acid

15-lipoxygenase

15-HPETE

Metabolites

Lipoxin A

Lipoxin B

11-HPETE

Metabolites

Cyclooxygenase

PGG_2

PG endoperoxide synthetase

PGH_2

Endoperoxide reductase

PGD_2

$PGF_{2\alpha}$

E isomeras

Hydroxylation

PGE_2

Hydroxylation

Other hydroxylated PGs

H_2O

PGA_2

PGC_2

PGB_2

TXA isomerase

TXA_2

H_2O

TXB_2

PG endoperoxide isomerase

PGI_2

H_2O

6-keto-$PGF_{1\alpha}$

5-lipoxygenase

5-HPETE

Metabolites

LTA_3

Metabolites

GSH

LTC_4

11-trans-LTC_4

LTD_4

LTE_4

TABLE 16-1 Principal Classes of the Classical Prostaglandins and Their Actions[a]

Structure	Compound	Effect on blood pressure	Effect on nonvascular smooth muscle (uterus/intestine)
↑ [increasing polarity]			
(ring with 9-OH, 11-OH; 1 R-COOH; 20 R-CH_3)	PGF (PO_4 soluble)	Transient increase	Very active
(ring with 9-keto, 11-OH; 1 R-COOH; 20 R-CH_3)	PGE (ether soluble)	Decrease	Very active
(ring with 9-keto, 10,11 double bond; 1 R-COOH; 20 R-CH_3)	PGA	Decrease	Inactive
↓ [increasing lipophilicity]			

[a] Reproduced in part from J. B. Lee, *in* "Principles of Endocrinology" (R. H. Williams, ed.) 5th ed., p 855. Saunders, Philadelphia, Pennsylvania, 1975.

series (PGA) is the least polar or hydrophilic, is the most lipophilic, and has no hydroxyl groups substituted on the five-membered ring, but rather a ketone group and an unsaturation (Table 16-1). These classes of PGs have characteristic actions depending on the five-membered-ring substituents, as shown in Table 16-1.

Inspection of Table 16-1 provides an easy means to relate structure with activity. Since PGF is the most water-soluble, it has hydroxyl groups on both C-9 and C-11. As will be seen later, PGF plays a role in the termination of pregnancy. This association will make it possible to remember that the F series is active in muscular contraction (i.e., as in uterine contractions of parturition).

The PGE series is intermediate between PGF and PGA. Consequently, it shares both activities of those groups as well as their characteristic ring substituents. The structure, therefore, can be deduced as one in which the ring has both a hydroxyl (typical of PGF) and a ketone (typical of PGA) and has the activities of both stimulating muscular contraction and decreasing blood pressure. Finally, PGA is the least polar (a ketone and a double bond in the ring) and has activity opposite that of PGF; it reduces blood pressure and has little effect on nonvascular muscle contraction.

In designating the PGs, two kinds of subscripts may appear at the end of the abbreviation (e.g., $PGF_{2\alpha}$). The subscript, which is a number, denotes the number of double bonds in the structure. The Greek letter, α, following the subscript number refers to the steric position of the C-9 substituent (α indicates extension behind the plane of the ring, away from the reader). This becomes apparent in the structure of $PGF_{2\alpha}$. If 2 is added to this number (4 in the case of $PGF_{2\alpha}$), we get the number of double bonds that must have been present in the fatty acid precursor of $PGF_{2\alpha}$, that is,

FIGURE 16-3 Routes of synthesis to some of the PGs, PGI_2, TX, LT, and LP from arachidonic acid. Note that all of the classes of PG relatives are synthesized through a cyclic endoperoxide intermediate except for the LTs and LPs and related structures, which are essentially modified fatty acids. A major activity of cyclooxygenase (COX2) is inducing inflammation and is repressed by glucocorticoids at the transcriptional level. Nonsteroidal anti-inflammatory drugs (aspirin and indomethacin) inhibit COX2 by transacetylation of the catalytic center of the enzyme. Whereas steroids should also shut down LT and LP formation in addition to PG (COX inhibition) by inhibition of phospholipase A_2 mediated by induction of lipocortin/annexin I, nonsteroidal drugs (NSAID) are not expected to affect LT and LP since they operate mainly on COX.

eicosatetraenoic acid (20:4ω6) or arachidonic acid. The information in parentheses, in this case (20:4ω6), presents information about the fatty acid. Thus, 20 is the number of carbons in the structure, 4 is the number of double bonds, and ω6 indicates that there are six methyls in the chain from the end until a double bond appears (Figure 16-4).

D. General Aspects of PGs

The synthesis of PGs and relatives from membrane lipids, release of the finished products into the cytoplasm, and actions are summarized by Figure 16-5. Thus, when PG synthesis is stimulated in a cell, a fatty acid precursor, usually arachidonic acid, is released from the membrane lipids by phospholipase A_2. It is acted on by the membrane PG synthetase complex, which produces a cyclic endoperoxide intermediate (Figure 16-4). This intermediate can give rise to PGE_2, $PGF_{2\alpha}$, and PGI_2 in cells that contain PGI_2 synthetase or to TX and TXA_2 in cells that contain TX synthetase (Figure 16-3). When homo-γ-linolenic acid (20:3) (Figure 16-4) is the parent fatty acid, a different cyclic endoperoxide is produced. This intermediate can give rise to PGE_1 and $PGF_{1\alpha}$. Thus, the PGs or relatives occurring in a given cell depend on the precursor and especially upon the enzymes present in the biosynthetic pathways.

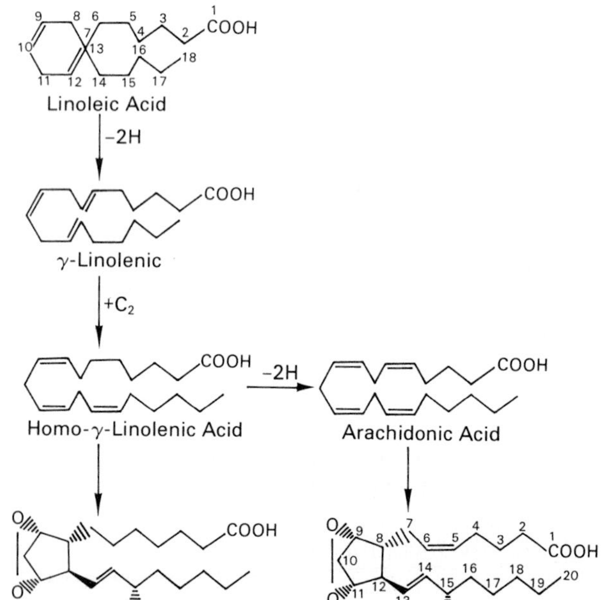

Linoleic Acid

−2H

γ-Linolenic

$+C_2$

Homo-γ-Linolenic Acid —(−2H)→ Arachidonic Acid

PGG_1 (Cyclic Endoperoxide from Homo-γ-Linolenic Acid)

PGG_2 (Cyclic Endoperoxide from Arachidonic Acid)

FIGURE 16-4 Interconversions of fatty acids leading to the production of major endoperoxide intermediates in the synthesis of PGs.

II. CHEMISTRY

A. Conformation of PGs and Thromboxanes (TX)

The conformational analysis of $PGF_{2\alpha}$ has been accomplished by the use of X-ray diffraction techniques. Two independent structures appear as familiar hairpin configurations, as shown in Figure 16-6. In contrast, TX molecules as represented by TXB_2 (Figure 16-3) do not adopt the PG hairpin configuration, as shown in Figure 16-7. These configurations specify obvious differences between the architectures of the PG and TX receptors. The side chains of these structures probably wobble considerably in solution or in the gel phase.

B. Phospholipase A_2 Enzymes

Phospholipase A_2 is responsible for the release of arachidonic acid from cell membrane phospholipids. Accordingly, its activity and regulation are important aspects of the generation of prostaglandins and their relatives in various situations, including the protective effects of PGs, as well as in inflammation and others. This enzyme is an obvious therapeutic target although progress in this area has been difficult. Significantly, the secreted enzyme, especially the enzyme found in synovial fluid (s), appears to be a gene product different from the enzyme in cellular cytosol (c). They also differ in catalytic requirements, as shown in Table 16-2. The synovial fluid enzyme consists of 128 amino acids while the cytosolic enzyme is 749 amino acids long, and there is no homology between the two proteins except for a His/Asp pair, which is catalytic in the case of synovial (s) PLA_2. Figure 16-8 shows molecular models of porcine pancreatic PLA_2, and the position of the Ca^{2+} ion is indicated in the model. Table 16-3 summarizes physiological processes affected by the phospholipase A_2 enzymes.

III. BIOCHEMISTRY

A. Biosynthesis

The biosynthetic mechanisms are reviewed in Figure 16-3. Succeeding reactions are catalyzed by glutathione-requiring isomerase, which is a GSH S-transferase, demonstrated to have $\Delta^{5,4}$-3-ketosteroid isomerase activity and requiring glutathione anion as a catalytic cofactor. The action of aspirin (acetylsalicylic

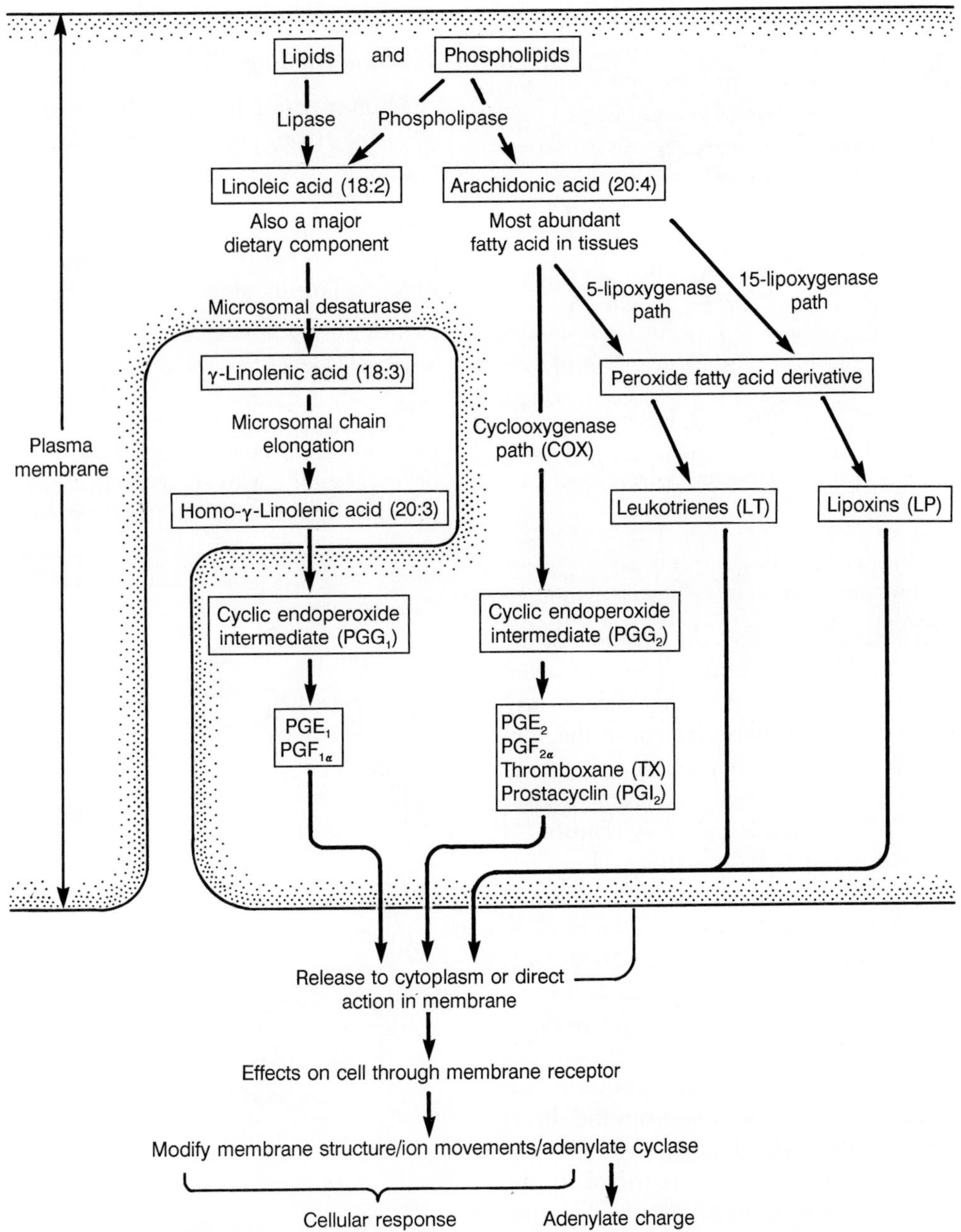

FIGURE 16-5 Overview of PG and its relatives' syntheses and actions.

acid) or indomethacin is to acetylate the PG endoperoxide synthetase. This results in loss of the cyclooxygenase activity, but not the peroxidase activity. Apparently one acetylation per molecule of 68,000 Da is sufficient to complete the reaction and involves the hydroxyl group of a serine in the N-terminal portion (Figure 16-9).

A parallel set of reactions can be developed for substrates other than arachidonic acid. These reactions, pictured in Figure 16-10, account for PGE_1, $PGF_{1\alpha}$, PGA_1, and PGB_1.

The leukotrienes (LT) derive from arachidonic acid, but do not proceed through a cyclic endoperoxide. Instead, this class of compounds derives by modification of arachidonic acid through the 5-lipoxygenase pathway or from other fatty acids, as shown in Figure 16-11. The three-dimensional structure of 15-lipoxygenase from soybean has been solved, as shown in

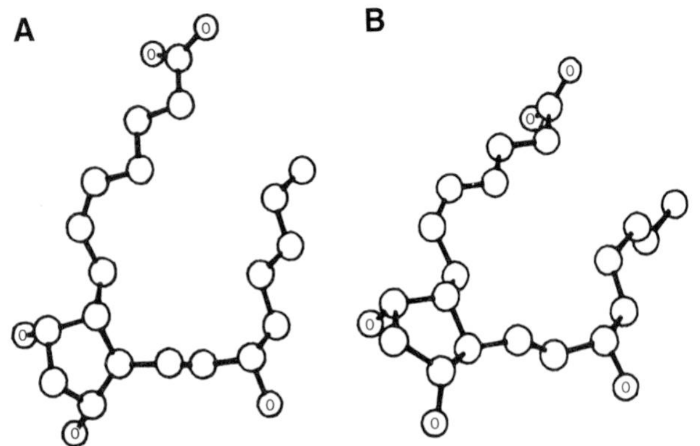

FIGURE 16-6 The two conformers of $PGF_{2\alpha}$. The views are the projections on the planes defined by atoms C-12, C-15, and O-15. Reproduced from Langs, D. A., Erman, M., and Detita, G. T. (1977). *Science* **197,** 1003–1005. Copyright 1977 by the AAAS.

Figure 16-12. The model indicates the position of the iron atom. Close to the iron atom are two adjacent unoccupied positions, which are thought to be the location of interaction of the 1,4-diene system of the substrate fatty acid with molecular oxygen during catalysis. The lipoxins (LP) derive directly from arachidonic acid through the 15-lipoxygenase pathway (Figure 16-13). Little is known about lipoxin actions. However, LPA dilates the microvasculature, and both LPA and LPB inhibit the cytotoxic effects of natural killer cells.

As can be seen from Figure 16-11, there are at least seven known LTs. Importantly, LTC_1, LTC_4, and LTC_5 contain glutathione, and LTD_4 contains cysteinylglycine. The D series contains sulfur amino acid residues as well. None of these very active compounds arise via a cycloendoperoxide, but all arise from the direct peroxidation of the fatty acid substrate on C-5. These compounds are very active bronchoconstrictors in the lungs and airways and are of importance in the etiology of asthma.

B. Metabolism

The PGs are metabolized in many positions, for example, PGE_2 (Figure 16–14A). TXA_2, the major active TX in the human platelet, is metabolized to the less active TXB_2 (Figure 16-14B). PGI_2 is metabolized to 6-keto-$PGF_{1\alpha}$ (Figure 16-14C). In the cases of PGI_2 and TXA_2, metabolism concerns the opening of an oxygen-containing ring. Futher metabolic steps are indicated as for the PGs.

C. Binding Proteins and Receptors

1. Serum Binding

Human serum has been shown to bind PGs, presumably by strong noncovalent interactions. PGs bind tightly in inverse relation to their polarity. Thus, the order of tightness of binding is PGA > PGE_1 > $PGF_{2\alpha}$. Since it is known that PGA_1 is cleared slowly in comparison to other PGs, its strong interaction with serum may explain its slow turnover. Specific binding proteins have not been reported.

2. Receptors

There appear to be three types of receptors located in various cell membranes. These receptors either raise or lower cellular levels of cyclic AMP or stimulate the phosphatidylinositol metabolic pathway (Table 16-4).

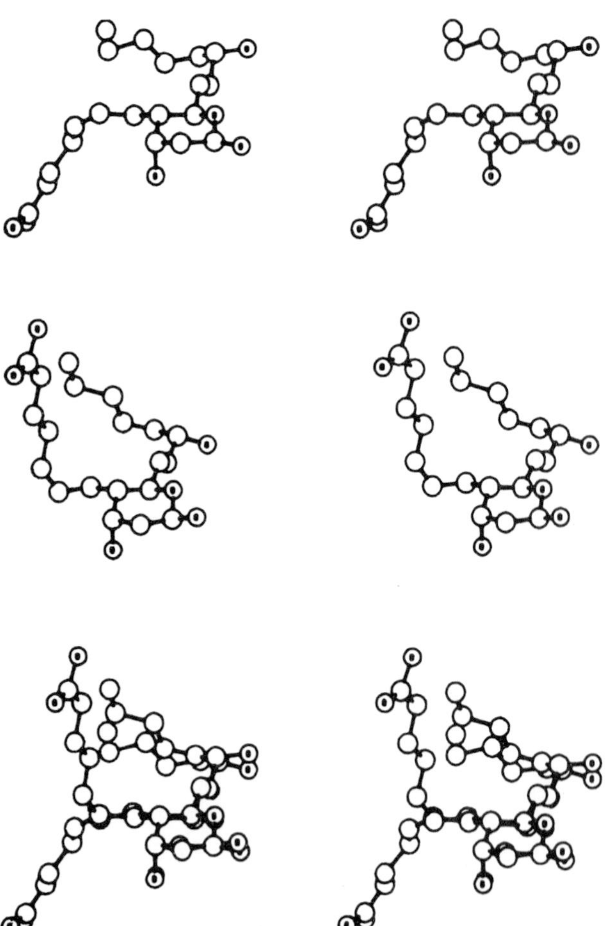

FIGURE 16-7 Stereograms of the α-tail (top) and β-tail (middle) conformers of TXB_2 and a composite (bottom) of the two conformers. Reproduced from Langs, D. A., Fortier, S., Erman, M. G., and Detita, G. T. (1979) *Nature (London)* **281,** 237–238. Copyright 1979 Macmillan Journals Limited.

TABLE 16-2 Characteristics of Human s- and $cPLA_2$ Enzymes[a]

	$sPLA_2$	$cPLA_2$
Original source	Human synovial fluid	Human monocyte cytosol (U937 cell line)
Mass	18 kDa	85 kDa (predicted from cDNA) 100–110 kDa (SDS–PAGE)
Localization	Extracellular fluids	Cytosol
Ca^{2+} requirement for maximal activity *in vitro*	Millimolar	Micromolar
Substrate specificity	All phospholipids	Phospholipids containing arachidonic acid
Regulation	Ca^{2+} levels	Ca^{2+} levels, binding domain for Ca^{2+} and lipids, membrane translocation, phosphorylation
Effect of *p*-bromophenacyl bromide	Inhibition and alkylation at active site His residue	Insensitive
Role of His/Asp pair	Catalytic	Unknown

[a] Reproduced with permission from Glaser, K.B., Mobilio, D., Chang, J.Y., and Senko, N. (1993). Phospholipase A_2 enzymes: Regulation and inhibition. *Trends Pharmacol. Sci.* **14,** 92–98.

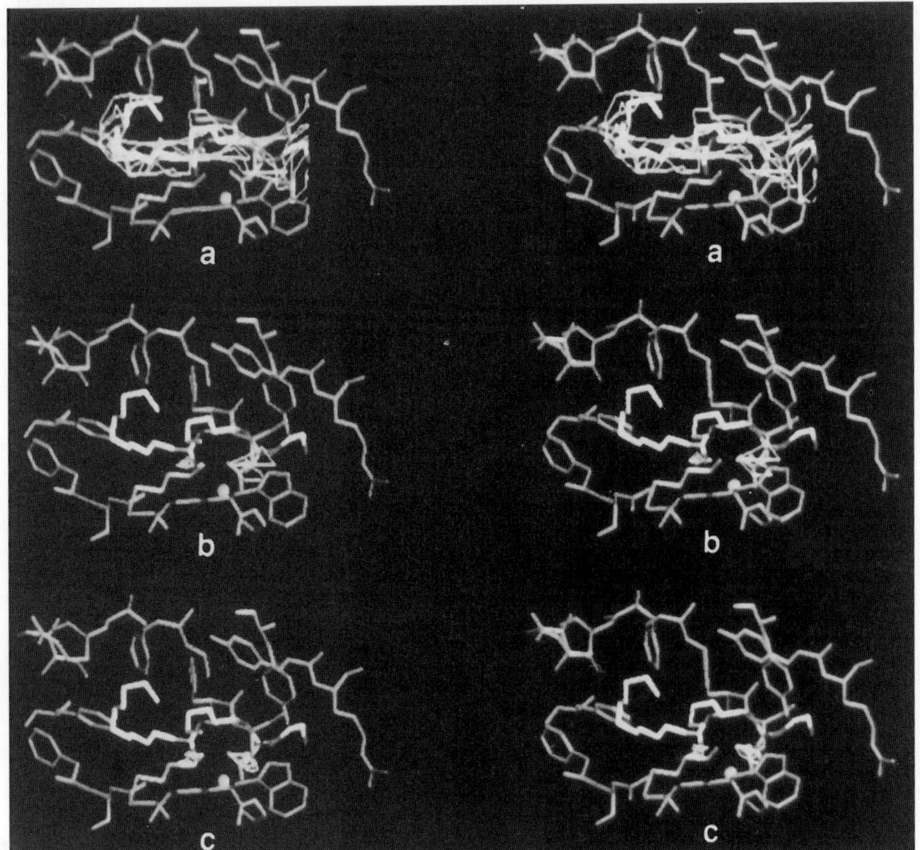

FIGURE 16-8 Stereopair representations of an inhibitor binding at the active site of porcine pancreatic PLA_2. The sphere is essential Ca^{2+} ion. These contours show the most favorable low-energy positions, calculated with the GRID program, for three different probes on the inhibitor: (a) CH_2, (b) carbonyl oxygen, (c) *O*. The original figure is in color and certain details are lost in black and white reproduction. Reproduced with permission from Glaser, K. B., Mobilio, D., Chang, J. Y., and Senko, N. (1993). Phospholipase A2 enzymes: Regulation and inhibition. *Trends Pharmacol. Sci.* **14,** 92–98.

TABLE 16-3 Physiological and Pathological Processes in which PLA_2 or a Product of PLA_2 Activity Has Been Implicated[a,b]

Physiological
Fat digestion
Lung surfactant metabolism
Cell membrane homeostasis
Insulin release
Sperm maturation–penetration
Lipoprotein metabolism
Hemostasis and blood clotting
Pathological
Rheumatoid arthritis
Psoriasis
Inflammation
Septic shock–endotoxin reactions
Premature labour
Renal disease
Reye's syndrome
Hypertension
Hyperthermia
Diabetic ketosis
Respiratory distress
Pancreatitis
Inflammatory bowel disease

[a] Most of the studies characterized specifically $sPLA_2$.

[b] Reproduced with permission from Glaser, K.B., Mobilio, D., Chang, J.Y., and Senko, N. (1993). Phospholipase A2 enzymes: Regulation and inhibition. *Trends Pharmacol. Sci.* **14,** 92–98.

Thus, there are receptors specific for PGD_2, PGE_2, $PGF_{2\alpha}$, PGI_2, and TXA_2 that have been studied. Some of these ligands, because of instability, are used in the radioactive form to measure receptor binding. Levels of PG producing a biological response correlate with the data of binding to the membrane receptor, as expected. Prostanoid receptors are so-called to include thromboxane receptors.

Two prostanoid receptors have now been cloned (TP and EP_3 receptors). When the cloned receptors are transfected into COS cells, they become functional. This family of receptors undoubtedly is linked to G proteins and, from study of the TP receptor, has seven membrane-spanning domains. The distribution of prostanoid receptors in various tissues and cells is shown is Table 16-5.

IV. BIOLOGICAL ACTIONS

A few biologically important systems will be discussed that emphasize the importance of prostaglandin relatives. At this point it is useful to examine the signal transduction of prostanoid receptors. Receptors involved in the contraction of smooth muscle, EP_1, EP_3, FP, and TP receptors, generally require an elevation of Ca^{2+} free in the cytoplasm. Some of these systems may involve the opening of Ca^{2+} channels operated by the activated receptor. Cytoplasmic Ca^{2+} levels can increase by an influx of Ca^{2+} through opened Ca^{2+} channels or by the mobilization of intracellular stores through the activation of the phosphatidylinositol system and the generation of IP_3 or both. It is possible that some receptors, EP_3 for example, can be coupled to different transducing systems in different tissues.

A. $PGF_{2\alpha}$ as an Agent for Therapeutic Abortion

At the termination of a normal pregnancy, parturition is started by an increased availability of fetal free cortisol generated by programmed changes in the fetal hypothalamus. The nature of the "clock" is unknown. With increased ACTH and a fall in the level of serum

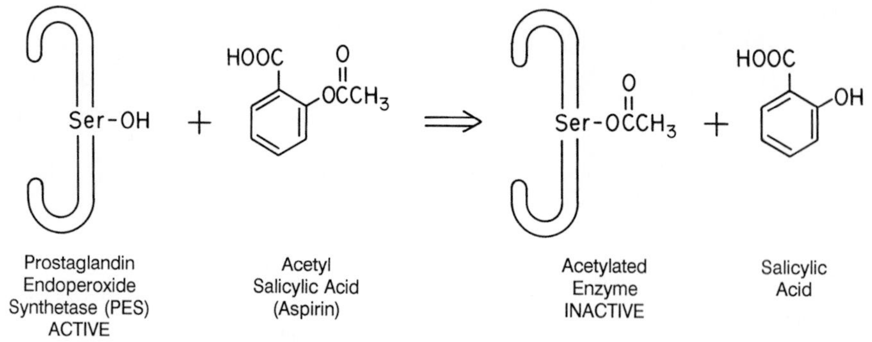

Prostaglandin Endoperoxide Synthetase (PES) ACTIVE + Acetyl Salicylic Acid (Aspirin) ⟹ Acetylated Enzyme INACTIVE + Salicylic Acid

FIGURE 16-9 Proposed mechanism for the inactivation of the cyclooxygenase component of prostaglandin endoperoxide synthesis by aspirin. Reproduced with permission from Gurr, M. I., and Harwood, J. L. (1991). "Lipid Biochemistry, An Introduction," 4th ed., p. 104, Chapman and Hall, London.

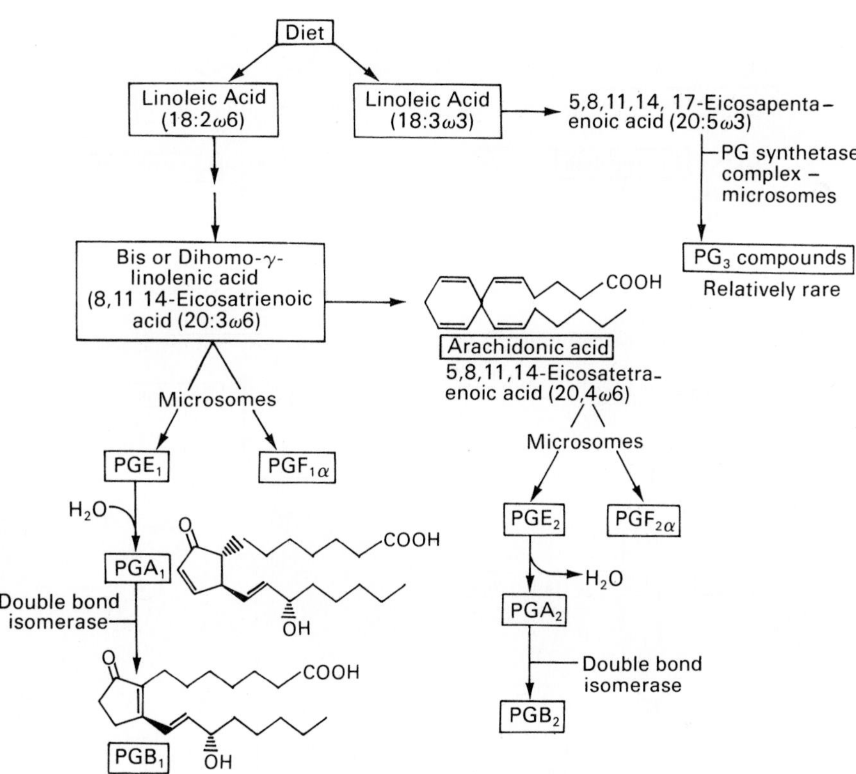

FIGURE 16-10 Overview of the biosynthesis of abundant and less abundant PGs.

transcortin, the circulating level of free cortisol is increased sharply. This leads to a marked decrease in placental production of progesterone and a stimulation of estradiol synthesis, $PGF_{2\alpha}$ synthesis, and oxytocin release, all of which bring about contractions of the uterus to expel the fetus. $PGF_{2\alpha}$ has been used as an agent for therapeutic abortion by intraamniotic injection, sometimes in combination with oxytocin. The uterine myometrium appears to contain $PGF_{2\alpha}$ receptors. The interaction of $PGF_{2\alpha}$ with receptor generates muscular contraction through increased levels of Ca^{2+} in the cytoplasm. This is expected to occur by the stimulation of phosphatidylinositol metabolism and the generation of IP_3. PGE analogues are often used during late first and early second trimester pregnancies before vacuum aspiration. Combination of prostaglandin with an antiprogestin can lower the effective concentration of prostaglandin, thus lowering gastrointestinal side effects and uterine pain.

B. Pancreas

Some experiments performed with isolated, perfused rat pancreas suggest that PGE_2, in the micromolar range, can stimulate the release of glucagon and insulin. Since glucagon release precedes that of insulin, the latter may be released secondarily by glucagon rather than in direct response to PGE_2. The effective concentrations of PGE_2 are in a range close to that observed in extracts of rat pancreas. It seems possible that PGE may be a transducer for several known secretagogues of glucagon.

C. Ectopic Bone Resorption

A transplantable mouse fibrosarcoma secretes a potent bone resorption-stimulating factor, which has been identified as PGE_2. Animals bearing this tumor have elevated levels of serum PGE_2 and Ca^{2+}, which can be lowered by the administration of inhibitors of PG synthetase, such as indomethacin.

D. Autonomic Nervous System

Stimulation of sympathetic nerves causes the release of PGEs. PGEs inhibit adrenergic transmission by depressing responses to the adrenergic transmitter at a postjunctional level. PGEs may also inhibit transmitter release from nerve terminals. Opposite effects are seen with PGFs, which often heighten the responses of effectors to norepinephrine. $PGF_{2\alpha}$ can facilitate the release of epinephrine from the adrenal medulla.

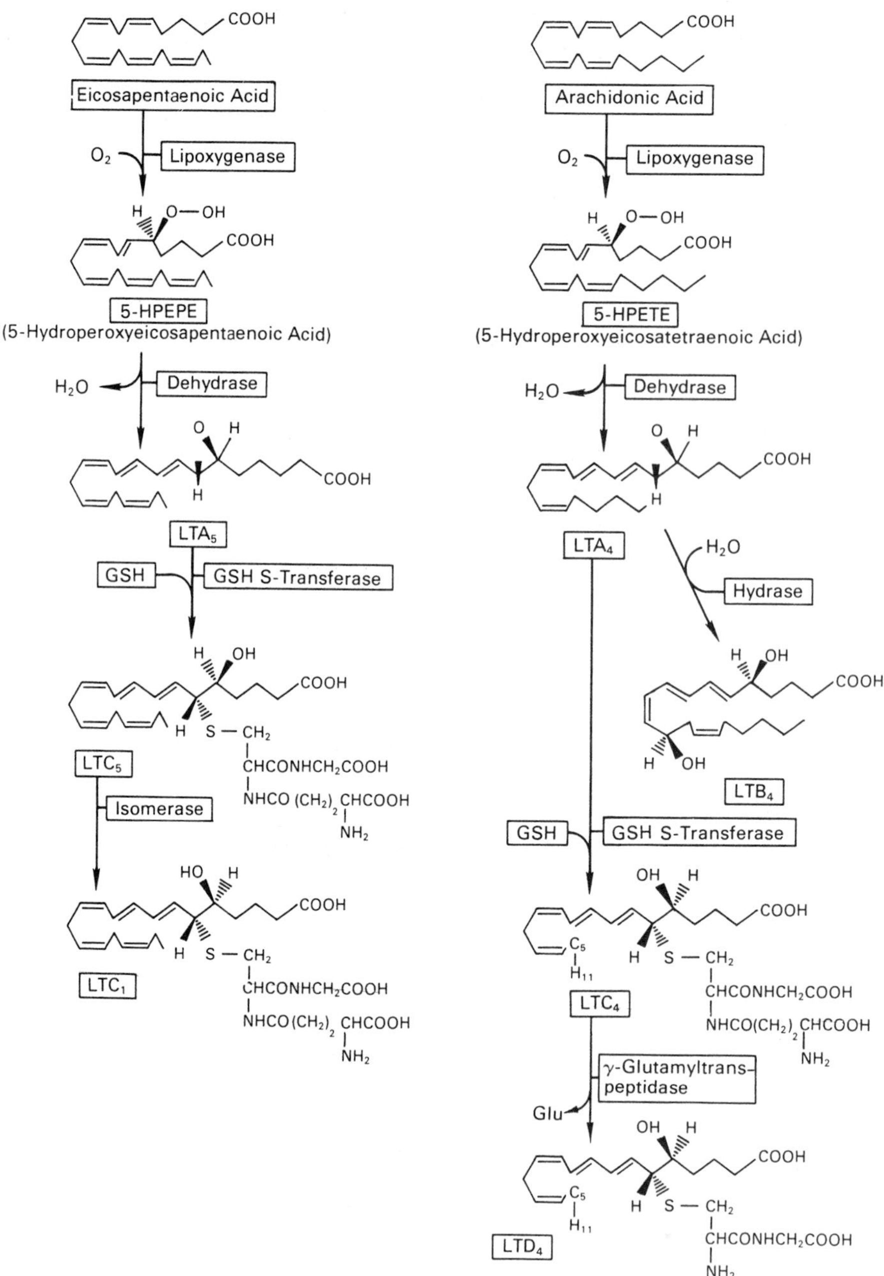

FIGURE 16-11 Biosynthesis of LT. Subscripts are numbers of double bonds.

E. PGI_2 in Fibroblasts

Reports with 3T3 mouse fibroblasts or human foreskin fibroblasts show that these cells synthesize PGI_2 (Figure 16-15). The stimulation of cyclic AMP in human fibroblasts is produced by PGs in the order $PGI_2 > PGH_2 = PGE_2 \gg PGE_2$, while PGD_2 is inactive. The stimulation by PGH_2 occurred through its transformation to PGI_2.

F. PGs and the Pain Mechanism

The pain mechanism is essential for survival, since acute pain is a warning mechanism for threatening conditions. Chronic pain is a more complicated subject, but both conditions may be grouped together for the purposes of our discussion. Relatively little is known about the pain mechanism.

The nociceptors, when excited by potentially harm-

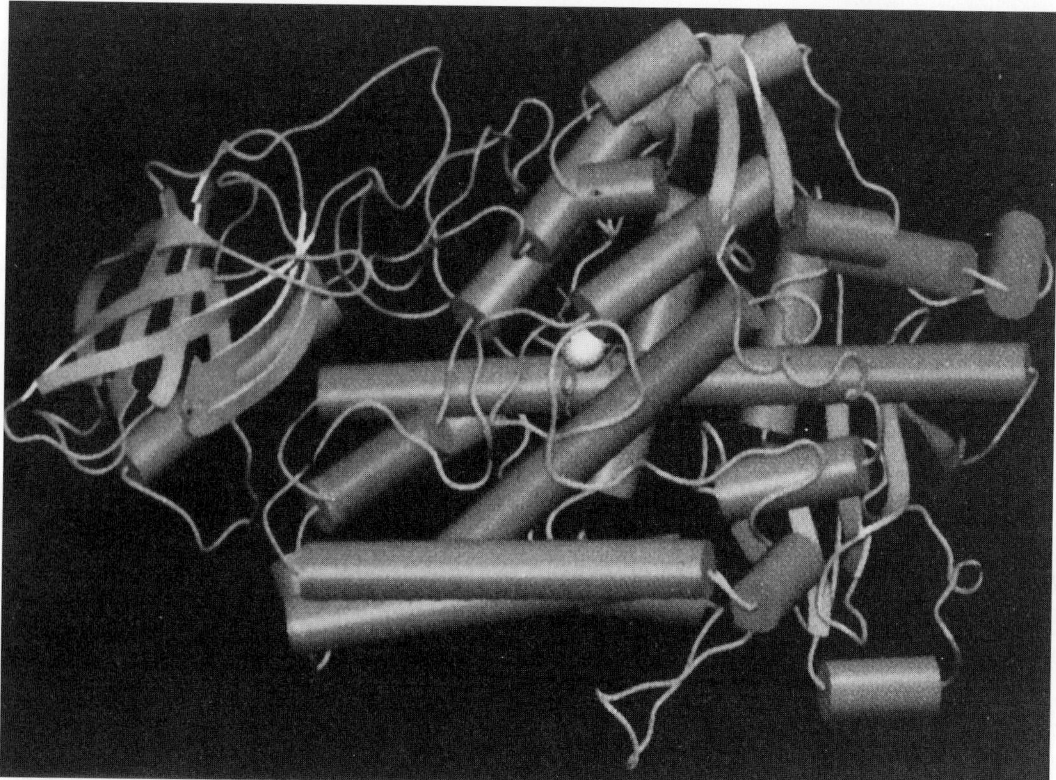

FIGURE 16-12 Schematic diagram of the three-dimensional structure of soybean lipoxygenase-1. The α-helices are represented by cylinders, the strands in the β-sheets by arrows, the coils by narrow rods, and the iron by a ball. Only three of the four iron ligands are shown. Domain I consisting of a single β-barrel is on the left. Domain II contains the iron, all of the helices in the structure, and two small β-sheet structures that lie on the surface of the enzyme. Reproduced as in black and white from a colored figure from Boyington, J. C., Gaffney, B. J., and Amzel, L. M. (1993). The three-dimensional structure of an arachidonic acid 15-lipoxygenase. *Science* **260,** 1482–1486.

ful stimuli, cause pain by way of their afferent nerve fibers (Figure 16-16). Probably every organ in the body contains these receptors. There are two classes of nociceptive afferents: those found among thin myelinated Aδ fibers and those among the nonmyelinated C fibers. The Aδ fibers are associated with sharp, focused pain, whereas the C fibers are associated with dull, burning, diffuse pain.

Chemical substances excite nociceptors or sensitize them to other stimuli (Figure 16-16A), resulting in the generation of pain. Endogenous substances in this category are PGE and bradykinin. These are referred to as "algesic substances" or agents producing pain. Substance P is considered in this category as it is thought to be a transmitter of the pain signal being released from the nociceptor (afferent) nerve terminal. Nociceptor firing occurs at levels of stimuli below those producing a frank sensation of pain; thus, these impulses probably summate spatially and temporally in the central nervous system before pain is perceived. Nociceptor firing rate and perception of pain both increase with the strength of the stimulus. PGs and other algesics apparently can cause pain when applied to the skin of humans. PGE_2 can lower the threshold of nociceptors to other stimuli, such as heat. It seems clear that PGEs operating through a receptor to stimulate adenylate cyclase and the level of cyclic AMP have a direct relation to the generation of pain. A speculative overall scheme tying together a number of observations is presented in Figure 16-17.

Pain is produced by an irritant, trauma, or nervous stimulation causing the release of fatty acids from membrane phospholipids, and PGEs are formed by the PG synthetase system. These dissolve in a cellular membrane and bind to the PGE receptor, which causes the stimulation of adenylate cyclase to form elevated amounts of cyclic AMP. It is not clear whether cyclic AMP is actually required for the production of pain or

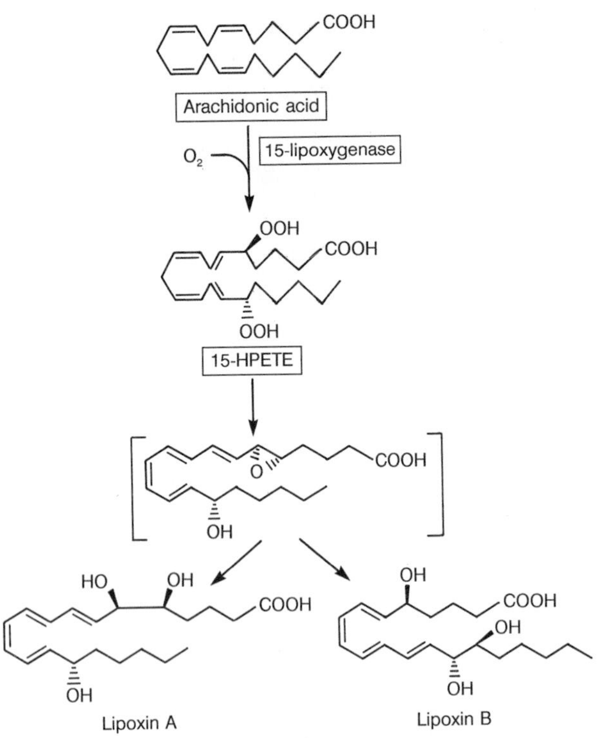

FIGURE 16-13 Formation of lipoxins from arachidonic acid.

the release of substance P, or whether PGE merely changes the properties of the nociceptor membrane, which results in increased firing. The elevated cyclic AMP, if involved, somehow produces pain, which, if persistent, results in a stress response via the hypothalamus (CRH) to release ACTH and then cortisol from the adrenal gland. In addition to ACTH, β-lipotropin is formed in equivalent amounts, which is processed to enkephalin via β-endorphin, and enkephalin binds to a cell membrane receptor, which may compete with the PGE receptor in some way or produce a second messenger opposing the action of cyclic AMP.

Elevated levels of circulating cortisol may produce a polypeptide, called "lipocortin" or annexin I (~40,000 Da), in the same cells that originally produced the elevated PGEs. This polypeptide is an inhibitor of phospholipase A_2 and thus quenches the production of PGEs leading to pain in the first place. This is expected to be a slower process than the analgesia produced by enkephalin. In addition, there are direct nervous connections between the hypothalamus and the adrenal medulla, which lead to the release of epinephrine and enkephalins in large amounts. The function of medulla enkephalin is presently unknown; however, it may have a positive action in counteracting pain and stress by peripheral actions, but its access to the central nervous system is unlikely in view of the blood–brain barrier.

G. Antihypertensive Renal Effects

In response to some stimuli (e.g., fright), catecholamines are secreted, causing renin to be released from the kidney juxtaglomerular apparatus. Renin stimulates the production of angiotensin II (Figure 15-7), and blood pressure elevation ensues from enhanced aldosterone production, Na^+ reabsorption, and constrictive effects on the vasculature. In response to the increased pressure, there is increased blood flow in the renal medulla, resulting in the release of PGA_2 or PGE_2 (see Chapter 15).

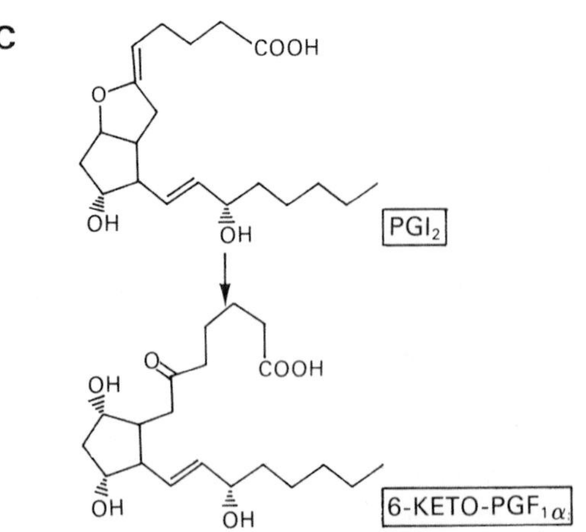

FIGURE 16-14 (A) Positions in which PGs are metabolized. (B) Metabolism of TX. (C) Metabolism of PGI_2.

TABLE 16-4 Prostanoid Receptors[a]

Receptor	Subtype	Ligand	Selective agonist	Selective antagonist	Transduction mechanism
DP		$[^3H]PGD_2$	BW245C	BWA868C	↑cAMP
EP	EP_1	$[^3H]PGE_2$		AH6809 SC-19220	IP_3/DG
	EP_2		Butaprost AH13205		↑cAMP
	EP_3		Enprostil GR63799		↓cAMP (IP_3/DG?)
					IP_3/DG
FP		$[^3H]PGF_{2\alpha}$	Fluprostenol Cloprostenol		IP_3/DG
IP		(PGI_2) $[^3H]$Iloprost	Iloprost Cicaprost		↑cAMP
TP		(TXA_2) $[^3H]$U-46619	U-46619	Vapiprost (GR32191)	IP_3/DG
		$[^3H]$SQ29548		BAYu3405	

[a] ↓↑ cAMP: inhibition or stimulation of adenylate cyclase. IP_3/DG: stimulation of phosphoinositide metabolism. Reproduced with permission from Coleman, R. A., and Humphrey, P. P. A. (1993). Prostanoid receptors: Their function and classification. *In* "Therapeutic Applications of Prostaglandins" (J. R. Vane and J. O'Grady, eds.), p. 15. Edward Arnold, London.

H. The Platelet System

PGI_2 and TXA_2 generally produce opposite reactions in the platelet. TXA_2 produced in the platelet stimulates platelet aggregation, while PGI_2 produced in the blood vessel wall inhibits platelet aggregation. It is not absolutely certain that TXA_2 is the active principle within the platelet, but there is good evidence supporting this conclusion. PGI_2 also relaxes coronary arteries whereas TXA_2 constricts them; PGI_2 lowers blood pressure and TXA_2 raises blood pressure.

The role of TXs and PGI_2 in platelet aggregation is suggested by Figure 16-18. Wounding and exposure of collagen in the vascular wall are sufficient to cause platelet aggregation. Arachidonate is released from the platelet membrane and TXA_2 is produced. TXA_2 causes the release of Ca^{2+} from intracellular storage sites, and the elevated Ca^{2+} inhibits adenylate cyclase, which is normally stimulated by circulating PGI_2 synthesized in the blood vessel walls. The elevated Ca^{2+} also stimulates the release of dense granules from the platelet that are rich in ADP and serotonin. ADP combines with the platelet, causing its surface to become sticky and the platelets to aggregate. The effect of ADP on platelets can be seen in Figure 16-19.

A hemostatic plug of platelets is formed, which aids in the repair process. The sequence of these events is diagrammed in Figure 16-20. Under normal conditions when there is no stimulus to set off the production of TXA_2 in the platelet, the circulating PGI_2 continuously stimulates an elevated level of cyclic AMP in the platelet, which inhibits the mobilization of internal stores of Ca^{2+} and inhibits the activities of phospholipase A_2 and cyclooxygenase, although these latter conclusions remain to be elaborated. The overall activities in the platelet are shown in Figure 16-21.

V. LEUKOTRIENES (LTs)

A. Background

These substances hold great promise for progress in diseases such as asthma. The LTs were identified as the principles of the slow-reacting substance of anaphylaxis (SRS or SRSA). The LTs are closely related to immediate hypersensitivity reactions. SRS causes slow contractile response of smooth muscle *in vitro* and has been isolated from the lung of pigs and humans. SRS-A from an asthmatic causes long-

TABLE 16-5 Distribution of Prostanoid Receptors in Various Tissues and Cell Types[a]

	Response	
Tissue–cell type	Stimulation[b]	Inhibition[c]
Smooth muscle		
Vascular	TP, (EP_3)	DP, EP_2, IP
Respiratory	TP, (EP_1)	EP_2
Gastrointestinal	EP_1, EP_3, FP, TP	EP_2 (DP), (IP)
Uterine	EP_1, EP_3, FP, TP	(DP), (EP_2) (IP)
Ocular	EP_1, FP, TP	EP_2
Inflammatory cells		
Monocytes		DP, EP_2
Granulocytes		DP, EP_2
Lymphocytes		EP_2
Neurons		
Autonomic		EP_3 (DP), (IP)
Afferent sensory	EP_2, IP	
Myofibroblasts	TP	
Platelets	EP_3, TP	
Adipocytes		EP_3
Corpus luteum	FP	
Epithelial cells	EP_3	EP_2

[a] Receptor types commonly found in each cell type or, in parentheses, found only in those from particular species. This table is reproduced by permission from Coleman, R. A., and Humphrey, P. P. A. (1993). Prostanoid receptors: Their function and classification. *In:* "Therapeutic Applications of Prostaglandins" (J.R. Vane and J. O'Grady, eds.), p. 21. Edward Arnold, London.

[b] Contraction of smooth muscle and myofibroblasts, activation of inflammatory cells to release mediators, depolarization of neurons, aggregation of platelets, luteolysis of the corpus luteum, and secretion from epithelial cells.

[c] Relaxation of smooth muscle, inhibition of mediator release from inflammatory cells, hyperpolarization of neurons, inhibition of aggregation by platelets, lysolysis by adipocytes, and secretion by epithelial cells.

lasting bronchoconstriction and is extremely potent. SRS-A has been shown to be formed by the incorporation of radiolabeled arachidonate and ^{35}S. SRS compounds have been characterized as LTs (see Figure 16-11). LTA, LTB, and LTC were isolated from a murine mastocytoma cell line and from rabbit polymorphonuclear leukocytes. LTs are derivatives of arachidonic acid or eicosapentaenoic acid, and many contain glutathione or cysteine.

B. LT and Asthma

Asthma is a complex disease resulting in part from narrowing of the airways. Mediators having various roles in this syndrome are derived from mast cells in the bronchus. These mediators act either directly on airway tissues or by indirect mechanisms, which include recruitment of inflammatory cells. Some mediators are preformed within the mast cell, whereas others are generated from the plasma membrane. The latter category includes the LTs, which may play a substantial role in asthma. The two classes of mediators are summarized in Figure 16-22. One of the major secretions of the mast cell is histamine, which, experimentally, causes a rapid and transient increase in airway resistance resulting from the action of histamine on bronchial smooth muscle. Other secretions include ECF-A (eosinophil chemotactic factor of anaphylaxis), NCF (neutrophil chemotactic factor of anaphylaxis), and those listed in Figure 16-22. The precise roles of the enzymes secreted have not been clarified.

The membrane-derived factors are produced from arachidonic acid, which is liberated by the action of phospholipase A_2 on phosphatidylcholine and other

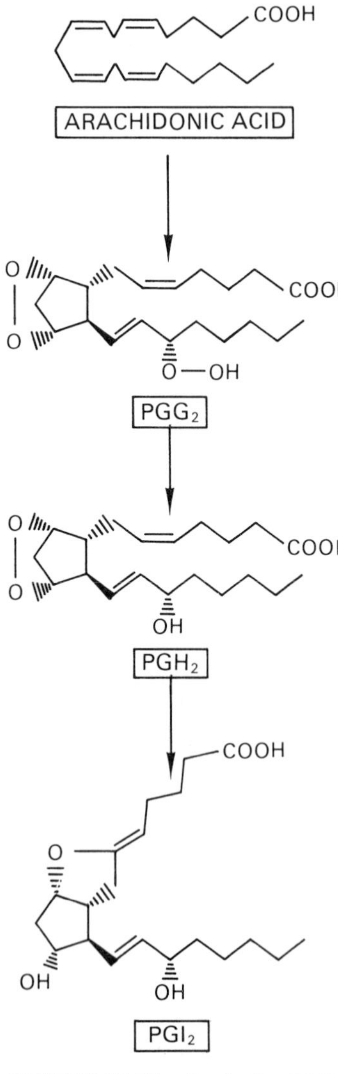

FIGURE 16-15 Synthesis of PGI_2.

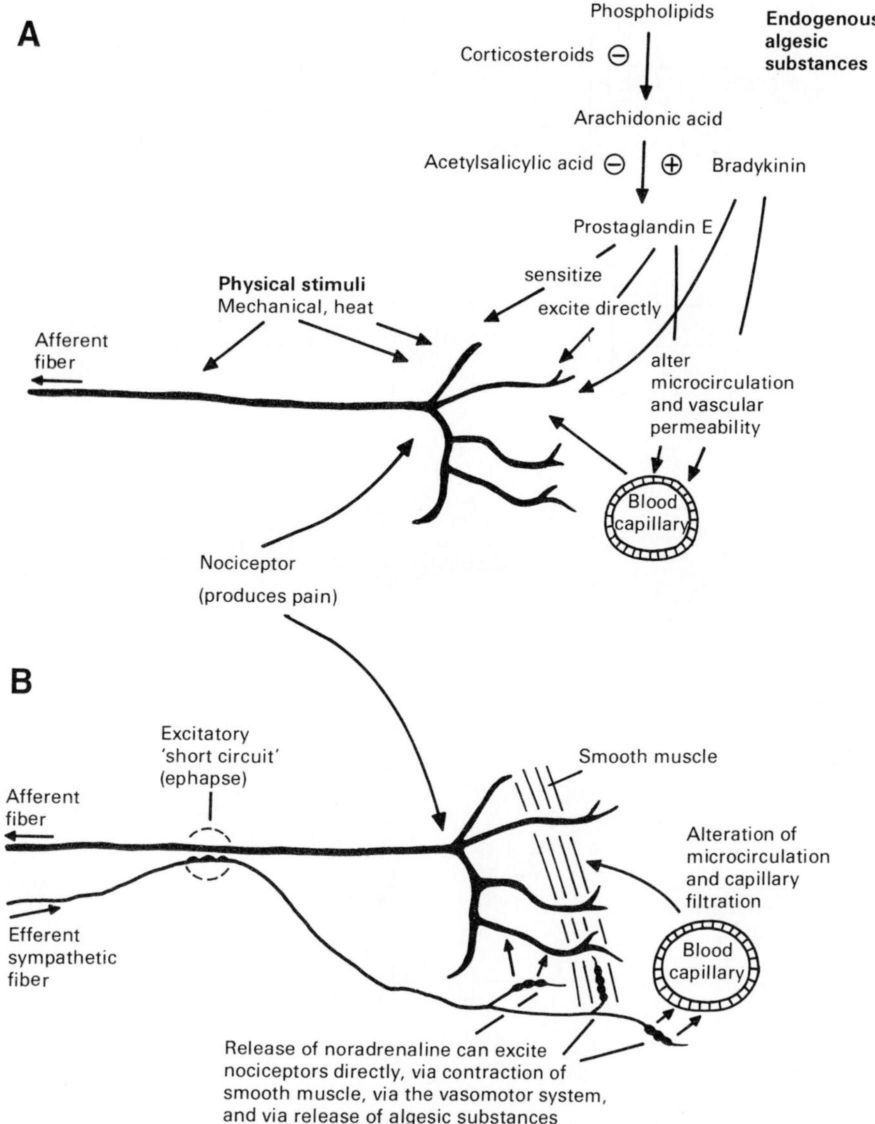

FIGURE 16-16 Pain transmission and the nociceptor. (A) Diagram showing the nociceptor and its environment. The nociceptor, represented anatomically as the terminal arborization of an afferent fiber, has its excitability influenced by physical stimuli and endogenous algesic substances. The action of these substances can also be indirect, operating on the microcirculation. Vanous substances, such as bradykinin, facilitate PG synthesis and can intensify the action of algesic substances. Analgesics can inhibit PGE synthesis, explaining part of their actions. (B) Diagram of the nociceptor and its environment and possible sympathetic mechanisms that may influence the excitability of the nociceptors. Sympathetic efferents can stimulate nociceptors when regulation is disturbed. Reproduced from Zimmerman, M. (1981). *Triangle Sandoz J. Med. Sci.* **20,** 7. Copyright 1981, Pergamon, Ltd., London/NY.

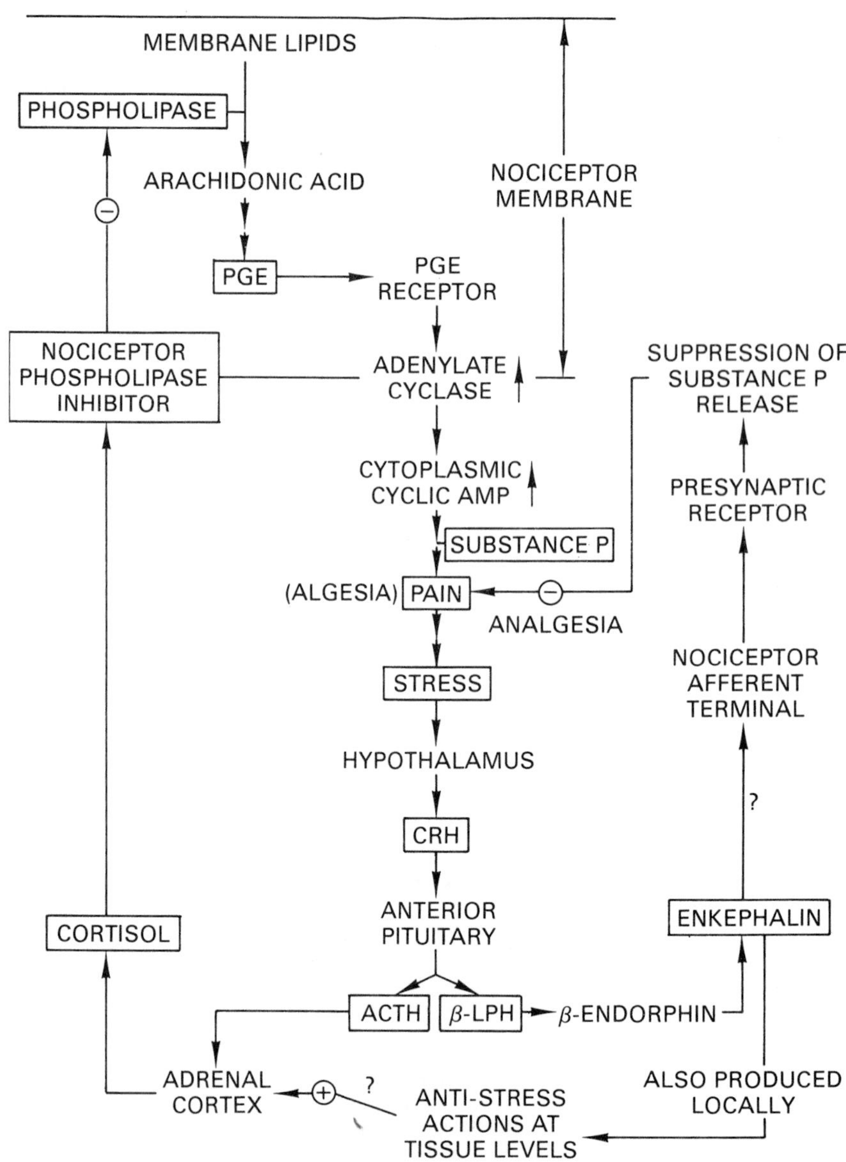

FIGURE 16-17 Highly speculative model on the actions of PGEs versus enkephalin and glucocorticoids on pain production.

phospholipids. When an allergen interacts with IgE in the cell membrane (Figure 16-23), cell membrane methyltransferases are activated and catalyze the conversion of phosphatidylethanolamine to phosphatidylcholine. This produces changes in the membrane and in the orientation of phospholipids and results in opening of the calcium ion gate, or constitution of the gate as the case may be, so that calcium is taken up from the extracellular space. This series of events is similar to other receptor systems (see Chapter 1). The increased level of calcium-activated membrane phospholipase A_2 results in the release of arachidonic acid from membrane phospholipid precursors. The cyclooxygenase-mediated metabolism of arachidonic acid would give rise to PGs, PGI_2, or TX, whereas lipoxygenase-mediated events would form LT and monohydroxy fatty acids.

The LTs are of special interest because they possess the activities of the slow-reacting substance of anaphylaxis. The precursor to the LTs is hydroperoxyeicosatetraenoic acid (5-HPETE) (Figure 16-11). The LTs contain a 5-hydroxy group, three conjugated double bonds, and other polar substituents such as glutathione (GSH) (as in LTC4), as shown in Figure 16-11. The most potent LTs have one cis and two trans double bonds in the triene portion of the molecule (C-7). In most *in vitro* systems involving potency of smooth muscle contraction, the order of active substances is

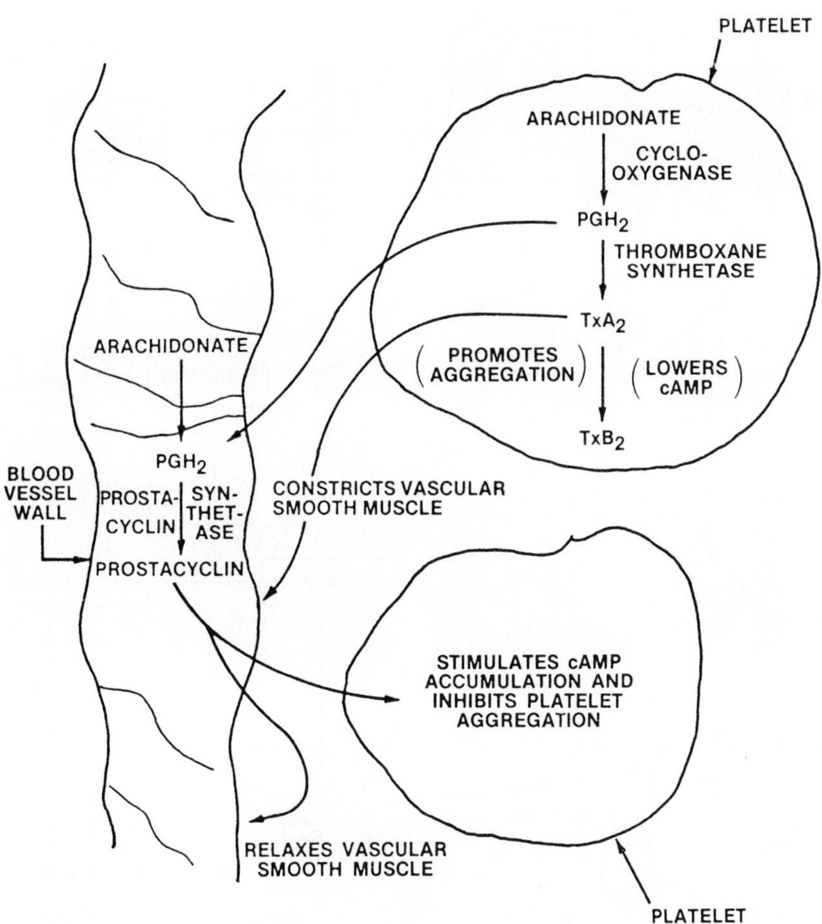

FIGURE 16-18 Model of human platelet homeostasis. Reproduced from Gorman, R. R. (1979). Modulation of human platelet function by prostacyclin and thromboxane A_2. *Fed. Proc. Fed. Am. Soc. Exp. Biol.* **38,** 83.

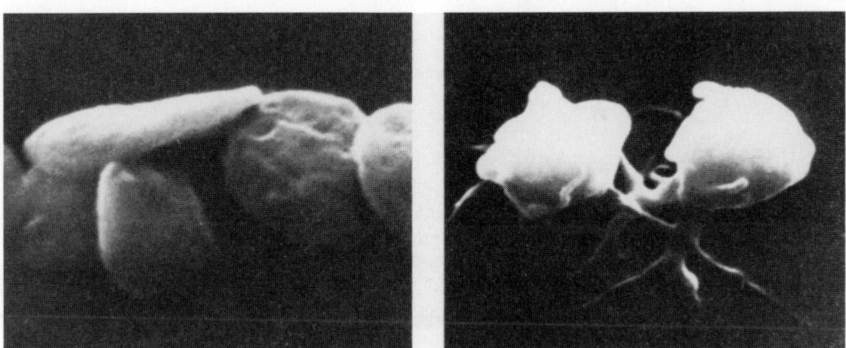

FIGURE 16-19 Platelets change shape when exposed to ADP, which is liberated from platelet dense granules. Platelets are smooth flat disks shown in the electron micrograph at the left. Treatment with ADP causes long processes to protrude from the platelets, which swell to become irregular "spiny spheres." The processes then adhere to a surface or to other platelets, shown on the right. Reproduced from Zucker, M. B. (1980). *Sci. Am.* **242,** 86–103.

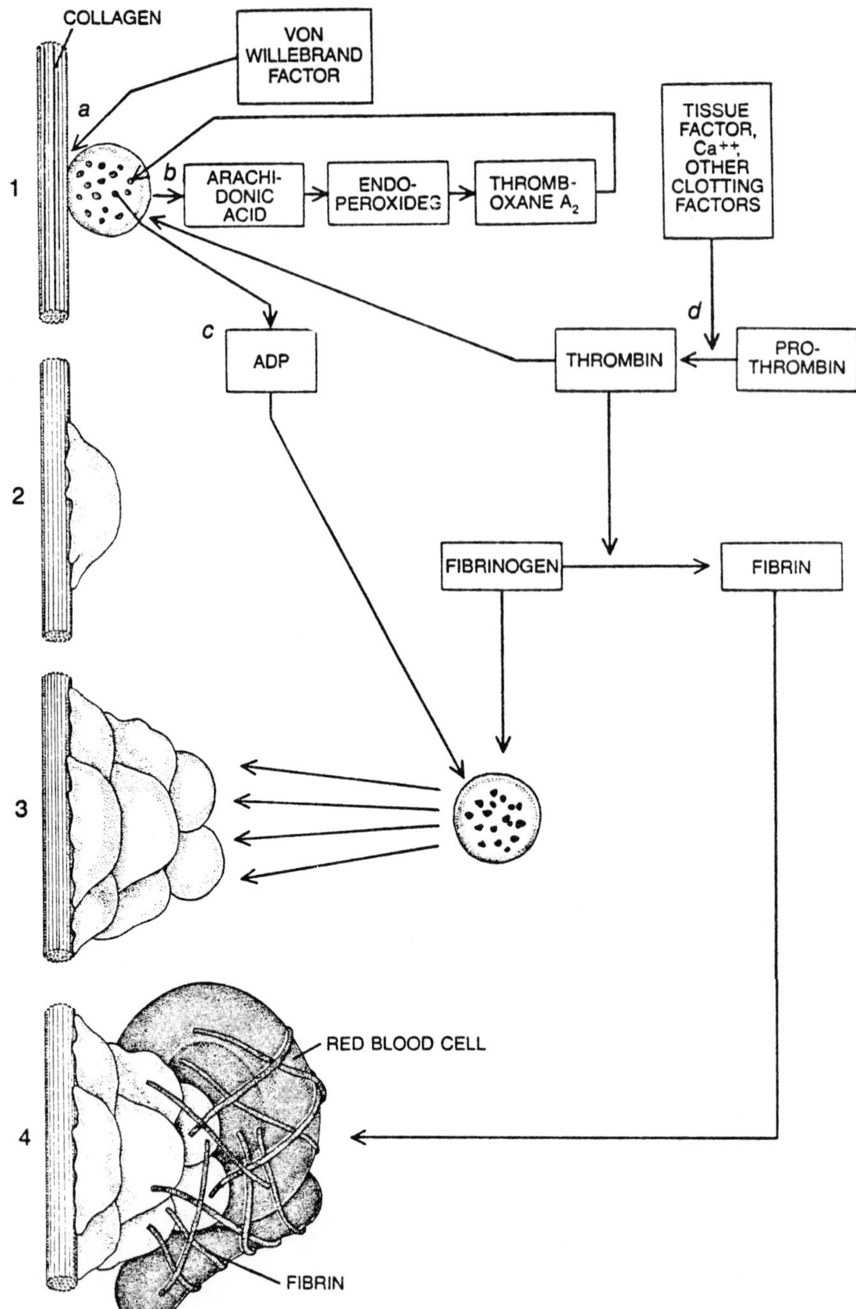

FIGURE 16-20 Complex interactions of platelets, their constituents, and other substances involved in the formation of the hemostatic plug and of thrombi are still not understood. Some are diagrammed here. Contact of a platelet with collagen (1) in the presence of von Willebrand factor (a) initiates a pathway (b) that stimulates the secretion of ADP from the dense granules (c). The adhering platelet changes shape, spreads out along the collagen, and degranulates (2). Meanwhile, a number of steps involving tissue factor, calcium ions, and other clotting factors convert prothrombin in the plasma into thrombin (d), which is also formed on the platelet surface. Thrombin also stimulates secretion and converts fibrinogen (from the plasma and platelets) into fibrin. Platelets aggregate under the influence of collagen, ADP, and thrombin (3). Strands of fibrin reinforce the plug. The process may stop at this stage or may go on to form a larger thrombus with trapped red blood cells (4). Reproduced from Zucker, M. B. (1980). *Sci. Am.* **242**, 86–103.

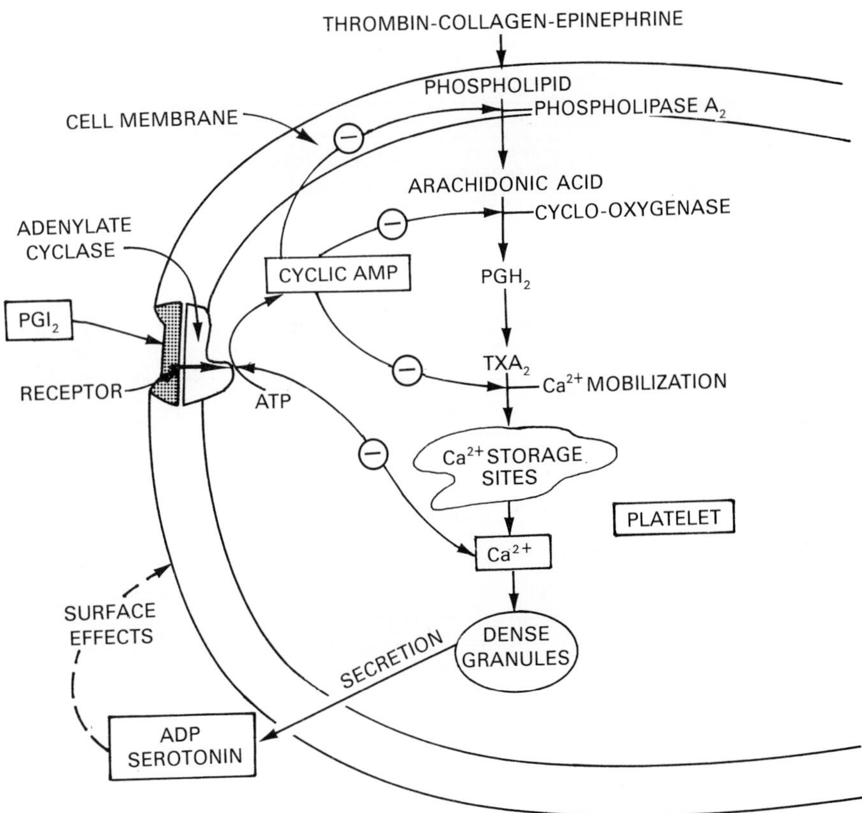

FIGURE 16-21 Interrelationships between platelet cell membrane stimulators (thrombin, collagen, epinephrine), TX (TXA), Ca^{2+}, PGI_2, and adenylate cyclase.

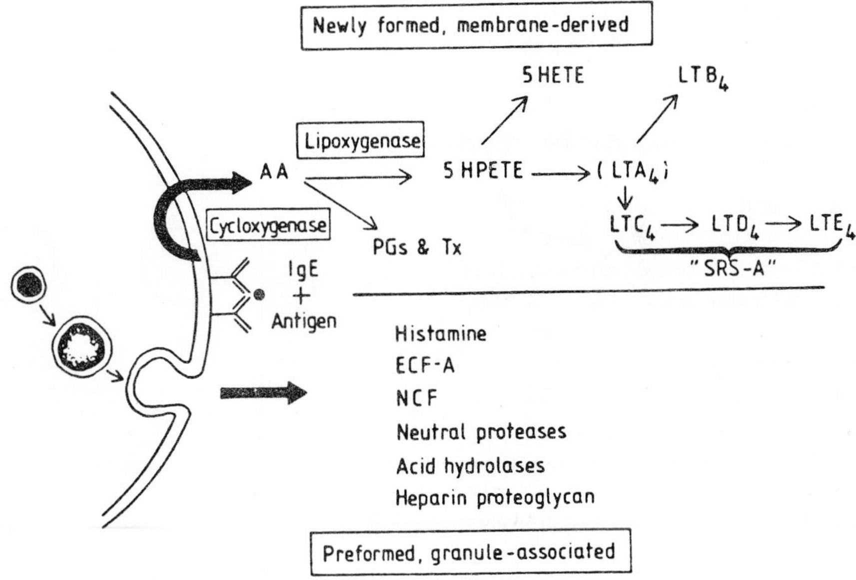

FIGURE 16-22 Mast cell-derived pharmacological mediators of hypersensitivity. Reproduced from Kay, A. B. (1982). *Eur. J. Respir. Dis.* **63** (Suppl. 122), 9–16. Copyright 1982 Munksgaard International Publishers Ltd., Copenhagen, Denmark.

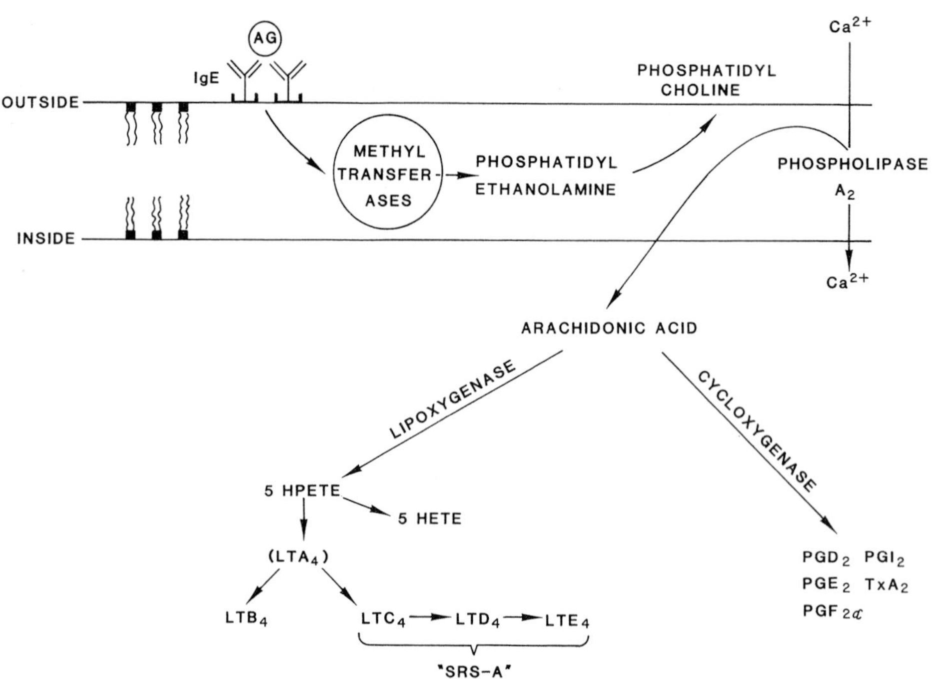

FIGURE 16-23 Membrane events and the release of free arachidonic acid. Reproduced from Kay, A. B. (1982). *Eur. J. Respir. Dis.* **63** (Suppl. 122), 9–16. Copyright 1982 Munksgaard International Publishers Ltd., Copenhagen, Denmark.

$LTD_4 > LTC_4 > LTE_4$. The polar LTs, LTB_4 especially (Figure 16-11), are potent chemotactic factors.

The other PGs and TXA_2 also have effects on smooth muscle. PGD_2, $PGF_{2\alpha}$, and TXA_2 constrict, but PGE_2 has a slight dilatory effect. PGD_2, PGE_2, and PGI_2 add to the effects on the permeability of histamine and bradykinin. Thus, while histamine gives a transient contraction of human lung smooth muscle *in vitro,* LTC_4 and LTD_4 cause prolonged responses and may be critical agents in asthma.

Thus, bronchial obstruction can be arbitrarily divided into three phases: a rapid spasmogenic phase, a late sustained phase, and a subacute inflammatory phase. These phases might occur in sequence following the initial release of agents from the mast cell resulting from the interaction of allergen and IgE or other nonimmunological stimuli, such as infection, exercise, or complement activation. These phases are represented in Figure 16-24. The rapid phase is probably histamine-mediated. A more severe stage of the reaction undoubtedly is related to the LTs. The late reaction may be associated with the reactivation of mast cells. LTs probably play a role in the late reaction as well since there is prolonged contraction of bronchial smooth muscle. Late reactions are inhibited by glucocorticoids, which act to prevent the release of arachidonic acid by inducing the synthesis of a peptide (lipocortin/annexin I) inhibitor of phospholipase A_2, although the mechanism needs further clarification. This is the same reaction that was discussed in connection with prostaglandin-induced pain (Figure 16-17). Corticosteroid-resistant asthma is a very serious problem in which LTs may

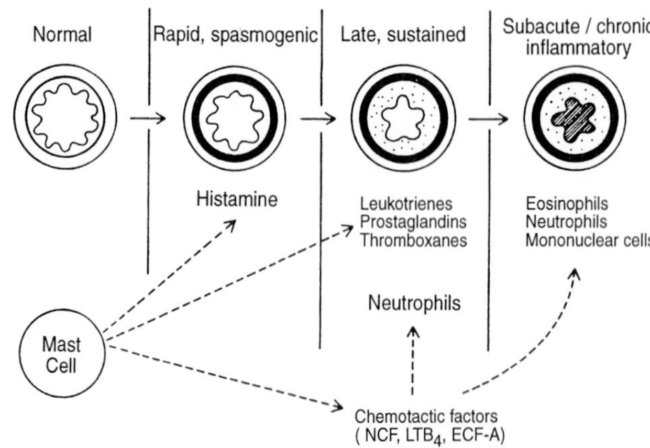

FIGURE 16-24 Mast cell mediators and airway obstruction in bronchial asthma. Reproduced from Kay, A. B. (1982). *Eur. J. Respir. Dis.* **63** (Suppl. 122), 9–16. Copyright 1982 Munksgaard International Publishers Ltd., Copenhagen, Denmark.

play an influential part. Prostaglandin relatives thus are important in producing the alterations in asthma. The cysteinyl LTs have special importance in the production of bronchospasm and bronchial hyperresponsiveness, and PGE_2 is involved in the bronchospastic and inflammatory responses. The contribution of other PG relatives is smaller than those mediators listed earlier. Not to be forgotten is a considerable variation from one patient to the next. An outline of the reactions occurring in asthma is shown in Figure 16-25. Glucocorticoids are a main therapy for asthma, as seen from this figure, in that they inhibit phospholipase A_2, presumably through the induction of lipocortin/annexin I, which inhibits the enzyme directly or indirectly. Glucocorticoids may also induce the cell death of eosinophils, which flood the lungs and account for some of the inflammatory problems. Little is known about the therapeutic effects of glucocorticoids on eosinophils, but glucocorticoids probably induce the eosinophil cell death mechanism and probably interfere with GM-CSF production/utilization needed for eosinophil proliferation.

VI. CLINICAL ASPECTS

A. Abortion

There is good evidence that PGs participate in various aspects of the ovarian cycle. In particular, $PGF_{2\alpha}$ has been implicated as one of the factors effective in causing terminal contractions on the uterine myometrium. The levels of $PGF_{2\alpha}$ in amniotic fluid align with the progress of labor during the birth process. Administered PG in-

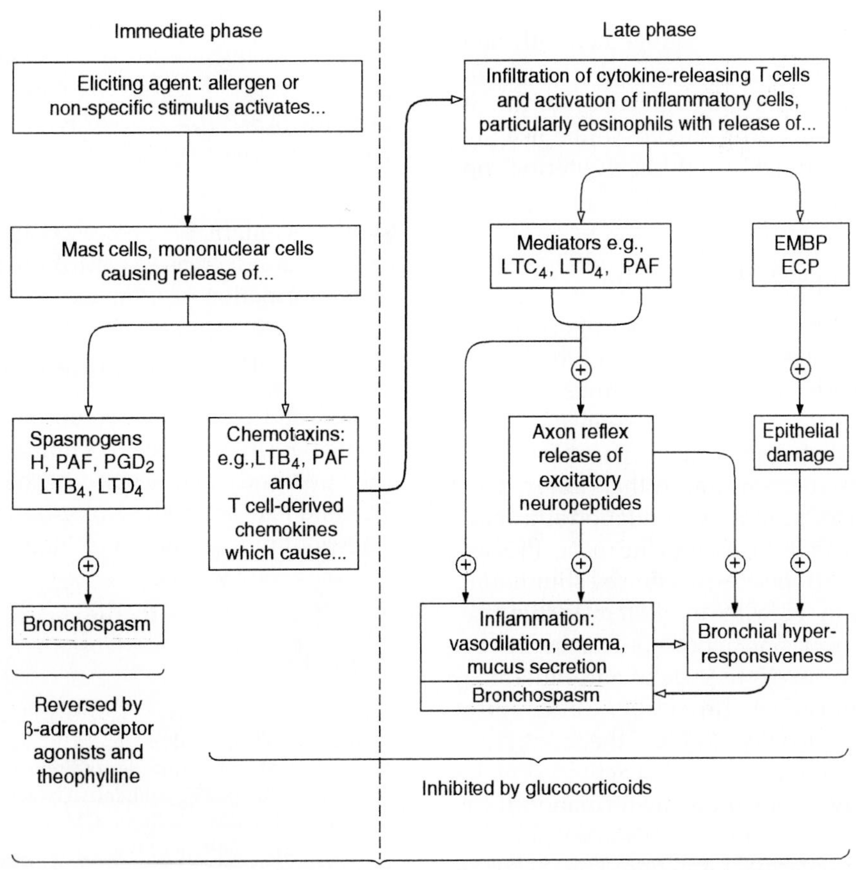

FIGURE 16-25 Outline of the reactions thought to occur in asthma. Abbreviations: H, histamine; PAF, platelet-activating factor; LTB_4, LTC_4, and LTD_4, leukotrienes B_4, C_4, and D_4; EMBP, eosinophil major basic protein; ECP, eosinophil cationic protein. Note that not all asthmatic subjects respond to cromolyn sodium or nedocromil sodium and that theophylline is only a second-line drug. Reproduced with permission from Rang, H. P., Dale, M. M., Ritter, J. M., and Gardner, P. (1995). "Pharmacology." Churchill-Livingstone, New York.

hibitors such as aspirin or indomethacin are known to delay the onset of labor. Administration of PG, often in the form of $PGF_{2\alpha}$, can produce side effects and should not be given to patients with a history of asthma. PGs are generally useful as abortifacients after the eighth week of pregnancy. In the second trimester, PGE_2 and $PGF_{2\alpha}$ can be given by the intrauterine route and other PGs can be given intramuscularly or vaginally to promote the termination of pregnancy.

B. Erectile Dysfunction

PGE_1 has been used successfully as a cure for erectile dysfunction. Sometimes prolonged erection was a problem after use of PGE_1. Self-injection is used with minimal side effects. Some patients may be sensitive to intracavernosal injections of PGE_1 and require adjustment of the dose. Topical application rather than injection may prove useful.

C. Gastroduodenal Ulceration

Misoprostol (PGE_1) can be used together with nonsteroidal anti-inflammatories to prevent gastric and duodenal ulcers. Diarrhea is a complication of the intake of misoprostol. This drug should not be used by pregnant women because of its effects on uterine contraction and abortion.

D. Myocardial Infarction

PGs play important roles in a number of organ systems, including the central nervous system, blood platelets (as described in this chapter), smooth muscles of the respiratory tract (also described here in connection with asthma), peripheral nervous system, gastrointestinal tract, and cardiovascular system. In the kidney, PGs, possibly mainly PGAs, produce vasodilation and accelerate the removal of sodium ion into the urine. PGs apparently can act in the opposite direction by stimulating the renin–angiotensin–aldosterone system. In this case, as in many others, opposing actions of PGs occur as a homeostatic mechanism. Nonsteroidal anti-inflammatory drugs, if administered chronically, can compromise the hypotensive activity of PGs in the kidney.

There is some reason to believe that activated platelet activity may play a role in plaque formation if the reaction is overresponsive and can also contribute to some instances of myocardial infarction. Thus, TX production by platelets would be balanced by PGI_2 production. "Sticky" platelets could have a negative effect on the development of myocardial infarction over the long term, and some claims have been made that chronic aspirin users could have a lower incidence of myocardial infarction, the explanation being that TX

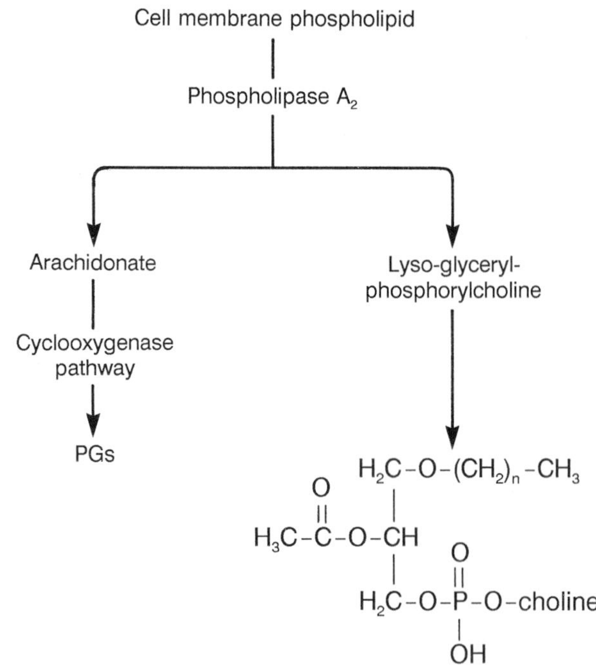

FIGURE 16-26 Generation of PGs and PAF at the site of embryo implantation.

synthesis might be decreased in platelets and there would be a decreased contribution of sticky platelets to the formation of infarcts.

E. Implantation of Embryo

PGs and platelet-activating factor (PAF) appear to participate in the process of human implantation. PG_2's are apparently synthesized at the implant action site. PGs and PAF are both synthesized from cell membrane phospholipid, as shown in Figure 16-26.

References

A. Books

Gurr, M. I., and Harwood, J. L. (1991). "Lipid Biochemistry." Chapman and Hall, London.

Lee, J. B. (1975). *In* "Principles of Endocrinology" (R. H. Williams, ed.), 5th ed., p. 855. Saunders, Philadelphia.

Rang, H. P., Dale, M. M., Ritter, J. M., and Gardner, P. (1995). "Pharmacology." Churchill Livingstone, New York.

Vane, J. R., and O'Grady, J., eds. (1993). "Therapeutic Applications of Prostaglandins" Edward Arnold, London.

B. Review Articles

Boyington, J. C., Gaffney, B. J., and Amzel, L. M. (1993). The three-dimensional structure of an arachidonic acid 15-lipoxygenase. *Science* **260,** 1482–1486.

Fenn, G. C., and Robinson, G. C. (1991). Misoprostol—A logical therapeutic approach to gastroduodenal mucosal injury induced by non-steroidal anti-inflammatory drugs? *J. Clin. Pharm. Ther.* **16,** 385–409.

Glaser, K. B., Mobilio, D., Chang, J. Y., and Senko, N. (1993). Phospholipase A2 enzymes: regulation and inhibition. *Trends Pharmacol. Sci.* **14,** 92–98.

Hammarstrom, S. (1984). The leukotrienes. *In* "Biochemical Actions of Hormones" (G. Litwack, ed.), Vol. 11, pp. 1–23. Academic Press, New York.

Schror, K. (1993). The effect of prostaglandins and thromboxane A_2 on coronary vessel tone-mechanisms of action and therapeutic implications. *Eur. Heart J.* **14** (Suppl.), 34–41.

Thien, F. C. K., and Walters, E. H. (1995). Eicosanoids and Asthma: An update. *Prostaglandins Leukotrienes and Essential Fatty Acids* **52,** 271–288.

C. Research Papers

Diczfalusy, U., and Alexson, S. E. H. (1990). Identification of metabolites from peroxisomal beta-oxidation of prostaglandins. *J. Lipid Res.* **31,** 307–313.

Gorman, R. R. (1979). Modulation of human platelet function by prostacyclin and thromboxane A_2. *Fed. Am. Soc. Exp. Biol. Fed. Proc.* **38,** 83.

Hirata, M., Hayashi, Y., Ushikubi, F., *et al.* (1991). Cloning and expression of cDNA for a human thromboxane A_2 receptor. *Nature* **349,** 617–620.

Kay, A. B. (1982). *Eur. J. Respir. Dis.* **63** (Supple. 122), 9–16.

Langs, D. A., Erman, M., and Detita, G. T. (1977). *Science* **197,** 1003–1005.

Langs, D. A., Fortier, S., Erman, M. G., and Detita, G. T. (1979). *Nature (London)* **281,** 237–238.

Lawrence, R. A., Jones, R. L., and Wilson, N. H. (1992). Characterization of receptors involved in the direct and indirect actions of prostaglandins E and I on the guinea pig ileum. *Br. J. Pharmacol.* **105,** 271–278.

Shenker, A., Goldsmith, P., Unson, C. G., and Spiegel, A. M. (1991). The G protein coupled to the thromboxane A_2 receptor in human platelets is a member of the novel Gq family. *J. Biol. Chem.* **266,** 9309–9313.

Strong, P., Coleman, R. A., and Humphrey, P. P. A. (1992). Prostanoid-induced inhibition of lipolysis in rat isolated adipocytes: probably involvement of EP_3 receptors. *Prostaglandins* **43,** 559–566.

Sugimoto, Y., Namba, T., Honda, A., *et al.* (1992). Cloning and expression of cDNA for mouse prostaglandin E receptor EP_3 subtype. *J. Biol. Chem.* **267,** 6463–6466.

Zimmerman, M. (1981). *Triangle Sandoz J. Med. Sci.* **20,** 7.

Zucker, M. B. (1980). *Sci. Am.* **242,** 86–103.

CHAPTER

17

Thymus Hormones

I. INTRODUCTION

A. Background

The thymus gland and various polypeptide hormones produced by thymus cells play a critical role in the functioning of the immune system. A few major polypeptide hormones elaborated by thymus cells are needed for growth and immunoglobulin production by antibody-producing cells. Here, the object will be to describe these hormones and their actions. Although the thymus epithelial cell is a rich source of thymus hormones, some of these substances are found in other cells of the body, so that the thymus is not always the exclusive source.

In addition to the central role of the thymus in the immune mechanism, the gland may produce components that regulate the secretion of hormones by other glands, for example, the release of prolactin from lactotrophs of the anterior pituitary. Thymus hormones also regulate the secretion of certain releasing hormones from the hypothalamus. However, if the thymus produces hormones that regulate the secretion of anterior pituitary hormones, then it is possible that feedback regulation could exist, which, in the case of prolactin, might lead to a partial explanation for its pronounced increase in stress (see Chapter 10). Regulation of hypothalamic secretions is important in the development of the sex organs because thymosin, a thymus hormone, is required at birth for the full release of GnRH, which, in turn, triggers the release of the gonadal hormones. Although this process often is not considered until puberty, the availability of GnRH released through the action of thymosin may have a role in the development of the sexual organs. Thus, babies with no thymus present at birth fail to develop normal gonads owing to the lack of thymosin from the thymus gland.

Peripheral T cells originate in the thymus. T cells are continually released that are fully mature and competent to function in antigen recognition. The precursor of the thymus during fetal and embryonic life develops functionally by immigration of precursor cells from bone marrow into the thymus, generation of T lineage-specific surface molecules, expression of the T-cell receptor, and finally diversification into $CD4^+$ or $CD8^+$ classes with effector functions. Some categories of T cells will express T-cell receptor (TCR) with α/β-subunits or TCR-$\gamma\delta$, the CD3 complex, and the CD4

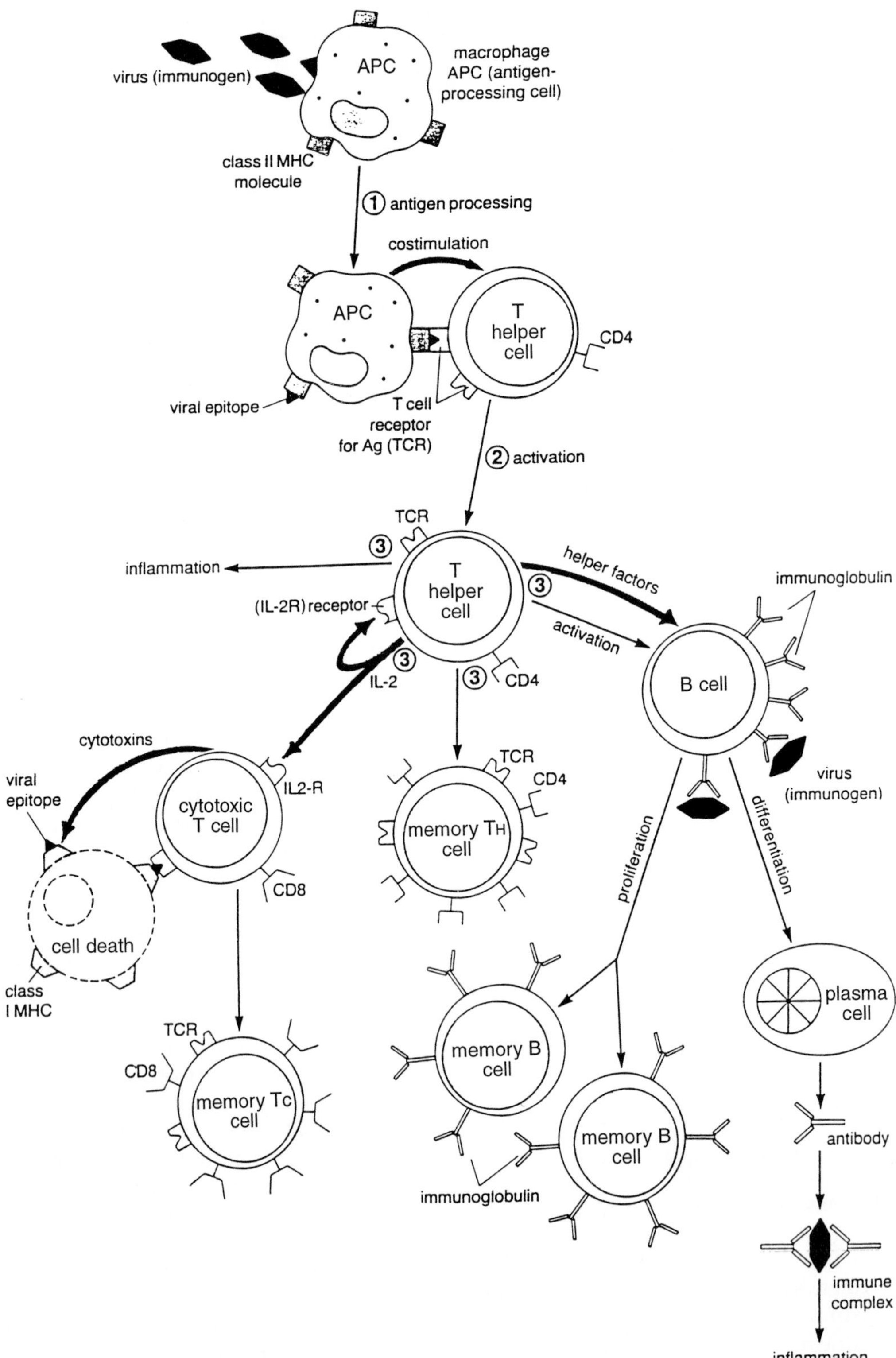

FIGURE 17-1 Sequence of events in a prototypic immune response. See text for explanation. Reproduced from Stites, D. P., Terr, A. I., and Parslow, T. G. (1994). "Basic and Clinical Immunology," 8th ed., p. 43. Appleton & Lange, Norwalk, CT.

and CD8 molecules. Immature cells are localized within the cortex of the thymus. The more mature cells are located in the medulla, and they leave the thymus to become peripheral T cells.

In the development of the thymus gland, involution occurs. Babies are born with a large thymus, which gradually becomes smaller in proportion to the development of the adrenal gland and the increasing secretion of cortisol. Involution occurs in cortex cells that are programmed for cell death initiated by glucocorticoids by a process of apoptosis (see Chapter 10), in which the dying cells are altered on their surface so as to be engulfed by macrophages without the development of inflammation. Early on, the large thymus in newborns was considered to be an anomaly since adults had a much smaller organ resulting from the maturation process regulated by cortisol. By the early 1900s it became recognized that tumors of the thymus gland were linked to the disease *myasthenia gravis*, and later removal of the tumor-bearing thymus became a successful therapy for this disease.

Because the thymus is the sole source of T lymphocytes, it plays a central role in the development of the immune system. Thymus cells produce hormones that function to activate and proliferate peripheral cells of the immune system, ensuring a surveillance function in the prevention of disease. Degeneration or loss of this system places the organism at risk for the development of cancers and infectious processes common in senescence, which seem to align with lowered functioning of the thymus. Thus, the size of the thymus is largest during the development of the immune system in early childhood. Thymus hormones that maintain a functioning immune system are being examined for use in therapy in diseases where thymus function is decreased.

B. Immune System

There are approximately 10^6–10^8 antibody specificities referring to different antigenic structures (epitopes), and a specific antigenic molecule will interact with and activate only a small proportion (<0.1%) of the antibody-producing cells. B cells are precursors of antibody-producing plasma cells, and T cells are thymus-derived cells that produce a number of stimulatory agents for B cells. The ability of a B-cell-derived immune (memory) cell to produce antibodies depends on a number of stimulatory factors as well as transpositions of variable genetic elements of the immunoglobulin structure, a subject of intense concentration in molecular genetics. Stimulatory activities are derived from thymus-derived cells (T cells) usually considered to be helper T cells (Figure 17-1).

The thymus contains T-cell lymphocytes, which in turn produce factors that regulate the expression of immunoglobulins by B-cell lymphocytes. T cells enter circulation from the thymus, as do B cells from bone marrow. Phagocytic cells (macrophage, antigen-processing cell, APC) are also involved in the immune process.

When an antigen (a foreign substance) enters the organism, B cells produce immunoglobulins with helper T-cell action. Antibodies on the surface of T cells with the same specificity as those produced by the activated B cells bind the antigen, and the complex is then transferred to a macrophage cell that ingests and destroys the antigen. This scheme (Figure 17-1) also depicts the factors contributed by helper T cells that stimulate antibody production. This chapter concentrates on a family of peptide hormones that regulate the proliferation and maturation of precursor lymphocytes into immunologically competent cells.

When an antigen is recognized by a T cell, a signal transduction mechanism involving the T-cell receptor comes into play. This series of events is shown in Figure 17-2. This figure shows the signal transduction events

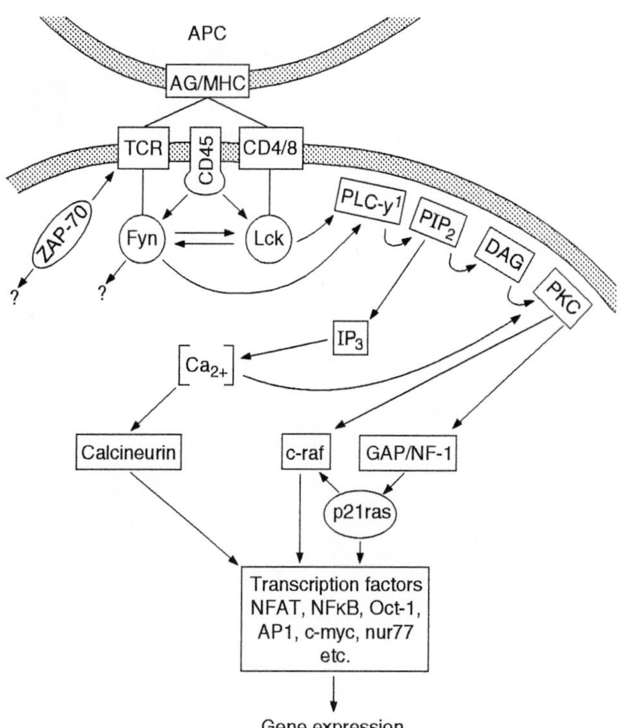

FIGURE 17-2 Simplified schematic representation of signal transduction events that occur in T cells following antigen recognition. Reproduced with permission from Kisielow, P. (1995). Apoptosis in intrathymic T-cell development. *In* "Apoptosis and the Immune Response" (C. D. Gregory, ed.), pp. 13–53. Wiley-Liss, New York.

initiated by the $\alpha\beta$ T-cell receptor. Tyrosine phosphorylation is called into play, which includes the ζ-chain of the T-cell receptor complex. For reference, a diagrammatic representation of the T-cell receptor signal transduction is given in Figure 17-2. Phospholipase $C_{\gamma 1}$ is also phosphorylated, reflecting the elevated activities of Fyn and Lck tyrosine kinases. The ζ-chain of the T-cell receptor then associates with the ZAP-70 tyrosine kinase (Figure 17-2). Activation of phospholipase C leads to hydrolysis of phosphatidylinositol 4,5-biphosphate (PIP_2), generating 1,2-diacylglycerol (DAG) and inositol triphosphate (IP_3), which lead to the activation of protein kinase C (PKC). Intracellular Ca^{2+} levels are increased by IP_3. Potential PKC targets include $p21^{ras}$, its GTPase activity controlling GAP and NF1 proteins, and raf and MAP kinases. Calcium increases activate calcineurin, a Ca^{2+} calmodulin-dependent protein phosphatase. NFAT, a transcriptional regulator of IL-2 gene expression, is a target of calcineurin that can activate NFAT by dephosphorylation. NF-κB and AP-1 can be targets of the T-cell receptor signal transduction pathway. Oct-1, c-*myc*, and nur 77 may also be involved.

II. ANATOMICAL AND MORPHOLOGICAL RELATIONSHIPS

The anatomy and histology of the thymus gland are pictured in Figure 17-3. The gland lies above the heart, as shown in Figure 17-3A, and is very much larger in

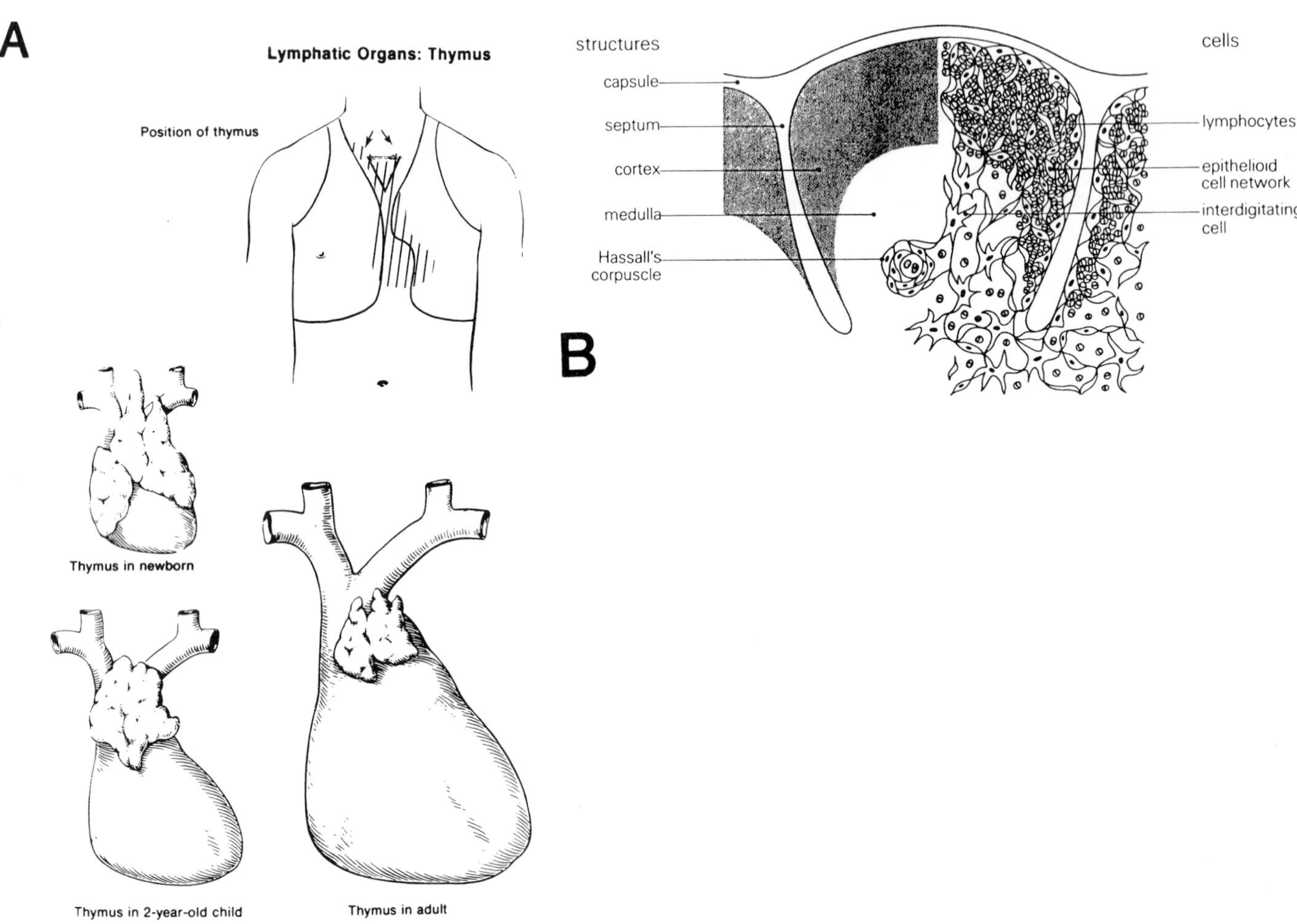

FIGURE 17-3 Anatomy and histology of the thymus gland. (A) Locations and comparative sizes at different stages of growth. Reproduced from Kahle, W., Leonhardt, H., and Platzer, W. (1978). "Color Atlas and Textbook of Human Anatomy," Vol. 2, p. 93. Year Book Medical Publishers, Chicago, IL. (B, p. 686) Histological organization of the thymus. The thymus is encapsulated and divided into lobules by septa. Densely packed dividing lymphocytes form a network of epithelioid cells in the cortex to the medulla. There are fewer lymphocytes in the medulla that contain more bone marrow-derived interdigitating cells. There is a close association of the developing lymphocytes with epithelial and interdigitating cells. Functions of the corpuscle structures are unknown. Reproduced from Roitt, I. M., Brostoff, J., and Male, D. K. (1985). "Immunology," p. 14.3. Gower Medical Pub. Ltd.

the newborn than in the adult. Shrinkage of the gland with age to adult status is the result of increasing function of the adrenal cortex with higher circulating levels of glucocorticoid hormones. Thymocytes are sensitive to the action of glucocorticoids mediated by the glucocorticoid receptor, and, as a result, the cells involute by a process of programmed cell death (specifically, apoptosis) under hormonal action. The histology of the thymus is shown in Figure 17-3B. Besides a capsule, there are two main layers of cells: the cortex and the medulla. Thymocytes are similar to lymphocytes of other lymphoid organs and to those in the circulation.

The thymus arises from an epithelial outgrowth of the third and fourth branchial pouches to form the thymus gland. The human fetal thymus is predominantly lymphoid, a pattern that is preserved throughout childhood. Later the lymphoid component decreases, particularly in the cortex, under the influence of endogenous adrenal glucocorticoids (see Chapter 10). The lymphoid cells either divide and die in the thymus gland or populate the peripheral lymphatic organs. Lobules constitute the histological cores of the thymus gland. The lobular structure is not evident until the third intrauterine month (Figure 17-4). Up to that time, the thymus appears as an epithelial organ. The periphery of the lobes is bounded by a capsule that originates the connective tissue septa. Each lobule is composed of an outer darkly staining cortex and an inner paler staining medulla, as visualized with hematoxylin and eosin (Figure 17-4). The thymus gland proportion to body weight is greatest at birth, but its absolute mass is greatest at puberty. Subsequently, there is involution and the thymic parenchyma is replaced by fat. This organ is the primary source of T lymphocytes.

III. CELL BIOLOGY

The immune system as a whole may be visualized as consisting of two major components: cellular and humoral immunity. Both respond to foreign proteins or other antigens. The γ-globulin fraction of plasma antibodies constitutes the humoral mechanism, whereas cellular immunity refers to lymphocyte products (lymphokines, e.g., the interleukins). The humoral mechanism protects against bacterial infection, whereas cellular immunity generates delayed allergic reactions and rejections of foreign tissue transplants or tumor cells. It neutralizes infections by viruses, fungi, and some bacteria. The development of these systems is outlined in Figure 17-5. Precursor cells destined to become lymphocytes originate in the yolk sac and find their way into the developing fetus. Those populating the thymus mature under the stimulation of thymic hormones into T-cell lymphocytes responsible for cellular immunity.

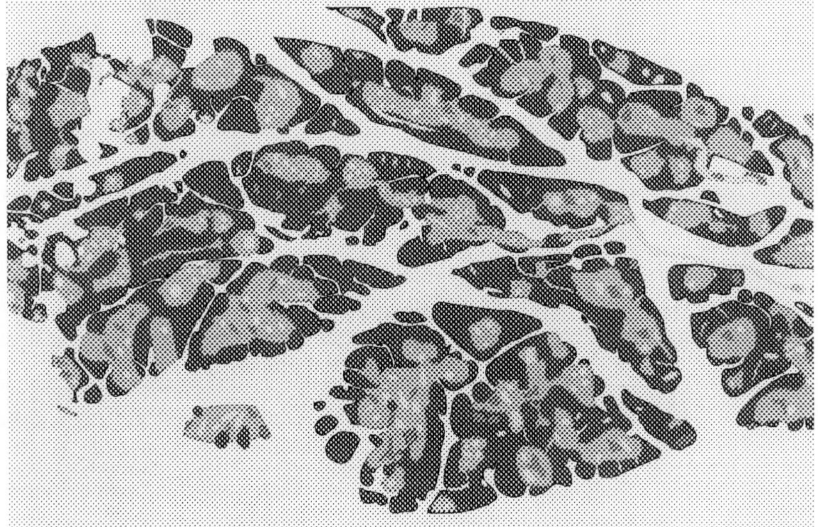

FIGURE 17-4 Normal child's thymus showing lobular structure. The thymus is derived from the entoderm of the third and fourth branchial arches with a probable contribution from the cervical sinus. Essentially the thymus is an epithelial organ and may be viewed as a foregut derivative. The characteristic lobular structure shown here appears in the third intrauterine month. Before then the rudimentary thymus is an epithelial organ. A capsule binds the periphery of the lobule, from which the connective tissue septa are derived. Each lobule is composed of an outer darkly stained cortex and an inner paler staining medulla, which is continuous throughout each half of the gland.

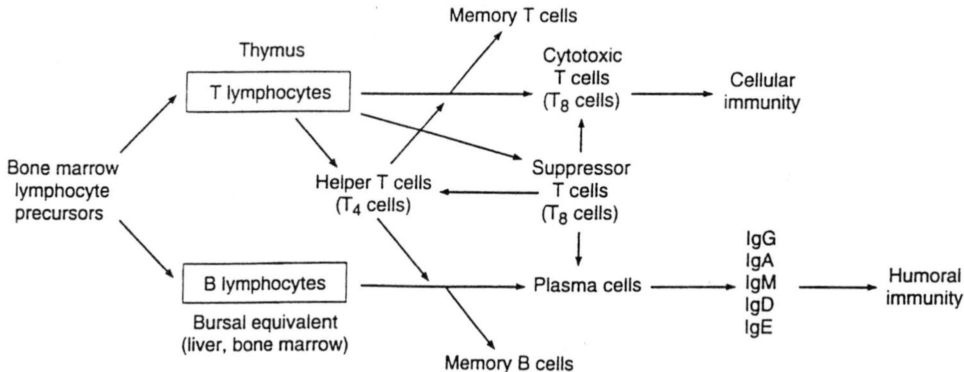

FIGURE 17-5 Development of the immune system. Reproduced with permission from Ganong, W. F. (1995). "Review of Medical Physiology." Appleton and Lange, Norwalk, CN.

IV. CHEMISTRY AND BIOCHEMISTRY

A. Introduction

There are as many as six peptides of the thymus called thymosins. These are described in Table 17-1 and appear to be acidic peptides in the molecular weight range 800–15,000. These "hormones" are produced in the thymus gland and control the development of the thymic-dependent lymphoid system as the thymus participates in the process of immune regulation. Among peptides isolated from the thymus, thymosin α_1, thymosin β_4, and polypeptide β_1 have been characterized fully and sequenced. The primary sequences of thymosin α_1, thymosin β_4, thymopoietin II, prothymosin α, and *facteur thymique serique* (FTS), now known as thymulin, are shown in Figure 17-6. The thymosins modulate immunological responses *in vitro* and *in vivo* as well as regulate the differentiation of thymic-dependent lymphocytes. Polypeptide β_1 is homologous with ubiquitin (see later) and the N-terminal 74 amino acids of a nonhistone chromosomal protein that resembles a histone.

Ubiquitin was discovered as a thymus polypeptide, but it was soon learned that it was present in a wide variety of tissues, hence the name, ubiquitin. Although this protein occurs in chromatin often conjugated with histone, it has been characterized in another context, being identical to the heat-stable polypeptide of a proteolytic system in which it plays an important role in cellular protein breakdown. As a consequence of these findings, ubiquitin is not considered to be a thymic hormone.

B. Thymosin α_1

Thymosin α_1 was the initial polypeptide to be isolated among the fractions obtained from bovine thymus. It is very active in the process of T-cell immunity and modulates the expression of terminal deoxynucleotidyltransferase. The N-terminal amino acid residue is blocked by an acetyl group. Inspection of Figure 17-6 indicates homology with thymosin α_2, but not with the other well-characterized hormones and factors. There are apparently no known homologies with

TABLE 17-1 Physical and Chemical Properties of Thymic Factors[a]

Compound	MW	Chemical properties	Origin
Thymosin fraction 5	1000–15,000	Family of heat-stable, acidic polypeptides	Bovine thymus
Thymosin α_1	3108	28 amino acid residues, p*I* 4.2	Bovine thymus + synthetic
Thymosin α_7			
Thymosin β_4	~5000		
Thymopoietin	5562	49 amino acid residues, heat-stable, p*I* 5.2	Bovine thymus + synthetic
Thymopoietin pentapeptide		Amino acid residues, 32–36 in thymopoietin	Synthetic
Thymic humoral factor	3200	31 amino acid residues, heat-labile, p*I* 5.7–5.9	Bovine thymus
Serum thymic factor (thymulin)	857	9 amino acid residues, EAYSQGGSN	Mouse, pig serum

[a] This table is taken, in part, from Table 3 of Wara, D. W. (1981). Thymic hormones and the immune system. *Adv. Pediatr.* **28,** 236.

Thymosin α_1	AcSDAAVDTSSEITTKDLKEKKEVVEEAEN
Thymosin α_2	AcSDAAVDTSSEITTKDLKEKKEVVEEAE$^{D}_{N}$GREAPAN
Prothymosin α	xSDAAVDTSSEITTKDLKEKKEVVEEAE$^{D}_{N}$GRDAPA$^{D}_{N}$GNAQNEENGEQEADN EVDEEEEEGGGEE(BGGZZZZZZZZZZZ)NGDEDEEAEAAPTGKRVAEDE DDDVETKKQKKTDEDD
Thymosin β_4	AcSDKPDMAEIEKFDKSKLKKTETQEKNPUPSKGTIEQEKQAGES
Thymopoietin II	SQFLEDPSVLTKEKLKSELVANNVTLPAGEQRKDVYVQLYLQTLTAVKR
Thymulin (Serum thymic factor)	DAYSQGGSN

FIGURE 17-6 Sequences of thymic hormones. Abbreviations: Ac, acetyl; x, unknown; B, Asp or Asn; Z, Glu or Gln. Underlined sequence represents a nuclear translocation signal.

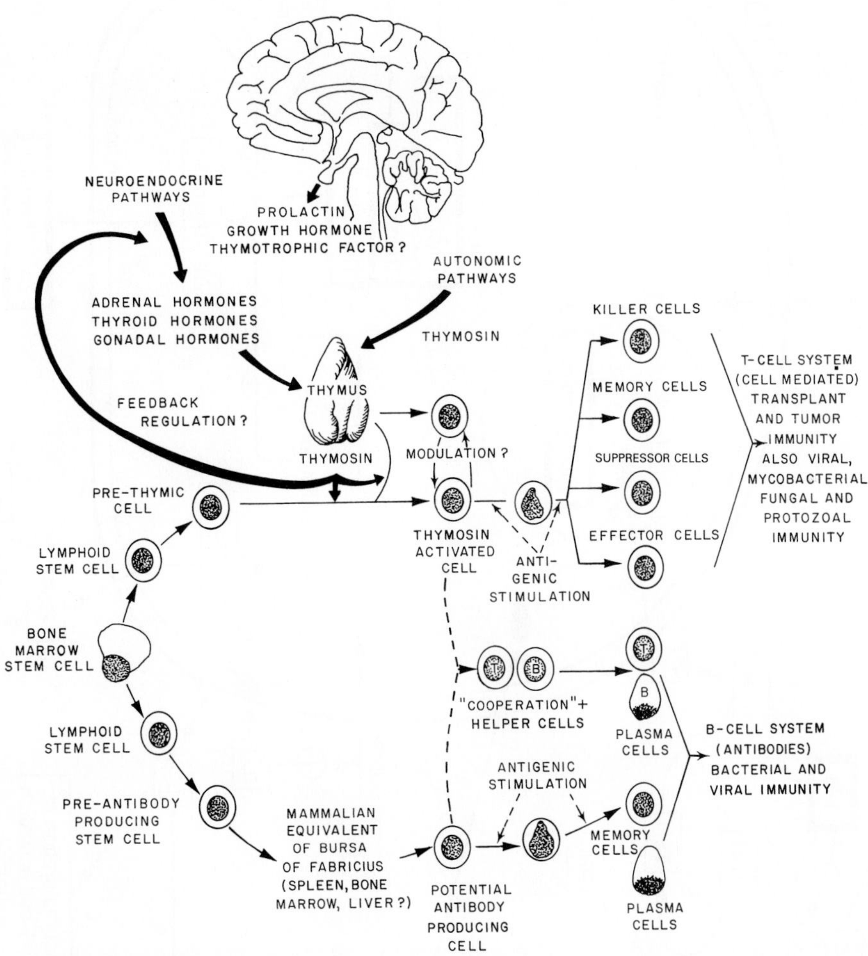

FIGURE 17-7 Hypothetical model illustrating the major interactions between the central nervous system and the neuroendocrine thymus. From Goldstein, A. L., Low, T. L. K., Thurman, G. B., Zatz, M. M., Hall, N., Chen, J., Hu, S.-K., Naylor, P. B., and McClure, J. E. (1981). *Recent Prog. Horm. Res.* **37,** 369–415.

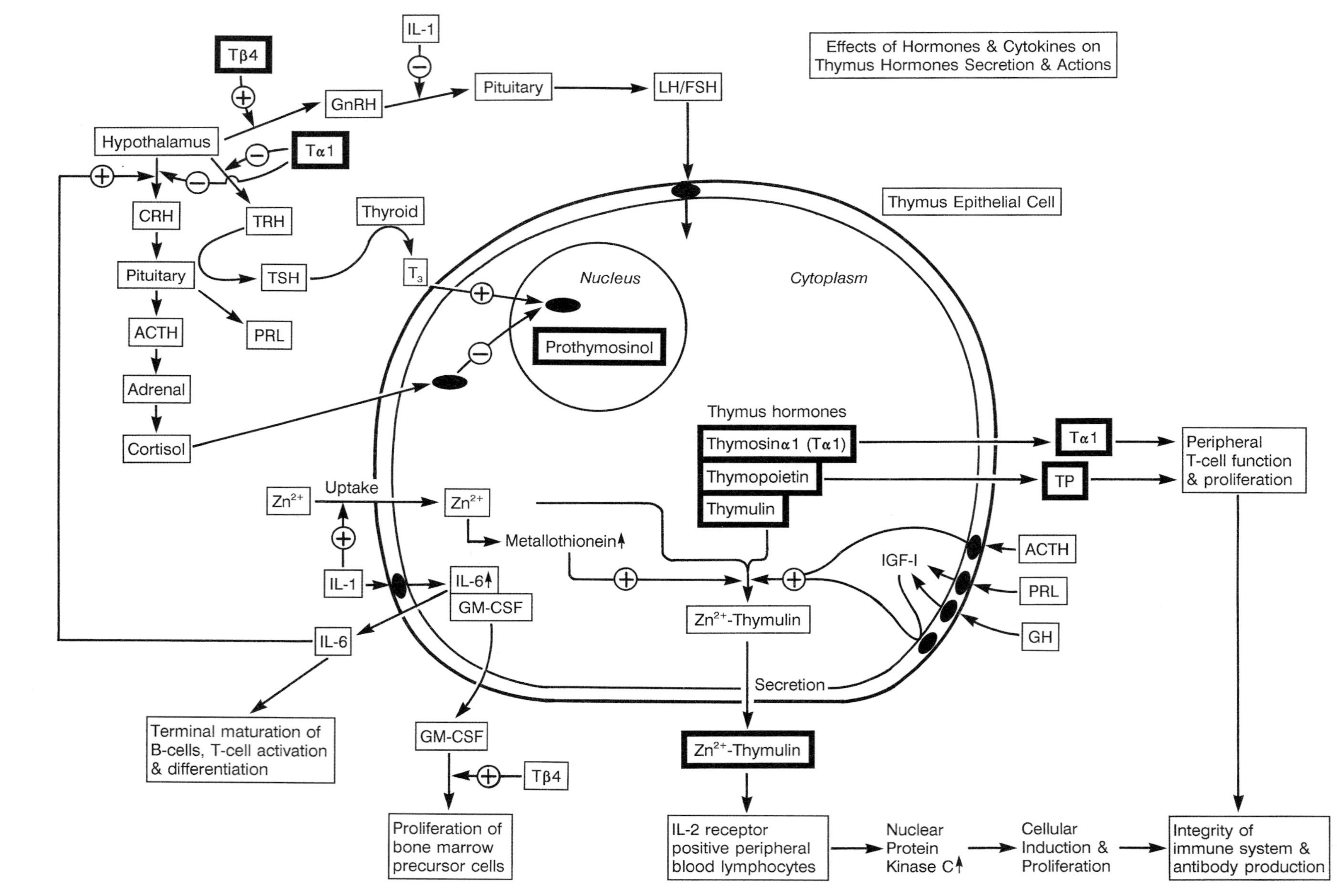
Effects of Hormones & Cytokines on
Thymus Hormones Secretion & Actions
Tβ4
IL-1
GnRH
Pituitary
LH/FSH
Hypothalamus
Tα1
CRH
TRH
Thyroid
Pituitary
TSH
T3
PRL
ACTH
Adrenal
Cortisol
Thymus Epithelial Cell
Nucleus
Cytoplasm
Prothymosinol
Thymus hormones
Thymosinα1 (Tα1)
Thymopoietin
Thymulin
Tα1
TP
Peripheral
T-cell function
& proliferation
Zn2+
Uptake
Zn2+
Metallothionein↑
IL-1
IL-6↑
GM-CSF
IL-6
ACTH
IGF-I
PRL
GH
Zn2+-Thymulin
Secretion
Zn2+-Thymulin
Terminal maturation of
B-cells, T-cell activation
& differentiation
GM-CSF
Tβ4
Proliferation of
bone marrow
precursor cells
IL-2 receptor
positive peripheral
blood lymphocytes
Nuclear
Protein
Kinase C↑
Cellular
Induction &
Proliferation
Integrity of
immune system &
antibody production

any of the primary sequences of thymosin α_1. Prothymosin α is a highly acidic protein lacking an amino-terminal signal peptide. It contains the nuclear translocation signal TKKQKKT, which may direct its localization to the nucleus, causing it to differ from the other thymic hormones. It also appears to be phosphorylated by casein kinase 2. Contrary to earlier indications, thymosin α_1 appears to be a native peptide in several tissues rather than an uncontrolled proteolytic product of prothymosin α, and the two polypeptides appear to be in separate compartments of the thymus epithelial cell, with prothymosin α in the nucleus and thymosin α_1 in the cytoplasm.

As shown in Figure 17-6, four thymic peptides appear to be unrelated with regard to sequence homology. The multiplicity of peptides produced by the same tissue with different chemical structures suggests the possibility that, although they share some common activities, they may also elicit highly individual functions in regulating the immune system. Interestingly, limited homologies do show up with rabbit skeletal muscle tropomyosin α-chain, 50 S ribosomal protein L7 from *Escherichia coli*, troponin I from rabbit skeletal muscle, and bovine prothrombin. Thymosin α_1 has been chemically synthesized, and the product is equally active as the biological material. Thymosin α_1 mRNA has been translated in a wheat germ system by using antiserum to this protein. These results suggest that a larger peptide of molecular weight 16,000 is a primary translation product, which is subsequently degraded to the native hormone (molecular weight 3108). The oligodeoxynucleotide encoding thymosin α_1 has been synthesized, inserted into a plasmid, and cloned in *E. coli*.

FIGURE 17-8 Effects of hormones and cytokines on thymus hormone secretion and actions. Three major thymus hormones, thymosin, thymopoietin, and thymulin, are thought to reside in the cytoplasm of the thymus epithelial cell. Some evidence suggests that prothymosin α resides in the nucleus and contains a nuclear translocation signal, TKKQKKT. The three cytoplasmic hormones are secreted into blood circulation. Thymulin is secreted as a 1:1 complex with Zn^{2+} dependent upon Zn^{2+} uptake by the cell. This is stimulated by extracellular IL-1. Internal Zn^{2+} further stimulates metallothionein, which catalyzes the combination of Zn^{2+} with thymulin, and then secretion of the complex occurs. Zn^{2+}-thymulin acts on IL-2 receptor-positive peripheral blood lymphocytes, causing stimulation of protein kinase C and cellular proliferation, which ensures antibody production. Similar effects are exerted by thymosin α_1 (Tα_1) and thymopoietin (TP), which act on peripheral T cells causing them to proliferate. The thymus interacts with the hypothalamo–pituitary axis, as shown in the upper left of the figure. Thymosin β_4 (Tβ_4) stimulates the release of GnRH from the hypothalamus, while thymosin α_1 (Tα_1) inhibits the release of CRH and TRH from the hypothalamus. Interleukin 6 from the thymus epithelial cell also affects the release of hormones from the hypothalamus, stimulating the release of CRH. IL-1 inhibits the release of gonadal hormones from the anterior pituitary under the influence of GnRH. Cortisol from the adrenal is a classical involutor of certain thymus cells interacting with the cytoplasmic corticosteroid receptor and ultimate transcriptional effects. Triiodothyronine (T_3), acting on a nuclear receptor, has positive effects on the thymus cell. Some thymus hormones are synthesized in other cells of the body, probably in smaller amounts than occur in the thymus cell. Thymus hormones play a major role in preserving the functioning of the immune system. Thymus function is generally thought to decline in aging, probably underlying the increasing prevalence of diseases in old age. Tβ_4, thymosin β_4.

C. Thymosin β_4

Thymosin β_4 has a molecular weight of 4982 and also has its N-terminal amino acid residue blocked by an acetyl group (Figure 17-6). It is the predominant thymosin in mammalian cells. This protein induces the expression of terminal deoxynucleotidyltransferase in transferase-negative mouse bone marrow cells *in vitro* and *in vivo* and increases the activity of this enzyme in hydrocortisone immune-suppressed mice *in vivo*. Thymosin β_4 therefore acts on lymphoid stem cells and in the early maturation process of thymus-dependent lymphocytes.

The primary sequence of thymosin β_4, like that of thymosin α_1, does not show homology with the sequences of other known proteins. However, there is an internal sequence duplication. Although it would be attractive to consider thymosin β_4 as a discrete hormone of the thymus, it appears in many tissues. No pro- or prepropeptide is formed as its precursor, and in mammalian tissues it is usually accompanied by a related peptide, thymosin β_{10}. Thymosin β_4 and β_{10} probably are not secretory peptide hormones.

Thymosin β_3 is another peptide completely related to thymosin β_4, but differs at the C-terminal end and has a molecular weight of 5500. It may also be a proteolytic fragment of thymosin β_4. Thymosins β_9, $\beta_{met\ 9}$, and β_{10} are minor variants of thymosin β_4, and, like thymosin β_4, they bind to ATP G actin. These proteins bind to monomeric actin and inhibit its polymerization. Part of the binding site of thymosin β_9 has been shown to reside in the carboxy terminus. The N-terminal tetrapeptide of thymosin β_4 is Ac-Ser-Asp-Lys-Pro and, by itself, affects cell cycle regulation by inhibiting *in vivo* entry of cell populations into the S phase; it also inhibits calmodulin-dependent phosphodiesterase. On the other hand, the 11–19 fragment of thymosin β_4 (11-KFDKSKLKK-19) elevates phosphodiesterase activity even in the absence of Ca^{2+} or calmodulin. The N-terminal tetrapeptide of thymosin β_4 Ac-SDKP, is formed from one-step enzymatic processing of thymosin β_4, and both peptides have been detected

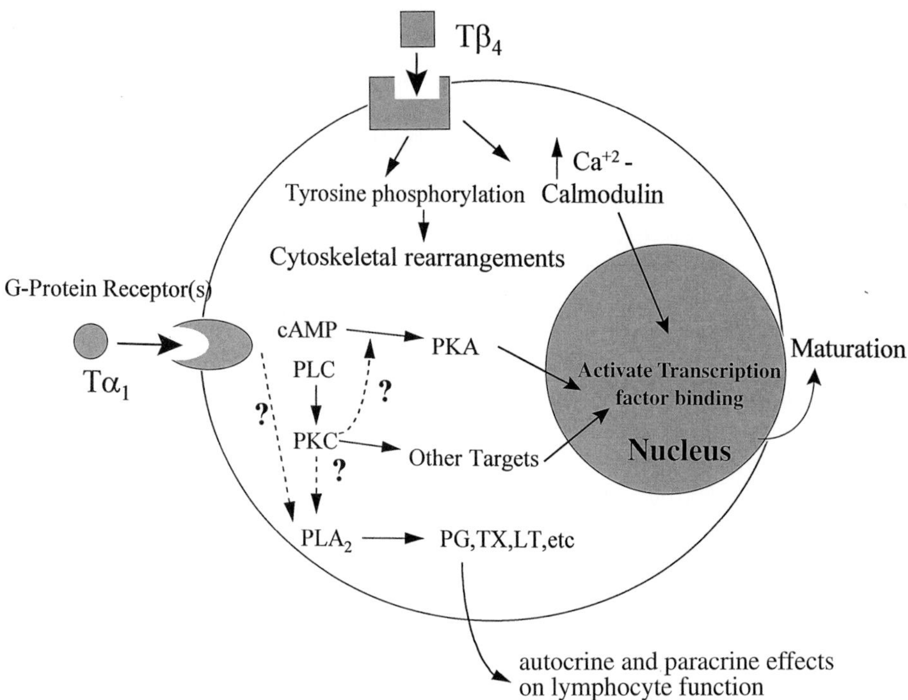

FIGURE 17-9 Proposed mechanism of $T\alpha_1$ and $T\beta_4$ action. $T\alpha_1$ activity has been associated with activation of cAMP/PKA- and PLC/PKC-dependent pathways. These second messenger pathways are typically G-protein receptor linked. $T\alpha_1$ activity involves the production of prostaglandins and the activation of phospholipase A_2 (PLA_2). It is currently unknown whether the activation of PLA_2 is directly through a receptor-mediated event or distal to the activation of PKC. The downstream targets of both PKC and PKA are transcription factors involved in cellular activation. The production of arachidonic acid metabolites have also been found to have significant effects on immune cell function. Although a large body of evidence has defined $T\beta_4$ as a major intracellular actin sequestering protein, exogenous $T\beta_4$ action has also been associated with the activation of tyrosine phosphorylation and increases in cytosolic calcium increases with subsequent alterations in cellular actin content (courtesy of Dr. Allan Goldstein, George Washington University, 1996).

in bone marrow cells. Incubation of [^{3}H]thymosin β_4 with intact or lysed bone marrow cells generated [^{3}H]Ac-SDKP, and the tetrapeptide was not degraded further. Thus, Ac-SDKP appears to be a new regulator of the hemopoietic system.

The mRNA for thymosin β_4 increases in fibroblasts induced to proliferate at the G1–S transition even in the presence of cycloheximide, but not in synchronized cycling cells. Thus, thymosin β_4 is regulated by cell proliferation, but it is not a cell cycle-regulated gene.

Thymosin β_4 synergizes with GM-CSF in myelopoiesis. There is a protein that binds thymosin β_4, which is an elastase inhibitor. It is related to the serpin (serine protease inhibitor) family and plasminogen activator inhibitor 2. Thus, the physiological role of elastase inhibitors may be the thymosin β_4-regulated rearrangement of leukocyte cytoskeleton components (see Figure 17.9).

D. Thymosin Fraction 5

This fraction has been found to cause the release of ACTH from CRH-insensitive mouse pituitary cells in culture without decreasing the intracellular ACTH stores. Apparently Ca^{2+} uptake is required for this activity.

E. Thymic Humoral Factor (THF-γ_2)

A single injection to immunodeficient aging C57BL mice raised the frequency of mitogen-responsive T cells in the thymus and spleen cell populations to levels equivalent to those of young mice. A summary of the production and secretion of thymus hormones, their actions, and their interactions with other organs, such as the hypothalamus and anterior pituitary is shown in Figure 17-8.

The thymus has been found to contain the natriuretic peptides BNP, CNP, and ANP. They are synthesized in the thymus and produced by different types of thymus cells. The mRNA of ANP is elevated in parallel to the extent of involution caused by dexamethasone, and BNP protein is elevated 2-fold 4 days after dexamethasone, while ANP protein is elevated 10-fold in the same time period after dexamethasone. The thymic natriuretic system may be involved in communication between the immune system and the neuroendocrine system.

Glucocorticoids have now been shown to induce protein kinase C_E, a Ca^{2+} independent kinase, by *de novo* synthesis in immature but not in mature T cells, aligning with the cell type sensitive to apoptosis induced by this hormone. The increase in protein kinase C_E and apoptosis induced by glucocorticoid are inhibited by actinomycin D, cycloheximide, and RU-486, a specific inhibitor of glucocorticoid receptor action. Induced protein kinase C_E translocates from the cytoplasm to the nuclear membrane, thus implicating the activity of this kinase in the glucocorticoid-induced apoptosis of immature T cells.

The thymus controls certain liver functions. Adult thymectomized rats are either observed as controls or treated with thymus extracts or sex hormones. In males, thymectomy causes reductions in liver microsomal cytochrome P450 and aminopyrine-*N*-demethylase activities. In addition, there is a fall in hypothalamic GnRH, plasma LH, and plasma testosterone. These levels can be restored by treatment with testosterone proprionate. Thymectomy of the female results in increased liver malondialdehyde and decreases in liver superoxide dismutase, glutathione, fluidity, and Ca^{2+} uptake in liver microsomal and mitochrondrial membranes. There are also lowered levels of hypothalamic GnRH and plasma estradiol. Treatment with thymic extract reverses all of these changes in the liver, implicating the action of thymic hormones on the maintenance of a variety of liver systems through a thymus neuroendocrine liver pathway.

Thymosin binds to a membrane receptor to exert its activity. This receptor may be coupled to adenylate cyclase in the absence of ligand (thymosin). Upon thymosin interaction with receptor, phosphodiesterase becomes activated, probably involving cadmodulin, and the cellular levels of cyclic AMP may fall by the phosphodiesterase-catalyzed conversion of cyclic AMP to AMP. Rather little information is available on the thymosin mechanism of action.

TABLE 17-2 Immunodeficiency Diseases with Thymic Alterations[a]

Syndrome	Thymus	Humoral immunity	Cellular immunity	Immunoglobulins	Lymphocytes	Heredity	Additional findings
DiGeorge syndrome	Absent	Normal	Deficient	Normal	Normal/low		Absence of parathyroid; anomalies of heart and great vessels
Nezelof syndrome (congenital thymic dysplasia)	Hypoplasia	Normal or slightly deficient	Deficient	Normal or slightly deficient	Reduced	Autosomal recessive	
Reticular dysgenesis	Hypoplasia	Deficient	Deficient	Deficient	Reduced		Leukopenia
Agammaglobulinemia (Swiss type)	Hypoplasia	Severely deficient	Absent or deficient	Deficient	Markedly reduced	Autosomal recessive	Dyschondroplasia, ulcerative colitis
Thymic alymphoplasia	Hypoplasia	Deficient	Absent or deficient	Deficient	Reduced	Sex-linked recessive, male only	
Louis–Bar syndrome (ataxia telangiectasia)	Hypoplasia	Slightly deficient	Deficient	Normal or slight IgA deficiency	Variable	Autosomal recessive	Cerebellar ataxia, telangiectasia, lymphoma, dyschondroplasia, endocrine anomalies, neutropenia
Agammaglobulinemia with thymoma	Spindle cell tumor	Deficient	Deficient	Deficient	Reduced		

[a] Reproduced from Willich, E. and Walter, E. (1992). Developmental abnormalities of the thymus. *In* "The Thymus" (E. Walter, E. Willich, and W. R. Webb, eds.), pp. 63–69. Springer-Verlag, Berlin.

F. Thymopoietin

Three distinct thymopoietins have been characterized that are splice variants involving similar N-termini, but different C-termini with molecular weights of 75, 51, and 39 kDa. mRNAs are expressed in various tissues but are most abundant in adult thymus and fetal liver. These proteins may have unique functions and have subcellular locations in different compartments. It is of interest that expression of human myotonin protein kinase cDNA in COS cells encodes a protein with a thymopoietin-like domain. A thymopoietin-related peptide has been cloned from a thymus cDNA library and contains the active site of thymopoietin (amino acids 32–36). This protein and thymopoietin differ in only one amino acid residue from bovine thymopoietin and in five residues from bovine thymopoietin I. Thymopoietin related peptide is a novel peptide counterpart of bovine thymopoietin.

V. BIOLOGICAL AND MOLECULAR ACTIONS

While thymosin peptides appear to act before and during the prothymocyte state, thymosin α_1 or prothymosin α acts at early as well as late stages of thymocyte maturation. It seems likely that thymosin β_3 is a proteolytic fragment of thymosin β_4.

By immunofluorescent microscopic methods, thymosin α_1 has been located in the cytoplasm of thymic epithelial cells of the medulla and in cells covering the cortical surface, whereas by the same methodology β_3 and β_4 were detected in cells covering the cortical surface. This finding aligns well with the known ability of β_3 in inducing terminal deoxynucleotide transferase expression and with the knowledge that cells containing this enzyme predominate in the thymic cortex. The thymic medulla, on the other hand, is terminal deoxynucleotide transferase negative and the action of thymosin α_1 is to suppress the expression of this enzyme.

Recent work implicates the autonomic and neuroendocrine systems in modulating the immune system. The thymus appears to play an integral role in the central nervous system regulation of the immune system. Figure 17-7 summarizes the interactions between the central nervous system and the neuroendocrine thymus. Apparently cholinergic receptors occur on the surface of thymosin-producing epithelial cells. Cholinergic and β-adrenergic receptors occur on the surface of T lymphocytes. Detectable levels of γ-aminobutyric acid, a known neurotransmitter, have been observed in the thymus. In addition to neural regulation, various hormones are now believed to regulate thymosin production. Growth hormone increases levels of thymosin α_1. TSH may also have an influence. The role of glucocorticoids on the thymus (Chapter 10) is well known and there may be specific interactions between glucocorticoids and thymosin production.

The thymus has been found to contain the natriuretic peptides BNP, CNP, and ANP. They are synthesized in the thymus and produced by different types of thymus cells. The mRNA of ANP is elevated in parallel to the extent of involution caused by dexamethasone and BNP protein is elevated 2-fold 4 days after dexamethasone while ANP protein is elevated 10-fold in the same time period after dexamethasone. The thymic natriuretic system may be involved in communication between the immune system and the neuroendocrine system.

Glucocorticoids have now been shown to induce protein kinase C_E, a Ca^{2+}-independent kinase, by *de novo* synthesis in immature but not in mature T cells aligning with the cell type sensitive to apoptosis induced by this hormone. The increase in protein kinase C_E and apoptosis induced by glucocorticoid are inhibited by actinomycin D, cycloheximide, and RU486, a specific inhibitor of glucocorticoid receptor action. Induced protein kinase C_E translocates from the cytoplasm to the nuclear membrane, thus implicating the activity of this kinase in the glucocorticoid-induced apoptosis of immature T cells.

The thymus controls certain liver functions. Adult thymectomized rats are either observed as controls or treated with thymus extracts or sex hormones. In males, thymectomy causes reductions of liver microsomal cytochrome P450 and aminopyrine-*N*-demethylase activities. In addition, there is a fall in hypothalamic GnRH, plasma LH, and plasma testosterone. These levels can be restored by treatment with testosterone proprionate. Thymectomy of the female results in increased liver malondialdehyde and decreases in liver superoxide dismutase, glutathione, fluidity, and Ca^{2+} uptake in liver microsomal and mitochrondrial membranes. There are also lowered levels of hypothalamic GnRH and plasma estradiol. Treatment with thymic extract reverses all of these changes in the liver, implicating the action of thymic hormones on the maintenance of a variety of liver systems through a thymus neuroendocrine liver pathway.

Thymosin binds to a membrane receptor to exert its activity. The thymosin α_1 ($T\alpha_1$) receptor may be coupled to adenylate cyclase and then to the protein kinase A pathway as well as the phospholipase C-protein kinase C pathway (Figure 17-9). On the other hand, the receptor for thymosin β_4 ($T\beta_4$) may be linked through tyrosine phosphorylation to coskeletal rearrangements as well as to the Ca^{2+}/calmodulin system resulting in enhancement of transcription factor binding in the nucleus (Figure 17-9). The ideas expressed in Figure 17-9 are speculative.

VI. CLINICAL ASPECTS

Clearly the involvement of thymosin in the immune system makes it a potentially crucial agent in diseases of immunodeficiency and in several types of cancers. Clinical studies have demonstrated prolonged survival of patients with immunodeficient diseases who have been treated with thymosin.

Clinical importance may be attached to a deficiency of thymic functions or to enlargement of the thymus above the normal size. If the thymus is not functioning properly, there will be problems with the T-cell part of the immune system. Low numbers of lymphocytes would be found in the absence of the thymus, resulting in severe immunodeficiency. Thymic peptides are frequently administered in cases of T-cell immunodeficiency.

A major cause of thymic enlargement can be neoplasms of the immune system. Thus, in childhood, acute lymphoblastic leukemias or malignant non-Hodgkin's lymphomas can be found. *Myasthenia gravis* can also be the consequence of a thymus neoplasm. Some immunodeficiency diseases are listed in Table 17-2.

References

A. Books

Goldstein, A. L. (1984). "Thymic Hormones and Lymphokines: Basic Chemistry and Clinical Applications." Plenum Press, New York and London.

Goss, J. A., and Flye, M. W. (1993). "The Thymus—Regulator of Cellular Immunity." R. G. Landes Company, Austin, TX.

Gregory, C. D. (1995). "Apoptosis and the Immune Response." Wiley-Liss Publications, New York, Chichester, Brisbane, Toronto, Singapore.

Willich, E., Walter, E, Otto, H. F., and Hofmann, W. J. (1992). "The Thymus: Diagnostic Imaging, Functions, and Pathologic Anatomy." Springer-Verlag, Berlin, Heidelberg, New York, London, Paris, Tokyo, Hong Kong, Barcelona, Budapest.

B. Review Articles

Debatin, M. (1992). Immunologic function of the thymus. *In* "The Thymus: Diagnostic Imaging, Functions, and Pathologic Anatomy" (E. Walter, E. Willich, and W. R. Webb, eds.), pp. 15–21. Springer-Verlag, Berlin.

Hofmann, W. J., and Otto, H. F. (1992). Anatomy and embryology of the thymus. *In* "The Thymus: Diagnostic Imaging, Functions, and Pathologic Anatomy" (E. Walter, E. Willich, and W. R. Webb, eds.), pp. 5–13. Springer-Verlag, Berlin.

Sater, D., and Nachmias, V. T. (1994). Beta thymosins as actin binding peptides. *Bioessays* **16,** 473–479.

Willich, E., Walter, E., Otto, H. F., and Hofmann, W. J. (1992). Developmental abnormalities of the thymus. *In* "The Thymus: Diagnostic Imaging, Functions, and Pathologic Anatomy" (E. Walter, E. Willich, and W. R. Webb, eds.), pp. 63–69. Springer-Verlag, Berlin.

C. Research Papers

Goldstein, A. L., Low, T. L. K., Thurman, G. B., Zatz, M. M., Hall, N., Chen, J., Hu, S.-K., Naylor, P. B., and McClure, J. E. (1981). Current status of thymosin and other hormones of the thymus gland. *Recent Prog. Hormone Reg.* **37,** 369–415.

Goya, R. G., Castro, M. G., Hannah, M. J., Sosa, Y. E., and Lowry, P. J. (1993). Thymosin peptides stimulate corticotropin release by a calcium-dependent mechanism. *Neuroendocrinology* **57,** 230–235.

Manrow, R. E., Sburlati, A. R., Hanover, J. A., and Berger, S. L. (1991). Nuclear target of prothymosin alpha. *J. Biol. Chem.* **266,** 3916–3924.

Wara, D. W. (1981). Thymic hormones and the immune system. *Adv. Pediatr.* **28,** 236.

CHAPTER

18

Pineal Hormones

I. INTRODUCTION

A. General Comments

The pineal (so named for its pinecone shape) gland is rather unique with respect to the magnitude of change characterizing its evolutionary development. In amphibians it is primarily a photoreceptive organ, but in higher forms it has evolved from this state to a gland producing hormones that receives light information by way of the lateral eyes and sympathetic nerves. Pineal products include melatonin, other methoxyindoles, serotonin, and acetylserotonin (see Figure 18-1). Although vasotocin has been reported to be a pineal product, it is not clear whether significant amounts of vasotocin are actually synthesized in the pineal gland. Production of the methoxyindoles is dependent on light or darkness, and these hormones have been shown to regulate seasonal reproductive activities.

There is some difference of opinion concerning the primacy of the pineal hormones with regard to physiological action. Melatonin is currently thought to be the principal active substance produced by the pineal, and it can replace most pineal functions after pinealectomy. Secretion of melatonin from the pineal is generated by norepinephrine (NEP), which is released from neighboring neurons. This activity in releasing NEP and the functioning of the pineal gland are essentially turned on in darkness and suppressed by environmental light. The dark–light conditions are transmitted through the eyes through the suprachiasmic nuclei, whose signals are suppressed in the light and activated by the absence of this inhibition in the dark.

Of necessity the phenomenon of rhythmicity arises in connection with the pineal relating to its sensitivity to light and darkness. This directly translates to the development and seasonal cyclicity of the gonads and may also have effects on the pulsatile release of anterior pituitary hormones (see Chapter 5).

B. Evolutionary Aspects

The profundity of evolutionary alterations in the structure of this organ deserves some comment. In lower forms, the pineal organ functions as a light receptor and in this definition is often referred to as the "third vertebrate eye." In 1959, it was recognized that the ultrastructure of the receptor cells in the parietal eye of the lizard resembles that of retinal cones, and the most highly differentiated sensory elements of this type occur in the pineal vesicle of the lamprey, the frontal organ of the frog, and the parietal eye of the lizard. The similarities between the neuronal organiza-

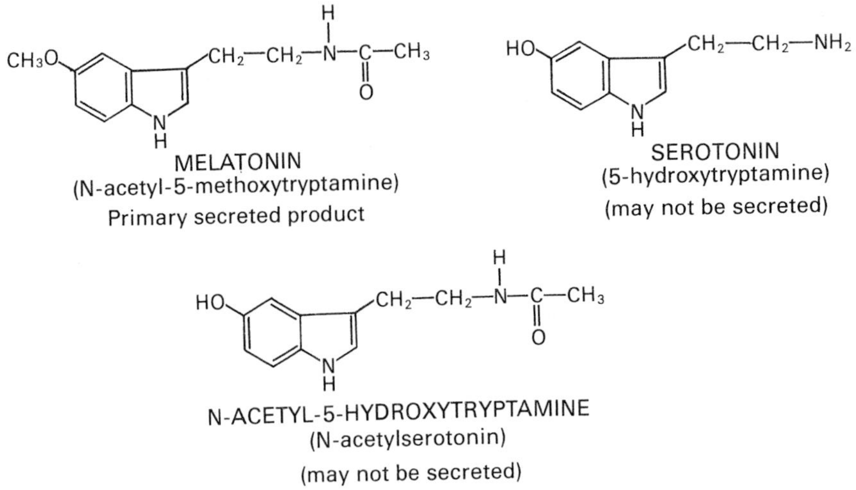

CH_3O — indole — $CH_2{-}CH_2{-}N(H){-}C({=}O){-}CH_3$

MELATONIN
(N-acetyl-5-methoxytryptamine)
Primary secreted product

HO — indole — $CH_2{-}CH_2{-}NH_2$

SEROTONIN
(5-hydroxytryptamine)
(may not be secreted)

HO — indole — $CH_2{-}CH_2{-}N(H){-}C({=}O){-}CH_3$

N-ACETYL-5-HYDROXYTRYPTAMINE
(N-acetylserotonin)
(may not be secreted)

FIGURE 18-1 Structures of pineal chemicals.

tion of the frog's pineal complex (epiphysis) and retina are evident in Figure 18-2. There appears to be a photolabile substance absorbing maximally at 560–580 nm in the frog or around 500 nm in other species. Presumably light energy is converted to electrical energy and hormonal output.

The evolution of this organ consists first of photoreceptor cells, which gradually become replaced by secretory rudimentary photoreceptor cells and pinealocytes (Figure 18-3). Sympathetic noradrenergic innervation becomes more pronounced, so that light information coming in through the lateral eyes and sympathetic efferents replaces the direct response to light characteristic of pinealocytes.

Ultimately the effects of pineal messages involve color change, endocrine control of reproduction, phototactic and locomotor reactions, detection of polarized light, and rhythmic phenomena. Clearly, the pineal gland is an important component of the photoneuroendocrine system.

Wurtman and Axelrod (1968) describe the pineal as a "neuroendocrine transducer which secretes a hormone, melatonin, in response to norepinephrine release from its sympathetic nerves, especially nocturnally, when the nerves are firing most frequently, and whose hormone provides the brain and possibly other organs with a time signal that cues other time-dependent physiological processes such as gonadal maturation, gonadal cyclicity, and perhaps sensitivity to environmental stimuli."

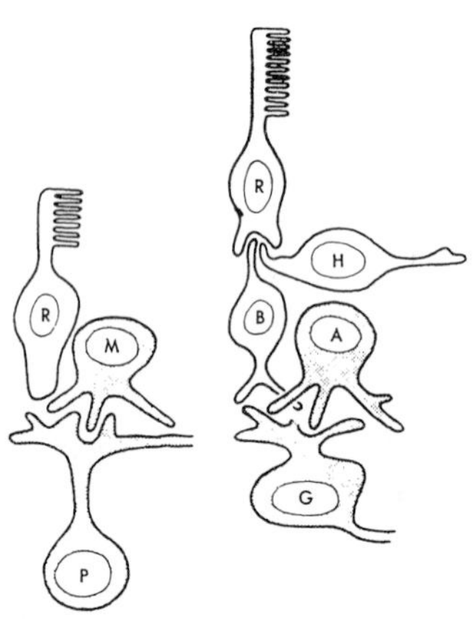

FIGURE 18-2 Neuronal organization of the pineal complex of the frog in comparison to the neuronal circuitry in the retina. Epiphysis: R, photoreceptor cell; M, multipolar nerve cell (interneuron); P, neuron contributing to the pineal tract. Retina: R, receptor cell; B, bipolar cell; H, horizontal cell; A, amacrine cell; G, ganglion cell of the optic nerve. A, G, M, and P are acetylcholinesterase-positive elements. Reproduced from Oksche A. (1980). The pineal organ—a component of photoneuroendocrine systems: Evolution, structure, function. *In* "Hormones, Adaptation and Evolution" (S. Ishii, T. Hirano, and M. Wada, eds.), p. 129. Jpn. Sci. Soc. Press, Tokyo.

II. DEVELOPMENT

The human pineal gland starts as an evagination of the roof of the third ventricle in the second month of gestation. The pineal anlage, the ependyna, a lobular structure, fuses into a solid and becomes invested with vascular connective tissue, giving a lobular appearance to the pineal parenchyma. The pineal gland then changes from a vertical to a horizontal organ and becomes innervated. In neonates, the pineal is shown to consist of two cell types, type I and type II. Type I cells

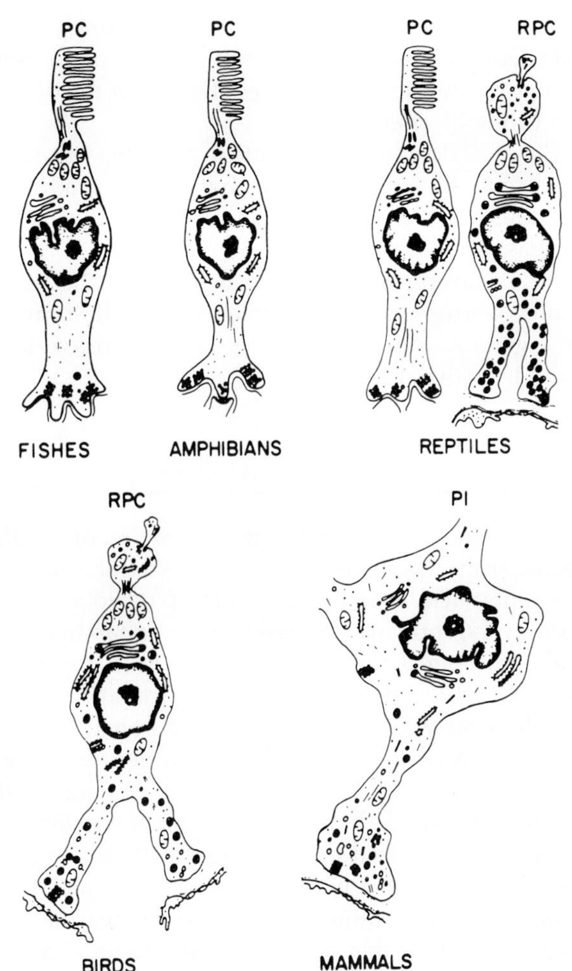

FIGURE 18-3 Vertebrate pineal cells showing evolution from a photoreceptor cell to a pinealocyte. Diagram of typical vertebrate pineal cells. Abbreviations: PC, photoreceptor cell; RPC, rudimentary photoreceptor cell; PI, pinealocyte. Reproduced from Hansen J. T., and Karasek, M., (1982). Neuron or endocrine cell? The pinealocyte as a paraneuron. *In* "The pineal and its hormones" (R. J. Reiter, ed.), pp. 1–9. Alan R. Liss, Inc., New York.

predominate at birth, are often pigmented, and are positive for S-100 proteins and negative for neuron-specific enolase. Type II cells contain relatively large amounts of neuron-specific enolase and are negative for nulanin and S-100 proteins. After 1 year there are few type I cells, so that the pineal undergoes considerable development during early childhood.

It has been observed that serum melatonin levels decline during continued development, so that there is an apparent relationship between pineal gland function and development in general. During gestation, maternal melatonin crosses the placenta, and after birth, maternal melatonin is passed to the infant through the milk. Full circadian rhythm of these amines probably is not established until well after the third month of life, and the pineal does not reach adult size until the fourth year of life.

III. ANATOMY AND CELL BIOLOGY

In humans, the pineal gland is located in the brain grossly between the thalamus and the mesencephalon, as shown in Figure 18-4. In adults the pineal is a flat, cone-shaped structure with dimensions of about 5–8 mm long and 3–5 mm wide and weighing around 120 mg. It is located at the posterior border of the third ventricle above the roof of the diencephalon, being connected to it by a short stalk. The pineal is covered by a layer called the *pia mater.* From the pia mater, connective tissue septa with blood vessels and unmyelinated nerve fibers enter the pineal to surround the cords of cells and follicles (alveoli) to form irregular lobules. Its location with respect to light stimuli is more clearly visualized in Figure 18-5. The chain of events with regard to the photoeffect is shown by light entering one of the eyes, and the signal is carried to the suprachiasmic nuclei of the hypothalamus, to the interomediolateral cell column of the spinal cord, then to the superior cervical ganglia, and finally to the post-ganglionic sympathetic neurons, probably signaling the pineal via norepinephrine release.

In infants, the pineal is large and many cells are arranged in alveoli. Before puberty, deposits of calcium, magnesium phosphate, and carbonate appear, which usually render the pineal opaque to X rays.

There are several cell types in the adult pineal organ, but the major ones are pinealocytes and interstitial cells. The pineal is developed maximally at about 7 years of age. As mentioned, salts invest the capsule and septa. The organ is supplied with many blood vessels and both myelinated and unmyelinated nerve fibers. Within the organ the capillaries are thin and fenestrated, but this may not apply to all species. The nerve fibers derive from the sympathetic autonomic nervous system. Nerve terminals may end directly on pineal cells. The neurosecretory granules containing norepinephrine are 40 nm in diameter. Serotonin is also present in the nerve endings, as well as within the pinealocytes.

The pinealocytes have cytoplasmic processes that end in swellings. The other major cell type in the pineal organ is the interstitial cell, which is irregular in shape and scattered among the pinealocytes. These cells resemble astrocytes, and some consider them to be neuroglia.

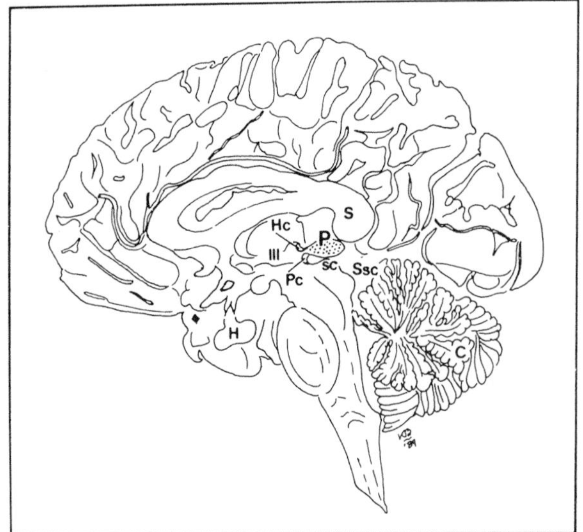

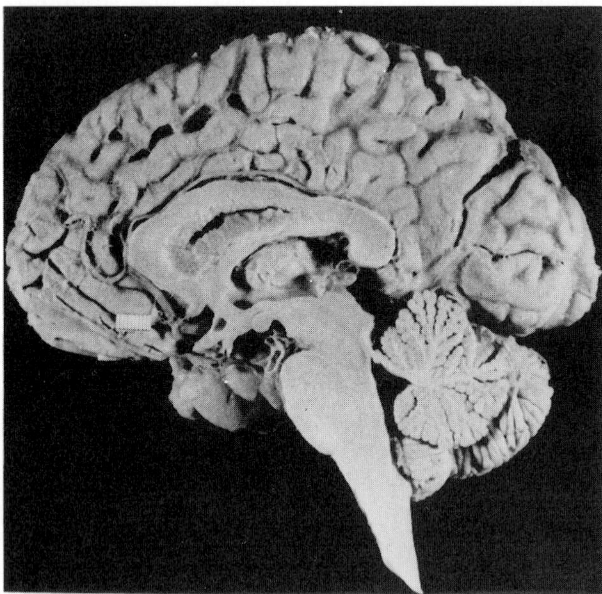

FIGURE 18-4 Medial surface of the right hemisphere of the human brain. Abbreviations: C, cerebellum; H, hypophysis; Hc, habenular commissure; P, pineal gland; Pc, posterior commissure; S, splenium of the corpus callosum; SC, superior colliculus; Ssc, superior subarachnoid cistern; III, third ventricle. Scale in the bottom half shows 1-mm divisions. Reproduced with permission from Shafii, M., and Shafii, S. L. (eds.) (1990). "Biological Rhythms, Mood Disorders, Light, Therapy, and the Pineal Gland." American Psychiatric Press, Inc., Washington, D.C.

IV. CHEMISTRY

A major hormone secreted by the pinealocyte is melatonin, shown in Figure 18-1. Serotonin and *N*-acetylserotonin also are important in the pineal, but they may not be secreted. Both melatonin and serotonin are derived from tryptophan, with the characteristic feature of all of these compounds being the indole ring (Figure 18-1). Other hydroxy- and methoxyindoles are present in the pinealocyte, as will be evident from the discussion of the biochemistry.

The pinealocyte may secrete a peptide, vasotocin, which may have important activities in reproductive functions and which is related in structure to oxytocin and vasopressin (Figure 4-4).

Two neurophysins have been found in the pineal gland. Neurophysins are proteins derived from the same gene product as either vasopressin or oxytocin in the hypothalamus.

V. BIOCHEMISTRY

Tryptophan is the precursor of melatonin. After uptake from the extracellular blood vascular system, this amino acid is converted, in the pinealocyte, to 5-hydroxytryptophan catalyzed by tryptophan hydroxylase. Serotonin is the product of the next step in the pathway achieved by the action of aromatic L-amino acid decarboxylase on 5-hydroxytryptophan.

Serotonin is the trivial name for 5-hydroxytryptamine. The serotonin concentration remains high in the pineal during daylight hours as a result of the signaling mechanism reviewed in Figure 18-8, but falls in darkness when serotonin either is converted to melatonin or falls in concentration for some other reason. Two enzymes achieve this conversion: serotonin *N*-acetyltransferase, which converts serotonin, using acetyl-coenzyme A, to *N*-acetylserotonin, and hydroxyindole-*O*-methyltransferase (HIOMT), which catalyzes the transfer of a methyl group from *S*-adenosylmethionine (SAM) to the 5-hydroxyl of *N*-acetylserotonin. The product of this last reaction is melatonin (5-methoxy-*N*-acetyltryptamine). During darkness, there is increased release of norepinephrine from the sympathetic neurons terminating on the pinealocytes (Figure 18-6).

Melatonin circulates in the blood and is metabolized by the liver. Apparently, some or all of the melatonin is secreted either into the blood directly or into the cerebrospinal fluid before entering the bloodstream. It is unclear exactly how the metabolism of melatonin is partitioned between the liver and central nervous system, if the latter plays a role in this activity at all. Various effects of melatonin have been ascribed to occur in the central nervous system and perhaps elsewhere.

The biosynthesis of melatonin is shown in Figure 18-7. In darkness, norepinephrine is secreted from adrenergic neurons ending on pinealocytes. The norepinephrine binds to the β-adrenergic receptor (see Chap-

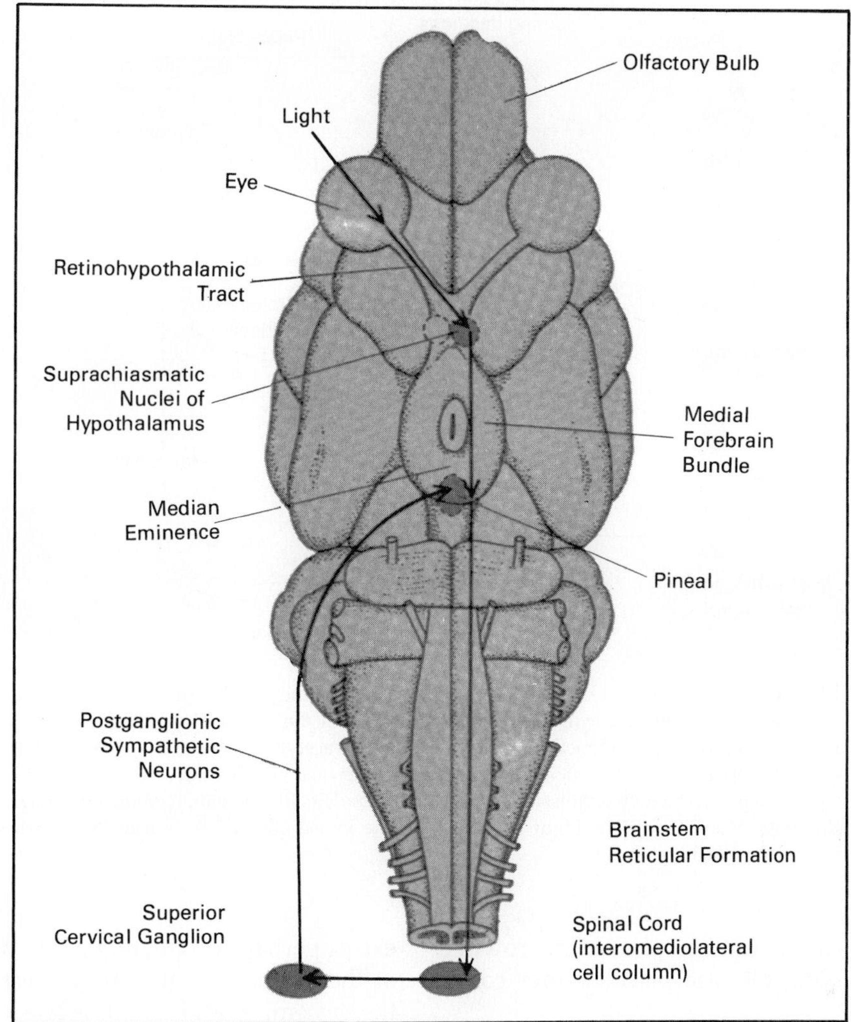

FIGURE 18-5 Anatomy of the pineal organ with respect to light stimulus. Light is transmitted from the retina along several brain pathways, which merge at the superior cervical ganglia of the sympathetic nervous system, as do all neural impulses of central nervous system origin that travel to the pineal. These impulses affect the synthesis of melatonin through the mediation of norepinephrine release and its stimulation of β-adrenergic receptors on the pineal cell membranes. This conversion of neural input identified the pineal as a neuroendocrine transducer. Reproduced from Wurtman, R. J. (1980). "Neuroendocrinology," (D. T. Krieger and J. C. Hughes, eds.), p. 103. Sinauer Associates, Sunderland, MA. courtesy of Nancy Lou (Gahan) Makris.

ter 11) on the pinealocyte membrane (may be a synapse) and results in the stimulation of intracellular cyclic AMP levels. The elevation of cyclic AMP probably operates as described elsewhere (Figure 1-37) and stimulates protein kinase activity, which in turn causes the phosphorylation of specific proteins and results in the stimulation of the synthesis of *N*-acetyltransferase.

In cultured rat pineal glands during the induction of serotonin-*N*-acetyltransferase by catecholamine, a nuclear protein is phosphorylated in the early stages of enzyme induction. This protein is also phosphorylated when the glands are treated directly with dibutyryl cyclic AMP, overstepping catecholamine ligand and its receptor. Increased levels of this enzyme lead to the conversion of serotonin to melatonin and its secretion from the cell probably into the bloodstream. Environmental light, perceived via the retinas, diminishes the flow of impulses along the pineal's sympathetic nerves, thus decreasing the release of norepinephrine onto pinealocytes. This decreases the activities of serotonin-*N*-acetyltransferase and HIOMT, suppressing the synthesis and secretion of melatonin. This pineal mech-

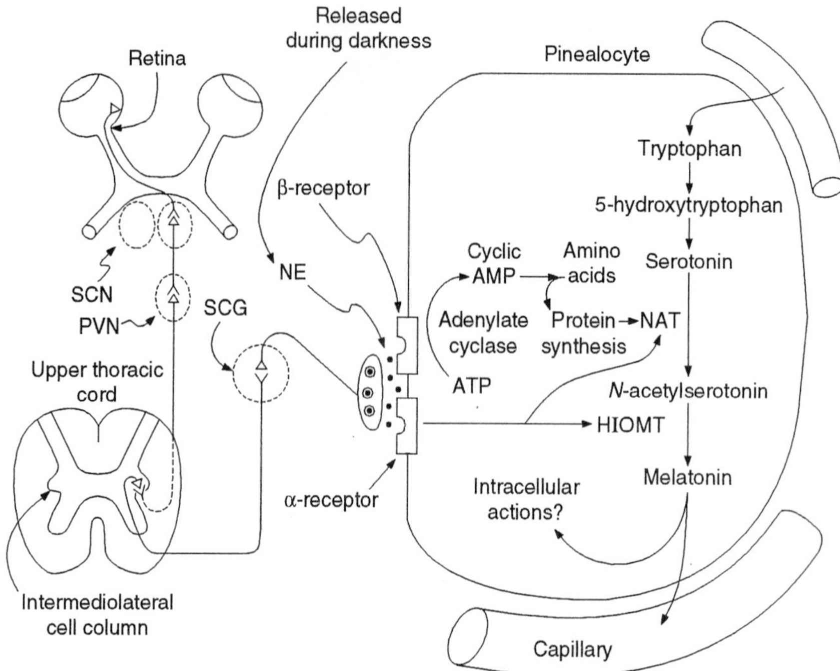

FIGURE 18-6 Neural connections between the eyes and the pineal gland as demonstrated in various mammals and synthesis of melatonin within the pineal gland. Abbreviations: HIOMT, hydroxyindole-*O*-methyltransferase; NAT, *N*-acetyltransferase; NE, norepinephrine, PVN, paraventricular nucleus; SCG, superior cervical ganglia; SCN suprachiasmatic nucleii. Reproduced with permission from Shafii, M., and Shafii, S. L. (eds.) (1990). "Biological Rhythms, Mood Disorders, Light Therapy, and the Pineal Gland." American Psychiatric Press, Inc., Washington, DC.

anism may mediate the stimulation of ovarian growth that occurs when young rats are placed under constant light.

The metabolic reactions involved in the conversion of tryptophan to melatonin are shown in Figure 18-7. The regulation of *N*-acetyltransferase is more elaborate than has been described up to this point. As mentioned before, the increased elaboration of norepinephrine, stimulated by the absence of light, results in the enhanced synthesis of *N*-acetyltransferase and coenzyme A. An *N*-acetyltransferase-inactivating substance (NIS) may be present. The NIS presumably would act when the concentration of acetyl-CoA falls (light) to further reduce the production of melatonin. γ-Aminobutyric acid (GABA) could have a modulating effect on norepinephrine increases in *N*-acetyltransferase protein. Cyclic GMP may have some effect in this system, although the dominant effect of norepinephrine operates through cyclic AMP.

Although lighting is the synchronizer of this process, experiments have been done with animals housed in continuous darkness. Under these conditions, melatonin synthesis and secretion continue to exhibit circadian rhythms, which require sympathetic nerves innervating the pineal. Obviously another cycling signal exists, perhaps originating in the suprachiasmic nuclei of the brain. Estrogens may exert some influence on the system, modulating the production of melatonin, although the evidence is not strong. Melatonin clearly acts on the brain, but it may also act on the pituitary and other organs. In all species examined, melatonin has been found to reach peak concentrations in darkness in CSF, blood, and urine. Figure 18-8 shows the relative levels of pineal-produced substances as well as the concentrations of melatonin in plasma and urinary metabolites as a function of the hourly light–dark cycle.

Melatonin is cleared in the urine for the most part by way of the liver, which contains a microsomal enzyme catalyzing 6-hydroxylation. Following hydroxylation, some of the product becomes esterified with sulfate and is excreted in that form (Figure 18-9).

VI. BIOLOGICAL AND MOLECULAR ACTIONS

A. General Comments

It is becoming clear that melatonin operates on the brain. For example, melatonin is known to produce drowsiness in humans. The mechanism of melatonin

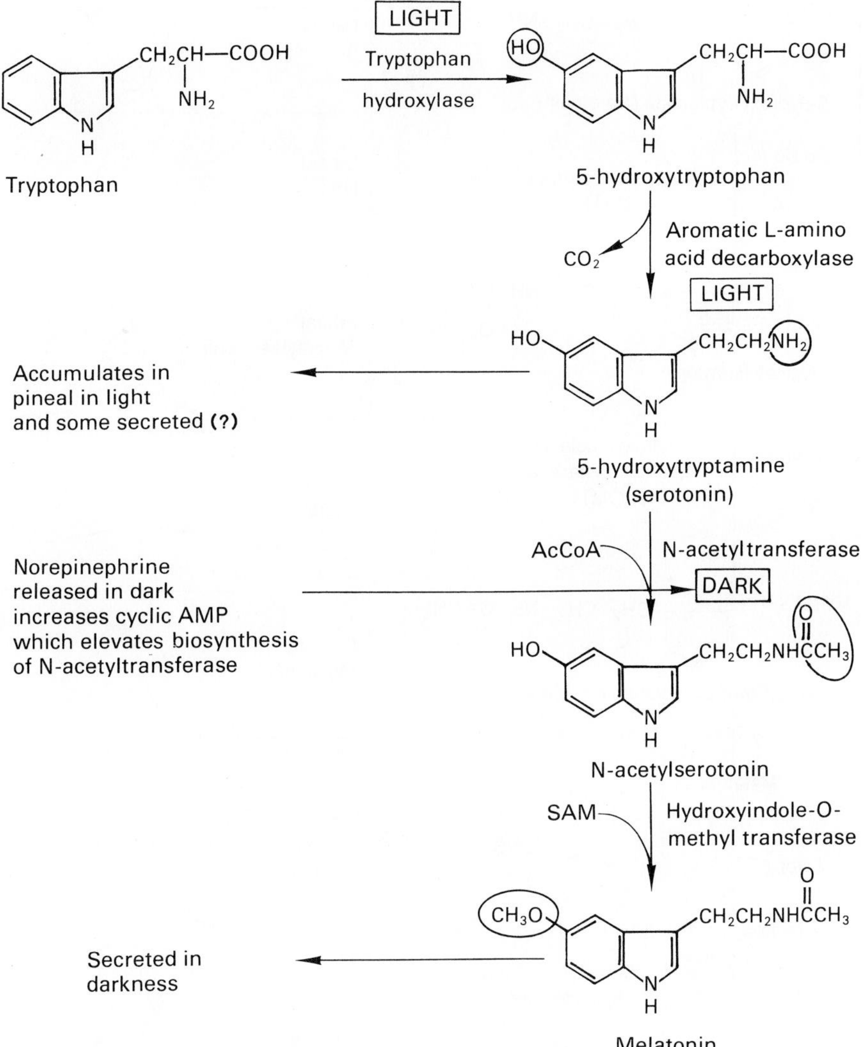

FIGURE 18-7 Reactions converting tryptophan to melatonin in pinealocytes. Circles highlight the group altered in the reaction.

secretion by pinealocytes is not understood. It is possible that this very lipid-soluble hormone leaves the cell by free diffusion, as pointed out by Wurtman (1980), in relation to its accumulation by the cell. In other words, the pinealocyte probably does not behave like a neuronal cell utilizing the process of exocytosis of neurosecretions.

B. Melatonin Receptor

Specific melatonin-binding sites have been defined by using [2-^{125}I] iodomelatonin as a radioactive ligand. High-affinity sites have been found in suprachiasmic nuclei and *pars tuberalis.* The levels of the melatonin receptor(s) in these structures are inversely related to the plasma concentration of melatonin. Consequently, the level of melatonin receptors is increased during the day. The light–dark cycle appears to be the regulatory factor. The density of melatonin receptors decreases during the night, even in pinealectomized animals. One hour of light can reverse the nighttime decrease in the suprachiasmic nuclei. There is evidence, however, that the mechanisms that regulate these changes in receptor activity in the two structures are different. In the *pars distalis,* for example, inhibition of Ca^{2+} influx and membrane potential are important mediators of melatonin effects.

Evidence has been presented for the presence of melatonin receptors in rat liver nuclei, leading to the conclusion that melatonin has an intracellular action. The signal transduction mechanism of melatonin receptors, having a K_D value in the picomolar range (one

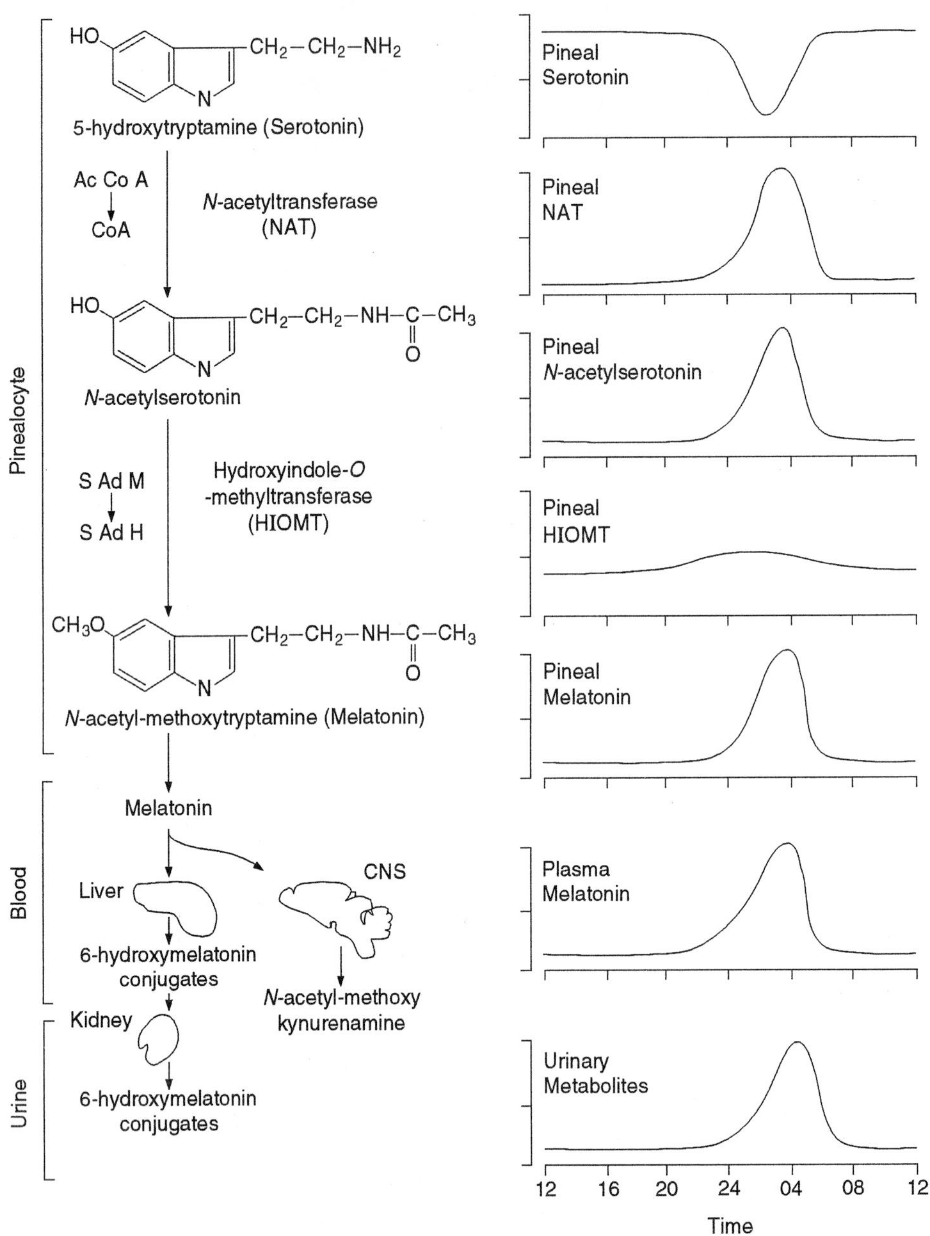

FIGURE 18-8 Rhythms in various pineal, blood, and urinary constituents known to exist in mammals, including humans. Reproduced with permission from Shafii, M., and Shafii, S. L. (eds.) (1990). Biological Rhythms, Mood Disorders, Light Therapy and the Pineal Gland.'' American Psychiatric Press, Inc., Washington, DC.

reported K_D value is $6 \times 10^{-11}\,M$), is thought to involve a *pertussis* toxin-sensitive G protein coupled to the membrane melatonin receptor, which, in turn, interacts with adenylate cyclase in an inhibitory manner to cause a dimunition of intracellular cyclic AMP.

High-affinity melatonin receptors have now been cloned and there appear to be two major subtypes. The cloning has been done from sheep and human species. The existence of more than one type of melatonin receptor was inferred from biphasic dose–response curves obtained with various ligands for the melatonin receptor. Ligand binding appears to involve hydrogen bonding between the 5-methoxy (Figure 18-1) and amide groups of melatonin and complementary amino acid residues and electrostatic interactions between an aromatic amino acid of the receptor-binding site and

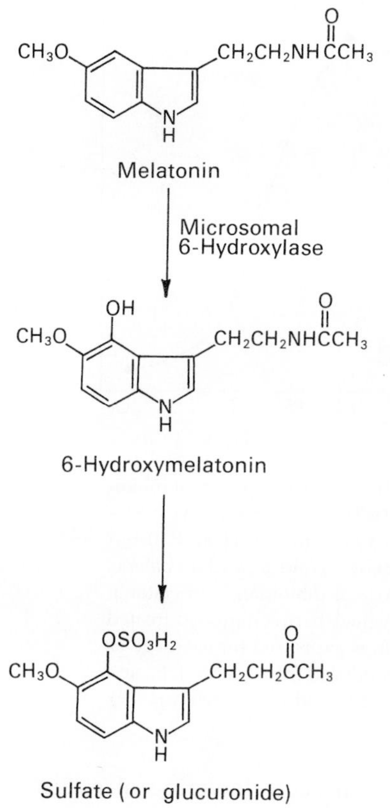

FIGURE 18-9 Metabolism of melatonin.

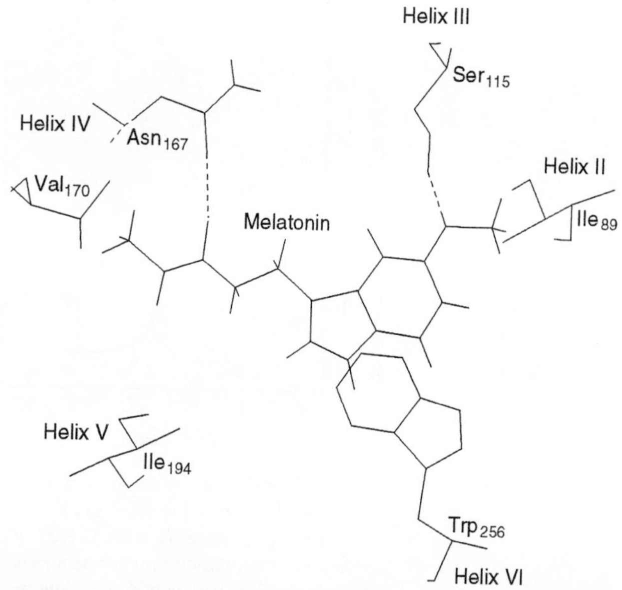

FIGURE 18-10 Postulated melatonin receptor-binding site model. This model shows two possible hydrogen-bonded interactions between melatonin and serine and asparagine residues in transmembrane helices III and IV of the receptor. Potential hydrogen bonds are denoted by dashed lines. Reproduced with permission from Sugden, D., Chong, N. W. S., and Lewis, D. F. V. (1995). Structural requirements at the melatonin receptor. *Br. J. Pharmacol.* **114**, 616–623.

the indole ring of melatonin. The interaction of melatonin with the melatonin receptor ligand-binding site is shown in Figure 18-10.

C. Actions of Melatonin on Anterior Pituitary

1. *LH and FSH*

Melatonin inhibits the secretion of LH and FSH induced by GnRH. In particular, this has been demonstrated with anterior pituitary cells obtained from 1- to 20-day-old rats. The inhibition by melatonin was effective in the range 0.1–1.0 n*M* when the cells were obtained from animals that were 5, 10, or 15 days of age. Suppression of anterior pituitary hormone release ranged to 50% of the full response to GnRH, as shown in Figure 18-11. Melatonin interferes either with the GnRH stimulation of anterior pituitary hormone release or with the secretion process, probably through an indirect mechanism. It has been shown in cultured pituitary glands from prepubertal rats that melatonin inhibits GnRH-induced LH secretion. Considering that GnRH has a signal transduction system involving phosphatidylinositol cycle activation, whereas melatonin signals the inhibition of adenylate kinase with a lowered cyclic AMP concentration, some more distant system must be affected by melatonin. Interestingly, pinealectomy results in an increase in ornithine decarboxylase activity in rat anterior pituitary, and this enhanced activity persists for at least several weeks. Other reports that pinealectomy results in increased mitotic activity in the rat adenohypophysis align with the effect on ornithine decarboxylase activity, which usually reflects mitotic activity. Whether this is related to the release of LH and FSH in response to GnRH is unclear. Cells from 20- to 30-day-old animals no longer showed the response to melatonin, presumably contributing to the regulation of development of the reproductive system. In the rat, puberty occurs 2–3 weeks after melatonin's ability to inhibit the pituitary gland disappears. In humans, peak nocturnal plasma melatonin levels fall by about 75% between ages 7 and 12, when concurrently measured plasma LH is rising (Waldhauser *et al.*, 1984). This suggests that melatonin may have a role in timing human puberty.

2. *Regulation of Growth Hormone Release*

It has been suggested for some time that melatonin could inhibit the release of growth hormone from the anterior pituitary. Other work suggests that melatonin accomplishes this by stimulating the release of somato-

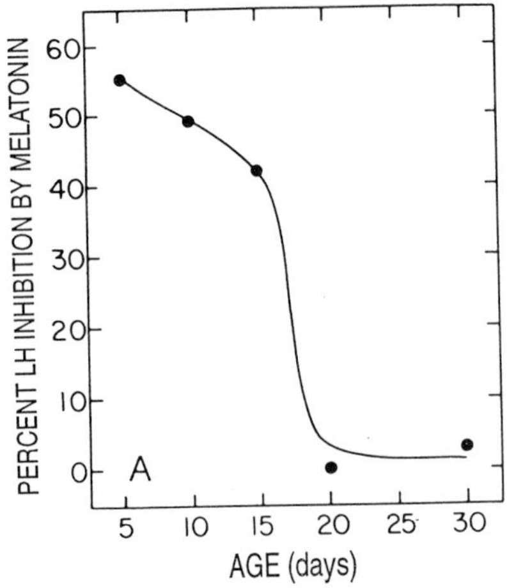

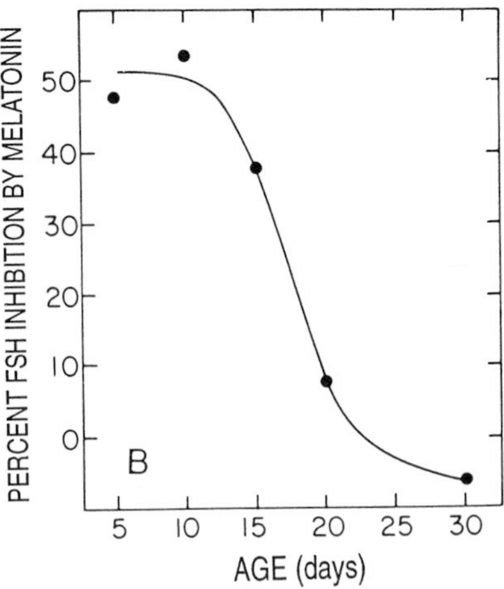

FIGURE 18-11 (A) Relationship between animal age and percentage of maximal inhibition by melatonin of pituitary LH response to LHRH (GnRH). The difference in LH values between groups treated with LHRH alone and groups treated with LHRH and 10 n*M* melatonin is expressed for each as a percentage of stimulation over control by LHRH alone. (B) Relationship between animal age and percentage of maximal inhibition by melatonin of pituitary FSH response to LHRH. The difference in FSH values between groups treated with LHRH alone and groups treated with 10 n*M* melatonin is expressed for each age as a percentage of stimulation over control by LHRH alone. Reproduced from Martin J. E., and Sattler, C. (1979). *Endocrinology* (*Baltimore*) **105**, 1007–1012. © by The Endocrine Society (1979).

statin, which has been observed by using explants of rat medial basal hypothalamus. The action of somatostatin on inhibiting growth hormone release is described in Figure 5-23. Interestingly, melatonin concentrations effective in stimulating somatostatin release were in the range of 10–100 n*M* for explants of hypothalamus.

Relatively little is known about the specific actions of Arg-vasotocin, and little or no information is available concerning an Arg-vasotocin receptor. In fact, some believe that vasotocin may not be present in the pineal at all.

VII. CLINICAL ASPECTS

Most inferences concerning pineal hormones and diseases have come from studies with experimental animals. Thus, pinealectomy at an early time in development in some species can lead to premature ovarian function, and treatment with melatonin can suppress ovarian function. As described earlier, the amounts of melatonin secreted nocturnally decline just prior to and during pubescence. Moreover, an acute oral dose (1–3 mg/kg) given to young adults causes prolactin to be secreted. A number of different kinds of tumors can affect the function of the pineal gland, usually resulting in central nervous system disturbances and either premature or delayed gonadal functioning, especially when the tumors occur in the young.

There may be a point of interaction above the pituitary between melatonin and ACTH-corticosteroid secretion. An acute lowering of cortisol secretion in humans (by metyrapone, for instance) stimulates melatonin secretion. Melatonin also appears to inhibit the release of CRH, thus impinging on the hypothalamic–pituitary–adrenal axis.

When melatonin is low in serum, more CRH is released, culminating in increased cortisol secretion. Increased cortisol levels are associated with depression. Interestingly, in bipolar disease it has been reported that serum melatonin would allow for higher levels of cortisol. In manic states the level of melatonin is high, which should depress release of CRH and generate lower levels of cortisol. Consequently, melatonin is closely related to the functioning of the hypothalamic–pituitary–adrenal axis and associated psychiatric manifestations (Figure 18-12).

However, there are intermediate situations in which depressed patients also produce normal levels of melatonin, and there are apparently cases of depression, such as the "winter blues" that are more dependent on biorhythm phasing, of which melatonin production can be a marker. Although complex, it is likely that melatonin production will have significance for certain types of depression and for the chronicity of sleep–wake cycles and may play a role in others; melatonin may also be important in this latter type where environmental bright light is a factor. There also seems to be the possibility that, in some cases of depression, high levels of melatonin are produced, but for some reason may not be utilized appropriately (in the hypo-

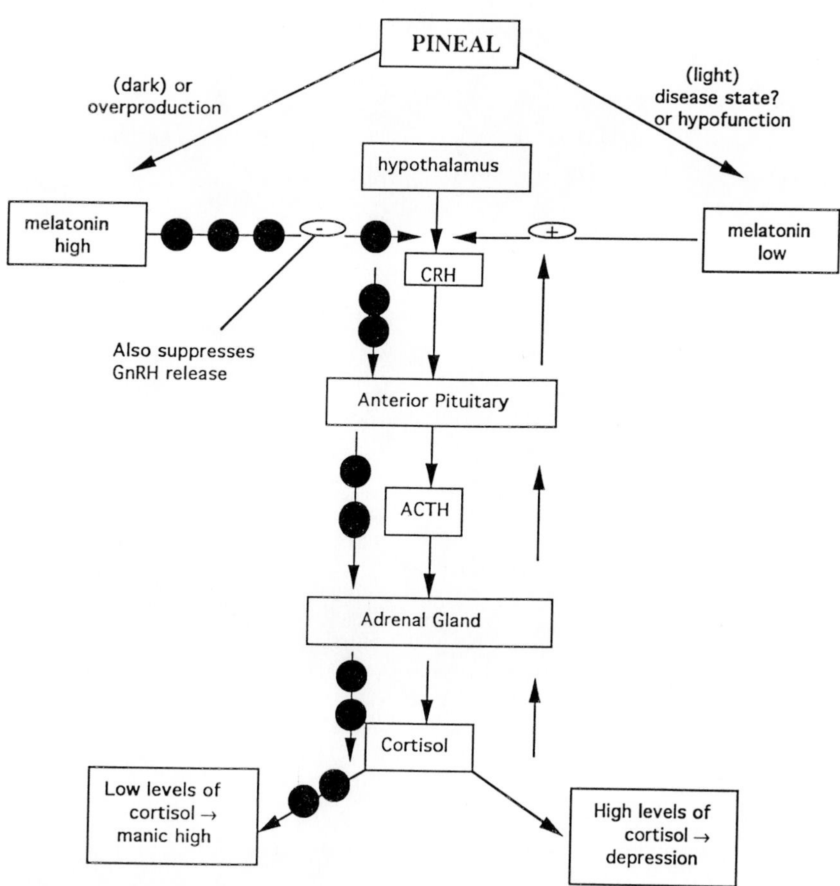

FIGURE 18-12 Melatonin and the hypothalamo-pituitary-adrenal axis in manic-depression states.

thalamus). This condition undoubtedly would benefit from more knowledge of the possible melatonin "receptor" and of whether the receptor could assume various states of functioning possibly through mutations.

References

A. Books

Shafii, M., and Shafii, S. L., eds. (1990). "Biological Rhythms, Mood Disorders, Light Therapy And The Pineal Gland." American Psychiatric Press, Inc., Washington, DC.

Wurtman, R. J. (1980). "Neuroendocrinology" (D. T. Krieger and J. C. Hughes, eds.). Sinauer Associates, Sunderland, MA.

B. Review Articles

Arendt, J. (1995). The pineal gland: basic physiology and clinical implications. *Endocrinology* (L. J. DeGroot, ed.) **1,** 432–444.

Dubocovich, M. L. (1995). Melatonin receptors; are there multiple subtypes? *Trends Pharmacol. Sci.* **16,** 50–56.

Pierpaoli, W., and, Jesnikov, V. A. (1994). The pineal aging clock. Evidence, models, mechanisms, interventions in the aging clock. *Ann. N.Y. Acad. Sci.* (W. Pierpaoli, W. Regelson, and N. Fabris, eds.) **719,** 461–473.

Waldhauser, F., Frisch, H., Walhauser, M., Weiszenbacher, G., Zeitlhuber, U., and Wurtman, R. J. (1984). Fall in nocturnal serum melatonin during prepuberty and pubescence. *The Lancet,* Feb. 18, 362–365.

Wurtman, R. J., and Axelrod, J. (1968). The formation, metabolism and physiologic effects of melatonin. *Advances in Pharmacology,* **6,** 141–151.

C. Research Papers

Ebisawa, T., Karne, S., Lerner, M. R., and Reppert, S. M. (1994). Expression cloning of a high-affinity melatonin receptor from xenopus dermal melanophores. *Proc. Natl. Acad. Sci. USA* **91,** 6133–6137.

Martin, J. E., and Sattler, C. (1979). Developmental loss of the acute inhibitory effect of melatonin on the *in vitro* pituitary luteinizing hormone and follicle-stimulating hormone responses to luteinizing hormone-releasing hormone. *Endocrinology (Baltimore)* **105,** 1007–1012.

Reppert, S. M., Weaver, D. R., and Ebisawa, T. (1994). Cloning and characterication of a mammalian melatonin receptor that mediates reproductive and circadian responses. *Neuron* **13,** 1177–1185.

Sugden, D., Chong, N. W. S., and Lewis, D. F. V. (1995). Structural requirements of the melatonin receptor. *Br. J. Pharmacol.* **114,** 618–623.

CHAPTER

19

Cell Growth Factors

I. GROWTH FACTOR SIGNAL TRANSDUCTION

In recent years there has been an explosion of information on the continuation of the signal generated by the interaction of growth factor with cell membrane growth factor receptor. Thus, the hormone–growth factor interacts with the receptor, and this interaction often produces a conformational rearrangement in receptor and receptor-associated proteins so that the message is carried by some initial system. This system can be a part of the receptor itself, which may contain a protein kinase catalytic center that catalyzes the phosphorylation of specific amino acid residues, such as tyrosine residues if it is a tyrosine protein kinase, resulting in autophosphorylation of the receptor on the cytoplasmic side of the receptor molecule. This could result in an activation leading to other interactions. Likewise, the receptor may dimerize upon interacting with ligand, and the dimerized receptor could undergo further interactions in the cell membrane and in the cell. The receptor may interact in specific ways with G proteins or enzymes of the phosphatidylinositol cycle to produce other changes, or the receptor could be interacting with an ion channel that could become activated following ligand–receptor interaction. These general comments set the stage for this section.

A number of growth factor receptors are members of a protein tyrosine kinase receptor family, and others have specific characteristics concerning their structure. These topics are summarized in Table 19-1.

Similarities between some groups of ligands indicate that they may have derived from the same or similar ancestors. Examples are insulin, IGF-I, and IGF-II. This group shares homology to the extent that there can be crossover binding to their receptors. For example, both insulin and IGF-II can bind to the IGF-I receptor, although much greater amounts of IGF-II and insulin are needed to activate the receptor. However, it appears that blockage of IGF-I by competition with a similar but inactive peptide, while greatly inhibiting the growth of certain cells in culture, has a much smaller effect on intact animals. Perhaps this is indicative of the growth factors (e.g., insulin or IGF-II) that are present in great enough quantity to overcome the effect of the inhibitor. Another group of hormones sharing a common ancestor is LH, FSH, hCG, ACTH, and MSH, and this has been discussed in Chapter 5. There are also similarities between receptors. The receptors for IGF-I and insulin are very similar, and receptors for IGF-I, insulin, epidermal growth factor

TABLE 19–1 Single-Transmembrane-Segment Receptors[a]

Protein Tyrosine Kinase Receptor Family
Epidermal growth factor
Neu-differentiation factor (human EGF receptor 2)
Insulin
Insulin-like growth factor-I
Fibroblast growth factors
Platelet-derived growth factors A and B
Colony-stimulating factor-1 (macrophage colony-stimulating factor)
Nerve growth factors (neurotrophins)
Hepatocyte growth factor (scatter factor)
Guanylate Cylase Receptor Family
Atrial natriuretic peptide types A, B, and C
Serine–Threonine Kinase Receptor Family
Transforming growth factor β, types I and II
Activin
Inhibin
Multisubunit Receptor Family
Growth hormone
Prolactin
Placental lactogen
Erythropoietin
Interleukin-2, -3, -4, -5, -6, -7
Granulocyte macrophage colony-stimulating factor
Granulocyte colony-stimulating factor
Interferon-α, -β, and -γ
Tumor necrosis factor p75
Leukemia inhibitory factor
Oncostatin
Ciliary neurotrophic factor
Tumor Necrosis Factor–Nerve Growth Factor Receptor Family
Tumor necrosis factor p55
Nerve growth factor p75
Phosphotyrosine Phosphatase Receptor Family
Ligands unknown
Plasma Protein Receptor Family
Low-density lipoprotein
Transferrin
Asialoglycoproteins
Polymeric immunoglobulin A–immunoglobulin M
Mannose 6-phosphate–insulin-like growth factor-II
Urokinase plasminogen activator
α_2-Macroglobulin

[a] Reproduced with permission from Gammeltoft, S., and Kahn, C. R. (1995). Hormone signaling via membrane receptors. *In* "Endocrinology" (L. J. DeGroot, ed.), 3rd ed., Vol. 1, pp. 17–65. W. B. Saunders Co., Philadelphia, PA.

(EGF), and platelet-derived growth factor (PDGF) all have tyrosine kinase domains on the cytoplasmic portion of the receptors. Receptors for PRL, GH hematopoietic growth factors, and cytokines have similar domains and interact with the Janus (JAK) cytoplasmic tyrosine kinases. Receptors that interact with G proteins often contain seven transmembrane domains, and the activation of enzymes by G-protein subunits is often similar.

II. GROWTH HORMONE (GH) AND SOMATOMEDINS (IGFs)

Although this subject overlaps that of Chapter 5 on anterior pituitary hormones, it is a reasonable starting point since IGF-I is a major growth factor. The relationship between growth hormone and IGF-I is shown in Figure 19-1.

Here it is shown that GH is released through the action of hypothalamic growth hormone-releasing hormone (GRH), as described earlier in Chapter 3. The GH released into circulation has two effects: One effect is to directly act on lipid and fat metabolism, analogously to cortisol action, in a manner to oppose the actions of insulin and IGF-I, as shown by its ability to directly increase blood sugar (Figure 19-1). The other action of GH is on liver and other organs to generate IGF-I, which, like insulin, produces skeletal growth as well as tissue growth. IGF-I is active as a negative feedback agent on the hypothalamus and causes the production of somatostatin (GIH), which inhibits the release of GH from the anterior pituitary, and it also acts at the level of the pituitary to inhibit expression of the GH gene in response to GRH. When GH causes IGF-I to be produced, IGF-I-binding proteins are also synthesized in the liver and other tissues, and both IGF-I and binding proteins are secreted into the bloodstream. Consequently, IGF-I (and other somatomedins) is complexed with binding proteins in the circulation. There are six IGF-binding proteins (IGFBPs) in humans and each is produced by a separate gene. These binding proteins contain cysteine-rich domains that are involved in the interaction with IGFs. IGFBPs I and II have R-G-D sequences that may bind to cell surface integrin receptors. Consequently, IGFs circulate in the bound form, a condition that allows for relatively high concentration and stability. In some cases the binding proteins interfere with the action of IGF at its cell membrane receptor, and in other cases, certain binding proteins enhance the growth effects of IGFs upon cells. Apparently, the extent to which a specific binding protein is phosphorylated may determine its activity with regard to the binding of IGF to its receptor. IGF-I can be dissociated from an IGF-I-BP by the action of serum endoproteases and also by heparin-like substances. Properties of human IGFBPs are given in Table 19-2.

III. INTERACTION OF IGFs WITH RECEPTORS

Not only do IGFs and insulin have high homologies, but the receptors for IGF-I and insulin are also homologous. Thus, IGF-I receptor binds IGF-I and, at higher levels, insulin, and it also carries binding sites for IGF-

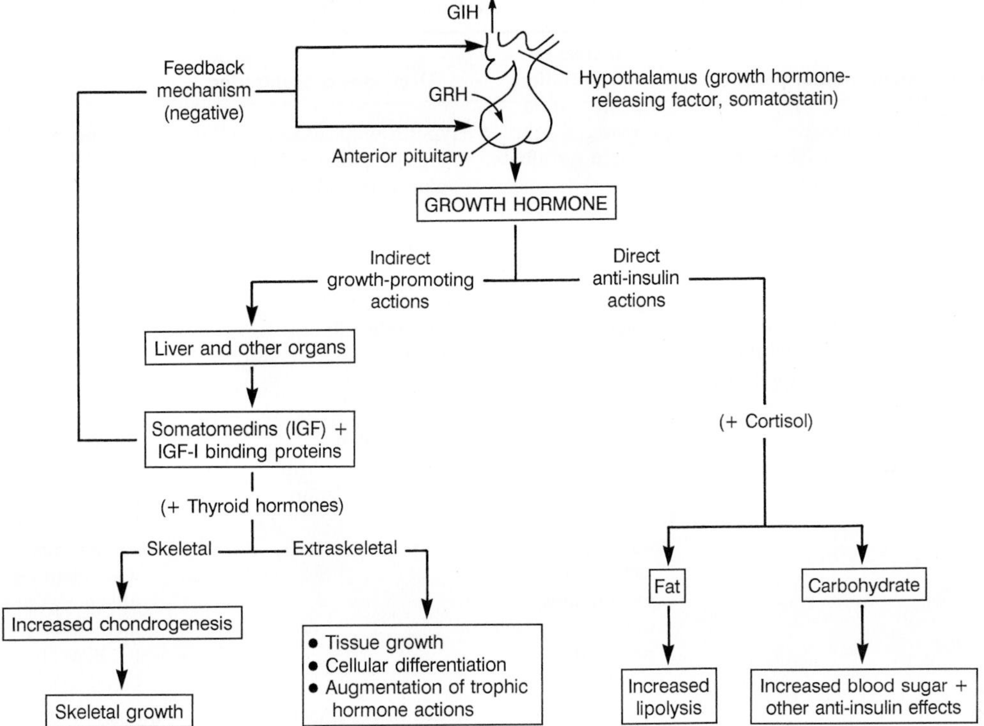

FIGURE 19-1 Somatomedin hypothesis of growth hormone (GH) action. The direct actions of GH include diabetogenic and lipolytic actions and stimulatory action on several hepatic enzymes. These direct actions are antagonistic to insulin and often synergistic with cortisol. The anabolic and growth-promoting actions of GH are mediated through the somatomedins. IGF-I participates in the negative feedback on GH secretion at the hypothalamic level by stimulating somatostatin production and at the pituitary level by directly blocking the effect of GH-releasing hormone on the expression of the GH gene. Reproduced with permission from Russell, W. E., and Van Wyk, J. J. (1995). Peptide growth factors. *In* "Endocrinology" (L. J. DeGroot, ed.), 3rd ed., Vol. 1, pp. 2590–3623. W.B. Saunders Co., Philadelphia, PA.

II, which are apparently different from the sites to which IGF-I binds. The β-subunits of both receptors contain a tyrosine kinase and ligand binding triggers the activity of the tyrosine kinase, resulting in autophosphorylation on tyrosine residues. Other proteins may be phosphorylated on tyrosine residues by the receptor kinase. IGF-I receptors are present in most tissues: the hormone exerts its effect in concert with other hormones produced in specialized tissues, and the IGF-I receptor is down-regulated after ligand binding in a typical process characteristic of many membrane receptors. IGF-II receptors bind IGF-II nearly exclusively and interact poorly with IGF-I. On the other hand, as has already been pointed out, IGF-II can bind to the IGF-I receptor and this interaction is important in humans, who, in contrast to some rodents, produce IGF-II. The IGF-II receptor is identical to the mannose 6-phosphate receptor and contains separate domains for binding IGF-II and mannose 6-phosphate. Interestingly, the mannose 6-phosphate-binding domain activates the TGF-β precursor by removing mannose, and this domain has other activities as well. The interaction of IGF-II with the IGF-II receptor may lead to the uptake of extracellular calcium possibly mediated by a G protein.

IV. SIGNAL TRANSDUCTION OF RECEPTORS FOR GH, IGFs, AND INSULIN

The growth hormone receptor and its immediate signaling partners are shown in Figure 19-2.

The human GH receptor binds human GH exclusively and does not interact with other species of GH. The structure of the extracellular domain of the GH receptor has been determined by X-ray crystallography and when studied in the presence of ligand it was shown that the extracellular domain forms a dimer that interacts with one GH molecule. The structure of the hormone-binding domain of the GH receptor is shown in Figure 19-3.

Dimerization is required for binding of GH and cellular responses.

Figure 19-4 shows a signal transduction scheme for cytokine receptor, growth factor receptor, and insulin

TABLE 19-2 Properties of IGF-Binding Proteins (IGFBPs)[a]

IGF-binding protein	Major tissues producing	Increases concentration	Decreases concentration	Other comments
IGFBP-1	Liver, kidney reproductive tissues: prostate, uterus, decidua	Decreased insulin or insulin sensitivity, fasting, diabetes, insulin resistance, pregnancy, hypopituitarism	Hyperinsulinism: feeding, glucose infusion, GH, glucocorticoids, partial hepatectomy increases liver level	Independent of GH; may transport IGFs out of vascular space
IGFBP-2	Fetal tissues, but declines after birth; CNS and cerebral spinal fluid, higher in lymph than serum	Hypophysectomy Diabetes	GH Glucocorticoids Insulin	May transport IGFs out of vascular space
IGFBP-3	Serum	GH treatment Acromegaly Adolescence Infusion of IGF-1 Various growth factors, *in vitro* feeding after nutrient restriction	GH deficiency Stimulators of cAMP Nutrient restriction	IGF-I or IGF-II is linked to other subunits, one of which is IGFBP-3; this complex prevents loss of IGFs from intravascular compartment; serum levels dependent on GH and nutritional status; IGF-I may mediate GH induction of IGFBP-3'; helpful in diagnosing GH deficiency
IGFBP-4	Liver Plasma Many other organs	cAMP	Reduction of cAMP	Binds to IGF-I and IGF-II equally; blocks effects of IGF-I in cell culture; may modulate effects of IGFs in CNS
IGFBP-5	Kidney Bone Endocrine tissues			Binds strongly to extracellular matrix and may serve as storage compartment for IGF-I when bound to extracellular matrix
IGFBP-6	All tissues		Acromegaly	10-fold higher affinity for IGF-II than for IGF-I; regulation may resemble that of IGFBP-1 and -2

[a] Information for this table was derived from Russell, W. E., and Van Wyk, J. J. (1995). Peptide growth factors. *In* "Endocrinology" (L. J. DeGroot, ed.), 3rd ed., Vol. 1, pp. 2590–3623. W. B. Saunders Co., Philadelphia, PA.

receptor. While this scheme may describe signal transduction for EGF, PDGF, FGF, NGF, and insulin, it is not clear whether this pathway utilizes IGF-I.

It would be surprising if IGF-I signal transduction operated by a different pathway, given that IGF-I can produce insulin effects by binding to the insulin receptor and insulin can produce IGF-I-like effects by binding to the IGF-I receptor. Figure 19-4 shows one pathway for receptor-linked tyrosine kinases and starts with autophosphorylation on tyrosine residues after ligand binding. Other proteins may also be phosphorylated, as in the case of the insulin receptor phosphorylating the insulin receptor substrate-1 (IRS-1). GRB2 associates with phosphorylated tyrosine residues in the receptor kinase. This is followed by the action of SOS complexed to GRB2 on membrane-bound *ras* by GDP–GTP exchange and *ras* activates raf kinase, which starts a phosphorylation cascade involving the four kinases shown in the figure. The end of this process is the effect on growth and metabolism.

V. EPIDERMAL GROWTH FACTOR (EGF) FAMILY

This growth factor is part of a family of growth factors that includes, besides EGF, TGF-α, amphiregulin, heparin-binding EGF, and cripto. The linear

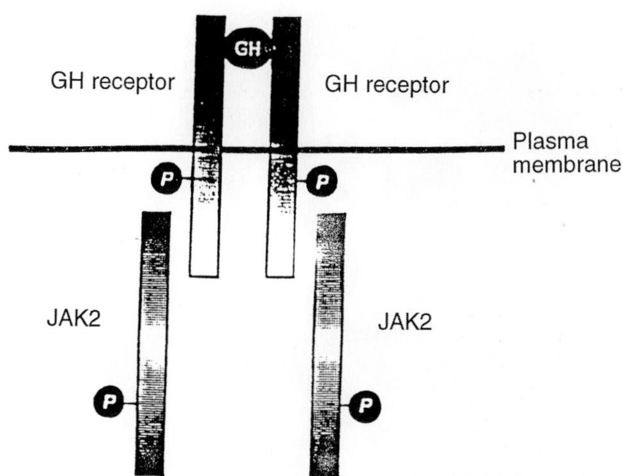

FIGURE 19-2 Structure of single-transmembrane-segment receptors that form homologous dimers; the growth hormone receptor. One GH molecule binds to identical binding sites on the extracellular domains of two GH receptors that form a homodimer. The intracellular domains associate with a tyrosine kinase of the Janus family of kinases. An encircled P denotes a potential tyrosine phosphorylation site on the intrecellular portion of GH receptor. Reproduced with permission from Gammeltoft, S. and Kahn, C. R. (1995). Hormone signaling via membrane receptors. *In* "Endocrinology" (L. J. DeGroot, ed.), 3rd ed., Vol. 1, pp. 17–65. W.B. Saunders Co., Philadelphia, PA.

sequences of these hormones are shown in Figure 19-5. The signal transduction process of the EGF receptor can be formulated as shown in Figure 19-6. While EGF promotes mitosis in cells of mesodermal and ectodermal origin, TFG-α confers the transformed phenotype on normal cells. TGF-α has substantial homology to EGF and both growth factors operate through the EGF receptors. TGF-α may be the fetal ligand for EGF receptor. Amphiregulin stimulates the growth of normal fibroblasts and keratinocytes while it inhibits the growth of several breast carcinoma lines. It has a 43-amino-acid extension at its terminus compared to EGF, and this extension may contain a nuclear translocation signal. Heparin-binding EGF also has an N-terminal extension and acts through EGF-R. While the gene for cripto has been cloned, little is known about its signal transduction.

The smallest form of active TGF-α contains 50 amino acids and is homologous to EGF. As shown in Figure 19-5, a presentation of six cysteine residues occurs in this family and provides the similar three-loop structures of EGF and TGF-α that explain the ability of both structures to bind to the EGF receptor. There is no evidence for a specific TGF-α receptor or for the activities of TGF-α on epithelial and mesenchymal cell proliferation; migration and differentiation appear to be due to the mediation of the TGF-α–EGFR interaction.

The 50-amino acid form of TGF-α is released from a membrane precursor, as shown in Figure 19-7. Two forms of this 159- or 160-amino-acid pro-TGF-α are coexpressed due to the microheterogeneity of mRNA splicing. The cysteine-rich cytoplasmic domain is highly conserved. The 50-amino-acid TGF-α is released from the precursor by proteases that cleave AV bonds at both termini (black arrows). Sometimes the processing is incomplete and larger forms of TGF-α are released that are still biologically active. This type of processing to a mature hormone from a membrane precursor is common throughout the EGF family. EGF is matured in a similar fashion, although the membrane precursor is larger than pro-TGF-α. Presumably the release of mature soluble growth factor from the precursor membrane form is a regulated event.

VI. TRANSFORMING GROWTH FACTOR-β (TGF-β) FAMILY

The crystal structure of TGF-β has been determined, and a computer model is shown in Figure 19-8. TGF-β either stimulates or inhibits cell growth or affects cellular functions distinct from mitosis. Among the activities of TGF-β are inhibition of proliferation of mesenchymal, epithelial, endothelial, and transformed cells, stimulation of proliferation of these cell types, control of extracellular matrix interactions, suppression of immune function, regulation of embryogenesis and cellular differentiation, fibroblast transformation, inhibition of adrenal steroidogenesis, and regulation of the biosynthesis and action of FSH. These represent many diverse activities, including several that are distinct from the effects on cellular proliferation. There are three isoforms in the mammal (TGF-β1–3), a chicken form (TGF-β4), and a frog form (TGF-β5). There are other more distant members of this family. The members of this family are related by sequence homology but, for the most part, act through separate receptors, although some of these receptors, as expected, may share some homology. TGF-β is found in the α-granules of platelets, and some species, such as pig, chicken, and mouse, contain isoforms. Aspects of the structure of TGF-β are described in Figure 19-8.

Members of the TGF-β family are pictured in Figure 19-9. Among the family members are inhibins, activins, Müllerian inhibitory substance (MIS), bone morphogenetic protein, Vg1 and Vgr-1 from xenopus, and DPP-C in *Drosophila* development. Inhibins, activins, and Müllerian inhibitory substance are discussed in Chapters 12 and 13.

TGF-β is generated from an inactive precursor protein. The precursors are glycosylated and substituted

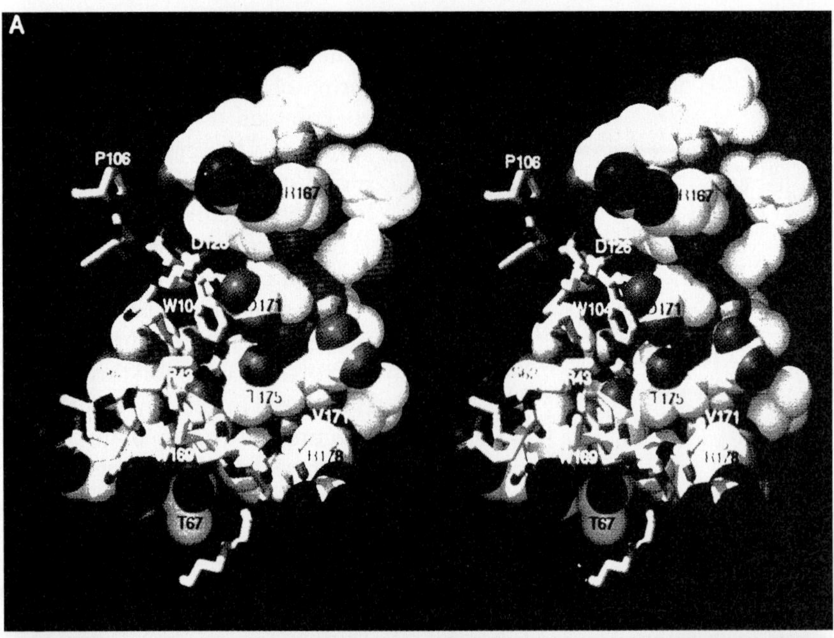

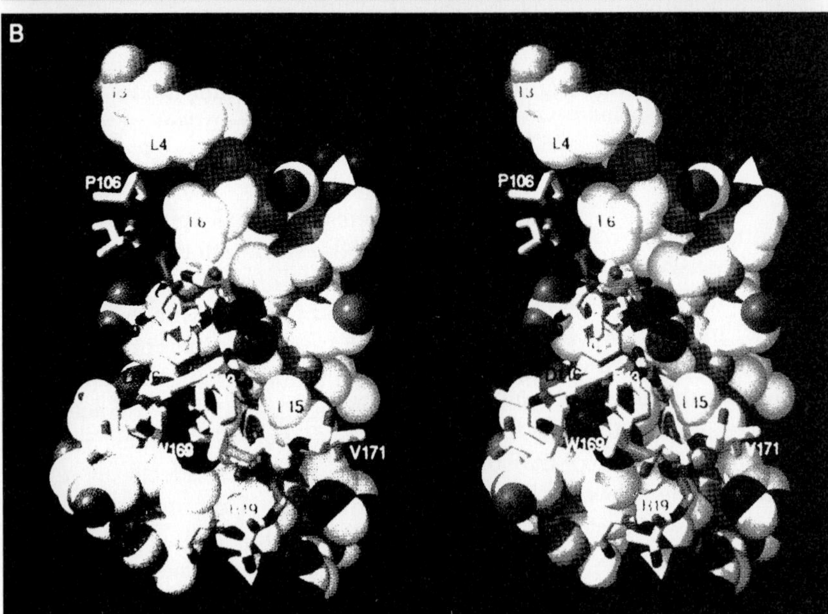

FIGURE 19-3 Close-up of interfaces between hormone and receptors: (A) binding site I; (B) binding site II. The hGH is represented by a space-filling model and the receptors by a stick model. The hGH backbone atoms are cyan, side chain carbons are white, and side chain oxygens and nitrogens are red and blue, respectively. The receptor carbon atoms are in yellow, with red oxygens and blue nitrogens. Selected residues are labeled. This figure is reproduced in black and white from a colored illustration with permission from de Vos, A. M., Ultsch, M., and Mossiakoff, A. (1992). Human growth hormone and extracellular domain of its receptor: Crystal structure of the complex. *Science* **255,** 306–212.

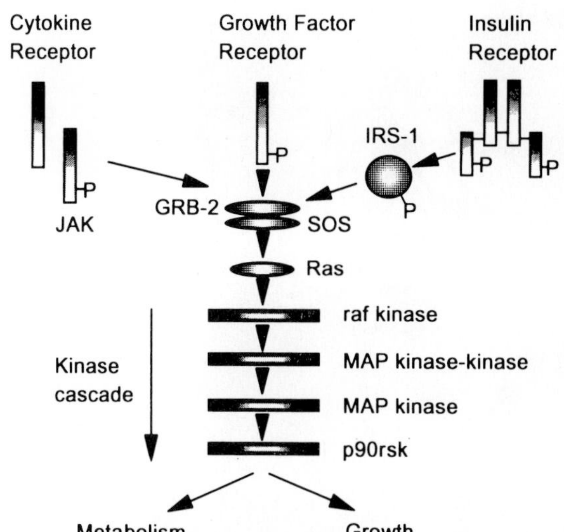

FIGURE 19-4 Signal pathway of receptor-activated tyrosine kinase receptors. Activation of receptor-linked tyrosine kinases by cytokines, growth factors, and insulin induces receptor autophosphorylation. The complex of GRB2 and the son of sevenless (SOS) guanine nucleotide-releasing protein is recruited to the plasma membrane by the receptor via SH2 domain–phosphotyrosine interaction. SOS activates raf kinase, which initiates a phosphorylation cascade, mitogen-activated protein (MAP) kinase-kinase, MAP-kinase, and $p90^{rsk}$, leading to cellular effects on metabolism and cell growth. Reproduced with permission from Gammeltoft, S., and Kahn, C. R. (1995). Hormone signaling via membrane receptors. *In* "Endocrinology" (L. J., DeGroot, ed.), 3rd ed., Vol. 1, pp. 17–65. W.B. Saunders Co., Philadelphia, PA.

with mannose phosphate. Dimerization follows to a 110-kDa inactive precursor. The precursor has to be activated in a process that is not well understood, but one that may involve proteases. TGF-β in serum is bound to α_2-macroglobulin, which is cleared by the liver; however, the latent complex of TGF-β has an extended half-life and is probably activated by proteases and binding to the IGF-II receptor through substituent mannose 6-phosphate residues. This complex between latent TGF-β and IGF-II receptor may be the best conformational form to allow the activating proteases to attack the latent TGF-β. Latency is apparently a function of the content of sialyl substituents. Binding of latent TGF-β to IGF-II receptor could also facilitate endocytosis of the complex to avail the latent TGF-β to the lysosomes and an activating environment. The active TGF-β is a 12.5-kDa peptide in the form of a 25-kDa homodimer linked by disulfide bonds.

There is a membrane-binding protein that forms a 300-kDa complex with TGF-β. The function of the binding protein is unknown except that it contains large amounts of heparin sulfate and chondroitin sulfate. This complex may represent a storage form of TGF-β. The signal-transducing form of TGF-β appears to involve the growth factor in an heteromeric complex of type I and type II TGF-β receptors, which form ligand–receptor complexes of about 65 and 130 kDa, respectively, generating a heterodimer of about 200 kDa. The direct effects of TGF-β through its heterodimeric receptor complex appear to be reversible forms of inhibition of cell growth directed at the G_1 phase of the cell cycle. TGF-β may also interfere with the expression of c-*myc*, which is a growth factor for many cells. Also, TGF-β may affect c-*myc* expression together with pRB, the protein product of the retinoblastoma gene. The stimulatory effects of this growth factor are more indirect than the negative effects and may be accounted for by the ability of TGF-β to induce other growth factors, for example, the β-chain of PDGF. In some cases, TGF-β has been shown to be a signal for inducing apoptosis. Although this mechanism is not well understood, it is becoming clear that cascades of cysteine proteases are involved in apoptosis, and these proteases, once activated, catalyze the cleavage of important proteins in the cell and may also account for the activation of other destructive enzymes, such as nucleases. Which particular proteases are activated by TGF-β will prove to be of interest.

VII. PLATELET-DERIVED GROWTH FACTOR (PDGF)

PDGF consists of two chains, A and B, which are separate gene products joined by a disulfide bond. Consequently, three forms of PDGF exist: PDGF-AA, PDGF-AB, and PDGF-BB. The molecular weight of the dimer is about 30 kDa, the A chain being 12 kDa and the B chain 18 kDa. Human platelets contain all three isoforms. PDGF-AA has been found in the context of the osteosarcoma cell-derived growth factor (ODGF), and PDGF-BB has been found in the context of $p28^{sis}$ secreted by simian sarcoma virus-infected cells. PDGF-AA is a secreted form of the growth factor, whereas PDGF-BB is associated with intracellular compartments.

There are two PDGF receptors, α and β. Both receptors have five extracellular Ig-like domains and intracellular tyrosine kinase domains (Figure 19-10). When the dimeric ligand binds to a receptor, a receptor dimer is formed that is capable of activating the signal transduction process, including autophosphorylation of intracellular tyrosine residues as an initial step. It is of interest that both receptors α and β induce signal transduction, but only the β-receptor is involved in chemotaxis. The α-receptor binds both the A and B chains of PDGF, but the β-receptor binds only the B chain of the growth factor. Thus, receptor dimer $\alpha\alpha$ binds PDGF-

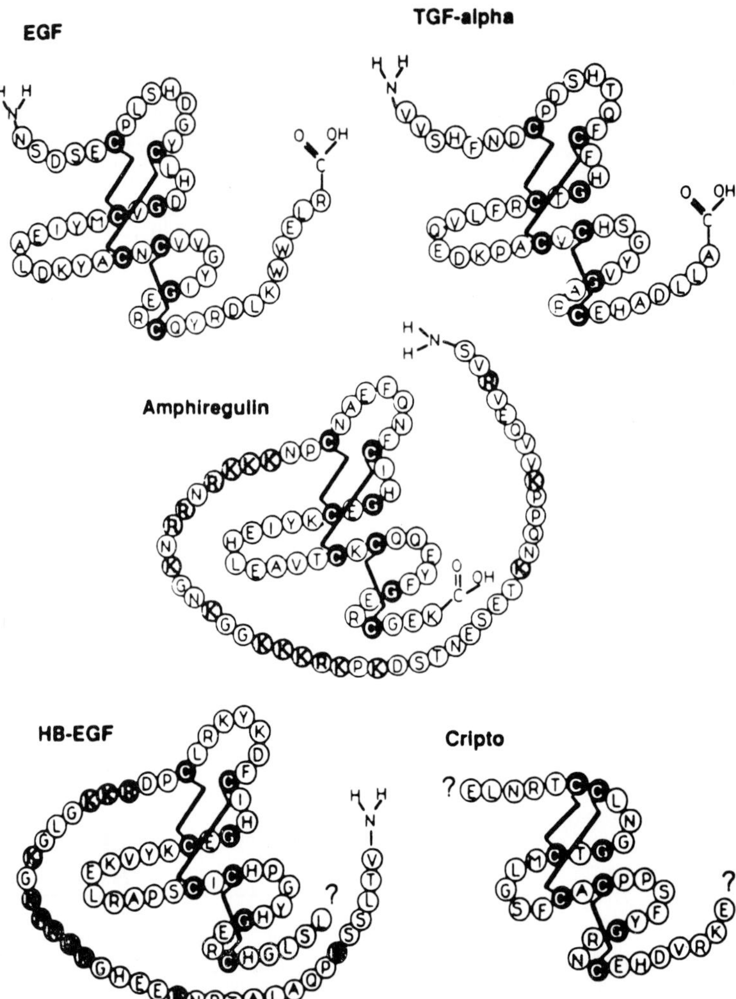

FIGURE 19-5 Members of the EGF superfamily. Cysteine bridges are indicated by solid black lines. Conserved amino acids are shown in black, and positively charged residues in the N-terminal extensions of HB-EGF and amphiregulin are indicated by shaded circles. All members of the EGF family of mitogens contain a highly conserved 36- to 40-amino acid motif, containing six cysteines that link internally to produce three peptide loops. In addition, all members of the family are synthesized embedded within larger precursor proteins. All three forms of TGF-α are capable of activating EGF–TGF-α receptors. Reproduced with permission from Russell, W. E., and Van Wyk, J. J. (1995). Peptide growth factors. *In* "Endocrinology" (L. J., DeGroot, ed.), 3rd ed., Vol. 1, pp. 2590–3623. W. B. Saunders Co., Philadelphia, PA.

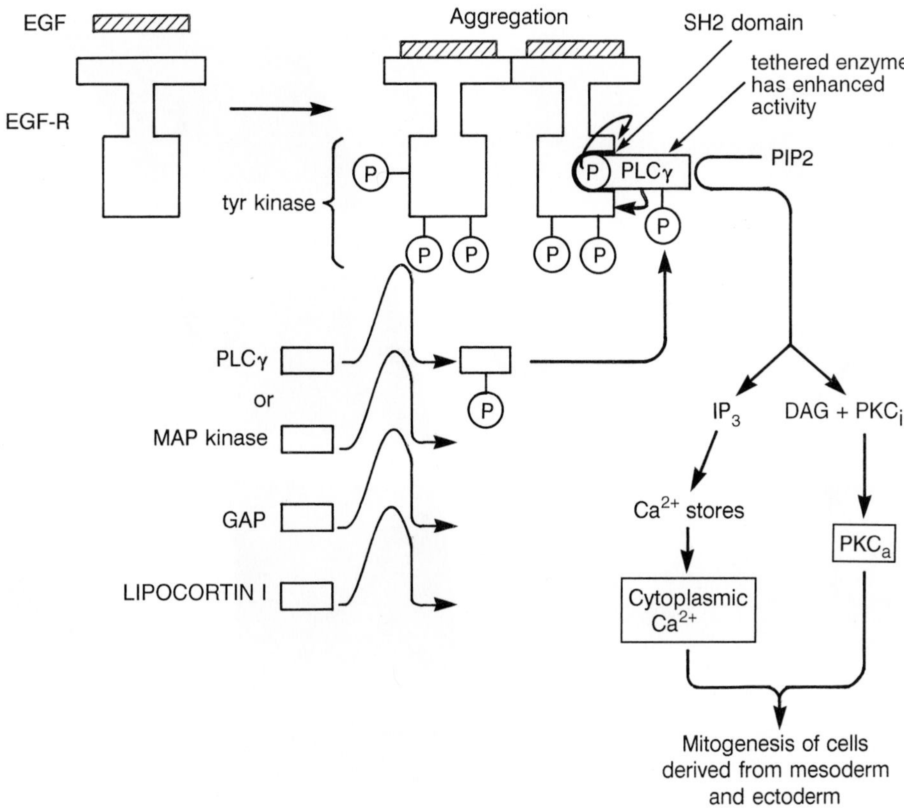

FIGURE 19-6 Signal transduction of EGF-R.

AA while receptor dimers $\alpha\alpha$ and $\alpha\beta$ bind PDGF-AB, and all of the receptor dimers, including $\beta\beta$, bind PDGF-BB. The PDGF–receptor complex is internalized and degraded, and the signal transduction process set off by the growth factor–receptor complex has already been indicated (Figure 19-4).

PDGF promotes cell division and does this by inducing a state of competence to undergo DNA synthesis. Other growth factors in the PDGF-stimulated cells are involved in progression through the G_1 phase of the cell cycle on the way to the S phase. As mentioned earlier, PDGF also functions in smooth muscle cells and fibroblasts as a chemoattractant, although other growth factors involved in progression are not required for PDGF to exert chemoattractant activity. PDGF is also involved in wound healing, and it is responsible for the appearance of macrophages in wounds where fibroblasts and endothelial cells replicate as part of the wound-healing process under the influence of PDGF. PDGF may also be involved in atherosclerosis, where initial injury to the intima can lead to platelet aggregation and degranulation. In this process, PDGF is released and smooth muscle cells migrate to the site and cause the sequestering of lipids. PDGF may have other important activities.

VIII. FIBROBLAST GROWTH FACTOR (FGF)

There are two separate forms of FGF, the basic FGF (bFGF) and the acidic FGF (aFGF), and these factors stimulate cell division in cell lines derived from neuroectoderm and mesoderm. The angiogenesis factor appears to be FGF, and FGF causes the proliferation of vascular endothelial cells. *int*-2, present in tumors, is related to FGF. *hst*-1 and kFGR are FGF-4. FGF-4 and -5 are related proteins, both oncogenes, and FGF-6 is related to *hst*.

bFGF is a 16-kDa, acid-labile protein, and aFGF is 15 kDa and has homology to bFGF. FGFs have affinity for heparin, which enhances the biological activity of FGF and increases its stability. FGFs are extracellular and may be secreted from the cell by Ca^{2+}-dependent exocytosis. Outside the cell, bFGF associates with negatively charged proteoglycans. Thus, FGFs seem to be stored in the extracellular matrix. Receptors for FGF are *flg* and *bek*. They contain three IgG-like domains in the extracellular region; they have single transmembrane domains and tyrosine kinase in the cytoplasmic domains. Either receptor binds both forms of FGF. There are a large number of FGF receptor isoforms

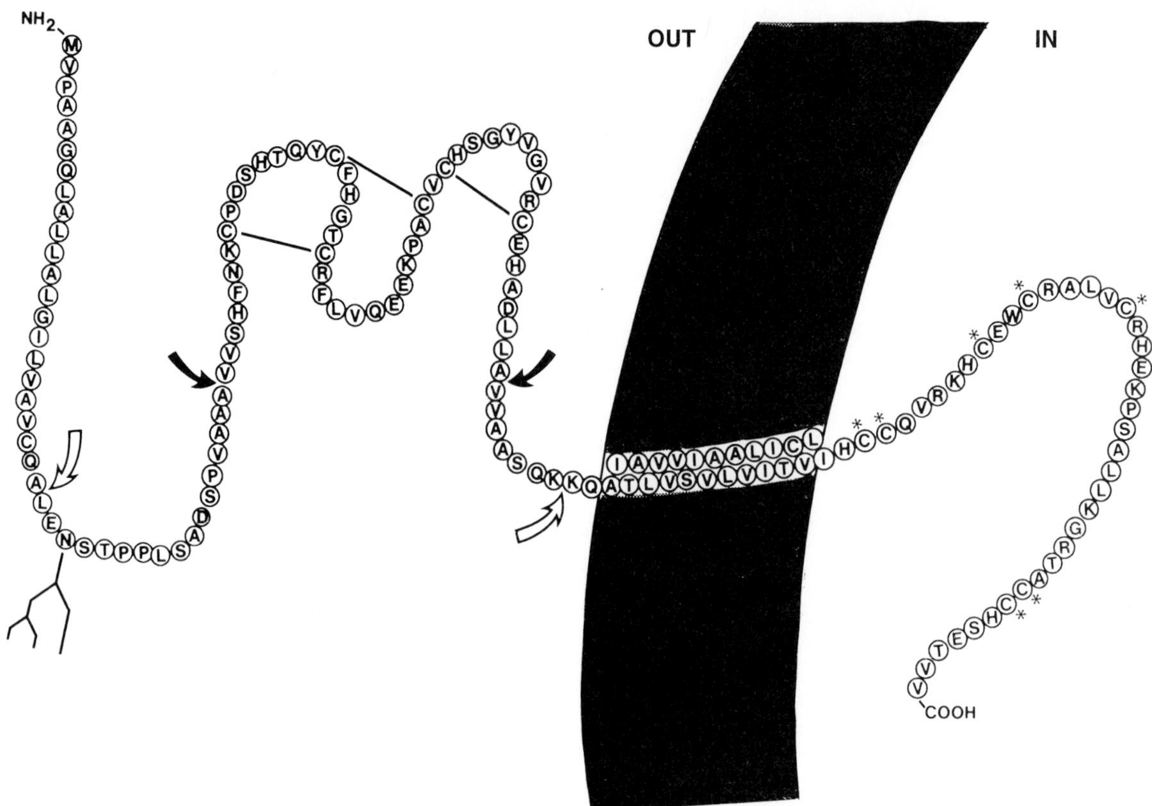

FIGURE 19-7 Complete amino acid sequence of the 159-amino acid pro-TGF-α is shown as embedded in the plasma membrane. A consensus site for N-linked glycosylation is indicated, and cysteine residues within the cytoplasmic domain are marked with asterisks. Black arrows indicate sites of proteolytic cleavage that release the mature, 50-amino acid form. The open arrow near the amino terminus marks the apparent signal peptide cleavage site between amino acids 22 and 23. The open arrow immediately external to the plasma membrane marks a lysine–lysine bond that could serve as a cleavage site for a trypsin-like protease. Reproduced with permission from Lee, D. C., Luetteke, N. C., Qiu, T. H., Chen, X., and Berkowitz, E. A. (1993). Transforming growth factor-alpha. *In* "Growth Factors in Perinatal Development" (R. C. Tsang, J. A. Lemons, and W. F. Balistreri, eds.), pp. 21–38. Raven Press, NY.

deriving from splice variants of four different genes. FGF resembles PDGF in its biological activity, and its signal transduction process is similar to that of other growth factor receptors, including PDGF (Figure 19-4).

IX. NERVE GROWTH FACTOR (NGF)

NGF was discovered in the mouse submaxillary gland. It has a molecular weight of 13,259 and exists in a complex of 140 kDa. The complex contains two molecules of NGF (β-subunits) together with two α- and two γ-subunits (the γ-subunit is arginine esterase). Apparently the identity of the α-subunits remains unclear. Although NGF is not a mitogen, it prevents apoptosis of neurons. A now classical demonstration of the induction of apoptosis is the withdrawal of NGF, usually in chick embryos, that triggers programmed cell death. This is the case during development, but dependence on NGF for neuronal survival does not occur in the adult. NGF causes a rat pheochromocytoma, PC-12, to differentiate into sympathetic neurons. Experimentally, NGF can also act as a chemoattractant for neutrophilic leukocytes. The structure of the complex containing NGF is shown in Figure 19-11.

Other growth factors with homology to NGF do exist. Examples of these are neurotrophin-3 (NT-3) and brain-derived neurotrophic factor (BDNF). Two other factors fall into this class of NGF homologues, but they have not been well described. Ciliary neurotrophic factor (CNTF) is a survival factor for chick ciliary ganglia parasympathetic neurons; however, it is different from neurotrophins NT-3 and BDNF. CNTF is expressed throughout the central nervous system and the peripheral nervous system and is a survival factor for neurons in these regions at various developmental stages. It is

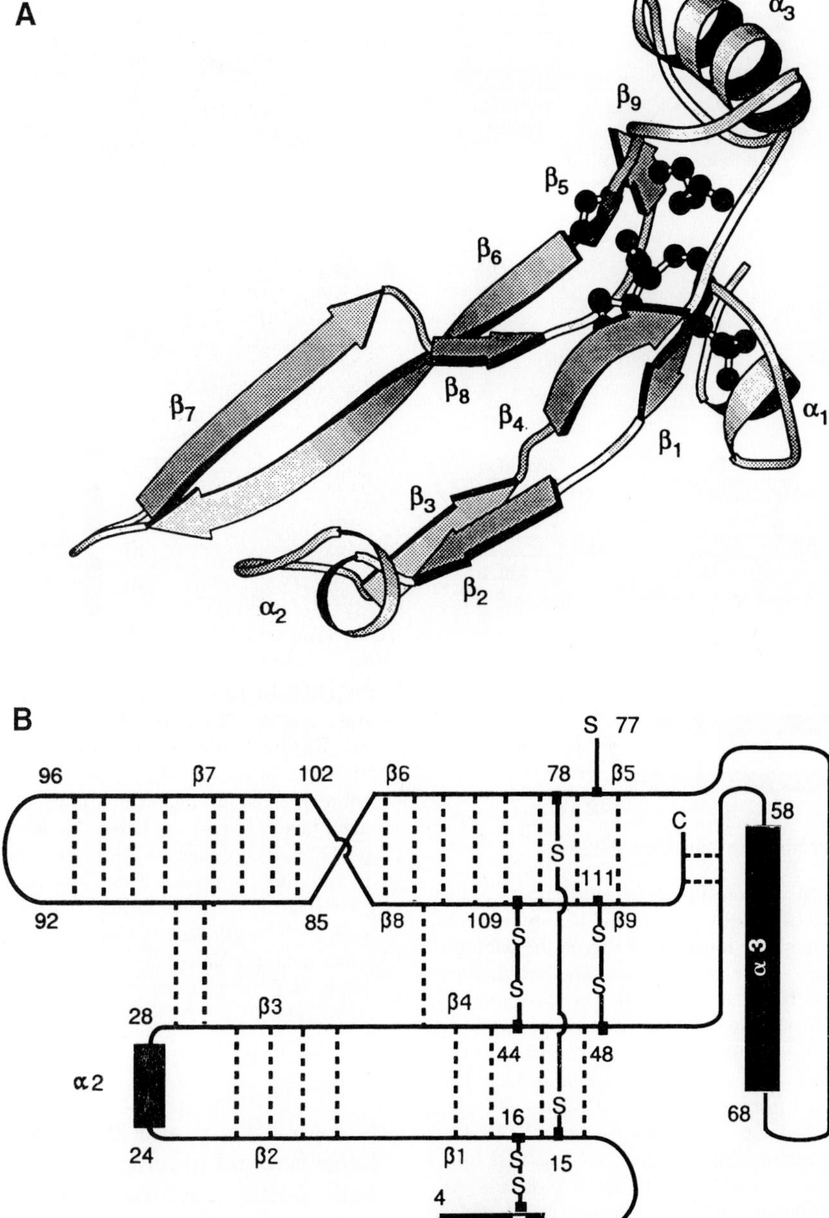

FIGURE 19-8 Crystal structure of TGF-β. (A) Topology diagram of a TGF-β2 subunit. The α-helices are labeled as α_1, α_2, and α_3, and peptide strands in β-sheets are labeled from β_1 through β_9. The residues involved in the regular secondary structure are as follows: α_1, residues 4–8; α_2, 24–28; α_3, 58–68; β_1, 15–18; β_2, 20–23; β_3, 37–40; β_4, 42–46; β_5, 77–80; β_6, 82–91; β_7, 96–102; β_8, 104–106; and β_9, 109–112. (B) Schematic drawing of the primary and secondary structure of a TGF-β2 subunit. Hydrogen bonds in the β-strands and loops are indicated by dashed lines. The analogy to a left hand can be seen. The heel (helix α3) is to the right and the fingers (β-strands) are to the left with the third and fourth fingers twisted. Reproduced with permission from Daopin, A., Piez, K. A., Ogawa, Y., and Davies, D. R. (1992). Crystal structure of transforming growth factor-β-2: An unusual fold for the superfamily. *Science* **257**, 369–373.

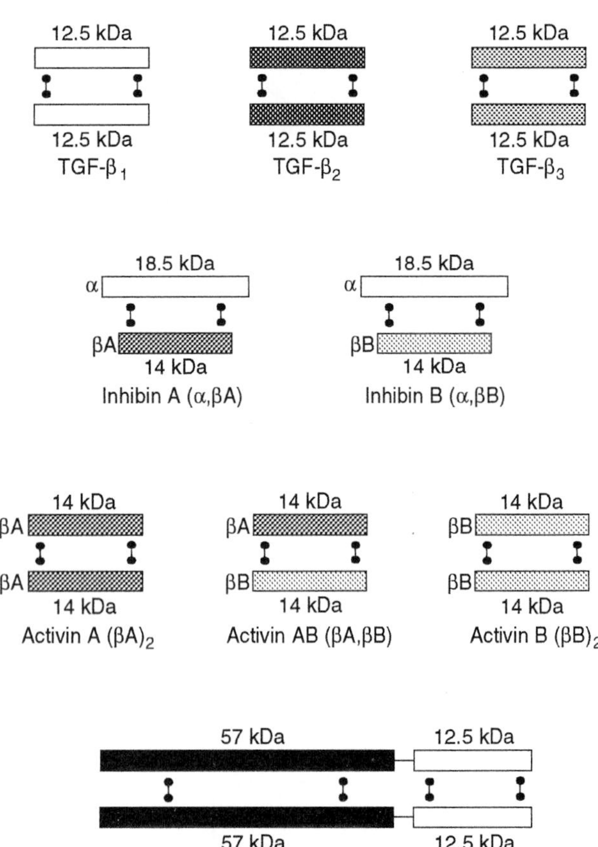

FIGURE 19-9 Structural relationships between members of the transforming growth factor-β (TGF-β) family. TGF-β_{1-3} are homodimers with different subunits. The homology between the subunits is greater than 80%. The α-subunits of inhibins and activins are unlike any other structures in the TGF-β family. The β-subunits of inhibin A and inhibin B (designated β_A and β_B) are approximately 30% homologous to TGF-β subunits. The activins consist of either homodimers or heterodimers of the β_A and/or β_B-subunits. Members of this family were isolated from intact cells and tissues as larger prohormones, which, in the case of the three patriarchal TGF-β's, remain inactive until the mature dimers shown are liberated from their latent form by exposure to acidic conditions or proteases. An exception to this is the Müllerian inhibiting substance (MIS), which was isolated as a much larger active dimer, each element of which is 70 kDa. The smaller homodimer, consisting of identical 12.5-Da subunits, is cleaved from the precursor by plasmin digestion and found to retain the same biological activity as the larger form of MIS. Reproduced with permission from Russell, W. E., and Van Wyk, J. J. (1995). Peptide growth factors. *In* "Endocrinology" (L. J., DeGroot, ed.), 3rd ed., Vol. 1, pp. 2590–3623. W.B. Saunders Co., Philadelphia, PA.

also a factor that prevents the death of motor neurons. CNTF seems to belong to a family of hematopoietic cytokines, as shown in Figure 19-12. These cytokines include granulocyte colony-stimulating factor (G-CSF), oncostatin M (OSM), leukemia inhibitory factor (LIF), and interleukin (Il-6).

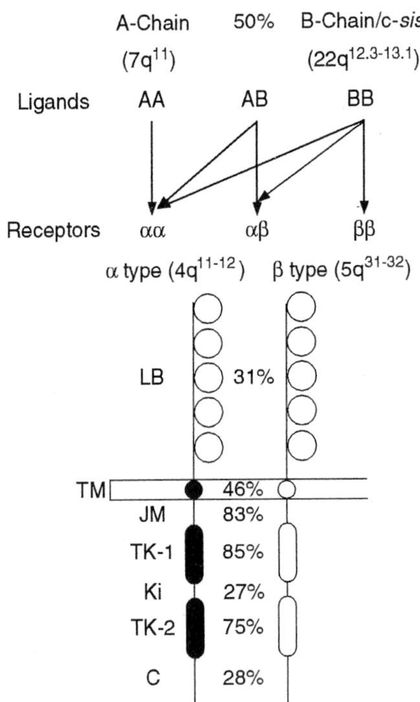

FIGURE 19-10 Biological relationship between three PDGF isoforms and two PDGF receptors. Arrows show the binding affinities for each dimerized receptor. The amino acid sequence homology (%) between α and β-PDGF receptors is shown in each domain. Abbreviations: LB, ligand binding; TM, transmembrane; JM, juxta membrane; TK, tyrosine kinase; Ki, kinase insert; C, carboxy terminal. Human chromosomal location is shown in parentheses. Reproduced with permission from Matsui, T. (1994). Platelet derived growth factor system in human oncogenesis and its signaling pathways. *In* "Growth Factors: Cell Growth, Morphogenesis, and Transformation." (T. Nakamura and K. Matsumoto, eds.), Gann Monograph on Cancer Research pp. 42, 1–12, Japan Scientific Societies Press, Tokyo, CRC Press, Boca Raton/Ann Arbor/London/Tokyo.

NGF binds to two receptor types, one of which is high affinity and the other is low affinity. The latter is found in lymphocytes and macrophages, and both forms may derive from a single gene. The tyrosine kinase receptor, *trk*, is found in neural tissues

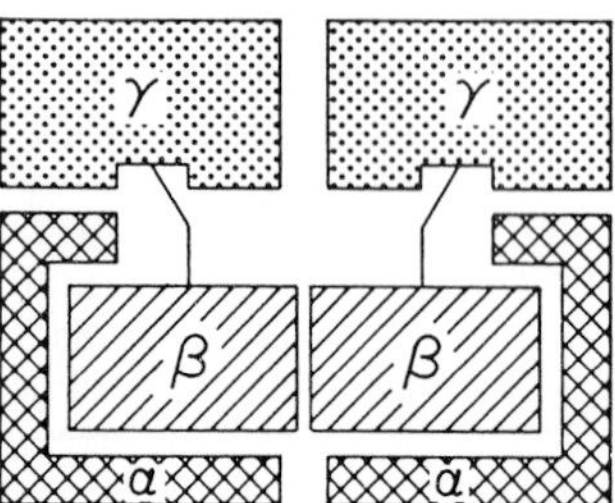

FIGURE 19-11 Mouse 7 S nerve growth factor (NGF). The β-subunit represents the monomeric NGF. Reproduced with permission from Underwood, L. E., and van Wyk, J. J. (1985). *In* "Textbook of Endocrinology" (R. H. Williams, ed.) 7th ed., p. 172. W.B. Saunders Co., Philadelphia, PA.

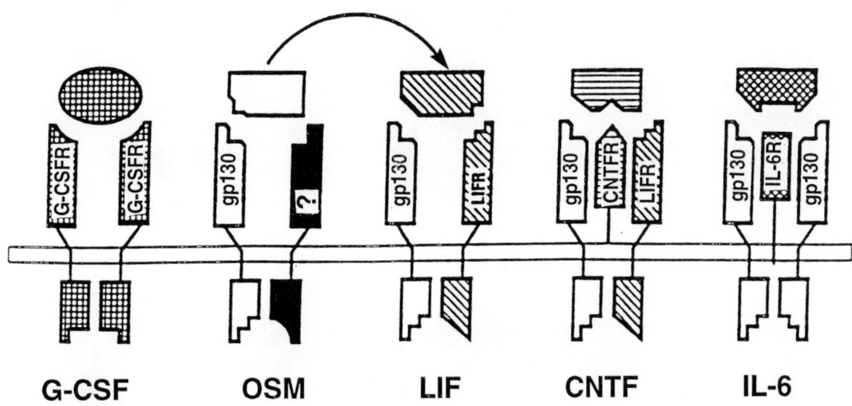

FIGURE 19-12 Diagram to illustrate similarities between the ligand–receptor complexes of the ciliary neurotrophic factor (CNTF) and related cytokines. Abbreviations: G-CSF, granulocyte colony-stimulating factor; OSM, oncostatin M; LIF, leukemia inhibitory factor; and IL-6, interleukin-6. The initial step in signal transduction is binding of the respective ligands to a low-affinity binding site, the expression of which determines cellular specificity. The low-affinity binding sites for the CNTF and IL-6 receptors are soluble, extracellular α-subunits that attach to the cell membrane through glycosophosphatidylinositol (GPI) linkages. A similar α-subunit has not been identified for LIF, OSM, and G-CSF, and the initial low-affinity binding is thought to be to one of the transmembrane subunits. Initial binding of the ligand is rapidly followed by dimerization of two transmembrane β-subunits. Dimerization creates high-affinity binding sites for their respective ligands and permits signal transduction. The binding subunit of OSM has not yet been identified. Note that the gp-130 subunit is utilized by CNTF, IL-6, LIF, and OSM, whereas the subunit initially described as the LIF receptor (LIFR) is shared with CNTF. The arrow signifies that OSM can also bind to the LIF receptor. Reproduced with permission from Russell, W. E., and Van Wyk, J. J. (1995). Peptide growth factors. *In* "Endocrinology" (L. J., DeGroot, ed.), 3rd ed., Vol. 1, pp. 2590–3623. W.B. Saunders Co., Philadelphia, PA.

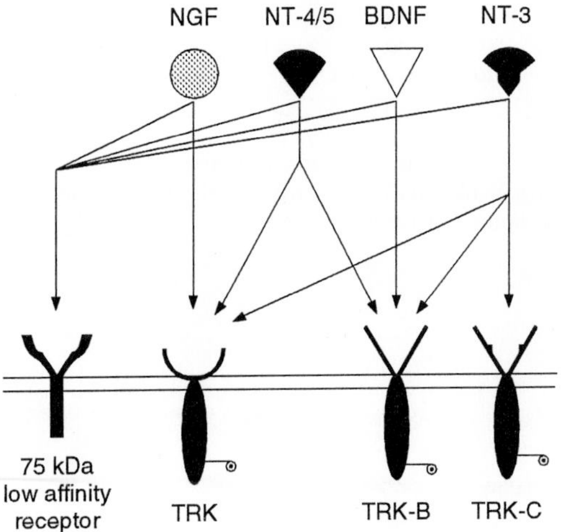

FIGURE 19-13 Neurotrophin–neurotrophin receptor interactions. Reproduced with permission from Maness, L. M., Kastin, A. J., Weber, J. T., Banks, W. A., Beckman, B. S., and Zadina, J. E. (1994). The neurotrophins and their receptors: Structure, function, and neuropathology. *Neurosci. Biobehav. Rev.*, **18,** 143–159.

and binds NGF as well as NF-3 and BDNF. Binding of neurotrophic factors to *trk* receptors is summarized in Figure 19-13. Furthermore, the signal transduction pathway following interaction with the *trk* receptor involves several components, including PI_3k, SRC, RAS, PLC-γ-1, and downstream pathways. These are summarized in Figure 19-15. Thyroid hormone positively regulates the levels of NGF in the brain during development and is required for normal development.

Hematopoietic growth factors are discussed in Chapter 15. Other growth factors relating to the gastrointestinal tract are discussed in Chapter 8.

X. HEPATOCYTE GROWTH FACTOR (HGF, SCATTER FACTOR)

HGF is a disulfide-linked heterodimer of 69-kDa α- and 34-kDa β-subunits derived by cleavage of an 85-kDa precursor protein. The precursor protein has four double-loop structures (kringle domains). The cleavage site for the release of α- and β-subunits of HGF is identical to that in plasminogen for the genera-

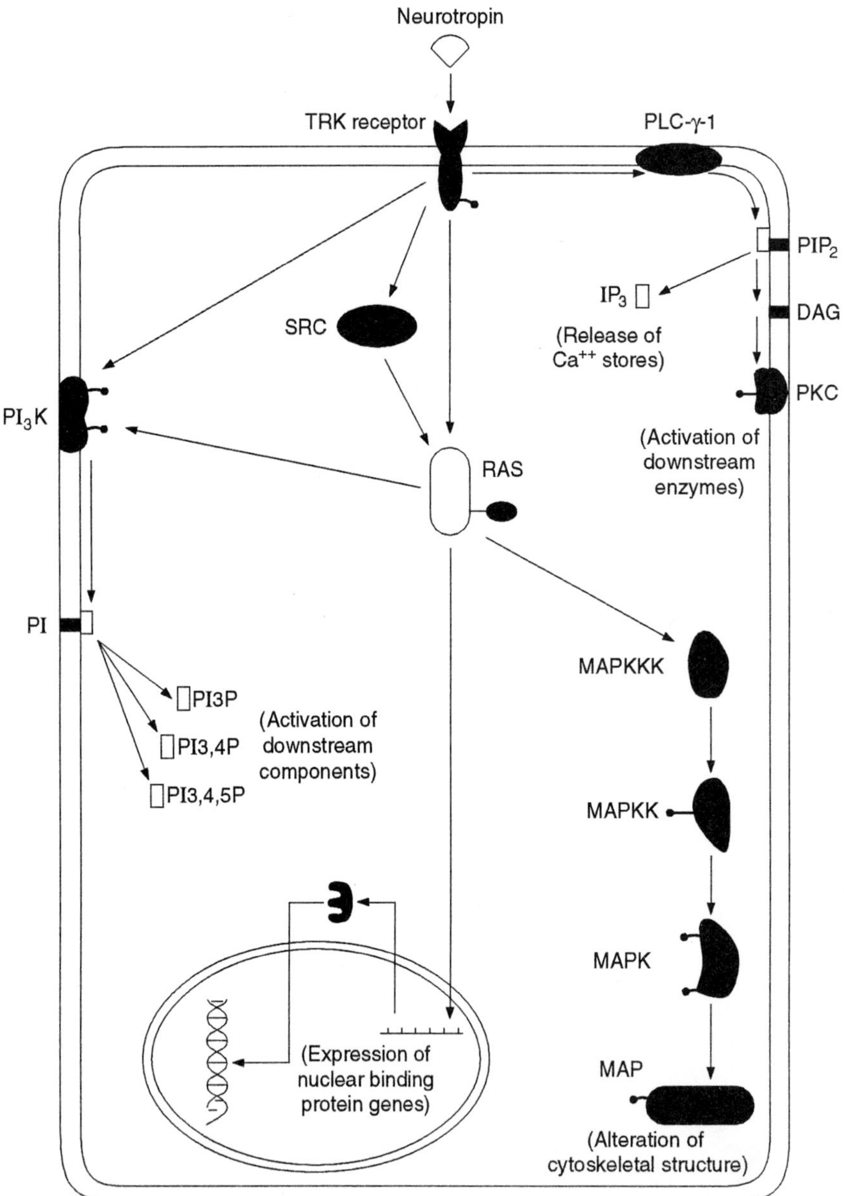

FIGURE 19-14 Putative neurotrophin signal transduction mechanisms. Reproduced with permission from Maness, L. M., Kastin, A. J., Weber, J. T., Banks, W. A., Beckman, B. S. and Zadina, J. E. (1994). The neurotrophins and their receptors: Structure, function, and neuropathology. *Neurosci. Biobehav. Rev.* **18,** 143–159.

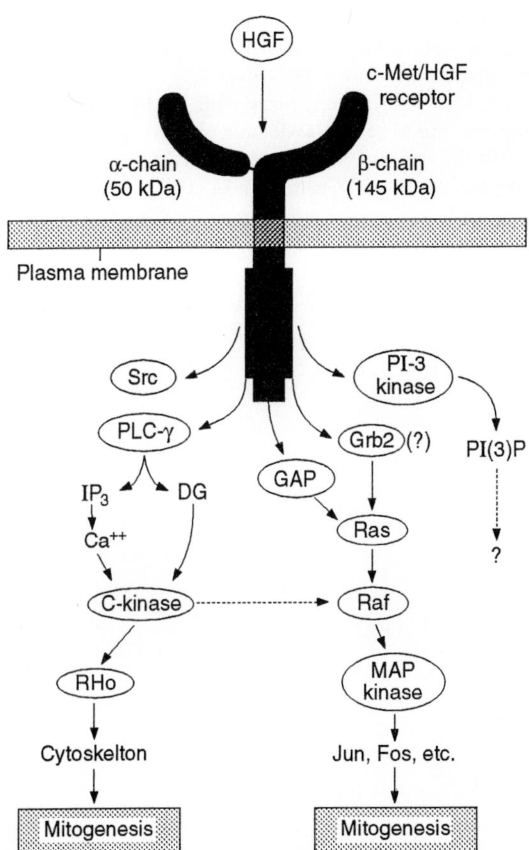

FIGURE 19-15 Schematic structure of c-*met*–HGF receptor and signal transduction pathways from the receptor. Reproduced with permission from Matsumoto, K., and Nakamura, T. (1994). Pleiotropic roles of HGF in mitogenesis, morphogenesis, and organ regeneration. *In* "Growth Factors: Cell Growth, Morphogenesis, and Transformation" (T. Nakamura and K., Matsumoto, eds.), Gann Monograph on Cancer Research, Vol. 42, pp. 91–112, Japan Scientific Societies Press, Tokyo, CRC Press, Boca Raton/Ann Arbor/London/Tokyo.

tion of plasmin through cleavage by plasminogen activator. Curiously, the precursor of HGF is as active as the heterodimer of α- and β-subunits.

The HGF receptor is the c-*met* protooncogene protein product, which is a membrane tyrosine kinase receptor. The kinase is in the β-chain of the receptor on the cytoplasmic side of the cell membrane. HGF stimulates growth in a wide variety of tissues, in spite of its name, besides hepatocytes, including melanoma cells, melanocytes, keratinocytes, renal tubule cells, and mammary epithelia. Within an hour of partial hepatectomy, there is a 15-fold increase in serum HGF levels.

The signal transduction pathways for HGF are indicated in Figure 19-15.

References

A. Books

DeGroot, L. J. (1995). "Endocrinology," 3rd ed., Vol. 1. W. B. Saunders Co., Philadelphia, PA.

B. Review Articles

Daughaday, W. H. (1995). Growth hormone, insulin-like growth factors, and acromegaly. *In* "Endocrinology" (L. J. DeGroot, ed.), 3rd. ed., Vol. 1, pp. 303–329. W. B. Saunders Co., Philadelphia, PA.

D'Ercole, A. J. (1993). Growth factors and development. *In* "Growth Factors in Perinatal Development" (R. C. Tsang, J. A. Lemons, and W. F. Balistreri, eds.), pp. 1–20. Raven Press, Ltd., New York.

Gammeltoft, S., and Kahn, C. R. (1995). Hormone signaling via membrane receptors. *In* "Endocrinology" (L. J. DeGroot, ed.), 3rd. ed., Vol. 1, pp. 17–65. W. B. Saunders Co., Philadelphia, PA.

Gill, G. N., Bertics, P. J., Thompson, D. M., Weber, W., and Cochet, C. (1985). Structure and regulation of the epidermal growth factor receptor. *In* "Cancer Cells" (J. Feramisco, B. Ozannie, and C. Stiles, eds.), pp. 11–18. Cold Spring Harbor Laboratory Press, Cold Spring Harbor, NY.

Hsuan, J. J. (1989). Transforming growth factors β. *Br. Med. Bull.* **45,** 425–437.

Lee, D. C., Luetteke, N. C., Qiu, T. H., Chen, X., and Berkowitz, E. A. (1993). Transforming growth factor-alpha. *In* "Growth Factors in Perinatal Development" (R. C. Tsang, J. A. Lemons, and W. F. Balistreri, eds.), pp. 21–38. Raven Press, New York.

LeRoith, D., Adamo, M., Werner, H., and Roberts, C. T., Jr. (1995). Molecular and cellular biology of the insulin-like growth factors. *In* "Molecular Endocrinology" (B. D. Weintraub, ed.). Raven Press, Ltd., New York.

Maness, L. M., Kastin, A. J., Weber, J. T., Banks, W. A., Beckman, B. S., and Zadina, J. E. (1994). The neurotrophins and their receptors: structure, function, and neuropathology. *Neurosci. Biobehav. Rev.* **18,** 143–159.

Matsui, T. (1994). Platelet derived growth factor system in human oncogenesis and its signaling pathways. *In* "Growth Factors: Cell Growth, Morphogenesis, and Transformation" (T. Nakamura, and K. Matsumoto, eds.), Gann Monograph on Cancer Research 42, pp. 1–12. Japan Scientific Societies Press, Tokyo, CRC Press, Boca Raton, FL, Ann Arbor, MI, London, Tokyo.

Matsumoto, K., and Nakamura, T. (1994). Pleiotropic roles of HGF in mitogenesis, morphogenesis, and organ regeneration. *In* "Growth Factors: Cell Growth, Morphogenesis, and Transformation" (T. Nakamura, and K. Matsumoto, eds.), Gann Monograph on Cancer Research 42, pp. 91–112. Japan Scientific Societies Press, Tokyo, CRC Press, Boca Raton, FL, Ann Arbor, MI, London, Tokyo.

McDonald, N., Murray-Rust, J., and Blundell, T. (1989). Structure-function relationships of growth factors and their receptors. *Br. Med. Bull.* **45,** 554–569.

Russell, W. E., and Van Wyk, J. J. (1995). Peptide growth factors. *In* "Endocrinology" (L. J. DeGroot, ed.), 3rd ed. Vol. 1, pp. 2590–3623. W. B. Saunders Co., Philadelphia, PA.

Sun, P. D., and Davies, D. R. (1995). The cysteine-knot growth-factor superfamily. *Annu. Rev. Biophys. Biomol. Struct.* **24,** 269–291.

Waterfield, M. D. (1989). Growth factor receptors. *Br. Med. Bull.* **45,** 541–553.

C. Research Papers

Cunningham, B. C., Ultsch, M., de Vos, A. M., Mulkerrin, M. G., Clauser, K. R., and Wells, J. A. (1991). Dimerization of the extracellular domain of the human growth hormone receptor by a single hormone molecule. *Science* **254,** 821–825.

Daopin, A., Piez, K. A., Ogawa, Y., and Davies, D. R. (1992). Crystal structure of transforming growth factor-β-2: An unusual fold for the superfamily. *Science* **257,** 369–373.

de Vos, A. M., Ultsch, M., and Mossiakoff, A. (1992). Human growth hormone and extracellular domain of its receptor: Crystal structure of the complex. *Science* **255,** 306–212.

Schlunegger, M. P., and Grütter, M. G. (1992). An unusual feature revealed by the crystal structure at 2.2 Å resolution of human transforming growth factor-β-2. *Nature* **358,** 430–434.

CHAPTER

20

Hormones and Cancer

I. INTRODUCTION

A cancer cell is a cell that has escaped the controls that maintain its normal differentiated function within the homeostatic mechanism of the body. This definition probably is applicable to most cancer cells and therefore emphasizes the role of signal transduction in the loss of organismic control. Thus, cancer may be viewed as an aberration of the signal transduction process or, putting it in simpler terms, it derives from a cell that has escaped or lost the ability to respond to controls from outside (or inside) the cell. This often takes the form of developing a mechanism in which a growth signal is permanently turned on, in contrast to the normal situation in which the cell only responds when there is a timely signal, turning on a specific receptor within or on the surface of the cell. As it happens, some of the oncogenes that are expressed by cancer cells are parts of cell membrane growth hormone receptors that are ligand independent, constitutively activated, and positively signal the cells to divide.

There are several aspects of hormones related to the cancer problem: (1) cancer is sometimes generated through the action of hormonally active substances, for example, the effect of diethylstilbestrol, a potent estrogenic substance; (2) the development of cancer is often connected with the cellular expression of oncogenes [many oncogenes function in signal transduction by replacing the normal function of a receptor, transducing protein, or transactivating factor(s)]; (3) many tumors inappropriately produce (ectopic production) biologically active levels of hormones that create dysfunctions of the signal transduction process and produce abnormal effects; (4) hormones and antihormones are used to treat certain types of cancer; and (5) the status of hormone receptors, for example, in breast cancer, can form the basis of staging the cancer and its capacity for therapy. These topics will be considered in this chapter.

II. RELATION OF SOME HORMONES TO CARCINOGENS AND DEVELOPMENT OF CANCER FROM INAPPROPRIATE HORMONAL TREATMENT

Some time ago it was recognized that certain well-known carcinogenic substances structurally resembled steroid hormones. These carcinogens had been shown to produce tumors in experimental animals. In this context, the structures of benzo[*a*]pyrene and 7,12-dimethylbenz[*a*]anthracene can be compared to the structure of estradiol (Figure 20-1). The structural similarities are even more apparent when the molecules are aligned and overlaid as shown in Figure 20-2. It thus is not surprising that steroid hormones can cause cancer if they are administered inappropriately or in high amounts.

A typical example of the carcinogenicity of hormones is tumor production in humans caused by treatment with diethylstilbestrol (DES), which is an artificial estrogen. The structure of DES is shown in Figure 20-3. In this figure, the chemical structure of DES is shown in the lower left-hand corner, and the three-dimensional models of its various forms are shown in the body of the figure. Notice that the A ring is superimposable with that of estradiol, and this may be, at least in part, the ring recognized by the estradiol receptor, so that DES is a strong artificial estrogen.

In the early 1970s it was noted in the medical literature that there was an increase in the occurrence of clear cell adenocarcinoma of the vagina, usually in teenaged women. It became apparent that this was due to DES treatment of the mothers of these young women where the drug had been administered for complications in pregnancy. DES provoked malignancies derived from usually absent Müllerian tissue present in the vagina after birth, which appeared to be a direct consequence of the treatment of the mother with DES. Apparently, persistent congenital Müllerian epithelium may undergo further differentiation in the presence of hormones, and the occurrence of Müllerian epithelium is otherwise rare. Consequently, physicians believed that the generation of cancerous tissue could have been derived from the persistent Müllerian glandular tissue under the influence of DES from the maternal environment. As many as 2–4 million pregnant women may have been treated with DES, so that about 2 million daughters and 2 million sons were gestated *in utero* under conditions of DES treatment of the mother. Whereas there have been important incidences in cancer in the female offspring, there have been few effects in the male offspring.

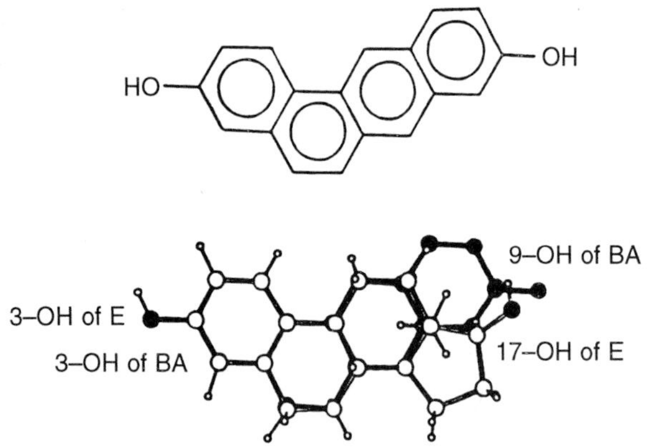

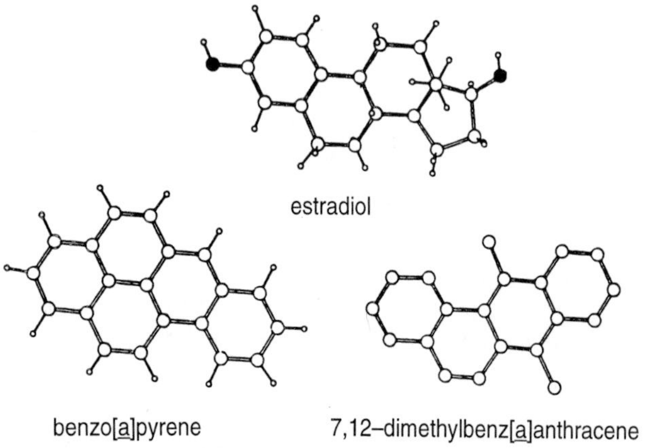

FIGURE 20-1 Top views looking down at the ring structures of estradiol and benzo[*a*]pyrene and dimethylbenz[*a*]anthracene. In both cases, the A, B, and C rings of the hormone seem to be superimposable over the analogous rings of the carcinogens. Reproduced with permission in part from Glusker J. P. (1979). *Biochem. Act. Hormones,* **6**, 121–204.

FIGURE 20-2 Structure of estradiol (E, white bonds) overlaid on the structure of 3,9-dihydroxybenz[*a*]anthracene (BA, black bonds). Although the crystal structure of benzanthracene has not been determined, its structure is drawn from structural data on benzo[*a*]pyrene and some phenols. The hydroxyl groups on O-3 of both compounds have been aligned. The proximity of the other hydroxy groups in each compound is evident. Reproduced with permission from Glusker, J. P. (1987). *Biochem. Act. Hormones,* **6**, 121–204.

III. HORMONE-RELATED TREATMENT OF CANCER

Clearly, many cancers are related to the status of hormones in the body, for example, breast cancer may or may not utilize estrogens as growth factors, prostate cancer cells may depend on a supply of androgens, etc.; these dependencies usually are a feature of the presence of the cognate receptor in the tumor cell, and hormone independence becomes a feature of tumor cells that no longer express the appropriate receptor. Thus, there are two approaches to cancer treatments involving hormones: in one, the gland responsible for the offensive secretion of a hormone may be removed (gland extirpation; this is also a route to remove a tumor of a specific gland), and in the other, hormones and their antagonists may be used as chemotherapeutic agents. A third avenue of treatment is becoming important, and this relies on agents that block the activity

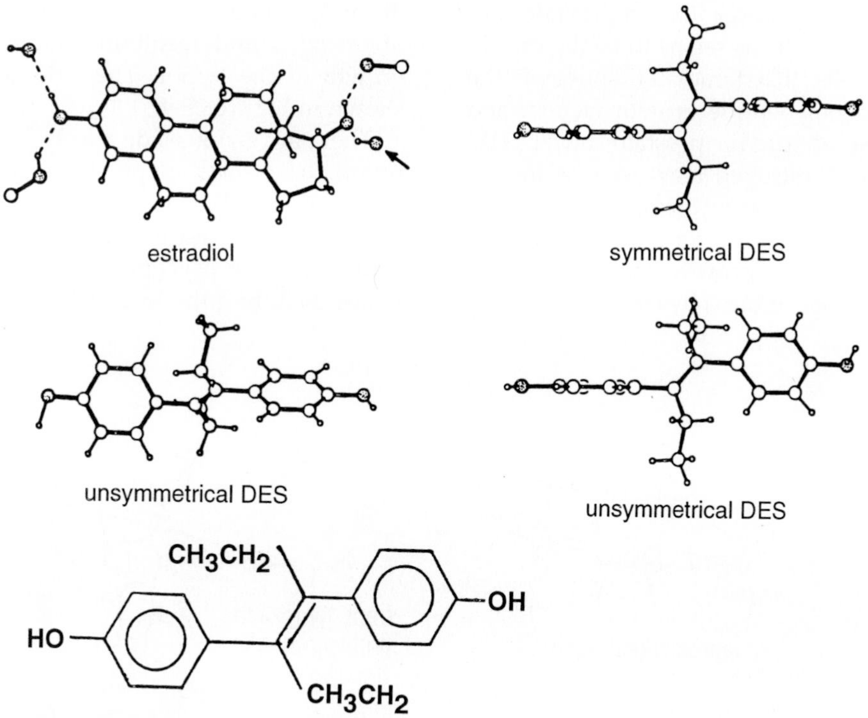

FIGURE 20-3 Chemical and three-dimensional structures of diethylstilbestrol compared to estradiol. The chemical structure of DES is shown in the lower left-hand part of the figure. The various forms of DES are shown in their three-dimensional models. The arrow points to the oxygen atom of the water molecule that is H-bonded to O-17 and has been suggested to be part of the unit that binds to the receptor (the other part may involve the A ring, as mentioned before). If this is so, O---O distances for estradiol and DES are both 12.1 Å. Reproduced with permission in part from Glusker, J. P. (1979). *Biochem. Act. Hormones*, **6**, 121–204.

of a hormone usually at the receptor level. In this case, a mimic of an active site of the ligand, if it is a peptide, can be synthesized and chemically stabilized. This agent will bind to the receptor and often block the further action of the receptor; this approach can be effective against growth factors that are required for rapidly growing cancer cells, but much less deleterious to mature normal tissue cells that either do not divide or do so at a very slow rate. This is a variation of the well-known approach of using a competitive inhibitor of ligands for steroid receptors when the steroid is required for growth of the tumor, which is often the case for estrogen receptor-positive tumors requiring estrogen for growth or androgen-positive prostate cancer cells requiring androgens for growth. Some of the cell types in these heterogeneous tumors will not express steroid receptors, and their growth will be steroid independent.

Removal of glands that produce hormones required for the growth of tumors is still sometimes used, most often with the ovary, testis, adrenal and pituitary. In some cases, gland removal leads to a marked regression of the hormone-dependent tumor, for example, in breast cancer when the tumor has been demonstrated to be estrogen receptor positive and progesterone receptor positive (indicating that estrogen is required for growth and the estrogen receptor is functional in inducing the synthesis of the progesterone receptor). Thus, ovariectomy can be used to treat estrogen receptor- and progesterone receptor-positive mammary tumors. The endogenous levels of estrogen fall shortly after ovariectomy, and a response can be expected in certain involved tissues, such as soft tissues, lymph nodes, and lung metastases, but metastases to liver and brain do not respond.

Removal of the testes is sometimes done for prostate cancer, but the results of this maneuver are not much better than for treatment with estrogens and in some cases antiestrogens (tamoxifen). Tamoxifen is being used more frequently for the treatment of both mammary cancer and prostate cancer. It is an antiestrogen, and in this regard its beneficial effects may be more understandable in breast cancer than in prostate cancer. However, it is possible, as seems to be the case in mammary tumor cells, that tamoxifen shuts off the supply of autocoid polypeptide growth factors, and this action may also be true for prostate cancer cells. In the mammary cell, estrogen may account for the increase in growth by stimulating the release of growth factors such as tumor growth factor α and insulin-like growth factor, and negative growth factors like tumor growth factor β may be inhibited by estrogen. Clearly, the action of tamoxifen can interfere with these effects of estrogen and result in the inhibition of cellular growth of the tumor. The structure of tamoxifen is shown in Figure 20-4.

There are serious side effects of giving estrogens to retard the growth of prostate cancers, such as fluid retention, vomiting, gynecomastia, and cardiovascular complications; these effects are not seen following castration, but the psychological effects of castration for males including the loss of libido and impotence are

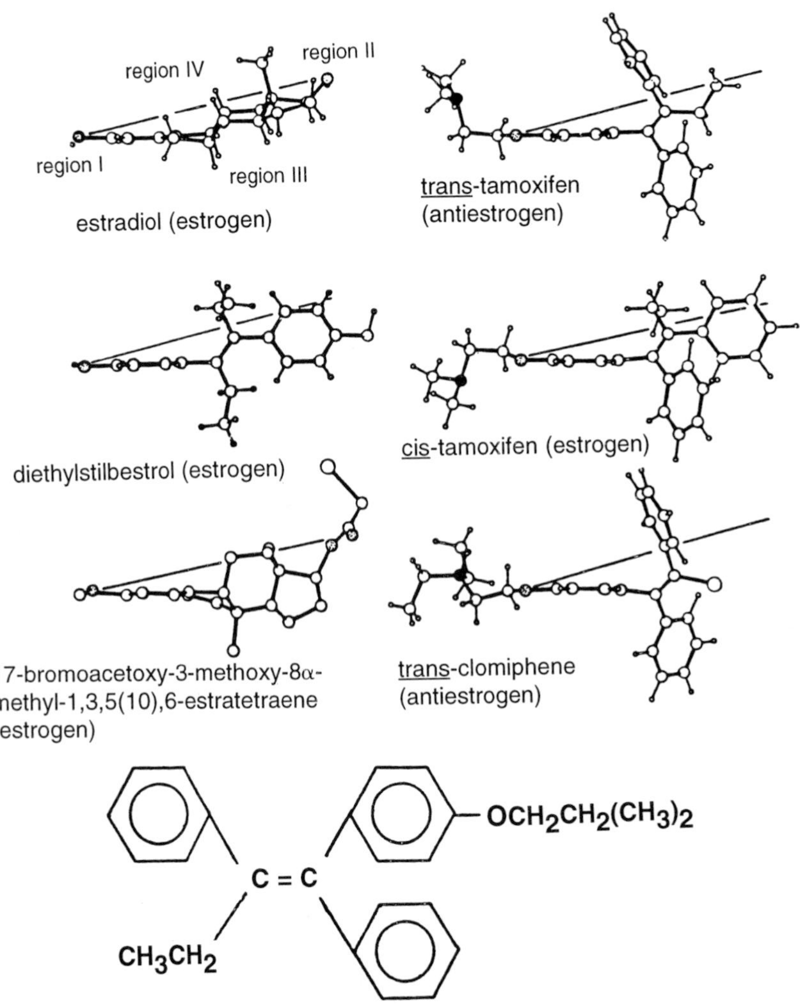

FIGURE 20-4 Structures of some estrogens and antiestrogens. The chemical formula for tamoxifen is at the bottom of the figure, while the other structures shown are taken from three-dimensional information. In estradiol, O-3 and O-17 are connected by a line that is repeated in the other compounds shown. Numbered regions have been defined in the literature (for more information, see the reference). The phenyl groups project into region IV in the two antiestrogens. *cis*-Tamoxifen: $R_1 = C_2H_5$; $R_2 = R_4 = C_6H_5$; $R_3 = (C_6H_4)O(CH_2)_2N(CH_3)$ *trans*-Tamoxifen: $R_1 = R_4 = C_6H_5$; $R_2 = C_2H_5$; $R_3 = (C_6H_4)O(CH_2)_2$-$N(CH_3)_2$ *trans*-Clomiphene: $R_1 = R_4 = C_6H_5$; $R_2 = C_1$; $R_3 = (C_6H_4)O(CH_2)_2N(C_2H_5)_2$. $(R_1)(R_3)C = C\ (R_2)(R_4)$. Reproduced in part with permission from Glusker J. P. (1979). *Biochem. Act. Hormones*, **6**, 121–204.

severe. When the tumor is localized and encapsulated, the prostate is usually removed and there are new techniques that preserve some of the functions of the prostate. However, if the tumor has spread, endocrine therapy usually is used, involving estrogens or tamoxifen as long as the tumor remains endocrine dependent. Eventually, this dependence may disappear, the tumor becomes hormone independent, and the disease becomes more difficult to manage.

Analogues of GnRH have been shown to decrease endogenous testosterone similarly to castration and without the side effects of estrogen treatment. These analogues no doubt will eventually replace some of the more common types of therapy. The mode of action of estrogen treatment in males for prostate cancer is that the hormone inhibits the release of LH from the anterior pituitary, increases the synthesis of the sex hormone-binding globulin, which binds more free testosterone and effectively inactivates it, and decreases the level of circulating testosterone by directly inhibiting testosterone synthesis in the Leydig cell. The inhibition of the conversion of testosterone to dihydrotestosterone occurs by inhibition of the responsible enzyme testosterone 5α-reductase, in the prostate (and in other androgen targets), dihydrotestosterone being the form of the hormone that the androgen receptor prefers to bind.

Treatment with hormones has complications. Although we have been discussing the use of estrogens in prostate cancer and the use of antiestrogens in mammary cancer, there is much discussion about the supposed beneficial use of estrogens to counteract or lessen the severity of the effects of menopause. We have already noted the effects of DES, an estrogen, in its danger for the production of cancer in the female fetus if the mother was treated with the hormone during pregnancy. In postmenopausal women taking estrogen, the chances of getting endometrial cancer increases 5–10-fold over women not taking estrogen. This risk seems to be dose and duration dependent, but the risk would have been obviated if a hysterectomy had been performed, beforehand. Other tumors, such as those of the breast, cervix, vagina, or liver, can occur after prolonged treatment as concluded from test results in animals. Also, women taking oral contraceptives, which usually consist of a mixture of an estrogen and a progestin, run a greater risk of blood clots in various parts of the body. Thus, the beneficial effects of treatment with sex hormones have to be carefully weighed against the risks.

In addition to the cases presented, there are also occasions when other hormone-secreting glands are removed. Although adrenalectomies were performed in the past, at present drugs can be used that interfere with the synthesis of steroid hormones and accomplish the same effect. Where breast cancer tissue is estrogen dependent, dehydroepiandrosterone, a weak androgen from the adrenal *zona reticularis,* is also a precursor of a potent estrogen after conversion by the aromatase enzyme system in the mammary tumor cell. Since dehydroepiandrosterone is secreted in large quantity, this represents a significant source of growth factor for the tumor and in previous times adrenalectomy would be combined with ovariectomy as a treatment for breast cancer. Sometimes hypophysectomy is done as a treatment of cancer patients since the trophic hormones derive from this source and would accomplish an effect similar to end organ removal, such as ovariectomy and adrenalectomy. Needless to say, there is the problem of end organ hormone replacement, such as with cortisol and others, to maintain essential functions.

IV. HORMONE RECEPTOR STATUS OF BREAST CANCERS

Breast cancer is diagnosed at a level of about 100,000–120,000 new cases per year in the United States, and 40,000 die each year from this most common cause of cancer in women. For some time breast cancers have been classified on the basis of receptor measurements from biopsy specimens. Thus, in general, these cancers fall into three categories: estrogen receptor (ER) positive and progesterone receptor (PR) positive; ER positive and PR negative; and ER negative and PR negative. These measurements show that in those cancers with normal levels of ER and PR the cells should be responsive to the growth-stimulating effects of estrogen. Furthermore, the positive expression of PR indicates that the ER is functional in that PR is a phenotypic product of ER action. Such tumors are more amenable to treatment and therapy with antiestrogens and other maneuvers. The progression of tumor status from hormone dependent to hormone independent (ER-, PR-) is a bad sign, and a hormone-independent tumor is virtually impossible to control. A similar situation is obtained in prostatic cancer with regard to the hormone (dihydrotestosterone) dependence of the tumor for growth. As long as these tumors are hormone dependent, they can be treated with estrogen inhibitors (tamoxifen) in the case of mammary cancer or with androgen inhibitors in the case of prostatic cancer. Some ideas concerning the actions of estrogen and tamoxifen on ER-positive breast cancer cells are shown in Figure 20-5. As was mentioned before, tamoxifen is also being used in prostatic cancer and is probably effective due to its ability to inhibit the production of

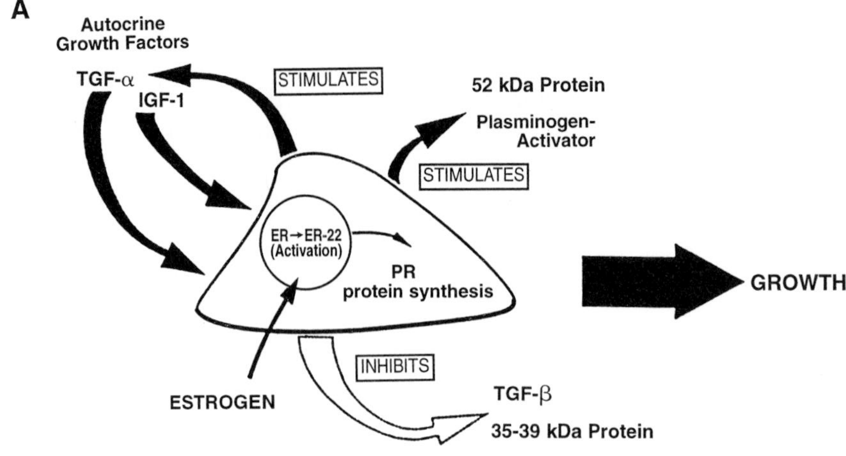

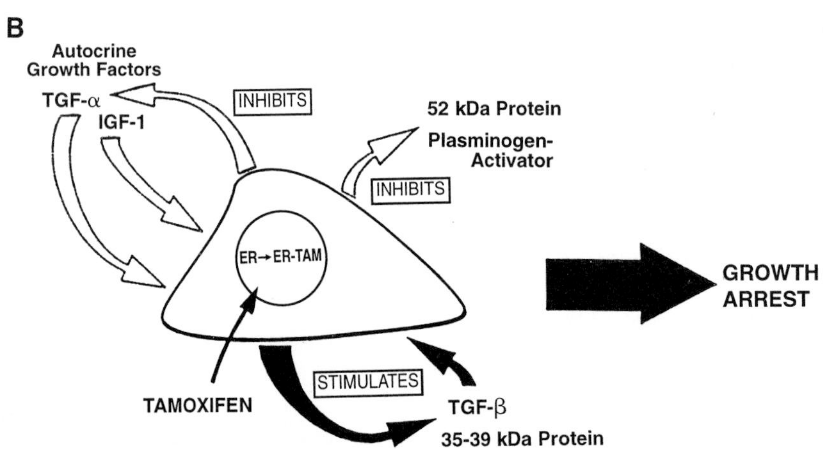

FIGURE 20-5 (A) Estrogen-mediated effects on ER+ breast cancer cells. (B) Inhibition of estrogen-stimulated growth in breast cancer cells by tamoxifen. These figures were reproduced with permission from Jordan, V. C. (1989). "Endocrine Therapy of Breast Cancer" (F. Cavalli, ed.), Vol. III. Springer-Verlag, Berlin.

peptidic growth factor autocoids, and it may increase the levels of certain growth-inhibitory factors.

Biopsy of mammary cancer and subsequent measurement of steroid receptor levels can thus stage the tumor in its progression from hormone dependence to partial hormone dependence to final hormone independence when chemotherapy is often without effect. Unfortunately, the indications of measurements of steroid receptor levels in breast cancer are not always absolute. Not all of the ER+ tumors in postmenopausal women are capable of responding to hormonal therapy, and it appears that the ability of receptor to bind steroid does not ensure the complete functioning of the receptor within the cell, such as the ability to translocate to the nucleus. On the other hand, the nucleus may contain null ERs that associate with the nucleus in the absence of steroid and may represent a refractory subpopulation. There have also been indications in the literature of other mutations in the ER, and these may form a subgroup within the ER+ population of breast cancers. Clearly, more research into this area of defects in ER function is needed to sharpen prognostic ability. Some work has shown that the cellular protooncogene *erb*B-2 (HER 2/*neu*, this encodes for part of the EGF receptor) is overexpressed by some breast cancers, and its expression has been associated with low likelihood of survival in about one-fifth of a study group.

Another finding is that normal human mammary cells produce a substance(s) that specifically inhibits the growth of mammary tissue. This protein (47, 65 kDa) has been purified and an antibody has been prepared. The protein inhibits the growth of five transformed human mammary cell lines without affecting transformed lines derived from tissues other than mammary cells. This is an interesting development and will be even more so if means are found to induce this activity in cancer cells.

Thus, receptor measurements may offer a more specific evaluation of a subset of patients. The continued refinement of the presence of steroid receptors and their mutants as correlated to their specific functions, together with the measurement of new indicators such as cellular protooncogenes, will greatly enhance prognosis and treatment of this disease.

V. ECTOPIC PRODUCTION OF HORMONES BY TUMOR CELLS

There are many examples of tumors producing hormones that are unrelated to the tissue of origin. This only underscores the dedifferentiated state of the cancer cell. Apparently there is no clear understanding of the reason for the production of a specific unexpected hormone by a given tumor cell, except that rearrangements of genes may occur and the rearranged gene may insert next to a highly expressing sequence and be overproduced. Thus, the term "ectopic" is used, which refers to the generation of a hormone away from the normal site of its production.

Various tumors have been shown to secrete ACTH and to cause hypercortisolism, even when the tumor is undetectably small for many years. The secretion of this hormone ectopically occurs mainly with bronchial carcinoid tumors and to a smaller degree with pheochromocytomas, thymic carcinoids, and islet cell tumors. The carcinoid tumors can be very difficult to locate. This condition can lead to Cushing's syndrome, characterized by overproduction of cortisol from the adrenal gland as a result of the uncontrolled levels of ACTH or its stimulating hormone, CRH, derived directly from ectopic production of the hormone. These tumors can also secrete proopiomelanocortin-related peptides.

The ectopic production of parathyroid hormone and concomitant hypercalcemia are somewhat rare. However, they have been shown to occur in a case of small cell carcinoma of the lung and in a case of ovarian carcinoma. Computed tomography of the abdomen revealed a right ovarian mass 5–8 cm in size. When the cancer was removed, the level of PTH declined. Molecular analysis revealed rearrangement of the upstream region of the PTH gene, as shown in Figure 20-6.

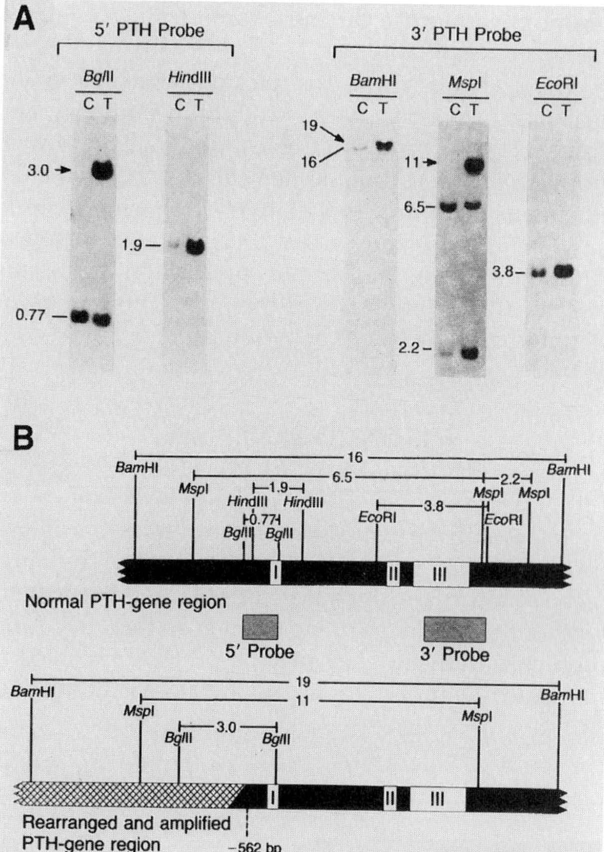

FIGURE 20-6 PTH gene region in the ovarian carcinoma. (A) shows the results of Southern blot analysis of the PTH gene region in the ovarian carcinoma. DNA (10 μg) from the peripheral blood leukocyte control (C) and from the ovarian tumor (T) was digested with the indicated restriction endonucleases and analyzed. The bands normally expected are indicated by dashes, while the abnormal, rearranged bands are indicated by arrows. Band sizes are in kb corresponding to the linear map in (B). (B) shows restriction maps of the normal PTH region (top) and the rearranged and amplified PTH gene region in the ovarian tumor (bottom). PTH exons I, II, and III are represented by open bars, introns normally present and flanking regions are represented by solid bars, and the cross-hatched bar represents the DNA placed upstream of PTH by the rearrangement. Shaded rectangles represent the fragments of normal PTH used as probes. At the bottom, −562 bp marks the upstream *Hin*dIII site aligned above it in (A), which is the 3′ border of the 150-bp BglII–*Hin*dIII segment that contains the precise breakpoint. Reproduced with permission from Nussbaum, S. R., Gaz, R. D., and Arnold, A. (1990). *New Eng. J. Med.* **19,** 1324–1328.

A case was reported of acromegaly from ectopic growth hormone-releasing hormone (GHRH) secreted to an extent 1000 times greater than normal by a bronchial carcinoid tumor. In this case, plasma growth hormone levels were increased. There was metastatic disease demonstrated by the production of GHRH and GH by tumor tissue in the lung, submaxillary gland, and breast. The condition was treated with chemotherapy (nitrosourea, CCNU, and 5-fluorouracil in repeated cycles) and the tumors regressed. This is a rare syndrome and also unusual in its positive response to chemotherapy since carcinoid tumors are not known for this susceptibility.

These are but a few examples out of a great many in which hormones can be produced by tumors, and these ectopic hormones are able to abrogate the normal hormonal homeostatic mechanism. In general, the descriptions given may be relatively rare, and it seems possible that in many cases of ectopic production of hormones there is aberrant production of tissue-specific transcription factors that, in the normal tissue counterpart, would not be produced.

VI. ONCOGENES AND HORMONAL FUNCTIONS

In general, the expression of two or more oncogenes is required for a cell to become cancerous. The oncogene products are functional parts of receptors, usually without the ligand-binding domain so that they function constitutively; they may be proteins involved in transduction processes or they are DNA-binding proteins that could operate at the transcriptional level to cause the overproduction or suppression of specific genes. Thus, in many cancer cells tumor suppressor functions are repressed, and the signal transduction process, often starting with a mitogenic signaling receptor system, is permanently turned on so that the growth of the cell is out of control and is permanently signaled to divide. Moreover, other factors may be turned on, e.g., proteolytic functions that may allow the growing tumor to spread and become metastatic. Some oncogenes may be expressed that interfere with the normal process of cell death and render a cell susceptible to changes leading to the formation of a tumor cell. A new set of genes that stimulate programmed cell death or apoptosis may function as tumor suppressors.

Many of the receptor-related alterations occurring as a result of oncogene expression involve the growth factor receptors. Growth factors with oncogenic potential are listed in Table 20-1. It can be appreciated that this list is still growing. In addition to the growth factors that can form the basis for oncogene activity, the receptors for growth factors themselves can also form

TABLE 20–1 Growth Factors with Oncogenic Potential[a]

E_{max} name(s)	E_{max} mature factor size	E_{max} cell source	E_{max} target cell	E_{max} receptor
PDGF	32 kDa (16-kDa B chain; 14–18-kDa A chain), + CHO	Blood platelets, endothelial cells, placenta	Mesenchymal cells smooth muscle, placental trophoblast	185-kDa tyrosine kinase
EGF	6 kDa (53 aa)	Submaxillary gland, Brunner's gland, possibly parietal cells	Wide variety of epithelial and mesenchymal cells	c-*erb*B gene; 170-kDa tyrosine kinase
TGFα	5.6 kDa (50 aa)	Transformed cells, placenta, embryos	Same as EGF	Binds to EGF receptor, may have separate one
CSF-1	70 kDa (2 × 35 kDa); 60% CHO	Mouse L cells	Macrophage progenitors	c-*fms* protooncogene; 170-kDa tyrosine kinase
CSF-2 (granulocyte macrophage CSF)	15–28 kDa (127 aa) (1–50% CHO)	Endotoxin-induced lung; placenta	Macrophage and granulocyte progenitors	Unknown
Multi-CSF (IL-3)	28 kDa (134 aa) (50% CHO)	T lymphocytes	Eosinophil, mast cell, granulocyte, macrophage progenitors, T lymphocytes	Unknown
IGF-I	7 kDa (70 aa)	Adult liver and (somatomedin C)	Epithelial, mesenchymal	450-kDa complex (2 α-chains of 130 kDa; 2 β-chains of 85 kDa
IGF-II	180 aa	Fetal liver, placenta (somatomedin A)	Epithelial, mesenchymal	Single polypeptide chain of 260 kDa
IL-1		Mononuclear phagocytes	Wide variety; fibroblasts, T cells, endothelial cells	
IL-2 153 aa (human)	15 kDa (133 aa); some CHO	T helper cells	Cytotoxic T lymphocytes	55 kDa (33-kDa protein + 22-kDa CHO)
FGF	Unknown	Brain, pituitary, chondrosarcoma	Endothelial, fibroblasts	Unknown
β-NGF	307 aa	Submaxillary gland	Sympathetic and sensory neurons	130 kDa (possibly kinase)

[a] This table is reproduced with permission in part from Burck, K. B., Liu, E. T., and Larrick, J. W. (1988). "Oncogenes, An Introduction to the Concept of Cancer Genes," p. 158. Springer-Verlag, NY.

TABLE 20–2 Nuclear Oncogenes[a]

Protooncogene or tumor suppressor gene	Protein molecular weight
Thyroid hormone receptor (*erb*A)	52,000
Transcription Factor AP-1 Components	
c-*jun*	47,000
jun-B	40,000
jun-D	40,000
c-*fos*	62,000
fra-1	38,000
fos-B	45,000
Myc family	
c-*myc*	62,000–64,000
L-*myc*	60,000–66,000
N-*myc*	62,000–64,000
Other nuclear oncogenes	
myb	75,000
ets-1	51,000
ets-2	56,000
ski	60,000
rel	68,000
Tumor suppressor genes	
Rb	110,000
p53	53,000

[a] Reproduced from Cooper, G. M. (1990). "Oncogenes," p. 238. Jones and Bartlett Publ., Boston.

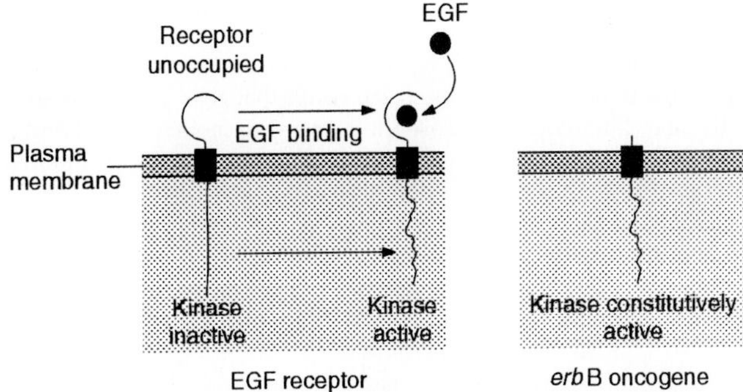

FIGURE 20-8 Mechanism of *erb*B oncogene activation. The tyrosine kinase activity of the EGF receptor (*erb*B protooncogene) is regulated by EGF binding; in contrast, the *erb*B oncogene tyrosine kinase is constitutively active. Reproduced with permission from Cooper G. M. (1990). "Oncogenes," p. 183. Jones and Bartlett Publishers, Boston.

the basis of oncogene activity. Most of the receptors appear to be tyrosine kinases and have the potential enzymatic activity built into their structures. A list of these tyrosine kinases appears in Table 20-2, and the family of tyrosine kinase receptors, some of which are oncogenes, is indicated in Figure 20-7.

As an example, the overexpression of *erb*B, which was pointed out in a previous discussion of breast cancer, represents a gene encoding information for a

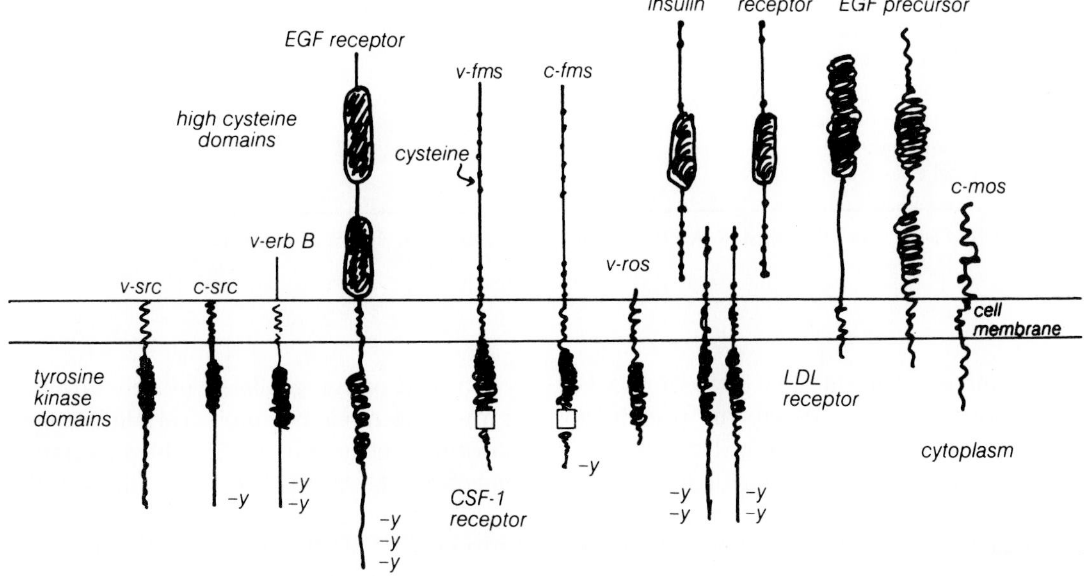

FIGURE 20-7 Family of cell surface receptors having tyrosine kinase activity. The positions of tyrosines (y) are possible phosphorylation sites. Domains sharing similar function–sequences include the cytoplasmic tyrosine kinase domains and the extracellular "high cysteine" domains. The location of cysteine (.) is also noted. Reproduced with permission from Burck, K. B., Liu, E. T., and Larrick, J. W. (1988). "Oncogenes, An Introduction to the Concept of Cancer Genes," p. 166. Springer-Verlag, New York.

TABLE 20–3 Some Genes Involved in Human Cancers

Genes known as proto-oncogenes code for proteins that stimulate cell division; mutated forms, called oncogenes, can cause the stimulatory proteins to be overactive, with the result that cells proliferate excessively. Tumor suppressor genes code for proteins that inhibit cell division. Mutations can cause the proteins to be inactivated and may thus deprive cells of needed restraints on proliferation. Investigators are still trying to decipher the specific functions of many tumor suppressor genes.

ONCOGENES	
Genes for growth factors or their receptors	
PDGF	Codes for platelet-derived growth factor. Involved in glioma (a brain cancer)
erb-B	Codes for the receptor for epidermal growth factor. Involved in glioblastoma (a brain cancer) and breast cancer
erb-B2	Also called HER-2 or neu. Codes for a growth factor receptor. Involved in breast, salivary gland and ovarian cancers
RET	Codes for a growth factor receptor. Involved in thyroid cancer
Genes for cytoplasmic relays in stimulatory signaling pathways	
Ki-ras	Involved in lung, ovarian, colon and pancreatic cancers
N-ras	Involved in leukemias
Genes for transcription factors that activate growth-promoting genes	
c-myc	Involved in leukemias and breast, stomach and lung cancers
N-myc	Involved in neuroblastoma (a nerve cell cancer) and glioblastoma
L-myc	Involved in lung cancer
Genes for other kinds of molecules	
Bcl-2	Codes for a protein that normally blocks cell suicide. Involved in follicular B cell lymphoma
Bcl-1	Also called PRAD1. Codes for cyclin D1, a stimulatory component of the cell cycle clock. Involved in breast, head and neck cancers
MDM2	
TUMOR SUPPRESSOR GENES	
Genes for proteins in the cytoplasm	
APC	Involved in colon and stomach cancers
DPC4	Codes for a relay molecule in a signaling pathway that inhibits cell division. Involved in pancreatic cancer
NF-1	Codes for a protein that inhibits a stimulatory (Ras) protein. Involved in neurofibroma and pheochromocytoma (cancers of the peripheral nervous system) and myeloid leukemia
NF-2	Involved in meningioma and ependymoma (brain cancers) and schwannoma (affecting the wrapping around peripheral nerves)
Genes for proteins in the nucleus	
MTS1	Codes for the p16 protein, a braking component of the cell cycle clock. Involved in a wide range of cancers
RB	Codes for the pRB protein, a master brake of the cell cycle. Involved in retinoblastoma and bone, bladder, small cell lung and breast cancer
p53	Codes for the p53 protein, which can halt cell division and induce abnormal cells to kill themselves. Involved in a wide range of cancers
WT1	Involved in Wilms' tumor of the kidney
Genes for proteins whose cellular location is not yet clear	
BRCA1	Involved in breast and ovarian cancers
BRCA2	Involved in breast cancer
VHL	Involved in renal cell cancer

Reproduced with permission from Weinberg, R. A. (1996). How cancer arises. *Sci. Am.*, September, 62–70.

protein that resembles a truncated EGF receptor lacking the EGF-binding domain. When this protein is overexpressed and inserted into the cell membrane, the cell is driven to mitosis without having to wait for external molecules of EGF to bind to a normal receptor containing a ligand-binding domain; instead, the receptor is permanently turned on and the cell divides at a rapid rate, having lost its homeostatic control mechanism. This situation is represented in Figure 20-8.

Instead of being related in structure to membrane growth hormone receptors, certain oncogenes encode information for proteins involved in transducing signals from the membrane. An example of this is the *ras* family of oncogenes. *Ras* functions in every way like a normal GTP-binding protein in the signal transduction pathway of many membrane receptors, usually resulting in the activation of adenylate cyclase as typified by the adrenergic receptors. However, *ras* proteins in mammalian cells do not seem to interact with adenylate cyclase, so that the target for these oncogenes is not clear. On the other hand, the GAP (GTPase-activating protein) does interact with *ras* and greatly stimulates the activity of GTPase, so that GAP may be a negative regulator of *ras* activity.

Nevertheless, members of the *ras* family are frequently identified in various human solid tumors, usually involving a single base substitution of the normal cellular constituent. Some believe that because of the relatively high incidence of *ras* oncogenes in lesions that may progress to malignancy, *ras* may play a role in the initiation of cancer. The oncogene has been found to contain several point mutations that alter the functioning of the p21 product's GTPase activity. Further work is needed to clarify the abnormal activities associated with the *ras* oncogene in mammalian cells.

There are also several nuclear oncogenes that presumably function in various ways at the transcriptional level. Some of these are listed in Table 20-2. Some oncogenes that are associated with specific functions and with specific types of cancers are listed in Table 20-3.

There is much active research on these proteins, and in many cases their modes of action are not yet known.

VII. CONCLUSION

In this brief chapter, we have discussed the relationship of hormones to cancer. Because cancer represents an escape from normal cellular regulation in terms of signal transduction, it is not surprising that hormones and their receptors are intimately related to carcinogenesis. In fact, carcinogens might be considered as external or xenobiotic hormones. There is great resemblance between the structures of some carcinogens and normal hormones. One of the chief mechanisms for escape of the normal signal transduction controls is the overexpression of oncogenes that are functional parts of receptors without the ligand-binding domain, a perfect solution by the aberrant cell to shortcutting the external signal. No doubt, new oncogenes will be-discovered that affect all levels of the signal transduction pathway, involving receptor processing and transcriptional controls.

References

A. Books

Bagdade, J. D., ed. (1990, 1991). "The Yearbook of Endocrinology." Mosby Year Book, St. Louis.

Burck, K. B., Liu, E. T., and Larrick, J. W. (1988). "Oncogenes, An Introduction to The Concept of Cancer Genes." Springer-Verlag, New York.

Cooper, G. M. (1990). "Oncogenes." Jones and Bartlett Publishers, Boston.

Lupulescu, A. (1990). "Hormones and Vitamins in Cancer Treatment." CRC Press, Boca Raton, FL.

"Physicians' Desk Reference," 46th ed (1992). Medical Economics Data.

Young, R. C., ed. (1990). "The Yearbook of Oncology." Mosby Year Book, St. Louis.

B. Review Articles

Burke, L. (1981). Diethylstilbestrol: effect of in utero exposure. *In* "Current Problems of Obstetrics and Gynecology," (J. M. Levinthal, ed.), pp. 1–58. Year Book Medical Publishers, Inc., Chicago.

Glusker, J. P. (1987). Structural aspects of steroid hormones and carcinogenic polycyclic aromatic hydrocarbons. *In* "Biochemical Actions of Hormones" (G. Litwack, ed.), Vol. 6, pp. 121–204. Academic Press, Inc., New York.

Jordan, V. C. (1989). Resistance to antioestrogen therapy: a challenge for the future. *In* Cavallo, F., ed., "Endocrine Therapy of Breast Cancer" (F. Cavallo, ed.), pp. 51–60. Springer-Verlag, Berlin.

Weinberg, R. A. (1996). How cancer arises. *Sci. Am.* September, 62–70.

C. Research Papers

Ervin, P. R., Jr., Kaminski, M. S., Cody, R. L., and Wicha, M. S. (1989). Production of mammostatin, a tissue specific growth inhibitor, by normal human mammary cells. *Science* **244,** 1585–1587.

Herbst, A. L., Ulfelder, H., and Poskanzer, D. C. (1971). Adenocarcinoma of the vagina: association with maternal stilbestrol therapy with tumor appearance in young women. *New Engl. J. Med.* **284,** 878.

Nussbaum, S. R., Gaz, R. D., and Arnold, A. (1990). Hypercalcemia and ectopic secretion of parathyroid hormone by an ovarian carcinoma with rearrangement of the gene for parathyroid hormone. *N. Engl. J. Med.* **19,** 1324–1328.

Paik, S., Hazan, R., Fisher, E. R., Sass, R. E., Fisher, B., Redmond, C., Schlessinger, J., Lippman, M. E., and King, C. R. (1990). Pathologic findings from the national surgical adjuvant breast and bowel project: prognostic significance of erbB-2 protein overexpression in primary breast cancer. *J. Clin. Oncol.* **8,** 103–112.

Raam, S., Pappas, R. N., and Tamura, H. (1988). Defective estrogen receptors in human mammary cancers: their significance in defining hormone dependence. *J. Natl. Cancer Inst.* **80,** 756–761.

Slaman, D. J., and Clark, G. M. (1988). Amplification of c-erbB-2 and aggressive human breast tumors? *Science* **240,** 1795–1798.

APPENDIX

A

Compilation of Most Known Hormones in Higher Mammals and Humans[a]

Trivial name and class	Abbreviation	Source	Action
Amino acid-derived hormones			
Adrenals			
Epinephrine (adrenaline)	EP	Adrenal medulla (CNS)	Glycogenolysis in liver, increases blood pressure
Intestine			
Histamine		Gut, CNS, mast cells, many tissues	Gastric secretion; may affect CNS
Bursa of Fabricius (chickens)			
Bursin		Bursa of Fabricius	A tripeptide inducing differentiation of avian and mammalian B-precursor cells
Nervous system			
Acetylcholine		Neurons	Variety of activities in nervous system, innervates adrenal medulla
Dopamine (also belived to be PIF)		CNS	Inhibits PRL release (and other actions)
γ-Aminobutyric acid	GABA	CNS	Neurotransmitter; inhibits release of CRF and PRL
Norepinephrine	NEP	CNS neurons	Neurotransmitter; increases blood pressure
Serotonin		CNS neurons, gut	Affects smooth muscles, nerves, stimulates release of GH, TSH, ACTH (CRF), and inhibits LH release
Pineal gland			
Acetylserotonin			Affects GH release from anterior pituitary
Melatonin			Inhibits GH release from anterior pituitary; affects reproductive functions
Octopamine			Inhibition of monoamine oxidase
Thyroid/parathyroid gland			
Thyroxine (T_4) and triiodothyrone (T_3)		Thyroid gland	Increases oxidation rates in tissues
Autoimmune antibodies acting like unregulated hormones			
Autoimmune anti-insulin		Human	Produces diabetes-like syndrome
Long-acting thyroid stimulator	LATS	Human	Stimulates thyroid-like TSH, but is regulated by thyroid hormone

(*continues*)

Appendix A—*Continued*

Trivial name and class	Abbreviation	Source	Action
Fatty acid (arachidonic acid)-derived hormones			
Blood			
Prostacyclin	PGI	Vascular endothelium, blood	Prevents aggregation of platelets
Thromboxane B_2	TXB_2	Platelets	Metabolite of TXA_2
Lung			
Leukotriene C_4	LTC_4		Long-acting bronchoconstrictor
Leukotriene D_4	LTD_4		Long-acting bronchoconstrictor
Leukotriene B_4	LTB_4		
Leukotriene A_4	LTA_4		Intermediate en route to LTC_4/LTD_4
Leukotriene E_4	LTE_4		Long-acting bronchoconstrictor
Lipoxin A	LPA		Inflammatory
Lipoxin B	LPB		Inflammatory
Prostaglandin $E_1 + E_2$	PGE_1 or PGE_2	Wide variety of cells	Stimulates cyclic AMP
Prostaglandin $F_{1\alpha} + F_{2\alpha}$	$PGF_{1\alpha}$ or $PGF_{2\alpha}$	Wide variety of cells	Active in dissolution of *corpus luteum,* ovulation + parturition contractions
Prostaglandin A_2	PGA_2	Kidney	Hypotensive effect
Thromboxane A_2	TXA_2	Platelets, lung, etc.	Causes platelet aggregation
Polypeptides/Proteins			
Adipose tissue			
Leptin		Fat cells	Loss of fat; satiety factor
Adrenals			
Met-enkephalin + Leu-enkephalin		Adrenal medulla and CNS cells	Analgesic actions in CNS; other unknown effects
Blood			
Angiotensin II		Blood, lungs, brain, many tissues	*Zona glomerulosa* cells of adrenal cortex to stimulate synthesis + release of aldosterone
Bradykinin		Plasma, gut, other tissues	Vasodilator; lowers blood pressure
Bone marrow			
Granulocyte-macrophage colony stimulatory factor	GM-CSF	Bone marrow	Stimulates production of granulocytes and macrophages
Granulocyte-colony stimulatory factor	G-CSF	Bone marrow	Stimulates production of granulocytes
Macrophage-colony stimulatory factor	M-CSF	Bone marrow	Stimulates production of macrophages
Interferon-γ	IFNγ, M	Bone marrow	
Interleukin-1	IL-1	Bone marrow	Produces a wide array of effects in the immune responses and induces other cytokines
Interleukin-2	IL-2	Bone marrow	Stimulates proliferation and maturation of $T_{1/2}$ cells and production by $T_{1/2}$ cells of cell cytokines that control the differentiation and proliferation of B cells
Interleukin-3	IL-3 (multi-CSF)	Bone marrow	Acts on the pluripotent hematopoietic stem cell to promote the development of many blood cells
Interleukin-4	IL-4	Bone marrow	Promotes proliferation of B cells
Interleukin-5	IL-5	Bone marrow	Promotes proliferation of B cells
Interleukin-6	IL-6	Bone marrow	Activates stromal bone marrow cells to produce colony-stimulating factors and acts as a pyrogen (i.e., raises body temperature), promotes proliferation of B&T cells
Interleukin-7	IL-7	Bone marrow	Promotes proliferation of B-cell progenitors
Interleukin-8	IL-8	Bone marrow	Activates macrophages
Interleukin-9	IL-9	Bone marrow	Enhances most cell growth
Interleukin-10	IL-10	Bone marrow	Inhibits clonal expansion of T-cells
Interleukin-11	IL-11	Bone marrow	
Interleukin-12	IL-12	Bone marrow	
Interleukin-13	Meg CSF	Bone marrow	

Appendix A—*Continued*

Trivial name and class	Abbreviation	Source	Action
Stem cell factor	SCF	Bone marrow	
Leukemia inhibitory factor	LIF	Bone marrow	
Platelet-Derived growth factor	PDGF	Platelets	Similar to FGF
Heart			
Atrial natriuretic factor (also known as atriopeptin)	ANF	Atria	Blood pressure lowering; stimulates renal sodium excretion; increases GFR and urine volume
Hypothalamus			
Arg-vasotocin		Hypothalamus and pineal gland	Regulates reproductive glands
Corticotropic-releasing hormone	CRH	Hypothalamus	Releases ACTH + β-endorphin in anterior pituitary
Gonadotropic-releasing hormone	GnRH	Hypothalamus, distributed in CNS; milk, gonadal cells containing GnRH receptors	Releases FSH and LH in anterior pituitary
Growth hormone release inhibiting hormone (somatostatin)	GIH	Hypothalamus, extrahypothalamic brain, spinal cord, pancreas, stomach, and intestine	Inhibits release of GH and TSH in anterior pituitary; regulates pancreatic hormones
Growth hormone releasing hormone	GRH		Releases GH in anterior pituitary
Melanotropin release inhibiting factor	MIF		Prevents release of MSH in anterior pituitary, probably not in man
Melanotropin releasing factor	MRF		Releases MSH in anterior pituitary probably not in man
Neurotensin		Hypothalamus, intestine (mucosa)	May have neurotransmitter actions; in pharmacological amounts, has several effects on gut
Prolactin releasing hormone	PRH or TRH		Releases PRL in anterior pituitary
Prolactin release inhibiting hormone	PIF		Inhibits prolactin release
Substance P	SP	Hypothalamus and CNS, intestine	Transmits pain and other functions, increases smooth muscle contractions of GI tract
Thyrotropic releasing hormone	TRH	Hypothalamus, extrahypothalamic brain, spinal cord, and brain stem	Releases TSH and PRL in anterior pituitary
Gastrointestinal			
ACTH			↑ Cortisol release
α-Melanocyte stimulating hormone	αMSH	Intermediary pituitary	↑ Melanin release
Beta-endorphin		Anterior/intermediate pituitary	↓ Intestinal transit, analgesic action in CNS
Bombesin		Nerves + endocrine cells of gut	Hypothermic hormone; increases gastrin + gastric acid secretion; many other actions
Cholecystokinin (pancreozymin)	CCK		Stimulates gallbladder contraction + bile flow; enhances secretion of pancreatic enzymes
Dynorphins			↓ Intestinal transit
Enkephalin			↓ Intestinal transit
Enteroglucagon (different from pancreatic glucagon, glucagon-like immuno-reactivity)	GLI	Gut, L cells of ileum and colon, brain	Increases insulin release
Galanin			↑ Plasma glucose, ↑ fundus contraction
Gastrin		G cells in midpyloric, glands in stomach antrum	Increases secretion of gastric acid + pepsin + many other effects
Gastrin inhibitory polypeptide	GIP		↓ Gastric secretion, ↑ intestinal secretion
Gastrin releasing peptide	GRP		↑ Release gastrin
Glicentin			↑ Hepatic glucose, ↓ acid secretion
Growth hormone releasing hormone	GRH		↑ Release of growth hormone
Motilin		Duodenum (jejunum), pineal, pituitary	Acts on GI tract to alter motility; stimulates contraction of fundus + antrum + decreases gastric emptying
Neuromedin B			↓ Acid secretion

(*continues*)

Appendix A—*Continued*

Trivial name and class	Abbreviation	Source	Action
Neuromedin K			Contraction gastrointestinal smooth muscle
Neuropeptide Y	NPY		↑ Vasoconstriction
Neurotensin			↓ Acid secretion, ↓ gastric motor activity
Oxyntomodulin			↑ Insulin release, ↓ acid secretion
Pancreastatin			↓ Islet somatostatin release
Pancreatic polypeptide	PP		↓ Pancreatic secretion
Peptide histidine isoleucine	PHI		↑ Pancreatic secretion
Peptide YY	PYY		↓ Pancreatic secretion, ↓ gallbladder contraction
Secretin		Duodenum when pH of its contents is less than 4.5	Stimulates pancreatic acinar cells to release bicarbonate + water which are transported to duodenum to elevate pH
Somatostatin			↓ Acid secretion, ↓ pancreatic function
Substance K			Contraction gastrointestinal smooth muscle
Substance P			Contraction gastrointestinal smooth muscle
Vasointestinal peptide	VIP	GI tract, hypothalamus, + elsewhere	Neurotransmitters in peripheral autonomic nervous system; relaxes smooth muscles of circulation; increases secretion of water and electrolytes from pancreas and gut
Kidney			
Erythropoietin	EPO	Kidney	Acts on bone marrow to induce terminal differentiation + initiation of hemoglobin synthesis
Liver			
Somatomedins (insulin-like growth factors)	IGF-I IGF-II	Liver, muscle, kidney, + other tissue	Cartilage sulfation; somatic cell growth, insulin-like effect
Lung			
Eosinophil chemotactic factor of anaphylaxis		Lung mast cells	After release, selective chemoattractants for eosinophils
Ovaries			
Relaxin		*Corpus luteum*	Increases during gestation (may inhibit myometrial contractions)
Pancreas			
Glucagon		A cells	Glycogenolysis in liver; increases cyclic AMP
Insulin		B cells	Glucose utilization in liver; promotes synthesis of glycogen
Pancreatic polypeptide	PP	Pancreatic islets (peripheral cells of)	Has a number of effects on gut in pharmacological amounts
Proinsulin		B cells	Precursor to insulin
Pituitary			
Adrenocorticotrophic hormone	ACTH	Adenohypophysis	Stimulates synthesis and release of cortisol and dehydroepiandrosterone from adrenal cortex
Arg-vasopressin	AVP or ADH	Posterior pituitary	Increases water reabsorption in kidney
β-Endorphin		*Pars intermedia* and anterior pituitary	Analgesic actions in CNS
Fibroblast growth factor acid FGF basic FGF	FGF aFGF bFGF		Stimulates proliferation of cells derived from endoderm and mesoderm in presence of serum
Follicle-stimulating hormone	FSH	Adenohypophysis	Stimulates development of ovarian follicle and secretion of estrogen; stimulates seminal tubules and spermatogenesis

Appendix A—*Continued*

Trivial name and class	Abbreviation	Source	Action
Growth hormone (somatomammotropin or somatotropin)	GH	Adenohypophysis	Somatic cell growth mediated by somatomedins, hyperglycemia, liver steroid metabolism, bone sulfation reactions
Lipotropin	LPH	Adenohypophysis	Fat mobilization; source of opioid peptides
Luteinizing hormone (formerly "interstitial cell regulating hormone" in reference to male)	LH	Adenohypophysis	Stimulates Leydig (interstitial) cell development in male + production of testosterone; stimulates corpus luteum and its production of progesterone in female
Melanocyte-stimulating hormone	MSH	Adenohypophysis *pars intermedia*	CNS functions (e.g., in memory retention) and skin-darkening reaction
Ovarian growth factor			Prolongs ovarian cell survival
Oxytocin	OT	Posterior pituitary, hypothalamus	Lactating mammary gland, milk letdown; uterine contraction at parturition
Prolactin	PRL	Adenohypophysis	Synthesis of milk constituents in mammary gland; stimulates testosterone production; secondary growth hormone effects in liver (e.g., as hyperglycemic acid); mammary gland secretory cell differentiation
Thyroid-stimulating hormone	TSH	Adenohypophysis	Stimulates thyroid gland follicles to secrete thyroid hormone
Placenta			
Human chorionic gonadotropin	hCG		LH-like functions, maintains progesterone productivity during pregnancy
Human placental lactogen	hPL		Acts like PRL and like GH because of large amount of hPL produced
Transforming growth factor α	TGFα		Binds to EGF receptor
Transforming growth factor β	TGFβ		Confers transformed phenotype on normal cells
Salivary gland			
Epidermal growth factor (formerly urogastrone)	EGF		Stimulates proliferation of cells of ectodermal and mesodermal origin with serum; inhibits gastric secretion
Nerve Growth Factor	NGF		Stimulates neuronal growth
Skin			
Alytesin		Amphibian skin	Stimulates gastric acid secretion
Cerulein		Frog skin	Similar to CCK + gastrins
Litorin		Amphibian skin	Stimulates gastric secretion
Ranatensin		Amphibian skin	Stimulates gastric secretion
Submaxillary gland			
Nerve growth factor	NGF		Differentiation and growth of embryonic dorsal root ganglia
Testes			
Antimüllerian hormone		Fetal Sertoli cells of the testes	Mediates involution of the Müllerian ducts
Inhibin		Seminiferous tubule (and ovary)	Negative feedback inhibitors of FSH secretion from anterior pituitary
Thymus			
Thymic humoral factor	THF		Activates adenylate cyclase in thymus + spleen cells
Thymopoietin I and II, α-thymosin			Stimulates phagocytes; stimulates differentiation of precursors into immune competent T cells
Thyroid/parathyroid			
Calcitonin	CT	Parafollicular C cells of thyroid gland	Lowers serum calcium
Parathyroid hormone	PTH	Parathyroid glands	Stimulates bone resorption; elevates serum Ca^{2+}, stimulates phosphate excretion by the kidney

(*continues*)

Appendix A—*Continued*

Trivial name and class	Abbreviation	Source	Action
Vascular endothelium			
Endothelins	E-1,E-2 E-3	Endothelial cells	Vasoconstriction
Steroid hormones			
Adrenals			
Aldosterone and 11-deoxy-corticosterone	DOC	*Zona glomerulosa* of adrenal Cortex	Salt retention in kidney
Cortisol (hydrocortisone)		*Zona fasciculata* and *zona reticularis* of adrenal cortex	Antistress hormone; carbohydrate metabolism; circulating glucose increased; liver glycogen increased; depresses immune system; anti-inflammatory agent
Dehydroepiandrosterone	DHEA	*Zona reticularis* of adrenal cortex	Weak androgen; major secretion of fetal adrenal cortex; can be converted to estrogen; may have other unknown actions
Brain			
Catecholestrogens			Stimulates catecholamine receptors in CNS
Kidney			
1,25-Dihydroxyvitamin D_3 (1,25-dihydroxycholecalciferol)			Stimulates intestinal Ca^{2+} absorption, stimulates a Ca^{2+} binding protein in a variety of tissues, especially in intestine, and many other responses
24R,25-Dihydroxyvitamin D_3 (24R, 25-dihydroxycholecalciferol)			Has receptor in chondrocytes
Ovaries			
17β-Estradiol and estriol		Ovarian follicle (corpus luteum)	Uterine endometrium development, female tissues
Progesterone		Corpus luteum, placenta (ovarian follicle)	Breast development; uterine endometrium development
Testes			
Dihydrotestosterone	DHT	Seminiferous tubule + other male tissues (e.g., prostate)	Conversion product of testosterone which binds to androgen receptor
Testosterone		Leydig cells (interstitial cells of testis) (adrenal)	Spermatogenesis/male characteristics
Second messenger substances			
Arachidonic acid		All cell membranes	Precursor of prostaglandins/ leukotrienes/lipoxins
Ca^{2+} (and calmodulin)		All cells	Required for secretory activity; enzyme regulator
Cyclic AMP		Many cells	Protein phosphorylation
Cyclic GMP		Many cells	
Other			
Nitric oxide	NO	Vascular endothelium	Vasodilation
Insect hormones			
Ecdysone (steroid)			Stimulates molting
Juvenile hormone (hydrocarbon terpenoid)	JH		Controls molting
Plant hormones			
Auxins (indoleacetic acid)		All higher plants and some lower plants	Stimulates extension growth and cell division in cambrium and foot
Giberellins (diterpenoid cells)		Widespread plant regulators	Stimulates extension growth
Kinins (N-substituted)			Promotes cell division

[a] This table is composed of most of the known hormones of higher animals and man; in addition, the key insect hormones and plant hormones are listed at the end of the table. The first level presents hormones by structure (amino acid-derived hormones, autoimmune antibodies acting like unregulated hormones, fatty acid-derived hormones, polypeptide hormones, steroid hormones, second messenger substances, insect hormones, and plant hormones). Each of these categories is then subdivided, where appropriate, by organ; finally, the individual hormones in a category are entered alphabetically. ($\uparrow$) Indicates an increase; ($\downarrow$) indicates a decrease.

APPENDIX

B

Human Blood Concentrations of Major Hormones[a]

Hormone	Fluid[b]	Units	Method or instrument
Aldosterone			
Standing (normal-salt diet)	S, P	4–31 ng/dl	Immunoassay
Recumbent (normal-salt diet)	S, P	<16 ng/dl	Immunoassay
Normal-salt diet (100–180 meq of sodium)	U	6–25 μg/day	Immunoassay
Low-salt diet (10 meq of sodium)	U	17–44 μg/day	Immunoassay
High-salt diet	U	0–6 μg/day	Immunoassay
Androstenedione	S	60–260 ng/dl	Immunoassay
Antidiuretic hormone (arginine vasopressin)	P	1.0–13.3 pg/ml	Immunoassay
Calcitonin	S		Immunoassay
Female		0–20 pg/ml	
Male		0–28 pg/ml	
Catecholamines			
Dopamine	U	65–400 μg/day	Liquid chromatography
	P	0–30 pg/ml	Liquid chromatography
Epinephrine	U	1.7–22.4 μg/day	Liquid chromatography
Supine	P	0–110 pg/ml	Liquid chromatography
Standing	P	0–140 pg/ml	Liquid chromatography
Norepinephrine	U	12.1–85.5 μg/day	Liquid chromatography
Supine	P	70–750 pg/ml	Liquid chromatography
Standing	P	200–1700 pg/ml	Liquid chromatography
Chorionic gonadotropin (hCG) (nonpregnant)	S	<10 mIU/ml	Immunoassay
Corticotropin (ACTH)	P	6.0–76.0 pg/ml	Immunoassay
Cortisol	P		Immunoassay
Fasting, 8 AM–12 noon		5.0–25.0 μg/dl	
12 noon–8 PM		5.0–15.0 μg/dl	
8 PM–8 AM		0.0–10.0 μg/dl	
Cortisol, free	U	20–70 μg/day	Immunoassay
C peptide	S	0.30–3.70 μg/liter	Immunoassay
11-Deoxycortisol (after metyrapone)	P	>7.5 μg/dl	Immunoassay
1,25-Dihydroxyvitamin D	S	16–42 pg/ml	Immunoassay
Erythropoietin	S	<19 mU/ml	Immunoassay
Estradiol	S, P		Immunoassay
Female			
Premenopausal adult		23–361 pg/ml	
Postmenopausal		<30 pg/ml	
Prepubertal		<20 pg/ml	

(continues)

Appendix B—*Continued*

Hormone	Fluid[b]	Units	Method or instrument
Male		<50 pg/ml	
Gastrin	P	0–200 pg/ml	Immunoassay
Growth hormone	P	2.0–6.0 ng/ml	Immunoassay
Hemoglobin A_{1c}	P	3.8–6.4%	Liquid chromatography
Homovanillic acid	U	0.0–15.0 mg/day	Liquid chromatography
17-Hydroxycorticosteroids	U		Colorimetry
Female		2.0–6.0 mg/day	
Male		3.0–10.0 mg/day	
5-Hydroxyindoleacetic acid (lower in women than in men)	U	2–9 mg/day	Colorimetry
17-Hydroxyprogesterone	S		Immunoassay
Female			
Prepubertal		0.20–0.54 μg/liter	
Follicular		0.02–0.80 μg/liter	
Luteal		0.90–3.04 μg/liter	
Postmenopausal		<0.45 μg/liter	
Male			
Prepubertal		0.12–0.30 μg/liter	
Adult		0.20–1.80 μg/liter	
25-Hydroxyvitamin D	S	8–55 ng/ml	Immunoassay
Insulin	S	0–29 μU/ml	Immunoassay
Male		5.0–23.0 mg/day	
17-Ketogenic steroids	U		Colorimetry
Female		3.0–15.0 mg/day	
17-Ketosteroids	U		Colorimetry
Female and male ≤10 yr old		0.1–3.0 mg/day	
Female and male 11–14 yr old		2.0–7.0 mg/day	
Female ≥15 yr old		5.0–15.0 mg/day	
Male ≥15 yr old		9.0–22.0 mg/day	
Metanephrines, total	U	0.0–0.90 mg/day	Spectrophotometry
Parathyroid hormone	P	10–60 pg/ml	Immunoassay
Parathyroid-related protein	P	<1.5 pmol/liter	Immunoassay
Pregnanediol	U		Gas chromatography
Female		0.2–6.0 mg/day	
Follicular phase		0.1–1.3 mg/day	
Luteal phase		1.2–9.5 mg/day	
Pregnancy		Gestation period dependent	
Male		0.2–1.2 mg/day	
Pregnanetriol	U	0.5–2.0 mg/day	Gas chromatography
Prolactin	S		Immunoassay
Female		0–15 ng/ml	
Male		0–10 ng/ml	
Renin activity	P		Immunoassay
Normal salt intake			
Recumbent 6 hr		0.5–1.6 ng/ml/hr	
Upright 4 hr		1.9–3.6 ng/ml/hr	
Low salt intake			
Recumbent 6 hr		2.2–4.4 ng/ml/hr	
Upright 4 hr		4.0–8.1 ng/ml/hr	
Upright 4 hr, with diuretic		6.8–15.0 ng/ml/hr	
Somatomedin C	P		Immunoassay
Female			
Preadolescent		60.8–724.5 ng/ml	
Adolescent		112.5–450.0 ng/ml	
Adult		141.8–389.3 ng/ml	
Male			
Preadolescent		65.5–841.5 ng/ml	
Adolescent		83.3–378.0 ng/ml	
Adult		54.0–328.5 ng/ml	
Testosterone, total, morning sample	P		Immunoassay
Female		20–90 ng/dl	
Male, adult		300–1100 ng/dl	

Appendix B—*Continued*

Hormone	Fluid[b]	Units	Method or instrument
Testosterone, unbound, morning sample	P		Equilibrium dialysis
Female, adult		0.09–1.29 ng/dl	
Male, adult		3.06–24.0 ng/dl	
Thyroglobulin	S	0–60 ng/ml	Immunoassay
Thyroid-hormone-binding index		0.83–1.17	Charcoal resin
Thyroid-stimulating hormone	S	0.5–5.0 μU/ml	Immunoassay
Thyroxine free	S	0.8–2.7 ng dl	Direct equilibrium dialysis
Thyroxine-binding globulin	S	Age and sex dependent	Immunoassay
Thyroxine free index		4.6–11.2	Calculation
Thyroxine, total (T_4)	S	4–12 μg/dl	Immunoassay
Triiodothyronine, total (T_3)	S	75–195 ng/dl	Immunoassay
Vanilmandelic acid (VMA)	U	1.4–6.5 mg/day	Liquid chromatography

[a] Abstracted from C. D. Jordan, J. G. Flood, M. LaPosata, and K. B. Lewandrowski (1992) Normal Reference laboratory values. *New England J. Medicine* 327, 718–724.

[b] S denotes serum, P plasma, and U urine.

APPENDIX

C

Clinically Relevant Endocrine Disorders[a]

Disease	Description
Acromegaly	Inappropriate and continued secretion of growth hormone by a tumor of pituitary cells which leads to soft tissue swelling and hypertrophy of the skeletal extremities (usually in the third or fourth decade).
Addison's disease	Adrenocortical insufficiency resulting from a deficient production of glucocorticoids and/or mineralocorticoids due to a destruction of the adrenal cortex.
Bartter's syndrome	Characterized by increased angiotensin, renin, aldosterone, secretion, hypokalemia, alkalosis, and hyperplasia of the renal juxtaglomerular cells, but with normal blood pressure; the disease, which is possibly an autosomal-dominant disorder, usually appears in late infancy or early childhood; the primary defect may be a lesion in chloride reabsorption in the kidney loop of Henle.
Celiac disease	An intestinal malabsorption disorder occurring in some individuals who may lack an enzyme necessary for the hydrolysis of *N*-glutamyl peptides in the small intestine. As a consequence, the affected individual is intolerant of some proteins—usually those derived from wheat, oats, barley, or rye; the disease is also referred to as gluten-sensitive enteropathy.
Chiari-Frommel syndrome	The occurrence of galactorrhea (non-nursing-related lactation) and amenorrhea (absence of expected menstrual periods) during the postpartum period; the disease may be due to the presence of a prolactin-secreting tumor.
Cretinism	Characterized by a permanent neurological and skeletal retardation and results from an inadequate output of thyroid hormone during uterine and neonatal life; may be caused by iodine deficiency, thyroid hypoplasia, genetic enzyme defects, or excessive maternal intake of goitrogens.
Cushing's disease	Hypercortisolism resulting from the presence of small pituitary tumors which secrete ACTH leading to excess production of cortisol by the adrenals.
Cushing's syndrome	The circumstance of glucocorticoid excess without specification of the specific etiology; it may result from endogenous causes but is more commonly iatrogenic.
Diabetes insipidus	A deficient secretion of vasopressin which is manifested clinically as *diabetes insipidus;* it is a disorder characterized by the excretion of an increased volume of dilute urine.
Diabetes mellitus	A disease characterized by a chronic disorder of intermediary metabolism due to a relative lack of insulin which is characterized by hyperglycemia in both the postprandial and fasting state (see also types I and II *diabetes mellitus*).
Diabetes mellitus (insulin-dependent or type I diabetes)	The form of diabetes which appears in the second and third decade of life and is characterized by a destruction of the pancreas B cells; this form of the disease is normally treated with daily administration of insulin.
Diabetes mellitus (insulin-independent or type II diabetes)	The form of diabetes arising after the fourth decade, usually in obese individuals; this form of the disease does not normally require treatment with insulin.

(*continues*)

Appendix C—*Continued*

Disease	Description
Empty sella syndrome	Empty sella is a term that describes *sellae* that fill with air during pneumoencephalography. It is frequently associated with the flattening of the pituitary gland. The etiology of the disease is unknown; the pituitary function is usually normal.
Fanconi syndrome	A renal tubular defect in the absorption of a variety of substances including H_2O, phosphate, sodium, bicarbonate, and amino acids; frequently an osteomalacic bone disease and a distal tubular acidosis may accompany the disease.
Feminization	Feminization of males, usually as manifested by enlargement of the breasts (gynecomastia) which can be attributed to an increase in estrogen levels relative to the prevailing androgen levels.
Froehlich's syndrome	A condition usually caused by craniopharyngioma (a tumor of the hypothalamus) which results in a combination of obesity and hypogonadism; sometimes termed adiposogenital dystrophy
Galactorrhea	The persistent discharge from the breast of a fluid that resembles milk and that occurs in the absence of parturition or else persists postpartum (4–6 months) after the cessation of nursing.
Gigantism	This condition appears in the first year of life and is characterized by a rapid weight and height gain; affected children usually have a large head and mental retardation; to date no specific endocrine abnormalites have been detected.
Goiter	Goiter may be defined as a thyroid gland that is twice its normal size; endemic goiter is the major thyroid disease throughout the world. Goiter is frequently associated with a dietary iodine deficiency; in instances of sporadic goiter it may occur as a consequence of a congenital defect in thyroid hormone synthesis.
Grave's disease	An autoimmune disease characterized by the presence in serum of a long-acting thyroid stimulator (LATS) that is an antibody for the receptor for TSH. Grave's disease is the most common cause of thyrotoxicosis.
Gynecomastia	Abnormal breast enlargement which may occur in males during puberty.
Hartnup's disease	An intestinal transport disorder. The condition may be diagnosed by the massive urinary excretion of monoamino-monocarboxylic amino acids; frequently there are pellagra-like rashes after exposure to sunlight as well as attacks of cerebellar ataxia.
Hermaphroditism	True hermaphroditism is defined as the presence of both testicular and ovarian tissue in the same individual; pseudohermaphroditism is a discrepancy between gonadal and somatic sex.
Hirsutism	An increase in facial hair in women which is beyond that cosmetically acceptable; this condition may be associated with a number of masculinizing disorders including Cushing's syndrome, congenital adrenal hyperplasia, and polycystic ovary syndrome.
Hyperaldosteronism	An inappropriate secretion of aldosterone. It can occur as a primary adrenal problem (e.g., adrenal tumor) or can be secondary to other metabolic derangements that stimulate its release; it is often characterized by inappropriately high levels of plasma renin.
Hyperparathyroidism	Inappropriately high secretion of PTH leading to hypercalcemia. Frequently associated with the hyperparathyroidism is a metabolic bone disease characterized by excessive bone calcium reabsorption; frequently attributable to an adenoma of the parathyroid gland.
Hypoparathyroidism	Inappropriately low secretion of PTH leading to hypocalcemia; the disease is either idiopathic or iatrogenically induced.
Hypophosphatasia	An autosomal recessive trait characterized by elevated serum and urine inorganic pyrophosphate, a low serum alkaline phosphatase, and frequently hypercalcemia; may be related to a dysfunction of the osteoblasts.
Klinefelter's syndrome	Typically characterized by male hypogonadism; the presence of extra X chromosomes is likely the fundamental underlying etiological factor. It is characterized by varying degrees of decreased Leydig cell function and seminiferous tubule failure.
Milk-alkali syndrome	Affected subjects have hypercalcemia, nephrocalcinosis, soft tissue calcification, renal impairment, alkalosis, and hyperphosphatemia; the syndrome can result as a consequence of an excessive dietary intake of milk and other absorbable alkali (e.g., Na_2CO_3 or $NaHCO_3$); it is uncommon today.
Myxedema	Hypothyroidism clinically manifested by the presence of a mucinous edema; the disease may appear at any time throughout life and is attributable to disorders of the thyroid gland or to pituitary insufficiency.
Nelson's syndrome	A pituitary adenoma occurring in 10% of patients with Cushing's disease; afflicted subjects have a severe skin pigmentation.
Osteomalacia	A bone disease in adults characterized by a failure of the skeletal osteoid to calcify; it is usually caused by an absence of adequate access to vitamin D.
Polycystic ovary syndrome	A complex of varying symptoms ranging from amenorrhea to anovulatory bleeding often associated with obesity and hirsutism. The term denotes an absence of ovulation in association with continuous stimulation of the ovary by disproportionately high levels of LH.

Appendix C—*Continued*

Disease	Description
Pseudohypoparathyroidism	A familial disorder characterized by hypocalcemia, increased circulating levels of PTH and a peripheral unresponsiveness to the hormone; afflicted individuals frequently are of short stature, with mental retardation and short metacarpals and/or metatarsals.
Rickets	A failure in the child of the skeletal osteoid to calcify; it is usually caused by an absence of adequate amounts of vitamin D; it is characterized by a bowing of the femur, tibia, and fibulas.
Turner's syndrome	A condition present in females with a 45, XO chromosome pattern (i.e., complete absence of the X chromosome). The XO individual is typically short with a thick neck and trunk and no obvious secondary sex characteristics.
Waterhouse-Friderichsen syndrome	Acute adrenal insufficiency resulting from severe systemic infection by *meningococcus* characterized by a high fever, meningeal irritation, and vascular collapse.
Werner's syndrome	Multiple endocrine neoplasia, type 1, caused by an autosomal recessive inheritance. It is often characterized by a severe testicular atrophy and a mild insulin-resistant diabetes.
Zollinger-Ellison syndrome	Tumors of the pancreas which result in excessive secretion of gastrin; the afflicted subject has recurrent duodenal ulcers and diarrhea caused by hypersecretion of gastric acid.

[a] This list was abstracted from the United States National Library of Medicine—Medical Subject Headings—Tree Structures—1996, pp. 294–298. U.S. Department of Commerce, Washington, DC, 1996. The diseases were included in the National Library of Medicine table on the basis of frequency of publication of papers about the given disease topics.

APPENDIX

D

The Genetic Code[a]

First position (5′ end)	Second position U	C	A	G	Third position (3′ end)
	Phe	Ser	Tyr	Cys	U
U	Phe	Ser	Tyr	Cys	C
	Leu	Ser	stop	stop	A
	Leu	Ser	stop	Trp	G
	Leu	Pro	His	Arg	U
C	Leu	Pro	His	Arg	C
	Leu	Pro	Gln	Arg	A
	Leu	Pro	Gln	Arg	G
	Ile	Thr	Asn	Ser	U
A	Ile	Thr	Asn	Ser	C
	Ile	Thr	Lys	Arg	A
	Met	Thr	Lys	Arg	G
	Val	Ala	Asp	Gly	U
G	Val	Ala	Asp	Gly	C
	Val	Ala	Glu	Gly	A
	Val	Ala	Glu	Gly	G

[a] The genetic code is expressed in terms of triplet sequences as they would appear in a natural messenger RNA molecule. The sequence of each codon is read from left to right, starting with the 5′ end and going toward the 3′ end. The codons UAA, UAG, and UGA are "stop" signals that result in chain termination and are used to indicate the end of a polypeptide chain. No separate codon is required to start the polypeptide chain, since the first amino acid is always methionine (Met).

APPENDIX

E

Amino Acid Abbreviations

Amino acid	Three letter	Single letter
Alanine	Ala	A
Arginine	Arg	R
Asparagine	Asn	N
Aspartic acid	Asp	D
Cysteine	Cys	C
Glutamic acid	Glu	E
Glutamine	Gln	Q
Glycine	Gly	G
Histidine	His	H
Isoleucine	Ile	I
Leucine	Leu	L
Lysine	Lys	K
Methionine	Met	M
Phenylalanine	Phe	F
Proline	Pro	P
Serine	Ser	S
Threonine	Thr	T
Tryptophan	Trp	W
Tyrosine	Tyr	Y
Valine	Val	V

APPENDIX

F

Units of Measurement in Biological Systems

The Systemé International d'Unités (SI units), which is now generally accepted as the universal scientific system of measure, is based on the meter and kilogram (m, kg) rather than the centimeter and gram of the older cgs system. It has seven basic units of measurement:

Quantity	*SI unit*	*Symbol*
Amount of substance	mole	mol
Mass	kilogram	kg
Time	second	s
Length	meter	m
Electric current	ampere	A
Thermodynamic temperature	kelvin	K
Light intensity	candela	cd

Several important units employed in biological systems which are derived from SI units include the following:

Quantity to be measured	*Definition in terms of basic units*	*SI unit*	*Symbol*
Volume	10^{-3} m^3	liter	L
Frequency	s^{-1}	hertz	Hz
Pressure	kg m^{-1} s^{-2}	pascal	Pa
Energy[a]	kg m^2 s^{-1}	joule	J

[a] One calorie is the energy required to heat 1 cm^3 of water from 14.5 to 15.5°C. One Calorie, which is equivalent to 1000 calories or 1.0 kilocalorie, is the standard unit employed nutritionally. To convert calories into joules, multiply by the conversion factor of 4.185.

For a complete description of SI units see the IUPAC Manual of Symbols and Terminology for Physicochemical Quantities and Units (1970). *Pure Appl. Chem.* **21,** 3–44.

Other units employed in biological systems include the following:

Unit	*Abbreviation*	*Comment*
Mole	mol	
Becquerel	Bq	Units of radioactivity[a]
Curie	Ci	
Calorie	cal	
Kilocalorie	kcal or Cal	
Svedburg	S	(1 S = 10^{-13} sec)

[a] 1 becquerel = 1 dps or 60 dpm, while 1 Ci = 3.7×10^{10} Bq.

The following prefixes and symbols are used to indicate multiples or decimal fractions of the SI units.

Prefix	*Factor*	*Symbol*
exa	10^{18}	E
peta	10^{15}	P
tera	10^{12}	T
giga	10^{9}	G
mega	10^{6}	M
kilo	10^{3}	k
hecto	10^{2}	h
deca	10^{1}	da
deci	10^{-1}	d
centi	10^{-2}	c
milli	10^{-3}	m
micro	10^{-6}	μ
nano	10^{-9}	n
pico	10^{-12}	p
femto	10^{-15}	f
atto	10^{-18}	a

Index

B

C

H

I

J

R

S

T

U

V